# TAXONOMY OF VASCULAR PLANTS

# TAXONOMY OF VASCULAR PLANTS

*by George H. M. Lawrence*
PROFESSOR OF BOTANY AT THE BAILEY HORTORIUM
CORNELL UNIVERSITY

MACMILLAN PUBLISHING CO., INC.
*New York*

Copyright, Macmillan Publishing Co., Inc., 1951

All rights reserved. No part of this book may be reproduced or transmitted in any form or by any means, electronic or mechanical, including photocopying, recording or by any information storage and retrieval system, without permission in writing from the Publisher.

MACMILLAN PUBLISHING CO., INC.
866 THIRD AVENUE, NEW YORK 10022

COLLIER MACMILLAN CANADA, LTD.

PRINTED IN THE UNITED STATES OF AMERICA

PRINTING   20 21 22 23   YEAR   7 8 9

C. A. ARNOLD, *An introduction to paleobotany*. Copyright 1947 by McGraw-Hill Book Company, Inc., New York.

W. W. ATWOOD, *The physiographic provinces of North America*. Copyright 1940 by Ginn and Company, Boston.

L. H. BAILEY, *Manual of cultivated plants*, revised edition. Copyright 1924 and 1949 by Liberty H. Bailey. Reprinted by permission of The Macmillan Company, New York.

S. A. CAIN, *Foundations of plant geography*. Copyright 1944 by Harper & Brothers, New York.

A. J. EAMES, *Morphology of vascular plants*. Copyright 1936 by McGraw-Hill Book Company, Inc., New York.

R. A. GOOD, *The geography of flowering plants*, 1947. Reprinted by permission of Longmans Green & Company, Limited, London.

J. HUTCHINSON, *British flowering plants*, 1948. Reprinted by permission of P. R. Gawthorn, Limited, London.

J. HUXLEY, *Evolution the modern synthesis*. Copyright 1943 by Julian S. Huxley. Reprinted by permission of Harper & Brothers, New York.

J. HUXLEY [Ed.], *The new systematics*, 1940. Reprinted by permission of The Clarendon Press, Oxford.

G. L. STEBBINS, JR., *Variation and evolution in plants*. Copyright 1950 by Columbia University Press, New York.

O. D. VON ENGELN, *Geomorphology*. Copyright 1942 by The Macmillan Company, New York.

# PREFACE

Taxonomic botany is an advanced subject that deals not merely with the identification and naming of plants but also with their classification. Any approach to an understanding of plant taxonomy as thus broadly conceived requires considerable previous botanical training in related fields, especially those of morphology, anatomy, cytology, genetics, paleobotany, and geomorphology. It cannot be presumed that every student may have the opportunity or time to acquire as a prerequisite even an elementary knowledge of these related sciences, nor should it be presumed that any single text need provide the material for such a background. However, any textbook devoted to the subject of taxonomic botany must make considerable use of the data available from these related topics, and must provide a sufficient explanation of their character and significance. This book has been prepared as a text to meet these needs as required by beginning and intermediate students of systematic botany. Its level of presentation is sufficiently elementary not to presuppose formal training in all of the allied botanical sciences, yet it does open channels to further study of them by those more advanced students who may have had the benefit of a considerable botanical background.

This text represents several marked departures from existing works. The subject matter is divided formally into two parts. Part One presents the more academic and theoretical considerations of taxonomy, supplemented by explanatory chapters of the practical fundamentals found to be essential to a minimum working knowledge of the science. Part Two is composed of a systematic enumeration of the 264 families of vascular plants known to grow as indigens or exotics in North America north of Mexico, together with technical descriptions of each and pertinent discussions and other features as mentioned below. Supplementing this second part is an appendix providing an illustrated glossary of the botanical terms used throughout the text. The division of the text into these two parts is deliberate and has been effected in order to bring into one part (the first) those subjects that are topics of more detailed study and discussion, and from whose bibliographies may be made assigned readings. The separation of the treatment of the individual families into the second part was effected in the recognition that ordinarily the student is assigned one or a few families for study at any one time and in a sequence that may not necessarily follow that of this text. Adoption of this view made it possible to set the treatment of the second part in a smaller type face, thereby permitting the inclusion of this relatively large number of families within a single volume. A second departure incorporated in Part Two is in the form of detailed discussions of the phylogenetic and related considerations of each family, together with bibliographies to the taxonomic, morphological, and cytogenetic literature pertinent to that family.

The text is an outgrowth of the author's experience as a research taxonomist and as a teacher. The first 8 chapters deal with the theoretical and basic principles prerequisite to a comprehension of the scope of the science, and to an ability to make intelligent use of its literature. The ninth chapter provides an explanation of the subject of plant nomenclature, supported by an annotated abridgement of the rules by which it operates. This annotated abridgement of the rules has been found to be a necessary adjunct to laboratory exercises concerned with the solution of nomenclatural problems encountered in most taxonomic work. Chapters X through XIII deal with the practical considerations attendant on field or herbarium studies, and serve as sources of basic information concerning procedures, techniques, and general curatorial problems. The last chapter of Part One comprises a considered and classified compilation of the literature of systematic botany, selected to indicate the scope of the literature and, by means of annotations, the particular use and reference value of the more significant items that it includes. The bibliographies terminating each chapter of Part One, and those at the close of the treatments of each family accounted for in Part Two, make no attempt to represent completeness and contain, for the most part, those references pertinent to American taxonomy and believed to be helpful to the student desiring to explore a particular phase of the subject beyond the limited scope of this text.

That the text may not be unreasonably provincial in its scope or range of usefulness, more families of plants are treated in Part Two than is usual. It is left to the instructor to select for detailed study, from this extensive enumeration, those families most pertinent to his particular needs. This treatment increases appreciably the utility of the text as a reference work and serves also to remind the student that there are many more families of vascular plants than the half hundred or so that may be studied in detail in the average taxonomy course. It may be of interest to know that 87 per cent of the families treated in Part Two have representatives indigenous to or well established without cultivation in continental United States and that 70 per cent of the total contain indigens occurring within the limits of the Gray's Manual area.

The arrangement of the families follows that of Engler and Diels. The selection of this system is not because of any belief that it is the phylogenetically correct system, but rather because it is the only available system that has been devised to account in detail for the flora of the world. Taxonomy is employed throughout the civilized world, and any system devised to meet the needs of those classifying plants must be adapted to studies of plants from the world viewpoint. Furthermore, the Engler system was selected because most current American floras and manuals are based on it, as also are all of the larger American herbaria. There are more recent systems of classifying plants that more closely approximate our current concepts of true relationships of vascular plants, but of the two currently receiving most favor one (that of Bessey) suffers currently from need of revision in the light of knowledge gained since 1915, and the other (that of Tippo) has been applied only to the larger units of classification of the world flora.

It will be noted in the second part of the text that neither floral diagrams nor formulas are provided. These devices do serve as aids in fixing family characteristics in the minds of some students and are available in existing taxonomy texts. It is believed that their use, and the degree of complexity to which they may be developed, is a matter to be left to the judgment of each instructor and not properly a component of a text of this character. The diagnostic drawings augmenting the descriptions of most families illustrate characteristic features of important members and serve as visual aids to instructor and student.

The illustrations and diagrams of Part One and of Appendix II have all been prepared especially for this work. Illustrations have been provided to show diagnostic and technical characters of each of the 264 families treated in Part Two, and are represented by 263 figures. Of this number, 179 have been made available through the generosity of the author and the publisher of Bailey's Manual, and are figures that had been prepared earlier for that work under my direction. The balance of the figures have been drawn since then for this text.

The glossary, comprising Appendix II, has been adapted from that in Bailey's *Manual of cultivated plants* (ed. 2, 1949) subject to modifications, amplifications, and a few corrections. The attempt has been made to account for every term employed in this text Terms considered essential to the working vocabulary of beginning students are preceded by an asterisk, and the majority of these terms are illustrated in the figures accompanying the glossary.

In preparing descriptions of plant families, effort has been made to insure that characteristics of comparable structures of broadly related families are given and in this way the critical comparisons of such structures between these families are made possible. Like most introductory texts, little or no new and original material is presented in either Parts One or Two, it being especially true of taxonomy that before the complexities of advanced systematics can be assimilated effectively, the basic fundamentals, concepts, and working vocabulary must be mastered.

In the preparation of the chapters of Part One, assistance has been enlisted of and freely given by a number of renowned authorities. The manuscript copy of Part One has been read by Dr. R. C. Rollins, Director of the Gray Herbarium, Harvard University and by Dr. J. M. Fogg, Jr., Vice Provost, University of Pennsylvania, each of whom made many helpful criticisms and suggestions. That of the first 5 chapters was read critically by Professor A. J. Eames of Cornell University; Chapter V on phylogeny by Dr. W. B. Turrill, Keeper of the Herbarium, Royal Botanic Gardens, Kew; Chapter VIII on biosystematics by Dr. H. G. Baker, University of Leeds, by the trio of Drs. Jens Clausen, David Keck, and William Hiesey of the Carnegie Institution of Washington, and by Professor H. H. Smith of Cornell University; proof of Chapter IX on nomenclature was read in manuscript by Dr. T. A. Sprague and in proof by Professor J. Lanjouw; and Chapter XIV on taxonomic literature by W. T. Stearn, Librarian, Lindley Library, London. Proof of this chapter was read by Dr. A. Becherer, Conservatoire de Botanique, Geneva, well known for his knowledge of European botanical literature. From each of these, many valuable suggestions have been received and I gratefully acknowledge their assistance and express my sincere thanks for their generosity and helpfulness. The treatments of many of the families of Part Two have been read critically by botanists who have monographed or made special studies of them and these include (together with the families reviewed): Abbe, E. C. (Betulaceae, Fagaceae, Leitneriaceae); Allen, C. K. (Lauraceae); Bailey, L. H. (Palmae); Ball, C. R. (Salicaceae); Benson, L. (Ranunculaceae); Camp, W. H. (Ericaceae); Clausen, R. T. (Crassulaceae); Clover, E. (Cactaceae); Constance, L. (Hydrophyllaceae, Umbelliferae); Cowan, R. S. (Rutaceae); Cronquist, A. (Compositae, Sapindaceae, Simaroubaceae); Cutler, H. C. (Ephedraceae); Dillon, G. and Schweinfurth, C. (Orchidaceae); Epling, C. (Labiatae); Fassett,

N. C. (Podostemaceae); Foster, R. C. (Iridaceae); Gleason, H. A. (Melastomaceae); Holm, R. W. (Asclepiadaceae); Johnston, I. M. (Boraginaceae); Kearney, T. H. (Malvaceae); Kobuski, C. E. (Theaceae); Laubengayer, R. A. (Polygonaceae); Leonard, E. C. (Acanthaceae); McVaugh, R. (Campanulaceae); Maguire, B. (Caryophyllaceae, Guttiferae, Hypericaceae); Manning, W. E. (Juglandaceae); Mason, H. L. (Polemoniaceae); Moldenke, H. N. (Eriocaulaceae, Verbenaceae); Moore, H. E. Jr. (Amaryllidaceae, Commelinaceae, Geraniaceae); Morton, C. V. (Gesneriaceae, Solanaceae); Moseley, M. F. Jr. (Casuarinaceae); Munz, P. A. (Hydrocaryaceae, Onagraceae); Pennell, F. (Scrophulariaceae); Perry, L. M. (Myrtaceae); Rollins, R. C. (Cruciferae); Smith, A. C. (Araliaceae, Cercidiphyllaceae, Eupteleaceae, Hippocrataceae, Illiciaceae, Magnoliaceae, Myristicaceae, Schisandraceae, Tetracentraceae, Trochodendraceae); Smith, F. N. (Santalaceae); Smith, L. B. (Bromeliaceae); Swallen, J. R. (Gramineae); Tippo, O. (Eucommiaceae, Urticales); Tryon, R. M. (Pteridophyta); Uhl, C. H. (Crassulaceae); Uhl, N. W. (Helobiae); Wheeler, L. C. (Euphorbiaceae); Woodson, R. E. Jr. (Apocynaceae, Asclepiadaceae).

To each of these authorities of special interests I acknowledge freely the very material assistance received, and while most or all of their constructive criticisms were incorporated in the respective treatments, I assume full responsibility for the presentation in this text. In addition to these, I wish to acknowledge the cooperation of the botanical illustrator, Miss Marion E. Ruff, whose meticulous care and skill are reflected in the drawings prepared by her. Material assistance and encouragement have been received from my associates at the Bailey Hortorium, notably W. J. Dress, H. E. Moore Jr., E. Z. Bailey, and L. H. Bailey, to all of whom I express my sincere thanks and appreciation for their friendly cooperation. Numerous other botanists have been consulted on many aspects of material incorporated in this text and have been most generous in their response. To each of them I express my gratitude for their helpful opinions and counsel. Revised drafts of the manuscript were typed by my wife, Miriam B. Lawrence, without whose patience and understanding this book would not have been completed.

G. H. M. L.

*Ithaca, N. Y.*
*July, 1951*

# CONTENTS

## PART I
## Principles and Practices of Plant Taxonomy

### CHAPTER I
Introduction . . . . . . . . . . . . . . . 3

### CHAPTER II
Taxonomy and its Significance . . . . . . . . . 6
Interrelationships with allied sciences; problems in taxonomy; opportunities.

### CHAPTER III
History of Classification . . . . . . . . . . . 13
Period I. classifications based on habit; Period II. artificial systems based on numerical classifications; Period III. systems based on form relationships; Period IV. systems based on phylogeny; other contemporary systems.

### CHAPTER IV
Principles of Taxonomy . . . . . . . . . . . 42
Major categories of classification; minor categories of classification; infraspecific categories; morphological criteria.

### CHAPTER V
Phylogenetic Considerations . . . . . . . . . 92
Significance to taxonomy; diversity of phyletic concepts; contributions to phylogenetic knowledge; phylogeny and the higher categories.

### CHAPTER VI
Current Systems of Classification . . . . . . . . 114
Bentham and Hooker; Engler and Prantl; Rendle; Wettstein; Pulle; Skottsberg; Bessey; Hallier; Hutchinson.

## CHAPTER VII
The Geography of Vascular Plants . . . . . . . . . 141

Genesis and evolution; geography and its relationship to plant distribution; vegetation areas of the earth; physiographic areas of North America; taxonomic significance of physiographic areas; phytogeography—static vs. dynamic; principles of dynamic geography; importance of phytogeography to taxonomy.

## CHAPTER VIII
Biosystematics and Cytogenetics . . . . . . . . . 169

Biosystematics and modern taxonomy; mechanics of evolution; biosystematic categories; methods in experimental taxonomy; importance to taxonomic research and interpretation; apomixis; limitations of cytogenetic criteria.

## CHAPTER IX
Plant Nomenclature . . . . . . . . . . . . 192

Beginnings of organized nomenclature; codes of nomenclature—Paris, Rochester, Vienna, American; International Rules of Botanical Nomenclature; primary and other provisions of the Rules; units of classification.

## CHAPTER X
Plant Identification . . . . . . . . . . . . 223

Taxonomic literature; herbaria.

## CHAPTER XI
Field and Herbarium Techniques . . . . . . . . . 234

Collecting procedures; preparation of specimens; preservation of specimens; housing of bulky materials; herbarium cases; type specimens.

## CHAPTER XII
Monographs and Revisions . . . . . . . . . . 263

Procedures.

## CHAPTER XIII
Floristics . . . . . . . . . . . . . . . 275

Types of floristic study; procedures.

## CHAPTER XIV

Literature of Taxonomic Botany . . . . . . . . . . 284

General taxonomic indexes; world floras; regional floras and manuals; monographs and revisions; bibliographies, catalogues, and review serials; periodicals; glossaries and dictionaries; cultivated and economic plants; maps and cartography; miscellaneous reference works.

## PART II

### Selected Families of Vascular Plants

Introduction . . . . . . . . . . . . . . . 333

Pteridophyta . . . . . . . . . . . . . . . 334

Embryophyta Siphonogama . . . . . . . . . . 354

    Gymnospermae . . . . . . . . . . . 355

    Angiospermae . . . . . . . . . . . . 370

        Monocotyledoneae . . . . . . . . . 371

        Dicotyledoneae . . . . . . . . . . . 438

## APPENDIX I

Suggested syllabus for elementary course . . . . . . . 733

## APPENDIX II

Illustrated glossary of botanical terms . . . . . . . . 737

Index . . . . . . . . . . . . . . . . . 777

# PART I

# PRINCIPLES
# AND PRACTICES OF
# PLANT TAXONOMY

# CHAPTER I

# INTRODUCTION

Taxonomy is a science that includes identification, nomenclature, and classification of objects, and is usually restricted to objects of biological origin; when limited to plants, it is often referred to as systematic botany. In this text the taxonomy of vascular plants includes the systematics of the taxa [1] known as pteridophytes, gymnosperms, and angiosperms. Taxonomy, as a biological science, has been defined variously by leading authorities, and the above statement is not to be construed as a formal definition but rather as an indication of the scope of the subject. Some earlier botanists restricted the term taxonomy to account for the principles underlying a system of classification. In contrast, systematics was treated by them as pertaining to the classification of plants within a particular nomenclatural system. There may be valid cause for recognizing these distinctions in concept, but if they were to be accepted there would remain the need for a single collective term to account for the several functions represented by identification, nomenclature, and classification. A plant taxonomist is recognized universally as one who identifies, names, and classifies plants; taxonomy as a workable term must likewise embrace these same functions.

The terms taxonomy and systematic botany are used interchangeably throughout this text. Plant taxonomy as a science is treated in its orthodox sense, that is, a science based fundamentally on morphology with the support of all interrelated sciences. This text treats descriptive taxonomy largely as of historical interest and, while recognizing the biological objectivity of the newer systematics and the potentials of their goals, accepts their tenets only in so far as they are compatible with the objectives outlined below.

Identification is the determination of a taxon as being identical with or similar to another and already known element; the determination may or may not be arrived at by the aid of literature or by comparison with

[1] Explanation and discussion of the terms taxon and taxa are provided on p. 53.

plants of known identity. In some instances, the plant may be determined to be new to science, a situation that results after elimination of the possibilities of its being like any known element. No names need be involved in the process of identifying a plant. For example, suppose that three related plants are at hand and are determined to represent three separate species; then suppose that a fourth plant of the same relationships comes to hand and by examination is found to be similar to the second of the first three; the recognition of this fourth plant as similar to another is its identification—the new plant has been identified as being like another known plant.

Nomenclature is concerned with the determination of the correct name of a known plant according to a nomenclatural system. Once the plant has been identified it becomes necessary that it have a scientific name that, in effect, it may have a handle by which to be designated. The naming of plants is a subject of international importance. It is a function of taxonomy that is regulated by what are known as the International Rules of Botanical Nomenclature. These rules direct the procedures to be followed for the determination of the name to be applied to a particular plant, or to be followed in situations requiring the selection of a name for a new plant.

Classification is the placing of a plant (or group of plants) in groups or categories according to a particular plan or sequence and in conformity with a nomenclatural system. Every species is classified as a member of a particular genus, every genus belongs to a particular family, the family to an order, the order to a class, and so on. In actual practice, classification deals more with the placing of a plant group in its proper place within the selected schema than the placing of an individual plant in one of several minor categories. The theory of evolution postulates that plants now living are descendants of ancestral types; this means that there are genetical relationships of all degrees of proximity among living plants just as there are similar relationships among vertebrate and other animals. The principles of modern plant classification recognize that these relationships exist, and these principles have been formulated in an attempt to bring related plant groups together in so far as knowledge of them permits. This knowledge, assembled for the purpose of arriving at a satisfactory classification, must be drawn from all available sources—many of which are beyond the limits of taxonomy. Among these sources are those phases of science that deal with plant morphology, anatomy, cytology, genetics, paleobotany, phytogeography, ecology, geomorphology, and biochemistry (serology). If a plant is a variant of a species, it is not difficult to determine the species with which it is affiliated. However, the higher the category, the greater is the difficulty of determining the relationship of its components to other taxa of the same level, and the problems of classification become increasingly complex. In ordinary practice, categories above that of the family are seldom considered in taxonomic studies of plant relationships (classification). The classification of the higher as well as the lower categories is a subject somewhat apart from taxonomy in the descriptive or classical sense, but by modern concepts is closely integrated with it. The classification of taxa is called phylogeny. The main interests of phylogenetic studies are the composition, disposition, and classification of taxa especially in the categories of family, order, class, and larger groups. The results of such studies are of vital importance to the taxonomist. The significance of phylogeny is receiving increasing attention by most taxonomists, and greater recognition is being given to it especially at the lower levels of classification.

# CHAPTER II

# TAXONOMY AND ITS SIGNIFICANCE

Taxonomic study has among its objectives the learning of the kinds of plants on the earth and their names, of their distinctions and their affinities, their distributions and habitat characteristics, and the correlation of these facets of knowledge with pertinent scientific data contributed by research activities of related fields of botanical endeavor. In the beginnings of the science of taxonomy, small fragments of plants were collected, and they, together with the scant notations on their labels, provided the basis of cursory herbarium studies and of initial records of the floras of large areas. It is now recognized that competent taxonomic work is the product of a knowledge of the plant as it grows naturally, of study of an adequate series of specimens representing it and its immediate allies, and of a synthesis of these data with other available pertinent data collated from related fields. The information accumulated from these studies is fundamental to the scientific knowledge of the inventory of the earth's plant resources.

A secondary objective of taxonomy is the assemblage of knowledge gained. This is usually in the form of treatises useful to fellow scientists and to civilization in general. The mere acquisition of the knowledge, the mere possession of the inventories, is sterile unless it is made available to others; only then is it of use and an aid to the progress of civilization. This conversion of taxonomic data from a status of sterility to fertility is accomplished in many ways. Floras are published to account for the plants of a given area; manuals are prepared that the plants of an area may be the more readily identified and named; revisions and monographs are published that one may know the extent and delimitations of a particular group and its components; distributional studies are published that others may know of range extensions, corrections, and interrelationships of the taxa within an area. In addition to these, there are built up great collections of pressed specimens of the plants that serve as the basis of the scientific studies and publications, collections that become the cornerstones of the work published, for they are the proof and the evidence of the identity of the material concerned. All these products of taxonomic research add to the resources available to the scientist. They are essential to any study of the natural resources of an area, to studies of land potentials, to evaluations of resources of raw materials possibly suited to man's needs in a multiplicity of activities (as, for example, forest products, medicines, food, ornamentals, agricultural crops, and industry). As long as world populations increase and areas of low population density exist, man will demand an increasing quantity of biological data concerning those areas of low population density—data that will serve as factors influencing human migrations. For purposes of comparison, he will demand also similar current information on the more densely populated areas.

A third, and scientifically basic, objective is the demonstration of the tremendous diversity of the plant world and its relation to man's understanding of evolution. An organized reconstruction of the plant kingdom as a whole can be made only after the inventory of its components (the plants) has been assembled. When this has been done, the charting of the degree and character of variation will demonstrate its diversity, and these data can then be integrated with other facets of evolutionary knowledge to produce a more accurate phylogenetic schema.

## Interrelationships with allied sciences

Taxonomy is dependent on many other sciences and they in turn are equally dependent on it. A taxonomist must have a knowledge of morphology; he must know not only the gross morphology of the plants with which he works, but if he is to comprehend the relationships of these plants he must often be conversant with studies of their embryology, floral anatomy, ontogenetical development, and teratological variations. Modern systematists place considerable value on the importance of cytogenetic findings as criteria in delimiting the species and its elements; data of this character have proved to be of inestimable value in demonstrating the presence and taxonomic significance of exceptional chromosomal situations and of breeding behavior over successive generations. In this connection attention has been focused on the significance of the correlation of these studies with the responses of genetically uniform plants when grown simultaneously in an assortment of environments. By means of these correlations, the less significant environmental characters may be segregated from the more fundamental characters of genetic origin. This segregation has resulted in a shift of emphasis from the viewpoint that the species is of morphological distinction only to that of its being a biological unit of combined morphological and genetical distinctions. In addition to an appreciation and understanding of the contributory value of morphological, anatomical, and cytogenetical findings, modern taxonomic studies reflect the significance of distributional patterns and of more detailed data concerning the extent of normal variation and its causes. All these wider viewpoints demonstrate the increasing dependence of taxonomy on the findings of related sciences; the product of modern taxonomic research is rapidly becoming one of synthesis rather than of individual conclusions.

**Problems in taxonomy**

Taxonomy is one of the older botanical sciences. Its development followed closely that of the exploration of the earth's surface. Prior to the explorations into the New World, man's knowledge of plants was restricted for the most part to the plants of the greater Mediterranean area and extensions from it. Communications with new lands accompanied exploration and colonization. Collections of natural resources were sent back to homelands. As the number of new plants received from these outlying areas multiplied and the information about them increased, the older systems for their classification and naming became inadequate. These were replaced by successively more adequate and at the same time more complex systems. During the eighteenth and early nineteenth centuries the studies we now call taxonomic dominated the field of botanical activity. However, it was not long before interest in the subject was surpassed by interest in the new and related fields. This was particularly true in those fields opened by Darwin's and Wallace's theories of evolution, by DeVries' theories of origin by mutation, by Mendel's laws of heredity, and by rapid developmental improvement in equipment and techniques, especially the microscope, whereby could be observed the nature of chromatin material within nuclei and more recently perhaps even the genes themselves. A half century ago many biological workers believed that most of the vascular plants of the world were then known, that the taxonomist occupied himself with efforts to differentiate between tweedledee and tweedledum, and that when not so busied was annoying others and inflating his ego by changing names, splitting species, or merely "working over old hay." Recent decades have witnessed a revival of interest in the science; a revival engendered in part by renewed explorations, by the recognition that taxonomic groups are biological entities and not merely morphological aggregates, by a re-evaluation of phylogenetic criteria in which wholly new concepts of group relationships have materialized, by extended field studies correlating morphological variations with environmental and distributional factors, and by a realization of the significance of synthesis of all these related data toward the resolution of the problems of systematics in the world's vascular flora.

The vascular flora of the earth is not at all so well known as the many savants of a half century ago believed. Vast areas of South America, Africa, the large island groups that comprise Oceania, Australasia, and much of western Asia are recognized today as botanically little known. To bring the problem closer home, and despite the existence of published floras for some of the areas, the lands of Central America and of Mexico have not been well explored for their plants; every year finds literally scores of species wholly new to science being described from collections made in these regions. Domestically, the components of the vascular flora of the southeastern region of the United States are not adequately known, those of Alaska are only beginning to become known, and much territory in western Canada remains to be surveyed for its plants. In all these areas the problems of floristics are many and challenging.

The taxonomist is interested in the problems associated with the distributions of plants, for distributional data may be related closely to the migrations of plants. Knowledge of plant distribution is pertinent to the determination of geographic areas of origins of species, of genera, and often of families—all factors that are important in determining matters of genetical relationships. These studies in distribution and geography bring taxonomy into the field of phytogeography, the inquiry into why a group occupies the area that it does, how long it has been there, how rapidly it is migrating, and what evolutionary trends it is showing. Studies with this wider viewpoint represent a synthesis of ecologic, genetic, and taxonomic aspects leading to a better understanding of a series of common problems.

The subject of speciation has always presented a problem to the taxonomist. The literature that has accumulated in the attempt to answer the question of "what is a species?" is voluminous, and by no means are the answers in accord. Long has it been said that species are judgments, which is another way of saying that there can be as many definitions and concepts of a species as there are taxonomists. The newer systematics, based on the premise that species are not judgments, but rather are biological units that have evolved from a series of often identifiable ancestors, has postulated certain hypotheses by which the smaller taxa of plants (genera, species, and their subdivisions) may be circumscribed by a combination of genetical, ecological, and morphological criteria. A species is considered, by supporters of these views, to be an objective definitive unit. However, analysis of their work fails to reveal any application of complete objectivity, and the taxa they would label as species do represent judgments to a degree. Nonetheless, their efforts toward objectivity have produced noteworthy milestones in taxonomic progress. The application of their criteria has placed taxa of more or less equivalent biological significance in their respective categories. These categories take on a more definite meaning, they have a biological significance, and plants assigned to them can be expected to follow prescribed patterns of behavior; large and unwieldy groups can be reorganized into biologically related taxa, and order can be expected slowly but definitely to emerge from confusion. The application of these principles of biosystematics requires infinite time, care, and patience, and may represent the long-term and ultimate solution to many taxonomic problems concerning genera known to be genetically less stable than others. In the interim until such data are available, the older and more orthodox principles of taxonomy will remain in effect; they must remain in effect because there is as yet no substitute for them. The scores of thousands of plants recognized today as species have been established by the application of existing and established principles of taxonomy. These principles are employed by the biosystematists in selecting the taxa with which they work and by which they measure—to a degree—the progress of their research. The use of orthodox and conventional principles is the only means whereby immediately useful and applicable results of taxonomic research can come to fruition. The problems awaiting the test of the principles of biosystematics are many, some are more acute than others, and in them the opportunities for basic taxonomic research are unlimited.

### Opportunities

The resolution of the accumulated knowledge of the earth's flora is far from perfect, and will become more perfect only by, first, exploration of its areas, collection of its components, their study and classification, and second, by the publishing of competently prepared floras, manuals, revisions, and monographs of elements of its flora. Large areas of our continents have no published accounting of their floras, others have none less than 50 to 100 years old, and the probability is that as many more plants have been discovered for these latter areas since the time of their publication as were accounted in them. In focusing attention on taxonomic problems that are close to home, it should be noted there is no flora of Mexico, nor of Canada, nor even of the United States. Domestically, we do have floras accounting for the flowering plants of many of our states; many of these are old, others were based on herbarium collections and unsupported by careful field observations, some states have no flora published within the last century, and a few have no published flora at all. The problems in these fields of taxonomic activity are many and acute, and they are within the reach of every student of taxonomy. In this connection it should be noted also that some of our currently best local or county floras have been the result of assiduous field studies and collections by amateur botanists. Similar problems await resolution by the trained taxonomist who is most interested in monographic work. There is a great need for competently prepared revisions and monographs of a large number of genera and families.

In addition to these problems open to and currently requiring the abilities of the trained and amateur taxonomist who is interested in the indigenous flora, there are equally important and abundant problems open to the trained taxonomist who will work with cultivated plants. Too often the botanist will avoid the exotic or lowly cultivated plant, deprecating it as "rubbish" or not deigning to attempt to identify and name it because, not knowing where it was native, he does not know what flora or manual would account for it; another would consider it to be a monstrosity, a hybrid, or a clonal selection and hence beyond the pale with respect to his abilities. Despite and because of these ill-conceived defenses by many taxonomists the challenge to identify cultivated plants is paramount. Here is a fertile and too-long neglected field for taxonomists to enter. Few taxonomists have any degree of intimate acquaintance with the world's flora, but rather are inclined to become specialists in the flora of a particular region or for a particular group; it is an easier and surer path. The cultivated flora has no geographic boundaries; it has been brought in by man from all of the continents. Surprisingly enough, of the scientifically named species current in the domestic trade, the great majority of them can be reasonably well aligned with their indigenous counterparts. To be sure, vernacularly named variants representing all manner of genetic potpourri and devoid of all vestiges of apparent relationship to indigenous ancestors, are found in some groups (such as chrysanthemums, wheat, corn, apples) but their generic affinities are readily recognized and their speciation—if any—is of minor concern. The solution to the problem of the systematics of cultivated plants would appear to lie with the production of more and better world monographs and revisions of families and genera, and by increased attention on the part of trained taxonomists to the needs of the economic botanist and horticulturist.

# CHAPTER III

# HISTORY OF CLASSIFICATION

The contemporary status of any facet of science is better comprehended and oriented by the student who has a knowledge of its beginnings and development. The history of plant classification is a fascinating subject because one learns not only of the men responsible for it and of their contributions, but also that from all these contributions there is being evolved a classification of plants based on biological facts. From this historical survey one should expect to learn not only of the progress of the science but also of the persons who effected that progress. By acquainting oneself with these men as persons one finds that their scientific contributions become of more lasting interest.

Many different classifications of plants have been proposed. They are recognizable as being or approaching one of three types: artificial, natural, and phylogenetic. An artificial system classifies organisms for convenience, primarily as an aid to identification, and usually by means of one or a few characters. A natural system reflects the situation as it is believed to exist in nature and utilizes all information available at the time. A phylogenetic system classifies organisms according to their evolutionary sequence, it reflects genetic relationships, and it enables one to determine at a glance the ancestors or derivatives (when present) of any taxon. The present state of man's knowledge of nature is too scant to enable one to construct a phylogenetic classification, and the so-called phylogenetic systems represent approaches toward an objective and in reality are mixed and are formed by the combination of natural and phylogenetic evidence.

Artificial systems are not commonly in current use. The classical example is the so-called sexual system of Linnaeus, by which plants are classified by the number and arrangement of sexual parts. A similar usage of artificial is found in the well-known artificial keys where taxa are segregated by pairs of single or few characters and the sequence of the taxa in the key is usually wholly unrelated to their natural or phyletic groupings.

Natural systems of classifications (or those alleged to be natural) have existed since before the time of Linnaeus, and the philosophy on which they were based remained dominant until displaced by those based on modern concepts of evolution founded on the works of Darwin and his successors. The philosophy of pre-Darwinian naturalists was essentially that a natural system and a Divine plan were the same, for, accepting the dogma of Special Creation, they believed that a Divine power created all nature according to a perfect plan which man could discover to a degree. To them, a natural system was one in accord with nature, and was one which represented life as it occurred in nature.

Organized systems of classification fall readily into one of four types and periods. The earliest systems were based on the habit of the plant; these were replaced by a widely adopted system based on numerical and sexual parts of the plant, a system abandoned in favor of others in which relationships of form provided the central focal point. The more recent and contemporary systems are those based on presumed phylogenetic relationships. Only those of the last period pretend to be phylogenetic systems or to be systems based on the synthesis of all discoverable evidence.

### Period I. Classifications based on habit

The systems of classifications propounded by the Greeks, and by herbalists and botanists for a period extending over 10 centuries, were based primarily on the habit of plants. Trees, herbs, vines, and so on constituted major groups of plants. Systems of this character were dominant during a period from about 300 B.C. to the middle of the eighteenth century, a period during which philosophers, herbalists, and botanists devised numerous, often crude, systems of classification considered by them to represent natural affinities.

*THEOPHRASTUS* (370–285 B.C.), a student of Aristotle and known as the Father of Botany, reflected the philosophy of his teacher and of Plato, Aristotle's teacher, when he classified all plants on the basis of form or texture: trees, shrubs, undershrubs, and herbs, and distinguished between annual, biennial, and perennial duration. Theophrastus also differentiated between centripetal (indeterminate) and centrifugal (determinate) inflorescences, recognized differences in ovary position, and in polypetalous and gamopetalous corollas. Although he brought plants together by these groupings, making possible more generalized and intelligent discussion, he recognized only vaguely relationships among them, and the groups he established were strictly artificial. In his *Historia plantarum* he roughly classified and described about 480 kinds of plants.

*ALBERTUS MAGNUS* (1193–1280), Bishop of Ratisbon and sometimes known as Albert of Bollstädt, was a scholar who is believed to have been the first to recognize, on the basis of stem structure, the differences between monocotyledonous and dicotyledonous plants—an unusual observation in view of the crude lenses then available. In other major respects he accepted the classification of Theophrastus.

*OTTO BRUNFELS* (1464–1534) was one of the first of the group of renowned herbalists who described and to some degree illustrated the plants of the world then known. Particularly were they interested in the purported medical values and domestic uses of plants, assumed attributes that were greatly confused by superstitions, misconceptions, and sorcerous folklore. Brunfels, a German, was first a Carthusian monk, then a schoolmaster, a Protestant theologian, and ultimately a physician. He produced one of the first illustrated herbals (comprising largely material from the works of Theophrastus, Dioscorides, and Pliny) and is to be remembered particularly as the first person to recognize the *Perfecti* and *Imperfecti*, groups of plants based on the presence and absence of flowers, respectively, and as observable by the naked eye when the plant or flowering branch was held at arm's length.

The herbalists as a group are important for their contributions to the descriptive phases of systematics. Many genera commemorate their names, as for example, *Brunfelsia* for Brunfels; *Fuchsia*, for the Bavarian physician Leonard Fuchs (1501–1566); *Lobelia* for the Dutch herbalist Mathias de L'Obel (1538–1616); *Gerardia* for the English barber, surgeon, and botanist John Gerard (1545–1612); and *Clusia* for the Flemish botanist Charles L'Ecluse (1526–1609).

*JEROME BOCK* (1498–1554), who wrote also under the Latinization of his name Hieronymus Tragus, was also an herbalist. He was in successive periods a schoolteacher, Lutheran minister, and a physician who indulged in botanical studies as a hobby. In his herbal he classified plants as herbs, shrubs, and trees, but endeavored to bring together related plants within these categories, for in the preface of his herbal he stated: [1] "I have placed together, yet kept distinct, all plants which are related and connected, or otherwise resemble one another and are compared, and have given up the former old rule or arrangement . . . For

---

[1] English translation from the German, in Arber, A., *Herbals*, p. 166.

the arrangement of plants by [these older methods] occasions much disparity and error." In addition to originality in arrangement, Bock, unlike some of his plagiarizing predecessors and contemporaries, based his plant descriptions largely on personal and critical observation. In addition he provided notes on the natural distribution of many plants, and denied the validity of much of the folklore and superstition associated with presumed virtues of plants.

*ANDREA CESALPINO* (1519–1603) was an Italian botanist and physician who has been referred to as the first plant taxonomist. In his one important botanical work, *De plantis* (1583), he included an introductory book or chapter that provided the basis for his classification of about 1500 plants. Cesalpino was an Aristotelian scientist in that his conclusions were based on reasoning rather than an analysis by observation. Teleology was accepted by him and treated as of major importance. He believed leaves to have been provided for the protection of buds, flowers, or fruit, denied the existence of sex in flowers,[2] treated the pith of dicots as a homologue of the spinal column of vertebrate animals, and contended that plants had a nutritive soul. Taxonomically he classified plants first on the basis of habit, dividing all into trees and herbs and subdividing them on the type of fruit and seed produced. Second, he recognized and used as classifying characters, those of the ovary being superior or inferior, the presence or absence of bulbs (fleshy rootstocks) or of milky vs. watery sap, and the number of locules in an ovary. Cesalpino wrote his opinions in narrative form and they were not arranged in any outline or synopsis. This may account in part for their not having been adopted by his contemporaries or immediate successors. On the other hand, the views of Cesalpino did influence the thinking of such later men as Tournefort, Ray, and Linnaeus.

*JEAN [JOHANN] BAUHIN* (1541–1631), a French and Swiss physician, is important for his excellently illustrated *Historia plantarum universalis* published posthumously (1650) in three volumes by his son-in-law J. H. Cherler. This was a comprehensive work and dealt with the synonymy of about 5000 plants. In addition, it included descriptions, which for the first time in botanical history were good diagnoses of the species he treated. Earlier his brother Gaspard [Casper] (1560–1624) published a *Pinax* (1623), a work containing names and synonyms of about 6000 species and one that remained dominant for over a century.

---

[2] It has been alleged that the existence of sexuality in plants was denied by the Church, and that the Vatican disciplined communicants who would admit it, but no evidence has been found in support of this.

Like his near-contemporaries, the herbalists, Gaspard Bauhin classified plants on the basis of texture and form. He is to be remembered also as one of the first to distinguish nomenclaturally between species and genera. To many of the plants classified and described by him, he gave a generic and specific (trivial) epithet. This binary nomenclature, with which Linnaeus is usually credited, was founded, therefore, by Bauhin more than a century before its use by Linnaeus in his renowned *Species plantarum*.

*JOSEPH PITTON de TOURNEFORT* (1656–1708) continued the long-standing form classification of plants and devised his own modifications of earlier systems. He divided flowering plants into 2 large categories, trees and herbs. Each of these was subdivided into groups based on characters such as flowers petal-bearing or nonpetal-bearing, flowers simple or compound (groups akin to the polypetalous and gamopetalous subdivisions of later authors), and flowers irregular or regular. His system was adopted widely throughout Europe. It prevailed in France until superseded by that of de Jussieu (about 1780) and in the rest of western Europe until displaced by that of Linnaeus (about 1760). This system exerted great influence but was considerably inferior to that published a decade later by John Ray.

In addition to developing a system of plant classification, Tournefort is known as the father of the modern genus concept. Many of his generic names were validated by later botanists and remain well known to the present. They include (to mention only a few) *Salix, Populus, Fagus, Betula, Lathyrus, Acer,* and *Verbena*. Although not the first to use generic names (cf. under Bauhin), Tournefort was the first to set the genera apart, treating the genus as the smallest practical unit of classification, and considering the species as variants of the genus.

*JOHN RAY* (1628–1705) was an English philosopher, theologian, and naturalist. He proposed a system of classification, long before that of Linnaeus, which took the best from such predecessors as Albertus, Cesalpino, Malpighi, and Grew and grouped plants together on the basis of relationship of form. As presented in its last revision (*Methodus plantarum,* 1703) Ray proposed a classification accounting for nearly 18,000 species, and under the main divisions of woody and herbaceous plants, he recognized the taxa of monocots and of dicots, the classes based on fruit type (cone-bearing, nut-bearing, bacciferous, pomiferous, pruniferous, and siliquous), and subdivided these on the basis of leaf and flower characters. It was a system based on form and gross morphology of plant structures, and in many respects was superior to the artificial Linnaean system that came later. Ray's classification was the direct antecedent of the system later devised by Bernard de Jussieu. It was the basis of the philosophical approach used by Linnaeus in his *Critica botanica* and his *Philosophia botanica* and (as pointed out by Svenson, 1945) "the technical approach of *Species plantarum,* the *methodus,* is modeled also on the work of Ray."

**Period II. Artificial systems based on numerical classifications**

This period was characterized by systems of classification that were designed deliberately to be artificial and to serve solely as aids to identification. In this period of plant classification, the botanists did not follow the classification of form that had been prevalent since the era of Aristotle, yet in arriving at their conclusions of relationships they were influenced by the type of reasoning so forcefully put forward centuries earlier by Aristotle.

*CAROLUS LINNAEUS* (1707–1778), or Carl Linné (Fig. 1) as he was known until he reached middle age, is regarded as the father of taxonomic botany and zoology, and is considered by many to have been the most prodigious systematist of all time. Linnaeus, his works, and extant collections have been and will continue to be the hub from which all serious taxonomic research at the lowest (species) level must emanate. Because of past and present great importance of this titan to systematic botany, every student should become reasonably conversant with his life, his works, and the important works of his students.

Linnaeus was born May 23, 1707, in Råshult, southern Sweden. Early in life Carl developed a germ of interest in plants. At the age of twenty he entered the University of Lund, but transferred to the University of Uppsala a year later. While a student, and under the guidance of Dr. Rudbeck (for whom he later named the genus *Rudbeckia*), he published (1729) his first paper on the sexuality of plants. The favorable publicity accorded this paper was responsible to a large degree for his appointment as a demonstrator in botany at Uppsala. The following year, promoted to the rank of Docent, he published (under the Latinized title of *Hortus uplandicus*) an enumeration of the plants in the Uppsala Botanical Garden. In this listing the plants were arranged according to the system of Tournefort. However, the number of plants in the garden soon exceeded in kinds the number that could be classified readily by Tournefort's system, causing Linnaeus to publish a new edition of his *Hortus uplandicus*. In this new edition the plants were classified according to a system of his own devising—his so-called sexual system. For the next 7 years he studied primarily those plants not available to him at Uppsala.

The period of 1737 through 1739 was exceedingly important in the life of Linnaeus. In the spring of 1737 he went to Germany and in June to Holland where, after a very short term of residence, he received his M.D. degree from the University of Haderwijk. The following month he visited Dr. Gronovius in Leyden, a physician and naturalist of much

**Fig. 1.** CAROLUS LINNAEUS, 1707–1778. Sketch by M. E. Ruff, from photograph, courtesy of the New York Botanical Garden.

means and affluence. Gronovius was so favorably impressed by Linnaeus that he offered to publish immediately Linnaeus' manuscript of his later famous *Systema naturae,* a work providing the foundation for the classification of all plants, animals, and minerals. On his return journey to Sweden in August, he stopped in Amsterdam to meet a Dr. Burman, a physician, who persuaded Linnaeus to remain with him long enough to work over and identify a collection of plants received from Ceylon. While here Linnaeus met Mr. George Clifford, Director of the Dutch East India Company and one of the very wealthy men of Europe. Clifford, a hypochondriac, was a patient of Dr. Burman, and on Burman's advice hired Linnaeus to serve as his personal physician and also to identify all the plants growing on his vast estates at Hartecamp, Netherlands, not far from Leyden. For Linnaeus, the living conditions at Clifford's were the best of his life and not only was he, for the first time in his life, completely free from all financial worries, but was able also to have published, at Clifford's expense, several important manuscripts. It was by these fortuitous circumstances that late in 1737 Linnaeus' now well-known *Genera plantarum* and *Flora lapponica* were published. While these works were in press, Clifford sent Linnaeus to visit the botanists and botanic gardens of England. On his return he spent 9 months preparing the manuscript for a sumptuous work, *Hortus Cliffortianus,* in which were named and described many temperate and tropical plants grown by Clifford. This work is important to present-day systematists because in it were illustrated and amply described many plants treated very briefly by Linnaeus in his later works, including also many plants received by Clifford from America. In April of 1738 Linnaeus again prepared to return to his homeland, making the trip via Paris, where he visited the de Jussieu brothers and went on a week-end collecting excursion with Bernard de Jussieu.

No other period of Linnaeus' life was as productive of botanical work as the approximately three years spent in Holland. During this period he published 14 treatises, many of them prepared or written during earlier years in Sweden. Most of these were the first editions of the books that today make him famous. Only one new work, his *Species plantarum,* was published after this time. During this stay in Holland he did not learn to speak Dutch. Although he was president of a select organization of Dutch scientists, and a lecturer on many occasions, he consistently delivered his talks in Latin—the language then well understood and used professionally by all academically educated people.

On his return to Stockholm, Linnaeus established a very large medical practice, treating among others the Queen of Sweden, and was appointed physician to the Admiralty. Later Linnaeus was appointed Professor of Practical Medicine at Uppsala. This appointment gave him the academic prestige he needed, the opportunity to teach botany, his preferred subject, to conduct field trips, and to direct the administration of the botanic garden. He held this position until 1775, when he was retired at his own request on a small pension. During this extended period his classes were well attended, and it is recounted that on days of field trips the class would often number 200 or more. These field trips were well organized. According to Jackson's account (1923), Linnaeus was always at the head of the troop, followed by an "Annotator" to take dictation as directed, "another was Fiscal who had supervision of the discipline of the troop, . . . others were marksmen to shoot birds, etc." At the conclusion of the trip "they marched back to town, the Professor at their

# HISTORY OF CLASSIFICATION

head, with French horns, kettledrums and banners, to the botanic garden where repeated 'Vivat Linnaeus' closed the day's enjoyment." After the first 20 years of this appointment, Linnaeus' popularity as a teacher waned appreciably, and no longer did he continue to attract the number and quality of students. He died January 10, 1778 after an illness of two

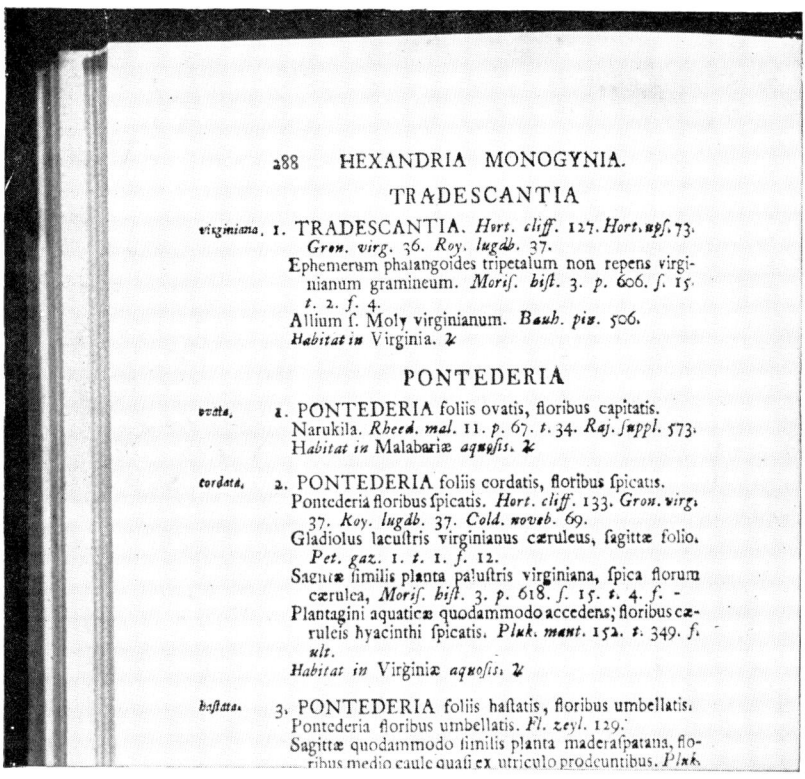

Fig. 2. Portion of page from Linnaeus' *Species Plantarum*.

years. He was buried during an impressive candlelight service 12 days later and, at his own request, his body was "unshaven, unwashed, unclad, enveloped with a sheet." Interment was in a crypt in the Cathedral at Uppsala.

Any analysis of Linnaeus' works must take into account the conditions of the times in which they were produced. They can be compared with botanical works of our time no more than one can compare modes of travel of that era with those available in this one. Linnaeus' so-called sexual system of classification was revolutionary and probably exerted more influence than any other of his contributions. The strength of the system lay in its relative simplicity, in the fact that it was a schema whereby plants could be arranged and could be found again. That a plant unknown to the average botanist could be identified and named by the system, and not merely classified by it, was an innovation of importance. As Jackson (1923) has stated,

> It was put forward at a time when such a plan was most required; when earnest searchers for two centuries had amassed so many plants of various forms ... for the disentangling of which no thoughtful and practicable plan had until then been laid down. Now when all this material threatened to overwhelm the builder, the sexual system was produced, by which plants could easily be examined and thus determined.

In brief, Linnaeus' sexual system provided 24 classes for all plants: the plants were sorted into most of these classes on the basis of number, union, and length of stamens. The classes were subdivided into orders on the basis of the number of styles in each flower.[3] The system was presented first in the second edition of his *Hortus uplandicus* (1732) and was expanded and served as the basis for his *Genera plantarum* (1737). This latter work is of importance in modern taxonomy as a source of descriptions of 935 genera. It was published in five editions (that of 1754 being

[3] An outline of the classes of the Linnaean system of classification is given as an illustration of the mechanics by which it operated, together with a few examples showing the disposition of some typical American plants.

Klass 1. *Monandria.* Stamens one.
  *Lemna, Scirpus*
Klass 2. *Diandria.* Stamens two.
  *Veronica, Salvia*
Klass 3. *Triandria.* Stamens three.
  *Iris, Sisyrinchium*
Klass 4. *Tetrandria.* Stamens four.
  *Mentha, Ulmus, Cornus*
Klass 5. *Pentandria.* Stamens five.
  *Primula, Myosotis*
Klass 6. *Hexandria.* Stamens six.
  *Rumex, Alisma, Berberis*
Klass 7. *Heptandria.* Stamens seven.
  *Aesculus*
Klass 8. *Octandria.* Stamens eight.
  *Fagopyrum*
Klass 9. *Enneandria.* Stamens nine.
  *Rheum, Ranunculus*
Klass 10. *Decandria.* Stamens ten.
  *Acer, Kalmia*
Klass 11. *Dodecandria.* Stamens 11–19.
  *Euphorbia, Calla*
Klass 12. *Icosandria.* Stamens twenty or more, epipetalous.
  *Rosa, Rubus, Spiraea*
Klass 13. *Polyandria.* Stamens twenty or more, attached to axis.
  *Tilia, Papaver, Nymphaea*
Klass 14. *Didynamia.* Stamens didynamous.
  *Monarda, Linaria, Linnaea*
Klass 15. *Tetradynamia.*
  Members of the Cruciferae
Klass 16. *Monadelphia.*
  Malvaceae, Geraniaceae
Klass 17. *Diadelphia.*
  *Lathyrus, Trifolium*
Klass 18. *Polyadelphia.*
  *Hypericum*
Klass 19. *Syngenesia.*
  *Lobelia, Viola,* Compositae
Klass 20. *Gynandria.*
  *Aristolochia,* Orchidaceae
Klass 21. *Monoecia.*
  *Typha, Quercus, Thuja*
Klass 22. *Dioecia.*
  *Salix, Urtica, Juniperus*
Klass 23. *Polygamia.*
  *Empetrum,* many Compositae
Klass 24. *Cryptogamia.*
  Algae, fungi, mosses, ferns

of nomenclatural significance today) plus two supplements called *Mantissae,* and in all a total of 1336 genera of plants were diagnosed. The *Species plantarum* published by Linnaeus in 1753, and chosen by modern botanists as a starting point of present-day nomenclature, has become the work most important in the systematics of vascular plants. Linnaeus intended it to serve as a reference work to the plants of the world, classified artificially for ease of identification according to the Linnaean (or sexual) system. In a sense, it was "a revision of Bauhin's *Pinax* in which the unit of classification was to be the species." In it (see Fig. 2) Linnaeus listed plants by his classes (see footnote 4) giving for each species (1) its generic name; (2) its trivial name (or specific epithet, which, when combined with the generic name, formed the binomial); (3) a specific phrase name (considered by Linnaeus as its specific name) in the form of a polynomial descriptive phrase, intended to serve as a "definition of the species . . . corresponding to the distinctions in dichotomous keys of our current manuals" (Svenson, p. 276), a polynomial that was not a description by Linnaeus' standards nor was it based on herbarium specimens before him; (4) abbreviated references to previous publications, specimens in his own herbarium, and citations of figures of the species; and (5) the region of nativity of the species. The 2-volume work (paged consecutively) was a "recension of previously published work (of comparatively recent authors) in which species were for the first time differentiated." It represented the first assignment of a binomial to practically every species of plant, although Linnaeus himself did not intend this to be a primary function of the work.[4] In this work varieties were first distinguished (by Greek-letter prefixes) from species. Linnaeus was not the first to employ binomials, for Bauhin (1596) did so but without basic philosophy concerning them, and Rivinus proposed (1690) that no plant name should consist of more than two words. In recognition of this and of earlier works, Linnaeus was knighted (1753) and later (1761) granted a patent of nobility, from which latter date he was known as Carl von Linné.

[4] For a thorough analysis of Linnaeus' handling of material and literature and for his use of phrase names and synonymy, see Svenson (1945), who emphasized that Linnaeus looked upon his *Species plantarum* "as a compendium without descriptions," and that "binomial nomenclature was not intended by Linnaeus to supersede the polynomial specific name." Svenson made it clear that "the term 'specific name' applies only to the phrase-name" (i.e., the polynomial following the trivial name), and that "the term 'specific epithet' is sometimes used in place of the Linnaean 'trivial name'. . . . The term 'diagnosis' . . . has been adopted by some as a substitute [term] for the Linnaean specific phrase-name, but in modern usage it may lack entirely the differential character of the Linnaean specific phrase-name." For a second review and analysis of Linnaeus' concepts, see Croizat (1945).

Linnaeus' system of classification was artificial. It is commonly titled a sexual system, but this is a misnomer; it was not fundamentally a sexual system, for the emphasis was largely on the numerical relationships of the sex organs (see footnote 3). It was a system based on differences rather than on similarities, and one so artificial that related elements often fell in widely separated classes. Linnaeus recognized this weakness, acknowledged that his system was designed as an aid to identification, and attempted to devise a system based on more natural relationships. In the sixth edition of his *Genera plantarum* (1764) he appended a list of 58 natural orders together with their included genera. Greene (1909) presented numerous examples to support his view that Linnaeus believed in a natural order of Creation, not represented by his sexual system, and had a clear concept of a natural classification. Linnaeus acknowledged (1731) that "all species number their origin first from the hand of the Omnipotent Creator . . .", but accepted the continuity of the species by sexual reproduction, for he wrote (1757) ". . . this hypothesis [transmutation of species] endured until the time of Harvey, who dared to assert that all life came from the egg, and was similar to the mother . . ."[5] As Copeland (1940) has written "it is not he [Linnaeus], but the smaller men who followed him, who may be said to have delayed the development of the natural system by an uncritical acceptance of his artificial system."

Linnaeus' importance in the history of taxonomy rests not only upon his researches and published works, but also upon the enthusiasm he instilled in his many students. Several of these became important botanists in their own right and were responsible also, by their extensive collections in foreign lands, for adding appreciably to the number and range of plants known to Linnaeus.

*PEHR [PETER] KALM* (1716–1779), a Swede, was one of Linnaeus' more important students. He had already demonstrated his ability as a collector and traveler by expeditions in Finland and Russia when the Universities of Åbo and Uppsala sent him on a trip to America. At the time of his return, Linnaeus had been confined for some time to his bed as a result of a heart attack but, on learning of Kalm's arrival with many bundles of American plants, he left his bed and transferred his concern from his heart to Kalm's collections.

*FREDRICK HASSELQUIST* (1722–1752), one of Linnaeus' favorite students, spent 2 years collecting in the Levant area and died of

[5] Quotations from translations from the Latin by Svenson 1945, pp. 280–281).

fever in Bagda (in or near Smyrna) early in 1752. It was from these collections that Linnaeus learned firsthand of many plants indigenous to Palestine, Arabia, Egypt, Syria, and Smyrna. However, the flora of Egypt, especially that of the Cairo area, became better known to Linnaeus through the collections and notes of a Finnish student, Pehr Forskål (1732–1761) who, once garbed as a peasant to escape persecution from unfriendly Bedouins, found about 100 new species and 30 new genera. Forsskål died in Arabia of malaria, and his collections and manuscripts were sent back to Copenhagen, headquarters of the Danish expedition of which he was a member.

Perhaps none of Linnaeus' students contributed more to his knowledge of plants of other parts of the world than did Carl Peter Thunberg (1743–1828), himself the author of two floras and many scientific writings. At a time when the ports of Japan were closed to all but the Dutch, Thunberg obtained a post as surgeon in the East India Company's service, an appointment that made it possible for him to become one of the first Occidentals to make anything approaching extensive collections of Japanese plants. Prior to his trip and residence at an isolated Japanese port, he spent three years collecting in the Cape of Good Hope area of South Africa and there found 300 species of plants that were new to science. In 1784 he succeeded Linnaeus' son as Professor of Botany at Uppsala. Traveling in foreign lands, especially in the tropics, was exceedingly hazardous, and diseases took their toll of Linnaeus' students, for no less than six of those whose trips he sponsored died in foreign lands. These students of Linnaeus, in addition to increasing the knowledge of plants, added to the fame and importance of their teacher by employing his systems of classification and nomenclature in their own published works.

By 1760 the Linnaean system was adopted to a large extent in Holland and Germany, and was coming into prominence in England where it was advocated by Dillenius and Sloane, leading English botanists of that time. It was never received favorably in France, and the only French student Linnaeus ever had (Henri Missa) left Sweden in a pique with his teacher and was not heard from afterwards. The French botanists held to the Tournefortian system until it was replaced by that of de Jussieu.

After Linnaeus' death his collections came into the hands of his eldest son, Carl, who also was a botanist and who was appointed to the chair at the university formerly occupied by his father. On the death of this son (1783), the collection reverted to Linnaeus' widow and daughters. Their sole interest in it was that it be disposed of to the highest cash bidder.

After lengthy negotiations the collections (including the herbarium) were sold to the wealthy and later famous English botanist, James Edward Smith, for the sum of 1000 guineas. These collections were packed in 26 large cases and shipped from Stockholm in September, 1784. The Linnaean collections (including the herbarium) were purchased from Smith's estate by the Linnean Society of London (of which Smith was a founder) for 3000 guineas and continue to be administered by that organization.[6] It has been alleged that, in an effort to prevent transfer of this nationally important collection, the King of Sweden dispatched a warship to follow it and bring it back. Despite anecdotes in the literature, and often presented as statements of fact, there is no evidence to establish any bases of fact for this myth.

The Linnaean system of classification remained dominant in many botanical centers for nearly a generation following Linnaeus' death. The last important work employing the system was the fourth edition of the *Species plantarum,* completely rewritten, and greatly enlarged under the editorship of Carl Ludwig Willdenow (1765–1812), Professor of Natural History at the University of Berlin and Director of the Berlin Botanical Gardens. This edition, of four tomes usually bound in nine volumes, was the last and is its most comprehensive. Willdenow made no significant contributions to or modification of the Linnaean system. In America the Linnaean system was employed by the early and immigrating botanists (Schweinitz, Muhlenberg, Pursh, and Thomas Walter) and was adhered to tenaciously by Amos Eaton as late as 1840, long after it had been rejected by such men as Nuttall, Torrey, and Gray.

**Period III. Systems based on form relationships.**

The second half of the eighteenth century was a period when great numbers of living plants, seeds of plants, and collections of prepared specimens were coming into Europe's botanical centers from all of the world's continents. A large proportion of these were of species new to science, each to be given a name, to be described, and placed in a classificatory system. With the increase in knowledge of the world's flora came the realization (about 1800) that there were greater natural affinities between plants than the so-called "sexual system" of Linnaeus would

---

[6] By aid of a grant of £2000 from the Carnegie Foundation, the extant botanical, zoological, entomological, and geological collections and manuscripts of Linnaeus were photographed on microfilm in 1939. One set of prints was sent to this country and deposited at the Arnold Arboretum, Jamaica Plain, Mass. Negatives were prepared from it and a second set of prints was made for general reference, with another deposited at the Chicago Museum of Natural History. These microfilms of Linnaeus' herbarium specimens are of value to students needing superficial knowledge of Linnaean types.

indicate. This realization was not derived from theorizing and logic, but was made cogent by man's increased knowledge and understanding of the organography and functioning of plants. Due acknowledgment must be made to the development of optics, to the awareness of the biological significance of sex organs in plants, and to an understanding of the rudiments of floral morphology. It was near the close of the eighteenth century and at the beginning of the next that revolutionary changes in systems of plant classification first were apparent. The new systems were called natural systems and are so in the sense that they reflected man's understanding of nature at that time. Plants were placed together because there existed a correlation of characters in common. The epoch-making work of Charles Darwin and Alfred Russel Wallace had not appeared. Modern evolutionary theories were unknown, and the concept of consanguinity of relationships was still vague (its underlying forces had been studied but not recognized by such men as Lamarck, Erasmus Darwin, Chambers, and others). The systems of this period were natural in that they served the desire of the human mind for true order; but better than Tournefort, Ray, and others, they also served the practical need of a classification by functioning adequately as an aid to identification. They are not phylogenetic systems and, in the sense that the latter are natural to the extent of allegedly indicating "blood" or genetical relationships, these earlier systems are esoteric. However, this in no way detracted from their utility, and most of the great floras of the world have been based on them.

*MICHEL ADANSON* (1727–1806) was a French botanist, a member of the faculty of l'Academie des Sciences at the Sorbonne, Paris, and an early plant explorer of the tropics. His major contributions included the rejection of all artificial classifications for the natural, and the descriptions of taxa more or less equivalent to modern orders and families (the latter a unit suggested earlier by Ray); all set forth in his 2-volume work, *Familles des plantes* (1763).[7]

*JEAN B. A. P. M. de LAMARCK* (1744–1829) was a French biologist best known to taxonomists for his *Flore françoise* (1778), written in the form of an artificial key to provide a means of identifying the plants of France. In the introduction of this work he laid down the principles of his concept of a natural classification: (1) the determina-

[7] Early botanists, following Linnaeus, designated as *Classis* (Class) the category now known as an order, and as *Ordo* (Order) that now known as a family. While there were some exceptions (by Adanson, de Candolle), this usage of class and order prevailed until the beginning of the twentieth century.

tion of which plant precedes another in a natural series; (2) the rules for the natural grouping of species; and (3) the treatment of orders and families. Lamarck is known also for his theory (known as *Lamarckism*) that changes in environment caused changes in the structure of organisms (such as new uses of parts or organs, or the loss of parts or organs through disuse), and that these changes were inherited by the offspring.

*DE JUSSIEU.* An apothecary of Lyon, France, named de Jussieu had three sons: Antoine (1686–1758), Bernard (1699–1776), and Joseph (1704–1779). All three became botanists; the two elder brothers remaining in France, the youngest spending many years in South America where he went insane after losing botanical collections made and accumulated over a period of five years. Antoine and Bernard both studied under the great French teacher-botanist Pierre Magnol (1638–1715) at the University of Montpellier. Antoine succeeded Tournefort as Director de Jardin des Plantes, Paris. Later he added his brother Bernard to the staff. In 1759 Bernard de Jussieu rearranged the plants in the garden at La Trianon, Versailles, according to a system more or less of his own devising, but with many similarities with the natural classification in Linnaeus' *Fragmenta methodi naturalis* and in Ray's *Methodus plantarum*. Here for the first time there was assembled a system that was not on the Aristotelian concept of habit that Tournefort adhered to, nor one that was artificial, as was the system of Linnaeus. De Jussieu divided the flowering plants into groups on the basis of monocot vs. dicot, ovary position, presence or absence of petals, and fusion or distinctness of petals. He never published his system and in 1763 he called in his 15-year-old nephew, Antoine Laurent de Jussieu (1748–1836) to work with him.

Ten years later Antoine Laurent de Jussieu published his first paper proposing a new classification of plants. It represented an improved version of his uncle's system and was in the form of a memoir dealing with the relationships within the Ranunculaceae. It initiated an era of natural systems, and the beginning of a period of classification that witnessed many modifications of this one system. During the following year de Jussieu published an *Exposition d'un nouvel ordre de plantes* in which he proposed to classify all plants on the basis of: acotyledonae, monocotyledoneae, and dicotyledoneae; the latter he divided into five groups based on corolla characters, calling them apetalae, petalae, monopetalae, polypetalae, and diclinae. This preliminary proposal of a new basis for classification, published in 1774, came to fruition in 1789 (the year of the French Revolution) in his *Genera plantarum*. In this work, flowering plants were divided into fifteen classes along the lines outlined above, and these classes were subdivided into 100 orders (*Ordines naturales*), each of which was clearly differentiated, named, and provided with a description. A large number of his orders (i.e., families) are to be found intact in the most modern of present-day classifications. The related taxa of Palmae, Liliaceae, Amaryllidaceae, and Iridaceae were grouped together, the Monopetalae contained those families now known as the gamopetalae, while his class Diclines irregulares (an unnatural assemblage) contained the Coniferae, Amentiferae, Urticaceae, Cucurbitaceae, and Euphorbiae. All except the first and last of his 15 classes are exclusively angiosperms, and it is significant of his keen perception that, while their sequence has been altered by reshufflings, most of these taxa have been accepted as valid by a majority of systematists for a century and half after de Jussieu. Following publication of this major work, de Jussieu was occupied primarily with the preparation of monographic studies. In addition to these activities he was the founder (in 1793) of one of Europe's great museums, the Musée d'Histoire Naturelle de Paris. In 1826 he resigned his professorship in favor of his son, Adrien de Jussieu.

*DE CANDOLLE.* Three generations of de Candolles have contributed much to the science of systematic botany. The first important botanist of this great family was Augustin Pyrame de Candolle (1778–1841). He was born in Geneva, Switzerland, and received his botanical training in Paris under Desfontaines. During his lifetime he made many botanical contributions to the fields of physiology, morphology, and taxonomy, but his taxonomic works are the more outstanding. His first important botanical work was accomplished soon after arrival in Paris; it consisted of preparing a descriptive text to accompany a series of folio color plates of succulents drawn for L'Héritier (Charles Louis L'Héritier de Brutelle) by the famous botanical artist, Redouté. While still in Paris de Candolle prepared a new and revised edition of Lamarck's *Flore françoise*. From 1808 to 1816 he was Professor of Botany at Montpellier where he published his important *Théorie élementaire . . .*, wherein he contended that anatomy and not physiology must be the sole basis of classification. The last 25 years of his life were spent in Geneva where he continued his monumental work, the *Prodromus systematis naturalis regni vegetabilis,* in which he proposed to classify and describe every species of seed plant then known to science. This was a stupendous undertaking, and was in reality a species plantarum projected on a more gigantic scale than any previous botanical work. De Candolle himself wrote and produced the first 7 volumes, with the succeeding 10 volumes written by specialists and published after his death under the editorship of his son, Alphonse de Candolle (1806–1893).

As an individual, de Candolle was very much aware of his own importance and greatness. His zeal, industry, and enthusiasm far exceeded that of any of his contemporaries. De Candolle proposed a system of classification that was followed in all his taxonomic works. It was a system that was similar in many respects to, and an elaboration of, that of de Jussieu, but retrogresses from the latter in that the ferns were treated as coordinate with the monocots. It has been considered to be superior to the de Jussieu system in that the dicots were subdivided into more basic primary groups of apparent closer affinities. For example, they were divided first into two groups on the basis of presence or absence of corolla, those with corollas subdivided on the basis of gamopetally vs. polypetally, and the latter again divided on the character of ovary position. Aside from these considerations, de Candolle accounted for 161 families of plants as compared with the 100 families known to and recognized by de Jussieu. His importance as a systematist was increased by the production of his *Prodromus*, and by his nearly 100 monographs. By the expansion of de Jussieu's system he demonstrated the inadequacy of the Linnaean classification to the extent that the latter could not meet the challenge and by 1840 was superseded.

During this first half of the nineteenth century there was considerable activity in other parts of Europe in the development of systems of classification, all representing modifications and elaborations of the de Jussieu system. In England, Robert Brown (1773–1858), a contemporary of de Candolle, devised no system of classification, but contributed nonetheless to better understandings of floral morphology and to classification problems. He demonstrated that the gymnosperms were a group apart from the angiosperms and that they were characterized by the presence of naked ovules, and so paved the way for Hofmeister's later formal designations of the two taxa. He was the first to explain the floral morphology and pollination of the Asclepiadaceae, and added much to the knowledge of the Orchidaceae. He demonstrated the nature of the cyathium in Euphorbiaceae, the morphology and probable derivation of the grass flower, and the floral morphology of the Polygalaceae. Each of these contributions produced a better understanding of the members of these families as concerns their biology and classification.

During the period of 1825 to 1845, no less than 24 systems of classification were proposed. They represented in the main only minor improvements or elaborations over the system of de Jussieu and, aside from the major contributions of de Candolle and of Brown, gave little indication of deep analysis of basic considerations. Many of these modifications represented primarily the adaptations of de Candolle's works into other languages, accompanied or not by changes in the names given to the higher taxa within the classification. Notable among these are the systems of Endlicher, Brongniart, and Lindley. Endlicher (1805-1849), a Viennese botanist, divided the plant kingdom into the thallophytes (algae, fungi, and lichens) and cormophytes (mosses, ferns, and seed plants). His is a system once widely used in Europe but not adopted by British or American botanists. It is embodied in his *Genera plantarum* (1836-1840) wherein were treated 6835 genera (6235 were of vascular plants) and was a signficant contribution of its time. Brongniart (1770-1847) was a Frenchman who, with considerable foresight, did not consider the taxon often referred to as the Apetalae to be one of closely related elements but distributed the species among those petaliferous families to which they appeared to be more closely related. He called seed plants Phanerogamae and, following Linnaeus, placed all lower groups in the Cryptogamae. In England, John Lindley (1799-1865) proposed a system based on the better features of those of his predecessors. It was an important revision, not so much for its context and organization as for the fact that it was the first comprehensive natural system to have been published in English and thus made readily available to large numbers of English-speaking peoples. The Lindley system was the first to be accepted widely in Great Britain and America as a successor to that of Linnaeus.

*BENTHAM AND HOOKER.* The system presented by these two botanists was published as a 3-volume work in Latin, titled *Genera plantarum,* and was a work that climaxed the period under discussion. George Bentham (1800-1884), an Englishman of moderate independent income, was an amateur but well-trained botanist until almost middle age, after which time he gave the subject of systematic botany all his attention. In addition to being a most critical, discriminating, and analytical systematist he also was an accomplished linguist and Latinist. Prior to his joint publication with Hooker of the *Genera plantarum,* Bentham published world monographs of the families Labiatae, Ericaceae, Polemoniaceae, Scrophulariaceae, and Polygonaceae. In addition to these he was the author of a 7-volume flora of Australia. Sir Joseph Dalton Hooker (1817-1911). son of the botanist Sir William J. Hooker, was more the plant explorer and plant geographer than was Bentham. In addition to botanical research, a considerable part of Hooker's time was devoted to administrative duties associated with the directorship of the Royal Botanical Gardens at Kew. About 1857 Bentham and Hooker agreed to produce jointly a *Genera plantarum* (1862–1883). This work comprised the names and descriptions of all genera of seed plants then known, classified according to the system proposed by these two botanists. About two-thirds of the contents of the work was written by Bentham, a task requiring a period of nearly 25 years of concentrated effort.

The Bentham and Hooker system was patterned directly on that of de Candolle (a close friend and associate of Bentham), but its text differed from that of the de Candolle system and from any work published by de Candolle, in that every genus was studied anew from material in British and Continental herbaria. Full and complete descriptions were prepared from studies and dissections of the plants themselves and did not represent a compilation made from existing literature. Almost every large genus was subdivided into subgenera and/or sections, each of which was named, diagnosed, and delineated, together with the assignment of important species belonging to them. In this work, the category intermediate between that of class and of order (i.e., the family of contemporary systems) was termed a Cohort, a classificatory unit comparable to the order of more modern systems, and represented a unit of closely related families (the name Cohors was first used by Endlicher for a taxon of higher level). The *Genera plantarum* was accepted from the first throughout the British Empire and in the United States and was adopted to a lesser extent by some Continental botanists. In this country the classification remained dominant until shortly before the turn of the present century and is currently retained by many British botanists and in British herbaria.

There was a marked constancy in these systems, from that of de Jussieu to that of Bentham and Hooker. All were predicated on the dogma of the constancy and immutabiilty of species. True, the later systems were superior to that of de Jussieu, but even the detailed and meticulous work of Bentham and Hooker represented improvements only in degree. Plants were treated by the authors of all these systems as inanimate objects. By way of analogy, it could be said that as the Linnaean system might have grouped all 4-legged articles together (irrespective of whether they were tables, chairs, chests, or commodes) so did the natural systems of this third period group articles of a kind together (as one would group tables, chairs, chests, and commodes—irrespective of the number of legs each might have). It is interesting to know that the publication of Darwin's theories of evolution and origin of species coincided with the time of production of the first volume of Bentham and Hooker's *Genera plantarum* and that Hooker then favored a complete reorganization of their classification but was deterred from effecting it by Bentham, who did not then accept the essentials of Darwin's work, although he did do so about a decade later. It was the publication of Wallace's and of Darwin's theories that automatically closed this period in the history of classification systems.

### Period IV. Systems based on phylogeny [8]

The rapid spread and acceptance of the theories of Darwin crystallized the dissatisfaction that was associated with the de Candollean system. Sachs was one of the first to abandon the de Candollean system, substituting for it in 1868 a composite system that never was widely accepted. Most of the systems of this fourth period are predicated on the theories of descent and evolution. By these theories, it is now universally accepted by biologists that existing forms of life are the product of evolutionary processes. The classification systems of this period have attempted to classify plants from the simple to the complex (recognizing some seemingly simple conditions to represent reductions from more complex ancestral conditions), and most of these systems have endeavored to establish the genetic and ancestral relationships. The facts that several different systems have been produced, and that a system clearly accounting for the true phylogenetical relationships of plants remains yet to be produced, are evidence that phylogenetic taxonomy of vascular plants is yet in its early stages of progress and will achieve its objectives only as additional facts concerning evolutionary origins and developments of existing plants become known.

*AUGUST WILHELM EICHLER* (1839–1887) in 1875 proposed the rudiments of the first system based on an approach to genetical relationships between plants. This was not a phylogenetic system in the modern sense, but Eichler did accept the concept of evolution. In 1883 he elaborated his earlier treatise into a unified system accounting for all major groups of the entire plant kingdom. It was a system that gradually replaced that of de Candolle in almost all but British and American botanical circles, where the influence of Bentham and Hooker remained dominant. The ultimate widespread acceptance of the basic

[8] See Chapter VI for explanation and analysis of current systems of classification.

tenets of Eichler's system makes desirable a familiarity with its significant features.

Eichler divided the plant kingdom into two subgroups: Cryptogamae and Phanerogamae. The latter contained the seed plants and the former the ferns, bryophytes, hepatics, fungi, and algae. The Cryptogamae were separated into three divisions: Thallophytes, Bryophytes, and Pteridophytes. Eichler treated the Algae as separate from the Fungi and divided the former into the four well-known groups of Cyanophyceae, Chlorophyceae, Phaeophyceae, and Rhodophyceae. The bryophytes were divided into two classes, the Hepaticae and Musci, while the pteridophytes were separated into three classes, Equisetineae, Lycopodineae, and Filicineae. The seed plants (Phanerogamae) were divided for the first time into the two major taxa of Angiospermae and Gymnospermae with the former composed of two classes, the Monocotyledoneae and the Dicotyledoneae. Eichler's classification of the Phanerogamae was predicated on the premise that plants now complex in their reproductive organization represented peaks in their evolutional development; or, as expressed by Hutchinson (1948) ". . . plants without petals . . . were usually regarded as representing a more primitive type than those with well-developed petals . . ." Eichler did not consider tenable the data then available to him (and that later served the basis of an opposing premise), that some of the so-called higher plants (as *Typha,* the cattail) might be simple by the loss or reduction of ancestral structures, a premise whose adoption would have altered materially his alignment of major groups of plants.

*ADOLPH ENGLER* (1844–1930) in 1892 published, as part of a guide to the plants in the Breslau botanic garden, a classification that was based on that of Eichler, and which was so well publicized as to be adopted by a majority of botanists of the world soon after the turn of the present century. Engler's system differed from that of Eichler more in matters of detail and in the nomenclature of major categories than in the basic philosophy or fundamental concepts on which the categories were established. Many of Engler's modifications of Eichler's system show the influence of earlier proposals of Braun, of Brongniart, and of Sachs. The seed plants (termed Embryophyta Siphonogama by Engler) were divided into Gymnospermae and Angiospermae, the latter into two classes, Monocotyledoneae and Dicotyledoneae and the dicots into the subclasses Archichlamideae (composed of the Choripetalae with separate petals, and Apetalae without petals) and Metachlamydeae (the corolla gamopetalous petals united). By this system each subclass is subdivided into orders and they are composed of presumably related families. In their accounting of the principles for a systematic arrangement of the angiosperms, Engler and Diels (1936) listed conditions accepted as primitive and contrasted each with the presumed derived condition. Among the objectionable features was the acceptance by Engler of the dichlamydeous flowers (perianth of two series, as calyx and corolla) as derived from monochlamydeous flowers (perianth of a single series or whorl), the derivation of all parietal placentation in syncarpous ovaries from axile placentation, of free-central placentation from parietal, and the interpretation of the majority of simple unisexual flowers as primitive.

One reason for the widespread adoption of the Engler system by botanists was that Engler and his associate, Prantl, applied their system to the plants of the world and by their 20-volume work, *Die natürlichen Pflanzenfamilien* (1887–1899), provided a means for the identification of all of the known genera of plants from algae to the most advanced seed plants. This was an illustrated work, with modern keys, and was responsible perhaps more than anything else for "putting across" the system. A second and much more detailed edition of this work, under the successive editorships of several outstanding German systematists (Engler, Engelmann, Diels, *et al.*) was begun in Berlin in 1924. Only minor modifications and changes were effected in his system, primarily in Engler and Gilg's (and later by Engler and Diels) *Syllabus der Pflanzenfamilien*, a one-volume work published in many editions that gave their arrangement of the classes, orders, and families of plants. The latest and eleventh edition of this work was published in 1936.

Engler considered the monocots to be more primitive than the dicots, the orchids more highly developed than the grasses and, among the dicots the so-called Amentiferae (willows, birches, oaks, and walnuts) together with other families whose flowers are devoid of perianth, to represent primitive types from which petaliferous elements were evolved. These views are not currently accepted *in toto* by many schools of taxonomic thought. The current dominance of the Engler system in many areas is due largely to the impetus given it by the detailed, far-reaching, all-embracive influence of Engler's many publications. In addition to those mentioned above, these include encyclopedic works written by leading German taxonomists under Engler's editorship.

*RICHARD VON WETTSTEIN* (1862–1931), an Austrian botanist, published in 1901 *Handbuch der systematischen Botanik* and in the posthumous 2-volume fourth edition (1930–1935) presented his latest views on the phylogeny of plants. While patterning the basic structure of his system somewhat along that of Engler's, Wettstein rearranged the relative positions of many dicot families and presented views concurred with by most contemporary phylogenists. In general his system was a much better phylogenetic classification than was Engler's. (See Chapter VI for explanation and analysis.)

*CHARLES E. BESSEY* (1845–1915), a student of Asa Gray and long established at the University of Nebraska, was the first American to make a major contribution to the knowledge of plant relationships and classification and the first to represent a classification to be truly phylogenetic. He did not accept the Eichler-Engler hypotheses. In his early years as a botanist he was considerably influenced by the upheaval in scientific thought occasioned by the argumentative subject of species origin and evolution as propounded by both Wallace and Darwin. In general, Bessey's was the system of Bentham and Hooker realigned according to evolutionary principles, with the cohorts of the latter called orders, and in most cases each was accompanied by new names, and the orders were retermed families. After several revisions, Bessey's system appeared in its last form in 1915 with the orders and families arranged as indicated in Chapter VI.

*HANS HALLIER* (1868–1932) published a phylogenetic classification that was based on many of the same phyletic principles as that by Bessey. It differed primarily in his having taken greater cognizance of the then current researches in paleobotany, anatomy, serology, and ontogeny than had Bessey and was more of a synthesis of recent (and frequently untested) findings than other contemporary classifications. Hallier rejected Engler's concept of the primitive flower, adopting instead, as did Bessey, the strobiloid type of flower. His treatment of the monocots may be less critical than that of the dicots, and his classification, despite innumerable examples of his keen perception scattered throughout it, is not in accord with recent findings to nearly the extent of the Bessey classification. (See Chapter VI for full discussion.)

*JOHN HUTCHINSON* (born 1884) formerly of the Royal Botanic Gardens, Kew, England, and a leading contemporary exponent of a phylogenetic system of classification, has concerned himself primarily with the angiosperms, publishing his classification of them in a 2-volume work, *The families of flowering plants* (1926, 1934) and as revised in his *British flowering plants* (1948). Hutchinson's classification has closer affinities with those of Bentham and Hooker and of Bessey than with that of

Engler, but differs from any of these by several fundamental theses. In his latest revision (1948) Hutchinson outlined the principles of classification adopted by him (see Chapter VI).

The system of Hutchinson is the most recent to be presented in appreciable detail, but to date this detail has been accorded only to the seed-bearing plants and of them primarily to the angiosperms. He has published no recent conclusions on the phylogenies within the gymnosperms and none at all concerning groups below the seed plants. Many taxonomists await Hutchinson's amplification and discussion, substantiated by references to factual data, of his principles of phylogeny. Many also await his presentation of reasons accounting for his classification of taxa in the categories of family and above.

### Other contemporary systems [9]

Numerous other phylogenists have devised and proposed systems of classification other than those discussed above. Notable among them are the systems proposed by Rendle, by Mez, and by Tippo.

*ALFRED BARTON RENDLE* (1865–1938), Keeper of the Department of Botany, British Museum of Natural History, London, from 1906 to 1930, is known not only for his studies of the Gramineae, Orchidaceae, and Naiadaceae, for his leadership associated with international nomenclatural legislation, but also for his 2-volume *Classification of flowering plants* (1904, 1925). The revision of the first volume was published in 1930. Rendle's system, basically that of Engler and Prantl, is one of convenience rather than one of modern phylogenetic significance. It treats the amentiferous and apetalous plants as among the primitive dicots, places the grasses subordinate to the lilies, and does not accept the phylogenetic considerations that treat the Ranales as primitive to other dicots (for details, see Chapter VI).

*KARL CHRISTIAN MEZ* (1866–1944), Professor of Botany, University of Koenigsberg, Germany, presented a paper in 1926 (revised and amplified in 1936) recounting his theory that relationships between the larger groups of genetically related plants could be determined by study and analysis of their protein reactions. This physiological approach, sometimes known as the serum diagnosis, consisted of mixing an extracted plant protein with serum, either in animal or *in vitro* and, following the formation of antibodies in the inoculated serum, of adding a pro-

---

[9] See Chapter VI for an accounting of additional contemporary systems, including those by Pulle, Skottsberg, *et al.*

tein extract prepared from the plant whose relationship is being studied. If a precipitation occurred when this second protein extract was added to the serum, a genetical relationship was believed to be indicated to exist between the plants involved. The proximity of the relationship was considered to be indicated by the abundance and character of the precipitate and the degree of dilution of the serum and extract at which precipitation would occur. The relationships indicated by this technique concurred to a degree with those proposed by Bessey and by Hutchinson. The utilization of data derived from serum diagnoses may contribute important corroboratory evidence to the formation of an ultimate system of classification.

*OSWALD TIPPO,* of the University of Illinois, published an outline (1942) of a projected classification based on ". . . the newest, sound developments in all the branches of plant phylogeny and in which the various groups are named in such a way as to indicate rank or degree of affinity and in which the various group names are brought into conformity with the classificatory system used by zoologists" (see Fig. 17). The system was not developed by Tippo, who "claims no great degree of originality for his classification for it is a compilation, a synthesis of several proposals." The classification of the nonvascular plants followed that of Smith (1938) and the basic arrangement of vascular plants followed that proposed by Eames (1936). The concepts and indicated relationships were in part the result of greater recognition and synthesis of paleobotanical data than are reflected in other phylogenetic systems. This emphasized the dependence on vegetative characters such as plant form, branching, and vascular anatomy in arriving at the natural groupings of the larger categories.

The Engler system remains dominant in the great herbaria of this country, but largely because it is the latest to account in detail for the entire plant kingdom. The more nearly phylogenetic systems of Bessey, of Hutchinson, and of Tippo are thought to be more accurate in their arrangement and more sound in the tenets on which they are founded, but—as pointed out in the Preface to this text—no one of them has been applied to the world flora (even of vascular plants) to a degree comparable with that of the Engler system. The student should recognize that the Pteridophyta themselves are not a homogeneous group bound by any one or two sets of characters; that there is no distinct line of demarcation between gymnosperms and angiosperms, or between monocotyledons and dicotyledons; and that a truly phylogenetic classification will be reticulate in character and will not lend itself to listing taxa in straight lines. Tippo's skeleton outline clearly indicates the probable major groupings to be expected in the system of the near future.

This review of the systems of classification demonstrates that taxonomic efforts devoted to the development of a more nearly perfect classification of plants have followed in the wake of those men whose interests have shifted from generalization to interests of specialization. Linnaeus, the greatest cataloguer and classifier of all time, was not only a systematist but a capable naturalist, a biological investigator, and a practicing physician of note; A. P. de Candolle was a physiologist, and a nomenclaturist, as well as a producer of extensive monographs; Engler was as much a plant geographer and botanical administrator as he was a taxonomist; Bessey, doubtless of necessity, devoted more efforts to teaching than to phylogenetic studies; much of the recent work of Hutchinson, aside from his studies on the flora of west tropical Africa, has been directed toward phylogenetic studies. It is probable that the phylogenist of the future will be so concerned with his own efforts as to have little concern for the identity of plants *per se*. The average taxonomic botanist is concerned today more with studies associated with floristics and monography of minor units (families and genera and their components) than he is with coping with the fundamental problems of phylogeny of major taxa. Underlying causes for these changes are indicated in Chapter V. Chronology forces the premature closing of the treatment of this last and current period of systems of plant classification. The period has not ended. Research in the fields of paleobotany, comparative anatomy, serology, cytogenetics, and morphology continues and, with the accumulation and synthesis of data from these and other interrelated branches of botanical science, a new and more nearly perfect system of classification will be evolved.

*LITERATURE*

ADANSON, M. Familles des plantes. 2 vols. Paris, 1763. [Ed. 2, 1847, by Adanson. A. and Payer, J., contains a biographical account].
ARBER, A. Herbals, their origin and evolution. London, 1912; ed. 2, 1928.
BAUHIN, C. [Phytopinax] seu enumeratic plantarum. . . Basel, 1596.
———. [Pinax] theatri botanici. Basel, 1623.
BESSEY, C. E. Evolution and classification. Contr. Bot. Dept. Univ. Nebraska, N. S. 7: 1894.
———. Phylogenetic taxonomy of flowering plants. Ann. Mo. Bot. Gard. 2: 1–155, 1915.
BRONGNIART, A. T. Essai d'une classification naturelle des champignons. Paris, 1825.
BRONGNIART, M. A. Historical notice of Antoine Laurent de Jussieu. Mag. Zool. Bot. 2: 293–308, 1838. [Engl. transl.].
BRUNFELS, O. Herbarium vivae icones. Argentorati, 3 tomes in 1530.

CESALPINO, A. De plantis libri. Florentiae, 1583.
CHRISTENSEN, C. Index to Pehr Forsskål. Flora aegyptiaco-arabica 1775 with a revision of Herbarium Forskallii contained in the botanical museum of the University of Copenhagen. Dansk. Bot. Arkiv 4: no. 3, 1922.
COPELAND, H. F. The phylogeny of the angiosperms. Madroño, 5: 209–218, 1940.
CROIZAT, L. History and nomenclature of the higher units of classification. Bull. Torrey Bot. Club, 72: 52–75, 1945.
DAUDIN, H. De Linné à Jussieu, methodes de la classification et idée de serie. Études d'histoire des sciences naturelles, 1: 1926.
DE CANDOLLE, A. P. Théorie élémentaire. Paris, 1813.
———. Prodromus systematis naturalis regni vegetabilis. 17 vols. and 4 index vols. Paris, 1824–1873.
EAMES, A. J. Morphology of vascular plants—lower groups. New York, 1936.
EICHLER, A. B. Blüthendiagramme construirt und erlautert, 2 vols. Leipzig, 1875–1878.
ENDLICHER, S. L. Genera plantarum. Vienna, 1836–1850.
ENGLER, A. Führer durch den königlichen botanischen Garten der Universität zu Breslau. Breslau, 1886.
———. [Ed.] Das Pflanzenreich. Leipzig, 1900 and continuing.
ENGLER, A. and DIELS, L. Syllabus der Pflanzenfamilien. Ed. 11. Berlin, 1936.
ENGLER, A. and GILG, E. Syllabus der Pflanzenfamilien. Eds. 9 and 10. Berlin, 1924.
ENGLER, A. and PRANTL, H. Die natürlichen Pflanzenfamilien, 20 vols. Leipzig, 1897–1915; ed. 2, [not completed], 1924–1942.

GREENE, E. L. Landmarks of botanical history. New York, 1909.
———. Linnaeus as an evolutionist. Proc. Wash. Acad. Sci. 11: 17–26, 1909.
———. Carolus Linnaeus. Philadelphia, 1912.
HALLIER, H. Phylogenetic studies of flowering plants. New Phytologist, 5: 151–162, 1905.
HARVEY-GIBSON, R. J. Outlines of the history of botany. London, 1919.
HAWKS, E. Pioneers of plant study. London, 1928.
HUTCHINSON, J. Contributions toward a phylogenetic classification of flowering plants. Kew Bull. 1924: 114–134.
———. The families of flowering plants. 2 vols. London, 1926, 1934.
———. British flowering plants. London, 1948.
JACKSON, B. D. Linnaeus. London, 1923.
JUSSIEU, A. L. DE. Genera plantarum. Paris, 1789.
KLEIN, L. M. The botany of Albertus Magnus. The Month, 48: 382–396, 1883.
LINDLEY, JOHN. Introduction to the natural system of botany. London, 1830.
———. The vegetable kingdom, etc. ed. 3. London, 1853.
LINNAEUS, C. Hortus uplandicus. Stockholm, 1730.
———. Systema naturae. Lugd. Bat., 1735; ed. 2, Stockholm, 1740; ed. 6, 1748; ed. 10, 2 vols., 1758–1759; ed. 12, 3 vols., 1766–1767 [intermediate editions not cited were not edited by Linnaeus].
———. Flora lapponica. Amsterdam, 1737.
———. Hortus cliffortianus. Amsterdam, 1737.
———. Genera plantarum. Lugd. Bat., 1737; ed. 2, 1742; ed. 5, 1754; ed. 6, 1764.
———. Species plantarum. 2 vols. Stockholm, 1753; ed. 2, 2 vols., 1762–1763; ed. 3, 2 vols., 1764; ed. 4, 6 vols., 1797–1830 [edited by C. L. Willdenow]; ed. 6, 2 vols. [continued by H. F. Link et al.], 1831–1833.

MEZ, C. Die Bedeutung der Serodiagnostik für die stammesgeschichtliche Forschung. Bot. Arch. 16: 1–23, 1926.
——. Morphologie und Serodiagnostik . . . Bot. Arch. 38: 86–104, 1936.
MILLER, PHILIP. The gardener's dictionary. London, 1734; ed. 8, 1768.
MOBIUS, M. Geschichte der Botanik von den ersten Anfängen bis zur Gegenwart. Jena, 1937.
POOL, R. J. A brief sketch of the life and work of Charles Edwin Bessey. Amer. Journ. Bot. 2: 505–518, 1915.
RAVEN, C. E. John Ray, naturalist; his life and work. New York, 1942.
——. English naturalists from Necker to Ray: a study of the making of the modern world. Cambridge, England, 1948.
RAY, J. Historia plantarum, 2 vols. London, 1686–1688.
RENDLE, A. B. Classification of flowering plants. Cambridge, England. Vol. 1, 1904; vol. 2, 1925; ed. 2, vol. 1, 1930.
——. George Clifford's herbarium and the "Hortus Cliffortianus." Journ. Bot. 61: 114–116, 1923.
SACHS, J. History of botany (1530–1860). London, 1890. [Transl. by H. E. F. Garnsey; rev. by I. B. Balfour.]
SARGENT, C. S. Scientific papers of Asa Gray. 2 vols. Boston, 1889.
SMITH, G. M. Cryptogamic botany. Vol. 1. New York, 1938.
SPRAGUE, T. A. The herbal of Otto Brunfels. Journ. Linn. Soc. Bot. (Lond.) 48: 79–124, 1928.
——. Botanical terms in Albertus Magnus. Kew Bull. pp. 440–459, 1933.
—— and M. S. The herbal of Valerius Cordus. Journ. Linn. Soc. Bot. (Lond.) 52: 1–113, 1939.
—— and NELMES, E. The herbal of Leonhart Fuchs. *Op. cit.* 48: 545–642, 1931.
SVENSON, H. K. On the descriptive method of Linnaeus. Rhodora, 47: 273–302, 363–388, 1945.
Theophrasti de historia et causis plantaru . . . T. Gaza interprete. Parisiis, 1529.
THUNBERG, C. P. Flora Japonica, etc. Lipsiae, 1784.
——. Flora capensis, etc. Vol. 1, Upsaliae, 1807–1813. Ed. nova. 2 vols. Hafniae, 1818.
TIPPO, O. A modern classification of the plant kingdom. Chron. Bot. 7: 203–206, 1942.
TOURNEFORT, J. P. DE. Elemens de botanique . . . 3 vols. Paris, 1694.
——. Institutiones rei herbariae. 3 vols. Paris, 1700.
WETTSTEIN, RICHARD. Handbuch der systematischen Botanik. Leipzig, 1901; ed. 4, 2 vols., 1935.

CHAPTER IV

# PRINCIPLES OF TAXONOMY

Taxonomy is a functional science. The direction, character, and extent of its functions are guided by principles that have developed with the increase in knowledge of the plants themselves. The formulation of these principles began with the period of descriptive taxonomy which functioned on a scientific level in the nineteenth century, and only recently has waned and may become secondary in importance. This descriptive period began with the works of Tournefort, de Jussieu, and Linnaeus. Most taxonomic work of the period was based on observations of the similarities and differences of usually gross morphological characters of the plants concerned. Plants were described and classified on the basis of these characters. The describing of plants from newly explored areas was then a major function of taxonomy. Succeeding these pioneers were such leaders as Robert Brown, the Hookers (William and his son Joseph), John Lindley, and George Bentham, all of England; the three generations of de Candolles of Geneva and Paris, Edmund Boissier of Geneva, Carl Willdenow and Curt von Sprengel of Berlin, Eduard von Regel of Leningrad, and Asa Gray of Cambridge (Mass.).

During this period these and other systematists published various principles of taxonomic procedure. The first of these was by Linnaeus in his *Critica botanica*.[1] Others were by Adanson, John Lindley, de Candolle, and Sir Joseph Hooker. Many of the criteria established by these early botanists have withstood the test of time and are accepted by contemporary workers. More important, however, is the recognition by present-day taxonomists that gross morphological characters are not always adequate to provide reliable means of differentiation between features that are of major significance from those of minor significance, or to serve necessarily in the determination of genetical relationships between taxa. This recognition has resulted in the need for a re-evaluation of all taxonomic work conducted by these earlier devotees of the descriptive

---

[1] For the latest English translation from the Latin, see that by Sir Arthur Hort.

method, a re-evaluation that will take into consideration, in addition to the morphological criteria, all other scientific data pertinent to the situation and as contributed by allied botanical sciences. Present-day taxonomy is based on the primary importance of morphological distinctness and affinity, but it is influenced appreciably by the findings of the cytologist, geneticist, anatomist, and others.

The principles of taxonomy are concerned primarily with the criteria employed. However, the intelligent discussion of these criteria presupposes an understanding of the units of classification. For this reason it is appropriate that an explanation and discussion of the latter precede other objectives of this chapter.

Taxonomy is based on the hypothesis that genetical relationships exist between plants, that present-day plants are, through successive generations, the offspring of ancestral plants that may or may not now be extant. It is based also on the assumption that there has occurred, during the developmental epochs of the earth's history, an evolution of plant characteristics to the extent that surviving plants often were of increased structural complexity and genetic organization over their ancestors. As these evolutionary processes progressed, the much-removed offspring of ancestors that once were extant and were closely related have become less closely related—to the degree that today the determination of plant relationships is almost wholly a subject based on postulations, hypotheses, and tenuous conjectures. Because of the predication of the science of taxonomy on these concepts of genetic relationship, it is desirable to place plants in categories that are indicative of their presumed genetic affinities. Each category represents a group of plants; as species, genera, families, etc. No one of these is subject to precise definition, their delimitation or circumscription varies, and each is subjective in character.[2] Groups such as Angiospermae, Monocotyledoneae, or Rosales are of considerable magnitude. They are, respectively, examples of the categories of subdivision, class, and order. Each of these is composed of many and diverse kinds of plants, and although each is an example drawn from a different level of classification, collectively they may be referred to as major categories. The more familiar groups, represented by such examples as *Petunia* (a genus), regal lily (a species), or cabbage (a variety) are of relatively small magnitude and by comparison are of minor categories. A minor category may be considered one whose name is also a part of the name of the particular plant. A major category is

[2] For a historical account of the development of higher categories of classification and considerations of the diversified nomenclature given them, see Croizat (1945) and Just (1945).

any one of the higher categories whose names are not a part of the name of the plants belonging to them. The major categories of vascular plants have been reasonably well established in recent decades and their circumscriptions are the special studies of the phylogenist rather than of the ordinary taxonomic worker. The minor categories have been the subject of more intense study by a greater assemblage of workers and for this reason have been interpreted more diversely. Each of the more important categories is discussed below individually, with those of major magnitude preceding those of narrower limits.

The categories of taxonomic groups and the terms denoting them are prescribed by Chapter II of the International Rules of Botanical Nomenclature (ed. 3), of which Article 13 states:

The definition of each of these categories varies, up to a certain point, according to individual opinion and the state of the science; but their relative order, sanctioned by custom, must not be altered. No classification is admissible which contains such alterations.

To the extent that the above-quoted article applies, the subject of classification units is one of nomenclature. The definition and general circumscription of the categories is a subject of classification and is a vital part of the principles of taxonomy.

### Major categories of classification

The plant kingdom is divided into *divisions*. Zoologists divide the animal kingdom into *phyla*, and for purposes of uniformity of category names between the two branches, some biologists and botanists have adopted for plants the term phylum in place of division. The term division is prescribed by the rules of nomenclature to represent the category of highest magnitude within the plant kingdom, whereas phylum is not mentioned. The number of divisions into which the plant kingdom is divided varies with different systems of classification. By some authors it was considered to be four; by the latest revision of the widely current Engler system it was considered to be 12. Among vascular plants the Spermatophyta (seed plants) may represent a division.[3] It is not possible to define a division concisely or with precision. Divisions of vascular plants are relatively few in number. They are taxa that are distinguished by characters that are common only to the constituent elements of each division. For example, the Spermatophyta are characterized, in part, by the possession of a sporophyte generation that is dominant over the much

---

[3] By the Engler system, this division bears the less familiar name, Embryophyta Siphonogama.

reduced gametophyte generation, by the presence of ovules, and by the production of reproductive structures called seeds. The characters employed are often of reproductive, morphological, or internal anatomical structures. For the most part, it is true that these selected characters are believed to have been fundamental to early ancestors and that their presence throughout an extended sequence of generations undoubtedly was responsible in part for survival of present-day representatives. It must be recognized also, since these are biological material and all possessing the germ plasm that is life, that in the far distant past, the components of one division had one or more ancestors in common with the components of another division. The existence of these relationships means that the boundaries or criteria by which one division is separated from another division are not so sharp, clear-cut, or inviolate of exception as one may be led to believe. Exceptions do exist and may be observed for almost every divisional character. For this reason the division is characterized by an aggregate of characters rather than by any single infallible one. Similar situations exist in all major categories of plants.

It is provided by the rules of nomenclature that any category of plants may be divided into subordinate categories intermediate between it and that of next lower rank. This is accomplished by adding the prefix *sub* to the name of the higher category. By this provision, a division may be composed of two or more subdivisions, or (when the intermediate category is not required) a division may be composed of two or more classes. The division Spermatophyta was divided into two *subdivisions,* the Gymnospermae and the Angiospermae. On the other hand, the division Pteridophyta was subdivided, not into subdivisions, but directly into categories called classes. This is because fewer kinds of categories are recognized within the Pteridophyta than within the Spermatophyta.

A division (or subdivision) is composed of *classes*. The class is the next full category subordinate in rank to the division. The names applied to classes are Latin names, as are those of all taxonomic categories, and ordinarily have the ending *-eae*. The two classes of the subdivision Angiospermae are the Monocotyledoneae and Dicotyledoneae. The Gymnospermae are not subdivided into classes by most authors, but are treated as composed of a number of distinct orders, a category explained below. In the case of the Dicotyledoneae the number of kinds of plants is so great that most phylogenists have treated them as composed of several *subclasses*. There has been no accord in the selection of names applied to these subclasses or in the selection of taxonomic bases on which they have been established.

Each class (or subclass, when present) is subdivided into *orders*. The order is the category next in line and subordinate to that of class.[4] The Latin names of orders conventionally have the ending *-ales*, as Rosales or Cycadales. However, the names of some orders of long standing, and which were given to the taxa before the existence of currently accepted rules of nomenclature, do not terminate in *-ales* but rather have the ending *-ae*. These names have priority and nomenclaturally are allowed. Examples of them are found in the names Glumiflorae and Tubiflorae (the orders to which belong respectively the grasses and asters). An order possesses a degree of phylogenetic unity that is determinable with greater assurance than that of the higher taxa of division or class. The relationships of its components and their eligibility for inclusion within the category can be ascertained by and established on more definite criteria and characteristics.

In some cases it has been found desirable to treat large orders as comprised of *suborders*. It is customary, and in accordance with the rules, to terminate the names of these suborders with the ending *-ineae* as Malvineae, a suborder of the Malvales.

An order of plants is comprised of one or more *families*. The category of *family* is the smallest of the major categories and, of them, is the most frequently encountered in ordinary taxonomic studies. The Latin names of all but 8 of the families of vascular plants are terminated by the conventional ending *-aceae*, as Pinaceae, Rosaceae, or Ranunculaceae. Certain family names sanctioned by long usage do not end in *-aceae* and have irregular terminations that are allowed as exceptions to the rule. These families are: Palmae, Gramineae, Cruciferae, Leguminosae, Guttiferae, Umbelliferae, Labiatae, and Compositae. However, the Rules (ed. 3) authorize the substitution of an alternative name ending in *-aceae,* as Poaceae for Gramineae.

The family usually represents a more natural unit than any of the higher categories. This is true because usually more is known about the components of a family, and correlations between a greater number of characters usually exist. For example, families such as the grass family, the sedge family, the mustard family, and innumerable others, are readily recognized as natural taxa whose respective members have definite characters bonding them together. Not all families, or taxa conservatively

---

[4] Earlier botanists employed the term order for the category now named family. This former interpretation was used in the Bentham and Hooker system of classification and is encountered in some current floras based on this system. The International Rules of Nomenclature (ed. 3) prescribe its use as indicated above and make its use illegitimate in lieu of the term family.

recognized as such by many botanists, are natural. Examples of such unnaturalness are to be found, among flowering plants, in the Englerian concept of the Saxifragaceae and Onagraceae and, in the ferns, in the conservative concept of the Polypodiaceae. Each of these families is an unnatural taxon and, as currently and conservatively broadly delimited, must have been derived from heterogeneous and relatively unrelated ancestors; that is, they are presumed to be of polyphyletic origins. The solution to the problems presented by these unnatural families may rest in their division into smaller phylogenetically homogeneous families. The natural family is believed to be one whose members were derived from common ancestral stock (i.e., of monophyletic origin). The families of higher plants are separated from one another by characters generally inherent in the reproductive structures. They are characters usually associable with features such as inflorescence type, ovary position, placentation type, pistil and carpel number, ovule type, embryology, such androecial conditions as monandry, diandry, and syngenesism, and the disposition of sexes as in dioecism and monoecism. The increase in knowledge of the origins, interrelationships, and ancestral types of these features will gradually result in the resolution of many of the problems concerned with the more accurate determination of family circumscription. A family is not of any particular size; it may be comprised of one genus or of 100 genera. Its claim to this rank lies with the degree and constancy by which it differs from other families within the same order or suborder.

When a family is large and comprised of many components, it is often found desirable to divide it into phyletic units called *subfamilies*. These bear Latin names that usually are terminated by the ending *-oideae*. One subfamily of the Rosaceae is the Rosoideae, another of the same family is the Pomoideae. Large subfamilies are sometimes subdivided into *tribes*, which are subordinate phyletic groups whose Latin names have the ending *-eae*. One tribe of the Compositae is the Astereae, another the Inuleae. In some cases a family is divided directly into tribes. Occasionally it is desired to recognize phyletic groups subordinate to the tribe; these are called *subtribes*. They are designated by Latin names ending in *-inae*.

### Minor categories of classification

A minor category of classification is considered here to be one whose name becomes a part of the name of the plant. It may be a genus, a species, or any one of the several categories subordinate in rank to the species. For reasons of phyletic clarity and convenience in classification, the category of genus is sometimes subdivided into subgenera, sections, subsections, and series. These latter are phyletic groups whose Latinized names do not enter into the name of the plants concerned and which basically are not of the same functional significance as is the category of genus, species, or variety. As mentioned at the beginning of this chapter, more workers are concerned with the minor categories of classification than are concerned with the major ones. Because of this, more data are available concerning the components of these units, and more diversity of opinion and concept exists concerning their delimitations. In general, the concepts of these minor units are represented by two or more schools of thought, and it is desirable to present the views of these as objectively as possible. For details other than those provided below the student is referred to articles cited in the bibliographic references at the close of this chapter.

The *genus* is subordinate to the family. Each family is comprised of one or more genera. The generic name of a plant is the first of the two words comprising a binomial as, for example, in the binomial *Quercus alba* the generic name is *Quercus*. The Latin names of genera are substantives (or adjectives used as such), are always capitalized, are always in the singular number, and may be taken from any source whatsoever or composed in an entirely arbitrary manner. They have no uniform endings.

The genus is a category of long standing. It is highly probable that genera were recognized as groups of plants of common affinity before science came of age, and before written languages existed. Among the early herbalists, Oscar Brunfels is credited with having had the clearest concept of the category of genus. A little over a century later, in 1716, there was published (posthumously) a monumental work by the French botanist Joseph Pitton Tournefort, *Institutiones rei herbariae*, wherein he developed his thesis that the fundamental category of classification was the genus, and that plants having in common two or three characters of reproductive structures were usually to be treated as members of the same genus. Basically the Linnaean concept of the genus (1737 *et seq.*) was in accord with that of Tournefort, and in this respect it is significant that of our common indigenous plants, more of them are in genera named and circumscribed by these two men than have been named subsequently by other botanists.

Any consideration of the concepts of the generic category must recognize that the philosophies of these early and great men were of a period when the theory of Special Creation prevailed, a doctrine accompanied by the corollary that life forms were immutable. It was during this period of dogmatism that basic tenets were established by which much of our currently available taxonomic literature was written. This literature was based on a system "organized on a basis of similarities, having as its fundamental principle a doctrine based on the thesis that a community of similar morphological structures indicates relationship" (Camp, 1940, p. 382). The genera of Linnaeus, Rafinesque, Hooker, Gray, Torrey, and of later botanists were based largely on the belief that a genus is a category whose components (i.e., species) have more characters in common with each other than they do with the components of other genera within the same family. This concept of the genus is supported by many contemporary taxonomists. An inherent weakness of this concept is that in circumscribing genera, often within a given family, it is not possible to treat selected characters as possessing equal value for all genera of the same family. It very often happens that characters that are adequate to separate some genera within a family are insufficiently stable to separate even species within another genus of the same family. Despite this shortcoming, the concept has provided an expedient solution to a practical need: it has served and will continue to serve a purpose.

Many botanists believe that the genus is more than a taxonomic category. If the theory of evolutionary descent via the transfer of mutable germ plasm through successive generations is accepted, then just as it is the goal that taxa represented by the higher categories be arranged within a natural and phyletic system of classification indicative of genetic relationships, so also is it desirable that the category of genus be treated as a phyletic unit; a category circumscribed and disposed to indicate the phyletic relationships of it and its components with other similarly established genera and their components. This second concept treats the genus as a biological category, and taxonomic studies of genera that accept this concept take into account not merely the morphological similarities by which it is conventionally recognized, but also the origins, migrations, genetic, cytologic, physiologic, and ecologic behavior, and geologic history associated with its components. The re-evaluation of existing genera by the tenets of such a concept must result in some genera being divided into segregates, others being combined with what were thought to have been distinct genera, and still others maintaining their *status quo*. Genera established by adherents of the first concept are the products of descriptive taxonomy, whereas those redefined according to the second concept are the products of a modern phyletic taxonomy.[5]

---

[5] For discussion of the delimitations of genera, see Moran (1942).

A genus may be divided into *subgenera* and they in turn into *sections, subsections,* and *series,* or the genus may be treated as composed of sections without the intercalation of subgenera. Some genera are not subdivided, but comprise only a group of very homogeneous species. The names of generic subdivisions are usually adjectives in the plural number and agree in gender with the generic name.

The *species* long has been considered to be the basic unit of all taxonomic work. This view had its origins in earliest civilizations and the species was the category on which the theory of Special Creation founded its beliefs (note that *Homo sapiens* is a species). This theory postulated that all the kinds (i.e., species) of plants and animals were created in their present form, that the number of species then on earth was the same number that had been there since the beginning of time, and that having these species man proceeded to devise classifications whereby they were grouped into genera, the genera into families, the families into orders, and so on. According to the theory of evolutionary development, no single category is a basic phyletic unit, for the category of species is no more fundamental to a phylogenetic schema than is any other category, and all categories must be accepted as somewhat artificial and all considered as interlocking links in a 3-dimensional composite of relationships. However, despite this, it must be recognized that nomenclaturally the species is the category on which the binomial system has been established. It is the category that has received more attention by biologists than all others combined.

**What is a species?** Botanists of every generation have attempted to answer this question, one for which there may be no single answer. An approach to an answer was provided by Camp and Gilly (1943, pp. 380–381), who wrote that:

> There are even some among us who have advocated that we discard the concept of a species altogether. Therefore, the question which the systematist should seek first to answer is not: Upon what criteria should the concept of a species-unit be based? Rather, he must enquire: Does the species-unit deserve to be a fundamental philosophical concept? This, perhaps fortunately for his own peace of mind, has long ago been decided for him.
>
> The concept of species or *kind,* as a unit, has become so firmly entrenched in the mind of man—so much a part of his awareness, so necessary to his basic philosophy—that it remains only for the systematist to interpret this unit . . .

There are many schools of thought that have attempted to produce interpretations of the unit, but the number of botanists holding to different views is diminishing. The time honored answer to the question of "What is a species?" has been that a species is a concept, that it is the product of each individual's judgment. The modern taxonomist requires more data in the jelling of a species concept for, in addition to morphological distinctions, inquiry is made of the character and extent of morphological variation within the populations that collectively are treated as a species; to these may be added considerations of distribution that may or may not take into account the geologic history of the areas concerned. This concept was undoubtedly of major importance to du Rietz (1930), who postulated that species were "the smallest natural populations permanently separated from each other by a distinct discontinuity in the series of biotypes." This view stresses the importance of morphological continuity transmissible from one generation of components of the species to another. Lamprecht (1949), a cytogeneticist, traced the developments of and changes in species concepts, classifying them to be identified with one of the following five periods of taxonomic research: "(1) descriptions without taxonomic systems, (2) the period of development of artificial and natural systems, (3) systems in the light of evolution, (4) the period beginning with the detection of speciation by addition of genomes, and (5) that beginning with the detection of the genic basis of the species barrier." From this view it is clear that most current taxonomic work comes within his third period, and that the biosystematist is now working in the last two periods.

Any consideration of solutions to the species problem must take into account the current views that (1) all populations tend to vary and that no two are ever exactly alike, (2) that some of these variations are adaptive and are of survival value, (3) that forces of nature result in the extinction of some individuals while others survive the same forces, (4) that some of the variations displayed by individuals within a population must be hereditary if successive generations are to be modified from ancestral conditions, and (5) that the environment of the individuals must not be static lest the course of evolution be checked by the forces of natural selection. These views collectively incorporate the principles of the theories of natural selection and of evolution. If they are pertinent to the background and evolution of the individual and likewise, through aggregates of individuals, of the species, it is apparent that there must be many kinds of species. These kinds of species are biologically different taxa that have arisen by means of different selective and genetic mechanisms.[6] The species is perpetuated by one of a relatively few types of reproduction. Sexual reproduction plays the dominant role, but within

---

[6] For discussions of the kinds of species and their genetic origins, organizations, and behaviors, see Camp and Gilly (1943) and Lamprecht (1949).

this type of reproduction the "fertility relations between the individuals or between intraspecific groups will certainly affect the type of the population" (Babcock, 1947, I, 35). In addition to these, the genetic constitutions of populations vary, for, as shown by chromosome complements, the plants may be diploids, or they may be various kinds of polyploids. Plants of different genetic constitution behave differently, the progeny may or may not be like their parents, and by virtue of the potentials inherent in these internal genetic conditions, plants may be sexually compatible or incompatible, migrate or regress, or may survive or perish. It has only recently been recognized that asexual reproduction also may be responsible for kinds of species. It has become established that certain populations often treated as species do not reproduce sexually but rather by asexual devices, many of which are obscure and superficially not apparent. This may be accomplished by various types of apomixis such as parthenogenesis or simple vegetative reproduction of clonal elements.[7] These different kinds of species may or may not possess gross morphological characters by which they may be differentiated. The recognition by Clausen, Keck, and Hiesey (1939) of the significance of these genetic factors in the speciation of plants resulted in a subclassification of species which has gained considerable support and is presented in detail in Chapter VIII.

From this it is apparent that the problem of speciation is not simple. The taxonomist is faced with the need of acknowledging that biologically there are different kinds of species and that each species represents a kind of population. Accompanying this is the recognition that these genetic systems called species must be resolvable into the existing functional binomial system of nomenclature.

The existence of different and sometimes seemingly divergent views on the subject of what is a species need not be confusing. It is important to know that they exist. It is desirable to know the principles of thought involved in the more important of them so that they may be rationalized and evaluated. In arriving at an opinion, it should be remembered that in most instances there are yet insufficient data by which to determine the speciation status by the standards of the biosystematist, and that for purposes of expediency and practicality the taxonomist must give to the populations under study "(1) a circumscription which is not only biologically as sound as possible, but (2) which also is in accord with an effective system of nomenclature. Furthermore, the interpretation of these

---

[7] For an analytical review of literature on the role of apomixis in species formation, see Chapter VIII, pp. 184–185, and Stebbins (1941), and Gustafsson (1946–47).

## PRINCIPLES OF TAXONOMY

items must be balanced; there must be no undue emphasis on one above the other, otherwise a bifurcation of concept will result leading to chaos in systematics." (Camp and Gilly, 1943, p. 381.)

Irrespective of the nature of the concept by which species are circumscribed, the unit must fit into the binomial system of nomenclature. It must have a Latin name and that Latin name, composed of two words—the generic name and the specific name—is the binomial. The specific epithet is the second word of a binomial. The regulations governing the subject of binary names are presented in Chapter IX.

In addition to the terms applied to these major and minor categories, there has been proposed the term *taxon* (plural, *taxa*) as being more adequate and specific than such ambiguous terms as entity or taxonomic group.[8] The term entity is unsatisfactory since no taxonomic or biologic category above that of the individual can strictly be an entity, and the term group is inadequate since it is a collective term not properly applicable to an individual plant. The term taxon has the merit of possessing a single meaning, unfettered by vagueness or ambiguity. It is short yet descriptive. It is believed that its acceptance will add clarity and precision to biological literature.

### Infraspecific categories

Any category below the rank of species is an infraspecific category. It is a variant of the species.[9] Taxonomically, the origin of species variants has been the subject of two basic philosophies. According to the first, botanists of the nineteenth century and earlier consistently in their taxonomic works would establish a species. Then if it was determined that a variant of that species existed, the variant was named and described as if it were an appendage to the species. This philosophy

---

[8] The term *taxon* was presented at the Utrecht symposium on nomenclature, convening during the summer of 1948, by H. J. Lam, who formally proposed "to indicate taxonomic groups of any rank with the term *taxon* (plural: *taxa*). This term was first introduced by A. Meyer in his: Logik der Morphologie im Rahmen einer Logik der gesamten Biologie, 1926, p. 127; cf. also pp. 133, 241." It is believed that Meyer did not mean to introduce the term taxon to be used for any systematic unit. He introduced it to distinguish a minor category in taxonomy from a major category of phylogeny that Haeckel (in *Generelle Morphologie* 1:2. 1866) had designated phylon. At the International Botanical Congress, held in Stockholm, 1950, it was voted to use the term taxon throughout the next edition of the Rules of Botanical Nomenclature, wherever appropriate.

[9] The designation infraspecific category is introduced to avoid the ambiguity occasioned by the use of subspecific category, since an infraspecific category clearly is any of several categories subordinate to that of species, whereas a subspecific category may be used in this sense or may be used more strictly to designate the category of subspecies.

The term variant is used here in the literal sense, for any element differing from the typical, and is not to be confused with the term variety; the latter being an English translation of the Latin *varietas*, a technical term used to designate a particular category subordinate to the rank of species.

stemmed directly from the theory of Special Creation, for the variant was not considered to be a genetic part of the species, but rather was an orphan element apart by itself that did not merit species rank but was attached to that species which it most closely resembled. Such was the philosophy of men from the time of Linnaeus to Asa Gray and later. Full comprehension of the dynamic forces inherent in the theories of evolution resulted in the second and more recent philosophy that biologically a species comprises the element represented by the initial binomial and all subsequently described variants ascribed to that binomial. In other words, by this later philosophy, it was merely by chance that one plant (or population) served to be the basis (nomenclatural type) of the new binomial (i.e., species). If it had happened that one of the elements later treated taxonomically and nomenclaturally as a variant of a binomial had been discovered first, then the second element would have been the typical element of the species (i.e., it would have been the initial element first to have been designated by a binomial) and the first would have become a variant of it. That is, if the variant and the typical element of a species are of the same rank they are presumed to be biological equivalents, one is no more the species than is the other. Properly speaking in technical treatments, the description of the species as represented by the binomial would then be of a scope that would account for the characteristics of all varietal elements of it and not of the typical element alone. If the species is thus broadly circumscribed, the typical element is to be contrasted with successively described variants and treated nomenclaturally in a definite category subordinate to the species and coordinate with other included elements of the same infraspecific rank. When more than one infraspecific taxon is present, a trinomial form of nomenclature is followed, as, for example, *Carex aquatilis* var. *aquatilis* for the typical element, and *Carex aquatilis* var. *altior* for an infraspecific taxon of it.

The rules of nomenclature provide for the infraspecific categories of subspecies, varietas, subvarietas, forma, forma biologica, forma specialis, and individuum. Of these the three most commonly recognized are the subspecies, variety (varietas), and forma. The situation exists here, as it does with respect to the category of species, that there are differences of opinion as to what is a subspecies, a variety, or a forma.

The *subspecies* has been given many definitions, but of them the following may serve to indicate the basic differences of concept: (1) that they are baby species or species of small magnitude that are distinguished by less obvious or less significant morphological features than are more obvious species within the same genus; (2) that they are major morphological variations of a species that have geographic distributions of their own, which are distinct from the area occupied by other subspecies of the same species; or (3) that a subspecies is the category to which should be referred those elements which by possession of satisfactory geographic, ecologic, and morphologic characters are suspected to be counterparts of the ecotype (a biologically significant element determinable only after analysis by slow and tedious experimental techniques).[10] The use of the category subspecies is receiving increasing favor by taxonomists, but in each case it must be determined which of the three basic types of concept was adopted by the author. For example, the subspecies of the Englerian school was a category of major morphological distinction with or without disjunctive distribution. By earlier European botanists (Link, Sprengel, *et al.*) it was reserved for horticultural elements, by early devotees of the American Code it was "applied indiscriminately to anything below the rank of species" (Weatherby, p. 161), and by some contemporary biosystematists it is employed for an unproved but suspected ecotype. (For definition and discussion of the term ecotype, cf. pp. 176–177.)

The *variety* (Latin, *varietas*) has been used as a category to designate as many or more concepts as has that of subspecies. Horticulturists have used it indiscriminately for any variant of the species; botanists have considered it to be (1) a morphological variant of the species without regard for distribution, (2) a morphological variant having its own geographical distribution, (3) a morphological variant sharing an area in common with one or more other varieties of the same species, and (4) a variant representing only a color or habit phase. From this it is clear that the same plant may be designated a subspecies by one botanist and a variety by another, or that the variety of one author is placed in the category of *forma* by another author. This lack of unanimity of concept is disconcerting, but it is a factor to be recognized in any appraisal of taxonomic literature. In this regard, it is not especially important that agreement exist if by even diverse modes of evaluation the same pattern of relationship is reached. There is no historical basis for priority of usage of either the term subspecies or variety. As stated by Fosberg (1942),

The solution seems actually simple enough, if one recognizes that there are many types of evolutionary process in operation, producing many kinds of species, and

---

[10] For details concerning these and other concepts of the subspecies category, cf. references in bibliography to papers by R. T. Clausen, Du Rietz, Fernald, Fosberg, Pennell, Weatherby.

that intraspecific units may be incipient species in various stages of development. These stages may be at least roughly indicated by the categories in which the groups are placed. Each taxonomist may take up the system of categories set up in the International Rules and apply it to the groups of plants with which he is working in the way that, in his judgment, best expresses the relationships of the groups of individuals concerned. The Rules require only that the order of the categories be not disturbed, and that each plant be placed in a species, genus, family, order, class, division and kingdom. All other categories are to be used at the discretion of the worker. In this way the system will retain the flexibility that is absolutely essential to make it fit the wide variety of evolutionary situations to which it must apply. Discarding any of the categories, whether from reasons of historical confusion or personal prejudice, impairs this flexibility.

The *forma* is commonly the smallest category used in ordinary taxonomic works. It is generally applied to trivial variations occurring among individuals of any population. Such variations as represented in corolla color, fruit color, habitat response are, by this concept, placed in the category of *forma*. Some botanists consider a *forma* to be any variant that occurs sporadically in a species population irrespective of the degree of morphological variation or constancy, the significant criteria being that of no geographical discontinuity. The *forma* of this latter concept is equivalent to the variety of many botanists.

*Lesser categories* below that of *forma* appear in the literature, primarily in European floras and in horticultural works. The category termed a *race* is employed in some floras in lieu of or subordinate to the term *forma*.[11] In horticultural work, the term *clone* [12] is applied universally to individual plants that are propagated by asexual (vegetative) means. The original definition of the term inadvertently has been considerably amplified and emended by some geneticists and biosystematists to include also those indigenous plants that reproduce asexually by apomixis.

### Morphological criteria

All taxonomists are agreed that the differences between plants, and the similarities that plants may possess in common, are measurable to a large degree by the morphological characters of those plants. A morphological character, to a taxonomist, is one inherent in or manifested by

---

[11] By some authors of biosystematic works the term race or ecological race is used in lieu of the term ecotype. This usage may be unfortunate and seemingly may add to the ambiguities of terminology, since the term ecotype (although of varying definition) is now well established in the literature and need not be abandoned for etymological reasons nor simplified to facilitate understanding. The term race has been used in other senses by geneticists and its use as a taxonomic category by some European systematists is not to be ignored.

[12] For discussions of the origin and spelling of the term and its application, see Stearn, W T. (1947).

a structural component of the plant. In general it is concerned with the organography of the plant, with characters that usually are discernible with the aid of no more than a good hand lens. The more recent recognitions by taxonomists of the significance of the related sciences of anatomy, cytology, and comparative morphology have resulted in an expansion of the concept as to what is a morphological character and to the extreme that some taxonomists now consider the number of chromosomes to represent as significant a morphological character as that provided by the number of stamens. The value of the morphological character is measured by its constancy. The more constant the character, the greater is the reliability that can be placed upon it. In this regard it must be remembered that a character that is reasonably constant among the members of one group may be a weak character among members of another group. For example, among the families of some dicotyledonous plants the character of ovary position (inferior vs. superior) is adequate to separate families and orders, but within some monocotyledonous plants the two types of ovary position are to be encountered within a family (e.g., Bromeliaceae), or even within a single genus (e.g., *Bomarea*).

***Vegetative vs. reproductive structures.*** The vegetative structures of plants (i.e., those of leaves, stems, buds, and habit of growth) are readily apparent to the eye and have an appeal to the beginning taxonomist. Their observation generally does not require the use of a microscope or high-powered lens nor the employment of razor blade or dissecting needles. However, their use as taxonomic criteria is limited for two basic reasons: the total number of vegetative characters available is few as compared with the number of species of vascular plants, and too often the vegetative characters are not particularly constant.

The number of species of woody plants of any temperate flora is usually fewer than the number of species of herbaceous plants of the same flora. For this reason, most genera, and sometimes the species, of these woody plants may be separated largely and almost exclusively by vegetative characters. Keys to woody plants are popular for this reason. However, even among the genera of woody plants, vegetative characters lack the necessary variability of type, stability, and constancy necessary for the identification of species. Examples are abundant in such temperate genera as *Crataegus, Prunus, Quercus, Rhododendron, Salix*, and *Vaccinium*. As a rule, the greater the number of species in any genus of woody plants the more difficult is their identification by means of vegetative characters.

In the preparation of keys [13] to woody plants it frequently is desirable to use vegetative characters to as great an extent as practicable. It has been found from experience, in dealing with plants of temperate regions, that certain vegetative characters are generally more constant than are others. (In certain genera and families, however, there are notable exceptions.) Those characters known to be more reliable than others are listed below in sequence of approximate descending constancy:

> Leaves needle- or scalelike vs. leaves broad and laminate
> Venation parallel vs. venation reticulate (netted)
> Leaves compound vs. leaves simple
> Plants erect trees or shrubs vs. plants vinelike
> Stipules present vs. stipules absent
> Leaves lobed or divided vs. leaves neither lobed nor divided
> Leaves opposite or whorled vs. leaves alternate
> Leaf margins entire vs. leaf margins variously toothed
> Leaves persistent and coriaceous vs. leaves deciduous
> Spines or prickles present vs. spines or prickles absent
> Leaf base tapering (cuneate) vs. leaf base broad (cordate, obtuse, truncate)
> Leaves with 1 main mid-vein vs. leaves with 3 primary veins

Succeeding the above would be characters of leaf form, leaf apex, the leaves petioled or sessile, and pith characters of the twig or stem. Characters of vesture (pubescence), surface, coloration, milky vs. watery sap, and sizes of plant or leaves are generally useful only in the separation of species or their variants.

Reproductive structures are those associated with the flower and fruit. The number of organs in these structures, and the number of components of these organs, are many times that available from vegetative structures. For this reason the number of mathematical combinations and recombinations of reproductive characters is infinitely greater and ordinarily exceeds the number of species (or smaller units) to be differentiated. In addition to being more abundant, reproductive characters in general are much more nearly constant than are vegetative characters. The reliability of any single reproductive character selected for taxonomic use must be determined individually, however, for each particular group of plants.

**Ligneous vs. herbaceous characters.** In recent decades increased emphasis has been placed on the phyletic significance of the woody as opposed to the herbaceous character of plants. It is believed by many botanists that among the higher vascular plants, those of woody habit were the antecedents of those of herbaceous habit. Hutchinson, in the latest version of his classification (1948), has made this hypothesis the

---

[13] Cf. pp. 225–228 for discussion of keys, their preparation, and use.

basis of his two classes (*Lignosae* and *Herbaceae*) of the *Dicotyledoneae*. In doing so he warned

... against taking the view that he has reverted to the old practice of putting all the woody plants into one group and all the herbaceous plants into another. . . . What has been done is to put all those families which are *predominately woody*, and the more *primitive* genera of which are *woody*, into the woody phylum, and conversely the same has been done with the *herbaceous* families.

In other words, the *woody* division consists of *trees and shrubs plus herbs which are related to trees and shrubs*, and the *herbaceous* division consists of *herbs plus some clearly related woody plants*.

In addition to the significance placed on these criteria by Hutchinson, the more elemental distinction between woody and herbaceous plants has practical value in artificially separating the minor units of classification. In many cases, the demarcation between the two criteria is not sharp. Some plants that appear to be herbaceous are actually woody in that their stems in addition to being woody in texture are of perennial duration (e.g., *Gaultheria procumbens, Epigaea repens,* or *Vinca minor*). Other plants are basally woody but have stems that for the most part are herbaceous. Such plants are often referred to as being suffrutescent in character (e.g., *Pachysandra terminalis, Arabis caucasica,* or *Aralia hispida*).

**Inflorescences.** An inflorescence conventionally is considered to be the arrangement of flowers on the floral axis. By Linnaeus (1751), it was said to be a mode of flowering; that is, he considered it to be a phenomenon or function and not a structural component of the plant. The present-day classification of inflorescences, although utilizing to an extensive degree the terminology of Linnaeus, is based on the relatively unscientific observations of Link (1807) as modified by Roeper (1826).[14] These botanists considered the inflorescence to be a structural entity of the plant, subject to subdivision into definite classes and types for use as taxonomic criteria. The basic classification of the inflorescence by Roeper (into a series of determinate and indeterminate types) is, with some amplifications and modifications, that used today in taxonomic and less technical descriptions of flowering plants.[15] As pointed out by Rickett (p. 189), inflorescences are branch systems and are "all related

[14] See Rickett, H. W. (1944), for a detailed analysis of the subject of inflorescence classification, and from whose review much of the data for this section of the chapter were obtained. See also Croizat (1944) for views treating the inflorescence as an aggregate of vegetative and floriferous axes.

[15] A determinate inflorescence, simply expressed, is one in which the terminal flower of the axis opens first with the remaining flowers of that axis opening successively from top to bottom. An indeterminate inflorescence is one in which the lowermost flowers of an axis open first and the terminal flowers open last, with the successive opening of flowers accompanied by an elongation of the axis.

in the sense of being derivable from one underlying type." It is conceded generally that the primitive inflorescence or its antecedent was a system of branches each of which was terminated by a flower. Every individual flower is a terminal unit. It is borne on the end of an axis, and it matters not whether that axis conventionally be termed a peduncle, pedicel, receptacle, torus, or stem. The structure represented by any one of these terms is technically a floral axis. It is important to an understanding of the use of inflorescences in the classification of these plants to know what type of inflorescence is primitive, and of the possible developmental sequence of other types from the primitive. Unfortunately there is yet insufficient knowledge concerning the phyletic relationships of the families of flowering plants to permit more than speculation on the origins of inflorescences. However, speculations based on what are believed to be the best available data do provide rational approaches to a solution of the problem.

Three theories have been proposed to account for the evolution of present-day inflorescences from the presumed primitive condition of a branch system of terminal flowers. The first postulates the primitive type to have been a panicle, the second that it was the single terminal flower, and the third that it was a dichasium. The first theory has been held by many investigators, including Nägeli (1883), Čelakovský (1892), and Pilger (1922). By acceptance of the panicle as primitive, it is easy to explain the derivation of all other types from it as was done in part by Goebel (1931). Pilger supported his views in defense of the theory with analogies which he conceded do not exist among living woody angiosperms. The second theory, supported by Parkin (1914), proposes the development of multiflowered inflorescences from the solitary flower by addition of lateral flowers (and branches of flowers) by axillary growth. Any consideration of this theory must take into account the generally accepted view, discussed later in this chapter, that branch systems are a fundamental part of the plant organization, and that the flower is probably made up of highly modified branch systems. Acceptance of the belief that foliar organs are highly modified branch systems adds logic to a theory that the primitive inflorescence also was a branch system and that most existing examples of solitary flowers represent examples of reduction and suppression. The third theory, that the dichasium (Fig. 3) may represent the primitive type, has been accepted by Rickett (1944).

In its simplest form the primitive dichasium consisted of three flowers disposed in a cluster formed on a single peduncle by a dichotomous branching immediately beneath a terminal flower (Fig. 3q). From this simple dichasium there may be developed a more complex dichasium that is formed by "a repetition of the same apparent dichotomy in each lateral branch." In nature, the dichasium may be simple or compound, ample or restricted, and it would be hazardous to speculate as to which form preceded the other in ancestral types. There is no clear paleobotanical evidence to support the view that the dichasium was ancestral to the panicle, neither can the latter be established as the more primitive. Either one may have been derived from the other. There is some fossil evidence that in a few rare instances the solitary flower may represent a primitive situation. It has been shown (Florin, 1948) that primitive and now extinct conifers possessed a paniculate inflorescence, a fact that should not, however, be construed as evidence indicative of gymnospermous ancestry of the angiosperms. There is a theory held by many botanists that the Pteridospermae (seed plants with fernlike foliage) provide the ancestral group best possessing angiospermlike morphology; in this taxon the ovules, borne on complex fernlike leaves, suggest a paniculate arrangement. Irrespective of the relative primitiveness of the panicle and the dichasium, the views of Rickett on the possible evolution of inflorescence types, as outlined below, are most helpful and provide a logical understanding of possible relationships.

The dichasium was postulated by Rickett to have given rise, by only slight structural modifications, to inflorescence types such as bostryches, cincinni, verticels, cymes, helicoid cymes, scorpioid cymes, and by apparent elimination of internodes it may have been an antecedent to types such as the umbel, capitulum, and corymb (Fig. 3). The obvious structural organization of complex and congested inflorescences indicates that many of them may have developed from various combinations and aggregations of dichasia. In this regard, it is now believed that the seemingly simple raceme and spike represent types that are reduced, along one of two or more evolutionary lines, from more complex ancestral types of dichasia or panicles. That is, the raceme or the spike may have been evolved through steps involving the reduction of individual dichasia to as little as a single flower accompanied or not by similar reduction of leaves to subtending bracts. It has been established that, in many cases, the presumed distinction of an inflorescence being determinate (i.e., centripetal) or indeterminate (centrifugal) is without validity and that ultimately in a new classification of inflorescence types, this criterion will have less value than is currently accorded it. In certain inflorescence types it is a valid characteristic, while in others it is unreliable. For example, the umbel of members of the Umbelliferae is an indeterminate type of inflorescence whose outer flowers open before the inner ones. However, in many of the Amaryllidaceae (as in *Allium* [16]) the umbel is a determinate inflorescence as indicated by the central flower opening prior to the outer flowers. Structurally, the inflorescence is an umbel in either family, but phyletically the two kinds of umbel are unrelated and represent reductions by two different evolutionary lines (Fig. 3).

A derivative and advanced group of inflorescence types over those considered to have more direct affinity with the dichasium is that known as the *monochasium*. A monochasium is an inflorescence resulting from a dichasium in which one branch of each dichotomy continues to develop while the other branch is suppressed completely (e.g., aborts). The result is a sympodial axis composed of a series of superposed axes and is sometimes referred to as a sympodial dichotomy (Fig. 3). This type of inflorescence is phyletically a complex inflorescence, since it is made up of a series of branch systems. It is the "scorpioid cyme" of botanical floras and manuals and is characteristic of many boraginaceous genera (*Myosotis, Heliotropium, Symphytum*) as well as of representatives of the Polemoniaceae and Hydrophyllaceae.[17] In these taxa its character and origin are usually obvious by macroscopic inspection. However, reduction often has advanced to a degree where the monochasium is not as apparent in all cases as in the so-called scorpioid cyme. That is, in some groups, the sympodial false axis (actually composed of determinate units) may be so refined in appearance as to seem to be a simple and true axis of indeterminate growth; such an inflorescence is the raceme of *Claytonia*, an inflorescence that morphologically is as much a helicoid cyme as is that of *Myosotis*. In other genera, the monochasium has been reduced to an umbel as in the florists' geranium (*Pelargonium*). The umbel of *Allium*, mentioned above, is of similar origin.

From this brief review of possible origins of inflorescences, it may be speculated that either the panicle or the dichasium (rarely the solitary flower) is the primitive type of inflorescence, and that from them have been evolved all other types of inflorescences. That is, the solitary flower, the panicle, and the dichasium all may be primitive types.[18] By this spec-

---

[16] For discussion of the family to which *Allium* belongs (Liliaceae vs. Amaryllidaceae), cf. Part II, p. 415.
[17] Note from study of Fig. 3, that the so-called scorpioid cyme here referred to in quotes is properly a helicoid cyme (helicos, meaning snaillike) and that the true scorpioid cyme is an inflorescence type not commonly encountered.
[18] The pteridosperm leaves bear ovules in ways that would provide a background and historical basis for this: the ovule terminal and solitary on huge fronds; the ovules terminal on few or all pinnules; or the ovules marginal and terminal throughout the frond.

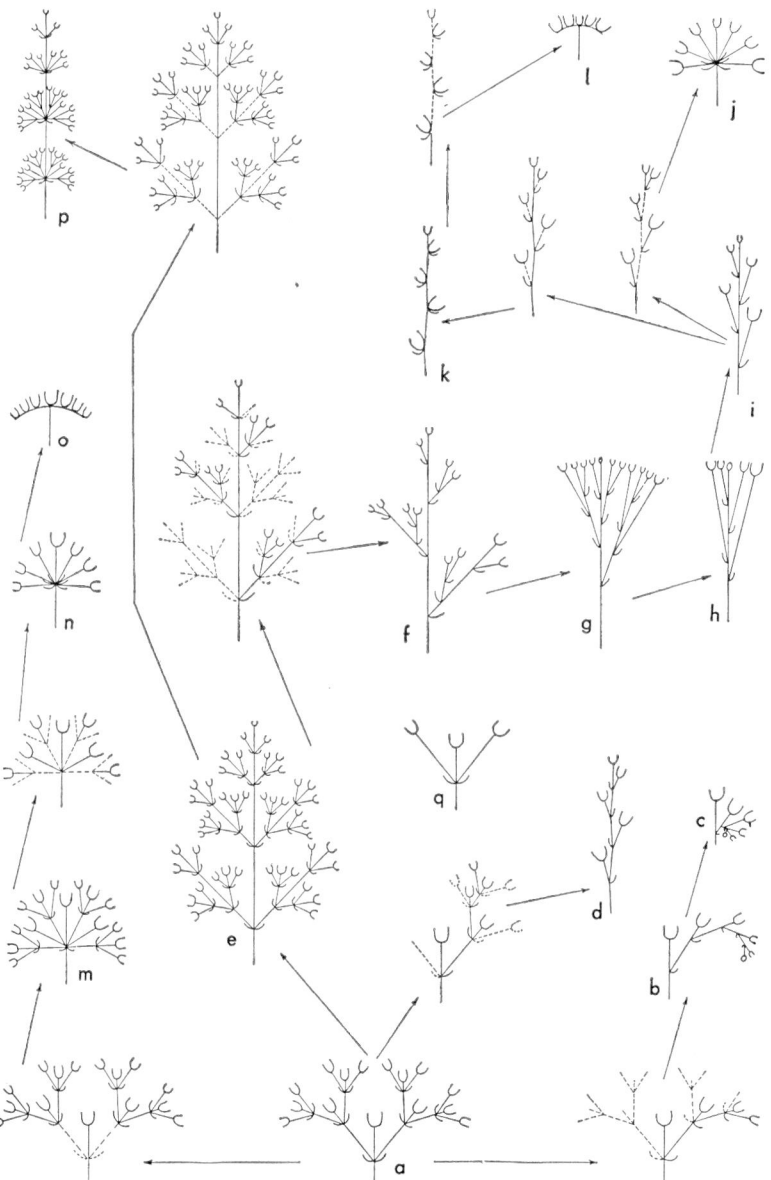

**Fig. 3.** Schematic diagrams of hypothetical evolution of inflorescence types: a, compound dichasium; b, helicoid cyme; c, cincinnus; d, scorpioid cyme; e, thyrse; f, panicle; g, compound corymb; h, simple corymb; i, raceme; j, indeterminate umbel; k, spike; l, indeterminate head; m, cyme; n, determinate umbel; o, determinate head; p, verticillate inflorescence; q, simple dichasium.

ulation, it is indicative that usually the conventional raceme is more primitive than the spike but that both are advanced over the panicle, that while the panicle may have been primitive, some present-day examples of the panicle may have been derived from a thyrse, and that the numerous cymose types of inflorescences are perhaps more primitive than most of those mentioned above. Accordingly, the modern or present-day inflorescence represented by a solitary flower in almost all plants is a product of reduction, a reduction that may have resulted from any one of many multiflowered inflorescence types, whose origin may be indicated only by comparative studies of inflorescences of related taxa.

The significance of the inflorescence as a taxonomic criterion has been established. In some families it is of significance in recognizing entire families (Umbelliferae, Cornaceae, Compositae, Labiatae, Gramineae, Amaryllidaceae [*sensu* Hutchinson]) and in others in recognizing tribes or genera (*Galium, Smilax, Malus, Philadelphus*). The consideration of phyletic significance of the inflorescence cannot be evaluated apart from other criteria, but in dealing with minor units of flowering plants often it may be helpful in establishing their position in a particular schema or classification.

***Reproductive elements and their taxonomic significance.*** Taxonomists, since and before the time of Linnaeus, have relied heavily on the value of characters inherent in the reproductive parts of vascular plants to serve as criteria of particular significance in their classifications. In the pteridophytes reproductive characters are associated in general with the sporangia and sex organs of the gametophyte; in the gymnosperms they are those of the pollen-producing and fruiting structures; and in the angiosperms are those of both the flowers and the fruiting structures. This last taxon is the most complex of all, and phylogenetically is the most highly developed and advanced. For these reasons great diversity and complexity of reproductive characters are typical of its members. Since the objectives of taxonomy include identification and classification, it is necessary to understand the composition of the flower, of the theories purporting to account for the origin of its components, and of the presumed evolutionary levels represented by present-day types of these components. The disciplines of morphology, anatomy, and paleobotany have contributed the data and the conclusions responsible for these understandings.

Prerequisite to any basic understanding of the flower is an appreciation of the view that vascular plants structurally are evolved from or composed of different types of branch systems. As applied to flowering plants, and on the evidence adduced from the fossil record that terminal reproductive structures were present prior to the evolution of foliaceous leaves, it is probable that at least 3 types of branch systems have served as antecedents of present-day leaflike structures (Wilson, 1941). These 3 types are:

(1) a branch or cluster of branches bearing terminal sporangia from which the stamen is descended; (2) branches which became flattened (but were never primarily photosynthetic) to become the ovule-bearing carpels; (3) sterile branches or branch systems which became the leaves.

This postulate should be kept in mind when considering the conventional and very generally accepted theory (Eames, 1931) that "the flower morphologically is a determinate stem with appendages, and these appendages are homologous with leaves." In other words, flowers are homologous with leafy stems; the pedicel and receptacle of the flower are homologs of the stem axis, and the sepals, petals, stamens, and pistil(s) are homologs of the leaves (i.e., the latter are appendages of the stem). If the quotation from Wilson is accepted, it becomes clear that although foliaceous leaves may be appendages of stems they may be presumed to have been evolved by the dorsiventral compression and lateral fusion of photosynthetic branches.[19] Similar origins were postulated by Wilson to account for the leaflike appendages represented by the stamens and by the carpels. Recognition that the foliar units are fundamentally modified branch systems brings the contentions of some botanists that floral parts are of appendicular origin into complete harmony with the contention of others that they are of stem or branch origin. Statements to the effect that foliage leaves and floral elements are homologs does not mean that they are necessarily derived one from the other.[20]

The flower typically is composed of the following 4 types of components (as arranged from the bottom of the floral axis upwards): *sepals,* collectively the calyx; *petals,* collectively the corolla; *stamens,* collectively the androecium; and *carpels* (the basic unit of the pistil), col-

---

[19] By "dorsiventral compression and lateral fusion of photosynthetic branches" is meant the flattening of presumably green terete secondary stems in a single plane with their posterior (inner) surfaces in a ventral (uppermost) position, and furthermore, that by marginal connation of adjoining edges a green leaflike organ evolved. By this view, any leaf is a derivative of a group of branches and because of this origin represents a highly modified branch system.

[20] The views presented here to account for the nature and origin of the angiospermous flower are believed to represent opinions of the majority. However, there are other and contradictory views on these matters, and for an introduction to them the reader is referred to the papers by Arber (1937), Bancroft (1935), and Ozenda (1949), wherein the opposing views are reviewed but not supported. Of these, perhaps the most widely publicized was Saunders' *theory of carpel polymorphism,* early refuted by Eames (1931) and currently rejected by most investigators.

lectively the gynoecium. The calyx and corolla comprise the perianth (floral envelope, perigone); neither is essential to the functioning of the flower as a reproductive structure.

**Perianth.** The perianth is an accessory part of the flower more or less enveloping the organs of reproduction, and is classically treated as composed of an outer cycle or whorl (the calyx) and an inner cycle or whorl (the corolla). This explanation is one of extreme simplicity. The origin of the perianth of the modern flower is usually explained by one of two hypotheses.

By one interpretation, the flower is assumed to have primitively been provided with a perianth of 2 cycles (one the calyx and the other the corolla). If it is assumed that upon phylogenetic reduction the corolla inevitably drops out first and the flower then becomes apetalous, the remaining perianth is one of sepals. On the other hand, if it is accepted that reduction may affect each cycle equally, and that the corolla is capable of assuming a sepaloid form (or the sepals a petaloid form), then a perianth is to be considered composed of units of undeterminable phyletic origin with its components termed *tepals* (as in many monocots).

The second and more widely accepted interpretation treats the ancient flower to have been an axis bearing many foliaceous bracts in a spiral arrangement, and assumes that the apical ones were fertile (mega- and microsporophylls) and the lower ones were sterile. These sterile bracts were similar in appearance, with the upper ones gradiently resembling more closely the microsporophylls than they did the lowermost sterile bracts. Because of this foliaceous similarity they are designated tepals. From this primitive ancient condition, and in conformance with the laws of phyllotaxy interacting with progressive spatial restrictions (resulting in apparent vertical compression of the axis), there was evolved the modern flower. This has a perianth that is usually dicyclic (and sometimes tricyclic) and pentamerous, tetramerous, trimerous, or dimerous. By his view, all parts of the perianth are homologous, and phylogenetically are tepals. The terms calyx and corolla become descriptive terms of taxonomic importance, but may lack phylogenetic significance. Differentiation into calyx and corolla may and probably did take place at many phylogenetic levels in angiosperm evolution, and there is ample evidence of petaloid calyces and of sepaloid corollas. Generally speaking, characters of the perianth are of greater significance to plant identification than to classification. Some apparent corollas have been derived from the androecium (Aizoaceae), others represent petaloid calyces (many ranaliaceous plants).

**Sepals.** The sepals compose the outermost series of the components of the typical flower and collectively comprise the *calyx*. Most sepals are supplied by three vascular strands (traces or bundles), as are also most leaves, and possess anatomy like that of the leaves of the plant on which they appear. For these reasons, they have been considered morphologically to be bracts and to have been derived from foliaceous leaves. Generally they are green in color (photosynthetic) and bear a resemblance to bracts or leaves. They may be distinct [21] (polysepalous or aposepalous) or connate (gamosepalous or synsepalous) or in some families be reduced to modified hairs or scales and referred to collectively as a pappus.[22] Sometimes they are highly colored and petallike (*Delphinium, Helleborus, Mirabilis*).

**Petals.** The petals compose the second series, or inner envelope, of the perianth (i.e., the corolla). Morphologically and anatomically they are more often like sterile stamens than they are like leaves. Evidence for this view is based on the presence (in most petals) of a single vascular strand, a characteristic also of most stamens. Usually the petal is colored other than green (nonphotosynthetic) and contributes to the showiness of the flower. It has been considered that corollas of distinct petals are more primitive than those whose petals are marginally connate, and this situation has been accorded much significance in classification schemas. However, Hutchinson (1948), who accepted polypetaly (apopetaly) as more primitive than gamopetaly (sympetaly), rejected the view that these two criteria were of sufficient phyletic importance to separate orders of plants into two subclasses (Polypetalae and Gamopetalae). The criteria were treated by him to be of minor phyletic importance. Terms used in describing the shape, texture, and disposition of perianth parts are the same as are used in describing similar parts of leaves.

**Stamens.** The stamen is the male reproductive organ of the flower. Collectively the stamens comprise the *androecium*. The stamen is conventionally treated as composed of an *anther* (the pollen-producing element) and a *filament* (the stalk). The more recent views of Wilson (1941), based on anatomical studies of the androecium of the Parietales and

---

[21] See footnote 38 (p. 84) for explanation of the terms *distinct* and *connate*.

[22] Many taxonomists assume the pappus of the Compositae flower to represent in all genera a reduced and modified calyx. Studies by Koch (1930) of the floral anatomy in the Compositae indicated that the calyx and corolla of some ancestral types had fused to form the corolla of some present-day members. In those members of the family where this has been suggested the pappus is presumed to be bracteate in origin. This latter view also provides plausible explanation for the origin of the apparent double pappus encountered in many genera of the family, in which situations the inner one may or may not be a modified calyx and the outer one perhaps derived by modification of bracts.

Malvales, differ in some respects from the perhaps better-known classical theories of androecial origins as presented originally by Goethe and by de Candolle (1813). The older theories treated the stamen as a sporophyll, a fertile homolog of the foliaceous leaf, whereas by Wilson's view it was treated as a direct derivative of a primitive branch system bearing terminal sporangia, that is, a branch system that did not pass through any stage of dorsiventral compression in its phyletic development. In amplifying this premise, Wilson further considered that

> A hypothetical primitive stamen may be conceived, in which a relatively long shank, or axis, itself an arm or limb of a dichotomy, is terminated by a system of dichotomously-divided branches, each ultimate branchlet bearing a single sporangium. By reduction of the two most remote dichotomizing branches, with the resultant fusion of pairs of sporangia, an anther of two synangia each composed of two sporangia may be derived. . . . The evolution of anthers, two-celled at maturity, but four-celled in ontogeny, probably took place very early in the evolution of the Angiosperms.
>
> . . . it may be concluded that in those flowers with numerous stamens, . . . the stamens represent not primary organs or sporophylls of the flower, but the apices of branch systems, the bases, or many branches of which, have dropped out in the course of evolution. The presence of a few and fixed number of stamens is then to be regarded as a modification of the condition obtaining in the flower with numerous stamens, each stamen representing the reduction of an entire branch system.

This view of the origin of the stamen is pertinent also to any consideration of the origin of the petal of most petaliferous families, since the petals are generally conceded to have been evolved from stamens. The stamens and petals of most flowers are supplied by a single trace, but in members of several families (including the Magnoliaceae and Lauraceae) each stamen is supplied with 3 traces, evidence of a primitive staminal condition in these representatives. It was postulated by Wilson that the staminal tube present in monadelphy (monadelphous stamens being those whose filaments are fused laterally into a cylinder, as in the hollyhock) is the "result of fusion of several branch systems" and is indicative of a primitive androecial condition.

If the above hypotheses are accepted, monadelphous stamens are among the more primitive, diadelphous stamens slightly more advanced, while those groups having stamens clustered in fascicles are more primitive than distinct stamens. Among the groups with androecia of distinct stamens, those with stamens numerous would be considered more primitive than those with stamens few and definite in number, and androecia reduced to a single stamen would be the most advanced of all types.

It is significant that angiosperm families now generally conceded to be the more primitive (Magnoliaceae, Ranunculaceae) have numerous stamens that are spirally arranged. Furthermore, since the stamens are distinct and free, it is to be presumed (in accordance with Wilson's findings) that they represent emerging branch tips of complicated branch systems. The spiral arrangement of floral parts has been accepted as a condition more primitive than that of whorled arrangement. A study of androecial types indicates that the views of Wilson are not necessarily contradictory to the classical concept that spirally arranged stamens are more primitive than whorled stamens. Wilson has attempted to account for situations where there exists a reduction in the number of stamens (per unit branch of the androecium). This reduction would appear to be an independent line of advance from that represented by the reduction from the spiral to the whorled arrangement. There is anatomical evidence that (1) some present-day stamens, superficially simple, are individually the remnants of much reduced (and usually dichotomous) branch systems (e.g., in the Melastomaceae), (2) within some orders of plants monadelphy preceded polyandry (e.g., Malvales), and (3) that in some admittedly primitive orders the androecium of present-day plants is represented by tips of emerged branch tips disposed in a high phyllotaxy (i.e., spirally arranged, as the stamens in Magnoliales and Ranales). The fossil record produces no known barrier to the view that each of these 3 manifestations may have arisen independently. Each of the 3 may represent lines of androecial evolution whose phyletic significance can be evaluated only in relation to other phyletic criteria. A taxon accepted to be advanced, as determined by an aggregate of characters, may possess an androecium of a presumed primitive type (as in Malvaceae).

Relatively few families are characterized by having spiral stamen arrangements (Fig. 4 Aa), and in the majority of families the stamens are whorled (Figs. 4 Ab-Ad). Families with a generally fixed number of whorled stamens usually can be divided into those whose stamens are in two whorls or are in a single whorl. Presence of a single whorl is considered the more advanced condition because, in most instances, the anatomical evidence has shown that the second (and inner) whorl of stamens that were present in their ancestral types has since been lost by suppression or by reduction. In some taxa, the remnants of this second whorl can be recognized as staminodes or nectaries, or as additional petals. The order or sequence in which stamens develop is of taxonomic importance, especially in an androecium of many stamens. In most families development is centripetal (the outer ones developing before the inner, sometimes described as acropetal) but in others it is centrifugal (basipetal).[23] This distinction indicates relationships between groups of one type or the other, and the opinion is held (Corner, 1946) that "the primitive massive, centrifugal androecium must have been derived from the usual centripetal state."

In addition to the presumed phyletic significance and origin of stamens as discussed above, there are numerous other characteristics associated with them that are of significance especially in the taxonomy of minor

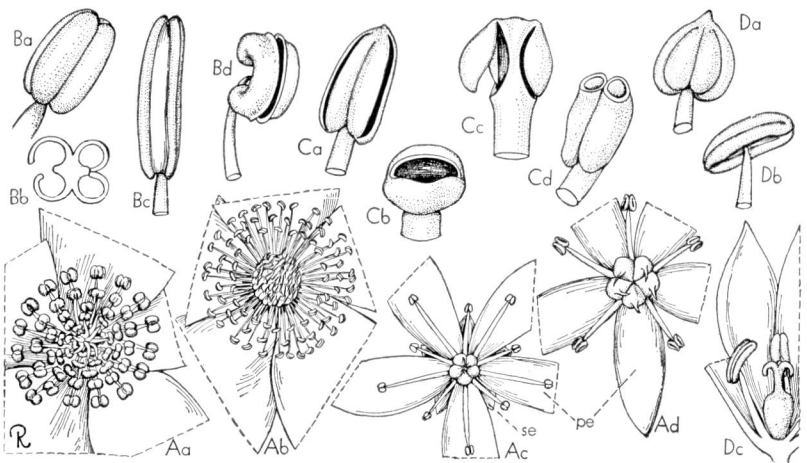

**Fig. 4.** Androecium and stamen types. A, androecium types: Aa, stamens spirally arranged; Ab, stamens in three whorls; Ac, stamens in two whorls, those of outer whorl the shorter (diplostemonous); Ad, stamens in a single whorl (dehiscence introrse). B, anther cell number: Ba, cells four at anthesis; Bb, same, cross-section; Bc, cells two at anthesis; Bd, cell one at anthesis. C, anther dehiscence types: Ca, longitudinal; Cb, transverse; Cc, valvular flaps; Cd, apical pores. D, anther position: Da, basifixed; Db, dorsifixed and versatile; Dc, dorsifixed, not versatile (dehiscence extrorse).

taxa. The anther of the typical stamen at maturity (anthesis) is composed of two *thecae* (actually each theca at time of anthesis represents two sporangia, the theca often termed a cell or chamber) separated from one another by an extension of the filament passing between them (connective). Fundamentally the anther is of 4 thecae, but the septum between the two adjoining ones situated on the same side of the connective usually disappears during the late ontogenetic development of the anther and prior to anthesis, although in a few taxa (as Lauraceae) the septa persist and the anthers are then 4-thecate at anthesis. In some families (Malvaceae, some Bombacaceae) the anther has a single theca, a situation

---

[23] Corner (1946) cited the following families as characterized in part by having centrifugal androecia: Actinidiaceae, Aizoaceae, Bixaceae, Cactaceae, Capparidaceae, Dilleniaceae, Hypericaceae, Loasaceae, Lecythidaceae, Malvaceae, Paeoniaceae, Theaceae, Tiliaceae.

shown by Wilson to be due to "the fusion of two neighboring sporangia into single synangia." The anther is fundamentally an organ that is terminal on the filament and when so it is said to be *basifixed*. A more advanced situation is present when the anther appears to be attached dorsally to the filament and is then said to be *dorsifixed,* and when dorsifixed it is sometimes *versatile* (Fig. 4 D). The anther cell (theca) dehisces typically by a vertical or longitudinal split, but in more advanced situations the dehiscence may be transverse (*Hibiscus, Elatine*) or by a pore (Polygalaceae, most Ericaceae). In some families the connective between the thecae becomes highly developed and appendaged, even to the extent (in Cannaceae and Marantaceae) of being petaloid and showy. The androecium is said to be *syngenesious* when its anthers are connate marginally. The anthers of syngenesious androecia ordinarily (in Campanulaceae spp., Compositae) form a cylinder through which the style or stigmas pass, but sometimes (in some Gesneriaceae) the syngenesious anthers are dissociated and remote from the style (see Fig. 317e).

In the development of the anther there are produced in the pollen sac many usually rounded cells termed *pollen mother cells.* Four *pollen grains* usually are formed from each of these. Pollen grains are of different morphological types and are receiving increasing recognition as a source of valuable taxonomic characters. The outer membranous coat of the pollen grain is cuticular in composition and is termed the *exine*. Often it is formed into various kinds of sculptured surfaces (furrowed, papillose, honeycombed, echinate, etc.) that are characteristic of particular taxa. Pollen grains that are free from one another are said to be granular. Sometimes the pollen grains are in clusters of 4, *tetrads* (as in many Ericaceae), but in at least two families (most Asclepiadaceae, Orchidaceae) the pollen of each anther sac is agglutinated into usually waxy to firm masses called *pollinia*. Pollen grains are often of value in distinguishing possible or suspected hybrids from nonhybrids, since the pollen of some hybrids is nonviable and is shrunken and nonturgid in appearance.[24]

A conspicuous modification of the androecium is the reduction of one or more stamens to sterile structures termed staminodes. These may

---

[24] Viable fresh pollen grains (of nonhybrid origin) contain an abundance of chromatin material. Smears or smashes of such pollen are stained readily with aceto-carmine solution or aniline (cotton) blue, and an indication of possible hybridity may be demonstrated if the stained smears of pollen of suspected hybrids fail to take on any of the stain. The technique and interpretation of the findings require skill and experience. A magnification of × 120 is generally adequate in the application of this preliminary test. It is emphasized that this test may only indicate possible hybridity and results should never be interpreted as presumptive of hybridity.

appear as a filament minus an anther (*Penstemon*), as expanded petal-like structures (members of the Aizoaceae, in *Canna*), knobbed glandular processes (many orchids), or in the form of a nectary (Loasaceae).

Some groups of plants are pollinated almost exclusively by wind and are described as being *anemophilous*. Others are pollinated largely by animal vectors, particularly by insects, and are said to be *entomophilous*. There has been considerable discussion over which of these conditions is the more primitive, and some of the basis for divergent views may be due to the probability that some of the present-day anemophilous plants perhaps have evolved from ancestors that were entomophilous, and that the latter in turn were from anemophilous prototypes. It is probably not possible to say that one of these conditions of itself is more primitive than the other, but the presence of one or the other may indicate relationships in particular cases.

**Carpels.** The carpel is the basic foliar unit of the female reproductive organ of the angiosperm flower. It is an ovule-bearing structure evolved from what is currently and generally accepted to have been a nonphotosynthetic foliar leaflike appendage, and as such was a megasporophyll. This nongreen leaflike appendage is believed generally to have had a development parallel with that of the foliage leaf. It is not believed to have been derived from a foliage leaf. When solitary in the flower, the folded carpel and its stigmatic surface are better known taxonomically as the ovary and stigma. These two structures (ovary and stigma), plus the style (or styles) when present, comprise the *pistil*. The term carpel sometimes is used in taxonomic works in instances where the term pistil is the more common.[25] The two terms are not always synonymous. The ovary of a pistil may be unicarpellate or it may have been formed by the

---

[25] There is no unanimity of opinion concerning the definition of the term pistil. Some taxonomists, and many morphologists, consider the term a source of confusion and abandon it. In its place they apply the term carpel to all unicarpellate organs and gynoecium to all multicarpellate female elements within a single flower (the latter use, irrespective of the number of free ovaries present). Basis for this view lies in the contention that the term pistil has been applied variously during the years of descriptive botany. These reasons for rejection of the term pistil are considered untenable and use of the term is defended, on the grounds that to a taxonomist (working primarily with macroscopic organs) 3 basic terms are needed to discuss the essential features of the female element: (1) the megasporophyll or foliar element, (2) the gross structure represented by a single ovary with its style (or styles) and stigma (or stigmas), irrespective of the number of foliar elements entering into its composition and of the freedom or fusion of those elements, and (3) the female element or household of the flower *in toto*, irrespective of the number of free and distinct ovaries involved. The 3 terms that meet these taxonomic needs are, respectively: carpel, pistil, and gynoecium. As defined above, the term pistil is neither ambiguous nor confusing. It is used by current taxonomic workers and was used (among others) by Linnaeus, Ludwig, Mirbel, St. Hilaire, de Candolle, Robert Brown, Hooker, and Asa Gray.

union of more than one carpel. Another term applied to the pistil is *gynoecium,* a collective term for the female element of the flower irrespective of whether it is represented by one or by many pistils (Fig. 5 a-c).

The gynoecium provides many taxonomic characters of major significance. In order to have a clearer comprehension of the importance of the gynoecium and of the phyletic values ascribed to its components, it is desirable to know something of the morphological theories proposed to account for its origin and types. The most deeply entrenched and widely

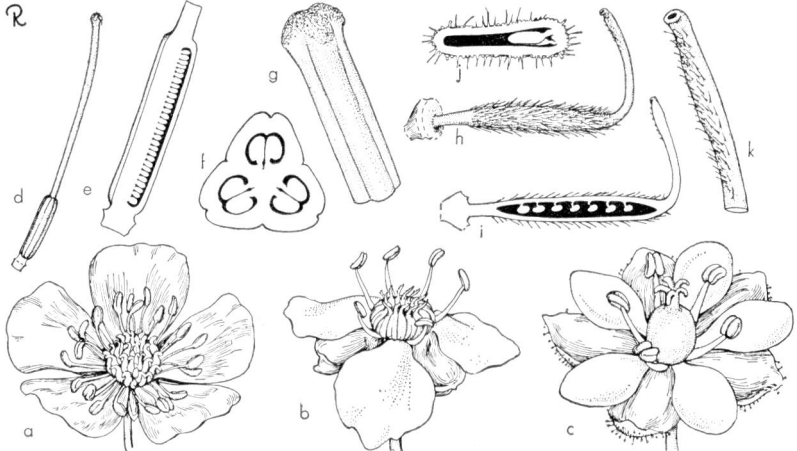

**Fig. 5.** Gynoecial types: a, pistils many and spiralled, the gynoecium apocarpous; b, pistils many and whorled (cyclic), the gynoecium apocarpous; c, pistil solitary, the gynoecium syncarpous; d, pistil one, ovary compound; e, same, ovary vertical section; f, same, ovary cross-section; g, same, style apex and 3-lobed stigma; h, pistil one, ovary simple; i, same, vertical section; j, same, ovary cross-section; k, same, style-tip and simple stigma.

accepted theory for the origin of the carpel is the Candollean or appendicular theory. It was first proposed by Goethe (1790),[26] amplified by de Candolle (1827), revised and supported by van Tieghem (1871), and more recently defended on the basis of detailed anatomical studies by Eames (1931), Arber (1937), and others.[27] The fundamentals of this theory are that the antecedent of the carpel (and of which no living examples exist) was a nonphotosynthetic foliar appendage, probably a leaflike palmately 3-veined dorsiventral structure supplied with 3 vascular strands or traces. Functionally it is believed to have been originally

[26] By Goethe's concept, the primitive carpel was held to be a green photosynthetic leaf bearing marginal ovules, a view not now accepted.

[27] Three current theories held in opposition to this view are to the effect that: (1) carpels have originated from the floral axis and are axillary organs, (2) all floral organs are structures that always have existed, as organs *sui generis,* and (3) carpels do not exist at all (acarpy), a theory advocated by Thompson.

an open flat megasporophyll bearing ovules on its margins (Fig. 6a), and to have folded lengthwise with the ovules inside and margins fused (Fig. 6c). The resultant structure is a unicarpellate, 1-loculed ovary. It is the product of a single carpel, and is termed a *simple ovary*. Only one carpel (i.e., only one foliar unit or megasporophyll) contributed to its

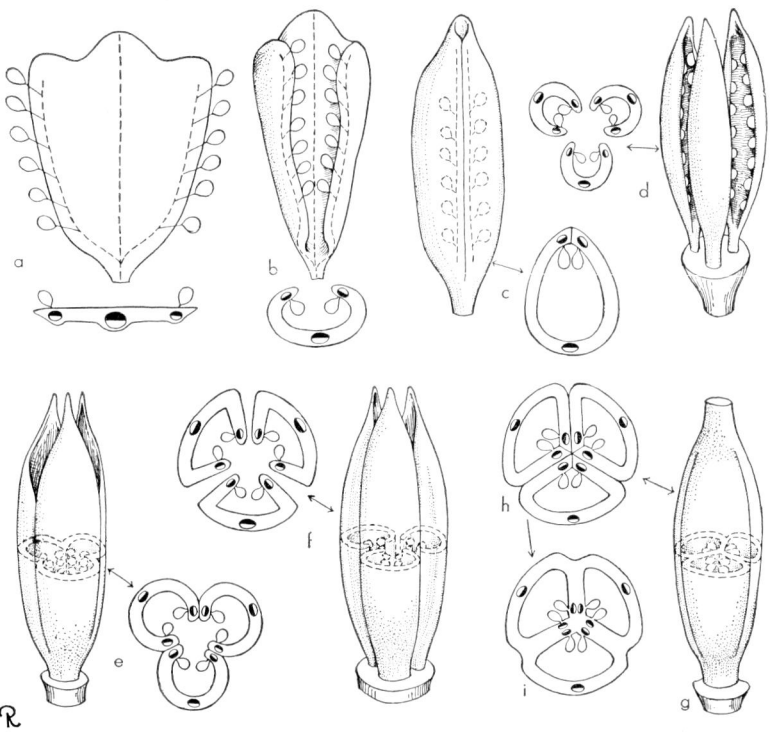

**Fig. 6.** Hypothetical evolution of simple and compound ovary. a, three-lobed carpel with submarginal ovules; b, same, somewhat involute; c, simple ovary derived from "b" by infolding of ovules and connation of ventral margins; d, axis bearing three involute open carpels; e, compound ovary derived from "d" by connation of edges of adjoining carpels; f, axis with three open carpels with adjoining sides more or less parallel; g, compound ovary derived from "f" by connation of adjoining sides and margins; h, cross-section of "g" (hypothetical); i, cross-section of "g" (actual) showing loss of carpellary demarcation in the three septa. (Note: vascular strands shown with xylem elements blackened.)

formation. In the cross-sectional view of a simple ovary (Fig. 6c) note the locations of the midrib and the two marginal veins of the carpel, observing that the latter are in a relatively close proximity to one another and furthermore that the two rows of ovules (indicated diagrammatically in the expanded carpel) are so close together within the simple ovary as to appear almost as if in a single row. The zone or area occupied by this row of ovules is the *placenta*. Examples of this type of ovary are to be found in the Leguminosae and many Ranunculaceae. Unlike the situation in animal anatomy, the angiosperm placenta is not a particular kind of tissue. It is the place where the ovules are attached.[28] A simple ovary, inasmuch as it is the product of a single carpel, always has a single placenta. Furthermore, the single placenta of a simple ovary is *parietal* in position (i.e., it always is attached to the side of the locule wall, or to an intrusion of the side wall into the locule).[29]

**Placentation types.** In many plant families the gynoecium is composed of several to many simple pistils (i.e., pistils each with a simple ovary) as in most of the Ranunculaceae. This type of gynoecium is currently believed to represent a primitive phyletic condition. It is from such a gynoecium, or from types ancestral to it, that the *compound ovary* is believed to have developed. A compound ovary is any ovary comprised of two or more carpels. If, perhaps by compression and fusion of adjoining ovarian tissues, several simple pistils are connate (grown together), the result is a single pistil of several carpels. Figure 6g–h is a diagram of the ovary of such a pistil. Note that fusion is incomplete in that the midrib and marginal veins of each of the 3 carpels are indicated. The placental areas of each of the 3 original simple ovaries are shown to be converged about an imaginary central axis. By the fusion of the 3 sets of walls of as many adjoining simple ovaries, the resultant compound ovary is one of 3 locules, each with a row of ovules in the axil formed by the septa (Fig. 6i). The placentation of this type of compound ovary is said to be *axile,* since the ovules are on the central axis which has been formed by the fusion of the carpel margins.[30] Any ovary having axile placenta-

[28] In some plants (as the tomato, *Lycopersicon,* or in *Epigaea*) the cells of the placental area proliferate and form a prominent tissue mass that extrudes into the locule and is properly referred to as the placenta or placental zone. See Fig. 273.

[29] Some botanists designate this placental condition as *marginal* or ventral, pointing out correctly that it differs from the parietal placentation of a syncarpous (i.e., compound) ovary in the placenta representing two connate margins of the same carpel and not the connate margins of two adjacent carpels. This distinction is phyletically important but is not always apparent from macroscopic examination common to taxonomic study (for example, only by knowledge of floral anatomy or comparative morphology can one identify a unilocular ovary with a single placenta and seemingly with marginal placentation, to be a tricarpellate ovary derived by reduction from an ancestral form that was trilocular with axile placentation). Parietal is accepted here as a descriptive term, applicable to any placenta situated on the ovary wall or an intrusion of that wall.

[30] Axile placentation is the result of the union of carpel margins at the center of the ovary. The term axile indicates merely that the ovules are situated in the axil of two adjoining septa. It should be understood clearly that in axile placentation no true axis or stem is present in the central column, that there is no stelar tissue involved (in *Pyrus* and *Cydonia* the stem tip may extend a short distance up between the carpels). It is believed that the ovules of all angiospermous ovaries are always of foliar and never of cauline origin.

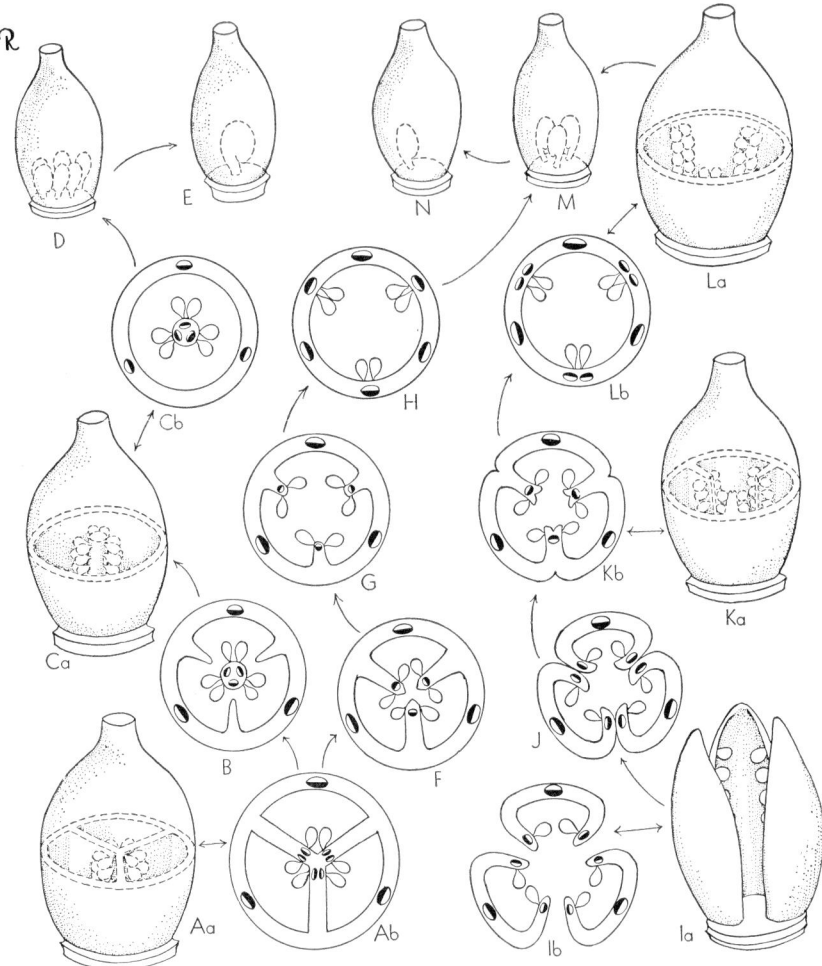

**Fig. 7.** Presumed evolutionary development of ovary and placentation types. A: trilocular ovary with axile placentation, derived as shown in Fig. 6; Aa, schematic view of ovary; Ab, cross-section of same showing carpellary vascularization (xylem elements blackened). B, intermediate stage between "Ab" and "C," note partial loss of septa and retention of ventral vascular strands (each adjoining pair fused resulting in a reduction of six strands to three). C: unilocular ovary with free-central placentation, derived from "Aa"; Ca, schematic view of ovary; Cb, cross-section of same, differs from "B" only in complete loss of septation. D, compound ovary with basal placentation, derived from "Ca" by reduction of central placentae. E, compound ovary with single basal ovule, derived from "D" by ovule reduction. F, cross-section of unilocular tricarpellate ovary with parietal placentation, derived from "Ab." G, same as "F" but septa reduced. H, advanced stage of "G" (no placental intrusion). I: hypothetical primitive situation of axis with three open carpels; Ia, schematic view; Ib, cross-section of same. J, compound ovary (unilocular, tricarpellate) with parietal placentation, derived by connation of adjoining carpel margins. K: compound ovary with parietal placentation, derived from "J"; Ka,

tion (irrespective of the number of locules) is always a compound ovary because it is one that has been formed by the connation of two or more carpels. It is composed of as many carpels as it has placentae, and each carpel is usually represented by a locule.

A simple ovary typically has *parietal* placentation (Fig. 6c). This same apparent placental type may occur in a compound ovary and by one of at least two evolutionary lines of development. Presumably it could have arisen from a gynoecium of open and incurved carpels standing in close proximity to one another and which became fused together marginally. Figure 7Ia indicates diagramatically such a situation involving 3 carpels. A single pistil results from the marginal connation of these carpels. It is a pistil that has a compound ovary of 1 locule and 3 parietal placenta. Note that each placental area is formed from the union of two adjacent carpel margins, each bearing a row of ovules and that the midrib of each carpel stands about midway between the placentae. The second line of evolutionary development believed responsible for some examples of parietal placentation in multicarpellate ovaries is that derived from an ovary having axile placentation. Figure 7A–7H illustrates this situation. In this hypothetical instance, a 3-carpelled and 3-loculed ovary gives rise to a 3-carpelled unilocular ovary by separation of the 3 central placental zones and their recession to or toward the ovary walls. The number of carpels remains the same, but the placental position is changed from axile to parietal, and when the 3 locules open into each other a single locule results. It is usually true that an ovary with parietal placentation, irrespective of the number of carpels, is an ovary of a single locule.[31] In many compound ovaries of this type, the placentae are not restricted to the inner periphery of the ovary wall but are on intrusions of the ovary wall (Fig. 7G) or (as in *Coptis* or *Gentiana* spp.) are on the inward-projecting free margins of the carpel or of adjoining carpels. By some authors this placental condition is termed falsely parietal, but macro-

---

[31] One exception to this occurs in the ovary of members of the family Cruciferae, where the ovary is normally 2-loculed and 4-carpelled with the ovules attached parietally along one or both peripheral margins of the commissural sides of two of the carpels. (See Fig. 152.)

---

schematic view of ovary; Kb, cross-section of same (note union of adjoining ventral carpellary strands and compare xylem orientation with that shown in "F," the latter presumed the more advanced condition). L: compound ovary with parietal placentation, derived from "K" (note absence of intruding carpellary margins); La, schematic view of ovary; Lb, cross-section of same (compare normal xylem orientation of ventral strands with reversed positions in "H," the retention of each pair of strands in "Lb" a less advanced situation than the fusion of them as shown in "H"). M, compound ovary with basal placentation, derived by placental reduction of "H" or of "La." N, compound ovary with single basal ovule, derived from "M" by ovule reduction.

scopic examination of a given ovary seldom affords a means of determining clearly between truly parietal and falsely parietal. The distinction is of morphological and phyletic significance, but is of little taxonomic use in ordinary identification procedures.

A third basic type of placentation within a compound ovary is that known as *free-central* placentation (see Fig. 7Ca). It is generally agreed that this type has been evolved from the axile type of placentation by the persistence of the central column and its placentae and the disappearance of the partitions, or septa. There is evidence that, in a few genera having free-central placentation (e.g., *Primula*), the axis on which the ovules are borne is comprised of carpellary tissue into which extends stem tissue for a short distance, the latter representing the extension of the stele from the subtending pedicel. An ovary with free-central placentation always has a single locule, but is composed of two or more carpels and hence is always a compound ovary.[32] However, since the placental zones are closely appressed laterally about the axis and usually without any line of demarcation between them, it is necessary to resort to other means, as indicated below, to determine or approximate the carpellary number.

The reduction of the central axis in the free-central type of placentation to a nubbin terminated by a single ovule produces a type designated as *basal placentation,* and taxonomically this comprises a fourth type of placentation.[33] In its extreme form, none of the central axis remains and the ovule appears as if on the floor of the locule. In other instances, the situation where the number of apparently basal ovules is more than one (usually 2 to 4) represents a condition believed to have been derived from axile placentation types or, in unilocular ovaries this probably has come through the free-central stage (cf. Figs. 7D and 7E). Taxonomists, in descriptions and keys, generally refer to them as basal when no axile condition is macroscopically apparent. A uniovulate placentation in some instances has been derived from parietal placentation of a multicarpellate ovary (Gramineae, Compositae) or of a unicarpellate ovary (as in some Ranunculaceae or Rosaceae). Because of these two derivations of a uniloculate 1-ovuled ovary, it is apparent that although the uniovulate

---

[32] The central column consists of a placental mass of proliferated tissue composed of as many fused placentae as there are carpels. The number of placentae contributing to this mass is usually not evident, but is strikingly so in some genera of Caryophyllaceae and in a few genera of Primulaceae.

[33] A fifth type is that known as *lamellate placentation,* occurring when the placental area and its ovules covers one or more broad, flattened, and usually lamellate surfaces which may be peripheral or represent platelike septa (as in *Papaver, Nuphar, Nymphaea,* and *Butomus*).

ovary appears to be simple, it may represent a highly advanced condition and may be a simple or a compound ovary (cf. Fig. 7N). The pendulous ovule, characteristic of some uniovulate ovaries, usually represents a reduction from parietal placentation but may have been derived from axile placentation (as in some Caprifoliaceae).

Taxonomically, these types of placentation provide characters of considerable significance and utility. Very often they are indicators of phyletic relationship between and within major groups. They are characters that frequently are employed in analytical keys, especially in keys to families. It is important to be able to differentiate and recognize the different types, and to have an intelligent concept as to their relative level of development from the primitive simple ovary with its parietal (or marginal) placentation, to the extreme degree of reduction represented by the uniloculate 1-ovuled simple or compound ovary. Occasionally it is desirable to know the number of carpels represented in an ovary. Accurate determination of carpellary number frequently requires microscopic study of a series of microtome sections, and in many instances may even then be determined only by comparative studies. However, the number of styles, stigmas, or stigma branches or lobes of a pistil often serve as indicators of the number of carpels present. In general, if the number of any one of these features is two or more, it may be assumed that the ovary bearing them is a compound rather than a simple ovary. An ovary of two or more complete locules is always compound. An ovary of a single locule is simple only if it has a single parietal zone of placentation and is comprised of a single carpel.

**Ovary position.** The position of an ovary is referred to as being superior, inferior, or half inferior. A *superior* ovary is one that is situated above the point of attachment of perianth and androecium. An ovary whose position is superior is considered to represent the primitive position. An *inferior* ovary is one situated below the apparent point of attachment of the perianth and androecium. An ovary whose position is *half inferior* (or subinferior) is more or less intermediate between these two. (For diagrams of these types, see Fig 8.)

There are two theories to account for the evolution of the inferior from the superior ovary, and in some instances the situation is explained only by accepting views from each of them. In a typical angiosperm flower the 5 whorls or series of components (calyx, corolla, two whorls of stamens, and the gynoecium) are borne on an axis called the receptacle or torus (Fig. 8A). The older and perhaps until recently the more widely accepted theory proposed to account for the derivation of the inferior ovary from the superior is the *receptacular theory,* by which it was concluded that the inferior ovary is embedded in or surrounded by a tube or cup of receptacular tissue. By this theory it was held also that in those flowers whose ovary is superior, but surrounded by a cuplike or cylinderlike tube called the *hypanthium* (more commonly and inaccurately termed the calyx tube and by some morphologists called the floral tube) (Fig. 8B and F), the hypanthium itself is of receptacular tissue (Fig. 8Ca). Anatomical studies of a wide range of materials by Eames (1931), Jackson (1934), and MacDaniels (1937) indicated that, for the majority of

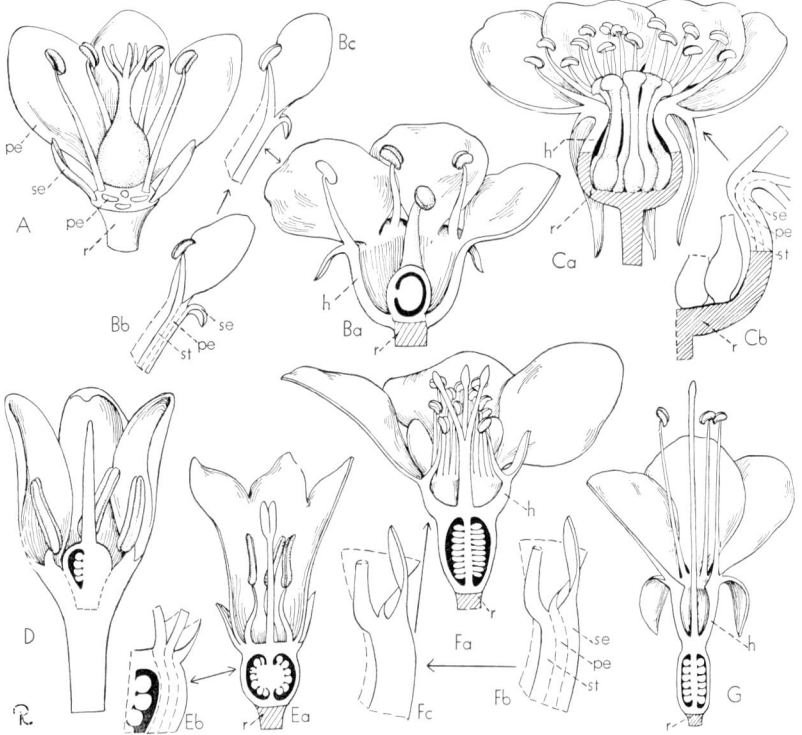

**Fig. 8.** Ovary position. A, ovary superior, perianth and stamens hypogynous; B, ovary superior, perianth segments and stamens perigynous; Ba, vertical section; Bb, detail of Ba to show composition of hypanthium; Bc, same, minus lines of adnation of perianth and androecium; C, ovary superior, perianth and stamens perigynous, hypanthium of "Rosa" type; Ca, vertical section; Cb, detail to show relation of receptacular cup to hypanthium and composition of latter; D, ovary half-inferior, stamens perigynous; E, ovary inferior, perianth and stamens epigynous; Ea, vertical section of flower; Eb, detail of same to show adnation of hypanthium to ovary; F, ovary inferior, hypanthium present, perianth and stamens epigynous; Fa, vertical section of flower; Fb, detail to show components of hypanthium; Fc, section of hypanthium; G, ovary inferior, stamens exserted, hypanthium present. (h hypanthium, pe petal, r receptacle, se sepal, st stamen.)

flowers possessing inferior ovaries, or ovaries within an hypanthium, the tissues previously believed to have been receptacular in origin are formed from the fusion of the foliar or appendicular units, as represented by perianth parts and stamens (Fig. 8E). This fusion represents a phylogenetic, and not an ontogenetic, phenomenon.

The second, and more widely accepted, of the two theories is known as the *appendicular theory*. By this view, it is held that the hypanthium that may surround the superior ovary, and the tissues that are adnate to the inferior ovary, are homologous in most cases and are components of the flower (i.e., they are appendicular in origin) and are not of the axis, receptacle, or torus. In most instances (an exception being the Santalaceae and Calycanthaceae), no stelar tissues are found in the hypanthium, irrespective of whether it be free or adnate to the ovary. In the genus *Rosa* it has been established that only the upper part of the hypanthium is appendicular in origin, for it was shown by Jackson (1934) that the basal half or less has resulted from the peripheral extrusion of the apex of the floral axis (receptacle).[34] There is ample evidence that the inferior ovary of present-day plants represents two lines of evolutionary development: (1) the usual situation represented by adnation of foliar elements (calyx, corolla, androecium) to the ovary wall and (2) the uncommon condition represented by a depression and peripheral extrusion of the receptacle with the latter surrounding and adnate to the ovary. On the basis of recent studies, made independently by several workers, it is now generally accepted that while the inferior ovary represents an advanced condition derived from the superior ovary, "epigynous and perigynous flowers differ in no essential respect from the hypogynous condition; the inferior ovary is, in most cases at least, appendicular in origin, not receptacular." (Wilson and Just, p. 103.)

Most phylogenists have placed much emphasis on the phyletic significance of ovary position, and to the extent that families have been classified in orders largely on the basis of this character. It is also true that the orders composed of families that are characterized by both types of ovary position have been grouped together in most classifications. In general, the ovary position provides a character of considerable reliability, although intergradation occurs between all three ovary position types in a few genera and families.

**Flower types.** Modern taxonomy demands an understanding of the morphological composition of the flower. Descriptive taxonomic litera-

---

[34] For a review of the pertinent literature, see Wilson and Just (1939) and Douglas (1944).

ture is predicated on an understanding of the various types of flowers as indicated by gross morphology, and a knowledge of flower types is essential to satisfactory use of keys and descriptions available in floras and manuals of vascular plants. Many of the floral conditions indicated below are of little phyletic significance, and often are less reliable than is indicated by their extensive use by many authors. In addition to

**Fig. 9.** Flower types and arrangement: a, flower staminate (perianth uniseriate); b, flower pistillate (perianth uniseriate); c, flower bisexual (perianth biseriate and corolla polypetalous); d, flowers unisexual (plant monoecious); e, flower actinomorphic or regular; f, flower zygomorphic; g, flower irregular; h, flower gamopetalous.

knowing these floral types, one should endeavor to grasp some comprehension of their relative constancy and reliability for use in the characterization or differentiation of taxonomic units.

**Distribution of sexes.** A flower represented by the two perianth series, the androecium, and the gynoecium, is a *complete flower*; one lacking the perianth series but possessing both sex elements is an *incomplete flower*, but since it possesses both kinds of sex elements it is termed a *perfect flower* and (with or without the perianth) often is referred to as *bisexual* or *hermaphroditic*. A flower lacking the organs of one sex is an *imperfect flower* and is *unisexual* (Fig. 9a, b). Irrespective of the perianth situation, a unisexual flower possessing only an androecium is a *staminate flower*,[35] while one possessing only a gynoecium is a *pistillate flower*.

Some species of plants characterized by unisexual flowers have the staminate flowers restricted to one plant and the pistillate flowers restricted to another. Such plants are *dioecious* (e.g., *Salix, Myrica, Garrya*, most spp. of *Fraxinus*). In other unisexually flowered taxa the pistillate flowers and the staminate flowers are on the same plant (Fig. 9d). Such a plant is *monoecious* (e.g., *Zea, Betula, Pinus, Euphorbia*).[36] Some plants are characterized by the presence of both bisexual and unisexual flowers in the same inflorescence or on different parts of the plant. Such inflorescences or plants are said to be *polygamous* (e.g., many members of the Compositae). A derived condition is that existing when a plant is functionally dioecious, but has scattered throughout its inflorescences a few perfect flowers (e.g., some spp. of *Ilex, Celastrus, Acer, Rhus*). Such plants are said to be *polygamodioecious*.

The primitive type of flower is postulated by most modern morphologists to have been a complete flower, that is, it was composed of a perianth, androecium, and gynoecium. Any situation where one of these floral components is missing is considered to represent a derived or advanced condition. For example, a case of phyletic advancement is indicated by the suppression or reduction of parts, as when stamens are represented by staminodes (which may be nectiferous, petaliferous, or filamentous) or when the gynoecial elements are represented by pistillodes. The reduction may have taken place within all of the androecial elements (in which case the flower is unisexual and pistillate) or to have affected only some of the stamens of the flower. Likewise, it may have affected some or all of the carpellary elements of the gynoecium. In a few cases, both sex organs have become suppressed or reduced to a point of being wholly nonexistent or at least nonfunctional, with the floral axis terminated only by the perianth parts. Such a structure morphologically is not a flower,

---

[35] In some works (e.g., Gray's *Manual*, ed. 7) staminate flowers are designated neutral since they are incapable of producing seed. However, as explained on p. 84, the term neutral is reserved for flowers lacking any sex element.

[36] Many authors have misapplied the terms dioecious and monoecious to flowers rather than restricting them to plants. However, if one holds a single unisexual flower in hand, it is not possible to determine from examination of it whether dioecism or monoecism exists; one must know the character of all the unisexual flowers on the plant to determine which condition exists—hence the term applies to the plant, or to the taxon, and not to the flower.

since it contains no reproductive elements, but in descriptive taxonomic literature is termed a *neutral* or *sterile flower* (e.g., some of the flowers of some spp. of *Hydrangea, Viburnum,* or Compositae).

**Disposition of perianth parts.** The arrangement of perianth parts in a spiral and without sharp demarcation between calyx and corolla usually is considered the most primitive (e.g., *Magnolia, Nymphaea, Nelumbo*). If one looks down on an open flower of this type, one notes a symmetry of perianth parts, and it is apparent that such a perianth can be bisected in two or more planes and into similar halves (as, for example, one can cut a pie into two equal halves from any selected point of the circumference). Such a perianth is *actinomorphic* or *regular*.[37] Flowers with perianth series in whorls rather than spirals may, with equal precision of application of the terms, have actinomorphic or regular perianths (Figs. 9c, 9e). Sometimes the corolla may be actinomorphic and the calyx not so, or the reciprocal condition may be present. In other plants the perianth parts, or segments, are so disposed that the floral envelopes may be bisected only along one diameter or in one plane (e.g., in Orchidaceae, papilionaceous Leguminosae, most Labiatae). These perianths are *zygomorphic* (Fig. 9f). In a few groups the perianth, or its parts, is of such disposition that it is not possible to bisect it into two like parts, and it is said then to be *irregular* (Fig. 9g). Generally speaking, the term irregular is more often applicable to flowers than to perianths (e.g., in *Canna* or Zingiberaceae).

**Fusion and modification of perianth parts.** The primitive situation of perianth parts has until recently been accepted to be that of distinct and free [38] from one another. Most contemporary systems of angiosperm classification accept plants with perianth parts free and distinct to be more primitive than those with perianth parts of each series

---

[37] The terms actinomorphic and zygomorphic (or their synonyms) are also often applied to flowers instead of only to the perianth. Unfortunately many authors, without precise application of terms, have described flowers as actinomorphic (or otherwise) when only the perianth (or only the corolla) was intended.

[38] The student should differentiate between the terms distinct and free as used taxonomically. The term *distinct* is employed to indicate complete separation of like parts (as one petal distinct and not connate to another petal), whereas *free* indicates complete freedom from union of adjoining, but unlike, parts (as the stamens free and not adnate to the petals).

In addition to the above, and in the interest of precision of expression, the student should understand the differences between the terms connate, coherent, adnate, and adherent. *Connate* means the fusion of like parts (as petal to petal); *coherent* means the meeting in close proximity by cohesion (viscidity, vesture, etc.) or otherwise of like parts but lacking fusion of their tissues (as pollen grains sometimes coherent); *adnate* refers to the fusion of unlike parts (as stamens to petals); *adhesion* refers to the coming into close contact of unlike parts but lacking fusion of their tissues (as the indehiscent husk of *Juglans* adherent to the nut).

connate. Corollas composed of distinct petals are *polypetalous* (choripetalous, apopetalous) (Fig. 9c), those whose petals are to any degree marginally connate are *gamopetalous* (sympetalous) (Fig. 9h). The gamopetalous condition may be very obvious (e.g., in *Aster, Convolvulus, Digitalis*) or superficially not present (e.g., in *Armeria, Chionanthus*) and in cases of deeply segmented corollas the condition may be determined only after careful dissection and study.

In general, most botanists accept the view that a perianth composed of free and distinct parts (as of sepals, petals or of tepals) is more primitive than one in which such parts are partially or completely connate or adnate (thus resulting in synsepally, sympetally, or the production of an hypanthium). There is abundant evidence indicating that taxa whose flowers exhibit fusion of perianth parts arose independently (i.e., by polyphyletic origins) in widely separated families and orders. A majority of morphologists, and many botanists, currently reject those phyletic views that would group all dicot orders whose families possess gamopetalous corollas (as, for example, the taxon Sympetalae). There is evidence also that some corollas now composed of distinct petals, or of segments connate only at the base, have been derived from ancestors whose corollas were conspicuously or completely gamopetalous (as in some Onagraceae, Plumbaginaceae). Likewise the vascular anatomy of some of these seemingly near-polypetalous corollas shows that their segments contain marginal vascular strands that clearly once belonged to adjoining lobes or segments of the same perianth series; this is evidence that in such situations the corolla segmentation probably was preceded by a more completely gamopetalous condition. Perianths having these complex vascular situations are believed generally to be more advanced phyletically than are those exhibiting simple but complete connation of parts. For these reasons it should be borne in mind that while the absence of fusion of perianth parts is generally to be construed to represent a primitive situation, it may in some taxa represent a very advanced phyletic situation.

***Fruits.*** A fruit may be defined as the product of the ripened ovary or pistil of a flower and may be composed in part also of accessory floral or vegetative parts. It is the seed-bearing (or containing) organ of a plant, and in this sense the ripened female strobilus (or cone) of a conifer may be referred to as a fruit. Fruits are important in the classification, delimitation, and identification of seed plants because generally they provide characters very reliable in the characterization of genera and of families. Many kinds and types of fruits exist. An appreciable number of classifications have been proposed for their systematic organization;[39] all are empirical and artificial. None has been devised during much of the last century, and none is based on the fundamental morphology of the structures concerned. The basic weakness of these classifications is their artificiality, and since they have been devised without due regard for the morphology of the ovaries and gynoecia from which the fruit is produced, the definitions of their categories or types invariably are violated by an abundance of exceptions. Until a more satisfactory classification has been devised, contemporary taxonomists are obliged to continue to employ the currently used conventional classification that follows the basic schema presented in Asa Gray's 6th edition of *Structural botany* or in *Gray's lessons in botany*. A synopsis adapted from that published by Gray (1879) and accounting for the more commonly encountered fruit types is given below. In making reference to it, the omnipresence of exceptions must be acknowledged.

Fruits simple, the product of a single pistil.
  Fleshy and usually indehiscent.
    Texture homogeneous, fleshy throughout........................*Berry*
    Texture heterogeneous.
      Fruit exterior a firm, hard, or leathery rind.
        Septae present, several to many......................*Hesperidium*
        Septae absent ................................................*Pepo*
      Fruit exterior soft.
        Center of fruit with a single "stone" ......................*Drupe*
        Center of fruit with papery or cartilaginous carpels ............*Pome*
  Dry fruits.
    Fruit indehiscent, usually 1–2-seeded.
      Winged ..................................................*Samara*
      Wingless.
        Pericarp thin.
          The pericarp adnate to the seed ........................*Achene*
          The pericarp loose and free from seed.....................*Utricle*
        Pericarp thick and hard, sometimes bony.
          Fruit small, from a 1-loculed ovary .....................*Achene*
          Fruit usually large, from a 2-more-loculed ovary ..............*Nut*
    Fruit usually dehiscent, 1-many-seeded.
      Product of a unicarpellate ovary.
        Dehiscing by ventral suture only .........................*Follicle*
        Dehiscing by two longitudinal or transverse sutures.
          Sutures longitudinal .................................*Legume*
          Sutures transverse ..................................*Loment*
      Product of a bi- or multicarpellate ovary.
        Fruit splitting into 1-seeded halves .....................*Schizocarp*
        Fruit splitting and releasing seeds.
          Dehiscence circumscissle ................................*Pyxis*

[39] For references to fruit classifications, cf. entries under Dickson, Gray, Rendle, Masters, and Winkler.

Dehiscence longitudinal ................................*Capsule*
(Silicles and siliques are specialized types of capsules characteristic of the Cruciferae family.)
Fruits compound, the product of 2 or more pistils.
The product of several pistils of a single gynoecium connate or coherent, and usually fleshy ........................*Aggregate fruit*
(An *accessory fruit* is a type of aggregate fruit in which the conspicuous and often fleshy part of the fruit is of nonovarian origin.)
The product of several gynoecia aggregated in one mass......*Multiple fruit*

**Seeds.** Seeds are fertilized mature ovules and each contains an embryo. An abundance of taxonomic and phylogenetic characters are provided by the seeds of plants. Generally speaking, those of general taxonomic use are the more superficial and associated with the seed coat (usually derived from the outer integument of the ovule, sometimes termed testa). Outgrowths such as wings (as in the pine, trumpet vine, or maple) or the coma (a tuft of hair as in milkweed, cotton) are examples of this character. In other groups, the presence and character of pits, sutures, sculpturing, and surface configurations are of significant value taxonomically (Fig. 322). Of more fundamental classificatory value are the internal characters afforded by seeds of many plant groups; particularly is this true of the presence or absence of endosperm (albumen), shape and position of the embryo, and the character, number, and arrangement of the cotyledons. In this regard, Martin (1946) devised a phylogenetic schema of seed plants based on the comparative internal morphology of seeds that compares very favorably with modern concepts based on combinations of other structures.

**Embryology.** The taxonomic value of embryological data has been ignored by the systematist until recently, or has been considered useful only to the phylogenist dealing with the classification of taxa in the higher categories. In this regard it has been emphasized by Maheshwari (1945, 1950) and by Just (1946) that the taxonomist should be cognizant of the significance of these data and recognize their value and limitations. Embryological characters found to be of taxonomic and phyletic significance include those of the male and female gametophyte as well as of the embryo. Just (1946) proposed a set of symbols whereby the attributes of each of the pertinent characters might be designated in condensed form. In focusing attention on the use of these data by the taxonomist he stated (p. 354):

The accepted orders of flowering plants may well be delimited differently, if embryological data are applied consistently, and groups which have so far not been investigated are studied embryologically. Therefore, the use of embryological data will be greatest as supplementary evidence in the improvement of our systems of classification, particularly with regard to the determination of the correct position and affinities of families and orders, whereas their application in purely descriptive works such as floras and manuals as well as in teaching elementary plant taxonomy may never prove feasible.

Embryological data need not be accorded more recognition than other taxonomically valuable characters. They do, however, deserve their rightful place among the others, a position they have not yet attained in the eyes of all botanists.

Much earlier, Rutgers (1923) proposed the use of formulae by which to designate embryo-sac and embryo conditions, but made no attempt to include the range of gametophytic characters of both sexes that were taken into account by Just. The recent contribution by Johansen (1950) should do much to enlighten the taxonomist of the value of embryological data, and provides a ready and classified reference to these data and to the pertinent literature.

**Karyology.** Cytology and cytogenetics have provided many data concerning the number, morphology, and behavior of chromosomes within the nucleus of the gametophyte that are of taxonomic significance. For a discussion of this see Chapter VIII.

*LITERATURE:*

ANDERSON, E. The concept of the genus. II. A survey of modern opinion. Bull. Torrey Bot. Club, 67: 363–369, 1940.

ARBER, A. The interpretation of the flower: a study of some aspects of morphological thought. Biol. Rev. 12: 157–184, 1937.

BABCOCK, E. B. Systematics, cytogenetics and evolution in *Crepis*. Bot. Rev. 8: 139–190, 1942.

BANCROFT, H. A review of researches concerning floral morphology. Bot. Rev. 1: 77–99, 1935.

BARTLETT, H. H. The concept of the genus. I. History of the generic concept in botany. Bull. Torrey Bot. Club, 67: 349–362, 1940.

BRIQUET, J., ed. International rules of botanical nomenclature. ed. 3. Jena, 1935.

BROWN, W. H. The bearing of nectaries on the phylogeny of the flowering plant. Proc. Amer. Philosoph. Soc. 79: 549–594, 1938.

BUCHHOLZ, J. T. Seeds. Sci. Monthly, 38: 367–369, 1934.

CAMP, W. H. The concept of the genus. IV. Our changing generic concepts. Bull. Torrey Bot. Club, 67: 381–389, 1940.

CAMP, W. H. and GILLY, C. L. Floral abnormalities in *Linaria vulgaris,* with notes on a method by which new genera may arise. Torreya, 41: 33–42, 1941.

——— . The structure and origin of species. Brittonia, 4: 325–385, 1943.

ČELAKOVSKY, L. F. Gedanken über eine zeitgemässe Reform der Theorie der Blütenstände. Bot. Jahrb. Engler, 16: 33–51, 1892.

CHUTE, H. M. The morphology and anatomy of the achene. Amer. Journ. Bot. 17: 703–723, 1930.

CLAUSEN, J., KECK, D. and HIESEY, W. The concept of species based on experiment. Amer. Journ. Bot. 26: 103–106, 1939.

CLAUSEN, R. T. The terms "subspecies" and "variety." Rhodora, 43: 157–167, 1941.

CORNER, E. J. H. Centrifugal stamens. Journ. Arnold Arb. 27: 423–437, 1946.

COULTER, J. M. Evolution of sex in plants. Chicago, 1914.

CROIZAT, L. The concept of inflorescence. Bull. Torrey Bot. Club, 70: 496–509, 1943.
———. History and nomenclature of the higher units of classification. Bull. Torrey Bot. Club, 72: 52–75, 1945.
DE CANDOLLE, A. P. Théorie élémentaire de la botanique. Paris, 1813.
———. Organographie végétale. Vol. 1. Paris, 1827.
DICKSON, A. Suggestions on fruit classification. Journ. Bot. 9: 309–312, 1871.
DOBZHANSKY, T. Genetics and the origin of species. ed. 2. New York, 1941.
DOUGLAS, G. E. The inferior ovary. Bot. Rev. 10: 125–186, 1944.
DU RIETZ, G. E. The fundamental units of biological taxonomy. Svensk Bot. Tidskr. 24: 333–428, 1930.
EAMES, A. J. The role of flower anatomy in the determination of angiosperm phylogeny. Proc. Int. Cong. Plant Sci. 1926 (Ithaca, N. Y.) 1: 423–427, 1929.
———. The vascular anatomy of the flower with refutation of the theory of carpel polymorphism. Amer. Journ. Bot. 18: 147–188, 1931.
EGLER, F. E. The fructus and the fruit. Chron. Bot. 7: 391–395, 1943.
ERDTMAN, G. An introduction to pollen analysis. Waltham, Mass., 1943.
FAEGRI, K. Some fundamental problems of taxonomy and phylogenetics. Bot. Rev. 3: 400–423, 1937.
FLORIN, R. Die Coniferen des Obercarbons und der unteren Perms. Paleontographica, 85, Abt. B, pt. 2, Lief I-V, 1938–1940.
FOSBERG, F. R. Subspecies and variety. Rhodora, 44: 153–157, 1942.
GOEBEL, K. I. E. Blütenbildung und Sprossgestaltung. Anthokladian und Infloreszenzen. 1931. [Suppl. 2 to Organographie der Pflanzen, ed. 3.]
GOETHE, J. W. VON. Versuch die Metamorphose der Pflanzen zu erklären. Gotha, 1790.
GOODSPEED, T. H. and BRADLEY, M. T. Amphidiploidy. Bot. Rev. 8: 271–316, 1942.
GRAY, A. Structural botany. ed. 6. New York, 1879.
———. Gray's lessons in botany. rev. ed. New York, 1887.
GREENMAN, J. M. The concept of the genus. III. Genera from the standpoint of morphology. Bull. Torrey Bot. Club, 67: 371–374, 1940.
GUSTAFSSON, Å. Apomixis in the higher plants. Lunds Univ. Arsskrift, N.F. 42(3) and 43(2,12), 1946–1947. (For review, cf. Stebbins, G. L. in Evolution, 3: 98–101, 1949.)
HALL, H. M. The taxonomic treatment of units smaller than species. Proc. Int Cong. Plant Sci. 1926 (Ithaca, N. Y.) 2: 1461–1468, 1929.
HOGBEN, L. Problems of the origins of species. In Huxley, J. The new systematics, 269–286, Oxford, 1940.
HORT, A. F. (translator). The "Critica botanica" of Linnaeus. Rev. by M. L. Green. London, 1938.
HUNT, K. W. A study of the style and stigma, with reference to the nature of the carpel. Amer. Journ. Bot. 24: 288–295, 1937.
HUTCHINSON, J. British flowering plants. London, 1948.
HUXLEY, J. S. Species formation and geographical isolation. Proc. Linn. Soc. (Lond.) 150: 253–264, 1938.
JACKSON, G. The morphology of flowers of *Rosa* and certain closely related genera. Amer. Journ. Bot. 21: 453–466, 1934.
JOHANSEN, D. A. Plant embryology. 305 pp. Waltham, Mass., 1950.
JONES, S. G. Introduction to floral mechanisms. 274 pp. London and Glasgow, 1939.
JUST, T. The morphology of the flower. Bot. Rev. 5: 115–131, 1939.
———. The proper designation of vascular plants. Bot. Rev. 11: 299–309, 1945.
———. The use of embryological formulas in plant taxonomy. Bull. Torrey Bot. Club, 73: 351–355, 1946.

———. The relative value of taxonomic characters. Amer. Midl. Nat. 36: 291–297, 1946.
KAUSSMANN, B. Vergleichende Untersuchungen über die blattnatur des Kelch-, Blumen-, und Staubblätter. Bot. Arch. (Leipzig), 42: 503–572, 1941.
LAMPRECHT, H. Systematik auf genetischer und zytologischer Grundlage. Agr. Hort. Genetica, 7: 1–26, 1949.
LINK, J. H. F. Grundleheren der Anatomie und Physiologie der Pflanzen. 304 pp. Berlin, 1807. Nachträge, 84 pp., 1809.
LINNAEUS, C. Fundamenta botanica . . . Amsterdam, 1736.
———. Philosophia botanica . . . ; and by Hort, A., 1938. Stockholm, 1751. (Eng. transl. by Rose, H., 472 pp. London, 1775.)
LOTSY, J. P. On the species of the taxonomist in its relation to evolution. Genetica, 13: 1–16, 1931.
MACDANIELS, L. H. The morphology of the apple and other pome fruits. Cornell Univ. Agr. Exp. Sta. Mem. 230, 1940.
MAHESHWARI, P. The place of angiosperm embryology in research and teaching. Journ. Indian Bot. Soc. 24: 25–41, 1945.
———. An introduction to embryology of angiosperms. 453 pp. New York, 1950.
MASTERS, M. T. Classification of fruits. Nature, 5: 6, 1871.
METCALFE, C. R. and CHALK, L. Anatomy of the dicotyledons; leaves, stem, and wood in relation to taxonomy, with notes on economic uses. 2 vols. Oxford, 1950.
MORAN, R. Delimitation of genera and subfamilies in the Crassulaceae. Desert Plt. Life, 14: 125–128, 1942.
NÄGELI, C. W. VON. Mechanisch-physiologische Theorie der Abstammungslehre, 822 pp. München, 1884.
OZENDA, P. Researches sur les dicotylédones apocarpiques. 183 pp. Paris, 1949. [Reprinted from, Publ. Lab. de l'Ecole Normale Supérieure, Ser. Biol. vol. 2, 1949.]
PARKEN, J. The classification of flowering plants. Northwest Sci. 20: 18–27, 1946.
PILGER, R. K. F. Über Verzweigung und Blütenstandsbildung bei den Holzgewachsen. Bibl. Bot. 23: 1–38, 1922.
POPE, M. A. Pollen morphology as an index to plant relationship—I. Morphology of pollen. Bot. Gaz. 80: 63–73, 1925.
RICKETT, H. W. The classification of inflorescences. Bot. Rev. 10: 187–231, 1944.
ROEPER, J. A. C. Observations sur la nature des fleurs et des inflorescences. Seringe Mél. Bot. 2: 71–114, 1826. [Also a Latin translation in Linnaea, 1: 433–466, 1826.]
RUTGERS, F. L. Reliquae Treubianae III. Embryosac and embryo of *Moringa oleifera* Lam. Ann. Jard. Bot. Buitenzorg, 33: 1–66, 1923.
SHERFF, E. E. The concept of the genus. IV. The delimitations of genera from the conservative point of view. Bull. Torrey Bot. Club, 67: 375–380, 1940.
STEARN, W. T. The use of the term "clone." Journ. Roy. Hort. Soc. 74: 41–47, 1947.
STEBBINS, G. L. JR. Apomixis in the angiosperms. Bot. Rev. 7: 507–542, 1941.
———. The role of isolation in the differentiation of plant species. Biol. Symp. 6: 217–233, 1942.
———. The genetic approach to problems of rare and endemic species. Madroño. 6: 241–258, 1942.
TOURNEFORT, J. P. DE. Institutiones rei herbariae. Paris, 1700; ed. 3, 1719. [This work essentially a Latin edition of his Elemens de botanique. An English translation in 2 vols., probably by J. Martyn, London, 1719–1730.]
VAN TIEGHEM, P. Recherches sur la structure du pistil et sur l'anatomie compareé de la fleur. Mém. sav. étrang. à l'inst. II, 21: 1–261, 1871.

WEATHERBY, C. A. Subspecies. Rhodora, 44: 157–167, 1942.
WILSON, C. L. The phylogeny of the stamen. Amer. Journ. Bot. 24: 686–699, 1937.
———. The evolution of the stamen. Chron. Bot. 6: 245, 1941.
———. The telome theory and the origin of the stamen. Amer. Journ. Bot. 29: 759–765, 1942.
WILSON, C. L. and JUST, T. The morphology of the flower. Bot. Rev. 5: 97–131, 1939.
WINKLER, H. Versuch eines "natürlichen" Systems der Früchte. Beitr. Biol. Pflanzen, 26: 201–220, 1939.
———. Zur Einigung und Weiterführung in der Frage des Fruchtsystems. Op. cit. 27: 92–130, 1940.
WODEHOUSE, R. P. Pollen grains, their structure, identification and significance in science and medicine. 574 pp. New York, 1935.
———. Evolution of pollen grains. Bot. Rev. 2: 67–84, 1936.

## CHAPTER V

# PHYLOGENETIC CONSIDERATIONS

*Phylogeny* is the evolutionary history of a taxon, and attempts to account for its origin and development. It is a function of taxonomy, by acceptance of a broad definition of the latter term. The term phylogeny is the antonym of *ontogeny*.[1] A primary objective of phylogenetic studies in botany is the determination of origins and relationships of all taxa of both extinct and present-day plants and the classification of them according to a system that will indicate their genetical or "blood" relationships. A truly phylogenetic classification does not now exist; it is doubtful if it ever will exist, but there is reason to believe that, with the acquisition of many more data and by the synthesis of all available data, a more satisfactory classification than any now known may be produced. It should be made clear that this phylogenetic system of the future undoubtedly will be 3-dimensional (or more) and reticulate in character and that it will be too complex in organization to be of practical use in the everyday classification of plants.

### Significance to taxonomy

Phylogeny deals with the evolutionary history of all taxa, from those in the category of division or phylum down to the species and their subdivisions. It is a function of taxonomic research at all levels of classification. A goal of phylogenetic research is the production of a phylogenetic system of classification. In its complete (and probably unattainable) form this phylogenetic system would enable one to determine the ancestor of a plant at any stage of its evolutionary development; it would show

---

[1] Ontogeny differs from phylogeny in that it accounts for the life history of the individual plant from its development from the zygote to the production of its own gametes. Ontogenetical studies deal also with the development of structures from the stage of primordial initiation to full maturity; phylogenetical studies of individual structures within a plant deal with the comparison of their evolutionary changes through successive generations from time of origin to the present.

92

the genetic and time relationship of any one taxon to another. In this regard it should be noted that there is a distinction between phylogeny and genealogy, for while the former deals with the evolution of the taxon the latter deals with the ancestry of the individual. Likewise, there is a distinction between phylogenetic and taxonomic classification, for (as pointed out by Turrill, 1942, p. 685) "taxonomy is based on characters, phylogeny on changes of characters." From this, it develops that there is a distinction between a complex 3-dimensional phylogenetic classification as outlined above and the existing so-called phylogenetic classifications.[2] Present-day classifications may be to the phylogenetic classification of the future what an artificial key is to a synopsis.[3] Sprague (in Huxley's *New systematics,* p. 441) summed up the situation clearly when he stated that, "in making an artificial classification there is arbitrary selection of characters, no attempt being made to arrive at groups exhibiting a maximum correlation of characters. In attempting to build a natural classification the units . . . are arranged in various ways until such maximum correlation is obtained."

It is probable that phylogenetic studies at the level of genus and below have been of greater significance and utility to the taxonomist than have those of the major groups. The reason for this is that the methods of the phylogenist who deals with higher taxa have been limited more or less to those that deal with paleobotanical and morphological research, whereas the taxonomist endeavoring to learn the phylogenetic relation-

[2] Distinctions between the 3 classification types (artificial, natural, and phylogenetic) have been given (Chapter III, p. 13). In using these terms, one should note that the term natural classification has been used by many contemporary botanists as synonymous with phylogenetic classification. In this text, the term natural is restricted to the classifications based on form relationships. Turrill (1942) emphasized the confusion and ambiguity that has resulted from the indiscriminate use of the term "natural" when applied to both pre-Darwinian and post-Darwinian classifications, and has advocated that it be abolished with regard to classification, and the term "general" be substituted for it. Turrill pointed out that none of the present-day so-called phylogenetic classifications is truly phylogenetic, but is only presumed to be so, or is phylogenetic only in so far as available evidence allows. For this reason he extended the application of the term "general" to cover also all modern post-Darwinian systems of classification, and reserved the term "phylogenetic classification" for a system to be developed at a future time when all the facts of evolution (now lacking) have been discovered. The validity of Turrill's views is patent, but their application is deferred for the present because (1) the term "natural" is deeply entrenched in biological literature and thinking, its original usage from before the time of Linnaeus through the nineteenth century is clear and consistent, and since the ambiguities are of relatively recent origin it is to be hoped that they may yet be eliminated if contemporary biologists will be more precise and restrictive in their use of the term; and (2) the term "phylogenetic" implies a classification based on evolutionary sequences and genetic relationships, and since this has been the underlying principle of the more recent modern classifications, then they are phylogenetic in principle as contrasted with the earlier natural classifications.

[3] For an explanation of the distinction between an artificial key and a synopsis, see Chapter X, p. 225.

ships of the components of families, genera, and species has used these methods augmented by cytogenetic and serological research.

It may be asked why, if a present-day classification meets our practical needs, we should strive to piece together a complex phyletic system. The basis of all evolutionary theories is the belief that living organisms may have progressed (though with many digressions) from primitive to more advanced forms. Only a fraction of the total of these forms is known, but the size of the fraction is not known. The living forms about us (species, genera, families, etc.) represent only the tips of branches, end products of evolutionary processes. This being true, the evolutionist—or phylogenist—will never be content until the theory of evolution has been proved as a law of nature. This he proposes to accomplish by the discovery and fitting of all evolutionary stages of life into one biological cosmos. A major segment of that cosmos will be a valid phylogenetic classification complete in all details and free from gaps and missing links. A phylogenetic system of classification for plants would provide the answer to questions of their origin, to their modes of evolution, to problems of monophyleticism vs. polyphyleticism, the identity of primitive and advanced characters, etc. It would result in a single stable classification of relationships.

## Diversity of phyletic concepts

There is little unanimity of current opinion on phylogenetic matters. This situation is attested to by the diversity of opinion represented by the presumedly phylogenetic classifications of Bessey, Hutchinson, Wettstein, Pulle, Skottsberg, *et al.*[4] The reason for this is the lack of factual data. Botanists do not know enough about the vascular plants of the past (except perhaps of the ferns and conifers) to be able to distinguish with certainty between characters indicative of primitive conditions and those indicative of advanced conditions. The construction of the ultimate phylogenetic classification must be based on established facts regarding the characteristics of ancestors of every taxon level. These ancestors existed in remote geologic time, and because of their relative simplicity their characters are said to be *primitive,* while those of their present-day descendants are said to be *advanced*. The primitive characters of contemporary, and presumedly phylogenetic, systems of angiosperm classification are not often based on paleobotanical evidence illustrative of an-

---

[4] The omission of Engler's system from this listing is deliberate, since Engler did not consider his system to be a phylogenetic classification in the broad sense of the concept (for explanation, see Chapter VI).

cestral conditions, and for the most part their primitiveness may be a matter of personal opinion or of judgments based on circumstantial evidence. Many of the so-called primitive characters in published lists (including those in this text) are alleged to be primitive because they occur in members of primitive taxa, and the taxa are primitive because they have primitive characters. It is difficult to find, by objective methods, devices to break this cyclic reasoning, and, among the angiosperms at least, there is inadequate paleobotanical evidence to support one view and to reject the other. For example, Eichler, Wettstein, Rendle, and others considered the unisexual apetalous cyclic flower to be primitive, whereas Bessey, Hutchinson, and others treated it as advanced and considered the bisexual polypetalous flower with spiral arrangement of parts to be primitive. The meager paleobotanical evidence of earliest angiosperms (in point of geologic age) is of plant structures identified with taxa representative of both views. Although it is not concrete evidence from the paleobotanical record, but is to a large extent circumstantial and based on assumptions, the evidence that favors the views of Bessey is stronger than that which favors the views of Eichler.[5]

The lack of paleobotanical evidence hampers and perhaps retards current phylogenetic studies of the angiosperms, but this situation does not mean that the phylogenist must "rest on his oars" and await new paleobotanical data. Rather, he must continue to make the best use possible of all other available evidence. Bailey (1949) expressed this view when he said that "it should be emphasized . . . that diversified investigations of surviving angiosperms provide the only available means at present

[5] Studies by Sporne (1948, 1949) have emphasized the utility of biometrical correlations between floral and vegetative characters in assessing the relative advancement of dicotyledonous families. In arriving at his correlations, he defined a primitive character as "one which, possessed by some present-day families, was also possessed by their ancestors," and a primitive family as "a present-day family which has retained a relatively large number of primitive characters and which has diverged very little from the ancestral." By statistical analysis dealing with presence or absence of 12 allegedly primitive characters throughout 259 dicot families, Sporne arrived at an "advancement index (%)" for each family. The characters selected to be most primitive were:

| | |
|---|---|
| Trees or shrubs | Petals free |
| Leaves glandular | Stamens pleiomerous |
| Leaves alternate | Carpels pleiomerous |
| Leaves stipulate | Seeds arillate |
| Flowers unisexual | Seeds with two integuments |
| Flowers actinomorphic | Seeds with integument bundles |

As a result, the most primitive dicot families were concluded to be the Flacourtiaceae, Anonaceae, Magnoliaceae, Myristicaceae, and Euphorbiaceae, while the most advanced included the Labiatae, Valerianaceae, Dipsacaceae, Phrymaceae (the Compositae standing about two-thirds the distance up the scale). The amentiferous families were scattered but averaged about midway along the scale of advancement with the Fagaceae the most primitive and Garryaceae the most advanced.

of morphologically characterizing this great group of the vascular plants . . ." It has been pointed out by phylogenists that there is strong, and seemingly irrefutable, evidence among living plants that certain basic characters or conditions are primitive and that others are derived from them. For example, in leaves of vascular plants as well as in the reproductive parts, the primitive arrangement (in most angiosperms) is in a spiral (supported by the fossil record). Furthermore, in the angiosperms, the anatomy shows that most so-called whorled or cyclic arrangements of floral parts are vertically compressed spirals, and likewise many leaf whorls are not true whorls but are compressed spirals. This sequence is presumed to be evidence that the spiral arrangement is more primitive than is the cyclic. In some apetalous flowers (as in Salicaceae, Juglandaceae, Urticaceae, Gramineae, some Centrospermae, and others) vestigial vascular systems are present that occupy positions anatomically homologous with those in petaloid flowers that lead to perianth parts. This is presented as evidence that apetaly is an advanced condition, derived from ancestors whose flowers had a perianth. Similarly, evidence from living plants supports the view that apocarpous gynoecia (unicarpellate ovaries) are primitive and that syncarpous gynoecia are advanced. Furthermore, these several allegedly primitive characters occur in combination in some families as a positive correlation of high value, a situation that strengthens the view that those families are more primitive than are others represented by similar correlations of lower value or of negative value.

The diversity of phyletic concepts is due also to the lack of synthesis of all available data. In all fairness to Hutchinson, his system cannot be criticized in this regard since he did not accompany his presentation with reasons for the alignment of most of the taxa. However, there is little to indicate that Bessey accepted much evidence other than that provided by gross and comparative morphology in the development of his system (and the tenets on which it was based) from that of Bentham and Hooker. Hallier and Wettstein both drew on the paleobotanical, serological, and anatomical data then available (in addition to morphological considerations), but many data which have since been accumulated are in contradiction to their opinions. Another reason for these diversities is that some phylogenists have given too little attention to the phyletic significance of pollen grain and starch grain morphology, to the physiological bases of serology and allied physicochemical relationships, and to the relationship of genic constitutions to segregation and establishment of major taxa.

The problems of phylogeny are so complex that only by considered analysis and synthesis of all possible evidence will there be anything approaching harmony of opinion on the subject.

The incompatibilities of some data with other data are responsible in part for the diversity of phyletic views. These resolve themselves into incompatibilities of interpretation of facts, and incompatibilities between bits of circumstantial evidence that too often are treated as facts. Evolutionary patterns laid down in conformance with morphological findings often are at variance with those based on physiological findings; in many instances the fossil record does not support segments of an allegedly phylogenetic classification based on the comparative morphology of extant taxa; in some instances, paleobotanists have alleged that entomophilous flowering plants existed during certain geologic ages that paleontologists have considered to be devoid of insect pollinators; embryological situations presumed by some authorities to be highly advanced occur in flowering plants treated by other botanists as primitive (as in some of the Amentiferae); and primitive characters of wood anatomy sometimes are at variance with allegedly primitive morphological characters of the same taxon. Existence of these conflicting situations should not be construed to imply that a relatively primitive extant flowering plant must possess primitive characters in all its parts, for it is accepted generally that a plant may be primitive in one or more respects and advanced in others. The basis of the conflict often rests on the bias of the phylogenist who may reject (or not even consider) characters contributed by disciplines in which he himself is neither well versed nor a specialist.

The existence of these diversities of phyletic concept should not be discouraging, nor be taken to indicate that all is confusion, but rather should be viewed as part of a healthy situation. The recognition of the causes of the diversities should serve to demonstrate to the phylogenist the need of exhuming more paleobotanical materials, and of carefully integrating with them the findings, based on living material, of research from all fields of biology. The formulation of phylogenetic classifications demands the teamwork and collaboration of botanists of all disciplines and the considered evaluation of data without bias.

### Contributions to phylogenetic knowledge

*Paleobotany* must be the foundation of phylogeny. The lack of solid and factual foundations for phylogenetic studies is a direct reflection of our knowledge of the paleobotanical history of vascular plants, especially of the angiosperms. The situation was expressed well by Turrill (1942, p. 508) when he wrote,

> The great diversity of opinion in published accounts of plant phylogeny suggests . . . that the available data are still too few for the construction of a valid general phylogenetic scheme . . . the paucity of relevant paleobotanical data in most groups of plants (partial exceptions are the Pteridophyta and Gymnospermae) is a major cause of uncertainty as to whether or not proposed series are phylogenetic, and, if they be, in which direction they should be read.

One of the requisites of phylogeny is the determination of origins of taxa, not only the origins of those in the highest categories but also those of levels such as class, order, and family. Present-day taxa of these latter categories among the angiosperms originated (presumably) millions of years ago, yet (*fide* Thomas, 1936) what little is known of them from the fossil record is at variance with views of their origin as accepted by most phylogenists. Some phylogenists, lacking necessary data from the paleobotanist, have fabricated missing links to make plausible their otherwise unfounded views (as Arber and Parkin's hypothetical Hemiangiospermae, 1907).

*Anatomy* is a source of evidence available from both paleobotanical and living material and is of considerable value to the phylogenist. Once the fossil record has clarified the issue of primitive vs. advanced characters, and the issue has been settled as to what type of stem anatomy preceded another, the existing accumulation of anatomical evidence from both vegetative and reproductive parts will be of critical importance. In the interim, much assistance is provided by these facts establishing relationships or probable affinities between taxa in the lower categories, especially below that of the order. Knowledge of the phyletic relationships within the angiosperms, the monocotyledons, and numerous families of dicotyledons, has been advanced by evidence of this type (see papers by Bailey, Bailey and Sinnott, Chalk, Cheadle, Eames, Heimsch, Metcalf and Chalk, Record, and Tippo). Similar and equally significant advancements have accrued from studies of the inflorescence and floral anatomy.

*Morphology* has dominated phylogenetic research for over a century. Initially it was restricted to gross morphology, and studies of it were followed by a period of intense ontogenetical investigation (an approach to phylogeny defended by Sahni, 1925, and by Lam, 1948, reviewed by deBeer, 1936, and currently de-emphasized by many botanists), and lately superseded by the widespread recognition of the significance of investigations of embryology and floral anatomy. Noteworthy also are the phyletic studies of seeds, especially of embryo and endosperm (Martin,

1947). One reason for this dominant position of morphology is that morphological studies permit ready determination and correlation of characters; another is that they can be subjected to comparative analysis to a greater degree than can characters from most other studies. In the phyletic studies of minor categories, it seems probable that much can be learned by the application of quantitative and biometrical methods to an analysis of morphological data (see papers by Anderson, 1936, Anderson and Abbe, 1934, Fassett, 1941, Epling, 1942, *et al.*). In the determination of phyletic positions within the higher categories, one objective of both anatomical and morphological studies has been evaluation and utilization of characters believed to be the more conservative.[6] Another objective has been the determination of the direction from which to read a series of transitional morphological situations. In other words, the solution of the basic problem of determining which end of the series is primitive and which is advanced rarely can be based on indisputable evidence.

*Cytology* and cytotaxonomy are studies which to date have been of most value in the phyletic resolution of taxa below the level of genus. Even then, conditions often have been so uniform that they have contributed little to the understanding of relationships within some genera (*Rhododendron,* many gymnosperms, cacti, *et al.*). Cytological data of themselves are inconclusive in determining or strongly indicating phyletic relationships of major categories. An example of this is afforded by one assemblage of data designed to indicate that the Magnoliales may have been derived from wide crosses between different groups of gymnosperms (Anderson, 1934). Since the Magnoliales are phyletically old, these presumed crosses did not occur between present-day gymnosperms, but between ancient ancestors of them. Postulations or hypotheses of what might have happened genetically in previous geological ages generally are more in the realm of philosophy than of science.

*Phytogeography,* particularly as correlated with the *morphology of the earth,* undoubtedly will contribute increasingly to the accuracy of phyletic investigations. Too little is known yet about the early land formations of the earth, but with resolution of this hiatus the early migrations of

---

[6] By conservatism of characters is meant their persistence within the plant over an extensive period of evolutionary development. For example, anatomists generally hold to the view that characters of stem anatomy are very conservative, since they vary little or not at all among species of a given genus, rarely between genera of a given family (when they do that genus is suspected by the anatomist of being taxonomically out of place), and that members of many families possess common or similar characters of stem anatomy. The morphologists also generally contend, for example, that the vascular anatomy leading to gynoecia and androecia are modified more slowly than the gross gynoecial or androecial structures, and therefore by their conservatism the vascular conditions serve to indicate probable ancestral situations.

plants will be understood better and there must then be harmony beween the phyletic arrangement of major categories and the evolution of their distributions. As was true for cytogenetical evidence, so is it true also for phytogeography that evidence contributed by studies of it currently is more important in the phylogenetic classification of taxa at and below the level of the family.

*Physiology* has produced criteria that have been by-passed by most phylogenists, and especially by taxonomists dealing with the minor categories. The contributions of Chester, Reichert, Molisch, and, to a lesser degree, of Mez attest to the basic value of serological and physicochemical investigations to a better understanding of relationships. The classical serological studies by Molisch and others are in surprising agreement in many respects with several phylogenetic classifications based primarily on morphological evidence. Molisch contributed much to our knowledge of distributions of chemical products among plant genera and families and concluded (to quote Turrill, 1942, p. 503) that, "while we are only at the beginning of phytochemical knowledge, the phylogenetic value of phytochemistry is already considerable. Especially can the chemistry of plant substances and their distribution suggest the correctness or otherwise of phylogenetic schemes based on morphological or other criteria." It is probable that further phytochemical studies may aid considerably in confirming or rejecting the transfer of genera from one family to another, or of a family from one order to another. Reichert's work on starches (1919) should be re-examined and evaluated in conjunction with phyletic studies at all levels. Studies of pharmacognosy have provided much information on the presence of such organic materials in plant tissues as alkaloids, glucosides, resins, oleoresins, volatile oils, etc. These are believed in some instances to be indicators of phylogenetic relationships, and further investigations are needed in this direction.

The contributions from the physiological approach are yet in the primitive or initial stages. This in no way lessens their potential value or significance, but rather serves to emphasize the need of revitalized activity by the physiologist, together with a broader outlook on the basic problems. The collaborating physiologist must admit the possibility that parallel or convergent evolution is present in his field of interest as it has been demonstrated to exist in the disciplines of morphology, cytology, and genetics. The fact that end products of biochemical processes occur in different genera does not of itself mean that those genera are more closely related to each other than to genera not possessing them. Furthermore, it is not merely the chemical end product that is of significance, but also the processes by which it was produced. The end product, measured by serological or biochemical tests, may be compared to the processes or materials that produce it as the gross morphology of the plant is to the vegetative and reproductive anatomy of that plant. There is unlimited opportunity for the biochemically trained physiologist who will work in concert with the morphologist, anatomist, and taxonomist on problems of phylogeny of vascular plants at all taxal levels.

## Phylogeny and the higher categories

*Pteridophytes.* It has been established by numerous researchers that the Pteridophyta are not a phylogenetic taxon (Eames, Andrews, Arnold, Copeland, Wettstein). The available evidence makes it clear that phylogenetically they must be thought of and treated not as one but as 3 or 4 taxa, as: the psilopsids, descended probably from the Psilophytales of lower Devonian time; the lycopsids (including the Lycopodiaceae, Isoetaceae, and Selaginellaceae) descended perhaps from such ancestral stocks as the lepidodendrids of the early Carboniferous; the sphenopsids (scouring rushes) derived perhaps from the calamites and sphenophylls and perhaps the Hyeniales of the lower Devonian; and a fourth taxon, represented by those ferns comprising the Filicales, whose ancestral stocks are traceable back into the Permian (the upper Paleozoic). The paleobotanical evidence in support of these views is more adequate than that known for most other major categories of plants. These major taxa are among the oldest of land plants. Also, they are the oldest of vascular plants, for nearly three-quarters of their evolutionary history had elapsed by the time modern flowering plants were first known and they were old when the cycads and ginkgoes became established. Some of these ancient ferns (as anemias, marattias, angiopteris) are known by living descendants today, and others by descendants of close affinity (as the fossil *Osmundites* and its contemporary counterpart *Osmunda*).

The availability of this paleobotanical material has made it possible to reconstruct many of the evolutionary channels of development that lead to present-day vascular cryptogams. It has made it clear that the ferns and their allies are of a polyphyletic origin, that they have had no known common ancestor, and that each of the 4 contemporary major taxa of vascular cryptogams (psilopsids, sphenopsids, lycopsids, and Filicales) may be as different from the others as the club mosses or scouring rushes are from buttercups or orchids. The evolutionary picture of these plants, well known though it may be in a relative sense, is far from complete. Many gaps and extensive voids remain to be filled in the almost crude phylogenetic structure now available; parts of that structure have been constructed from weak or even hypothetical evidence, and nothing definite is known of the character or identity of fern antecedents.

**Gymnosperms.** These are the cycads, ginkgo, taxads, conifers, Gnetales, and some extinct taxa. Counterparts of present-day cycads and the ginkgo date back through the Triassic and into the Permian (Upper Carboniferous), while modern conifers were dominant in the Cenozoic and extended doubtfully into the Upper Cretaceous. Prior to this, there were transition conifers that were climax vegetational types from the Upper Cretaceous back through the Triassic. Conifers ancestral to these lived during the Paleozoic. For much of our knowledge of the Paleozoic conifers we are indebted to the researches by Florin (1938–1945) that have contributed also to a clearer understanding of relationships within modern conifers. These early conifers were trees and shrubs, appeared perhaps more like modern Araucarias than other extant types, with small flattened leaves in spirals, loose stroboloid inflorescences that were intermediate in organization between those of the Cordaites and the present-day pines. The better understanding of these plants and their descendants led Florin to the conclusion (1948) that the taxads are not a part of the Coniferae, but are a separate and equivalent taxon.

Evidence produced during the past few years has supported the views of some phylogenists that the Cordaites, or plants ancestral to them, may have been the principal ancestors of the conifers. The Cordaites are an extinct taxon, of greater antiquity than the conifers, of unknown origin, and are responsible in part for the views of some paleobotanists that the origin of the gymnosperms probably extended back into the early Devonian epoch. In any case, the identity of the early ancestors of the conifers becomes speculative, and no fossil plant remains have been discovered that definitely can be said to be ancestral to the cycads or to the ginkgo.

The knowledge of phylogeny within the gymnosperms has advanced appreciably with the increase in knowledge of their paleobotanical background. Within the conifers it has become increasingly clear that the several families are of a more remote relationship to one another than was formerly believed and that the family Pinaceae of older works deserves to be treated taxonomically as an order or suborder, and its tribes raised to the family level. No ancestors are known for the Gnetales, Welwitschiales, or Ephedrales, and none of these orders is closely allied to other gymnosperms. Like the ferns, the gymnosperms are certainly of polyphyletic origin, and there is no evidence to suggest a common ancestor for the cycads, ginkgo, taxads, and the conifers. Similarly, there is no evidence to suggest phyletic relationships between the pteridophytes and the gymnosperms; the available evidence suggests that the latter are not derived from the vascular cryptogams.

***Angiosperms.*** The paleobotanical record for the seed plants is so inadequate that phylogenists have had no solid foundation on which to construct a classification of phyletic relationships. Much of the meager paleobotanical evidence is of vegetative parts, and almost none of the reproductive material is derived from flowering plants more primitive than can be found among modern angiosperms. The lack of this material is responsible to a large extent for the diversity of opinion as to what types of structures are indicative of primitive conditions and what are clearly advanced. Plants of ranalian affinities and those of amentiferous affinities are among the oldest of paleobotanical material. From the evidence available, it would seem that the angiosperms "blossomed forth" with a sudden surge in abundance and variety in the late Cretaceous, for fossil material of trees scarcely differentiable from those now growing about us is obtained from strata of that epoch.[7] Some of this material dates back into the Jurassic, and it is becoming more clear that the angiosperms originated earlier than was indicated by evidence of a quarter century ago.

The origin of the angiosperms is not known. Many theories (some more philosophical than scientific) have been presented to account for their origin; some are supported by circumstantial evidence that is convincing; others are based on more tenuous evidence, but enjoy current acceptance; and a few are so lacking in credibility as to be accepted only by the gullible. There seems not to be a majority of opinion in favor of one view over another, but if the admittedly inadequate and partially circumstantial evidence in support of the pteridosperm theory is considered satisfactory then that theory is the most difficult to refute. The pteridosperms (seed-bearing plants with fernlike foliage) had their origins in the Devonian and were dominant during most of the Carboniferous. They produced pollen from microsporophylls and seeds from their naked ovules. The plants were monoecious or dioecious and the probably apetalous solitary megasporangia or ovules were variously disposed (often in panicles). They are the most primitive of known seed plants, but the phyletic gap between them and the earliest true flowering plants of the

---

[7] For an analysis of reasons accounting for this gap in the paleobotanical record, see Just (1948, pp. 97–56).

Jurassic is exceedingly great. In recounting theories of angiosperm origin Andrews (1947) has said,

> . . . the pteridosperms seem to present the only possible fossils to which we may look as a starting point for the flowering plants. The seed pod of the latter constitutes their most distinctive feature, and the cupule which encloses the seed (or seeds) in the pteridosperms is the most likely precursor of the flowering-plant seed pod. It has not been proved that this is actually the case, but since it is the most plausible and convincing evidence that we have to go on, then it is justifiable to use it at least as a working hypothesis until more evidence is accumulated either to support it or to disprove it.

A similar view was taken by Arnold (1947), who stressed also that the angiosperms must have had "a longer and more extensive pre-Cretaceous history than so far has been revealed by the fossil record and that the scarcity of fossils is due to their predominatingly upland habits where the remains were not readily buried and preserved." Darrah (1939) conceded the pteridosperms to be probable ancestors of the cycads, but held that "the relationship of the seed-ferns to groups other than the true ferns and cycads, is entirely debatable." He was of the opinion that the angiosperms arose "without shadow of doubt from some gymnospermous stock."

It is the opinion of other paleobotanists and phylogenists that the angiosperms have been derived from the gymnosperms or from stocks ancestral to them. In 1907 Arber and Parkin endeavored to establish this theory by the fabrication of a hypothetical connecting link that they called the Hemiangiospermae, in which the reproductive structure was constructed according to the plan of the cycadeoid flower; a structure assigned a perianth of many distinct parts, open foliaceous carpels with marginal megasporangia, and an androecium of many stamens, all arranged in spirals. There is no evidence in the fossil record that such a structure ever existed, and without it their theory of angiosperm origin from the gymnosperms via cycadaceous ancestors has no substance. Despite this absence of fact, the theory was accepted by Bessey, by Hutchinson, and others. Hagerup postulated a diphyletic origin for the angiosperms with one line extending from the Filicales through the cycads to the Ranales (the Polycarpicae) and the second from the lycopods through the Cordaites, conifers, and Gnetales to such angiosperm taxa as the Centrospermae and Personatae.[8] These views were rejected in part by Chaudefaud (1946), who pointed out the incompatibility of their basic tenets with morphological conditions known to exist in the ovules of pteridosperms, cycads, and some modern flowering plants. The signifi-

[8] For a summary and analysis of these and other contemporary views, see Just (1948).

cance of the Gnetales to the solution of problems of angiosperm origins has been emphasized by Markgraf (1930), who held the angiosperms to be a taxon derived from gymnosperms, and one that branched off from ancestors of modern gymnosperms so long ago as to be of no particular relationship to present-day conifers. Fagerlind (1947) reinterpreted the morphology of the *Gnetum* flower, demonstrated it to be homologous with those of *Ephedra* and *Welwitschia,* and considered the 3 taxa to have had a common ancestor which was derived from the same stock as the single ancestor postulated to have given rise to the Pro-angiosperms (from which all modern angiosperms evolved polyphyletically).

A few phylogenists have held that the angiosperms may have originated from the Caytoniales, a rather recently discovered group of the middle Jurassic. However, as reviewed by Arnold, these plants have now been shown to have "undoubted affinity with the pteridosperms, and they are now classified as Mesozoic remnants of that group. The gap between the Caytoniales and the flowering plants is probably as great as that between any of the other vascular plants and the angiosperms."

As recently as 1946 it has been suggested by Lemesle that certain ranalian families were descended from the Bennettitales. This was based, according to Bailey (1949), on a misinterpretation of the phyletic significance of tracheary elements of the pertinent taxa together with Lemesle's failure to recognize the existence and significance of parallel and convergent evolution of anatomical features. It was pointed out further by Bailey that "such sweeping generalizations, based upon limited and inadequate data, . . . have raised uncertainties and doubts in the minds of many botanists regarding the value and reliability of anatomical evidence in the study of phylogeny."

An entirely different approach to the problem of angiosperm origin was published by Sahni (1920), redefined, extended, and given impetus by Lam (1948) and repudiated in essence by Bailey (1949, p. 64 *ff.*). These views presented by Lam (together with his classification based on them) are becoming known as the Stachyosporous theory, and in brief, he recognized the seed plants to be composed of two taxa: Phyllospermae and Stachyosporae. The Phyllospermae have their megasporangia (ovules) enclosed in foliar carpels and contain most of the apocarpous dicots (Polycarpicae) and their derivatives, i.e., the ranalian taxa. The Stachyosporae have their megasporangia "protected by sterile organs (e.g., sterilized microsporangiosphores; pseudocarpels)" and contain the Monochlamydeae of Engler and "perhaps also some Monocotyledons and Sympetalae." Lam has emphasized the need for rejection of the Angiospermae as a starting point for morphological interpretations of all other vascular plants and substituted in their place the facts known about the psilopsids, pteropsids, and the Protoangiospermae, and in arriving at his conclusions has made much of ontogenetical and teratological evidence. (For a critique reviewing and rejecting the Stachyosporous theory, cf. Eames, 1951.)

Little credence has been given to the conjecture presented by Anderson (1934) that the angiosperms may be hybrids of widely divergent gymnosperms, but Turrill (1942) has indicated that "it may be that there is more truth than most botanists seem to accord to Lotsy's view . . . that hybridization is the key to evolution." Similarly, it has been held by Lawson (1930) that the high ratio of pollen sterility present in endemic species in the two dominant dicot families of the Australian flora (Myrtaceae and Proteaceae) is evidence of their hybrid origin, and this plus other evidence has led him to believe that hybridization and natural selection have been the major factors in the evolution of the angiosperms. Any favorable consideration of these views requires also reconsideration of Goldschmidt's opinion that major taxa of the angiosperms may have arisen directly by mutation.

The determination of whether the earliest seed plants, and especially the earliest angiosperms, were woody or herbaceous (or if each type individually may have been ancestral) is basic to any phyletic arrangement of them. The fossil record is not sufficiently complete to solve or to contribute materially to a solution of the problem. Morphologists have supported both views. Arber upheld the herbaceous ancestry for the angiosperms and contended that the woody habit is "an expression of racial senile degeneration"; evidence was amassed from many sources by Eames (1911), Sinnott and Bailey (1914), and by Sinnott (1916) in support of the primitiveness of the woody habit over the herbaceous. Their views have been accepted in principle, or subscribed to independently, by an increasing number of botanists, including Arnold, Bessey, Darrah, Hallier, Hutchinson, and others.[9] Acceptance of the view that the woody habit preceded the herbaceous demands acceptance also of the view that the angiosperms are a polyphyletic taxon, for, as Sinnott and Bailey pointed out, "it is quite evident, therefore, that whichever of these two classes [i.e., woody or herbaceous angiosperms] is the more recent it must have arisen quite independently many different times, and from numerous ancient stocks." Evidence is accruing steadily from all sources

---

[9] For a review of the evidence in defense of each of the two views, see Darrah, pp. 168–172.

that supports this view, but not all phylogenists accept polyphyleticism for the angiosperms. Both Bessey and Hutchinson have rejected it in favor of a monophyletic origin, but others (Lotsy, Hallier, Eames) pointed out that from evidence now available there can be no answer to the question of origin of many taxa (e.g., Casuarinaceae, Salicaceae, Leitneriaceae, Proteaceae, to mention a few) other than a polyphyletic one; they are families that stand alone and far apart from others with no known ancestral relationships or interrelationships. The absence of known origins for the angiosperms reduces most phylogenetic views close to the level of philosophical speculation.

The role of vegetative anatomy has been prominent in providing valuable evidence for phylogenetic studies of especially the woody angiosperms,[10] with much of the research in this country having been done by Bailey and his students at Harvard. It has been pointed out by Tippo (1946) that extension of these studies to the herbaceous dicots will "not only uncover new phylogenetic sequences but . . . [will] extend the lines of specialization already established [for the woody dicots] into this relatively virgin territory." The work by Cheadle (1942) and his students on monocot stem anatomy has laid the foundations from which valuable phylogenetic results may be expected.

Opinions have differed as to whether the monocotyledonous plants are more advanced or more primitive than the dicotyledonous plants. If the view is accepted that the woody habit preceded the herbaceous, then, since the vast majority of ancient and modern monocots are herbaceous, it would point to the monocots having been derived from one or more dicot ancestors. Botanists accepting this view treat the dicots as the more primitive. Mrs. Arber has held that the monocots preceded the dicots. Current criteria from morphological and anatomical studies of both vegetative and reproductive structures would place the monocots as probably having been derived from the dicots. There is evidence for the belief that the monocots are more likely to be monophyletic in origin than is true for the dicots.

The subject of the phylogeny of vascular plants is not one to give the student a sense of satisfaction, for neither a cursory nor an exploratory study of it engenders the feeling of having grasped or comprehended a segment of scientific knowledge. This regrettable sense of insecurity is due in part to the speculative nature of many of its findings or conclusions, speculations by many botanists—based on the same or different evidence—and of great diversity. The insecurity will persist as long as ex-

[10] For a critical review and bibliography of this subject, see Tippo (1946).

tensive gaps exist in our knowledge of the paleontological record. In the interim, factual evidence of direct relation to the phylogeny of the vascular plants is accumulating. New allegedly phyletic classifications will be produced, but it cannot be accepted that the taxonomist will reject yesterday's classification for that of today nor be prepared to set aside today's for tomorrow's when it arrives. These new and successive classifications have their place; they are to be studied, to be tested, and to be subjected to critical analysis in the light of the latest available evidence. It is a responsibility of the originator of each to insure that each part of the new classification be accompanied by adequate discussion of its establishment on facts. It should be remembered that the phylogenist is dependent on taxonomy to accomplish his objectives, but that the taxonomist can identify, name, and classify his plants satisfactorily without phylogeny. Turrill has pointed out that, "the study of phylogeny is, indeed, not only justifiable as a deduction from the general theory of evolution, but in its own inherent right is a subject of great biological importance and interest. Historically, classification preceded, and must precede, investigations of phylogeny."

Much of this chapter, perforce, has treated the subject of phylogeny as applied to the major taxa of vascular plants, and in this more attention has been focused on the inadequacies of the evidence, the subjectiveness of the phylogenist's approach, and the complexities of the problem. It has been emphasized that acquisition of additional paleobotanical evidence is of paramount importance to the solution of problems associated with the origins of flowering plants. The paleontological record is the best available evidence of the nature of plants of the past, and while it is probably true that other types of evidence are subordinate to it, one must remember that evidence of any single type is inadequate of itself. The phylogenetic classification must be a synthesis of evidence obtained from all sources.

Most of these phylogenetic considerations serve to acquaint one with the situation rather than to instruct on the subject. The solution of phyletic problems as associated with the higher levels of classification is not one to confront the average student of taxonomy. However, there are many problems of phylogeny that do confront the taxonomist. They deal with phylogeny of and within the family, the genus, and the species. There are many approaches to solutions of the origin and evolution of these taxa, and they confront anyone engaged in revisionary or monographic research. The increase in exploration of these approaches promises to make more correlations available to the resolution of phyletic relationships and affinities. The birth of this stage of phylogeny will engender more confidence in the classification produced.

*LITERATURE:*

ANDERSON, E. Origin of the angiosperms. Nature, 133: 462, 1934.
——. The species problem in *Iris*. Ann. Mo. Bot. Gard. 23: 457–509, 1936.
ANDERSON, E. and ABBE, E. C. A quantitative comparison of specific and generic differences in the Betulaceae. Journ. Arnold Arb. 15: 43–49, 1934.
ANDREWS, HENRY N. JR. Ancient plants and the world they lived in. 279 pp. Ithaca, N. Y., 1947.
——. Some evolutionary trends in the pteridosperms. Bot. Gaz. 110: 13–31, 1948.
ARBER, A. The tree habit in angiosperms: its origin and meaning. New Phytologist, 27: 69–84, 1928.
——. The interpretation of the flower, a study of some aspects of morphological thought. Biol. Rev. 12: 157–184, 1937.
ARBER, E. A. N. and PARKIN, J. On the origin of angiosperms. Journ. Linn. Soc. Bot. (Lond.) 38: 28–80, 1907.
ARNOLD, C. A. An introduction to paleobotany. 433 pp. New York, 1947.
——. Classification of gymnosperms from the viewpoint of paleobotany. Bot. Gaz. 110: 2–12, 1948.
BAILEY, I. W. Origin of the angiosperms: need for a broadened outlook. Journ. Arnold Arb. 30: 64–70, 1949.
BANCROFT, H. A review of researches concerning floral morphology. Bot. Rev. 1: 77–99, 1935.
BEMMEL, A. C. V. VAN. Modern systematics. Chron. Nat. 104: 97–99, 1948.
BESSEY, C. E. The point of divergence of monocotyledons and dicotyledons. Bot. Gaz. 22: 229–232, 1895.
——. Phylogeny and taxonomy of the angiosperms. Bot. Gaz. 24: 145–178, 1897.
——. Revisions of some plant phyla. Univ. (of Nebraska) Studies, 14: 37–109, 1914.
——. The phylogenetic taxonomy of flowering plants. Ann. Mo. Bot. Gard. 2: 109–164, 1915.
BOWER, F. O. Primitive land plants. Cambridge, England, 1935.
BROWNE, I. M. P. Some views on the morphology and phylogeny of the leafy vascular sporophyte. Bot. Rev. 1: 383–404, 1935.
BUCHHOLZ, J. T. Generic and a sub-generic distribution of the Coniferales. Bot. Gaz. 110: 89–91, 1948.
CAMPBELL, D. H. The phylogeny of the angiosperms. Bull. Torrey Bot. Club, 55: 479–497, 1928.
——. The phylogeny of monocotyledons. Ann. Bot. 44: 311–331, 1930.
——. The evolution of land plants. 731 pp. Stanford, Calif., 1940.
CHAUDEFAUD, M. L'Origine et l'évolution de l'ovule des Phanérogames. La rev. sci. 84: 502–509, 1946.
CHALK, L. The phylogenetic value of certain anatomic features of dicotyledonous plants. Ann. Bot. n.s. 1: 429–437, 1937.
CHAMBERLAIN, C. J. The gymnosperms. Bot. Rev. 1: 183–209, 1939.
CHEADLE, V. I. The role of anatomy in phylogenetic studies of the Monocotyledoneae. Chron. Bot. 7: 253, 254, 1942.
CHESTER, K. S. A critique of plant serology Parts I, II. Quart. Rev. Biology, 12: 19–46, 165–190, 1937.

COCKERELL, T. D. A. The origin of the higher flowering plants. Science, 81: 458–459, 1935.
CONZATTI, C. El origen probable de las monocotiledoneas. Proc. 8th Amer. Sci. Cong. 3: 197, 1942.
COPELAND, H. F. The phylogeny of the angiosperms. Madroño, 5: 209–218, 1940.
CROW, W. B. Phylogeny and the natural system. Journ. Genetics, 17: 85–155, 1926.
DARRAH, W. C. Principles of paleobotany. 239 pp. Leyden, 1939.
DAVY, J. BURTT. On the primary groups of dicotyledons. Ann. Bot. N. S. 1: 429–437, 1937.
DEBEER, C. R. Embryology and evolution. 116 pp. Oxford, 1936.
DU RIETZ, G. E. The fundamental units of vegetation. Proc. Int. Cong. Plant Sci. 1926 (Ithaca, N. Y.) 1: 623–627, 1929.
EAMES, A. J. On the origin of the herbaceous type in the angiosperms. Ann. Bot. 25: 215–224, 1911.
———. The role of flower anatomy in the determination of angiosperm phylogeny. Proc. Int. Cong. Plant Sci. 1926 (Ithaca, N. Y.) 1: 423–427, 1929.
———. Morphology of vascular plants—lower groups. 433 pp. New York, 1936.
———. Again: The new morphology. New Phytologist, vol. 50, 1951. [In press]
EPLING, C. The American species of *Scutellaria*. Univ. Calif. Publ. Bot. 20: 1–146, 1942.
ERDTMAN, G. Suggestions for the classification of fossil and recent pollen grains and spores. Svensk. Bot. Tidsk. 41: 104–114, 1947.
FAEGRI, K. Some fundamental problems of taxonomy and phylogenetics. Bot. Rev. 3: 400–423, 1937.
FAGERLIND, F. Strobilus und Blüte von *Gnetum* und die Möglichkeit, aus ihrer Struktur den Blütenbau der Angiospermen zu deuten. Arkiv Bot. 33A (8): 1–57, 1947.
FASSETT, N. C. Mass collections: *Rubus odoratus* and *Rubus parviflorus*. Ann. Mo. Bot. Gard. 28: 299–369, 1941.
FLORIN, R. Die Koniferen des Oberkarbons und des unteren Perms. 1–8. Paleontographica, 85B: 1–729, 1938–1945.
FLORY, W. S. Chromosome numbers and phylogeny in the gymnosperms. Journ. Arnold Arb. 17: 83–89, 1936.
FOSTER, A. S. Leaf differentiation in angiosperms. Bot. Rev. 2: 349–372, 1936.
———. Phylogenetic and ontogenetic interpretations of the cataphyll. Amer. Journ. Bot. 18: 243–249, 1931.
FRENGUELLI, J. El origen de las angiospermas. Bull. Soc. Argentina de Bot. 1: 169–208, 1946.
GILBERT, S. G. Evolutionary significance of ring porosity in woody angiosperms. Bot. Gaz. 102: 105–120, 1940.
GILMOUR, J. S. L. and TURRILL, W. B. The aim and scope of taxonomy. Chron. Bot. 6: 217–219, 1941.
GOLDSCHMIDT, R. The material basis of evolution. 436 pp. New York, 1940.
GREGOIRE, V. La valeur morpholologique des carpels dans les angiosperms. Bull. Classe Sci. 17: 1286–1302, 1931.
GUNDERSON, A. Flower buds and phylogeny of dicotyledons. Bull. Torrey Bot. Club, 66: 287–295, 1939.
———. The classification of dicotyledons. Torreya, 39: 108–110, 1939.
———. Flower structure and the classification of dicotyledons. Brooklyn Bot. Gard. Rec. 30: 93–98, 1941.
———. Flower forms and groups of dicotyledons. Bull. Torrey Bot. Club, 70: 510–516, 1943.

HAAS, O. and SIMPSON, G. G. Analysis of some phylogenetic terms with attempts at redefinition. Proc. Amer. Philosoph. Soc. 90: 319–349, 1946.
HAGERUP, O. Zur Abstammung einiger Angiospermen durch Gnetales und Coniferae. Danske Vidensk. Biol. Medd. 11: 4, 1934; 13: 6, 1936; 14: 4, 1938; 15: 2, 1939.
HALLIER, H. Provisional scheme of the natural (phylogenetic) system of flowering plants. New Phytologist, 4: 151–162, 1905.
———. L'origine et la système phylétique des angiospermes exposés à l'aide de leur arbre généalogique. Arch. Néerl. Sci. Exact. et Nat. Ser. III B, 1: 146–234, 1912.
HARRIS, T. M. The ancestry of the angiosperms. Proc. Sixth Int. Bot. Cong. Amsterdam, 2: 230–231, 1935.
HILL, A. W. The morphology and seedling structure of geophilous species of *Peperomia*, together with some views on the origin of monocotyledons. Ann. Bot. 20: 395–427, 1906.
HUTCHINSON, J. Contributions towards a phylogenetic classification of flowering plants. Kew Bull., 1923: 65–89, 241–261; 1924: 49–66, 114–134.
JEPSEN, G. L., SIMPSON, G. G., and MAYR, E. eds. Genetics, paleontology, and evolution. Princeton, N. J., 1949.
JUST, T. Gymnosperms and the origin of angiosperms. Bot. Gaz. 110: 91–103, 1948.
———. Some aspects of plant morphology in evolution. In Jepsen, et al. Genetics, paleontology, and evolution. Princeton, N. J., 1949.
KOZO-POLJANSKI, B. On some "third" conceptions in floral morphology. New Phytologist, 35: 479–492, 1936.
LAM, H. J. Studies in phylogeny. Blumea, 3: 114–158, 1938.
———. Classification and the new morphology. Acta Biotheoretica, 8: 107–154, 1948.
———. A new system of the Cormophyta. Blumea, 6: 282–289, 1948.
———. Stachyospory and phyllospory as factors in the natural system of the Cormophyta. Svensk Bot. Tidskr. 44: 517–534, 1950.
LEMESLE, R. Les divers types de fibres à ponctuations aréolées chez les dicotylédones apocarpiques le plus archaiques et leur rôle dans la phylogénie. Ann. sci. nat. bot. et biol. végétale, 7: 19–40, 1946.
LEWIS, D. The evolution of sex in flowering plants. Biol. Rev. 17: 46–67, 1942.
LOTSY, P. Phylogeny of plants. Bot. Gaz. 49: 460–461, 1910.
———. Evolution considered in the light of hybridization. 55 pp. Christchurch, New Zealand, 1925.
MAHESHWARI, P. A critical review of the types of embryo-sacs in angiosperms. New Phytologist, 36: 359–417, 1937.
———. An introduction to the embryology of angiosperms. 453 pp. New York, 1950.
MARKGRAF, F. Monographie der Gattung *Gnetum*. Bull. Jardin Bot. Buitenzorg, Ser. 3, 10: 407–511, 1930.
MARTIN, A. C. The comparative internal morphology of seeds. Amer. Midl. Nat. 36: 513–660, 1946.
MATTHEWS, J. R. Floral morphology and its bearing on the classification of angiosperms. Trans. Bot. Soc. (Edinb.) 23: 60–82, 1941.
MAURITZON, J. Die Bedeutung der embryologischen Forschung für das natürliche System der Pflanzen. Lunds Univ. Arsskr. N. F. II, 35 (15): 1–70, 1939.
MAYR, E. Systematics and the origin of species. New York, 1942.
METCALFE, C. R. The systematic anatomy of the vegetative organs of the angiosperms. Biol. Rev. 21: 159–172, 1946.

POPE, M. A. Pollen morphology as an index to plant relationship. Bot. Gaz. 80: 63–73, 1925.
PULLE, A. Remarks on the system of the Spermatophytes. In Compendium van de Terminologie Nomenclatuur en Systematiek der Zaadplanten. S. 134–139, Utrecht, 1937.
REICHERT, E. T. A biochemic basis for the study of problems of taxonomy, heredity, evolution, etc. with special reference to the starches. Washington D. C., 1919.
SAHNI, B. On the structure and affinities of *Acmopyle Pancheri* Pilger. Trans. Roy. Philosoph. Soc. (Lond.), Ser. B. 210: 253–310, 1920.
———. Ontogeny of vascular plants and the theory of recapitulation. Journ. Indian Bot. Soc. 4: 202–216, 1925.
SARGANT, E. The reconstruction of a race of primitive angiosperms. Ann. Bot. 22: 121–186, 1908.
SCHAFFNER, J. H. The importance of phylogenetic taxonomy in systematic botany. Ecology, 19: 296–300, 1938.
SCHNARF, K. Die Bedeutung der embryologischen Forschung für das natürliche System der Pflanzen. Biol. Gen. 271–288, 1933.
SCHUBERT, C. The evolution of primitive plants from the geologist's point of view. New Phytologist, 19: 272–275, 1921.
SINNOTT, E. W. Investigations on the phylogeny of the Angiosperms. Amer. Journ. Bot. 1: 303–322, 1914.
———. The evolution of herbs. Science, 44: 291–298, 1916.
SINNOTT, E. W. and BAILEY, I. W. Investigations on the phylogeny of the angiosperms. Part 3, Amer. Journ. Bot. 1: 441–453, 1914; Part 4, Ann. Bot. 28: 547–600, 1914; Part 5, Amer. Journ. Bot. 2: 1–22, 1915.
———. The evolution of herbaceous plants and its bearing on certain problems of geology and climatology. Journ. Geol. 23: 289–306, 1915.
SPORNE, K. R. Correlation and classification in Dicotyledons. Proc. Linn. Soc. (Lond.) 160: 40–58, 1948.
———. A new approach to the problem of the primitive flower. New Phytologist, 48: 260–276, 1949.
SPRAGUE, T. A. A discussion on phylogeny and taxonomy. Proc. Linn. Soc. (Lond.) 152nd Session, 243–250, 1940.
———. Taxonomic botany, with special reference to the angiosperms. In Huxley, J., The new systematics, pp. 435–454, 1940.
STEBBINS, G. L. JR. Cytological characteristics associated with the different growth habits in the dicotyledons. Amer. Journ. Bot. 25: 189–197, 1938.

———. Paleobotany and the origin of the Angiosperms. Bot. Rev. 2: 397–418, 1936.
TIPPO, O. A modern classification of the plant kingdom. Chron. Bot. 7: 203–206, 1942.
———. The role of wood anatomy in phylogeny. Amer. Midl. Nat. 36: 362–372, 1946.
TURRILL, W. B. The expansion of taxonomy with special reference to the Spermatophyta. Biol. Rev. 13: 342–373, 1938.
———. Taxonomy and phylogeny I–III. Bot. Rev. 8: 247–270, 473–532, 655–707, 1942.

VESTAL, P. A. Wood anatomy as an aid to classification and phylogeny. Chron. Bot. 6: 53, 54, 1940.
WALTON, J. An introduction to the study of fossil plants. London, 1940.
WEEVERS, TH. The relation between taxonomy and chemistry of plants. Blumea, 5(2): 412–422, 1943.
WERNHAM, H. F. Floral evolution. New Phytologist, 10: 78–83, 109–120, 145-159, 217–226, 293–307, 1911.
WIELAND, G. T. Origin of angiosperms. Nature, 131: 360–361, 1933.
WODEHOUSE, R. P. The phylogenetic value of pollen-grain characters. Ann. Bot. 42: 891–934, 1928.
———. Evolution of pollen grains. Bot. Rev. 2: 67–84, 1936.

# CHAPTER VI

# CURRENT SYSTEMS OF CLASSIFICATION

Botanists throughout the world currently are using one of five basic systems of plant classification, or modifications of them.[1] It may seem at first that this is an unnecessarily large number when each is a more or less systematic presentation of the same families and genera, but examination of the situations usually indicates that there is reason for the variety. Systems are chosen for use for one of several reasons, including (1) the precedent established by large herbaria that tend to dominate the practices of subordinate groups under their influence, (2) the influence of standard floras and manuals, and (3) the degree to which a particular system may lend itself to the needs of the botanist. Specimens are arranged within a herbarium according to a particular system, and when that herbarium is large the change from one system to another may be a stupendous undertaking and scarcely justified unless there is some approach to permanence. Floras of large areas are prepared by professional taxonomists who usually are associated with a center of taxonomic research. The classification accepted at that center is usually employed in the flora, and through the widespread influence of the flora and of students trained at that center, the classification often dominates the area. These two reasons, individually or in concert, may account for the general use of the Bentham and Hooker classification in the British Commonwealth, the Engler system in the eastern half of the United States, and the Bessey system in the north central part of this country. In other instances, progressive and alert systematists have recognized the phyletic inadequacies of these systems, and for pedagogical reasons have adopted the Hutchinson system or modernized versions of older systems. The trend in this country is increasing in the direction of abandonment of the Engler system by the teaching taxonomist, while the Engler system is re-

[1] They are the systems of Bentham and Hooker, Engler, Bessey, Hutchinson, and Tippo. The latter has not been developed to date below the level of Class, and in other countries the systems by Pulle and by Skottsberg are accorded equal if not greater acclaim.

tained in the newer comprehensive floras and manuals and by curators of herbaria. If it could be demonstrated that these newer systems, or the modifications of older systems, were significantly closer to actual phylogeny, there is little doubt that a majority of herbarium curators and authors of major floras and manuals would accept them without regard for the labor involved. However, as emphasized in the previous chapter on phylogeny, there is no assurance that any classification system now known (even in skeleton form) represents the actual phylogenetic situation—especially with regard to the flowering plants. Until a system has been developed that approaches finality, there may be small gain in progressing by laborious hops from one stage to the next. On the other hand, one is expected to rearrange taxa within a basic classification as new phyletic evidence is produced.

The text of this chapter attempts to state the underlying principles of the basic classification systems of Bentham and Hooker, Hallier, Engler, Bessey, Hutchinson, and Tippo; to indicate the arrangement of at least the orders; and to point out strong and weak features of each. In addition to these systems, recognition is given to major variations of them as presented by Wettstein, Rendle, Pulle, and Skottsberg.

### Bentham and Hooker [2]

Many of the latest floras authored by British botanists have the plants arranged according to the Bentham and Hooker system. Because of this, and because of the marked similarity between it and that of Bessey, and to a lesser degree of Hutchinson, it is desirable to outline its major categories. The Phanerogams or seed plants were classified as follows:

I. Dicotyledons
  A. Polypetalae (corollas of separate petals)
    Series    I. Thalamiflorae (stamens hypogynous and usually many, no disc present)
        Ranales, Parietales, Polygalineae, Caryophyllineae, Guttiferales, Malvales
    Series    II. Disciflorae (stamens hypogynous, disc present)
        Geraniales, Olacales, Celastrales, Sapindales
    Series    III. Calyciflorae (stamens perigynous or epigynous, ovary mostly inferior)
        Rosales, Myrtales, Passiflorales, Ficoidales, Umbellales
  B. Gamopetalae (corolla of partially or completely connate petals)
    Series    I. Inferae (ovary inferior)
        Rubiales, Asterales, Campanales
    Series    II. Heteromerae (ovary superior, androecium of 1 or 2 series, carpels mostly more than 2)
        Ericales, Primulales, Ebenales

[2] For historical review of the system, see Chapter III, pp. 31–33.

　　　　Series　III. Bicarpellatae (ovary superior, androecium of 1 series. carpels 2)
　　　　　　　Gentianales, Polemoniales, Personales, Lamiales
　　C. Monochlamydeae (flowers apetalous) [3]
　　　　Series　　I. Curvembryeae (embryo coiled, ovule mostly 1)
　　　　　　　Nyctagineae, Chenopodiaceae
　　　　Series　 II. Multiovulatae aquaticae (several seeded, immersed aquatics)
　　　　　　　Podostemaceae
　　　　Series　III. Multiovulatae terrestres
　　　　　　　Nepenthaceae, Aristolochiaceae
　　　　Series　IV. Microembryeae (embryo minute in endosperm)
　　　　　　　Piperaceae, Myristicaceae
　　　　Series　 V. Daphnales (ovary unicarpellate, uniovulate)
　　　　　　　Laurineae, Proteaceae, Elaeagnaceae
　　　　Series　VI. Achlamydosporeae (ovary usually inferior, unilocular, ovules 1–3)
　　　　　　　Santalaceae, Loranthaceae
　　　　Series VII. Unisexuales (flowers unisexual)
　　　　　　　Euphorbiaceae, Platanaceae
　　　　Series VIII. Ordines anomali (of uncertain relationships, nearer to VII than anything else)
　　　　　　　Salicineae, Empetraceae, Ceratophyllaceae
I. Gymnospermae
　　Gnetaceae, Coniferae, Cycadaceae
II. Monocotyledons
　　　Series　　I. Microspermae (ovary inferior, seeds minute)
　　　　　　　Orchidaceae, Burmanniaceae
　　　Series　 II. Epigynae (ovary usually inferior, seeds large)
　　　　　　　Bromeliaceae, Irideae, Amaryllideae
　　　Series　III. Coronarieae (ovary superior, perianth colored)
　　　　　　　Liliaceae, Pontederiaceae
　　　Series　IV. Calycineae (ovary superior, perianth greenish)
　　　　　　　Juncaceae, Palmae
　　　Series　 V. Nudiflorae (perianth mostly none, seed albuminous)
　　　　　　　Pandanaceae, Typhaceae, Aroideae
　　　Series　VI. Apocarpae (pistils more than 1 and distinct)
　　　　　　　Alismaceae, Naiadaceae
　　　Series VII. Glumaceae (perianth reduced, scaly bracts present and conspicuous)
　　　　　　　Gramineae, Cyperaceae

　　The *Genera plantarum* was produced jointly by the 2 authors, and is a classification of only the seed plants. The seed plants were considered by them to number about 97,205 species distributed among the major taxa and categories as indicated by Table 1. The Bentham and Hooker system was in essence a refinement of those by de Candolle and by Lindley, which in turn were based directly on that of de Jussieu. The refinements are to be noted in the Polypetalae, where the new series Disciflorae are interpolated between the Thalamiflorae and the Calyciflorae, and in the

[3] No cohorts (i.e., orders) were designated for groups of families in the remainder of the series, and in lieu of this one or two representative families are cited as examples.

revision of the classification of those apetalous taxa that comprised the Monochlamydeae. The gymnosperms were treated as a third taxon collateral with and placed between the dicots and monocots. While this may appear to be inconsistent with current knowledge of the subdivision (and much of which was then known) it was an advance over the treatment of them by de Candolle, who had distributed the conifers among well-defined taxa of dicots.

Table I. Distribution of taxa in Bentham and Hooker's *Genera plantarum*

|  | ORDERS (Families) | GENERA | SPECIES (Estimated) |
|---|---|---|---|
| Dicotyledons |  |  |  |
| Polypetalae | 82 | 2,610 | 31,874 |
| Gamopetalae | 45 | 2,619 | 34,556 |
| Monochlamydeae | 36 | 801 | 11,784 |
| Gymnospermae | 3 | 44 | 415 |
| Monocotyledons | 34 | 1,495 | 18,516 |
| Total | 200 | 7,569 | 97,205 |

One of the most valuable contributions of this work is the descriptions of taxa of all levels, for, as Bentham stated (1883), "The descriptions in the 'Flora' are drawn up from the actual examination of specimens. . . . Nothing in my work is merely copied, except in a very few cases where the material at my disposal was insufficient, and where I have specially referred to my authority." This originality has elevated

Fig. 10. Sir Joseph Dalton Hooker, 1817–1911.
Sketch by M. E. Ruff, from photograph, courtesy of the New York Botanical Garden.

the work to a level of its own, since all subsequent classification systems are compilations from the literature. Because the generic descriptions were based on direct observations of one or the other of the authors, they became models of accuracy. The greater part of the work was by Bentham, who devoted nearly 27 years to his share of the contribution, and Table II indicates the division of authorship by taxa (data from Bentham, 1883).

Table II. Authorship of taxa in Bentham and Hooker's *Genera plantarum*

|  | G. BENTHAM | J. D. HOOKER |
|---|---|---|
| Dicotyledons |  |  |
| Polypetalae |  |  |
| Thalamiflorae | All others | Cruciferae |
|  |  | Capparideae |
|  |  | Resedaceae |
| Disciflorae | Lineae | All others |
|  | Humiriaceae |  |
|  | Geraniaceae |  |
|  | Olacineae |  |
| Calyciflorae | Leguminosae | All others |
|  | Myrtaceae |  |
|  | Umbelliferae |  |
|  | Araliaceae |  |
| Gamopetalae |  |  |
| Inferae | Asterales | Rubiales (most) |
|  | Campanales |  |
| Heteromerae | Ebenales (most) | Ericales |
|  |  | Primulales (most) |
|  |  | Sapotaceae (some) |
| Bicarpellatae | All |  |
| Monochlamydeae | All others | Nyctagineae to Batideae |
|  |  | Nepenthaceae, Cytinaceae |
|  |  | Balanophoraceae |
| Gymnospermeae | All |  |
| Monocotyledons | Orchideae | Palmae |
|  | Gramineae | Nudiflorae, Apocarpae |
|  | All others |  |

## Engler and Prantl

Adolph Engler and Karl A. E. Prantl are names associated with the system developed originally by Eichler (about 1875) and modified by Engler and his associates (see Chapter III). It is accepted generally as the first of several allegedly phylogenetic systems, but study of Engler's explanatory remarks (1897) indicates that while he distinguished between primitive and advanced conditions and accepted to a degree the principle of simplicity by reduction, he did not consider that his system was a truly phylogenetic classification. This was pointed out by Turrill (1942, p. 671), who wrote "Engler's system does not claim to be phylogenetic in the complete sense but to show in its sequence of groups progressive complexity of structure, apart from accepted subordinate reductions." In this connection, Engler stated [translation by Uline, 1898], "The sequence of series and

Fig. 11. ADOLPH ENGLER, 1844–1930. Sketch by M. E. Ruff, from photograph, courtesy of the New York Botanical Garden.

families is treated with special reference to the progressive steps which are manifested in floral structure, fruit and seed development, and differentiation of tissue." In other words, Engler attempted to devise a system that had the utility and practicality of a natural system based on form relationships and one that was compatible with evolutionary principles. He considered the angiosperms to have had a polyphyletic origin from an unknown and hypothetical taxon of extinct gymnosperms, and that "a great number of parallel series came into existence from the beginning . . ." Some of Engler's evolutionary concepts were colored by an admixture of Lamarckism, as evidenced by his statement that "in those series in which wind-pollination prevailed . . . a highly developed corolla, having no particular value, stood no chance of transmission to the generation following." The Engler classification remained essentially unchanged, aside from position changes of a few small families, in all eleven editions of his *Syllabus*. The sequence followed in the eleventh (and last) edition, prepared by Diels (1936), is that followed in Part II of this text.

Many evaluations of the merits and demerits of the Engler system have been published. Phylogenists who deprecate its basic tenets (as presented on pp. 34–35) sometimes fail to realize that the system was not conceived to be phylogenetic in the modern sense. There is a logical sequence of form relationships in the arrangement of its larger taxa; the steps of seemingly simple structures placed at the lower phyletic levels, superposed by continuing steps of seemingly more complex structures is purposeful albeit superficial. This concept of a structure, polyphyletic in origin, of parallel series of steps is borne out by Engler's groupings of flowering plants on the premises that evolutionary lines progressed from apetaly to polypetaly and gamopetaly, apocarpy to syncarpy, hypogyny to epigyny, and actinomorphy to zygomorphy. The system is not defensible as a phylogenetic classification, but because of the vast literature that has adopted its principles it has made an impress for half a century that will not rapidly be displaced.

### Rendle

Alfred Barton Rendle (1865–1938) published a classification that was patterned after the Engler system, but differed from it in a few more or less minor features. Like Engler, he treated the monocots as more primitive than the dicots, but the Palmae were treated as a separate order, the Principes, considered to be the primitive family of the Spadiciflorae with the Aroideae (i.e., Araceae) and Lemnaceae advanced taxa in the same order. In a number of instances Rendle omitted families or orders (as the Aponogetonaceae, Triuridales, Butomaceae, Cyclanthaceae, *et al.*). In the disposition of the dicots, Rendle's system diverges more markedly from that of Engler, and like the latter he made no claim that it was strictly phylogenetic but rather that the orders were "grouped in three grades which correspond to grades of differentiation in the floral structure." Plants in the first grade (Monochlamydeae) included orders with a simple type of flower and he noted that "some may be reduced forms, the affinities of which are to be sought in the higher grades . . . [with others] representing lines of development from earlier extinct groups." In the second grade (Dialypetalae) the orders were arranged in ascending sequence, starting with the Ranales with many spirally arranged parts to the Umbelliflorae with cyclic and reduced parts. The third grade (Sympetalae) comprised orders that represented "still higher grades of floral development which have sprung from various dialypetalous groups." Rendle postulated polyphyletic origins for components of each of these grades. He placed emphasis on wind vs. insect pollination as phyletic criteria, and considered that generally the woody habit was more primitive than the herbaceous. Rendle was cautious in placing many taxa, noting that the evidence for many of his views was sometimes circumstantial or even conjectural. One of the more significant merits of the two-volume work lies in the clarity and completeness of descriptions, the recording of exceptions to the rule, the discussions of relationships, and the historical presentation of earlier classifications.

### Wettstein

Richard von Wettstein (1862–1931) originated a system that was based on many of the primary dicta of the Engler system, but was the more nearly phylogenetic classification. Like Engler, Wettstein believed that among the angiosperms the bisexual and perianthed flower had evolved from ancestors whose flowers were unisexual and naked. He treated the angiosperms as descended from the gymnosperms via the Gnetales, or ancestral stocks closely allied to them, and believed that the vascular cryptogams were ancestral to the gymnosperms. In his arrangement of angiosperm families and orders he departed further from the Engler system than did Rendle, demonstrated greater acumen, and took cognizance of a larger range of data (including serological) than did either Engler or Rendle. It is considered generally that among those systems accepting apetaly as a primitive condition, the system of Wettstein more closely approaches phyletic realities than does that of Engler or Rendle. Wettstein differed from Engler in considering the dicots more primitive than the monocots (the latter to have been derived from ranalian stocks). In general, Wettstein considered functionless reproductive organs or parts as evidence of reduction, survival by possession of adaptive modifications (as in Sarraceniaceae) to indicate advancement, woody plants to be more primitive than herbaceous, many-flowered inflorescences to be more primitive than few or solitary flowered types, and the spiral arrangement of parts to have preceded the cyclic.

The system has never been accepted widely in this country. However, it is significant to note that in addition to treating the monocots as derived from ranalian ancestors, Wettstein's treatment of the monocot families departs markedly from Engler's in that the polycarpic Heliobiae are considered primitive, with the ancestors of Alismaceae and Butomaceae interpreted to have been ancestral also to the monocots; the grasses and sedges were placed in separate orders; and the Spadiciflorae (palms and aroids) and Pandanales (bur reed and cattails) were treated as the most advanced of all monocots, preceded by the Gynandrae (orchids). Similar, but less conspicuous divergences are to be noted in the dicots, where the Cactaceae are allied with the Aizoaceae in the Centrospermae, and the Tricoccae (Euphorbiaceae, *et al.*) placed between the Centrospermae and Hamamelidales; and the Parietales (violets, sundews, begonias, etc.) transferred to a position beside or slightly advanced over the Rhoeadales (poppies), and these followed by the Guttiferales and Rosales. The Myrtales are moved to a position intermediate between the Rosales and Columniferae (mallows), with the latter followed by the Gruinales (Geraniales of Engler) and Terebinthales. Wettstein's interpretation of relationships within the Sympetalae does not differ significantly from that of Engler. On the whole, Wettstein's classification is one of considerable merit, and many of his conclusions of phyletic relationships have been adopted in subsequent classifications.

## Pulle

August A. Pulle (born 1878) of the Utrecht Botanical Museum and author of the current *Flora of Surinam,* published (1938) a modification of the Engler classification of seed plants. He accepted the Spermatophyta to be a division, but rejected the Englerian concept that it was composed of 2 subdivisions (Gymnospermae and Angiospermae), treating it instead to contain 4 collateral subdivisions (Pteridospermae, Gymnospermae, Chlamydospermae, and Angiospermae). The last were divided into the monocots and dicots, with the orders of the latter placed in 8 series. In a revision (1950), he merged the Cactales with the Centrospermae, moved the Proteales from a position near the Santalales to one treating them as derivative of the Thymelaeales and derived the latter from the Myrtales; the Terebinthales were broken up into the Malpighiales, Polygales, Rutales, Sapindales, and Balsaminales; and the highly advanced series composed of the Tubiflorae and its satellite orders was held to be disjunctive from the others with no definite line of derivation indicated. The organization of the revised system of Pulle is as follows:

Pteridospermae (a subdivision of 2 extinct families)
Gymnospermae
  Cycadinae (Class)
  Bennettitinae (extinct)
  Cordaitinae (extinct)
  Ginkyoinae
  Coniferae
    (Araucariales, Podocarpales, Pinales, Cupressales, and Taxales)

Chlamydospermae
   Gnetales (Ephedraceae and Gnetaceae), Welwitchiales
Angiospermae
   Monocotyledoneae
      Spadiciflorae, Pandanales, Helobiae, Triuridales, Farinosae, Liliiflorae, Cyperales, Glumiflorae, Scitamineae, Gynandrae
   Dicotyledoneae
      Series I
         Casuarinales, Piperales, Salicales, Garryales, Leitneriales, Juglandales, Julianales, Myricales, Balanopsidales, Hydrostachyales, Fagales, Urticales, Centrospermae, Polygonales, Plumbaginales, Primulales
      Series II
         Santalales, Balanophorales
      Series III
         Hamamelidales, Ranales, Ebenales
      Series IV
         Aristolochiales, Rosales, Podostemales, Myrtales, Thymelaeales, Proteales
      Series V
         Rhoeadales, Batidales, Sarraceniales, Parietales, Cucurbitales, Guttiferales, Diapensiales, Ericales, Campanulatae
      Series VI (derived from the Rosales)
         Pandales, Malvales, Tricoccae, Geraniales, Malpighiales, Polygales, Rutales, Sapindales, Balsaminales, Rhamnales
      Series VII (derived from the Sapindales)
         Celastrales, Umbelliflorae, Rubiales
      Series VIII
         Ligustrales, Contortae, Tubiflorae, Plantaginales, Callitrichales, Hippuridales

The classification by Pulle of the gymnosperms (including his Chlamydospermae) is in closer accord with modern phyletic opinion than is that by most other authors, and in his explanation of his system he suggested that perhaps the seven families of the old Coniferae may represent 7 rather than 5 orders.

The angiosperm classification, while patterned (for the dicots) after that of Wettstein, was original in a number of respects. Pulle has developed the growing opinions that the Monochlamydeae and the Sympetalae are not natural taxa, and because of the presumed polyphyletic origin of the latter has rearranged its component families as indicated above. In presenting the outline of his early classification Pulle wrote (1937):

> Whilst I suppose that most taxonomists will agree with me as regards the connection *Ranales-Rosales-Myrtales,* some of them may object to the position and the connections of the *Tricoccae* and many will even think the connections *Ranales-Ebenales, Rhoeadales-Batidales, Tubiflorae-Callitrichales* and *Tubiflorae-Hippuridales* rather doubtful. As regards the latter two orders we have to keep in mind that placing them with the *Tricoccae* or with the *Myrtales,* has satisfied nobody and that a number of their characters rather point to a relationship with the *Tubiflorae.* So we do better to accept them as strongly reduced forms at the end of a series than to give them a place to which they certainly have no right.

## Skottsberg

Carl Skottsberg (born 1880) Professor of Botany at Göteborg, Sweden, devised for the plant kingdom a modification of the Engler classification, utilizing some concepts embodied in Wettstein's system, that served the basis of his *Växternas Liv* (vol. 5, 1940). His treatment of the pteridophytes and gymnosperms was conventional except that the class Gnetinae was made a collateral of the Coniferae and comprised the unifamilial orders Ephedrales, Welwitschiales, and Gnetales. Of the angiosperms, he interpreted the monocots to have been evolved from an unknown primitive dicot, and in both taxa treated apocarpy and polycarpy as primitive, with syncarpy and monocarpy as advanced (an exception being the Casuarinaceae, which he placed at the beginning of the dicots but not as ancestral to other dicots). The apetalous families were viewed to have been of polymorphic origin and were redistributed from the positions accorded them by Engler, Wettstein, and Pulle. Skottsberg's disposition of the dicot orders was as follows:

Choripetalae
  Verticillatae (Casuarinaceae)
  Polycarpicae (Ranales), Aristolochiales, Nepenthales, Sarraceniales, Piperales, Polygonales, Centrospermae, Cactales, Rhoeadales, Parietales, Cucurbitales, Guttiferales, Rosales, Podostemonales, Hydrostachyales, Myrtales, Hamamelidales, Salicales, Garryales, Myricales, Balanopsidales, Leitneriales, Juglandales, Julianales, Fagales, Batidales, Urticales, Proteales, Santalales, Balanophorales, Columniferae (Malvales), Tricoccae, Callitrichales, Gruinales (Geraniales), Pandales, Terebinthales, Polygalales, Sapindales, Celastrales, Rhamnales, Umbelliflorae
Sympetalae
  Plumbaginales, Primulales, Bicornes, Diospyrales, Tubiflorae, Contortae, Ligustrales, Plantaginales, Rubiales, Campanulatae, Synandrae

Skottsberg differed from both Engler and from Wettstein and adopted some of the views of the Benthamian-Bessey school, as indicated by the transfer of the amentiferae to a position following the Rosales, and differed from Pulle in retaining the Primulales in the Sympetalae. However, he did not follow the more current views of those anatomists and morphologists who distribute the amentiferous families (or orders) among the various petaliferous families to which they are alleged to have closer affinities than they do (with some exceptions) to each other. This classification is, in many respects, more balanced than others reviewed to this point, and has the merit over that by Pulle, of presenting a seemingly more realistic revision of the monocots as well as of most of the dicots.

### Bessey

Charles E. Bessey in 1893 read a paper presenting the nucleus of what was to become his well-known classification; this was amplified in 1897, and his final treatment published in 1915. The Bessey classification was original in many respects, particularly in its basic philosophy. On the other hand, Bessey accepted some of the dispositions made in the earlier classification of Bentham and Hooker and in the Engler and Prantl

Fig. 12. CHARLES EDWIN BESSEY, 1845–1915. Sketch by M. E. Ruff, from photograph, courtesy Brooklyn Botanical Garden.

classification. Bessey gave emphasis to this when he wrote (1897) that his classification "does not differ as much from the two older systems as they differ from one another. . . . Bringing together the results of the studies of these matters . . . we find it possible to make such modifications of the two systems as will give us an arrangement which fairly agrees with the present state of our knowledge."

Bessey considered the seed plants to have had a polyphyletic origin and to be composed of three separate phyla of which he dealt only with the Anthophyta (angiosperms). Each of these phyla was considered by him to have been derived separately from vascular cryptogams. He derived the Anthophyta from his Cycadophyta and the latter from implied bennettitalian ancestry. He divided the angiosperms conventionally into the Oppositifoliae (dicots) and Alternifoliae (monocots). His classification of the components of these two taxa was based on evidence obtained

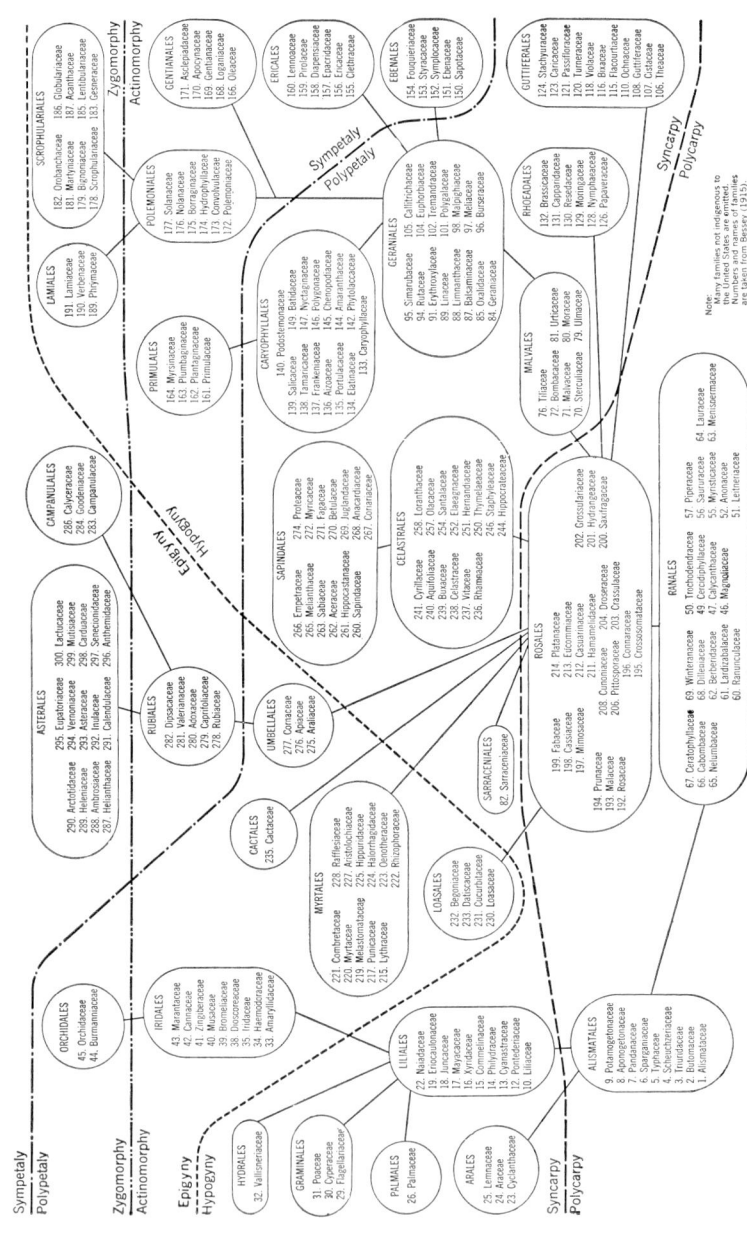

**Fig. 13.** Classification of the angiosperms, after Bessey, 1915. Arranged by J. F. Cornman.

from 3 lines of investigation: historical (paleobotanical), ontogenetical (including embryology), and the morphological studies of homologies. Bessey subscribed to the strobiloid theory of origin of the angiosperm flower. This postulated that the flower originated from a vegetative shoot in which some of the spirally arranged phyllomes (dorsiventrally compressed foliaceous branch systems) were modified and functioned as reproductive organs (the upper sterile ones developing into perianth parts, the adjoining distal phyllomes bearing microsporangia and developing into the androecium (stamens), and the uppermost ones bearing megasporangia (marginal or submarginal ovules) and by their infolding as distinct organs became a gynoecium of many distinct pistils. From this primitive strobiloid flower he concluded (1897) that a modern angiosperm flower "in which the phyllomes are the least modified must be regarded as primitive, . . . the simple pistil developed from a single phyllome is primitive . . . [and] the several-seeded, compound ovary must be lower [in evolutionary position] and the compound ovary with but one seed must be higher. . . . So too when all parts of the flower are separate it is a primitive condition, and when they are united it is a derived structure." To these tenets he added two more, based on the direction of fusion of parts: by one the fusion of parts was lateral or transverse (across the flower) and involved adnation of unlike parts (as in the formation of a hypanthium or in the development of an inferior ovary), by the other the fusion was horizontal and usually involved connations of like parts (as in syncarpy, monadelphy, gamopetaly) and both types of fusion would be in the most highly advanced taxa.[4]

The ranalian plants or their ancestors were considered the primitive angiosperms. One branch gave rise to the monocots, with the dicots interpreted to have bifurcated early into two primary evolutionary lines, the rosalian line (Cotyloideae) characterized principally by transverse adnations of unlike parts, and the ranalian line (Strobiloideae) characterized by vertical connations of like parts (see Fig. 13).

Disposition of angiosperm families according to the Bessey classification:[5]

Class, Alternifoliae (Monocotyledoneae)
  Subclass, Strobiloideae
    Order Alismatales
      1, Alismataceae; 2, Butomaceae; 3, Triuridaceae; 4, Scheuchzeriaceae;

---

[4] Bessey (1915) also enumerated 28 dicta on which his classification was predicated.
[5] Numbers preceding family names are those employed by Bessey (1915) and, while providing a linear sequence for bookkeeping purposes, do not of necessity imply a phyletic sequence within the order. Spellings of family names are taken directly from Bessey.

5, Typhaceae; 6, Sparganiaceae; 7, Pandanaceae; 8, Aponogetonaceae; 9, Potamogetonaceae

Order Liliales

10, Liliaceae; 11, Stemonaceae; 12, Pontederiaceae; 13, Cyanastraceae; 14, Philydraceae; 15, Commelinaceae; 16, Xyridaceae; 17, Mayacaceae; 18, Juncaceae; 19, Eriocaulonaceae; 20, Thurniaceae; 21, Rapataceae; 22, Naiadaceae

Order Arales

23, Cyclanthaceae; 24, Araceae; 25, Lemnaceae.

Order Palmales

26, Palmaceae

Order Graminales

27, Restionaceae; 28, Centrolepidaceae; 29, Flagellariaceae; 30, Cyperaceae; 31, Poaceae

Subclass, Cotyloideae

Order Hydrales

32, Vallisneriaceae

Order Iridales

33, Amaryllidaceae; 34, Haemodoraceae; 35, Iridaceae; 36, Velloziaceae; 37, Taccaceae; 38, Dioscoreaceae; 39, Bromeliaceae; 40, Musaceae; 41, Zingiberaceae; 42, Cannaceae; 43, Marantaceae

Order Orchidales

44, Burmanniaceae; 45, Orchidaceae

Class, Oppositifoliae (Dicotyledoneae)

Subclass, Strobiloideae

Superorder Apopetalae-Polycarpellatae

Order Ranales

46, Magnoliaceae; 47, Calycanthaceae; 48, Monimiaceae; 49, Cercidiphyllaceae; 50, Trochodendraceae; 51, Leitneriaceae; 52, Anonaceae; 53, Lactoridaceae; 54, Gomortegaceae; 55, Myristicaceae; 56, Saururaceae; 57, Piperaceae; 58, Lacistemaceae; 59, Chloranthaceae; 60, Ranunculaceae; 61, Lardizabalaceae; 62, Berberidaceae; 63, Menispermaceae; 64, Lauraceae; 65, Nelumbaceae; 66, Cabombaceae; 67, Ceratophyllaceae; 68, Dilleniaceae; 69, Winteranaceae.

Order Malvales

70, Sterculiaceae; 71, Malvaceae; 72, Bombaceae; 73, Scytopetalaceae; 74, Chlaenaceae; 75, Gonystylaceae; 76, Tiliaceae; 77, Elaeocarpaceae; 78, Balanopsidaceae; 79, Ulmaceae; 80, Moraceae; 81, Urticaceae

Order Sarraceniales

82, Sarraceniaceae; 83, Nepenthaceae

Order Geraniales

84, Geraniaceae; 85, Oxalidaceae; 86, Tropaeolaceae; 87, Balsaminaceae; 88, Limnanthaceae; 89, Linaceae; 90, Humiriaceae; 91, Erythroxylaceae; 92, Zygophyllaceae; 93, Cneoraceae; 94, Rutaceae; 95, Simarubaceae; 96, Burseraceae; 97, Meliaceae; 98, Malpighiaceae; 99, Trigoniaceae; 100, Vochysiaceae; 101, Polygalaceae; 102, Tremandraceae; 103, Dichapetalaceae; 104, Euphorbiaceae; 105, Callitrichaceae

Order Guttiferales

106, Theaceae; 107, Cistaceae; 108, Guttiferaceae; 109, Eucryphiaceae; 110, Ochnaceae; 111, Dipterocarpaceae; 112, Caryocaraceae; 113, Quiinaceae; 114, Marcgraviaceae; 115, Flacourtiaceae; 116, Bixaceae; 117, Cochlospermaceae; 118, Violaceae; 119, Malesherbiaceae; 120, Turneraceae;

121, Passifloraceae; 122, Achariaceae; 123, Caricaceae; 124, Stachyuraceae; 125, Koeberliniaceae
Order Rhoeadales
126, Papaveraceae; 127, Tovariaceae; 128, Nymphaeaceae; 129, Moringaceae; 130, Resedaceae; 131, Capparidaceae; 132, Brassicaceae
Order Caryophyllales
133, Caryophyllaceae; 134, Elatinaceae; 135, Portulacaceae; 136, Aizoaceae; 137, Frankeniaceae; 138, Tamaricaceae; 139, Salicaceae; 140, Podostemonaceae; 141, Hydrostachydaceae; 142, Phytolaccaceae; 143, Basellaceae; 144, Amaranthaceae; 145, Chenopodiaceae; 146, Polygonaceae; 147, Nyctaginaceae; 148, Cynocrambaceae; 149, Batidaceae
Superorder Sympetalae-Polycarpellatae
Order Ebenales
150, Sapotaceae; 151, Ebenaceae; 152, Symplocaceae; 153, Styracaceae; 154, Fouquieriaceae
Order Ericales
155, Clethraceae; 156, Ericaceae; 157, Epacridaceae; 158, Diapensiaceae; 159, Pirolaceae; 160, Lennoaceae
Order Primulales
161, Primulaceae; 162, Plantaginaceae; 163, Plumbaginaceae; 164, Myrsinaceae; 165, Theophrastaceae
Superorder Sympetalae-Dicarpellatae
Order Gentianales
166, Oleaceae; 167, Salvadoraceae; 168, Loganiaceae; 169, Gentianaceae; 170, Apocynaceae; 171, Asclepiadaceae
Order Polemoniales
172, Polemoniaceae; 173, Convolvulaceae; 174, Hydrophyllaceae; 175, Borraginaceae; 176, Nolanaceae; 177, Solanaceae
Order Scrophulariales
178, Scrophulariaceae; 179, Bignoniaceae; 180, Pedaliaceae; 181, Martyniaceae; 182, Orobanchaceae; 183, Gesneriaceae; 184, Columelliaceae; 185, Lentibulariaceae; 186, Globulariaceae; 187, Acanthaceae
Order Lamiales
188, Myoporaceae; 189, Phrymaceae; 190, Verbenaceae; 191, Lamiaceae
Subclass Cotyloideae
Superorder Apopetalae
Order Rosales
192, Rosaceae; 193, Malaceae; 194, Prunaceae; 195, Crossosomataceae; 196, Connaraceae; 197, Mimosaceae; 198, Cassiaceae; 199, Fabaceae; 200, Saxifragaceae; 201, Hydrangeaceae; 202, Grossulariaceae; 203, Crassulaceae; 204, Droseraceae; 205, Cephalotaceae; 206, Pittosporaceae; 207, Brunelliaceae; 208, Cunoniaceae; 209, Myrothamnaceae; 210, Bruniaceae; 211, Hamamelidaceae; 212, Casuarinaceae; 213, Eucommiaceae; 214, Platanaceae
Order Myrtales
215, Lythraceae; 216, Sonnerataceae; 217, Punicaceae; 218, Lecythidaceae; 219, Melastomataceae; 220, Myrtaceae; 221, Combretaceae; 222, Rhizophoraceae; 223, Oenotheraceae; 224, Halorrhagidaceae; 225, Hippuridaceae; 226, Cynomoriaceae; 227, Aristolochiaceae; 228, Rafflesiaceae; 229, Hydnororaceae
Order Loasales
230, Loasaceae; 231, Cucurbitaceae; 232, Begoniaceae; 233, Datiscaceae; 234, Ancistrocladaceae

Order Cactales
235, Cactaceae
Order Celastrales
236, Rhamnaceae; 237, Vitaceae; 238, Celastraceae; 239, Buxaceae; 240, Aquifoliaceae; 241, Cyrillaceae; 242, Pentaphyllaceae; 243, Corynocarpaceae; 244, Hippocrataceae; 245, Stackhousiaceae; 246, Staphyleaceae; 247, Geissolomataceae; 248, Penaeaceae; 249, Oliniaceae; 250, Thymelaeaceae; 251, Hernandiaceae; 252, Elaeagnaceae; 253, Myzodendraceae; 254, Santalaceae; 255, Opiliaceae; 256, Grubbiaceae; 257, Olacaceae; 258, Loranthaceae; 259, Balanophoraceae
Order Sapindales
260, Sapindaceae; 261, Hippocastanaceae; 262, Aceraceae; 263, Sabiaceae; 264, Icacinaceae; 265, Melianthaceae; 266, Empetraceae; 267, Coriariaceae; 268, Anacardiaceae; 269, Juglandaceae; 270, Betulaceae; 271, Fagaceae; 272, Myricaceae; 273, Julianaceae; 274, Proteaceae
Order Umbellales
275, Araliaceae; 276, Apiaceae (Umbelliferae); 277, Cornaceae
Superorder Sympetalae
Order Rubiales
278, Rubiaceae; 279, Caprifoliaceae; 280, Adoxaceae; 281, Valerianaceae; 282, Dipsacaceae
Order Campanulales
283, Campanulaceae (incl. *Lobelia*); 284, Goodeniaceae; 285, Stylidaceae; 286, Calyceraceae
Order Asterales
287, Helianthaceae; 288, Ambrosiaceae; 289, Heleniaceae; 290, Arctotidaceae; 291, Calendulaceae; 292, Inulaceae; 293, Asteraceae; 294, Vernoniaceae; 295, Eupatoriaceae; 296, Anthemidaceae; 297, Senecionidaceae; 298, Carduaceae; 299, Mutisiaceae; 300, Lactucaceae

The Bessey system has received increasing support in recent years, and while no large herbaria arrange their material in accordance with its classification, it has been employed in a few manuals and floras and does lend itself to pedagogical requirements better than its predecessors. Much more information of a phylogenetic nature is now available than was known a quarter century ago, with the result that his system as of 1915 is scarcely more correct than several others, and by some botanists is held to be less phylogenetic than that of Skottsberg. Information is currently available, and more is being added, to enable the synthesis of a system of classification superior to that of Bessey. There is good reason to believe that an improved and modified Bessey system, expanded to account for the entire plant kingdom, will be formulated within the next decade or two.

### Hallier

Hans Hallier (1868–1932), of Hamburg and Leyden and a student of Haeckel, presented a system of classification designed to be phylogenetic and based primarily on study of herbarium material and the literature, but taking cognizance of ontogenetical observations and available paleobotanical evidence.[6] He considered the seed plants to comprise 2 phyla, each monophyletic, and derived the angiosperms from an unknown and extinct tribe of cycads situated near or descended from the Bennettitales, admitting affinities with the Marattiales. The conifers (with which were merged the ginkgo and the taxads) were derived from the Cycadaceae or from stocks immediately ancestral to them.

The monocots were treated as more advanced than the dicots and to have been derived from extinct (and presumably unknown?) stocks ancestral to the Lardizabalaceae. In the classification of both the dicots and monocots Hallier followed independently the dicta established by Bessey that recognized the strobiloid type of flower as primitive, and held polycarpy and the spiral arrangement of parts to be primitive over syncarpy and cyclic arrangements. Some of his conceptual differences from the dicta set forth by Bessey may be ascribed to his emphasis on characters as (cf. 1912 reference) (a) branching sparse but robust as opposed to deliquescent types, (b) entire simple leaves that are coriaceous and have pinnate venation as being more primitive than other types, and (c) flowers large, solitary, and terminal as opposed to compound inflorescences of small or minute flowers. He placed more significance on ovule morphology and position than have most other phylogenists. He had little patience with the Engler classification, and demonstrated the inadequacies of the one by Bentham and Hooker. It is clear that while his classification has some features in common with that of Bessey, the two developed independently.

The dicots were considered older and more primitive than the monocots, with all dicots derived through the orders of his Protérogènes (which included the Magnoliaceae). As a taxon, the dicots were interpreted to be composed of 4 primary subdivisions containing 29 orders:[7] (1) the Protérogènes (Ranales [excluding Magnoliaceae], Nepenthales, Caryophyllales, Piperales, and Aristolochiales); (2) the Anonophylales (Anonales [including Magnoliaceae], Hamamelidales, Umbellales, Campanulales, Columnales [Malvales]); (3) the Rhodophyles (Cruciales [Capparidales], Gruinales [Geraniales], Aesculales [Sapindales and Leguminosae], Terebinthales, Proteales, Rosales, Rhamnales); and (4) the

[6] Hallier published numerous papers on his classification, including two major revisions. Material incorporated in the treatment of this chapter has been taken largely from his paper of 1912, and it should be noted that the phylogenetic ideas and the classification presented in this paper are much more detailed and very different from those in the more commonly cited and brief account of his classification (in English) of 1905.

[7] In many instances Hallier assigned endings to ordinal names not the oldest nor in conformity with the Rules, and these have been altered here to the conventional -ales.

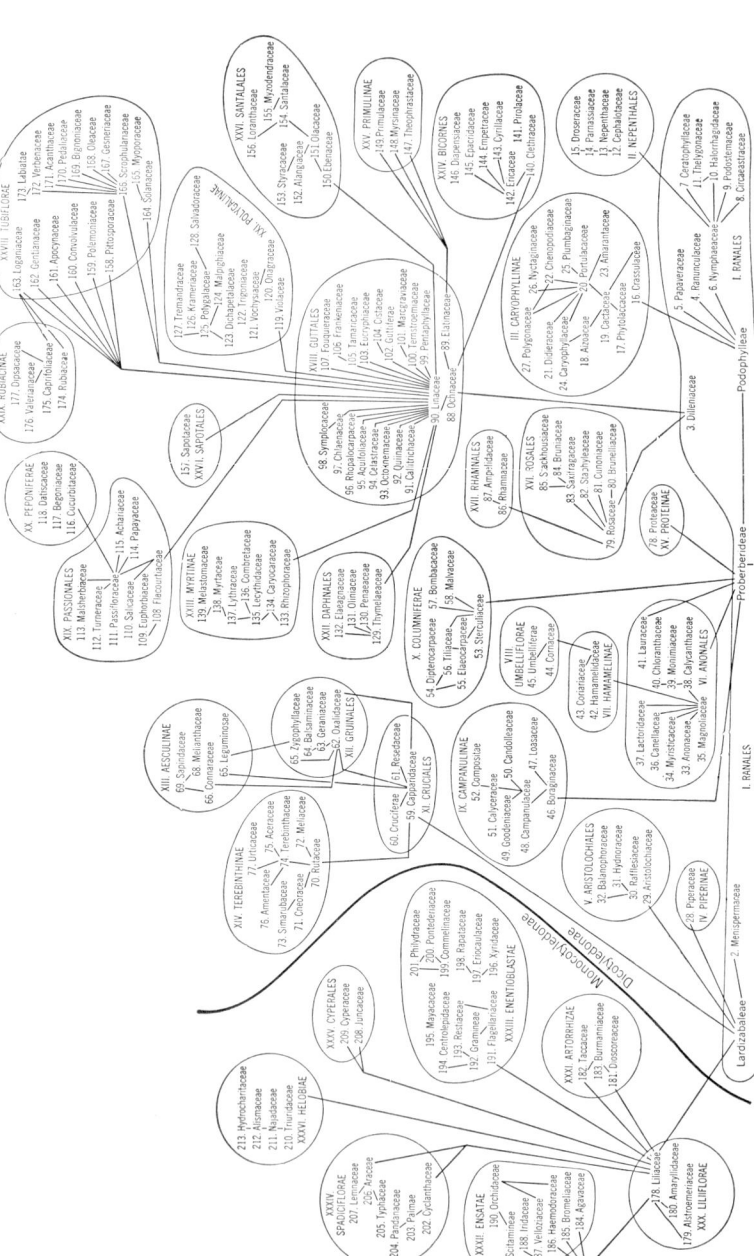

**Fig. 14.** Hallier's classification of the angiosperms, 1912. Solid lines connecting orders and families indicate possible or probable lines of evolution or alliance.

Ochnigènes (Guttales, Passionales, Péponifères [cucurbits, begonias, datiscas], Polygales, Daphnales, Myrtales, Bicornes, Primulales, Santales, Sapotales, Tubiflores, Rubiales). In this classification (see Fig. 14) he treated most of the amentiferae (including the Casuarinaceae) as "reduced descendants of Pistacia-like Terebinthaceae and not descendants of Hamamelidaceae" with the Salicaceae as reduced descendants of Flacourtiaceae, and the Juglandaceae as derived from anacardiaceous ancestors. The Cactaceae were allied with the Aizoaceae, the Plumbaginaceae (but not the Primulales) with the Caryophyllaceae, the Hamamelidales derived from the Magnoliaceae, the Umbelliflorae from the Hamamelidaceae, and the Cucurbitaceae from the Passifloraceae.

The treatment of the monocots as projected, but incomplete, in his 1905 version gave promise of being in closer agreement with present opinions than was the more complete and somewhat scrambled assemblage in the 1912 revision. It is significant to note, however, that he recognized the Amaryllidaceae to be an unnatural taxon and subdivided it into the Agavaceae, Alstroemeriaceae, and the Amaryllidaceae (derived from the Allioideae of Liliaceae). Likewise he separated the Gramineae and Cyperaceae into different orders with several other families associated with each (see Fig. 14).

The value of the Hallier classification lies not so much with the overall arrangement as with the insight displayed in the realignment of genera and families from that of the previous systems of Engler and of Bentham and Hooker, and for the influence his work has had on subsequent phyletic proposals. It is a classification that must be consulted for the ideas expressed, and frequently one finds situations indicative of possible or probable relationship that others have overlooked.

### Hutchinson

John Hutchinson, the author of the classification bearing his name, presented his first comprehensive version of his presumed phylogenetic arrangement in his 2-volume classic *Families of flowering plants* (1926–34). Later he published his revised concepts in his book *British flowering plants* (1948),[8] from which has been drawn much of the material presented in the following treatment.

The gymnosperms were treated as a lineal monophyletic series in ascending sequence from the primitive Cycadaceae, Ginkgoaceae, Taxaceae, Pinaceae, to the Cupressaceae (although it was conceded that the

---

[8] For a brief summary of Hutchinson and his classification, see that in Chapter III, pp. 36–37.

Araucarieae "may have been evolved independently"), and no cognizance seems to have been taken of studies on their phylogeny as published during the last two decades.

Hutchinson's primary contributions have been his phylogenetic studies of the angiosperms, a group he considered monophyletic and whose origin he was content to leave in the hypothetical proangiosperms of

Fig. 15. JOHN HUTCHINSON, 1884–
Sketch by M. E. Ruff, from photograph, courtesy J. Hutchinson.

Arber and Parkin. His basic phyletic principles paralleled those of Bessey, with a few very significant exceptions. In his initial subdivision of the angiosperms the herbaceous representatives (Ranales) formed the stem of one evolutionary line (the Herbaceae) and the woody representatives (Magnoliales) formed the stem of the second line (the Lignosae). In defending this basis for his major cleavage, Hutchinson wrote (1948, p. 7),

> In addition, he [the author] now has the courage of his convictions with respect to the recognition of two main phyla in the Dicotyledons, the one *fundamentally and predominately woody*, the other *fundamentally and predominately herbaceous*, and the families have been rearranged in accordance with this view, in the belief that they have in nature developed on parallel lines.

In refutation of earlier criticisms that he had followed Aristotelian classifications by placing the woody plants in one group and the herbs in another, Hutchinson replied (pp. 10–11),

> What has been done is to put all those families which are *predominately woody* into the woody phylum, and conversely the same has been done with the *herbaceous* families.

A second departure in Hutchinson's phyletic dicta from those of Bessey is to be noted in the former's de-emphasis of the significance of the gamopetalous taxa as having been derived monophyletically or diphyletically from the polypetalous taxa.

The revised classification was predicated on 22 principles that are comparable to Bessey's dicta, and these appear below in condensed form:

1. Evolution is both upwards and downwards, the former tending toward preservation . . . and the latter to their reduction and suppression [of characters] . . .
2. Evolution does not necessarily involve all organs at the same time; . . .
3. Broadly speaking, trees and shrubs are more primitive than herbs in any one family or genus . . .
4. Trees and shrubs are older than climbers in any one family or genus.
5. Perennials are older than biennials and annuals; . . .
6. Aquatic flowering plants are derived from terrestrial ancestors, and epiphytes, saprophytes, and parasites are more recent than plants of normal habit; . . .
7. Dicotyledons are more primitive than monocots.
8. Spiral arrangement is more primitive than cyclic.
9. Simple leaves are usually more primitive than compound leaves.
10. Unisexual flowers are more advanced than bisexual; dioecious plants are more recent than monoecious.
11. The solitary flower is more primitive than the inflorescence . . .
12. Aestivation types are evolved from contorted to imbricate to valvate.
13. Apetalous flowers are derived from petaliferous flowers.
14. Polypetaly is more primitive than gamopetaly.
15. Actinomorphy is more primitive than zygomorphy.
16. Hypogyny is usually more primitive than perigyny and epigyny is the most advanced.
17. Apocarpy is more primitive than syncarpy.
18. A gynoecium of many pistils preceded one of few pistils.
19. Seeds with endosperm and small embryo are older than seeds without endosperm and a large embryo; . . .
20. Numerous stamens, in general, indicate greater primitiveness than does an androecium of a few stamens (exception, Malvaceae).
21. Separate anthers, in general, indicate greater primitiveness than does an androecium of either fused anthers or filaments.
22. Aggregate fruits are more highly evolved than single fruits; as a rule the capsule precedes the berry or drupe.

A comparison of these principles with the dicta of Bessey shows that, of the above, numbers 1–3, 7–10, 13–21 are enumerated also by Bessey with the same context and, for the most part, in the same sequence.

There are many examples of concurrence by Hutchinson with the earlier views of Bessey and of Hallier. His arrangement of the angiosperm orders and many of the families is indicated in Fig. 16.

There has not been any appreciable adoption to date of this classification in this country, although the revision of the monocots has been accepted in many quarters.[9] This is due largely to his fundamental thesis that the dicots have evolved in two directions, one from the herbaceous Ranales and the other from the woody Magnoliales. A second barrier to widespread adoption of the Hutchinson system in this country is the insistence on a monophyletic origin for the seed plants in general and the angiosperms in particular. Furthermore, the failure of its author (to date) to provide explanatory notes setting forth reasons for the disposition of the orders (and in some instances, the families) has caused many discerning botanists to defer acceptance until these deficiencies are corrected. It would appear that much relatively recent morphological, anatomical, and cytological evidence may not have been given due consideration by Hutchinson in the arrival at an appreciable number of his conclusions. On the other hand, it must be pointed out that in the opinion of most taxonomists he has contributed a very real service by his careful and critical appraisal of family and ordinal limitations, and in his excellent artificial keys to the families of dicots (1926) and to the genera of most monocots (1934). The classification has been a greater stimulant to phyletic thinking during the past decade or two than any other similar contribution.

This brief survey and partial analysis of contemporary classifications has attempted to fulfill 3 objectives: an explanation of the phyletic principles on which each was based, an outline of each, by enumeration or diagram of the disposition of the major taxa, and to provide through the mesh of an assortment of schemas an indication of current trends in phyletic thinking. In connection with this third objective, attention is directed to one or two phylogenetic considerations. Initially, there is the problem of deciding whether the views of Engler, which supported the primitiveness of the apetalous flower over that of the polypetalous strobiloid type, are valid or invalid. In this country, at least, they are rejected by a majority who would treat the strobiloid type of flower as the more primitive; a view reflected in the classifications of Bessey, Hallier, Hutchinson, and Skottsberg. The basic phyletic views of the dicot families by Bessey has suffered from less adverse critisism over half a century

---

[9] For review of Hutchinson's views on the Liliiflorae, see discussions following the Liliaceae and Amaryllidaceae in Part II.

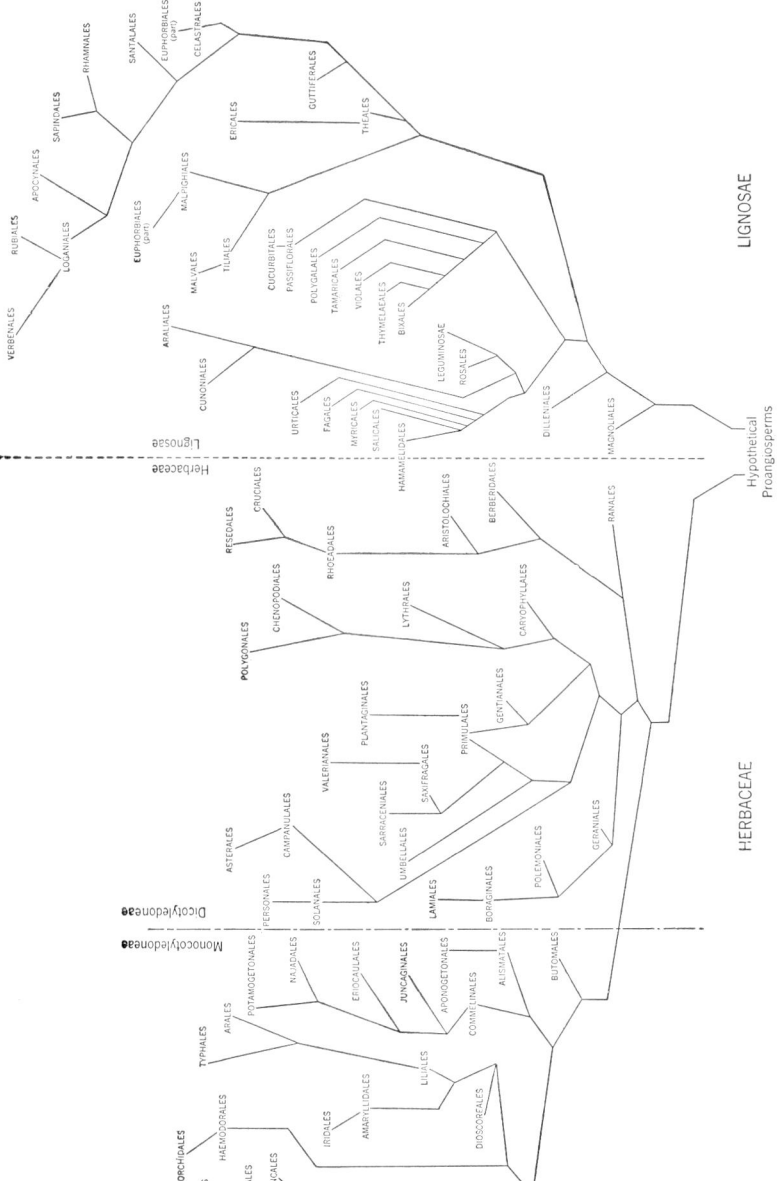

**Fig. 16.** A classification of the angiosperms. Adapted from Hutchinson, *British Flowering Plants*, P. R. Gawthorn, Ltd., 1948.

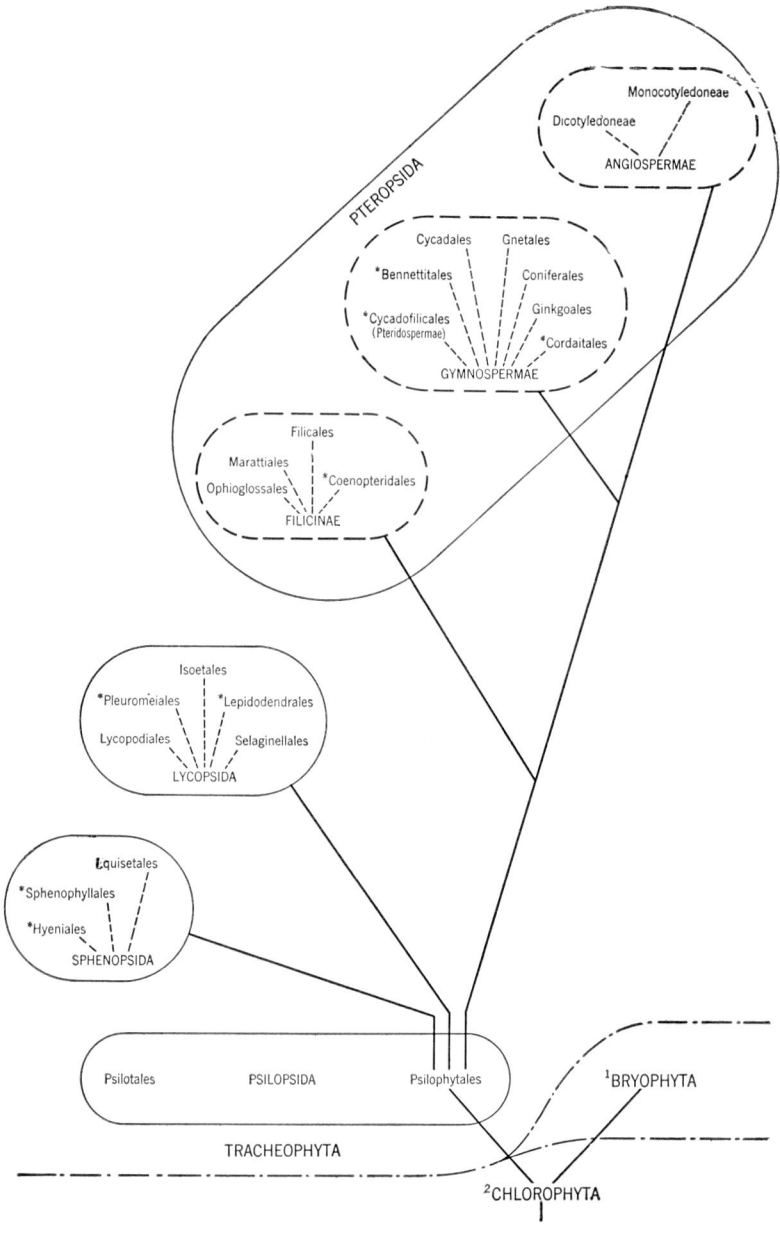

**Fig. 17.** Classification of the angiosperms, adapted from Tippo (1942). Extinct taxa marked by an asterisk. The Bryophyta and Chlorophyta (algae) are non-vascular cryptogams.

than have those of Hutchinson over two decades. There has been much new evidence of phylogenetic significance published since the last revision of the Bessey classification (1915), as emphasized in most of the discussions in the family treatment of Part II of this text. The skeletal system proposed and never elaborated by Tippo (Fig. 17), on the basis of earlier work by Smith and by Eames (see p. 38), has been received favorably by many. From the contributions of the phylogenists whose classifications have been reviewed in this chapter, and as augmented by the vast quantity of material evidence that has been made available in recent years, it would seem that a classification utilizing evidence of the highest correlations and greatest compatibility may yet be produced within the span of the present generation. There is no promise that such a classification will phylogenetically be appreciably closer to the ultimate than others, but there is assurance that such a classification should be in closer harmony with currently accepted information.

*LITERATURE:*

BENTHAM, G. On the recent progress and present state of systematic botany. Rept. 44th Meeting Brit. Assoc. Adv. Sci. 27–54, 1874; and [with few deletions, edited by Asa Gray] in: Amer. Journ. Sci. Ser. III, 9: 288–294, 346–355, 1875.

———. On the joint and separate work of the authors of Bentham and Hooker's "Genera Plantarum." Journ. Linn. Soc. Bot. (Lond) 20: 304–308, 1884.

BENTHAM, G. and HOOKER, J. D. Genera Plantarum. 3 vols. London, 1862–1883.

BESSEY, C. E. Evolution and classification. Bot. Gaz. 18: 329–332, 1893.

———. The point of divergence of monocotyledons and dicotyledons. Bot. Gaz. 22: 229–232, 1895 [A discussion of Engler's classification.]

———. Phylogeny and taxonomy of the angiosperms. Bot. Gaz. 24: 145–178, 1897.

———. The phylogenetic taxonomy of flowering plants. Ann. Mo. Bot. Gard. 2: 109–164, 1915.

ENGLER, A. Übersicht über die Unterabteilungen, Klassen, Reihen, Unterreihen und Familien der Embryophyta siphonogama. Natürliche Pflanzenfamilien, Nachtr. II-IV, 1897.

———. The groups of angiosperms. Bot. Gaz. 25: 338–352, 1897. [Abstr. of English transl. by E. B. Uline.]

ENGLER, A. and DIELS, L. Syllabus der Pflanzenfamilien. Aufl. 11. Berlin, 1936.

HALLIER, H. Provisional scheme of the natural (phylogenetic) system of flowering plants. New Phytologist, 4: 151–162, 1905.

———. On the origin of angiosperms. Bot. Gaz. 45: 196–198, 1908.

———. L'origine et le système phylétique des angiospermes. Arch. Néerl. Sci. Exact. Nat. Ser. III B, 1: 146–234, 1912.

HUTCHINSON, J. The families of flowering plants. I, Dicotyledons, 328 pp., 1926. II, Monocotyledons, 243 pp. London, 1934.

———. A botanist in Southern Africa. London, 1946.

———. British flowering plants. 374 pp. London, 1948.

PULLE, A. A. Remarks on the system of the spermatophytes. Med. Bot. Mus. Herb. Rijks-Univ. Utrecht, 43: 1–17, 1937.

———. Compendium van de Terminologie, Nomenclatuur en Systematiek der Zaadplanten. 338 pp. Utrecht, 1938; ed. 2, 370 pp. Utrecht, 1950.

———. The classification of the spermatophytes. Chron. Bot. 4: 109–113, 1938.

RENDLE, A. B. The classification of flowering plants. Vol. 1, Gymnosperms and Monocotyledons, 403 pp. Cambridge, 1904. Vol. 2, Dicotyledons, 636 pp. Cambridge, 1925. [The second edition of vol. 1 (1930) entailed only typographical changes and a brief Appendix.]

SENN, G. Die Grundlagen des Hallierschen Angiospermensystems. Beih. Bot. Centralbl. 18: 129–156, 1904.

SKOTTSBERG, C. Växternas Liv. Vol. 5, 735 pp. Stockholm, 1940.

TIPPO, O. A modern classification of the plant kingdom. Chron. Bot. 7: 203–206, 1942.

TURRILL, W. B. Taxonomy and phylogeny. III. Bot. Rev. 8: 655–707, 1942.

WEITSTEIN, R. Handbuch der systematischen Botanik. ed. 4. 1152 pp. Leipzig and Wien, 1935.

# CHAPTER VII

# THE GEOGRAPHY OF VASCULAR PLANTS

Systematic botany in its restricted sense treats the classification, identification, and naming of plants. However, and as already discussed in the introductory chapter, plant classification is predicated on the accepted tenets of evolutionary development of the plants themselves, of their environment, and of the relationships of their past and present distributions to the past and present geography of the earth. A taxonomist is not satisfied merely to know a plant, or to know its name. In order that classification may be the more accurate, a knowledge of genetic relationships is of equal importance. The full significance of interrelationships of plants can be appreciated only after it is understood that related plants have, or have had through their ancestors, common or interrelated geographic distributions. Presentation of the subject of present-day plant distributions necessitates an understanding of factors affecting them. This is as true of distributions over several continents as it is of distributions within the limits of a single continent. Theories concerning these distributions should account for the present existence of cospecific and cogeneric taxa now growing naturally on separate continents, should account for restrictions of family groups to particular continents or combinations of continents, and should explain also the distribution of prehistoric floras as represented by the fossil record. Several theories attempting to satisfy these requirements, but based on very different postulates and premises, have been presented from time to time by various authors. In order to understand these theories one should understand the composition of past and present world floras. This is aided by an appreciation of the past and present physiography of the earth's surface and of the theories pertaining to its presumed evolutionary development.

This chapter presents this background of distribution of existing floras in a most elementary manner. The subject of plant geography, although an integral part of systematics, is a broad field of study in itself. It requires in particular a knowledge of living and extinct plants, of physical geography, and of geomorphology. Presented here are only the basic rudiments of the subject, sufficient only to demonstrate the significance of plant geography to any taxonomic study and to provide an introduction to the pertinent current literature. Supplemental reading should be done, particularly in the works of Good, of Cain, and of Atwood.

### Genesis and evolution

An appreciation of theories concerning the origin of the earth, the formation and distribution of its surface masses, and the evolutionary changes that are believed to have occurred through the eras of time contributes to a proper perspective of plant geography. The subject of these initial views is geology, but it has direct bearing on the distribution of plants and on problems in taxonomy.[1]

To the geologist, the earth is not a solid but an elastic mass. Its inner core (centrosphere) is postulated to be a crystalline sphere about 4,000 miles in diameter. This is surrounded by two concentric zones of vitreous material (mesosphere and asthenosphere) that, combined, are about 1850 miles thick. These are resilient and yield to very low shearing stresses. Enveloping all of these is the so-called earth's crust (lithosphere) which is a rock shell about 50 miles thick. The strength of the earth is said (von Engeln, p. 23) to be derived from the force of gravity, the presumed viscosity of its interior, and the rigidity of its outer shell. However, even this outer shell is not solid or inflexible.

The apparent surface of the earth is comprised of continental platforms (land masses) and ocean basins (water masses). The transition from one to the other is abrupt. Geologically, the continental platforms are slabs composed principally of a granite rock that is less dense than the basalt rock composing the floor of ocean basins, or present there under a thin layer of granitic rock. Furthermore, these platforms (composed of *sial*) are actually not immobile but float in the underlying layer of denser basaltic rock (termed the *sima*). The viscosity of the sima is so great that only over periods of perhaps millions of years is there an appreciable movement of the slabs of sial.

It was not until the present century that a definite theory, based on the above concepts of the earth's surface, was presented. This was done by Wegener (1915) and is known as the theory of continental drift, or as the Wegener theory. The concept was not original with Wegener, but

---

[1] Much of the information that follows in this section has been condensed or adapted from von Engeln, *Geomorphology* (1942), and Good, *The Geography of the flowering plants* (1947).

because of the impetus given it by his publications, his name is now intimately linked with it. The essentials of the theory as summarized by Good (p. 323) are these:

> During the earlier part of the Paleozoic epoch the continents were all joined together into one huge land mass or Pangaea, and subsequently they separated and drifted apart until they have come to reach the positions they now occupy . . . This movement centered on Africa, which, with the main part of continental Asia, has retained its original position more or less unchanged. The theory also envisages a movement or wandering of the poles, thus accounting for considerable alterations in the distribution of the climatic zones.

The presentation of this theory in Wegener's concise form resulted in much discussion on the part of geologists, physiographers, and biologists. Modifications and refinements of the theory have been proposed by du Toit (1937), and reference should be made to his work for a synthesis of them. Du Toit's views differed from those of Wegener primarily in that he postulated the presence in an earlier era of two primordial continents, a northern one (Laurasia) and a southern one (Gondwana), whereas Wegener accepted one initial continent (Pangaea). Du Toit further considered that Gondwana broke up into 5 lesser masses comprising the 4 present southern continents and India, which then drifted to their present locations. Perhaps the chief barrier to more general acceptance of the theory is that an adequate force to bring about the drifting of these continental slabs has not been demonstrated. In general, however, the basic tenets of the theory are becoming more widely accepted, although accord is by no means general. Campbell (1942) expressed the opinion that du Toit's theory "would best explain most of the problems in the geographic distribution of the floras of the Southern Hemisphere." [2]

The sima, or basalt layer underlying the continental slabs, has plasticity as well as viscosity. For this reason it is understandable that not only may the continental masses have moved horizontally, but that, as weight of the masses increased or decreased, they also moved vertically. This latter type of movement also has been effected by the tremendous weight of ice sheets, and vertical movement has resulted in depression by tilting or by settling, or in elevation changes of reverse character. Any of these modifications would result in changes of shore lines and in changes in areas of land surface exposed to the atmosphere. As noted above, there has been found to be a sharp demarcation between continental platforms and ocean basins. The transition from continental levels to ocean bot-

[2] For views of geologists that generally favor the theory of continental drift, see references to papers by Rastall (1929), Baker (1932), and Holland (1944).

toms is, geologically speaking, quite abrupt. As von Engeln has pointed out (p. 25),

> Over 21 per cent of the earth's surface area occurs between sea level and 3300 feet elevation; over 23 per cent between 13,000 and 16,000 feet below sea level. The percentage distribution at the other intervals is in every instance much lower These two levels may be regarded as representative of the altitudes of the continental platforms and the ocean floor respectively.

The significance to a plant geographer of the existence of this distinct disjunctivity between continental platforms and ocean basin is that there are differences in character of oceanic borders. For example, there is a sharp drop from the continental mass to deep waters, along most of the coast of western North and South America. A similar situation occurs off the east coast of the Japan-Philippine Island archipelagos and extends southeastward to the north shores of New Guinea. In other words, the Pacific Ocean extends over an exceedingly deep ocean basin. A comparable situation does not exist under the Atlantic Ocean. Along most of the coast line of the eastern shores of North and South America, there extends seaward for varying distances up to a few hundred miles a submerged shelf that for the most part has less than 600 feet of water over it. Along its easterly edge this shelf drops off rather abruptly to depths of several thousand feet. This submerged shelf is an integral part of the continents. There is every reason to believe that it has not always been submerged. Elevation of these now submerged continental platforms by approximately 600 feet would bring the continents of Europe and North America almost together in their northern limits and considerably narrow the span of ocean separating the land masses to the south.

Another theory, perhaps more widely supported a half century ago than at present, is that the continents—particularly northeastern North America and western Europe—were connected by land bridges. It is a theory unsupported by direct evidence. It arose at a time prior to acceptance of the theory of evolution, and before the possibility of continental movement was conceived. It was the only proposal then available which explained the diverse distributions of life forms. This theory has been reviewed in detail by du Toit, who concluded, in contrasting it with the continental drift theory, that "the differences between the two doctrines are indeed fundamental and the acceptance of the one must largely exclude the other."[3] A documented summary of these two theories has been presented by Good (pp. 323-331).

---

[3] A review of geologists' opinions in support and in rejection of this theory is that by Gregory (1929).

Geologic eras, their characteristics and floras, are a subject most pertinent to paleobotany, but also are fundamental to any rational concepts concerning contemporary plant distributions. There is increasing accord on the subject of the age of the earth, but it is of such astronomical figures as almost to be beyond comprehension by most biologists. Present information places it in the vicinity of a few billions of years.

This span, too great for ordinary comprehension, is divided into 5 eras: Archeozoic, Proterozoic, Paleozoic (or Primary), Mesozoic (or Secondary), and Cenozoic. It is agreed quite generally that the first two eras account for more than half of all geologic time. No fossil remains of vascular plants have been found in rocks believed formed during these first two eras. Life did exist then and was represented largely by primitive marine forms. The third era, the Paleozoic, saw the development of fishes and fernlike plants, and accounts for about 30 per cent of all geologic time. Succeeding it, the Mesozoic era accounts for about 11 per cent, and the Cenozoic for about 4 per cent of all time.

Each of the last 3 eras is divided into periods (Fig. 18). In discussing the evolution of plants and associating their development with that of the earth, its geography and even its fauna, it is more usual to refer to these periods than to the eras. The Paleozoic is divided into 7 periods called respectively, and beginning with the oldest, Cambrian, Ordovician (Champlanian), Silurian, Devonian, Lower Carboniferous (Mississippian), Upper Carboniferous (Pennsylvanian), and Permian. The Mesozoic era is divided into 3 periods, Triassic, Jurassic, and Cretaceous; the Cenozoic into 5 periods (by some authors termed epochs), Eocene, Oligocene, Miocene, Pliocene, and Pleistocene, followed by the recent or postglacial period, in which man now lives, often designated as the Holocene.

The earliest vascular plants were in existence in the late Silurian, although now extinct. They were the Psilophytales, a group reaching its zenith in the mid-Devonian and disappearing early in the Carboniferous. Coincidental with their diminishing importance in the upper Devonian was the beginning of the Lepidodendrid type of Lycopsida, a group dominant along with the Calamites in the Carboniferous and not extending beyond the Permian. These periods of the Paleozoic were characterized by such tree-fernlike plants that formed great forests, whose spores or pollen were produced in great quantity and were the source of much of our present-day coals and petroleums. During the latter half of the Paleozoic the gymnosperms became conspicuous and formed large forest areas.

The Mesozoic saw the slow, gradual increase in the true ferns we

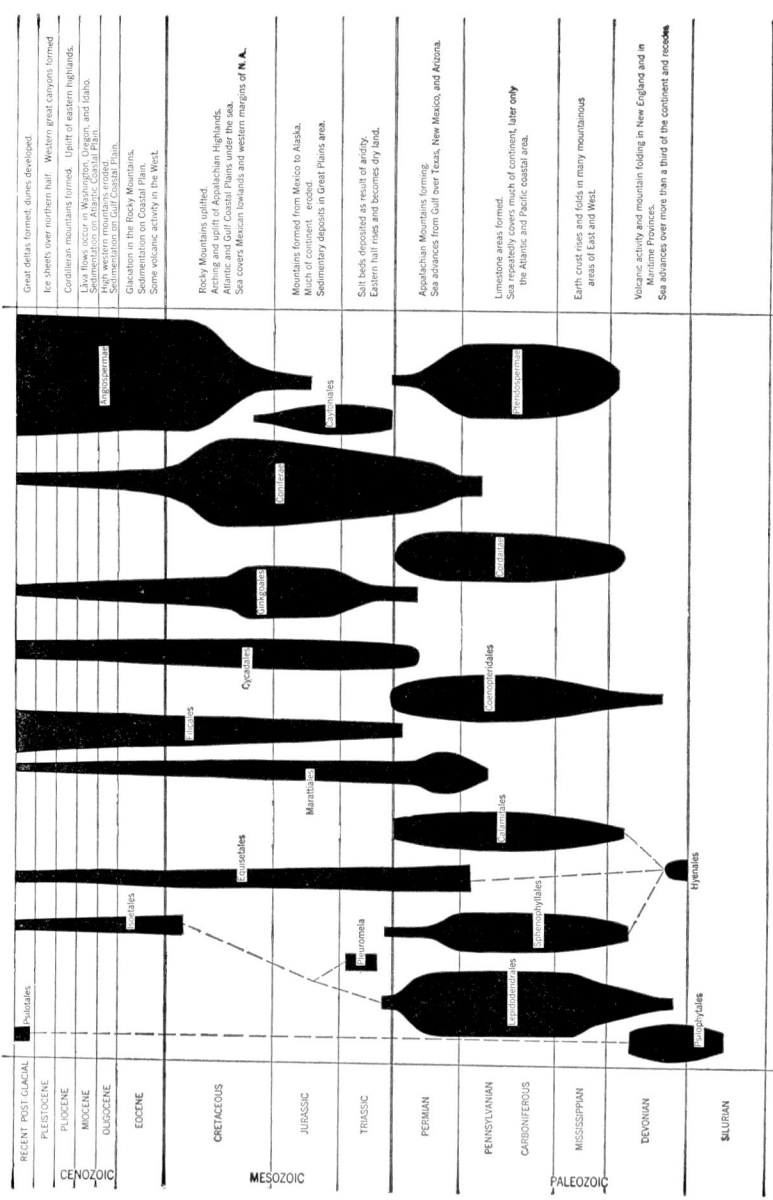

**Fig. 18.** Chart of presumed evolutionary development of plant life during the geologic periods.

know today (the Filicales), the zenith of the gymnosperm, and more especially, of the cycad floras, and (in the Cretaceous) the definite existence of the angiosperms or flowering plants. By the end of the Cretaceous these were surely a conspicuous element of the flora. It was during this era that reptiles reached their peak of development, particularly in the late Jurassic.

Botanically, there is little difference between the composition of the dominant flora of the Tertiary and of the Quaternary. In both cases the flowering plants were dominant over gymnosperms and ferns. Significant changes took place over the earth, however, in climates and in physiography. The Rocky Mountains were elevated to present prominence (Pliocene), the Pyrenees were pushed upward (Oligocene), and the Alps of central Europe elevated (Pliocene). The great epoch of glaciation over much of the earth's northern surface was during the Pleistocene, and following this, civilization had its recognizable beginnings.

## Geography and its relationship to plant distribution

The physical factors making or maintaining continents, oceans, mountains, and deserts are a prime factor in delimiting the places where plants grow. Physical geography is directly and indirectly such an integral part of plant geography that the latter can be studied only in terms of dependence on and subjection to the influence of the former.

The earth's surface is comprised of six continental land masses (North America, South America, Eurasia, Africa, Australia, and Antarctica) and the oceans surrounding each. In addition to these there is a very large number of lesser island land masses. The distribution and size of the latter cause them to be significant to phytogeographic problems. Such islands or archipelagos include Greenland, Iceland, British Isles, the Canaries, the West Indies, Madagascar, the Seychelles, the Aleutians, Japan, Formosa, the Philippines, Malaya, New Caledonia, New Guinea, New Zealand, Tasmania, and the Hawaiian Islands.

The relative positions of continents and islands are of particular importance to students of vascular floras. A study of Fig. 19 shows the earth's surface to be mostly in the northern hemisphere. It is significant to studies of plant distributions that the more northern latitudes of North America and Eurasia form an almost continuous belt of land around the globe. In studying Fig. 19 in this regard, it must be remembered that it is the outline of the continental platforms that is significant, and not necessarily the boundaries of that portion which today is above the ocean's surface. It should be noted that the ocean basins become broader and their water

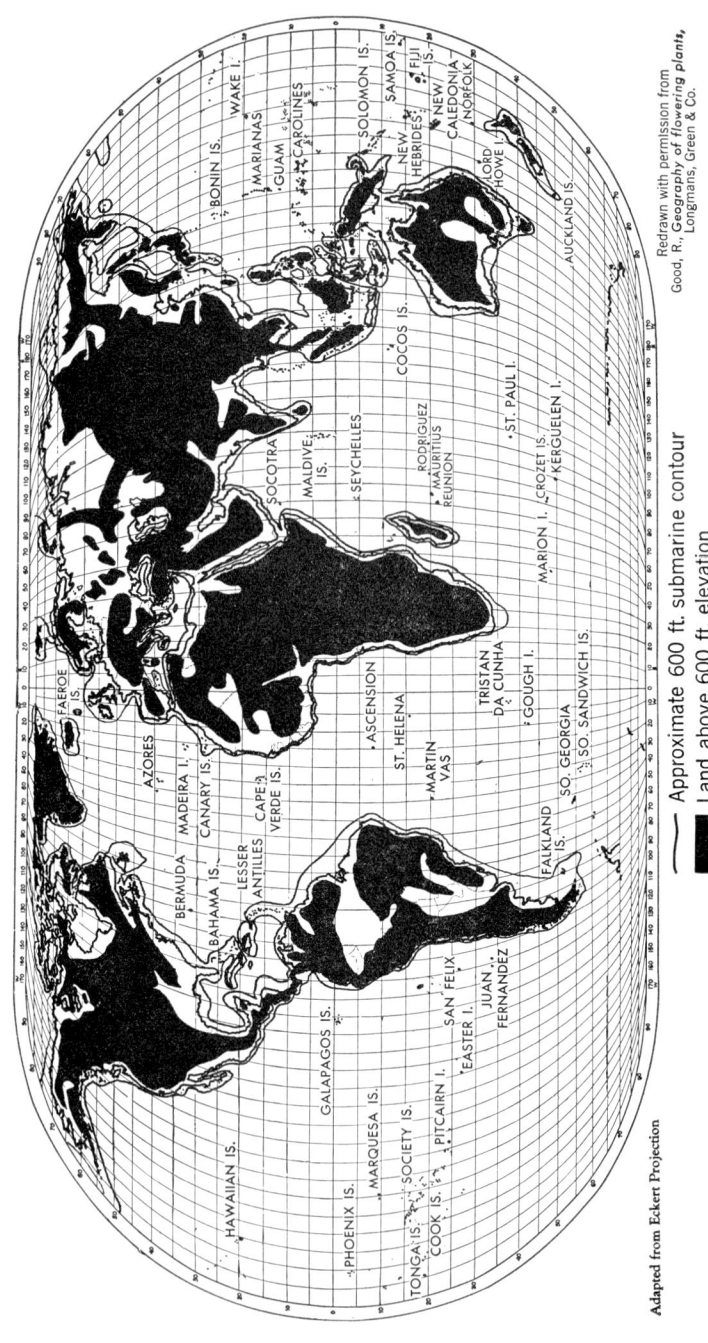

**Fig. 19.** World map of contemporary continents and major insular groups. Areas 600 ft. or more above sea level shown in solid black, the 600 ft. submerged land curve shown as a line outside present continental and insular limits.

surfaces larger as the temperate regions of the southern hemisphere are reached. The obvious corollary to this is that the land masses below the equator of South America, Africa, and intermediate islands are far more remote from one another than are similar masses at an equivalent northern latitude. Figure 19 also shows that many islands are an integral part of an adjoining continental platform and are islands only because their height exceeds the depth of the water over the continental shelf of which they are a part. Islands of this character include Newfoundland, some of the West Indies, Greenland, the British Isles, and the many thousands of islands from northern Japan through the Philippines and Malaya. These are continental islands. Other islands are noted to project from the floors of the ocean basins and are oceanic islands. The latter are more common in the Pacific than in the Atlantic Ocean, and include, in the Pacific, the Bonins, the Mariannas, the Galapagos, and the Hawaiian Islands, and, in the Atlantic, the Bermudas, the Azores, Cape Verde, and Ascension. The island of Madagascar along the east coast of Africa is likewise an oceanic island.

The presence of mountains on all the continental land masses is a feature of importance to the phytogeographer. Particularly is this true when the mountains form chains or ranges of great elevation and length. Historically, all existing ranges are of relatively recent geologic origin (see col. 3 of Fig. 18). Because of the land connections linking some continents, and because of the geologic history of the earth, only three great mountain systems are recognized on the earth's surface. These are distinguished as the western American system, composed of the Cordilleran ranges of North and South America; the Old World Eurasian-Australasian system composed of the Pyrenees, Alps, and Caucasus of Europe, the Sino-Himalayas, the central Asian plateau, and the mountains of Malaya; and the island mountains of Africa. In addition to the physical importance of mountains, their climatic importance is also of concern to the botanist, for, as elevation increases, the climate becomes cooler—a factor materially affecting distributions of plants on mountains in equatorial zones.

There are many factors responsible for the major distributions of plants. They may be *inherent* or *geographic*. Inherent factors include those that deal with the evolution and relative immobility of the individual. Evolution takes into consideration the development of the plant over periods of geologic time, and integrally associated with it is the organism's genetic constitution. The individual immobility of the growing plant is of significance, since the existence also of a mobile or motile phase in its life history is essential to establishment of a range, and in most cases to perpetuation of the race. Geographic factors are of two major sorts: *barrier factors* and *climatic factors*. A barrier is considered by Good as an area that is of a character that "cannot be crossed by a spreading species in the ordinary processes of its dispersal." Since barrier factors may be of greater significance to world flora distributions than climatic factors, they will be discussed first, and the subject of climatic factors reserved for discussion later in the chapter.

The disposition of continental masses and their probable movements during geologic times is correlated directly with a barrier factor known as ocean expanses. Bodies of water that are of considerable expanse (as are the oceans), uninterrupted by even small islandic masses, provide one of the most effective barriers to plant migration. When large bodies of water are dotted with islands, their effectiveness as barriers may be reduced, and for this reason many islands (even though small in area) are of considerable significance in accounting for some present-day plant distributions. The presence of mountain ranges represents a form of barrier insurmountable by many plants. The reason for this is that at the higher elevations of a range, environmental conditions usually are so different from those usual for the plants of lower elevations as to be functionally lethal. Furthermore, mountain ranges, by their height and extent, materially influence the precipitation of rainfall by increasing it on the windward and decreasing it on the leeward side. A third barrier factor is that provided by large desert areas where, because of climatic conditions, the range extension of all but the extreme xerophytic plants is arrested. In all these types of barriers it is to be remembered that in no case are they absolute barriers to plant migrations. Their significance lies in the fact that they bar range extensions over or across them of the great majority of plants. The climatic factors to be discussed subsequently include temperature, moisture relations, light, and wind. In addition to these there are edaphic and biotic factors whose significance to plant geography has been the subject of many books.

### Vegetation areas of the earth

There are basically 4 vegetation zones on either side of the equator: arctic or alpine, temperate, subtropical, and tropical. Due to the elevation of certain land masses within the temperate zone, nearly one-fifth of its area is occupied by arctic or alpine plants, and only three-fourths by temperate plants. Similar situations exist in the floras of the subtropical and tropical zones. In the subtropics, where snow-covered mountains

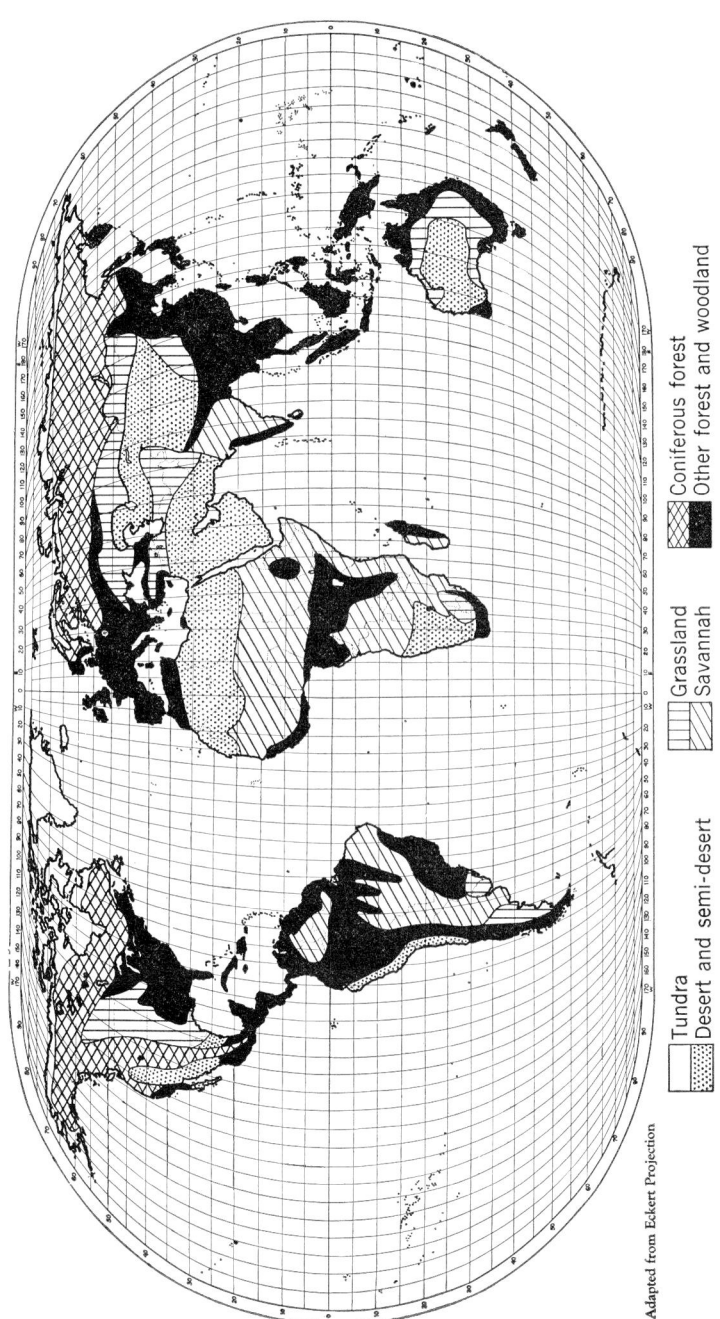

Fig. 20. World map showing distribution of major vegetation types, simplified from Brockmann-Jerosch. Redrawn from Good, *The Geography of the flowering plants*, by permission of Longmans, Green and Company.

Adapted from Eckert Projection

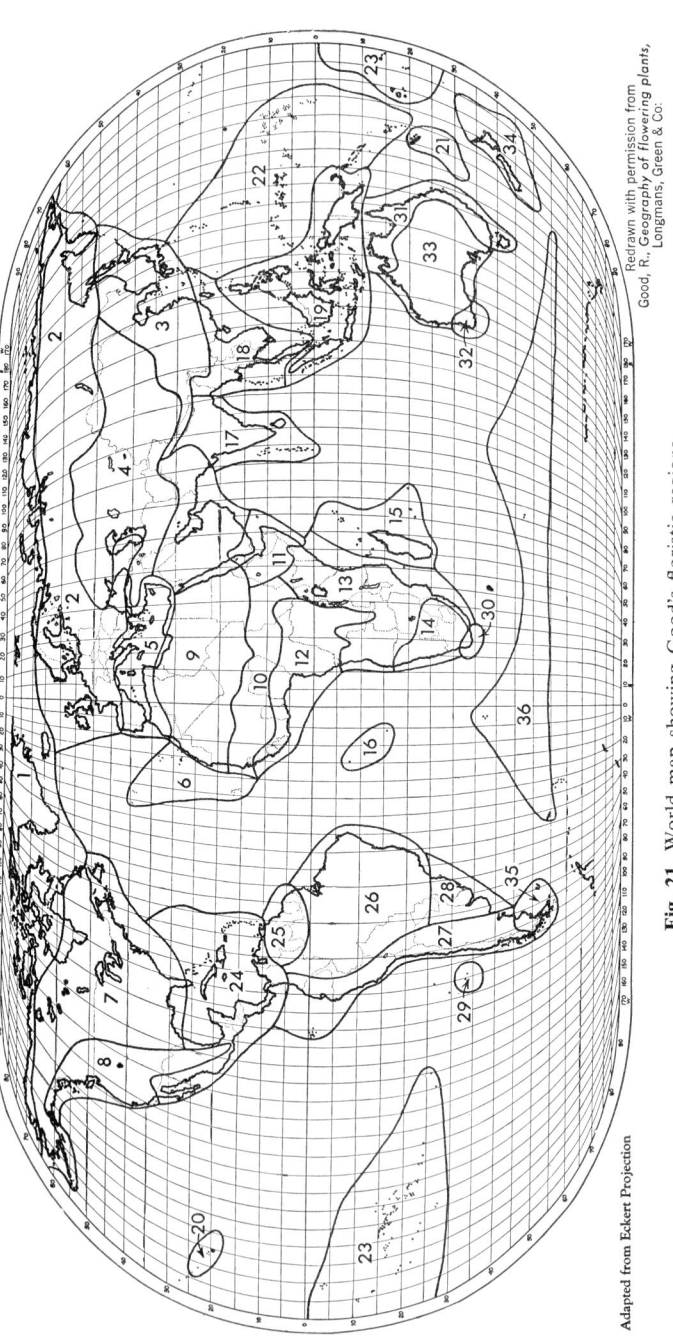

**Fig. 21.** World map showing Good's floristic regions.

1. Arctic and Subarctic
2. Euro-Siberian
3. Sino-Japanese
4. W. and C. Asiatic
5. Mediterranean
6. Macaronesian Transition
7. Atlantic North American
8. Pacific North American
9. African-Indian Desert
10. Sudanese Park Steppe
11. N.E. African Highland
12. West African Forest
13. E. African Steppe
14. South African Transition African Island
15. N.E. African Highland
16. Ascension and St. Helena
17. Indian
18. Continental S. E. Asiatic
19. Malayan Archipelago
20. Hawaiian
21. New Caledonia
22. Melanesian and Micronesian
23. Polynesian
24. Caribbean
25. Venezuela-Guiana
26. Brazilian
27. Andine
28. Pampas
29. Juan Fernandez
30. Cape
31. N. and E. Australian
32. S. W. Australian
33. C. Australian
34. New Zealand
35. Patagonian
36. S. Temp. Oceanic Islands

Adapted from Eckert Projection

Redrawn with permission from Good, R., *Geography of flowering plants*, Longmans, Green & Co.

occur, about one-tenth of the area supports arctic floras and nearly twice this much supports temperate zone plants. Another limitation associated with this system is that it fails to give consideration to the influence of thermal ocean currents, as the Gulf Stream on the British Isles. For these reasons, any classification as enumerated above is too general to be of significant value in systematic studies or in understanding relationships between floras. An elaboration of this classification was published by Hansen (1920), who subdivided the earth into 7 zones on either side of the equatorial zone, delimiting them by definite boundaries of latitude; the equatorial zone being 15° on either side of the equator, the tropical zones extending from 15° to 23.5°, the subtropical zones from 23.5° to 34.0°, the warm temperate zones 34.0° to 45°, the cold temperate zones 45° to 58°, the subarctic zones 58° to 66.5°, the arctic zones 66.5° to 72°, and the polar zones extending from 72° to the poles. This elaboration is subject to the same limitations as the more general one described above in that altitudinal variations support plants of the colder zones and the latitudinal zone is not representative of a particular floristic type.

The more biological systems of subdividing the earth's land surface into vegetational areas recognize the importance of floristic regions. Many such systems have been proposed, some based on geography, some on dominance of family or generic units, and others on physiographic areas. Schouw (1823) divided the earth's surface into 25 kingdoms representing for the most part geographic areas. Good (1947) divided the area into kingdoms, regions, and provinces, the 5 kingdoms being akin to latitudinal zones, the 36 regions accounting for the major floristic or physiographic areas with them, and the numerous provinces often representing areas of notable endemism or having politicogeographic boundaries. Since the classification by Good is the basis for discussions throughout this work it is represented here, in part, by Fig. 21. It does not pretend to be based on the physiographic provinces of the world, since that would demand a treatment too detailed for inclusion within a single volume. The merit of his system over that of Campbell (1926) or of Engler and Drude (1896) lies in its graphicness and in the fact that recognition is given to actual distributions even though they may not conform to orthodox latitudinal zones. It is functional in that it serves also to correlate phylogenetic studies with phytogeography.

### Physiographic areas of North America

Since this text is expected primarily to serve the needs of students in North America, the physiographic areas of this continent are given consideration. A physiographic area, province, or region is a natural unit possessing a more or less common topography; it is an area throughout which the geographic conditions are much the same. The most recent work devoted to the subject is that of Atwood, much of whose geographic data have been drawn on in the preparation of this section of the chapter.[4]

Physiographic areas are of particular significance to taxonomy. A physiographic area is a natural unit; the environmental conditions existing within it possess a greater uniformity than exists in the combined environmental conditions of it and an adjoining or disjunctive area. For this reason any given species is more likely to succeed and extend its range within a single area than it is to expand into adjoining or to succeed in disjunctive areas. In general it is true that each physiographic area supports a flora peculiar to itself. A manual or flora published to account for the plants of a physiographic area is of greater biological significance than one accounting for the plants of a political area. Likewise, it may be of more biological significance to know of the physiographic area (rather than the county) in which a plant grows. In making notes accompanying collections, or to be incorporated on herbarium labels, both physiographic and political locations may be given to advantage.

As indicated by Fig. 22, Atwood divided the North American continent into nine physiographic provinces. Most of these were subdivided into divisions or subprovinces—units delimited by Atwood in his individual treatments of them. The nine provinces and their characteristics follow.

THE ATLANTIC AND GULF COASTAL PLAIN

A lowland area bordering and gently sloping to the sea, extending from Cape Cod to New York City, through Trenton, N. J., to Philadelphia, Baltimore, Richmond, Augusta, Columbus, Ga., and west to Montgomery, Ala., whence it loops north to Illinois, back across southeastern Oklahoma, Texas, and along the Gulf of Mexico to Yucatan. It is bordered throughout its length by adjoining provinces of materially higher elevations. The continental shelf (see Fig. 19) is fundamentally a part of it. That part of the inner boundary extending from the lower Hudson River Valley southwestward to central Alabama is known as the *fall line.* The name indicates that this junction of adjoining provinces with the coastal plain is also a junction between the hard rocks underlying the areas of higher elevation

---

[4] Another treatise on physiographic areas (of the United States) is by Fenneman (1938). However, the work by Atwood has been selected for reference here since it provides a concise account for all the North American continent, with uniformity of treatment, and gives consideration to climate and rainfall—important to phytogeography —factors too often omitted by Fenneman. Students of Pacific coast floras should be familiar also with the physiographic subdivisions proposed by Munz and Keck (1949).

Another and dynamic approach to the problem of subdividing the continent was that by Dice (1943), who distinguished 28 biotic provinces.

# THE GEOGRAPHY OF VASCULAR PLANTS 155

and the soft and younger sedimentary rocks of the plain. Because of this, falls or rapids are produced at the zone where streams pass from the hard to softer rocks in their seaward descent. Botanists familiar with these falls or rapids are quick to recognize floristic changes that exist in areas above and below them; i.e., floristic changes between the 2 physiographic provinces.

**Fig. 22.** Map of North America showing Atwood's Physiographic Provinces. Adapted from *Goode's Series of Base Maps*, Henry M. Leppard, Editor. Copyright 1937 by the University of Chicago. Redrawn from Atwood, *The Physiographic Provinces of North America*, by permission of Ginn and Co.

## THE APPALACHIAN HIGHLANDS

This province is composed of most of the bolder relief features in the eastern part of North America; an association of mountain ranges, deeply dissected plateaus, and rolling uplands and, in New England and the Maritime Provinces of eastern Canada, a coastal area. It is divided by the Hudson River Valley into two rather natural divisions: the New England-Acadian, or northeastern, division and the southwestern division.

## THE LAURENTIAN UPLAND

This province comprises a vast and geologically ancient area extending from Baffin Land in the Arctic Ocean west to beyond Geat Bear Lake, south to Lake Winnipeg and northern Wisconsin, and east to the St. Lawrence River and Labrador. The Adirondack Mountains of New York State are a part of this region.

Most of the southern third of this area of about 1,000,000 square miles supports coniferous forests. A reduction in annual precipitation (to about 10 inches) and progressively shorter growing seasons (to about 4 weeks) becomes more restricting as polar regions are approached, and about a third of the area in the north is a tundra vegetation composed primarily of herbaceous plants and very dwarf shrubs.

## THE CENTRAL LOWLANDS

These are the gently rolling prairies of the continent, the central plains, the wheat, corn, cotton, and farm belts extending from Ohio northwest to central Saskatchewan, from there south through the Dakotas to Texas and joining the Gulf coastal plan to the south. Atwood divided it into 6 subprovinces (see Fig. 22), each delimited by individual physiographic features.

The annual rainfall over all but the northwestern part of this province is about 20 to 30 inches, with most of the precipitation occurring during the summer season. The northern and more recently glaciated areas contain abundant lakes, excellent soil, muck bog, and swampland. Forest areas are few and when present are of the deciduous type. The southern subprovinces contain few lakes, have poor soil, and even less forested area.

## THE INTERIOR HIGHLANDS

This is a small province situated between the southern end of the central lowlands, west of the Mississippi Valley and north of the Gulf coastal plain. It comprises the Ozark plateau, the St. François, Boston, Ouachita, Arbuckle, and Wichita mountains, and the Arkansas Valley lowland. Some of these arise in the Osage plains of the central lowlands, but physiographically belong in this province.

The area is one of low mountains and plateaus of mostly sedimentary rock formations. The area is also primarily one of deciduous forests and supports a more varied and richer flora than adjoining provinces.

## THE GREAT PLAINS

This province is a broad belt about 4600 miles long and 400 miles wide, extending from the Rio Grande in the south to almost the Arctic Ocean in the north, and from the Rocky Mountains in the west to the central lowlands in the east. Although seemingly flat, the area slopes to the eastward, with its elevation of about 1500 feet at its juncture with the central lowlands being nearly 4000 feet lower than that of its western margin. The eastern boundary is not well defined, but its western edge ends abruptly at the base of the Rocky Mountains.

At its eastern boundary the annual precipitation is about 20 inches, but the volume decreases progressively westward until it drops to 10 inches or less. This low rainfall is a limiting factor for vegetation, grassland and chaparral being the dominant vegetation in the south and central parts, with forests appearing in central and southern Canadian zones and the tundra forming the climax vegetation in the north.

## THE CORDILLERAN RANGES

The mountainous belt of the western section of the North American continent is known as the cordillera. It is treated as comprised of three physiographic provinces: the Rocky Mountains, the cordilleran plateaus, and the Pacific Borderlands.

*The Rocky Mountain province* extends from just north of the Bering Strait via the Brooks Range and Canadian Rockies to northwestern Wyoming. Here the mountain ranges are interrupted by a large plateau, not a part of the province, and continue disjunctively via the Wasatch Range of Utah, through Colorado, New Mexico, and westernmost Texas. In Mexico the province is represented by the three extensive Sierra Madre ranges. There are many smaller ranges included in the Rocky Mountain province, notably, the Jefferson and Gallatin ranges in Montana, and the Absaroka and Big Horn ranges in Wyoming, the Black Hills of South Dakota, and the Front Range and San Juan Mountains of Colorado. Geologically the mountains of this province are believed to have had their origin in the late Mesozoic era, chiefly by anticlinal folding.

The vegetation, below timberline, is largely forests or open ranges. In the Brooks Range of Alaska and northern Canadian Rockies the annual rainfall is only 10 to 20 inches. This is true also of much of the southern third of the province. Parts of it in Alberta, Idaho, and Montana receive 20 to 50 inches of rain annually. Conifers comprise the dominant forest trees. In this area are found the highest mountains in the continent. Changes in elevation are often abrupt, especially in the eastern part, and isolated pockets and valleys are of significance in plant distributions.

The *cordilleran plateaus* comprise a second portion of the cordilleran ranges. It is a disjunctive province of 6 somewhat arbitrarily delimited subprovinces (see Atwood) and all form a chain or series of plateaus extending from the Bering Sea into south central Mexico (see Fig. 22).

## THE PACIFIC BORDERLANDS

The physical features extending north and south along the western margin of the continent, from the Aleutians and the Alaska peninsula to Baja California consist of two chains of geologically young and rugged mountains between which is a broken series of lowland troughs. Collectively these comprise the Pacific borderlands. Atwood divided it into 10 subdivisions.

## Taxonomic significance of physiographic areas

An elementary knowledge of physical geography and of the major physiographic features of this continent is essential to an intelligent understanding of existing floristic and monographic studies.

In addition to the general data given above for each of the major physiographic regions of North America, the student should know, for any region whose flora is made the subject of intensive studies, such essentials as: the physical geology of the area; its geological history; past and present watersheds and drainage basins; soil types and their physicochemical characteristics; climatic conditions.

Armed with these data, and with a field knowledge of the flora of an area (as substituted by adequate collections of exsiccatae of that flora), one has some of the fundamentals essential to studies concerning endemism, migrations, disjunctive and continuous distributions, and vicarious forms and finally is in a position to understand better the character of speciation of elements comprising the flora in question. In most detailed floristic studies knowledge of pertinent facts concerning the subdivisions of the physiographic province may be of as much or greater significance than that for the province as a unit. Nonetheless, if the study is to be treated in adequate perspective, its provincialness must not be narrowed to the exclusion of recognizing the significance of the physiographic province and relationship of that province to the continent and to the earth as a whole.

**Phytogeography—static vs. dynamic**

Phytogeographic studies have been, until relatively very recent times, essentially of a static character. Pioneered by the work of the German botanists Grisebach, Engler, and Drude, and represented more recently by the more general works of Raunkiaer, of Campbell and, to a lesser degree, of Good, they have been descriptive in character. It is to these works that the student should refer to learn of the major floristic areas of the world, their limits, the composition of their flora, and their interrelationships with each other and with physiographic features, climates, soils, and civilizations. Descriptive plant geography is of major significance to taxonomic and ecologic studies. A knowledge of the distributions of families, genera, and species of plants adds to one's understanding of relationships of these groups. However, informative and fascinating as this phase of the subject may be, it is rarely of itself a means to an end. For that one must turn to the dynamic aspects of phytogeography.

Dynamic plant geography is another means of denoting interpretive plant geography. It represents the synthesis and integration of accumulated and ascertainable facts of the related sciences of cytology, genetics, paleobotany, ecology, evolution, taxonomy, comparative morphology, and phylogeny. In the words of Cain (1944) it

searches for causes of distributional phenomena, both modern and historical, it finds its explanations in the material which is more particularly the province of special sciences.

Interpretive plant geography is a borderline science which depends for its materials and some of its concepts upon the more specialized sciences, and it derives its distinction as a field of study from synthesis and integration. Interpretive plant geography is a second phase that follows naturally after descriptive plant geography.

THE GEOGRAPHY OF VASCULAR PLANTS 159

This relatively new approach to phytogeography could not have arisen prior to the postulations of the theory of evolution by Darwin and by Wallace. Undoubtedly it had its nebulous and then unrecognized beginnings in the works of Hall and Clements (1925) in California, of Turesson in Denmark, and of Julian Huxley in his *New systematics*. It has been stimulated, to mention only a few authors, by the more recent work of Erdtman on pollen analysis, of du Rietz on boreal distributions, and the contemporary classics of Babcock and of Clausen, Keck, and Hiesey. The findings of most of these and of scores of other workers are synthesized and presented with critical analysis by Cain. Brissenden's recent translation from the Russian of Wulff's *An introduction to historical plant geography* (1943) is an easily read treatise presenting a somewhat more challenging but critical analysis of the subject from a different approach.

**Principles of dynamic geography**

Plant geography is a subject that has developed with and because of progress in basic biological science. Commencing with the broad basic tenets published by Good (1931), revised and emended by Mason (1936), and brought to the present by Cain (1944), a set of principles now is available for study and analysis. Those of Good were developed into a comprehensive thesis, and are embodied in his most recent work (1947). Cain's acceptance of both Good's initial work, largely as augmented by the findings of Mason, was represented by his amplifications of them in his *Foundations of plant geography*. Cain's 13 basic principles, accompanied by copious illustrative data drawn from his own field studies and the analyses of hundreds of papers by other authors, provided the bulk of the subject matter of his book. These basic principles as first proposed by Good, emended by Mason, and amplified by Cain are:

A. Principles concerning the environment:
   1. Climatic control is primary.
   2. Climate has varied in the past.
   3. The relations of land and sea have varied in the past.
   4. Edaphic control is secondary.
   5. Biotic factors are also of importance.
   6. The environment is holocoenotic.
B. Principles concerning plant responses:
    7. Ranges of plants are limited by tolerances.
    8. Tolerances have a genetic basis.
    9. Different ontogenetic phases have different tolerances.
C. Principles concerning the migration of floras and climaxes:
    10. Great migrations have taken place.
    11. Migrations result from transport and establishment.

D. Principles concerning the perpetuation and evolution of floras and climaxes:
   12. Perpetuation depends upon migration and evolution.
   13. Evolution of floras depends upon migration, evolution, and environmental selection.

From a study of Cain's treatment of these principles, it is apparent that 4 of his chief amplifications lie in the incorporation in them of (1) the vitalness of the genetic constitution to evolutionary processes as well as to distribution and tolerance, (2) the significance of biotic factors in controlling distribution, (3) that responsible factors act mutually and as a concerted force (holocoenotically) rather than independently or separately, and (4) that the individual plant demonstrates varied tolerances to climatic and edaphic factors during its reproductive cycle.

The basic tenets of Good's theory were reviewed by Wulff (1943, pp. 138–144), who accepted them in principle, pointing out that Good failed to take into consideration the two factors accounting for ". . . the process of divergence which a species undergoes during the course of its migrations and, . . . the movements of floras on mountain slopes from one altitudinal belt to another." In his latest treatment (1947) Good incorporated several provisions in his theories and illustrations to satisfy these situations.

A study of the literature embodying these principles brings one almost immediately to references to subsidiary theories, some now relegated to history, others with supporters as numerous as disclaimers, and at least one now quite universally accepted. Because of their significance and the extent to which they are referred to in the literature they are summarized below.

*Willis' age and area theory* (1922) is basically the hypothesis that the longer a species has existed the greater will be its area of distribution, or, conversely, the area occupied by a species is directly proportional to its age. To quote from Willis (1922, p. 83),

if a species enter the country and give rise casually to new (endemic) species, then, if the country be divided into equal zones, it will generally occur that the endemic species occupy the zones in numbers increasing from the outer margins to some point near the centre at which the parent entered.

Following the same reasoning, he postulated also that the genus with the most species is the oldest and that with a few or a single (monotypic) species is the youngest. No allowance was made for a genus being an ancient relic represented by a few or a solitary remaining species.

His presentation of the theory was documented by data collected during a period of over 20 years, during which time he benefited from ample field studies in Ceylon and in the tropics of Asia and of South America. His ideas were criticized severely on the basis that the theory, while applicable to the tropical examples studied by him, was not serviceable as a general rule by which to determine the relative age of any area.[5] It was readily refutable by the fossil record which showed that plants of restricted distribution today (or not even known now to exist in the wild, as *Ginkgo*) were once widely distributed, even in areas whose present climate would not now be favorable to them. Studies of contemporary north temperate floras indicated that size of the area of distribution of a genus or species depended more on favorable dispersal factors and adaptability to climatic and edaphic (i.e., soil) factors than on age. The present consensus is that in theory Willis' ideas are valid, but that in practice the presence and interaction of numerous other factors (internal and external) preclude its functioning as a general feature.

*Guppy's theory of differentiation* (1906) is of academic interest, since it is now negated by the preponderance of cytogenetic data accumulated since its advocacy. It postulates that the currently larger groups of plants (families, orders, etc.) were the first to appear, and that with time they became differentiated into successively smaller units and ultimately into many species of each genus, and races of each species. His theory was supported by considerable evidence, but was based on the unsupported premise that there existed

for terrestrial plants an era when uniformity in environment was the rule—an era, one might imagine, of great atmospheric humidity, when persistent cloud-coverings blanketed the globe and when the same equitable temperature everywhere prevailed . . .

From this reasoning he concluded that a uniform environment would produce conditions resulting in uniform evolutionary changes and ultimately approach the evolution of a uniform flora. Recognizing that such a uniform flora does not exist, he considered the present-day multiplicity of kinds of plants to have resulted from contemporary changes toward a diversity of climates, accompanied by radical evolutionary responses by all organisms.

*Good's theory of tolerance* (1931) was retained by him (1947) as an entity apart from his principles of plant distribution. It was the subject of favorable discussion by Wulff (1944) and has been incorporated in essence by Cain in his 13 principles of plant geography. The implication

---

[5] For a critique of Willis' theory, see references to papers by Berry, Fernald, Gleason, and Sinnott (1924), representing collectively a symposium on the subject.

of this theory (a collection of truisms), as expressed by Good (1931), was that,

> a species is able to occupy only those parts of the world where the external conditions are within those of its range of tolerance. This total area which a species can occupy in virtue of its tolerance is conveniently termed its "potential area." The size of the potential area will tend to vary with change in external conditions.

The theory was predicated on the subsidiary hypothesis which postulated that,

> ... environmental change has, at least during the more recent past, been more rapid than change in tolerance or morphology, or, in other words, by the view that progressive adaptation to external change has not had time to occur *in situ*.

Evidence was advanced to demonstrate that most plants do not possess unlimited tolerance nor capacity to respond by adaptation to changing environments as rapidly as these changes occur. This is the nucleus of the theory, since, as Wulff pointed out (p. 141),

> ... if both tolerance and climate changed at an equal rate, there would be no plant movement and the entire theory would collapse. But usually this does not occur, since evolutionary changes in a species ... proceed considerably more slowly than changes in habitat conditions. The movements of floras known to us suffice to show that these changes are not simultaneous. They may coincide in time and rate only in case there occurs a mutation affecting the tolerance of a species.

The theory has been substantiated cytologically by examples from the Leguminosae provided by Senn (1943) and gave evidence to Cain's conclusions that (pp. 470–471) "polyploidy is of importance in the theory of specific tolerances because it results in changed reaction norms and thus allows the population to occupy a different area in which the ecological or climatic conditions deviate from those to which the diploids were adapted."

### Importance of phytogeography to taxonomy

Phytogeography is rapidly becoming a science of its own. The number of botanists devoting their research activities to it is increasing. This situation is apparent particularly in the field of dynamic plant geography, where the studies are conceded to be of paramount importance to basic taxonomic work.

One of the most fundamental of all taxonomic problems is the determination of the limits of species—the nomenclatural unit of classification. These limits vary for every category (genus, a family, or a larger unit). Taxonomy is passing through a closing era of descriptive taxonomy and into an opening era of dynamic systematics. It finds all appreciative and progressive systematists synthesizing and integrating with the morphological studies entailed in plant classification the results of studies of the related scientific phenomena inherent in, or manifested by, plants. To this end the findings of dynamic plant geography contribute significantly.

The applications of techniques and employment of methods demonstrated as essential to solutions of dynamic phytogeographic problems may, and often do, provide the monographic and the floristic taxonomist also with partial or complete answers to his problems. They may be of particular aid in the solution of problems concerned with the determination of the existence and significance of conditions indicated by the following topics.

**Migrations and evolutions of floras.** A *migration* of any element of a flora represents an accomplished fact. It is a situation of the moment and not one in process of occurring. The movement of a population through two or more sexual generations is one of *dispersal*. Dissemination is a precursor of migration. A migration has occurred only after the seed or spores of an individual have been disseminated, followed by their germination, successful competition, and the development of the new individual as a part of the flora in the new territory. Dissemination is subject to many factors, such as (1) physical characteristics of the parent plant, or of its fruit or seed, that may inhibit or facilitate dispersal, or (2) presence of barriers that present physical obstruction to dispersal, or of barriers such as climatic or edaphic conditions unfavorable for germination or successful growth and competition by the new individual. Plant dispersal is taking place constantly. That is to say, at any given moment countless numbers of migrations have occurred, and the disseminations, germinations, and growth then taking place are all degrees of development enabling new migrations to come into being. These migrations represent new ranges or occupation of new area, and also the evolution of a changing flora.

**Discontinuous distribution.** When the total area occupied by a species population includes regions on which the species does not exist, that species is said to have a discontinuous distribution. This condition is true, provided the unoccupied area is greater than the ordinary dispersal range of the species. These discontinuous or disjunctive distributions may be of minor or major character, depending on the remoteness of one population from another. They may be environmental, with the areas of absence representing regions whose climate, soil, or topography are unfavorable to successful invasion by the species. Major discontinuities exist, for example, when the same genus or species occurs on two widely separated parts of a continent, or of the earth, with no representation existing between. These discontinuous areas are considered to be the product of destruction of formerly intervening populations, or of their migration during geologic time, as caused by climatic, geologic, or physiographic changes. Instances of major disjunctions are many, and long have been of great interest to botanists because of the problems presented by their present distributions. It is of interest to note that of the approximately 300 families of vascular plants, over two-thirds contain one or more major disjuncts occurring on two or more continents. Two terms are encountered occasionally in recent botanical literature that are concerned with this general subject: polytopy and polyphylesis. *Polytopy* is the situation represented by a major discontinuity of distribution. When an element (usually species, genus, or family) occurs in two or more discrete areas, it is said to be polytopes. The most generally accepted hypothesis accounting for the existence of these polytopes is based on the premise that they are genetically closely related. However, some botanists hold that they have arisen independently and polyphyletically. *Polyphylesis* is the situation represented by a polyphyletic origin (from two or more ancestral lines).

**Endemism.** The situation known as endemism is one comprised of an element having one restricted region or area of distribution. The phytogeographer and associated taxonomists recognize two types of endemism. One is where the element is a young species or genus which may not yet have attained its maximum area as determined by its dispersal barriers. This is termed, in the strict sense, an *endemic*. The second type is one in which the element is an old or relic one, now occupying a contracting and much smaller area than before, an element that is surviving but not contributing to flora evolution—it is termed an *epibiotic*. Endemism is of significance to the taxonomist since he is vitally concerned with the history of the flora. Regions that are of considerable geologic antiquity, that have not been subject to major climatic revolutions, and that long have been physiographically isolated contribute a greater number of endemics and epibiotics than do regions not so characterized.[6]

**Centers of area.** In detailed analyses of plant distribution the taxonomist is interested in more than learning of distributional limits and causes thereof. He needs to know of other related conditions, such as the center of origin of the taxon, the center of its variation (development), and the center of frequency of individuals comprising it. If the taxon is actively

---

[6] For supplementary reading on endemism, see references to papers by Fernald (1924, 1931), Mason (1946), Sinnott (1917) Stebbins (1942), and Turesson (1925).

contributing to the evolution of the flora, it may be that knowledge of centers of dispersal are of significance, and conversely, if the situation is one of survival, there are centers of refuge to be determined. The determination of these centers is made with the aid of paleobotanical data, from which distributions of earlier and now nonexistent populations can be learned, by study of present distributional patterns and of phylogenetic relationships, and by determination of detailed climatic and geological histories of the area.

*Senescence.* There has been much speculation on the part of botanists on the presumed correlation of the vigor and aggressiveness of taxa (species, genera, etc.) with their juvenility, and of their senescence with antiquity. It has been theorized that genera, after evolvement, pass through a cycle of phylogenetic history commencing with monotypism, expansive speciation, and migration resulting in polytopism, followed by a decline, survival, epibioticism, and finally extinction. Cytological data in support of such theories have been projected, and by subjective analysis the speculations possess a disarming degree of rationality. The declining arc of this cycle is termed the senescence of the element. However, many questions have arisen that the concept of senescence does not answer. It is a hypothesis not yet sufficiently accepted by many critical plant geographers to be classed as a theory.

*Cytogenetic criteria.* The role of genetics and cytology in aiding the solution of basic taxonomic problems is largely the subject of Chapter VIII of this text. However, it is pertinent to considerations of dynamic phytogeography to point out that the determination of the cytological situations, such as chromosome counts, behavior, and morphology, and of others, such as polyploidy, amphiploidy, apomixis, and hybridity, are frequently of major importance in solving problems associated with speciation, evolution, and regression. Through the works of Anderson, Stebbins, Clausen, Turesson, Babcock, and Senn (to mention only a few), the value and ultimate necessity of such data in arriving at the solution of problems concerned with plant geography and with the new systematics have been conclusively demonstrated.

*LITERATURE:*

ADAMS, C. C. Post-glacial origin and migrations of the life of northeastern United States. Journ. Geog., 1: 303–310, 352–357, 1902.
ATWOOD, W. W. The physiographic provinces of North America. Boston, 1940.
BABCOCK, E. B. Cyto-genetics and the species-concept. Amer. Nat., 65: 5–18, 1931.
———. Systematics, cytogenetics and evolution in *Crepis*. Bot. Rev., 8: 139–190, 1942.
BAKER, H. B. The Atlantic Rift and its meaning. Privately printed. Ann Arbor, 1932

BROOKS, C. E. P. Climate through the ages. London, 1926.
CAIN, S. A. Foundations of plant geography. New York, 1944.
———. Pollen analysis as a paleo-ecological research method. Bot. Rev. 5: 627–654, 1939.
CAMPBELL, D. H. Outline of plant geography. London, 1926.
———. Continental drift and plant distribution. Science, 95: 69–70, 1942.
CLAUSEN, J., KECK, D. D., and HIESEY, W. M. Experimental studies on the nature of species. I. Effect of varied environments on western American plants. Carnegie Inst. Wash. Pub. 520: 1–452, 1940.
DARLINGTON, C. D. Taxonomic species and genetic systems. In Huxley, The new systematics (pp. 137–160). Oxford, 1940.
DAVIDSON, J. A. The polygonal graph simultaneous portrayal of several variables in population analysis. Madroño, 9: 105–110, 1947.
DICE, L. R. The biotic provinces of North America. 78 pp. Ann Arbor, 1943.
DIELS, L. Pflanzengeographie (Sammlung Göschen 389). Aufl. 2. Berlin, 1918.
DOBZHANSKY, T. Genetics and the origin of species. ed. 2. New York, 1941.
DRUDE, O. Atlas der Pflanzenverbreitung. (Berghaus Phys. Atlas) Gotha, 1887.
DU RIETZ, G. E. Factors controlling the distribution of species in vegetation. Proc. Int. Cong. Plant Sci., 1926. (Ithaca, New York) 1: 673–675, 1929.
———. The fundamental units of biological taxonomy. Svensk Bot. Tidskr. 24: 333–428, 1930.
———. Life-forms of terrestrial flowering plants. I. Acta Phytogeogr. Suecica. 13: 215–282, 1940.
DU TOIT, A. L. Our wandering continents. Edinburgh and London, 1937.
ENGLER, A., and DRUDE, O. Vegetation der Erde. Aufl. 1. Leipzig, 1896.
ERDTMAN, G. An introduction to pollen analysis. Waltham, Mass., 1943.
FENNEMAN, N. M. Physiography of eastern United States. New York, 1938.
FERNALD, M. L. The geographic affinities of the vascular floras of New England, the Maritime Provinces and Newfoundland. Amer. Journ. Bot. 5: 219–236, 1918.
———. Isolation and endemism in northeastern America and their relation to the age-and-area hypothesis. Amer. Journ. Bot. 11: 558–572, 1924.
———. Persistence of plants in unglaciated areas of boreal America. Mem. Amer. Acad. Arts and Sci., 15: 295–317, 1925.
———. The antiquity and dispersal of vascular plants. Quart. Rev. Biol., 1: 212–245, 1926.
———. Some relationships of the floras of the Northern Hemisphere. Proc. Int. Cong. Plant Sci., 1926. (Ithaca, New York) 2: 1487–1507, 1929.
———. Specific segregations and identities in some floras of eastern North America. Rhodora, 33: 25–63, 1931.
GLEASON, H. A. Species and area. Ecology, 6: 66–74, 1925.
GOOD, R. A summary of discontinuous generic distribution in the Angiosperms. New Phytologist, 26: 249–259, 1927.
———. A theory of plant geography. New Phytologist, 30: 149–171, 1931.
———. The geography of the flowering plants. 403 pp. London, 1947.
GRAEBNER, P. Lehrbuch der allgemeinen Pflanzengeographie. Aufl. 2. Leipzig, 1929.
GRAY, A. Address of Professor Asa Gray, ex-president of the association. (Consists of a comparative study of the floras of Eastern North America and of Eastern Asia.) Proc. Amer. Assoc. Adv. Sci. 21: 1–31, 1873.
———. Diagnostic characters of new species of phaenogamous plants, collected in Japan by Charles Wright, with observations upon the relations of the Japanese flora to that of North America, and other parts of the northern temperate zone. Mem. Amer. Acad. Arts and Sci. 6: 377–452, 1859.

GREENMAN, J. M. The age-and-area hypothesis with special reference to the flora of tropical America. Amer. Journ. Bot., 12: 189–193, 1925.
GREGORY, J. W. The geological history of the Atlantic Ocean. Quart. Journ. Geol. Soc. 85: lxvii-cxxii, 1929.
———. The geological history of the Pacific Ocean. Quart. Journ. Geol. Soc. 86: lxxii-cxxxvi, 1930.
GUILLAUMIN, A. Les régions floristiques du Pacifique. Proc. 3rd Pan-Pacific Sci. Cong., 1926.
GUPPY, H. B. Observations of a naturalist in the Pacific between 1896 and 1899. 2 vols. London, 1903–06.
HALL, H. M. The taxonomic treatment of units smaller than species. Proc. Int. Cong. Plant Sci. 1926. (Ithaca, N. Y.) 2: 1461–1468, 1929.
HOLLAND, SIR T. H. The theory of continental drift. Proc. Linn. Soc. (Lond.), 1942–43: 112–119, 1944.
HOLMES, A. The age of the earth London, 1937.
HULTÉN, E. Outline of the history of arctic and boreal biota during the Quaternary period. Stockholm, 1937.
HUXLEY, J. [ed.]. The new systematics. Oxford, 1940.
JOLY, J. The surface history of the earth. Oxford, 1925.
JUST, T. Geology and plant distribution. Ecological Monogr. 17: 127–137, 1947.
KÜCHLER, A. W. A geographic system of vegetation. Geogr. Rev. 37: 233–240, 1974.
MASON, H. L. The principles of geographic distribution as applied to floral analysis. Madroño, 3: 181–190, 1936.
———. The edaphic factor in narrow endemism. I, II. Madroño, 8: 209–226; 241–257, 1946.
———. Evolution of certain floristic associations in western North America. Ecological Monogr. 17: 201–210, 1947.
MUNZ, P. A. and KECK, D. D. California plant communities. El Aliso, 2: 87–105, 1949.
OOSTING, H. J. The study of plant communities. 389 pp. San Francisco, Calif., 1948.
RASTALL, R. H. On continental drift and cognate subjects. Geol. Magazine, 66: 447–456, 1929.
RAUNKIAER, C. (ed. Tansley, A. G.). The life forms of plants and statistical plant geography. Oxford, 1934.
RIDLEY, H. N. The dispersal of plants throughout the world. Ashford, England, 1930.
SCHUCHERT, C. Historical geology of the Antillean Caribbean region. New York, 1935.
SCHUCHERT, C. and DUNBAR, C. O. Historical geology, ed. 4, New York, 1941.
SCHOUW, J. F. Grundzüge einer allgemeinen Pflanzengeographie. Berlin, 1823.
SEARS, P. B. Glacial and postglacial vegetation. Bot. Rev. 1: 38–51, 1935.
SENN, H. A. The relation of anatomy and cytology to the classification of the Leguminosae. Chron. Bot. 7: 306–308, 1943.
SINNOTT, E. W. The 'Age and Area' hypothesis and the problem of endemism. Ann. Bot. 31: 209–216, 1917.
———. Age and area and the history of species. Journ. Bot. 11: 573–578, 1924.
STEBBINS, G. L. JR. The role of isolating mechanisms in the differentiation of plant species. Biol. Symposia, 6: 217–233, 1942.
———. The genetic approach to problems of rare and endemic species. Madroño, 6: 241–258, 1942.
THOMAS, H. H. Palaeobotany and the origin of the Angiosperms. Bot. Rev. 2: 397–418, 1936.

THORNTHWAITE, C. W. The climates of North America according to a new classification. Geogr. Rev. 21: 633–655, 1931.
TURESSON, G. The plant species in relation to habitat and climate. Hereditas, 6: 147–236, 1925.
TURRILL, W. B. Principles of plant geography. Kew Bull. 1939: 208, 237.
VON ENGELN, O. D. Geomorphology, systematic and regional. New York, 1942.
WEGENER, A. (trans. Skerl, J. G. A.) The origin of continents and oceans. London, 1924.
WILLIS, J. C. Age and area. A study in geographical distribution and origin of species. Cambridge, England, 1922.
———. Some conceptions about geographical distribution and origin of species. Proc. Linn. Soc. (Lond.) 150: 162–167, 1938.
———. The course of evolution by differentiation or divergent mutation rather than by selection. Cambridge, England, 1940.
WOODSON, R. E. JR. Notes on the "historical factor" in plant geography. Contr. Gray Herb. 165: 12–25, 1947.
WULFF, E. V. Introduction to the historical geography of plants. Bull. Appl. Bot. Gen. and Plant Breeding. Supp. 52, Leningrad, 1932.
———. (trans. Brissenden, E.) An introduction to historical plant geography. Waltham, Mass., 1943.

## CHAPTER VIII

# BIOSYSTEMATICS AND CYTOGENETICS

Biosystematics,[1] in the unrestricted sense, is a phase of botanical research that endeavors, by study of living populations, to delimit the natural biotic units, and to classify them objectively as taxa of different orders of magnitude. This necessitates use of data from the fields of ecology, genetics, cytology, morphology, phytogeography, and physiology, particularly as observed from plants grown under artificial and natural conditions of environment. The biosystematist strives to determine objectively whether a taxon (or a population) belongs in the category of genus, species, subspecies, etc. In this approach emphasis is placed on cytogenetics and cytotaxonomy supplemented by the classical approaches of morphology, ecology, and phytogeography. Cytotaxonomy is the integration of cytology and taxonomy in the effort better to understand and to resolve problems of plant relationships, and cytogenetics is the combining of cytological and genetical techniques in the effort to arrive at the solution of a problem.

A first step in biosystematic investigations is a thorough sampling of the taxon (it may or may not be a species) and its populations and the cytological study of the chromosomes of many populations within geographic races, species, genera, and so on. Differences in chromosome number, their morphology, and behavior at meiosis usually indicate genetic differences of taxonomic significance.

A second step includes the determination of the ability of the different populations to hybridize, and a study of the vigor and fertility of the

---

[1] Biosystematics is not the only term available; *experimental taxonomy* is another term for essentially (although not originally) the same discipline, and covers transplant studies and cytogenetics, plus considerations of the classical aspects of taxonomy; *genonomy* (the law of race or offspring) was proposed by Epling (1943) as a more appropriate term to connote the application of laws of blood relationship, but has not been adopted by biosystematists; *genecology* (a combined form of gene ecology, in the sense of race ecology) is used by Turesson and his followers, but this term has not been accepted in America where it is confused euphonically with the medical term gynecology (often pronounced in this country with a soft g but elsewhere with a hard g).

hybrids. This discloses the presence or absence of breeding barriers between groups, and is of taxonomic importance as indicating the natural limits of the taxa of various levels or order.

A third step studies the homologies of the chromosomes in the hybrids, as determined at meiosis. These are an important indicator of the degree of genetic relationship in the material.

Information obtained from these 3 steps is compared with the data obtained from comparative morphology and geographical distribution. The resultant classification of the taxa to which it is applied (within the category of genus and taxa of lower level) has an increased objectivity over one obtained through a consideration of morphology and distribution alone but may not in all instances be an acceptable substitute for the classification resulting from the synthesis of data obtained from all sources.

### Biosystematics and modern taxonomy

From the generalized definition of biosystematics given above and the resolution of its objectives, it becomes clear that it has the same goal as does modern taxonomy, and that it differs from modern taxonomy only in points of emphasis and technique. That is, the two are basically methodological variants of the single discipline, taxonomy. The explanations given below are presented to clarify these two methods of approach and to provide a background for the discussions of biosystematics that follow.

Taxonomy was originally almost exclusively a descriptive science, and the taxonomist arrived at conclusions on the basis of gross morphological characters evident from the specimen, either as it grew in the field or as it appeared on the herbarium sheet. That is known as *descriptive taxonomy* (termed *alpha taxonomy* by Turrill). The primary objective of the descriptive taxonomist was to identify and name plants. With the increase in knowledge of the plants as they grew in the field, the descriptive taxonomist realized that other factors, now known as the ecological and phytogeographical, were correlated with the morphological characters selected arbitrarily as the distinguishing criteria, and the later descriptive taxonomists have taken these factors into consideration. Examples of descriptive taxonomists would include almost all those of the nineteenth century, and such recent Americans as N. L. Britton, E. L. Greene, P. A. Rydberg, J. K. Small, Marcus Jones, B. L. Robinson, *et al*. In arriving at their taxonomic conclusions these recent botanists did not utilize direct evidence available from the disciplines of cytogenetics, cytotaxonomy, comparative anatomy, embryology, and others.

Taxonomists of the alpha school have diminished in number, and many now recognize the limitations of the approach. In their stead there are now many who practice the conviction that the most reliable taxonomic conclusions result from the synthesis of pertinent information drawn from as many related disciplines as possible. Modern taxonomists seek all available pertinent evidence and make use of it. However, few modern taxonomists possess the combined training of a cytologist, geneticist, morphologist, and anatomist, and few presume to be proficient technicians in all these and related disciplines. In their research few personally make their own morphological and anatomical preparations, or conduct extensive crossing programs or transplant studies, but an increasing number initiate their own cytological investigations. On the other hand, they do seek and utilize evidence from such of these studies as have been made, independently or in collaboration, by other specialists.

The modern taxonomist endeavors to review all known evidence, and takes it into consideration in arriving at a conclusion, whereas the biosystematist procures evidence by new research (alone or by teamwork) from the disciplines of cytogenetics, ecology (including transplant studies), phytogeography, physiology, and morphology, and derives conclusions therefrom. The former endeavors to arrive at the best solution possible with that which is available within a reasonable period of time, as measured by practical limitations. The latter plans a program unconditioned by the time element or presence of practicabilities to obtain specific and requisite evidence before attempting to arrive at conclusions. Both accept the classificatory units whose rank and sequence are fixed by international legislation. The modern taxonomist utilizes evidence from every discipline without being required by strict definition and delimitation of those units to place emphasis on any type of evidence; but the biosystematist has established, in addition to these, one or more sets of classificatory units based primarily on ecological and genetical criteria (giving consideration also to correlated morphological differences), and has so defined their limits as to make evidence from the disciplines of ecology, cytology, and genetics a prerequisite to classification and to treat as supportive evidence that derived from other disciplines. The distinctions between modern taxonomists and biosystematists are neither remote nor fixed, for not only do the two approaches converge toward a common point, but they are connected by reticulations of increasing frequency. This is represented by the utilization of biosystematic methods by the modern taxonomist, and by many individuals conducting research projects in each of these channels.

In general, there is a somewhat greater objectivity associated with the procedures and resolution of classificatory units of the biosystematist than with those of the modern taxonomist. The experimental taxonomist strives to arrive at an over-all picture of biological relationships by attempting to correlate the relationships evidenced by ecology, cytogenetics, morphology (gross and microscopic), phytogeography, and physiology. There are advantages and limitations to each of these approaches to a common goal, but it is significant that the findings of taxonomy (classical as well as modern) and of biosystematics are of mutual importance, and that each contributes to the other to build a stronger single discipline.

## Mechanics of evolution

In biology, evolution is a doctrine that holds that species of living organisms had a common descent, through a series of modifications in successive generations, brought about through selection (acting on variability) in the direction of adaptation. It is a doctrine that holds that higher forms of life have been derived from more primitive forms. Theories of evolution and genetics make no attempt to account for the origin of life, but many have been presented to account for the mechanisms by which life forms have evolved. One of the earliest was Lamarckism, which maintained that evolutionary changes are caused directly by changes in environment and by degrees of use or disuse on the part of the individual. This was followed by Darwinism and biometry, and neo-Darwinism.

Darwinism is a theory accounting for the evolution of organisms by the transmission of variations, in ratios proportionate to the number of individuals produced during successive generations, thereby causing a spread in variation frequency (i.e., a few of some variants and an abundance of others) which "involves steady structural change in the group as a whole, and this change is evolution" (Simpson, 1945). The most notable of Darwin's contributions, a subsidiary to his concept of evolution, was his theory of natural selection. This was based on 3 observable facts and two deductions.[2] Of the former, the first is the tendency of all organisms to increase numerically in a geometric ratio; the second,

[2] Adapted from Huxley, J. *Evolution, the modern synthesis.* London, 1943, pp. 14–26.

the fact that despite this tendency toward a progressive increment, the numbers of a given species actually remain more or less constant; and the third, the fact that all species vary appreciably in nature. His initial deduction, based on the first two facts, was that there must exist in nature a competition for survival or a struggle for existence. His second deduction was (to quote Huxley) that,

since there is a struggle for existence among individuals, and since these individuals are not all alike, some of the variation among them will be advantageous in the struggle for survival, others unfavourable. Consequently a higher proportion of individuals with favourable variations will on the average survive, a higher proportion of those with unfavourable variation will die or fail to reproduce themselves. . . . Thus natural selection will act constantly to improve and to maintain adjustment of animals and plants to their surroundings. . . .

Darwin's theories of evolution have been accepted in many respects and are the basis of most contemporary evolutionary concepts. Subsequent studies demonstrated limitations in his deductions. He did not realize that recombinations of many variations are but nonheritable modifications induced by environment, nor that existing genetic elements may produce new inheritable variations. Darwinism was temporarily overshadowed by the mutation theory of de Vries, by Mendelism, and surpassed by the work on pure lines by Johannsen, but it has since been revived in a form now known as neo-Darwinism.

Mendelism, based on laws established by Gregor Mendel and rediscovered at the dawn of the present century, recognizes the principle of particulate inheritance, but in its original conception gave no explanation of its physical basis. With Mendelism and the recognition of genes (the units or bearers of heredity), genetics as a science came into existence. Almost immediately the chromosomes were recognized as the carriers of the genes. It had been pointed out by Mendel and elaborated on by others that the inheritance of any one unit is independent of that of any other unit (sometimes referred to as the particulate nature of inheritance). It is also recognized that a plant may be homozygous (have received like genes from each parent) or heterozygous (formed by the union of chromosomes of unlike genes, or by mutation in a homozygote), and certain characters may be genetically dominant (be brought to expression by the single occurrence of a gene) while their alternatives are recessive (requiring the presence of a gene from each parent to come to expression). Finally, the distribution in the progeny of the segregating dominant and recessive characteristics may be predicted in accordance with ratios resulting from chance segregation and recombination.

*Mutations* (sometimes designated "sports") are believed generally to be the product of intrinsic changes in the substance or structure of the chromosomes. They may be responsible for minor or major differences in plants. One authority (Goldschmidt, 1940) is of the opinion that families of plants and animals may have arisen *de novo* by macromutation and that connecting links may never have existed between them and their presumed relatives or ancestors. This view is not generally accepted. The heritable variations in plants are now generally believed to have arisen by mutation, to have been diffused by crossing, and maintained by selection. *Modifications* are the nonheritable responses of a given heredity to different environments, while mutations are heritable changes in the germ plasm or the chromosomes. Heritable and nonheritable variation may be differentiated only by experimentation. Mutations alter the nature of genes while recombinations juggle existing genes (cf. Huxley, 1943, p. 21).

Cytological phenomena of particular significance to biosystematics and to the mechanics of evolution are associated with problems of hybridity.[3] Cytologically, a *hybrid* is the product of the union of two unlike gametes. There are many kinds of hybrids. The taxonomist may recognize intergeneric hybrids (as between two genera of orchids or crucifers), interspecific hybrids (between two species), intraspecific hybrids (between subdivisions of a species). The constitution of artificial hybrids is known more precisely than that of spontaneous hybrids from the wild, the parents of which are usually deduced. The degree of fertility of the hybrid may give some indication of the degree of genetic relationship between its parents. In general, hybrids between the taxonomically less closely related species (as determined by comparative morphology, ecology, etc.) of a genus tend to be sterile or of low fertility, whereas hybrids between taxonomically more closely related species or infraspecific taxa tend to be more fertile. Thus there is correlation between crossability, hybrid fertility, and taxonomic relationship.

*Polyploidy* is a condition existing when the somatic chromosome complement of a vascular plant is composed of more than two sets of the monoploid (haploid) or *n*-number of the taxon or genus (resulting in triploids, tetraploids, etc.). Evidence indicates that most natural polyploids are of hybrid origin, and while they may be accepted taxonomically as a valid species, their origin may establish most of them phyletically as

---

[3] For discussion of the relationship of these and other cytological phenomena to taxonomy and evolution, see papers by Allan (1937 and 1949), Anderson (1937), Baker (1947 and 1950), Heilborn (1929), Lotsy and Goddijn (1928), and W. W. Smith (1933 and 1936).

taxa of higher advancement than diploid species of the same genus.[4] For reasons indicated below, it is helpful in taxonomic studies to know whether a particular polyploid is an autoploid or amphiploid.

*Autoploidy* is a type of polyploidy in which (*fide* Clausen, Keck, and Hiesey, 1945) each of the chromosome sets has been derived from the same species. Since in autoploids each set of chromosomes is present more than twice, their chromosomes may conjugate not only in 2's but in 3's, or 4's (multivalent formations). This may lead to irregular distribution of chromosomes to the gametes and therefore to sterility to a greater or lesser extent. It is unknown whether in the course of great periods of time, an autoploid may overcome this multivalent formation or not and therefore become functionally diploid—but this may be true in some existing autoploids of great age.

Taxonomically autoploids are difficult to distinguish from their ancestral diploids and are the least distinct of all taxa. Often they are not classified as distinct taxa or only as *formae*. However, it is held by some that when autoploids are morphologically and ecologically distinct from their diploid progenitors, biologically they should be considered distinct species, for the numerical difference in chromosome number adds a genetic barrier to the morphological and ecological distinctness. This is true because some sterility may result if the autoploid is crossed with a diploid. (There are recorded cases where tetraploids have crossed with diploids and produced fertile triploids, but these are unusual situations.)

*Amphiploidy* (amphipolyploidy, amphidiploidy) is a particular type of polyploidy (allopolyploidy) characterized by the addition of both sets of chromosomes from each of two species. For example, the amphiploid *Primula kewensis* has 18 pairs of chromosomes and is a fertile hybrid, having 9 pairs of chromosomes from *P. floribunda* and 9 pairs from *P. verticillata*. The amphiploid is commonly an interspecific hybrid and is most likely to succeed if the parent species are related sufficiently to produce a vigorous $F_1$ and yet remotely enough to prevent pairing between those of chromosomes coming from different sources. Such pairing would result in a breakdown of the balance between their combined chromosome sets. The amphiploid will be stable when the parental chromosomes are so different that those of one parent do not synapse with those of the other. Amphiploidy is an evolutionary process respon-

---

[4] In this regard it should be remembered that while some polyploid species may be hybrids, they are genetically stable hybrids of usually considerable age. That is, they are for the most part not $F_1$ or $F_2$ generation hybrids, nor necessarily hybrids that have arisen during the current century or centuries, but may even have originated in a previous geologic age.

sible for the formation of many species, and sometimes genera, in nature; examples of natural species supposed to have arisen by this means include *Phleum pratense, Iris versicolor, Poa annua, Spartina Townsendii, Rumex Acetosella, Brassica Napus, Brassica juncea, Prunus domestica, Nicotiana Tabacum,* and *Galeopsis Tetrahit*.[5]

### Biosystematic categories

The major objective of biosystematic studies is to arrive at a better understanding of the natural relationships of plants, particularly those of the rank of genus and below. This also is an objective of orthodox taxonomy. Based on the data of genetics, cytology, ecology, and morphology, the biosystematist has developed a classification for experimentally investigated natural taxa. These categories are not intended as substitutes for the units used in classical or practical taxonomy, and they are not necessarily the equivalent of these, although they may be counterparts of them. It is not proposed that they have status in nomenclatural matters. Each provides a single-word term for a biosystematic situation, and in no case should the term be applied to a plant or a population unless the situation for which the term stands has been proved experimentally to exist for the particular taxon. The biosystematic categories represent evolutionary nodes, and many populations occupy positions intermediate between these nodes. They are nodes in the sense that each biosystematic category represents a step or a level in the evolutionary scale of differentiation from that of a local population to that of a genus (e.g., from the local population to the comparium). The names of these units are appearing with increasing frequency in taxonomic literature, and it is incumbent on the taxonomist to understand them and to appreciate their significance. None of these terms should be applied to a plant or a population unless its right to the category has been established and recorded. The four most widely accepted categories of the biosystematist are, in order of ascending phyletic value, ecotype, ecospecies, cenospecies, and comparium.[6]

The *ecotype* is the basic unit in biosystematics. It is accepted as a phyletic unit "adapted to a particular environment but capable of producing fully fertile hybrids with other ecotypes of the same ecospecies."

---

[5] See Goodspeed and Bradley (1942) for a review of amphiploidy and for an enumeration of 124 known amphiploids (designated by parental formulas rather than by binomials), and Clausen, Keck, and Hiesey (1945). See Stebbins (1947) for discussion of types of allopolyploidy.

[6] For explanations of concepts involved in these categories see papers by Clausen, Keck, and Hiesey, Gregor, and Turesson, as cited in the references at the end of this chapter. For a discussion of other categories of species, see Camp and Gilly (1943).

The ecotypes of one species are not isolated by genetic barriers and remain genetically distinct only because they thrive in ecologically different environments. One ecotype differs from another of the same ecospecies by many genes and the unit is somewhat parallel with, but not necessarily identical to, the geographic variety or subspecies of taxonomists. Some ecotypes are the equivalents of geographic subspecies, but more than one ecotype may be included in these, particularly such ecotypes as are physiologically but not morphologically distinct. The term ecotype was proposed first by Turesson (1922) for an "ecological unit to cover the product arising as a result of the genotypical response of an ecospecies to a particular habitat." Turesson later (1929) emphasized the genetic crossability between ecotypes. The more comprehensive definition of the ecotype by Gregor et al. (1936) described it as

a population distinguished by morphological and physiological characters, most frequently of a quantitative nature; interfertile with other ecotypes of the ecospecies, but prevented from freely exchanging genes by ecological barriers.

The determination of interfertility among ecotypes of the same ecospecies is usually established by controlled tests on plants transferred from the wild to the experimental ground.[7]

The *ecospecies* was defined first by Turesson as a group of plants comprised of one or more ecotypes, within the cenospecies, whose members are able to interchange their genes without detriment to the offspring.[8] Related ecospecies are usually separated by incomplete genetic barriers which, in addition to ecological barriers, are adequate to preclude free interchange of genes with any other ecospecies. When ecospecies of one cenospecies are crossed, the resultant hybrids are either partially sterile, or, if fertile, they produce many weaklings in the $F_2$ generation (as slow-growing dwarfs, individuals highly susceptible to diseases against which parents enjoyed immunity, and teratological misfits). Such weaklings are unable to compete, and fail to reproduce. A few such hybrid segregates may possess sufficient vigor to survive. These may be reabsorbed by interbreeding into one or the other parental ecospecies. Related ecospecies generally inhabit different but often contiguous ecological or geographical areas, thus retaining a relative genetic purity. In general, the ecospecies

[7] Different kinds of ecotypes are recognized by the biosystematist as, *edaphic, climatic,* and *biotic ecotypes*. Some authors, in an effort to be more understandable and without any intention of abandoning the term ecotype, have designated these as edaphic, climatic, and biotic races.

[8] It has been pointed out that even ecotypes cannot exchange genes without detriment to the offspring if the environment favors one ecotype over the other. This is one of the basic reasons why genetic studies of them must be made in the garden.

approximates the conventional and conservative taxonomic species (referred to by some botanists as the Linnaean species, but the species of Linnaeus was often more inclusive).

The *cenospecies* is a group of plants representing one or more ecospecies "of common evolutionary origin, so far as morphological, cytological, and experimental facts indicate." Cenospecies of the same comparium are separated by genetic barriers so nearly absolute that all the hybrids between them are sterile unless amphiploidy (amphidiploidy) occurs.[9] For this reason, distinct cenospecies may exist in a single environment without genetic intermixing. Rather often the cenospecies parallel the taxonomic sections or subsections of the genus. It is noted that ecological separation forms no part of the definition. As was true for the ecotype and ecospecies, the identification of a phyletic unit as a cenospecies must be based on genetic experiment to determine the degree of fertility (if any) of the $F_1$ generation.

The *comparium* is the biosystematic unit that often is comparable to the genus. It is composed of one or more cenospecies that are able to intercross. Distinct comparia are unable to intercross, and complete genetic incompatibility prevails between them. There are, however, numerous taxa, accepted by the orthodox systematist as genera, that may contain two or more comparia (e.g., in the Leguminosae). In some families the accepted and conventional genera are not the equivalents of comparia or even of cenospecies (e.g., some Crassulaceae, Orchidaceae, Cruciferae).

Discussions of biosystematic categories and concepts involve also the use of the genetic terms of genotype, biotype, and phenotype. As explained by Stebbins (1950, pp. 36–37),

... each individual organism has a genotype and a phenotype. The *genotype* is the sum total of all the genes present in the individual. ... All the first-generation progeny of a cross between two completely homozygous individuals have exactly similar genotypes no matter how different are the two parents. But depending on the degree of heterozygosity of one or both parents of any mating, whether within a population, between varieties, or between species, the genotypes of the offspring will differ from each other to a greater or lesser degree.

Knowledge of the existence of similar genotypes, or of the nature of differences between genotypes, is of fundamental importance to bio-

[9] It has been pointed out by Winge (1917), by Anderson (1937), and by Goodspeed and Bradley (1942) that amphiploidy involving distinct species produces a unit that genetically is a new species (or unit of higher rank) and should be given a binomial (or name of appropriate higher rank). Clausen *et al.* (1945, p. 149) expressed a similar view, noting that the amphiploid is ". . . a genetic species from its inception, and the taxonomist may anticipate from the nature of its origin that morphological characters may be found to distinguish it as a taxonomic species as well."

systematic studies. The relationship between genotypes and biotypes was made clear by Stebbins (l.c.) who wrote,

The *biotype* consists of all the individuals having the same genotype . . . The biotype in cross-fertilized organisms usually consists of a single individual. But in self-fertilized plants, the individuals may become completely homozygous and produce by selfing a progeny of individuals all within the same genotype, and therefore belong to the same biotype.

The *phenotype* is the form or appearance of an individual, and represents the result of external factors (as growing conditions) on its genotype. Thus, two individuals of the same genotype may appear to be different (i.e., they have two phenotypes) if each has grown in an environment different from the other. Conversely, two plants may have the same phenotype but have different genotypes.

## Methods in experimental taxonomy

The techniques of experimental taxonomy comprise methods of testing to determine to which of the above-described biosystematic units a population belongs. The most reliable evidence and conclusions are obtained from application of all methods: those of orthodox taxonomy, as well as those of cytology and genetics, combined with cultivation in uniform and in varied environments. Ordinarily the methods of testing in uniform and in contrasting environments are applied simultaneously to a number of populations of the same and of different taxonomic rank within a genus. When dealing with clonal divisions of perennials, they are known to some botanists as transplant studies because the plants are moved from their native habitat and are propagated and grown in new environments. The work had its beginnings with the classical studies of Alexis Jordan, Kerner von Marilaun, and Bonnier, was later much expanded by Turesson, and since has been employed by Hall, Gregor, Clausen *et al.,* Marsden-Jones, Turrill, Babcock, and many others.

*Growth in uniform environment* provides a means of studying the variability in heredity. Population samples of the same taxonomic category (e.g., species, subspecies) are procured from different environments and are grown in a common experimental plot or under controlled environmental conditions.[10] This permits comparison of behavior of plants of unlike heredity in the uniform environment, thereby distinguishing between hereditary variation and environmental modification.

[1] For an account of the latter, see Clausen, Keck, and Hiesey (1948).

*Growth in varied environments*, as afforded by a series of field stations or artificially controlled environments, is of equal importance as a means of observing the interplay between heredity and environment. For this purpose, the individuals may be cloned and distributed to two or more markedly different environments. In each of the environments the cloned plants may then be subjected to observations such as seasonal reactions, and their performances measured. Conclusions obtained from the varied environment tests greatly help in plotting the range of environmental tolerances of individuals, races, and species.

*Cytogenetic analysis* may involve the 3 following types or techniques. (1) Cytological studies are made of as many populations as practical to determine the possible correlations between visible chromosome differences and differences in external morphology and geographic distribution within the species complex. (2) Selected forms are crossed and the hybrid progeny grown to determine the fertility of the $F_1$ generation, the fertility and vigor of the $F_2$ generation, and to analyze the genotypes of related ecotypes and ecospecies. (3) The chromosomal homologies of these natural units are analyzed through studies of the chromosome pairing in their hybrids.

*Crossing programs to test the presence or absence of sterility barriers* are a part of the biosystematic study. The crossing test is applied to any taxa suspected of possessing these barriers, and in instances where the barrier is found not to be absolute the $F_1$ and $F_2$ generations must be studied statistically to determine their vigor and fertility. Such tests are conducted also with plants of populations presumed to be of close affinity to determine the genetic mechanism that may be allowing them to retain distinctness. The crossing tests are applied also to geographical or ecological extremes to determine if such extremes are separated by genetic barriers.[11]

In any and all of these studies it is essential that complete and accurate records be maintained. Not only should there be records of cytogenetic and ecologic data, but these should be accompanied by photographs and adequate herbarium specimens to serve as permanent

---

[11] Students of biosystematics should be cognizant of the work of Laibach (1929), Tukey (1935), Dobzhansky (1937), and Cooper and Brink (1942), that deal with isolating mechanisms and with the endosperm as a barrier to interspecific hybridization in some taxa of flowering plants, as evidenced by the viability of excised embryos of such hybrids when grown in pure culture. This knowledge permits broadening the limits of interspecific hybridization although the existence of such barriers prevents natural intercrossings.

documentary records of the material. Both herbarium specimens and cytological preparations are important records for future reference.

### Importance to taxonomic research and interpretation

*Aid in delimiting taxa of infrageneric categories.* Cytogenetic and cytotaxonomic studies are being applied with increasing frequency to problems of generic phylogeny. Taxonomists occasionally arrive at divergent views on the relationships of species within a large genus; some believing the genus to be represented by a number of polymorphic species, others that it contains a large number of species with intergrading variants, and that it is composed of a number of sections or subsections, and still others treating it as composed of several segregate genera. Genera of this character are usually of wide distribution and possess a few to several centers of distribution. The apparent conflict of views of various taxonomists working with the same genus can often be clarified by application of a combination of cytogenetic and cytotaxonomic studies.

For example, *Nicotiana*, a genus of about 60 species, has been studied in detail by Goodspeed and others, and by correlating the findings of distributional patterns and morphological differences with studies of chromosome counts and breeding behaviors, it has been demonstrated that the genus is composed of 3 natural subgenera and 11 sections. "The members of these 'genetic groups' were considered to possess distinctive morphological, distributional, and cytological characters sufficiently in common, on the one hand, to demonstrate that phylogenetic relationships within the individual groups were relatively intimate, and on the other, to set apart the groups themselves one from another." (Goodspeed, 1945, p. 577.)

Similar studies have been made, although perhaps to a lesser extent, by Cleland on *Oenothera*, by Babcock and Jenkins on *Crepis*, by Manton on Cruciferae, by McKelvey and Sax on *Yucca* and *Agave*, and by Senn on Leguminosae. Studies of this character are exceedingly helpful in determining the origins and evolutionary trends in a genus, in providing better evidence of primitiveness and advancement within the taxon, and in better enabling the taxonomist to group the species in natural relationships.

*Aid in delimiting species.* The results of biosystematic studies have provided the taxonomist with new data and have revitalized the interest and enthusiasm of taxonomic workers, especially in understanding the species as a biological unit. Biosystematic studies have been in progress for over a quarter century, and while the taxonomist may not accept their approach to his problems as sole means to an end, he recognizes the value of and utilizes their contributions.

The species (i.e., the ecospecies), as conceived by the biosystematist, is a group of interbreeding or potentially interbreeding individuals reproductively isolated from other groups of individuals. It is a unit delimited primarily by genetical criteria and secondarily by criteria derived from ecological and morphological evidence.

Botanists who consider the species a biological unit and who accept the genetical concept of it find the techniques of the biosystematist to be of considerable importance in the delimitation of species and of subordinate taxa. Similarly, these same techniques are a primary means of differentiating ecotypes from ecospecies,[12] and at the same time they identify some taxa as hybrids and others as only biological units of lower classification levels (including apomicts) frequently not deserving nomenclatorial recognition. The biosystematic approach is helpful in the solution of phylogenetic problems, utilizing the knowledge that fertility is usually a more certain indicator of close relationships than is nonfertility.[13] As pointed out and explained by Turrill (1940), these approaches are useful also in the determination of the "degree of plasticity of genotypes, . . . the occurrence and constancy of correlation of characters, . . . the occurrence and nature of sterility barriers, . . . the evaluation of characters, . . . the recognition of hybrids, [and] . . . the phylogeny of species."

***Aid in determining relationships.*** Most biosystematic studies have been restricted to the cenospecies and lower units, but they are of value also to studies of the higher categories. Members of one comparium are more likely to be related to one another than to members of other comparia, and this relationship may enable the biosystematist to delimit more objectively those genera whose circumscriptions have been diversely interpreted by taxonomists. In this connection it must be recognized that evolutionarily premature blocking of interbreeding through embryo-

---

[12] For application, see among others Jenkin's (1933) report on work with grasses.

[13] Other techniques, scarcely biosystematic in scope, have been developed by which numerical values, based on biometrical analysis of cytogenetic data, have been used to weight morphological characters employed in the delimitation of species and infraspecific units. In practice this generally has resulted in the circumscription of ultraconservative "macrospecies." A recent application of this technique is that by Roberty (1950) which represents an amplification of its use by Maillefer (1944), who in turn credited Kreyer (1930) as the originator.

endosperm interference may divide a phyletic genus into several comparia. In such instances the comparium is not the equivalent of the genus. For this reason, the crossing criterion may offer little or no aid to the determination of generic limits in such a family as the Leguminosae. Conversely, the absence of sterility barriers between currently accepted genera of some families (Gramineae, Orchidaceae, Polypodiaceae, some Rosaceae, etc.) point to a close evolutionary relationship between such genera. In many families, such as Polygonaceae, Aizoaceae, Palmae, Vitaceae, Bignoniaceae, etc., cytogenetic data are scanty or lacking and must be obtained before the utility of biosystematics in the delimitation of their genera can be ascertained.

## Apomixis

Apomixis is a term for many types of reproduction in which there is no fusion of male and female gametes. In many seed plants offspring are produced from seeds that have developed from ovules containing unreduced unfertilized eggs or from somatic cells associated with the egg. The resultant plant is called an *apomict*. Since it is produced without the inclusion in its germ plasm of the chromosomes or genes from a paternal sex cell, it is identical cytogenetically with the mother plant (unless it be a haploid apomict.) This reproductive phenomenon is genetically homologous with such artificial methods of asexual reproduction as propagation by cuttings, graftage, budding, or division.

Many types of apomicts occur.[14] They may arise from sporophytic tissue without involving formation of a morphological gametophyte and egg cell (i.e., from cells of the nucellus or integument), or from the female gametophyte and egg cell, and yet possess an unreduced chromosome number. Explanation of the causes of apomixis lies in the field of cytogenetics and not of taxonomy, but it is important that the taxonomist have an appreciation of them because a knowledge of apomixis is of increasing importance in understanding speciation in many difficult genera. Gustafsson has defended the view (originally presented by Ernst, 1918) that in nearly every group of higher plants the majority of experimentally proven apomicts are definitely or probably of ancestral hybrid origin. It is known that hybridization makes possible recombinations of mutant forms and sets up genetic situations favorable to the establishment of apomictic cycles. Apomixis may be facultative or obligative, and, while

[14] For detailed explanations of these, as well as a detailed treatment of apomixis in higher plants and an enumeration of known apomicts, see Gustafsson (1946–1947).

evidence is lacking, it has been suggested that the former type may precede the latter in evolutionary sequence within a genus.[15]

It is known that genera containing apomicts usually are very polymorphic, having a large number of taxa, and usually have a large number of species assigned to them by the descriptive taxonomist. Widespread polymorphism within an apomictic genus has been ascribed by Gustafsson to (1) hybridization and segregation, (2) polyploidization and haploidization, (3) mutation, (4) autosegregation (as associated with apomixis), and (5) elimination of intermediate biotypes. The presence of apomixis in these genera makes possible the survival of hybrids and hybrid derivatives that ordinarily would be eliminated otherwise because of their sterility. The perpetuation of these derivatives in nature, often associated with particular distributions, sometimes results in the descriptive taxonomist's treating each biotype or apomictic hybrid as a morphologically distinct and seed-producing species. As pointed out by Stebbins (1949), "in a sexual species biotypes equally different in morphological characteristics are passed over because they are connected by innumerable intermediate and recombination types." The presence of apomixis cannot be established from the herbarium specimen (although pollen-sterility studies may be suggestive) nor from field observation of living material. It can be established only by carefully conducted cytogenetic investigation. A basic technique is to emasculate flowers, bag them to prevent cross-pollination, and determine if seed are set and, if so, are viable. This procedure is accompanied by cytological studies of the chromosomal situation and to determine from it the type of the apomictic process. The results obtained by the bagging technique require careful checking before asexuality can be presumed to exist, for, as was true in Guayule (*Parthenium*), pollination may be required without being followed by fertilization, and the plants are pseudogamous rather than asexual propagules.

The taxonomist has been faced by the problem of how to name and classify plants established to be apomicts. The apomictic clone is an asexual propagule and by neither morphological nor biological criteria may it be

---

[15] A *facultative apomict* (as in *Festuca ovina*) reproduces itself both by normal sexual reproduction and by repetition of the asexual process. It is designated sometimes as an amphiapomict. That is, if the flower is pollinated at the right time some of the ovules may be fertilized and sexual reproduction occur, or if unpollinated or if no gametic union takes place, there follows asexual reproduction by apomictic processes.

An *obligative apomict* (spp. of *Rubus*) reproduces only by asexual means. Its inability to reproduce sexually under any conditions is believed by some to be due generally to complete failure at meiosis to produce viable gametes; this often occurs in hybrids that are the progeny of parents with a high and a low chromosome number respectively.

considered homologous with a sexually reproducing species. It should not be designated by a binomial. Turesson (1926) designated facultative apomicts of *Festuca ovina* by a ternary epithet preceded by the abbreviated designator *aapm* (amphiapomict). Babcock and Stebbins (1938) found *Crepis* to contain primitive sexual species (diploids) and apomictic agamospecies (a large, variable complex), and designated the components of the latter by ternary names, as *formae apomictae* of the primary species. However, in some instances the primitive sexual species is not now extant, yet apomicts inferred to have been derived from it have been demonstrated to exist. To provide also for this contingency, Gustafsson (1947) proposed that apomicts (apomictic strains) be designated by binomials and be called *microspecies* (abbreviated, *msp.*), a taxon that "may, or may not, contain several close biotypes." His nomenclatural treatment of them is in accordance with the taxonomic practice of European taxonomists dealing with apomictic complexes, but injects a source of confusion into botanical nomenclature by using the binomial for 2 concepts of a species. There is no provision in the Rules requiring use of the designator *msp* to distinguish one kind of a species from another. The nomenclature of apomicts was the subject of much discussion at the Stockholm Botanical Congress (1950). It was voted there to add a new article to the Rules to provide for this contingency. This article, now in effect, reads as follows:

Taxa which are apomicts may, if so desired, be treated in the following manner: (1) if they are considered of specific rank, by the interpolation of the abbreviation *ap.* between the generic name and the epithet; (2) if they are considered to be of infraspecific rank, by the interpolation of the abbreviation *ap.* between the categories.

The application of this rule presents no difficulties. If, in the hypothetical case of a *Rubus,* a taxon first described as *Rubus novus* later is established to be an apomict, this biological situation may be indicated by designating the taxon as *Rubus* ap. *novus*. In the event it is held to be only a form of a species having an older binomial (for example, *Rubus altus*), it could be designated as *Rubus altus* f. ap. *novus*. The Rules do not provide a new category for apomicts. The interpolation of the abbreviation *ap.* does not of itself constitute the making of a new name or combination of names (i.e., no new author citation is involved). However, if a known apomict having only a binomial is being reduced in rank to a lower category for the first time, the publication of that change in rank must be made in accordance with the directives of the Rules (cf. p. 201) and the publication of a new name or of a new combination may be required.

## Limitations of cytogenetic criteria

A presentation of the limitations of cytogenetic criteria, in so far as they apply to taxonomy, demands recognition of equally significant limitations of taxonomic criteria as they apply to taxonomy and especially to phylogeny. The weakness of the taxonomic approach has been emphasized in the chapters on the principles of taxonomy and phylogenetic criteria, and just as there is a responsibility to point out those inherent in previous and current taxonomic thinking, so should the student of taxonomy be informed of and be able to recognize the more significant shortcomings and inadequacies of published cytogenetic findings. This is particularly true since both biosystematics and modern taxonomy are predicated on the significance of cytogenetic evidence.

The classificatory units of ecotype, ecospecies, cenospecies, and comparium by definition are circumscribed to an appreciable extent by cytogenetic criteria, and they are not biologically or taxonomically the exact equivalents of such pre-existing taxa as genera, species, subspecies, etc., established by international legislation. A major limitation is imposed on the quantitative productivity of the biosystematist by the time required to obtain the cytogenetic evidence necessary to permit his classification of a plant or a population. This may be offset, to be sure, by the qualitative level of the result, but it is a limitation as measured by the demands placed on taxonomy. This limitation is one reason why the biosystematic approach is of greater importance to the determination of the phyletic classification of the individuals and populations than it is to their identification. In other words, because of this time-factor limitation, both biosystematics and modern taxonomy are indispensable to one another and neither is likely to absorb the other.

The biosystematist depends on the methods of the taxonomist to identify and name the organisms with which he works, and the taxonomist will rely on the biosystematist to help him ascertain the probable biological category in which each organism belongs. Each must decide whether the organism belongs in an approximate equivalent taxonomic category (as the comparium is to the genus) or if, for practical considerations, the best interests will be served by treating it otherwise. For example, the literature indicates that all taxonomists accept as taxonomically valid the tropical orchidaceous genera *Cattleya, Brassavola*, and *Laelia*, yet no genetic barriers exist between some or all of their species (the trigeneric hybrid *Brassocattlaelia* is represented by many hybrids designated by binomials). It is doubtful if any modern taxonomist would combine these 3 genera under a single generic name, although by the standards of the biosystematist no one of the 3 genera appears to approach the genetic distinctness of a comparium.[16] The geneticist, while passing no judgment on their validity as genera, might say that the situation (represented by readily interbreeding genera) suggests that evolutionarily these genera are very closely related. If one were to start with plants representing species of these genera, transplanted from nature, and were to determine their biological status by biosystematic methods, it would require about 15 years to produce a flowering $F_2$ and to arrive at a conclusion as to its phyletic classification. Similar studies with many woody plants would require longer periods; a major limitation to biosystematic investigations.

Cytogenetic data are only as reliable as the identification of the material used in the study. Scores of examples can be cited of chromosome numbers of indigenous material of species, published without the identity of the plants involved having been documented by herbarium specimens.[17] The discerning taxonomist is increasingly reluctant to cite or to rely on chromosome numbers published in atlases or other compilatory lists unless it is known that the identity of the specimens used for the study may be verified.[18] Unfortunately the great majority of cytological findings known in some quarters as unreliable have not been so publicized, and the average taxonomist is unaware of the situation. For this reason it is most important that adequate herbarium specimens be made

---

[16] The reverse situation, already pointed out, exists among some taxa treated by the conservative taxonomist as genera, and by the splitter as several genera, for here the biosystematist may find (as in the Leguminosae, *Astragalus* for example) that some of the so-called species of the taxonomist react cytogenetically as do comparia. This application of cytogenetics would indicate that no one of these apparent comparia could be treated taxonomically as the equivalents of genera.

[17] It is known that many plants in botanic gardens and arboreta are not the species (and sometimes not the genera) that their label names indicate, yet both the taxonomist and the cytogeneticist use this material (or worse, seeds from it that on germination are found to be inadvertently produced hybrids of which it is the seed parent) for their studies without attempting either to verify the identification of the plants or to make herbarium specimens of those used in their researches from which future reidentifications may be made.

[18] One example selected from many serves to illustrate the need of herbarium preparations of material used in cytological studies. In 1924 Longley published cytological studies in *Rubus* (Amer. Journ. Bot. 11: 249–282) that were based on collections then growing at the Arnold Arboretum but that since have been destroyed; species names in the paper were taken from garden labels, no herbarium specimens were reported to have been made of the plants, and it is a matter of record that the identity of those particular *Rubus* collections has been recognized by several competent authorities to have been badly scrambled. Despite this situation, documented in the record, taxonomists continue to cite Longley's paper as valid evidence of cytological conditions alleged to exist in species represented by the reported binomials. If adequate herbarium material had been prepared and was extant, some of this cytological evidence might be salvaged.

of all plants used in biosystematic studies and that they be cited in published results.[19]

In addition to the above limitations, the student should note also that other botanical evidence is to be included in future biosystematic studies if they are to reflect a synthesis of all available evidence. Information valuable to the resolution of generic limits is to be found in the studies of comparative embryology as proposed by Just (1946). Biochemical studies have shown that serological techniques may be of greater significance in resolving relationships in minor taxa (as genera and subspecies) than of many higher taxa, and there is evidence that exploration of the physiological functions responsible for the serological end products may be more significant as an aid to determining natural relationships. Most published biosystematic work has dealt with the more polymorphic taxa (plants whose speciation in many instances has not been resolved satisfactorily by any previous taxonomic methods), and highly valuable contributions have resulted. It is to be expected that the resolution of many problems of generic delimitation among less polymorphic taxa will be approached by the application of biosystematic analysis.[20] There is promise that cytogenetical and biosystematical studies will do much toward providing scientific answers to these and other problems of classification, and certainly the phylogenetic classification of taxa at the level of genus and below cannot be answered without benefit of their findings. Modern taxonomy will not approach the ultimate in progress unless it utilizes to the full the evidence provided by the biosystematist.

*LITERATURE:*

ALLAN, H. H. Wild species-hybrids in the phanerogams. I–II. Bot. Rev. 3: 593–615, 1937; *op. cit.* 15: 77–105, 1949.

ANDERSON, E. Cytology in its relation to taxonomy. Bot. Rev. 3: 335–350, 1937.

———. Supra-specific variation in nature and in classification. Amer. Nat. 71: 223–235, 1937.

ANDERSON, E. and OWNBEY, R. P. The genetic coefficients of specific difference. Ann. Mo. Bot. Gard. 26: 325–348, 1939.

BABCOCK, E. B. Genetic evolutionary processes. Proc. Nat. Acad. Sci. 20: 510–515, 1934.

———. Systematics, cytogenetics, and evolution in *Crepis*. Bot. Rev. 8: 139–190, 1942.

———. The genus *Crepis*. The taxonomy, phylogeny, distribution and evolution of *Crepis*. 197 pp. Berkeley, Calif., 1947.

[19] For discussion and proposed solution of this problem, see Rattenbury (1948).
[20] Examples of the need for biosystematic studies on taxa of this type may be illustrated by the current unsatisfactory taxonomic situation that prevails between *Agave* and *Furcraea, Yucca* and *Clistoyucca* and *Samuela* and *Beschorneria,* and the *Malus-Pyrus-Sorbus* complex.

BABCOCK, E. B. and JENKINS, J. A. Chromosomes and phylogeny in *Crepis*. Univ. Calif. Publ. Bot. 18: 241–292, 1943.

BAKER, H. G. Criteria of hybridity. Nature, 159: 221–223, 1947.

———. Gene-flow between interfertile plant-forms in nature. Biol. Rev. 1950 (in press).

BRUUN, H. G. Cytological studies in *Primula*, with special reference to the relation between karyology and taxonomy of the genus. Symb. Bot. Upsal. 1: 1–239, 1932.

CAMP, W. H. and GILLY, C. L. The structure and origin of species. Brittonia, 4: 323–385, 1943.

CLAUSEN, J., KECK, D. D. and HIESEY, W. M. The concept of species based on experiment. Amer. Journ. Bot. 26: 103–106, 1939.

———. Experimental studies on the nature of species. I, Effect of varied environments on western North American plants. Carnegie Inst. Wash. Publ. 520: 1–452, 1940; II, Plant evolution through amphiploidy and autoploidy, with examples from the Madiinae. *Op. cit.* 564: 1–174, 1945; III, Environmental responses of climatic races of *Achillea. Op. cit.* 581: 1–129, 1948.

CLELAND, R. E. The problem of species in *Oenothera*. Amer. Nat. 78: 5–28, 1944.

COOPER, D. C. and BRINK, R. A. The endosperm as a barrier to interspecific hybridization in flowering plants. Science, 95: 75–76, 1942.

DOBZHANSKY, T. Genetic nature of species differences. Amer. Nat. 71: 404–420, 1937.

———. Genetics and the origin of species. ed. 2. 446 pp. New York, 1941.

EPLING, C. Taxonomy and genonomy. Science, 98: 515–516, 1943.

ERLANSON, E. W. Experimental data for a revision of the North American wild roses. Bot. Gaz. 96: 197–259, 1934.

FAEGRI, K. Some fundamental problems of taxonomy and phylogenetics. Bot. Rev. 3: 400–423, 1937.

GOLDSCHMIDT, R. The material basis of evolution. 436 pp. New Haven, 1940.

GOODSPEED, T. H. Cytotaxonomy of *Nicotiana*. Bot. Rev. 11: 533–592, 1945.

———. On the evolution of the genus *Nicotiana*. Proc. Nat. Acad. 33: 158–171, 1947.

GOODSPEED, T. H. and BRADLEY, M. V. Amphidiploidy. Bot. Rev. 8: 271–316, 1942.

GREGOR, J. W. Experimental delimitation of species. New Phytologist, 30: 204–217, 1931.

———. The ecotype. Biol. Rev. 19: 20–30, 1944.

———. Presidential Address: Some reflections on intraspecific ecological variation and its classification. Trans. Bot. Soc. (Edinb.) 34: 377–391, 1946.

GREGOR, J. W., DAVEY, V. McM., and LANG, J. M. S. Experimental taxonomy. I. New Phytologist, 35: 323–350, 1936.

GUSTAFSSON, Å. Apomixis in higher plants. I–III Lunds Univ. Arsskr. N. F. 42, no. 3; 43, nos. 3 and 12, 370 pp., 1946–47.

HEILBORN, O. Chromosome numbers and taxonomy. Proc. Int. Cong. Plant Sci. 1926 (Ithaca, N. Y.) 1: 307–310, 1929.

HUXLEY, J. S. Clines: an auxiliary taxonomic principle. Nature, 142: 219–220, 1938.

——— [Ed.]. The new systematics, 583 pp. London, 1940.

———. Evolution the modern synthesis. 645 pp. New York and London, 1943.

JENKIN, T. J. Interspecific and intergeneric hybrids in herbage grasses. Initial crosses. Journ. Genetics, 28: 205–264, 1933.

JUST, T. The use of embryological formulas in plant taxonomy. Bull. Torrey Bot. Club, 73: 351–355, 1946.

KOSTOFF, D. Cytogenetics of the genus *Nicotiana*. Karyosystematics, genetics, cytology, cytogenetics, and phylesis of tobaccos. 1070 pp., Sofia, Bulgaria, 1943.

KREYER, G. K. *Valeriana officinalis* L. in Europa und im Kaukasus. Bull. Appl. Bot. Leningrad, 23: 3–260, 1930.

LAIBACH, F. Ectogenesis in plants. Methods and genetic possibilities of propagating embryos otherwise dying in the seed. Journ. Heredity, 20: 200–208, 1929.

LOTSY, J. P. and GODDIJN, W. A. Voyages and exploration to judge of the bearing of hybridization upon evolution. Genetica, 10: 1–315, 1928.

MAILLEFER, A. Etude du *Valeriana officinalis* L. et des especes affines. Mem. Soc. Vaudoise Sci. Nat. 8: 277–340, 1946.

MANTON, I. Introduction to the general cytology of the Cruciferae. Ann. Bot. 46: 509–556, 1932.

MARSDEN-JONES, E. M. The genetics of *Geum intermedium* Willd. haud Ehrh., and its back-crosses. Journ. Genetics, 23: 377–395, 1930.

MAYR, E. Systematics and the origin of species. 334 pp. New York, 1942.

RATTENBURY, J. A. Chromosome number publication. Madroño, 9: 257–258, 1948.

ROBERTY, G. Gossypiorum revisiones tentamen. Candollea, 13: 9–165, 1950.

RUTTLE, M. L. Cytological and embryological studies on the genus *Mentha*. Gartenbauwissenschaft, 4: 428–468, 1931.

SENN, H. A Chromosome number relationships in the Leguminosae. Bibliographia Genetica, 12: 175–337, 1938.

SIMPSON, G. G. Tempo and mode in evolution. 237 pp. New York, 1944.

———. Tempo and mode in evolution. Trans. N. Y. Acad. Sci. ser. II, 8: 45–60, 1945.

SMITH, W. W. Some aspects of the bearing of cytology on taxonomy. Proc. Linn. Soc. (Lond.) 145: 151–181, 1933.

———. Problems in classification of plants. Journ. Roy. Hort. Soc. 61: 77–90, 117–134, 1936.

STEBBINS, G. L. JR. The significance of polyploidy in plant evolution. Amer. Nat. 74: 54–66, 1940.

———. Apomixis in the angiosperms. Bot. Rev. 7: 507–542, 1941.

———. Polyploid complexes in relation to ecology and the history of floras. In Ecological aspects of evolution. Amer. Nat. 76: 36–45, 1942.

———. The concept of genetic homogeneity as an explanation for the existence and behavior of rare and endemic species. Chron. Bot. 7: 252–253, 1942.

———. Types of polyploids: their classification and significance. In Advances in genetics, pp. 403–429. New York, 1947.

———. Asexual reproduction in relation to plant evolution. Evolution, 3: 98–101, 1949. [A review of Gustafsson's 'Apomixis in higher plants.']

———. Variation and evolution in plants. 643 pp. New York, 1950.

TUKEY, H. B. Artificial culture methods for isolated embryos of deciduous fruits. Proc. Amer. Soc. Hort. Sci. 32: 313–322, 1935.

TURESSON, G. The genotypical response of the plant species to the habitat. Hereditas, 3: 211–350, 1922.

———. The plant species in relation to habitat and climate. Hereditas, 6: 141–236, 1925.

———. Studien über *Festuca ovina* L. I. Hereditas, 8: 161–206, 1926, II, *op. cit.* 13: 177–184, 1930. III, *op. cit.* 15: 13–16, 1931.

———. The selective effects of climate upon the plant species. Hereditas, 14: 99–152, 1930.

TURRILL, W. B. Experimental attacks on species problems. Chron. Bot. 7: 281–283, 1942.

———. The ecotype concept; a consideration with appreciation and criticism, especially of recent trends. New Phytologist, 45: 34–43, 1946.

WARBURG, E. F. Taxonomy and relationship in the Geraniales in the light of their cytology. New Phytologist, 37: 130–159, 189–210, 1938.

WINGE, Ø. The genetic aspect of the species problem. Proc. Linn. Soc. (Lond.), Session 150 (4): 231–238, 1938.

## CHAPTER IX

## PLANT NOMENCLATURE

Nomenclature [1] is allied to taxonomy in that it deals with the determination of the correct name to be applied to a known taxon or to a known plant. Once a plant has been identified, the correct name must be given it. This is a function of nomenclature. As illustrated in the Introduction, identification is one thing, nomenclature is another.

Plants are objects and some of them have always been useful to man. For this reason, prehistoric man must have talked about plants, probably differentiated the useful from the useless, and in so doing he must have given them names. There is no record of the beginnings of plant nomenclature, for they are as obscure as the records of earliest civilizations. The first books dealing with plants are taken direct from early Greek or Latin manuscripts, or are ancient translations into those classical languages. The plant names in them likewise are generally in Greek or Latin, but some are unaltered barbarous names used by the then uncivilized tribes.

The scientific names of plants are now based on the Latin language, known universally by scholars and students throughout the world. Descriptions of plants are not written in the classical Latin of Cicero or of Horace, but in the "lingua franca" spoken and written by scholars during the Middle Ages. This Latin was a development of the popular or "vulgar" Latin spoken by ordinary people in classical times, and was written by all but a handful of poets and literary men (cf. Bodmer, pp. 309–348). Latin is specific and exact in its meaning. By its preciseness and conciseness it is particularly pertinent to the needs of descriptive phases of the natural sciences. The Latin language employs the Roman

[1] Webster's Unabridged Dictionary, ed. 2 (1948), gives the preferred pronunciation of nomenclature to be nō-měn-clā-tūre, with the accent on the first and third syllables. In discussing matters of nomenclature, distinction should be made between the words nomenclatural and nomenclatorial. When the subject is nomenclature, the adjective *nomenclatural*, or the adverb *nomenclaturally*, is used. When it is a book or work dealing with names and their synonyms (i.e., to a *nomenclator*, as Steudel's *Nomenclator*, or the original *Index Kewensis*) it may be referred to as a *nomenclatorial* work.

alphabet, and for this reason the use of Latin words in plant names obviates the confusion that would result if these names were written indiscriminately in the characters of other alphabets as, for example, Greek, Chinese, Hebrew, or Sanscrit.

Objections have been raised against employment of Latin names for plants. Efforts have been made to develop an English nomenclature based on the use of English rather than Latin names. The current edition of *Standardized plant names* (1940) is an example of an attempt in this direction; in it every plant was given a so-called "common name" which, in most cases, represents not a commonly known name but an anglicization of the Latin name. It has been proposed (Kelsey, 1945) that these "common names" be employed for all horticultural and economic plants, that they be established and made permanent, and that governmental legislation be enacted to provide an agency to enforce their employment. There are many basic problems that confront the advocate of a system comprised only of English names, or of names based on any spoken nationalistic tongue. In the first place, it is difficult to ascertain which of several English names for a given plant may have been employed first, and secondly, these names by themselves are of little use in identification, inasmuch as there are no treatments of the world flora based on them.

Many English names have often been applied to the same plant. For example, the poisonous shrub *Rhus Vernix* is known variously as poison sumac, poison dogwood, poison elder, swamp sumac, etc. There are innumerable instances where the same English name has been applied to wholly different plants: in some regions a cowslip may be the marsh marigold (*Caltha palustris*), while in others it is a primrose (*Primula veris*); the purple loosestrife is *Lythrum Salicaria* (a member of the Lythraceae), but the yellow loosestrife is *Lysimachia vulgaris* (a member of the Primulaceae); likewise, when is a pine a pine, for in the vernacular a pine may be a species of *Pinus, Araucaria, Agathis, Callitris,* or *Casuarina*. English names may indicate no close affinities at all, for while the oaks may be associated with the genus *Quercus*, the tanbark oak is a *Lithocarpus*, the poison oak a *Rhus*, the silk oak a *Grevillea*, and Jerusalem oak a *Chenopodium*.

It is not possible to learn the scientific name of most plants if only their vernacular names are known, for all botanical works by which plants are classified are based universally on Latin names. Another barrier to the exclusive use of common, or anglicized, names is that their employment is restricted essentially to the people of one language. An acceptable system of nomenclature must be international in character. A rose has a different vernacular name in the tongue of every people by whom it is known. Multiply the hundreds of vernacular names of cultivated or native roses by the number of languages to which they are common, and a concept is gained of the extent of nomenclatural chaos that would result should such a system be applied, to the exclusion of Latin names, to all plants.

The use of scientific names, rather than of common or vernacular names, has much to commend it. Scientific names do indicate generic and usually genetic relationships; hence they are of biological significance. They are international in scope and are common to all tongues, and they are relatively unambiguous. It has been said that scientific names are difficult to pronounce and to remember, that because they are polysyllabic they are confusing, and that to the non-Latinist they are without meaning. The puerility of such contentions becomes apparent when one listens to youth conversing freely about superheterodynes, polyclinal cleavages, and television, or the modern mother and housewife discussing riboflavin, schizophrenia, conjunctivitis, or even paradichlorobenzene.

## Beginnings of organized nomenclature

Before the middle of the eighteenth century the names of plants commonly were polynomials. That is, they are composed of several words in a series, constituting a more or less terse description of a plant. It was a cumbersome system and, with additions of hundreds of plants new to science, was superseded by the binomial system, which was used to a degree by Rivinus and later was employed by Linnaeus (in his *Species plantarum,* 1753). The binomial system postulates that the name of every species of plant consists only of two words (for example, that of the white oak is *Quercus alba*): in this binomial, the first word (*Quercus*) designates the genus to which the plant belongs, and the second (*alba*) a particular species of that genus. This second word of a plant's name is a specific epithet; the two words in combination comprise the species name and form a binomial, and the binomial is a binary epithet.

The advent of increasing numbers of new plants from areas of recent exploration caused concern over methods of naming these novelties, and elemental rules were proposed to serve as guides to plantsmen. The rules of nomenclature as we know them today had their beginning in Linnaeus' *Critica botanica* (1737), a work comprising an elaboration of the Aphorisms of his *Fundamenta botanica* (1736). The former was written primarily to explain the taxonomic and nomenclatural principles employed by Linnaeus in his *Genera plantarum* and *Hortus Cliffortianus*. The Aphorisms constituted his Principles of Nomenclature and many of them are paralleled closely by contemporary rules of nomenclature.[2] Later, in his *Philosophia botanica* (1751), Linnaeus amplified the views expressed in his Aphorisms and Principles and established the real beginnings of a sound nomenclature for plants.

In the days of Linnaeus, by common accord, no two genera could have the same generic name, and no two species within a given genus could have the same species name. It was accepted also that when a genus was divided into two or more genera, the original generic name must be retained for one of them; that when a variety was recognized, it must be associated nomenclaturally with that species of which it was a variant and not be given a binomial name apart from the parent species. However, except for Linnaeus' Principles of Nomenclature, there were no generally accepted rules governing the naming of plants. In a general way, priority of publication was given recognition; most authors after the time of Linnaeus endeavored to avoid assigning different names to identical species or identical names to different species, and when such situations did arise there was a tendency to employ the earlier name in subsequent works. However, the number of species and genera new to science increased so rapidly during the early part of the nineteenth century that confusion in nomenclature did exist. For many years there was no index from which botanists of one country or district might learn of names that had been employed elsewhere by other botanists.

The appearance in 1813 of Augustin de Candolle's *Théorie élémentaire de la botanique* provided the first significant work since the publications of Linnaeus, on explaining matters of organography and nomenclature. On pp. 221–257 of the first volume, he gave explicit instructions on nomenclatural procedures, many of them taken from Linnaeus, and pointed out the fallacies of proposals and practices of other botanists of that generation. Here, for what probably was the first time, was given a complete and detailed set of rules on plant nomenclature; rules that definitely were the forerunner of the *Lois de la nomenclature botanique* proposed a half-century later by his son, Alphonse de Candolle.

The first important post-Linnaean index to the names of flowering plants was Steudel's *Nomenclator botanicus* (1821). This work comprised a list of the Latin names of all flowering plants then known, to-

[2] See Sir Arthur Hort's English translation of the *Critica botanica*, 1938.

gether with their synonyms. A second edition appeared in 1840. The Nomenclator was used universally by botanists of Europe and America, and was the forerunner, by more than half a century, of the now current *Index Kewensis.*

In many countries the nomenclatural patterns and practices were set by botanists of considerable prestige, influence, and veneration. Names employed by them gained ascendancy even though their application may have been influenced unduly by national or personal jealousies. As time passed, the need for unification of procedure became increasingly apparent, and the Swiss botanist Alphonse de Candolle headed the movement for an assembly of botanists to study the situation and to initiate corrective measures.

### Codes of nomenclature. Paris Code, 1867

The first organized efforts toward standardization and legislation of nomenclatural practices were at the First International Botanical Congress, meeting in Paris in August, 1867. About 150 European and American botanists were invited to attend this Congress. In advance of their arrival de Candolle sent to each for preliminary study a copy of his *Lois de la nomenclature botanique* (*Laws of botanical nomenclature*). At the Congress these laws were scheduled for discussion, and after a few days of relatively uncritical study and slight revision it was resolved that, "these Laws, as adopted by this Assembly, shall be recommended as the best Guide for Nomenclature in the Vegetable Kingdom." By these "de Candolle Rules" (as the rules comprising the Paris Code of 1867 are often known) the starting point for all plant nomenclature was with Linnaeus (no date or work of Linnaeus was specified for this starting point); the rule of priority was treated as fundamental, and no provision was made for exceptions to it; considerable attention was given to the matter of author citation and to the terms to be applied to the categories of plants; and the requirements for valid publication and for acceptance and rejection of names were established. The rules of the Paris Code represented an excellent beginning in the right direction, but their application revealed numerous inherent deficiencies. The need for modification and for revision became increasingly evident, with the result that scarcely a decade had elapsed before various schools of thought were putting into practice individual interpretations and "rules" that were not a part of the Paris Code. The international character of the code was soon dissipated.

One of the early major divergences from the Paris Code was the to the Vienna Rules. For this reason, the Vienna Rules did not gain world-wide acceptance.

The Fourth International Congress met in 1910 in Brussels. Its only significant action, in regard to plant nomenclature, was the establishment of different starting points for priority of names of nonvascular plants, the recognition of the value of the "type concept" as proposed by the Rochester Code (but not its incorporation into the Rules), and clarification of phraseology in several of the Vienna Rules.

### American Code, 1907

The proposals of the Rochester Code formed the basis of the American Code, and the two are not significantly different except that the latter was modified in some respects to agree with the Vienna Code, although its proponents refused to accept either the principle of *nomina generica conservanda* or of requiring Latin diagnoses to accompany names of new taxa. On the other hand, the American Code was predicated on the acceptance of the type concept, and provided also that a binomial may not be used again for a plant in any way if it has been employed previously for another plant, even though the previous use may have been illegitimate. This revision of the Rochester Code was published in 1907 under the name American Code (cf. Amer. Nomen. Comm., 1907). Its sponsorship and circle of adherents remained essentially unchanged, and was mainly that which supported the Rochester Code movement.

The committee on nomenclature of the Botanical Society of America, acting under pressure from supporters of the American Code, formulated a new code in 1918 and, in an effort to enlist the support of more American botanists, called it the Type-Basis Code (cf. Hitchcock, 1919). This code differed from the American Code only in minor respects, making the doubtfully significant concession to the effect that, at a later date, a list of *nomina generica conservanda* might be added if, in the opinions of its advocates, such a list were to become desirable. No such list ever was prepared, and the code never was adopted by the Society.

### International Rules of Botanical Nomenclature, 1930

It was not until the close of the Cambridge (England) Congress [4] of 1930 that accord and harmony existed among major botanical factions. Initial steps toward this goal were taken in Europe in the early 1920's

---

[4] Sometimes erroneously referred to as the "London Congress." The latter is properly known as the Imperial Botanical Conference and convened in London in 1924. (Cf. Rendle 1924, Sprague 1929).

(cf. Sprague, 1924) and further progress was effected at the Ithaca (N. Y.) Congress of 1926. At the latter congress no fundamental nomenclatural matters were brought to a vote, but these were subjects of discussion, and were assigned to a permanent committee for continued study.

At the Cambridge Congress, every effort was made to harmonize the basic differences between the Vienna Rules and the American Code. In preparation for this, credit must be accorded to T. A. Sprague and M. L. Green for their success in convincing leaders at botanical centers on the Continent and in Britain, and to A. S. Hitchcock, leader of the American Code bloc, for his reconciliation of colleagues and associates to the necessity for compromise (cf. Sprague, 1920). As a result the new rules legislated at Cambridge were a product of conciliation and accord by parties of both factions. For the first time in botanical history, a code of nomenclature came into being that was international in function as well as in name.

## International Rules of Botanical Nomenclature

The third edition of the International Rules of Botanical Nomenclature resulted from the decisions reached at the Cambridge Congress. At the Sixth International Botanical Congress, Amsterdam (1935), the Rules were again a subject of discussion, and only one major change was legislated. It provided for the advance from January 1, 1932, to January 1, 1935, as the date after which all diagnoses of plants new to science (excepting the Bacteria) must be in Latin. Other changes consisted primarily of clarification of phraseology of rules and recommendations. At the Amsterdam Congress, there was an organized movement for the establishment of a selected list of *nomina specifica conservanda*. This principle was rejected by the Congress by a vote of 208 to 61.

The third edition of the Rules contained an enumeration of the *nomina generica conservanda* as of 1935. In the supplement to that edition (Sprague, 1948) was given an enumeration of the *nomina generica conservanda proposita* (generic names proposed for conservation), and additional names were proposed for conservation subsequently. By authority granted by the sixth Congress (cf. Proc. Int. Bot. Congr. 1: 359, 1936) the Special Committee for Phanerogamae and Pteridophyta published an enumeration of additional *nomina generica conservanda (Pteridophyta* and *Phanerogamae)*. This included the ferns and seed plants proposed for conservation in the third edition of the Rules, plus about 300 additional genera proposed subsequent to 1935 (cf. Sprague, 1940). A full account of the action taken at the 1935 Congress is available in the Proceedings (1936) and has been summarized briefly by Sprague (Science, 1936, Kew Bull., 1936).[5]

The Rules of Nomenclature are divided into chapters and sections, and these are subdivided into articles and recommendations. Botanists who follow the Rules accept the articles as mandatory and binding, they are "the law"; the recommendations are provided for guidance, that more uniform practices may result. At the end of the Rules, several appendices are provided; each is devoted to a special topic and is a source of instruction, guidance, and supplementary information. Here will be found such subjects as typification, classified lists of conserved names, and directives pertaining to the nomenclature of cultivated plants.

At the time of the present writing, the third edition of the Rules, as modified at Amsterdam, is current but is to be replaced by a new and fourth edition based on legislation enacted at the Seventh International Botanical Congress (Stockholm).

The Seventh International Botanical Congress was held in Stockholm in 1950, and the Rules of Nomenclature were the subject of lengthy and considered discussions (cf. Lanjouw, 1950). Many minor changes and refinements were legislated and will be incorporated in the next edition. Notable among the results effected at the nomenclature sessions of this 1950 Congress are the following modifications or changes to the third edition of the Rules:

The introduction into the Rules of the term *taxon* (pl. *taxa*) as the designator of any taxonomic entity or group (cf. p. 53 for explanation of the term).

The elevation of a former recommendation to the mandatory level of article, to read: "Names of orders are taken from that of one of their principal families, with the ending *-ales* . . For names of taxa above the rank of family, the rule of priority is not compulsory."

Complete revision of the section on the names of hybrids and other special categories. It was provided that names of *nothomorphs* (different hybrid forms of the same parentage, i.e., of pleomorphic hybrids) "may be designated by an epithet preceded by a binary name of the taxon and the term nothomorph (abbreviated as *nm.*) in the same way as subdivisions of species are classed under the binary name of the species."

Addition of a new article in this section for the nomenclatural treatment of apomicts. It reads: "Taxa which are apomicts may, if so desired, be treated in the following manner: (1) if they are considered of specific rank, by the interpolation of the abbreviation *ap.* between the generic name and the epithet; (2) if they are considered to be of infraspecific rank, by the interpolation of the abbreviation *ap.* between the categories."

A complete revision of Art. 18 (ed. 3, of the Rules), dealing with the subject of selecting and designating types, resulted in the establishment of 5 articles and 2 recommendations in an effort to more adequately deal with this difficult subject.

[5] An unofficial edition of the rules of nomenclature has been published (see Camp, et al., 1948) in which were brought together all approved *nomina conservanda*.

Revision of the article dealing with the names of cultivated plants (Art. 35, ed. 3). Henceforth it is to be associated with a new Appendix VII.

The clarification and amplification of conditions of valid publication of names involved many minor changes.

The provision of the Rules (Art. 63, ed. 3) allowing the designation of certain names as *nomina dubia* was rejected.

A new article which provides in effect that a taxon of any rank below that of species, which includes the type of the species, is to be designated by a repetition of the specific epithet and without the citation of an author's name. This retroactive Rule requires rejection of such subdivisional epithets as *typicus, genuinus,* or *originarius.*

An amendment to Art. 70 of the Rules (ed. 3) by which it is now directed that "the use of the terminations *i* or *ae,* instead of *ii* or *iae* . . . is treated as an orthographic error" and is to be corrected.

The Congress rejected the principle of *nomina specifica conservanda* by a ballot of 320 to 40 votes (cf. p. 200 for similar action taken on this subject in 1935).

A rephrasing of a former recommendation which dealt with the capitalization of specific names. (Rec. XLII, of ed. 3) and now reads: "All specific and trivial names or epithets should be written with a small initial letter, although writers desiring to use capital initial letters for particular names or epithets may do so . . ."

A new appendix was prepared and approved for the determination of types.

These and other changes to the Rules made at this Congress became effective immediately, and botanists knowing of them should not await publication of a new edition of the Rules before following them.

## Primary and other provisions of the Rules

The advent of a fourth edition of the Rules, in which the numbers of the articles and the recommendations of the current (third) edition will be displaced by new numbers, makes it impractical to discuss the individual directives by number. For this reason the context of the Rules is presented below by chapters, sections, and subsections. All discussions below are based on the Rules as they stand as a result of legislation enacted at the Stockholm Congress. This digest of the Rules is in no sense a substitute for them, and the student must consult the official editions of the Rules for all nomenclatural directives. The carefully selected examples that are provided as illustrations of the application of each article are especially helpful and contribute materially to a clearer understanding of each article. Quotations of articles in the following text are based on the forthcoming edition of the Rules, but have no official sanction or standing.

CHAPTER I. *General Considerations and Guiding Principles*

These few articles provide in essence a preamble to the Rules, and (1) establish the need for Rules; (2) declare the independence of botanical from zoological nomenclature; (3) provide for the basis of scientific names on the Latin language; (4) delimit nomenclature to dealing with terms of rank and name; and (5) specify that the Rules apply to all forms of plant life.

CHAPTER II. *Categories of Taxa, and the Terms Denoting Them*

This is composed of 4 articles by which are specified the names of the categories, from species through division, and their order of sequence.

The units of classification (i.e., the categories of taxonomic groups, such as orders, families, genera, and species) constitute a separate section treated on page 217 of this text, where their nomenclatural status is explained. For a discussion of concepts, circumscription, and taxonomic status of these categories see pp. 44–56.

CHAPTER III. *Names of Taxa*

This chapter contains the major portion of the Rules, and is divided into 15 sections.

SECTION 1. *The principle of priority*

This section provides that: The purpose of giving a name to a taxon is not to indicate the characters or the history of the taxon, but to supply a means of referring to it.

Each taxon with a given circumscription, taxonomic position, and rank can bear only one valid name, the earliest that is in accordance with the Rules of Nomenclature.

SECTION 2. *The type method*

The five articles of this section provide that names of plants are based on nomenclatural types. The nomenclatural type of a species (or unit of lower rank) of a vascular plant is usually an herbarium specimen. Some material does not lend itself well to preservation of characters in dried form (such as succulents), in which case the herbarium sheet of parts that can be so preserved (as flowers) may be augmented by critical and diagnostic drawings and photographs. Some species, in the absence of herbarium material, may be typified by a description or a figure prepared by one not necessarily the original author. Great care is given to type specimens. They are the record for the future as well as the present. Should there be disagreement between an author's diagnosis of a plant and the characters of that plant as shown by the type specimen, those of the specimen are considered to be correct. So great is the value attached to type specimens that in many institutions they are preserved in special envelopes, folders, cases, or vaults. Many institutions have a policy not to loan types.

This Rule provides that, whereas the type of a species or taxon of lower rank is an individual specimen, the type of a genus, subgenus, or section is a designated species, that of a family, subfamily, or tribe, is a genus and that of an order or suborder is a family.

In a taxonomic work concerned with old and usually large genera (as in some Linnaean genera) it occasionally happens that the taxonomist divides the genus into two or more genera. The question arises as to which of these genera should retain the original name. Until recently there was an option of two courses that might be followed. One was the so-called "residue method," whereby the generic name was applied to the taxon of the original genus remaining after the segregates separated from it had been named. The other alternative was the "type method," whereby the generic name was applied to include the so-called "type species," that is, the species the original author had in mind when establishing the genus. In working with contemporarily established genera there is usually little question as to which is the type species of a generic name since it is usually indicated when the genus is established. However, the type species of long-established genera often is not known and the genus as conceived and established by many early botanists was a composite concept, based on not one but several species. To early botanists, the modern concept of a type species was unknown. In an effort to incorporate the good points of both of these methods, and to eliminate their shortcomings, the "standard method" was devised (Sprague, 1926); it represents the modern type concept. This method proposes to fix permanently the application of generic names by the acceptance of "standard species," but leaves the selection of the standard species for each genus to be decided on the merits of the individual case, that serious changes in nomenclature may be avoided. A list of standard species (*species lectotypicae propositae*) was prepared for the Linnaean genera (Appendix II of the Rules) and presented for adoption at the Amsterdam Congress, at which time a resolution recommending their adoption by botanists was passed (Sprague, 1936, p. 187). A new revision of an appendix comprising a "guide for the determination of types," has been approved for inclusion in the fourth edition of the Rules.

Any discussion of types and typification raises questions concerning definitions of terms applied to types. As a group, botanists have not been precise or consistent in their choice of terms, and a degree of confusion has resulted. Much has been written on the subject (cf. Frizzell, 1933, for bibliography), but the more recent and most lucid paper on the subject is that by Blake (1943).

The subject was acted on at the Stockholm Congress (1950) and the following kinds of types were officially designated:

A *holotype* is the one specimen or other element used by the author of the name, or designated by him, as the nomenclatural type (i.e., the element to which the name of the taxon is permanently attached).

A *lectotype* is a specimen or other element selected from the original material to serve as the nomenclatural type, when the holotype was not designated at the time of publication, or when the holotype is missing.

A *neotype* is a specimen selected to serve as the nomenclatural type of a taxon in a situation when all material on which the taxon was based is missing.

A *paratype* is a specimen cited with the original description, other than the holotype.

An *isotype* is a specimen believed to be a duplicate of the holotype.

A *syntype* is one of two or more specimens or elements used by an author when no holotype was designated, or in lieu of a holotype, or when one of two or more specimens were designated simultaneously as the type.

In addition to the above, it is accepted generally that a *cotype* is a second specimen from the same plant from which the holotype was collected. The term *type,* used alone and unqualified, generally refers to the holotype. A *topotype* is a specimen collected at the same station as was the type (i.e., from the type locality). There are other terms ending with the suffix *-type,* but for the most part they are concerned with ecological or genetical situations. For explanation of the nomenclatural designations to be given to the typical element of a species, or taxon in a lower category, see page 202.

SECTION 3. *Limitation of the principle of priority; publication, starting points, conservation of names*

Four articles comprise this section, and they provide that names have no claim to recognition by botanists unless validly published (see also Sect. 6).

Dates for starting points in the nomenclature of various taxa as follows:

1753. Algae, Lichens, Hepaticae (with Linnaeus' *Species plantarum,* ed. 1). Vascular plants (Pteridophyta and Phanerogamae) as of May 1st, with exceptions as provided by *nomina generica conservanda* and explained below.

1801. Some Fungi, all Muscineae (mosses).

1821–1832. Other Fungi.

*Nomina generica conservanda:* certain names, not the oldest, applied to some genera, are conserved over older names because these later names have been in common use for long periods. The rejected name is designated a *nomen rejiciendum.*

The Rule, as legislated at the Stockholm Congress, amplifies this article to apply to "the nomenclature of genera and categories of higher rank." This authorizes the conservation of names in the categories of genus, family, and order. At the same time, the principle favoring the conservation of species names was defeated by a ratio of 8 votes to 1.

An enumeration of conserved generic names was included in the unofficial edition of the Rules (edited by Camp, *et al.,* 1948). For the statistically minded, there are now 237 conserved generic names for nonvascular plants and 787 conserved generic names for vascular plants. These figures will be increased by nearly a fifth, if proposals now awaiting disposal are given favorable action.

SECTION 4. *Nomenclature of the taxa according to their categories*

SUBSECT. 1. *Names of taxa above rank of family.* Names of orders are taken from that of one of their principal families, with the ending *-ales.* Suborders are designated in a similar manner, with the ending *-ineae.* For names of taxa above the rank of family, the rule of priority is not compulsory.

At the Stockholm Congress it was provided (as a new recommendation) that the endings of divisions should be *-phyta,* of subdivisions or subphyla *-phytina,* and of classes and subclasses (of vascular plants) *-opsida* and *-idae,* respectively.

SUBSECT. 2. *Names of families and subfamilies, tribes, and subtribes.* Names of families or subfamilies are taken from the name of an existing or former genus of the taxon, and end in *-aceae* and *-oideae,* respectively.

The name Rosaceae is taken from its included generic name *Rosa,* Caryophyllaceae from a pre-Linnaean (and now nonexistent) generic name *Caryophyllus.* The names of 8 families are excepted, but each will have an accepted alternative name (to be conserved, for use by anyone who elects the alternative name) ending in *-aceae*; either name may be used. These families, together with their suggested alternative names, are: Palmae (Arecaceae), Gramineae (Poaceae), Cruciferae (Brassicaceae), Leguminosae (Fabaceae), Guttiferae (Clusiaceae), Umbelliferae (Ammiaceae), Labiatae (Lamiaceae), Compositae (Asteraceae). For further nomenclatural notes on family names, cf. Sprague, 1921.

The names of tribes end in *-eae* and of subtribes in *-inae.*

SUBSECT. 3. *Names of genera and subdivisions of genera.* The name of a genus, a subgenus, or a section may be taken from any source and may be composed in any arbitrary manner. It is a substantive and is written with an initial capital.

SUBSECT. 4. *Names of species (binary names).* Names of species are binary combinations consisting of the name of the genus followed by a single specific epithet. If the epithet consists of two words, these must either be united or joined by a hyphen. Symbols forming part of specific epithets proposed by Linnaeus must be transcribed.

SUBSECT. 5. *Names of taxa below the rank of species (ternary names).* Epithets of subspecies and varieties are formed like those of species and follow them in order, beginning with those of the highest rank.

Two subdivisions of the same species, even if they are of different rank, cannot bear the same subdivisional epithet, unless they are based on the same type.

SUBSECT. 6. *Names of hybrids and some special categories.* This comprises a group of articles that deal with names of hybrids, nothomorphs (pleomorphic hybrids), apomicts, and clones. (Cf. also pp. 185, 202.)

SUBSECT. 7. *Names of plants of horticultural origin.* Plants brought into cultivation from the wild and which differ in no fundamental way from the parent stocks bear the same names as are applied to the same species and subdivisions of species in nature.

Plants arising in cultivation through hybridization, mutation, or other processes which tend to establish recognizable differences from the parent stocks, receive epithets preferably in a common language ("fancy" epithets) markedly different from the Latin epithets of species or varieties. Detailed regulations for the nomenclature of plants in cultivation will appear in a separate appendix to the Rules (ed. 4).

SECTION 5. *Conditions and dates of effective publication.*

A group of 3 articles dealing with the publication of plant names, by which it is provided that:

a. Publication is effected . . . by distribution, by sale, by exchange . . . of printed matter; the deposit of manuscripts in libraries . . . of microfilm or similar reproduction . . . does not constitute publication.

b. Publication formerly could be effected by distribution of an exsiccata (dried specimen) accompanied by the original diagnosis of it. A new article, added to the Rules at Stockholm, legislates that as of January 1, 1952, onwards, this practice no longer constitutes effective publication.

SECTION 6. *Conditions and dates of valid publication of names*

A validly published name is one which has been (1) effectively published, and (2) accompanied by a description of the taxon or by a direct or indirect reference to a previously and effectively published description of it. In essence, while *effective publication* of a name deals with the mechanics of its distribution, *valid publication* deals with both distribution of the name and with preparation of the textual matter prior to distribution.

The 2 examples that follow are of effectively published names, and illustrate valid publication:

a. *Phalaris arundinacea* L., Sp. Pl. 55, 1753. As indicated by the citation following the italicized name, the binomial was published by Linnaeus in his *Species plantarum,* page 55, in 1753. The name was accompanied by a diagnosis. Publication was effective and valid.

b. *Digitaria sanguinalis* (L.) Scop., Fl. Carn., ed. 2, 1:52, 1772.
*Panicum sanguinale* L., Sp. Pl., 57, 1753.

The name *Digitaria sanguinalis* was made by Scopoli and (as stated in the abbreviated citation) was effectively and validly published by him in his second edition of the *Flora carniolica* (on page 52 of volume 1, in 1772). Scopoli did not accompany his name with a critical diagnosis of the species, but he did make a reference to a previously and effectively published description of the plant as given by Linnaeus (under the generic name of *Panicum*) in the latter's *Species plantarum,* page 57, 1753. The listing of the name of the plant on which the new name is based (a basonym), together with the citation of the name, its author, and reference to the source of original publication, is sufficient to meet the requirements of the Rule. An adequate diagnosis must accompany a new name when no validly published synonym exists, and the name then is that of a taxon new to science.

There are many technical aspects to this topic. Notable among them are articles that provide that:

a. From January 1, 1935, names of new taxa of recent plants, the Bacteria excepted, are considered as validly published only when they are accompanied by a Latin diagnosis.

Descriptions of new taxa published prior to January 1, 1935, are to be treated as valid even though published in any modern language, including Japanese, Russian, or other languages in which non-Roman alphabets were employed. This Rule requires that all taxa (including the Algae), whether below the rank of species or of higher category, must have their diagnoses written in Latin if the date of publication is January 1, 1935, or later.

b. A name of a taxon is not validly published when it is cited merely as a synonym.

c. The name of a species or of a subdivision of a species is not validly published unless it is accompanied by (1) a description of the taxon; or (2) the citation of a previously and effectively published description of the taxon under another name; or (3) a plate or figure with analyses showing essential characters; but this applies only to plates or figures published before January 1, 1908.

Examples illustrative of the first two of these conditions were given above. The second condition may require further explanation. For example, a plant may have been described originally as a species of a different genus, and if it is being transferred subsequently to a new generic name, the transfer will result in its having a different binomial. In many instances the plant may have been treated originally as a variety of a species and given an adequate description in the category of *varietas*. Later, should the plant be raised from the category of *varietas* to that of species, it is given a binomial. The description accompanying the name when it was first established as a variety may have been such that it is not necessary to redescribe the plant, and reference to that earlier description is then sufficient to validate the publication of the binomial.

d. The date of a name or of an epithet is that of its valid publication. For purposes of priority, however, only legitimate names and epithets published in legitimate combinations are taken into consideration.

Distinction is made between a "name" and an "epithet." A name is the scientific appellation applied to a member of any taxon; as *Pinus* is the scientific name of a particular genus, *Pinus nigra* the name of a species of that genus, and *Pinus nigra* var. *caramanica* the name of a variety of that species. An epithet is the ultimate designation of a member of a group; as *nigra* is a specific epithet, and *caramanica* is a varietal epithet.

A legitimate name or epithet is one that is in strict accordance with the Rules. It is important to remember that unless a name has been published validly, it is an illegitimate name, and that illegitimate names are not to be given consideration in matters of priority.

e. New names published on or after January 1, 1952, must be accompanied by a clear indication of the category to which the plant belongs; otherwise the name is to be treated as invalid.

SECTION 7. *Citation of authors' names and of literature for purposes of precision*

For the indication of the name (unitary, binary, or ternary) of a group to be accurate and complete, and in order that the date may be readily verified, it is necessary to cite the author who first published the name in question.

The lily family was named by Adanson. The scientific name of this family, together with the citation of its author's name, is: Liliaceae Adanson.

The lily genus was named by Linnaeus and its scientific name together with its author citation is *Lilium* Linnaeus. The American Turk's-cap lily was named by Linnaeus *Lilium superbum,* and because he was the author of the binomial, his name follows the binary name (viz. *Lilium superbum* Linnaeus, and commonly written *Lilium superbum* L.).

The name of the author is not a part of the name of the plant. To differentiate them, it is customary to indicate the scientific name of the plant in print by means of an italic type face. The author's name is provided for purposes of precision and to identify a particular plant name with the description of that plant as published by that author. By means of author citations, it is recognizable at a glance that *Bergenia* Moench is a genus based on a different plant from that of *Bergenia* Necker.

When a name has been proposed but not published by one author and is subsequently validly published and ascribed to him by another author who supplied the description, the name of the latter author must be appended to the citation with the connecting word *ex*.

For example, Robert Brown named a specimen *Capparis lasiantha*. He prepared no description to accompany his name, but recognized the specimen to represent a new species and penned the name on the herbarium sheet. Later, de Candolle, concurring with Brown, published the new species using Brown's name for it. De Condolle studied Brown's specimen and from it prepared the diagnosis to accompany his publication of the new name. The correct and complete name and author citation is *Capparis lasiantha* R. Br. *ex* DC.

When a genus or a taxon of lower rank is altered in rank but retains its name or epithet, the author who first used the name legitimately must be cited in parentheses, followed by the name of the author who effected the alteration of rank.

An example illustrates the application of this Rule.

The German botanist Carl Willdenow named the American yellow ladyslipper *Cypripedium pubescens,* and its name and author citation is *C. pubescens* Willd. Recently Correll concluded that the plant was only a variety of a similar Eurasian species known by the name of *Cypripe-*

*dium Calceolus* L. The new name for the American plant then became *Cypripedium Calceolus* L. var. *pubescens*. This is an example of a plant (*C. pubescens*) reduced from the category of species to that of *varietas* ("altered in rank"). Because there existed no earlier name for the plant in the new category the plant retains its former binary epithet (i.e., Correll elected to retain the epithet *pubescens* for it). By the Rule, the name of the original author (Willdenow) must be cited in parentheses and the name of the author who made the change (Correll) must follow. Accordingly, the name of the plant with its complete author citation is *Cypripedium Calceolus* L. var. *pubescens* (Willd.) Correll.

A recommendation of the Rules states that,

> Authors' names put after names of plants are abbreviated unless they are very short.

The abbreviations used for authors' names are more or less established by usage, and efforts are made to insure that they are clear and understandable. Most manuals and the larger floras provide an alphabetical list of abbreviations employed for authors' names. Sometimes the initial of the first name is used to avoid confusion (i.e., R. Br. for Robert Brown and Wm. Br. for William Brown).[6] When father and son are both authors of names, the letter *f* may follow the name of the son (e.g., Hook. f. for Sir J. D. Hooker, son of the botanist William Hooker) and represents the Latin word *filius*, son.

SECTION 8. *Retention of names or epithets of taxa which are remodelled or divided*

A group of three articles that provide in effect that
1. The changing of diagnostic characters of a taxon does not warrant changing the name of that taxon.
2. When a genus is divided into two or more genera, the generic name must be retained for that genus containing the type species of the original entity.
3. When a species is divided into two or more species, the specific epithet must be retained for one of them, or (if it has not been retained) must be re-established When a particular specimen was originally designated as the type, the specific epithet must be retained for the species including that specimen. When no type was designated, a type must be chosen according to the regulations given (Appendix I of the Rules).

The same rule applies to subdivisions of species, for example, to a subspecies divided into two or more subspecies, or to a variety divided into two or more varieties.

The sugar maple was first described by Humphrey Marshall, who named it *Acer saccharum*. Later François Michaux considered it com-

---
[6] Other common and standard abbreviations include L. (Linnaeus), DC. (de Candolle), HBK. (Humboldt, Bonpland, and Kunth), BSP. (Britton, Sterns, and Poggenberg), R&P (Ruiz Lopez and Pavon), T&G (Torrey and Gray), and W&K (Waldstein and Kitaibel).

posed of two taxa, each of which he treated as species. In agreement with the above Rule, Michaux retained the name *Acer saccharum* Marsh. for the taxon he believed Marshall to have so named, and named the species he separated from it *Acer nigrum* Michx. f.

SECTION 9. *Retention of names or epithets of taxa below the rank of genus on transference to another genus or species*

a. When a species is transferred to another genus (or placed under another generic name for the same genus), without change of rank, the specific epithet must be retained or (if it has not been retained) must be re-established, unless one of the following obstacles exists: (1) that the resulting binary name is a later homonym, or a tautonym; (2) that there is available an earlier validly published specific epithet.

When the specific epithet, on transference to another generic name, has been applied erroneously in its new position to a different species, the combination must be retained for the plant on which the epithet was originally based.

The application of this Rule is best explained by a few illustrative examples.

In 1753 Linnaeus described what is now known to have been a hemlock, under the name *Pinus canadensis*. Carrière recognized that it was a hemlock and not a pine, and transferred the species to the genus *Tsuga* (the genus of hemlocks). This involved no change of rank, and the original epithet of *canadensis* was retained, resulting in the combination *Tsuga canadensis* (L.) Carr.

The French botanist du Roi placed the American larch in the genus *Pinus* and named and described it as *Pinus laricina*. Michaux correctly recognized it to be a larch and named it *Larix americana*. The use of the epithet *americana* is contrary to the above Rule, for Michaux did not retain the original specific epithet of *laricina*. Later Karl Koch corrected the situation by rejecting Michaux's binomial and re-establishing du Roi's earlier epithet, making the combination of *Larix laricina* (du Roi) Koch. Michaux's binomial is an illegitimate name and that of Koch is the valid name of the American larch.

Aside from the technicalities, exceptions, and recommendations that are a part of this Rule, its essential feature is that when a species is transferred from one genus to another without any change in rank (i.e., when it remains a species in either case), the epithet given it for the first time in the category of species is the epithet that must be used for it at all times unless prevented by the so-called "homonym Rule" (p. 213), or the tautonym Rule (p. 215):

b. The name of a variety, or other subdivision of a species, must be given the same consideration as would the epithet of a species name.

A plant treated as a variety (or other infraspecific taxon) ordinarily must be given the oldest varietal name for it in that category, and when transferred without change of rank from one species or genus to another must retain that epithet unless there are legitimate reasons for rejecting it.

There is little cause for confusion or misunderstanding in the application of this Rule. However, it should be remembered that a provision of the American Code carried the matter of priority to a greater extreme and directed that when combined names were of the same date, the priority of page within a given work, or priority of position of two names on a page, should be considered, and the name on the earlier page or appearing first on a page had priority over the other name. Scientific papers written before acceptance of the Cambridge edition of the Rules may reflect the principle of priority as based on the American Code provision. Contentions in support of the American Code rule have no standing in contemporary nomenclatural practices and no one should be confused or misled by them.

SECTION 10. *Choice of names when two taxa of the same rank are united*

a. When 2 or more taxa of the same rank are united the oldest legitimate epithet is retained. If the names or epithets are of the same date, the author who unites the taxa has the right of choosing one of them. The author who first adopts one of them, definitely treating another as a synonym or referring it to a subordinate taxon, must be followed.

SECTION 11. *Choice of names when the rank of a taxon is changed*

a. When a tribe becomes a family, when a subgenus or section becomes a genus, when a subdivision of a species becomes a species, or when the reverse of these changes takes place, and in general when a taxon changes its rank, the earliest legitimate name or epithet given to the taxon in its new rank is valid, unless that name or the resulting combination is a later homonym.

It was observed that when a name was transferred *without change of rank* the epithet remained unchanged in the new category, *if* it were the oldest available name. However, when on transfer *a group changes its rank*, the oldest epithet *in the new category* must be employed.

For example, in southeastern Europe there are, among others, 3 thrifts that belong to the genus *Armeria*. For many years they were known as *A. canescens* (Host) Ebel (1827), *A. majellensis* Boiss. (1848), and *A. majellensis* var. *brachyphylla* Boiss. (1879). In a recent taxonomic treatment of the genus the last two names were treated as a single variant of *A. canescens*. By application of this article, quoted above, when a species (in this case *Armeria majellensis*) becomes a variety, the oldest name in the new category must be employed. Because *A. majellensis* and

*A. majellensis* var. *brachyphylla* were treated as a single variant of *A. canescens,* and hence were combined under a single ternary name, that ternary name must be the oldest available in the category, namely *brachyphylla* (1879). As a result, the 3 taxa were interpreted to comprise a single species of two varieties, one to be represented by the typical element of the binomial *A. canescens* and the other by the ternary epithet, var. *brachyphylla.* The name *A. majellensis* became a synonym of *A. canescens* var. *brachyphylla.*

The example cited above illustrates the application of the article when the change of rank is downward, that is, from the category of species to the lower category of variety. The same conditions apply when the change is upward, that is, if the variety is raised to a species or a species to a genus. In any of these cases, the first available epithet given the plant *in the new category* (rank) is its valid epithet.

SECTION 12. *Rejection of names*

a. A name or epithet must not be rejected, changed, or modified merely because it is badly chosen, or disagreeable, or because another is more preferable or better known.

b. A name must be rejected if it is illegitimate . . .

A name is illegitimate in the following cases: (1) if it was nomenclaturally superfluous when published;[7] (2) if published in contravention of specified earlier Articles, if its author did not adopt the earliest legitimate epithet available for the taxon with its particular circumscription, position, and rank; (3) if it is a later homonym; (4) if it is a rejectable generic name; (5) if it is a specific name published in a work where binary nomenclature for species was not consistently employed, or if the name is a tautonym.

c. A name of a taxon is illegitimate and must be rejected if it is a *later homonym,* that is, if it duplicates a name previously and validly published for a taxon of the same rank based on a different type. Even if the earlier homonym is illegitimate, or is generally treated as a synonym on taxonomic grounds, the later homonym must be rejected.

This Rule, sometimes known as the "homonym rule," was incorporated in the Rules in its present form at the Cambridge Congress and represents one of the features adopted from the American Code. Prior to 1930, according to the Vienna Rules, if the earlier homonym was an illegitimate name, it was not given consideration. However, in accordance with the Rule quoted above, consideration must be given to both legitimate and illegitimate earlier homonyms. The two examples that follow illustrate the application of this Rule.

[7] Students are likely to overlook the significance of Sect. 1 of this article by failing to recognize a superfluous name. An excellent example of the application of this rule was provided and explained by M. L. Sprague (Kew Bull. 1933, pp. 152–154) wherein the name *Silene Cucubalus* was shown to be the valid name of a plant often and more commonly known by the illegitimate names *S. inflata, S. vulgaris,* and *S. latifolia.*

The French botanists, Franchet and Savatier, described as new a species of Japanese maple in 1879 and named it *Acer parviflorum*. However, unknown to them and nearly a century earlier, this same specific epithet had been given to a wholly different species of maple by the German botanist Ehrhart. The later use of the binomial by Franchet and Savatier created a later homonym (i.e., the use of one name for two different species). It so happens that Ehrhart's name is treated universally as a synonym of Lamarck's *Acer spicatum* and accordingly is an illegitimate name (since it was superfluous when published). By the "homonym rule," even illegitimate names must be given consideration in such matters. For this reason, Rehder corrected the situation by giving Franchet and Savatier's species the new name of *Acer brevilobum*. The synonym and citations for this are as follows:

*Acer brevilobum* Rehder, Journ. Arnold Arb. 19: 85, 1938.
    *Acer parviflorum* Franch. & Sav., Enum. Fl. Jap. 2: 321, 1879; non Ehrh., Beitr. Naturk. 4: 25, 1779.

A somewhat more complex example is provided by a nomenclatural situation in the coniferous genus *Thuja*. In 1824 Lambert named and described a little-known species of the genus as *T. plicata*. Later (1847) Endlicher gave the same binomial to a different but well-known species of *Thuja*. Nothing was done to correct the situation. In 1868 Hoopes reduced Lambert's little-known plant to varietal status and named it *Thuja occidentalis* var. *plicata* (Lamb.) Hoopes. Apparently unaware of Hoopes' combination, Masters in 1897 treated the Endlicher plant, then an important forest and horticultural subject, as *T. occidentalis* var. *plicata* (Endl.) Mast. Masters' combination was a later homonym and as such it was an illegitimate name. Rehder corrected this in 1939 by renaming Endlicher's plant *Thuja occidentalis* var. *Mastersii*.

The directive of this article, that a name is illegitimate and must be rejected "even if the earlier homonym is illegitimate," has resulted in numerous name changes.

    d. A name of a taxon must be rejected if, owing to its use with different meanings, it becomes a permanent source of confusion or error.

A number of names, mostly binomials, have been treated as *nomina ambigua*. One concerns the red oak of eastern United States, and the case will serve to illustrate application of the Rule.

Linnaeus, in 1753, described an American oak as *Quercus rubra*. As indicated by the synonymy he gave for it, several oaks now recognized as representing other species were combined under his name. As now interpreted, *Q. rubra* L. was based on at least two different species; one is the red oak of northeastern North America, and the other the so-called Spanish oak of southeastern United States. It was not known to which of these species his binomial should apply, and botanical practice of the past is of little help, for in 1771 du Roi applied the Linnaean name to the red oak, calling the southern Spanish oak *Q. falcata* Michx. In 1915 Sargent (after a study of Linnaean material and references) endeavored to establish the name *Q. rubra* L. for the southern Spanish oak and *Q. borealis* Michx. f. for the northern red oak. Confusion resulted, some people following Sargent's interpretation and others retaining the older interpretation of du Roi. Because of the ambiguity associated with Linnaeus' name of *Q. rubra*, Rehder has proposed that it be treated as a *nomen ambiguum* and, following his treatment of the problem, most contemporary botanists now apply the name *Q. falcata* Michx. to the southern Spanish oak and *Q. borealis* Michx. f. to the northern red oak with name *Q. rubra* L. rejected completely from use for any plant.

e. A name of a taxon must be rejected if the characters were derived from two or more entirely discordant elements, unless it is possible to select one of these elements as a satisfactory type of the name.

One example of application of this Rule is that associated with the genus *Actinotinus*, established by Oliver in 1888 on a specimen derived from the two genera *Viburnum* and *Aesculus*. Oliver's type specimen was received from a native Chinese collector who had inserted the inflorescence of a *Viburnum* into the terminal bud of an *Aesculus*. The name *Actinotinus* is a *nomen confusum* and must be rejected.

f. Generic names are illegitimate and must be rejected if they (1) are words not intended as names; (2) coincide with a morphological term; (3) are unitary designations to species; or (4) consist of two separate and unhyphenated words.

g. Specific epithets are illegitimate in the following special cases and must be rejected: (1) when they are merely words not intended as names; (2) when they are merely ordinal adjectives being used for enumeration; (3) when they exactly repeat the generic name with or without the addition of a transcribed symbol (*tautonym*); (4) when they were published in works in which the Linnaean system of binary nomenclature for species was not consistently employed.

In (3) above, the term tautonym applies to any binomial where the binary name repeats the generic name. This practice was legitimate under the American Code where, for example, many authors subscribing to this code placed the white pine of northeastern United States (*Pinus Strobus* L.) in *Strobus*. The retention of the oldest binary epithet (*Strobus*) for the plant resulted in the combination, made by Dr. J. K. Small, of *Strobus Strobus* (L.) Small. This binomial is a tautonym and as such must be rejected as an illegitimate name. (See Moldenke, 1932, for an extensive list of examples.)

> h. The original spelling of a name must be retained except in cases of typographic or unintentional orthographic error.

The Rules provide many examples to illustrate this article and its several recommendations. An example of a typographic error is *Rosa Pisartii* Carr., whose correction to *Rosa Pisardii* is allowed.

An example of an orthographic error is illustrated by the specific epithet *sinensis*, an orthographic variant of *chinensis*, and both names may not be employed legitimately for two species in the same genus.

The forming of specific epithets when taken from the name of a person is part of an Article (Art. 70, ed. 3); the important consideration being that "when the name ends in a consonant, the letters *ii* are added (thus *Ramondii* from Ramond), except when the name ends in *-er* when *i* is added (thus *Kerneri* from Kerner)." Names ending in vowels are terminated in the genitive case by the single *i*.

A new note to this article reads:

> The use of the terminations *i* or *ae* instead of *ii* or *iae* . . . is treated as an unintentional orthographic error . . . Thus, the original spelling of the name *Pinus Griffithi* is to be corrected to read *Pinus Griffithii*.

A recommendation reads:

> All specific and trivial names or epithets should be written with a small initial letter, although writers desiring to use capital initial letters for particular names or epithets may do so when these are derived directly from the names of persons (or deities), or are vernacular (or barbaric) names, or are previously published (including pre-Linnaean and invalid) unmodified generic names.

This recommendation advocates the decapitalization of all specific and infraspecific epithets. At the same time it recognizes the prerogative of those who capitalize names when taken from the names of persons (*Pinus Coulteri,* in honor of William Coulter), when taken from another genus (*Dianthus Caryophyllus*, the epithet *Caryophyllus* being a pre-Linnaean generic name) or when a vernacular or barbaric name (*Schinus Molle,* the epithet *Molle* being a Peruvian vernacular name). There is not complete accord among botanists regarding the capitalization of specific names. Papers have been published in favor of it (Bailey, 1946; Fernald, 1947, 1949) and against it (Blake, 1940; Beetle, 1944; Steere, 1945; Polunin, 1950). The current editorial policy of the Kew Bulletin on decapitalizing species names and the announced decapitalization of them in Supplement XI of *Index Kewensis* point to the ultimate exit of capitalization.

## Units of Classification

The Rules of Nomenclature prescribe the categories into which plants should be classified. These categories constitute the units of classification. Their sequence and order of importance is fixed by the Rules. Changes from this sequence are not allowable. The units of classification are arranged below regressively, in descending order from units of greatest magnitude to units of least magnitude. Illustrative examples are indicated parenthetically. See pp. 44–56 of Chapter IV, for pertinent taxonomic considerations.

Kingdom (Vegetable)
   Division (*Thallophyta, Bryophyta, Pteridophyta,* and *Spermatophyta,* or as given by Engler and Prantl, *Embryophyta Siphonogama*)
     Subdivision (*Gymnospermae* and *Angiospermae*)
       Class (*Monocotyledoneae* and *Dicotyledoneae*)
         Subclass (*Archichlamydeae*)
           Order (*Rosales;* unless conserved, the names of orders terminate in *-ales.*)
             Suborder (*Rosineae;* the recommended terminal designation for suborders is *-ineae.*)
               Family (*Rosaceae;* families have the ending of *-aceae,* with exceptions as noted.)
                 Subfamily (*Rosoideae;* the terminal ending for subfamilies is *-oideae.*)
                   Tribe (*Roseae;* the termination for names of tribes is *-eae.*)
                     Subtribe (*Rosinae;* the authorized termination for subtribes is *-inae.*)
                       Genus (*Rosa;* generic names may be composed in any arbitrary manner.)
                         Subgenus (*Eurosa;* names of subgenera are usually substantives resembling the names of genera. A new Article of the Rules provides that the subgenus containing the type species of a generic name must bear that name unaltered if no earlier legitimate name is available.)
                           Section (*Gallicanae;* sections and subsections are subdivisions into which subgenera are sometimes divided.)
                             Species (*Rosa gallica;* Article 27 (ed. 3) and its recommendations applies to the names of species.)
                               Categories below the rank of species, in descending order of importance are:
                               Subspecies (ssp.)
                               Varietas (var.)
                               Subvarietas (subvar.)
                               Forma (f.)
                               Clone (cl.)

The progress of the development of rules of nomenclature is represented, with one exception, by a direct lineal growth based on trial, error, and correction; it was the development from the Paris to the Stockholm Codes. The exception was the revolt from the orthodox by the group of American botanists who conceived the American Code of nomenclature.

To comprehend this leftist movement fully, one should recognize first that the membership of the rightist element was entrenched and steeped to a high degree with the botanical traditions of revered preceptors, of botanically endowed heritage, and of near-sentimental conservatisms;

Fig. 23. NATHANIEL LORD BRITTON, 1859–1934. Sketch by M. E. Ruff, from photographs, courtesy the New York Botanical Garden.

all laudable sentiments but characteristics that sometimes obscured the objectivity that makes for science. The leftist American faction was headed by N. L. Britton, a scientist trained originally as a geologist, not imbued with idolatry of botanical masters nor steeped in botanical lore and heritage. He entered the field of taxonomic botany unencumbered by sentimental prejudice or bias. Britton's ideas of absolute priority were logical, if neither sound nor practical. He refused to accept the more widespread view that nomenclature has a practical function; that of giving a concise reference to a particular concept (i.e., a name to a taxon). Priority is only a means to an end and, as others have pointed out, "the happy idea of *nomina generica conservanda* was the saving of a situation." Britton's ideas of the type concept, embodied from the first in his nomenclatural proposals, provided the beginnings of an objective basis for future taxonomic work. However, his views on typification also were lacking in practicalities and it remained for A. S. Hitchcock, his colleague in these matters, later to remedy the defects by setting forth the basis of a natural typification in place of Britton's mechanical procedure. Hitchcock's views were taken up by Sprague, through whom the European botanists were led to accept them. The values of Britton's leftist movement ultimately were recognized and with the close of the Cambridge Congress there came to all taxonomists the realization of an international unity.

Today there is unity, but not harmony, as concerns nomenclatural legislation, trends, and practices among plant taxonomists (cf. Core, 1945; Corner, 1939; Martin, 1945; Little, 1948; Smith, 1949; Camp, 1950; and Gilmour, 1950). Those lacking scientific perspective, and concerned by temporarily discomforting changes in plant names, would freeze the nomenclature of plants as it now stands and allow no further name changes; others, in the "interest of stability," would advocate *nomina specifica conservanda*; and some botanists would divorce the subject of nomenclature of horticultural subjects from that of indigenous plants, ignoring the fact that the two groups of plants are inseparable, since most species of cultivated plants have counterparts in the wild, or that additional plants constantly are being introduced into cultivation. As noted earlier, the matter of capitalization of binary and ternary names is not treated with accord. The trend toward decapitalization is gaining momentum. There is no uniformity in practice in the delimitation and choice of subspecific categories, and while not a part *per se* of plant nomenclature, the vacillations and fluctuations in concept of these categories as encountered in the literature do affect the stability of plant names. It has been urged that application of the rules be tempered by judgment (Gleason, 1947). The entire question of the influence of the experimental method in taxonomy on the nomenclature of the future is replete with dynamic potentials, and is responsible in part for the introduction into the Rules of the categories nothomorph, apomict, and clone. The influence of increasing cytological and genetical findings surely will be reflected to a greater extent in future nomenclatural regulations and practices.

From all of this it must be concluded that continuous progress has been made in the development of effective nomenclatural legislation. This progress has not yet reached its zenith. It is doubtful if it ever will. The Rules of Nomenclature epitomize one facet of botanical science, and as nature is not static, neither can the nomenclature of its components be so and continue to remain on a level of scientific objectivity.

LITERATURE:

A.A.A.S., BOT. CLUB, NOMENCL. COMMISSION. Rochester code of nomenclature. Bull. Torrey Bot. Club, 19: 290–292, 1892, and Bot. Gaz. 34: 287–288, 1892.

———. Code of botanical nomenclature. Bull. Torrey Bot. Club, 31: 249–290, 1904.

———. American code of nomenclature. Op. cit. 34: 167–178, 1907.

AIRY-SHAW, H. K. Typification of new names derived from persons or places. Kew Bull. 1: 35–39, 1947.

BAILEY, L. H. Some present needs in systematic botany. Proc. Amer. Philosoph Soc. 54: 58–65, 1915.

———. Various cultigens, and transfers in nomenclature. Gentes Herb. 1: 113 115, 1923.

———. Statements on the systematic study of variables. Proc. Int. Cong. of Plant Sci. 1926 (Ithaca, N. Y.) 2: 1427–1433, 1929.

———. How plants get their names. New York, 1933.

———. Species-names with capital letters. Gentes Herb. 7: 168–174, 1946.

BEETLE, A. A. Specific decapitalization. Chron. Bot. 7: 380–381, 1943.

BLAKE, S. F. Decapitalization of specific names. Amer. Journ. Bot. 27: suppl. 22. 1940.

———. Cotype, syntype, and other terms referring to type material. Rhodora, 45: 481–485, 1943.

BODMER, F. The loom of language. (Edited and arranged by L. Hogben.) London, 1944.

BRIQUET, J. Texte synoptique des documents destinés à servir de base aux débats du Congrès Internationale de nomenclature botanique de Vienne, 1905. Berlin, 1905.

BRIQUET, J. and RENDLE, A. B. [eds] International Rules of Botanical Nomenclature, ed. 3. Jena, 1935.

CAMP, W. H. The names of plants in cultivation. Nat. Hort. Mag. 27: 83–86, 1948.

CAMP, W. H., RICKETT, H. W. and WEATHERBY, C. A. [eds.] International Rules of Botanical Nomenclature. [Unofficial special edition.] Brittonia, 6: 1–120, 1947.

CLAUSEN, R. T. On the citation of authorities for botanical names. Science, 88: 299–300, 1938.

CORE, E. L. On the need for revision of the International code of botanical nomenclature. Castanea, 10: 116–119, 1945.

CORNER, E. J. H. Notes on the systematy and distribution of Malayan Phanerogams. Garden's Bull. (Straits Settlements) 10: 1–81, 1939 [See p. 81 for opinions in opposition to *nomina specifica conservanda.*]

DE CANDOLLE, ALPHONSE. Lois de la nomenclature botanique, Genève, 1867. [For English translation, see Weddell, H. A., Laws of botanical nomenclature, together with an historical introduction and a commentary. London, 1868.]

EWAN, J. Isotype versus co-type as designators for duplicate type. Chron. Bot. 7: 8–9, 1942.

FERNALD, M. L. Some spermatophytes of eastern North America. Rhodora, 42: 239–246, 1940. [Views on concepts of categories of subspecies, varietas, and forma.]

———. *Sedum Rosea*, not *S. roseum*. Rhodora, 49: 79–81, 1947. [A defense of capitalization of binary names as provided by the Rules.]

———. A most useful series of illustrations. Rhodora, 51: 31–32, 1949. [Supporting capitalization of specific names.]
FRIZZELL, D. L. Terminology of types. Amer. Midl. Nat. 14: 637–668, 1935.
GILMOUR, J. S. L. The conservation of species names. Särtryck ur Botaniska Notiser, 1950: 340–343. [A paper opposing *nomina specifica conservanda,* but proposing legislation for *nomina excludenda,* thereby arriving at the same objective of nomenclatural stability without establishment of lists of conserved names.]
GLEASON, H. A. The preservation of well known binomials. Phytologia, 2: 201–212, 1947.
GREEN, M. L. History of plant nomenclature. Kew Bull. 1927: 403–415.
HANSON, H. C. Codes of nomenclature and botanical congresses. Amer. Bot. 31: 114–120, 1925.
HIGGINS, V. The naming of plants. London, 1937.
HITCHCOCK, A. S. Type-basis code of botanical nomenclature. Science, 49: 333–336, 1919.
———. Methods of descriptive systematic botany. New York, 1925.
———. A basis for agreement on nomenclature at the Ithaca Congress. Amer. Journ. Bot. 13: 291–300, 1926.
———. The relation of nomenclature to taxonomy. Proc. Int. Cong. Plant Sci. 1926 (Ithaca, N. Y.) 2: 1434–1439, 1929.
HORT, ARTHUR. The "Critica Botanica" of Linnaeus. [Transl. from the Latin.] London, 1938.
JACKSON, B. D. [ed.] Index Kewensis. 4 vols. London, 1893–95. Supplements I–X, London, 1900–1940.
———. History of the compilation of the Index Kewensis. Journ. Roy. Hort. Soc., 49: 224–229, 1924.
KELSEY, H. P. Present-day plant name confusion. Horticulture (Boston, Mass.), 23: 519, 1945.
KELSEY, H. P. and DAYTON, W. A. [eds.] Standardized plant names. ed. 2. 675 pp. Harrisburg, Pa., 1942.
LANJOUW, J. Botanical nomenclature and taxonomy; a symposium . . . at Utrecht, the Netherlands, June 14–19, 1948. 87 pp. Waltham, Mass., 1950.
———. (ed.) Synopsis of proposals concerning the International rules of botanical nomenclature, submitted to the Seventh International Botanical Congress, Stockholm, 1950. 255 pp. Utrecht, 1950.
LINNAEUS, C. Fundamenta Botanica . . . Amsterdam, 1735.
———. Critica Botanica. Leyden, 1737.
———. Ph losophica botanica. 362 pp. Stockholm, 1751. (Eng. transl. by Rose, H., 1775.)
———. Species Plantarum. Holmiae, 1753.
LITTLE, E. L. A proposal to stabilize plant names. Phytologia, 2: 451–456, 1948.
MARTIN, A. C. Instability of scientific names in plants. Amer. Midl. Nat. 34: 799–800, 1945.
MOLDENKE, H. N. A discussion of tautonyms. Bull. Torrey Bot. Club, 59: 139–156, 1932.
POLUNIN, N. Specific and trivial decapitalization. Bull. Torrey Bot. Club, 77: 214–221, 1950.
REHDER, A. New species, varieties, and combinations from the herbarium and the collections of the Arnold Arboretum. Journ. Arnold Arb. 1: 44–60, 1919.
———. The varietal categories in botanical nomenclature and their historical development. Journ. Arnold Arb. 8: 56–68, 1927.
RENDLE, A. B. Rules of nomenclature. [In Brook, F. T. (ed.) Report of the pro-

ceedings of the imperial botanical conference, pp. 301–307.] London, 1924. [cf. also, Journ. Bot. 62: 79–81, 243–244, 1924.]

RICKETT, H. W. Linnaeus' rules of nomenclature. Torreya, 41: 188–191, 1941.

———. Citation of author's names in taxonomy. Bull. Torrey Bot. Club, 75: 172–174, 1948.

ROBERTY, G. Proposition sur la nomenclature des groupements systématiques de rang inférieur à l'espèce. Candollea, 10: 293–344, 1946.

SMITH, A. C. The principle of priority in biological nomenclature. Chron. Bot. 9: 114–119, 1945.

———. A legislated nomenclature for species of plants? Amer. Journ. Bot. 36: 624–626, 1949.

SPRAGUE, T. A. Plant nomenclature: some suggestions. Journ. Bot. 59: 153–160, 1920. [cf. also, op. cit. pp. 289–297, for separate reactions from A. Rehder, J. Groves, and N. L. Britton; op. cit. 60: 129–139, 1921, for reply by Sprague; op. cit. pp. 256–263 for rejoinder by J. H. Barnhart, and pp. 313–318 for separate supplementary views by Sprague and A. S. Hitchcock.]

———. The nomenclature of plant families. Journ. Bot. 60: 69–73, 1921.

———. Suggestions for a World-Code of plant nomenclature. Science, 57: 207, 1923.

———. Proposed changes in the International Rules. Journ. Bot. 62: 196–198, 1924.

———. Standard-species. Kew Bull. 1926: 96–100.

———. The gender of generic names a vindication of Art. 72 (2). Kew Bull. 1935: 545–557.

———. Principal decisions concerning nomenclature made by the Sixth International Congress. Kew Bull. 1936: 185–188.

———. Proposed additions and amendments to the Int. Rules of Bot. Nomen. Kew Bull. 1939: 317–339.

——— [ed.] International botanical congress, Cambridge (England) 1930. Nomenclature. Proposals by British botanists. 203 pp. London, 1929.

———. Additional *nomina generica conservanda* (Pteridophyta and Phanerogamae). Kew Bull. 1940: 81–134.

——— [ed.] Supplement, International Rules of botanical nomenclature, embodying the alterations made at Amsterdam in 1935. 28 pp. Cheltenham, 1948.

STEERE, WM. C. Decapitalization of specific names of Bryophytes. Bryologist, 48: 38–41, 1945.

STEUDEL, E. G. Nomenclator botanicus. Stuttgart, 200 pp. 1821; ed. 2, 2 vols. 1840–41.

STOUT, A. B. The nomenclature of cultivated plants. Amer. Journ. Bot. 27: 339–347, 1940.

SVENSON, H. K. On the descriptive method of Linnaeus. Rhodora, 47: 273–302, 363–388, 1945. [reprinted in Contrib. Brooklyn Inst. of Arts & Sci., Brooklyn Bot. Gard., No. 103, 1945.]

WEATHERBY, C. A. Changes in botanical names. Amer. Midl. Nat. 35: 795, 1946.

———. Valid and legitimate names—and *Thalictrum polycarpum* S. Wats. Madroño, 7: 83–85, 1943.

# CHAPTER X

# PLANT IDENTIFICATION

The identification of an unknown plant, as explained early in the text, is its determination as being identical with or similar to another and already known plant. However, it is customary in taxonomic practice to combine the identification of a plant with the determination of the correct name to be applied to that plant. The distinction between identification and nomenclature was explained in the Introduction. Plant identification is accomplished generally by means of one or more of several different methods or combinations of methods. No one method can be said to be better than another, and selection depends on the individual situation.

**Taxonomic literature**

The literature devoted to the systematics of the world's plants is written in all modern languages, and has been accumulating for two centuries or more. An adequate knowledge of it is possessed by only a relatively few bibliophilic specialists, most of whom are botanical librarians. The subject is of such great importance to systematic botany that Chapter XIV of this text is devoted to it. The following discussion deals primarily with the utilization of taxonomic literature and the material in herbaria.

When the unknown plant is of a known locality, the usual procedure is to refer to a book or treatment accounting for the plants of that region. This is usually a flora or manual, and, if the latter, contains both analytical keys and descriptions. When the literature of a particular area or region is not known, reference should be made to Blake and Atwood's *Guide to the floras of the world*.[1]

With the appropriate manual and the unknown plant at hand the next step is to make use of the manual. Generally speaking, the first use of the manual is for determination of the family to which the unknown belongs, a step that is accomplished by use of the artificial analytical key

[1] Part I of this work (1942) accounts for the floras of the New World, Africa, Australia, and islands of the Pacific and Indian Oceans. If the unknown plant is from outside these limits, see entries under *Floras* in Chapter XIV.

to families (see pp. 225–228 for discussion of keys, synopses, and their uses). Knowing the name of the family, one turns to the section where it is treated and there, by means of the key to genera, repeats the procedure to determine its generic name. After this, and by use of the key to species, the specific identity of the unknown is learned. This results in knowing not only the identity of the plant, but also in learning the binomial credited to it by the author of the manual. For many reasons the identity and name of a plant obtained solely by use of a manual may be incorrect, incomplete, or both. When identifying any unknown by aid of keys in a manual, always check the description of each group arrived at by means of the key (family, genus, and species) to insure that there is a reasonable agreement between the characters observed in the unknown plant and those provided in the description of the plant it is presumed to be. When a marked variance exists between these two (plant and description), a misidentification probably exists.

A second method of identifying an unknown is the utilization of the latest floras and check lists of the particular region. Most floras of a state, county, or smaller political unit lack keys and descriptions. However, they comprise an index to the plants known for the locality and generally provide other pertinent habitat, distributional, and frequency data. An accounting of these items eliminates all plants that do not occur in the area and reduces to a working minimum the number of possibilities that may account for the identity of the unknown. By process of elimination, an unknown of established generic identity usually can be aligned with one of a few species, and identification completed by comparison of its characters with those in the descriptions in any standard work accounting for plants of the area.

The third method, and one of the most reliable if the material represented by the unknown is reasonably complete, is identification by means of the latest monographic or revisionary work accounting for the particular family, genus, or section, represented by the unknown. Identification by this method presupposes knowledge of the family and usually the generic name of the unknown. If the generic name is known, the next problem is to learn if a monographic treatment is available. There are indexes to this subject,[2] but none is of recent publication as concerns monographic works restricted to American or New World plants. It is in part to compensate for this deficiency that bibliographies are provided at the close of the treatment of each family in Part II of this text.

[2] See Chapter XIV, for entries under Bibliographies, catalogues, and review serials, especially those by Jackson, Merrill and Walker, Pritzel, and Rehder (1949).

*Keys* are devices useful in identifying an unknown. They represent one type of taxonomic literature. A key is an artificial analytical device or arrangement whereby a choice is provided between two contradictory propositions resulting in the acceptance of one and the rejection of the other. A key may be short and limited to a single pair of contradictory propositions (a *couplet*), or it may be composed of an extensive series of these. A *synopsis* is a device, usually in the format of a key, presenting graphically the technical characters which in general or in the aggregate differentiate taxa. It is not designed ordinarily for use in identification. For examples, see those on p. 372 and pp. 439–442.

1. Tendrils simple, not forked or branched.
   2. Pistillate flowers each with 2 small staminodes...............*Bryonopsis*
   2. Pistillate flowers without staminodes.
      3. Staminate flowers subtended by a shield-shaped bract....*Mormordica*
      3. Staminate flowers not subtended by a bract................*Cucumis*
1. Tendrils forked or branched.
   4. Leaves lobed for more than halfway to midrib.
      5. Plant monoecious; flowers bright yellow...................*Citrullus*
      5. Plant dioecious; flowers greenish..........................*Abobra*
   4. Leaves entire or lobed to much less than halfway to midrib.
      6. Staminate flowers with elongated hypanthium; flowers white..*Lagenaria*
      6. Staminate flowers with hypanthium short or none; flowers yellow
                                                                    *Benincasa*

**Fig. 24.** Example of a yoked or indented key.

The currently conventional and most acceptable type of key is the *dichotomous key*, a type usually of one of two formats. In any key, each statement of a couplet is termed a *lead*. In most taxonomic literature originating in this country the collateral leads of a given couplet are arranged in yokes, and each lead is identified by a letter or figure. Each successive subordinate yoke is indented under the one preceding it. An example of a dichotomous key with yoked or indented leads is given in Fig. 24. The second arrangement of a dichotomous key has bracketed or parallel leads. The 2 leads of each couplet always are together. Using the same context as in Fig. 24, such a key would be organized as shown in Fig. 25.

There are advantages and disadvantages to each of these types of dichotomous keys. The yoked type has the advantage of grouping similar elements in such a manner that they can be grasped visually as groups. However, it is apparent from the above example that in extended keys of this type there is a sloping and shortening of lines to the right with a resultant loss of economy of page space. In the parallel or bracketed type, the advantages of the indented format are lost and conversely the disadvantages are offset, because there is no opportunity in the bracketed type to group blocks of leads visually with elements having one or more characters in common. In this type of format, however, all leads are of approximately the same line length and produce a maximum of efficiency of page space. Some authors, employing this second format, do not indent alternate couplets, but instead bring all out to a common margin at the left of the page.

```
1. Tendrils simple, not forked or branched.................................2
1. Tendrils forked or branched.........................................4
    2. Pistillate flowers each with 3 small staminodia...............Bryonopsis
    2. Pistillate flowers without staminodia................................3
3. Staminate flowers subtended by a shield-shaped bract............Mormodica
3. Staminate flowers not subtended by a bract.....................Cucumis
    4. Leaves lobed for more than halfway to midrib........................5
    4. Leaves entire or lobed much less than halfway to midrib .. . .. . ....6
5. Plant monoecious; flowers bright yellow.........................Citrullus
5. Plant dioecious; flowers greenish................................Abobra
    6. Staminate flowers with elongated hypanthium; flowers white....Lagenaria
    6. Staminate flowers with hypanthium short or none; flowers yellow.Benincasa
```

**Fig. 25.** Example of a bracketed key.

There are differences of opinion and practice regarding the designators of couplet leads. In the examples above the leads are numbered. Until recently, the more prevalent practice has been to use lettered leads, which, had they been substituted above for the numbers would have been as in Fig. 26.

**Fig. 26.** Examples of the use of letters and numerals in leads of keys.

Convention, more than any other factor, seems to be responsible for the continued use of lettered leads. They have few advantages over numbered leads and have the following disadvantages:

1. In extended and lengthy keys there are not enough letters in the alphabet to accommodate the necessary couplets, with the result that authors resort to other alternatives, as continuing with letters of the lower

case (a, aa; b, bb; etc.) or with Greek letters or lower-case Roman numerals.

2. The presence of often many couplets, each designated by the same character or letter, is a source of confusion, especially when referring back to a lead (some keys have a dozen or more couplets in a single key, each designated by the same pair of letters).

3. Use of lettered couplets makes it very difficult to refer quickly and precisely to a particular lead in a key. This is of special pertinence when referring to couplets or leads in writing, and more so when using keys with lettered leads in classroom exercises.

In the numbered key, each couplet is numbered consecutively and no two couplets of a given key receive the same number. This provides a direct reference to each couplet. Some authors distinguish the first lead of a numbered couplet from the second by adding the letters a and b, respectively, to each lead of the couplet number (Fig. 26).

In the construction of the indented or yoked key, it usually happens that one section contains a smaller number of taxa than the other (note that only 3 genera appear in the first division of the key in Fig. 24, whereas 4 appear in the second). It is generally desirable to have the smaller division within a key precede the larger. This facilitates finding the collateral or second lead of the divided couplet and keeps the two in closer proximity to one another. There are objections to this feature, considered by some to more than offset the mechanical advantage, for in such an artificial arrangement it is not possible to have the elements of the key appear in any sort of phylogenetic sequence. If the key is to serve merely as an aid to identification, this disadvantage is of no significance, but if it is to precede a systematic and phylogenetic accounting of a group of elements it will result in their appearing in a different sequence from that by which they will be arranged in the text that follows. The awkwardness is partially countered by numbering the elements in the key in the sequence by which they will appear in the text.

Some rules to be considered in construction of any key are:

1. Insure that the key is always strictly dichotomous.
2. Select characters that are in opposition to one another so that the two leads of each couplet comprise two contradictory propositions, one of which will fit the situation and the other not apply.
3. In so far as possible phrase leads to read as positive statements, especially the initial lead of the couplet.
4. The initial word of each lead of the couplet should be identical. That is, if the first lead of a couplet starts with the word "flowers," the second lead of the same couplet should begin "flowers." This facilitates orientation of leads of any one couplet.

5. Two consecutive couplets should not each begin with the same word. This potential source of confusion may be by-passed if the repeated initial word of the second couplet is preceded by the article "the."
6. Avoid use of overlapping limits in variation, or generalities in opposing leads of the couplet. For example, it is bad form to phrase leads as in the following:
    1. Inflorescence a raceme; pedicels 4–6 cm long.
    1. Inflorescence a raceme or panicle; pedicels 6–10 cm long.
    or
    1. Flowers on long peduncles; leaves very broad.
    1. Flowers on short peduncles; leaves narrower.

    In the first example the characters overlap and the leads do not represent absolutely contradictory statements. In the second example, the characters are given as generalities and lack definiteness: who can say how much is long or short, or is broad, or narrower than broad?
7. Use macroscopic morphological characters in so far as possible in separation of groups (family, genus, species, etc.). That is, avoid separation on the basis of disjunctive geographic distributions, because one may not always know the source of an unknown; cytological data of themselves are little help in identifying a plant, chromosome numbers biologically may be significant but cannot be determined from a herbarium specimen, and are no help in a key.
8. In keys to dioecious plants, it is helpful to provide two separate keys, one utilizing characters for the staminate plant and the other for characters of the pistillate plant. The sample key provided in Fig. 24 illustrates the difficulties encountered when this rule is violated, for often when identifying dioecious plants one has material of one sex or the other, but rarely has material of both sexes at hand at one time.

## Herbaria

A herbarium is a collection of plant specimens that usually have been dried and pressed, are arranged in the sequence of an accepted classification, and are available for reference or other scientific study. Some kinds of plants do not lend themselves readily to drying techniques without significant loss of diagnostic features (as in the fleshy members of the Aizoaceae and some Cactaceae) and may be preserved in liquid instead of pressed and dried; others have parts so bulky as not to be suited for pressing flat, and are dried without pressing and are stored in special boxes (as palm inflorescences, cones and branches of many gymnosperms, and many dry fruits).[3]

For the last hundred years it has been the usual practice to mount each specimen, together with a label bearing pertinent data, on a sheet of high-quality paper. These sheets are stored in specially designed cases known as herbarium cases. Before the era of mounting each collection on individual sheets, botanists pasted or sewed them in bound volumes of plain pages. The herbaria of the herbalists, and of almost all botanists as late as 1820 or later, were preserved in this manner. Figure 27 shows

---

[3] See Chapter XIII for preparation and preservation of herbarium specimens.

Fig. 27. One page of a volume of an early herbarium, prepared in 1821.

one volume of this earlier era, as contrasted with Fig. 28, which shows an open case of a modern herbarium.

The modern herbarium is far removed in basic concept from the collections of scraps that were called herbaria two centuries ago. It is by no means a collection of "hay," as some scientists may disparagingly allude to it, and in the words of A. C. Smith (1948, p. 584),

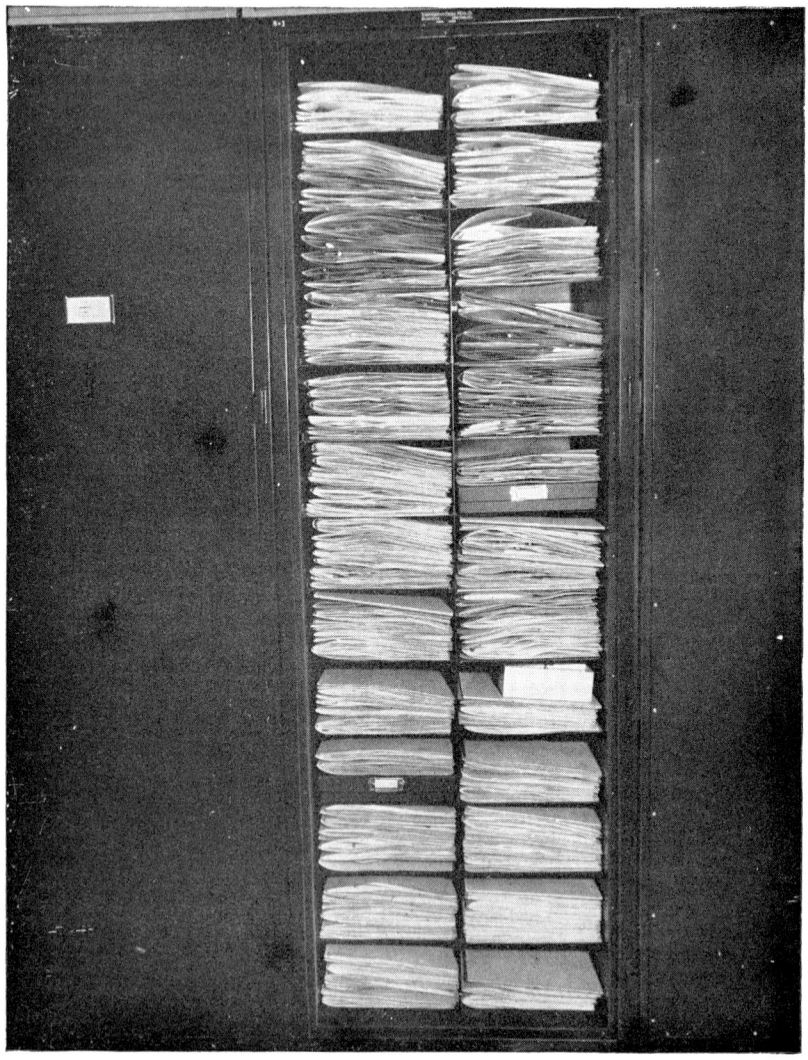

**Fig. 28.** Open steel case of modern herbarium.

## PLANT IDENTIFICATION

The unattainable goal of systematic botanists is to record the distribution of all plants and to classify them in a system which will, in some manner, depict their phylogenetic history. The primary purpose of our great herbaria is thus to present a picture, by means of representative specimens, of the composition of the modern plant world. These herbaria must be the chief basis for future monographic and phytogeographical studies. But they also serve many others beside the taxonomist; the economic botanist, the ethnobotanist, the morphologist, the geneticist, and students of many other disciplines seek much of their basic data in the collections and publications of systematic botanists.

There are many herbaria located throughout the world that individually and collectively are of inestimable value to botanical workers and to all who depend on their services. The list below cites some of these herbaria in sequence of descending size, as based on the number of specimens in each.[4]

| | |
|---|---|
| Royal Botanical Gardens, Kew | 5,000,000 |
| British Museum (Natural History) | 4,000,000 |
| Museum Natural History, Paris | 3,500,000 |
| V. L. Komarov Botanical Institute of the Academy of Sciences of U.S.S.R. | 3,000,000(?) |
| Conservatoire et Jardin Botaniques de Genève | 3,000,000 |
| Royal Botanic Garden, Edinburgh | 1,500,000 |
| National Herbarium, Melbourne, Australia | 1,500,000 |
| United States Herbaria: [5] | |
| U. S. National Herbarium, Washington, D. C. | 2,250,000 |
| New York Botanical Garden | 2,241,685 |
| Missouri Botanical Garden | 1,500,000 |
| Gray Herbarium, Cambridge, Mass. | 1,325,000 |
| Chicago Natural History Museum (Field Museum) | 1,246,000 |
| Academy of Natural Sciences of Philadelphia | 1,000,000 |
| Farlow Herbarium, Cambridge, Mass. | 982,171 |
| University of California | 864,928 |
| Arnold Arboretum, Jamaica Plain, Mass.[6] | 630,000 |
| Bureau of Plant Industry, Beltsville, Md. | 430,000 |

[4] One of the 3 largest herbaria of the world (containing about 4,000,000 specimens) was that located in Berlin. It was almost completely destroyed by allied bombs and fire in the winter of 1943. By this loss science has been irretrievably deprived of vast collections of plants from many parts of the world, and of the type specimens of perhaps a few thousands of species. One historically and nomenclaturally famous collection, that of Willdenow, was saved from destruction, reported to have been removed to "regions to the east" (cf., Babcock, 1948), and since has been returned to Berlin.

[5] Much of the information, and all statistical data pertinent to herbaria of the United States, have been taken from Jones and Meadows (1948). Other statistical data have been based on the paper by Gager (1938).

[6] The accounting by Jones and Meadows lists the several herbaria of Harvard University separately, since they are maintained in all respects as separate functions of the University. However, it should be noted that if the figures of these herbaria (Gray Herbarium, general spermatophytes; Farlow Herbarium, cryptogamic collections; Arnold Arboretum, ligneous spermatophytes; New England Botanical Club herbarium (general) and containing 197,708 specimens; and the Oakes Ames Orchid Herbarium, orchidaceous collections numbering 65,000 specimens) are combined, they total nearly 3,200,000 specimens, or more than in the herbaria of any single American institution.

In addition to the herbaria listed above, there are many other collections throughout the world that individually number a million specimens or more. Among them are the herbaria at Uppsala, Zurich, Brussels, Florence, and Vienna. Important herbaria located elsewhere include those situated at Cape Town, Pretoria, Leyden, Buenos Aires and Tucumán (Argentina), Buitenzorg (Straits Settlements), Calcutta, Canton, Brisbane, and Wellington.

There are about 170 organized herbaria in the United States, of which 74 contain 50,000 specimens or more each. The oldest domestic herbaria of particular note include (with dates of founding) those at the Academy of Natural Sciences, Philadelphia (1812), University of Michigan (1838), California Academy of Sciences (1853), University of Missouri (1856), Missouri Botanical Garden (1857), Gray Herbarium (1864) and U. S. National Herbarium (1868).

Students and scholars engaged in revisionary or floristic studies that involve unusually careful study on the identification of plants always have occasion to study the nomenclatural type specimens of some or many of the taxa concerned. From *Index Kewensis*, or other source, the name of the original describer of the plant is known. From the original description the identity of the collector of the type specimen may usually be ascertained. The next problem is to learn where the collections of that collector (or, in the case of long-established binomials, of the botanist who described it as new) are located. In some cases they may not all be at one institution and then the quest is to find the specimen, if extant, among the several collections. There are guides to the locations of collections of the more important taxonomists, and these should be consulted before making extensive individual searches.[7]

*LITERATURE:*

BLAKE, S. F. and ATWOOD, A. C. Geographical guide to floras of the world. Part I. U.S.D.A. Misc. Publ. 401, Washington, D. C., 1942.
DEAM, C. C. Flora of Indiana. Indianapolis, 1940.
DOLE, E. J. [ed.]. An annotated list of the ferns and seed plants of the State of Vermont. ed. 3 [Burlington?], 1937.
GRASSL, C. O. An international system of botanical districts. Bull. Torrey Bot. Club, 63: 519–524, 1936.
HANES, C. R. and F. N. Flora of Kalamazoo county, Michigan: vascular plants. Schoolcraft, Mich., 1947.
HITCHCOCK, A. S. Methods of descriptive systematic botany. New York, 1925. [Chapters VII, VIII, XVIII.]
JONES, G. N. A botanical survey of the Olympic Peninsula, Washington. Univ. Washington Pub. Biol. 5: 1936.

[7] See pp. 328–329 of Chapter XIV for references to literature dealing with locations of botanical collections.

———. Flora of Illinois. Notre Dame, Ind., 1945.
JONES, G. N. and MEADOWS, E. Principal institutional herbaria of the United States. Amer. Midl. Nat. 40: 724–740, 1948.
REEVES, R. G. and BAIN, D. C. A flora of South Central Texas. College Station, Texas, 1946.
STONE, W. The plants of southern New Jersey, with special reference to the flora of the pine barrens and the geographic distribution of the species. Ann. Rept. New Jersey State Mus. 1910: 23–828, 1911.

CHAPTER XI

# FIELD AND HERBARIUM TECHNIQUES

**Collecting procedures**

There are at least 3 ways of handling fresh plant material for processing into herbarium specimens. When time and carrying facilities permit, the most satisfactory method is to press each plant as it is collected. This is done in a field press made up of a pair of collecting frames, a few blotters, and folded sheets of newsprint. A second method is to accumulate the material in a metal collecting can or *vasculum* (pl., *vascula*.) These are available from biological supply houses, but professional botanists generally prefer sizes larger than those available from commercial houses, and have cans made to specification by local metalsmiths (Fig. 29-b). In actual use, the vasculum is first lined with a few thicknesses of well-moistened newsprint to retard wilting of specimens, and experience has shown that plants keep better in a full vasculum than in one that is only partly filled.[1] Plants should be pressed as soon as opportunity permits, but if the full vasculum is stored in a cool place most kinds may be held over night without a serious loss of quality. The third method, used more in the tropical rain forests than in temperate regions, is to carry collected specimens in a rucksack. They are then pressed as soon as possible after return to camp headquarters.

Certain items of equipment are indispensable to plant collecting, particularly a collecting pick, a strong knife or a machete, and a pair of pruning shears. The collecting pick is essential for digging up rhizomes, deep-seated bulbs or corms, and the roots of most herbaceous plants. Picks are available generally from biological supply houses, but an excellent and less expensive substitute is a bricklayer's hammer available from any hardware store or mail-order house. The use of a belt-supported sheath to carry the pick is a helpful accessory (Fig. 29-c). The pruning

---

[1] Most experienced field botanists who use vascula find that the vasculum should be painted white to reflect the sun's rays and also as an aid in quickly locating it when cached or lying among vegetation.

# FIELD AND HERBARIUM TECHNIQUES

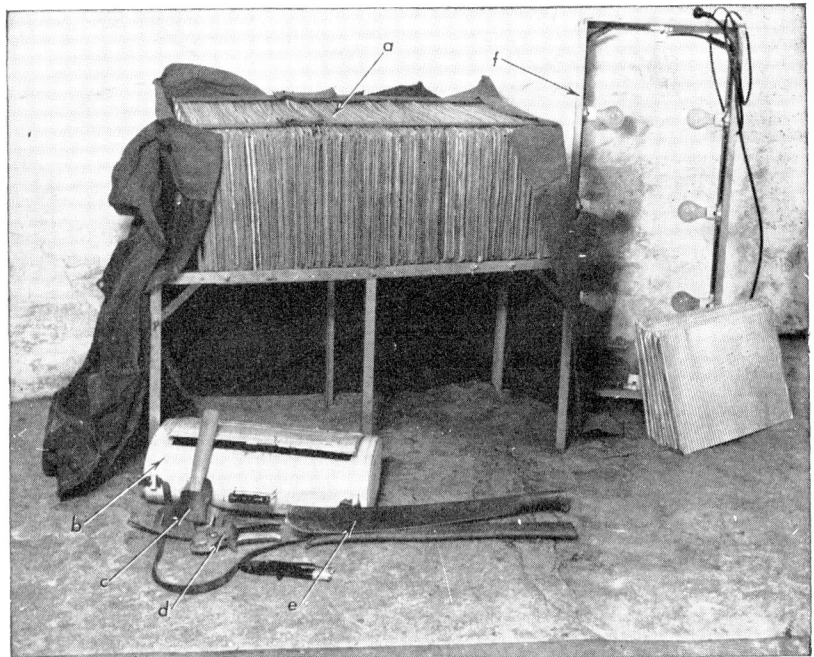

**Fig. 29.** Collecting, pressing, and drying equipment: a, field press of corrugates and blotters, on collapsible aluminum drying rack, with protective canvas around back and sides; b, vasculum; c, pick and sheath; d, pruning shears; e, machete and sheath; f, collapsible drying frame with incandescent bulbs, aluminum corrugates at right.

shears are especially useful in cutting woody material to pressing size. In addition to these items, it will be found that a garden rake or potato digger is of use when collecting submerged aquatics from a boat.

### Preparation of specimens

*Plant presses* are of several types, the selection of which depends on the use to be made of them and on the drying techniques to be used. The efficiency of the press is determined largely by its ability to hold the material under a constant and firm pressure, to dry the specimen to a degree short of crispness, and to retain the colors of all parts in so far as possible. Conventionally, most presses comprise a pair of wood or metal frames, blotters, pressing paper, and straps or strong cord.[2] The specimen to be pressed is arranged within the folded sheet of pressing paper that has been placed on a blotter, and another placed over it. If the plants are

[2] See Pool, R. J., *Flowers and flowering plants,* ed. 2. pp. 360–361, 1941, for details of making a plant press from plywood boards.

to be dried with the aid of artificial heat, a sheet of corrugated material is used between each pressing paper and its specimen (under some conditions described below, the blotters may then be omitted), otherwise no corrugates are used and the press is built up by an alternation of blotter-pressing paper-blotter, etc. The press frames are on the top and bottom of the press, and it is then "locked up" by means of straps or stout cord. Ordinarily the straps are drawn as tightly as possible in order to flatten all parts and to bring as much of each specimen as possible in contact with the moisture-removing blotter. Straps used on plant presses may be of leather or of webbing (trunk straps). If heat is to be used in the drying process, the trunk straps are to be preferred, since leather dries out and is then short-lived. Some botanists prefer a sash cord to straps, for the cords are of less bulk and they interfere less with the movement of air through the press when the material is to be dried by means of artificial heat. Pressing frames are usually constructed of slats of oak, ash, or hickory (about ¾ in. wide and ¼ in. thick) riveted together (copper or aluminum rivets preferred) and are generally 12 by 18 inches.

Blotters designed for the purpose, and of the standard size of 12 by 18 in., are available from biological supply houses in a wide range of weights and qualities. Botanists differ in their preferences for blotting materials, and the most costly types are not of necessity the most efficient. Other things being equal, one should select the quality that will absorb a given volume of water in the shortest period of time. This quality of blotter will not only absorb moisture from the plant quicker than other types, but also it can be dried more quickly for repeated use. Thickness of the blotting material is largely a matter of individual choice, although the stiffer heavy feltlike grades are the least efficient.

Corrugates, often referred to as ventilators, are used in presses when plants are dried by means of artificial heat. A corrugate is a sheet of pasteboard or thin metal, with fluted ducts. As used in plant presses, they are 12 by 18 in., and are cut with the ducts extending across the sheet (not lengthwise). The function is to provide air passages through the press for movement of dry heated air. This air comes in contact with the moist surfaces of blotters or paper and hastens drying of the specimen. The use of corrugates was advocated as early as 1910 by Collins, who favored the single-faced type in which one surface of the sheet was covered and the other had exposed ridges and grooves. Later, Ricker (1913) pointed out advantages of a double-faced corrugate, which had both surfaces covered. As early as 1926 it was shown by Stevens that corrugates made of aluminum had certain advantages not available in those made of pasteboard, but they never became popular, due to unavailability through commercial sources, until recently (cf. Camp, 1946). These aluminum corrugates are not faced and may be nested for storage.

Single-faced pasteboard corrugates are particularly efficient when used in connection with one or two blotters (in the sequence of blotter-corrugate-specimen-blotter-etc., or blotter-corrugate-blotter-specimen-blotter-corrugate, etc.) but have the disadvantage of remaining efficient for only a relatively short time. It has been estimated (Camp, p. 240) that this type of corrugate is ready for discard after having been used for an average of 10 times. The effective life of any corrugate is determined by the freedom of the ducts or grooves from closure (as caused by denting or crushing from the specimen, or by blunting of the edges from mechanical damage in handling). Air cannot flow through a closed duct, and as the number of closed ducts increases, the efficiency of the sheet is reduced. The portions of the specimen lying against closed ducts become cooked and are of inferior quality. Double-faced corrugates are used with or without blotters in plant presses, depending on the character of the material. They have a much longer effective life, but are less pliant and do not produce as uniform a pressure over all parts of the specimen. This latter disadvantage is offset to a degree by the use of blotters or of padding material as described on p. 240. The use of aluminum corrugates has met with mixed reactions, and it is becoming clear that while they are of value for use in certain situations, they are not likely to replace pasteboard corrugates. Some of their advantages and disadvantages are as follows:

| *Advantages* | *Disadvantages* |
| --- | --- |
| Nest easily, facilitating packing and reducing bulk. | Heavier than pasteboard. |
| | Sharp edges cut hands. |
| Not affected by wetting and not readily dented as are pasteboards. | Difficult to handle in large presses because of slipping. |
| Permit possibly more rapid drying. | Eight to ten times as costly as pasteboard. |
| Permanent, or of greater effective life than pasteboard. | Subject to denting by woody stems. |
| | Rigidity of grooves may mark specimens of many kinds of plants. |

Presses with aluminum corrugates cannot have their straps drawn as tightly as can those with pasteboard corrugates, and the former require more frequent taking up of slack as the drying of the plants progresses. The advantage of the metal ones over those of pasteboard seems to lie in their being unaffected by water and hence of value for collecting on expeditions where the hazard of soaking of packs or presses is omnipresent, or where high humidity or frequent rains appreciably shorten the life of pasteboard corrugates.

Papers used for pressing plants are usually salvaged from old newspapers. One recommended procedure is to tear a standard newspaper sheet in half, fold it in half, and cut or tear off one end so that the length of the sheet never exceeds 16 inches (the amount to be torn off is gauged readily by the practiced eye).[3] Some botanists prefer to use unprinted newsprint (cut to 22 by 16 in. and folded once), since it offers a blank surface on which to record collection data. Others believe (perhaps without evidence) that the printed newsprint may serve during periods of temporary storage as a repellent to insects. Paper heavier than commercial newsprint is unsatisfactory since it is less pliant and retards the drying process.

*Pressing plant materials* requires careful attention to detail if quality specimens are to result. In the first place, it is necessary to select specimens judiciously, and in this regard the following points should be observed:

Select specimens that are free from evidence of insect feeding, rust infections, and other obvious pathological symptoms.

Ordinarily avoid the depauperate individuals.

Ensure that the specimen is either in flowering or fruiting condition. Sterile material is generally worthless.

If specimen is herbaceous, always include enough of the underground parts to show their character.

Arrangement of the specimen within the pressing paper is an important step in preparing herbarium specimens. When the specimen is fresh, its parts are usually pliant and can be arranged easily. Skill in specimen arrangement comes with experience, and many collectors take justifiable pride in their work. There are many aspects of plant pressing that are learned best from experience and by working with collectors of experience. However, the following items are offered as suggestions to the inexperienced:

Greatest efficiency is obtained from the pressing paper and the press only when the maximum surface of the paper is covered by the specimen or specimens.

Ordinarily, a specimen should be restricted to a single folded sheet of pressing paper; some large-foliaged materials are exceptions.

When individual plants are very small, many may be pressed in a single pressing paper.

Herbaceous specimens longer than 16 in. may be accommodated *in toto* by folding in a V-shaped or N-shaped manner.

---

[3] The reduction of pressing paper size to 16 inches is necessary to avoid making specimens too long to be mounted on standard herbarium paper (11½ by 16½ in.).

# FIELD AND HERBARIUM TECHNIQUES 239

Avoid breaking brittle stems at point of fold by means of moderate maceration at the fold before bending.

Springy stems may be held in place after folding by slipping slitted slips of paper over the bent ends.

Prune specimens judiciously to prevent appreciable overlapping of parts, but always leave a basal portion (petiole segment, or flower pedicel) to indicate location of pruned part.

Whenever possible, arrange one or more leaves (or parts of leaves) with lower side uppermost.

When leaves are pinnately compound, it may be necessary to excise all except one leaf. If the remaining leaf is too large for the sheet, it may be split lengthwise (or the leaflets removed from one side) providing the terminal leaflet (when present) remains attached and is not mutilated.

Large palmately compound leaves usually may be split in half lengthwise and one half discarded.

Very large and tall herbs, if of excurrent growth, may be split lengthwise along the stem (leaving the inflorescence intact or not) and one half discarded. Another choice is to press a section from the bottom, another from the middle, and a third from the top; indicating this in the notations.

In some plants, only a single leaf and the inflorescence may be accommodated within a sheet. Always note leaf arrangement in such instances and include the entire petiole and portion of stem from which it was produced.

Plants with gamopetalous corollas should have a few flowers pressed separately and some of these split open and spread out before pressing.

All roots, or other underground parts, should be washed free of soil before pressing whenever possible.

*Special aids for better arrangement* of specimens or preserving the identity of parts of them have been devised or recognized to be of value and include:

Metal bars, about ¾ by 15 in., and ⅛ in. thick, are helpful in holding and folding grass and sedge materials that are considerably longer than the pressing sheet.[4]

Flowers with deliquescent corollas or perianths often stick permanently to the pressing sheet, or are so thin as to be possible of removal only with great difficulty (especially true of iridaceous and commelinaceous flowers). The difficulty is minimized by placing a *single layer* of absorbent cleansing tissue beneath and over the flower before pressing. With care, this tissue may be removed by peeling after the specimen is dry.

Wet strips of newspaper about 1 by 4 in. are of aid in holding down obstreperous parts, when a collector is alone and needs an extra pair of hands. These strips dry in the press and are readily removed from the specimen.

Pads of newsprint (of a few to several thicknesses of paper) are

---

[4] In the absence of bars, zigzag folds of stems may be held temporarily by sliding a slotted strip of stout paper over the V, thus leaving both hands free to make successive folds.

very useful when placed around bulky portions of some specimens, for they aid in keeping nearby foliage flat by assuring a more uniform pressure on it while in press. Some botanists have advocated use of pieces of ¾-inch sponge rubber for this purpose: either is equally efficient, paper is the cheaper. Fosberg (1939) described the use of pads made of cotton batting placed between layers of newsprint. These are inexpensive, easily made, and result in the preparation of excellent specimens, especially of bulky plants.

Aquatic plants frequently are filmy or somewhat filamentous and are difficult to arrange on the sheet when removed from the water. The technique generally used is to tear some unprinted pressing papers into fourths, float the paper on the water until completely wetted, float the specimen nearby, slide over the wet and slightly submersed paper, lift the paper carefully (sloping it to facilitate water runoff) and, keeping the specimen in place on it, place the wet sheet and its specimen within a regular pressing sheet. After drying, the mucilage frequently produced from the specimen may cause it to adhere to the quarter sheet, and no effort is made to remove the aquatic plant from it.

*Keeping wet material* without its spoiling is sometimes a problem faced by collectors working in tropical regions, or under emergency situations when lacking adequate drying facilities. Two techniques have been found useful in these cases, but the results are inferior to those from the usual methods of processing material. In either case, the objective is to keep the material from decomposing after it has been collected and arranged in pressing papers, until such time as it can be dried by usual procedures. Schultes (1947) favored the following technique:

> Specimens are pressed between blotters for about 24 hours. The press is then opened, and each specimen is dipped in a tray containing a solution of 2 parts of commercial 40 per cent formaldehyde and 3 parts of water. After a few seconds of immersion, the specimens are replaced in the pressing papers without draining, and the sheets of specimens piled on top of one another without blotters or corrugates between them. Pressure is then applied for a few hours and the bundle then is enclosed in an airtight package (cellophane or Pliofilm that may be sealed, or several thicknesses of rubberized cloth). The bundle is wrapped and shipped to a place where the specimens are removed and dried conventionally. Plants may be held wet under these conditions for a month or more, and if drying must be delayed further, preservation of specimens may be maintained by opening and redipping the specimens in the same formaldehyde-water mixture.

One disadvantage in the use of the above technique is that the preserving solution causes drying, blackening, and cracking of the preparator's skin under conditions involving any appreciable contact with it. It is recommended that rubber gloves be worn if the technique is used extensively. Another disadvantage is the requirement of carrying with one a gallon or more of the preserving solution. Specimens made by this means are of inferior quality as concerns color retention and prettiness, but are adequate for the necessary scientific studies made of them. A modification of this technique was proposed by Fosberg (1947) whereby a preserving solution of one part formaldehyde and two parts of 70 per cent alcohol were substituted for the Schultes solution. The advantages were better wetting, penetration, and more satisfactory control of decomposition. Application by means of a flat 2-inch brush was advocated in place of dipping.

Hodge (1947) found the substitution of a 40–50 per cent solution of alcohol as the preservative in place of the formaldehyde-water mixture to be satisfactory, and pointed out that in the tropics, any crude form of alcohol, or any similarly spirituous liquor (as rum) was equally effective, much cheaper, and more easily obtained than was formaldehyde. Moore (1950) favors use of hydroxyquinoline sulfate as a preservative in this technique (for use, see p. 255).

*Drying techniques* are of two types: those accomplished without heat, and those with the aid of artificial heat. The majority of American collectors now dry their specimens with the aid of heat. Either technique can produce specimens of poor quality, and because the drying process is much accelerated when heat is used, greater care must be observed during all its stages if quality specimens are to result.

Drying without heat was universal until about the advent of the present century. The procedure is somewhat as follows. Plants are placed in pressing papers between the blotters of a conventional field press. No corrugates are employed. The press is locked up for about 24 hours; this is known as the "sweating" period. It is then opened, and as blotters are removed each pressing sheet is turned back, the specimen examined, and parts rearranged as the situation demands. This rearranging is an important step, for at this time the parts of most plants are somewhat flaccid and have lost their natural spring (or as one botanist expressed it, they now are "tamed"), and it is a simple matter to place the leaves in desired positions, turn back corolla lobes, etc. The difference between a poorly and a well-arranged specimen usually results from the attention given it at this stage of the process. After rearranging, the folded sheet is lifted onto a fresh dry blotter and covered by another dry blotter. This is repeated for every specimen until all have been examined, rearranged as necessary, and placed between dry blotters. The new pile of blotters and specimens is then locked up in the press and allowed to stand for another

24 to 36 hours, when the process of replacing wet (or damp) blotters with dry ones is repeated. A third change of blotters follows, and the length of the period between this and the previous change will be determined in part by the kind of material being dried, but ordinarily it is after 2 to 3 days. Except for fleshy and succulent material, which may require much longer time, most specimens are completely cured by this technique in about a week. Among the disadvantages of the method are the following:

> A minimum of about a week is required for completion of drying.
> Blotters must be changed 3 or 4 times; every wet blotter removed must be dried, usually by placing in the sun.
> A large number of blotters must be at hand; the number of specimens that can be processed per collector is small as compared to drying by aid of heat.
> The labor expenditure per specimen is excessively high.
> The possibility of impairment of specimens by mold and other fungus growth becomes a major consideration when used in many tropical climates. Larvae, invariably present within flowers of many plants, continue to eat and destroy plant parts while drying progresses.

Drying with the aid of artificial heat is the prevalent method. It is accomplished by means of heated dry air passing up and through the ducts of the corrugated sheet. The efficacy of the method depends entirely on the presence of open ducts throughout the entire sheet of corrugated material. Every closed duct impairs the quality of the specimen. Those opposed to this technique (see Fernald, 1945) have pointed out that specimens dried over heat (1) lose any waxy bloom or glaucescence that may have been present, (2) become brittle during drying, (3) do not retain the coloration that is present in specimens dried without heat, and (4) are often permanently marked by ridges of the corrugates. Of these objections, only the first is valid, and it concerns a character that is best accounted for by notations on the label accompanying the specimen. A specimen is not brittle when properly dried over heat, it should retain better coloration of foliage and of flowers than when prepared by any slower method, and it should not show the markings of corrugate ridges if ventilators are used properly. Irrespective of the heat source, the basic procedures to be followed for this method of drying are:

> Specimens are "sweated" in the field press for about 24 hours.
> They are then opened, examined individually, rearranged, and each sheet transferred to the drying press.
> The drying press is locked up with less pressure than that exerted on the plants when in the field press, and the press is placed over the heat source. (If in the open, a fire-resistant canvas skirt is placed around the press and extended to below the heat source, see Fig. 29.)

The press is tightened after 6 hours of drying, tightened again after 12 hours and at the same time is turned over so that the cool side faces the heat.

After 24 hours the press is removed from the heat source, opened, and all folders of dry specimens removed; drying is continued for wet specimens.

The total time over the heat will vary with the intensity of the heat source, from a minimum of 12 hours (considered too rapid drying by most collectors) to a day and a half or 2 days. The choice of heat sources will depend on the situation. Ordinarily electricity is the simplest and most readily controlled, and the use of incandescent bulbs (Fig. 29-f), resistance coils, or Calrod units all have been proved satisfactory. Resistance coils are the most hazardous since specimen or blotter fragments may fall on them and ignite, providing an incendiary source that may burn the press. Hot steam pipes are an excellent source when the heat is constant and especially if a circulation of hot air can be directed upward through the press. Drying plants by artificial heat in the field presents numerous problems. Kerosene lanterns have been used very effectively (7 lanterns are sufficient for a press 4 feet long, and when burned continuously require about 5 gallons of fuel a week). It should be remembered that, to prevent their smoking, wicks need to be trimmed at least twice daily when being burned continuously, and a supply of wicks for replacements becomes a staple necessity. Gasoline pressure cookstoves provide a steady, smokeless, high heat and are well adapted to plant press drying. It is necessary in connection with their use (unless specially modified) for the collector to arise several times during the night to pump up pressure in the gasoline tank, otherwise the flame goes out. Either of these heat sources is a potential source of danger, by loss of the press from fire, and caution must be used at all times when they are employed.[5]

*Mounting specimens* is a phase of specimen preparation that has been developed from a number of different approaches. In this country specimens are mounted on sheets of standard size herbarium paper (11½ by 16½ in.). After mounting, they are stored in special cases built to accommodate sheets of this size. The quality of paper varies according to the need of the herbarium. Curators of most herbaria recognize that their collections are permanent accessions, and mount the specimens on the longest-lasting and most durable paper available for its weight, a 100 per cent rag paper. Paper of this quality is expensive, and cheaper quality papers (of lower rag content) are available. Herbarium papers in a selection of qualities are available from biological supply sources. In

---

[5] For accounts of drying plants by artificial heat, see references to papers by Camp, Fernald, Gates, Lundell, Lunell, MacDaniels, Maillefer, Smith, Steyermark.

Fig. 30. Mounted herbarium specimens.

addition to the selection of paper, there remains the choice of a means by which the specimen is to be affixed to it.

Most herbaria use a glue or paste to fasten specimens to the sheets. One or two curators have used wire staples, but the tendency of staples to rust is an objectionable feature. The specimens of at least one large American herbarium are fastened solely by means of narrow strips of adhesive linen, and many curators employ a combination of paste-glue and gummed linen strips. Use of any type of adhesive-coated cellophane tape for the purpose of affixing specimens to herbarium sheets is to be discouraged. It is a practice used largely by uninformed amateurs who may be unaware that such tapes will discolor, and of the relatively short life of the adhesive coating. The adhesives probably used more than all others are Special "A" Tin Paste and Improved Process Glue.[6] The former is an inverted starch paste and the latter is a fish glue. The paste has been found reasonably satisfactory for all materials except those with large coriaceous or highly cutinized leaves; for the latter, a mixture of 3 parts of the paste with one part of the glue has been found to be more satisfactory (Merrill, 1926).

There are 3 techniques most commonly used in mounting specimens with paste or glue. The glass plate method, preferred at many institutions, requires the use of a piece of (preferably) plate glass at least 14 by 20 in. The paste is spread thinly over most of the surface with a flat 2-inch brush. The specimen is removed from the pressing sheet, placed face upward on the prepared plate, with all parts of the lower side in contact with the paste. It is then lifted, with the aid of forceps, and transferred to the sheet of mounting paper. The pressing sheet of newsprint is placed over the specimen, firmed, and taken off and discarded—this removes all excess paste from edges of leaves and flowers. The label is pasted on the lower right- or left-hand corner of the sheet. After one or two specimens have been mounted, it will be necessary to reapply fresh paste to the plate, and at intervals the accumulated dried paste and bits of debris must be scraped off with a spatula or other flexible blade, and fresh paste applied. The plate should be washed and set to dry after each mounting period.

The second, and alternate, technique requires no glass plate. The specimen is laid on the pressing paper, lower side uppermost, and with the aid of a sash brush, small amounts of paste are brushed directly on

---

[6] Both products are manufactured by the Russia Cement Co., Gloucester, Mass. They are available from that source in 1-gallon cans and larger volumes. They keep indefinitely when covered, and require no thinning or heating before use.

major portions of the specimen. It and the label are then affixed to the herbarium sheet as noted above. Some curators consider the second method to be more rapid, less messy, and much more economical of paste. The advantage of the glass plate method lies in assurance of coverage by paste over all parts of the lower side of the specimen, this resulting in a better and more permanent contact with the paper.

A third technique of applying paste is not generally used, except in connection with mounting aquatics or other flimsy subjects. The specimen to be mounted is laid, lower side uppermost, on a piece of cheesecloth (usually 2 to 3 thicknesses). It is sprayed with a diluted solution of paste by means of an atomizer, and then flipped over onto the sheet of herbarium paper. The diluted paste generally is adequate to affix the very light weight and thin texture of such specimens.

Adhesive strips of linen may be used to affix a specimen to the sheet without the aid of any additional adhesive material, but the practice is not employed extensively and the disadvantages seemingly more than offset the advantages. Supporters of the method point out that it permits removal of a specimen from the sheet by slitting the strips, making the plant available for critical study of both sides. When study has been completed, the specimen must be remounted on a fresh sheet if neatness of material is a factor. The disadvantages are the lack of permanence, the additional time required, and the disconcerting appearance of the sheets bearing a half dozen or more strips. On the other hand, it is often necessary to use strips of gummed linen tape as additional aids to hold heavy or largely woody specimens to the sheet. These strips (about 1½ inches long) should be as narrow as practicable, varying from $\frac{1}{16}$ to $\frac{3}{16}$ inch wide, and their use should be restricted to a minimum.

More recently it has been found (Archer, 1950) that specimens may be affixed to the sheet neatly and readily by application of thin bands of quick-drying liquid plastic. In urging further experimentation with this technique, Archer emphasized the importance of using a permanent type of plastic (i.e., cellulose acetate, or other duPont formulae specified by him) that would not become yellow with age nor crack with usage. It is likely that this use of a liquid plastic material may provide a supplementary technique of special value for woody specimens or others ordinarily requiring the use of gummed linen strips.

*Herbarium labels* are an important part of any permanent plant specimen, irrespective of whether the specimen is pressed and mounted, preserved in fluid, or stored dry in boxes. The purpose of the label is to provide the person using the specimen with pertinent data not apparent from the material. Consideration should be given to the design and format of the printed form of any label. To most people, the name and source of the specimen constitute the information first desired from examination of its label. This is to be considered in designing any label, and to avoid selection of type faces that are distractingly large or bold. Ordinarily a

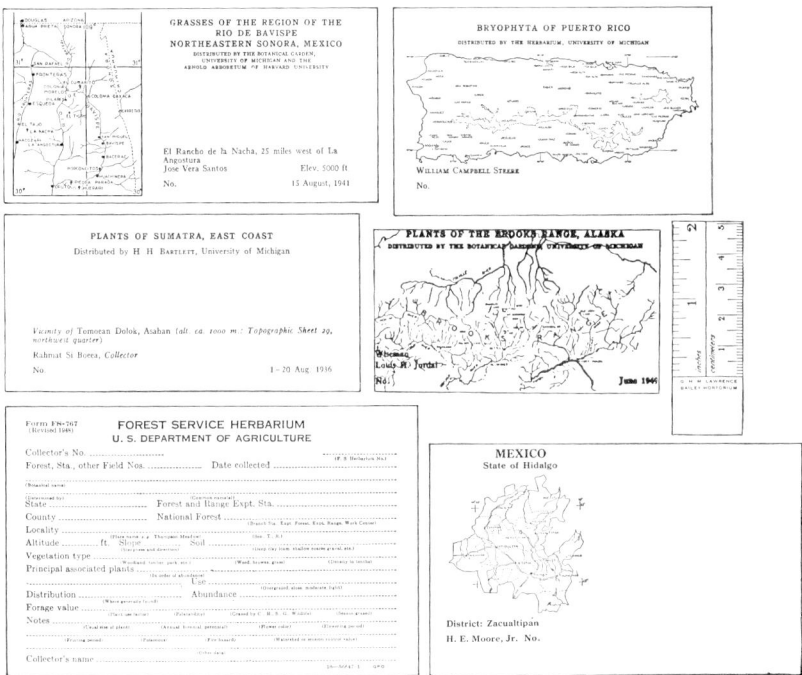

Fig. 31. Examples of herbarium labels. Outline maps printed in black when at end of label and in faint color when covering label (data then typed in black over the map). Scale at right center used when photographing type specimens and is retained on sheet as permanent record (negative number and date are added).

10-point type face is as large as is needed for the title of the label, and faces of 8-point or 6-point are used for the remaining data. The label should be large enough to accommodate the data to be placed on it. Very large labels are not only offensive in appearance, but demand space often needed for the specimen. Under no circumstances should a label be so large as to require folding. Printed labels that designate the data to be provided and the spaces in which they are to be inserted, are wasteful of space and are an affront to the competent and intelligent collector. For ordinary purposes, most labels are of sizes approximating 4½ by 2¾ in. (varying ½ in. in either dimension).

Botanists engaged in floristic studies of an area should give consideration to including a reproduction of a condensed outline map of the area (or a political section of it) on one side of the label, especially if the area is one not represented in detail on the maps of any domestically published atlas. In practice, a colored dot is placed on the map to indicate the approximate location where the specimen was collected (see Fig. 31 for samples of labels used by various botanists). The making of labels by the mimeographing process is to be discouraged unless well done and impressed on a high-quality bond paper. Few mimeographed labels have characters that are sharp and clear, and too often they are run off on the usual short-lived sulfite mimeograph paper which yellows or disintegrates with age.

Data on labels should be typed, preferably on a machine equipped with elite type characters, since this produces 12 characters to the inch instead of the 10 characters of pica type.[7] Duplication of labels by means of carbon copies of the typed data is not satisfactory because the copies smudge and have a secondhand appearance. Their use often produces the illusion that the original copy went with a better specimen of the same collection. Data written on labels in longhand are always acceptable, but care must be taken that it be very legible, especially as concerns place names. It should be remembered that handwritten labels must be legible and comprehensible to persons not familiar with the English language or with local geographical names.

### Preservation of specimens

Herbarium specimens are subject to serious damage by being eaten by various insects. Notable among these predators is the tobacco or herbarium beetle (*Lasioderma serricorne*) or the drugstore beetle (*Stegobium paniceum*) and the more minute and colorless book louse (*Atropos divinatoria*). These insects may complete their life cycle within dried specimens, attacking material of all ages, and chewing through the sheets to pass from one specimen to another. The herbarium beetle, the worst offender of the three, completes its life cycle in 70 to 90 days, with larval stage occupying 35 to 50 days. Large collections of herbarium specimens are known to have been completely destroyed by insect infestations (especially by the larvae) during periods of unattended storage. Constant vigilance against ravages by insects must be maintained if specimens are to be preserved. Preservation of herbarium collections from insect damage is accomplished most effectively by combined use of

[7] See Chapters XII and XIII and Walker (1942) for label data.

insecticides and repellents. There is no known means of treating herbarium specimens whereby they are permanently free from insect attack and at the same time readily available for use.

*Insecticides* kill the insect either by contact or by being eaten. Contact insecticides used in herbarium management include cyanide gas, paradichlorobenzene (PDB), carbon disulfide gas, a mixture of ethylene dichloride and carbon tetrachloride, or DDT. Placing specimens in a chamber which is heated to a degree lethal to the insect is another means of killing. Digestive poisons that have been used for this purpose include salts of mercury or arsenic.

The first 4 of the above mentioned insecticides are functionally gases, and to be effective the specimens must be in an enclosed chamber (a special airtight compartment or a reasonably airtight metal herbarium case). Most institutions using them fumigate the entire herbarium at regular intervals of 1, 2, or 3 years—depending on the degree of control found to exist. Furthermore, every parcel or bundle of plants that comes into the herbarium must be fumigated before the specimens are mounted or distributed throughout the collection. New accessions are a prime source of reinfestation. This is particularly true of material received on a temporary loan basis; it too must be fumigated on receipt if all sources of infestation are to be checked. Specimens collected on expeditions and awaiting determination and naming should be fumigated on receipt. These freshly collected specimens often contain actively foraging larvae of other insects which should be killed as soon as possible. There is no residual or permanent gain from any form of gas poisoning. Its effective period is that when present as a gas in high concentration.

The most effective insecticide is cyanide gas. It is exceedingly poisonous to all forms of life and may be used only with special precautions and generally by specially trained personnel.[8] It is now used exclusively at the Kew, the world's largest herbarium, under procedures described in detail by Ballard (1938). Its use demands either (1) a specially constructed airtight chamber into which the specimens are placed, or (2) closure and sealing of the building in which the herbarium is situated. When accomplished by the use of a special chamber, the cost of fumigation is less than one dollar per thousand specimens, but if all spaces in a building are to be fumigated, the cost is usually far in excess of this amount.

[8] Any contemplated use of cyanide gas for the control of insect pests in herbarium specimens should be preceded by an investigation of local public health ordinances. Many municipalities forbid use of the gas within jurisdictional areas, except under supervision and cognizance of public health officers.

Paradichlorobenzene, the common insecticide for clothes moths, is effective in killing herbarium insects if used in sufficient quantities. The crystals vaporize more rapidly at temperatures of 110°-130° than at normal room temperature, and specimens should remain in the airtight chamber for periods of 3 to 4 days at such a temperature, or for longer periods at lower temperatures. This material is more commonly used in most herbaria as a repellent than as an insecticide.[9]

Carbon disulfide (carbon bisulfide) is probably the most commonly employed fumigant in herbaria. Ordinarily it is purchased in cartons of 5-pound bottles or 1-pound cans. The technical grade is as effective and much more economical than the more refined C.P. quality. Carbon disulfide is a very volatile, highly inflammable liquid. It produces a gas at room temperatures of 50°-60° F that is characterized by a disagreeable sulfurous odor. The danger of fire or serious explosion attendant on the use of this material cannot be overemphasized.[10] The gas is of a highly explosive character, and the material should never be used in a room having an open flame, sparks, or even high temperatures. Fumes are readily ignited by hot electric resistance coils. The gas is heavier than air, and is an efficient and economical insecticide. Ordinarily, in herbaria housed in steel cases, it is handled for fumigating by placing a metal tray (pie plates are ideal) containing about 6 to 8 fluid ounces of the liquid on the top of each tier of shelves, closing the door tightly, and leaving it closed for 2 to 3 days. The usual procedure is to treat all cases at one time and to close the room for the fumigating period. No herbarium case is gastight, and seepage of the gas will occur. For this reason, it should be used at a period when either the room can be sealed efficiently or the entire building vacated of personnel. It must be remembered that the fire hazard is present during all this period. Herbarium specimens in ordinary wooden or pasteboard cases or boxes must be transferred to a special case or chamber for adequate fumigation. Adequate concentration of either cyanide gas or of carbon disulfide will kill the eggs of the herbarium beetle; however, these concentrations do not always reach all specimens, and complete control is not achieved in most cases. In many instances this also may be because of undue leakage of the gas from the steel cases or to an inadequate volume of liquid used for the case. Other things being equal, the more bulky the plant material in the case, the more space to be filled with gas and the greater is the volume of liquid required.

[9] For papers on the use of paradichlorobenzene in the herbarium, see Martin (1925) and Merrill (1948).
[10] The protection offered by many fire insurance policies is void for losses or damages attributable to use of this chemical.

Entomologists now advocate substitution of a mixture of 3 parts (by volume) of ethylene dichloride with one part of carbon tetrachloride for the carbon disulfide. This mixture is slightly more expensive but is of equal toxicity as an insecticide; it volatilizes readily at room temperatures *and is noninflammable.* Ethylene dichloride is an inflammable liquid, but ceases to be so when mixed with the incombustible carbon tetrachloride at the ratio of 3 to one. The mixture may be stored indefinitely when tightly stoppered. In fumigating practices, a 10-15 per cent increase in volume of fumigant is recommended over that calculated for carbon disulfide.

Bichloride of mercury (corrosive sublimate) is a stomach poison that is highly toxic and generally lethal to all forms of animal life that ingest it. There are two or more techniques commonly used for the application of this material to herbarium specimens. The crystals or powder is dissolved in a solution of 95 per cent alcohol in quantity sufficient to produce a supersaturated solution at room temperature. This is the stock solution. It should be kept in a glass-stoppered bottle clearly labeled as to its contents and marked POISON.[11] The solution used to poison the specimens is prepared by adding one part of stock solution to 9 parts of alcohol. According to one technique, this poisoning solution is poured into a glass or nonmetallic tray and each plant specimen, before mounting, is dipped in it, returned to its pressing sheet, and dried between blotters for 24 hours. Following this, it is ready for mounting. The second technique uses the same poisoning solution, but applied by means of a flat brush instead of by dipping the specimen. This modification is more economical of material, is generally a quicker procedure, and is less likely to result in any undesirable whitish residue of the chemical on the plant parts. The material should not be sprayed with an airbrush, unless the latter is of a noncorrosive metal. Ordinarily bichloride of mercury solution is not applied to specimens already mounted on herbarium sheets, because if the plant material is of recent collection the alcohol will dissolve some of the pigments and leave an unsightly discolored residue on the sheet, and furthermore, the chemical itself usually will produce a brownish or blackish mercuric stain when in contact with paper. Com-

---

[11] This stock solution is generally made up with ethyl alcohol, but where the cost of this solvent prohibits its use, methyl (wood) alcohol or isopropyl alcohol may be used. The poisonous character of this material is not apparent immediately on ingestion, and when consumed into the human system it causes a gradual and finally complete cessation of function of the kidneys. It is a cumulative poison, and repeated intake of minute quantities is additive and injurious. Death is not immediate, but it is certain. The material is a strong irritant to mucous membranes and one should not rub the eyes at any time while handling it or diluted solutions of it.

plete control of insect infestation cannot be insured by poisoning with this material. Nonetheless, it is one of the most effective stomach poisons known for the purpose and does provide sufficient control to justify its use. However, the metalic salt must be eaten before the insect is affected by it, and since it is applied only to the surface of the plant it is sometimes escaped entirely by those larvae that are working within a stem or within a bulky dense inflorescence. Penetration of the liquid, even by dipping of the specimen, is not appreciable in fleshy, succulent, or cutinized parts of plants, in the heads of many of the Compositae, or in waxy, thick, coriaceous foliage. *Bichloride of mercury is not a permanent poison.* As years pass, the material becomes inert and is then worthless as an insecticide. It is alleged by some botanists to be responsible for dermatitis symptoms, others object to it because of the possibility of internal poisoning from handling the treated parts (no case of poisoning from this source has been found reported in the record), and additional objections exist on the grounds of its lack of complete protection against insect damage due to its loss of potency.

DDT (dichloro diphenyl trichloroethane) has been advocated as an insecticide to be dusted on the plants as soon as they are dried and as removed from the press (Howard, 1947). It was found that a light sprinkling (as with a salt shaker) of 100 per cent DDT killed all insects present on or in the plant at time of collection, and concentrations of 25 to 50 per cent DDT have been found by other botanists to be equally effective. The material, dissolved in one of several solvents, has been used in place of bichloride of mercury as a presumably permanent contact poison for herbarium specimens. However, it is now known that currently available preparations of DDT do not provide a permanent insecticide, and that the toxic value of it dissipates within one or two years after application. Investigations on the use of DDT liquid applications to the interior of the herbarium cases are being conducted, since it is much simpler to spray such interiors at regular intervals of about a year than to treat individual specimens with any nonpermanent material.

Heat has been mentioned as an efficient means of controlling insect infestations. The subject was discussed by O'Neill (1938) and in at least one American herbarium (University of Montreal) each steel herbarium case has been equipped with a thermostatically controlled electric heating unit by means of which the spaces within the case may be heated to temperatures lethal to eggs, beetles, and larvae. O'Neill demonstrated that a temperature of 140° F was adequate to kill all stages of insect development within a few minutes. However, because of the poor conductivity of the volume of paper present in a well-filled herbarium case, it is necessary to raise the air temperature within the case to 170° F for a period of 4 to 5 hours. There is no evidence to show that this heating increases the brittleness of the specimens. The construction of a special heating chamber has been found useful for application of this method, but the cost of handling specimens is high.

*Repellents* are chemicals that deter infestation by insects but do not function necessarily as agents that are toxic or lethal to the insect. The two principal repellents are naphthalene compounds (naphtha flakes) and paradichlorobenzene (PDB). The report by Merrill (1948) on the efficacy of repellents indicated that the placing of a small (2 or 3 oz. size) muslin bag filled with a mixture of two parts naphthalene and one part paradichlorobenzene was placed in each case and pasteboard carton containing specimens. These bags are refilled with the mixture once a year, and have been found to be effective as repellents when used in cases known to be free from initial insect infestation at the time of insertion.

### Housing of bulky materials

Not all plant materials, because of their excessive bulk and weight, are suited for mounting directly on herbarium sheets. Other materials, especially fruits and some vegetative parts, are of too succulent or fleshy a texture to retain, when pressed and treated in the conventional manner, any of the characters that are so diagnostic when fresh. All of these cases demand special consideration, that the necessary morphological features may be preserved as satisfactorily as possible.

*Dry treatments* involve the use of boxes, folders, and envelopes. Material of large fruiting structures (cones, capsules, woody palm inflorescences, etc.) are usually placed in boxes designed to fit into herbarium case pigeonholes. These boxes are of strong pasteboard construction, covered with a buckram cloth, and have a cover lipped on all sides, with a label holder and handle affixed to the front end. Bailey (1946) has pointed out the utility of a folder of heavy white ledger paper for use in the housing of unmounted palm specimens. The same type of folder is equally suited for coniferous material whose leaves are persistent on drying. Small dry fruits too bulky to remain directly on the specimen affixed to a sheet are preserved in stout clasp-type envelopes on which are pasted labels that are duplicates of those on the conventionally pressed specimens.

*Liquid preservation* is employed by curators of some herbaria, especially for such succulent material as found in the Aizoaceae, Cactaceae, and Euphorbiaceae, or for material temporarily preserved for demonstration or dissection purposes in the classroom. There are many disadvantages to use of liquids as a means of permanent preservation, primarily the bulk, weight, and cost of adequate containers and preserving fluids. In addition, attention must be given to periodic replacement of the preserving fluid. Most materials can be preserved adequately, or their diagnostic characters shown by use of other methods. In the case of succulents, these other methods usually involve the preservation of the outer shell, which is skinned away from the softer fleshy central portion, and by sectioning and by drying. Prior to this, the plant should be photographed (black and white, not in color except as a supplementary photograph) in sufficient detail to show the form and such other significant details as may otherwise be lost. A metric scale or other object should be included in the photograph to indicate sizes. Flowers of succulents usually are excised and pressed conventionally, and either mounted on sheets or placed in pockets or envelopes together with the more bulky dried portions. Photographs should be kept in the herbarium together with the dried material.

There are many formulas available for preservation of plant materials in liquid. Each is recommended usually to meet the needs of a particular use for the material. The time-honored liquid preservative for general taxonomic use is a 5 per cent aqueous solution of the commercial formaldehyde. This provides a very economical but only moderately satisfactory preservative. Formaldehyde solutions prevent decomposition of the material, but otherwise they have little to commend their use over solutions described below. Specimens preserved in formaldehyde become unduly limp and flaccid, and gamopetalous flowers particularly fail to retain characteristic forms. The solution causes a hardening of the skin even from contact with the small amounts remaining with specimens that have been washed in clear water. The odor is unnecessarily offensive, especially to beginning students, and the fumes of the diluted solution irritate mucous membranes of nasal passages and eyes of sensitive individuals. Many schools have ceased to use formaldehyde "pickle" for any material preserved for classroom use, and by using inoffensive preservatives or techniques have stimulated increased interest in dissection of the better preserved specimens.

If it is anticipated that morphological study is to be made of parts of the preserved material, a solution of 70 per cent alcohol is to be used. Material that is to be subjected only to gross morphological examination may be preserved in a solution of 50 per cent alcohol. This is adequate to prevent decomposition, and usually hardens floral parts more than does the 5 per cent solution of commercial formaldehyde. Material that has been preserved in a mixture of 50 or 60 per cent alcohol is much more pleasant and easy to handle. Alcohol (ethyl or isopropyl) is much more satisfactory than is formaldehyde for "pickling" specimens for class and laboratory dissection. If the "pickled" material is somewhat fleshy, or if zygomorphic corolla form is to be retained, a 70 per cent alcohol solution is more satisfactory.

Oxyquinoline sulfate (or hydroxyquinoline sulfate), in 1-2 per cent aqueous solution has been found to be a general preservative to be used as a substitute for either alcohol or formaldehyde (Swingle, 1930). The concentration of the solution is not a critical factor; a 0.1 per cent solution checks bacterial decomposition in most plants, and solutions of 5 per cent of the material have no deleterious effect on preserved material. The solution is noncorrosive to the skin, is practically odorless, it does not evaporate any more rapidly than does water, and it is inexpensive. If spilled, or lost by leakage, no injurious effects to fabrics result (although it should be kept from ferrous metals). Swingle reported that materials preserved in it were killed and adequately fixed for macroscopic morphological study. It is particularly useful for field "pickling," since small envelopes or capsules of measured quantities of powder may be taken on a trip and mixed with water as needed (cf. Moore, 1950). Material preserved in this solution does not retain most colors (greens persist the longest) and in general corolla tissues soften as do those preserved in formaldehyde solutions.

Color retention and preservation of flowers by means of formaldehyde fumes have been found satisfactory for some kinds of material, and because of the ease of preparation and economy, the use of this technique is particularly adapted to classroom needs. Proposed by Harrington (1947) and somewhat modified since then, the technique consists of placing a pad of absorbent cotton in the bottom of a wide-mouthed glass. This is then wet with commercial formaldehyde (undiluted) and the cotton covered with a protective perforated cellophane disk. Fresh flowers are placed on the disk and the container is sealed tightly until the flowers are to be used. Some flowers retain their shape and color for many months, although those of certain families (Compositae, Commelinaceae, and Iridaceae) are for the most part notable exceptions that do not respond favorably to the method.

Preservation of colorless parasitic materials (as *Monotropa*) that ordinarily blacken when placed in the usual preserving fluids was reported by Nieuwland and Salvin (1928) to be accomplished by the following procedure,

> Place specimen in a test tube of 95 per cent ethyl alcohol; add 0.5 g of sodium sulfite, and 0.5 cc of concentrated hydrochloric acid.
> Seal tightly and shake.
> Allow to stand one week.
> Siphon or pipette off the alcohol, replace with xylol, and seal airtight.

Retention of green coloration is frequently desired in pickled material prepared for class use. Several formulas have been found satisfactory, but one of the simplest to prepare and use is a modification of that by Keefe (1926), as follows:

> 90.0 cc 50 per cent ethyl alcohol (isopropyl alcohol may be used)
> 5.0 cc commercial formaldehyde
> 2.5 cc glycerin
> 2.5 cc glacial acetic acid
> 20.0 g cupric chloride (for normally yellow-green foliage use half this quantity)
> 2.5 g uranium nitrate

Fresh material is placed directly in the solution and left until needed. Flowers placed in this solution will not retain any of their original colors. Herbarium specimens made from material taken from this solution are reported to retain their green foliage color indefinitely.

*Quick-freeze* equipment has been found by the author and others (cf. Harrington, 1950; Rollins, 1950) to provide an excellent means of preserving material for class dissection purposes. Most flowers and fruits packaged in moisture-tight cellophane type wraps or in glass jars keep their color, form, and texture indefinitely when stored at temperatures of 10°F and lower. Material taken from the freezer retains its color and much of its structure for periods of 2 to 4 hours. Success has been experienced with flowers of *Tradescantia, Eichhornia, Hypericum, Iris, Ipomoea,* and others of fugacious or deliquescent character. *Monotropa* does not blacken until an hour or more after thawing. A freezer of 20 cu ft capacity, carefully packed, will hold enough material for a class of about twenty students. For lists of names of plants well suited to this type of preservation, see Rollins (1950) pp. 294–297.

*Special techniques* pertinent to the needs of particular families are the problems of every collector. Before making extensive collections for herbarium purposes, the collector should study the very competent papers by Johnston (1939), Fosberg (1939), Fogg (1940), Ricker (1913), and Archer (1945). Techniques designed to meet the needs of special groups have been recounted by Bailey (1946) for the palms, Macdougal

(1947) for cacti, Sharp (1935) for gymnosperms, and Verdoorn (1945) for succulents.

All taxonomic research involves preparation and study of floral dissections from herbarium material. When working on revisionary studies it is desirable to keep these dissections for ready reference and restudy. Several techniques have been found satisfactory for this purpose. According to one, the specimen is dissected on a glass slide (1 by 3 in. or 2 by 3 in.) after having been boiled a minute or less in an aqueous solution of 10 per cent glycerin. On completion of the dissection, the specimen is covered with a few drops of warm glycerin, and the slide is stored on a flat tray (usually in a small, dustproof case) after having been labeled to identify it with the herbarium specimen from which it was removed. These dissections may be preserved indefinitely by occasional addition of a drop of a solution of one per cent oxyquinoline sulfate. They never dry out completely and may be dissected further at any time as necessary.

An account of a different treatment of such dissections has been published by Quisumbing (1931), wherein fresh flowers, or the dried flowers removed from a herbarium sheet, are boiled in water, dissected, studied, and transferred to a glass slide on which has been placed a sufficient number of drops of sodium silicate (common water glass) to almost float the dissection. The preparation is covered quickly with a cover slip (transparent plastic may be used for large material) and allowed to dry. Best results are obtained if all surplus moisture is removed by teasing with pledges of absorbent tissue before transferring to the permanent slide mount. If it is necessary later to remove the cover slip for further dissection, the medium can be softened by boiling in water "for an hour or so and then left in an evaporating dish for a longer period."

### Herbarium cases

The modern herbarium case is of welded and reinforced steel construction, with 2 or 3 tiers of pigeonholes (the number depending on whether the case has a single door or double doors) each usually 19 in. deep, 13 in. wide, and 8 in. high. These cases when empty weigh 350 to 700 pounds each. They are not fireproof, but are dust-tight and are designed to be sufficiently gas-tight for most fumigation requirements (Fig. 28).

Small collections may be housed temporarily in specially designed cardboard cartons that may be stacked in tiers to heights of six feet or more without damage to boxes or contents (Merrill, 1926). These cases are used in many larger herbaria to house overflow collections or to alleviate other congested conditions (Fig. 32). They have the advantage of economy and are reasonably convenient. They are not at all dustproof, and insects have free access to the contents. Fumigation of specimens housed in them may be accomplished either by removal of material to a fumigation chamber or by general fumigation of the room they are in.[12]

Fig. 32. Temporary herbarium storage boxes.

Plants are arranged in the herbarium according to a selected classification. If the collection comprises a local flora, or is restricted to indigenous plants of the United States, the selection of the classification is not of major importance. Most American herbaria are arranged according to either the Engler or the Bessey system.[13] If the herbarium is to include specimens from all parts of the earth, it is necessary for the present to

---

[12] These cartons are available currently from the Starr Corrugated Box Co., Maspeth, L. I., N. Y., at an approximate cost (in 1950) of $25.00 per hundred. Knobs or pulls for the door flaps are an additional item.

[13] See pp. 127–130 for sequence of American families of flowering plants according to the Bessey system.

follow either the Engler system or that of Bentham and Hooker—
unless the tenets of the Hutchinson system are accepted.

Herbarium specimens, after being mounted, are sorted by family and
by genus. Those of each genus are placed within a folded genus cover.
In most herbaria, when there are several sheets of a given species, they
are placed in a genus cover apart from the miscellaneous species of that
genus. By convention, the genus cover has the name of the genus written
or printed on the lower left edge, and when the cover is restricted to
specimens of a single species, the genus initial and species name are
placed on the lower right margin. Curators following the Engler system
usually precede the generic name on the lower left by the number assigned to that genus by Dalla Torre and Harms.[14] In small herbaria, it
is customary to file the genera and species in alphabetical sequence under
each family. In the large herbaria of this country the genera are filed in
phyletic sequence according to Dalla Torre and Harms. In such instances,
there usually is an index card to genera filed in the pigeonhole where
each family starts. This card lists the genera alphabetically and following each name is the Dalla Torre and Harms sequence number for
the genus, or there is indicated the number of the herbarium case in
which each genus is situated. The species of some large genera (as *Carex,
Prunus, Senecio, Rhododendron,* etc.) may be filed also according to
a particular phyletic sequence, and in such instances a similar index card
accompanies the cover of the first species of the genus. The basis for the
phyletic sequence of species is usually that given in a monograph of
the group.

Most large herbaria segregate specimens also by the continent to
which the material is indigenous. Usually this fact is indicated by appropriate index tabs or names on the front edge of each genus cover, or
by using genus covers of different colors for each geographic region. By
these systems, if a species is represented by specimens from both eastern
United States and other specimens from continental Europe, those of
each region will be in separate genus covers and all the genus covers of
a single species will be together. Likewise, in large collections, the specimens of a given species within a single cover may be arranged in a predetermined geographic sequence. In the United States this is usually
from north to south and east to west. The particular sequence is readily
discernible from inspection of the sheets within any selected cover. Anyone making use of the facilities of a herbarium employing this practice

[14] Dalla Torre, C. G. and Harms, H. [eds.]. Genera siphonogamarum. Berlin, 1900–1907. For description of this work, see p. 287 of this text.

should take cognizance of it and ensure that he does not disarrange the specimens by careless or indiscriminate shuffling. The principal advantage of geographical arrangements of specimens within a cover is that it enables one to observe quickly the probable general distribution of the species, and also to ascertain readily whether or not a specimen from a particular place is included in the collection.

### Type specimens

Type specimens are the specimens on which the name of an element has been based. In the event of a discrepancy between various descriptions of the element, recourse is had to the type specimen, and the situation existing with it often clarifies the matter. For this reason, type specimens (or types, as they are usually termed) are among the most valuable in any herbarium. There can be only a single type specimen for each nomenclatural element. Because of their value, types are given special care by curators of herbaria. In some herbaria, they are kept in separate and special cases. This precludes unnecessary handling and permits more adequate inspection for possible insect infestation. Usually the sheet bearing the type is placed within a protective cover of some sort, or within a large envelope.[15] If the herbarium is not in a reasonably fireproof structure, the types should be removed from it and given the benefit of the most protective housing available. Evidence of the cogency of taking these precautions lies in the fact that 7 American herbaria have been destroyed by fire, in whole or in part, during the last 50 years.[16] A greater and more tragic loss of literally thousands of type specimens resulted from the burning of the great herbarium at Berlin in 1943, as a result of allied bombing raids. Type specimens should not be used or handled any more than is essential, and curators of many herbaria do not permit their being sent out on loan to other botanists or institutions.

*LITERATURE:*

ARCHER, W. A. Collecting data and specimens for study of economic plants. U.S. Dept. of Agr. Misc. Publ. 568: 1–52, 1945.

———. New plastic aid in mounting herbarium sheets. Rhodora, 52: 298–299, 1950.

[15] One practice of increasing frequency is to place the type specimen within a protective folder of genus-cover material whose folded lower edge has been overprinted with a broad red band. This conspicuous edge facilitates locating a type when it has been included with covers of other specimens.

[16] According to data provided by Jones and Meadows (1948), these were the herbaria at the following institutions (together with date of loss): California Academy of Sciences and the Univ. of Idaho 1906, Earlham College 1924, Clemson College 1925, Univ. of Tennessee 1934, Tuskegee Institute 1940, and Univ. of Kentucky 1948.

BAILEY, L. H. The palm herbarium with remarks on certain taxonomic practices. Gentes. Herb. 7: 153–180, 1946.
BAKER, G. E. Freezing laboratory materials for plant science. Science, 109: 525, 1949.
BALLARD, F. Herbarium specimens and gas poisoning. Kew Bull. 1938, 387–389.
BENSON, L. Notes on taxonomic techniques. Torreya, 39: 73–75, 1939.
BLAKE, S. F. Better herbarium specimens. Rhodora, 37: 19, 1935.
CAMP, W. H. On the use of artificial heat in the preparation of herbarium specimens. Bull. Torrey Bot. Club, 73: 235–243, 1946.
COLEY, M., and WEATHERBY, C. A. Wild flower preservation, a collector's guide. 197 pp. New York, 1915.
COLLINS, F. J. The use of corrugated paper boards in drying plants. Rhodora, 12: 221–224, 1910.
―――. Better herbarium specimens. Rhodora, 34: 247–249, 1932.
FASSETT, N. C. Herbarium technique. Rhodora, 51: 59–60, 1949.
FERNALD, M. L. Injury to herbarium specimens by extreme heat. Rhodora, 47: 258–260, 1945.
FESSENDEN, G. R. Preservation of Agricultural specimens in plastics. U.S. Dept. of Agr. Misc. Publ. 679: 1–78, 1949.
FOGG, J. M. Suggestions for collectors. Rhodora, 42: 145–157, 1940.
FOSBERG, F. R. Plant collecting manual for field anthropologists. Philadelphia, 1939.
―――. The herbarium. Sci. Monthly, 63: 429–434, 1946.
―――. Formaldehyde in plant collecting. Science, 106: 250–251, 1947.
GATES, B. An electrical drier for herbarium specimens. Rhodora, 52: 129–134, 1950.
GLEASON, H. A. Annotations on herbarium sheets. Rhodora, 35: 41–43, 1933.
HARRINGTON, H. D. Preserving plants in formaldehyde fumes. Turtox News, 25: 238, 1947.
―――. Preserving flowers by freezing. Turtox News, 28: 51, 1950.
HARRINGTON, H. D. and SMITH, A. C. Methods of preserving and arranging herbarium specimens. Journ. N. Y. Bot. Gard. 31: 112–125, 1930.
HODGE, W. H. The use of alcohol in plant collecting. Rhodora, 49: 207–210, 1947.
HORR, W. H. A rapid plant dryer. Trans. Kans. Acad. Sci. 50(2): 191–193, 1947.
HOWARD, RICHARD A. The use of DDT in the preparation of botanical specimens. Rhodora, 49: 286–288, 1947.
JOHNSTON, I. M. The preparation of botanical specimens for the herbarium. Arnold Arbor., Harvard Univ., 1939.
JONES, G. N. and MEADOWS, E. Principal institutional herbaria of the United States. Amer. Midl. Nat. 40: 724–740, 1948.
LUNDELL, C. L. A useful method for drying plant specimens in the field. Wrightia, 1: 161, 162. 1946.
MACDANIELS, L. H. A portable plant drier for tropical climates. Amer. Journ. Bot. 17: 669–670, 1930.
MACDOUGALL, T. A method for pressing cactus flowers. Cactus and Succ. Journ. 19: 188, 1947.
MAILLEFER, A. Les herborisations et la dessication des plantes pour herbiers. Bull. Soc. Vaudoise Sci. Nat. 62: 421–429, 1944.
MARTIN, G. W. Paradichlorobenzene in the herbarium. Bot. Gaz. 79: 450, 1925.
MERRILL, E. D. An efficient and economical herbarium paste. Torreya, 26: 63–65, 1926.
―――. An economical herbarium case. Torreya, 26: 50–54, 1926.
―――. On the technique of inserting published data in the herbarium. Journ. Arnold Arb. 18: 173–182, 1937.

———. On the control of destructive insects in the herbarium. Journ. Arnold Arb. 29: 103–110, 1948.
MILLSPAUGH, C. F. Herbarium organization. Field Mus. Nat. Hist. Publ. Mus. Tech. Ser. 1: 1–18, 1925.
MOORE, H. E. JR. A substitute for formaldehyde and alcohol in plant collecting. Rhodora, 52: 123–124, 1950.
NICHOLS, G. E. and ST. JOHN, H. Pressing plants with double-face corrugated paper boards. Rhodora, 20: 153–160, 1918.
NIEUWLAND, J. A. and SLAVIN, A. D. Preservation of *Monotropa* and similar plants without discoloration. Proc. Indiana Acad. Sci. 38: 103–104, 1928.
O'NEILL, H. Heat as an insecticide in the herbarium. Rhodora, 40: 1–4, 1938.
PIERCE, W. D. Retention of plant colors. Science, 84: 253–254, 1936.
QUISUMBING, E. Water glass as a medium for permanently mounting dissections of herbarium material. Torreya, 31: 45–47, 1931.
RICKER, P. L. Directions for collecting plants. U.S. Dept. of Agr. Bur. Pl. Ind. Circ. 126: 27–35, 1913.
ROBINSON, B. L. Insecticides used at the Gray Herbarium. Rhodora, 5: 237–247, 1903.
ROLLINS, R. C. Deep-freezing flowers for laboratory instruction in systematic botany. Rhodora, 52: 289–297, 1950.
SALISBURY, E. The collection, preservation and interchange of biological material. Roy. Soc. Empire Sci. Cong. Rpt. 1946 (2): 206–211, 1948.
SCHULTES, R. E. The use of formaldehyde in plant collecting. Rhodora, 49: 54–60, 1947.
SCULLY, F. J. Preservation of plant material in natural colors. Rhodora, 39: 16–19, 1937.
SHARP, A. J. An improvement in method of preparing gymnosperms for the herbarium. Rhodora, 37: 257–268, 1935.
SMITH, G. G. A drying cabinet for the herbarium. Journ. So. African Bot. 12: 43–45, 1946.
STEVENS, F. L. Corrugated aluminum sheets for the botanist's press. Bot. Gaz. 82: 104–106, 1926.
STEYERMARK, JULIAN A. Notes on drying plants. Rhodora, 49: 220–227, 1947.
SWINGLE, C. F. Oxyquinoline sulphate as a preservative for plant tissues. Bot. Gaz. 90: 333–334, 1930.
TURRILL, W. B. Taxonomy in the seed-bearing plants. Kew Bull. 1950: 453–461. [Largely an account of herbarium management at Kew.]
VERDOORN, I. C. On the genus *Aloe:*—Preparation of herbarium material at Pretoria. Chron. Bot. 9: 150, 151, 1945.
WALKER, E. Recording localities on specimen labels. Chron. Bot. 7: 70–71, 1942.
WHERRY, E. T. A plastic spray coating for herbarium specimens. Bartonia, 25: 86, 1949.

## CHAPTER XII

## MONOGRAPHS AND REVISIONS

The preparation required for the production of monographs and revisions often appears unduly formidable to one who is beginning in the field of taxonomic research.[1] This chapter proposes only to suggest some of the basic considerations preparatory to initiating the investigation on which it may be expected to base the monograph or revision, and to suggest steps to be followed, together with suggested sources of information.[2]

It must be emphasized initially that research, designed for ultimate publication as a revision, or as a monograph, demands a measured degree of competency. Competency is gained only by study, guidance, and experience. All scientific research must be executed with the utmost care, and with an awareness of the responsibility assumed by the investigator to all who would place confidence in the results of that research. Integrity of effort is paramount in importance. Objectivity must prevail without bias. In the field of taxonomy, a monographer becomes a specialist of the components of chosen taxa, and his opinions and judgments are generally respected, although not necessarily concurred with. Ordinarily, once a trained systematist begins monographic work, he selects as his subject of special interest one or more taxa not already under active investigation. Likewise, other systematists select other taxa for their studies and each is left much to his own devices. Anyone versed in current taxonomic literature can name a score of workers, each of whom is known as a specialist in a particular taxon, and becomes known as an authority on the taxonomy of those taxa. The knowledge and experience gained from breadth and depth of study in the relatively restricted sphere that is the monographer's cannot help but engender sound judgment of

---

[1] A discussion of what constitutes a monograph and a revision is provided in p. 304 of this text. Bibliographic citations of monographs and revisions are provided in Part II of this work, where they terminate the treatment of each family.

[2] Much valuable information, not included here, is to be found in Chapter X of Hitchcock's *Methods of descriptive systematic botany* (1925).

components of one's special interest. Each monographer usually becomes an authority on the systematics of particular taxa, for at least his own generation. There is attendant on this authority the added responsibility of knowing that published findings are likely to be accepted or unchallenged for a generation or more, and that periods of this length may elapse before the conclusions are subjected to serious study by another equally or more competent person. The time to comprehend the signficance of this responsibility, and to acquire by diligence the background necessary to assume it, is when one first contemplates initiating a research project that may result in a monograph or revision.

There are at least two approaches to taxonomic problems. The conventional one is based on the study of the selected taxon as represented by herbarium material and on studies of the ranges of variations of the plants as they occur in the field under habitat and climatic conditions. To these observations is added knowledge gained from study of the pertinent writings of past and present workers in taxonomic fields and, more especially, in the related fields of morphology, anatomy, cytology, genetics, physiology, and ecology. A second approach is that utilized by the biosystematist, whereby the data concerning cytogenetical aspects, ecological factors, and the biological status of speciation are determined *de novo* and simultaneously with, and as a prerequisite to, studies of a morphological nature. No one can deny the advantages offered by the second approach, but associated with them are counterbalancing disadvantages that cannot be ignored.[3]

## Procedures

No one procedure for conducting a taxonomic investigation of a revisionary nature may be prescribed as being better than another, and no one procedure will fit all situations. The final course must be determined largely by the particular problem and situation. It will be influenced by the scope of the problem, the facilities available, and the extent of research published by predecessors and contemporaries.

*Scope of the problem selected* will be governed largely by (1) the immediate need of the sought-for solutions, (2) maximum time available for the completion of initial studies, and (3) availability of materials, both plant and bibliographic, with which to work. It is assumed with reasonableness that one starting on a monographic or revisionary problem for the first time would not contemplate selection of a taxon larger than a genus (or a subdivision thereof). The size of the taxon

[3] For discussion of these and related matters see Chapter VIII.

selected is of secondary importance in so far as its selection for fundamental scientific research is concerned, and one cannot say with objectivity that a genus acceptable for an assigned problem in taxonomic research shall be composed of a stipulated number of species, for who can foretell how many species the forthcoming revision may accept? For example, the genus of pussytoes or ladies' tobacco is *Antennaria*; by most botanists it has been considered to be composed in North America of several scores of species, yet a recent monographer might conclude it to be composed of fewer than a dozen species. It may be that the biosystematic approach is better to determine the true situation. In large genera, the revision may be restricted to one or more subgenera or sections. It is more satisfactory to limit the problem to a study of a small and presumed phyletic taxon occurring over much of a major physiographic province or subdivision, than to a major taxon occurring within a small politically bounded area. That is, a more satisfactory understanding of the relationships of the plants is to be expected if the study is one of a genus, subgenus, or section occurring in the Atlantic coastal plain (or subdivision thereof), than if the study is of a family as it may occur, for example, in a county of Ohio. The size of the taxon should be small enough to permit a thorough taxonomic investigation of its components.

In determining a tentative scope for the problem, a preliminary survey should be made of:

1. Pertinent existing research—how recent, by whom, on the basis of what kinds of data.
2. Opportunities for studying material at some of the larger herbaria prior to making requests for loans of herbarium specimens.
3. The number of binomials involved (not necessarily the number of species)—who published them and where, on the basis of collections by whom, what types are available and where.
4. Field study—extent now proposed, minimum time and travel considered indispensable.
5. Possibilities of collaboration with one or more other researchers for purposes of obtaining supplementary data on members of the selected group as concerns their morphology, cytology, genetics, and ecology.

*Preliminary investigations* are necessary to learn (1) whether others are engaged in taxonomic study on the same problem, a point best discovered by inquiry of curators of the larger herbaria who are in close touch with studies being made of material in their collections; (2) of explorations by others in areas where the group is established, since by this means valuable new and often unidentified material may become available; and (3) of the taxonomic literature of the taxon. Guidance in matters pertaining to the literature is often especially helpful to one commencing systematic studies.

Before commencing revisionary work, one should be conversant with all of the floras, monographs, and revisions that have been published since 1753 accounting for plants of the general area selected (obviously this requirement must be tempered if the revision is to be world-wide in scope). Some taxonomists have found it very helpful to copy the treatments on microfilm (35 mm) and blow up from them two copies of every page containing original descriptions (one to be cut up and filed alphabetically for reference, the other available for phyletic shuffling). From these works one becomes familiar with the taxa recognized by earlier botanists. A second step is to learn of all the binomials that have been applied to members of the genus, restricting this when possible to plants of the area in question. This information is obtainable for American plants, with the least effort, from the *Gray Herbarium Card Index*.[4] From this same source, one may learn of the infraspecific names that have been applied to American species of any genus of vascular plants. In the absence of availability of this card index, reference should be made to *Index Kewensis* and its Supplements (see p. 286 for description and scope). The third step is to obtain, from the citation accompanying the names in these indexes, the original description for each. These should be copied with meticulous accuracy, or reproduced photographically, and be available for reference at all times during the course of the investigation. From the original description may be ascertained usually the locality of original collection (in species a century or more old this is usually disappointingly vague), the date of collection, and the name of the collector (often indicated as the collector of the specimen on which the name was based, and in cases of early works not indicated at all). The locating of many type specimens calls for considerable search, and for some early collectors and botanists no collections are extant.[5]

In addition to these several types of literature, one should study, in advance of extended field explorations, the pertinent literature dealing with the physiography, geology, and ecology of the region over which the plants of the study occur. Likewise, valuable information is to be had from phytogeographic treatments of these and other plants in this area. An initial source for locating this literature is the current abstracting and

---

[4] For a discussion of this work, its character, scope, and of institutions in the U. S. possessing complete sets of it, see p. 287.

[5] For references to literature on locations of type collections, see p. 328.

review series, such as *Biological abstracts* (see p. 307). As the study progresses, there will be the need to know of morphological, cytological, and genetical studies that may have been made of these same taxa.

It is well within the scope of preliminary investigations to examine and study the herbarium material available in the nearest large herbaria, with a view toward becoming familiar with the range of speciation within the selected taxon, the variation of taxa treated as species by other workers, and the making of preliminary tests of the constancy of characters employed extensively in manuals and earlier revisionary accounts. In addition to the suggestions indicated above, one should investigate the characters by which the taxon is conventionally distinguished from others of the same classification level. The validity of the bases for distinction should be challenged and should be the subject of further investigation if found weak or wanting.

In the course of these introductory reviews of literature and studies of material, one should acquire an understanding of the extent and general composition of the taxon as it occurs throughout its range (this may be world-wide or over only a single continent). It is not sufficient to know the plants of a taxon as they occur within one or more physiographic provinces, if this knowledge is gained to the exclusion of knowing of other elements elsewhere in the range of the group. Too often, revisions have been made of taxa as they occur, for example, in eastern North America, without due consideration and study of their relatives in other boreal areas of the world.[6] Revisionary studies must take into consideration the present distributions in the light of the geological and paleobotanical record. This requires an investigation of possible migrations, determinations of areas of origin for components of the group, and the presumed origin of the group itself. Investigations of this nature often are clarified by the results of pollen studies of lesser taxa within the group (Erdtman, 1943 *et seq.*), and from an examination of such reliable cytological analyses and genetical data as may be available. It is necessary to integrate taxonomic findings, in so far as possible, with those resulting from studies of the same or closely related elements via the approaches of morphology, anatomy, cytogenetics, paleobotany, and related fields. In the synthesis of these data, the taxonomist should be prepared to point out and defend incompatibilities.

Preliminary studies make possible the determination of whether or not

---

[6] For references to excellent background material on the subject of interrelationships of boreal floras, see those papers by Fernald, Gray, Hultén, and Wulff, cited in the bibliography to Chapter VII.

real problems, in a taxonomic sense, are present. The nature of the problems may be represented by polymorphism, presumed or demonstrated hybridity, apomixis, speciation, sectional or subgeneric delimitations, geographic races or variants, or the discernment of previously unobserved or untested criteria for delimiting species or infraspecific elements.

*Field studies* are desirable for an understanding of the relationships of any group of plants, and should preface and accompany the herbarium and bibliographic phases of the study. When it is not possible to study the plants in the field, effort should be made to grow and study as many kinds as possible in one or more test gardens, obtaining seeds and plants for such purpose from indigenous sources in so far as possible.

Studies made in the field should cover both the season of flowering and of fruiting, and collections should be made of material in all stages of development. In planning field studies, it is necessary to know of the available maps of the area,[7] both for use in plotting distributions and in planning itineraries. During the period in the field, efforts should be made to relocate type localities, in so far as possible, and to endeavor to locate original stations. There is value in collections from type localities, not only because they may comprise topotypes (whether they do or not depends on the accuracy with which the station is relocated) but also because by careful radiating search from the area of the nomenclatural type, the fringes and centers of distribution of plants identifiable with that type may be determined. The possession of soil and geologic maps when in the field often provides immediate explanations for obvious breaks and limits in distribution, or may make it possible to predict distributional limits more accurately. Notes should be made of precise locations of collections and of habitat types and of soil characteristics not evident from the soil maps. Topographic maps are available for most areas of this country, but for areas not treated by them it is desirable to record altitudes of stations.

Airplane reconnaissance photographs, available for much of the area of the United States, of Canada, and for an appreciable belt of Mexico, are of limited value to taxonomic field studies since it is difficult to determine variations in altitude from them and generally only the ecological aspects of the flora are readily discernible. They do have value when they are of areas of difficult accessibility or remoteness from highways, since they often reveal foot and truck trails, and isolated areas of aquatic and semiaquatic habitats are generally apparent (see p. 325 for sources).

[7] See the section on maps in Chapter XIV.

Opportunity should be taken while in the field to make detailed studies of characters often lost in the dried specimen. Such characters include:

> Coloration of foliage and floral parts, especially glaucescence, corolla vernation, corolla color changes associated with flower development, anther color before and after dehiscence, gynoecial coloration.
> Viscidity of parts, especially nectariferous organs.
> Pollinating agencies and vectors, and time of day of pollination.
> Rootstock characters (too often overlooked).
> Texture of foliage and perianth parts.
> Color and character of freshly matured fruits.

A primary function of field study is the determination of natural variation within and between populations. Field collecting involves not merely an alertness for the unusual (this type of material is represented already in undue abundance in herbaria, and is mistaken too often by the uninitiated as representative), but rather an effort to have the specimens represent a range of the natural variations without an undue emphasis on any one phase. In making field collections, one should assign a different collection number to specimens from different colonies within the same general area. Abundance of material permitting, it is well to prepare 4 or 5 sets of duplicate collections to be distributed, on completion of the study, to important herbaria.

*Mass collections* is the term given to the technique whereby specimens are made from many individuals of a series of interbreeding colonies of a species. These collections, when properly made, provide "a record of a population as well as of the individuals which make up that population and . . . give the facts about variation which can be obtained from populations but not from individuals." (Anderson, 1941, p. 287.) Mass collections, in some cases, may provide a means of determining relationships with greater accuracy, especially as concerns the elements below the level of a species and the groups to which they belong. They have been found to be helpful in determining whether a variation is one associated with or restricted to colonies, to habitats, or to regions. The knowledge of these correlations is particularly pertinent to determining if a population represents a subspecies, variety, or forma.

In making mass collections, attention should be given to two considerations: the part of a plant to collect, and the selection of a random sample of the population. Taxonomic experience in the selected genus usually will provide the answer to the first consideration, and one will select for collection those parts that are significant for their characters. As Anderson has indicated (pp. 289–290), individual judgment must be relied on in the determination of what constitutes a random sample. When the subject is a small plant, the entire plant is collected. If the subject is large, the collection may be only of an inflorescence, a leaf, or fruits, or other critical portions. Thirty to 50 samples should be made of a population, and these, when properly selected, constitute the material for study and permanent herbarium record from which 3 different kinds of information may be obtained: [8] (1) frequency of variation; (2) discontinuity of variation; (3) correlation between variables.

The preparation and study of mass collections is particularly pertinent to problems associated with revisionary studies (cf. use by Erickson, Schery, Cutler, Fassett), but is equally valuable for floristic investigations. The material is easily incorporated with other conventional herbarium specimens and is not as space-consuming as might be anticipated. It is the practice of some taxonomists to keep the mass collections of any one population within a single genus cover, and the data accounted for by a single label affixed to the cover.

*Herbarium studies* are usually conducted concurrently with field work or, in many instances, during seasons when field work is unproductive. Opinions on approaches to the problem are numerous and diverse; those given below merely suggest those satisfactory to some workers.

After one has acquired some field experience, and has benefited from the study of one or two ample herbarium collections, knowledge of the group may be sufficiently adequate for a study of a large series of herbarium material. For this purpose, material is obtained on loan from herbaria rich in representatives of the genus for the area in question. In borrowing material from another herbarium, it must be remembered that the material is irreplaceable, and that a responsibility is assumed on accepting it for study. This responsibility is threefold: (1) to the institution making the loan possible; (2) to the institution to which the loan was made (loans are effected generally between institutions rather than from an institution to an individual); and (3) to the tyro who is in no position to acquire an unsavory reputation on the occasion of the first loans made for his use. On receipt of loaned material, a careful count should be made of the number of sheets before returning the receipt that usually accompanies the specimens. It is a practice of many taxonomists to place a lightly penciled code letter on a corner of every sheet (identifying that sheet with its institutional source) and to accompany the letter with a serial number assigned in sequence, one for each sheet. This permits, on completion of study of borrowed material, a quick and accurate sorting of the sheets received, and enables one to

[8] For details of analysis of these data, see Anderson (1941 and 1943).

reassemble them in the same order in which they arrived. At the time of reassembly, the penciled code letter and number should be erased very lightly without defacement of the sheet.

Borrowed and local material is usually sorted into folders of kinds, preparatory to any detailed study of each sheet, and the labeled folders are filed systematically. When not in actual use they should be kept in a case safe from mechanical injury or the incursion of insects. In working over large assembled collections for the first time, and preparatory to initial sortings into folders of presumed like kinds, some workers find that greater objectivity is effected if labels on the sheets have been covered by clipped slips of paper, since by this means one is freed from the tendency to be influenced by opinions of others.

As a result of the extensive reading that will have been done, and by virtue of knowledge gained from both the field experience and the preliminary herbarium studies, one will have views on apparent or presumed relationships, similarities, and dissimilarities evident in the assemblage of herbarium specimens. This is the time to test these views. If all preconceived views measure up to the facts, it is possible that the methods used may lack in objectivity. Presumably problems of identification will arise. They may involve selection or the determination of characters (here the analysis of earlier-prepared mass collections should be of help); they may involve irregular distributional patterns; and there may be problems of presumed hybridity, or of a correlation of overlapping of characters with similar overlapping of distributions, the problems then being to determine whether the overlapping of characters is due primarily to a lack of a dispersal barrier or to a lack of a genetic barrier. After the plants have been identified, there remains the question of determining the categories to which they belong and an evaluation of the criteria by which these categories shall be delimited. Subsequent to this, there is the matter of devising one or more keys adequate to separate on morphological characters each of the recognized taxa, irrespective of the categories in which they are placed. These keys should be subjected to the test of satisfactory use by taxonomists other than the originator, and preferably by means of specimens whose labels are covered at the time of the testing.[9] In groups where fruit characters are equally as important as flower characters, the effort should be made to have a key based on fruits and another based primarily or exclusively (conditions permitting) on flowering and vegetative structures.

Resolution of these problems brings one to the point of investigating

[9] See Chapter X, for keys and their preparation (pp. 225–228).

the nomenclatural situations of each element to be given a name. This presupposes a thorough understanding of the principal features of the current edition of the Rules of Nomenclature and is not a subject for discussion at this point. Once the plants have been identified, the categories in which they are to be placed determined, and names for each taxon decided upon, there is the task of preparing adequate and uniform descriptions of each nomenclatural taxon. Descriptions must be prepared with care; data should be obtained of the detailed characteristics of the type specimen of each element, and all measurements must be made with precision, with enough specimens involved to insure complete ranges of dimensional variations of all structures.

The remaining task, after nomenclatural and taxonomic problems have been treated, is annotation of all specimens assembled for the study and recording each specimen so annotated. Annotation slips are most convenient for this purpose. Each should bear the full name and author citation for the plant, the name (not merely initials) of the person responsible for the annotation, and the date. The record of every sheet so annotated should include (so far as the data are available): name of the collector, collection number (when missing, the accession number may be substituted, provided that notation is made to this effect), place of collection (state or province, county, town), habitat, date of collection, and herbarium source or sources of the particular specimen (see p. 274 for reference to standard abbreviations of world herbaria). Some of these data will be needed for inclusion in the manuscript of the projected revision or monograph.

In preparing the accumulated data for presentation, it is usually desirable to incorporate some of them in an illustrative form, making use of the more graphic formats that serve as visual aids. These may be in the form of (1) charts or graphs designed to show trends of variations or ratios of constancy between structures or organs, (2) illustrations of organs, floral dissections, type specimens or portions thereof, and especially illustrations of new species or other elements, and (3) maps of geographic distributions based on the material included in the study. Examples of the use of graphs and charts are illustrated in the works of Anderson (1936) and Epling (1942, 1944). Illustrations of the technical features may be represented by photographic reproduction, or by means of pen-and-ink drawings. Supporters of the use of photographs of technical structures contend that the medium presents the situation without possibility of interpretation by the artist, whereas those preferring the pen-and-ink drawings point out that in the photographic rendition the features of particular structures often are obscured by less relevant aspects—as vesture or overlapping parts, features readily de-emphasized in the line drawing. There is little question but that a well and accurately executed drawing is superior to and more satisfactory than the best photographic reproduction. Maps prepared to show distributions of elements are usually based on one or more types of the standard outline maps as discussed on p. 323 of Chapter XIV.

In all except the most condensed revisions, it is customary to include with the treatment of each species or infraspecific unit, a listing of the herbarium specimens examined by the author and identified as that taxon. Familiarity with the current literature will provide abundant examples of the many methods of citing these specimens (sometimes restricted to citation of specimens of large exsiccatae or to types), and the selection of a particular format will be conditioned to a large degree by the publication space available to the author. Irrespective of the format, it is the practice to indicate the name of the herbarium where each specimen is known to be deposited by a standard abbreviation. These abbreviations, selected and compiled by Lanjouw (1939, 1941). have been adopted internationally and should be followed by students requiring them.

With these data, the investigator may be ready to assemble the results into final form for publication.[10] The details to be incorporated, format to be used, and general editorial style to be followed will be subject to the policies of the particular serial or journal to whose editor one expects to submit the finished paper for publication. The best procedure is to observe papers of similar character in that journal. A few references are included among those cited below to serve as sources of generally accepted guidance in the preparation of scientific papers.

*LITERATURE:*

ANDERSON, E. The species problem in *Iris*. Ann. Mo. Bot. Gard. 23: 457–509, 1936. [See especially, Figs. 5, 6, 8–14.]

———. The technique and use of mass collections in plant taxonomy. Ann. Mo. Bot. Gard. 28: 287–292, 1941.

ANDERSON, E. and TURRILL, W. B. Biometrical studies on herbarium material. Nature, 136: 986–987, 1935.

BLAKE, S. F. and ATWOOD, A. C. Geographical guide to the floras of the world. An annotated list with special reference to useful plants and common names. Part I, U. S. Dept. of Agr., Misc. Publ. 401. Washington, D. C., 1942.

CUTLER, H. C. Monograph of the North American species of the genus *Ephedra*. Ann. Mo. Bot. Gard. 26: 373–427, 1939.

[10] Additional and valuable suggestions by Blake and Atwood (1942, pp. 8–9) should be consulted during the course of preparation of any revisionary or monographic treatment.

EPLING, C. The American species of *Scutellaria*. Univ. Calif. Publ. Bot. 20: 1–146, 1942. [See especially Figs. 3, 4, 7, 8.]

———. The living mosaic. 26 pp. Berkeley, Calif., 1944.

ERICKSON, R. O. Mass collections: *Camassia scilloides*. Ann. Mo. Bot. Gard. 29: 293–298, 1941.

FASSETT, N. C. Mass collections: *Rubus odoratus* and *Rubus parviflorus*. Ann. Mo. Bot. Gard. 28: 292–374, 1941.

HITCHCOCK, A. S. Methods of descriptive systematic botany. New York, 1925.

JOUGHIN, G. L. Basic reference forms. A guide to established practice in bibliography, quotation, footnotes, and thesis format. New York, 1941.

LANJOUW, J. On the standardization of herbarium abbreviations. Chron. Bot. 5: 142–150, 1939; 6: 377–378, 1941.

SCHERY, R. W. Monograph of *Malvaviscus*. Ann. Mo. Bot. Gard. 29: 183–244, 1942.

TRELEASE, S. F. The scientific paper, how to prepare it, how to write it. ed. 4. Baltimore, 1947.

UNIVERSITY OF CHICAGO PRESS. A manual of style. ed. 11. New York, 1949.

# CHAPTER XIII

# FLORISTICS

## Types of floristic study

Floristic studies are taxonomic studies of a flora, or of a major segment of a flora, of a given area. They may range in extent from a compiled check list of vascular plants of a small politically bounded area to a thorough taxonomic or biosystematic analysis of the components of the vascular flora of a continent. Because of this wide scope, the subject of floristic study appeals to the interests of both the amateur and the professional botanist. It is a field of interest in which the amateur botanist has taken an active and contributory part. The amateur botanist should be encouraged to continue to investigate such problems, for they provide an avocation that can produce important contributions to the knowledge of any major flora. Among professional botanists, there are many who have not the time, facilities, or interest required for revisionary studies, but do find that study of the flora of a particular area provides an opportunity for welcomed field activity and the knowledge that valuable data are being recorded. From this it is clear that here is a type of botanical activity that may be conducted for scientific gain which can be limited in scope to meet the potentials and the needs of any serious-minded investigator.

As is pointed out in Chapter XIV, there is a distinction between a flora of an area or region and a manual of the same region. A flora is an inventory of the plants of a definite area. This inventory is usually authenticated by citations of herbarium specimens and of localities or stations where each element is known to have occurred. It is customary to arrange the plants treated in a flora according to a recognized system of classification. A manual is a book that provides means of identifying and naming a plant. In addition to providing the data included in a flora, it always includes keys to plants of the rank of species and above and descriptions of all taxa that are recognized nomenclaturally.

*Compilatory lists* are of two types: those based on existing accounts collated from the literature, and those based on herbarium specimens but without the latter being subjected to critical study by the compiler of the list; a third type is that represented by a combination of the other two. They serve the purpose of meeting an interim need, or of accounting for the plants of an area until a more authentic list or flora has been produced. A compilatory flora can be of aid to anyone in an area not represented already by a modern manual or an adequate flora; it serves also as an index, or check list, to one collecting material for herbarium specimens designed to be representative of the flora of an area, or to one endeavoring to know the approximate composition of a flora. A flora of this character is primitive as compared with other types, and when completed it adds no more to the scientific understanding of the plants of the area than was known previously; admittedly it is only a compilation of existing data from many sources. On the other hand, if the compilation has been thorough, this primitive flora may be a more accurate compendium of the scientific knowledge of the plants of the area than is available from any but the most recent and most critical manual of the region.

In the preparation of a list based on both the literature and on herbarium specimens, the following outline may provide suggestions for procedure.

1. Prepare a preliminary list from the latest manual of the region. Restrict the list to those taxa ascertainable from the manual to be indigenous to the area selected.
2. Survey the literature covering the period since the manual of the region was published for reports concerning:
    a. New range extensions into or through the area
    b. New stations for plants within the area
    c. New taxa (e.g., new species or less than specific entities) described for the first time and from localities within the area
    d. Nomenclatural changes affecting plants of the area
    e. Taxonomic status of plants within the area

For each of the above items, one should consult existing works of floristic, revisionary, and monographic character. Much supplementary material may be had from these sources to be incorporated into the initial preliminary list prepared from the older manual. The material should be arranged systematically, at least by families. The next step will be to compare the information in the list with specimens of the plants concerned. In doing this one should be alert: (1) to list all locations of stations within the area where each nomenclatural element has been collected, with a record of the name of the collector and date (collector's names may be abbreviated); (2) to note, in leafing through folders, if there are other nomenclatural elements present among the collections that are not noted in the prepared list (new stations are often discovered in this way).

If a compilatory list is to be published, its character should be indicated clearly. This may be accomplished by the title or subtitle (e.g., "A preliminary study of the flora . . ." or "A tentative accounting of the flora of . . ." or by means of an adequate subtitle as, "A flora of . . ., based on a compilation of existing records"). Some botanists are of the opinion that compilatory lists should not be published, but rather that they should function as the basis of a more thorough study that ultimately will result in an original work of considerable reliability.

*Field study* of a particular area is the basic approach to a thorough inventory of the plants involved. The determination of the size of the area is subject to many of the factors considered in Chapter XII, except that in floristic investigations custom has sanctioned the prevalent practice of coinciding the limits of the study with those of political boundaries. Some of the better and recent floras (Deam, 1940; Hanes, 1947; Dole, 1937) are restricted solely to political areas without regard for physiographic provinces or their subdivisions; others (Jones, 1945; Reeves and Bain, 1946) designed to meet teaching needs ignore either political or physiographic areas; and those of a third group (Jones, 1936; Stone, 1910) account only for the plants of or within physiographic or hydrographic regions. Greater unity and stronger interrelationships are present among the plants of a flora based on a physiographic area than among those based on other types. On the other hand, practical considerations may dictate that the boundaries of a projected flora coincide with county or state lines. It is sometimes desirable to prepare a flora of an area within a physiographic and/or hydrographic region, and for this purpose the square degree may be more acceptable than a politically bounded unit (cf. Grassl, 1936). A square degree is an approximate rectangle bounded by one degree of latitude and one degree of longitude. Such an area is especially satisfactory when the focal point or geographic center of the flora coincides with the center of the rectangle.

Floras based on field studies are significant contributions, and users of them know that the author of the flora knew the plants involved as living organisms and as subjects whose forms represented responses to a multitude of factors, and that he did not treat them as piles of dried specimens drawn from pigeonholes, reshuffled, reannotated, and returned. Every floristic study based on field observations deserves to be given careful and painstaking attention. The suggestions that follow are provided as guides to indicate some of the more significant phases to be investigated.[1]

## Procedures

Field collections are the core of any modern floristic study. Once the limits of the selected area have been established, the primary phase of the study is to become thoroughly familiar with the plants of that area. The best evidence of knowledge of the plants of an area is the collection of specimens that have been prepared of them. As much time as possible should be spent in the field, and collections of the plants should be made during all weeks of the growing season in order that material will be had, in fruit as well as in flower, of every taxon believed to be a part of the flora. It is not unusual for several years to elapse during the course of these field studies. In temperate regions it is important that the region be visited during the winter season and notes made of floristic features pertaining to the persistence of leaves and of fruits, the significant bud characters that may be of value in identification procedures, and likewise provide the opportunity to observe late-blooming and very early-blooming elements as well as sporadic out-of-season blooming during occasional cycles of unseasonal activity. Possession and subsequent incorporation of these diverse data add to the thoroughness of the study. In making collections, particularly if the area is one whose plants are not well represented in herbaria, one should always prepare several sets—unless in particular cases the material is of rarity, for in no case should zest for specimens eradicate a station—for distribution to interested botanists or institutions. Particular care must be taken to insure that all parts of the area are visited and that the collections are representative of all habitat types.

In preparing herbarium specimens, close attention should be given to recording all necessary and pertinent data concerning characters of the specimen that may not be apparent after drying.[2] There are two ways of identifying the specimen with the field notes. One is to affix a consecutively numbered small tag to each specimen as it is collected and to assign the same number to notations made at the time of collection. The second is to press the specimen as it is collected and write the number and data on the sheet in which it is pressed, transferring the

---

[1] See Blake and Atwood, pp. 7–8, for other suggestions to be considered in the preparation of any flora. Many of those suggestions may seem to be elemental, but they were mentioned because of omission from many existing floras. The more significant of them are quoted in Chapter XIV (pp. 288–303) of this text.

[2] See Chapter XII, p. 269, for a listing of some of the characters to be noted.

# FLORISTICS

data to the notebook later. The disadvantage of the first system is that if the notebook is lost the unlabeled collections may be worthless; the disadvantage of the second system is that it presupposes immediate pressing in the field and involves making the notations twice. Observations of other features than those of the reproductive structures, should be noted, particularly of the following.

Texture; if herbaceous, indicate duration: annual, biennial, perennial.

Habit of plant: erect, prostrate, decumbent, twining, etc.

Sizes of plants (when larger than can be accommodated *in toto* on the herbarium sheet). Use metric system for sizes, and when of trees give diameter in centimeters or decimeters at breast height (abbreviated d.b.h.); figures given should be of the particular tree from which the specimen was taken.

Fragrance or other odors: of flowers, fruits, crushed herbage.

Pollinating agents: diptera, hymenoptera, lepidoptera, carrion beetles, etc.

Station location. Keep accurate record of place where each specimen was collected.

Frequency. Arrive, at the start of the study, on concepts of classifying frequency on basis of abundant, common, frequent, infrequent, rare, very rare, etc.

Habitat type: acid bog, beech-hemlock forest, salt marsh, etc.

Other plants of the association. Name a few genera.

Soil type (estimated) in which the plant grows: sandy loam, podsol, etc.

Soil $p$H, readily determinable by simple field techniques.

Exposure to sun: full, partial, or in moderate or dense shade.

Direction of habitat exposure: north slope, etc.

Elevation, determined from topographic map when available.

Vernacular names. Inquire of same from local residents of long standing, even though you may know one or more accepted vernacular names for the element.

In the accumulation of data, it is often helpful to devise a form for use in organizing information from notes of various collections. The data compiled on any one form should be accompanied by the number or numbers of the collections from which they were taken.

Most published floras include only the names of plants known to be native and cite for them those localities or stations where they are indigenous. The determination of whether a plant is indigenous or not is not always simple. Categories accepted by many authors to account for the various degrees of establishment of nonindigenous plants include:

*Naturalized:* a plant fully established, reproducing, migrating, and expanding in its area of occurrence.

*Introduced:* a plant brought in deliberately by man and growing in the area without cultivation (true of many weeds).

*Adventive:* a plant that enters the area by any means (particularly natural dissemination or inadvertent dissemination by man) and grows but is unable to meet competition by continued natural reproduction.

Suspicion should be raised against new stations of exotics (indigens of other regions) situated near places of habitation, cemeteries, dumps, railroad yards, or highways of urban localities. Such plants may be adventive or introduced, and in time some may become naturalized. If plants belonging to any of these categories are included in a published flora, their nativity status and degree of influence in the floristic picture should be indicated clearly.

Floristic studies usually require several seasons of investigation in the field, and during this period the number of collections may reach several thousands and occupy considerable space. For this reason the specimens should be sorted by collection number, or phyletically according to an accepted classification, and stored in containers or cases.[3] One of the first steps, after the plants have been brought in from field activities, is to prepare labels for each collection number. Every specimen must have a label, and all labels for a given collection number should be replicates of one another (see pp. 246–248 for information on labels and their preparation).

*Identification of specimens* is accomplished either by working out the identities by the usual taxonomic procedures or by arranging to send sets of selected unknowns to specialists who agree to report on their determinations in return for having the specimens sent them. Ordinarily both procedures are followed, the investigator completing as many of the determinations as possible and soliciting the aid of specialists (monographers of particular groups) for the remainder.

Suggested approaches to problems of plant identification are given in Chapter X, and are pertinent to the problems of determining material collected during the progress of a floristic study. Unless the plants of the selected area are initially completely foreign to the investigator, it is probable that an appreciable number are identifiable as to family status on sight with a small number recognizable as to genus, and some of the collections as to probable specific indentity. Any of these spot determinations should be treated as tentative until verified. Material not known is worked out to family status by following the treatment in available manuals or such comprehensive works as Engler and Prantl, or Bentham and Hooker. Knowing the family, one may usually ascertain the genus by use of the same references, or by monographic works treating those families. Every effort should be made to consult the latest literature pertaining to each group. In instances of conflicting taxonomic views, it

---

[3] Chapter XI, pp. 248–253, for instructions for control of insects while in storage. Freshly collected material must be cared for in this regard, especially if it is to be stored for several months or more.

will be necessary to review the evidence and arrive at one's own opinions. In the final manuscript of the flora, one should explain taxonomic or nomenclatural departures from the views of specialists.

*Herbarium studies* may be integrated with the preliminary determination of the field collections, particularly if the investigator is affiliated with or situated near one of the larger herbaria. There are at least 3 phases of herbarium study involved in the preparation of a flora. They are (1) verification of preliminary determinations made of the field collections, (2) determination of the collection records of plants of the area as evidenced by material already in the herbarium, and (3) examination of authentic material for the verification of names applicable to the collections involved. These procedures are explained in Chapter XII.

Examination of types, isotypes, or other acceptably authentic material is an essential part of any reliable floristic study. In each instance, notation should be made of whether or not the type has been seen, for much more confidence may be placed in a determination arrived at by comparison with type material than in one not so authenticated. The location of type specimens is often a problem of major proportions.[4] In those cases where the type collections are in foreign herbaria, it occasionally happens that photographs of them are available in one or another of the American herbaria.

*Existing literature* constitutes a third channel of investigation in any floristic study. The literature to be studied usually is scattered, and there is no single index to it, although use of the references suggested on pp. 305–308 should be helpful in this regard.

The first step in the investigation of the literature is to learn of the floristic studies made of the area by previous investigators. This may be in the form of direct evidence, as manuals of the region or more especially earlier floras. Each of these previous reports must be analyzed critically. All statements of previous records must be held in a tentative status until each has been challenged and verified. The verification is made by examination of the herbarium specimen on which the earlier record was based. It occasionally happens that there is no specimen extant to substantiate the earlier report. If the current field studies fail to verify the alleged existence of the element, the name should be included in the new flora only if a definite indicator (as an asterisk) precedes the name. By this means it is marked as an unverified inclusion, and an explanatory

---

[4] See Chapter XIV, pp. 328–329, for references to literature pertaining to the locations of type collections.

note should accompany the entry. Published reports in the literature of occurrence must be substantiated by herbarium specimens or else be viewed with extreme scepticism. The same critical view should be taken of all reports of range extension into the area and of new taxa described from collections from within the area. In other words, the investigator should account only for those plants of which he has examined specimens known to have been collected within the area. The integrity of the flora rests to a large extent on the adoption of this view, and inclusion of unverifiable entries serves to lessen the value of any floristic work.

Mention has been made above of the use of existing monographs and revisions. These usually serve two primary purposes: (1) when competently prepared, they represent (as of their publication date) the best source for taxonomic and nomenclatural data of the group and in general merit greater confidence than the more compilatory manuals; (2) modern monographic works invariably provide citations of specimens examined, and the investigator of a floristic work will profit from a study of these, selecting all that are of collections made from within his area. Not only do these collections provide specific examples, but duplicates of the specimens seen by the monographer often are available to the investigator and serve as reasonable counterparts for purposes of identification.

In the preparation of any flora, there is a need for study of other floristic works that were concerned with the floristics of adjoining areas, or areas from within the same physiographic province. These floristic works may be in the format of a flora, or may comprise less formal papers that deal with taxonomic or distributional problems of plants within the region. Much valuable material, and considered opinions of taxonomic scholars, are too often buried in these more general articles. For them, the investigator should scan the pages of those botanical serials known to include taxonomic papers of a particular area (as *Madroño, Rhodora, Castanea, American Midland Naturalist, Wrightia, Bulletin of the Torrey Botanical Club*, etc.).

There are other sources to be investigated in searching for literature pertinent to the study. Prominent among them are the several indexes to taxonomic literature. One of the best indexes for nomenclatural data is the *Gray Herbarium Card Index* (see p. 287) which, while restricted to plants of North and South America, is the only index to names and synonymies of plants in categories below the rank of species. The *Torrey Club Card Index* is invaluable as an author index to all botanical literature, and is available at large botanical libraries. During the last half century more literature concerning the taxonomy of the plants of northeastern North America has appeared in the pages of *Rhodora* than in any other comparable serial. For this reason, the 50-year index to this work (now in preparation) will be of considerable help to students of the flora of this region.

*LITERATURE:*

FOGG, J. M. JR. Some methods applied to a state flora survey. Contr. Gray Herb. 165: 121–132, 1947.

HITCHCOCK, A. S. Methods of descriptive taxonomy. New York, 1925. [Chapters VII, XVIII.]

CHAPTER XIV

# LITERATURE OF TAXONOMIC BOTANY

Taxonomy is fundamentally a descriptive and highly documented science. For this reason its literature is voluminous and constitutes so vital a part of its structure that, irrespective of whether the problem is one of identification of an unknown plant, solution of a nomenclatural puzzle, or a monographic or floristic study, acquaintanceship must be made with the more important publications of the subject.

This chapter treats the more pertinent items of taxonomic literature exclusive of monographs and revisions, many of which are cited under the families described in Part II of this book. There is no complete nor even adequate bibliography dealing exclusively with taxonomic literature; such a work would occupy many volumes. This complete literature is to be found in volumes devoted to individual subjects, in pamphlets, and in articles published in periodicals. Because every nation has its taxonomists, speaking and writing in every modern language, the texts of these books and articles are in as many languages. This international character of taxonomic literature must be recognized, and while most of the entries selected for inclusion in the bibliographies that follow are in the English language, one must appreciate that some of the most important and valuable taxonomic publications are in other languages, and that linguistic limitations must not be a barrier to a knowledge of this literature. Many of the entries in the bibliographies listed below are annotated, so that anyone not thoroughly familiar with a particular work may learn something of its scope, utility, and significance. These entries are arranged according to the following classifications:

General taxonomic indexes (p. 285)
World floras and manuals
  North America and West Indies (p. 291)
  Mexico and Central America (p. 296)
  South America (p. 297)
  Europe (p. 298)
  Asia (p. 300)

284

Africa (p. 301)
Australasia (p. 302)
Monographs and revisions (p. 304)
Bibliographies, catalogues, and review serials (p. 305)
Periodicals
  New World (p. 310)
  Old World (p. 313)
  Cultivated plants (p. 316)
Glossaries and dictionaries (p. 317)
Cultivated and economic plants (p. 319)
References to miscellaneous topics
  Maps and cartography (p. 323)
  Biographical references (p. 326)
  Dates of publication (p. 327)
  Locations of type specimens (p. 328)
  Directories and addresses (p. 329)
  Color charts (p. 329)
  Outstanding botanical libraries (p. 330)

## General taxonomic indexes

The following are the important indexes to vascular plants. They are indexes of plant names and not to literature concerning the plants. Indexes serve as an aid to locating quickly the source of original

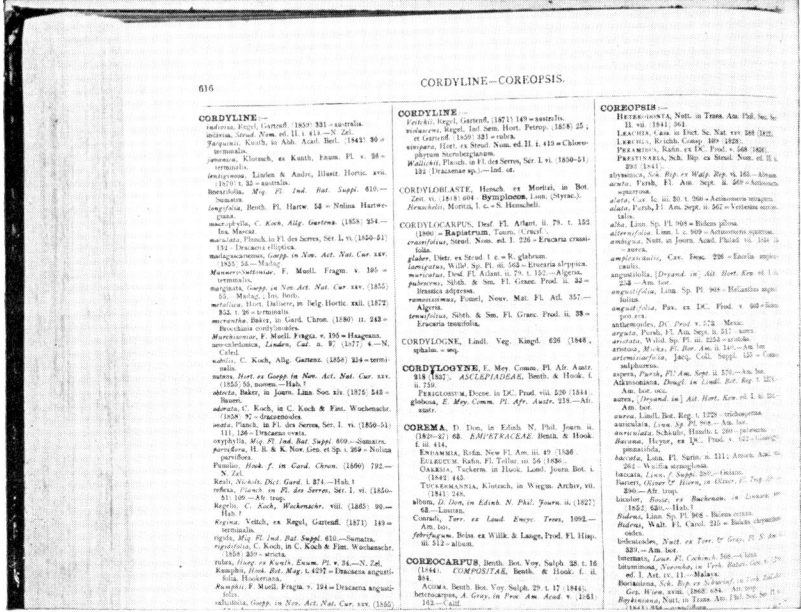

Fig. 33. Portion of page of *Index Kewensis*. By permission of the Clarendon Press, Oxford

publication of a name, to learn if a particular name has been applied to a plant, or to what order, family, subfamily, or tribe, a plant of a given name may belong. One index is to the sources of illustrations of plants. These indexes are the nucleus of any significant taxonomic library, and it is incumbent on the student of taxonomy to know of their availability and importance.

Index Kewensis plantarum phanerogamarum. 2 vols., 10 suppl. Oxford, 1893–1947.

This work is the cornerstone to the literature on the systematics of flowering plants. The compilation of the original work was made possible by a gift of money from Charles Darwin. The Clarendon Press at Oxford undertook publication in return for the copyright. It was compiled at the Royal Botanic Gardens by B. Daydon Jackson and his clerical assistants under the direction of J. D. Hooker. It was published in 1893–1895 in 2 volumes (sometimes bound in 4) followed by supplements regularly published at 5-year intervals. The original work consisted of an alphabetical enumeration of the genera of seed plants (gymnosperms and angiosperms) published from the time of Linnaeus to the year 1885. Under each generic name was given, in alphabetical sequence, every species epithet known to have been published for a member of that genus; this was followed by the name of the author of the combination, the place of publication of that name (often much condensed and abbreviated), and an indication of the native country of the plant (Fig. 33). The date of publication was not given. Species epithets treated by the editorial staff

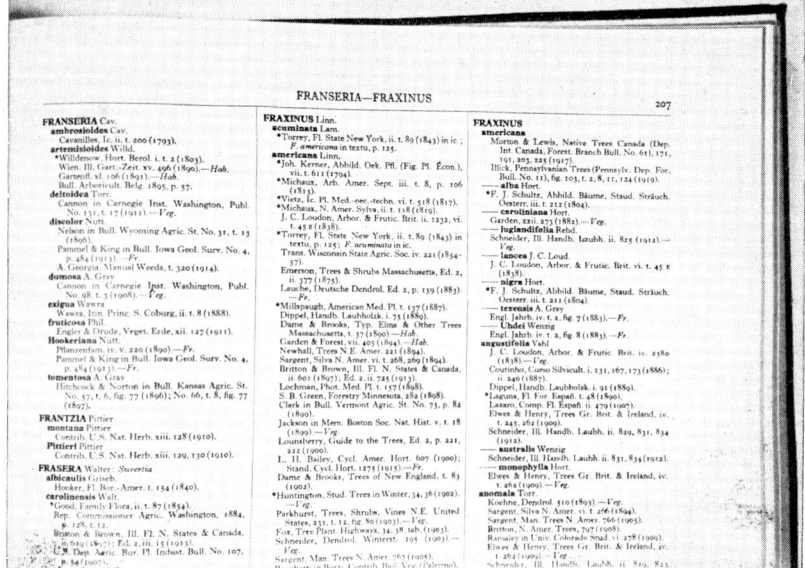

Fig. 34. Portion of page of *Index Londinensis*. By permission of the Clarendon Press, Oxford.

as synonyms of presumed valid names were set in italics, followed by the presumed valid name of the species concerned. Many illegitimate names were included, but without indication of their illegitimacy. Supplements have been published up to 1940, and in the later issues changes of editorial policy have resulted in the date of publication having been indicated, double author citation employed, and the practice of indicating synonomy discontinued. Entries in later supplements usually have indicated if the name is that of a new species or a transfer from an existing name, and if the binomial is illegitimate. *Index Kewensis is the reference employed to determine the source of the original publication of a generic name or binomial of a seed plant.* It does not account for names of ferns nor of any plants in divisions subordinate to the spermatophytes, nor for names of plants as placed in categories below that of species. For a short history of its compilation see Journ. Roy. Hort. Soc. 49: 224–229, 1924.

Index filicum. Hafniae [i.e., Copenhagen], 1906, with supplements to 1933.

This work is somewhat comparable to *Index Kewensis* in that it provides references to the sources of original publication of generic and specific names applied to the Filicineae (true ferns). It does not include the so-called "fern allies." The original work and supplements were edited by Carl Christensen.

Gray herbarium card index. Cambridge, Mass.

A card index issued quarterly to subscribers, accounting for all new names and new combinations applied in any category to the flowering plants and pteridophytes of the western hemisphere. In so far as concerns generic names and binomials, this index duplicates *Index Kewensis*, but it is invaluable to students of New World plants since it accounts for Latin names given to vascular plants since 1885, irrespective of the category and irrespective of whether published in a New World or Old World publication. Approximately 260,000 cards have been issued to date.[1]

Genera siphonogamarum. Berlin, 1900–1907.

This work, edited by C. G. Dalla Torre and H. Harms, accounts in one volume for the names published for families and genera of spermatophytes. Orders and families are arranged essentially according to the Engler system, and under each family name are given the names of subfamilies, tribes, etc., together with the names of genera assigned to each. Under each genus is indicated the number of species described in it. With the names in each category are given the appropriate author citation, source and date of publication, and synonyms. A 284-page index concludes the work, making it the more valuable as a ready reference from which to learn the family to which any generic name of a seed plant may belong. The names of genera are arranged systematically and are numbered consecutively throughout the work from 1 (*Cycas*) to 9629 (*Thamnoseris*). These numbers are used by the curators of many herbaria as indexing guides for the insertion and locating of material arranged according to the Engler system.

---

[1] The following list of North American institutions has been provided by the Director of the Gray Herbarium (1949) as comprising those of this continent possessing complete sets of this index: Arnold Arboretum, Carnegie Museum (Pittsburgh), Central Experimental Farm (Ottawa), Chicago Museum of Natural History, Dudley Herbarium (Stanford), Gray Herbarium, Missouri Botanical Garden, Montreal Botanical Garden, New York Botanical Garden, Southern Methodist Univ., State Coll. of Washington (Pullman), U. S. Dept. Agr. (Beltsville), U. S. National Herbarium (Washington), Univ. Calif. (Berkeley, Los Angeles), Univ. Illinois, Univ. Michigan, Univ. Minnesota, Univ. Oklahoma, Univ. Texas, Univ. Washington (Seattle), Univ. Wisconsin.

Index Londinensis to illustrations of flowering plants, ferns, and fern allies. Oxford, 1920–1931, with one 2-volume supplement, 1941.

A 6-volume work providing an alphabetical index, by genus and species, to the illustrations of flowering plants and ferns from 1753 to end of 1920 (the supplement covering the period 1921–1935). It accounts for illustrations of all Latin-named plants in categories below the rank of species as well as those given binomials (Fig. 34). An illustration is listed under the name used in the particular work cited, and there is no assurance that the illustration is necessarily authentic as to name. The work was compiled at the Royal Botanic Gardens, Kew, by Otto Stapf and his clerical assistants under the auspices of the Royal Horticultural Society of London.

### World floras

A flora is a work devoted to the plants of a particular region, and also usually is restricted to a major segment of the plant kingdom (as vascular plants, flowering plants, etc.). The more complete flora is one that accounts for all the vascular or seed plants. In any flora these plants are arranged according to one or another of the available systems (Engler, Bessey, Hutchinson, etc.), giving for each plant the complete scientific name, author citation, reference to source of original publication, synonymy, and geographic distribution within the area in question. There may or may not be keys to the entries treated, and descriptions of each kind of plant may or may not be included. In floras of relatively small and restricted areas the keys and descriptions are frequently omitted, the flora then serving more as an authenticated inventory of the plants within the area, rather than as a direct means of identification of those plants. The latter is more the function of a manual than of a flora.

In the preparation of a flora, irrespective of the size of the area to be treated, the student is enjoined to take cognizance of the suggestions of Blake and Atwood (pp. 8–9) wherein it is recommended that

. . . the essential features of even the barest list of plants include a title accurately describing its contents; the most definite possible statement of the geographical area covered . . .; and a statement of the material on which the list is based. . . . Additional items that add greatly to the value of a flora include accounts of the topography, hydrography, climate, geology, and soils; botanical explorations and list of collectors . . .; list of herbaria in which specimens are deposited; . . . ecology, phytogeography, life zones, endemic species; notes on native and cultivated useful plants; local vernacular names; lists of doubtful and excluded species, . . .; a gazetteer of localities . . . supplemented, if possible, by a map; a list of botanical names first published in the work; and an index.

Few floras include all these features, but it is clearly apparent that omission of any not only detracts from completeness but lessens the scientific value of the contribution.

There is no single world flora that accounts for every species, even of the spermatophytes, on the earth. The herbarium material on which to base such a flora does not exist. The so-called world floras (e.g., Bentham and Hooker's *Genera plantarum,* Engler and Prantl's *Die natürlichen Pflanzenfamilien*), for the most part, do not treat units below the category of genus, and some of them (Hutchinson's *Families of flowering plants,* vol. 1) do not account for units below the category of family.

In addition to the works cited below there are the much earlier works of Linnaeus, Lindley, and de Candolle that, to anyone but the taxonomist engaged in research, are now more of classical than of practical interest.

BAILLON, H. Histoire des plantes. 13 vols. Paris, 1867–1895.
   A comprehensive work treating all families and genera of vascular plants. Copiously illustrated with figures of pertinent reproductive structures. No keys provided. Extensive bibliographic references in footnotes. Text in French.

BAILLON, H. The natural history of plants. 8 vols. London, 1871–1888.
   This is an English translation of the first of 8 volumes of the French work, cited above and does not include the 5 volumes on the monocotyledons.

BENTHAM, G. and HOOKER, J. D. Genera plantarum. 3 vols. London, 1862–1883.
   This is one of the great reference works of all time devoted to the genera of seed plants. Its primary claim to greatness is that the descriptions of genera treated in it were prepared from studies of the plants themselves. These descriptions were not a product of compilation. The genera were grouped under families, the families under cohorts (orders in present-day terminology); all in accordance with the Bentham and Hooker system of classification. There are no keys in the work; major taxa are separated by means of synoptical devices. Large genera are divided into subgenera and sections, and important species of each are usually indicated. The text is in Latin. See pp. 115–118 of Chapter VI of this book for more detailed analysis.

DE CANDOLLE, A. P., A., and C. Prodromus systematis naturalis regni vegetabilis. 17 vols. With Buek's index, vols. 1–4. Paris, 1824–1873
   This was the first work of its scope since Willdenow's edition of *Species plantarum,* attempting to account for all species of seed plants. It was prepared before modern keys were generally employed, and the later volumes made use of synopses (cf. Chapter X, p. 225) in lieu of keys. The work treats only the dicots. The first 7 volumes were written by Augustin Pyramus de Candolle, and the remaining volumes, edited by his son Alphonse, were written for the most part by selected monographers. In all, 35 botanists participated. "While its title of *Prodromus* (forerunner) designates it merely as a preliminary survey of the vegetable kingdom to be followed by more elaborate monographs, nevertheless it contains the last comprehensive revision of many an important genus." For the dates of publication, see W. T. Stearn's paper in Candollea, 8: 1–4, 1939.

DE CANDOLLE, ALPHONSE and CASIMIR, eds. Monographiae phanerogamarum, 7 vols. Paris, 1879–1891.
   A work designed to provide monographs to families, and to serve as a successor to the *Prodromus.* The following families are treated: vol. 1, Smilacées, Restiaceae, Meliaceae; vol. 2, Araceae; vol. 3, Philydraceae, Alismaceae. Butomaceae, Juncaginaceae, Commelinaceae, Cucurbitaceae; vol. 4, Burseraceae, Anacardiaceae, Pontederiaceae; vol. 5, Cyrtandreae [Gesneriaceae], Ampélidées [Vitaceae]; vol. 6, Andropogoneae [Gramineae]; vol. 7, Melastomacées.

ENGLER, A. and DIELS, L. Syllabus der Pflanzenfamilien. Berlin, 11th ed., 1936.
   This is a syllabus of the plant families and is, in a sense, a condensation of the Engler system of classification into a single volume. It contains a synoptical key to all the families of plants (exclusive of Bacteria), indicates the various subsidiary categories to which the families belong, provides synonymies, author citations, and brief family descriptions. The volume is well illustrated with diagrams and drawings of technical characters. The text is in German.

ENGLER, A. and PRANTL, K. Die natürlichen Pflanzenfamilien. 23 vols. Leipzig, 1887–1915; ed. 2, 8 vols. 1924– [Incomplete, 1942].
   This is, in effect, a glorified "genera plantarum." It is written in German, well illustrated, and is documented by selected bibliographies. The attempt was made to provide dichotomous keys and adequate descriptions to all families of plants (except the Bacteria) from the Algae through the Compositae, arranging them in accordance with the Engler system of classification. For each family was given a summary of knowledge concerning the embryology, morphology, and anatomy of vegetative and reproductive structures. A bibliography of pertinent and outstanding taxonomic papers supplements these data. The second edition promises to be a most important taxonomic work, and publication, suspended toward the close of World War II, is reported soon to be resumed. The monographic treatments of the families were prepared largely by specialists under the editorship of Engler, Prantl, and Diels.

HUTCHINSON, J. The families of flowering plants. 2 vols. London, 1926 and 1934.
   Volume I treats the dicotyledonous plants and volume II the monocotyledonous plants. The treatment of the dicots is preceded by a statement of Hutchinson's classification and a key to the families. Each family is accompanied by a description and the names of prominent genera. Volume II is similar, but includes also a key to the genera of most families.

### Regional floras and manuals

Floras are regional in scope. In the following account no distinction is made between a flora and a manual; in fact most of the works listed are manuals rather than floras. The student should know of the more important floras and manuals of the earth's continents. Anyone engaged in taxonomic work, as a primary or secondary interest, is called on frequently to identify plants, and when these are unknown indigenous plants they are identified usually by the aid of a manual or flora of the area from which the plants originated. Occasionally a collection of plants is received in exchange or by purchase, and their identification is checked by reference to treatments in the pertinent floras. Knowledge of the more important floras is invaluable to the traveler interested in the plants of an area otherwise little known to him. Anyone pursuing taxonomic work in the field of cultivated plants must be well versed in the larger regional floras of the world, since cultivated plants may come from any part of the world.

There is great need for modern competently prepared floras. This is true not only for much of North America, but to an even greater degree for almost all other continents, with the possible exception of parts of

Europe and a few colonized areas. In the United States there is need for preparation of local floras, preferably of physiographic rather than of political areas. Most of the available comprehensive floras of other countries exclusive of much of Europe are over 50 years old, and knowledge of their plants has been amplified manyfold since then. See Chapter XIII for details.

In addition to floras and manuals for restricted areas, the student should be cognizant of the existence of the many papers available concerning the floristics and phytogeography of those same areas and their components.

The most complete index to the literature comprising the floras and manuals of the world is that projected by Blake and Atwood.[2] Of this work Part I has been published. It accounts for general and local floras devoted to Africa, Australia, North America, South America, and islands of the Atlantic, Pacific, and Indian oceans. Because of the completeness and utility of this publication the treatment that follows accounts, in so far as the New World is concerned, only for the larger and perhaps most frequently used floras. Many data incorporated into the explanatory notes preceding the following enumerations of floras have been adapted from Blake and Atwood.

**North America**

There is no complete flora for the continent. Neither are there complete floras yet available for all of Alaska or Canada. Only about one-third of the area of the latter has its plants accounted for by modern lists. Only 3 of the 8 Central American countries (including Mexico) possess modern floras. Mexico, the largest, and richest in number of species, has no complete flora, although much has been published on it. Greenland's flora is comparatively well known. All 48 states of the United States are covered by local or regional floras, although no separate state floras or lists are known for 6 states. Of the West Indies, perhaps the plants of Cuba are the least adequately accounted for.

A selected annotated list of floras considered especially helpful in the identification of vascular plants of this region follows.

BRITTON, N. L., *et al.* North American flora. Published by the New York Botanical Garden, 1905 and continuing.
    This flora, projected to comprise 34 volumes, is still in the course of preparation. When complete, it will provide data concerning every accepted species

[2] BLAKE, S. F. and ATWOOD, A. C. Geographical guide to the floras of the world. An annotated list with special reference to useful plants and common plant names. Part I, U. S. Dept. Agr., Misc. Publ. 401. 1942.

of plant known to be indigenous to North America, from the bacteria and algae through the seed plants. For the units of each category there are dichotonious keys, terse descriptions, synonymy, reference to habitat, distributional ranges, and designation of a type. There are no illustrations, annotations, or taxonomic or nomenclatural discussions. North America is treated as comprising the continent, Greenland, and the West Indies (Trinidad, Tobago, and Curaçao are not included). Each family or genus has been prepared by a monographer following a uniform editorial policy. Parts published prior to 1935 were in accord with the so-called American Code of Nomenclature; subsequent parts follow the International Rules, ed. 3. To date, the treatments of the following orders or families of vascular plants have been completed in whole or in part: Ophioglossaceae to Cyatheaceae, Typhaceae to Gramineae, Cyperaceae, Bromeliaceae, Chenopodiaceae, Amaranthaceae, Saxifragaceae, Rosaceae, Leguminosae, Geraniales through Polygalales, Umbelliferae, Ericales, Apocynaceae, Rubiaceae, Compositae.

Fig. 35. ASA GRAY (1810–1888). Foremost American botanist of his day, founder of the Gray Herbarium at Harvard University, and author of one of the most renowned manuals of American botany. Sketch by M. E. Ruff, from photograph, courtesy L. H. Bailey.

## Alaska and Canada

ANDERSON, J. P. Flora of Alaska and adjacent parts of Canada. An illustrated descriptive text of all vascular plants known to occur within the region covered. Iowa State College Journal of Sci. 18: 137–175, 381–445 (1943); 19: 133–205 (1945); 20: 213–257, 297–347 (1946); 21: 363–423 (1947); 23: 137–187 (1949).

A descriptive flora, with keys, to the vascular plants of Alaska. Original illustrations of critical characters of most species are provided, and for each species are given the essential synonymy, distribution, and an English vernacular name. Arrangement follows the Engler system, and the work is complete through Plumbaginaceae [as of 1949].

HULTÉN, E. Flora of Alaska and Yukon. Lunds univ. årssk. N. F. Avd. II 37: 1–27 (1941); 38: 129–412 (1942); 39: 413–467 (1943); 40: 569–795 (1944); 41: 797–978 (1945); 42: 981–1066 (1946); 43: 1067–1200 (1947); 44: 1201–1341 (1948); 45: 1345–1481, 1949.

An annotated flora with keys to species, but with no keys to families or genera. No descriptions included. Full synonymy and local and general distributional data provided (including maps for each species). Based on the Engler system. Complete to Campanulaceae.

HULTÉN, E. The flora of the Aleutian Islands and westernmost Alaska peninsula, with notes on the flora of Commander Islands. 397 pp. Stockholm, 1937.

A copiously annotated flora, with synonymy, with detailed general distributional ranges. No keys or descriptions are provided for the 479 species treated. A bibliography on Alaskan flora is included.

HULTÉN, E. Flora of Kamtchatka and the adjacent islands. Kungl. Svenska Vetenskakad. Handl. Ser. 3, Bd. 5 (1926–1928) and 8 (1929-1930).

MARIE-VICTORIN, FRÈRE. [Conrad Kirouac] Flore laurentienne. 917 pp. Montreal, 1935.

An annotated and illustrated descriptive flora, with keys, accounting for the plants of southern Quebec. Text in French.

LOUIS-MARIE, PÈRE. Flore-manual de la province de Quebec. 319 pp. Montreal, 1931. (Contr. 23, Institut agricole d'Oka).

An elementary flora, in the form of keys to the vascular plants. Text in French.

## Western United States and Canada

ABRAMS, LEROY. An illustrated flora of the Pacific states, Washington, Oregon and California. 2 vols. [Incomplete]. Palo Alto, Calif., 1923, 1944.

A descriptive flora to be completed in 4 volumes. Volumes 1 and 2, Filicales through Krameriaceae, have been published. A revised edition of Volume 1 was published in 1940, differing from the original edition in that the nomenclature was revised to conform with the International Rules (ed. 3).

COULTER, J. M. New manual of botany of the central Rocky Mountains (vascular plants). . . . rev. by Aven Nelson. New York, 1909.

JEPSON, W. L. A manual of the flowering plants of California. Berkeley, 1923-1925.

A descriptive flora with keys accounting for 4019 species of vascular plants.

JEPSON, W. L. A flora of California. 3 vols. [Incomplete]. San Francisco, Berkeley, 1909-1943.

A descriptive flora of the seed plants, with bibliographic references, localities, illustrations, and keys.

KEARNEY, T. H. and PEEBLES, R. H. Flowering plants and ferns of Arizona. U. S. Dept. Agr. Misc. Publ. 423. 1942.

An analytical flora accounting for about 3200 species, with keys to elements of all categories. Descriptions are given only for genera; distributional data, synonymies, and habitats are given for elements in all categories.

MUNZ, P. A. A manual of southern California botany. Claremont, Calif., 1935.

A descriptive flora with keys, accounting for the vascular plants of California, from Los Angeles County southward.

PECK, M. E. A manual of the higher plants of Oregon. Portland, Ore., 1941.

PIPER, C. V. Flora of the state of Washington. Contr. U. S. Nat. Herb. vol. 11. 1906.

RYDBERG, P. A. Flora of the Rocky Mountains and adjacent plains, Colorado, Utah, Wyoming, Idaho, Montana, Saskatchewan, Alberta and neighboring parts of Nebraska, South Dakota, North Dakota and British Columbia. 2d ed. New York, 1922.

This is a descriptive flora with dichotomous keys accounting for 6029 binomials of pteridophytes and spermatophytes. The essential difference between first and second editions is that the latter contains 33 pages of additions and corrections.

WOOTON, E. O. and STANDLEY, P. C. Flora of New Mexico. Contr. U. S. Nat. Herb. vol. 19. 1915.

### Central and midwestern United States

DEAM, C. C. Flora of Indiana. Indianapolis, 1940.

An annotated flora with keys to the species of Indiana plants. No descriptions of species are included, although many valuable floristic and bibliographical data are provided. The treatment of various groups contributed by specialists. Many supplements have been published as Indiana Plant Distribution Records in the Proceedings of the Indiana Academy of Science.

REEVES, R. G. and BAIN, D. C. A flora of south central Texas. College Station, Texas, 1946.

Keys and descriptions to families, genera, and species; arrangement according to the Bessey system "with a few modifications." The area covered by the flora is within the Southern Post Oak area of the coastal plain.

Fig. 36. PER AXEL RYDBERG (1860–1931). Trained originally as an engineer, ardent field collector in western and mid-western United States, and author of several floras of those regions. Sketch by M. E. Ruff, from photograph, courtesy Brooklyn Botanical Garden.

RYDBERG, P. A. Flora of the prairies and plains of central North America New York, 1932.

A descriptive flora, with keys, accounting for approximately 4000 species of vascular plants of Kansas, Nebraska, Iowa, Minnesota, South Dakota, North Dakota, southern Manitoba, and southeastern Saskatchewan.

SCHAFFNER, J. H. Field manual of the flora of Ohio and adjacent territory. Columbus, 1928.

A manual of the vascular plants in the form of keys. Includes a bibliography and glossary, but without descriptions.

STEMAN, T. R. and MYERS, W. S. Oklahoma flora. Oklahoma City, 1937.

An annotated descriptive flora, with keys, to the vascular plants of Oklahoma. Gramineae and Cyperaceae omitted.

TIDESTROM, I. Flora of Utah and Nevada. Contr. U. S. Nat. Herb. vol. 25. 665 pp. 1925.
   A systematic enumeration of about 3700 species of vascular plants with keys but without descriptions. Treatments of several groups contributed by various authors.

## Northeastern United States

BRITTON, N. L. Manual of the flora of the northern states and Canada. 3d ed., rev. and enl. New York, 1907.
   A descriptive flora of the vascular plants of the area, with keys.
BRITTON, N. L. and BROWN, A. An illustrated flora of the northern United States, Canada, and British possessions from Newfoundland to the parallel of the southern boundary of southern Virginia and from the Atlantic westward to the 102nd meridian. 2d ed. rev. and enl. 3 vols. New York, 1913.
   A descriptive flora with keys to approximately 4600 species of vascular plants and a figure illustrating each species. In this work each species is given a vernacular name, which frequently is only an anglicization of the Latin name.

**Fig. 37.** MERRITT LYNDON FERNALD (1873-1950).
For over a quarter of a century Director of the Gray Herbarium and Editor of *Rhodora*, best known as an authority on the vascular flora of primarily the coastal regions of northeastern North America and as author of the latest (8th) edition of Gray's *Manual of Botany*. Sketch by M. E. Ruff, from photograph, courtesy M. L. Fernald.

FERNALD, M. L. Gray's manual of botany. ed. 8. Boston and New York, 1950.
   A manual of the pteridophytes and spermatophytes of the central and northeastern United States and adjacent Canada. This edition is essentially a new work, embodying over 40 years of revisionary and floristic studies by the late Professor Fernald. New and dichotomous keys provided throughout to 1133 genera and 5523 species.

## Southeastern United States

SMALL, J. K. Manual of the southeastern flora; giving descriptions of the seed plants growing naturally in Florida, Alabama, Mississippi, eastern Louisiana, Tennessee, North Carolina, South Carolina, and Georgia. New York, 1933.

A descriptive flora with keys to approximately 5500 binomials of spermatophytes (pteridophytes not included). The flower characters of each genus are indicated by illustrative figures.

Fig. 38. JOHN KUNKEL SMALL (1869–1938). Authority on the higher plants of southeastern United States, indefatigable field collector throughout that region, and author of the most recent comprehensive work on its flora. Sketch by M. E. Ruff, from photograph, courtesy Brooklyn Botanical Garden.

## Mexico and Central America

CONZATTI, C. Flora taxonomica mexicana (plantas vasculares). Vol. 1 (1939) Pteridophytes; vol. 2 (1943–1947) Monocotyledonae. Mexico, 1939–1947 [Incomplete].

A briefly descriptive flora with keys to families, genera, and species of Mexican vascular plants. Text in Spanish.

HEMSLEY, W. B. Biologia Centrali-Americana. Botany. 5 vols. 1879–1888 [1915].

The botanical section of the comprehensive work edited by F. Ducane Godman and O. Salvin. Localities and synonymy but no keys; only new species are described.

STANDLEY, P. C. Trees and shrubs of Mexico. Contr. U. S. Nat. Herb. vol. 23. 1920–1926.

A briefly descriptive flora of approximately 5700 species of native woody vascular plants with keys. Treatment of several families contributed by specialists.

STANDLEY, P. C. Flora of the Panama Canal zone. Contr. U. S. Nat. Herb. vol. 27. 1928.

A briefly descriptive flora of the spermatophytes, with keys.

STANDLEY, P. C. The flora of Barro Colorado Island, Panama. Contr. Arnold Arb. 5. 1933

STANDLEY, P. C. Flora of Costa Rica. Publ. Field Mus. Nat. Hist. Bot. 391, 392, 420, 429, Bot. Ser. vol. 18. 1616 pp. 1937–1938.

An enumeration of approximately 6000 spermatophytes with brief descriptions of genera and species of the dicotyledons.

STANDLEY, P. C. and RECORD, S. J. The forests and flora of British Honduras. Pub. Field Mus. Nat. Hist. 350. Bot. Ser. vol. 12. 432 pp. Chicago, 1936.

A descriptive flora to approximately 2100 species of vascular plants. Contains keys and descriptions to the species of the larger genera and to the genera of the larger families.

STANDLEY, P. C. and STEYERMARK, J. A. Flora of Guatemala. Fieldiana: Botany, 24: pts. 4–6, 1949 [Incomplete].

A flora providing keys, descriptions, synonymy, vernacular names, and geographic distributions; based on the Engler classification, complete to date for families Ulmaceae through Saurauiaceae.

WOODSON, R. E. JR. and SCHERY, R. W. Flora of Panama. In Ann. Mo. Bot. Gard. 30(2), 1942; 31(1), 1944; 32(1), 1945; 33(1), 1946; 33(4), 1946; 35(1), 1948; 36(1), 1949. [Incomplete.]

A flora providing keys and descriptions to families, genera, and species, together with synonymy, geographic distribution, and citations of specimens examined. Based on the Engler classification. Complete to date for Cycadaceae through Orchidaceae and Lauraceae through Cruciferae.

## West Indies

BRITTON, N. L. and MILLSPAUGH, C. F. The Bahaman flora. 695 pp. New York, 1920.

A descriptive flora with keys, essential synonymy, and notes of general distribution. Accounts for all plant life from algae through spermatophytes.

FAWCETT, W. and RENDLE, A. B. Flora of Jamaica, containing descriptions of the flowering plants known from the island. Vols. 1, 3–5, 7. London, 1910–1936 [Incomplete].

A descriptive flora, with keys, synonymy, distributional data, citation of exsiccatae, etc.

GRISEBACH, A. H. R. Flora of the British West Indian islands. 789 pp. London 1864.

Actual publication dates are 1859–1864 (cf. Jackson, B. D. in Journ. Bot. 30: 347, 1892). Despite its antiquity, the only general flora of the West Indies (exclusive of Cuba, Puerto Rico, and Hispaniola). A descriptive flora, without keys.

## South America

As pointed out by Blake and Atwood, of the 13 South American countries (plus Patagonia) only 4 have recent and essentially complete lists of vascular plants (Patagonia, Surinam, Uruguay, and Venezuela), and none of these is a flora or manual in the modern sense. The floras now in preparation for Brazil, Peru, and Argentina are of great promise. Many countries of the continent have no partial flora, manual, or list that is less than 100 years old.

DESCOLE, H., *et al.* Genera et species plantarum argentinarum. Buenos Aires, 1943-continuing.

A sumptuous work projected to occupy many tomes, accounting for the flora of Argentina. The arrangement of families follows no established system. A descriptive flora with keys to genera and species, synonymies, distributional data (with maps) and full page plates illustrating each species. Many plates hand-colored. Text in Spanish. Inasmuch as the work is incomplete and the family arrangement unorthodox, the families treated to date are listed here: Tome I (1943), Zygophyllaceae, Cactaceae, Euphorbiaceae; Tome II (1944), Asclepiadaceae, Valerianaceae; Tome III (1945), Centrolepidaceae, Mayacaceae, Xyridaceae, Eriocaulaceae, Bromeliaceae; Tome IV (1948), Cyperaceae.

GAY, C. Historia fisica y politica de Chile . . . botanica flora chilena. 8 vols. and atlas of 103 col. pl. Paris, Santiago, 1845–1854.

A descriptive flora of Chilean plants. Text in Spanish except for Latin diagnoses. For dates of publication, see Johnston, I. M. in Darwiniana, 5: 154–165, 1941.

MACBRIDE, J. F. Flora of Peru. Pub. Field. Mus. Nat. Hist. Bot. 1936–1949 [In process of completion].

An annotated list of spermatophytes, with keys, descriptions, citation of exsiccatae. Treatments of several families by specialists. Text in English.

MACLOSKIE, G., et al. Reports of the Princeton University expeditions to Patagonia. vol. 8, botany. 1896–1899.

The Flora Patagonica, an annotated list of vascular plants, with keys, brief descriptions, bibliography, and floristic data.

MARTIUS, K. F. P. VON, et al. Flora Brasiliensis . . . 15 vol. in 40. 3805 pl. Berlin, 1840–1906.

One of the most sumptuous of floras (in excess of 20,000 pages) accounting for 22,767 species. A costly and now rare work; one of the few comprehensive floras of Brazilian plants.

PITTIER, H. F. Genera plantarum Venezuelensium. Clave analitica de los generos de plantas hoy conocidos en Venezuela. 354 pp. Caracas, 1939.

Key to the genera of vascular plants of Venezuela. Text in Spanish.

PULLE, A. A. Flora of Surinam (Dutch Guiana). 4 vols. Amsterdam, 1932–1940 [Incomplete].

A descriptive flora of phanerogams. No keys. Text in English.

REICHE, K. F. Flora de Chile. 6 vols. Santiago, 1896–1911.

A descriptive flora, with keys. Lacks most of the "Apetalae," and parts available are not altogether trustworthy.

## Europe

There are several hundred floras and manuals accounting for the vascular plants of Europe, but few have been published during the last half century. Cited below are those believed more frequently referred to in checking identities, affinities, and nomenclature of most European plants.

BENTHAM, G. Handbook of the British flora; a description of the flowering plants and ferns indigenous to, or naturalized in the British Isles. For the use of beginners and amateurs. ed. 7, rev. by A. B. Rendle. 606 pp. Ashford, Kent, 1930.

A descriptive flora, with keys, descriptions, and distributional ranges. The introduction includes explanations of gross morphology of structures, notes on classification, herbarium techniques, glossary, and key to the families.

BUTCHER, R. W. and STRUDWICK, F. E. Further illustrations of British plants. Ashford, Kent, 1946.
   A supplementary handbook to Fitch and Smith, with 485 new figures with descriptions.
COSTE, H. Flore descriptive et illustrée de la France. 3 vols. Paris, 1900–1906.
   A descriptive flora to the vascular plants of France and Corsica. Keys to species provided. A figure is provided to illustrate each species. Text in French. Dates of publication and contents of the 16 fascicles are stated in a note following the preface to vol. 3.
COUTINHO, A. X. P. Flora de Portugal (plantas vasculares). Paris, Lisbon, 1913. ed. 2, 1939.
   A standard flora in the form of keys. For Supplemento (154 pp.) see Bol. Soc. Brot. Ser. II, vol. 10, 1935.
FIORI, A. Nuova flora analitica d'Italia. 2 vols. Firenze, 1923–1929.
   A conservative flora of the vascular plants of Italy, employing an excessively broad concept of species. Full and complete descriptive keys are provided for genera and species. No other descriptions included. Text in Italian. It is a companion work to the same author's Iconographia florae italicae. ed. 2. 1921.
FITCH, W. H. and SMITH, W. G. Illustrations of the British flora; a series of wood engravings, with dissections, of British plants. ed. 5. Ashford, Kent, 1931.
   An illustrated companion to Bentham's Handbook of the British flora, providing 1321 figures of as many species.
HAYEK, A. Prodromus florae peninsularae Balcanicae. 3 vols. Berlin, 1923–1933. [Beihefte 30 of Fedde, Repert. Sp. Nov.].
   A descriptive flora of the vascular plants of southeastern Europe. All groups are keyed. The treatment is modern and conservative. Text and keys in Latin. For dates of publication, see Stearn, W. T. in Journ. Soc. Bibl. Nat. Hist. 1: 117–119, 1937.
HEGI, G. Illustrierte Flora von Mittel-Europa. 13 vols. Munich, 1906–1931. 2nd ed., vol. 1, 1936; vol. 2, 1939 (incomplete).
   A very complete conservative classical work, accounting for the vascular plants of central Europe. The text, well illustrated, is in German. Descriptions, keys, synonymy, and copious bibliography are provided. Treatment of many families by specialists. For dates of publication, see Becherer, A. in Candollea, 5: 342–344, 1934.
HUTCHINSON, J. British flowering plants. London, 1948.
   This book deals primarily with the "evolution and classification of families and genera, with notes on their distribution" while data ordinarily expected in a flora are treated as of secondary importance. It is primarily a treatise (much revised over that in the author's *The families of flowering plants*) of Hutchinson's system of classification, using British material in so far as possible as illustrative of its organization. Keys are provided to the families and genera of British plants, but no formal descriptions are included, and distributions are sketchily recounted.
KOMAROV, V. L., *et al.* Flora U. R. S. S. (Flora Unionis Rerumpublicarum Sovieticarum Socialisticarum). 15 vols. Moscow, 1934–date [Incomplete].
   This work is a comprehensive flora of Russia. All families, genera, and species are described, and keys are provided. The work is illustrated and the entire text (except for Latin names of plants) is in Russian. For lists and maps, whereby anyone ignorant of Russian can nevertheless translate the geographic and other abbreviations of the Flora U. R. S. S. into English equivalents and thereby ascertain the general distribution of species described in this work, see Stearn W. T. in New Phytologist, 46: 61–67, 1947.

LAGERBERG, T. and HOLMBOE, J. Våre ville planter. 5 vols. Oslo, 1937–1940 [not completed].
    This is the Norwegian edition of the Swedish, *Vilda våxter i Norden* by T. Lagerberg (Stockholm 1937 *et seq.*). Many line drawings and color photographs. Text in Norwegian.
LID, J. Norsk flora. Oslo, 1944.
    A 1-volume handbook accounting, with keys and descriptions, for the vascular plants of Norway. About 360 figures of line drawings provide habit sketches of the majority of species treated. Text in Norwegian.
LINDMAN, C. A. M. Svensk fanerogamflora. ed. 2. Stockholm, 1926.
    A standard flora of Sweden, well illustrated. Keys are provided. The classification follows the Linnaean "sexual system."
POST, G. E. Flora of Syria and Palestine. ed. 2. 2 vols. London, 1932–1933. (The 2nd ed. published posthumously and edited by T. E. Dinsmore.)
SAMPAIO, G. Flore portugaise. ed. 2. Porto, Portugal, 1947.
    A flora of 792 pages and 693 figures accounting for the flowering plants of Portugal. The taxonomy is conservative and the nomenclature follows the Vienna Rules of 1905. The species are accounted by descriptive dichotomous keys following the Engler classification. Synonymy very incomplete.
URSING, B. Svenska våxter i text och bild. Stockholm, 1947.
    A field manual accounting for the species of Swedish flowering plants; habit sketches of the majority are in color. The key to the families is based on the Linnaean "sexual system" of classification following Swedish tradition; conventional artificial keys are provided to genera and species. Text in Swedish.
WILLKOMM, H. M. and LANGE, J. Prodromus florae Hispanicae. 3 vols. Stuttgart, 1861–1880. [Supplementum.] 1893.
    This remains a standard work on plants of Spain. For dates of publication, see Wiltshear in Journ. Bot. 53: 371, 1915.

## Asia

Considering its area and vast range of climatic zones, Asia is probably the most deficient in floras of any great land mass. Japan has no modern flora in a Roman alphabet. China has no national flora, and few regional floras. The existing flora of India is mostly more than 50 years old. The plants of much of western Asia (especially Afghanistan, Iran, Iraq, Oman, and greater Arabia) are not well known, or so little known that they are not accounted in any flora less than a century old. However, despite this lack of formal descriptive floras, the last half century has witnessed extensive collecting in the near and middle east, whose results appear in the enumerations published by Stapf, Bornmüller, Handel-Mazzetti, P. H. Davis, Nabelek, Rechinger, *et al.,* in European botanical periodicals.

BOISSIER, E. Flora orientalis sive enumeratis plantarum in oriente a Graecia et Aegypto ad Indiae fines hucusque observatarum. 5 vols. Basel and Genève, 1867–1884. Supplementum, 1888.
    An important flora for the vascular plants of the area. Synopses and descriptions provided for plants in all categories. Text in Latin.

GAMBLE, J. S. and FISHER, C. E. C. Flora of the Presidency of Madras. 11 parts, usually in 3 vols. London, 1915-1938.
HOOKER, J. D. Flora of British India. 7 vols. London, 1876-1897.
    A descriptive flora of the species of vascular plants of British India. Modified synoptical type keys provided.
KANJILAL, U. N., *et al.* Flora of Assam. 5 vols. Shillong, 1934-1940 [incomplete to date].
    This flora was intended originally as a forest flora of the region, but many herbaceous plants are included. Descriptions and keys are provided. Text in English.

## Africa

Exploitation by colonization has been responsible for a moderate understanding of the plants of Africa, but even this is so inadequate that of the 49 geographical divisions recognized by Blake and Atwood there are more or less complete floras, containing keys or descriptions, of only 7 (Algeria, Anglo-Egyptian Sudan, Angola, Egypt, Swaziland and Transvaal, and Tunisia). However, certain large areas are covered by more or less comprehensive floras, only 1 of which is modern. Blake and Atwood cite 23 countries, colonies, and protectorates of Africa that are entirely without general floras or lists limited to their area.

A very brief listing of African floras follows.

ADAMSON, R. S. and SLATER, T. M. Flora of the Cape peninsula. 889 pp. Cape Town, 1950.
    A descriptive flora, with keys, synonymies, and distributions of the vascular plants of the area. Based on the Engler classification.
BATTANDIER, J. A. and TRABUT, L. Flore analytique et synoptique de l'Algérie et de la Tunisie. 406 pp. Alger, 1902 [1904].
    A flora, in the form of keys, accounting for 3316 species of vascular plants. Text in French.
BATTISCOMBE, E. Trees and shrubs of Kenya Colony. 201 pp. Nairobi, 1936.
    An annotated descriptive list of woody plants.
BEWS, J. W. An introduction to the flora of Natal and Zululand. 248 pp. Pietermaritzburg, 1921.
    An annotated list of 3786 species of seed plants, including keys to families and genera.
BURTT DAVY, J. A manual of the flowering plants and ferns of the Transvaal with Swaziland, South Africa. 529 pp. London, 1926-1932 [Not completed].
    An annotated briefly descriptive flora, with keys, to the vascular plants. Arrangement follows Hutchinson system.
EXELL, A. W. Catalogue of the vascular plants of S. Tomé. 428 pp. London, 1944.
    A catalogue of special importance for its critical synonymy and citations of the literature, for many of the plants of this Gulf of Guinea island occur also in tropical West Africa. No keys or descriptions included.
HARVEY, W. H., SONDER, W., *et al.* Flora capensis: being a systematic description of the plants of Cape Colony, Caffraria, and Port Natal (and neighboring territories). 7 vols. Dublin, 1859-1865; London, 1896-1933.
    For history of this work, see Thistleton-Dyer in Kew Bull. 1925: 289-293, and for dates of publication see Marshall, H. S. in Journ. Soc. Bibl. Nat. Hist.

1: 196. 1939. A detailed descriptive flora with synoptical keys. Many data of earlier volumes now of considerable antiquity.

HUTCHINSON, J. and DALZIEL, J. M. Flora of west tropical Africa. The British west African colonies, British Cameroons, the French and Portuguese colonies south of the Tropic of Cancer to Lake Chad, and Ferdinando Po. 2 vols. London, 1927–1929.

A modern illustrated flora with descriptions and keys, citation of exsiccatae, and ranges of distribution.

MUSCHLER, R. A manual flora of Egypt. 2 vols. Berlin, 1912.

A descriptive flora, without keys, of 1503 species of vascular plants. Text in English. The work is not critical, and information in it is subject to challenge and verification.

OLIVER, D. et al. Flora of tropical Africa. 10 vols. London, 1868–1937 [Not completed].

A descriptive flora of seed plants with keys, synonymy, and distributional ranges. For bibliographic notes see Blake and Atwood (p. 63).

PHILLIPS, E. P. The genera of South African flowering plants. 702 pp. Cape Town, 1926. Botanical Survey of South Africa. Memoir 10.

Descriptions and keys of South African plants.

ROBYNS, W. editor. Flore du Congo Belge et du Ruanda-Urundi. Vol. 1. Bruxelles, 1948 [Incomplete].

A critical descriptive flora, with keys, synonymies, and distributions, published under the aegis of the Institut National pour l'étude agronomique du Congo Belge. Based on the Engler classification. Volume I includes Gymnospermae through Polygonaceae. Text in French.

TÄCKHOLM, V. and G. and DRAR, M. Flora of Egypt. 2 vols. Cairo, Egypt, 1941–1950 [Incomplete].

A descriptive flora based on the Engler classification, with keys, synonymies, and distributions. Many exotic plants and those of economic importance included. Volume I treats the ferns, gymnosperms, and monocots (through the grasses). Volume II treats the Cyperaceae through the Juncaceae.

## Australasia

Following the delimitations established by Blake and Atwood, Australasia is restricted to include only Australia, New Zealand, and Tasmania. As compared with other regions, this is provided with several comparatively modern descriptive floras or lists, and no major area is devoid of an accounting of its plants.

BAILEY, F. M. The Queensland flora. 6 vols. Brisbane, 1899–1902.

A descriptive flora of vascular plants, with keys, vernacular names, and distributional ranges. A 66-page general index to scientific names was published separately in 1905.

BENTHAM, G. Flora Australiensis: a description of the plants of the Australian territory. 7 vols. London, 1863–1878.

A descriptive flora of Australia and Tasmania with keys, synonymies, and distributional ranges. For dates of publication, see Marshall, H. S. in Journ. Soc. Bibl. Nat. Hist. 1: 71, 1937.

BLACK, J. M. Flora of South Australia. 746 pp. Adelaide, 1922–1929.

A descriptive flora, with keys and glossary, of 2430 vascular plants (Orchidaceae by R. S. Rogers).

CHEESEMAN, T. F. Manual of the New Zealand flora. ed. 2, rev. and enl. 1163 pp. Wellington, 1925.
: A descriptive flora of vascular plants with keys, synonymies, distributional ranges, and Mauri vernacular names.
CHEESEMAN, T. F. and HEMSLEY, W. B. Illustrations of the New Zealand flora. 2 vols. 251 pl. Wellington, 1914.
DIXON, W. A. The plants of New South Wales; an analytical key to the flowering plants, except the grasses and ferns of the state, set out in an original method, with an up-to-date list of native and introduced flora. 322 pp. Sydney, 1906.
EWART, A. J. Flora of Victoria. 1257 pp. [Melbourne], 1930.
: An annotated descriptive flora with keys and distributional notes.
EWART, A. J. and DAVIES, O. B. The flora of the Northern Territory . . . with appendices by J. H. Maiden, *et al.* 387 pp. Melbourne, 1917.
: An annotated flora with keys or brief synoptical diagnoses, and distributional notes.
KIRK, T. The forest flora of New Zealand and the outlying islands. 345 pp. 142 pl. 1889.
: A descriptive flora, without keys; the illustrations are excellent.
MOORE, C. and BETCHE, E. Handbook of the flora of New South Wales. 582 pp. Sydney, 1893.
: A description of the flowering plants and ferns indigenous to New South Wales.
RODWAY, L. The Tasmanian flora. 320 pp. Hobart, 1903.
: A descriptive flora with keys, distributional ranges, and glossary.
TATE, R. A handbook of the flora of extratropical South Australia, containing the flowering plants and ferns. Adelaide, 1890.
: About 1900 vascular plants are accounted for by keys. The treatment accounts for the vascular plants of an area as far north as Tropic of Capricorn.

## Pacific insular floras

BROWN, F. B. H. and E. D. W. Flora of southeastern Polynesia. Bishop Museum Bull. 84, 89, 1931–1935.
CHRISTOPHERSON, E. Flowering plants of Samoa. Bishop Museum Bull. 128, 1938.
DEGENER, O. Flora Hawaiiensis, or the new illustrated flora of the Hawaiian Islands. Book I-III. Honolulu, 1932–1938 [Incomplete].
: A descriptive flora, loose-leaf, unpaged, keys unpublished to date, each species described and illustrated.
HILLEBRAND, W. Flora of the Hawaiian Islands; a description of their phanerogams and vascular cryptogams. London and New York, 1888.
: A descriptive flora, with keys to genera and species, accounting for about 1000 species. Vernacular names are included.
MERRILL, E. D. An enumeration of Philippine flowering plants. 4 vols. Manila, 1922–1926.
ROCK, J. F. C. The indigenous trees of the Hawaiian Islands. Honolulu, 1913.
: A descriptive account, without keys, of the indigenous trees of the islands.
STEENIS, C. G. G. J. VAN, ed. Flora Malesiana. Batavia, 1948. [Incomplete.]
: A new flora, comprising an illustrated systematic account of the plants of Malaysia, i.e., the area extending from Sumatra to New Guinea and from Luzon to Timor and New Guinea, including Indonesia, the Malay peninsula, the Philippine Islands, the Moluccas, Celebes, and Borneo. It includes keys, descriptions, references, and the commonly cultivated as well as wild plants. Text in English.

## Monographs and Revisions

A taxonomic monograph of a group of plants is a comprehensive treatise representing an analysis and synthesis of existing taxonomic knowledge of that taxon, plus the results of original research in so far as systematics are concerned. In other words, it is "as complete an account as can be made at a given time of any one family, tribe, or genus, 'nothing being neglected which is necessary for a perfect knowledge of it.'" The usual subject of a taxonomic monograph is the genus or the family. The investigations, on which the monograph is based, generally account for all elements of the group in question; that is, all species if of a genus, or all genera and species if of a family. A monograph is generally worldwide in its scope and application, and is not limited to the representatives of a geographical area; it reviews and evaluates all taxonomic treatments that have been made of that taxon; it should be based whenever possible on an extensive field knowledge of the plants concerned, or failing that, on their behavior under cultivation and, in all cases, on a study of the collections of the larger herbaria of the world or of the area accounted for; it is predicated almost without exception on studies of the available type specimens of all nomenclatural elements involved; it often embraces a study of the phylogeny of the taxon and considers the phylogenetic relationships of the taxon with other related taxa; it takes cognizance of, and synthesizes in so far as possible, all cytological, genetical, morphological, anatomical, paleobotanical, and ecological studies made of members of the taxon by its author, co-workers, or others. All elements of the treatise are accounted for by dichotomous keys, full synonymies, complete descriptions, precise designations of types, together with notes as to where the types are deposited, citations of specimens examined, distributional ranges (supplemented by maps of the same), notes on habitats, and discussions of taxonomic and nomenclatorial considerations as may be appropriate. Monographs of this scope are few in number. The term monograph as used loosely, notably by horticulturists little acquainted with taxonomic literature or procedures, covers not only such works, but any systematic work devoted to a single taxon.

A taxonomic revision differs from a monograph primarily in degree of scope and completeness. Often it accounts for only a section of a genus or for the elements as restricted to a continent or smaller geographical area. Many revisions make no attempt to review all previous work on the taxon or to take cognizance of the interrelated sciences of cytotaxonomy, genetics, ecology, etc. It is not uncommon for revisions to be based solely on herbarium studies; they may thus be provisional and introductory to more thorough treatments.

There are no modern bibliographies devoted to the subjects of monographs and revisions of the world. However, the bibliographies by Pritzel, Jackson, Rehder, and Merrill and Walker do list several thousand monographs and revisions (cf. pp. 306–307). No bibliographies of monographs or revisions are given in this chapter, but a selection of those dealing with important genera (and of the family when available) is given at the end of the description of each family accounted for in Part II of this text. A list of so-called "monographic" works on genera of hardy herbaceous plants was published by Stearn, W. T. in Journ. Roy. Hort. Soc. 67: 296–303, 1942.

## Bibliographies, catalogues, and review serials

The literature of systematic botany is of such magnitude that no one individual can be thoroughly conversant with more than a small part of it, few taxonomic institutions have complete card catalogues of it, and there is no one book or set of books accounting for it. Despite its vastness and the great number of items composing it, the student of taxonomy must learn how to find his way about this literature. Familiarity with available works of reference will save time when searching for information about a particular problem. The 3 principal sources are published bibliographies, catalogues of outstanding botanical libraries, and periodicals containing reviews of current literature. Each source is discussed below, and annotated references to selections of each type are included.

*Important bibliographies* of taxonomic literature are cited below. They are works that attempt to account for all books and scientific literature published during a given period on the subject of the particular bibliography. Most bibliographies include a list of abbreviations employed for the many series whose papers are cited. Such a list, together with the full title of the periodical, is often helpful when endeavoring to locate a serial in a library card catalogue.

BAY, J. C. Bibliographies of botany. In, Progressus rei botanicae, 3: 331–456. 1910.
    An accounting of the bibliographies published up to 1910 in all fields of botany. A bibliography of bibliographies.

Bibliogr. Agr. BIBLIOGRAPHY OF AGRICULTURE. (U.S. Dept. of Agr. Library) U.S. Dept. of Agr. Misc. Publ. 337. 1942.
    A monthly periodical accounting for all subjects of broad agricultural character. A section of each issue treats taxonomic papers.

Cat. Sci. Papers CATALOGUE OF SCIENTIFIC PAPERS. Royal Society of London. Vols. 1–6 (period 1800–1863), 7–8 (1864–1873), 10–11 (1874–1883), 12 (1800–1883, omitted in vols. 1–11), 13–19 (1884–1900).

An index, by authors, to titles of scientific papers in serials and periodicals of the world for the period 1800–1900, in volumes indicated. For each entry are given name of author, title, source, publication date.

Int. Cat. Sci. Lit. INTERNATIONAL CATALOGUE OF SCIENTIFIC LITERATURE. M, BOTANY. Royal Society of London. 1902–1916.

Volumes issued annually, each indexing by author and subject heading all original botanical contributions irrespective of whether a book, pamphlet, or periodical. The series accounts for literature of the period 1900–1913.

JACKSON, B. D. Guide to the literature of botany; being a classified selection of botanical works, including nearly 6000 titles not given in Pritzel's Thesaurus. 626 pp. London, 1881.

MERRILL, E. D. and WALKER, E. H. A bibliography of eastern Asiatic botany. Arnold Arboretum of Harvard Univ. 719 pp. Jamaica Plain, Mass., 1938.

An accounting of over 21,000 author entries of all known works concerned to the least degree with plants of eastern Asia, arranged by authors and titles, with an excellent subject index. A valued innovation is a reference list of serial abbreviations, with full title of the serial and pertinent bibliographic notes accompanying each. The majority of world-important botanical serials are accounted for.

Plt. Sci. Lit. PLANT SCIENCE LITERATURE. U. S. Dept. of Agr., Bur. Plt. Ind. Library. 1930–1940. (Superseded by Bibliography of Agriculture.)

A mimeographed monthly periodical of limited distribution, but available in most university and institutional libraries.

PRITZEL, G. A. Thesaurus literaturae botanicae, etc. 547 pp. 1847–1851; ed. 2, 576 pp. 1872.

One of the fundamental bibliographic references for early botanical literature, the second edition much the more valuable. A reprint of ed. 2 was issued in 1924, and another in 1949.

REHDER, A. The Bradley bibliography. A guide to the literature of the woody plants of the world published before the beginning of the twentieth century. Compiled at the Arnold Arboretum of Harvard University under the direction of Charles Sprague Sargent. 5 vols. 1911–1918.

Accounts for all publications, as issued through 1900, on woody plants, including books, pamphlets, and articles in periodicals, arranged by subjects and authors.

―――. Bibliography of cultivated trees and shrubs. 825 pp. Jamaica Plain, Mass., 1949.

A bibliography citing the author and place of publication of all Latin names of taxa treated in the author's *Manual of cultivated trees and shrubs,* ed. 2, and of their synonyms, with the nomenclature brought up to date as of 1947. The names of a majority of the species of woody plants of temperate North America are included.

*Catalogues* cited below account for the books of special libraries rich in botanical titles, and are of especial value in taxonomic studies. It is often necessary to know the full name of a particular author, to know the unabridged and exact title of a work, to know when it was published, or when a particular edition was issued. These data are usually available from such catalogues.

Catalogue of the library of the British museum (natural history). 5 vols. London, 1903–1915. Supplement. 3 vols. 1922–1940.

This catalogue (including the works accounted for in the supplement) accounts for books, maps, periodicals, and pamphlets in the fields of natural history up to 1920 (through I), to 1930 (letter J–O) and to 1938 (letters P–Z). Entries are alphabetical by author, with cross-indexing employed to an appreciable extent.

Catalogue of the library of the Massachusetts horticultural society. 587 pp. Cambridge, Mass., 1918.

An author-subject catalogue of about 22,000 volumes, exclusive of the collection of about 11,000 nursery and seed trade catalogues.

Catalogue of the library of the royal botanic gardens, Kew (up to 1898). Bull. of Misc. Inf. [later as Kew Bull.]. Additional Series III, 790 pp. London, 1899. Supplement (1898–1915), 433 pp. London, 1919.

A convenient catalogue in 1 volume accounting for general works, travels periodicals, serials, and manuscripts; arranged within each of these groups alphabetically. The majority of taxonomic works (through 1897) are listed, and the catalogue is a source of complete author name, unabridged titles, size and number of volumes, and dates of publication. The supplement lists entries received, and usually published, during the period 1898–1915.

THE LINDLEY LIBRARY. Catalogue of books, pamphlets, manuscripts, and drawings. 487 pp. London, 1927.

A catalogue enumerating the holdings of the Royal Horticultural Society's Lindley Library up to 1926.

*Review serials* are periodicals, usually issued at regular intervals, that provide either (1) a bibliography of current literature of a particular subject, (2) an abstract of papers or books in special fields, (3) reviews of titles of current literature, or (4) any combination of these functions. In evaluating their treatments it is well to remember that an abstract is a brief factual summary of a paper, frequently prepared by its author, whereas a review is an often critical appraisal and evaluation of the paper, and is by a person other than the original author. Most major botanical periodicals provide bibliographies (annotated or otherwise) of current literature pertinent to interests of subscribers. This is especially true of such serials as

*Bulletin of the Torrey Botanical Club*
*Bulletin de la société botanique de France*
*Nuovo giornale botanico italiano*
*Lingan Science Journal*
*Acta phytotaxonomica et geobotanica*

Other periodicals are devoted in whole or major part to the subject of abstracts or summaries of botanical works and papers and are accounted for below with annotations.

Biol. Abstr. BIOLOGICAL ABSTRACTS; A COMPREHENSIVE ABSTRACTING AND INDEXING JOURNAL OF THE WORLD'S LITERATURE IN THEORETICAL AND APPLIED BIOLOGY, EXCLUSIVE OF CLINICAL MEDICINE. Philadelphia, 1926+.

Bot. Abstr. BOTANICAL ABSTRACTS; A MONTHLY SERIAL FURNISHING ABSTRACTS IN THE INTERNATIONAL FIELD OF BOTANY IN ITS BROADEST SENSE. Baltimore, 1918–1926.

Bot. Centralbl. [Bot. Zbl.] BOTANISCHES CENTRALBLATT; REFERIERENDES ORGAN FÜR DAS GESAMTGEBIET DER BOTANIK. Jena, 1880+.

    Sometimes catalogued as Botanisches Zentralblatt. Abstracts of world botanical literature, arranged by subject groups and indexed by authors. Text in German.

Bot. Jahrb. Engler [Bot. Jb.] BOTANISCHE JAHRBÜCHER FÜR SYSTEMATIK, PFLANZENGESCHICHTE UND PFLANZENGEOGRAPHIE. Founded by A. Engler. Leipzig, Stuttgart, 1880+.

    A general taxonomic periodical with a review of literature (abstracts) terminating each volume. Text in German.

Just's Bot. Jahrb. [Just's Jber.] JUST'S BOTANISCHER JAHRESBERICHT. Vols. 1–10 as Botanischer Jahresbericht. 1873+.

    Abstracts or summaries of world botanical literature. Arranged by major subject groups, indexed by subject and author. Text in German.

## Periodicals

A periodical, or serial, is a publication appearing usually at regular intervals. Each issue is called a number, or sometimes is termed a fascicle (*Heft* in German). Collectively these numbers or fascicles comprise a volume (*Band* in German). In the case of periodicals appearing at regular intervals—biweekly, monthly, or quarterly—a volume usually comprises the issues of a calendar year. Scientific periodicals usually are sponsored either by a scientific organization, such as a learned society, or an educational or nonprofit research institution, such as a university or museum.

Sponsors of periodicals often publish more than one serial. A society may publish a monthly serial to provide a source of publication for a variety of relatively short papers contributed by its members, together with records of its own proceedings. Such a periodical is usually entitled Journal, Annal, Bulletin, or Proceeding. The same society also may publish at much less frequent intervals another serial to account for longer and more monumental works, often by a single author, such as monographs or floras. These are often entitled Memoirs or Transactions. To illustrate these situations, two publications of the Torrey Botanical Club (New York) are the *Bulletin of the Torrey Botanical Club* and *Memoirs of the Torrey Botanical Club*.

The titles of some periodicals are long, and in citing them it is customary to abbreviate or condense them. Unfortunately, there is no uniformity of practice in this country as regards their abbreviation, and to one not familiar with the periodical literature this multiplicity of abbreviations may be confusing. A very excellent listing of a large number of the botanical periodicals of the world and their abbreviations is to be found in Merrill and Walker's *A bibliography of eastern Asiatic botany* (pp.

xiii-lii), and in Rehder's *Bibliography of cultivated trees and shrubs.* Another of more restricted scope is that by Schwarten and Rickett (Bull. Torrey Bot. Club, 74: 348-356, 1947).

American students of taxonomy should know that there are American periodicals devoted in whole or in part to papers dealing with systematic botany; the advanced student of necessity should have an acquaintanceship with these. It should be appreciated also that the American serials on this subject constitute but a small fraction of the world's periodical literature of taxonomic publications. There are nearly 1000 different periodicals regularly containing articles on systematic botany. Some of these periodicals, containing articles pertinent to taxonomy of American plants, are published outside the New World and in languages other than English.

The number of botanical serials is so great that probably no library contains complete sets of them all. When a local library does not have a desired serial, or a particular volume of a serial, it is necessary (1) to borrow that volume, (2) to visit a library known to possess it, or (3) to obtain a photographic copy such as a microfilm or photostat. In determining what libraries have the desired volume, reference is made to the current edition of the *Union list of serials,* a standard reference work possessed by almost all leading civic and institutional libraries. In this work, all periodicals of the world catalogued by American libraries are listed alphabetically. Under each listing are given, in abbreviated form, the names of all American libraries known to have the serial and, when the set is incomplete, the volumes available at the particular library. A British publication of the same plan is *A world list of scientific periodicals,* published in the years 1900-1933 (Oxford Univ. Press, London). This work enumerates 24,029 periodicals with standard abbreviations, and states in which of the main British libraries they can be found.

Library card catalogues usually cite full titles when indexing periodicals. However, botanists generally use the abbreviated form of the title in citations and bibliographies, and for this reason it is necessary to know the full title as catalogued. Because of this, there are given below 3 annotated lists, each with the usual abbreviated title, and the full title, of the more significant serials that deal in whole or in part with the literature of systematic botany. The first list is that of American periodicals. The second is a short selected list of those Old World periodicals believed most frequently consulted by American students of taxonomy. The third is of periodicals important in studies of the systematics of cultivated plants.

In the listings that follow, abbreviations given for periodicals are for the most part those used by Merrill and Walker.[3] Libraries differ widely in the method of cataloguing titles of periodicals and too often fail to cross-index titles under the various possible headings. In general, if a periodical is published by a municipal or government institution (as a national museum) it is catalogued under the name of the sponsoring country or city; likewise, if by a society, academy, or educational institution, it will be catalogued under the title of the organization, which may in turn be placed under the name of the city where it has its seat. For the purpose of uniformity in listing the periodicals that follow, the abbreviated form is given, followed by the full name of each. When this name is known to be at variance with the listing in catalogues of our larger libraries an alternative title also may be indicated. Students may find Russian periodicals especially confusing, since not only have titles changed completely with changes from a tsarist to a soviet regime and the official language of scientific publications changed from French to Russian, but names of cities also have changed (as St. Petersburg to Leningrad, etc.). It is sometimes necessary to enlist the aid of library personnel in locating titles of such periodicals in a given card catalogue.

The list of New World serials of taxonomic botany enumerated below makes no claim for completeness. Some are national in scope, others contain taxonomic articles primarily of regional interest. Most of the entries are currently published, and for these the first year of issue is cited and the plus sign (+) is given after the date. Other serials are no longer published, a fact indicated by the date of last publication. Forty-eight of these periodicals are North American, and 6 are Central or South American in origin.

Amer. Fern Journ. [Amer. Fern J.] AMERICAN FERN JOURNAL. 1910+.
    Published by the American Fern Society.
Amer. Journ. Bot. [Amer. J. Bot.] AMERICAN JOURNAL OF BOTANY. 1914+.
    Published by the Botanical Society of America.
Amer. Midl. Nat. AMERICAN MIDLAND NATURALIST; DEVOTED TO NATURAL HISTORY, PRIMARILY THAT OF THE PRAIRIE STATES. Notre Dame, Ind., 1909+.
Anal. Inst. Biol. [Mexico]. [An. Inst. Biol. Univ. Méx.] ANALES DEL INSTITUTO DE BIOLOGÍA, UNIVERSIDAD DE MÉXICO. Mexico, 1929+.

---

[3] British botanists for the most part follow the abbreviations suggested in the British World List of Scientific Periodicals (indicated here in square brackets when differing from American usage).

It is to be hoped that a set of abbreviations may soon be published that have been agreed upon by a committee representing international interests. These abbreviations "should shorten any abbreviated word by at least 3 letters, they should be clear and definite, and they should be intelligible to the inexperienced person of a nationality and language other than that represented by the abbreviated title." Most bibliophiles are agreed that 1-word periodical names (as Brittonia) should not be abbreviated.

Ann. Mo. Bot. Gard. [Ann. Mo. bot. G'dn.] ANNALS OF THE MISSOURI BOTANICAL GARDEN. St. Louis, Mo., 1914+.
  The Annual Report of the Director of this garden issued 1889-1911 contains many important taxonomic papers.
Bartonia. BARTONIA. PROCEEDINGS OF THE PHILADELPHIA BOTANICAL CLUB. 1908+. (Publication suspended during 1915–1923.)
Bishop Museum Bull. [Bull. Bishop Mus., Honolulu] BULLETIN OF THE BERNICE P. BISHOP MUSEUM OF POLYNESIAN ETHNOLOGY AND NATURAL HISTORY. Honolulu, 1922+.
Bishop Museum Mem. [Mem. Bishop Mus., Honolulu] MEMOIR OF THE BERNICE P. BISHOP MUSEUM OF POLYNESIAN ETHNOLOGY AND NATURAL HISTORY. Honolulu, 1899+.
Bot. Gaz. BOTANICAL GAZETTE. Crawfordsville, Ind., 1875+.
Bot. Rev. BOTANICAL REVIEW; INTERPRETING BOTANICAL PROGRESS. Lancaster, Pa., 1935+.
Brittonia. BRITTONIA; A SERIES OF BOTANICAL PAPERS. New York, N. Y., 1931+.
  Published by the New York Botanical Garden.
Bull. Biol. Soc. Wash. BULLETIN OF THE BIOLOGICAL SOCIETY OF WASHINGTON. Washington, D. C., 1918+.
Bull. Calif. Acad. Sci. BULLETIN OF THE CALIFORNIA ACADEMY OF SCIENCES. San Francisco, 1884–1887.
Bull. Mo. Bot. Gard. [Bull. Mo. Bot. G'dn.] BULLETIN OF THE MISSOURI BOTANICAL GARDEN. St. Louis, Mo., 1913+.
Bull. Torrey Bot. Club [Bull. Torrey Bot. Cl.] BULLETIN OF THE TORREY BOTANICAL CLUB. New York, N. Y., 1870+.
Caldasia. CALDASIA. Instituto de ciencias naturales, universidad. Colombia, 1941+.
Canadian Field-Nat. [Canad. Field Nat.] CANADIAN FIELD-NATURALIST. Ottawa, Ont., 1887+.
  Published by the Ottawa Field Naturalists' Club, and from 1887–1919 as the Ottawa Naturalist.
Castanea. CASTANEA. Morgantown, W. Va., 1936+.
  Published by the Southern Appalachian Botanical Club, and from 1936–1937 as the Journal Southern Appalachian Botanical Club.
Claytonia. CLAYTONIA. Lynchburg, Va., 1934–1939.
  Superseded by The Virginia Journal of Science.
Contr. Biol. Lab. Cath. Univ. Amer. CONTRIBUTIONS FROM THE BIOLOGICAL LABORATORY OF THE CATHOLIC UNIVERSITY OF AMERICA. Washington, D. C., 1915+.
Contr. Bot. Lab. Univ. Penn. [Contr. bot. Lab. Univ. Pa.] CONTRIBUTIONS FROM THE BOTANICAL LABORATORY AND THE MORRIS ARBORETUM OF THE UNIVERSITY OF PENNSYLVANIA. Philadelphia, 1892+.
Contr. Dudley Herb. CONTRIBUTIONS FROM THE DUDLEY HERBARIUM. Stanford University, Palo Alto, Calif., 1927+.
Contr. Gray Herb. [Contr. Gray Herb. Harv.] CONTRIBUTIONS FROM THE GRAY HERBARIUM, HARVARD UNIVERSITY. Cambridge, Mass., 1891+.
Contr. inst. bot. univ. Montréal. CONTRIBUTIONS DE L'INSTITUT BOTANIQUE DE L'UNIVERSITÉ DE MONTRÉAL. 1922+.
  Published as Contributions de le laboratoire de botanique, 1922–1938.
Contr. U. S. Nat. Herb. CONTRIBUTIONS FROM THE U. S. NATIONAL HERBARIUM. Smithsonian Institution, Washington, D. C. 1890+.
  Volumes 1–7 published by the U. S. Dept. of Agriculture.
Darwiniana. DARWINIANA. Buenos Aires, 1922+
  Published by the Instituto de botánica Darwinion, academia nacional de ciencias exactas, fisicas y naturales.

El Aliso. EL ALISO; A SERIES OF PAPERS ON THE NATIVE PLANTS OF CALIFORNIA. Rancho-Santa Ana Botanic Garden, Anaheim, Calif., 1948+.
Erythrea. ERYTHREA; A JOURNAL OF BOTANY, WEST AMERICAN AND GENERAL. Berkeley, Calif., 1893–1900, 1922–1938.
Field Mus. Bot. [Field Mus. Publ. Bot.] FIELD MUSEUM OF NATURAL HISTORY, BOTANICAL SERIES. Chicago, Ill., 1895–1943.

Published by the Chicago Natural History Museum, formerly the Field Columbian Museum, 1894–1908, and later the Field Museum of Natural History, 1908–1943.
Gentes Herb. [Gentes Herb., Ithaca] GENTES HERBARUM; OCCASIONAL PAPERS ON THE KINDS OF PLANTS. Ithaca, N. Y., 1920+

Published privately by L. H. Bailey 1920–1934, and subsequently by the Bailey Hortorium, Cornell University, 1935+.
Journ. Acad. Sci. Philadelphia. [J. Acad. nat. Sci. Philad.] JOURNAL OF THE ACADEMY OF NATURAL SCIENCES OF PHILADELPHIA. Ser. 1, 1817–1842; ser. 2, 1847–1918.
Journ. Arnold Arb. [J. Arnold Arb.] JOURNAL OF THE ARNOLD ARBORETUM. Harvard Univ., Jamaica Plain, Mass., 1919+.
Journ. Elisha Mitchell Soc. [J. Elisha Mitchell sci. soc.] JOURNAL OF THE ELISHA MITCHELL SCIENTIFIC SOCIETY. Chapel Hill., N. C., 1883+.
Journ. So. Appal. Bot. Club: see under Castanea.
Journ. Washington Acad. Sci. [J. Wash. Acad. Sci.] JOURNAL OF THE WASHINGTON ACADEMY OF SCIENCE. Washington, D. C., 1911+.
Leafl. West. Bot. LEAFLETS OF WESTERN BOTANY; A PUBLICATION ON THE EXOTIC FLORA OF CALIFORNIA AND ON THE NATIVE FLORA OF WESTERN NORTH AMERICA. San Francisco, 1932+.
Lilloa. LILLOA; REVISTA DE BOTÁNICA. Instituto Miguel Lillo, universidad nacional, Tucumán, 1937+.
Lloydia. LLOYDIA; A QUARTERLY JOURNAL OF BIOLOGICAL SCIENCE. Lloyd Library, Cincinnati, 1938+.
Madroño. MADROÑO. San Francisco, 1916+.

Published by the California Botanical Society.
Mem. Calif. Acad. Sci. MEMOIR OF THE CALIFORNIA ACADEMY OF SCIENCES. Vols. 1–5. San Francisco, 1868–1905.
Mem. Gray Herb. [Mem. Gray Herb. Harv.] MEMOIR FROM THE GRAY HERBARIUM, HARVARD UNIVERSITY. Cambridge, Mass., 1917+.
Mem. Torrey Bot. Club [Mem. Torrey Bot. Cl.] MEMOIR OF THE TORREY BOTANICAL CLUB. New York, N. Y., 1889+.
Muhlenbergia. MUHLENBERGIA. Vols. 1–10. Los Gatos, Calif., 1900–1915.
Occas. Papers Calif. Acad. Sci. OCCASIONAL PAPERS OF THE CALIFORNIA ACADEMY OF SCIENCES. San Francisco, 1890+. (Suspended 1906–1921).
Physis. PHYSIS. Buenos Aires, 1912+.

Published by the Sociedad argentina de ciencias naturales.
Phytologia. PHYTOLOGIA. New York, N. Y., 1935+.
Proc. Acad. Sci. Philadelphia. [Proc. Acad. nat. sci. Philad.] PROCEEDINGS OF THE ACADEMY OF NATURAL SCIENCES OF PHILADELPHIA. 1841+
Proc. Biol. Soc. Washington. PROCEEDINGS OF THE BIOLOGICAL SOCIETY OF WASHINGTON. Washington, D. C., 1880+.
Proc. Calif. Acad. Sci. PROCEEDINGS OF THE CALIFORNIA ACADEMY OF SCIENCES. San Francisco, ser. 1, 1854–1876; ser. 2, 1880–1896; ser. 3 (as Botany), 1897–1904; ser. 4, 1907+.
Proc. Washington Acad. Sci. [Pro. Wash. Acad. Sci.] PROCEEDINGS OF THE WASHINGTON ACADEMY OF SCIENCE. Washington, D. C., 1899–1911.

Rhodora. RHODORA. Boston, Mass., 1899+.
    Published by the New England Botanical Club.
Rodriguesia. RODRIGUESIA. Instituto de biologia vegetal; jardim botanico, estação biologica do Itatiaya. Rio de Janeiro, 1935+.
Torreya. TORREYA; A MONTHLY JOURNAL OF BOTANICAL NOTES AND NEWS. New York, N. Y., 1901–1945.
    Published by the Torrey Botanical Club, and incorporated in the Bull. Torrey Bot. Club in January, 1946, under a section by itself but without separate pagination.
Univ. Calif. Publ. Bot. UNIVERSITY OF CALIFORNIA PUBLICATIONS IN BOTANY. Berkeley, 1902+.
Va. Journ. Sci. VIRGINIA JOURNAL OF SCIENCE. Charlottesville, Va., 1940+
    Formerly titled CLAYTONIA.
Wrightia. WRIGHTIA. Southern Methodist University. Dallas, Texas. 1945+.

The list of Old World serials given below accounts only for those of major taxonomic significance, and contains relatively few titles selected from a lengthy bibliography. Fifty-eight are of European, 4 of Asiatic, and 5 of African origin.

Abh. Akad. Wiss. Berlin. [Abh. preuss. Akad. Wiss.] ABHANDLUNGEN PREUSSISCHE AKADEMIE DER WISSENSCHAFTEN. Berlin, 1804–1907.
Abh. Akad. Wiss., München. (Same as Abh. bayer. Akad. Wiss. Math.-nat., which see.)
Abh. bayer. Akad. Wiss. Math.-phys. ABHANDLUNGEN BAYERISCHE AKADEMIE DER WISSENSCHAFTEN. MATHEMATISCH-PHYSIKALISCHE KLASSE. Munich, 1829+.
Abh. zool.-bot. Gesell. Wien [Abh. zool.-bot. Ges. Wien] ABHANDLUNGEN ZOOLOGISCH-BOTANISCHE GESELLSCHAFT IN WIEN [Vienna], 1901+.
Acta horti Gotoburg. ACTA HORTI GOTOBURGENSIS. Botaniska trädgård, Gothenberg, Sweden, 1924+.
Acta horti Petrop. ACTA HORTI PETROPOLITANI . . . Trudy imperatorskago S. Petersburgskago Botanicheskago Sada. Leningrad, 1871–1931.
Acta univ. Lund. ACTA UNIVERSITATIS LUNDENSIS. University of Lund, Sweden, 1864–1904.
Allg. bot. Zeitschr. [Allg. bot. Z.] ALLGEMEINE BOTANISCHE ZEITSCHRIFT FÜR SYSTEMATIK, FLORISTIK, UND PFLANZENGEOGRAPHIE. Karlsruhe, 1895–1927 (suspended 1920–1925).
Anales soc. españ. hist. nat. [An. Soc. esp. Hist. nat.] ANALES DE SOCIEDAD ESPAÑOLA DE HISTORIA NATURAL. Madrid, 1872–1901.
Ann. Bolus Herb. ANNALS OF THE BOLUS HERBARIUM AND LIBRARY. University of Cape Town. Cape Town, 1914+.
Ann. mus. hist. nat. (Paris). ANNALES DU MUSÉUM NATIONAL D'HISTOIRE NATURELLE. Paris, 1802–1913.
Ann. naturhist. Mus. Wien. [Ann. naturh. Mus. Wien.] ANNALEN DES NATURHISTORISCHES HOFMUSEUM. Wien (Vienna), 1886+. (Sometimes catalogued as Königliches naturhistorisches Hofmuseum in Wien, Annalen.)
Ann. sci. nat. ANNALES DES SCIENCES NATURELLES. Paris, 1824–1833. Ser. II. Bot. 1834+ (in several series).
Beih. bot. Centralbl. [Beih. bot. Zbl.] BEIHEFTE BOTANISCHES CENTRALBLATT, 1891+. (Sometimes catalogued as Botanisches Zentralblatt, Beihefte.)
Beih. Repert. sp. nov. Fedde. BEIHEFTE, REPERTORIUM SPECIERUM NOVARUM REGNI

VEGETABILIS. Leipzig, 1911+. (Fedde in above abbreviation refers to Dr. F. Fedde, the founder.)
Ber. Akad. Wiss. Berlin. [Ber. preuss. Akad. Wiss.] BERICHTE PREUSSISCHER AKADEMIE DER WISSENSCHAFTEN. Berlin, 1836–1855.
Bibl. bot. [Bibl. bot. Stuttgart] BIBLIOTHECA BOTANICA. Stuttgart, 1886+.
Biol. Medd. Dansk. Vid. BIOLOGISKE MEDDELELSER DANSKE VIDENSKABERNES SELSKAB. Copenhagen, 1917+. Sometimes catalogued under (Kongelige) Danske Videnskabernes Selskab, Det, Biologiske Meddelelser.
Blumea. BLUMEA, TIJDSCHRIFT VOOR DE SYSTEMATIEK EN DE GEOGRAFIE DER PLANTEN. Rijksherbarium. Leiden, 1934+.
Boissiera. BOISSIERA. Conservatoire de botanique, Geneva, 1936+.
Bol. soc. brot. BOLETIM DA SOCIEDADE BROTERIANA. Coimbra, Portugal. Ser. 1, 1880–1920; ser. 2, 1922+.
Bol. soc. españ. hist. nat. [B. Soc. esp. Hist. nat.] BOLETIN DE SOCIEDAD ESPAÑOLA DE HISTORIA NATURAL. Madrid, 1901+.
Bot. Centralbl. [Bot. Zbl.] BOTANISCHES CENTRALBLATT. Jena and Dresden, 1880+.
Bothalia. BOTHALIA. National Herbarium, So. Africa. Pretoria, 1921+.
Bot. Jahrb. Engler. [Bot. Jb.] BOTANISCHE JAHRBÜCHER FÜR SYSTEMATIK, PFLANZENGESCHICHTE UND PFLANZENGEOGRAPHIE. Leipzig, 1880+.
Founded and edited by Adolph Engler. Each volume contains 1 or more separately numbered and paged appendices titled Beiblätter and Literaturbericht.
Bot. Mag. see Curtis's Bot. Mag., p. 317.
Bot. Mag. (Tokyo). [Bot. Mag., Tokyo] BOTANICAL MAGAZINE. Tokyo, 1887+.
Published by the Tokyo Botanical Society, and beginning with vol. 46 (1932) the Botanical Society of Japan. Text in Japanese and western languages.
Bot. Notiser. BOTANISKA NOTISER; LUNDS BOTANISKA FORENING. University of Lund. Lund, Sweden, 1839+.
Bot. Zeit. [Bot. Ztg.] BOTANISCHE ZEITUNG. Berlin, 1843–1910.
Bull. herb. Boissier. BULLETIN DE L'HERBIER BOISSIER. GENÈVE, Chambésy, Switzerland, 1893–1899; ser. 2, 1901–1908. (Continued as the Bulletin de la société botanique de Genève.)
Bull. mus. hist. nat. (Paris). BULLETIN DU MUSÉUM NATIONAL D'HISTOIRE NATURELLE. Paris, 1895–1928; ser. 2, 1929+.
Bull. soc. bot. Fr. BULLETIN DE LA SOCIÉTÉ BOTANIQUE DE FRANCE. Paris, 1854+.
Bull. soc. bot. Genève. BULLETIN DE LA SOCIÉTÉ BOTANIQUE DE GENÈVE. Geneva, 1878–1905; ser. 2, 1909+. (Title of Ser. 1, Bulletin des travaux de la société botanique de Genève.)
Bull. soc. hist. nat. Afr. nord. BULLETIN DE LA SOCIÉTÉ D'HISTOIRE NATURELLE DE L'AFRIQUE DU NORD. Algiers, 1909+.
Candollea. CANDOLLEA. Geneva, 1922+. (Supersedes Annuaire, Genève conservatoire et jardin botanique.)
Danske Vidensk. Biol. Medd. see under Biol. Medd. Dansk. Vid.
Danske Vidensk. Selsk. Skr. DANSKE VIDENSKABERNES SELSKABS SKRIFTER NATURVIDENSKABELIG OG MATHEMATISK AFDFLING. Copenhagen, 1824+ (in many series).
Denkschr. Akad. Wiss. Math.-nat. (Wien). DENKSCHRIFTEN DER AKADEMIE DER WISSENSCHAFTEN IN WIEN, MATHEMATISCH-NATURWISSENSCHAFTLICHE KLASSE. Wien [Vienna], 1850+.
Fedde Repert. see Repert. spec. nov. Fedde.
Fl. Pl. So. Afr. [Flower. Pl. S. Afr.] FLOWERING PLANTS OF SOUTH AFRICA. Johannesburg, 1920+.

Since 1947 the title has become Flowering Plants of Africa. A botanical periodical of hand-colored plates with descriptions of the flowering plants indigenous to Africa.

Flora. FLORA, ODER ALLGEMEINE BOTANISCHE ZEITUNG. Jena, 1818+.

Published by the Bayerische botanische Gesellschaft, Regensburg. During much of the period of issue, each volume contained a separately paged part known in various periods as, BEILAGE (1819–1830), BEIBLÄTTER (1832–1844), BESONDERE BEILAGE (1843–1844, 1850–1858), or REPERTORIUM (1864–1872).

Fortschr. Bot. [Fortschr. Bot. Berl.] FORTSCHRITTE DER BOTANIK. Berlin, 1931+.

Göteborgs Bot. Trad. GÖTEBORGS BOTANISKA TRÄDGÅRD. Same as Acta horti Goth. which see.

Hook. Bot. Misc. HOOKER'S BOTANICAL MISCELLANY. London, 1830–1833. Continued as Hooker's Journal of Botany.

Journ. Bot. [J. Bot. Lond.] JOURNAL OF BOTANY, BRITISH AND FOREIGN. London, 1863–1943.

Journ. Jap. Bot. [J. Jap. Bot.] JOURNAL OF JAPANESE BOTANY. Tokyo, 1916+.

Journ. Linn. Soc. Bot. (Lond.). [J. Linn. Soc. Bot.] JOURNAL OF THE LINNEAN SOCIETY. BOTANY. London, 1855+.

Journ. So. Afr. Bot. [J. S. Afr. Bot.] JOURNAL OF SOUTH AFRICAN BOTANY. Cape Town, 1935+.

K. bayer. Akad. Wiss. Math.-nat. (Same as Abh. bayer. Akad., etc., which see.)

Kew Bull. KEW BULLETIN. Royal Botanic Gardens. Kew, England, 1887+. (Formerly titled Bulletin of Miscellaneous Information.)

Konig. Danske Vidensk. Selsk. Skr. see under Danske, etc.

Leningrad Glav. bot. sad. Trudy. LENINGRAD GLAVNYI BOTANISCHESKII SADA TRUDY. (Same as ACTA HORTI PETROP., which see.)

Lignan Sci. Journ. [Lignan Sci. J.] LIGNAN UNIVERSITY, SCIENCE JOURNAL. Cantor, China, 1923+.

Linnaea. LINNAEA. EIN JOURNAL FÜR DIE BOTANIK IN IHREM GANZEN UMFANGE. Berlin, 1826–1866; n.s. 1867–1882.

Subtitle varies, merged with JAHRBUCH, UNIVERSITÄT BOTANISCHER GARTEN, BERLIN.

Magyar Bot. Lap. [Ung. bot. Bl.] MAGYAR BOTANIKAI LAPOK (UNGARISCHE BOTANISCHE BLÄTTER). Budapest, 1902+.

Med. Rijks Herb. Leiden. [Meded. Rijks-Herb.] MEDEDEELINGEN VAN'S RIJKS HERBARIUM. Leiden, 1910–1933.

Published by the Rijksuniversiteit, Leiden. Superseded by BLUMEA.

Meddel. Göteborgs Bot. Trad. MEDDELANDEN GÖTEBORGS BOTANISKA TRÄGÅRD Same as Acta horti Gotoburg, which see.

Mem. Acad. Sci. St. Petersb. MÉMOIRES DE L'ACADÉMIE IMPÉRIALE DES SCIENCES DE ST. PETERSBURG. Leningrad, 1728+.

Recent volumes published under the Russian title, Trudy Sl. Flora i systematika vyshikh rastenii.

Mem. soc. bot. Fr. MEMOIRES DE LA SOCIÉTÉ BOTANIQUE DE FRANCE. Paris, 1905+.

Monatsber. Akad. Wiss. Berlin [Mon. preuss. Akad. Wiss.] MONATSBERICHTE PREUSSISCHE AKADEMIE DER WISSENSCHAFTEN. Berlin, 1865–1881.

Notes Bot. Gard. Edinburgh. [Notes R. bot. G'dn. Edinb.] NOTES FROM THE ROYAL BOTANIC GARDEN, EDINBURGH. Edinburgh, 1900+.

Nuovo gior. bot. ital. NUOVO GIORNALE BOTANICO ITALIANO. Firenze, 1869–1893; n.s. 1894+.

Published by the Società botanica italiana.

Proc. Linn. Soc. (Lond.). PROCEEDINGS OF THE LINNEAN SOCIETY OF LONDON. London, 1838+.

Rep. Bot. Soc. Exch. Cl. REPORT OF THE BOTANICAL (SOCIETY AND) EXCHANGE CLUB OF THE BRITISH ISLES. Manchester, England, 1862–1947. Succeeded by Watsonia.
Repert. sp. nov. Fedde. [Repert. nov. Spec. Regn. Veg.] REPERTORIUM SPECIERUM NOVARUM REGNI VEGETABILIS; HERAUSGEGEBEN VON PROFESSOR DR. PHIL. FRIEDRICH FEDDE. Berlin-Dahlem and Leipzig, 1905+.
Repert. sp. nov. Fedde, Beih. REPERTORIUM SPECIERUM NOVARUM REGNI VEGETABILIS. HERAUSGEGEBEN VON PROFESSOR DR. PHIL. FRIEDRICH FEDDE. BEIHEFTE. Berlin-Dahlem and Leipzig, 1911+.
   This serial is distinct from the preceding.
Sitz. Akad. Wiss. Berlin. [Sitz. preuss. Akad. Wiss.] SITZUNGSBERICHTE PREUSSISCHE AKADEMIE DER WISSENSCHAFTEN. Berlin, 1882–1921.
Sitz. Akad. Wiss. Math.-nat. (Wien.). SITZUNGSBERICHTE DER AKADEMIE DER WISSENSCHAFTEN IN WIEN, MATHEMATISCH-NATURWISSENSCHAFTLICHE KLASSE. Wien [Vienna], 1848+.
Sitz. bayer. Akad. Wiss. Math.-nat. SITZUNGSBERICHTE BAYERISCHE AKADEMIE DER WISSENSCHAFTEN, MATHEMATISCH-PHYSIKALISCHE KLASSE. München, ser. 1, 1871–1913; ser. 2, 1914+.
Sunyatsenia. SUNYATSENIA. Journal of the Botanical Institute, College of Agriculture, Sun Yat Sen University, Canton, China, Canton, 1930+.
Svensk Bot. Tidskr. SVENSK BOTANISK TIDSKRIFT. Stockholm, 1907+.
Trans. Linn. Soc. Bot. (Lond.). TRANSACTIONS OF THE LINNEAN SOCIETY OF LONDON. BOTANY. London, 1875+.
Ung. bot. Bl. UNGARISCHE BOTANISCHE BLÄTTER. Budapest, 1902+.
   This is the German title of Magyar Botanikai Lapok.
Verh. Zool.-bot. Ges. Wien. VERHANDLUNGEN DER K. K. ZOOLOGISCH-BOTANISCHEN GESELLSCHAFT IN WIEN [Vienna]. Vienna, 1851+.
   The letters "K. K." (Kaiserlich-Königlichen) were included in the title only for vols. 8 (1858) to 67 (1918).
Watsonia. WATSONIA. JOURNAL OF THE BOTANICAL SOCIETY OF THE BRITISH ISLES. Oxford, 1949+.
   Successor to Rep. Bot. Soc. Exch. Cl.

It is with some reluctance that periodicals concerned primarily with systematics of cultivated or economic plants are here treated separately. The taxonomist who deals with the taxonomy of cultivated plants makes regular use of the serials listed above and in addition must be thoroughly cognizant of papers published in periodicals such as those listed below. There can be no valid distinction between taxonomy of indigenous and of cultivated plants. The systematist of indigenous plants merely makes less use of the world's literature than does the systematist of cultivated plants who must be conversant with all literature dealing with the kinds of plants irrespective of their immediate origins. Taxonomic literature restricted to cultivated plants is not voluminous, and periodicals that deal primarily with indigenous plants may treat cultivated plants also (as is true of Curtis's *Botanical Magazine* and *Gentes herbarum*). Serials most frequently used in taxonomic studies of cultivated plants are as follows (exclusive of those devoted wholly to special families or groups).

Addisonia. ADDISONIA. New York, 1916+.
Belg. hort. LA BELGIQUE HORTICOLE, JOURNAL DES JARDINS, DES SERRES ET DES VERGERS. Liége, 1850–1885.
Bot. Mag., see Curtis's Bot. Mag.
Bot. Reg. THE BOTANICAL REGISTER. London, 1815–1847.
    Sometimes known as Edwards' botanical register, in honor of its first editor.
Bull. Alpine Gard. Soc. [Bull. Alp. G'dn. Soc.] QUARTERLY BULLETIN OF THE ALPINE GARDEN SOCIETY OF GREAT BRITAIN. Wallington, England, 1930+.
Curtis's Bot. Mag. CURTIS'S BOTANICAL MAGAZINE. London, 1784+.
Deutsche Gärt.-Zeit. DEUTSCHE GÄRTNER-ZEITUNG. ZENTRALBLATT FÜR DAS GÄRTNERISCHE FORTBILDUNGSWESEN IN DEUTSCHLAND. Erfurt, 1877–1885. (Continued as Möller's deutsche Gärtner-Zeitung.)
Fl. des serres. FLORES DES SERRES ET DES JARDINS DE L'EUROPE. Ghent, 1845–1880.
Floral Mag. THE FLORAL MAGAZINE: COMPRISING FIGURES AND DESCRIPTIONS OF POPULAR GARDEN FLOWERS. London, 1860–1871; n.s. 1872–1881.
Gard. Chron. [G'dners. Chron.] THE GARDENERS' CHRONICLE. London, 1841–1873; ser. 2, 1874–1886; ser. 3, 1887+. (Not to be confused with the Gardeners' chronicle of America, a wholly different serial.)
Gartenflora. GARTENFLORA; ZEITSCHRIFT FÜR GARTEN UND BLUMENKUNDE. Berlin 1843–1851; n.s. 1852+.
HERBERTIA. YEARBOOK OF THE AMERICAN AMARYLLIS SOCIETY. Orlando, Fla., 1934+.
    The first 2 volumes known only as Yearbook of the American Amaryllis Society. The name of the society changed in 1944 to American Plant Life Society.
Hook. icones. [Hook. Icon. Plant.] HOOKER'S ICONES PLANTARUM. Royal Botanic Garden. Kew, England, 1837+.
Illustr. Gart.-Zeit. ILLUSTRIERTE GARTENZEITUNG; EINE MONATLICHE ZEITSCHRIFT FÜR GARTENBAU UND BLUMENZUCHT. Berlin, 1856–1887.
Journ. Roy. Hort. Soc. [J. R. Hort. Soc.] JOURNAL OF THE ROYAL HORTICULTURAL SOCIETY OF LONDON. London, 1845–1855; n.s. 1866+.
Kew Bull. KEW BULLETIN. Royal Botanic Gardens. Kew, England, 1887+. Formerly titled, Bulletin of Miscellaneous Information.
Lodd. Bot. Cab. LODDIGES' BOTANICAL CABINET; CONSISTING OF COLOURED DELINEATIONS OF PLANTS FROM ALL COUNTRIES . . . BY CONRAD LODDIGES & SONS. London, 1817–1833.
Möllers Gärtnerztg., see under Deutsch. Gärt.-Zeit.
Nat. Hort. Mag. NATIONAL HORTICULTURAL MAGAZINE. Washington, D. C., 1922+.
    Published by the American Horticultural Society.
Paxt. Mag. Bot. PAXTON'S MAGAZINE OF BOTANY, AND REGISTER OF FLOWERING PLANTS. London, 1834–1849.
Rev. hort. [Rev. hort., Paris] REVUE HORTICOLE. Paris, 1829+.
Rev. hort. belge. [Rev. Hortic., Brux.] REVUE DE L'HORTICULTURE BELGE ET ÉTRANGÈRE. Ghent, 1875–1914.

## Glossaries and dictionaries

A glossary is an alphabetical enumeration of terms together with an explanation of their meanings, whereas a dictionary is one of words pertinent to its subject matter. Almost all modern manuals, and many floras, include a glossary of the botanical terms employed. Several comprehensive and nearly all-inclusive glossaries have been published as separate works of definitive character. Those of particular value to English-speaking taxonomists are cited below. Another usually reliable source for the explanation of technical botanical terms is an unabridged dictionary such as Webster's, the Century, Funk and Wagnalls', or the Oxford Dictionary.

EASTWOOD, A. Excerpt from illustrated dictionary of botanical terms by John Lindley, Ph.D., F.R.S. 1848. Privately reprinted, San Francisco, 1938.

    A glossary comprising a 40-page excerpt from Book III of Lindley's Introduction to Botany (pp. 346–383), giving definitions and Latin equivalents for over 600 terms and illustrations of about 350 terms. The availability of Latin equivalents is of aid to the publishing systematist seeking correct equivalents of terms, especially for the gradations of color (over 85 latinized color terms given),[4] texture, vesture, variegation, etc.

GILBERT-CARTER, H. Glossary of the British flora. 79 pp. Cambridge, England, 1950.

    This glossary includes generic names and trivial epithets of Latin names of British plants, and is of especial value for the critical attention given to the pronunciation and meaning of each name.

GRAY, A. Structural botany, or organography on the basis of morphology. To which are added the principles of taxonomy and phytogeography and a glossary of botanical terms. 442 pp. New York, 1879.

    A standard reference, now out of print, of value for the thoroughness and clarity with which gross morphological features of vascular plants are presented.

GRAY, A. Gray's lessons in botany. The elements of botany for beginners and for schools. 226 pp. New York, 1887.

    One of the best, reasonably priced annotated glossaries of structural botany. Copiously illustrated with line drawings.

JACKSON, B. D. A glossary of botanic terms; with their derivation and accent. ed. 4. London, 1928.

    The standard English language botanical glossary. Latin equivalents accompany most of the several thousand terms defined.

WILLIS, J. C. A dictionary of flowering plants and ferns. ed. 6. Cambridge, England, 1931.

    A very valuable work accounting alphabetically for the families (with characters of each) and genera and for many terms relating to the morphology, ecology, and taxonomy of vascular plants.

Dictionaries of plants are often encyclopedic in scope and are so few in number that any accounting of them must not be limited to those written in English. Most of the data available in botanical dictionaries are to be found nowhere else, least of all in the usual unabridged dictionaries of common use. Most botanical dictionaries are of plant names and are sources for the etymology of Latin or vernacular names, for biographical data of persons for whom plants have been named, and for vernacular names in various languages. One dictionary cited below (Bail-

---

[4] See also reference to Dade's list of Latin names of colors, p. 329.

lon) is almost a botanical encyclopedia. Annotated titles of a few of the most significant dictionaries are given below.

BACKER, C. A. Verklarend woordenboek der Wetenschappelijke namen van de in Nederland en Nederlandsch-indie in het wild groeiende en in tuinen en parken gekweekte varens en hoogere planten. 664 pp. Groningen and Batavia, 1936.

An excellent dictionary of Latin plant names, giving the etymology of generic names and of most binary names, especially those derived from barbaric or personal names. Entries based on personal names give dates of birth and death of the commemorized person and a brief biographical note. Vernacular names not accounted for. Text in Dutch.

BAILLON, H. Dictionnaire de botanique. 4 vols. Paris, 1876–1892.

Reputed to be the most complete dictionary of the plant kingdom. Abundantly illustrated with figures, especially of flowers, floral dissections, and fruits. Text in French.

BEDEVIAN, A. K. Illustrated polyglottic dictionary of plant names in Latin, Arabic, Armenian, English, French, German, Italian, and Turkish languages, including economic, medicinal, poisonous and ornamental plants, and common weeds. Cairo, 1936.

GERTH VAN WIJK, H. L. A dictionary of plant names. 2 vols. The Hague, 1911–1916.

Volume 1 of this work includes bibliography; alphabetical list of Latin botanical names, with English, French, German, and Dutch vernacular names. Volume 2 consists of an index to vernacular names.

LYONS, A. B. Plant names scientific and popular, including in the case of each plant the correct botanical name in accordance with the reformed nomenclature, together with botanical and popular synonyms. ed. 2, 630 pp. Detroit, 1907.

An alphabetical listing of 2327 genera, with 1 or more species given for each, together with vernacular names and principal uses and an index to vernacular names.

WITTSTEIN, G. C. Etymologisch-botanisches Handwörterbuch. Enthaltend: die genaue Ableitung und Erklärung der Namen sämmtlicher botanischen Gattungen, Untergattungen und ihrer Synonyme. 952 pp. Ansbach, 1852.

A notoriously unreliable German reference work, the author of which "apparently would always rather follow his imagination than consult authorities," and which has led many later authors astray by its "many fantastic derivations." (See Airy-Shaw, in Gard. Chron. III, 106: 328, 1939; Sprague, in Kew Bull. 1928: 269.)

### Cultivated and Economic Plants

Taxonomic literature dealing with the identification of cultivated and economic plants is not voluminous, and sometimes it is difficult to differentiate between that which is popular or horticultural and that which is based on more scientific principles. This literature is usually published as separate books devoted to a special subject, such as manuals, encyclopedias, or as sumptuous monographs of horticultural subjects. A notable exception is the serial *Economic botany* (New York). In addition, a few of the leading horticultural serials cited in the previous section are sources of usually competent taxonomic treatments of cultivated plants.

ALEFELD, A. Landwirthschaftliche Flora; oder die nutzbaren kultivierten Garten- und Feldgewächse Mitteleuropas in allen ihren wilden und Kulturvarietäten für Landwirthe, Gärtner, Gartenfreunde und Botaniker insbesondere für landwirthschaftliche Lehranstalten. 360 pp. Berlin, 1866.

A systematic arrangement of economic plants. Accounts for all species, varieties, and formae of garden, field, and orchard crops, with author citation, synonymies, and descriptions, and synoptical keys to related taxa. This work is indispensable to the study of the systematics of these cultivated plants. Uses and vernacular names are usually given. Text in German.

BAILEY, L. H. Cyclopedia of American horticulture. 4 vols. New York, 1900–1902.

A copiously illustrated work treating of 2255 genera and about 8800 species of cultivated plants with descriptions, synoptical keys, and author citations. Bibliographic references are made to botanical monographs. A 6-volume edition (1906) contains one of the first keys in English to the families and genera of cultivated plants.

BAILEY, L. H. The standard cyclopedia of horticulture. 6 vols. New York, 1914–1917. [Later printings of ed. 2 bound in 3 vols.]

This cyclopedia represents a rewriting of the *Cyclopedia of American horticulture,* and despite references to "editions" dated as late as 1949 all are essentially identical with the original issue of 1917. A second edition was published (1922) which involved corrections of only a few typographical errors. All later so-called "editions" are a reprint of this second edition. In addition to the alphabetically arranged entries, Volume I contains a synopsis of the plant kingdom with the families of cultivated plants (described and illustrated) arranged according to the Engler classification, together with a key to the families and their cultivated genera based on the Bentham and Hooker classification. The work accounts for 3214 genera and 14,553 species of cultivated plants accompanied by author citation, synonymy, descriptions, references to published illustrations, and synoptical keys.

BAILEY, L. H. Manual of cultivated plants. ed. 2. 1116 pp. New York, 1949.

A manual for the identification of about 5300 of the commonly grown cultivated plants. Keys are provided. Single author citation is employed. A glossary and an enumeration of English equivalents of most noncapitalized specific epithets are included.

BAILEY, L. H. and E. Z. Hortus second. New York, 1941.

A one-volume concise dictionary of gardening, general horticulture, and cultivated plants in North America. Nearly 32,000 plant names are accounted for, and are arranged alphabetically, cross-referenced by Latin and common names (no author citations are employed), and each is followed by a brief description, synonymy, and an indication of place of nativity of each species treated. This work is of value as an index of the plants that are offered in the American trade, for the spelling of Latin names, the identification of the botanical family to which a plant belongs, and for the distinctive characteristics of families, genera, and species.

BEAN, W. J. Trees and shrubs hardy in the British Isles. ed. 4, in 3 vols. London, 1950.

An encyclopedic treatment, the genera and species arranged alphabetically, based on personal horticultural experience and literature compilations. Descriptions and author citations given for each entity. No keys included.

BELLAIR, G. and SAINT-LÉGER, L. [eds.]. Les plantes de serre. Paris, 1939.

An encyclopedic work accounting for all plants cultivated in Europe (as of 1938). Listings are alphabetical by Latin names. Single author citation is given. There are no keys. Common names, descriptions and brief cultural notes are provided. Text in French.

BERGMANS, J. B. Vaste planten en rotsheesters. ed. 2. Tegelen, Netherlands, 1939.

A garden dictionary accounting for all ordinary plants of Dutch (and American) gardens exclusive of annuals and those woody elements not a part of the garden landscape. Genera and species arranged alphabetically by Latin names, accompanied by author citation, synonymy, common names, descriptions, and detailed accounting of varieties. No keys included. Text in Dutch.

BOIS, D. Les plantes alimentaires chez tous les peuples et à travers les âges. 4 vols. Paris, 1927–1937.

This work provides descriptive accounts and illustrations of wild and cultivated edible plants, with history, uses, vernacular names, and bibliographies. The volumes treat the following groups: vol. 1, vegetables; vol. 2, fruits; vol. 3, spices and condiments; vol. 4, sources of beverages. Text in French.

BONSTEDT, C. Parey's Blumengärtnerei. 2 vols. Berlin, 1931–1932.

An illustrated encyclopedia of cultivated plants arranged according to the Engler classification. Nondichotomous keys to genera and species are provided, but the families are not keyed. All Latin names are accompanied by author citations. Text in German. Dates of publication, provided by W. T. Stearn are: I: 1–672, 1930; 673–940, 1931; II: 1–288, 1931; 289–792, 1932.

CHOUARD, P. and LAUMONNIER, E. [eds.]. Le bon jardinier. ed. 151. Paris, 1947.

An encyclopedic treatment, in French, accounting for the cultivated plants of France and their culture.

GREY, C. H. Hardy bulbs, including half-hardy bulbs and tuberous and fibrous-rooted plants. 3 vols. London, 1937–1938.

An illustrated botanical-horticultural work accounting for most bulbous monocots grown in English gardens. No keys are included. Plants are arranged alphabetically by families. For the most part, only Latin-named taxa are accounted for; double author citation is employed. Vol. I treats the Iridaceae; vol. II, the Amaryllidaceae, Commelinaceae, Haemodoraceae, Orchidaceae, and Scitamineae; and vol. III, the Liliaceae.

HEDRICK, U. P. Cyclopedia of hardy fruits. New York, ed. 1, 1922; ed. 2, 1938.

HILL, A. F. Economic botany. A textbook of useful plants and plant products. New York, 1937.

HOLLAND, J. H. Overseas plant products. 279 pp. London, 1937.

An annotated alphabetical enumeration (with bibliography) of economic plant products by trade names, with botanical names and localities, of "all the natural products of vegetable origin, imported on a commercial scale into the docks under the control of the Port of London Authority and into other ports, for landing and delivery to the consignees in the markets of the United Kingdom."

HYLANDER, N. Våra prydnadsväxters namm på svenska och latin. [The names of our ornamental plants in Swedish and Latin.] Stockholm, 1948.

A series of analyses of the nomenclatural problems of over 100 commonly cultivated garden subjects.

KIRK, J. W. C. A British garden flora. A classification and description of the genera of plants, trees and shrubs represented in the gardens of Great Britain, with keys for their identification. London, 1927.

An illustrated manual accounting, with keys, for the families and major genera of cultivated plants. Species are treated only incidentally, and without descriptions or keys.

MUENSCHER, W. C. Weeds. New York, 1935.

A treatise on dissemination and control of weeds followed by a systematic arrangement of families (Engler classification), a dichotomous key to the species, together with descriptions and illustrations of the species of weeds of northern United States and Canada.

NEAL, M. C. In gardens of Hawaii. Bernice P. Bishop Museum Special Publ. 40. Honolulu, 1948.

A manual of the cultivated ornamental plants of Hawaii. A work of sufficient scope to be pertinent to the systematics of ornamentals through the tropics. Keys are provided to the genera and to most species.

NICHOLSON, G. The illustrated dictionary of gardening. Vols. 1–4, London, 1884–1887; pages 251–608 of vol. 4 consisting of a supplement to the preceding vols. Supplement, vols. 1–2. London, 1900–1901.

All entries arranged alphabetically by genus. Latin names are unaccompanied by author citations and no keys are provided. First distributed to subscribers in parts; for date of publication of each, see volume 4, pp. 249–250.

POPENOE, W. Manual of tropical and subtropical fruits, excluding the banana, coconut, pineapple, citrus fruits, olive, and fig. New York, 1920.

A horticultural work with each entry accompanied by a Latin name, author citation, botanical description, and distribution. About 100 different tropical fruits are treated, many are illustrated.

REHDER, A. Manual of cultivated trees and shrubs hardy in North America exclusive of the subtropical warmer regions. ed. 2. New York, 1940.

A manual accounting for 486 genera and 2535 species of hardy woody plants. Only Latin-named taxa are included. Keys are provided to taxa of all categories through that of species. Double author citation is employed throughout, and all elements are briefly described. Dates of introduction to cultivation (not necessarily to cultivation in U. S.) included, degree of hardiness noted, and proximal dates of flowering and fruiting given.

REHDER, A. Bibliography of cultivated trees and shrubs. 825 pp. Jamaica Plain, Mass., 1949.

A bibliography of all scientific names appearing in the author's *Manual of cultivated trees and shrubs,* ed. 2, with nomenclature brought up to date as of 1947.

ROBBINS, W. W. The botany of crop plants. ed. 3. Philadelphia, 1936.

A text and reference well documented with bibliographies, treating the systematics, descriptive morphology, and economic importance of crop plants. The term crop is used in the broad sense to include crops of the field, garden, and orchard. Keys are provided to the important crop genera of some of the larger families, and species of a few genera.

SCHNEIDER, C. Illustriertes Handbuch der Laubholzkunde. 2 vols. Jena, 1904–1912.

A manual of the woody angiosperms (excl. Bambusae and Cactaceae) arranged according to the Engler classification, with each family, genus, and species keyed and described. Many drawings of floral and fruit details. Text in German.

SIEBERT, A. and VOSS, A. Vilmorin's Blumengärtnerei. 2 vols. Berlin, 1894–1896.

An illustrated encyclopedia of garden plants arranged according to the de Candolle classification. Keys are provided to genera and species and to groups of related genera. All taxa are described and authorities cited for Latin names.

WEHRHAHN, H. R. Die Gartenstauden. Ein Handbuch für Gärtner, Staudenzüchter und Gartenfreunde. 2 vols. Berlin, 1929–1931.

An illustrated manual of cultivated garden perennials, with keys, descriptions, author citations, and essential synonymy. Text in German.

## Maps and cartography

The subject of this section may seem remote or apart from that of the rest of the text; however, the taxonomic botanist knows the need of (1) an understanding of maps, the various projections, and the uses for which they were designed, (2) the kinds of maps available as sources of supplementary information, and (3) of the availability and sources of maps that treat any area of investigation. Likewise, the botanist who uses the treatises resulting from such investigation must have an appreciation of the maps employed in them.

One of the most clear, yet elementary, summaries of the kinds of maps, their projections, and their uses is that by Chamberlin (1947). A more technical accounting of the various map projections is by Deetz and Adams (1945), while the textbook by Raisz, *General cartography* (1948), is commended for its concise instructions relative to maps and map making. These 3 references are considered adequate to a general background for the intelligent selection, use, and comprehension of most maps used in conjunction with taxonomic studies.

In addition to knowing the kinds of maps and the particular uses for which each has been designed, it is desirable to know of the availability and sources of maps (exclusive of library collections) and of gazetteers. The information given below is suggestive of some sources, and is by no means bibliographic in character.

*Outline maps* are used extensively for plotting distribution of plants, their migrations, or of factors influencing these. Such maps are available in a wide assortment of sizes and in several different projections. Commercial sources include:

McKinley Publishing Co., Philadelphia, Pa.
University of Chicago Press, Chicago, Ill. (For the Goode Series of Outline Maps.)
Coast and Geodetic Survey, Washington, 25, D. C.
Superintendent of Documents, Washington 25, D. C. (For an outline map of U. S., with each state and county outline indicated.)

*Topographic maps* are published generally by federal governments or geographic societies. Those for areas of the United States are published and sold by the U. S. Geological Survey, and are available for nearly half the country. Index circulars, showing the areas for which maps may be had within each state, are available on application to the Director of the Geological Survey. These maps show, with local place names, all physical features, roads, trails, cities and towns, and state and county boundary lines. Most of them have been prepared at the scale of 1:62,500 (approx. 1 inch equalling 1 mile), but a few are at scales of 1:38,680 or 1:125,000.

New World areas south of the United States are represented by a series of maps known as the Map of Hispanic America, published by the American Geographical Society. These are available in sheets, prepared at the scale of 1:1,000,000 (approx. 1 inch equalling 15.8 miles). An index sheet is available from the publisher.

Maps of coastal and island areas of the United States and possessions (including islands of the Caribbean and much of the Pacific areas) are available as Nautical Charts (scale, 1:62,500) from the Coast and Geodetic Survey, Washington 25, D. C.

Maps of much of the earth's surface, especially those that were of or near areas of strategic importance during World War II, are available from the Army Map Service, War Dept., Washington 16, D. C. Similar maps, with overlays of air navigation data, are available from the Aeronautical Chart Service, U. S. Army Air Forces, Washington, D. C. The maps available from this second U. S. Army source are not duplicates of those available from the first and often account for entirely different areas.

A map of the world prepared at the scale of 1:1,000,000 is in preparation and much of it has been completed. For ease in handling, it is published in sheets, each sheet accounting for approximately 4 degrees of latitude and 6 degrees of longitude. These are available from C. S. Hammond & Co., 1 East 43rd St., New York 17, N. Y., and the 5 sheets representing the U. S. portion are available also from the Geological Survey, Washington, D. C.

A map for much of Eurasia is the British Council Map no. 1, Europe and the Middle East (1941), made by the Royal Geographical Society, London. This map is on a scale of 1:1,000,000 and shows the relief of the area from Iceland to the French Sudan in the south and to central China in the east.

*Road maps* of the county road systems are available at nominal charges for the counties of most states. Requests for information concerning them should be addressed to the highway department of the particular state. These maps are useful in field work, for plotting itineraries, and locations of individual collections. They have the advantage of usually being more up to date than most of the U. S. Geological Survey maps and are valuable adjuncts to the latter, both for road locations and for local place names.

*Geological maps* are essential to any detailed floristic or monographic research and, when available, are to be obtained directly from the Department of Geology and Mineral Resources (or the equivalent) of the particular state. Other geological maps are published by the U. S. Geological Survey, from which source lists of available maps may be obtained.

Soil maps and geological maps are equally important in any detailed floristic or monographic research. These maps show the type of soil profile under any area mapped, and are prepared usually at the scale of 1:62,500. In addition to soil profile data, they also show topographic features and such characteristics as locations of roads, railroads, and physiographic elements. These maps are produced by the Bureau of Plant Industry, Soils, and Agricultural Engineering, of the U. S. Dept. of Agriculture. A list of all available maps is included in Price List 46, Agricultural Chemistry and Soils and Fertilizers, issued by the Supt. of Documents, Washington 25, D. C.

*Air photographs* are not strictly classifiable as maps, but they may be used in taxonomic research as valued adjuncts to maps of various types. Much of the surface of the United States and Canada and several wide strips across Mexico have been covered photographically by either the U. S. Army Air Forces or for various agencies administered under the U. S. Dept. of Agriculture. Information concerning areas for which photographic coverage is available should be addressed to the U. S. Army Air Forces, Photographic Division, Washington, D. C. and to the Office of Information, U. S. Dept. of Agriculture, Washington, D. C.

*Gazetteers* are alphabetical listings of geographical place names. Generally they include names of physiographic features as well as of cities and towns. In a class allied to that of gazetters are *map indexes*. Most maps are not accompanied by indexes, but most of the maps published by the National Geographic Society (Washington, D. C.) have companion indexes that can be purchased with them. Maps accounting for most of the earth's surface (at scales of usually 1:4,000,000 and higher), together with indexes, are obtainable by purchase from this source. One of the most complete unabridged gazetteers for the world is Heilprin, A. and L. Lippincott's new gazetteer (1913) which provides information for the United States as of 1910, and for the remainder of the of the world as of 1905.

*LITERATURE:*

CHAMBERLIN, W. The round earth on flat paper: map projections used by cartographers. National Geographic Society, Washington, D. C., 1947.
DEETZ, C. H. and ADAMS, O. S. Elements of map projection with applications to

map and chart construction. U. S. Dept. of Commerce, Coast and Geodetic Survey Spec. Publ. 68, Washington, D. C., 1945.

RAISZ, E. General cartography. New York, 1948.

Government maps, and directions for obtaining them. Price list 53. Supt. of Documents, Washington, D. C.

**Miscellaneous reference works**

In addition to the entries cited above as fundamental to any library of taxonomic works there remain a number of supplementary items that do not fall within their limits. These are named below under several subheadings.

*Biographical references,* especially of botanists, are of considerable value. Sources for this information are standard encyclopedic biographical works that account for men of a given country; however, many men important as botanists or collectors are otherwise too obscure to have been listed in such general works. Mention has already been made (p. 319) of the usefulness of Backer, and of the limitations of Wittstein, as sources of brief biographical data and pertinent dates. The several editions of *American men of science* and of *Leaders in education* provide sources of most contemporary and neocontemporary American botanists. The value of the many editions of *Who's who* in America is of similar character, and the companion volume titled *Who was who* provides biographical sketches of Americans of note not now living. One of the better card files to biographical data of botanists of the world is at the library of the New York Botanical Garden. It was built up largely by the efforts of the late Dr. John H. Barnhart and is especially complete for data published up to about 1940. Many manuals and floras provide brief biographical data concerning authors of botanical names employed, or of important collectors of the area treated (cf. Gray's Man., ed. 8; Rehder, Man. Cult. Trees and Shrubs, ed. 2; Bailey, Man. Cult. Plts., ed. 2; Munz, Man. So. Calif. Bot.). Dates of birth and death are given when known in the Catalogue of the British Museum (Natural History) Library for authors of all entries. This is true also for entries in the U. S. Library of Congress Catalogue. When the year of death is known it is usually possible to find adequate biographical sketches in one or more botanical periodicals of the time.

In addition to the above, the following references are especially useful for biographical notes about European botanists: [5]

BRETSCHNEIDER, E. History of European botanical discoveries in China. 2 vols. London, 1898.

[5] The author is especially grateful to W. T. Stearn of the R. H. S. Lindley Library, London, for having provided these references.

BRITTEN, J. and BOULGER, G. S. A biographical index of deceased British and Irish botanists. ed. 2, [edited by A. B. Rendle] London, 1931.
BRIQUET, J. Biographies des botanistes à Genève. Ber. Schweiz. Bot. Ges. 50a, 1940.
BURNAT, E. Botanistes qui ont contribué à faire connaître la flore des Alpes-Maritimes. Bull. soc. bot. Fr. 30: cvii–cxxx. 1883. 2nd. ed. by Cavillier, F. in Mém. Riviera Scientifique, 5 [suppl.]: 1–95, 1941.
CHRISTENSEN, C. Den danske botaniks historie med tilhorende bibliografi. 2 vols. Copenhagen, 1924–1926.
COLMEIRO, M. La botánica y los botánicos de la Península Hispano-Lusitana; estudios bibliográphicos y biográficos. Madrid, 1858.
COSSON, E. Compendium florae Atlanticae, 1: 7–101. Paris, 1881.
HRYNIEWIECKI, B. Précis de l'histoire de la botanique en Pologne. Warsaw, 1933.
HYLANDER, N. Förteckning över Skandinaviens växter, utgiven av Lunds botaniska förening. 1, Karlväxter. ed. 3, Lund, 1941.
KROK, TH. O. B. N. Bibliotheca botanica suecana . . . Svensk botanisk litteratur från äldsta tiden t.o.m. 1918. Uppsala and Stockholm, 1925.
MAIWALD, V. Geschichte der Botanik in Böhmen. Vienna and Leipzig, 1904.
PRITZEL, G. A. Thesaurus Literaturae botanicae omnium gentium. ed. 2, Leipzig, 1872. Reprinted 1949.
SACCARDO, P. A. La botanica in Italia. Venice, 1895; seconda, 1901.
URBAN, I. Symbolae Antillanae seu fundamenta florae Indiae Occidentalis. Vols. 1, 2, 3, 5. Berlin, 1898–1908.

*Dates of publication* of taxonomic works are of great importance to the researcher who requires them in determining questions of name priority. The International Rules of Botanical Nomenclature are specific in defining the date of publication to be the exact date of release to the general public of the work in question. Ordinarily the date is given on the title page of a book; if in parts, the contents and dates of these are usually given on the verso of the title page. This date is usually accepted as date of publication unless established to be incorrect. It occasionally happens that a book or periodical bears a particular date on its title page but that the item was not released or available for distribution within the country of origin until a later date. This later date is the actual date of publication and is the date to be used in determining questions of priority. When a later date than that indicated (the work date) is known to be the correct date of publication, it is sometimes indicated in subsequent citations by enclosure in square brackets [ ]. It frequently happens that a work of several volumes is published over a period of years, and in works of taxonomic character it is essential to know the exact date when each volume was published. Likewise, each of the volumes of a single work may be published as separate parts, paged continuously for the volume. In such instances, it is necessary to know the contents and publication dates of each part. In contemporary publications, these data usually are indicated clearly, but in many cases, especially for works published a century or more ago, the exact dates of publication have been determined only after diligent and painstaking research. A bibliography accounting for papers dealing with the dates of publication of botanical works is given by Griffin et al.[6] Subsequent papers have been published on the subject (especially by H. S. Marshall and W. T. Stearn) and are to be sought in botanical periodicals, notably in Journ. Arnold Arb., Journ. Bot., Journ. Soc. Bibliog. Nat. Hist., and Rhodora.

*Locations of type specimens* are of concern to all taxonomists conducting any extensive research or making critical determinations of material. The incorporation of the type concept (cf. pp. 203–205) as fundamental to botanical nomenclature made the establishment of type specimens and knowledge of where they are deposited of paramount importance. There is no one work to which one may refer to learn where the types of a particular collector or author are located. In many instances the collections of any one person are to be found in not one, but several herbaria. In other instances the herbaria have been lost through one of several causes (insect depredations, neglect, fire, war damage, etc.). References to be consulted to learn of the locations of the majority of collections or of private herbaria include the following:

DE CANDOLLE, ALPHONSE. La Phytographie, ou l'art de décrire les végétaux considérés sous différents points de vue. 484 pp. Paris, 1880.

    Chapter XXX (pp. 391–462) contains an alphabetical enumeration of the authors and collectors of plants, with a notation on the herbaria in which their collections are deposited. This is one of the first, and currently valuable, lists of locations of type specimens.

[HITCHCOCK, A. S.] Location of type specimens. Mimeographed, 19 pp. 1934.

    An unannotated alphabetical list, prepared and distributed by the Committee on Nomenclature of the Botanical Society of America, indicating where type collections (cryptogamic and phanerogamic) of about 500 botanists of the world are located.

[HITCHCOCK, A. S.] Location of type specimens. List 2. Mimeographed, 30 pp. Washington, D. C., 1935.

    A revised and enlarged edition of the list issued in 1934, accounting for locations of types of about 600 botanists.

LASÈGUE, A. Musée botanique de M. Benjamin Delessert. Notices sur les collections de plantes et la bibliothèque qui le composent; contenant en outre des documents sur les principaux herbiers d'Europe et l'exposé des voyages entrepris dans l'intérêt de la botanique. 588 pp. Paris, 1845.

SAVAGE, S. A catalogue of the Linnaean herbarium. 225 pp. London, 1945.

    A catalogue of all specimens in the Linnaean herbarium together with detailed annotations concerning each specimen, indicating (when known) the date the specimen came into the herbarium, inscriptions on the labels by Linnaeus or eighteenth century authorities, and several plates of halftone reproductions of identified eighteenth century handwritings on herbarium

---

[6] Griffin, F. J., Sherbon, C. D., and Marshall, H. S. A catalogue of papers concerning the dates of publication of natural history books. Journ. Soc. Bibliog. Nat. Hist. 1: 1–30, 1936; first supplement, *op. cit.* 2:1–17, 1943.

sheets. The catalogue is of particular value when determining if a type specimen exists for a Linnaean name and further, if that specimen had been available to Linnaeus prior to publication of editions 1 or 2 of *Species plantarum* or had been added to the herbarium later by Linnaeus' son.

SHERBORN, C. D. Where is the . . . collection? An account of the various natural history collections which have come under the notice of the compiler. 148 pp. Cambridge, England, 1940.

A cataloguing, by author, of primarily zoological material, especially of British naturalists. A number of continental European botanists and their collections are included.

*Directories and addresses* of contemporary taxonomists serve several purposes. In addition to giving full names of individuals listed, they usually give such supplementary data as title, degrees, institution of affiliation, and special interests. Among the more current works of this category are the following:

INTERNATIONAL ADDRESS BOOK OF BOTANISTS; being a directory of individuals and scientific institutions, universities, societies, etc., in all parts of the world interested in the study of botany. London, 1931. [Prepared under direction of Diels, Merrill, and Chipp, as a committee appointed by Fifth International Botanical Congress.]

VERDOORN, F. Selected references on current research in plant taxonomy, ecology, and geography in Europe, Africa, Asia, and Australia. Chron. Bot. 6: 265–287, 298–311. 1941.

Arranged alphabetically by author, giving address and nature of research. New World botanists excluded.

VERDOORN, F. Botanical collectors in the Latin American countries (preliminary list.) Chron. Bot. 6: 171–172. 1941.

*Color charts* supply standards for comparison of living specimens and are essential for accurate color description. The following three charts are widely used: [7]

COLOR STANDARDS AND NOMENCLATURE. By R. Ridgway. Washington, D. C., 1912.
HORTICULTURAL COLOUR CHARTS, issued by the British Colour Council in collaboration with the Royal Horticultural Society. 2 vols. London, 1938, 1941.

The users of this chart will find it a convenience to number the 200 colors listed in the introductory text to vol. 2, using the margin, to attach the same numbers to the plates, and then to combine the plates of the 2 volumes into 1 numerical sequence, thus arranging the plates in accordance with this list. The use of this chart as a means of obtaining uniformity and precision in color descriptions by American horticulturists is advocated in Arnoldia 7: 41–52, 1947.

RÉPERTOIRE DE COULEURS . . . publié par la société française des chrysanthémistes et René Oberthür avec la collaboration de H. Dauthenay. Paris, 1905.

Reference should also be made to H. A. Dade, Colour Terminology in Biology (Imp. Mycol. Instit. Kew, Mycological Papers no. 6: 1943)

---

[7] This section on color charts contributed by W. T. Stearn.

by those who wish to describe colors in Latin or to translate Latin descriptions into English.

*Outstanding botanical libraries* in this country constitute a topic of practical importance to all students of taxonomy, for next in importance to knowing the literature is to know where it is available. No competent study is known to have been made of this subject and, aside from the situation with regard to periodicals, no known statistical studies are available on the relative size, strength, or importance of American libraries of botanical works. For this reason, the remarks that follow are based solely on the author's personal observations and are only of a generally informative nature.

There are four great centers of botanical literature in eastern United States and among them are housed the great majority of published botanical works.[8] They are the libraries of the following: Academy of Natural Sciences, Philadelphia; Harvard University (the libraries of the Gray Herbarium, Arnold Arboretum, Farlow Herbarium, and the Widener Library considered collectively), Cambridge, Mass.; New York Botanical Garden; and U. S. Dept. of Agriculture. Other American libraries of importance for their botanical titles in the taxonomy of higher plants include: the Library of Congress; University of California, Berkeley; Cornell University, Ithaca, N. Y.; Missouri Botanical Garden, St. Louis; New York Public Library, N. Y.; Iowa State University, Ames; and the Massachusetts Horticultural Society, Boston.

The subject of taxonomy, perhaps more than other fields of botanical science, demands an almost encyclopedic knowledge of the world's pertinent literature. In many biological subjects the current literature so outmodes the writings of a decade or more previous as to cause the older writings to be of only historic or academic interest. This situation does not exist in taxonomy, where systematic publications of a century or more ago may represent the latest information available, or findings must be considered along with those of contemporary writings as concerns a particular taxon. For this reason all taxonomists must know the important sources of taxonomic literature, especially of their own particular field of activity.

In this chapter the attempt has been made to account for the more significant bibliographic items pertinent to the general systematics of

---

[8] According to Wilson, L. R., Downs, R. B. and Tauber, M. F., A Survey of the Libraries of Cornell Univ., Ithaca, 1948, pp. 131–132, the ten leading collections of botanical periodicals in American University libraries (as of 1943) are at the following institutions, given in descending order: California (Berkeley), Cornell, Harvard, Iowa State, Minnesota, Illinois, Columbia, Wisconsin, Chicago, Michigan.

vascular plants. The items enumerated above comprise a nucleus of the most frequently used books, pamphlets, and periodicals. The student will have found wanting many titles pertinent to a specialized phase of the subject. These missing titles may be sought in available bibliographies of phytogeographical, genetical, ecological, cytological, and morphological subjects. Basic references dealing with these subjects are given in this text in footnotes or in the lists of selected references at the end of most chapters.

The material and data presented in this chapter are not to be studied at one sitting, nor to be read or assigned for reading as are other chapters herein. The material is intended to serve as a reference source. It is a classified annotated bibliography; a collation of elements of a greatly diversified literature or of sources to such literature.

# PART II

# SELECTED FAMILIES
# OF VASCULAR PLANTS

Students of taxonomy are expected to become conversant and familiar with the characteristics of the more important families of vascular plants of a particular or local flora, in addition to acquiring a knowledge of the theoretical considerations, basic guiding principles, and historical background of plant taxonomy. Since this textbook is designed to meet the needs of any area of the United States, treatments are provided of all families of plants having members known to be growing in the country, either as indigens or as exotics. Most courses in systematic botany include the pteridophytes and the gymnosperms among the plant groups studied, and for this reason they are included in this book. It is recognized that few courses in systematic botany are likely to account for all or even the majority of the families treated here, and it is expected that the instructor will select the treatment of such families for study as judgment and local conditions dictate.

The arrangement of the families that follow is that in the latest edition of Engler and Diels, *Syllabus der Pflanzenfamilien* (1936). Certain taxa elevated to the rank of family since that time are accepted here as families and are placed in the Englerian sequence. The selection of the Engler system of classification is no reflection of the author's personal preferences or convictions. It is a selection of necessity, for it is the most recent system to treat the flora (i.e., genera and species) of vascular plants from a world viewpoint. In further support of this selection it must be recognized that all the larger American herbaria are organized according to this classification, that most of the regional American floras are based on it, and that instructors in many schools continue to follow it. It is recognized that the Engler system is in many respects archaic, and in the light of our present knowledge is obsolete. It is recognized also that many courses of instruction in plant taxonomy are based on the systems of Bessey or of Hutchinson or on modifications of those systems. However, there is no classification that treats the families of vascular plants of the world which is based on either of the latter two systems. The Bessey system remains a provincial system restricted in a large degree to the classification of the flowering plants indigenous, primarily, to North America, and it has not been revised since 1915. The Hutchinson system, which has undergone 3 revisions to date, with more expected, treats only the angiosperms, and for this reason also is incomplete. Few students of phylogeny believe that either of these two latter systems represents the ultimate, or is yet even sufficiently close to it to represent a synthesis of the more recent findings. For these several reasons, the Engler arrangement seems to be the only one acceptable until one more nearly perfect has been developed for the vascular plants of the earth. For the benefit of the student, and for the instructor, the classification of most of the following families according to the systems of Bessey and of Hutchinson have been given in Chapter VI.

The treatments of each family that follow are made as uniform as individual situations permit. Characteristics of each family have been compiled from the literature. Every effort has been made to insure reliability of sources and to make the compilations as complete as the needs for them could be anticipated. In this regard, a summary of diagnostic characters separating closely related families has been included in most instances. Many of the treatments have been read critically by specialists, as acknowledged in the Preface. Illustrations of selected representatives of each family have been provided as visual aids to portray the critical and more significant features. The distributional ranges of each family are general and serve to indicate regions of occurrence, both global and continental. Disjunctive distributions of families are often indicated, but no attempt is made toward completeness in this regard. The size of the family, as to number of genera and species, is based either on recent monographs or considered estimates that endeavor to avoid extremes in conservatism or segregation. For the larger families it has been found useful to provide a synopsis of the subfamilies or tribes, giving their Latin names and synonyms as appropriate, and in doing so the treatment of Engler and Diels is followed unless superseded by more recent and authoritative monographs. References are made to important revisions or monographs as appropriate, especially when these are considered pertinent to students of American botany. Supplementing these technical data there is included usually a paragraph reviewing current and classical opinions concerning the phylogenetic position of the family, and in the case of large families, of their components. In succeeding paragraphs are given brief indications of the current economic importance of each family as may be contributed by specified genera and species. Reference to number of cultivated genera or species applies only to the situation existing in this country.

Omission of floral diagrams and formulas is deliberate. It is recognized that these devices have definite pedagogic value in the classroom, but it is believed that their use is associated largely with laboratory exercises, and this text makes no pretense to satisfy such a need.

## THE PLANT KINGDOM

According to the Engler system the plant kingdom is subdivided into 13 divisions, of which the first 11 include the plants more commonly known as the Thallophyta. The twelfth (Embryophyta Asiphonogama) is divided into 2 subdivisions (Bryophyta and Pteridophyta), while the thirteenth (Embryophyta Siphonogama) corresponds to the Spermatophyta. It was hypothesized by Engler and his associates that each successively higher group was evolved from the preceding lower group. At the present stage of botanical knowledge this classification is recognized as one of convenience. Much evidence has been presented in the last half century—especially by paleontological, anatomical, and morphological research—to refute the validity of the Engler system. Modern phylogenists agree that the evolution of the plant kingdom has not been as simple as the Engler system would indicate, and while no adequate system has yet been produced, it is reasonable to believe that the basic divisions of the ultimate classification of the plant kingdom may be along the lines proposed—but not validly published, taxonomically—by Eames (1936) and shown by Fig. 17. (For discussion of names of these taxa, see footnote 8, p. 355.)

The treatment of families in this text commences with those of the Pteridophyta (Embryophyta Asiphonogama, of Engler) and ends with the dicotyledonous family, Compositae.

### Division III. PTERIDOPHYTA

The pteridophytes are characterized by the presence of vascular tissue (phloem and xylem tissue of regular organization) in the sporophyte generation, with true roots present in all but a few aquatic members; alternation of a separate game-

# DIVISION III. PTERIDOPHYTA

tophyte generation disjunctive from the sporophyte generation clearly evident, the sporophyte generation being dominant; egg cells borne in archegonia and motile sperm cells produced from antheridia, both structures located on the thalluslike gametophytic plant. The minute, often chlorophyll-less, gametophyte gives rise to the large, usually foliaceous sporophyte. Spores of the latter germinate and give rise to the gametophyte and by this sequence complete the life cycle.

The division is composed of 5 classes: Articulatae (horsetails), Lycopodiinae (club mosses), Psilotinae, Isoetinae (quillworts), and Filicinae (true ferns). The first 4 are known widely by the misnomer "fern allies" and more recently have been designated the lycosphens by Wherry (1949). It is now apparent that there is much closer relationship between the Filicinae and the seed plants than between the former and the lycosphens. The latter represent families more primitive than the Filicinae, for, among other reasons, they are plants whose ancestors existed many ages earlier in geologic time (see Fig. 18).

Plants belonging to these classes may be separated as follows:

Vernation not circinate; lvs., when present, are minute or at most only quill-like.

    Sts. conspicuously jointed; lvs. whorled, forming a sheath at st. nodes....I. *Articulatae*

    Sts. not jointed; lvs. absent or if present not whorled and sheath-forming.

        Lvs. minute, in whorls, spirals or decussate, persistent............II. *Lycopodiinae*

        Lvs. not minute (or if so are caducous), when present are quill-like.

            Sts. dichotomously much-branched, leafless or essentially so; spores homosporous, in 3-loculed sporangia............................III. *Psilotinae*

            Sts. short and corm-like, with quill-like lvs.; spores dimorphic, in 1-loculed sporangia ................................................IV. *Isoetinae*

Vernation circinate; lvs. usually foliaceous, never scale or quill-like..........V. *Filicinae*

These spore-producing vascular plants (pteridophytes) are primarily of the tropics and subtropics, and are not so well represented in the flora of temperate North America as one might be inclined to believe. By conservative concepts, of the approximately 215 genera recognized for the world, only about 40 are indigenous within the area covered by this text, and of the approximately 9200 species for the world, fewer than 200 are native in North America north of Mexico.

*LITERATURE:*

BROUN, M. Index to North American ferns. 215 pp. Orleans, Mass., 1938 [Accounts for nomenclature of all categories of North American Pteridophytes.]

EAMES, A. J. Morphology of vascular plants. Lower groups. New York, 1936.

VERDOORN, F. Manual of Pteridology. The Hague, 1938.

## Class I. ARTICULATAE [1]

This class, known also as the Equisetinae, is represented among living plants by the single order Equisetales, composed solely of the genus *Equisetum*, of which many species are known. As a class, the Articulatae stand far apart from other vascular cryptogams and are derived from ancestors of great antiquity that were once of major dominance, in size of individuals and distribution, in the earth's flora. Notable among these ancestors were probably the Calamitaceae, and relatives of

---

[1] There is no index to Latin names of members of this or the succeeding 3 classes of pteridophytes that is comparable to *Index Kewensis* for the spermatophytes or *Index filicum* for the Filicinae. However, indexes are reported to be in course of preparation for names of members of this class and for the Lycopodiinae.

# 336 SELECTED FAMILIES OF VASCULAR PLANTS

the Sphenophyllales—giant horsetaillike plants dominant in the Paleozoic. Plants of this class are distinguished from those other classes by their rushlike, often branching axes, the stems jointed and mostly hollow; the leaves whorled at the nodes and mostly minute and toothlike; the reproductive bodies in conical or spikelike terminal cones with eusporangiate sporangia borne on the underside of peltate scales; and sperms multiciliate. Engler and Diels subdivided it into 4 subclasses, only 1 of which (Equisetales) contained an order of living plants.

## Order 1. EQUISETALES
### EQUISETACEAE. HORSETAIL OR SCOURING RUSH FAMILY

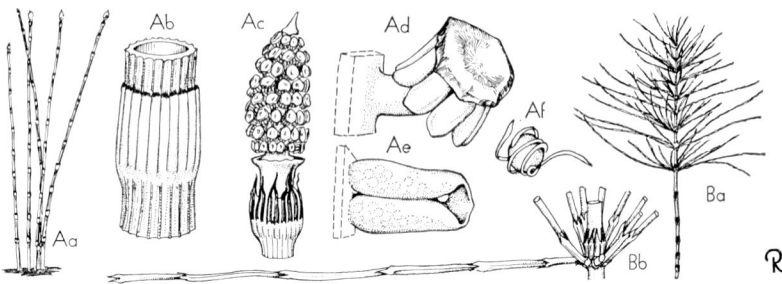

**Fig. 39.** EQUISETACEAE. A, *Equisetum hyemale*: Aa, fruiting stems, × ⅛; Ab, stem section, at node, ×3; Ac, strobilus, × 1½; Ad, sporangiophore, ×10; Ae, sporangium, × 20; Af, spore with elaters, × 80. B, *E. arvense*: Ba, sterile stem with branches, × ⅙; Bb, node, with leaves and lateral branch, × 1.

Plants annual or perennial, usually rhizomatous, the axis erect, surface often silicated, striated, or grooved, the stems with nodal joints at which appear whorls of connate scalelike leaves, the stems branched at the nodes or not, but when branched the branches whorled and as numerous as the leaves and alternate with them; stem hollow except at nodes and characterized by a usually large central canal (centrum) surrounded by a number of medium-sized ones each embedded in the outer tissue and under each of the external grooves (vallecular), connected by stomata, with a minute canal (carinal) situated under each ridge between the grooves; homosporous, the sporangia borne in terminal, sessile, or stalked cones (strobili); the latter are composed of numerous closely appressed peltate polygonal sporangiophores (sporophylls) projecting at right angles to the cone axis, the sporangia 5–10 per sporangiophore, arranged around under side of sporangiophore margin, eusporangiate, dehiscence by longitudinal slit; spores minute, numerous, chlorophyll-containing, and each spore provided with 4 hygroscopically sensitive spatulate elaters attached at a common point, and of importance in spore dissemination.

The class, order, and family are represented by the single genus *Equisetum*, comprised of about 25 species, and occurring on all the large land areas of the earth except Australia and New Zealand. The plants are primarily of wet places, but one species at least occurs in the driest of habitats. The majority of species occurs in the tropics and subtropics, with about 12 species more or less widely distributed over temperate North America.

Most phylogenists have considered the horsetails the most primitive of extant vascular plants, placing them below the Lycopodiinae in their schemas. Despite the lack of a definite bridging to ancestral fossil types, there is much evidence to demonstrate strong affinities with the Calamitales and Sphenophyllales of the Carboniferous, and to a lesser degree with the Hyenales of the Devonian. For these reasons and because, as Eames (1936) pointed out, the genus *Equisetum* possesses a complex branch system, an anatomy characterized by the presence of intercalary meristems and the advanced endarch type of xylem, and gametophytes with dorsiventral prothallia with advanced conditions existing in sex organ morphology, it seems clear that these are not primitive vascular cryptogams but are relics of probably highly developed ancestral types. By Eames they are separated as a major phylum, the Sphenopsida, a group considered by him as advanced over the Psilopsida and the Lycopsida.

The Equisetaceae are of diminishing economic importance. They were once much used for polishing woods and scouring utensils, and several species are noxious weeds in poorly drained soils. A few species have been advertised by domestic dealers in native plants.

*LITERATURE:*

SCHAFFNER, J. H. Geographic distribution of the species of *Equisetum* in relation to their phylogeny. Amer. Fern Journ. 20: 89–106, 1930.

## Class II. LYCOPODIINAE

A class of 4 orders, of which 2 are extinct. The class is characterized by the stems mostly elongate, creeping, and more or less indeterminate, the leaves scalelike to acicular, very short and mostly imbricated or crowded; roots dichotomous; sporangia solitary, subtended by a sporophyll, often condensed into terminal strobili, homosporous or heterosporous, eusporangiate in origin; sperms biciliate.

Phylogenetically the Lycopodiinae were considered by Engler and Diels to be advanced over the Articulatae (sphenopsid derivatives), and to be subordinate to the Psilotinae. Eames concluded that the Lycopsida were next in primitiveness to the Psilopsida, and that within the Lycopsida, the Lycopodiales were the more primitive order although "interrelationships among lycopsid forms are obscure." Wettstein and also Diels considered the Psilotinae as advanced over the Lycopodiinae but not to the degree reached by the Isoetinae.

## Order 2. LYCOPODIALES
### LYCOPODIACEAE. THE CLUB MOSS FAMILY

Low terrestrial or epiphytic plants, often mosslike in appearance, varying greatly in habit, but all with stems slender, mostly prostrate or in a few kinds erect or ascending, the branching basically dichotomous; roots mostly dichotomously branched; leaves very numerous, small, mostly in close spirals, whorls, or opposite pairs, often imbricated, rarely all basal (in *Phylloglossum*); reproduction by uniform spores produced from sporangia borne singly on upper side of leaves (sporophylls) near their base, in leaf axils, or on the stem immediately above a leaf, the sporophylls variable in size and when minute the axis usually much condensed with the numerous sporophylls forming a strobilus (or cone), the strobili sessile or peduncled; spores abundant, minute, variously reticulated or sculptured; vegetative reproduction by gemmae occurs in many species.

A family of 2 genera: *Lycopodium* with about 100 species, occurring over most of the earth (except the most arid areas) and abundant in subtropical and tropical forests, and *Phylloglossum,* a monotypic genus restricted to parts of Australia. Tropical species of *Lycopodium* are predominately epiphytes, and those of temperate and arctic regions are terrestrial. Twelve species are native to eastern United

States, and 7 extend westward to represent the genus on the Pacific coast region, with 3 extending southward into Texas.

Several species of *Lycopodium* (ground pine, creeping Jennie, crowfoot, princess pine) are gathered extensively commercially for decorative purposes during the Christmas season. Formerly the spores were used commercially by druggists as pill coatings, as well as in a number of official medicinal preparations.

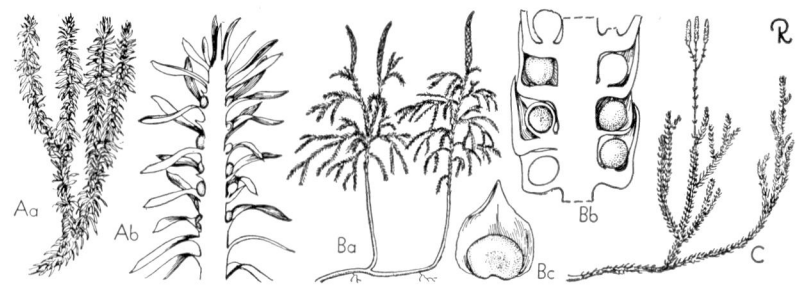

**Fig. 40.** LYCOPODIACEAE. A, *Lycopodium lucidulum*: Aa, fertile branch, × ⅙; Ab, branch tip, vertical section showing sporangia, × 1. B. *L. obscurum*: Ba, fertile branches, × ¼; Bb, portion of strobilus in vertical section, × 5; Bc, sporophyll and sporangium, × 5; C. *L. clavatum*: fertile and sterile branch, × ¼.

*LITERATURE:*

CLAUSEN, R. T. Hybrids of the eastern North American sub-species of *Lycopodium complanatum* and *L. tristachyum*. Amer. Fern Journ. 35: 9–20. 1945. [Cf. also *op. cit.* 36: 122. 1946.]
LLOYD, F. and UNDERWOOD, L. M. Review of the species of *Lycopodium* of North America. Bull. Torrey Bot. Club, 27: 147–168. 1900.
NESSEL, H. Beiträge zur Kenntnis der Gattung *Lycopodium*. Fedde, Rep. Spec. Nov. 39: 61–71. 1937.
UNDERWOOD, L. M. and LLOYD, F. E. Species of *Lycopodium* of the American tropics. Bull. Torrey. Bot. Club, 33: 101–124. 1906.
WILSON, L. R. The spores of the genus *Lycopodium* in the United States and Canada. Rhodora, 36: 13–19. 1934. [Includes key based on spores.]

## Order 3. SELAGINELLALES
## SELAGINELLACEAE. SELAGINELLA OR SMALL CLUB MOSS FAMILY

A monogeneric family of mosslike or lycopodiumlike plants, of small to moderate size; stems usually abundantly dichotomously branched; tufted, creeping, pendent, erect, or occasionally climbing; leaves arranged in spirals, decussate pairs or in 4 rows, characterized by presence of small membranous ligule sunken in pit at base of each leaf and on upper side, which develops early and shrivels prior to leaf maturity; roots mostly adventitious, dichotomous, often originating from supporting modified branches (rhizophores); plants heterosporous; microspores (small) and megaspores (large) in respective eusporangiate sporangia, borne in leaf axils; strobili (cones) formed in all species, lax or much condensed, terminal, usually 4-sided and sharply angled, each strobilus containing both mega- and microsporangia, the megasporangia (usually greenish-white) producing usually 4 large megaspores, the microsporangia (usually orange-red) producing several hundred minute microspores, the arrangement of sporophylls variable, but most commonly the megasporophylls are along the lower half of the strobilus; vegetative reproduction by bulbils, fragmentation or tip rooting common.

# DIVISION III. PTERIDOPHYTA

The single genus, *Selaginella*, is composed of about 600 species, of which 37 occur in the United States. The others are widely distributed, but mostly tropical, on all continents. Of the domestic species, 3 occur in the northeast, about 15 in the south and west into New Mexico, and 11 in the Pacific coast states.

The order is related to the Lycopodiales and is similar to it in many respects, but perhaps not so closely as once believed. It is readily distinguished by the presence of the ligule, heterosporous strobili, and morphologically by differences in the gametophytes. On the basis of characters of both sporophyte and gametophyte there is ample evidence to justify treating the Selaginellales as advanced over the Lycopodiales.

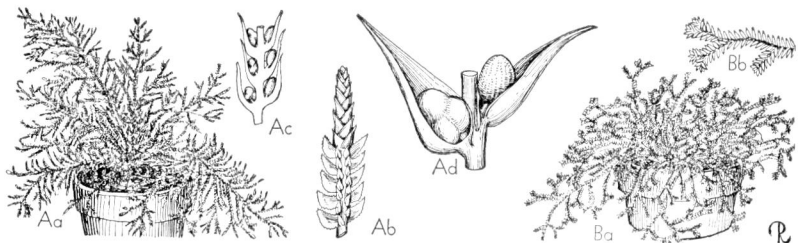

**Fig. 41.** SELAGINELLACEAE. A, *Selaginella pallescens*: Aa, habit of plant, × 1/10; Ab, branch tip, × 2; Ac, fertile branch, vertical section, × 3; Ad, same, showing megasporangium (left) and microsporangium right), × 10. B, *Selaginella Kraussiana*: Ba, habit of plant, × 1/10; Bb, branch tip, × 1. (From L. H. Bailey, *Manual of cultivated plants*, The Macmillan Company, 1949. Copyright 1924 and 1949 by Liberty H. Bailey.)

About 25 species of *Selaginella* are cultivated domestically as ornamentals, largely of tropical origins and grown as conservatory subjects. The highly publicized resurrection plant (*S. lepidophylla*) is shipped by the bale from Mexico and sold here as a novelty.

*LITERATURE:*

CLAUSEN, R. T. *Selaginella*, subgenus *Euselaginella*, in the southeastern United States Amer. Fern Journ. 36: 65–82. 1946.

REEVE, R. M. The spores of the genus *Selaginella* in north central and northeastern United States. Rhodora, 37: 342–345. 1935.

## Class III. PSILOTINAE

A taxon characterized in part by the uniformly dichotomously branched thallus-like sporophyte, lacking true roots or leaves, the sporangia (eusporangiate in origin) very large and 2-3-lobed; gametophytes somewhat resembling, but smaller than, the sporophyte; sperms multiciliate. A very primitive group of pteridophytes of relatively restricted distributions representing a class much isolated, phylogenetically, from other living plants. Only the single order Psilotales, represented by the one family, Psilotaceae.

## Order 4. PSILOTALES
### PSILOTACEAE. PSILOTUM FAMILY

Epiphytic, herbaceous perennial plants, erect or pendulous; rootless, but with slender dichotomous microrhizal rhizomes; leaves wanting, but in *Tmesipteris* the branch tip flattened laterally and leaflike in superficial appearance; reproduction by spores from homosporous sporangia, the sporangia larger than in other pteridophytes (2–4 mm across), conspicuous and borne solitary and terminal on branch tips, containing as many chambers as lobes, spores numerous; vegetative reproduction accomplished in *Psilotum* by means of gemmae developed on the rhizomes.

This is a small family represented by 2 genera and perhaps not more than 3 species. *Psilotum,* with 2 species, is frequent throughout tropics and subtropics, extending north, in hammocks, to Florida and the coast of South Carolina and in the Pacific to Hawaii. The monotypic *Tmesipteris* is mostly Australasian, extending north via Malaysia to the Philippine Islands.

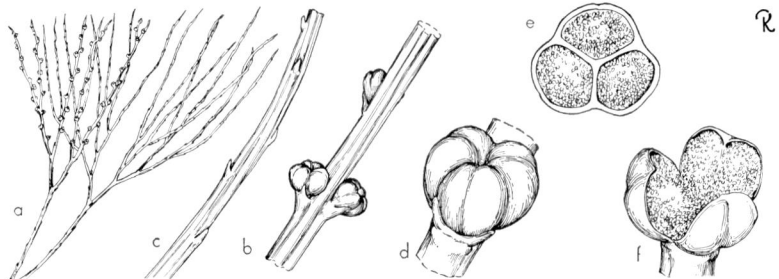

**Fig. 42.** PSILOTACEAE. *Psilotum nudum*: a, fertile and sterile branches, × ⅛; b, fertile branch node with sporangia, × 2; c, sterile branch portion with leaves, × 2; d, sporangium, × 6; e, same, cross-section, × 6; f, sporangium dehiscing, × 6.

There is no unanimity of opinion on the phylogenetic position of the Psilotales. Wettstein concurred with Engler and Diels in treating this order as lepidophytic in origin and as advanced over the Selaginellales, an interpretation extended further by considering also that they are advanced over the Isoetales. Eames (1936) treated them as the most primitive of extant vascular plants, pointing out that there is no evidence to support the hypothesis that the rootless and leafless character is a derived condition; on the contrary he considered the sporophyte to be thalluslike in many respects, the entire plant a continuous dichotomously branched axis that is scarcely differentiated. These situations, plus the presence of the primitive cauline and terminal sporangia and superficial antheridia that produce multiciliate sperms, were held to be largely responsible for his disposition of the order. A very different view was that of Chaudefaud (1950), who considered the Psilotales to have been ancestral to the Bryophyta.

*Psilotum* is grown to a limited extent in conservatories as a novelty.

*LITERATURE:*

CHAUDEFAUD, M. Les Psilotinées et l'évolution des Archigoniates. Bul. Soc. Bot. Fr. 97: 99–100, 1950.
EAMES, A. J. Morphology of Vascular Plants. New York. 1936. [Psilotaceae, pp. 71–98.]

### Class IV. ISOETINAE [2]

A class composed of the single family, Isoetaceae. Allied to the Selaginellales by presence of a ligule and existence of heterosporous condition, and to Lycopodiales in anatomical features of stem and leaves, but differs from both orders of that class by the general habit of plant, root system, morphology of stem, by possession of large chambered eusporangiate sporangia, and multiciliate sperm.

[2] The elevation of this group to the rank of class by Engler and Diels is not followed by most authors. Walton and Alston (in Verdoorn Man. Pteridology) treated it only as a family of Lycopodiales, while Eames maintained it as an order by itself, acknowledging it to be distantly related to the orders of club mosses.

## Order 5. ISOETALES
### ISOETACEAE. QUILLWORT FAMILY

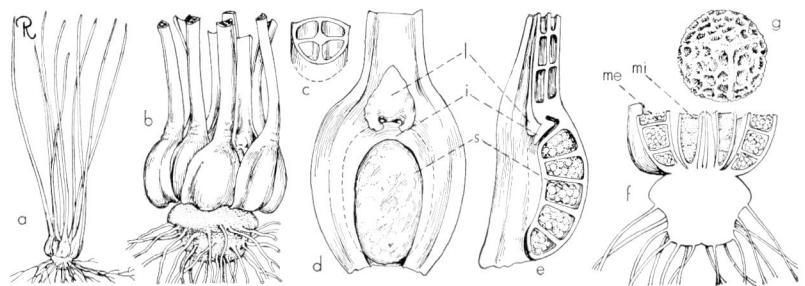

**Fig. 43.** ISOETACEAE. *Isoetes Engelmannii*: a, plant, habit, × ¼; b, corm, with attached leaf bases, × 1; c, leaf, cross-section, × 8; d, leaf base, central side, × 3; e, leaf base, vertical section, × 3; f, corm with portion of leaf bases, vertical section, × 1; g, megaspore, × 20. (i, velum; l, ligule; me, megaspores; mi, microspores; s, sporangium.)

Small herbaceous perennial aquatics or plants of wet places, amphibious or some species wholly submerged along lake or pond margins, grasslike or sedgelike in appearance; the tufted quill-like leaves arising from a thick flattened cormlike subterranean axis that becomes grooved or lobed with age; roots produced from basal portion of the axis; leaves usually a few centimeters (rarely to 10 dm) long, crowded on the obtusely elevated rim of axis, linear, tapering, terete or angled, traversed by 4 septate longitudinal air canals whose cross partitions bear a vascular strand (bast bundle) in center and often 4 or more in the periphery, the leaf base spoon-shaped with a persistent ligule present above the ventral concavity, each leaf a potential sporophyll; typically 3 sets of sporophylls present, the outermost megasporophylls enclosing microsporophylls with the innermost ones abortive, immature, or undeveloped (in a few spp. microsporophylls intermixed with megasporophylls); the sporangia, situated in the ventral concavity of each fertile sporophyll, are the largest of any living plant (4–7 mm long), their ventral side usually covered by a flap or fold (the velum), chambered by plates or bars (trabeculae) into several to numerous lacunae; microspores very numerous (150,000–1,000,000 in single microsporangium of most species), minute and of various surfaces; megaspores (gynospores of some authors) large, numerous (50–300 per megasporophyll), tetrahedal, the siliceous surface conspicuously sculptured, ridged, crested, or papillated; dehiscence effected only by decay of tissues or external disturbance; vegetative reproduction rare or nonexistent.

The family is composed of a single genus, *Isoetes*, having about 60 or more species of world-wide distribution exclusive of tropical regions, where they are rare or absent. About 20 species of *Isoetes* occur in the United States, with 7 in the northeast, 3 in the southeast, 8 along the Pacific coast area, and 3 in the plains areas extending southward into Mexico. Pfeiffer (1922) recognized the genus to be composed of 4 sections, based on the surface character of the megaspores. No members of the family are of significant economic importance.

The Isoetales are generally accepted as composing a class by themselves (Isoetinae) and *Isoetes* is the only extant remnant of a once dominant component of a part of the earth's flora. Among living plants they are most closely allied to the Lycopodiinae, although by Engler and Diels the Psilotinae was placed between these 2 classes. Eames considered the Isoetales to be more closely allied to the Selaginellales, pointing out in this regard (1936) that "there is little to tie the two groups together except the presence of a ligule." Phylogenists (Wettstein, Eames) consider the quillworts to be relics or derivatives of the lepidodendraceous Pleuromeiales, a dominant group of the Carboniferous that persisted into the Triassic, rather than derived from known ancestral stocks of either Lycopodiales or Selaginellales.

LITERATURE:

GRENDA, A. Über die systematische Stellung der Isoetaceen. Bot. Archiv. 16: 268–296, 1926.
PFEIFFER, N. E. Monograph of the Isoetaceae. Ann. Mo. Bot. Gard. 9: 79–232, 1922.

## Class V. FILICINAE

The Filicinae, or true ferns, represent the most dominant group of pteridophytes in our present-day fern flora. They are primarily terrestrial plants with members of a few families epiphytic and others limited to aquatic habitats. Members of the group are distinguished, as a class, from other pteridophytes by their conspicuous, typically large, foliaceous leaves that open usually from circinate vernation, and on which the 1-loculed sporangia are borne marginally or dorsally; the spores always microscopically minute, uniform, and (except for 2 families) homosporous.[3]

The Filicinae were divided by Engler and Diels (1936) into 2 subclasses, Eusporangiatae and Leptosporangiatae.[4] Each subclass was divided into 2 orders: Eusporangiatae into Ophioglossales and Marattiales; Leptosporangiatae into Eufilicales and Hydropteridales. The four orders of Filicinae are represented in temperate North America by about 12 families, 42 genera, and 110 species. All except 2 of the genera are of leptosporangiate ferns.

LITERATURE:

BOWER, F. O. The ferns (Filicales) treated comparatively with a view to their natural classification. 3 vols. Cambridge, England, 1923–1928.
CHRISTENSEN, C. Index Filicum (cf. p. 287 for annotated citation). 1905–1906. [Supplements issued through 1934.]
———. Filicinae. In Verdoorn, Manual of Pteridology. The Hague, 1938. [cf. pp. 522–550.]
COPELAND, E. B. Genera Filicum (Annales Cryptogamici et Phytopathologici, vol. 5). Chronica Botanica Co., Waltham, Mass., 1947.
EAMES, A. J. Morphology of vascular plants. Lower groups. New York, 1936.
ENGLER, A., and DIELS, L. Syllabus der Pflanzenfamilien. Aufl. 11. Berlin, 1936.

For additional references to illustrated works, regional treatments, and major monographs (to 1937) see Christensen (1938), pp. 525–527.

---

[3] Characters of the gametophyte (prothallium), while of major morphological and phylogenetical significance, are not of ordinary use taxonomically and are omitted from the descriptions that follow.

[4] In all ferns, the sporangia develop from one or more primordial cells known as initials. A eusporangiate fern is one whose sporangia have developed from a group of superficial initial cells that divided periclinally (i.e., into layers parallel to the surface) to form inner and outer layers of cells, with the inner becoming spore-producing and the outer remaining sterile and developing into the sporangium proper. A leptosporangiate fern is one whose sporangia developed individually, each from a single initial cell, which divided periclinally or obliquely, and of the 2 cells resulting from this division, only the outer one (by subsequent divisions and differentiations) gave rise to the sporangium and its contents. These sporangial origins are of fundamental importance to phylogenetic considerations of the pteridophytes; the eusporangiate type, occurring in the first 4 classes (and the first 2 orders of the fifth class), is the more primitive, and while because of its microscopic nature the character of sporangium origin is of no practical taxonomic use in the herbarium or field, its significance must be recognized and acknowledged.

# DIVISION III. PTERIDOPHYTA

## Subclass EUSPORANGIATAE

The eusporangiate ferns, comprising the primitive elements of the Filicinae, are represented by the orders containing the 2 families Ophioglossaceae and Marattiaceae. The eusporangiate character is common to all members of the 4 classes recounted to this point, and the members of this subclass are presumed to have been derived from ancestral stocks allied to the lycopsid and sphenopsid groups.

## Order 6. OPHIOGLOSSALES

Mostly terrestrial herbs, sometimes epiphytic; leaves solitary or few, not of circinate vernation, dimorphic or usually with a sterile foliaceous segment and fertile nonfoliaceous branch arising from a common stipe, the blade simple, lobed, or variously compound; fertile segment simple, racemose, or paniculate, the sporangia eusporangiate, large, naked, the walls more than 1 cell thick, without annulus, bivalvate and dehiscing by a slit, producing many thick-walled spores. Ophioglossaceae the only family.

### OPHIOGLOSSACEAE. GRAPEFERN FAMILY

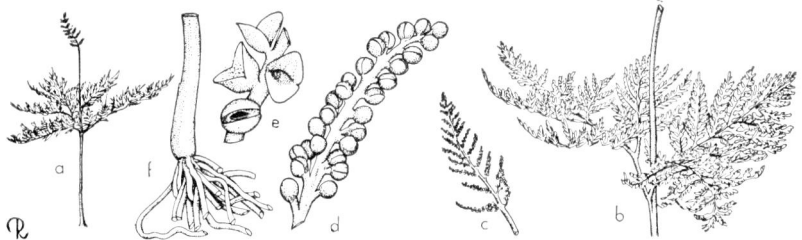

Fig. 44. OPHIOGLOSSACEAE. *Botrychium virginianum*: a, habit, × ½; b, sterile fronds, × ¼; c, fertile frond, × ¼; d, fertile pinna, × 2; e, sporangia, × 6; f, rootstock, × ½. (From L. H. Bailey, *Manual of cultivated plants*, The Macmillan Company, 1949. Copyright 1924 and 1949 by Liberty H. Bailey.)

A family with characters of the order, of 3 genera (*Botrychium, Ophioglossum,* and *Helminthostachys*) and about 60 species, of wide general distribution over the earth. *Botrychium* was treated by Clausen (1938) as composed of 23 species and distributed among 3 subgenera, with 11 species native to temperate North America. *Helminthostachys* is monotypic and is native in the Indo-Malayan and northeastern Australian regions. *Ophioglossum* (*fide* Clausen) has 28 species, placed by him in 4 subgenera, with 7 species considered indigenous within this country.

The ferns of the Ophioglossales are conceded to be among the most primitive of the Filicinae, but among themselves they are often highly specialized, with the evidences of specialization lying in features of reduction and simplification from more complex relatives and ancestors. The order has been considered a derivative or close ally of lycopodiaceous ancestral stocks, and in the gametophytes are found conditions that parallel those existing in members of the Lycopodiales and Psilotales. Within the family, *Botrychium* is accepted phylogenetically as the most primitive genus, and *Ophioglossum* as the most advanced.

*LITERATURE:*

CLAUSEN, R. T. A monograph of the Ophioglossaceae. Mem. Torrey Bot. Club, 19(2): 1–77, 1938.

## Order 7. MARATTIALES

Leaves circinate in vernation, attached to rhizome or a globose stem by a swollen joint, the pinnae jointed to rachis or stipe by swollen nodes, stipules present; sporangia eusporangiate (i.e., developed from a many-celled initial), the walls more than 1 cell thick opening basically by a ventral longitudinal split. Marattiaceae the only family.

Christensen (1938) recognized 2 families in this order: Angiopteridaceae and Marattiaceae. Copeland (1947) treated these as 2 subfamilies of Marattiaceae, while Engler and Diels subdivided them into 5 subfamilies. On the basis of morphological, anatomical, and phylogenetical considerations, Eames (1936) treated the order as composed of a single family.

### MARATTIACEAE. MARATTIA FAMILY

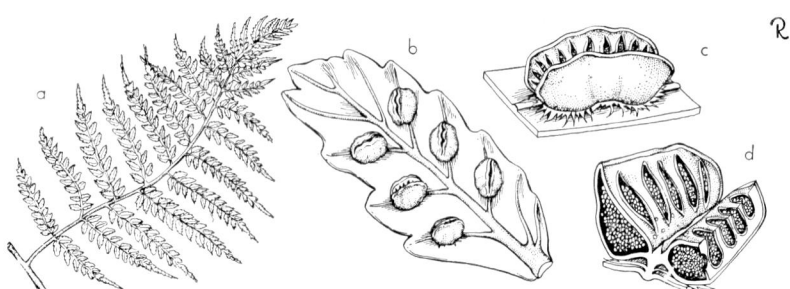

**Fig. 45.** MARATTIACEAE. *Marattia alata*: a, frond pinna, × ¼; b, pinnule with sori, × 3; c, sorus, × 7; d, sorus in sectional view, × 12. (c-d redrawn from Bauer.)

A family having characteristics of the order, composed of 6 genera, all except *Danaea* restricted primarily to tropical regions. None is indigenous to the regions accounted for in this text, but representatives of *Marattia* are cultivated domestically to a very limited extent in conservatories.

### Subclass LEPTOSPORANGIATAE

The leptosporangiate ferns, so named because of the character of the origin of the sporangia (cf. footnote 4), are the dominant ferns of temperate regions, and are accepted as advanced groups over the primitive eusporangiate genera of the Filicinae. The recognition of 2 distinct orders within this subclass is accepted by many taxonomists, but the validity of the Hydropteridales as an order has been challenged and rejected by recent morphological studies.

### Order 8. EUFILICALES

Herbaceous to arborescent, and terrestrial to aquatic ferns, varying widely in size and habit; leaves of circinate vernation, generally produced abundantly, occasionally dimorphic; sporangium leptosporangiate (i.e., developed from a single cell), the wall 1 cell in thickness, an annulus present. Treated by Engler and Diels as composed of 8 families, by Christensen (1938) of 15 families, by Copeland (1947) of 19 families, and by Holttum (1947) of 11 families.

*LITERATURE* (cf. also above, under Filicineae):

HOLTTUM, R. E. A revised classification of leptosporangiate ferns. Journ. Linn. Soc. Bot. 53: 123–158, 1947.

———— The classification of ferns. Biol. Rev. 24: 267–296. 1949.

## OSMUNDACEAE. OSMUNDA FAMILY

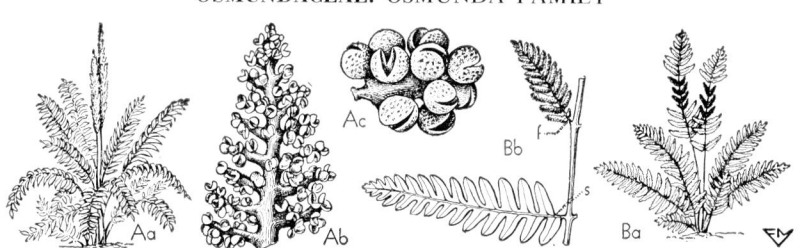

**Fig. 46.** OSMUNDACEAE. A, *Osmunda cinnamomea*: Aa, habit, much reduced; Ab, segments of fertile frond, × 3; Ac, sporangia, × 10. B, *Osmunda Claytoniana*: Ba, habit, much reduced; Bb, portion of frond with sterile (s) and fertile (f) segments, × ½. (From L. H. Bailey, *Manual of cultivated plants,* The Macmillan Company, 1949. Copyright 1924 and 1949 by Liberty H. Bailey.)

Terrestrial or subaquatic ferns of ordinary habit, rarely arborescent; leaves pinnately compound or decompound, uniform or (in *Osmunda*) dimorphic, the veins free and dichotomous, the stipe scaleless, somewhat sheathing at base, stipules or stipuloid structures present; indusia none, sori none; sporangia large, developing simultaneously, scattered on dorsal surface of leaf or on both sides of modified fertile portions of the leaf, thin-walled, short-stout-stalked, globose to pyriform, the annulus incomplete and imperfect, composed of thickened cells on one side near apex, dehiscence by vertical slit originating at annulus, extending over top to other side.

A family of 3 genera and about 20 species. *Osmunda,* with 12 species of temperate to tropical swampy regions, has 3 species extending in the aggregate from Newfoundland to Florida and west to the Northwest Territories of Canada southeast to Texas and into Mexico. The monotypic *Todea* is disjunctive in its distribution, occurring in Africa, Australia, and New Zealand. *Leptopteris* (6 species), with very thin filmy leaves, extends from New Zealand to Polynesia and Malaysia.

The family is primitive among ferns and while in some respects transitional between eusporangiate and leptosporangiate families, it does not serve as a bridge between them. Among its distinctive characters are the peculiar sporangial dehiscence with the annulus composed of a group of clustered, often lateral cells rather than the usual ring.

Economically *Osmunda* is of importance as the major source of fiber obtained from the roots and rhizomes and used in the culture of orchids and other epiphytes. The cinnamon fern (*O. cinnamomea*) and the interrupted fern (*O. Claytoniana*) are commonly cultivated.

*LITERATURE:*

BENEDICT, R. C. Osmundaceae. In North Amer. Flora, 16: 27–28, 1909.
FERNALD, M. L. Some varieties of the Amphigean species of *Osmunda.* Rhodora, 32: 71–76, 1930.

## SCHIZAEACEAE. CURLY GRASS OR CLIMBING FERN FAMILY

Terrestrial ferns of very diverse habit, some extremely small and grasslike, others climbing by leaves of indeterminate growth; leaves uniform or more commonly dimorphic; sori wanting or present, indusium none but a false indusium often evident, the sporangia solitary, marginal but sometimes seemingly dorsal, the annulus transverse, apical, complete; dehiscence by a vertical slit.

A family of 4 genera and about 160 species, mostly tropical, rare in temperate regions. Three species occur sporadically in this country as rarities, *Lygodium palmatum* (climbing fern) from Florida to New Hampshire and *Schizaea pusilla* (curly grass) from New Jersey to Nova Scotia and Ontario. *Lygodium japonicum* is escaped from cultivation and naturalized in parts of southeastern United States. A third genus *Anemia*, with erect dimorphic fronds, has 2 species indigenous to southern United States. *Schizaea Germanii* (*Actinostachys*) is a rarity of southern Florida. The family is distinguished from other leptosporangiate ferns by the solitary sporangia, each with a transverse apical annulus. The genera are so markedly distinct in habit that each has been raised individually to family rank by early authors.

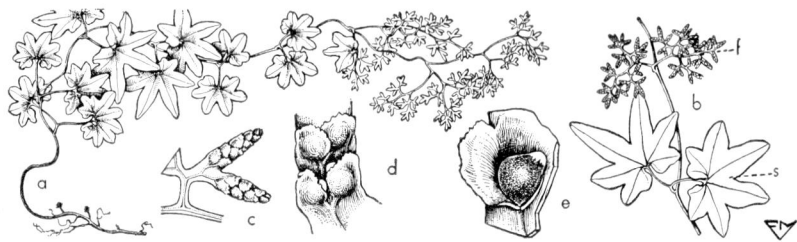

**Fig. 47.** SCHIZAEACEAE. *Lygodium palmatum*: a, plant, × ¼; b, section of frond showing sterile (s) and fertile (f) pinnae, × ½; c, fertile segments, × 2; d, sporangia covered by indusia, × 5; e, sporangium with indusium cut and opened back, × 10. (From L. H. Bailey, *Manual of cultivated plants,* The Macmillan Company, 1949. Copyright 1924 and 1949 by Liberty H. Bailey.)

The Schizaeaceae were considered by Eames (1936) and by Diels (1936) to be one of the 3 most primitive families of the order, a conclusion strongly supported by paleobotanical evidence and strengthened by the presence of characters interpreted to be primitive: among these, the large sporangia borne on naked leaf segments and the leaves with slow and long-continued growth. The family is further interpreted to have been derived from ancestral stocks that have given rise directly or indirectly to such families as the Dicksoniaceae, Marsileaceae, and Hymenophyllaceae.

*Lygodium* (climbing fern) is cultivated in subtropical regions outdoors on trellises, and in the north *L. palmatum* (the climbing fern) is much prized in ferneries as a rarity. *Schizaea* is offered occasionally by dealers in native plants, but is little cultivated and then as a novelty, whereas *Anemia* is not uncommon in conservatories or, in the south, in outdoor gardens.

*LITERATURE:*

MAXON, W. R. Schizaeaceae. In North Amer. Flora, 16: 31–53, 1909.
SELLING, O. H. Studies in recent and fossil species of *Schizaea*, with particular reference to their spore characters. Acta. Hort. Gotob. 16: 1–109, 1944 [Includes also an historical account of the genus and detailed bibliography.]

## GLEICHENIACEAE. GLEICHENIA FAMILY

Terrestrial ferns with long-creeping rhizomes; leaves pseudodichotomous and usually forking several times due to arrested growth of the main divisions that develop in succeeding seasons, the pinnae mostly coriaceous with veins free; sporangia few in dorsal sori, indusia wanting, the annulus complete, extending obliquely around the back and over the top of the pyriform sessile sporangium, dehiscence by a vertical slit along ventral side.

# DIVISION III. PTERIDOPHYTA

A family, considered by Diels as composed of a single genus (*Gleichenia*), by Christensen to be of 5, and by Copeland of 6 genera. About 130 species are recognized. They are mostly ferns of drier habitats (often weedy) in the tropics and subtropics of south temperate regions, with 2 species reaching Japan, a few others in Hawaii, and 1 extending from Brazil to the West Indies and north into Mexico with an isolated station on Mon Louis Island, Alabama. No representatives of the family are known now to be cultivated domestically.

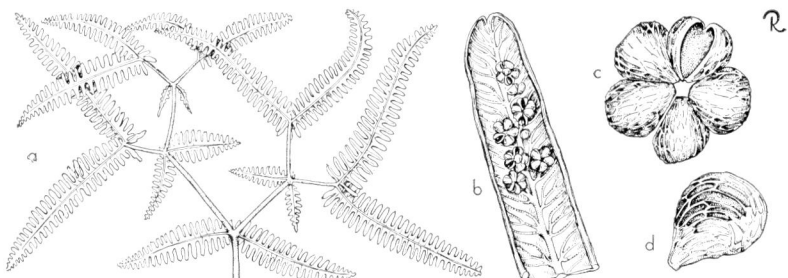

**Fig. 48.** GLEICHENIACEAE. *Gleichenia flexuosa*: a, portion of frond, × ¼; b, pinnule with sori, × 5; c, sorus, × 30; d, sporangium, × 40.

The Gleicheniaceae, along with the Osmundaceae and Schizaeaceae, were considered by Eames (1936) and by Diels (1936) to be one of the 3 most primitive families of the Eufilicales, and are generally acknowledged to represent ancestral stocks that gave rise to, among others, the Cyatheaceae. Evidences of primitiveness in the family are the dichotomously branched stems, the leaves of indeterminate growth, the nonindusiate simple and superficial sori, and the large sporangia producing numerous spores.

*LITERATURE:*
CHING, R. C. On the genus *Gleichenia*, Smith. Sunyatsenia, 5: 269–288, 1940.
MAXON, W. R. Gleicheniaceae. In North Amer. Flora, 16: 53–63, 1909.

## HYMENOPHYLLACEAE. FILMY FERN FAMILY

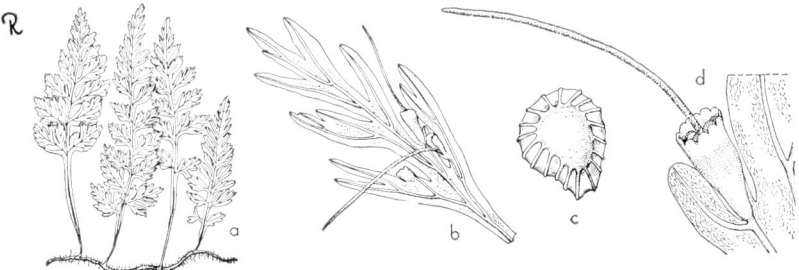

**Fig. 49.** HYMENOPHYLLACEAE. *Trichomanes Boschianum*: a, rhizome and fronds, × ⅜; b, fertile pinna, × 3; sporangium, × 50; d, fertile segment with sorus and indusium, × 6.

Small, delicate ferns; stem usually rhizomatous; leaves 1-celled thick between the veins, entire or dichotomously forked or 1–3 pinnate, rarely thalloid or orbicular, the veins dichotomously branched, the stomata absent (leaves absent in 2 genera);

sporangia in marginal sori that are raised on a slender columnar projection of the veinlet and enveloped by a cup-shaped indusium; sporangium thin-walled, with complete and horizontal (equatorial) annulus, dehiscence vertical or oblique; spores tetrahedal to globose.

Conventionally considered as composed of 2 genera (*Hymenophyllum* and *Trichomanes*) and 300–400 species. These are segregated into 33 genera by Copeland, whose treatment is gaining in recognition as the plants become better known. The species occur throughout the tropics and in New Zealand, with one species extending to central Europe and another (*Trichomanes Boschianum* Sturm) extending from Alabama to Kentucky and Illinois, where it occurs on moist sandstone cliffs. Almost all are plants of warm humid regions growing in at least partial shade on rocks, cliffs, and trees.

Species of both genera are cultivated as novelties and with difficulty.

*LITERATURE:*
COPELAND, E. B. Genera Hymenophyllacearum. Philip. Journ. Sci. 67: 1–110, 1938.

## CYATHEACEAE. CYATHEA FAMILY

Tree ferns, shrubby or, in some species, with stout, erect dictyostelic trunks to 15 ft or more high and 2–20 in. thick, the trunks branching in some cases and often clothed with matted adventitious roots or apically with hairs or scales; leaves generally large, to 15–20 ft long in some species, bipinnately divided, arranged spirally in a terminal widespreading crown, the stipe often densely chaffy-scaly or spiny at base, the base also sometimes with frondlike abortive pinnae (aphlebiae); venation mostly open, single or forking, never reticulate; sori superficial, dorsal on the veins, gradate, an indusium usually present (absent or vestigial in *Alsophila*), inferior, usually cup-shaped, or covering sorus from one side as a flaplike scale; sporangia small, on typically 4-celled stalks, thin-walled, the annulus complete, oblique, dehiscence transverse. See Fig. 50.

The family was considered by Engler and Diels (*sensu* Cyatheae), by Eames, and by Christensen as composed of 3 genera (*Alsophila, Hemitelia, Cyathea*), and by Copeland of 7 genera. Diels placed the number of species at 265, while by Christensen it was assessed to be about 700, and by Copeland in excess of 800. The family is restricted in distribution to tropical mountain forests from Mexico to Chile, Malaysia to Australasia and New Zealand, and Africa. Species of *Alsophila* and *Cyathea* are cultivated as tub and pot subjects. Pith from the trunks is used in India in making an intoxicating beverage. In New Zealand the pith was formerly used as a food product.

The Cyatheaceae were treated by Copeland (1947) as of probably antarctic origin with migrations having extended northward through western South America to Mexico, into Africa and likewise northward into Malaysia and the Pacific island groups. Eames (1936) considered the family to have been derived from ancestors clearly allied to the Gleicheniaceae.

## DICKSONIACEAE. DICKSONIA FAMILY

Tree ferns, or in some species, low-growing rhizomatous ferns, allied to, and by some authors (including Engler and Diels) treated as a subfamily of Cyatheaceae (other authors, including Bowers, reject existence of this alliance). Distinguished from Cyatheaceae by the marginal sori borne at vein tips and by the leaf bases hidden by an abundance of dense wool (no scales present). The indusium is cup-shaped or bivalvate.

Christensen, Wettstein, and Eames each recognized this as a distinct family.[5] The genera are 5 in number (2 are monotypic) and the species are about 30. Distribution is disjunctive with representatives occurring in eastern Asia, Malaysia, Hawaii, Central America, and the Juan Fernandez Islands. One genus (*Cibotium*) is of domestic economic importance as an ornamental tub or pot subject. In native localities the wool from stipe bases has been used for pillow stuffing.

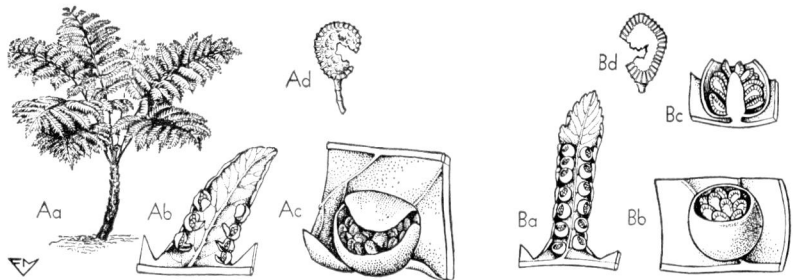

**Fig. 50.** DICKSONIACEAE. A, *Cibotium Schiedei*: Aa, plant, much reduced; Ab, frond segment with sori, × 2; Ac, sorus, × 10; Ad, sporangium, × 25. CYATHEACEAE. B, *Alsophila australis*: Ba, frond segment with sori, × 2; Bb, sorus, × 10; Bc, sorus, vertical section, × 10; Bd, sporangium, × 25. (From L. H. Bailey, *Manual of cultivated plants,* The Macmillan Company, 1949. Copyright 1924 and 1949 by Liberty H. Bailey.)

Eames has pointed out (1936) that the Dicksoniaceae are not morphologically nor phylogenetically closely allied to the Cyatheaceae. He concluded that the resemblances between them—arborescent habit, the small flat sporangia with nearly vertical annuli and mostly gradate sori—result from parallel advancement to the same level, and that they have evolved from schizaeaceous stock rather than the gleicheniaceous ancestors of the Cyatheaceae. Each family appears to represent developmental zeniths of its respective evolutionary line.

## POLYPODIACEAE. FERN FAMILY

Ferns of very diverse habit, rarely treelike; vegetative characters very variable, leaves usually monomorphic, but in some genera dimorphic, simple to decompound, the veins usually forking, open or reticulate; sporangia usually in sori; the sori mostly dorsal or marginal, naked or more often indusiate, generally mixed or gradient (sporangia of various degrees of development within each sorus); indusium (when present) inferior or superior; sporangia thin-walled, long- or short-stalked, the annulus incomplete, vertical, interrupted by the stalk, dehiscence transverse.

As treated here, the largest family of ferns, composed of about 170 genera and 7000 species, of wide distribution over most of the land areas of the earth, especially abundant in forests and humid areas, but occurring in almost all floristic areas or zones from desert to rain forest and from tropics to arctic or antarctic. Christensen (1938) recognized about 25 genera having in excess of 50 species each, of which those having North American representatives (together with total number of species for the genus) are *Dryopteris* (650 spp.), *Asplenium* (650 spp.), *Pteris* (250 spp.), *Polystichum* (225 spp.), *Adiantum* (200 spp.), *Athyrium* (180 spp.), *Cheilanthes*

---

[5] Copeland treated the components of this family as distinct from Cyatheaceae, but merged them into the large family (about 60 genera) of Pteridaceae (the latter not accepted by Engler and Diels nor in this text).

(130 spp.), *Blechnum* (180 spp.), *Phymatodes* (100 spp.), *Pellaea* (80 spp.), *Vittaria* (80 spp.), *Dennstaedtia* (70 spp.), *Notholaena* (60 spp.), and *Polypodium* (*sensu strictu*, 50 spp.). Many well-known genera of North America are represented by only 1 or a few indigenous species, as *Pteridium, Onoclea,* and *Camptosorus.*

**Fig. 51.** POLYPODIACEAE. A, *Nephrolepis exaltata*: Aa, pinna with sori, × 1; Ab, sorus, × 15; Ac, sporangium, × 25. B, *Blechnum*: pinna, × 1. C, *Asplenium bulbiferum*: Ca, pinna, × ½; Cb, sorus, × 5. D, *Pellaea viridis*: Da, pinna, × ½; Db, contiguous sori, ×2. E, *Platycerium bifurcatum*: habit showing fertile portion (f) of frond, × ⅙. F, *Phyllitis Scolopendrium*: Fa, distal end of frond with sori, × ½; Fb, sorus with double indusium, × 3. G, *Polystichum setiferum*: Ga, portion of rachis with fertile pinna, × ½; Gb, sorus, × 10, H, *Adiantum*: pinna with sori, × ½. I, *Davallia*: Ia, pinna with sori, × 2; Ib, sorus, × 10; Ic, sporangium, × 25. (From L. H. Bailey, *Manual of cultivated plants,* The Macmillan Company, 1949. Copyright 1924 and 1949 by Liberty H. Bailey.)

Pteridologists and morphologists are agreed that the Polypodiaceae, as constituted by Diels or Christensen, phylogenetically are not a natural family, but rather are an assemblage of extraordinarily diverse ferns that are held together largely by the possession in common of a vertical incomplete annulus associated with a lenticular long-stalked sporangium. The pertinent summary of the phylogenetic position of the members of this family as given by Eames (1936) is quoted below in part:

They serve as an excellent example of a polyphyletic group—a group which is tied together by the attainment of certain advanced characters which have been selected as a taxonomic basis. They stand definitely at the top among leptosporangiate ferns—an aggregation of the highest members of the several lines of this stock; they are the modern, the young ferns. The fossil record bears out this conclusion, which has been drawn from the evidence of comparative study, for no members of the family are known before the Mesozoic.

Recognition of the unnaturalness of this family resulted in recent reclassifications of it independently by several authors, notably Christensen (1938), Ching (1940), Dickason (1946). Holttum (1947, 1949), and Copeland (1947). Christensen retained the family in its conventional and conservative sense, but divided it into 15 subfamilies which he conceded were "perhaps better dealt with as families." Ching subdivided it into 32 families; Holttum considered its components to represent 5 families with one of these subdivided into 11 subfamilies; Dickason accepted Ching's 32 families but realigned them into 2 groups based on sorus position; and Copeland reclassified the Polypodiaceae as comprised of 9 families. For a review of these treatments see Weatherby (1948), whose personal opinions favored the treatment by Copeland. Since no 2 of these 5 reclassifications of Polypodiaceae were in agreement as to what constituted a family or as to the phylogenetic sequence of the elements of this assemblage, any claim to validity or acceptance of one over the others must await the test of time.

The ferns of this family are of little economic importance except for the approximately 225 species grown as ornamentals. Among the more important of the 44 genera having ornamentals reportedly cultivated in America are *Nephrolepis, Adiantum, Dryopteris, Platycerium, Polystichum, Polypodium, Dennstaedtia, Davallia, Pityrogramma,* and *Cyrtomium.*

*LITERATURE:*

CHING, R. C. On the natural classification of the family Polypodiaceae. Sunyatsenia, 5: 201–268, 1940.
CHRISTENSEN, C. 1938. [cf. entry under Filicinae.]
COPELAND, E. B. Comment on natural classification of the family Polypodiaeae. R. C. Ching. Sunyatsenia, 6: 159–177, 1941.
———. Genera Filicum. Waltham, Mass. 1947.
DICKASON, F. G. A phylogenetic study of the ferns of Burma. Ohio Journ. Sci. 46: 73–108, 1946.
EAMES, A. J. Morphology of vascular plants (pp. 274–277). New York, 1936.
HOLTTUM, R. E. 1947 and 1949. [cf. entries under Eufilicales.]
WEATHERBY, C. A. Reclassifications of the Polypodiaceae. Amer. Fern Journ. 38: 7–12, 1948.

## PARKERIACEAE. WATER FERN FAMILY [6]

Aquatic annual ferns, floating or rooting in substrate; rhizome short, erect; leaves pinnately decompound, mostly 1–2 ft long, glabrous, dimorphic, the sterile ones broad and laminate, less divided than the fertile ones, the veins anastomosing without included veinlets; leaves usually floating, proliferating new plants asexually in axils, the fertile ones more finely cut with narrower segments and revolute margins almost completely enclosing the scattered sporangia; no sori or indusia present; sporangia sessile on the veins, large, globose, thin-walled, the annulus broad and nearly complete to almost wanting, the dehiscence transverse.

A family of a single genus (*Ceratopteris*) containing perhaps seven species of mostly tropical distribution in Asia, Africa, and the New World, 2 species extending north into Florida and westward to Louisiana.

[6] In much of the literature this family passes under the later name of Ceratopteridaceae (Maxon, W. R. Pterid. Porto Rico, p. 379. 1926).

The Parkeriaceae were recognized by Benedict, Wettstein, Diels, Ching, and Copeland as a family. Other phylogenists (Bower, Christensen, and Holttum) have considered them either as a part of Polypodiaceae (Gymnogrammeoideae) or treated the single genus (*Ceratopteris*) as a component of one of the families segregated from Polypodiaceae (Holttum placed them in Adiantiaceae; Christensen in the sub-

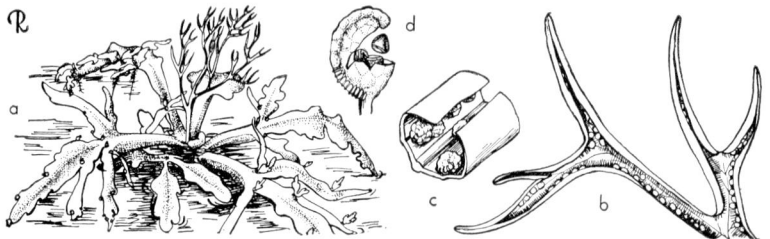

Fig. 52. PARKERIACEAE. *Ceratopteris thalictroides*: a, habit with sterile and fertile fronds, × ⅙; b, fertile frond segment, × 1; c, section of fertile frond showing sporangia, × 4; d, sporanguim, × 12. (From L. H. Bailey, *Manual of cultivated plants*, The Macmillan Company, 1949. Copyright 1924 and 1949 by Liberty H. Bailey.)

family Gymnogrammeoideae of Polypodiaceae, a view favored but not adopted by Copeland). Their claim to familial status apparently rests largely with their specialized sporophytic habit associated with their restriction to aquatic habitats, characters that readily permit macroscopic recognition, but are of little or no phylogenetic significance. However, it is apparent from analysis of its reproductive structures that *Ceratopteris* is closely allied with such genera as *Onychium* and *Cryptogramma,* and because of this, the views of those who do not retain Parkeriaceae as a separate family should be considered.

One species, *Ceratopteris siliquosa,* is grown as a food crop (for greens) in the Orient, and 2 or more species are cultivated here for ornament in pools and aquaria.

*LITERATURE:*

BENEDICT, R. C. The genus *Ceratopteris:* a preliminary revision. Bull. Torrey Bot. Club, 36: 463–476, 1909.

## Order 9. HYDROPTERIDALES

Aquatic leptosporangiate ferns, rooting in the substrate or free-floating, distinguished from the Eufilicales by the presence of sporocarps enclosing the usually heterosporous sporangia. The order was treated by Engler and Diels, Christensen, and by Copeland as composed of 2 families (Marsileaceae and Salviniaceae). Eames (1936) considered that heterospory arose independently within these families and that they are not closely related nor to be held together by this character: "the group Hydropteridineae is an unnatural one and should not be maintained . . . The Marsileaceae and the Salviniaceae are to be considered members of the Filicales, far advanced along the line of spore and gametophyte specialization and representing end products of two lines of elaboration."

## MARSILEACEAE. WATER CLOVER FAMILY

Aquatic or marsh plants with slender creeping rhizomes, growing in mud, the leaf with blades (when present) often floating on surface of water and petioles arising from rootstocks, the blades simple or with 2 to 4 pinnae, fan-shaped, the veins dichotomous and anastomosing at margin; plants monoecious, producing megasporangia and microsporangia; the sporocarps hard and bean-shaped, borne on the petioles laterally or at their bases, stalked, solitary or numerous. Morphologically, the sporocarp is a modified leaf segment, folded together, containing 2 rows of indusiated sori within. Megasporangia produce megaspores which on germination give rise to egg cells, while the microsporangia produce microspores that give rise to sperm-producing antheridia.

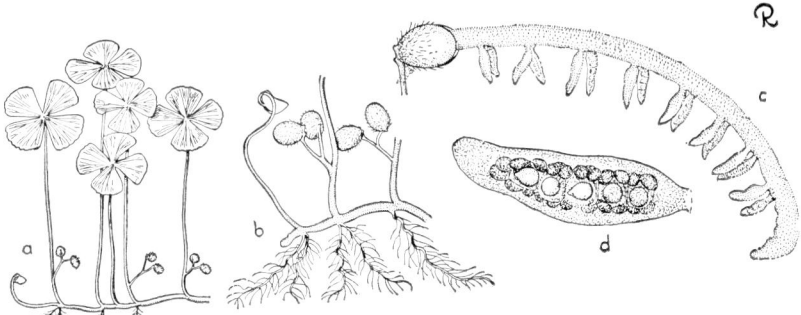

Fig. 53. MARSILEACEAE. *Marsilea quadrifolia*: a, portion of plant, × ⅓; b, sporocarps, × 1; c, germinating sporacarp, sori pendent from gelatinous tissue, × 1½; d, sorus, seen from below, showing large megaspores surrounded by smaller microspores, × 8.

A family of 3 genera and about 72 species. *Marsilea* (water clover, pepperwort) the largest with about 65 spp. of wide almost cosmopolitan distribution except for limited occurrence in North America, where 1 species is native from British Columbia to Texas and California, and naturalized east to Florida and Connecticut, 3 others mostly restricted to Texas, and a fifth naturalized in numerous localities. *Pilularia,* with 6 species of widely scattered distribution from Australia and New Zealand to Chile, Europe, and North America, has filiform or grasslike leaves. One species (*P. americana*) occurs in the southeastern United States from Georgia to Arkansas and also along the Pacific coast of North America from Oregon to southern California. The third genus, *Regnellidium,* with leaves of 2 opposite leaflets, is monotypic and known only from southern Brazil.

The family is of very limited economic importance. *Marsilea quadrifolia* is available in the trade as an aquatic novelty for pools and aquaria.

*LITERATURE:*

EAMES, A. J. Morphology of vascular plants. New York, 1936. [cf. Chapt. X for detailed account of morphology of the family and details of reproductive phenomena.]
STASON, M. The Marsileas of the western United States. Bull. Torrey Bot. Club, 53: 473–478, 1926.

## SALVINIACEAE. SALVINIA FAMILY

Small, floating aquatics not fernlike in appearance; stem small or wanting, rootless (in *Salvinia*) or with simple unbranched roots (in *Azolla*); the leaves small. not circinate in vernation, dimorphic and in whorls of 3 (in *Salvinia*) with 2 lateral, stiff-hairy, and floating, and the third submerged, dissected into 8–12 filiform hairy segments, or uniform and alternate (in *Azolla*) with each leaf papillate and divided into an upper aerial lobe and a lower submerged lobe; sporocarps in clusters or rows on submerged leaves (in *Salvinia*) or on the first leaf of a lateral branch (in *Azolla*), heterosporous, the sporangia borne on basal columns in the single cavity of the sporocarp, both megaspores and microspores present initially, but only one kind maturing, resulting in each sporocarp becoming entirely "male" or "female," uniform in size and shape (in *Salvinia*) or dimorphic (in *Azolla*) with the "male" sporocarp being much larger than the "female" and with numerous microsporangia, while the smaller megasporocarp contains a single megasporangium; the clusters of microsporangia within the sporocarp representing indusiated sori, annulus absent or only vestigial.

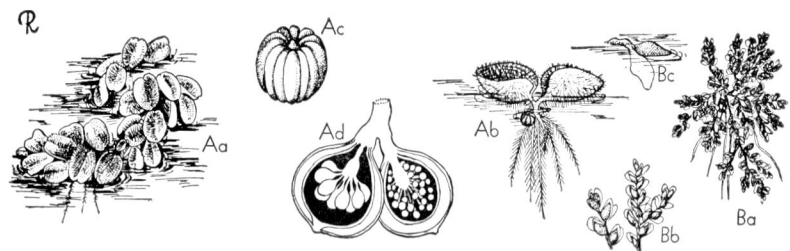

**Fig. 54.** SALVINIACEAE. A, *Salvinia rotundifolia*: Aa, habit of plant, × ½; Ab, single plant with sporocarp and submerged pinnatisect leaves, × 1; Ac, sporocarp, × 5; Ad, sporocarps, vertical section with megasporangia (left) and microsporangia (right), × 10. B, *Azolla filiculoides*: Ba, habit, × 1; Bb, sterile branch, × 2; Bc, leaf, × 8. (From L. H. Bailey, *Manual of cultivated plants,* The Macmillan Company, 1949. Copyright 1924 and 1949 by Liberty H. Bailey.)

A family of 2 genera, *Salvinia* and *Azolla,* comprising about 16 species occurring on all continents, but of greatest abundance in the tropics. Each genus is represented in the eastern and central United States by 1 native species, and 1 species of *Azolla* (*A. filiculoides*) is native to the Pacific coastal areas from Washington to southern California east to Arizona and south through Mexico, South America to the Strait of Magellan.

Christensen treated this group as composed of 2 families (Salviniaceae and Azollaceae) representing an independent order, the Salviniales. Engler and Diels, Eames, and Copeland considered it not sufficiently marked to justify such a major segregation. The genera are markedly different morphologically from one another and from other ferns, and there is considerable evidence to justify Christensen's treatment.

The family is of no major economic importance, although species of both genera are available in the trade for culture in pools and aquaria.

*LITERATURE:*

HERZOG, R. Ein Beitrӓg zur Systematik der Gattung *Salvinia.* Hedwigia, 74: 257–284, 1935.
SVENSON. H. K. The New World species of *Azolla.* Amer. Fern Journ. 34: 69–84, 1944.

### Division IV. EMBRYOPHYTA SIPHONOGAMA [7]

The seed plants; characterized by a very complex sporophytic generation and a much reduced gametophytic generation (the male gametophyte, present in early

---

[7] This division, in accordance with the terminology of Hallier's classification, is more commonly known by the older name Spermatophyta. It comprises the Phanerogamae of Eichler. It is probable that future classifications will treat the plants as components of the taxon Pteropsida.

## DIVISION IV. EMBRYOPHYTA SIPHONOGAMA

stages in the pollen grain and completing its development in the pollen tube is reduced to 2 or 3 cells in the angiosperms, and the female gametophyte [or embryo sac] reduced typically to as few as 8 cells), the male gametes usually eciliate (motile sperms in cycads and ginkgo), the megaspore never released from the megasporangium (ovule), the germinating pollen grain producing a pollen tube, the seed composed of or formed from the female gametophyte and integuments (usually 2, sometimes one or both lost by reduction); and containing an embryo sporophyte, megasporangium wall, and sometimes endospermous tissue.

This division has been differentiated classically from the Pteridophyta and others by formation of pollen tubes and production of seeds—criteria that have lost their validity as distinguishing characteristics due to discoveries nearly a half century ago that certain ancient fernlike plants (Pteridospermae) bore seeds. It is now recognized that the seed character is insufficient to set the angiosperms and gymnosperms apart from the ferns, club mosses, and horsetails. It is divided into 2 subdivisions: Gymnospermae and Angiospermae—the so-called coniferous plants and the flowering plants. Eames (1936) considered the Filicinae and Spermatophyta to comprise the taxon Pteropsida, a subphylum which together with the subphyla Sphenopsida, Lycopsida, and Psilopsida, comprised the Tracheophyta.[8]

### Subdivision I. GYMNOSPERMAE

By classical definition the seeds of members of this subdivision are produced from naked ovules, borne on the surface of the megasporophyll (cone scale or carpel); with these are the associated characters of absence of vessels in secondary wood, presence of resin canals, and flowers reduced to pollen sacs and ovules, usually arranged (in one or both sexes) in strobili. Exceptions are to be found to all of these characters in so far as distinction between the subdivisions are concerned.

The living gymnosperms comprise a major forest group of primarily temperate regions of both northern and southern hemispheres. Phylogenists consider them to be remnants of a once large and diverse group that was dominant in the Mesozoic, whose ancestors are known only fragmentarily from the fossil record. The orders comprising this subdivision are composed of markedly distinct and widely divergent groups that now are considered examples of more or less parallel development. This view displaces the older view of lineal development, whereby one family was interpreted to have been derived from another more primitive extant family. Knowledge of the status of their former dominance and characteristics of some now extinct members, together with homologies between reproductive structures of present and past representatives, has led to the belief that the conifers and ephedras may be closely related to the Cordaites of prehistoric times.

Living gymnosperms are classified by the Engler system into 4 orders (Cycadales,

---

[8] Just (The proper designation of vascular plants. Bot. Review 11: 299–309, 1945) has pointed out that the term Tracheophyta was first published by Sinnott in 1935 and is a *nomen nudum*. It has been used extensively in the last decade or more by morphologists, and scarcely at all by taxonomists. Another, and older, term proposed for the group by Pia (1931), was Stelophyta. This also is a *nomen nudum* and has not gained the acceptance by botanists that has been accorded Tracheophyta.

Most of the names proposed in recent times for these taxa are *nomina nuda* that require validation before use in any future classification. The distressing problem of determining priority for these names was resolved at the Stockholm Congress (1950) when it was voted that ". . . for names of taxa above the rank of family, the rule of priority is not compulsory." This gives complete freedom in the choice of names to be applied to these higher taxa, provided the selected name has been properly published, and the taxon for which it stands has been properly described.

Ginkgoales, Coniferae, and Gnetales), 12 families, and about 63 genera and 675 species. Characters distinguishing the orders are given in the following key:

Sperm cells ciliate, motile.
    Staminate sporophylls in strobili, the pistillate usually so; lvs. pinnately compound .................................................................... *Cycadales*
    Staminate sporophylls solitary or few and never in strobili; lvs. simple, dichotomously veined, flabellate ........................................................ *Ginkgoales*
Sperm cells eciliate, nonmotile.
    Staminate strobili simple; resin canals present .......................... *Coniferae*
    Staminate strobili compound; resin canals absent ........................ *Gnetales*

LITERATURE:

CHAMBERLAIN, C. J. Gymposperms, structure and evolution. Chicago, 1935.
———. Gymnosperms. Bot. Rev. 1: 183–209, 1935.
EICHLER, A. W. Coniferae, in Engler and Prantl, Die natürlichen Pflanzenfamilien, 2: 28–116, 1889.
HUTCHINSON, J. Contributions towards a phylogenetic classification of flowering plants. III. The genera of Gymnosperms. Kew Bull. 1924, pp. 49–66.
REHDER, A. Manual of cultivated trees and shrubs. ed. 2. New York, 1940. [Gymnosperms, pp. 1–71.]
WETTSTEIN, R. VON. Handbuch der systematischen Botanik. ed. 2. 1930. [Gymnospermae. pp. 453–587.]

## Order 10. CYCADALES
### CYCADACEAE. CYCAD FAMILY

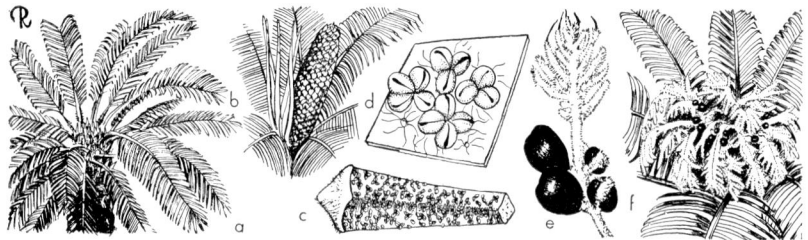

Fig. 55. CYCADACEAE. *Cycas revoluta*: a, habit of plant, much reduced; b, staminate strobilus, × 1/25; c, staminate sporophyll (lower side), × 3/8; d, staminate sporocarps, × 6; e, pistillate sporophyll bearing seed, × 1/6; f, fruiting crown, much reduced. (From L. H. Bailey, *Manual of cultivated plants,* The Macmillan Company, 1949. Copyright 1924 and 1949 by Liberty H. Bailey.)

Woody, more or less palmlike trees or shrubs; stems thick, tuberous, and mostly subterranean (*Zamia, Bowenia,* and *Stangeria*) or columnar and mostly unbranched (*Cycas, Dion, Ceratozamia*) growing to maximums of about 50 ft high in some genera, all with a very large pith, very slow growing (Dions 6–7 ft high are about 1000 years old, corresponding growth rates being common to most other arborescent genera); leaves alternate in close spirals that seem to be whorls and forming crowns at trunk apices (bipinnate in *Bowenia*), persistent (usually for 3–10 years), the leaf bases remaining after leaves drop, variable in size and number of leaflets (pinnae), leaf vernation erect, not circinate as in leptosporangiate ferns; fruiting structure a cone (except in ovulate plants of *Cycas,* where ovules are few, huge, and borne on petiole margins of pinnately compound open sporophyll), unisexual (the genera dioecious), very variable in size (some male cones a yard long, and some ovulate cones weighing to 100 pounds, or in others the individual ovules each the size of ducks' eggs and weighing several ounces), terminal on trunk or tuber (axillary in some species of *Macrozamia*); the male cone scales bearing abundance of scattered microsporangia, that produce very large motile sperm, sometimes grouped with an approach to the soral condition; in ovulate cones (excluding *Cycas*) the sporophylls are mostly weakly peltate, bearing marginal ovules, the sporophyll with its ovules comprising an open carpel; on fertilization the naked ovule develops into a usually drupelike seed, sometimes brightly colored.

A family of 9 genera and about 100 species confined to tropical and subtropical regions of the world, and there to a few restricted areas; centers of distribution in the New World are Mexico (*Dion, Ceratozamia*), and the West Indies (*Zamia, Microcycas*) to northern South America down the Andes to Chile and in localized areas of Alabama and Florida (*Zamia*); centers of distribution in the Old World are Australia (*Macrozamia, Bowenia, Cycas*) and South Africa (*Encephalartos, Stangeria*) with species of *Cycas* occurring sporadically from Australia to southern Japan (including also India and China) and 1 on Madagascar. The Cycadaceae are represented in our indigenous flora only by 4 species of *Zamia* that are restricted almost exclusively to Florida, with an additional station reported in adjoining Alabama.

The family is morphologically and taxonomically distinct from all other gymnosperms, as indicated by the compound leaves, the palmlike habit, and the very numerous scattered microsporangia.

It was once believed that the cycads were the link connecting the ferns with the seed plants. Current opinion, based largely on more complete and better known fossil evidence, is that the cycads are true gymnosperms with some fernlike characters, and that they represent a transitional stage. They are a relic group from which no group of greater advancement has evolved—hence the cycads are a link to no other group and stand as an offshoot from extinct near relatives (the seed ferns). They are not to be considered of close relationship to any other living gymnosperm.

The Cycadaceae were treated by Pilger (1926) as composed of 5 subfamilies: Cycadioideae, Stangerioideae, Bowenioideae, Dionoideae, and Zamioideae, with each of the first 4 represented by their respective type genera and the last by 5 genera.

Economically the family is of limited value. About 25 species (representing all genera except *Stangeria*) are cultivated domestically for ornament or as novelties in conservatories. Species of *Cycas* are among the more hardy of the family, and grow outdoors in southern California, Florida, and parts of the southwest and south. Leaves of *Cycas* are dried, dyed, and used for decoration. Seeds of *Cycas* and *Zamia* are reported to be edible. In the Orient the pith of *Cycas* species yields a sago starch used in making bread, and the Kaffir bread of the Hottentots is made from seeds of *Encephalartos*.

*LITERATURE:*

CHAMBERLAIN, C. J. Gymnosperms, structure and evolution. Chicago, 1935. [Cycadales, chapters IV–VIII.]

LAMB, SISTER MARY ALICE. Leaflets of Cycadaceae. Bot. Gaz. 76: 185–202, 1923. [Includes a vegetative key to the genera and most species.]

PILGER, R. Cycadaceae. In Engler and Prantl, Die natürlichen Pflanzenfamilien. ed. 2, 13: 44–82, 1926.

SCHUSTER, J. Cycadaceae. In Engler, Das Pflanzenreich. 99 (IV. 1): 1–168, 1932.

## Order 11. GINKGOALES
## GINKGOACEAE. GINKGO FAMILY

The ginkgo, or maidenhair tree (*Ginkgo biloba*), the only species of the 1 genus of the family and of the order. A much branched dioecious deciduous hardy tree, growing to 90 feet high; true vessels absent in the secondary wood, resin ducts present; leaves alternate (often seemingly whorled on very slow-growing lateral short shoots), flabellate, often bifid, dichotomously veined; staminate trees with stamens in pairs in a catkinlike bractless strobilus borne in leaf axils and bearing usually 2 anthers (microsporangia) on each stalk (sporophyll); ovulate trees with ovules borne abundantly in short shoots (spurs) in pedunculate pairs (one of each pair often aborting) or rarely racemose on the peduncle, each ovule subtended by a small raised collar (possibly the vestigial remains of the now nonexistent megasporophyll, the peduncle being stem tissue), fertilization by small motile sperm, the so-called fruit (actually a seed) plumlike and drupaceous, with a fleshy outer and a horny inner layer, putridly fetid with odor of rancid butter at time of falling, the embryo with 2 cotyledons.

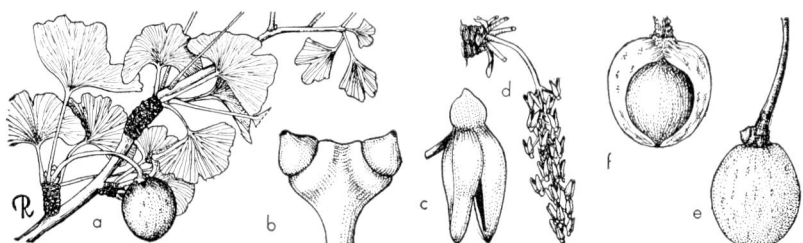

**Fig. 56.** GINKGOACEAE. *Ginkgo biloba*: a, fruiting branch, × ¼; b, pistillate strobilus, × 4; c, anther, × 8; d, staminate inflorescence, × 1; e, seed, × ½; f, seed, vertical section, × ½. (From L. H. Bailey, *Manual of cultivated plants,* The Macmillan Company, 1949. Copyright 1924 and 1949 by Liberty H. Bailey.)

The single species, not known in the wild, has long been cultivated in Chinese temple gardens where the tree is considered sacred. It was introduced in Europe in about 1730 and to this country in 1784. Trees (especially staminate, since the pistillate produce an abundance of malodorous fruit) are desirable for ornament in street plantings, where they are tolerant of smoke and low water supply. They withstand temperatures to −30°F.

The ginkgo is clearly a gymnosperm, but is without any near relatives in the group. Like the cycads, it is a relic of once wide-global distribution. The family is believed to have been represented by other genera when at its zenith, probably in the early Jurassic. It is a relic group, a link to no other living plants, and is considered to have been derived from other now extinct ancestors that were more closely related perhaps to stocks ancestral to the cycads.

Several scholars (notably Moule 1937, 1944 and Thommen 1949) have shown that etymologically the name *Ginkgo* is an orthographic error and that the correct translation of the Chinese characters used to designate the maidenhair tree should be "Ginkyo." The error originated with Kaempfer in 1712 and was introduced into formal botanical nomenclature by Linnaeus in 1771. Recently some botanists have accepted the views of Moule and of Thommen and have rejected the name *Ginkgo* as "an unintentional orthographic error," as allowed by the Rules, and have replaced it with *Ginkyo* (cf. Pulle 1946, Widder 1948).

*LITERATURE:*

CHAMBERLAIN, C. J. Gymnosperms, structure and evolution. Chicago, 1935. [Ginkgoales, Chapter X.]

MOULE, A. C. The name *Ginkgo biloba* and other names of the tree. T'oung Pao (Leiden), 33: 193–219, 1937.

―――. The name *Ginkgo.* Journ. Roy. Hort. Soc. 69: 166, 1944.

PILGER, R. Ginkgoaceae. In Engler and Prantl. Die natürlichen Pflanzenfamilien. ed. 2, 13: 98–109, 1926.
PULLE, A. A. Over de *Ginkgo* alias *Ginkyo*. Jaarboek Nederl. Dendrol. Ver. pp. 25–35, 1940–1946.
THOMMEN, E. Neues zur Schreibung des Namens *Ginkgo*. Ver. Nat. Ges. Basel, 60: 77–103, 1949.
WIDDER, F. Die Rechtschreibung des Namens *Ginkgo*. Phyton, 1: 47–52, 1948.

## Order 12. CONIFERAE [9]

Plants of this order differ as a group from those of the preceding 2 orders in having nonmotile sperms and simple, mostly acicular to linear or lanceolate leaves not dichotomously veined, the plants usually monoecious, and ovulate strobilus usually a woody cone (reduced to a single ovule and the fruit berrylike or arillate in a few genera).

The order was treated by Pilger (1926) as composed of 7 families, 48 genera (of which 15 are monotypic and mostly endemic), and about 520 species. Four of these families are represented by plants growing indigenously in this country, and all are represented here in cultivation. The families are not large, as compared with families of angiosperms where scores of genera have more species than are in this entire group of families, and this fact, plus the record of a relatively high number of monotypic endemic genera, attests to their antiquity.

The results of recent paleobotanical studies by Florin (1938–1944) have contributed a wealth of new material that promises to lead to re-evaluation of the phylogenetic relationships within the living gymnosperms, and particularly within the Coniferae. Strong fossil evidence was produced in support of his opinion that the Taxaceae and Cephalotaxaceae of Pilger are not true conifers, but represent a collateral order (Taxales) derived from nonstrobilate ancestral stocks. These recent views are not yet a part of conventional taxonomic concepts, but they indicate that changes are to be expected in the reclassification of some families here treated within the Gymnospermae. The tentative reclassification proposed by Buchholz (1934), wherein he divided the Coniferae into 2 suborders represented by 10 families, was never amplified nor later adopted by authors of taxonomic works.

*LITERATURE:*

ARBER, A. *et al.* "The origin of gymnosperms"; a discussion at the Linnean Society. New Phytologist 5: 68–76; 141–148, 1906.
BAILEY, L. H. The cultivated conifers. New York, 1933.
BEISSNER, L. Handbuch der Nadelholzkunde. ed. 3. Berlin, 1930. [Rev. by J. Fitschen.]
BUCHHOLZ, J. T. The classification of Coniferales. Trans. Ill. Acad. Sci. 25: 112–113, 1934.
DALLIMORE, W. and JACKSON, A. B. A handbook of Coniferae. ed. 2. London, 1931.
FITZPATRICK, H. M. Coniferae: keys to the genera and species. Scient. Proc. Roy. Dublin Soc. n.s. 19: 189–260, 1929.
FLORIN, R. Die Koniferen des Oberkarbons und des unteren Perms. Paleontographica, 85: 1–654, 1938–44.
PARDE, L. G. C. Les conifers. 294 pp. Paris, 1946.
PENHALLOW, D. P. A manual of the North American gymnosperms, exclusive of the Cycadales but together with certain exotic species. Boston, 1907.
PHILLIPS, E. W. J. The identification of coniferous woods by their microscopic structure. Journ. Linn. Soc. Bot. (Lond.) 52: 259–320, 1941.

---

[9] Engler and Gilg (1924), following the basic classification of Eichler, treated this order as composed of 2 families and 11 subfamilies or tribes: a treatment followed by authors of most American floras. However, more recent morphological, paleobotanical, and phylogenetical studies have produced evidence justifying the elevation of these subfamilies and tribes to the rank of family, a view accepted by Pilger (1926) and by Engler and Diels (1936).

PILGER, R. Gymnospermae. In Engler and Prantl, Die natürlichen Pflanzenfamilien. ed. 2. Berlin. 13: 1–447, 1926.
SAXTON, W. T. The classification of conifers. New Phytologist, 12: 242–262, 1913. [Provides excellent bibliography to 1913.]
SILVA TAROUCA, E. and SCHNEIDER, C. Unsere Freiland Nadelhòlzer. ed. 4. Wien, 1923.

## TAXACEAE. TAXUS OR YEW FAMILY

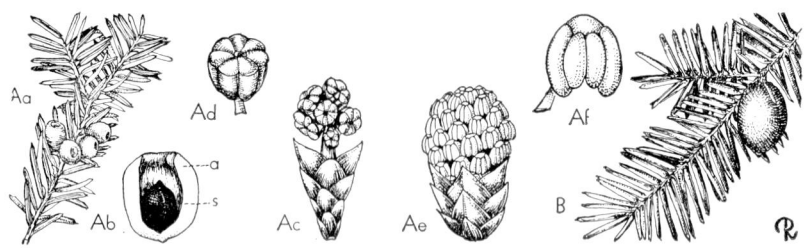

Fig. 57. TAXACEAE. Aa-Ad, *Taxus baccata*: Aa, fruiting branch, × ¼; Ab, aril (a) in section to show seed (s), × 1; Ac, male strobilus, × 4; Ad, stamen, × 8; Ae, *T. cuspidata*, male strobilus, × 8; Af, same, stamen, × 16. B, *Torreya californica*, branch with aril-covered seed, × ¼. (From L. H. Bailey, *Manual of cultivated plants,* The Macmillan Company, 1949. Copyright 1924 and 1949 by Liberty H. Bailey.)

Leaves persistent, alternate, often 2-ranked, linear to linear-lanceolate, with either pale green to tawny bands beneath, or with 2 glaucous lines beneath that are narrower than the alternating 3 green bands, mostly 1–3 cm long; plants dioecious or monoecious; staminate flowers solitary or in small axillary conelike clusters of 3–14 stamens, the anthers 3–9-celled, borne on peltate or apically thickened scales (microsporophylls); ovulate structure a usually solitary ovule, terminal (not in strobili), borne on a fleshy or rudimentary fertile sporophyll; seed dry and nutlike, surrounded or enveloped by a soft-fleshy usually highly colored arillus; embryo with 2 cotyledons.

A family of 3 genera and about 13 species, with 2 genera (*Torreya, Taxus*) occurring only in the northern hemisphere, and the monotypic *Austrotaxus,* an endemic of New Caledonia. *Torreya (Tumion)* has 3 species in China and Japan, one of limited distribution in western Florida, and another in California. *Taxus,* with 7 species for the genus (1 indigenous to North America), occurs sporadically over a wide disjunctive area of Europe, Asia, North Africa, and North and Central America.

The family is distinguished from other conifers by the peltate microsporophylls, the ovules of each flower solitary and terminal, and the seed subtended by or embedded in a fertile scale, the arillus. They are not a primitive family, the apparent gross simplicity of reproductive elements represents reduction, and their relationship to the Podocarpaceae is apparent. The Taxaceae were elevated to the rank of order by Pilger in 1916 (a view he rejected 10 years later), and accepted as the Taxales by Florin (1938, 1948) on the basis of fossil evidence that the members and their ancestral stocks were gymnosperms without cones.

Species of *Torreya* and *Taxus* are cultivated for ornament; the wood of the latter is used in cabinet work and for archery bows.

*LITERATURE:*
FLORIN, R. Die Koniferen des Oberkarbons und des unteren Perms. Paleontographica, 85: 1–654, 1938–44.

## DIVISION IV. EMBRYOPHYTA SIPHONOGAMA

———. On the morphology and relationship of the Taxaceae. Bot. Gaz. 110: 31–39, 1948.
PILGER, R. Die Taxales. Mitt. Deutsch. Dendr. Ges. 25: 1–28, 1916.
———. Taxaceae. In Engler and Prantl, Die natürlichen Pflanzenfamilien. ed. 2, Bd. 13: 199–211, 1926.

### PODOCARPACEAE. PODOCARPUS FAMILY

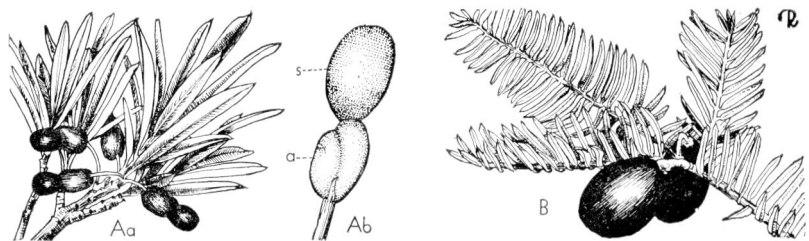

Fig. 58. PODOCARPACEAE. A, *Podocarpus macrophylla*: Aa, fruiting branch, × ¼; Ab, seed (s) and aril (a), × ½. CEPHALOTAXACEAE. B, *Cephalotaxus Harringtonia* var. *drupacea*: branch with aril-covered seeds, × ½. (From L. H. Bailey, *Manual of cultivated plants*, The Macmillan Company, 1949. Copyright 1924 and 1949 by Liberty H. Bailey.)

Trees to 100 ft or more high, or shrubby in some species; leaves persistent, alternate or opposite, or absent and represented by phylloclads, very variable from acicular to broadly lamellate; plants dioecious or monoecious; staminate flowers in terminal or axillary strobili, the stamens usually many, the anthers 2-celled; ovulate flower solitary, axillary or terminal, or in strobili with megasporophylls 1-ovuled and bracted; seed solitary; cotyledons 2.

A family of 7 genera and about 100 species, largely of the southern hemisphere, and none native in North America. The family was subdivided by Pilger into 3 subfamilies: Pherosphaeroideae, Podocarpoideae, and Phyllocladoideae, the first and last represented only by their respective type genera. Four genera (*Dacrydium, Phyllocladus, Podocarpus, Saxegothaea*) and 22 species are cultivated for ornament. *Podocarpus* is an important timber tree of Australasia.

*LITERATURE:*

BUCHHOLZ, J. T. and GRAY, N. E. A taxonomic revision of *Podocarpus*. I. The sections of the genus and their subdivisions with special reference to leaf anatomy. Journ. Arnold Arb. 29: 49–63; II. The American species of *Podocarpus*: sect. *Stachycarpus*, *op. cit.* 64–76; III. sect. *Eupodocarpus*, subsects. C & D, *op. cit.* 123–151, 1948. [Recognized 2 subgenera; *Stachycarpus* with 3 sects. and *Protopodocarpus* with 6 sects.]
PILGER, R. Podocarpaceae. In Engler and Prantl, Die natürlichen Pflanzenfamilien. ed. 2. 13: 211–249, 1926.
WASSCHER, J. The genus *Podocarpus* in the Netherlands Indies. Blumea, 4: 359–481, 1941.
WILDE, M. H. A new interpretation of coniferous cones. I. Podocarpaceae. Ann. Bot. 8: 1–41, 1944.

### ARAUCARIACEAE. ARAUCARIA FAMILY

Large trees to 140 ft high or more; branches more or less symmetrical, often whorled, the secondary branches mostly deciduous; leaves persistent, alternate, often 2-ranked, dimorphic, the juvenile larger and often differing from adult in form and arrangement, persisting until trees are large; trees dioecious or monoecious; staminate flowers in large axillary or terminal strobili, the microsporophylls many, spiraled, bracted, the microsporangia linear, many; ovulate strobili with the scales 1-ovuled; cone woody, very large, the scales deciduous at maturity; seed very large, with samaralike wing or wingless; cotyledons 2, rarely 4.

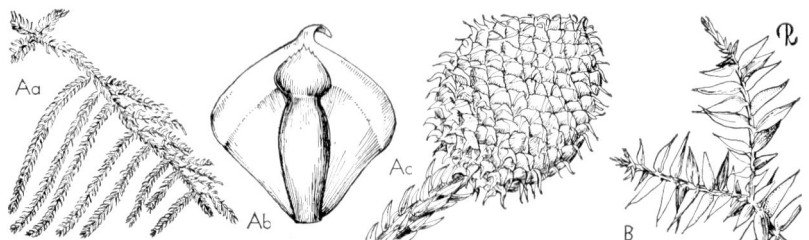

**Fig. 59.** ARAUCARIACEAE. A, *Araucaria excelsa*: Aa, branch with juvenile foliage, × ¼; Ab, cone-scale, lower side, with seed, × ½; Ac, cone, × ¼. B, *Araucaria Bidwillii*: juvenile branch, × ½. (From L. H. Bailey, *Manual of cultivated plants*, The Macmillan Company, 1949. Copyright 1924 and 1949 by Liberty H. Bailey.)

A family of 2 genera (*Araucaria, Agathis*) and about 32 species of the southern hemisphere, where they are important for their timber trees. About 12 species of both genera are cultivated domestically for ornament in Florida and California and some of *Araucaria* as conservatory subjects in the north.

*LITERATURE:*

BURLINGAME, L. L. The origin and relationships of the Araucarians. Bot. Gaz. 60: 89–114, 1915.

PILGER, R. Araucariaceae. In Engler and Prantl, Die natürlichen Pflanzenfamilien. ed. 2. 13: 249–266, 1926.

## CEPHALOTAXACEAE. PLUM-YEW FAMILY

Trees or shrubs; branches opposite; leaves persistent, uniformly small and linear, dense, spirally arranged and 2-ranked, with 2 glaucous lines beneath that are wider than the 3 alternating green bands; plants mostly dioecious, occasionally monoecious; staminate flowers in globose heads in leaf axils (*Cephalotaxus*) or in spicate strobili (*Amentotaxus*); ovulate structures in axils of scales at base of the twigs, consisting of several pairs of 2-ovuled megasporophylls; seeds 1 or 2, pediceled, large and drupelike (ripening the second season), about 1 in. long; embryo large, with 2 cotyledons. See Fig. 58.

A family of 2 genera and 6 species (*Amentotaxus* is monotypic and endemic to western China) of eastern Asia. Three species of *Cephalotaxus* are grown for ornament as far north as New York.

*LITERATURE:*

PILGER, R. Cephalotaxaceae. In Engler and Prantl, Die natürlichen Pflanzenfamilien. ed. 2. 13: 267–271, 1926.

---

**Fig. 60.** PINACEAE. A, *Larix leptolepis*: coning branch, × ½. B, *Pseudotsuga taxifolia*: coning branch, × ½. C, *Picea pungens*: coning branch, × ½. D, *Abies homolepis*: coning branch, × ½. E, *Pinus Strobus*: Ea, coning branch, × ½; Eb, fascicle of leaves, × ½; Ec, base of fascicle showing bracts, × 2; Ed, staminate strobilus, × 3; Ee, staminate sporophyll bearing pollen-sacs, × 10; Ef, pistillate strobilus, × 3; Eg, pistillate sporophyll distal edge, × 10; Eh, same, lower side showing two naked ovules, × 20. (From L. H. Bailey, *Manual of cultivated plants*, The Macmillan Company, 1949. Copyright 1924 and 1949 by Liberty H. Bailey.)

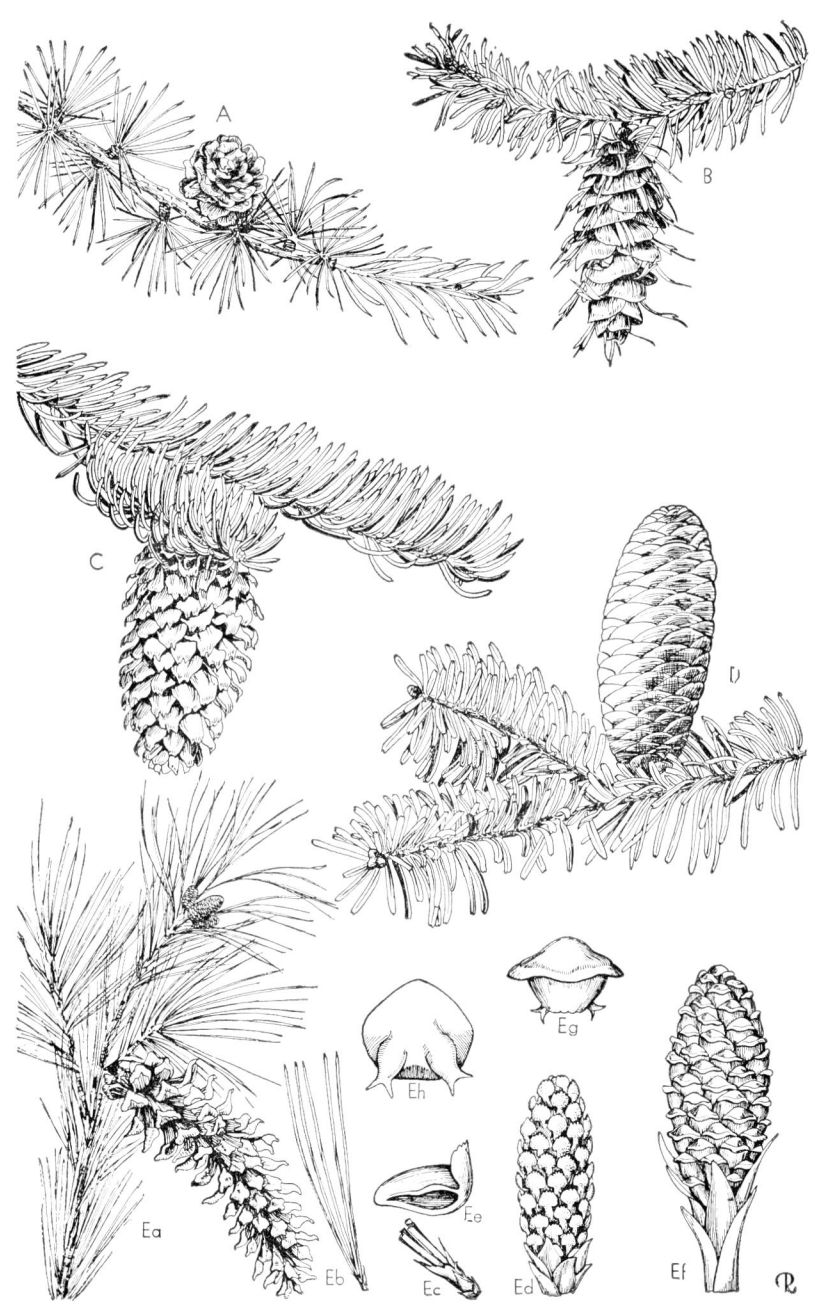

## PINACEAE. PINE FAMILY [10]

Trees, rarely shrubs; branches whorled or opposite or rarely alternate, the short-shoot (spur) characteristic of some genera and bearing leaves seemingly whorled or fascicled; leaves persistent or in a few genera deciduous, linear, in spirals of high phyllotaxies; plants monoecious; microsporangia in small herbaceous strobili, completely fused to the mostly many, spiraled scales, microsporophylls bearing mostly 2-6 sporangia on dorsal (lower) side, each anther typically 2-celled; ovulate strobili with scales in spirals and mostly numerous, each with typically 2 inverted ovules borne basally on ventral (upper) side, the scales subtended by small or large variously adnate bracts; cone woody, tightly closed until seed is ripe, the scales persistent (except in *Abies*); seeds winged, usually two on each scale, the embryo with 2-15 cotyledons.

A family of resinous woody plants, comprising 9 genera and about 210 species of wide distribution, especially throughout the temperate regions of the northern hemisphere. Two genera (*Keteleeria, Pseudolarix*) are restricted to China, with the latter the only monotypic genus in the family; 2 genera (*Pseudotsuga, Tsuga*) occur in both North America and eastern Asia; 1 genus (*Cedrus*) occurs in the Mediterranean region of Europe and north Africa and the western Himalayas of Asia; and 4 genera (*Abies, Picea, Larix, Pinus*) are dispersed over Eurasia and North America. The largest genera are *Pinus* (90 spp.), *Abies* (40 spp.), and *Picea* (40 spp.). Of the 7 genera occurring in North America, *Abies* (24 spp.) is generally of more northern distribution at higher altitudes, with all but a few species in the provinces of Alaska to California and Arizona, a few species occurring south to Guatemala, and 1 other in the upper northeastern part of the continent; *Pseudotsuga* (2 spp.) in the Pacific mountainous provinces; *Tsuga* (3 spp.) in the alpine and most northern of temperate areas across the continent extending down the Appalachian mountains to North Carolina; and *Pinus* (26 spp.), which is of the widest and most general distribution of all, occurring generally at lower latitudes and elevations in most provinces excepting the plains areas.

The Pinaceae are characterized by the cone scales flattened and distinct from the subtending bract, with the bract usually shorter than the scale (longer in *Pseudotsuga*), the 2 structures showing various degrees of adnation in the different genera. In all but *Larix* and *Pseudolarix* the leaves are persistent, while in all but *Abies* and *Keteleeria* the cones are drooping or pendulous, and in *Pseudolarix, Larix, Cedrus,* and *Pinus* the production of short shoots (lateral spurs or fascicles of leaves) and terminal twigs are characteristic features.

The family was treated by Pilger (1926) to be composed of 2 subfamilies; the Pinoideae with *Pinus* its only genus, and the Abietinoideae containing the remaining 8 genera.

The Pinaceae are of considerable economic importance for timber and pulpwood. They are important also as a source of naval stores (pitch, turpentine, and rosin) from *Pinus*, Venetian turpentine from *Larix*, Canada balsam from *Abies balsamea*, edible seeds from several species of *Pinus*, and numerous other products of lesser importance. About 165 species, representing all 9 genera, are cultivated domestically, and the number of horticultural forms and clones is legion.

*LITERATURE:*

FLOUS, F. Classification et évolution d'un groupe d'abiétinées. Trav. Lab. Forest. Toulouse, I: 211-286, 1936.

[10] By some authors this and the 2 families that follow are retained as a single family, Pinaceae, composed of 3 subfamilies: the Abietineae accounting for the Pinaceae as treated here, the Taxodiineae, and the Cupressineae.

FULLING, E. H. Identification by leaf structure, of the species of *Abies* cultivated in the United States. Bull. Torrey Bot. Club, 61: 497–524, 1934.
HARLOW, W. M. The identification of the pines of the United States, native and cultivated, by needle structure. Bull. N. Y. State Coll. Forestry (Syracuse), 4: 1–21, 1931.
LAMBERT, A. B. Description of the genus *Pinus*. London, 1803.
MARTINEZ, M. Las Pinaceas mexicanas. An. Inst. Biol. [Mexico], 16: 1–345, 1945.
PILGER, R. Pinaceae. In Engler and Prantl, Die natürlichen Pflanzenfamilien. ed. 2. 13: 271–342, 1926.
SHAW, G. R. The genus *Pinus*. Cambridge, Mass., 1914.

## TAXODIACEAE. TAXODIUM FAMILY

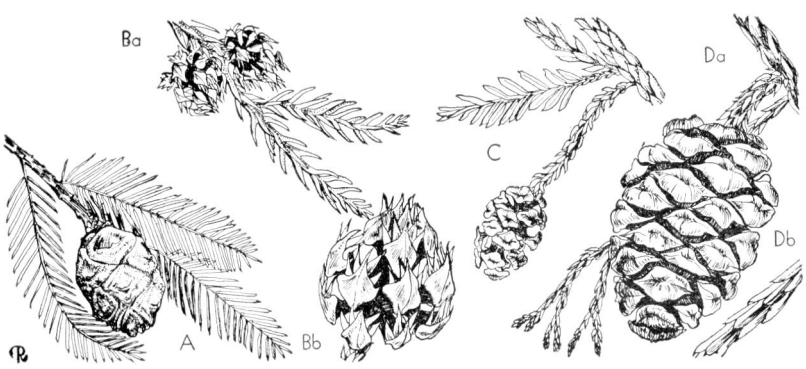

Fig. 61. TAXODIACEAE. A, *Taxodium distichum*: A, coning branch, × ½. B, *Cryptomeria japonica*: Ba, coning branch, × ½; Bb, cone, × 1. C, *Sequoia sempervirens*: coning branch, × ½. D, *Sequoiadendron giganteum*: Da, coning branch, × ½; Db, portion of leafy twig, × 1. (From L. H. Bailey, *Manual of cultivated plants*, The Macmillan Company, 1949. Copyright 1924 and 1949 by Liberty H. Bailey.)

Trees, rarely shrubs; lateral branchlets deciduous or persistent; leaves scalelike or needlelike to falcate, sometimes dimorphic (*Sequoia, Sciadopitys*), persistent or rarely deciduous (*Taxodium*), solitary and spiraled or the needlelike ones connate in pairs and in false whorls (*Sciadopitys*); plants monoecious; male strobili small, sometimes catkinlike, clustered in heads or in racemelike masses, terminal or axillary, the microsporangia (anther cells) 2–9; ovulate strobili terminal, the sporophylls bearing 2–9 erect or anatropous ovules, flat or peltate, the subtending bracts partially or completely adnate; cone woody or leathery-woody, more or less globose with persistent scales, the bracts not differentiable from scales, the seeds 2–9, wingless but with slight integumentary margin, the cotyledons 2–9.

A family of 10 genera and 16 species. Three genera (*Cryptomeria, Sciadopitys, Taiwania*) are endemic to Japan, and the last 2 named are monotypic (*Cryptomeria* also being so considered by some botanists), 2 are monotypic endemics of China (*Glyptostrobus, Metasequoia*), 2 others are monotypic endemics to southern Oregon and California (*Sequoia, Sequoiadendron*), and *Cunninghamia* has 1 species in China and another in Formosa. The only representative in the southern hemisphere is *Athrotaxis*, with 3 species in western Tasmania. *Taxodium* (3 spp.) occurs in North America from southern Delaware to Florida and Mexico, extending west into Illinois, Missouri, and Texas. This family is a heterogeneous assemblage whose members may best be distinguished from those of other families of conifers by the flat or peltate cone scales lacking distinct bracts and each scale producing 2–9 seeds; the dimorphic leaves provide another character of reliability when present. The family stands closer to Pinaceae and Cupressaceae than to others. Considerable interest centers on the genus *Metasequoia* since it was named and described from fossil remains 5 years before living trees were discovered (in 1946) in western China. The fossils of this conifer had been known for a century or more, but most paleobotanists had misidentified them as fossil sequoias. The segregation of *Sequoiadendron* from *Sequoia* has not been accepted by some botanists. (For a review of the situation cf. Dayton, 1943; Stebbins, 1948.) The 2 were accepted as generically distinct by Stebbins (1948), who contrasted them with *Metasequoia*.

Pilger (1926) treated the family as composed of 2 subfamilies; the Sciadopityoideae represented only by *Sciadopitys*, and the Taxodioideae containing the remaining genera. Buchholz (1934) considered each to be a separate family, a view likely to gain support as the plants are better known morphologically.

All the genera, represented by 14 of their combined species, are cultivated domestically for ornament; of these *Cunninghamia* (China fir), *Cryptomeria*, and *Sciadopitys* (umbrella pine) are the more prized for landscape uses. *Taxodium* is the source of commercial cypress lumber valued for its resistance to wood-rotting fungi and in demand for green house benches and racks. Similar uses are made of the redwood lumber of *Sequoia sempervirens*.

*LITERATURE:*

BUCHHOLZ, J. T. Generic segregation of the Sequoias. Amer Journ. Bot. 26: 535–538, 1939.
DAYTON, W. A. The names of the giant *Sequoia*. Leafl. West. Bot. 3: 209–219, 1943.
HU, H. and CHENG, W. On the new family Metasequoiaceae and on *Metasequoia glyptostroboides*, a living species of the genus *Metasequoia* found in Szechuan and Hupeh. Bull. Fan Memorial Inst. Biol. n.s. 1: 153–161, 1948.
PILGER, R. Taxodiaceae. In Engler and Prantl, Die natürlichen Pflanzenfamilien. ed. 2. 13: 342–360, 1926.
STEBBINS, G. L. JR. The chromosomes and relationships of *Metasequoia* and *Sequoia*. Science, 108: 95–98, 1948.

## CUPRESSACEAE. CYPRESS FAMILY

Trees or shrubs; leaves persistent, opposite or whorled, mostly small and scalelike, acicular or subulate, sometimes dimorphic (juvenile leaves larger and more slender than those of adult foliage); plants monoecious, or dioecious in some genera; male strobili small, terminal or axillary, or clusters of 2–24 stamens borne on lower side of margin of a broad and somewhat peltate microsporophyll, the anthers 2–6-celled, often in ternate whorls or in pairs; ovulate strobili terminal or lateral on short branches, the scales (megasporophylls) usually few (1–12), flattened and imbricate (of a single ternate whorl in *Fitzroya*), peltate (*Cupressus, Chamaecyparis*), or fleshy and connate (*Juniperus*), the ovules erect, 1–12 per scale; cone dry and woody with persistent opposite or whorled scales, or the fruit fleshy and berrylike (in *Juniperus*), small and rarely exceeding 1 in. long; seeds often with integumentary wings, the cotyledons mostly 2, rarely 5–6.

A family of world-wide distribution, composed of 15–16 genera and about 140 species with *Juniperus* (70 spp.), *Callitris* (20 spp.), and *Cupressus* (15 spp.) being the largest genera. Six genera are monotypic and mostly endemic: *Fitzroya*, southern Chile; *Tetraclinis*, southern Spain to north Africa; *Thujopsis*, Japan; *Callitropsis*, New Caledonia; *Diselma*, western Tasmania; and *Arceuthos*, southeastern Europe. *Fokenia* (3 spp.) is restricted to China, *Widdringtonia* (5 spp.) occurs in south and southeastern tropical Africa, and *Actinostrobus* (2 spp.) is restricted to western Australia. The remaining genera are of either widespread or disjunctive distributions: *Callitris* is Australasian; *Libocedrus* (9 spp.) occurs along Pacific coast of both North and South America, also in New Zealand, China, and Formosa; *Cupressus* grows in western North America south to Guatemala, Asia, and extends to eastern Europe; *Chamaecyparis* (6 spp.) is indigenous to Pacific and eastern North America, Japan, and Formosa; *Thuja* (6 spp.) is common to northern North America and eastern Asia; and *Juniperus* is widely spread over much of the temperate areas of the northern hemisphere. The characteristics distinguishing members of this family from those of other coniferous families are the leaves and cone scales opposite or whorled with the former small and (in adult foliage) usually scalelike, the ovules erect, and the cones all generally smaller than in other families.

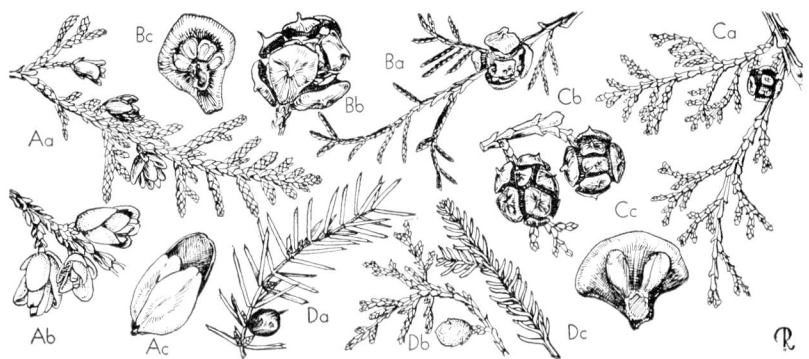

**Fig. 62.** CUPRESSACEAE. A, *Thuja occidentalis*: Aa, coning branch, × ½; Ab, cones, × 1; Ac, two-seeded cone-scale, × 2. B, *Cupressus macrocarpa*: Ba, coning branch, × ½; Bb, cone, × 1; Bc, four-seeded cone-scale, × 2. C, *Chamaecyparis pisifera*: Ca, coning branch, × ½; Cb, cones, × 1; Cc, two-seeded cone-scale, × 2. D, *Juniperus rigida*: coning branch, × ½; Db, *Juniperus virginiana*: coning branch, × ½; Dc, *Juniperus communis*: juvenile foliage, × ½. (From L. H. Bailey, *Manual of cultivated plants*, The Macmillan Company, 1949. Copyright 1924 and 1949 by Liberty H. Bailey.)

Domestically, the family is represented by 5 genera. *Juniperus* is widespread over much of the United States; *Cupressus* occurs only in the southwestern corner of the country, where it represents a northern extension from Mexico; *Chamaecyparis* is of disjunctive distribution, being represented by a single species in acid-bog habitats from Nova Scotia to Florida, and by 2 species on Pacific slopes from southern Alaska to California; *Thuja* likewise is a disjunct with 1 of its 4 species in northeastern America and another in northwestern America; *Libocedrus* is represented on this continent by 1 of its 8 species, it being here restricted to the arid transition zone from Oregon into Lower California.

The Cupressaceae were treated by Pilger as composed of 3 subfamilies: the Thujoideae containing those genera with woody flattened cone scales, the Cupressoideae with *Cupressus* and *Chamaecyparis* that have woody peltate cone scales and the Juniperoideae with fleshy cone scales and the cone berrylike.

Many genera (notably *Juniperus, Cupressus, Chamaecyparis, Thuja, Callitris*. and *Libocedrus*) contain important timber-producing species, and domestically all but *Callitris* are leading sources of wood for cabinet work, shingles, lead pencils, and construction purposes. All but 3 genera (*Fokenia, Diselma, Callitropsis*) representing about 70 species, are cultivated domestically for ornamental purposes. A volatile oil extracted from crushed Juniper berries is the principal flavoring ingredient of gin, and oil of cedar is obtained from *Thuja occidentalis*.

*LITERATURE:*

ANTOINE, F. Die Cupressineen-Gattungen. Wien, 1857.
MARTINEZ, M. Los *Juniperus* mexicanos. An. Inst. Biol. [Mexico], 17: 3–128, 1946.
——— Los *Cupressus* mexicanos. *op. cit.* 18: 71–149, 1947.
PILGER, R. Die Gattung *Juniperus.* L. In Mitt. Deutsch. Dendrol. Gesell. 43: 255–269, 1931.
——— Cupressaceae. In Engler and Prantl, Die natürlichen Pflanzenfamilien. ed. 2. 13: 361–403, 1926.
WOLF, C. and WAGNER, W. E. The New World cypresses (*Cupressus*). El Aliso, 1: 1–444, 1948.

## Order 13. GNETALES

The Gnetales differ from other gymnospermous orders by many characteristics, and may readily be distinguished by the compound staminate strobili, opposite or whorled leaves (found also in Cupressaceae), vessels in the secondary wood, and absence of resin canals. Only the naked ovule character keeps this order in the gymnosperms, and only Ephedraceae has a gymnospermous life history. They possess many characters common to the angiosperms and have been interpreted by some authors to be a partial connecting link between the two. However, it is of significance that no Gnetales have been found as far back as the Cretaceous, when angiosperms were already abundant, and there are important morphological considerations that lead some botanists to believe that the Gnetales are of themselves an isolated assemblage whose 3 genera are only remotely related to one another and far less so to either gymnosperms or angiosperms.

Markgraf, the latest monographer of the group (in Engler and Prantl, *Die natürlichen Pflanzenfamilien,* ed. 2, bd. 13, pp. 407–441, 1926), considered this order as composed of 3 families. However, so very marked and of such fundamental character are their many differences that each family deserves the rank of order. Only Ephedraceae are accounted for here, but the unigeneric (and monotypic) Welwitschiaceae and Gnetaceae are each as different from the other 2 families as are all the conifers from the cycads or from ginkgo. An opposite view was taken by C. J. Chamberlain, who, following the older taxonomic views, found it satisfactory to consider the plants as representing 3 genera of a single family.

The discovery of a new taxon (*Sarcopus aberrans*) in Indo China, alleged to be of this alliance, was reported by Gagnepain (Bull. Soc. Bot. Fr. 93: 313–320, 1946), who separated it from others of the Gnetales by its bisexual flowers whose female element is represented by a single naked ovule. He treated this new monotypic genus as a new family (Sarcopodaceae) and considered it to be the most advanced of the order. Further opinion on the taxonomic status of the plant must await availability of material for study by other investigators.

### EPHEDRACEAE. EPHEDRA FAMILY

Shrubs to about 7 ft high or much less, erect, decumbent or climbing, much branched, the branches usually green for several seasons; leaves mostly deciduous, opposite or whorled, more or less connate basally and usually reduced to membranous sheaths; plants dioecious or rarely monoecious; staminate flowers in compact subglobose to oblong, compound, usually axillary strobili, the latter usually opposite or in whorls of 3 or 4 at twig nodes, each strobilus bearing 2–8 opposite pairs of bracts along the primary axis, of these the lower 1 or 2 pairs sterile and the rest bear solitary male flowers, each flower borne on a short secondary axis arising between each fertile pair of bracts, the flower (sporangiophore) composed of 2 thin opposite scales (perianth of some authors) above which are 1–8 stamens

## DIVISION IV. EMBRYOPHYTA SIPHONOGAMA

with sessile or filamented anthers, the filaments often basally connate, the anthers dehiscing by terminal pores; ovulate flowers in an elongated acute strobilus, the strobili opposite or in whorls of 3–4 at branch nodes, each strobilus with usually many pairs of bracts along the axis (sometimes reduced to 2), the 4 or more lower pairs sterile, and bearing a usually terminal naked ovule (as many as 3 flowers may comprise a single strobilus), the ovule with 2 integuments, the outer integument represented by 4 basally coherent bracts (sometimes collectively referred to as the perianth), the inner of 2 similar bracts that elongate considerably at time of pollination and stylelike (tubullus of some authors) in appearance; seed with leathery integument, globose to cylindric, membranous or winged (due to bractlike integument) or forming a berrylike syncarp, often red in color, cotyledons 2.

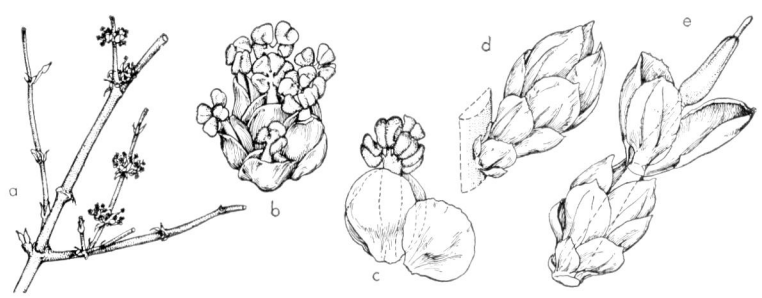

**Fig. 63.** EPHEDRACEAE. *Ephedra Torreyana*: a, staminate branch in flower, habit, × ¼; b, staminate inflorescence, × 4; c, staminate flower with bract, × 5; d, "pistillate" inflorescence, habit, × 3; e, same, expanded, × 3.

The order is represented in the United States by the 1 family and a single genus (*Ephedra*, of about 42 species). The genus is of wide sporadic distribution in arid regions of tropics and subtropics of northern and southern hemispheres. About 18 species occur in the Old World (France, Canary Islands, around the Mediterranean east to Persia, India, and China), and about 24 in the New World with 13–15 in North America and 9–11 in South America (Bolivia to Patagonia). The North American species (15) are most abundant in California, Arizona, and New Mexico, with ranges extending north and south into Mexico, and 4 species occurring as far east as Texas. The plants are readily distinguished by their evergreen horsetaillike stems, usually open and straggling habit, and small deciduous cones borne in stem axils.

Economically, the plants are of little importance aside from Asiatic species, from which the medicinal alkaloid ephedrine is obtained (about 1000 tons of the crude drug, baled plants, were imported in 1935). The plants are cultivated to a very limited extent (4 spp.) in this country, outdoors in favorable regions as a sand binder, or under glass as a novelty.

*LITERATURE:*

CHAMBERLAIN, C. J. Gymnosperms, structure and evolution. Chicago, 1935. [See, chapters XVII–XIX.]
CUTLER, H. C. Monograph of the North American species of the genus *Ephedra*. Ann Mo. Bot. Gard. 26: 373–427, 1939.
GROFF, G. W. and CLARK, G. W. The botany of *Ephedra*. Calif. Univ. Pubs. Bot. 14: 247–282, 1928.
MARKGRAF, F. Gnetales. In Engler and Prantl, Die natürlichen Pflanzenfamilien. ed. 2. 13: 407–441, 1926.

PEARSON, H. H. W. Gnetales. 194 pp. Cambridge Univ. Press, 1929.
THOMPSON, W. P. The morphology and affinities of *Gnetum*. Amer. Journ. Bot. 3: 135–184, 1916.

## Subdivision II. ANGIOSPERMAE [11]

The angiosperms, more commonly known as the flowering plants, are characterized as a group by the presence of vessels [12] in the stems; by the ovules enclosed within 1 or more carpellary sporophylls and comprising the ovary, this organ together with its terminal stigmatic zone (the 2 connected or not by a constricted stylar isthmus) constituting the basic element of the gynoecium, the pistil; the 1 or more microsporangia borne on a microsporophyll (designated as anther and filament respectively), comprising the basic element of the androecium, the stamen; the combination of members of one or both of these sex elements, accompanied or not by a perianth arising from a common axis, comprise the flower. Subsequent to pollination of the stigma and fertilization of the ovules, maturation of carpellary tissues occurs, with the ovules developing into seeds and the ovary wall into the fruit coat.[13]

There are exceptions to these characteristics of the angiosperms. Members of some families (as the Winteraceae, Tetracentraceae, Trochodendraceae) are believed never to have had vessels, while in others (Cactaceae) the vessels have been lost through specialization and suppression. In some angiosperms (*Platanus, Reseda*) the carpels are open and the ovules not completely enclosed within the ovary; likewise, in some gymnosperms (*Araucaria*, some Gnetales) the ovules are enclosed nearly as completely as in an angiosperm. In the case of *Caulophyllum* (an angiosperm) the ovary wall ruptures soon after fertilization, followed by extrusion of the ovule, and the berrylike so-called fruit is actually a naked seed. The presence of a flower, as a character to distinguish angiosperms from gymnosperms, is a matter of artificial definition since it is impossible to define the flower with morphological precision so as to exclude the cone or cluster of fertile sporophylls that characterizes the gymnosperms. Despite these exceptions, the taxonomic validity of the angiosperms as a group is not subject to serious challenge, and (aside from the Gnetales) there is never any question as to which subdivision a particular plant belongs.

The angiosperms are a vast group, providing the dominant vegetation of the earth's surface. Compared to other groups they are of modern development and are considered to be young. Their youth is not one of absolute age, for they are believed to have been in existence as early as the Paleozoic, but they are young in that it is only during and since the Mesozoic that their rapidly expanding develop-

---

[11] Bessey treated this subdivision as a phylum, the Anthophyta. The term angiosperm, from the Greek, literally means vessel seed, in allusion to the seed borne or produced within a vessel or ovary. The term anthophyte, also from the Greek, means flower plant.

[12] A vessel is composed of a series of nonliving water-conducting, typically thick-walled cells of the xylem, arranged or lying end to end, with the end walls of each cell perforated by large pores.

[13] The above constitutes an obviously technical circumscription of a flower. Reduced to its simplest elements, a flower may be represented by a solitary anther (as in some members of the Lemnaceae), by a single stamen (as in *Euphorbia*), or by a single pistil (as in *Ficus*) on an axis, but in most plants the flower is composed typically of 5 whorls or series of elements arranged (centrifugally) on an axis as follows: the gynoecium of 1 or more pistils, the androecium of typically 2 series (whorls) of stamens (1 series often suppressed), the corolla representing an inner series of bracts (petaloid), and the calyx representing the outermost series of bracts (sepaloid). For further discussion of the flower, see Chapter IV.

DIVISION IV. EMBRYOPHYTA SIPHONOGAMA 371

ment accelerated appreciably as to kind, distribution, and dominance. It is estimated that the angiosperms are composed of about 300 families, represented currently by about 200,000 species. Large areas of the earth's continents are not yet well known floristically, and it is reasonable to believe that, with more thorough exploration and investigation, the number of species will continue to increase.

See Chapter V for discussion of the phylogeny of the angiosperms and for references to the pertinent literature.

### Class I. MONOCOTYLEDONEAE [14]

Plants woody or, more commonly, herbaceous; stems with vascular bundles scattered throughout but not arranged in a single cylinder (i.e., not appearing as a ring in cross section); leaves usually parallel-veined with margins almost always entire; flowers, basically with parts in 3's or multiples of 3; seed embryos with a single cotyledon. Exceptions exist to each of these characters, but it is unusual to find any individual varying in more than 1 respect, and in the aggregate, these features may be considered adequate to characterize a monocot.

Phylogenetically, the monocots are now generally believed to have been derived from extinct and very primitive dicot stocks of ranalian ancestry. If so, they are more advanced phylogenetically than are the dicots. For this reason, the systems of Bessey and of Hutchinson placed the monocots after the dicots in lineal sequences, or as collaterals with and derived from them. Engler and associates considered the monocots the more primitive for reasons previously discussed in Chapter V. Their position in this text is in conformance with the Engler system.

The monocots have been treated in detail in recent times by Engler and associates, and by Hutchinson. They have been the subject of extensive morphological study by Arber *et al.* The treatment by Bessey (into 8 orders) lacked the critical study given by him to the dicots. Engler and Diels (1936) considered them as composed of 11 orders and 45 families, whereas Hutchinson (1934) considered the orders to number 26 and the families 68.

According to Engler the monocots of the southeastern United States would be grouped under 9 orders and 31 families, whereas for the same area Small (1933), using the same system with narrower concepts of categories, placed them in 16 orders and 49 families. In contrast to Small's concepts, the 8th edition of Gray's *Manual* treated the monocots of its area (northeastern United States and Canada) as representing 8 orders and 25 families, with the same plants placed in 8 orders and 27 families by Engler.

These contrasts indicate differences in taxonomic concept. Of the 11 monocot orders recognized by Engler, 9 are represented by plants native to this country, and of the 36 families treated below only 5 are represented solely by exotics. These 36 families account for about 2000 genera and 34,000 species in the world.

Each treatment of included monocot families contains considerations of phylogenetic position, as classified by Engler, by Hutchinson, and as indicated by other or more recent studies. In the case of monocot families, little comparison is made with the classification by Bessey, since it now seems clear that his major divisions within the class (on the basis of ovary position) resulted in an arrangement not supported by the morphological considerations and phylogenetic criteria of more recent time. As pointed out in the review and analysis of systems of classification,

---

[14] The name Monocotyledoneae is derived from the fact that embryos of members of this class have typically 1 cotyledon. For brevity in reference, the term monocotyledon and its adjectival form, monocotyledonous, are reduced in this treatment to monocot. Likewise, without loss of clarity or precision, the term dicotyledon (and dicotyledonous) is reduced to dicot, where appropriate.

Chapter VI, Hutchinson divided the monocots into 3 subphyla, and considered criteria based on inflorescences to be phylogenetically more fundamental than those based on ovary position. These criteria, together with those of carpellary conditions, floral anatomy, embryogeny, and vascular arrangements, are interpreted as more fundamental than such a relatively superficial character as that afforded by ovary position.

The following synopsis of the 10 orders of monocots is provided to indicate distinguishing characteristics. It does not represent a key by which individuals may be classified.

1. Perianth wanting, or reduced to bristles, bracts, scales (petaloid in few advanced genera of Helobiae).
    2. Fls. variously arranged but not in axils of dry chaffy bracts.
        3. Perianth represented by bristles or chaffy scales; leaves stiffly long and sword-shaped ............................................13. *Pandanales*
        3. Perianth usually of fleshy or herbaceous bracts; lvs. various or wanting; fr. usually drupaceous or baccate.
            4. Seeds without endosperm; pollen grains in triads; fr. 1-seeded..14. *Helobiae*
            4. Seeds with endosperm; pollen grains in diads or tetrads; fr. more than 1-seeded ........................................18. *Spathiflorae*
    2. Fls. in axils of dry, chaffy bracts ..........................15. *Glumiflorae*
1. Perianth generally present and in two series, the inner or both usually more or less petaloid (except in the order Synanthae).
    5. Plants typically woody; leaves generally decompound or palmately lobed, cleft, or divided.
        6. Carpels 3 in each ovary, each cell 1-ovuled ...................16. *Principes*
        6. Carpels 2–4, each cell many-ovuled ........................17. *Synanthae*
    5. Plants herbaceous, or if woody the leaves never compound nor flabellate.
        7. Seeds not minute, endosperm present.
            8. Endosperm mealy .....................................19. *Farinosae*
            8. Endosperm fleshy, horny, or cartilaginous.
                9. Functional stamens 3–6; seeds rarely arillate ..........20. *Liliiflorae*
                9. Functional stamens 1; seeds mostly arillate ..........21. *Scitamineae*
        7. Seeds very minute, lacking endosperm ...................22. *Microspermae*

*LITERATURE:*

ARBER, A. Monocotyledons. Cambridge, 1925.
BANCROFT, N. A review of literature concerning evolution of Monocotyledons. New Phytologist, 13: 285–303, 1914.
CAMPBELL, D. H. The phylogeny of monocotyledons. Ann. Bot. 44: 311–331, 1930.
ENGLER, A. and DIELS, L. Syllabus der Pflanzenfamilien. Aufl. 11. Berlin, 1936.
HUTCHINSON, J. The families of flowering plants. II. Monocotyledons. London, 1934.
RENDLE, A. B. The classification of flowering plants. I. Gymnosperms and monocotyledons. ed. 2. Cambridge, 1930.
WETTSTEIN, R. VON, Handbuch der systematischen Botanik. ed. 2. vol. 2. Leipzig, 1935.

## Order 14. PANDANALES

Leaves linear; flowers unisexual, the families of dioecious or monoecious plants; perianth of bristles or dry scales, gynoecium monopistillate, ovary 1-many-carpelled; pollen in diads, the stamens 1-many; fruit more or less nutlike; seeds with endosperm.

The order contains only the 3 families treated below. Hutchinson segregated the Pandanaceae from Typhaceae and Sparganiaceae as a separate order, interpreting it, on the basis of the coalescence of fruits into a syncarp, to be more advanced than the Typhaceae or Sparganiaceae. All 3 families now are believed to be simple by a reduction of reproductive parts and to have been derived from liliaceous stocks.

# DIVISION IV. EMBRYOPHYTA SIPHONOGAMA

## TYPHACEAE. CATTAIL FAMILY

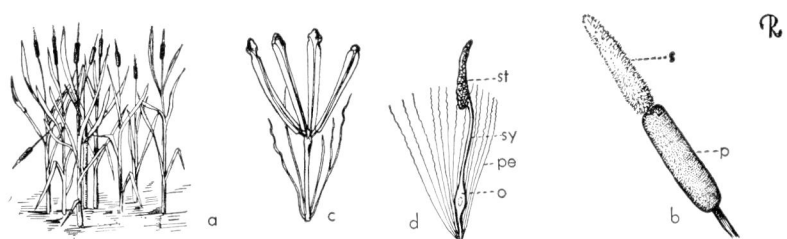

**Fig. 64.** TYPHACEAE. *Typha latifolia*: a, habit in fruit, much reduced; b, spike of staminate (s) and pistillate (p) flowers, × ⅛; c, staminate flower, × 5; d, pistillate flower with perianth (pe), ovary (o), style (sy) and stigma (st), × 10. (From L. H. Bailey, *Manual of cultivated plants,* The Macmillan Company, 1949. Copyright 1924 and 1949 by Liberty H. Bailey.

Perennial herbs of open marshes, rootstocks rhizomatous, creeping; leaves erect, long-linear, mostly basal, parallel-veined, sessile; plants monoecious; flowers unisexual, borne on a cylindrical spadix, the staminate above the pistillate and those of each sex subtended by 1 caducous bractlike spathe, the perianth represented by bristles; staminate flowers of 2–5 stamens variously monadelphous, the connate filaments bearing long silky hairs; pistillate flowers of 1 pistil, the ovary long-stipitate, unilocular, 1-carpelled with a single pendulous ovule, the style 1, usually filiform, the stigma 1, linear to spatulate, or rhomboidal, 1-sided, ovary stipe bearing many silky hairs; fruit a minute nutlet with persistent style; seed with mealy endosperm.

The single genus *Typha* contains about 15 species. The genus is of more or less cosmopolitan distribution throughout riparian and estuarian marshes of temperate and tropical regions of northern and southern hemispheres. The 2 species indigenous to North America and a third from Japan are cultivated for ornament. Leaves of both native species are used in weaving chair bottoms and matting. *Typha* is readily identified by its very long erect linear leaves and the dense spicate inflorescences, fuzzy-brown at maturity.

*LITERATURE:*

GRAEBNER, P. Typhaceae. In Engler, Das Pflanzenreich, 2 (IV. 1): 1–18, 1900.
HOTCHKISS, N. and DOZIER, H. L. Taxonomy and distribution of North American cattails. Amer. Midl. Nat. 41: 237–254, 1949.
ROSCOE, M. V. Cytological studies in the genus *Typha*. Bot. Gaz. 84: 392–406, 1927.

## PANDANACEAE. SCREW-PINE FAMILY

Trees or shrubs, erect or sometimes vines; stout aerial prop roots, when present, often actually support the trunk from the ground and are produced also from branches; leaves usually spiral, 4-ranked, congested with seemingly no internodes at branch tips, linear, sessile with sheathing base, tough, fibrous and leathery but not fleshy, strongly keeled and canaliculate, mostly spinulose along keel and margins; plants dioecious; flowers unisexual, paniculate or densely crowded, the perianth rudimentary or absent, the inflorescence enclosed initially by spathaceous or foliaceous bracts; staminate flowers scarcely distinguishable as such, the stamens numerous, densely packed or separated or in fasciculate clusters, scattered over a thyrsoid spadixlike axis, the filaments distinct or connate, the anthers 2-celled (the cells sometimes each once divided), basifixed, dehiscing by vertical slits; rudimentary ovary present and minute or absent; pistillate spadix simple, the flowers with or without hypogynous staminodes; pistils numerous, coherent in bundles or isolated, the ovary superior, unilocular, the ovules solitary or many, basal or parietal, anatropous, the style short or none, the stigma 1 and pistils sometimes united by stigmas; fruit a syncarp; each pistil developing into a drupaceous baccate or woody fruit (pulpy inside) cohering into multiple units; seeds small, endosperm fleshy, embryo minute.

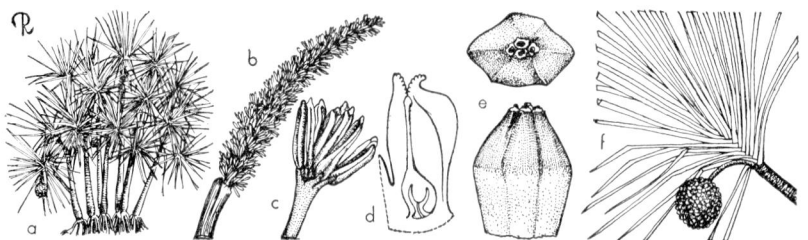

Fig. 65. PANDANACEAE. *Pandanus utilis*: a, habit of tree in fruit, much reduced; b, branch of staminate inflorescence, × ½; c, staminate flower, × 3; d, pistillate flower, vertical section, enlarged; e, drupe, side and top views, × ½; f, fruiting branch, much reduced. (From L. H. Bailey, *Manual of cultivated plants*, The Macmillan Company, 1949. Copyright 1924 and 1949 by Liberty H. Bailey.)

A family of 3 genera (*Pandanus, Freycinetia, Sararanya*) and perhaps 300 species, of tropical regions from Africa and Asia through Polynesia and Australasia, extending into the Hawaiian Islands. None is native to this country.

The family stands apart from the other 2 of the order by its woody often palm-like habit and more strongly by its large more or less fleshy syncarpous fruit. *Pandanus* (about 150 spp.), the screw pine, has 2–3 spp. cultivated in the open in southern subtropical United States and grown in the north as a foliage subject under glass. In regions of its nativity, the Pandanus leaves are an important source for thatch, matting, clothing, and assorted containers, while the immature fleshy pericarps of fruits of some species are a source of food.

*LITERATURE:*

MARTELLI, U. Enumerazione delle "Pandanaceae." Webbia, 3: 307–327, 1910; *op. cit.* 4: 1–105, 1913–14.

———. La distribuzione geografica delle Pandanaceae. Atti. Soc. Toscana Sci. Nat. Mem. 43: 190–209, 1933.

WARBURG, O. Pandanaceae. In Engler, Das Pflanzenreich, 3 (IV. 9): 1–97, 1900.

## SPARGANIACEAE. BUR-REED FAMILY

Aquatic perennial herbs, rhizomatous; stems leafy, simple or branched; leaves linear, alternate, sessile and sheathing, 2-ranked, erect or floating; plants monoecious; flowers unisexual, crowded into globose sessile or pedunculate clusters, the staminate above the pistillate, the perianth reduced to usually 3–6 minute membranous elongated or spatulate scales; staminate flowers of 3 or more stamens, the filaments mostly distinct, the anthers mostly oblong or cuneate, basifixed; pistillate flowers crowded, with perianth calyxlike, of 3–6 linear or spatulate scales; pistil with ovary superior, sessile, narrowed basally, mostly unilocular, the ovule 1 in each locule and basal or pendulous, the style simple or forked, the stigma simple and unilateral; fruit nutlike, indehiscent, with spongy exocarp and hard bony endocarp; seed with straight embryo in copious mealy endosperm.

# DIVISION IV. EMBRYOPHYTA SIPHONOGAMA

The family is composed of the single cosmopolitan *Sparganium*, a genus of about 20 species, limited largely to the temperate and frigid regions of northern and southern hemispheres. About 11–12 species are native to North America, 2 are common also to Eurasia, and 1 extends as far south as Florida. The family seems more closely related to the monoecious Typhaceae than to the dioecious Pandanaceae, as indicated by its herbaceous habit, aquatic associations, and characters of fruit. The disposition of the flowers in globose heads readily separates it from the cattails. The genus is rarely cultivated, and is of limited economic value as a source of food for wild life.

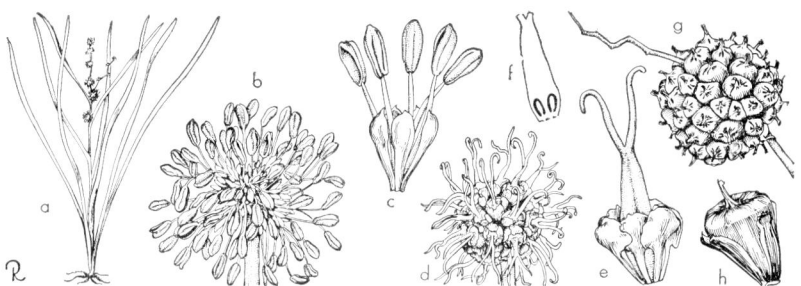

**Fig. 66.** SPARGANIACEAE. *Sparganium eurycarpum*: a, flowering plant, much reduced; b, staminate inflorescence, × 3; c, staminate flower, × 5; d, pistillate inflorescence, × 1½; e, pistillate flower, × 4; f, ovary, vertical section, × 4; g, fruit cluster, × ¾; h, fruit, × 1½.

*LITERATURE:*

FERNALD, M. L. Notes on *Sparganium*. Rhodora, 24: 26–34, 1922.
GRAEBNER, P. Sparganiaceae. In Engler, Das Pflanzenreich, 2 (IV. 10): 1–126, 1900.
RYDBERG, P. A. Sparganiaceae. In North Amer. Flora, 17: 5–10, 1903.

## Order 15. HELOBIAE [15]

Plants mostly of aquatic or marshy habitats, often almost completely submerged; sexual elements arranged into distinct mostly cyclic flowers; the flowers various and providing no characters diagnostic of the order as here defined. The complete or near absence of endosperm in the seed and the aquatic habit of most of the components provide the basic characters by which the 7 families are distinguished as a unit. In addition to these characters, plants of a majority of the genera possess minute scales (*squamulae intravaginales*) that are associated with the leaf bases. These are not axillary to the leaves with which each is associated, but as stated by Arber, probably are "appendages of the basal region of the leaf skin belonging to the next leaf above." Similar structures occur also in some of the Araceae and some Lemnaceae.

There is little doubt but that this is an unnatural order as defined by Engler. Hutchinson placed the taxa in 6 orders, and Small (following the Engler system) divided them into 3 orders. Hutchinson acknowledged that they possessed greater affinities to one another than to other monocots, and treated them as representing 1 of the 4 of his major branches of the monocots. Bessey placed most of the families in the Liliales (others were associated with Typhaceae), and elevated the

---

[15] Named Helobiae by Engler, this order is also known as the Najadales and as the Fluviales.

remaining families as a third order (on the basis of the ovary inferior) removed from those with a superior ovary. Rendle (1930) accepted Engler's concept of the order and placed it after the Pandanales, but did not consider it to have been derived from the latter, for he stated that it was "a group of orders [families] developing along its own lines."

Hutchinson's distribution of the taxa among 6 orders has much in its favor. However, he treated these 6 orders as derived from the Alismatales, with the Juncaginales ancestral to the Aponogetonales and the latter to the Najadales. The Potamogetonales were represented as an offshoot somewhat more primitive than the Juncaginales. The morphological studies of Cheadle (1942) and of Uhl (1947) discredited the likelihood of the Alismatales having been ancestral to other taxa of the Helobiae, and presented evidence to support the belief that the primitive families were the Aponogetonaceae, Scheuchzeriaceae, and Lilaeaceae, with the Najadaceae representing a probably highly advanced taxon.

## POTAMOGETONACEAE. POND WEED FAMILY

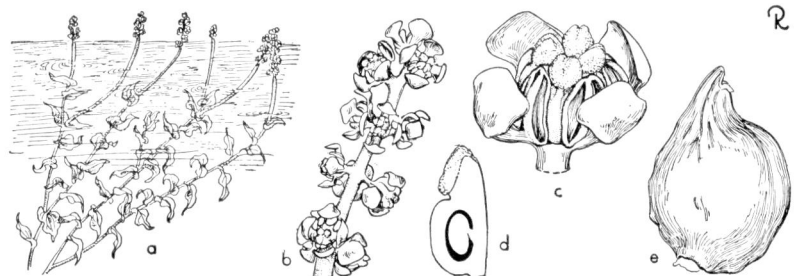

**Fig. 67.** POTAMOGETONACEAE. *Potamogeton Richardsonii*: a, flowering branches, × ¼; b, inflorescence, × 2; c, flower, habit, × 8; d, pistil, vertical section, × 10; e, fruit, × 8.

Aquatic, or very rarely marsh and bog perennials; stems often jointed and nodose, the lower nodes root-bearing and upper ones foliaceous; leaves sheathing basally, the sheath often apically ligulate, blades submersed or floating, 2-ranked; flowers bisexual or unisexual (plants then monoecious or less frequently dioecious), the parts arranged in 1–4-merous whorls; perianth (pseudoperianth or sepaloid connective) variable, of 4–6 distinct herbaceous valvate segments in one whorl, or membranous and tubular, or cup-shaped, or wanting (the perianth variously interpreted but not established to be bractlike in origin, adnate to the stamen and not a petaloid connective); stamens 1–4, the anthers 1–2-celled; the ovary unilocular and unicarpellate, the ovule 1, apical or parietal (*Zostera, Cymodocea*), pendulous; fruit an indehiscent nutlet or drupelet; seed without endosperm.[16]

A family of 8 genera and about 124 species of almost exclusively submersed aquatic habit, distributed widely over the globe and inhabiting oceanic coastal habitats, brackish tidal waters, and fresh waters of rivers, streams, lakes, ponds, and bogs. The largest genus, *Potamogeton*, common mostly to fresh-water habitats,

[16] The recent studies of Uhl (1947) on the floral morphology of flowers of genera within this family suggested that the so-called perianth parts are in fact individual bracts subtending and adnate to the stamens and that the flower is fundamentally an inflorescence composed of staminate flowers (each of a single stamen and monobracteate perianth and apetalous pistillate flowers). This view, first proposed by Kunth (1841), has been supported also by Miki (1937) *et al.*

has about 90 species of which perhaps 50 are North American, with the remainder distributed over much of the north temperate zone. The remaining genera are all small, and are common to shallow oceanic and brackish waters. Notable among them are *Zostera* (6 spp.) to which the salt-water eel grass belongs, *Ruppia* a monotypic cosmopolitan genus, *Zannichellia* (2 spp.) also cosmopolitan in distribution, *Phyllospadix* (2 spp.) limited to the Pacific coast of North America, *Cymodocea* (7 spp.) of warm pantropical waters with 1 species in Florida, and 2 genera (*Posidonia* and *Althenia*) not occurring in North American waters.

Some American authors (notably Britton, Fernald, and Rydberg), perhaps following the more conservative evaluations of Bentham and Hooker, have combined this and the succeeding family under the name Najadaceae. Engler and Diels (1936) accepted the family as composed of 5 tribes (Zostereae, Posidonieae, Potamogetoneae, Cymodoceae, and Zannichellieae). Hutchinson (1934) recognized no divisions within the family, but limited it to only *Potamogeton,* and each of the other genera was placed in separate unigeneric families. Taylor (1909) elevated 3 tribes of previous authors to the rank of family and treated the North American components as the Zannichelliaceae, Zosteraceae, and Cymodoceaceae.

The family is distinguished from related families of the order by the submerged aquatic habit, flowers with perianth rudimentary or none, stamens 1–4, and the gynoecium of 1–4 1-celled, 1-ovuled pistils. Following Taylor, Abrams placed *Zostera* and *Phyllospadix* in a family apart by itself (Zosteraceae) distinguished from the genera of his Potamogetonaceae by the anthers sessile, 2-ranked on the axis, the pollen filamentous, and the pistil with 2 slender stigmas; ribbonlike leaves are common to both these genera.

The relationships of the family to other families of the order are not clear, and authorities differ in interpretations. It would appear that the family stands in an intermediate position—more advanced than the Aponogetonaceae or Scheuchzeriaceae, but not developed to the degree shown by the Najadaceae or the Alismaceae.

The morphological findings by Uhl (1947) indicated that within the family the genera *Zostera* and *Phyllospadix* stand apart, and probably merit family recognition, a view also held by Hutchinson (1934) and by Miki (1937). *Ruppia* and *Potamogeton* were found to have affinities in common with *Zostera* and *Phyllospadix* to represent an advancement (by reduction) over them. Hutchinson's segregation of *Zannichellia* as a separate family (Zannichelliaceae) was supported by the findings of Uhl, who concluded that its ancestral stock was probably from both Potamogeton- and Zosteralike plants.

The plants of this family are of little value except in biological conservation activities where many genera are important sources of food for waterfowl, and provide protection to fish.

*LITERATURE:*

Ascherson, P. and Graebner, P. Potamogetonaceae. In Engler, Das Pflanzenreich, 2 (IV. 13), 1907.
Chrysler, M. A. The structure and relationships in Potamogetonaceae and allied families. Bot. Gaz. 44: 161–188, 1907.
Fernald, M. L. The linear-leaved North American species of *Potamogeton,* section *Axillares.* Mem. Amer. Acad. 17: 1–183, 1932.
Miki, S. The origin of *Najas* and *Potamogeton.* Bot. Mag. Tokyo, 51: 290–480, 1937.
Ogden, E. C. The broad-leaved species of *Potamogeton* in North America north of Mexico. Rhodora, 45: 57–105, 119–163, 171–214, 1943.
Taylor, N. Zannichelliaceae. In North Amer. Flora, 17: 13–27, 1909. [Includes treatments of *Zannichellia, Ruppia,* and *Potamogeton.*]
———. Zosteraceae. In North Amer. Flora, 17: 29–30, 1909. [Includes treatments of *Zostera* and *Phyllospadix.*]

———. Cymodoceaceae. In North Amer. Flora, 17: 31–32, 1909. [Includes treatments of *Cymodocea* and *Halodule*.]

UHL, N. W. Studies in the floral morphology and anatomy of certain members of the Helobiae. Thesis (Ph.D.) Cornell Univ. 1947.

## NAJADACEAE. NAJAS FAMILY

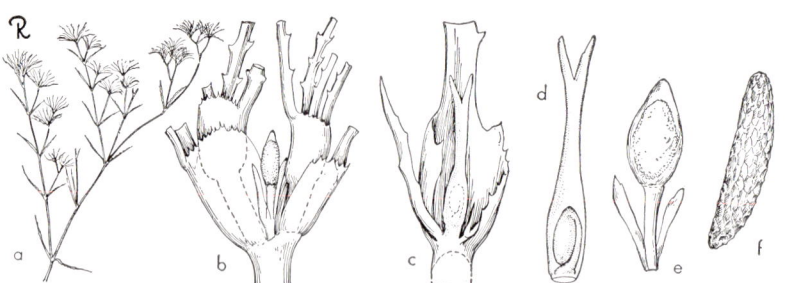

**Fig. 68.** NAJADACEAE. *Najas gracillima*: a, vegetative branch, × ⅙; b, branch axil with staminate flower, × 6; c, leaf with pistillate flower, habit, × 10; d, pistillate flower showing position of ovule), × 15; e, staminate flower, × 15; f, seed, × 10. (f redrawn from Fassett.)

Aquatic submerged monoecious or dioecious annuals of fresh or brackish water with much-branched nodal slender stems, rooting from lower nodes; leaves linear to linear-lanceolate, entire or toothed, sometimes with prominent auricles (stipules), subopposite or seemingly whorled, sessile with sheathing base and 2 minute scales within the sheath; flowers unisexual, minute, sessile, solitary or clustered in branch axils; staminate flowers each with 1 stamen (anther at first subsessile, the filament elongating during anthesis) enclosed in a minute, flask-shaped, membranous bract (often but incorrectly termed a spathe), 2-lipped at apex, the anther 1–4-celled, dehiscing vertically; pistillate flower without perianth or bract (spathe), or the latter may be present and when so it is hyaline and adherent to ovary; pistil 1, the ovary unilocular and unicarpellate, the ovule 1, basal, anatropous, the style 1, the stigmas 2–4 (commonly 3), subulate; fruit an achene, usually enclosed by foliaceous sheath; seed without endosperm.

A family of a single genus, *Najas,* composed of perhaps 40 species, of which 7 are native to and widely distributed throughout the United States. Of little or no economic importance. The family is distinguished from Potamogetonaceae by the annual duration, flowers always wholly submerged (not floating), and the ovule basal and erect.

Campbell (1897) interpreted the plants to be simple and to be among the most primitive of monocots and of angiosperms, postulating further that ancestral stocks were probably heterosporous Filicales. The phylogenetic position of the Najadaceae is becoming more clear: there is anatomical and morphological evidence to support the hypothesis that, by reduction and suppression of parts, it is a highly advanced family within the order, and that, contrary to its position in the Engler schema, it may stand at the top of the Heliobiae. Within the monocots it is undoubtedly a primitive family, but by no means the most primitive and probably not ancestral to more advanced types.

*LITERATURE:*

CAMPBELL, D. H. A morphological study of *Naias* and *Zannichellia*. Proc. Calif. Acad Sci. ser. 3 Bot. 1: 1–61, 1897.

CLAUSEN, R. T. Studies in the genus *Najas* in the northern United States. Rhodora, 38: 333–345, 1936.
RENDLE, A. B. Najadaceae. In Engler, Das Pflanzenreich, 2 (IV. 12): 1–21, 1901.
TAYLOR, N. Naiadaceae. In North Amer. Flora, 17: 33–25, 1909.

## APONOGETONACEAE. APONOGETON FAMILY

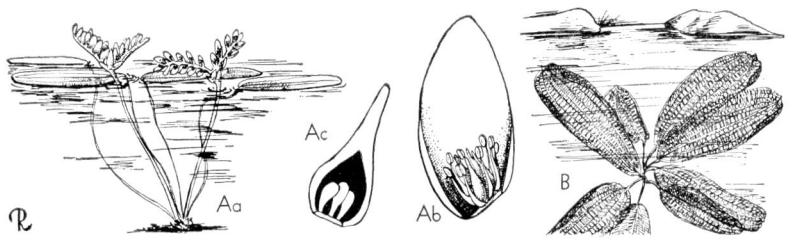

**Fig. 69.** APONOGETONACEAE. A, *Aponogeton distachyus*: Aa, habit, leaves floating, × ⅛; Ab, flower habit, × 3; Ac, pistil, vertical section, × 6. B, *Aponogeton fenestralis*: habit, leaves submerged, × ⅛. (From L. H. Bailey, *Manual of cultivated plants*, The Macmillan Company, 1949. Copyright 1924 and 1949 by Liberty H. Bailey.)

Aquatic perennials of fresh-water habitats; roots fibrous and from tuberous rhizomes; leaves mostly basal, floating or submerged, long-petioled, blades linear to oblong, reticulately veined with a few primary parallel veins and many transverse secondary veins, sometimes fenestrate; flowers usually bisexual (or unisexual by abortion), in simple or 2–8-branched spikelike inflorescences, ebracteate; perianth none or if present of 1–3 petaloid or spathaceous equal or unequal segments; stamens mostly 6 or more, persistent, hypogenous, the anthers 2-celled, dehiscing vertically, extrorse, on prominent free filaments; gynoecium of 3–6 sessile pistils; ovary unilocular, the ovules 2 or more, basal, anatropous, the style present, mostly slender-tapering, the stigma 1 and papillate; fruit a follicle; seed without endosperm.

A unigeneric family (*Aponogeton*) of about 22 species in tropical and south Africa and northeastern Australia to New Guinea. None is indigenous to the New World. The lattice-leaf plant of Madagascar (*Aponogeton fenestralis*) is highly prized by aquarium fanciers and the Cape pondweed (*A. distachyus*) is grown as a garden pool subject.

*LITERATURE:*

CAMUS, A. Le genre *Aponogeton* L.f. Bull. Soc. Bot. Fr. 70: 670–676, 1923.
KRAUSE, K. and ENGLER, A. Aponogetonaceae. In Engler, Das Pflanzenreich, 24 (IV. 13): 1–24, 1906.

## SCHEUCHZERIACEAE.[17] ARROW GRASS FAMILY

Marsh herbs, perennial, rushlike in appearance, the rhizome producing fibrous or sometimes tuberous roots; leaves basal or cauline, flat, linear, sheathing basally, ligulate; inflorescence a few-flowered spike or raceme; flowers bisexual or unisexual, bracteate (ebracteate in *Triglochin*), perianth actinomorphic, of 6 mostly similar and free segments in 2 series (rarely of 3 in 1 series); flowers (when perfect) with 6 free stamens, the anthers 2-celled, basifixed, dehiscing vertically, extrorse, the pistils 3–6, coherent or weakly connate basally and adaxially but separating at fruit maturity, the ovary 1-celled, the ovules 1-few, basal, erect, anatropous, the stigma 1 and usually sessile or on very short stout style, plumose or papillose; fruit a follicle; seeds without endosperm.

[17] Sometimes designated the Juncaginaceae.

A family of 3 genera of mostly wide distributions. *Triglochin,* with 12 species of temperate and subarctic regions of both hemispheres, is represented by 3 species in bogs and tidal marshes of North America; *Scheuchzeria,* a monotypic genus, is restricted to cold sphagnum bogs and wet shores of cooler regions of northern hemisphere; and *Tetroncium* is a monotypic genus of the American antarctic.

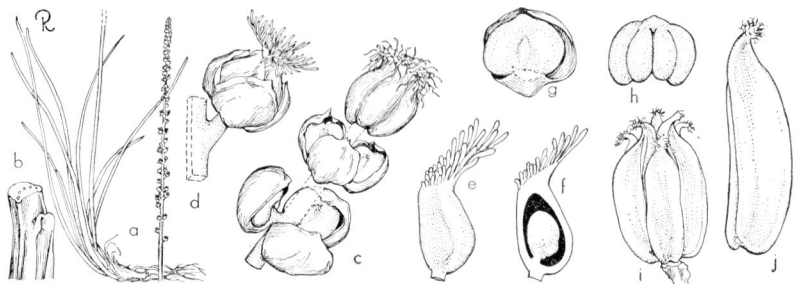

**Fig. 70.** SCHEUCHZERIACEAE. *Triglochin maritima*: a, plant with inflorescence, × ⅟₁₀; b, leaf, cross-section with ligule and portion of sheath, × 2; c, portion of inflorescence showing (below) staminate and (above) pistillate flowers, × 6; d, pistillate flower, habit, × 6; e, pistil, habit, × 15; f, pistil, vertical section, × 15; g, staminate flower, × 8; h, stamen, dorsal side, × 8; i, folliectum, × 5; j, follicle, × 8.

Engler and Diels interpreted the Scheuchzeriaceae to be composed of the 3 genera, *Scheuchzera, Triglochin,* and *Lilaea.* Hutchinson restricted the Scheuchzeriaceae to *Scheuchzeria,* recognized the Juncaginaceae as a separate family of 4 genera (one of which was *Triglochin*), and treated *Lilaea* as a distinct family, the Lilaeaceae. There is much evidence to support Hutchinson's views, and certainly *Lilaea* is sufficiently distinct, on grounds of gross morphology, from the other 2 to merit segregation into a family by itself—a view taken by Abrams in his recent flora of the Pacific states.

Hutchinson (1934), in addition to the realignment of genera as outlined above, transferred what he considered the Scheuchzeriaceae to the order Alismatales (together with Alismaceae and Petrosaviaceae). Recent studies of the floral anatomy of these and related genera (Uhl, 1947), indicated that *Scheuchzeria, Triglochin,* and *Lilaea* were closely related to one another and showed no close affinity to members of the Alismaceae. But Uhl also was of the opinion that *Scheuchzeria* is perhaps more closely related to *Aponogeton* than to other genera.

*LITERATURE:*

BRITTON, N. L. Scheuchzeriaceae. In North Amer. Flora, 17: 41–42, 1909.
BUCHENAU, F. Scheuchzeriaceae. In Engler, Das Pflanzenreich, 16 (IV. 14): 1–20, 1903.
UHL, N. W. Studies in the floral morphology and anatomy of certain members of the Helobiae. Thesis (Ph.D.). Cornell Univ. 1947.

## LILAEACEAE. FLOWERING QUILLWORT FAMILY [18]

Annual, aquatic, or marsh plants, stemless, the fibrous roots from a very short rhizome; leaves basal, crowded, alternate, linear, terete, and spongy, sheathing basally, the sheaths often with minute hyaline scales within; inflorescence a dense

[18] The name of this family is phonetically close to that of Liliaceae. Orthographically the names are distinct and each is nomenclaturally valid. Hutchinson (1934) proposed for this family the name Heterostylaceae "as an alternative name for anyone who may quite naturally object to the use of a family name so similar to that of Liliaceae." Adoption of such a proposal would be contrary to the existing rules of nomenclature.

spikelike raceme, the spikes of 2 types, basal (enclosed by the sheathing leaf bases) and comprised of pistillate flowers, or scapose and mixed often with perfect flowers in the central zone, the pistillate below and staminate above, frequently 2 lateral pistillate flowers being present at base of peduncle and sheathed by the leaf bases; flowers all without perianth, usually subtended by a bract (the bract considered by Hutchinson to represent a single perianth segment); basal spikes all with pistillate flowers of single sessile pistil, 1-loculed, the carpels probably 3, the ovule 1, basal, erect, anatropous, the style long and filamentous (to 10 cm in basal flowers) bearing a capitate stigma, the fruit a ribbed caryopsis; scapose spikes with pistillate flowers bracted or ebracteate, the pistil 1 and unilocular, the ovule 1, the lower pistils with long styles and the upper with shorter styles, the stigmas papillose; fruit a winged caryopsis; the staminate flowers (uppermost in the inflorescence) each composed of a single sessile 2-celled anther whose connective is dilated and bractlike, each stamen borne in axil of a minute bract; the perfect flowers mostly bracteate and consist of a staminate flower adjacent or closely appressed to and slightly below a pistillate flower.

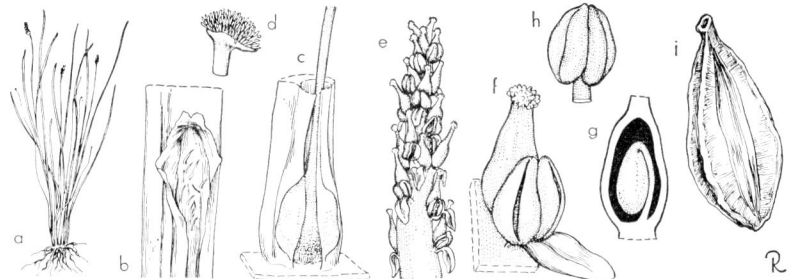

**Fig. 71.** LILAEACEAE. *Lilaea scilloides*: a, flowering plant, much reduced; b, leaf-sheath with hyaline scales, × 5; c, basal pistillate flower (lower portion), × 10; d, stigma of basal flower, × 10; e, scapose inflorescence, distal end, × 4; f, bisexual flower and its bract, × 12; g, ovary, vertical section, × 15; h, anther, × 12; i, fruit, × 10. (Drawn from material, courtesy H. L. Mason.)

A family of a single monotypic genus, *Lilaea scilloides* (*Heterostylus*), of alkaline lakes or muddy vernal waters in the Pacific area from British Columbia to Chile and Argentina. The family is of no known economic importance.

Taxonomically, *Lilaea* has been retained by most authors in the Scheuchzeriaceae, and was segregated as a separate unigeneric family by Small in 1909, a view concurred in, among phylogenists, only by Hutchinson (1934). Recent studies of floral morphology by Uhl (1947) indicated that the genus *Lilaea* probably was derived from *Triglochin*like ancestral types and is now sufficiently far removed from present-day *Triglochin* to justify its recognition as a family. Anatomical data presented by Uhl suggested that the so-called flowers of *Lilaea* are not true flowers, but to the contrary are

. . . reduced lateral branches of an inflorescence bearing unisexual flowers. The lowermost branches are each reduced to a single pistillate flower—the uppermost to a single staminate flower. Those which appear to represent perfect flowers are reduced to one staminate and one pistillate flower.

*LITERATURE:*

ABRAMS, L. Illustrated Flora of the Pacific States, rev. ed. Berkeley, Calif. 1940. [Lilaeaceae, 1: 95.]
TAYLOR, N. Lilaeaceae. In North Amer. Flora 17: 37, 1909.

## ALISMACEAE. WATER-PLANTAIN FAMILY [19]

**Fig. 72.** ALISMACEAE. *Sagittaria sagittifolia*: a, flowering plant, × 1/20; b, portion of inflorescence, × 1/2; c, staminate flower, × 1; d, pistillate flower, × 1; e, same, vertical section, less perianth, × 3; f, pistil, × 12; g, fruit, × 5. (From L. H. Bailey, *Manual of cultivated plants*, The Macmillan Company, 1949. Copyright 1924 and 1949 by Liberty H. Bailey.)

Perennial or annual marsh or aquatic herbs; plants cauline and erect or with leaves floating; roots fibrous from stout rhizome; leaves basal, long-petioled, sheathing basally, the blades very variable from linear to ovate or the bases sagittate or hastate, reticulately veined with few to many primary parallel veins converging apically and numerous close and parallel transverse veins; inflorescence a raceme or panicle; flowers bisexual or unisexual (in *Sagittaria*), pediceled, often in whorls, bracteate, the perianth of 6 free segments in 2 series, regular, vernation imbricate, the outer 3 green and sepallike, herbaceous and persistent, the inner 3 larger and petaloid, deciduous; stamens 6 or more (rarely 3), hypogynous, free, the anthers 2-celled, extrorse or dehiscing by lateral slits; gynoecium of numerous (6 or more) free pistils, spirally arranged or in advanced genera in a single whorl, the receptacle flat or convex; ovary superior, unilocular, the ovules solitary or infrequently several, basal or nearly so, the style mostly acicular and persistent, the stigma 1 and scarcely distinguishable from style; fruit an achene, rarely follicular and then dehiscing basally.

A family of 14 genera and about 55–60 species, of wide distribution, but mainly in temperate and tropical regions of the northern hemispheres; the plants mostly of fresh-water swamps and streams. The 5 genera native to North America (a possible sixth genus *Helianthium* here included in *Echinodorus*) are *Alisma* with 1 or 2 of its total of 6–8 species widely distributed in this country, *Machaerocarpus* a monotypic genus indigenous to California, *Echinodorus* with 2 of its 14–15 species occurring across the southern half of the country, *Lophotocarpus* an American genus of about 7 species of which 1 is restricted to west coast areas and 4 others to eastern and southern sections of the country, and *Sagittaria* primarily an American genus of about 40 species or less of which possibly 18 are within this country (the polymorphic character of several of the species makes the number ascribable to this genus subject to interpretation). Old World genera include *Elisma, Caldesia, Rautanenia, Burnatia, Ranalisma, Limnophyton, Wiesneria,* and *Damasonium*. Of these, *Sagittaria, Alisma,* and *Lophotocarpus* have the widest and most general distributions on nearly all continents.

The Alismaceae are distinguished from most other families of the order by the

---

[19] The name of the family was originally spelled Alismaceae, and the spelling Alismataceae was used for the name of the family as a *nomen conservandum*. However, Weatherby noted that if Haloragaceae may be derived from *Haloragis*, the name Alismaceae (from *Alisma*) also seems acceptable.

bisexual flowers having a 2-seriate perianth with the outer series clearly sepallike. They differ from the Butomaceae and Hydrocharitaceae in the ovary having 1 or more basal ovules (as opposed to parietal placentation) and the flowers usually arranged in whorls.

The family, together with the Butomaceae, was redefined recently by Pichon (1946), and the carpellary characters rejected as bases for separation. Pichon proposed that the Alismaceae be treated as composed of members having lactiferous ducts, leaves petioled and with expanded blades, petals mostly caducous, ovules campylotropous, and seeds with curved embryos. This emendation resulted in his transfer to Alismaceae of all members of the Butomaceae except *Butomus*, and segregation of the West Indian *Helianthium nymphaeifolium* from other species of *Helianthium* as a new monotypic genus, *Albidella*. Pichon recognized 2 tribes, Alismateae and Limnocharieae, separated on the carpellary differences accepted by authors retaining them as separate families. There appears to be much in favor of Pichon's realignment of genera in the 2 families, since it represented regrouping supported by macroscopic and microscopic morphological characters of the gynoecium and also by anatomical distinctions.

Phylogenetically the family is of considerable interest to botanists, since by many it is interpreted to represent one of the most primitive of extant monocot families. Hutchinson pointed out that the relatively recently discovered *Ranalisma* differed from ranunculaceous plants primarily only in possession of a single cotyledon and lack of endosperm. Cheadle (1942) found that vessels with simple or porous perforations were the rule in this family, an anatomical condition more advanced than that found in most other families of the order. The presence in the Alismaceae of bisexual flowers whose gynoecia have many pistils, whose androecia are of usually many stamens, mark it as more primitive than its position in the Engler system would indicate. However, since it possesses an advanced type of vascular anatomy whose presence is contradictory evidence against any hypothesis that would derive other families of the Helobiae from it, there is the probability that this and the next 2 families terminate a lateral branch from primitive elements of the order that has not given rise to more advanced families of this order.

The family is of limited economic importance. Rhizomes of *Sagittaria* constituted the wappata eaten by American Indians; those of Asiatic species are cultivated for food by the Chinese. Species of the genus are cultivated for ornament in pools and aquaria as also are species of *Alisma*.

*LITERATURE:*

BROWN, W. V. Cytological studies in the Alismaceae. Bot. Gaz. 108: 262–267, 1946.
CHEADLE, V. I. The occurrence and types of vessels in the various organs of the plant in the Monocotyledoneae. Amer. Journ. Bot. 29: 441–450, 1942.
ENGLER, A. Alismaceae. In Engler. Das Pflanzenreich, 16 (IV. 15): 1–66, 1903.
FERNALD, M. L. The North American representatives of *Alisma Plantago-aquatica*. Rhodora, 48: 86–96, 1946.
PICHON, M. Sur les Alismatacées et les Butomacées [includes *Albidella*, gen. nov., key to genera of redefined Alismaceae.] Not. Syst. [Paris] 12: 170–183, 1946.
SAMUELSSON, G. Die Arten der Gattung *Alisma*, L. Arkiv. f. Bot. 24a: 1–46, 1932.
SMALL, J. K. Alismaceae. In North Amer. Flora, 17: 43–62, 1909.
SMITH, J. G. North American species of *Sagittaria* and *Lophotocarpus*. Annual Rpt., Mo. Bot. Gard. 6: 27–64, 1895.

## BUTOMACEAE. FLOWERING RUSH FAMILY

Aquatic or marsh perennials, usually with milky juice; roots fibrous from a stout rhizome; leaves basal or cauline, ensiform or flat and dilated or petioled with orbicular to elliptic blades, reticulately veined with primary veins palmately parallel and transverse secondary veins or the leaves seemingly veinless; flowers solitary or in involucrate umbels, bisexual, the perianth of 6 free imbricated segments in 2 series, regular, the outer 3 sepallike (rarely colored), subherbaceous, the inner 3 larger, petallike, usually thin and deciduous; stamens mostly 6–9 or more (sometimes less in *Ostenia*), whorled, hypogynous, when numerous the outer often without anthers, the filaments free and flattened, the anthers 2-celled, basifixed, dehiscing by lateral slits; gynoecium of 6 or more pistils, distinct or only basally coherent, the ovary superior, unilocular, 1-carpelled, the ovules numerous, anatropous, scattered over inner surface with the placental region parietal and reticulately branched; fruit a follicle, each free or nearly so, dehiscing adaxially; seed without endosperm, the embryo horseshoe-shaped or straight.

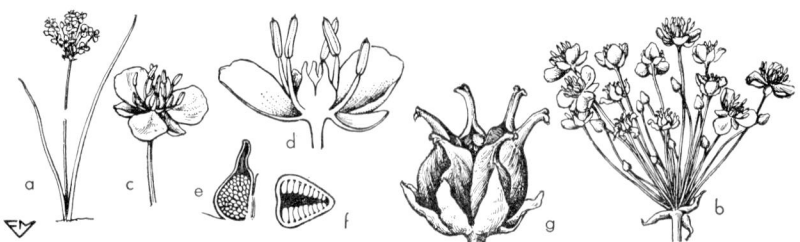

**Fig. 73.** BUTOMACEAE. *Butomus umbellatus*: a, plant (many leaves omitted), × 1/16; b. inflorescence, × 1/4; c, flower, × 1/2; d, flower, vertical section, × 1; e, pistil, vertical section, × 2; f, pistil, cross-section, × 3; g, follicles, × 2. (From L. H. Bailey, *Manual of cultivated plants,* The Macmillan Company, 1949. Copyright 1924 and 1949 by Liberty H. Bailey.)

A family of 6 genera and 9 species, represented in this country by *Butomus umbellatus* (a monotypic genus) now naturalized from Eurasia along the shores of the St. Lawrence River and Lake Champlain (N. Y.) and cultivated in water gardens for ornament. Two other genera of tropical America (*Limnocharis* with 2 species and *Hydrocleis* with 3 species) are cultivated as aquatic subjects. The 3 remaining genera are each monotypic; the first 2 are local endemics, *Ostenia* of Uruguay and *Elattosis* of Tonkin, while *Tenagocharis* is pantropical. Except as indicated above the family is domestically of no economic importance. *Butomus* has promise of value in wild-life conservation practices, and its rhizomes are baked and eaten in northern Asia.

The Butomaceae are readily distinguished from the Alismaceae, to which they seem closely related, by the numerous ovules and parietal placentation. (See under Alismaceae for a redefinition of the family by Pichon and realignment of all genera except *Butomus* under that family.)

This family, like the Alismaceae, is now considered to be among the primitive monocots, and the gynoecial situation of many pistils and ovules scattered over the inner wall of the carpel is perhaps an example of the more primitive type to be found among all monocots and dicots. Hutchinson treated it as a representative of ancestral progenitors of the liliaceous stocks, while the Bessey system seemingly ignored its phylogenetic significances. For the more recent phyletic views by Pichon (1946), see under Alismaceae above.

*LITERATURE:*

BUCHENAU, F. Butomaceae. In Engler, Das Pflanzenreich, 16 (IV. 16) 1903.
CORE, E. L. *Butomus umbellatus* in America. Ohio Journ. Sci. 41: 79–85, 1941.
NASH, G. V. Butomaceae. In North Amer. Flora, 17: 63–64, 1909.
PICHON, M. [See entry under Alismaceae.]

## HYDROCHARITACEAE. FROG'S-BIT FAMILY

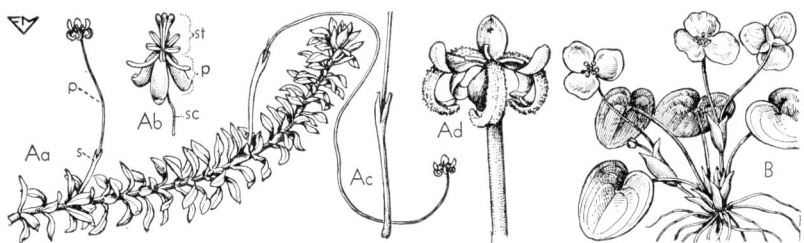

**Fig. 74.** HYDROCHARITACEAE. A, *Anacharis canadensis*: Aa, pistillate plant in flower showing spathe (s), perianth-tube (p), × 1; Ab, staminate flower showing stamens (st), perianth (p) and scape (sc), × 2; Ac, spathe of staminate flower; Ad, pistillate flower, × 4. B, *Hydrocharis Morsus-ranae*: flowering plant, × ¼. (From L. H. Bailey, *Manual of cultivated plants,* The Macmillan Company, 1949. Copyright 1924 and 1949 by Liberty H. Bailey.)

Partially or completely submerged aquatic herbs, rarely floating, of fresh- or salt-water habitats, roots terrestrial or floating; leaves basal and often crowded, or cauline (and then alternate, opposite or whorled), usually sessile, very variable in shape and size; inflorescence (or flower) subtended by a bifid spathaceous bract or by a pair of opposite bracts; flowers bisexual or more commonly unisexual (the plants then dioecious), actinomorphic, solitary or in umbels, when unisexual the pistillate flowers solitary and the staminate umbellate, the spathes sessile or long pedunculate (peduncles often spirally coiled); the perianth segments 6 (rarely 2), free, usually 2-seriate with 3 in each series, the outer often sepallike and green, valvate, the inner petaloid and imbricate or convolute; staminate flowers with stamens 3-many, the anthers 2-celled, dehiscing by parallel vertical slits, one or more rudimentary sterile ovaries (pistillodes) usually present; pistillate flowers unipistillate, the ovary inferior, unilocular with mostly 3-6 parietal placentae that are sometimes intruded into the loculus or the parietal zones ill defined and the ovules seemingly scattered over the surface, the style 1, usually divided into as many (mostly 3-6) branches as placentae, the ovules numerous on each placenta; staminodia often present; fruit berrylike, indehiscent, submerged; seeds many, without endosperm.

A family of 16 genera and 80-90 species, indigenous primarily to waters of the warmer regions of the world. Of the 5 genera represented in this country, 2 are of marine plants growing in restricted ranges of salt water only along the coast of Florida, and are pollinated under the surface of the water (*Thalassia,* with 1 species and *Halophila,* also with 1 species, floating, in coves and creeks); the remaining 3 genera are much more generally distributed, inhabit fresh water only, and are pollinated at or above the surface of the water (*Anacharis* [*Elodea, Philotria*], with 2 species, and *Vallisneria* and *Limnobium* each with a single indigenous species). For genera of other continents and their relationships, cf. Hutchinson, pp. 28-32.

The Hydrocharitaceae are closely related to the Butomaceae (both were placed in the order Butomales by Hutchinson), and differ from other related families by the same characters distinguishing the latter, and are recognizable from the Butomaceae by the inferior ovary, the spathe or paired floral bracts, and the unipistillate gynoecium. A contrary opinion was held by Miki (1937), who placed the family higher within the order and considered it the ancestor of Najadaceae, on the basis of an assumed inferior ovary in the latter. Studies by Uhl did not clearly support the view that the ovary of *Najas* was inferior, and it is difficult to believe that the Hydrocharitaceae with their markedly developed perianth and multicarpellate parietal placentation could be close to the Butomaceae-Alismaceae complex.

The family is interpreted to comprise 4 subfamilies. Of these, only the Vallisnerioideae are represented by plants native to this country. Several American authors have treated this taxon as the Vallisneriaceae (cf. Britton, Small, Rydberg, Abrams, and Pool). Hutchinson (following Dandy) treated the family as composed of the same elements that were placed in it by Engler, but subdivided them into 3 subfamilies, one of which was the Vallisnerioideae. The family was elevated by Small (1909) to the rank of order and subdivided into 2 families; Elodeaceae composed of *Halophila, Vallisneria,* and *Philotria* (a name given by Rafinesque in 1818 to the plant named *Anacharis* by Richard in 1811 and *Elodea* by Michaux in 1803, not Adanson in 1763), and Hydrocharitaceae composed of *Thalassia, Hydromystia* (indigenous from Puerto Rico southward), and *Limnobium*.

*LITERATURE:*

ERNST-SCHWARZENBACH, M. Zur Blütenbiologie einiger Hydrocharitaceen. Ber. Schw. Bot. Ges. 55: 33–69, 1945.
KAUSIK, S. B. Pollination and its influence on the behavior of the pistillate flower in *Vallisneria spiralis.* Amer. Journ. Bot. 26: 207–211, 1939.
MARIE-VICTORIN, (Frère) *L'Anacharis canadensis.* Histoire et solution d'un imbroglio taxonomique. Contr. Lab. Bot. Univ. Montreal, 18: 1–43, 1931.
———. Les Vallisnéries américaines. *Op. cit.* 46: 1–38, 1943.
RYDBERG, P. A. Elodeaceae. In North Amer. Flora, 17: 67–71, 1909.
———. Hydrocharitaceae. In North Amer. Flora, 17: 73–74, 1909.
SANTOS, J. K. Determination of sex in *Elodea.* Bot. Gaz. 77: 353–376, 1924.
SMALL, J. K. Hydrocharitales. In North Amer. Flora, 17: 65, 1909.
SVEDELIUS, N. On the different types of pollination in *Vallisneria spiralis.* Svensk. Bot. Tidskr. 26: 1–2, 1932.
WITMER, S. W. Morphology and cytology of *Vallisneria spiralis,* L. Amer. Midl. Nat. 18: 309–333, 1937.

### Order 16. GLUMIFLORAE [20]

The grasses and sedges; flowers very small and usually termed florets, arranged in spikelets, without a conventional perianth (sometimes represented by bristles or reduced to minute scales), in axils of dry chaffy bracts, stamens typically 3 or sometimes 6 or rarely more, the ovary superior and 2–3-carpellate; fruit a caryopsis or achene, and with endosperm present and abundant.

The plants placed in this order have been considered by most botanists to compose a natural assemblage, and one with which some phylogenists in the past included other families (i.e., Juncaceae, Centrolepidaceae, Thurniaceae, and Restionaceae). The taxonomic validity of placing both the grasses and the sedges into 1 order was rejected by Hutchinson (1934, 1948), who segregated them into separate orders (Graminales and Cyperales) and placed into a third and related order (Juncales) the other 4 families cited above. The results of morphological studies by Belk (1939) and by Blaser (1940, 1944) supported the segregation of the grasses and sedges into 2 separate orders, but failed to provide substantiating evidence to justify the transfer into this general association of the plants comprising the Juncales of Hutchinson.

It is generally agreed that the grasses and sedges are very advanced groups,

---

[20] Various names have been employed for this order, that of Glumiflorae being the oldest. Authors desirous of terminating all orders uniformly by the approved ending -ales have employed either Graminales or Poales.

whose seemingly simple inflorescence and floral structures represent drastic reductions from unknown ancestral types. It is also generally conceded that they have been evolved from primitive liliaceous ancestral stocks, stocks of unknown identities or contemporary relationships, although phylogenists have presented many hypotheses on the subject (Arber, Bessey, and Hutchinson considered these to be juncaceous in character). The 2 taxa (grasses and sedges) are not as closely related as was formerly believed to be the case, since it has been shown that they differ by the former having terminal flowers whose ovaries evolved from ancestral types having parietal placentation, and the latter by having axillary flowers whose ovaries evolved from types having free-central placentation; furthermore, the spikelet once thought to be a feature common to both is now not believed to be homologous and does not indicate a bond of phyletic relationship.

## GRAMINEAE. GRASS FAMILY [21]

Since the grass family is very large and one of gross morphological complexity, it has acquired a terminology peculiar to itself and deserves more detailed accounting than other families treated above. Alternative terms are indicated parenthetically.

Annual or perennial herbs, or rarely (among the bamboos) woody plants; roots fibrous, rhizomes present or absent; the stems (culms) erect, ascending, prostrate, or creeping, typically hollow or often solid (Andropogoneae, Maydeae), always closed at nodes, the latter often swollen, terete (rarely flattened or angled); leaves solitary at the nodes, 2-ranked (in a ½ phyllotaxy), usually parallel-veined, composed of 2 parts, the sheath enveloping the culm and overlapping or sometimes (as in *Bromus*) connate by its margins, and the blade, usually flat and generally linear to lanceolate, rarely (except in Bambuseae, *Zeugites, Pharus,* and others) with a constricted petiole (blades disarticulate from sheath in Bambuseae); along the inner margin of union of blade with sheath there is situated a variously shaped appendage, the *ligule,* typically membranous, occasionally hyaline or represented by a row of hairs (the ligule wanting in *Echinochloa* spp.); leaf epidermis within the family is basically of one of 2 types, (1) the festucoid type with minute unicellular hairs (cellules) and simple siliceous cells over the nerves, or (2) the panicoid type with minute 2-celled hairs and complex siliceous cells.

The basic inflorescence unit is a *spikelet* (described below); spikelets pedicelled or sessile, almost always aggregated terminally on primary culms or branches as one of 3 gross inflorescence types (spicate, racemose, paniculate); each spikelet composed of 1 or more florets and their subtending chaffy bracts. A spikelet usually contains few to several sessile flowers and is composed of an axis termed the *rachilla* (continuous or jointed) of several nodes, subtended by an exposed basal stalk, the *pedicel* (morphologically a peduncle); at the base of the spikelet are 2 bracts, the lower termed the *first glume,* and the next the *second glume* (empty glumes); the shape, texture, and venation of glumes provide good taxonomic characters, the first glume is frequently smaller than the second and sometimes so reduced in size as to be vestigial (as an obscure rim) or completely suppressed and wanting, while in a few genera both glumes have been lost by reduction. Above the glumes there is a very short section of the rachilla followed by the flower and its bracts, collectively, the floret and the number of florets varying from 1 to 50; each floret typically composed of usually 2 bracts, (1) the *lemma* or flowering glume, resembling the glumes, often greenish, keeled or

---

[21] According to the Rules the name of this family is Gramineae. Many authors substitute for it the name Poaceae, taking the latter name from *Poa,* the type genus of the family. Gramineae is a *nomen conservandum* and Poaceae is an allowable alternative name.

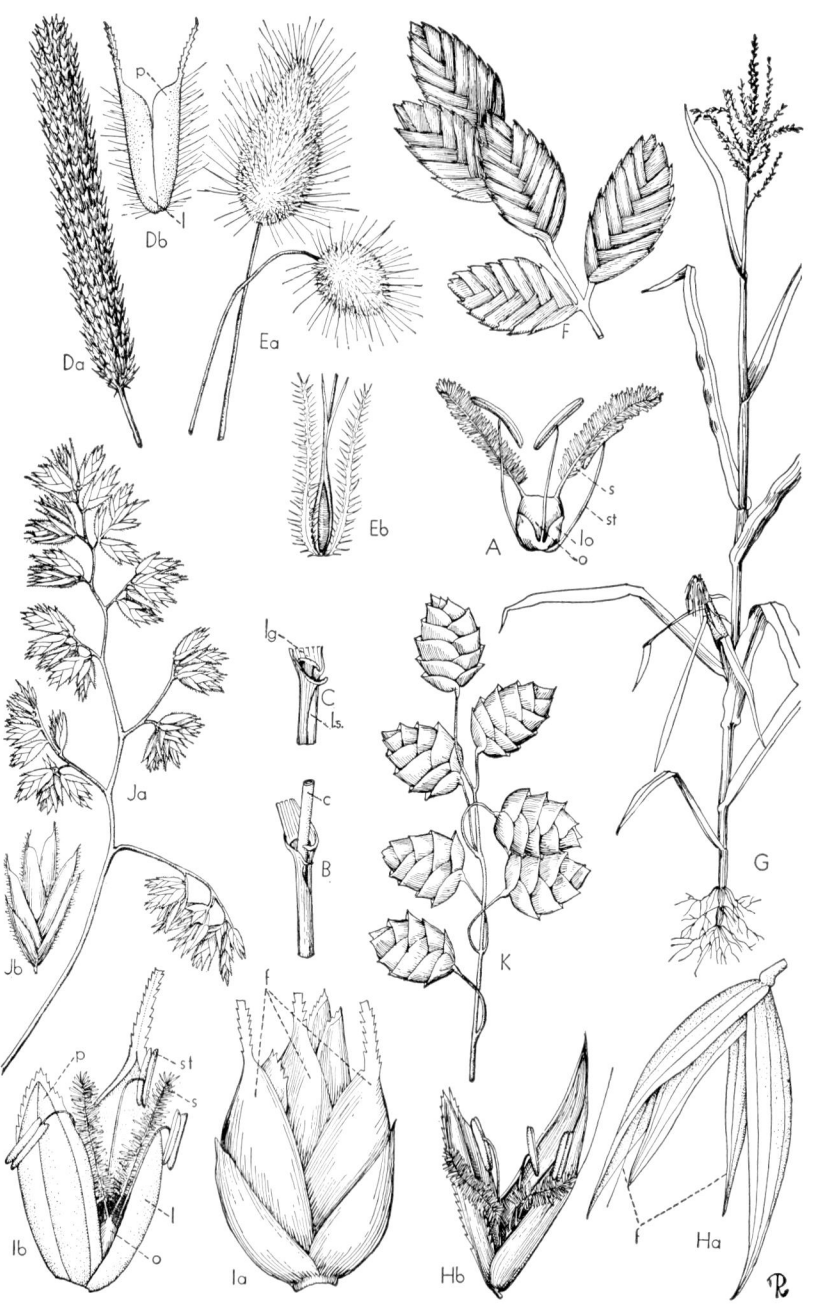

rounded, nerved, awned or awnless, fertile (enveloping a perfect or pistillate flower) or sterile (the flower staminate or none); and (2) the *palea* (or palet), the bract between the flower and the rachilla, usually 2-nerved or 2-keeled and its margins enclosing the flower, and in some grasses wholly enclosed by the lemma, and may be reduced to a minute veinless scale or be obsolete (morphologically the palea is a bracteole or prophyll). In the axil of the palea is the *flower,* unisexual or more frequently bisexual; normally chasmogamous but frequently cleistogamous in numerous genera; when bisexual, composed of a highly modified and greatly reduced perianth, represented by 3 *lodicules* in the Bambuseae and 2 in all other tribes, situated at the base of the flower outside the stamens (at anthesis, the increased turgidity of the lodicules expands the lemma and palea and is followed by exsertion of stamens and/or stigmas); the stamens, usually in a whorl of 3 (or 6 in 2 whorls in most bamboos and many Oryzeae), with occasionally 1 suppressed (in *Hierochloë*) or 2 suppressed and the functional stamen 1 (in *Cinna* and some spp. of *Festuca* and *Uniola*), rarely numerous (*Pariana,* of South America), hypogynous, anthers 2-celled, basifixed but so deeply sagittate as to appear and function as if versatile; the pistil with ovary superior, unilocular, 3-carpelled, the ovule 1 (anatropous, often adnate to adaxial side), styles usually 2 (3 in Bambuseae or the 2 are fused and appear as 1 in *Zea*), the stigmas papillate or more frequently plumose extensions of the styles; fruit mostly a caryopsis (the seed and adherent pericarp), rarely a nut or berry (in some Bambuseae), or utricle (*Sporobolus, Eleusine,* and others); seed with copious endosperm (the type of starch grain present has provided useful generic characters in studies of phylogeny).

Spikelets of grasses vary widely in different genera, particularly as to number of fertile florets in each and disposition of sexes within them. In several genera (as in *Zea, Coix,* the Olyreae, Phareae, and others) the spikelets are unisexual and the plants monoecious or dioecious (as in spp. of *Poa, Distichlis*).

The grass family, from point of view of number of individuals, is the largest and most widely distributed family of vascular plants. In the competition for dominance it becomes the vegetative climax in vast areas of low annual rainfall; notable among these are the plains and prairies of this continent, the savannas and pampas of South America, the steppes and plains of Eurasia, and the veldt of South Africa. Taxonomically, it is one of our largest families of seed plants. The number of genera in the world is a matter of diverse opinion, but it may be placed at between 450 and 525, varying in size from many that are monotypic to genera of 200 or more species each (as *Panicum,* 800± spp.; *Poa,* 200± spp.; *Paspalum,* 00± spp.; *Andropogon,* 200± spp.). The genera of grasses are generally classified by agrostologists (taxonomists who specialize in grasses) according to the schema of Haeckel who grouped them into 2 subfamilies (Festucoideae and Poacoideae) and 13 tribes.

All tribes are represented domestically by one or more of the 168 genera and 1220 species occurring in this country (of which 44 genera and 156 species are introduced). For taxonomic and phytogeographic considerations of these and other

---

**Fig. 75.** GRAMINEAE. A, typical grass floret: essential parts, × 4. B, culm showing upper portion of sheath and base of blade, × ½. C, upper portion of leaf-sheath with ligule, × ½. D, *Phleum pratense*: Da, inflorescence, × 1; Db, spikelet, × 8. E, *Lagurus ovatus*: Ea, inflorescence, × 1; Eb, spikelet, × 8. F, *Uniola latifolia*: portion of inflorescence, × 1. G, *Zea Mays*: × ¹⁄₁₀. H, *Avena sativa*: Ha, spikelet, × 4; Hb, floret, × 8. I, *Triticum aestivum*: Ia, spikelet, × 4; Ib, floret, × 8. J, *Dactylis glomerata*: Ja, inflorescence, × 1; Jb, spikelet, × 4. K, *Briza maxima*: inflorescence, × 1. (c culm, f floret, 1 lemma, lg ligule, lo lodicule, l.s. leaf-sheath, o ovary, p palea, s stigma, st stamen.) (From L. H. Bailey, *Manual of cultivated plants,* The Macmillan Company, 1949. Copyright 1924 and 1949 by Liberty H. Bailey.)

grasses the student is referred to one or more of the references to literature cited below.

Morphologically and taxonomically the grasses have been the subject of various and divergent opinions. The grass flowers, as known today, are unlike those of any other group, and only among the Cyperaceae are they elsewhere aggregated into spikeletlike structures. The spikelet is considered the basic floral unit in the family, and the gross inflorescences of grasses are composed of aggregates of spikelets rather than of clusters of separate flowers as in other orders. Within the spikelet, the grass flower is a highly developed structure whose morphology is complicated by the extensive suppression of parts and reduction of the conventional floral elements. The grass perianth has been interpreted variously, and it was largely due to the studies of Rowlee (1898) that the lodicules were accepted as vestigial perianth remains. More controvertible has been the question of carpel number in the pistil of members of this family, for by Haeckel (1883), Bews (1929), Rendle (1930), Diels (1936), and others the pistil was considered to be of a single carpel terminated by a 2–3-branched stigma, whereas Lotsy (1911), Weatherwax (1929), Arber (1934), Randolph (1936), and others have held that a tricarpellary ovary is present, a view supported by the detailed floral anatomy studies of Belk (1939). The latter found the gynoecium to be fundamentally a tricarpellate organ with 3 carpels joined edge to edge, and the single ovule of the ovary always to be attached to the posterior wall of the single locule. This shows that the present-day grass flower ovary with its 1 locule and 1 ovule has evolved from ancestral 3-carpelled stocks that had parietal placentation.

Within the family specialization has advanced along many lines. It is probable that the Bambuseae, the most primitive of living grasses, arose from forms more primitive than those that gave rise to the other tribes. The Festuceae and Hordeae are undoubtedly more primitive than other tribes with the Andropogoneae, Agrostideae, Paniceae, and Maydeae the most advanced or highly specialized. Hubbard (1948) considered the following characters to represent primitive conditions within the family.

Spikelets many-flowered
Flowers all fertile
Uppermost floret lateral on rachilla with axis produced beyond it
Glumes persistent at maturity
Florets exserted from glumes
Lemmas many-nerved
Lemmas herbaceous and leaflike
Lemmas awnless
Lodicules 6 or 3
Stamens 6, in 2 whorls
Stigmas 3

In addition to the above, he considered that most of the grasses here treated as comprising the Panicoideae originated within the tropical parts of the earth and migrated into temperate regions, whereas those of the Festucoideae (exclusive of the Bambuseae and Chlorideae) are more widespread in temperate regions and occur in the tropics only at high altitudes. He also considered the aquatic habitat of tribes such as the Oryzeae as an indicator of primitiveness and associated tribes occurring in more xerophytic habitats with more advanced phyletic conditions.

Economically the grasses are probably of greater importance than any other family of plants. From a world viewpoint their importance can be indicated by the following uses:

fodder for domestic animals (the 6 more important domestic genera): *Agrostis, Phleum, Dactylis, Sorghum, Setaria, Zea*

food for man (8 genera domestically important) rice (*Oryza*), corn (*Zea*), wheat, (*Triticum*), rye (*Secale*), oats (*Avena*), barley (*Hordeum*)
sugar and molasses (*Saccharum, Sorghum*)
cornstarch and by-products (*Zea*)
beverages
    sake (rice), whiskey (rye, barley, corn), rum (molasses, from sugar cane)
shelter
    thatch, bamboo framing, matting
industrial uses
    corn products (insulation materials)
    newsprint and other papers
    ethyl alcohol and derivatives
turf
    forage, decorative, sports areas (*Agrostis, Poa, Festuca, Stenotaphrum, Cynodon, Zoisia*)
ornamentals (over 80 genera cultivated domestically)

## LITERATURE:

ARBER, A. Studies in the Gramineae, I-X. Ann. Bot. in vols. 40–45, 1926–31.
———. The Gramineae: a study of cereal, bamboo, and grass. Cambridge, England, 1934
BELK, E. Studies in the anatomy and morphology of the spikelet and flower of the Gramineae. Thesis (Ph.D.) Cornell Univ., 1939.
BESSEY, E. A. The phylogeny of the grasses. Ann. Rep. Michigan Acad. Sci. 19: 239–245, 1917.
BEWS, J. W. The world's grasses. Their differentiation, distribution, economics, and ecology. London, 1929.
BROWN, W. V. A cytological study in the Gramineae. Amer. Journ. Bot. 35: 382–395, 1948.
CARRIER, L. The identification of grasses by their vegetative characters. U.S. Dept. Agr. Bull. 461, 1917.
CHASE, A. First book of grasses. New York, 1922. [A copiously illustrated book explaining the structure of grasses and their basic classification.]
———. Poaceae. North Amer. Flora, 17: 568–579, 1939.
CUTLER, H. C. and ANDERSON, E. A preliminary survey of the genus *Tripsacum*. Ann. Mo. Bot. Gard. 28: 249–269, 1941.
HAECKEL, E. Echte Gräser. Engler and Prantl, Die natürlichen Pflanzenfamilien, II, 2. 1887. [English transl. by Scribner & Southworth, 1890].
HITCHCOCK, A. S. A textbook of grasses. New York, 1914. [A very excellent book introducing the subject of agrostology; all tribes and genera represented by indigenous and cultivated plants fully described, and accompanied by keys, discussion, and detailed bibliography.]
———. Poaceae. North Amer. Flora, 17: 198–288, 1915; 299–354, 1931; 355–482, 1935; 483–542, 1937; 543–568, 1939.
———. Manual of grasses of the United States. U.S. Dept. Agr. Misc. Publ. 200. Washington, D.C., 1935. ed. 2 [Rev. by Agnes Chase], 1951.
———. The genera of grasses of the United States. Revision by Agnes Chase. U.S. Dept. Agr. Bull. 772, 1936.
HUBBARD, C. E. Gramineae. In Hutchinson, British flowering plants: evolution and classification. pp. 284–348. London, 1948.
NASH, G. V. Poaceae. North Amer. Flora, 17: 17–98, 1909; 99–196, 1912; 197–198, 1915.
RANDOLPH, L. F. Developmental morphology of the caryopsis in maize. Journ. Agr. Res. 53: 881–916, 1936.
ROWLEE, W. W. The morphological significance of the lodicules of grasses. Bot. Gaz. 15: 199–203, 1898.
SWALLEN, J. R. Poaceae. North Amer. Flora, 17: 579–638, 1939.
WEATHERWAX, P. The morphology of the spikelets of six genera of Oryzeae. Amer. Journ.

## CYPERACEAE. SEDGE FAMILY

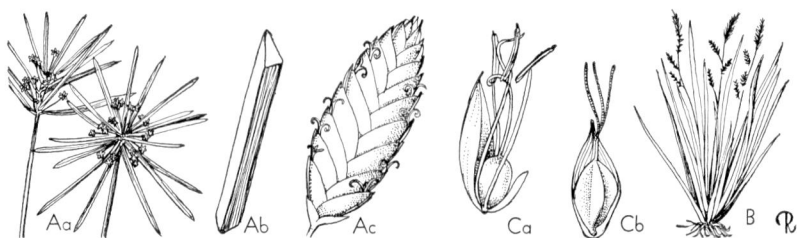

**Fig. 76.** CYPERACEAE. A, *Cyperus alternifolius*: Aa, habit flowering axis, side and top view, × ⅕; Ab, stem section, × ½; Ac, inflorescence, × 4. B, *Carex Morrowii*: habit, × ⅛. C, *Carex plantaginea*: Ca, flower, × 10; Cb, fruit, × 10. (From L. H. Bailey, *Manual of cultivated plants,* The Macmillan Company, 1949. Copyright 1924 and 1949 by Liberty H. Bailey.)

Perennial or infrequently annual grasslike or rushlike herbs, often of damp boggy, marshy, or riparian habitats; roots fibrous from a very short or elongated and creeping rhizome, the latter rarely tuberlike; stems (culms) mostly solid, often triquetrous, generally unbranched below the inflorescence, frequently leafless; leaves in basal tufts or basally cauline and then 3-ranked, in a ⅓ phyllotaxy (or infrequently a ⅜ phyllotaxy), of 2 parts—a grasslike blade and a closed (or rarely open) sheath, ligule usually absent; flowers very minute, subtended by chaffy bracts, bisexual or unisexual (when the latter the plants monoecious or very rarely dioecious), arranged in spikelets in spicate, racemose, paniculate, or umbellate inflorescence types, the latter frequently subtended by foliaceous bracts; flowers each in the axils of glumelike, closely imbricated bracts (scales, glumes), the perianth represented by hypogynous bristles or scales (rarely subpetaloid) of varying numbers in 2 whorls or completely suppressed and absent; stamens (in bisexual or staminate flowers) hypogynous, 1 to 6 and usually 3, the anthers basifixed, 2-celled, dehiscing by vertical slits, the filaments distinct; pistil (in bisexual or pistillate flowers) 1, the ovary superior, sometimes (in *Carex*) subtended and enveloped by a single posterior prophyll (perigynium, utriculus, or sac), unilocular, the ovule solitary, basal, erect, anatropous, the style 2–3-toothed or with 2–3 branches, when persistent and indurated, sometimes forming a beak on the achene; fruit nutlike (achene or nutlet), indehiscent, when from a 2-branched style pistil often more or less 2-sided or lenticular, and when from a 3-branched style often 3-sided, the achene sometimes enclosed in a sac (perigynium or utricle) or by a partially enveloping and connate glume; seed with albumen.

A very large family distributed throughout the world, but particularly abundant in the subarctic and temperate regions of northern and southern hemispheres. It was divided by Engler and Diels into 3 subfamilies (earlier Englerian classifications recognized only 2 subfamilies), 7 tribes, and about 72 genera (17 monotypic). Hutchinson recognized 7 tribes (some of which were reclassified from the Englerian arrangement) represented by 83 genera. The number of species for the world is about 3200, with genera having in excess of 100 spp. each including *Carex* (ca. 1100 spp.), *Cyperus* (ca. 700 spp.), *Scirpus* (ca. 200 spp.), *Fimbristylis* (ca. 125 spp.), *Rhynchospora* (ca. 200 spp.), *Eleocharis* (ca. 150 spp.), and *Scleria* (ca. 100 spp.). Eighteen genera occur in the northeastern section of North America, about 22 genera in the southeast, and 14 in the plains states, with Abrams crediting 13 genera and 236 species to the Pacific states. Of the indigenous genera, *Carex, Scirpus, Stenophyllus, Hemicarpha, Rhynchospora,* and *Eriophorum* (rare in the south) occur in almost every region of the country.

Members of the family generally are readily separable from the Gramineae by a combination of characters, viz., solid and often triquetrous stems, leaves in a ⅓ phyllotaxy with the leaf sheath generally closed, each flower usually subtended by a single bract (glume), and the absence of ligule and lodicules.

Morphological studies by Snell (1936) and Blaser (1940) have demonstrated that (1) primitive members of the family possessed a regular 6-parted perianth disposed in 2 whorls, which in some genera was reduced to bristles (the latter representing 3 or more veins of each sepal when in the outer whorl, or the single vein of each petal when in the inner whorl) or completely suppressed with no remnants, (2) the androecium was basically composed of 6 stamens in 2 series (a condition not now typical but occasionally observed as an anomaly in *Eleocharis*) with those of the inner series usually suppressed and 1 or 2 of the outer whorl suppressed in some genera or species, (3) the ovary may be either 3- or 2-carpelled, with always a single central basal ovule believed to have been derived from free-central placentation of ancestral stocks, and in some cases (*Scirpus*) the number of styles may not be a reliable index to carpel number, and (4) the character of a stylopodium in some genera (*Eleocharis, Scirpus*) as used in some taxonomic treatments is rejected as inconstant and of no taxonomic or phylogenetic significance; similarly, it was found that no reliability could be placed in the character of lenticular versus trigonal achenes, since intergradations between them were observed, and also that the degree of difference varied with external growth factors.

Some authors have treated the 3 subfamilies of Cyperaceae as independent families: Cyperaceae (restricted to the subfamily Scirpoideae), Rhynchosporaceae, and Caricaceae. Holttum (1948), largely on the basis of spikelet morphology studies, concluded that of the 3, the subfamily Scirpoideae was the most primitive, that possibly several lines led to the Rhynchosporoideae, and that from the latter (also by several lines) arose within the Caricoideae the genera of Cryptangieae, Sclerieae, and Cariceae with *Carex* the most highly developed genus in the family.

Phylogenetically the Cyperaceae have been allied closely to the Gramineae. Hutchinson separated the families into 2 separate orders (Cyperales and Graminales) of 1 division, considered the latter to be the more highly advanced, and treated both as having been derived from liliaceous ancestors via the Juncaceae complex. The studies by Snell and Blaser indicated that the Gramineae are not close allies of the Cyperaceae. It was pointed out by Blaser that (1) the superficial grasslike habit is of no phylogenetic significance and has appeared also in other unrelated families, (2) the spikelets of the Cyperaceae are not at all homologous with those of the grasses, since in the former they vary widely in organization within the family, whereas in the latter they are reasonably constant, (3) the basic placental condition of the Cyperaceae has been derived from an ancestral free-central type, whereas in the Gramineae it has been derived from a parietal type, and (4) the florets of the Gramineae (excluding bracts) are borne terminally, whereas in the Cyperaceae they are always axillary in position. From these considerations it is clear that the 2 families belong to 2 separate orders as treated by Hutchinson, and that they are not so closely related as he indicated them to be, but that there remains the "possible derivation of the Cyperaceae from small-flowered, few-seeded, hypogynous, Liliales with axile placentation . . . the transition having occurred through loss of septa and placental axis and may have occurred through some line not now recognizable" (Blaser, 1940).

Economically the family is of little importance. Pith from culms of *Cyperus Papyrus* was used by Egyptians as early as 2400 B.C. in papermaking and, aside from use as an ornamental aquatic, its importance is now only academic. *Cyperus alternifolius* (umbrella plant) is commonly grown in warm regions as an aquatic ornamental, and *C. esculentus* (chufa, groundnut), usually a weed, may be grown for its edible tuberous rhizomes.

*LITERATURE:*

BEETLE, A. A. Cyperaceae, Tribe 2, Scirpeae. In North Amer. Flora, 18: 479–504, 1947. [*Scirpus.*]

BLASER, H. W. The morphology of the flowers and inflorescences of the Cyperaceae. Thesis (Ph.D.) Cornell Univ., 1940.

———. Studies in the morphology of the Cyperaceae. I. Morphology of the flowers. A. Scirpoid genera. Amer. Journ. Bot. 28: 542–551, 1941. B. Rhynchosporoid genera. *op. cit.*, 832–838, 1941. II. The prophyll. *op. cit.*, 31: 53–64, 1944.

CLARKE, C. B. New genera and species of Cyperaceae. Kew Bull. Add. Series, 8: 1–196, 1908.

GALE, S. *Rhynchospora*, section *Eurhynchospora*, in Canada, the United States, and the West Indies. Rhodora. 46: 89–134, 159–197, 207–249, 255–278, 1944.

HOLTTUM, R. E. The spikelet in Cyperaceae. Bot. Rev. 14: 525–541, 1948.

MACKENZIE, K. K. Cyperaceae, Tribe 1, Cariceae. In North Amer. Flora, 18: 1–478, 1931–1935.

SNELL, R. S. Anatomy of the spikelets and flowers of *Carex, Kobresia,* and *Uncinis.* Bull. Torrey Bot. Club, 63: 277–295, 1936.

SVENSON, H. K. Monographic studies in the genus *Eleocharis.* Rhodora, 31: 121–135, 152–163, 167–191, 199–219, 224–242, 1929; II, *op. cit.*, 34: 193–203, 215–227, 1932; III, *op. cit.*, 377–389, 1934; IV, *op. cit.*, 39: 210–231, 236–273, 1937.

## Order 17. PRINCIPES [22]

The palms; woody plants; flowers 3-merous, small, mostly on compound spadices comprised of simple or branched rachillae, the inflorescence mostly enveloped by 1 or more large spathaceous bracts in bud; ovary superior, tricarpellate, ovules 3 or more with only 1 usually developing, superior; the fruit a berry or a drupe; seeds with endosperm. The Palmae are the only family of the order.

## PALMAE. PALM FAMILY [23]

Woody shrubs, vines, or trees; stems very short or seemingly none, or large-boled (mostly unbranched trees), or slender, long, and flexuous; leaves (fronds) in a terminal cluster (head or coma) in arborescent species or scattered alternately in climbing and some shrubby species, petioled, the base persisting or not and often spreading about or sheathing the trunk, in some genera much expanded and basally enveloping terminal bud and then forming a "crownshaft," the petiole (haft) smooth or margined, with teeth or prickles or the lower leaflets regressively smaller and modified into long spines; the blade simple and flabellate (fan palms) or pinnately compound (feather palms) with generally many leaflets (pinnae), or

---

[22] Botanists holding to the view that names of all orders terminate uniformly in -ales adopt the name Palmales or Arecales for this group.

[23] Palmae is the conserved name for the family. Some botanists hold that consistency is achieved only by naming each family after its type genus and adopt for this family the alternative name of Arecaceae.

---

**Fig. 77.** PALMAE. A, *Cocos nucifera*: Aa, tree in fruit, much reduced; Ab, flowering inflorescence and spathes, × 1/10; Ac, section of rachilla with pistillate and staminate flower, × 1/2; Ad, pistillate flower, vertical section, × 1; Ae, staminate flower, × 1; Af, fruit, × 1/8; Ag, same, longitudinal section, × 1/8; Ah, nut, × 1/6. B, *Washingtonia filifera*: Ba, tree, much reduced; Bb, flower, × 2; Bc, same, perianth expanded, × 2; Bd, fruit, × 1. C, *Thrinax microcarpa*: leaf, much reduced. D, *Chrysalidocarpus lutescens*: trunk apex showing flowering inflorescence, crownshaft and leaf-petioles, much reduced. (After L. H. Bailey, *Manual of cultivated plants*, The Macmillan Company, 1949.)

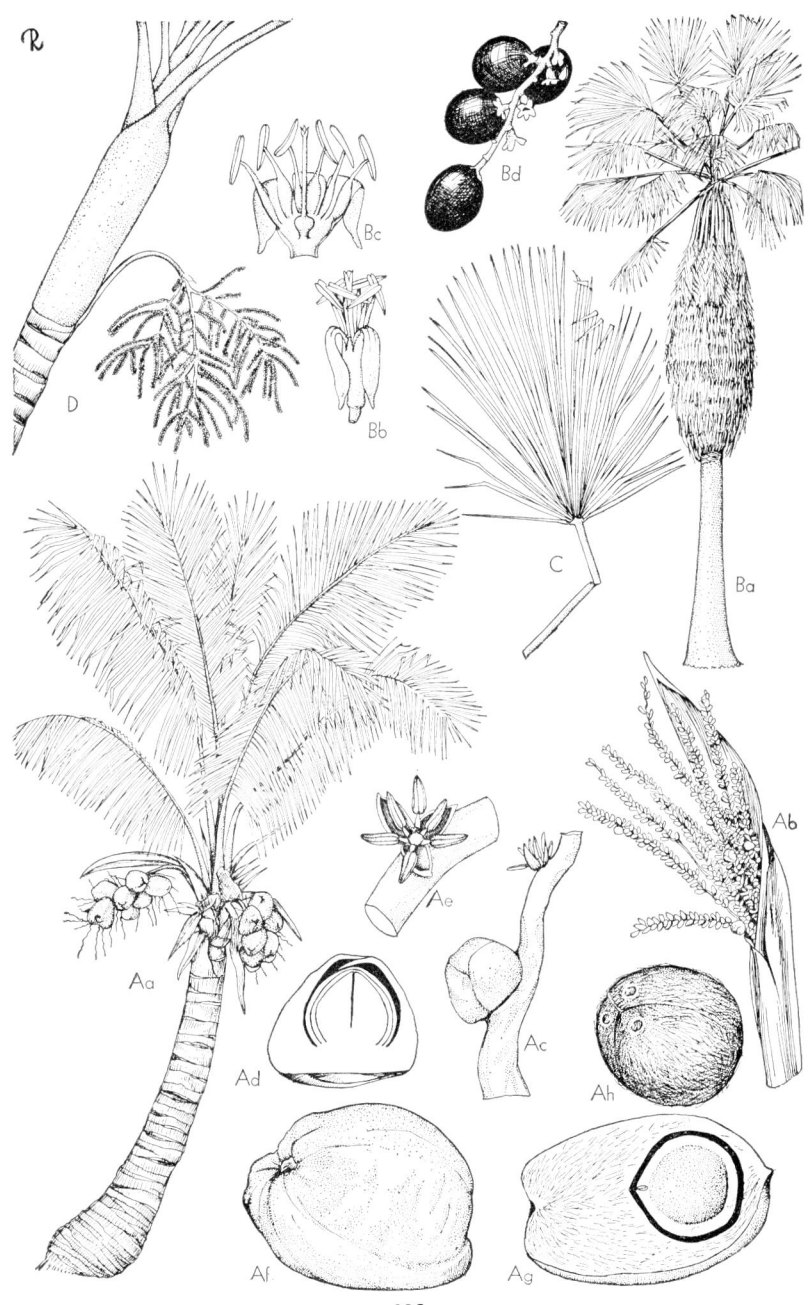

simple and pinnately veined; in leaves of many fan palms (*Thrinax*) a woody ligule (hastula) occurring, situated ventrally at junction of blade and petiole, in others (*Sabal*) the midrib (costa) extending well into the curved blade; inflorescence usually paniculate, mostly below (infrafoliar) or amongst the leaves (intrafoliar), or rarely above the crown of foliage (suprafoliar), sometimes termed a spadix (usually compound, branches termed rachillae, simple in Geonomeae spp. and others) whose base is a peduncle; the inflorescence subtended by 1 or more usually large bracts (spathes, when woody a *cymba*), and in some genera the cymba constricted basally into a handlelike claw (manubrium); flowers small, actinomorphic, sessile or very short-pedicelled, bisexual or more commonly unisexual (when so, the plants usually monoecious or in a few genera dioecious), perianth of 6 segments (vestigial in *Phytelephas*), in 2 series, free or connate, sepals generally imbricate or open in bud, petals mostly valvate in staminate flowers and imbricate in the pistillate, stamens mostly 6 (occasionally more in a few genera) in 2 whorls of 3 each, the anthers 2-celled, dehiscing by vertical slits, filaments distinct, usually short; pistil 1, or rarely 3 and basally connate, ovary superior (vestigial or none in staminate flowers), 1–3-celled (rarely more), the ovule usually solitary in each locule, basal or axile, erect or pendulous; fruit mostly a berry or drupe, the exocarp fleshy, fibrous or leathery, sometimes (in *Raffia*) of imbricated bracts; seed with endosperm, sometimes ruminate, the embryo small and its position within the seed often affording characters of generic value.

A large family of tropical and subtropical woody plants not yet well known taxonomically and, in the aggregate, in need of much systematic and morphological study. No recent world monographs are available, and existing treatments (Bentham and Hooker, Drude, Wendland, Beccari) are not in harmony with much independent research published since on individual genera and/or floristic groups (Burrett, Furtado, Bailey, *et al.*).

The family was divided by Bentham and Hooker into 8 tribes, and Hutchinson counted 210 genera. The number of valid species may be 4000 or more, and since the family is poorly represented in herbaria undoubtedly many genera and an appreciable number of species are yet unknown to science. Nine genera are native in the United States, represented in all by about 20 species. These include *Washingtonia* in southern California and Arizona, *Sabal* and *Serenoa* from coastal North Carolina into Florida and west along the coastal plain of the Gulf states into southeastern Texas, and native in Florida the genera *Thrinax, Cocothrinax, Paurotis, Pseudophoenix, Rhapidophyllum,* and *Roystonea.* The monotypic genus *Cocos* is extensively naturalized in Florida, but is not an indigen. Many genera, perhaps 95, are cultivated for ornament in southern Florida and California.

Domestically the family is of little direct importance, but from a world viewpoint it ranks next to the grasses in economic significance. Pantropically, its members provide food, shelter, clothing, and lesser necessities to life for entire populations of primitive peoples. To civilizations of temperate regions the palms are the sources of such products as copra, coconut and other oil, dates, coconut meat, rattan cane, raffia, ivory nuts, and carnauba wax.

*LITERATURE:*

BAILEY, L. H. [For many papers devoted to the systematics of American and cultivated palms, see Gentes Herbarum, vols, 2, 3, 4, 6, 7, and 8 (1930–1949).]

BOSCH, E. Blütenmorphologie und zytologische Untersuchungen an Palmen. Ber. Schw. Bot. Ges. 57: 37–100, 1947.

BURRET, M. [For titles of over 40 papers on American palms see Dahlgren cited below, pp. 289–290.]

DAHLGREN, B. E. Index of American palms. Field Mus. Nat. Hist. Bot. Studies, vol. 14 1936.

MARTIUS, C. F. P. VON. Historia naturalis palmarum. 3 vols. Leipzig, 1823–50.

## Order 18. SYNANTHAE [24]

An order of 1 family of palmlike monoecious plants characterized by the minute unisexual apetalous flowers congested on a more or less fleshy spadix that is enveloped by deciduous spathelike bracts.

### CYCLANTHACEAE. PANAMA-HAT-PALM FAMILY

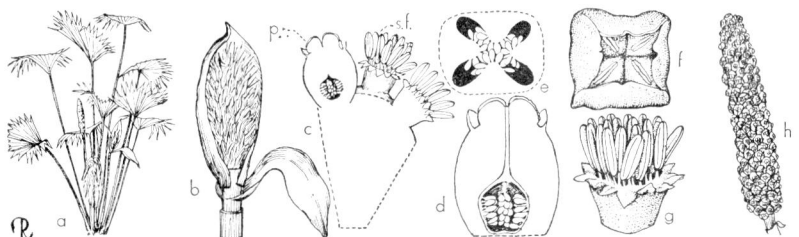

**Fig. 78.** CYCLANTHACEAE. *Carludovica palmata*: a, habit, much reduced; b, inflorescence, × ⅙; c, segment of flowering spadix, × 2 (p.f. pistillate flower, s.f. staminate flower); d, pistillate flower, vertical section, × 5; e, ovary, cross-section, × 10; f, pistillate flower, top view, × 4; g, staminate flower, × 5; h, fruiting spadix, × ⅙. (From L. H. Bailey, *Manual of cultivated plants*, The Macmillan Company, 1949. Copyright 1924 and 1949 by Liberty H. Bailey.)

Acaulescent palmlike monoecious perennial herbs, or suffrutescent climbers, mostly terrestrial or sometimes subepiphytic; leaves spiral or distichous, palmate, entire, flabellate, or 2-lobed or 2-divided, the lobes often much subdivided, the petiole basally sheathing; inflorescence a simple axillary spadix, peduncled, subtended and initially enveloped by 2 or more conspicuous caducous foliaceous or petalaceous spathes; flowers unisexual, densely crowded, the sexes in superposed whorls or in spiraled units of a pistillate flower surrounded by staminate flowers, or sexes alternating in spirals; staminate flowers with perianth none or represented by an obliquely toothed cuplike structure, the stamens numerous, the anthers 2-celled, dehiscing by longitudinal slits, the filaments basally connate and adherent or adnate to perianth when the latter is present; pistillate flowers with perianth none or with 4 connate perigynal lobes or free perianth segments (those of adjoining flowers sometimes connate, enlarged, and indurate in fruit), the staminodes 4, short or flexuously filiform, with or without rudimentary anthers, the pistil 1, the ovary 1-loculed, 2–4-carpelled, free or embedded in spadix, the ovules numerous, attached to walls on 2–4 parietal placentae or to apex of locule, anatropous, the placentation basically parietal, the style none or stoutly pyramidal, the stigmas 1–4, divergent; fruit a fleshy syncarp, of separate or connate berries frequently falling from spadix as layers of pulp; seeds numerous, with succulent testa, the endosperm present and copious, the embryo minute.

A family of the West Indies and tropical America not occurring in this country except as cultivated subjects. Composed of perhaps 6 genera and 45 or more species, and by some authors the genera treated as sections of *Carludovica* (cf. Gleason, 1929). Only *Carludovica* is domestically cultivated (infrequently) as a tropical novelty.

Phylogenetically the family (and order) is one of considerable specialization and advancement as indicated by the much reduced unisexual flowers and the

---

[24] Synanthae, the name of the order as employed by Engler and Diels (1936). For the same order some authorities use the name Cyclanthales.

inflorescence. Morphological studies by Harling (1946) supported the opinions of Engler and Diels (1936) that the family represented an order apart by itself, but stood between the palms and the aroids in advancement. Hutchinson also considered it to be an advanced derivative of the palms.

Economically important only as a source of material for Panama hats (obtained primarily from leaves of *Carludovica*), and to a lesser degree for ornamental and landscape value.

LITERATURE:

GLEASON, H. A. New or noteworthy monocotyledons from British Guiana. Bull. Torrey Bot. Club, 56: 1–8, 1929. [Comprises a revision of the genus *Carludovica*, resolving generic segregates into 6 sections on the broadly redefined genus.]

HARLING, G. Studien über den Blütenbau und die Embryologie der Familie Cyclanthaceae. Svensk. Bot. Tidskr. 40: 257–272, 1946.

## Order 19. SPATHIFLORAE

Mostly herbs or climbers; flowers minute, much reduced, on a thickened spadix that is subtended by a single large mostly herbaceous spathe, the perianth none or much reduced (not petaloid), the ovary usually superior, the fruit a berry. The families Araceae and Lemnaceae comprise the order.

## ARACEAE. ARUM FAMILY

Rhizomatous or tuberous, mostly herbaceous terrestrial (rarely aquatic) plants with sap milky, watery, or sharply pungent, calcium oxalate crystals generally present in the tissues, the plants sometimes epiphytic with aerial roots but commonly terrestrial; leaves simple or compound, solitary or few, basal or cauline, and alternate, petioled with membranous sheathing base, the blades ensiform and parallel-veined or variously lamellate with pinnate- or palmate-netted venation; inflorescence a simple spadix subtended by a sometimes caducous herbaceous spathe that is generally large and often brightly colored; flowers bisexual or unisexual (the plants then monoecious or rarely dioecious), small, often with fetid odor, the perianth rare in unisexual flowers and present in bisexual ones, of 4–6 free or connate segments; stamens 2, 4, or 8 (rarely solitary), hypogynous, the anthers mostly 2-celled, dehiscing by pores or slits, free or connate; pistil 1, the ovary superior or inferior (then embedded in spadix), 1-many-loculed, the ovules 1-many, the placentation basal, parietal, axile, or apical, the style and stigma present and variable or absent; fruit a berry; seeds mostly with endosperm.

A family of about 105 genera and 1400–1500 species, mostly tropical or subtropical in distribution but many of temperate regions. The outstanding authority on the systematics within the family was Engler, who divided it into 8 subfamilies (Pothoideae, Monsteroideae, Calloideae, Lasioideae, Philodendroideae, Calocasioideae, Aroideae, and Pistioideae) and 28 tribes. Hutchinson (1934) recognized 18 tribes, providing keys to these and to about 113 genera. Eight genera are indigenous to the United States: *Arisaema, Peltandra, Calla, Symplocarpus, Orontium,* and *Acorus* in much of the country east of the Rockies, with *Pistia* restricted to Florida and westward to Texas, and *Lysichiton* extending from Alaska southward to the Santa Cruz Mts. of California. Five of these genera are monotypic (*Peltandra, Orontium, Symplocarpus, Lysichiton,* and *Pistia*). From the world viewpoint, the largest genera include *Anthurium* (ca. 500 spp.), *Philodendron* (ca. 200 spp.), *Homalomena* (80 spp.), and *Arisaema* (ca. 60 spp.).

American representatives of the family are readily identified by the herbaceous character of the plant and by the inflorescence a spadix enveloped or subtended

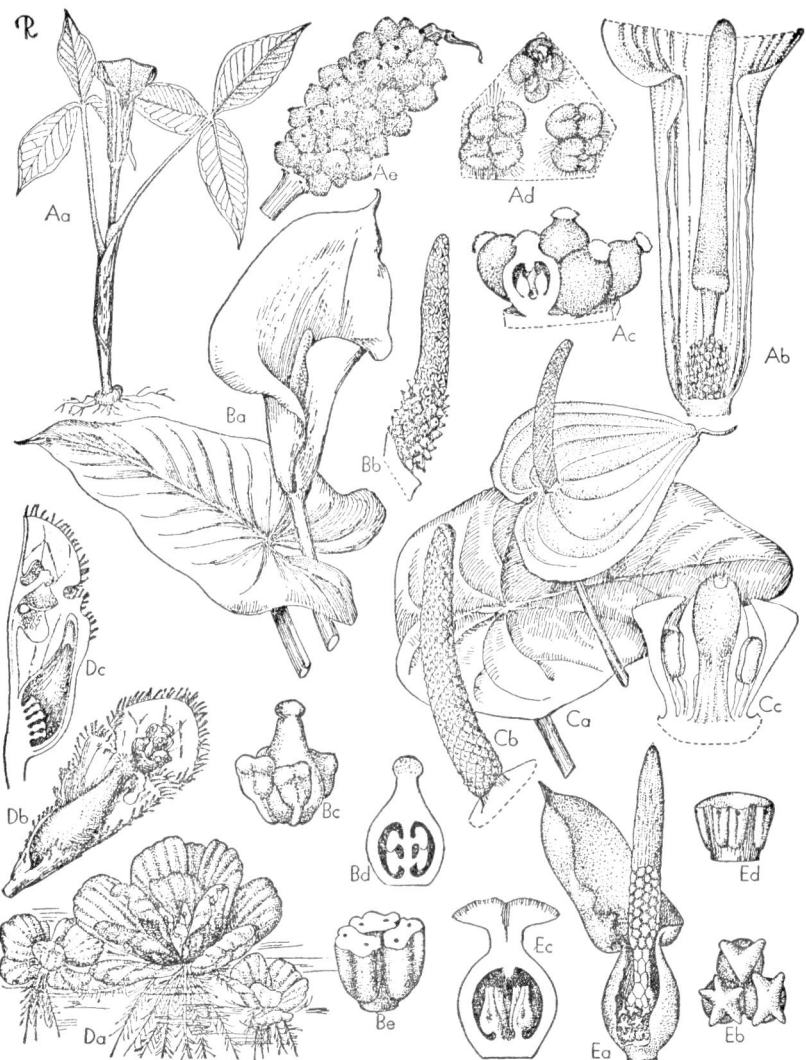

**Fig. 79.** ARACEAE. A, *Arisaema atrorubens*: Aa, plant in flower, × ⅙; Ab, spathe base showing spadix with pistillate flowers, × 1; Ac, fertile section of spadix with pistillate flowers, × 5; Ad, fertile section of spadix with staminate flowers, × 5; Ae, fruit-cluster, × ½. B, *Zantedeschia aethiopica*: Ba, leaf-blade and inflorescence, × ⅙; Bb, spadix, × ¼; Bc, pistillate flower with staminodia, × 8; Bd, same, vertical section, × 8; Be, staminate flower, × 8. C, *Anthurium Andraeanum*: Ca, leaf-blade and inflorescence, × ¼; Cb, spadix, × ½; Cc, flower, one perianth-segment and stamen removed, × 8. D, *Pistia Stratiotes*: Da, habit, × ½; Db, inflorescence, × 3; Dc, same, vertical section, × 3. E, *Alocasia Lowii*: Ea, inflorescence, spathe sectioned to show spadix, × ½; Eb, cluster of pistillate flowers, top view, × 3; Ec, pistillate flower, vertical section, × 6; Ed, staminate flower, × 4. (From L. H. Bailey, *Manual of cultivated plants*, The Macmillan Company, 1949. Copyright 1924 and 1949 by Liberty H. Bailey.)

by a single spathe. All indigenous genera, except *Arisaema*, are plants of bog or aquatic habitats and the former is a plant of moist woodlands. *Pistia* is a free-floating aquatic, and *Orontium* grows in water but roots in the substrate.

The phylogenetic position of the Araceae has been the subject of several interpretations. Engler (1920) considered them derivatives of the Palmae via the Cyclanthaceae. Wettstein (1935) considered them to be advanced over the orchids and as probably derived directly from Helobiae-Liliiflorae stocks. Bessey (1915) treated the palms and aroids to be of parallel or divergent development from liliaceous ancestors, whereas Hutchinson (1934) considered the family to have been derived from stocks ancestral to the Aspidistrae tribe of Liliaceae.

Economically the family is of little importance in the United States aside from members grown as ornamentals. In Old World tropics and subtropics the thickened rootstocks of *Colocasia* (taro) are a source of starchy food, while those of *Alocasia* are used similarly to a lesser extent. A score or more of horticultural variants of taro are grown for this purpose in Hawaii and other Pacific islands. The large "fruits" of *Monstera* are eaten in many tropical regions and prized for their delicate flavor. Representatives of about 35 genera are cultivated domestically for ornamental purposes. Notable among them is the calla of florists (*Zantedeschia*), not to be confused with the native *Calla* also grown along pool margins, and several genera whose members provide durable foliage plants for interior decoration, as *Dieffenbachia, Pothos, Philodendron, Scindapsus,* and *Aglaonema* Many species and garden variants of *Anthurium* are grown for both foliage and brightly colored, showy, waxy spathes. Plants of *Caladium* (Elephant's Ear) are grown for their multicolored foliage, and members of *Arisaema* (jack-in-the-pulpit), *Amorphophallus,* and *Helicodicerus* are grown in the open as curiosities. *Pistia* and *Orontium* are cultivated for ornament in pools, with the former also an aquarium subject.

*LITERATURE:*

ENGLER, A. Araceae. In de Candolle, Monogr. Phan. 2: 1–681, 1879.
ENGLER, A. and KRAUSE, K. Araceae. In Engler, Das Pflanzenreich 21 (IV. 23) 1–330, 1905; 37: (IV. 23) 1–160, 1908; 48: (IV. 23) 1–130, 1911; 55: (IV. 23) 1–134, 1912; 60: (IV. 23. Db) 1–143, 1913; 64: (IV. 23. Bc) 1–78, 1915; 71: (IV. 23 E) 1–139, 1920; 73: (IV. 23. F) 1–274, 1920; 74: (IV. 23. A) 1–71, 1920.
HUTCHINSON, J. The families of flowering plants. II. Monocotyledons, Araceae (pp. 117–124). London, 1934.
HUTTLESTON, D. G. Three subspecies of *Arisaema triphyllum*. Bull. Torrey Bot. Club, 76: 407–413, 1949.

## LEMNACEAE. DUCKWEED FAMILY

Floating or submerged perennial herbs; plants without roots or the roots reduc d to unbranched rhizoids; the plant body reduced to a small or minute oval, oblong flat or globose thallus, leafless, often purplish beneath, often reproducing asexually by buds, overwintering in temperate regions by production of buds that sink below the surface into substrate; monoecious; flowers unisexual, naked or initially enclosed by a membranous sheathing spathe; staminate flowers solitary or in pairs, of 1 or rarely 2 stamens, the anthers 1–2-celled, the filaments absent or if present filiform or fusiform; pistillate flowers solitary, the pistil 1, ovary unilocular, sessile, the ovules 1–7, basal, the style and stigma 1, simple; fruit a utricle, the seeds with or without endosperm.

The family is represented in fresh-water habitats throughout much of the world. Botanists are divided in opinion as to the number of genera: by Engler and Diels (1936) and by Wettstein (1935) it was considered to be 3 (*Lemna* with 8–10 spp., *Spirodela* with 3 spp., and *Wolffia* [25] with 12–14 spp.); by most American botanists it is placed at 4 (*Wolffiella* added as a generic segregate from *Wolffia*); whereas by Hutchinson (1934) it was treated as composed of only 2 genera (*Spirodela* merged with *Lemna,* and *Wolffiella* ignored). Acceptance of *Wolffiella* as a taxonomically valid generic entity is given greater justification by the recent discovery for the first time of flowering material (Mason, 1938) both in this country and in Argentina.

**Fig. 80.** LEMNACEAE. *Lemna minor*: a, flowering plants with rhizoids, × 2; b, plant in flower, × 8; c, sheath with inflorescence of three flowers, × 25; d, staminate flower, × 40; e, pistillate flower, × 40.

All 4 genera are indigenous throughout most of the country, with the family represented by 2 spp. of *Spirodela,* 7 of *Lemna,* 3 of *Wolffia,* and 1 of *Wolffiella.* Species of all genera are cultivated for ornament in pools and aquaria.

Most phylogenists are agreed that the family represents a degenerate offshoot of the Araceae with origins probably from *Pistia* or ancestral stocks of close affinity. Brooks (1940), on the basis of vegetative and floral anatomy and cytological studies, concluded that "the Lemnaceae have been derived from the Araceae, and that the Lemnaceae have a very reduced structure and show a reduction series from *Spirodela* through *Lemna* to *Wolffiella* and *Wolffia.*" An entirely different view was taken by Lawalrée (1945), who contended that the family was not of close affinity with any of the Araceae, but rather was a derivative of members of the order Helobiae. His principal basis for this was due to the presence of "squamulae intravaginales" in *Spirodela,* a situation common also to most members of the Helobiae.

Considerable interest has been evinced in the flowering and fruiting of species of Lemnaceae; both are rare phenomena and the latter the more so. Explanations accounting for the infrequency of these functions are wanting, and students are to be encouraged to search for reproducing material of these genera, and when it is found, to collect for critical study copious samples in liquid preservative.

*LITERATURE:*

BROOKS, J. S. The cytology and morphology of the Lemnaceae. Thesis (Ph.D.). Cornell Univ., 1940.
GOEBEL, K. Zur Organographie der Lemnaceen. Flora, 114: 278–305, 1921.
HEGELMAIER, F. Systematische Übersicht der Lemnaceen. Bot. Jahrb. 21: 268–305, 1895
LAWALRÉE, A. La position systématique des Lemnaceae et leur classification. Bull. Soc. [Roy.] Bot. Belgique, 77: 27–38, 1945.
MASON, H. L. The flowering of *Wolffiella lingulata* (Hegelm.) Hegelm. Madroño, 4· 241–251, 1938.

[25] Small (Manual of the southeastern flora) rejected the generic name *Wolffia* and accepted that of *Bruneria* in its place—a practice not followed elsewhere.

SAEGER, A. The flowering of Lemnaceae. Bull. Torrey Bot. Club, 56: 351–358, 1929.
THOMPSON, C. H. A revision of the American Lemnaceae occurring north of Mexico. Rep. Mo. Bot. Gard. 9: 21–42, 1898.

## Order 20. FARINOSAE

A large order of 13 relatively small families, of which only 6 come within the scope of this work; all having in common the single character of a mealy endosperm and characterized in general by a usually compound and usually superior ovary.

The order is not phylogentically a homogeneous taxon (Engler placed the 13 families belonging to it into 6 suborders), and the establishment of it on the endosperm character, as accepted by Engler, has since been rejected by Bessey, Hutchinson, Wettstein, and others. By these later authors, the families accounted for below were redistributed elsewhere among the monocots and as indicated in the discussion of each.

Families of the order not treated in this text include: Flagellariaceae, Restionaceae, Centrolepidaceae, Thurniaceae, Rapataceae, Cyanastraceae, Philydraceae.

### MAYACACEAE. BOG MOSS FAMILY

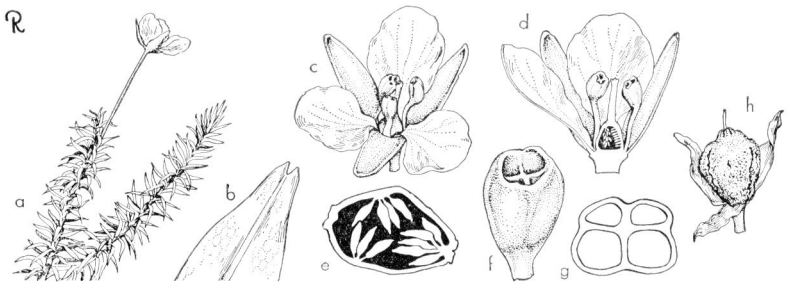

Fig. 81. MAYACACEAE. *Mayaca Aubletii*: a, fertile and sterile branch, × 1; b, leaf apex, × 30; c, flower, face view, × 4; d, flower, vertical section, × 4; e, ovary, cross-section, × 20; f, anther, × 15; g, anther, cross-section at anthesis, × 20; h, fruit, × 3.

Fresh-water aquatics, floating, or submerged and creeping; leaves filiform or linear, spirally alternate in high phyllotaxies, the blades sessile, 1-nerved, narrow, bidentate; flowers solitary and axillary on slender peduncles or aggregated near branch apices, bisexual, the perianth actinomorphic, of 2 series, 3-merous, the calyx of 3 free subvalvate sepals; the corolla about as long as calyx, white, violet, or rose, of 3 imbricate short-clawed obovate petals; stamens 3, alternate with the petals, hypogynous, the anthers 4-celled, dehiscing apically or subapically by pores or poricidal slits, basifixed, the filaments distinct, filiform; gynoecium of single pistil, the ovary superior, unilocular, 3-carpelled, the ovules several, biseriate, the placentae 3 and parietal, orthotropous, the style filiform, undivided, the stigma 1 and terminal, persistent on fruit until its dehiscence; fruit a triquetrous 3-valved capsule, dehiscing midway between placentae; seeds globose to ovoid, reticulated, with endosperm, the embryo small and apical.

A monogeneric family (*Mayaca*) represented by about 10 species with all but 1 (of tropical west Africa) in tropical or subtropical America. Two species (*M. fluviatilis, M. Aubletii*) occur infrequently to rarely along margins of pools and sluggish streams of the Atlantic and Gulf coastal plain from perhaps as far north as Virginia, south to Florida, and west to Texas. They are not of any known economic importance and none is cultivated.

Engler placed the family (along with the Restionaceae, Centrolepidaceae, Xyridaceae, and Eriocaulaceae) in his suborder Enantioblastae. Wettstein (1935) followed Hallier and elevated the suborder without change of name to the rank of order, and considered it as advanced over his Liliiflorae.

Phylogenetically the family is perhaps closer to the Flagellariaceae (not treated in this work) and Commelinaceae than to others of the Farinosae. Hutchinson (1934) recognized these 3 families as comprising the Commelinales, an order characterized by the leaves sheathing basally, the perianth of free parts in 2 whorls, differentiated into calyx and corolla. He considered them to have been derived from the Butomaceae and the Alismaceae, and credited them with having evolved from ancestral stock common also to the Xyridaceae, Eriocaulaceae, and Bromeliaceae.

*LITERATURE:*

PILGER, R. Mayacaceae. In Engler and Prantl, Die natürlichen Pflanzenfamilien. ed. 2, 15a: 33–35, 1930.

SMITH, A. C. Mayacaceae. In North Amer. Flora, 19: 1–2, 1937.

## XYRIDACEAE. YELLOW-EYED GRASS FAMILY

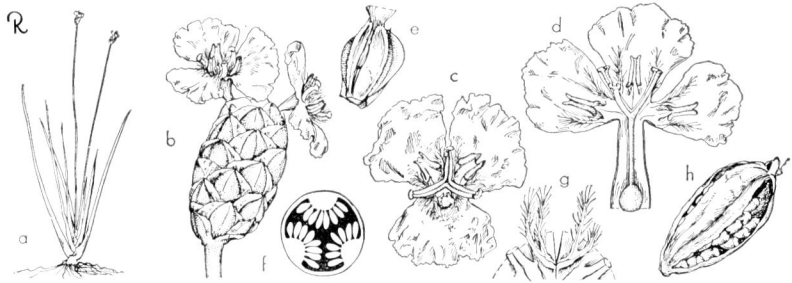

**Fig. 82.** XYRIDACEAE. *Xyris Congdonii*: a, plant in flower, × 1/20; b, inflorescence, × 1½; c, flower, face view, × 2; d, flower, perianth expanded, × 2; e, calyx, × 2; f, ovary, cross-section, × 10; g, staminodes, between two filament bases, × 2; h, fruit, × 5.

Small perennial or annual rushlike herbs; rootstock short and sometimes swollen and bulbous, roots fibrous; leaves mostly basal and tufted, sheathing, linear, terete or filiform; inflorescence a pedunculate, terminal, globose or cylindrical, bracteate head borne on a scape; flowers bisexual, zygomorphic, subtended by an involucre of stiff or coriaceous, imbricated bracts, each flower in the axil of an interfloral bract; perianth of 2 series, the calyx of 3 (or occasionally 2) sepals with the 2 lower (posterior) ones cymbiform, keeled, and chaffy, the inner sepal membranous and forming a hood over the corolla prior to anthesis, the corolla usually actinomorphic, marcescent, usually gamopetalous, the tube long or short and terminated by 3 uniform spreading lobes, usually yellow; stamens 3, often adherent to corolla, opposite corolla lobes and sometimes alternated with 3 (often bifid, plumose, or bearded with moniliform hairs) staminodes, the anthers 2-celled, extrorse, dehiscing by vertical slits, the filaments distinct and epipetalous, usually short and flattened; gynoecium of a single pistil, the ovary superior, unilocular (or imperfectly basally trilocular), 3-carpelled, the placentae 3 and parietal, or 1 and basal, or 3 and free-central, the ovules few to many (rarely reduced to 1) and orthotropous, the style usually 1 and long, or 3-branched, the stigmas 1–3 and linear; fruit an oblong 3-valved loculicidal capsule, enveloped by the persistent corolla tube; the seeds minute, usually apiculate, mostly striate, with copious mealy or fleshy endosperm and small apical embryo.

A family of 2 genera. *Xyris* (190 spp.) occurs from New Jersey southward into tropics and subtropics of North America and in tropics and subtropics of South America, Africa, and Australia. The genus is not known in Europe or continental Asia. *Abolboda* (9–10 spp.) is restricted primarily to tropical South America. About 19 species of *Xyris* (of the section *Euxyris,* characterized by the ovary 1-celled with 3 parietal placentae) occur in this country, extending from Newfoundland south along the Atlantic coast to Florida, westward to Texas, and inland along the St. Lawrence River into the Great Lakes to Lake Superior. The plants are to be found typically in sandy soils of shallow saline or brackish waters, fresh waters of bogs, open swamps, and low moist sandy grounds and roadside ditches. A few species are true aquatics and grow in swift-flowing streams.

The family is perhaps more closely related to the Mayaceae, Eriocaulaceae, and Commelinaceae than to others in the order. Hutchinson placed it (together with the Rapataceae, not in this treatment) in a separate order, the Xyridales, and considered them to represent, by reduction, an advancement over his Commelinales (cf. under Mayaceae), pointing out that the Xyridales were the "Compositae" of the monocots. Editors of the *North American flora* placed it in an order called Xyridales which, unlike the Xyridales of Hutchinson, was treated to comprise also the Mayaceae, Eriocaulaceae, Pontederiaceae, and Bromeliaceae.

Economically, the family is of little importance. Two species of *Xyris* are used in aquatic gardening and aquaria.

*LITERATURE:*

MALME, G. O. A. Xyridaceae. In Engler and Prantl, Die natürlichen Pflanzenfamilien. ed. 2, 15a: 35–38, 1930.

——. Xyridaceae. In North Amer. Flora, 19: 3–15, 1937.

## ERIOCAULACEAE. PIPEWORT FAMILY

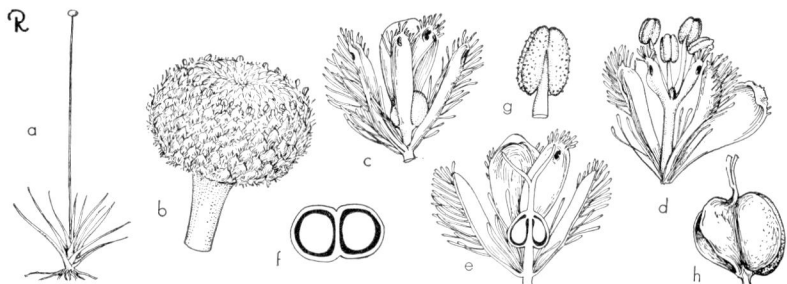

**Fig. 83.** ERIOCAULACEAE. *Eriocaulon compressum*: a, plant in flower, × ⅛; b, inflorescence, habit, × 2; c, pistillate flower, habit, × 6; d, staminate flower, habit, × 6; e, pistillate flower, vertical section, × 6; f, ovary, cross-section, × 15; g, anther, × 15; h, fruit, × 10.

Mostly small perennial herbs (rarely annuals) of cespitose habit, and of aquatic or marsh habitats, acaulescent or nearly so, roots fibrous; leaves mostly linear, basal, grasslike, usually crowded, often pellucid, sometimes membranous; plants usually monoecious or rarely dioecious; inflorescence a small head subtended by an involucre of bracts and borne capitately (the heads solitary or in umbellate aggregates) on an erect, slender, often scapose peduncle that usually much exceeds the leaves and is sheathed basally; flowers unisexual, regular or rarely irregular, small to minute, numerous, sessile to shortly pedicellate, on a variously shaped torus or receptacle, each flower usually borne in the axil of a scarious, scalelike, colorless or colored bract; when plants are monoecious the staminate and pistillate flowers either mixed within the inflorescence or the staminate in center and pistillate on the periphery; when dioecious the flowers of an entire head (and plant) of one sex or the other; perianth (perigonium) chaffy (scarious) or membranous, rarely hyaline, in 2 series but not differentiated into conventional calyx and petaloid corolla, but rather the 2 series distinct, 2–3-merous, the outer series (calyx) free or the segments coherent, the inner series (rarely absent) often stipitate and infundibular or cupular (by some authors this perianth is interpreted to be an involucre); stamens as many or twice as many as outer perianth segments (rarely fewer) and alternate with them, inserted on corolla (when present), the anthers 1–2-celled, introrse, dehiscing by vertical slits, filaments distinct and slender, the staminate flowers often with vestigial gynoecium (pistillodium) present; pistillate flower unipistillate, the ovary 2–3-loculed, superior, each locule with a single pendulous orthotropous ovule, the placentation axile, the style 1, terminal, with as many branches as ovary locules, each branch simple or forked, staminodes rarely present; fruit a membranous 2–3-celled, 2–3-seeded loculicidal capsule; embryo minute and apical.

A family of 12 genera and about 1110 species occurring primarily in bogs and along wet shores of the tropics, with a few members extending into temperate regions of both hemispheres. South America is the epicenter of distribution, with 1 genus occurring only in West Africa and Madagascar. Three genera occur in this country. *Eriocaulon* (about 370 species) is represented by 8 species which extend north from Florida to Newfoundland and west to Texas and Minnesota. *Syngonanthus*, having 177 species of tropical America, is represented by a single invading species (*S. flavidulus*) that extends northward along the pine barren areas of the coastal plain from Alabama to North Carolina. *Lachnocaulon*, a small tropical American genus of 10 species, extends northward along the coastal plain from Texas to Virginia. The family is not represented in western United States. The inflorescences of a few species of *Syngonanthus* are sold as "everlasting" flowers for decorations.

Engler and Diels (1936) treated these families (Mayacaceae, Xyridaceae, and Eriocaulaceae) along with others not represented in this country (Restionaceae and Centrolepidaceae) as comprising the suborder Enantioblastae (cf. discussion under Mayacaceae). Hutchinson placed them in 3 different but closely related orders, and considered the Eriocaulaceae as sufficiently apart from the others to be elevated to a separate order (Eriocaulales), and advanced over his Xyridales, on the basis of flowers always unisexual (an advancement over flowers often bisexual) and the ovules reduced to 1 per carpel.

*LITERATURE:*

MOLDENKE, H. N. Eriocaulaceae. In North Amer. Flora, 19: 17–50, 1937.

———. The known geographic distribution of the members of the Eriocaulaceae, together with a check-list of scientific names proposed in the group. Privately printed. New York, 1946.

———. The known geographic distribution of the members of the Verbenaceae, Avicenniaceae, Stilbaceae, Symphoricacaea, and Eriocaulaceae. Privately printed. New York, 1949.

RUHLAND, W. Eriocaulaceae. In Engler and Prantl, Die natürlichen Pflanzenfamilien. ed 2, 15a: 39–57, 1930.

## BROMELIACEAE. PINEAPPLE FAMILY

Mostly short-stemmed herbaceous epiphytes (terrestrial in Pitcairnioideae); leaves mostly basal and rosette-forming, rarely cauline, spirally alternate, mostly stiffly lorate, often spiny-serrate and troughlike (entire in Tillandsioideae and others), sometimes coarsely grasslike or linear to filiform and then flexuous, covered wholly or in part by minute peltate scalelike hairs appearing scurfy, often basally colored reddish or purplish, the base sheathing; inflorescence a terminal head, spike, raceme, or panicle, the flowers in axils of often brightly colored bracts (solitary and pseudolateral in *Tillandsia usneoides*); flowers bisexual or rarely functionally unisexual, actinomorphic or weakly zygomorphic, perianth of 2 series, the outer series of 3 herbaceous calyxlike segments (sepals) and the inner of 3 corollalike segments with petals free or variously connate, often brightly colored, in some groups with scales or prominent nectaries within; stamens 6, often borne on the corolla (not so in many Tillandsioideae), often at the base of the petals, free or adnate to them, distinct or the filaments basally connate, the anthers 2-celled, linear, usually versatile, dehiscing introrsely by vertical slits, distinct, the pollen grains providing good diagnostic characters (may be grooved, marked with pores or smooth); pistil 1, the ovary superior to inferior, 3-locules, with ovules usually numerous in each, the placentae axile and sometimes bifurcate, the style 1, the stigmas 3, sometimes spirally twisted; fruit a berry or capsule, often more or less enveloped by the persistent perianth, sometimes (in *Ananas*) syncarpous; seeds with copious mealy endosperm and small embryo, naked, sometimes winged or caudate, or with a plumose pappuslike appendage.

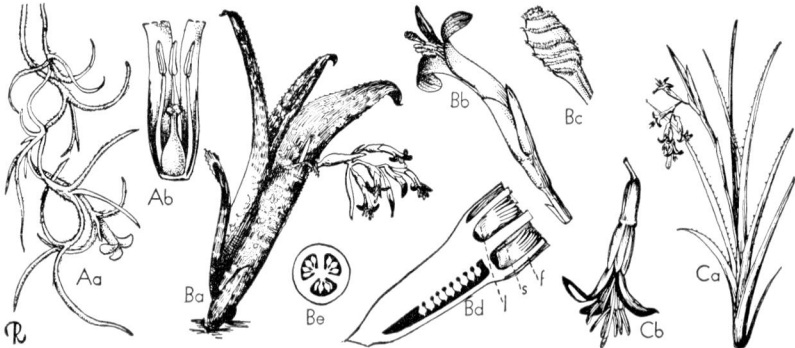

**Fig. 84.** BROMELIACEAE. A, *Tillandsia usneoides*: Aa, flowering branch, × ½; Ab, part of a flower, × 2. B, *Aechmea polystachya*: Ba, plant in flower, × ⅒; Bb, flower, × ½; Bc, stigma, × 2; Bd, ovary, vertical section, × 1; Be, ovary, cross-section, × ½. C, *Billbergia nutans*: Ca, plant in flower, × ⅒; Cb, flower, × ½. (From L. H. Bailey, *Manual of cultivated plants,* The Macmillan Company, 1949. Copyright 1924 and 1949 by Liberty H. Bailey.)

A family of 45–50 genera and about 1800–2000 species of tropical to warm temperate America, extending from eastern Virginia south and west to Texas and across northern Mexico southward to central Argentina and Chile. It is the largest family of seed plants almost wholly indigenous to the New World (except for *Pitcairnia Feliciana,* native in French West Africa). *Tillandsia* is the most widely distributed, extending south into Argentina and north into Virginia. Four genera are reported by Smith (1938) to be represented within the United States; *Hechtia* with 3 species in Texas, and the remainder largely in Florida—*Guzmania* with a single indigenous species, *Catopsis* with 2 species in southern Florida, and *Tillandsia* with 12 species of which 1 (*T. usneoides*) extends north into Virginia and westward along the Gulf of Mexico into Texas, another (*T. Baileyi*) is restricted to southern Texas, a third (*T. recurvata*) extends westward from Florida into Texas and occurs again in Arizona, and the remainder are limited to Florida or extend into more southerly extraterritorial ranges. Small (1933) treated *Tillandsia usneoides* and *T. recurvata* as monotypic genera (*Dendropogon usneoides* and *Diaphoranthema recurvata* respectively), a view rejected by Smith and other systematists.

The family was divided by Harms (1930) into 4 subfamilies. The 34 genera whose flowers have an inferior ovary and the fruit a berry were placed in the Bromelioideae; the remaining 25 genera (with flowers having a superior or half inferior ovary and capsular fruits) constituted the other subfamilies, with the genus *Navia*, comprised of terrestrial plants whose seeds are neither winged nor appendaged, treated as the Navioideae, but now considered an extreme of the Pitcairneae. The 12 genera whose seeds have long plumose appendages comprised the Tillandsioideae, and the remaining 12 genera, with seeds winged or non-plumosely tailed comprised the Pitcairnioideae. The Bromeliaceae are best distinguished by the presence of multicellular or stellate scales or hairs, and comprise a natural assemblage of genera characterized also by their often epiphytic habit, the usually stiff leaves often colored basally, the usually colored floral bracts, the herbaceous calyx, and the versatile anthers.

Phylogenetically the family was interpreted by Hutchinson (1934) to "represent the climax of a line of descent wherein the calyx and corolla have remained distinct or fairly distinct from each other, a feature retained from the Dicotyledonous stock." He treated it as related to, but more advanced than, the Commelinales, and separated it as an order by itself, the Bromeliales. Smith (1934) considered the strongest affinities of the family were with the Rapataceae (not treated in this text) and that both families probably arose from a common ancestral stock. He further considered the Bromeliaceae to be the most primitive members of the Farinosae. Within the family, *Puya* (Pitcairnioideae) was treated by Smith as probably the source of ancestral types from which the other subfamilies developed. Pittendrigh (1948) regarded the 3 subfamilies as separate derivatives of a common ancestor (presumably extinct and unknown), and indicated that while *Puya* may be the most primitive of living genera, the Tillandsioideae are not derived from it nor closely related to it.

Economically the Bromeliad family is of most importance for the edible pineapple (*Ananas comosus*), while the leaves of the same plant are the source of fiber used in weaving Pina cloth. The dried stems and leaves of Spanish moss (*Tillandsia usneoides*) are used in upholstery under the trade-name of "vegetable hair." Some species are important in tropical America as sources of cordage and fiber for fabrics (as caroá from *Neoglaziovia variegata* of Brazil, and pita floja from *Aechmea Magdalenae* of Central America to Colombia). About 20 genera are grown as ornamentals under glass and in the open in near frost-free regions; notable among them are *Billbergia, Aechmea, Vriesia, Nidularium, Pitcairnia, Puya, Guzmania,* and *Tillandsia.*

LITERATURE:

HARMS, H. Bromeliaceae. In Engler and Prantl, Die natürlichen Pflanzenfamilien. ed. 2. 15a: 65–159, 1930.
LINDSCHAU, M. Beiträge zur Zytologie der Bromeliaceae. Planta, 20: 506–530, 1933.
MEZ, C. Bromeliaceae. In Engler, Das Pflanzenreich. 103 (IV. 32): 32–100, 1935.
PITTENDRIGH, C. S. The Bromeliad-Anopheles-Malaria complex in Trinidad. I, The bromeliad flora. Evolution, 2: 58–89, 1948.
SCHULTES, R. E. *Aechmea Magdalenae* and its utilization as a fibre plant. Bot. Mus. Leafl. Harvard Univ. 9: 117–122, 1941.
SMITH, L. B. Geographical evidence on the lines of evolution in the Bromeliaceae. Bot. Jahrb. 66: 446–468, 1934.
———. Bromeliaceae. In North Amer. Flora, 19: 61–228, 1938.
———. Notes on the taxonomy of *Ananas* and *Pseudananas*. Harvard Univ. Bot. Mus. Leafl. 7: 73–81, 1939.

## COMMELINACEAE. SPIDERWORT FAMILY

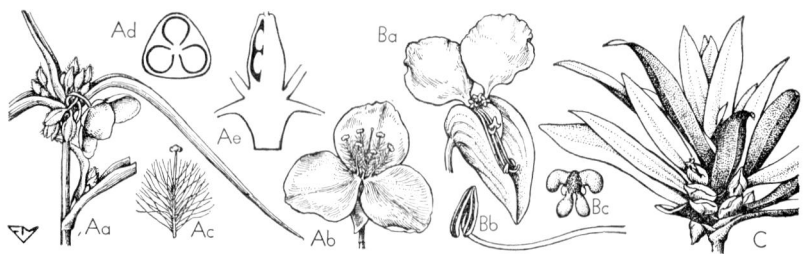

**Fig. 85.** COMMELINACEAE. A, *Tradescantia virginiana*: Aa, inflorescence, × ¼; Ab, flower, × ½; Ac, stamen, × 1; Ad, ovary, cross-section, × 5; Ae, ovary, vertical section, × 3. B, *Commelina coelestis*: Ba, flower, × 1; Bb, fertile stamen, × 4; Bc, staminodium, × 4. C, *Rhoeo discolor*: flowering plant, × ⅛. (From L. H. Bailey, *Manual of cultivated plants*, The Macmillan Company, 1949. Copyright 1924 and 1949 by Liberty H. Bailey.)

Succulent perennial or annual herbs, acaulescent or with nodose stems, roots fibrous or sometimes much thickened and tuberlike; leaves alternate, flat or trough-like, entire, parallel-veined, sheathing by a basal membranous and often closed sheath (petiole); inflorescence terminal, terminal and axillary, or less often axillary, a simple or compound helicoid cyme or thyrse, or the flower solitary, sometimes subtended by a boat-shaped (cymbiform) spathe or foliaceous bracts; flowers usually actinomorphic, or in some genera zygomorphic, bisexual, the perianth of 2 series, the outer one a green herbaceous calyx of 3 usually free and imbricated herbaceous sepals (rarely gamosepalous and rarely the sepals petaloid), the inner one a colored mostly ephemeral and deliquescent corolla of 3 free equal or unequal petals rarely united into a slender tube, the third petal sometimes much reduced; stamens typically 6 but often the fertile ones reduced to 3 by abortion (aborted stamens reduced to staminodes) or rarely 1 stamen functional (no staminodes in *Callisia*), hypogynous, the anthers with 2 parallel or divergent cells, dehiscing by longitudinal slits or (in *Dichorisandra*) by an apical pore, the filaments distinct (rarely are some connate), often bearded with colored moniliform hairs; pistil 1, the ovary superior, sessile or stipitate, 3-loculed (rarely 2), the ovules 1-few per cell and orthotropous, the placentation axile, style 1, stigma 1 and capitate or 3-fid; fruit a loculicidal capsule, sometimes enclosed by fleshy sepals, rarely fleshy and indehiscent (pergamentaceous); seeds with a punctiform to linear funicular scar, usually netted, muricate, or ridged (arillate in *Dichorisandra*), the endosperm copious and mealy, the embryo situated beneath a conspicuous disclike callosity (embryotega) on seed coat.

A family of largely tropical and subtropical plants, represented by 37 genera (9 monotypic) and about 600 species distributed over the warmer parts of the earth. Members of the subfamily Tradescantieae (flowers regular) are most abundant in the New World, whereas those of the subfamily Commelineae (flowers zygomorphic) are of greatest frequency in tropical Africa. The largest genera are *Aneilema* (70 spp.), *Murdannia* (50 spp.), *Cyanotis* (50 spp.), and *Tradescantia* (34 spp.). In the United States the family is represented by 4 genera with indigenous species (*Tradescantia* 20–22 spp., *Callisia* 1 sp., *Tripogandra* 3 spp., and *Commelina* 8–9 spp.) and 3 genera now naturalized in the southeast from Central America and the West Indies (*Rhoeo, Zebrina,* and *Aneilema*). Small's segregation of some southeastern species of *Tradescantia* under *Tradescantella* and *Cuthbertia* was taxonomically accepted by Woodson (1942), but for nomenclatural reasons the genera are treated under *Callisia* and *Tripogandra* respectively. The Commelinaceae are not represented among indigenous plants of the Pacific states and only *Tradescantia* extends indigenously into Canada.

Members of the family have been the subject of independent study and reclassification by 4 recent botanists (Brückner, 1930; Hutchinson, 1934; Woodson, 1942; and Pichon, 1946). No two arrived at the same conclusions or accepted the same criteria as bases for separation of groups within the family. In brief, Brückner based his major subdivisions on characters of perianth and androecia, Hutchinson considered those of inflorescence position and type to be of primary significance, Woodson (accounting only for the American elements) accepted inflorescence type as of first consideration and androecial conditions secondarily, and Pichon reclassified the genera largely on the bases of androecial, gynoecial, and perianth characters in that sequence. In the selection of one of these classifications over another, students should remember that the corollas of most members of the family are deliquescent and are poorly represented in most herbarium specimens (often lost entirely) and that the androecial characters presented by Brückner and Pichon may not be constant in all cases. For these reasons the bases employed by Hutchinson and Woodson seem to have the advantage of greater utility and constancy. It is not possible to select between these last two reclassifications, since Hutchinson attempted to account for all genera (pointing out need of further study in several cases) and Woodson has more thoroughly recounted the taxonomic status of the New World members of the family.

Most botanists are agreed that the family is represented by 2 tribes, Tradescantieae and Commelineae. Woodson distinguished the former by its "paired sessile scorpioid cymes which appear as a 2-sided unit superficially," whereas in the latter "the ultimate branches or units of the inflorescence are composed of individual scorpioid cymes which appear 1-sided superficially." Woodson further realigned several generic elements, merging *Murdannia* with *Aneilema* in the Commelineae and recognizing *Cuthbertia* as generically distinct from *Tradescantia*, but merging it and several other erstwhile segregates under the older Rafinesquian genus *Tripogandra*. Woodson further treated *Rectanthera* (*Spironema*) and *Tradescantella*, as congeneric with the Linnaean genus *Callisia*. Pichon both reclassified and redefined the family, separated from it the Australian *Cartonema* as Cartonemaceae, and considered the Commelinales of Hutchinson to be a natural grouping composed of the Mayacaceae, Cartonemaceae, Commelinaceae, and Flagellariaceae. He recognized Commelinaceae to be composed of 39 genera (4 described as new) which he placed in 10 tribes.

Phylogenetically the Commelinaceae have, in common with other members of the Farinosae, the characters of mealy endosperm and the perianth differentiation into corolla and calyx. Hutchinson treated them (along with the Mayacaceae and Flagellariaceae) as comprising the Commelinales, an order interpreted by him to have been derived from both the Butomales and Alismatales and considered advanced over each by the compound ovary (vice a gynoecium of several pistils). As noted under Bromeliaceae, Hutchinson considered the latter family to have been derived from his Commelinales.

Economically the family is of little importance except for a few members of 11–12 genera grown to a limited extent as garden ornamentals. Notable among these are species of *Tradescantia, Commelina, Rhoeo, Zebrina, Cyanotis, Tinantia,* and *Callisia* (*Spironema*).

LITERATURE:

ANDERSON, E. and WOODSON, R. E. The species of *Tradescantia* indigenous to the United States. Contr. Arnold Arb. 9: 1–132, 1935.

ANDERSON, E. and SAX, K. A cytological monograph of the American species of *Tradescantia*. Bot. Gaz. 97: 433–476, 1936.

BRÜCKNER, G. Beiträge zur Anatomie, Morphologie und Systematik der Commelinaceae. Engl. Bot. Jahrb. 61 (Beibl. 137): 1–70, 1926.

———. Commelinaceae. In Engler and Prantl, Die natürlichen Pflanzenfamilien. ed. 2, 15a: 159–181, 1930.

CLARKE, C. B. Commelinaceae. In de Candolle, Monogr. Phan. 3: 113–324, 1881.

HUTCHINSON, S. Commelinaceae, in The families of flowering plants. II. Monocotyledons. London, 1934. [pp. 53–57.]

PENNELL, F. W. The genus *Commelina* (Plumier) L. in the United States. Bull. Torrey Bot. Club, 43: 96–111, 1916.

PICHON, M. Sur les Commelinacées. Notulae Systematicae [Paris]. 12: 217–242, 1946. [Includes keys to tribes and genera and 3 new genera in the New World.]

THARP, B. C. *Commelinantia*, a new genus of Commelinaceae. Bull. Torrey Bot. Club, 49: 269–275, 1922. [cf. also *op. cit.*, 54: 337–340, 1927.]

WOODSON, R. E. Commentary on the North American genera of Commelinaceae. Ann. Mo. Bot. Gard. 29: 141–154, 1942.

## PONTEDERIACEAE. PICKEREL-WEED FAMILY

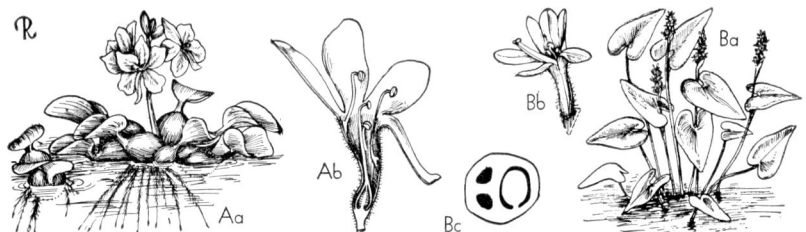

**Fig. 86.** PONTEDERIACEAE. A, *Eichhornia crassipes*: Aa, plant in flower, × ½12; Ab, flower, vertical section, × ½. B, *Pontederia cordata*: Ba, plant in flower, × ⅕15; Bb, flower, × 1; Bc, ovary, cross-section, × 10. (From L. H. Bailey, *Manual of cultivated plants,* The Macmillan Company, 1949. Copyright 1924 and 1949 by Liberty H. Bailey.)

Aquatic, mostly perennial herbs (annual only in *Hydrothrix*), floating or rooting in substrate; rootstock thick, short or creeping, with fibrous roots; stems very short or erect and unbranched, mostly enveloped by sheathing leaf bases; leaves usually in opposite pairs or in whorls of 3–4, with fleshy sheathing petioles or sessile and sheathing, the blades emersed or floating, lanceolate to broadly ovate or elliptic, sometimes completely reduced, and the leaf represented by a linear leaflike flattened petiole; inflorescence a raceme or panicle, usually subtended by a spathelike leaf sheath; flowers bisexual, mostly regular or in some genera zygomorphic, with (or more often without) subtending bracts, the perianth obscurely 2-seriate, of 6 imbricated segments, distinct or basally connate, all quite similar and petallike, sometimes twisted after anthesis and falling or the base thickened-persistent and enclosing the fruit; stamens typically 6, or reduced to 3 or 1, inserted on perianth, usually unequal or dissimilar, the anthers 2-celled, introrse, dehiscing by vertical slits or rarely poricidal, the filaments slender, free; pistil 1, the ovary superior, trilocular with axile placentae or unilocular with 3 parietal placentae, the ovules numerous, or solitary and pendulous, from each placenta, anatropous; the style 1, the stigma 1–6-lobed or 6-toothed or capitate; fruit a perfectly or incompletely 3-celled capsule or a 1-celled 1-seeded utricle; seed ribbed lengthwise, the endosperm copious and mealy, the embryo small, straight, and linear.

A family of 6–7 genera and about 28 species. All are fresh-water aquatic plants, distributed pantropically, a few genera extending laterally into the temperate zones of both hemispheres, absent in Europe and much of Australasia except in New Guinea and northern Australia. Of the 6 genera, *Hydrothrix* (monotypic) is known only in Brazil, *Reussia* (2 spp.) is South American and principally Brazilian, *Monochoria* (5 spp.) occurs in the Old World tropics extending north through Asia to Manchuria and south into Australia, *Heteranthera* (9–10 spp.) is of the American and African tropics, reaching north in the New World with 4 species in this country (1 species extending into Canada) from Florida to Ontario westward to California and Oregon (Alexander treated *H. dubia* as *Zosterella dubia*, a generic segregate not accepted by many American botanists, but one meriting serious consideration as a valid taxon), *Eichhornia* [incl. *Piaropus*] (5 spp.) is tropical American extending north probably to Florida and also in tropical Africa to Madagascar, and *Pontederia* (3 spp.) of the temperate and subtropic regions of the Americas has 2 species generally distributed over most of United States and southern Canada.

The family was divided by Schwartz (1930) into 3 tribes, the first 2 Eichhornieae and Heterantherae) with a trilocular ovary and the third (Pontederieae) with a unilocular ovary. It was included as a member of the Farinosae because of its endosperm and embryo character. Hutchinson placed the plants in the Liliales, stating (p. 106), ". . . they appear to me to be aquatic *Liliaceae*, tending towards the Aroid type, the spiciform inflorescence having a spathelike reduced leaf (leaf sheath)." Schwartz also treated the family as being closely related to the Liliaceae, but considered the characters of endosperm, the variability and reduction in the androecium, and the floral zygomorphy to be of sufficient importance to also justify placing it close to the Commelinaceae.

*Pontederia* (pickerel weed) and *Eichhornia* (water hyacinth) are grown to a limited extent as aquatic ornamentals in pools. *Eichhornia*, introduced as an ornamental, has become a serious pest clogging waterways in warm regions of this and other countries.

*LITERATURE:*

ALEXANDER, E. J. Pontederiaceae. In North Amer. Flora, 19: 51–60, 1937.
SCHWARTZ, O. Pontederiaceae. In Engler and Prantl, Die natürlichen Pflanzenfamilien. ed. 2. 15a: 181–188, 1930.

## Order 21. LILIIFLORAE [26]

An order characterized by Engler as with flowers typically 3-merous, the perianth 2-seriate and usually undifferentiated into calyx or corolla, the stamens 3–6, the ovary compound and 3-carpelled (rarely 2-carpellate), the seeds with endosperm.

The name Liliales has been applied to orders of various circumscriptions, and it is significant that their limits have been variously interpreted by Engler, Rendle, Wettstein, Bessey, and Hutchinson. Authors of American floras have usually followed one or another of these interpretations but in no case has there been consistency in this regard.

The order was treated by Engler and Diels to comprise 3 suborders and 8 families as follows (family names preceded by an asterisk are not treated in this text):

| Juncineae | Liliineae | | Iridineae |
|---|---|---|---|
| Juncaceae | Liliaceae | Amaryllidaceae | Iridaceae |
| | *Stemonaceae | *Velloziaceae | |
| | Haemodoraceae | *Taccaceae | |
| | | Dioscoreaceae | |

Hutchinson placed all the above families (except the Juncaceae) into his Corolliferae, a subphyllum that also included Engler's Scitamineae and Microspermae, and reclassified their genera as comprising 22 families and 7 orders as explained on p. 371. Engler's classification of families in this order was based largely on ovary

[26] Known also as the Liliales.

position, with the inferior ovary treated as derived from the superior. Hutchinson rejected this view with the statement that (p. 7) "the character of the *superior* or *inferior* ovary has often been stressed too much and has led to artificial classification." By this he rejected also the Bessey classification of liliaceous stocks (cf. p. 128). Continuing, Hutchinson stated further (p. 8) ". . . the type of inflorescence is of much more importance than the superior or inferior ovary, and the result is a nearer approximation of allied genera."

## JUNCACEAE. RUSH FAMILY

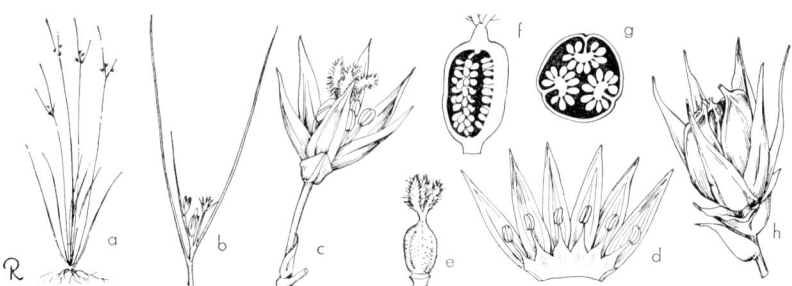

**Fig. 87.** JUNCACEAE. *Juncus tenuis*: a, plant in flower, × 1/10; b, inflorescence, × 1/2; c, flower, habit, × 5; d, perianth and androecium, expanded, × 4; e, pistil, habit, × 4; f, ovary, vertical section, × 9; g, ovary, cross-section, × 10; h, fruit with subtending bracts, × 6.

Rushlike or grasslike perennial or annual herbs (or rarely shrubs), often with hairy roots from an erect or horizontal rhizome; stems short; leaves mostly basal, tufted, usually linear or filiform, grasslike and flat or terete, sheathing basally or reduced to sheath only; inflorescence a panicle, corymb, or head or the flowers solitary; flowers bisexual or if unisexual the plants dioecious, regular, the perianth glumaceous or coriaceous (rarely scarious), biseriate, segments 3 in each series (or only 1 series present), not usually differentiated into calyx and corolla, often colored or greenish; stamens 6 or 3, opposite the perianth segments, distinct, the anthers 2-celled, introrse, dehiscing by vertical slits, basifixed, the pollen grains in tetrads; pistil 1, the ovary superior, unilocular with 3 parietal placentae, or trilocular with axile placentation or incompletely septate, the ovules 1 and basal, or many and biseriate on each placenta, anatropous; the style 1, very short to linear, or 3, the stigmas 3 and linear to lanceolate; fruit a 1–3-celled loculicidal capsule; seeds small, often tailed, with minute straight embryo enclosed by fleshy endosperm.

A family of 8 genera and about 315 species largely of temperate regions of the southern hemisphere. It is absent or rare in the tropics. Six genera, representing a total of only 10 species, are restricted to the southern hemisphere. The genus *Juncus* (225 spp.) is of world-wide distribution, with most species in the northern hemisphere and about 90 within the United States and Canada. The eighth genus, *Luzula* (80 spp.) likewise is largely one of the northern hemisphere, with about 10 species indigenous to this country.

The Juncaceae are a very old family, reportedly extending through the Tertiary and back into the Cretaceous. There are differences of opinion as to its ancient centers of origin; by Buchenau it was placed in the mountains of Eurasia, by Vierhapper in the Old World tropics, and by Weimarck it was considered as probably having spread northward from Antarctica. The family has been treated generally as allied to the Liliaceae, either as ancestral to it or as derived from it. Many authors are agreed that it is a degenerate reduced form of liliaceous stocks. Hutchinson placed it in an order by itself, treating it as advanced over his Liliales. He also treated it as one of 3 orders composing his class Glumiflorae and as a progenitor of his Cyperales and Graminales.

The Juncaceae are of little economic importance. A few species may be grown for ornament in locations adjoining aquatic habitats. Others are used locally in weaving matting, hats, chair seats; and the pith has been used for candlewicks.

*LITERATURE:*

BUCHENAU, F. Juncaceae. In Engler, Das Pflanzenreich, 25 (IV. 36): 1–284. 1906.
VIERHAPPER, F. In Engler and Prantl, Die natürlichen Pflanzenfamilien, ed. 2. Bd. 15a: 192–224, 1930.
WEIMARCK, H. Studies in Juncaceae, with special reference to the species in Ethiopia and the Cape. Svensk Bot. Tidskr. 40: 141–178, 1946.

## LILIACEAE. LILY FAMILY

Mostly perennial herbs, infrequently or only occasionally woody; rootstock a rhizome, bulb, corm, or tuber; stems erect or climbing, often modified into fleshy subterranean storage organs or cladophylls; leaves basal or cauline, alternate or whorled (opposite only in *Scolyopus*), mostly lamellate but sometimes reduced to scales or sheaths, sometimes fleshy or with prickly margins, occasionally fibrous, the venation mostly parallel (parallel-reticulate in few genera); inflorescence various; flowers bisexual (rarely unisexual and then the plants usually dioecious), regular or infrequently weakly zygomorphic, the perianth usually large and showy, mostly corollalike, in 2 series of 3 segments each (rarely 4, or more than 6), generally undifferentiated into corolla and calyx, and when so the segments termed tepals, the segments imbricate or the outer series valvate, sometimes connate into a tube; stamens 6 (rarely 3, 4, or up to 12), hypogynous or adnate to perianth, the filaments distinct or connate, the anthers 2-celled, extrorse or introrse, basifixed or versatile, dehiscing usually by vertical slits or rarely by a terminal pore; pistil 1, the ovary generally trilocular with axile placentation (rarely unilocular with 3 parietal placentae), usually superior, the ovules mostly numerous and biseriate, the style usually 1 (rarely 3), divided or trifid, the stigmas usually 3, or 1 with 3 lobes; fruit a septicidal or loculicidal capsule or a berry; seed with abundant endosperm.

A family of about 240 genera and 4000 species widely distributed over most of the vegetated land areas of the earth and especially abundant in the warm temperate and tropical regions. Except for some xerophytic representatives, members of the lily family do not form dominant or climax vegetations over areas of appreciable extent. The family was subdivided by Krause (1930) into 12 subfamilies and 35 tribes. Of the approximately 240 genera recognized by him, only about 55–60 are native to North America (exclusive of Mexico wherein occur about 12 additional genera) and these represent only about one-third of the tribes. The larger genera of the family include (with approximate number of species in each): *Asparagus* 300, *Smilax* 300, *Allium* 280, *Aloe* 180, *Scilla* 90, *Lilium* 70, *Kniphofia* 70, *Colchicum* 65, *Sanseviera* 60, *Tulipa* 50, *Fritillaria* 50, *Veratrum* 46, *Calochortus* 40, and *Trillium* 30.

The Liliaceae were defined within closer limits by Hutchinson (1934), who segregated most of the woody nonxerophytic members as the Ruscaceae, Philesiaceae, and Smilacaceae, most of the xerophytic Dracaenioideae that have basal or terminal fibrous leaves as the Agavaceae, the tribes Agapantheae, Allieae, and Gilliesieae that have umbellate and spathaceous inflorescences as a part of his Amaryllidaceae, the tribe Parideae characterized in part by a perianth of green calyx and showy corolla as the Trilliaceae, the saprophytic Petrosavieae as Petrosaviaceae, the Aus-

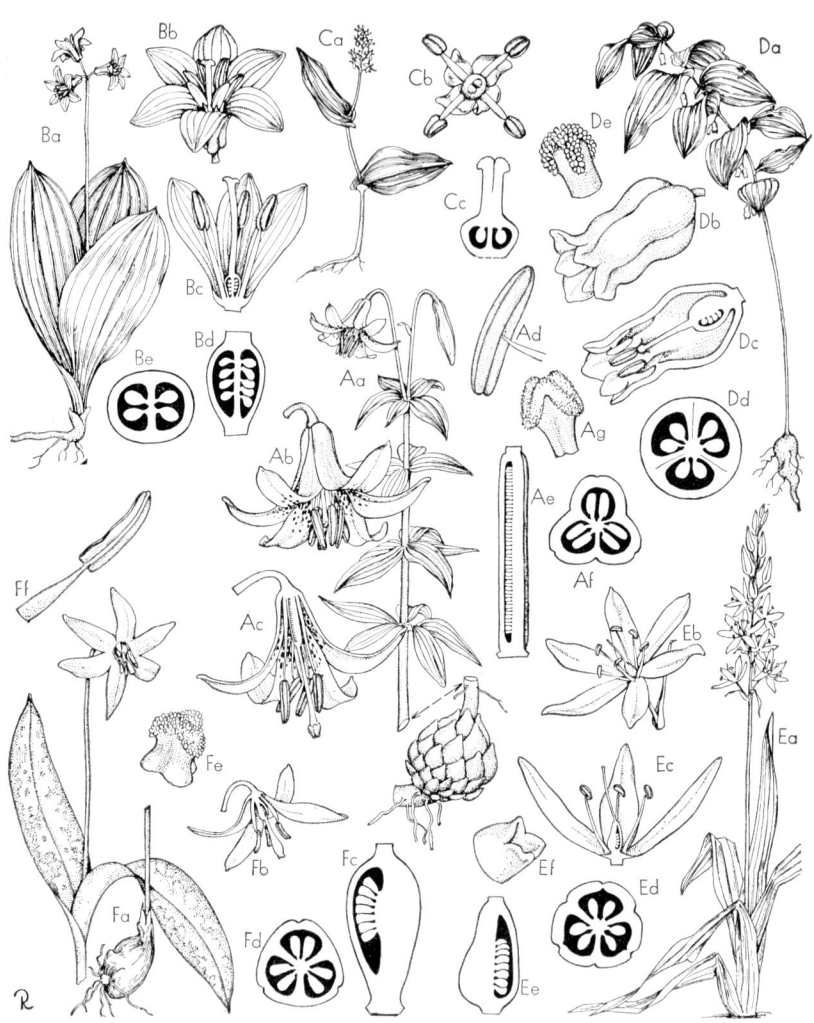

**Fig. 88.** LILIACEAE. A, *Lilium canadense*: Aa, plant in flower, with bulb, much reduced; Ab, flower, habit, × ⅓; Ac, same, vertical section, × ⅓; Ad, anther and distal end of filament, × 1½; Ae, ovary, vertical section, × 5; Af, same, cross-section, × 15; Ag, style tip with 3-lobed stigma, × 10. B, *Clintonia borealis*: Ba, plant in flower, × ¼; Bb, flower, habit, × 1; Bc, same, vertical section, × 1; Bd, ovary, vertical section, × 4; Be, same, cross-section, × 6. C, *Maianthemum canadense*: Ca, plant in flower, × ¼; Cb, flower, face view, × 3; Cc, pistil, vertical section, × 8. D, *Polygonatum pubescens*: Da, plant in flower, much reduced; Db, flower, habit, × 2; Dc, same, vertical section, × 2; Dd, ovary, cross-section, × 6; De, style tip with 3-lobed stigma, × 10. E, *Camassia Quamash*: Ea, plant in flower, × ⅛; Eb, flower, habit, × ⅓; Ec, same, vertical section, × ½; Ed, ovary, cross-section, × 3; Ee, same, vertical section, × 2; Ef, style tip with 3-lobed stigma, × 5. F, *Erythronium americanum*: Fa, plant in flower, × ½; Fb, flower, vertical section, × ½; Fc, ovary, vertical section, × 3; Fd, same, cross section, × 3; Fe, stigma, × 5; Ff, anther, with distal end of filament, × 3.

tralasian xerophytic linear and fibrous leaved Dasypogoneae, Lomandreae, and Calectasieae as the Xanthorrhoeaceae of his Agavales and the cormous Californian *Odontostomum* as part of his Tecophilaeaceae. By this reinterpretation essentially 4 subfamilies (Allioideae, Dracaenoideae, Luzuriagoideae, and Smilacoideae), 5 tribes (Petrosavieae, Dasypogoneae, Lomandreae, Calectasieae, and Parideae) and 2 subtribes (Odontostominae and Rusinae) were excluded by him from the Liliaceae of Krause and most authors. The remainder was treated as the Liliaceae and subdivided by him into 18 tribes composed of about 180 genera.

Phylogenetically the family is now believed to represent basic monocotyledonous stock from whose antecedents (i.e., ancestral liliaceous stocks) have evolved the great majority of present-day "petaloid monocot" families including such well-known representatives as the amaryllids, irids, palms, aroids, and orchids. Ancestrally its stocks are postulated possibly to have been derived from those of commelinaceous and butomaceous affinities. There is much morphological and anatomical evidence to indicate that the family, as classified by Krause, embraces a number of discordant taxa; the separation of families on the basis of ovary position may be less tenable than that based on inflorescence characters, and it is clear that several of the subfamilies and tribes treated by Krause as belonging to the Liliaceae are certainly as distinct from others in the family as the family is from other families of monocots.

The redefinition by Hutchinson of the Liliaceae and Amaryllidaceae on the basis of inflorescence character is probably more fundamental and stable than that based on ovary position, the latter a character that vacillates within several genera of these families between superior and inferior (*Ophiopogon, Bomarea, Hemerocallis*). The isolation of the xerophytic fibrous-leaved Dracaenoideae as one component of Agavaceae (together with another component similarly removed from the Amaryllidaceae) brings together a less firmly bonded group, morphologically, but one of common appearance, environment, and inflorescence type, a group as homogeneous biologically and no more artificial taxonomically than many others now differentiated within the Liliaceae of Krause. Anderson (1940) on the basis of studies of the floral anatomy and morphology of members of the Liliiflorae, and particularly the Liliaceae, favored the circumscription of the Liliaceae as maintained by Krause, but stated, with respect to Hutchinson's transfer of the umbellate spathaceous genera to Amaryllidaceae,

> The Agapantheae and Allieae have umbellate inflorescences and are closely related anatomically; consequently there may be some justification for the placing by Hutchinson of these forms in the Amaryllidaceae.

The transfer of the Agapantheae, Allieae, and Gilliesieae to the Amaryllidaceae has been accepted by some morphologists (Cheadle, 1942), supported by pollen studies (Maia, 1941), and its present adoption by many seems certain to gain support following gradual disintegration of what may well be a natural prejudice to change. For a discussion of the phylogeny of the Dracenoideae, see p. 413.

There is agreement that within the Liliaceae the rhizomatous groups are more primitive than are the bulbous, but it is probable also that within the former groups certain elements may have developed to levels as advanced as have the bulbous representatives. The Tofieldieae (Narthecieae, of Hutchinson) are undoubtedly the most primitive and have given rise to the Asphodeloideae from whose ancestral stocks most other elements of the family have evolved. It is also probable that the Tulipeae (including *Lilium*) stand well at the top of any phyletic schema of the family, but it is apparent that much study from all approaches is needed before any basic solution of the phylogeny within the group can be realized. At present, several hypotheses are on record; none is in accord with another.

Economically the lily family stands high in the number of its genera important as ornamentals with over 160 genera represented in the American trade. Tulips, lilies, hyacinths, scillas, autumn crocuses, and others constitute the bulk of the Dutch bulb trade. The day lilies (*Hemerocallis*) merit special attention. Asparagus is an important vegetable crop. The onion (*Allium*), if retained in this family, is of even greater commercial value. The red squill, used in rodent control measures, is from the bulbs of *Urginea*. Aloin, important in the drug trade, is obtained from large acreages of the genus *Aloe*.

LITERATURE:

ANDERSON, A. E. Some studies on the floral anatomy of the Liliales. Thesis (Ph.D.) Cornell Univ., 1940.
BERGER, A. Liliaceae-Asphodeloideae-Aloineae. In Engler, Das Pflanzenreich, 33 (IV. 38. III, ii): 1–347, 1908.
BUXBAUM, F. Die Entwicklungslinien der Liliodeae. Bot. Archiv. 38: 305–398, 1937.
CHEADLE, V. I. The occurrence and types of vessels in the various organs of the plant in the Monocotyledoneae. Amer. Journ. Bot. 29: 441–458, 1942.
ELWES, H. J. A monograph of the genus *Lilium*. 83 pp. 49 col. pl. London, 1877–1880. [For Suppl. see Grove.]
GATES, R. R. A systematic study of the North American genus *Trillium*, its variability, and its relationship to *Paris* and *Medeola*. Ann. Mo. Bot. Gard. 4: 43–92, 1917.
———. A systematic study of the North American Melanthaceae from the genetic standpoint. Journ. Linn. Soc. Bot. 44: 131–172, 1918.
———. A systematic analytical study of certain North American Convallariaceae, considered in regard to their origin through discontinuous variation. Ann. Bot. 32: 253–257, 1918.
GROVE, A. Supplement to Elwes' Monograph of *Lilium*. Pts. I–IV. 58 pp. 16 col. pl. London, 1933–1936. Pts. III and IV are by Grove, A. and Cotton, A. D.
JANKE, V. DE. Key to the Alliums of Europe. Herbertia, 11: 219–225, 1944. 1946.
KRAUSE, K. Liliaceae. In Engler and Prantl, Die natürlichen Pflanzenfamilien, ed. 2. Bd. 15a: 227–386, 1930.
MCKELVEY, S. D. Yuccas of the southwestern United States. Pt. 1: 1–150, 1938; Pt. 2: 1–192, 1947. Jamaica Plain, Mass.
MCKELVEY, S. D. and SAX, K. Taxonomic and cytological relationships of *Yucca* and *Agave*. Journ. Arnold Arb. 14: 76–81, 1933.
MAIA, L. D'O. Le grain de pollen dans l'identification et la classification des plantes. I. Sur la position systématique du genre *Allium*. Bull. Soc. Portugaise Sci. Nat. 13: 135–147, 1941.
MORTON, C. V. A check-list of Amaryllidaceae, tribe Allieae, in the United States. Herbertia, 7: 68–83, 1940.
OWNBEY, M. A monograph of the genus *Calochortus*. Ann. Mo. Bot. Gard. 27: 371–560, 1940.
OWNBEY, R. P. The liliaceous genus *Polygonatum* in North America. Ann. Mo. Bot. Gard. 31: 373–413, 1944.
SCHNARF, K. Ein Beitrag zur Kenntnis der Verbreitung des Aloins und ihrer systematischen Bedeutung. Oesterreich. Bot. Zeitschr. 93: 113–122, 1944. [A rearrangement of genera on the basis of Aloin content and fluorescent reactions.]
TRELEASE, W. The Yuccae. Rept. Mo. Bot. Gard. 13: 37–133, 1902.
———. The dersert group Nolineae. Proc. Amer. Philosph. Soc. 1: 404–442, 1911.
WATSON, S. A revision of the North American Liliaceae. Proc. Amer. Acad. Arts and Sci. 14: 213–288, 1879.

## HAEMODORACEAE. BLOODWORT FAMILY

Perennial herbs, often with orange to red sap; caulescent and usually stoloniterous, roots fibrous; leaves alternate, mostly basal, equitant, linear; inflorescence a simple or compound cyme, a raceme or panicle, often densely villous; flowers bisexual, regular or infrequently weakly zygomorphic, the perianth present and usually persistent, in 1 series, the 6 segments free or basally connate into a tube;

stamens 3 or less commonly 6, distinct, the anthers 2-celled, dehiscing by vertical slits, introrse; pistil 1, the ovary inferior, half inferior, or nearly superior, the locules and carpels 3, the placentation axile, the ovules solitary to numerous, the style 1 and usually filiform, the stigmas (or lobes) 3; fruit a loculicidal 3-valved capsule.

A family of 9 genera and about 35 species, largely confined to Australia (*Haemodorum* 20 spp.), South Africa (*Barberetta* 1 sp., *Dilatris* 2 spp., *Wachendorffia* 5 spp., *Pauridia* 1 sp.) and South America (*Hagenbachia* 1 sp., *Schiekia* 1 sp.) with 1 monotypic genus (*Lachnanthes*) extending northward in this country along the Atlantic coastal plain from Florida to Cape Cod.

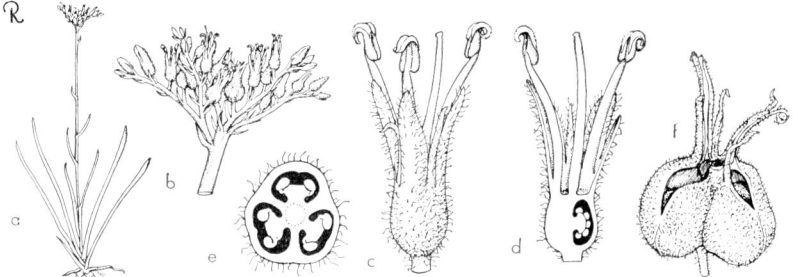

**Fig. 89.** HAEMODORACEAE. *Lachnanthes tinctoria*: a, plant in flower, × 1/10; b, inflorescence, × 1/2; c, flower, habit, × 4; d, flower, vertical section, × 4; e, ovary, cross-section, × 4; f, fruit, × 3.

There is disagreement among phylogenists as to how much or how little this family embraces. The above follows Pax's latest treatment (1930). By Bentham and Hooker and Gray's Manual (ed. 7) the monotypic *Lophiola* (New Jersey to Florida) was included within this family, but by Pax it was treated as a member of the tribe Conostylideae of his Amaryllidaceae. Hutchinson transferred the Conostylideae (which included *Lophiola*) to the Haemodoraceae with the observation that (p. 164) "thus constituted . . . the family is natural and homogeneous, not only in its *facies* but in its general distribution, which is predominantly austral, and mainly in Australia, with a few representatives in S. Africa and S. America." As reconstituted by Hutchinson, the family would be composed of 16 genera and about 93 species.

The family is of no known domestic economic importance.

*LITERATURE:*

PAX, F. Haemodoraceae. In Engler and Prantl, Die natürlichen Pflanzenfamilien, ed. 2. Bd. 15a: 386–391, 1930.

## AMARYLLIDACEAE. AMARYLLIS FAMILY [27]

Perennial mostly scapose plants; rootstock a rhizome, bulb, or corm; leaves mostly linear or lorate and basal (rarely cauline) or sometimes (in Agavoideae) fibrous, rigid and ensiform; inflorescence mostly umbellate, racemose, or paniculate, and sometimes reduced to a single flower; flowers bisexual, regular or infrequently zygomorphic, the perianth of 6 segments in 2 series, sometimes gamophyllous and

---

[27] Treated by Small (*Manual of the Flora of the Southeastern States*) as the Leucojaceae, a name given it by Batsch (1802). The conserved name Amaryllidaceae originated with Lindley (1836).

in some genera (*Narcissus*) bears a crown or corona;[28] stamens 6, inserted on the perianth, the anthers 2-celled, dehiscing by vertical slits or rarely by terminal pores (Galantheae), usually introrse or sometimes extrorse, versatile or occasionally basifixed, usually distinct but sometimes the filaments connected basally by a corolliform velamen (staminal corona);[29] pistil 1, the ovary inferior, or rarely half inferior to superior, trilocular (unilocular with basal placentation in *Calo-*

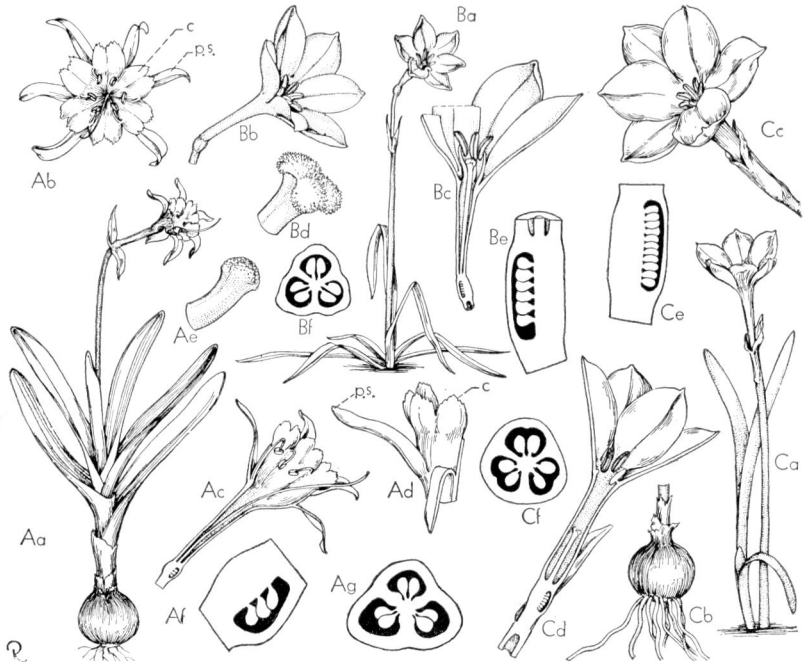

**Fig. 90.** AMARYLLIDACEAE. A, *Hymenocallis narcissiflora*: Aa, plant in flower, with bulb, × ⅛; Ab, flower, face view, × ⅙; Ac, same, vertical section, × ⅙; Ad, perianth section, showing corona, × ¼; Ae, style tip, with 3-lobed stigma, × 5; Af, ovary, vertical section, × 1; Ag, ovary, cross-section, × 1½. B, *Zephyranthes grandiflora*: Ba, plant in flower, less bulb, × ¼; Bb, flower, habit, × ½; Bc, same, vertical section, × ½; Bd, style tip and stigma, × 4; Be, ovary, vertical section, × 3; Bf, same, cross-section, × 3. C, *Cooperia pedunculata*: Ca, plant in flower, × ¼; Cb, bulb, × ¼; Cc, flower, habit, with spathe valve, × ½; Cd, same, vertical section, × ½; Ce, ovary, vertical section × 2; Cf, same, cross-section, × 3. (c corona, p.s. perianth segment.)

*stemma*), the placentation axile or the anatropous ovules paired and collateral and seemingly basal (ovule 1 in *Choananthus*), the style 1, the stigmas 3, or 1 and 3 lobed, or capitate; fruit usually a 3–celled capsule, or a berry (as in *Clivia, Cryptostephanus, Haemanthus*); seeds mostly several to many per locule, the endosperm present and fleshy.

[28] In the Zephyrantheae and Amarylleae there is a corona of perianth origin, usually consisting of scales or fimbriae (sometimes tubular as in *Placea*).

[29] In some genera the androecial corona is represented by an indistinct or distinct tooth at either side of the filament base (*Urceolinia, Vagaria*) or by membranes that are only shortly united (*Eurycles*).

As here defined (*sensu* Pax), the Amaryllidaceae are a somewhat heterogeneous family of 86 genera and about 1310 species, subdivided into 4 subfamilies (Amaryllidoideae, Agavoideae, Hypoxidoideae, Campynematoideae). The subfamily Amaryllidoideae (about 55 genera) is characterized by the plants bulbous, and the inflorescence an umbel (sometimes reduced to a single flower) subtended by 1 or more spathaceous bracts. It is generally accepted as composed of 2 tribes, the Amaryllideae (perianth lacking a corona) and the Narcisseae (perianth with a corona). The subfamily Agavoideae, containing the single tribe Agaveae, is composed of 7 genera of nonbulbous plants with the leaves fibrous and the flowers mostly small in large panicles or racemes. The subfamily Hypoxidoideae (22 genera) is composed of rhizomatous plants whose leaves are herbaceous and whose flowers are in cymes, racemes, or panicles. The 4 tribes are the Alstroemerieae, Hypoxideae, Conanthereae, and Conostylideae. Two genera, *Campynema* of Tasmania and *Campynemanthe* of New Caledonia, compose the subfamily Campynematoideae.

The classifications of Rendle, Wettstein, and Bessey were essentially in agreement with that of Pax and Hoffmann outlined above. However, that of Hutchinson (1934) represented a marked departure from these and one that is now favored by many authorities as representing phylogenetically a more realistic arrangement. Hutchinson restricted the Amaryllidaceae to those members having an umbellate inflorescence that is subtended by 1 or more spathaceous bracts. It is the subfamily Amaryllidoideae of Pax and Hoffmann plus those members of Liliaceae, (*sensu* Krause) possessing the same inflorescence characteristics (Agapantheae, Allieae, Gilliesieae). Within his Amaryllidaceae, Hutchinson elevated Pax's and Hoffmann's subtribes to the rank of tribe, recognizing each by the names accepted under the Engler system, except that 2 subtribes of the Narcisseae (Dentiferae and Phaedranassinae) are united and treated as a single tribe, the Eustephieae. The remaining 3 subfamilies are excluded from his concept of the Amaryllidaceae and treated as follows: the Agavoideae were elevated to the rank of family (Agavaceae, to which were added also the Yuccaae, Nolineae, and Dracaeneae from Liliaceae, *sensu* Krause); the tribe Alstroemerieae of Hypoxidoideae was raised to the Alstroemeriaceae; the tribes Conanthereae and Conostylideae of the same subfamily were incorporated, respectively, in the Tecophilaeaceae and Haemodoraceae; and the tribe Hypoxideae (together with the subfamily Campynematoideae) was segregated and elevated as the Hypoxidaceae.

The Amaryllidaceae (*sensu* Pax and Hoffmann) are widely distributed throughout the world, with the majority of their members occurring on the more level plains, plateaus, and steppe areas of the tropics and subtropics. Of the 86 genera, 3 (*Crinum, Hypoxis,* and *Amaryllis reginae*) are in the Old and New World (the first 2 are pantropical), 35 in tropical America, 22 in Africa, 12 in Australasia, 10 in the Mediterranean, and the remainder distributed in other regions. Only 1 genus (*Lycoris*) is restricted to eastern Asia, although perhaps 5 other genera have range extensions into the area. The largest genera, with the approximate number of species are: *Agave* (275), *Crinum* (130), *Bomarea* (120), *Hypoxis* (90), *Haemanthus* (60), *Amaryllis* [*Hippeastrum* of authors] (60), *Alstroemeria* (60), *Zephyranthes* (60), and *Narcissus* (43). Twenty-two genera are treated as monotypic.

The family has relatively few representatives in much of this country; they number 8 genera and about 66 species. Small (1933) recognized 7 genera (*Aletris, Hypoxis, Manfreda, Agave, Zephyranthes* [as *Atamasco*]*, Crinum, Hymenocallis*) represented by 38 binomials, as occurring in southeastern United States, and treated the family nomenclaturally as the Leucojaceae. Of these, *Aletris* with a half-inferior ovary, was placed in Liliaceae by Krause and by Hutchinson. Only *Hypoxis* is represented in the northeast and by a single species, although *Zephy* *ranthes Atamasco* occurs as far north as Virginia and *Manfreda virginica* (*Agave virginica*) to Virginia and Maryland. In the central states *Cooperia* extends from the southwest into Kansas, and *Hymenocallis* into Missouri and Illinois and *Habranthus texana* is restricted to Texas. In the southwest the Agavoideae are well represented, while 4 species of *Agave* are the only representatives of the family in the Pacific states (about 10 spp. of *Agave* occur in the southwest).

Economically the Amaryllidaceae contribute a large number of plants that are important to many activities. The agaves are primary sources of fiber used in cordage, patricularly of sisal, and henequen. Cuban and Mauritian hemp is made from leaves of the related *Furcraea*. In Latin American countries extensive acreages of Agave are grown as the source of the sugary exudate used as the basis for the distilled ginlike liquors mezcal and tequila. Pulque is a fermented beverage from the same source. Flour is made in Chile from roots of *Alstroemeria*. Aside from these products, the greatest domestic value of the family is among its ornamentals, which are represented by about 45 genera and perhaps 500 species. The better-known ornamentals are in *Narcissus, Amaryllis, Brunsvigia, Agave, Galanthus, Leucojum, Crinum, Nerine, Hymenocallis, Zephyranthes, Cyrtanthus,* and *Lycoris*.

*LITERATURE:*

BAKER, J. G. Handbook of the Amaryllideae, including the Alstroemerieae and Agaveae. 216 pp. London, 1888. [A monographic treatment with descriptions in English of all species known at that time.]

BELVAL, H. À propos des idées de Hutchinson sur les Amaryllidacées. Bull. Soc. Bot. Fr. 85: 486, 1938.

BRACKETT, A. Revision of the American species of *Hypoxis*. Contr. Gray Herb. 69: 120-155, 1923.

———. Some genera closely allied to *Hypoxis*. Op. cit. 69: 155–163, 1923.

CAVE, M. Sporogenesis and embryo sac development of *Hesperocallis* and *Leucocrinum* in relation to their systematic position. Amer. Journ. Bot. 35: 343–349, 1948.

FLORY, W. S. Cytotaxonomic notes on the genus *Habranthus*. Herbertia, 5: 151–153, 1938.

GRANICK, E. B. A karyosystematic study of the genus *Agave*. Amer. Journ. Bot. 31: 283–298, 1944.

HUME, H. H. The correlation of classification and distribution in *Zephyranthes*. Nat. Hort. Mag. 14: 258–275, 1935.

———. The genus *Cooperia*. Bull. Torrey. Bot. Club, 65: 79–87, 1938

HUTCHINSON, J. The families of flowering plants. II. Monocotyledons. New York, 1934. [Cf. Amaryllidaceae, pp. 128–135.]

McKELVEY, S. and SAX, C. Taxonomic and cytological relationships of *Yucca* and *Agave*. Journ. Arnold Arb. 14: 76–81, 1933.

MORAN, R. The Agavaceae. Desert Plant Life, 21: 64–69, 1949.

MORTON, C. V. A check-list of the bulbous Amaryllidaceae native to the United States. Year Book Amer. Amaryllis Soc. 2: 80–84, 1935.

PAX, F. and HOFFMANN, K. Amaryllidaceae. In Engler and Prantl, Die natürlichen Pflanzenfamilien, ed. 2, Bd. 15a: 391–430, 1930.

SEALY, J. R. *Zephyranthes, Pyrolirion, Habranthus,* and *Hippeastrum*. Journ. Roy. Hort. Soc. 62: 195–209, 1937.

TRAUB, H. P. The tribes of Amaryllidaceae. Herbertia, 5: 110–113, 1938.

TRAUB, H. P. and MOLDENKE, H. N. Amaryllidaceae: Tribe Amarylleae. Stanford, Calif. 1948, 1949.

TRAUB, H. P. and UPHOF, J. C. T. Tentative revision of the genus *Amaryllis* (Linn. ex parte) Uphof. Herbertia, 5: 114–131, 1938.

UPHOF, J. C. T. The history of nomenclature. *Amaryllis* (Linn.) Herb. and *Hippeastrum* Herb. Herbertia, 3: 101–109, 1935. [For refutation, cf. Sealy, J. R. in Kew Bull. 1939: 49–68, and for referee's opinion upholding Uphof's contentions, cf. Pam, A. in Journ. Roy Hort. Soc. 69: 102–103, 1944.]

———. Review of the genus *Habranthus*. Herbertia, 13: 93–97, 1946, 1948.

## DIOSCOREACEAE. YAM FAMILY [30]

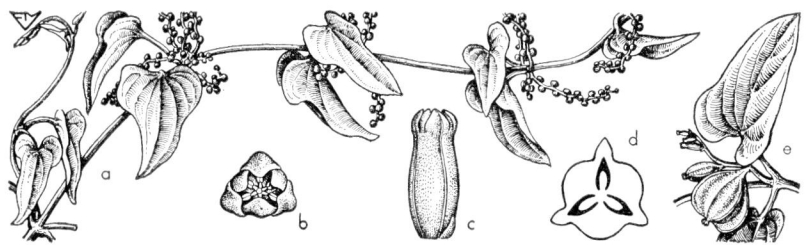

**Fig. 91.** DIOSCOREACEAE. *Dioscorea Batatas*: a, habit, staminate branch in bud, × ½; b, staminate flower, × 5; c, pistillate flower, × 5; d, ovary, cross-section, × 20; e, fruit, × ½. (From L. H. Bailey, *Manual of cultivated plants*, The Macmillan Company, 1949. Copyright 1924 and 1949 by Liberty H. Bailey.)

Climbing or twining herbs or shrubs, from a tuberous rhizome or much-thickened woody caudex; leaves alternate or rarely opposite, entire or palmately lobed or divided, mostly arrow-shaped with cordate bases, the venation mostly palmate-reticulate with each of several primary veins extending to apex, petiole usually jointed basally or twisted; inflorescence a spike, raceme, or panicle; flowers usually unisexual (bisexual in 4 Old World genera) and plants then generally dioecious, with regular, small, and inconspicuous flowers; perianth spreading or campanulate, of 6 parts in 2 series, usually basally connate and the structure then 6-lobed; staminate flowers with 6 stamens in 2 whorls with inner one sometimes reduced to 3 staminodia, the anthers basifixed, extrorse or introrse, 2-celled, the cells confluent or contiguous, dehiscing vertically, the connective occasionally much proliferated or appendaged, filaments distinct or briefly connate, a rudimentary pistil present or absent; pistillate flowers with single pistil, the ovary inferior, trilocular, the placentation axile, the ovules 2 to numerous in each cell, superposed and anatropous, the style 1 or more, the stigmas 3 or each 2-parted, the staminodes 2 or none; flowers when perfect have characters representing the combination of both sexes; fruit a capsule (often 3-winged) or berry; seeds sometimes winged, with endosperm.

A family of 10 genera and about 650 species divided into 2 tribes (Dioscoreae, dioecious plants with unisexual flowers, and Stenomerideae, plants with bisexual flowers) distributed widely throughout tropics and subtropics of the world and extending slightly into the north temperate zones. The largest genus is *Dioscorea* (ca. 650 spp.), most of whose species are tropical American. All but 3 of the other genera are of the Old World. Only *Dioscorea* occurs indigenously in this country, represented in the south by perhaps 6 species, 1 of which extends as far north as Ontario and Minnesota.

The Dioscoreaceae are interpreted by most American authors more conservatively to be of not more than 200 species. The most recent monographic treatment by Knuth (1930) recognized *Dioscorea* to be composed of about 600 species and distributed among 60 sections, and many species have been described subsequently. Hutchinson treated the Stenomerideae as 3 separate families, the Stenomeridaceae, Trichopodaceae, and Petermanniaceae.

Phylogenetically the family is derived from the Liliaceae and, except perhaps for *Petermannia* (which Hutchinson allied with the Alstroemereae) are a closely

[30] The name Dioscoreaceae has been conserved over other names, including Tamacaceae, used by J. K. Small (1933).

knit group. The netted-veined leaves, inferior ovaries, and usually unisexual flowers represent advancement over the liliaceous ancestors.

Economically the family is important as a source of several food plants, notably the yams of eastern commerce (*Dioscorea Batatas*), a very important article of food in the Far East. Shoots of the Mediterranean *Tamus communis* are eaten like asparagus. Both are cultivated in the southern regions of the country and another species of *Dioscorea* (the cinnamon vine) is grown for ornament.

*LITERATURE:*

KNUTH, R. Dioscoreaceae. In Engler, Das Pflanzenreich. 87 (IV. 43): 1–387, 1924.

———. Dioscoreaceae. In Engler and Prantl, Die natürlichen Pflanzenfamilien, ed. 2, Bd. 15a: 438–462, 1930.

PRAIN, D. and BURKILL, I. H. A synopsis of the dioscoreas of the Old World, Africa excluded, with descriptions of new species and of varieties. Journ. Asiatic Soc. Bengal, n.s. 10: 5–41, 1914. [An accounting of 107 spp. with keys, synonymies, and notes.]

SMITH, B. W. Notes on the cytology and distribution of the Dioscoreaceae. Bull. Torrey. Bot. Club, 64: 189–197, 1937.

## IRIDACEAE. IRIS FAMILY [31]

Perennial herbs or very rarely subshrubs (*Witsenia, Nivenia*), the roots produced from rhizomes, bulbs, or corms; stems solitary or several from the rootstock, or none and the peduncled flowers acaulescent from a corm; leaves mostly basal and numerous, generally equitant, mostly linear to ensiform, venation parallel; inflorescence generally racemose or paniculate, or the flowers solitary in a few genera; flowers subtended individually or in clusters by 2 spathelike bracts, usually showy, bisexual, regular or zygomorphic (sometimes those with flowers "regular" have curved perianth tubes), the perianth of 6 petaloid segments in 2 series undifferentiated into calyx and corolla (although those of the outer series usually distinguishable from those of the inner by color, size, or shape), all generally connate basally in a tube or the tube obsolete; stamens 3, representing the remaining outer whorl of an ancestrally biseriate androecium, situated opposite the outer perianth segments and often adnate to them, the anthers 2-celled, extrorse, dehiscing by vertical slits, mostly basifixed, the filaments usually distinct; pistil 1, the ovary inferior, trilocular with axile placentation or rarely (*Hermodactylus*) unilocular with 3 parietal placentae, the ovules few to many on each placenta (rarely solitary), anatropous, the style 1, often 3-branched, the stigmas 3, branches filiform to subulate or flabellate, entire or lobed or fimbriate, the style crests sometimes winged and petaloid, the stigmatic surface terminal or adaxial; fruit a loculicidal capsule dehiscing by 3 valves; seeds with copious endosperm, sometimes arillate, the embryo small.

A family of 58 genera and about 1500 species, distributed over much of the earth except in coldest regions; the center of distribution in Africa. About 33 genera are restricted to the Old World (28 to Africa), 14 to the New World, and 2 (*Libertia* and *Orthrosanus*) are disjuncts each occurring in Chile and Australasia. *Iris* is widely distributed over much of the temperate and subarctic regions of the northern hemisphere.

The family is represented in this country by indigenous species of 5 genera. *Iris* (represented by about 30 of the perhaps 225 known species) occurs over much of the continent and is especially abundant in the Mississippi delta region. *Sisyrinchium* (a large New World genus of 80 or more polymorphic species), is widely represented by 25 or more species, and 2 of its sections were recognized as genera by

---

[31] The name Iridaceae is conserved over other names for the family, including Ixiaceae

Fig. 92. IRIDACEAE. A, *Gladiolus tristis*: Aa, inflorescence, × ¼; Ab, flower, × ¼; Ac, same, vertical section, × ¼. B, *Crocus susianus*: plant with flower, vertical section, × ½. C, *Freesia refracta*: Ca, inflorescence, × ½; Cb, flower, vertical section, × ½; Cc, ovary, cross-section, × 5. D, *Iris Xiphium*: inflorescence, × ½. E, *Iris germanica*: flower, × ¼ (b beard, f fall, p.t. perianth tube, sp spathe, std standard, stg stigma-branch.) F, *Iris sibirica*: capsule, × ½. (From L. H. Bailey, *Manual of cultivated plants*, The Macmillan Company, 1949. Copyright 1924 and 1949 by Liberty H. Bailey.)

Abrams (*Olsynium* and *Hydastylus*). *Alophia* (*Herbertia,* Sweet not S. F. Gray), with 8 species distributed from Chile to Mexico, extends northward into southeastern and central Texas where it is represented on the plains and prairies by 3 species. *Calydorea* (a segregate of the South American *Nemastylis*) has 3 species from the eastern prairies of Texas north to Tennessee and Missouri. *Eustylis* is a monotypic genus of Texas and Louisiana. *Belamcanda chinensis,* a native of China, is naturalized in southeastern North America.

The treatment by Diels (1930) followed the earlier classification of Bentham and Hooker, who divided the family into the tribes Sisyrincheae, Ixieae, and Moraeeae.

Wettstein reclassified the genera into 3 subfamilies as follows: the Crocoideae, composed of the 2 genera *Crocus* and *Romulea* and characterized by the flowers solitary and acaulescent on a corm; the Iridoideae, composed of Diel's Moreeae and Sisyrincheae (except for the Crocinae); and the Ixioideae (Ixieae of Diels). This subdivision has been somewhat more widely accepted by followers of the Engler system than has that by Diels.

Hutchinson accepted the family essentially as delimited above, elevated the subtribes of Bentham and Hooker to the rank of tribe, subdivided the Irideae into 3 tribes (Irideae, Mariceae, Tigrideae) and added as a new tribe (Isophysideae) based on the monotypic Tasmanian *Isophysis* which differs from all other members of the family by its superior ovary, and which was treated by Diels as a subtribe of Liliaceae (as *Hewardia*). Hutchinson considered the Sisyrinchieae to be the most primitive tribe (next to the Isophysideae) and to represent the stock from which the Aristeae, Cipureae, Irideae, and Ixieae have been derived. He considered the Ixieae of Bentham and Hooker to be composed of 3 separate tribes: Ixieae, Gladioleae, and Antholyzeae, with the latter containing the most advanced members of the family.

Phylogenetically the family was treated by Diels, Wettstein, and by Bessey to have been derived from the Amaryllidaceae, a view influenced largely by the inferior ovary being common to both. Hutchinson discounted this view, which was emphasized by his transfer of the superior ovaried *Isophysis* from Liliaceae to Iridaceae, and pointed out that while the Iridaceae are similar to the Amaryllidaceae they have been "evolved separately from the Liliaceae and on different lines; rootstock usually a corm; inflorescence not umbelliform; some zygomorphy in the perianth." In addition to these considerations, Hutchinson considered the family to be sufficiently distinct and unrelated to other families to justify its separation as an order (Iridales) apart by itself.[32]

Economically the family is of greatest importance for its ornamentals. In addition to these, the rhizomes of several irises are commercial sources of orris root, used as an aromatic flavorant in dentifrices, and the stigmas of *Crocus sativa* are collected in commercial quantities for use in preparation of saffron dye. Domestically the iris and gladiolus are among the more popular of ornamental flowers, and over 10,000 vernacularly named clones are in the trade. Almost all genera of the family contain species of potential commercial value as ornamentals, but notable among them are *Crocus, Tigridia, Freesia, Ixia, Antholyza, Romulea, Neomarica, Moraea, Nemastylis, Eustylis, Belamcanda,* and *Sisyrinchium.*

LITERATURE:

ANDERSON, E. The problem of species in the northern blue-flags, *Iris versicolor* L. and *I. virginica* L. Ann. Mo. Bot. Gard. 15: 241–332, 1928.

———. The species problem in *Iris.* Ann. Mo. Bot. Gard. 23: 457–509, 1936.

---

[32] Hallier (1912) removed the Iridaceae from the Liliiflorae and placed them in his Ensatae, treating them as derived from the Liliaceae via the Velloziaceae.

BAKER, J. G. Handbook of the Irideae. 247 pp. London, 1892.
BICKNELL, E. P. Studies in *Sisyrinchium*. Bull. Torrey Bot. Club, 27: 373–387, 1900; *op. cit.* 28: 570–592, 1901; *op. cit.* 31: 374–391, 1904.
BRITTINGHAM, W. H. Cytological studies on some genera of the Iridaceae. Amer. Journ. Bot. 21: 77–82, 1934.
BROWN, N. E. The South African Iridaceae of Thunberg's herbarium. Journ. Linn. Soc. Lond. (Bot.) 48: 15–55, 1928.
DAUMAN, E. Die systematische Bedeutung des Blütennektariums der Gattung *Iris*. Bericht. Deutsch. Bot. Ges. 51: 157–164, 1933.
DIELS, L. Iridaceae. In Engler and Prantl, Die natürlichen Pflanzenfamilien, ed. 2, Bd. 15a: 463–505, 1930.
DYKES, W. R. The genus *Iris*, 245, pp. 48 col. pl., London, 1913.
FOSTER, R. C. Notes on nomenclature in Iridaceae. Contr. Gray Herb. 114: 37–50, 1936.
———. A cyto-taxonomic survey of the North American species of *Iris*. *op. cit.* 119: 3–80, 1937.
———. Studies in the Iridaceae. *op. cit.* 127: 33–48, 1939; *op. cit.* 136: 3–78, 1941; 155: 1–72, 1945.
HUTCHINSON, J. The families of flowering plants. II. Monocotyledons. London, 1934. [pp. 135–141.]
LEWIS, G. J. Iridaceae. New genera and species and miscellaneous notes. Journ. So. Afr. Bot. 7 (1): 19–59, 1941.
MAW, G. A monograph of the genus *Crocus*. 326 pp., 67 col. pl. London, 1886.
RANDOLPH, L. F. Chromosome numbers in native American and introduced species and cultivated varieties of *Iris*. Bull. Amer. Iris Soc. 52: 61–66, 1934.
RILEY, H. P. The problems of species in the Louisiana irises. Bull. Amer. Iris Soc. 74: 3–7, 1939.
VIOSCA, P. The irises of southeastern Louisiana. Bull. Amer. Iris Soc. 57: 3–56, 1935.

## Order 22. SCITAMINEAE [33]

A natural order of 4 tropical or subtropical families (Musaceae, Zingiberaceae, Cannaceae, Marantaceae) characterized by the functional stamens 1 or 5 (6 in *Ravenala madagascariensis*), the flowers usually zygomorphic, the leaves distichous or in spirals but the basal leaf sheath open and rarely closed, the ovary inferior, and the seeds with endosperm. Hutchinson divided the Musaceae into 3 families and counted the order to be composed of 6 families.

The order is generally accepted as comprising a well-knit assemblage of closely related, highly advanced monocots. It was considered by Engler to represent the ancestral stock of the Orchidaceae. Hutchinson treated it as the most advanced of his Calyciferae, or as he expressed it (p. 11), ". . . the climax of one line of development of the division in which the calyx and corolla have remained in separate whorls; often regarded as prototypes of Orchids, but here considered to be a parallel group . . ."

### MUSACEAE. BANANA FAMILY

Mostly large herbs, often treelike in appearance and then the caudex sometimes semiligneous, the stout stem unbranched and usually sheathed by the petioles; leaves large, alternate, sometimes distichous, entire (lacerations, when present, the result of mechanical injury), pinnately veined, convolute; inflorescence a spike or panicle or sometimes capitate, subtended by spathaceous bracts that may be large and coriaceous or semisucculent, and sometimes cymbiform; flowers bisexual or unisexual (when unisexual, the plants monoecious with staminate flowers within

---

[33] Authors designating orders by the ending -ales quite properly name this Scitaminales. On the basis of the presence of often arillate seeds in this order, it also has been referred to as the Arillatae. Hutchinson designated it the Zingiberales.

the upper bracts and pistillate flowers within the lower), each borne in the axil of a bract, irregular, the perianth present, of 6 parts in 2 series, the segments unequal in size and shape (sometimes even within the whorl), distinct or variously connate; stamens basically 6, one usually a staminodium and the other 5 fertile (*Ravenala* has 6 fertile stamens), the anthers 2-celled, narrowly linear, the cells parallel, dehiscing by vertical slits, the pollen granular, filaments distinct and filiform, the staminode (when present) a small and rudimentary stamen or petaloid; pistil 1, the ovary inferior, trilocular, the placentation axile, the ovules solitary and seemingly basal in each cell or numerous and clearly axile, anatropous, the style 1 and filiform, the stigmas usually 3 and each sometimes branched; fruit a 3-celled capsule or an elongated berry; seeds often arillate, with endosperm.

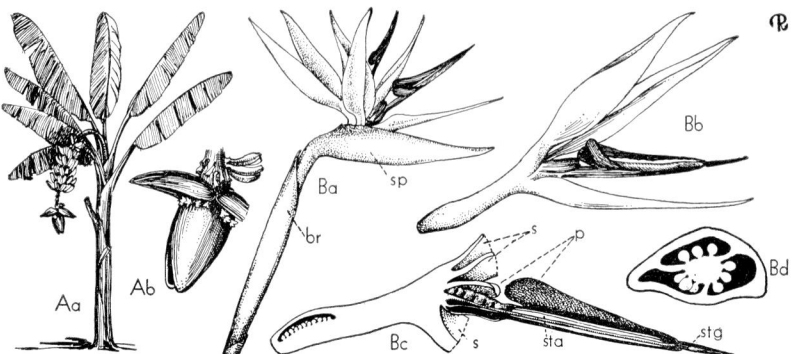

**Fig. 93.** MUSACEAE. A, *Musa paradisiaca*: Aa, plant in fruit and flower, much reduced; Ab, inflorescence, × 1/20. B, *Strelitzia Reginae*: Ba, inflorescence, × 1/6 (br bract, p petal, s sepal, sp spathe, sta stamen, stg stigma); Bb, single flower, × 3/8; Bc, same, vertical section (sepals partially removed), × 1/2; Bd, ovary, cross-section, × 2. (From L. H. Bailey, *Manual of cultivated plants,* The Macmillan Company, 1949. Copyright 1924 and 1949 by Liberty H. Bailey.)

A family of 5 genera and about 150 species of wide distribution in the tropics and present in the United States only as an escape or in cultivation.

Winkler (1930) treated the family as composed of 3 subfamilies, Stretlitzioideae, Musoideae, and Lowioideae. The Stretlitzioideae were divided into 3 tribes, the single genus in each being *Ravenala* (2 spp.), *Strelitzia* (5 spp.), and *Heliconia* (60 spp.). The subfamily Musoideae was treated as composed of a single unigeneric tribe to contain *Musa* (about 80 spp.) and the Lowioideae were likewise considered as represented by a single tribe and the 1 genus *Orchidanthera* (3 spp.). Hutchinson (1934) raised each subfamily to the rank of family (Musaceae, Strelitziaceae, Lowiaceae), and the genus *Ravenala* was subdivided into 2, *Ravenala* (Madagascar) and *Phenakospermum* (Guianas, Brazil).

Phylogenetically the Musaceae are undoubtedly the most primitive of the order, but one of the more highly advanced among monocots, as is evidenced by the reduced androecial conditions, inferior ovary, and the irregular flower. The presence of pinnately veined leaves is an advanced venation among monocots, but is of limited phylogenetic significance.

The family is of greater economic importance than others of the order, notably because of the significance of the banana (*Musa*) as a food and fiber plant. Many species of *Musa* are grown pantropically for their edible fruits, and the number of vernacular-named varieties extends into the hundreds. In some species of *Musa* the stem pith, apex of the immature flowering spike, and young shoots provide sources of food. *Musa textilis* is grown extensively in the Philippines and the Orient for the fiber from the sheathing leaf bases which is woven into Abaca cloth or used for cordage and known as Manila hemp. About 4 species of *Heliconia* are grown as ornamentals in the south or under glass. *Strelitzia reginae*, the bird-of-paradise flower, is an important commercial florist's crop in California and Hawaii, and is shipped to markets throughout the country. *Ravenala madagascariensis*, the traveler's-tree (traveler's-palm), is a flat-sided, distichously leaved ornamental grown in Florida and southern California as a landscape subject.

*LITERATURE:*

CHEESMAN, E. E. Classification of the bananas. I. The genus *Ensete* Horan. Kew Bull. 1947, 97–105; II. The genus *Musa* L. *op. cit.* 106–117; III. Critical notes on species. *op. cit.*, 1948, 11–28, 145–157.
DE WILDEMAN, E. Les bananiers, culture, exploitation, commerce, systématique du genre *Musa*. Ann. Mus. Col. Marseille, ser. 2, 10: 286–362, 1912.
SCHUMANN, K. *Musa*. In Engler, Das Pflanzenreich IV. 45: 13–28, 1900.
WINKLER, H. Musaceae. In Engler and Prantl, Die natürlichen Pflanzenfamilien, ed. 2, Bd. 15a: 505–541, 1930.

## ZINGIBERACEAE. GINGER FAMILY [34]

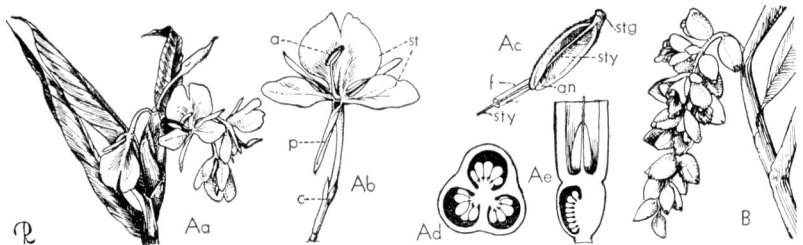

**Fig. 94.** ZINGIBERACEAE. A, *Hedychium coronarium*: Aa, flowering stem, × ⅙; Ab, flower habit (c calyx, p petal, st staminodia, a stamen, style and stigma), × ¼; Ac, anther and stigma (f filament, sty style, an anther-cell, stg stigma), × 1; Ad, ovary, cross-section, × 3; Ae, ovary, vertical section, × 2. B, *Alpinia speciosa*: inflorescence, × ⅙. (From L. H. Bailey, *Manual of cultivated plants*, The Macmillan Company, 1949. Copyright 1924 and 1949 by Liberty H. Bailey.)

Perennial herbs with creeping horizontal or tuberous rhizomes, rarely with fibrous roots; stems bracted and scapose or leafy and then very short or elongated; leaves basal or cauline, distichous, sheathing basally, alternate, sessile or petiolate, a ligule present at junction of blade with petiole or sheath, the blade mostly linear to elliptic and usually large, with closely parallel-pinnate venation with numerous veins oblique from midrib and strongly ascending; inflorescence a compact spike or open raceme, or the flowers solitary, each flower or cluster of flowers subtended by a usually conspicuous bract; flowers bisexual, irregular, the perianth of 6 parts, in 2 series, differentiated into a tubular 3-toothed or spathiform, somewhat herbaceous calyx and a tubular, unequally 3-lobed corolla (usually showy and of delicate texture) with posterior lobe usually the largest; fertile stamen 1, a large petaloid staminodium opposite it and a smaller staminodium may be present, the anther 2-celled, dehiscing vertically, the filament distinct and mostly slender and deeply

---

[34] The name Alpiniaceae was used for this family by J. K. Small, *Manual of the Southeastern Flora* (1933), and the naturalized genus *Alpinia* was treated nomenclaturally as *Languas*.

grooved; pistil 1, the ovary inferior, trilocular with axile placentation or unilocular with 3 parietal (rarely basal) placentae, the ovules many in each cell or placenta, the style 1, undivided, usually filiform and more or less enveloped in channel of filament of fertile stamen extending between the anther cells with the usually simple or capitate stigma protruding (in some genera the style not as above but 2-lipped or dentate); fruit a 3-valved, loculicidal capsule or fleshy, indehiscent, and berry-like; seeds with copious hard or mealy endosperm, the embryo straight.

A family of about 47 genera and 1400 species, distributed throughout the tropics and subtropics, but with *Costus* being pantropical and *Renealmia* occurring in both the New World and Old World. Of the remaining genera, 3 are restricted to Africa and 2 to Central and South America, 40 occurring only in Asia or extending southward to Australia. The family has no species indigenous in this country, but *Alpinia speciosa* is naturalized on hammocks in the southern Florida peninsula and *Hedychium coronatum* in marshes in eastern Georgia and along the Mississippi River below New Orleans, Louisiana. The larger genera of the family and their respective approximate number of species are: *Alpinia* (225), *Costus* (140), *Globba* (100), *Amomium* (90), *Zingiber* (80), *Renealmia* (70), *Curcuma* (54), *Bosenbergia* (50), and *Hedychium* (40).

The family was divided by Loesener (1930) into 2 subfamilies (Zingiberoideae and Costoideae) with the first subdivided into 3 tribes (Hedychieae, Globbeae, and Zingibereae) and the latter with no tribes and only 4 genera. Hutchinson rejected the subfamilies and treated the same 4 taxa as tribes. Distinguishing characters for the Zingiberaceae are the presence of aromatic oils, the ligule, the marked differentiation of the outer perianth series from the inner, the single stamen, and the large usually petaloid staminodium.

Economically the family is important as the source of ginger root (*Zingiber officinale*) used as a flavoring extract and as a condiment, for numerous fragrant oils used in perfumes, and for its ornamentals. In this country the more commonly cultivated ornamentals include: *Alpinia*, the shell ginger; *Hedychium*, the ginger lily or torch flower; *Elettaria*; *Cardamon*, whose seeds are used medicinally and as spice; *Roscoea* and *Curcuma*.

LITERATURE:

LOESENER, T. Zingiberaceae. In Engler and Prantl, Die natürlichen Pflanzenfamilien, ed. 2, Bd. 15a: 541–640, 1930.

## CANNACEAE. CANNA FAMILY

Large, coarse, perennial herbs, rootstock a tuberous rhizome; leaves cauline, large, and foliaceous, mostly oblong to broadly elliptic, pinnately veined with prominent midrib, a petiole present and sheathing the stem, no ligule present; inflorescence a raceme or panicle; flowers large and showy, bisexual, irregular, each subtended by a conspicuous bract and short-pediceled, the perianth differentiated into 2 series of 3 segments each, the outer series composed of 3 herbaceous more or less green to purple sepals which persist in fruit, the inner series of 3 similar but longer, erect, basally connate petals; androecium highly modified and comprising the showy part of the flower, the stamens 6 but sometimes reduced to 4, all briefly basally connate, petaloid, the 3 outer ones always sterile (one sometimes largest, reflexed, and termed the labellum), 2 of the inner more or less connate and the remaining fertile but petaloid stamen free, bearing a single 1-celled anther adnate to the petaloid margin near the apex; pistil 1, petaloid and usually winged (the two lateral weak wings sometimes interpreted as sterile stigmatic lobes), the ovary inferior, trilocular, multiovulate, the stigma 1 and represented by a stigmatic line on 1 apical margin: fruit a warty capsule dehiscing by collapse of the warty pericarp; seeds with very hard endosperm, small, numerous, subglobose, the embryo straight.

A family of the single genus, *Canna*. The species are 30–60 depending on the interpretation. Most authors concede it to be restricted to the New World tropics, with centers of distribution in the West Indies and Central America, but Eichler, Kranzlin, Engler, and Winkler have insisted that at least 1 and probably 3 species are indigenous to Africa and Asia. One species is native to the United States (*C. flaccida*) and extends northward along the swamps and marshes of the coastal plain from Florida into South Carolina. The West Indian *C. indica* is naturalized extensively along the coastal plain from Florida west to Texas.

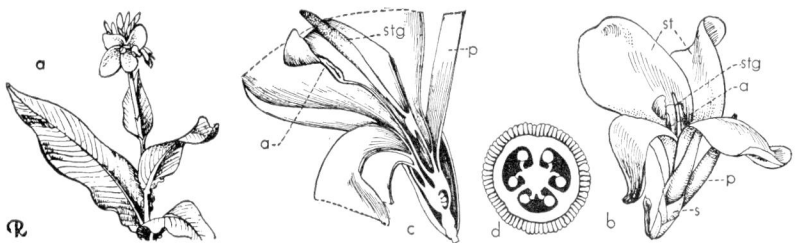

**Fig. 95.** CANNACEAE. *Canna generalis*: a, flowering plant, much reduced; b, flower (s sepal, p petal, a anther, stg stigma, st staminodia), × ¼; c, flower, vertical section, × ½; d, ovary, cross-section, × 4. (From L. H. Bailey, *Manual of cultivated plants*, The Macmillan Company, 1949. Copyright 1924 and 1949 by Liberty H. Bailey.)

Economically the family is important largely for the several species contributing to the ornamental canna employed prominently for tropical bedding effects, much of the material being of hybrid origin and treated as *C. generalis*. About 8 species are reported to be in the American trade.

*LITERATURE:*

BAILEY, L. H. *Canna*. Gentes Herb. 1: 118–120, 1923.
KRANZLIN, F. Cannaceen. In Engler, Das Pflanzenreich, 56 (IV. 47): 1–77, 1912.
WINKLER, H. Cannaceae. In Engler and Prantl, Die natürlichen Pflanzenfamilien, ed. 2, Bd. 15a: 640–654, 1930.

## MARANTACEAE. ARROWROOT FAMILY

Perennial herbs, stemmed or acaulescent, rhizomatous; leaves distichous, often imbricated, mostly basal or seemingly so, the blade usually linear, ovate, or oblong to elliptic, sometimes with one side straight and the other convex, the veins pinnate from the midrib and closely parallel, the petiole of 2 parts (a stalk and an open stem sheath), the petiole terete but also often winged and with a ligulate (or pulviniform) articulation at point of union with blade; inflorescence a spike or panicle, usually subtended and surrounded by spathaceous bracts, sometimes scapose and arising from the rhizomes; flowers bisexual, irregular, the perianth of 6 segments in 2 series, usually differentiated into a 3-sepaled calyx and an irregularly 3-lobed corolla that is basally tubular; androecium basically of 2 whorls, the outer whorl represented by 1 or 2 petaloid staminodes (sometimes absent), the inner by 1 fertile stamen and 1 or 2 (when two, often connate) petaloid staminodes, the fertile stamen usually petaloid with a 1-celled lateral or marginal anther whose filament may be apically distinct or fused with the petaloid portion, dehiscence by a vertical slit; pistil 1, the ovary inferior, trilocular (2 of the locules often sterile), the placentation axile, but the solitary ovule of each locule erect and appearing as if basal, the style 1, stout, flat and twisted, lobed, involute, or apically dilated, the stigma 1, terminal and sometimes truncate or a depressed foveolate stylar apex; fruit a loculicidal capsule or fleshy and berrylike but dehiscent; seeds often arillate, the endosperm copious, the embryo much incurved or folded.

The second largest family of the order, represented by 26 genera and about 350 species, all of the tropics and subtropics and primarily plants of moist or swampy forest habitats. The majority of the genera (19) are tropical American, with 7 native to Africa and 6 to Asia, the remainder being widely distributed in the Old World and only 1 (*Thalia*) occurring in both the Americas and the Old World (Africa). The only genus native to this country is *Thalia*, represented here by 2 of its 11 species; one, *T. dealbata*, of ponds, streams, and swamps of the coastal plain from Florida to South Carolina, extends westward to Texas and Missouri, and the second, *T. geniculata*, a predominately West Indian species, has its northern limits in Florida. One species of *Maranta* (*M. arundinacea*) is naturalized in southern Florida. The largest genera of the family are: *Calathea* (tropical South America, 130 spp.), *Ischnosiphon* (South America, 26 spp.), *Clinogyne* (Africa, 28 spp.), *Phrynium* (Old World tropics, 26 spp.), and *Maranta* (tropical America, 23 spp.).

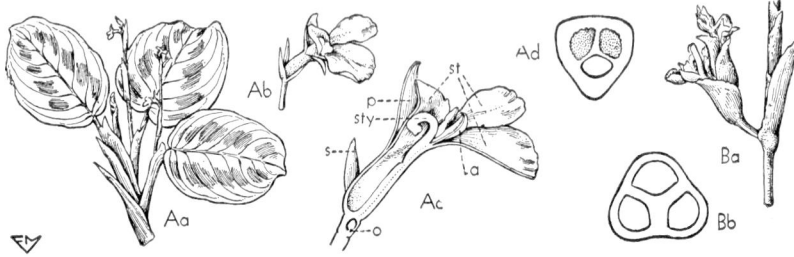

**Fig. 96.** MARANTACEAE. A, *Maranta bicolor*: Aa, flowering branch, × ⅙; Ab, flower, × 1; Ac, flower, vertical section (o ovary, s sepal, p petal, st staminodes, a anther, sty style), × 2; Ad, ovary, cross-section, × 15. B, *Calathea Lietzei*: Ba, inflorescence, × ½; Bb, ovary, cross-section, × 15. (From L. H. Bailey, *Manual of cultivated plants*, The Macmillan Company, 1949. Copyright 1924 and 1949 by Liberty H. Bailey.)

Distinguishing characters for the family are in the 3-sectioned leaves (sheath, petiole, blade) with the peculiar pulviniform ligule, the unusual androecial condition with its several staminodes and solitary 1-celled anther (a character common also to Cannaceae, but the latter family having multiovulate carpels), the 1-3-loculed ovary with a single, apparently basal ovule in each fertile locule, and the seed with curved or folded embryo.

Phylogenists agree that this is the most highly advanced family of the order, as evidenced by the highly complex androecium reduced to a single stamen and that to a 1-celled organ, by the reduction of the ovules to 1 in each locule, in many genera only 1 of the 3 locules ovule-bearing.

Economically the Marantaceae are important as a source of food products and for their ornamentals. West Indian arrowroot, or maranta starch, obtained from the rhizomes of *Maranta arundinacea*, is produced commercially in tropical America and is of importance in special diets requiring a readily digestible starch. Representatives of 6 genera are offered in the trade for ornamental uses. Among them are about 30 species of *Calathea* and 2-3 species of *Maranta* (much material sold as *Maranta* is actually *Calathea*), and *Thalia dealbata*, a popular garden aquatic.

*LITERATURE:*

LOESENER, T. Marantaceae. In Engler and Prantl, Die natürlichen Pflanzenfamilien, ed. 2, Bd. 15a: 654-693. 1930.

DIVISION IV. EMBRYOPHYTA SIPHONOGAMA

## Order 23. MICROSPERMAE [35]

The Microspermae are an order of 2 families (Burmanniaceae, Orchidaceae) allegedly characterized by the flowers usually irregular, bisexual, the ovary inferior and tricarpellate, the stamens adherent or adnate in whole or in part to the style, and by the very numerous minute seeds with an undifferentiated embryo and without endosperm or the latter exceedingly scant. There are exceptions to some of these characters, notably as concern stamens and seeds.

Advocates and supporters of the Engler system accept the order as the most highly developed taxon of monocots, a view that has lost support among many phylogenists. Hutchinson (1948) considered the grasses to be phylogenetically more advanced than the orchids. Wettstein (1935) likewise rejected the view that the orchids are the most advanced monocots, and treated his Spadiciflorae (palms, aroids, etc.) and Pandanales (Pandanaceae, Sparganiaceae, Typhaceae) as being the more advanced. Bessey (1915) concurred with Engler's view that the orchids are the most advanced monocots. There is no unanimity of opinion on the Englerian view that the Orchidaceae were derived from the Burmanniaceae or even that they are closely related, and only by the Engler system are the 2 families held to compose a single order. Hutchinson's views most closely approximated Engler's in this regard, for he accepted the Orchidaceae as derived from the Burmanniaceae, but elevated each to the rank of separate order and included 2 other small families in his Burmanniales. Engler's concept of this order closely paralleled that of Bentham and Hooker and differed only in that the Hydrocharitaceae, placed here by Bentham and Hooker, were excluded by Engler and aligned with the Heliobiae. Jonker, the most recent monographer of the Burmanniaceae, concluded that attempts to ally that family with the Orchidaceae were based on artificial or inconstant characters. The more recent consensus is that the orchids comprise an order by themselves, and that the Burmanniaceae are best treated as either an independent order or allied with the Liliiflorae.

### BURMANNIACEAE. BURMANNIA FAMILY

Annual or perennial often saprophytic herbs, usually small, slender and delicate, with stems generally unbranched; rhizomatous or tuberous; leaves numerous, scale-like, or the lowermost sometimes linear to lanceolate, mostly basal or the upper ones minute, alternate, venation parallel or indistinct; inflorescence racemose or cymose in a bifid cincinnus, or the flowers solitary and terminal; flowers bisexual, actinomorphic or irregular, mostly white or bright blue (rarely yellow); the perianth of 6 segments in 2 whorls, all segments connate basally into a single tube (often modified or very diverse in form) with the outer 3 lobes valvate in bud; stamens sometimes 3 or 6, inserted on the perianth tube, usually conspicuous and sometimes with an alate connective, the filaments short or scarcely apparent; pistil 1, the ovary inferior, trilocular with axile placentation or unilocular with 3 parietal (sometimes intruded) placentae, and when parietal the placentae sometimes meeting basally, the ovules very numerous and minute, the style 1, filiform (in Burmannnieae) or short cylindrical to conical (in Thismieae), with 3 branches or lobes bearing terminal stigmas; fruit a capsule dehiscing by longitudinal slits circumscissilely or loculicidally, often 3-winged, the perianth remains often persisting; seeds very numerous, minute, the endosperm scant, the embryo undifferentiated.

[35] This is the name employed for the order by Engler and followers. J. K. Small (1933) substituted for it the name Orchidales, but the latter generally has been restricted for use (as by Hutchinson, *et al.*) when the Orchidaceae were treated as comprising an order apart by themselves. Wettstein's Gynandreae has been used also in this latter and restricted sense in lieu of Orchidales.

The family was considered by Jonker (1938) as composed of 16 genera and about 125 species distributed widely throughout the tropics of the world. It is represented in this country by 3 genera and 5 species, most of which represent northern ranges of elements whose distributional centers are in the West Indies, Mexico, or southward. *Burmannia* has 3 indigenous species, one (*B. flava*) restricted to the southwestern tip of the Florida peninsula, and two (*B. biflora* and *B. capitata*) extending northward from southern Florida to North Carolina and west along the Gulf of Mexico to Houston, Texas. *Apteria* is represented only by *A. aphylla* which occurs over much of the same area (north to Georgia, west to Texas), and *Thismia americana* is a little-known, very rare endemic known only from an open prairie near Chicago, Illinois.

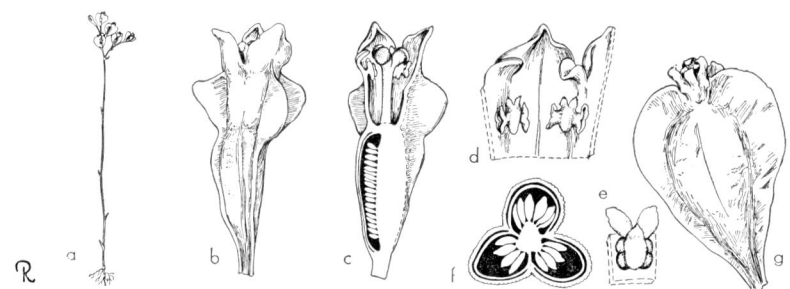

**Fig. 97.** BURMANNIACEAE. *Burmannia biflora*: a, plant in flower, × ½; b, flower, side view, × 7; c, flower, vertical section, × 7; d, perianth section with lobes and dehisced anthers, × 12; e, undehisced anther and appendages, × 10; f, ovary, cross-section, × 15; g, fruit, × 5.

The classification of Bentham and Hooker has been accepted until recently by most authors (Engler, Diels, *et al.*), wherein the family was divided into 3 tribes as follows: Euburmannieae with 3 stamens an actinomorphic perianth and the tube cylindrical, Thismieae with 6 pendant stamens an actinomorphic perianth and the tube oblong to obovoid, and Corsieae with 6 stamens and zygomorphic perianth. Schlechter (1921) monographed the Thismieae and treated them as composed of 10 genera. Hutchinson (1934) treated each of the 3 tribes as separate families. Wettstein (1935) treated the first 2 tribes as comprising the Burmanniaceae and the third as the Corsiaceae. Jonker (1938) concurred with Wettstein's disposition of the tribes and rejected Hutchinson's recognition of the Thismiaceae on the grounds that there was no sharp demarcation between them and the Euburmannieae, since the genus *Oxygyne* has characters common to both tribes and since the presence of an endosperm has been demonstrated in 2 species of *Thismia*.

The family has no members of domestic economic importance, although some members are reported to have been used (possibly as adulterants) in the composition of green tea.

*LITERATURE:*

ENGLER, A. Burmanniaceae. In Engler and Prantl, Die natürlichen Pflanzenfamilien, II (6): 44–51, 1889.

HUTCHINSON, J. Families of flowering plants. II. Monocotyledons. London, 1934. [pp. 175–179.]

JONKER, F. P. A monograph of the Burmanniaceae. Med. Bot. Mus. Rijksuniv. Utrecht, no. 51: 1–279, 1938.

SCHLECHTER, R. Thismieae. Notizbl. Bot. Gard. Berlin, 8 (71): 31–45, 1921.

## ORCHIDACEAE. ORCHID FAMILY

Perennial herbs; terrestrial, epiphytic, or saprophytic, sometimes vinelike, the terrestrial with fibrous or with thickened tuberous or cordlike roots, the epiphytic often with the leaf-bearing stem swollen to form a pseudobulb and often with aerial hanging cordlike roots covered by a layer of water-absorbing tissue (velamen), the saprophytic without chlorophyll; stems leafy or scapose, of sympodial or monopodial growth; leaves alternate or rarely opposite or whorled, simple, often distichous and sometimes closely imbricated, occasionally reduced to scales, membranous (and then sometimes plaited), coriaceous, or succulent, usually linear, lorate, or ovate, or orbicular, sheathing basally with the sheath generally closed and enveloping the stem: inflorescence a spike, raceme, or panicle, or else the flowers solitary; flowers bisexual, rarely unisexual (and then the plants either monoecious or dioecious), zygomorphic, always bracteate, sessile or pediceled (the ovary often not tumid at anthesis and then readily mistaken for a pedicel), mostly resupinate at anthesis due to a 180° twist in the ovary or pedicel during development; perianth typically of 6 segments in 2 series, the outer series of 3 sepals that are green or often colored and petaloid, all similar or the median more conspicuous in size or coloration, mostly imbricate, the inner series of 3 petals with the central (median) petal (the labellum or lip) usually larger, often very highly modified as to structure, shape and/or color and when resupinate stands in an abaxial position, this labellum frequently projected basally into a spur or sac with or without nectar within; ovary inferior, 3-carpelled, usually unilocular with 3 parietal biseriate placentae or in a few genera trilocular with axile placentation, the ovules very numerous, minute, anatropous, the style, stigmas, and stamens variously adnate into a single highly complex structure (the column or gynandrium); stamen 1 and terminal on the column or the stamens 2 and lateral (subtending the terminal stigma), each anther basically 2-celled and introrse, the pollen grains either granular (often in tetrads) or usually agglutinated into mealy, waxy, or bony masses (pollinia), the pollinia 2–8 per anther, the distal end of pollinium sometimes attenuated into a sterile filamentous strand (caudicle), free within each anther cell or more or less united to one another; staminodes often present as glandular or dentiform or ovate processes in 1–2 in number; stylar portion of column usually stout, the stigmas or stigma lobes basically 3, all or only 2 being fertile and functional (the 2 often confluent), when the latter situation exists, the 2 laterals fertile with the third and terminal stigmatic lobe modified into a small, sterile, proliferated outgrowth (rostellum) situated below the fertile and terminal solitary anther and between it and the usually depressed or concave fertile stigmas; in highly developed genera the rostellum becomes an integral part of the pollinium and then modified into a viscid disc or discs (viscidia); fruit a capsule, dehiscing by 3–6 hygroscopically sensitive valves which remain apically connate, the capsule opening medianly; seeds very abundant, minute, often fusiform, without endosperm, the embryo undifferentiated.

A very large family of about 450 genera and 10,000–15,000 species, of wide distribution over the earth in all hemispheres but most abundant in the tropics, where the majority of genera are epiphytes. Most of the genera of the temperate and all those of the arctic regions are terrestrial. When conservatively interpreted, the genera native to temperate and subtropical North America (exclusive of Mexico) are counted to be 43 and the species about 145. Of these genera, about a dozen are strictly epiphytic and restricted in this country to the southeastern United States. About 20 genera are present in northeastern North America, 36 genera in southeastern United States, 17 genera in the central plains and prairie states, and 9 genera

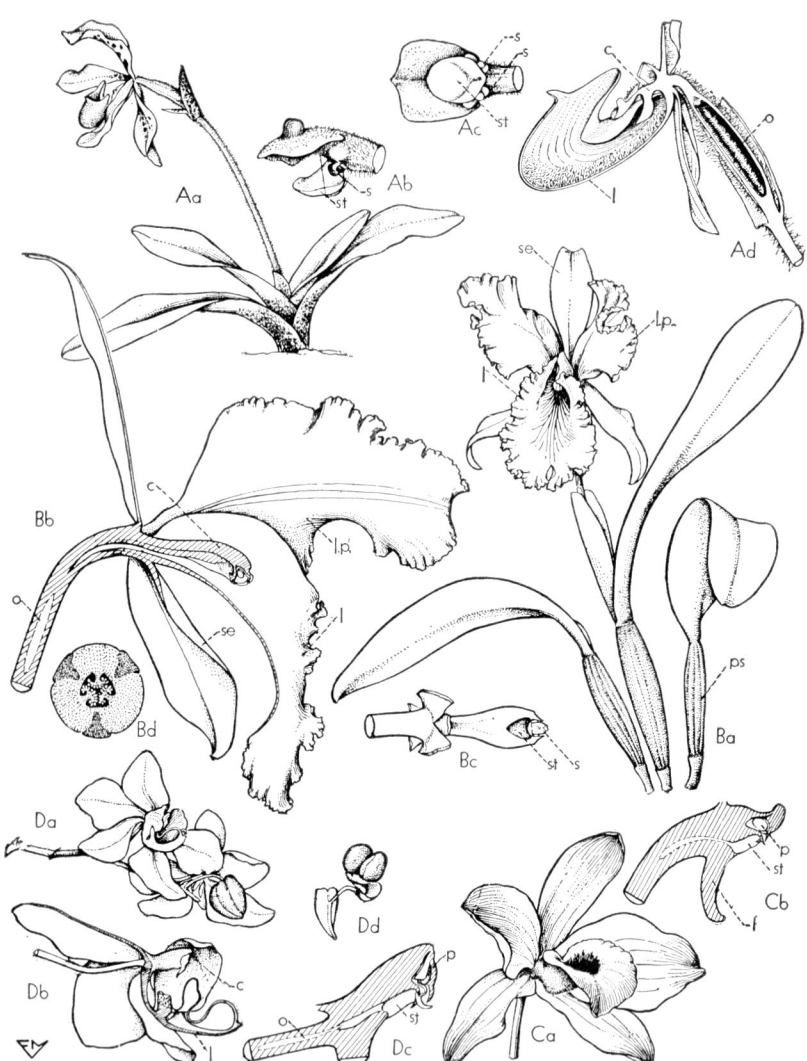

**Fig. 98.** ORCHIDACEAE. A, *Paphiopedilum insigne*: Aa, plant in flower, × ⅛; Ab, column, side view, × 1; Ac, same, from below, × 1; Ad, flower, vertical section, × ½. B, *Cattleya Lueddemanniana*: Ba, plant in flower, × ⅙; Bb, flower, vertical section, × ½; Bc, column from beneath, × ½; Bd, ovary, cross-section, × 2. C, Dendrobium nobile: Ca, flower, × ½; Cb, column, × 2. D, *Phalaenopsis amabilis*: Da, flowers, × ¼; Db, flower, vertical section, × ½; Dc, column, vertical section, × 1½; Dd, pollinia, × 3. (c column, f foot, l labellum, l.p. lateral petal, o ovary, ps pseudobulb, s stamen, se sepal, st stigma.) (From L. H. Bailey, *Manual of cultivated plants,* The Macmillan Company, 1949. Copyright 1924 and 1949 by Liberty H. Bailey.)

are credited to the Pacific coast states. The largest genera include *Spiranthes, Habenaria, Cypripedium, Listera* and *Epidendrum.*

The Orchidaceae, as described above, are more strictly circumscribed than earlier authors treated them, for the Diandrae are now considered to comprise a single tribe (Cypripediloideae or Cypripedilineae), whereas formerly they were considered to include also a second and more primitive tribe named the Apostasineae; views to this effect were held by Bentham and Hooker, Rolfe, Pfitzer, and others. However, Hutchinson, Schlechter, and other orchidologists considered the Apostasineae not to be orchids. They are a group of 3 Australasian monocot genera long considered to be relics of primitive orchids that were ancestral to the Cypripediloideae. They differ from orchids in possessing an actinomorphic perianth, and an androecium of 2 or 3 distinct stamens whose filaments are free or united at the base and to the style. There is no column, and the slender style is terminated by a 3-lobed stigma. By Hutchinson (1948) they were treated as the Apostasiaceae and as related to his Hypoxidaceae, and especially to the tropical genus *Curculigo*. In his discussion of them he stated, "these are not orchids in the true sense, although they may indicate the origin of the larger group."

The orchid flower is so highly specialized and modified as to require more explanation than that in the formal description above. It has undoubtedly arisen from liliaceous- or amaryllidaceouslike prototypes, and its complexity is more the product of adnation than of suppression of parts. In the flowers of most orchidaceous genera there is an outer series of 3 sepals (the abaxial referred to as the dorsal and the other two as laterals), and an inner series of 2 lateral petals and a labellum. It was once thought (Darwin, Brown, Rolfe) that the labellum represented a union of 2 stamens of the now absent outer whorl with a petal. Anatomical studies by Swamy (1948) showed that no vascular tissue of staminal origin was involved in the labellum, and that the latter was only one of the 3 members of the inner perianth whorl. All orchid flowers represent a reduction from the primitive ancestral type in that in the androecium (originally of 6 stamens in 2 whorls) 2 general conditions have developed: in the more primitive Diandrae, 2 of the outer stamens have been suppressed completely and the third is recognizable as a conspicuous staminode, while one of the inner whorl has been completely suppressed and the 2 remaining are the functional stamens situated on either side of the column posterior to the 3-lobed stigma; in the Monandrae 2 of the outer whorl of stamens have been suppressed and the third is present (fused with the style and stigmas) as the 1 functional anther situated terminally on the column,[36] while all 3 stamens of the inner whorl have been suppressed. The characters within the stigma and stamen of the Monandrae (the subfamily containing the great majority of the genera) are of particular significance taxonomically and morphologically. In this subfamily the stigma is situated in a usually depressed cavity (recognized by its glossy, viscid surface) immediately below the terminal anther, 2 stigmatic lobes or surfaces are present and often confluent; the third (absent in *Cephalanthera*) is modified into a nonfunctional lobe that secretes a quick-drying sticky material and is called the rostellum. In primitive monandrous genera the rostellum is quite distinct from the anther, but in higher genera the rostellum is inseparable from the anther in that it has become an integral part of the caudicles and viscid discs (viscidia) of the pollinia. The character of the anther is represented by an extensive series of forms, the recognition of which is important in identifying genera and species. The most primitive type is one of 2 cells containing granular pollen. This pollen cannot

---

[36] By some authors termed the "gynostegium" or "gynostemium." It has been considered by earlier authors (Oliver, Rendle, Willis) to represent an extension of the floral axis but studies by Swamy (1948) demonstrated it to be an appendicular structure containing the vascular strands normally associated with stamens and stigmas.

properly be called pollen grains but represents tetrads (daughter cells), and the contents of anther cells of this type are sometimes termed sectile pollinia. In more advanced genera the tetrads of pollen are agglutinated (by strands developed from sterile sporogenous material) into pollinia that are waxy or bony in texture. These pollinia are generally attached to the apical ends of the anther sac by slender stalks called caudicles. In the more advanced members the caudicles are terminated by a globose or disc-shaped gland. The number of pollinia varies among genera, the basic number being 2 (1 in each anther sac) but in other cases they may total 4 or 8, occasionally in different sized pairs (2 or 4 in each sac) in an anther. The treatment of the phylogeny of orchid genera by Rolfe is particularly helpful to an understanding of the gross morphology of the flowers of the many groups; the account of the evolution of the orchid flower by Godfery is predicated on direct ancestry from the Apostasineae, which he included within the Orchidaceae.

The most recent reclassification of the orchids, from a world viewpoint, was by Schlechter (1926) wherein he aligned the 610 genera recognized by him into 2 subfamilies, 4 tribes, and 81 subtribes. The latest classification of those orchids indigenous to United States and Canada was by Ames (1924), a conservative treatment agreeing in the main with that of Schlechter, but not recognizing the genera of the native orchids as so many as Schlechter held to be valid entities. The phyletic relationships of these genera and the subdivisions to which they belong are not indicated in many American floras or manuals, and for this reason the generic elements recognized by Ames are listed below according to Schlechter's schema. Characters of the categories are indicated within brackets. Well-known generic synonyms, or names of segregate genera appearing as valid in certain American floras, are placed in parentheses after their accepted name. Those generic names of the latter status are designated by an asterisk (*) if considered by Schlechter to be taxonomically valid.

Subfamily Diandrae [anthers 2 and lateral]
    Tribe Cypripediloideae (the only tribe of the subfamily)
        Cypripedium (*Calceolus, Criosanthes, Fissipes*)

Subfamily Monandrae [anther 1 and terminal]
    Division Basitonae [caudicle and viscidium arising from bases of pollinia; pollen always granular]
        Tribe Ophrydoideae (the only tribe of the division)
            Orchis (*Galeorchis*), Habenaria (*Blephariglottis\*, Coeloglossum\*, Gymnadenia\*, Gymnadeniopsis\*, Lysias, Perularia\*, Platanthera\**).

    Division Acrotonae [caudicle and viscidia arising from apices of pollinia; pollen various]
        Tribe Polychondreae [pollen granular; anther mostly persistent]
            Listera, Epipactis (*Serapias, Amesia*), Bletia (*Limodorum*), Triphora, Isotria, Pogonia, Vanilla, Cleistes, Arethusa, Calopogon, Prescottia, Cranichis, Ponthieva, Spiranthes (*Beadlea, Cyclopogon\*, Gyrostachys, Ibidium, Mesadenus\*, Neottia\*, Pelexia\**), Stenorrhynchus, Centrogenium, Goodyera (*Epipactis, Peramium*), Erythrodes (*Physurus\**), Tropidia.

        Tribe Kerosphaereae [pollen waxy or bony; anther commonly soon deciduous]
            Series Acranthae [inflorescence terminal or in axil of upper leaves]
                Pleurothallis, Malaxis (*Achroanthes*), Liparis, Tipularia, Calypso (*Cytherea*), Epidendrum (*Amphiglottis, Anacheilium, Auliza, Epicladium, Encyclia\*, Hormidium, Spathiger*), Basiphyllaea (*Carteria*), Polystachya.

            Series Pleuranthae [inflorescence lateral, from base of pseudobulb or axils of lower leaves or of the lower sheaths]
                Aplectrum, Hexalectris, Corallorhiza, Eulophia (*Platypus\*, Triorchos*), Cyrtopodium, Brassia, Ionopsis, Oncidium, Macradenia, Campylocentrum, Harrisella, Polyrrhiza.

DIVISION IV. EMBRYOPHYTA SIPHONOGAMA    437

The orchids of the British Isles were treated in excellent detail by Godfery (1933), whose phyletic arrangement of the 2 divisions of the Monandrae was based on a reversal of interpretation of the pertinent primitive versus advanced characters from that held by Schlechter (1926). Hutchinson (1934) accepted the views of Schlechter, but later (1948) rejected them in favor of those of Godfery "as they coincide more or less with the general principles on which the evolutionary system here put forward is based." The essential differences between the views of Schlechter and Godfery were that the latter considered genera, whose pollinia are attached by apical caudicles and viscidia, to be more primitive than those genera having them attached basally (i.e., that the Acrotonae are more primitive than the Basitonae). By this rearrangement, the more primitive genera of the Monandrae would include *Cephalanthera, Epipactis, Listera, Spiranthes,* and *Goodyera,* while the more advanced would be the genera *Habenaria, Orchis,* and the British *Ophrys.* This rearrangement of the 2 divisions in no way seriously affects the utility of the Schlechter schema, and would seem to make more reasonable the phyletic organization of its component genera.

Economically the orchids are important primarily for the ornamentals they contribute to the florists' industry and to horticulture. Capsules produced by members of the tropical genus *Vanilla* are the source from which natural vanilla extract is produced. Among the plants of greatest ornamental value are a few of terrestrial genera of temperate regions (*Cypripedium, Habenaria,* and *Bletilla*), the remainder being primarily epiphytic genera from the tropics or subtropics (*Paphiopedalum, Cattleya, Laelia, Phalaenopsis, Cymbidium, Odontoglossum, Dendrobium, Coelogyne,* and *Epidendrum,* with many other exotic genera cultivated by fanciers). For papers concerning culture of orchids from seed, consult Knudson, *et al.*

*LITERATURE:*

The literature restricted to the systematics of the Orchidaceae is voluminous. Unlike situations in many other families, it is not possible to restrict even a very selected accounting to entries concerning only North American elements. In addition to strictly taxonomic literature, the student should be cognizant also of the more important iconographic works and serials devoted to the subject of orchidology. The entries that follow comprise a minimum selection of titles.

AMES, O. The genus *Habenaria* in North America. Orchidaceae, fasc. 4, 1910.
―――. An enumeration of the orchids of the United States and Canada. 120 pp. Boston, 1924. [Includes key to genera, pp. 7–12. Now in process of revision.]
―――. The pollinia of orchids. Amer. Orchid Soc. Bull. 13: 190–194, 1944.
―――. Orchids in retrospect; a collection of essays on the Orchidaceae. 172 pp. Cambridge, Mass., 1948.
CORRELL, D. S. A contribution to our knowledge of the orchids of the south-eastern United States. Harvard Univ. Bot. Mus. Leafl. 8: 69–92, 1940.
―――. Orchids of North America. Waltham, Mass., 1950. [The only completely illustrated orchid flora of the United States and Canada.]
DAMMER, U. Kurze Uebersicht über die Gattung und Arten der Cypripedilinen. Orchis 1: 50–52, 57–60, 65–67, 76–79, 81–82, 1906–1907.
DARWIN, C. On the various contrivances by which British and foreign orchids are fertilised. London, 1862.
GODFERY, M. J. Monograph and iconograph of native British Orchidaceae. Cambridge, 1933.
HUTCHINSON, J. The families of flowering plants. II. Monocotyledons. London, 1934. [pp. 179–185.]
―――. British flowering plants. London, 1948. [pp. 272–280.]
KNUDSON, L. Physiological study of the symbiotic germination of orchid seeds. Bot. Gaz. 79: 345–380, 1925.
―――. Germination of seeds of *Vanilla.* Amer. Journ. Bot. 37: 241–247, 1950.

LINDLEY, J. Folia orchidacea. An enumeration of the known species of orchids. 300 pp. London, 1852–59.
PFITZER, E. Orchidaceae. In Engler and Prantl, Die natürlichen Pflanzenfamilien, II(6): 52–218, 1889.
ROLFE, R. A. The evolution of the Orchidaceae. Orchid Review 17: 129–132, 193–196, 289–292, 353–356, 1909; *op. cit.* 18: 33–36, 97–99, 129–132, 162–166, 289–294, 321–325, 1910.
SANDER, F. Reichenbachia; orchids illustrated and described. 2 vol. St. Albans (England), 1888–1890; ser. 2, 2 vol. 1892–94.
SCHLECHTER, R. Das System der Orchidaceen. Notizbl. Bot. Gart. und Mus. Berlin, 9(88): 563–591, 1926.
———. Die Orchideen, ihre Beschreibung, Kultur, und Züchtung: Handbuch für Orchideenliebhaber, Züchter, und Botaniker. ed. 2 [edited by E. M. Miethe], 960 pp., Berlin, 1927.
SCHUSTER, C. Orchidacearum Iconum index. Zusammenstellung der in der Literatur erscheinenden Tafeln und Textbildungen von Orchideen. Fedde, Repert. Spec. Nov. Beih. 60: 1–80, 1931; *op. cit.* 81–160, 1932; *op. cit.* 321–400, 1934. [An index to orchid illustrations that serves also as a very complete bibliography to pertinent literature.]
SWAMY, B. G. L. Vascular anatomy of orchid flowers. Harvard Univ. Bot. Mus. Leafl. 13: 61–95. 1948.
SERIALS:
Amer. Orchid Society Bull., Cambridge, Mass.
Australian Orchid Review, Sydney, Australia.
Pacific Orchid Society, Bull. Honolulu, T. H.
Journal des Orchidées. Paris. [6 vol., 1890–1896, edited by J. Linden.]
Lindenia: Iconographie des Orchidées, Gand & Bruxelles. [13 vol., 1885–1898, 814 col. pl.]
Orchid Digest, Berkeley, Calif.
Orchid Lore, Houston, Tex.
Orchid Review, London, England
Orchidaceae, Cambridge, Mass.
Orchidologia Zeylanica, Colombo, Ceylon
Orchis, Berlin (discontinued).
Orquidea, Rio de Janeiro, Brazil.
Philippine Orchid Review, Manila, Phil. Rep.

## Class II. DICOTYLEDONEAE [37]

Plants herbaceous or woody; stems with vascular elements arranged either in a hollow cylinder around the relatively small pith, or in bundles arranged in a single circle (as viewed in cross section), in woody dicot stems the sheath of cambium is situated close to the bark between the xylem and the phloem, and by means of cambial activity secondary growth layers are produced; leaves typically with netted venation of the palmate or pinnate type; flowers basically with parts numerous or in multiples of 4 or 5; seed embryos with typically 2 cotyledons.

The relative phyletic position of the monocots and dicots already has been presented (Chapter V, and under Monocotyledoneae, p. 371). The classification of the dicots has received varied treatment, even during the last few decades, and there is a lack of agreement as to the interrelationships of the taxa and as to their composition. Engler and Diels (1936) considered the dicot families to number 258, and distributed them among 44 orders (Reihen). In contrast to this view, Hutchinson (1926) considered the families to number 264 and the orders 76. Bessey (1915)

---

[37] The name Dicotyledoneae owes its origin to the typically 2 cotyledons present in embryos of members of this class. The character is not infallible, and examples of monocotyledony and polycotyledony are known within the taxon. For brevity, and without loss of clarity or precision, the term dicotyledon (and its adjectival form dicotyledonous) is reduced to dicot, where appropriate.

accepted 255 families distributed among 22 orders. On the basis of available data, it would appear that the number of families and orders more nearly approaches that recognized by Hutchinson, and that certainly the number of orders is greater than accepted by Engler or Bessey.

All except 6 of the dicot orders recognized by Engler are represented by species indigenous to this country, and 2 of those have species grown here as exotics. Recent studies have emphasized the taxonomic validity of certain families [38] not recognized by Engler but which are treated in this text as distinct. Of these 202 families of dicots, 154 are represented in this country by indigenous species and 48 others by exotic species that are regularly cultivated.

Effort has been made to provide for each family, in addition to carefully compiled technical descriptions, distributional notes, and data dealing with economic importance, a terse summary of available phyletic information. The disposition of the families by the several leading phylogenists and specialists has been indicated where pertinent, and especially where contemporary views of competent authority are at variance with those of Engler and his associates.

The distinguishing characteristics of the 44 orders of dicots are indicated in part by the analytical synopsis provided below. This synopsis does not represent an artificial key by which individual plants may be classified, since for the most part exceptions to the general situation are not indicated.

1. Flowers without a sepaloid or petaloid perianth (bracts may be present).
   2. Ovule with 20 or more embryo sacs.........................1. *Verticillatae*
   2. Ovule usually with only 1 embryo sac.
      3. Plants predominantly herbs; fls. mostly bisexual; infl. spicate......2. *Piperales*
      3. Plants predominantly woody; fls. mostly unisexual and the infl. of one or both sexes catkinlike.
         4. Staminate fl. a single naked stamen...................3. *Hydrostachyales*
         4. Staminate fl. of 2-more stamens, usually bracteate.
            5. Endosperm present; ovary superior.
               6. Fr. a capsule; seed comose........................4. *Salicales*
               6. Fr. a nut, drupe, or berry.
                  7. Ovary with 2-many ovules, each with 1 integument.
                     8. The ovary unilocular .......................5..*Garryales*
                     8. The ovary bilocular.....................7. *Balanopsidales*
                  7. Ovary with 1 ovule.
                     9. Ovule orthotropous, integument 1; style 2-branched ...............................6. *Myricales*
                     9. Ovule amphitropous, integuments 2; style simple ..................................8. *Leitneriales*
            5. Endosperm absent (except in 13); ovary inferior or superior.
              10. Ovary inferior.
                 11. Ovule solitary in ovary; lvs. pinnately compound.
                     12. Ovary 2-carpelled; plants monoecious......9. *Juglandales*
                     12. Ovary 3-carpelled; plants dioecious........10. *Julianiales*
                 11. Ovules 2 or more; lvs. simple.
                     13. Ovary 4-loculed; plants dioecious...........11. *Batidales*
                     13. Ovary 1-loculed; plants monoecious..........12. *Fagales*
              10. Ovary superior.
                 14. Stamens 4-12; endosperm present...............13. *Urticales*
                 14. Stamens numerous; endosperm lacking......14. *Podostemales*
1. Fls. with a perianth of sepals, or petals, or both.
   15. Petals distinct or mostly so, usually when not so are fused with sepals and stamens to form an hypanthium.
      16. Perianth parts generally alike (sepaloid or petaloid) and undifferentiated into calyx and corolla.

[38] Included here are the families Degeneriaceae, Eupteleaceae, Tetracentraceae, Winteraceae, Illiciaceae, Schisandraceae, Illecebraceae, Fumariaceae, and Hypericaceae.

17. Seed without endosperm............................15. *Proteales*
17. Seed with endosperm.
    18. Fls. mostly unisexual (sometimes bisexual in 17).
        19. Stamens opposite perianth parts and adnate to them ......................................16. *Santalales*
        19. Stamens alternate with perianth parts and free from them or numerous.
            20. Placentation axile or free-central, ovary usually inferior ............................17. *Aristolochiales*
            20. Placentation free-central, ovary superior..........................18. *Balanophorales*
    18. Fls. mostly bisexual..........................19. *Polygonales*
16. Perianth usually differentiated into calyx and corolla (corolla sometimes absent).
    21. Embryo coiled or curved, rarely straight; placentation basal or free-central.....................................20. *Centrospermae*
    21. Embryo usually straight; placentation usually axile or parietal.
        22. Floral parts usually spirally arranged; gynoecium mostly multipistillate, each pistil unicarpellate................21. *Ranales*
        22. Floral parts cyclic (androecium not always so); gynoecium mostly unipistillate and each pistil syncarpous (exceptions in 24).
            23. Lvs. tubular or with viscid glandular hairs, insectivorous ................................23. *Sarraceniales*
            23. Lvs. not insectivorous.
                24. Placentation basically parietal; floral envelopes never forming an hypanthium.................22. *Rhoeadales*
                24. Placentation predominantly axile or the ovules few or solitary in a uniloculate ovary; hypanthium often present.
                    25. Ovules orthotropous; plants dioecious; carpels 3, each with a single ovule..............25. *Pandales*
                    25. Ovules anatropous; plants not (or rarely) dioecious; carpels usually 1, or 4 or more.
                        26. Ovary predominantly inferior or surrounded by an hypanthium; pistils 1 to many, sometimes syncarpous; stamens mostly epigynous or perigynous......24. *Rosales*
                        26. Ovary inferior or superior, almost always syncarpous.
                            27. Stamens rarely more than twice as many as sepals, in 1 or 2 whorls, hypogynous.
                                28. The stamens twice as many as sepals and in 2 whorls, or in 1 whorl and opposite the sepals.
                                    29. Ovules pendulous with a ventral raphe and the micropyle up, or erect with a dorsal raphe and the micropyle down. ............26. *Geraniales*
                                    29. Ovules pendulous with a ventral raphe and the micropyle down, or erect with a ventral raphe and the micropyle up. ..............27. *Sapindales*
                                28. The stamens as many as the sepals and alternating with them (opposite the petals)...28. *Rhamnales*
                            27. Stamens usually numerous.

30. Ovary superior.
    31. Placentation usually axile; sepals usually valvate..29. *Malvales*
    31. Placentation usually parietal; sepals usually imbricate. ...............30. *Parietales*
30. Ovary usually inferior or surrounded by an hypanthium.
    32. Sepals and petals numerous, the series often not sharply differentiated; plants spiny, usually fleshy......31. *Opuntiales*
    32. Sepals and petals usually 4 or 5; plants not fleshy, rarely spiny.
        33. Ovules several to many in each locule (or if 1–2, the ovary superior and enveloped by an hypanthium)...32. *Myrtiflorae*
        33. Ovules 1–2 in each locule; ovary inferior; infl. cymose or umbellate..........33. *Umbelliflorae*

15. Petals connate (sometimes only very shortly so at base) and the corolla deciduous as a single unit.
    34. Ovary superior or generally so (inferior in one or few genera of nos. 35 and 38).
        35. Stamens free from corolla or adnate only at extreme base....35. *Ericales*
        35. Stamens borne on corolla.
            36. The stamens inserted at sinuses or connate into a tube..34. *Diapensiales*
            36. The stamens borne along corolla tube and not connate.
                37. Stamens opposite the corolla lobes, as many as or more than corolla lobes.
                    38. Ovary 1-loculed.
                        39. Placentation free-central, ovules usually many; style 1 .........  ....................36. *Primulales*
                        39. Placentation basal or the single ovule pendulous; styles or their branches 5............37. *Plumbaginales*
                    38. Ovary 2–5-loculed ..........................38. *Ebenales*
                37. Stamens alternate with corolla lobes, as many as the lobes or fewer.
                    40. The corolla herbaceous; fr. never a circumscissile capsule.
                        41. Corolla lobes usually convolute; ovary 2-carpelled, or ovaries 2 and each unicarpellate; stamens usually inserted at or near corolla base........39. *Contortae*
                        41. Corolla lobes imbricate; ovary 1 with 2–5 carpels; stamens usually adnate to the mostly elongated corolla tube ...........................40. *Tubiflorae*
                    40. The corolla scarious; fr. a circumscissile capsule ................................41. *Plantaginales*
    34. Ovary inferior.
        42. Stamens distinct, rarely coherent; lvs. opposite..............42. *Rubiales*
        42. Stamens coherent or variously connate; lvs. mostly alternate.
            43. Carpels 3, placentation predominantly parietal; fls. mostly unisexual; plants lianous herbs.......................43. *Cucurbitales*
            43. Carpels usually 2 or 5 (rarely 3), placentation axile or ovule 1 and basal or pendulous; plants infrequently lianous herbs ........................................44. *Campanulatae*

*LITERATURE:*

Most of the literature dealing with the phylogeny of the dicots has been accounted in Chapters V and VI. Throughout much of the treatment that follows, references are made to the respective treatments by Bessey, Small, Hallier, Hutchinson, Wettstein, and Rendle. These works are cited below and the citations are not repeated elsewhere in the following treatments of families.

BESSEY, C. E. Phylogenetic taxonomy of flowering plants. Ann. Mo. Bot. Gard. 2: 1–155, 1915.
HALLIER, H. L'origine et le système phylétique des angiospermes. Arch. Néerl. Sci. Exact. Nat. Ser. III B, 1: 146–234, 1912.
HUTCHINSON, J. The families of flowering plants. I. Dicotyledons. London, 1926.
———. British flowering plants; evolution and classification of families and genera with notes on their distribution. London, 1948.
RENDLE, A. B. The classification of flowering plants. II. Dicotyledons. ed. 1, reprinted with corrections. Cambridge, 1938.
SMALL, J. K. Manual of the southeastern flora. New York, 1935.
WETTSTEIN, R. VON, Handbuch der systematischen Botanik. ed. 2. Vol. 2. Leipzig, 1935.

## Order 1. VERTICILLATAE [39]

Characters as for the single family, Casuarinaceae. The phyletic position of the order is discussed under Casuarinaceae.

### CASUARINACEAE. CASUARINA FAMILY

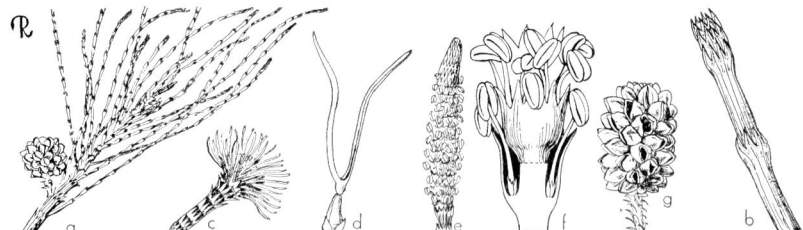

**Fig. 99.** CASUARINACEAE. *Casuarina equisetifolia*: a, fruiting branch, × ½; b, twig tip, × 3; c, pistillate inflorescence, enlarged; d, pistillate flower, × 15; e, staminate inflorescence, × 1; f, staminate flower, partly excised, × 10; g, fruit, × 1. (From L. H. Bailey, *Manual of cultivated plants*, The Macmillan Company, 1949. Copyright 1924 and 1949 by Liberty H. Bailey.)

Evergreen woody much-branched monoecious or dioecious trees or shrubs with jointed whorled and striate branches; leaves in whorls of 4–16, scalelike, usually linear to lanceolate, always connate basally, forming a sheath about the twig, and the leaf tips appearing only as teeth, internodes with striate grooves as many as leaves; flowers unisexual, without perianth; staminate flowers composed of a single stamen subtended by 4 obovate serrately margined bracteoles, in whorls, arising from within leaf sheaths along distal portion of twig (this fertile portion of twig sometimes said to be catkinlike), the anther basifixed, 4-celled, dehiscing by vertical slits; pistillate flowers capitate, each subtended by 1 bract and a pair of bracteoles; ovary superior, originally 2-loculed, but unilocular at anthesis by suppression, the ovules 2 on a single parietal placenta (1 ovule aborts), the carpels 2; style 1, very short, the stigmas 2 and linear; fruit a 1-seeded small-winged samara, enclosed by usually 2 woody bracteoles and a bract, all of which at fruit

---

[39] The name Casuarinales has been used for this order by Rendle, Small, *et al.*

maturity open like a capsule, many fruit and their bracteoles aggregated into a dry, woody, conelike, multiple fruit.

A unigeneric family (*Casuarina*) of the southern hemisphere and particularly of Australia, New Caledonia, to Malaya, and the Mascarene and other islands. Taxonomically, *Casuarina* is a poorly understood genus, and its species may number 40–50.

The family is distinguished readily from others by its equisetumlike jointed branches, whorled, minute leaves, and its woody, conelike fruit.

The apparently simple anemophilous flowers, the approach to a catkinlike staminate inflorescence, and the large rays in the wood resulted in the treatment of this family as the most primitive of dicots by Engler, and the retention of it within the Amentiferae by Rendle and by Wettstein, who interpreted it as the most primitive angiosperm and one derived from the Ephedraceae. It was recognized to be an advanced family by Bessey, by Hallier (as a member of his Amentacées), by Hutchinson, and by Tippo. Bessey treated it first (1897) as a member of the Sapindales and later (1915) as a leafless representative of and derived from the Hamamelidaceae within the Rosales; Hutchinson (1926) placed it at the top of the Amentiferae and derived the entire group from the Hamamelidales, and Tippo (1938) considered it, on the basis of anatomical evidence, to be a highly advanced derivative of the Hamamelidaceae. Moseley (1948), on the basis of anatomical evidence and floral morphology, concluded the family to be "moderately rather than highly specialized," and that the floral morphology indicated the family to have been derived by reduction from Hamamelidaceaelike ancestors—a view strongly supported by the anatomical evidence. An opposing view was expressed by Hjelmquist (1948), who, on the basis of the floral components and to the exclusion of anatomical evidence, held the family to be a primitive dicot but not a member of the Amentiferae. The preponderance of evidence gives more support to the views of Tippo and of Moseley than to those of Engler, Rendle, or Hjelmquist.

Economically the family is important as a timber tree in areas where indigenous (and there often known as the she-oak, a common name applied also to *Grevillea* of the Proteaceae) and several species are cultivated extensively in the warmer regions of this country as ornamentals.

*LITERATURE:*

HJELMQUIST, H. Studies on the floral morphology and phylogeny of the Amentiferae. Bot. Notiser, Suppl. 2: 1–171, 1948.

MOSELEY, M. F. JR. Comparative anatomy and phylogeny of the Casuarinaceae. Bot. Gaz. 110: 232–280, 1948.

POISSON, J. Recherches sur les Casuarina et en particulier ceux de la Nouvelle Calédonie. Paris, 1876. [Also in Nouv. Arch. Mus. Paris, I, 10.]

RENDLE, A. B. A new group of flowering plants. Nat. Sci. 1: 132–143, 1892.

TIPPO, O. Comparative anatomy of the Moraceae and their presumed allies. Bot. Gaz. 100: 1–99, 1938.

TREUB, M. Sur les Casuarinées et leur place dans le système naturel. Ann. Jard. Bot. Buitenzorg, 10: 145–231, 1891.

## Order 2. PIPERALES

An order composed of the Saururaceae, Piperaceae, and Chloranthaceae, with species of only the first 2 families indigenous to this country. Characteristics of the order are: plants predominantly herbaceous, leaves simple, the flowers in spikes or racemes, minute, mostly bisexual, and lacking perianth but often bracteate. The taxon has not been the subject of recent critical phyletic study; it was accepted by both Hallier and Hutchinson as a terminal and lateral offshoot of ranalian stocks.

## SAURURACEAE. LIZARD'S-TAIL FAMILY

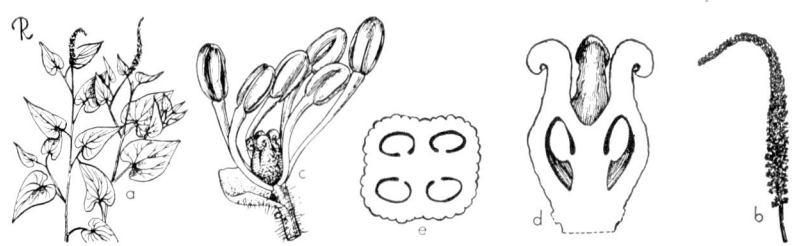

**Fig. 100.** SAURURACEAE. *Saururus cernuus*: a, flowering stems, × 1/10; b, inflorescence, × 1/4; c, flower, × 5; d, pistil, vertical section, × 18; e, ovary, cross-section, × 18. (From L. H. Bailey, *Manual of cultivated plants*, The Macmillan Company, 1949. Copyright 1924 and 1949 by Liberty H. Bailey.)

A small family of perennial herbs, usually of moist situations; stems erect; leaves broad, mostly ovate, alternate, simple, the stipules present and adnate to petiole; flowers bisexual, in dense slender peduncled spikes or racemes, without perianth, bracteate; stamens 6–8 or fewer by abortion, hypogynous or adnate to ovary and nearly epigynous, the filaments distinct, the anthers 2-celled, dehiscing by vertical slits; gynoecium superior to half-inferior, of 3–4 distinct 1-loculed, 1-carpelled pistils with parietal placentation, or the simple pistils basally connate into a single pistil with a compound 3–4-carpelled ovary and parietal placentae (axile in *Saururus*), the ovules usually 2–10 per carpel (often 1 in *Saururus*) and orthotropous, the styles and stigmas as many as carpels; fruit a semisucculent follicle or a fleshy capsule dehiscing by apical valves; seeds endospermous with small embryo.

Three genera (*Saururus, Anemopsis, Houttuynia*) and perhaps 4 species represent the family. One species of *Saururus* (*S. cernuus*) occurs in eastern North America and another in eastern Asia. *Anemopsis* is a monotypic genus (*A. californica*) indigenous to southwestern United States and northern Mexico. *Houttuynia* is represented by a single species occurring from the Himalayas to Japan.

The lizard's-tail family is allied to the Piperaceae, differing by its multicarpellate gynoecia, usually parietal placentation, and by the stem anatomy (vascular strands united into a single ring in Saururaceae and distinct in Piperaceae).

The family is of little economic importance, although dealers in native plants offer *Houttuynia, Anemopsis,* and *Saururus* as garden plants for wet soils.

*LITERATURE:*

HOLM, T. *Saururus cernuus* L., a morphological study. Amer. Journ. Sci. V, 12: 162–168, 1926.

## PIPERACEAE. PEPPER FAMILY

Erect or scandent herbs, shrubs or infrequently trees, evergreen when woody, the stems with vascular strands distinct and sometimes somewhat scattered as in monocots; frequently succulent when herbaceous, nodes often jointed or swollen; leaves alternate, rarely opposite or whorled, petiolate, entire, the stipules adnate to petiole when present, palmately or penninerved, often fleshy; flowers very minute, bracteate, usually bisexual or in some species unisexual, generally in dense fleshy spikes or the spikes umbellate; perianth absent; stamens 1–10 (primitively, probably 6 in 2 whorls of 3 each), hypogynous, the filaments usually distinct, the anthers each of 2 distinct or confluent cells (then appearing as 1-celled), dehiscence by longitudinal slits; pistil 1, the ovary superior, 1-loculed, the carpels 2–5 (primi-

# DIVISION IV. EMBRYOPHYTA SIPHONOGAMA

tive number probably 3), the ovule solitary basal orthotropous, the style 0–1, the stigmas 1–5 (often brushlike and lateral in *Peperomia*); fruit a small drupe (often described erroneously as a berry); seed small with endosperm and minute embryo.

A pantropical family of 10–12 genera, with the following in the New World: *Heckeria*, 8 spp. in South America; *Piper*, over 700 spp. in both hemispheres; *Nematanthera*, 2–4 spp. in the Guianas; *Verhuellia*, 2 spp. and *Symbryon* 1 sp., both of Cuba; and *Peperomia*, over 600 spp., in both hemispheres but predominantly from Mexico into South America. Four or 5 species of *Peperomia* (*Micropiper*, *Rhynchophorum*) are indigenous to rich woods and hammocks of peninsular Florida.

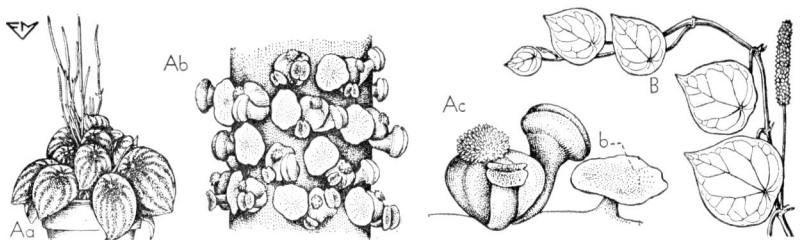

**Fig. 101.** PIPERACEAE. A, *Peperomia Sandersii* var. *argyreia*: Aa, plant in flower, × ⅙; Ab, section of inflorescence, × 10; Ac, flower and bract (b), × 25. B, *Piper nigrum*: branch with fruit, × ⅙. (From L. H. Bailey, *Manual of cultivated plants*, The Macmillan Company, 1949. Copyright 1924 and 1949 by Liberty H. Bailey.)

Members of the family are distinguished largely by the densely and minutely flowered spicate inflorescence, the naked flowers, and the 1-celled, 1-ovuled ovary. In many representatives the succulent herbage is distinctive.

Contrary to the views of Engler and of Rendle, it is now held generally that the family, while of undeterminate origin, is not one of the most primitive dicots, but probably is an independent and terminal offshoot of direct ranalian ancestry — a view supported by Hallier, Bessey, and Hutchinson.

Economically the family is important for the pepper of world spice markets: the ripened fruit of *Piper nigrum* of Java is the source of white pepper, while the unripe fruit of the same species is the source of black pepper. A narcotic beverage is produced in Oceania from roots of *Piper methysticum*. Species of *Macropiper*, *Piper*, and *Peperomia* are grown domestically as house plants for their foliage.

*LITERATURE:*

SMALL, J. K. The wild pepper-plants of continental United States. Journ. N. Y. Bot. Gard. 32: 210–223, 1931.

TRELEASE, W. The geography of the American peppers. Proc. Amer. Philos. Soc. 69: 309–327, 1930.

## CHLORANTHACEAE. CHLORANTHUS FAMILY

A small family of herbs, shrubs, or trees, evergreen when woody; leaves opposite, simple, minutely stipulate, usually aromatic, opposing petiole bases often meeting and connate; flowers bisexual or unisexual (the plants then monoecious), minute, bracteate, actinomorphic, in spikes, cymes, panicles, or heads; perianth none or rudimentary and caliciform in pistillate flowers, the bisexual flowers with a single stamen or with the stamens 3 and connate into a 3-lobed mass adnate basally to one side of the ovary, the anthers 2-celled (sometimes 1-celled) and dehiscing vertically, the pistil 1, with ovary inferior, 1-loculed, 1-carpelled, the ovule solitary pendulous and orthotropous, the stigma solitary and sessile or on a short style; when unisexual, the staminate flowers many on a pedunculed short-spicate axillary inflorescence, and the pistillate flowers few to several in a short raceme or spike, with the ovary usually enveloped by an adnate gamosepalous 3-toothed calyx; fruit a small ovoid to globose drupe; seed with abundant oily endosperm and a minute embryo.

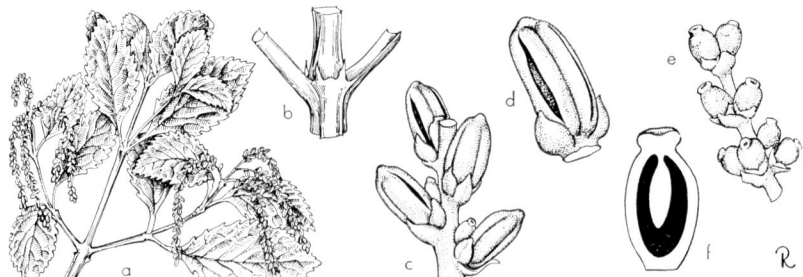

Fig. 102. CHLORANTHACEAE. *Ascarina lucida*: a, flowering branch, × ¼; b, stem node with stipules, × 2; c, staminate flowers, × 3; d, staminate flower, × 5; e, pistillate flowers, × 3; f, pistillate flower, vertical section, × 10. (c-f, adapted from Cheeseman.)

A tropical and subtropical family composed of 2 Old World genera (*Chloranthus* with about 12 spp., and *Ascarina* with 3 spp.) and a single New World genus (*Hedyosmum*, with about 25 spp. of tropical America and extending northward to Jalapa, Mexico). None of the family is indigenous to the United States. *Chloranthus glaber*, a low ornamental shrub, is grown in southern California for its foliage and bright red fruit.

The family is distinguished from others within the order by the opposite stipulate leaves, united petiole bases, and the inferior ovary with the solitary pendulous ovule.

*LITERATURE:*

CORDEMOY, C. JACOB DE. Monographie du groupe des Chloranthacées. Adansonia, 3: 280–310, 1863.

### Order 3. HYDROSTACHYALES

An order of a single unigeneric family, Hydrostachyaceae, of African dioecious aquatics.

### Order 4. SALICALES [40]

A distinctive order with the characteristics of its 1 family, Salicaceae. There is much evidence that the family has been derived from advanced ancestral stocks, but data are yet needed before the phyletic position of it or its ancestors can be indicated.

[40] The dicot orders Salicales through Fagales were designated by Eichler (1883) as a single order, Amentaceae. They were accepted as a phyletically homogeneous group of primitive dicots by Engler and by Rendle, both of whom rejected Eichler's name and considered them to represent several orders. Hutchinson transferred all these amentiferous orders (plus the Urticales) to a phyletic position that treats them as descendants of hamamelidaceous ancestors. The plants were restudied by Hjelmquist (1948) and treated by him as a natural assemblage of several orders comprising a single taxon, Amentiferae. For several decades a wide range of evidence has been accumulating in support of the view that these dicots are neither natural nor primitive taxa. Discussions of and references to this supporting literature are cited in the treatments of the several families that follow.

## SALICACEAE. WILLOW FAMILY

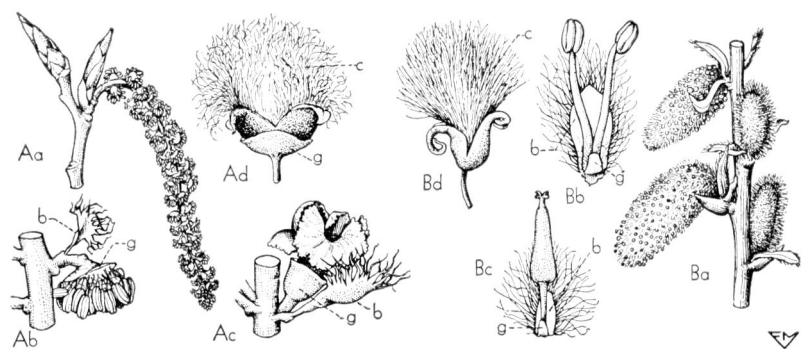

**Fig. 103.** SALICACEAE. A, *Populus balsamifera*: Aa, twig with staminate catkin, × ½; Ab, staminate catkin axis with flower, × 2; Ac, pistillate flower, × 4; Ad, capsule, × 2. B, *Salix fragilis*: Ba, staminate catkins, × 1 (catkins on left of twig at anthesis, others less mature); Bb, staminate flower, × 6; Bc, pistillate flower, × 6; Bd, capsule, × 2. (b bract, c seed coma, g gland.) (From L. H. Bailey, *Manual of cultivated plants*, The Macmillan Company, 1949. Copyright 1924 and 1949 by Liberty H. Bailey.)

Woody trees or shrubs (a few arctic species suffrutescent), dioecious (individuals rarely monoecious); leaves deciduous, simple, alternate or rarely subopposite (*S. purpurea*), stipulate (often caducously so), generally petiolate; flowers unisexual, those of each sex in dense erect to pendulous catkins, perianth absent or vestigial, each flower seemingly subtended by a fringed or hairy bract and a cupular disc or 1–2 nectiferous glands (when 2, one is above and one below the pedicel), appearing before or with the leaves; staminate flowers of 2-more stamens, the filaments distinct or (in some *Salix* spp.) basally connate (wholly connate in *S. sitchensis*), the anthers 2-celled, dehiscing vertically; pistillate flowers with pistil 1, sessile or short-stipitate (long-stipitate in *S. Bebbiana* and others), the ovary superior, 1-loculed, the placentae 2–4 and parietal, the carpels 2–4, the ovules anatropous, usually many with single integument (except in 2 spp. of *Populus*), the style 1 with 2–4 stigmas; fruit a 2–4-valved capsule; seeds comose, embryo straight, the endosperm little or none.

The family is composed of 2 genera (*Populus* with about 30–40 spp., *Salix* with about 300 spp.) of almost world-wide distribution (absent only in Australasia and the Malayan Archipelago, but 1 species in the Philippine Isls.), the most primitive species occurring in the tropics but with present centers of distribution in the north temperate and subarctic regions.[41]

Members of the family may be distinguished from catkin-bearing representatives of other families by the plants dioecious, by the flowers seemingly subtended by bracts and a cuplike disc or a gland, and by the comose seeds.

Morphological studies by Fisher (1928) on an extensive number of species of both genera concluded that (1) the "simplicity of the flowers . . . is largely due to extreme reduction and not to a retention of archaic features"; (2) the flowers once (in ancestral forms) possessed a perianth of certainly 1 and probably 2 series which is now represented by the cuplike gland of *Populus* and the usually finger-

---

[41] A third and monotypic taxon proposed as a genus by Nakai in 1920 under the name *Chosenia* was accepted as a genus by Wettstein, Komarov (who later rejected its validity as a genus), and Rehder. As a result of extensive morphologic and taxonomic studies, Hjelmquist (1948) rejected its claim to generic rank and included it in *Salix*.

like gland or glands of *Salix;* (3) the bract "does not subtend the flower . . . but is a leaf upon the pedicel of a formerly stalked ancient flower"; (4) the family is not allied with the Tamaricaceae as alleged by Bessey (1887) but is an advanced phyletic group of unknown ancestral relationships; and (5) that within the family, *Populus* is the more primitive. Hutchinson (1926) retained it with the other amentiferous families of presumed hamamelidaceous affinities and treated it as the most primitive of them. Hjelmquist (1948) was of the opinion that the cup or fingerlike gland characteristic of flowers in this family was "formed by the reduction of an undifferentiated bracteal envelope and that it is not quite appropriate to designate them as perianth." He included the Salicaceae in an order by itself and treated it as the most advanced within the taxon Amentiferae. There is evidence to indicate the Salicaceae to be a more highly specialized group than most or all other amentiferous families and the only one of an order that stands well apart from other orders of dicots.

Economically the family is important for the many species of both genera grown as ornamentals. *Salix* species are sources of wood made into charcoal and for withy twigs used in basketry.

*LITERATURE:*

ANDERSON, N. J. Salicineae. In de Candolle's Prodromus, 16(2): 190–331, 1868.
BALL, C. R. New or little known west American willows. Univ. Calif. Publ. Bot. 17: 399–434, 1934.
ERLANSON, E. W. and HERMANN, F. J. The morphology and cytology of perfect flowers in *Populus tremuloides* Michx. Papers Mich. Acad. Sci. 8: 97–110, 1928.
FISHER, M. J. The morphology and anatomy of flowers of Salicaceae. I. Amer. Journ. Bot. 15: 307–326, 1928.
HJELMQUIST, H. Studies on the floral morphology and phylogeny of the Amentiferae. Bot. Notiser, Suppl. 2: 1–171, 1948.
PETO, F. H. Cytology of poplar species and natural hybrids. Canadian Journ. Res. 16: 445–455, 1938.
SCHNEIDER, C. Über die systematische Gliederung der Gattung *Salix*. Österreichische Bot. Zeit. 65: 273–278, 1915.
———. A conspectus of Mexican, West Indian, Central and South American varieties of *Salix*. Bot. Gaz. 65: 1–41, 1917.
———. Notes on American Willows. Bot. Gaz. 66: 117–142, 318–353, 1918; *op. cit.* 67: 27–64, 309–396, 1919; Journ. Arnold Arb. 1: 1–32, 67–97, 147–171, 211–232, 1919; *op. cit.* 2: 1–25, 65–90, 185–204, 920; *op. cit.* 3: 61–125, 1921. [A revision of the genus in 12 articles, concluded by analytical keys, in Latin, to the American species.]
SUDWORTH, G. B. Poplars, principal tree willows and walnuts of the Rocky Mountain Region. U.S. Dept. of Agr. Tech. Bull. 420: 1–111, 1934.
WILKINSON, J. The cytology of *Salix* in relation to its taxonomy. Ann. Bot. II, 8: 269–284, 1944.

## Order 5. GARRYALES

An order with the characteristics of its only family, Garryaceae, and distinguished from other orders of dioecious amentiferous plants by its unilocular biovulate ovary, and baccate fruit. The relationship of the order is probably within or near the Umbelliflorae, and it is not at all a primitive taxon as interpreted by earlier botanists.

## GARRYACEAE. THE GARRYA FAMILY

A small unigeneric family of dioecious evergreen shrubs or less commonly trees; leaves opposite, simple, entire, leathery, the opposing petioles basally connate, estipulate; flowers unisexual, those of each sex disposed in pendulous silky-hairy catkinlike racemes; staminate flowers stalked, the perianth composed of 4 valvate decussate "sepals" or bracts, often apically coherent with stamens protruding between them; the stamens 4, distinct, alternate with the sepals, the anthers 2-celled, basifixed, dehiscing by vertical slits, a conical rudimentary pistil present; pistillate flowers in dense bracteate pendulous raceme, each flower sessile, without perianth, composed of a single pistil with ovary superior, 1-loculed, probably 2-carpelled with the placentation fundamentally parietal, the ovules anatropous, pendant from near the locule apex and with a single integument, the styles 2, distinct, subulate, spreading; fruit a berry (sometimes 1-seeded by abortion), globose to ovoid, terminated by the persistent styles; seed with a minute straight embryo and copious endosperm.

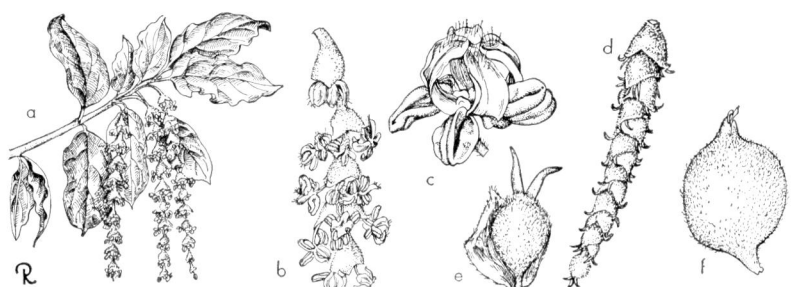

**Fig. 104.** GARRYACEAE. *Garrya elliptica*: a, branch with staminate inflorescences, × ⅓; b, portion of staminate inflorescence, × 1; c, staminate flower, × 5; d, pistillate inflorescence, × 1; e, pistillate flower, × 3; f, fruit, × 4.

*Garrya*, the only genus of the family, has about 15 species extending from California (4 spp.) through southwestern United States and Mexico (10 spp.) to the West Indies (1 sp. in Jamaica).

Some students (Hallock, Hjelmquist, Wettstein) of this and related families hold that the Garryaceae are not related to the amentiferous taxa, but belong to a separate order (Garryales) presumed to have affinities with the Umbelliferae. Other authors (incl. Jepson) have placed it near the Cornaceae, a family of similar alliance.

The family is characterized by the leaves opposite, estipulate, and with joined petiole bases, by the pendulous catkinlike unisexual inflorescences, and the 2-styled pistillate flowers.

Economically the Garryaceae are of little importance. A few species are cultivated as ornamental shrubs in southern California, being prized for their glossy evergreen foliage and reddish-purple drooping inflorescences.

*LITERATURE:*

BAILLON, H. Organogénie floral des *Garrya*. Compt. rendu de l'Assn. française. 1887, pp. 561–566. [Reprinted in Adansonia XII, tome 6.]
COULTER, J. M. and EVANS, W. H. *Garrya*. Bot. Gaz. 15: 93–97, 1890.
EASTWOOD, A. Notes on *Garrya*, with description of new species and key. Bot. Gaz. 36: 456–463, 1903.
HALLOCK, F. A. The relationship of *Garrya*. Ann. Bot. 44: 771–812, 1930.
HJELMQUIST, H. Studies on the floral morphology and phylogeny of Amentiferae. Bot. Notiser, Suppl. 2: 1–171, 1948.
WANGERIN, W. Garryaceae. In Engler, Das Pflanzenreich, 41 (IV, 56a): 1–17, 1910.

## Order 6. MYRICALES

An order containing only the Myricaceae, and distinguished from other amentiferous taxa by the unilocular bistylar ovary and the solitary orthotropous ovule. The phyletic position of the order is interpreted variously, but the combination of anatomical and morphological evidence favors views that treat its origin to be from stocks near or ancestral to the Hamamelidaceae or their relatives.

### MYRICACEAE. SWEET GALE FAMILY

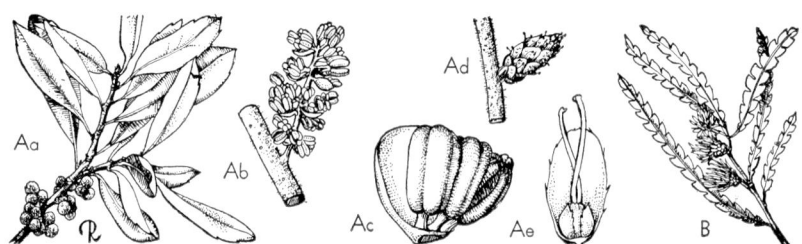

**Fig. 105.** MYRICACEAE. A, *Myrica pensylvanica*: Aa, habit in fruit, × ⅜; Ab, staminate inflorescence, × 2; Ac, staminate flower, × 8; Ad, pistillate inflorescence, × 2; Ae, pistillate flower, showing bract and 2 bracteoles, × 8. B, *Comptonia peregrina* var. *asplenifolia*, with young fruit, × ⅜. (From L. H. Bailey, *Manual of cultivated plants*, The Macmillan Company, 1949. Copyright 1924 and 1949 by Liberty H. Bailey)

Aromatic trees or shrubs, deciduous or evergreen, dioecious or monoecious (disposition of sexes very variable and the sex of flowers of a given plant or branch may vary from year to year) or sometimes individual flowers bisexual with both sex elements functional; leaves alternate, entire to pinnatifid, usually short-petioled, punctate resin-dotted, often coriaceous, estipulate (except in *Comptonia*); the unisexual inflorescence a densely flowered stiffish axillary spike, and when bisexual the staminate flowers subtend the pistillate; flowers mostly unisexual, devoid of perianth, usually bracteate; staminate flowers subtended by a single bract and with stamens 2–20 (usually 4–8), the filaments distinct or connate, the anthers basifixed, 2-celled, dehiscing by vertical slits; pistillate flower subtended by a bract and with or without bracteoles, the pistil 1, the ovary superior, 1-loculed, the ovule solitary basal (derived from parietal) and orthotropous with a single integument, the carpels 2, the style very short or not apparent, 2-branched, and stigmas 2; flowers when bisexual with central pistil and 3–4 stamens, bracted; fruit a small drupe enveloped or not by persisting bracteoles, often covered with a whitish waxy coating; seed with little or no endosperm, the embryo straight.

The number of genera composing the Myricaceae is unsettled. Most authors (including Engler, Asa Gray, Hutchinson) have treated it as unigeneric (*Myrica*, with about 40 spp.). By others (including Hjelmquist and Small), the 3 subgenera each have geen accorded generic rank (*Myrica, Gale,* and *Comptonia*), but most contemporary American authors (Fernald, Rehder, Abrams, Bailey) recognized only *Comptonia* (a monotypic genus) as generically distinct from *Myrica*. *Myrica* occurs in xerophytic or swampy areas over much of the northern hemisphere, and is represented in this country by about 6 species (2 restricted to the Pacific coast, 1 to Florida, 2 to Atlantic coastal plain and 1 cosmopolitan). *Comptonia peregrina* (*M. asplenifolia*), the sweet fern, westward from the sandy pinelands of the northern Atlantic coastal plain southward to North Carolina.

Members of the Myricaceae generally are recognizable by the aromatic fragrance of the foliage when crushed, by the yellow glandular dots on the leaves, and by the indehiscent waxy-coated 1-seeded fruits.

It was the opinion of Chevalier (1901) and of Hjelmquist (1948) that within the family the ancestors of the taxa *Gale, Myrica,* and *Comptonia* advanced from primitive to complex in that sequence. Hjelmquist treated the family as the most primitive of his Juglandales and retained it within the taxon Amentiferae. Hutchinson (1948) interpreted the family to be advanced over the Fagales and derived from stocks ancestral to the Hamamelidales. Youngken (1920) accepted the generic limits established by Chevalier, and made no phyletic study of the family.

Economically the family is important for the aromatic wax obtained from the fruit of several species and used in making bayberry candles, for tannic acid obtained from *M. Gale,* for the edible fruit of *M. Nagii,* and for the ornamental value associated with especially the pistillate plants of several species (*M. pensylvanica, M. cerifera,* and *M. Gale*).

*LITERATURE:*

CHEVALIER, A. Monographie des Myricacées; anatomie et histologie, organographie, classification et description des èspeces, distribution géographique. Mém. Soc. Sci. Nat. Cherbourg, 32: 85–340, 1901–02.
DAVEY, A. J. and GIBSON, C. M. Note on the distribution of sexes in *Myrica Gale.* New Phytologist, 16: 147–151, 1917.
STOKES, J. Cytological studies in the Myricaceae. Bot. Gaz. 99: 387–399, 1937.
YOUNGKEN, H. W. The comparative morphology, taxonomy, and distribution of Myricaceae of the eastern United States. Ann. Journ. Pharmacy, 87: 391–398, 1915.
———. The comparative morphology, taxonomy and distribution of the Myricaceae of the eastern United States. Contr. Bot. Lab. Univ. Pennsylvania, 4: 339–400, 1920.

## Order 8. LEITNERIALES [42]

A monotypic order (*Leitneria*) of very restricted distribution in this country and distinguished from others of this alliance by the unilocular and unistylar ovary having a single amphitropous ovule. Little is known about its probable phyletic affinities and for the present, like the Salicales, it stands apart from other dicot orders and clearly is a much reduced relic and not a primitive taxon.

### LEITNERIACEAE. CORK WOOD FAMILY

Deciduous dioecious shrubs or small trees; leaves alternate, simple, entire, estipulate, petiolate, subcoriaceous; cymules "flowers" (of authors) unisexual in catkinlike spikes before the leaves, the staminate inflorescence lax, composed of 40–50 large imbricated glandular-pubescent bracts each subtending a staminate cymule,[43] the pistillate inflorescence stiffly erect, composed of a primary axis bearing large primary bracts within each of which is a pair of much smaller secondary bracts, the 2 series (when fertile) subtending the pistillate cymule (many basal and some apical bracts are sterile); staminate cymule without perianth, stamens 3–12, filaments distinct, incurved, the anthers erect, basifixed, introrse, 2-celled, dehiscing longitudinally; pistillate cymule subtended by 3 bracts and a "perianth" of usually 4 (3–8) distinct bractlike tepals, the pistil 1, simple, the ovary superior, 1-loculed, the ovule 1, parietal and attached near apex, amphitropous, integuments 2, the

[42] Order 7, Balanopsidales (monotypic) of New Caledonia are not known to have representatives growing as exotics in this country.
[43] Studies by Abbe and Earle (1940) on the morphology and anatomy of the inflorescences have shown the "flowers" to be cymules, a condition not reflected by the gross morphology of the inflorescence, for this reason the term cymule is used in place of flower.

style 1 stoutly linear, constricted at union with ovary, the stigma 1; fruit a leathery compressed obovoid drupe, usually several aggregated in a cluster, each subtended by the persistent bract; seed with a large straight embryo and a thin fleshy endosperm.

The family is represented only by the single species *Leitneria floridana*, occurring in swamps from southern Missouri to Texas and Florida.

The plant is distinguished readily by its erect pistillate spikes as contrasted with the pendulous staminate aments, the simple superior ovary with parietal placentation, the nonaromatic foliage, and large embryo.

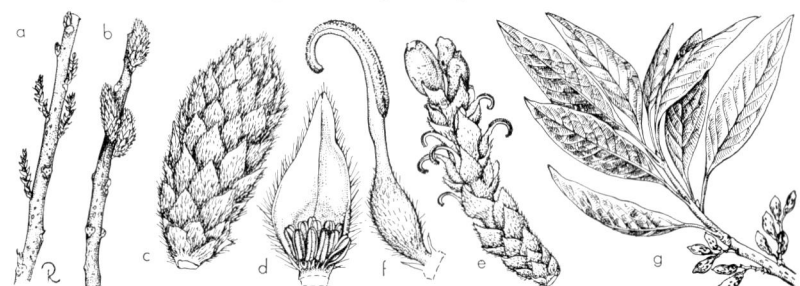

**Fig. 106.** LEITNERIACEAE. *Leitneria floridana*: a, branch with pistillate catkins, × ¼; b, branch with staminate catkins, × ¼; c, staminate catkin, × 1½; d, bract with stamens, × 5; e, pistillate catkin, × 1½; f, pistil, × 5; g, foliage branch, × ¼.

Engler and followers included the family with others having amentiferous inflorescences, Hutchinson (1926) following Heim (1891) derived it from the Rosales through the Hamamelidaceae. Bessey (1915) considered *Leitneria* to belong in the Ranales. Data provided by the critical studies of Abbe and Earle (1940) led them to the conclusion that the family should be placed in either the Rosales or Geraniales, with recognition of the need for more diagnostic evidence especially in the nature of cytological data and those to be obtained from study of the male gametophyte. More recently Hjelmquist (1948) included the family in the Amentiferae, and indicated that it represented "a line issuing from the type of Myricaceae." His conclusions were based on the floral morphology to the exclusion of consideration of contradicting anatomical evidence. There is little question but that the family represents an advanced taxon (as evidenced by reduction of floral parts and vegetative anatomy), but many more data are needed before its phyletic position safely can be indicated.

Economically the corkwood tree is of little importance except as a novelty and to a slight degree as an ornamental.

*LITERATURE:*

ABBE, E. C. and EARLE, T. T. Inflorescence, floral anatomy and morphology of *Leitneria floridana*. Bull. Torrey Bot. Club, 67: 173–193, 1940.

HEIM, F. Sur le genre *Leitneria* Chapm. Assn. Franc. pour l'avanc. des Sci. (Marseille), 231–232, 1891.

HJELMQUIST, H. Studies on the floral morphology and phylogeny of the Amentiferae. Bot. Notiser, Suppl. 2: 1–71, 1948.

## Order 9. JUGLANDALES

A monotypic order with the characters of its 1 family, Juglandaceae, and distinguished from others of the alliance by the plants predominantly monoecious, the flower solitary in the axil of a primary bract, the bicarpellate and inferior ovary with a solitary ovule, and the leaves pinnately compound. The phylogeny of the order is discussed in general under the family; however, there is evidence that the Rhoipteleaceae (here included in the Urticales) probably belongs in the Juglandales and may be ancestral to the Juglandaceae.

## JUGLANDACEAE. WALNUT FAMILY

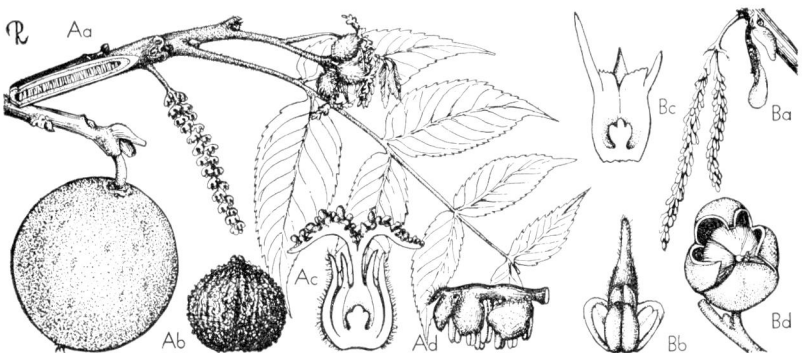

Fig. 107. JUGLANDACEAE. A, *Juglans nigra*: Aa, flowering branch, stem sectioned to show chambered pith, × ½; Ab, fruit and seed, × ½; Ac, pistillate flower, vertical section, × 2; Ad, staminate flower, × 2. B, *Carya*: Ba, staminate catkins, × ½; Bb, staminate flower, × 4; Bc, pistillate flower, vertical section, × 2; Bd, nut, the husk dehiscing, × ½. (From L. H. Bailey, *Manual of cultivated plants*, The Macmillan Company, 1949. Copyright 1924 and 1949 by Liberty H. Bailey.)

Deciduous monoecious trees or rarely shrubs (dioecious in *Engelhardtia* spp.); leaves alternate or rarely opposite (*Alfaroa, Engelhardtia* spp.), pinnately compound, usually resinous-dotted (lepidote) beneath and aromatic, estipulate; flowers unisexual, disposed in 1 of 4 basic inflorescence types: (1) a terminal paniculate cluster of several erect or pendulous staminate catkins, usually combined with a central largely or entirely pistillate catkin (as in spp. of *Platycarya, Alfaroa, Engelhardtia*); (2) a solitary terminal pendulous pistillate catkin (spp. of *Engelhardtia, Pterocarya*) which by reduction and specialization becomes a shorter few-flowered erect spike (*Juglans and Carya*) and thence to a 1-2-flowered spike (*Carya* and *Juglans*); (3) a 3-catkined staminate inflorescence lateral on old wood or at the base of new wood (spp. of *Engelhardtia* and *Carya*) derived by reduction from the more primitive staminate inflorescences of a 5-8-catkined panicle terminal on a short lateral branch (*Alfaroa*); (4) a solitary lateral staminate catkin, representing the most advanced type (*Juglans*, spp. of *Pterocarya*) resulting from reduction of the third type; the staminate flower possesses a floral envelope of basically 7 parts (a primary bract, 2 secondary bracts, and a perianth of 4 or fewer tepals), but sometimes 1 or more of the tepals may be absent or the perianth entirely absent (as in *Carya*), or the secondary bracts and the perianth may be absent (*Platycarya*); in some the primary bract is 3-lobed (*Engelhardtia, Alfaroa*) [44] the recep-

[44] The interpretation by Hjelmquist (1948) of the parts of the floral envelope differed from that of Manning given here for the staminate and pistillate flowers of certain genera. In the flowers of *Alfaroa* and *Engelhardtia* he considered the 2 lateral lobes of the 3-lobed bract as secondary bracts in both pistillate and staminate flowers, which means that there may be up to 6 tepals in the staminate flower, and in *Juglans* and *Pterocarya* he considered the secondary bracts absent in the staminate flower, with the tepals therefore 6 or fewer in number.

tacle attached to the bracts and bearing 3–100 stamens in 1 or more series, the filaments short, distinct, the anthers erect, basifixed, 2-celled, dehiscing longitudinally, a rudimentary pistil often present; the pistillate flower comprises a floral envelope and a pistil, the former consisting of a primary bract (sometimes 3-lobed), 2 secondary bracts, and a perianth of 4 tepals (1 or more secondary bracts and/or tepals may be lacking or modified) with the primary and secondary bracts adnate to the base of the ovary or to the whole ovary as are the tepals when present (in *Carya* the 4-toothed envelope is composed of a 3-cleft structure representing 3 secondary bracts or derived from the 2 connate secondary bracts, with the tepals modified to form a stigmatic disc), the pistil 1, the ovary inferior, 1-loculed above but 2–4-loculed below, normally 2-carpelled (sometimes 3), the ovule 1, erect in the center at the top of the incomplete partition although appearing basal in the young flower, orthotropous,[45] the styles 2 (or solitary with 2 lobes or branches); fruit drupelike nut with a dehiscent or indehiscent leathery or fibrous husk (the so-called exocarp, a structure derived from the involucre and perianth surrounding the ovary), or sometimes winged nutlet, incompletely 2–4-celled; the seed 2–4-lobed, the embryo very large, the endosperm none.

A family of 6 (or 7?) genera and about 60 species, mostly of the north temperate zone but with one distributional range extending through Central America, along the Andes to Argentina, and another from temperate Asia extending down to Java and New Guinea. In this country the family is represented by 4–6 species of *Juglans* (20 spp.), and about 14 species of *Carya* (*Hicoria*) (17 spp.). *Juglans* is the only genus of the family indigenous to the Pacific coast, where it is represented by a single species, and this genus with *Carya* and *Engelhardtia* are the only ones of the family to occur in both eastern and western hemispheres. The Juglandaceae are most abundant in eastern Asia and Atlantic North America.

The Juglandaceae are characterized by the presence of pinnately compound leaves, the flowers each subtended by a primary bract, apetalous, unisexual with the staminate generally lateral in pendulous (in *Juglans* and *Carya*) and the pistillate in terminal erect inflorescences (in U. S. representatives), by the distinctive floral envelope, the inferior ovary, the resultant usually leathery drupaceous nut or winged samara.

It has been the view of Bessey and Hutchinson that the Juglandaceae may have been evolved from one of the taxa in the Sapindales. Hallier included them in his Terebinthaceae as derived from the Rutaceae. Tippo noted that anatomically they are an advanced taxon and are not the simple family that Engler and Wettstein considered them. Because of their pinnate leaves, resinous condition, and floral morphology they may not be allied closely to the families in the Urticales, but may be well advanced over them. Heimsch (1942) stated that on the basis of wood anatomy the Juglandaceae are unrelated to the Anacardiaceae and the Julianaceae. Studies by Manning (1938, 1940, 1948, 1949) suggested that, on the basis of floral morphology, the genera *Engelhardtia* and *Alfaroa* may be the more primitive, with *Juglans* and *Carya* among the more advanced in the family. It is significant to note that, on the basis of wood anatomy studies, Heimsch and Wetmore (1939) arrived

---

[45] Macroscopic examination of the young ovary indicates the ovule to be basal and orthotropous; morphological studies by Benson and Welsford (1909) produced evidence that the ovule is parietal and anatropous; Nast (1935) disagreed somewhat with these earlier authors in her findings and interpretation of the floral anatomy and suggests that the ovule is of a modified axile type; Manning (1940) believed that the ovule is terminal at the apex and center of an incomplete partition, hence derived from an axile condition where the ovules were clustered near the center of a normal or incomplete partition (such as in the Fagaceae or in *Koelreuteria* of the Sapindaceae). This is further evidence that the family is phyletically advanced and not at all a primitive taxon

at conclusions regarding the phyletic relationships within the family that were in essential agreement with the floral morphology studies of Manning. Heimsch (1944) reported that the pollen morphology of members of the family supported his earlier conclusions as to the relationships of the genera. In his studies of the Amentiferae, Hjelmquist (1948) allied the Juglandaceae with and advanced over the Myricaceae (including both in the Juglandales).

Economically the Juglandaceae are of major importance. Wood of many species is of considerable value: lumber from the black walnut (*Juglans nigra*) and the Circassian walnut (*Juglans regia*) is in demand for cabinet and furniture making, and lumber from the hickory (*Carya* spp.) is sought for tool handles. The fruits of the English or Persian walnut (*Juglans regia*), black walnut, of species of hickory (*Carya*), and of the pecan (*Carya illinoensis*) are among the most valuable food nuts of the country. Species of *Juglans, Carya, Pterocarya,* and *Platycarya* are grown domestically for ornament, and the genus *Englehardtia* is a source of lumber highly valued in the orient.

*LITERATURE:*

BENSON, M. and WELSFORD, E. J. The morphology of the ovule and female flower of *Juglans regia* and a few allied genera. Ann. Bot. 23: 623–633, 1909.

HEIMSCH, C. *Alfaroa* pollen and generic relationships in the Juglandaceae. Amer. Journ. Bot. 31 (8): 3s, 1944.

———. Comparative anatomy of the secondary system in the "Gruinales" and "Terebinthales" of Wettstein, with reference to taxonomic grouping. Lilloa, 8: 83–198, 1942.

HEIMSCH, C. and WETMORE, R. H. The significance of wood anatomy in the taxonomy of the Juglandaceae. Amer. Journ. Bot. 26: 651–660, 1939.

HJELMQUIST, H. Studies on the floral morphology and phylogeny of the Amentiferae. Bot. Not. Suppl. 2 (1): 1–171, 1948.

LANGDON, L. M. Ontogenetic and anatomical studies of the flower and fruit of the Fagaceae and Juglandaceae. Bot. Gaz. 101: 301–327, 1939.

MANNING, W. E. The morphology of the flower of the Juglandaceae. I. The inflorescence. Amer. Journ. Bot. 25: 407–419, 1938. II. The pistillate flowers and fruit. *op. cit.* 27: 839–852, 1940. III. The staminate flowers. *op. cit.* 35: 606–621, 1948.

———. The genus *Alfaroa*. Bull. Torrey Bot. Club, 76: 196–209, 1949.

NAST, C. G. Morphological development of the fruit of *Juglans regia*. Hilgardia, 9: 345–381, 1935.

WITHNER, C. L. Stem anatomy and phylogeny of the Rhoipteleaceae. Amer. Journ. Bot. Bot. 28: 872–878, 1941.

### Order 11. BATIDALES [46]

A monotypic order with the characters of the family Batidaceae. One species of dioecious shrub with fleshy leaves, congested stout amentlike inflorences and the pistillate flowers with a 4-locular ovary. The evidence seems inadequate to justify close alliance of this taxon with any other family or order. Its ancestral stocks are not known and, like the Salicaceae, it stands alone, a morphologically much-reduced relic that is not the primitive dicot as interpreted by Engler and others.

### BATIDACEAE. BATIS FAMILY

Low straggling maritime dioecious shrubs with strong-scented spreading or prostrate stems; leaves opposite, simple, sessile, fleshy and semiterete, entire, stipulate; flowers unisexual, actinomorphic, minute, in axillary sessile bracteate catkinlike or

[46] Order 10, Julianales (one dioecious family) of Mexico and Peru, is not known to have representatives growing as exotics in this country. Current authorities, including Hjelmquist, are agreed that, despite the pinnately compound leaves and superficially amentiferous features, it is not allied to the Juglandaceae nor to other families of the Amentiferae.

conelike erect spikes, the pistillate spikes of 4–12 naked flowers, many, subtended by persistent imbricated bracts, the calyx shallowly 2-lipped campanulate and membranous; stamens 4 or 5, alternated with as many clawed and rhombately limbed staminodes (petals of some authors), the filaments distinct, the anthers dorsifixed, 2-celled, introrse, dehiscing vertically; pistillate flowers subtended by small deciduous nonimbricated bracts, lacking calyx and corolla, the pistil 1, the ovary superior, 4-loculed, each with a solitary basal anatropous ovule, the stigma 1 and cushionlike; fruit a berry, those of each inflorescence coherent or connate in a multiple structure; seeds with large spatulate embryo and no endosperm.

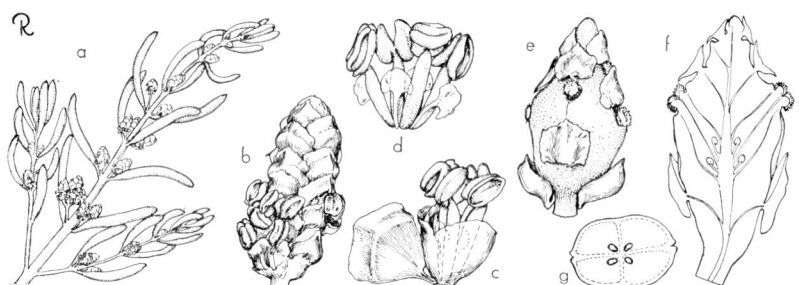

**Fig. 108.** BATIDACEAE. *Batis maritima*: a, flowering branch, × ½; b, staminate inflorescence, × 3; c, staminate flower, with bract, × 5; d, stamens and staminodes, × 5; e, pistillate inflorescence, × 5; f, same, vertical section, × 8; ovary, cross-section, × 5.

The family consists of the monotypic genus (*Batis maritima*) inhabiting littoral areas of tropical and subtropical regions of the New World and the Hawaiian Islands. It occurs in salt marshes on the Pacific coast from San Pedro, California, to the Galapagos Islands, and in similar habitats of the Atlantic coast from North Carolina to Brazil and the West Indies.

The Batidaceae are distinguished readily by the succulent subterete opposite leaves, the dioecious character associated with the conelike axillary inflorescences and the 4-loculed ovaries of the naked pistillate flowers.

The phyletic position of this family is not known. Bessey placed it at the end of his Caryophyllales, noting that "it very doubtfully belongs here." Hallier allied it with the Chenopodiaceae and Hutchinson included it with the Hamamelidaceae. The Batidaceae have little or no relationship to either the Juglandales or Fagales as interpreted by Engler and by Hjelmquist.

The family is of no known economic importance.

*LITERATURE:*

HJELMQUIST, H. Studies on the floral morphology and phylogeny of the Amentiferae. Bot. Notiser, Suppl. 2: 1–171, 1948.
WILSON, P. Batidaceae. North Amer. Flora, 21: 255, 1932.

### Order 12. FAGALES

An order considered by Engler to contain the 2 families, Betulaceae and Fagaceae, but subdivided by Hjelmquist into 2 orders (Fagales and Betulales) and retained by him within the Amentiferae. It is distinguished in part by the inferior unilocular ovary containing 2 or more ovules. Most American botanists are agreed that the taxon has been derived from stocks ancestral to the Hamamelidaceae and is advanced over that family, and evidence provided by studies of stem and floral anatomy and floral morphology is abundant to support the view that the order is not one of primitive dicots.

## BETULACEAE.[47] BIRCH FAMILY

Deciduous monoecious trees and shrubs; leaves alternate, simple, stipulate (stipules often deciduous) serrate margined; cymules unisexual (cf. footnote 48), the staminate in pendulous catkinlike inflorescences, the pistillate in short lateral or capitate pendent or erect strobiloid inflorescences; the inflorescence of either sex is composed theoretically of a primary axis on which are arranged spirally many much-condensed cymules (the "flowers" of most texts); each cymule (theoretically) is comprised of a basal primary bract from the axil of which arises a secondary axis bearing a pair of secondary bracts and terminated by a flower, with a tertiary axis arising from the axil of each secondary bract and terminated by a flower subtended by a pair of tertiary bracts;[48] the staminate inflorescence of all genera is composed of 3-flowered cymules, but both tertiary bracts have been lost in *Betula, Corylus, Carpinus, Ostrya,* one tertiary bract lost in *Alnus* and all bracts except the primary lost in *Ostryopsis;* a perianth of typically 4 minute tepals subtends the 1–4 stamens of each flower (as in *Alnus* and *Betula*) but are absent in flowers of *Carpinus, Corylus, Ostrya, Ostryopsis,* the stamens per cymule number 2–20, the filaments are short, distinct or basally connate, the anthers 2-celled with cells separate or connate, dehiscing longitudinally; in the pistillate inflorescence, the cymule is 2-flowered, except in *Betula,* which is 3-flowered, all bracts are present in *Carpinus, Ostrya, Ostryopsis,* both tertiary bracts lost in *Betula,* a single bract lost in *Corylus;* the tepals are present in *Carpinus, Ostrya, Corylus, Ostryopsis,* and are obsolescent in *Betula* and *Alnus;* in each pistillate flower the pistil 1, the ovary inferior or nude,[49] 2-loculed below and 1-loculed above the placental septum, each locule with 2 ovules or reduced to 1 by abortion, attached axially from locule apex, the styles 2 and linear, each with a single stigma, or a single and deeply 2-parted style; fruit a small nut or briefly winged samara, indehiscent, 1-celled, 1-seeded, the seed with large straight embryo, no endosperm.

A family of 6 genera and over 100 species (representing the tribes Betuleae and Coryleae), mostly in the northern hemisphere, of which the following 5 genera have species indigenous to this country: *Betula, Alnus, Corylus, Ostrya,* and *Carpinus.* About 12 of the 40 species of *Betula,* 10 of the 30 species of *Alnus,* 3 of the 15 species of *Corylus,* 1 of the 26 species of *Carpinus,* and 1 of the 7 species of *Ostrya* are indigenous to the more temperate regions of this country.

The family is distinguished from related families by the basically cymose grouping of the pistillate flowers on each bract, and the 2-celled inferior ovary.

---

[47] It was pointed out by Fernald (1945) and concurred with by Rehder (1946) that the name of Betulaceae for this family was antedated by the older and valid name of Corylaceae. The latter name has been adopted by these authors. Subsequently Little (1949) proposed that the name Betulaceae be conserved over all others. Some authors (Hutchinson, Hjelmquist, *et al.*) divide the family, treating it as composed of the Betulaceae (Betuleae) and Corylaceae (Coryleae).

[48] It must be remembered that this description is of a theoretical cymule, a type probably ancestral to any known existing types, and from which all living types differ in having lost (by suppression or condensation) the secondary and tertiary axes and/or bracts, so that when dissected the bracts appear to be superposed one on another, and as noted below may be fewer than the total complement of the typical cymule.

[49] In some members of the family (as in *Betula*) there is no external evidence of a perianth, hence it is not possible to diagnose macroscopically the type of ovary position and it is designated as *nude.*

**Fig. 109.** BETULACEAE. A, *Alnus rugosa*: Aa, twig with catkins, × ½; Ab, staminate flowers, × 3; Ac, single staminate flower, × 5. B, *Corylus Avellana*: fruit, × ½. C, *Carpinus caroliniana*: fruiting branch, × ½ D, *Betula pendula*: Da, twig with catkins, × ½; Db, pistillate flowers, × 10; Dc, staminate flowers, × 3. E, schematic diagram of theoretical 3-flowered pistillate cymule ancestral to modern types. F, *Carpinus japonica*: Fa, schematic diagram of pistillate cymule; Fb, pistillate cymule. G, *Betula Medwediewii*: Ga, schematic diagram of pistillate cymule; Gb, pistillate cymule. H, *Alnus crispa*: Ha, schematic diagram of pistillate cymule; Hb, pistillate cymule. ($A_1$, $A_2$, $A_3$ primary, secondary, and tertiary axes respectively; $B_1$, $B_2$, $B_3$ primary, secondary, and tertiary bracts respectively; broken lines indicate absence of parts. (A-D from L. H. Bailey, *Manual of cultivated plants,* The Macmillan Company, 1949. Copyright 1924 and 1949 by Liberty H. Bailey. E-H redrawn from Abbe, 1935, by M. E. Ruff.)

The phylogenetic position of the Betulaceae was considered by Hutchinson (1926), Tippo (1938) and others to be derived from hamamelidaceous stocks but to be more highly specialized than the Hamamelidaceae in all anatomical respects. Hjelmquist (1948) accepted it as representing 2 families belonging to the single order Betulales, which has arisen from stocks ancestral to the Fagales. There is considerable evidence to support the view that the reproductive structures are not simple but have flowers that are complex by reduction. Abbe (1935) has suggested that the present carpellary situation has been derived from a 3-carpellate ancestor —a situation that is presumed to have existed also for the Hamamelidaceae.

## DIVISION IV. EMBRYOPHYTA SIPHONOGAMA

Economically the family is important for the hardwood lumber obtained from the birches (*Betula*); oil of betula extracted from twigs of *Betula* spp. has the flavor and odor of wintergreen, the edible hazelnuts and filberts are produced from *Corylus* spp., high-grade charcoal is made from the wood of *Alnus* and *Betula*, birch beer is made or flavored from the sugary sap of *Betula* spp., and the very hard wood of *Ostrya* is prized for mallets and beetles.

### LITERATURE:

ABBE, E. C. Studies in the phylogeny of the Betulaceae. I. Floral and inflorescence anatomy and morphology. Bot. Gaz. 97: 1–67, 1935. II. Extremes in variation in the range of variation of floral and inflorescence anatomy. *op. cit.* 99: 369–431, 1938.

ANDERSON, E. and ABBE, E. C. A quantitative comparison of specific and generic differences in the Betulaceae. Journ. Arnold Arb. 15: 43–50, 1934.

BUTLER, B. T. Western American birches. Bull. Torrey Bot. Club, 36: 421–440, 1909.

FERNALD, M. L. Some North American Corylaceae (Betulaceae). I. Notes on *Betula* in eastern North America. Rhodora 47: 303–329, 1945. III. Eastern North American representatives of *Alnus incana*. *op. cit.* 333–361, 1945.

HJELMQUIST, H. Studies on the floral morphology and phylogeny of the Amentiferae. Bot. Notiser, Suppl. 2: 1–171, 1948.

REHDER, A. Notes on some cultivated trees and shrubs, III. Journ. Arnold Arb. 27: 169–170, 1946. [Gives reasons for accepting the name Corylaceae instead of Betulaceae.]

TIPPO, O. Comparative anatomy of the Moraceae and their presumed allies. Bot. Gaz. 100: 1–99, 1938.

WOODWORTH, R. H. Cytological studies on the Betulaceae, I–IV. Bot. Gaz. 87: 331–363, 1929; *op. cit.* 88: 383–399, 1929; *op. cit.* 89: 402–409, 1930; *op. cit.* 90: 108–115, 1930.

## FAGACEAE. BEECH FAMILY

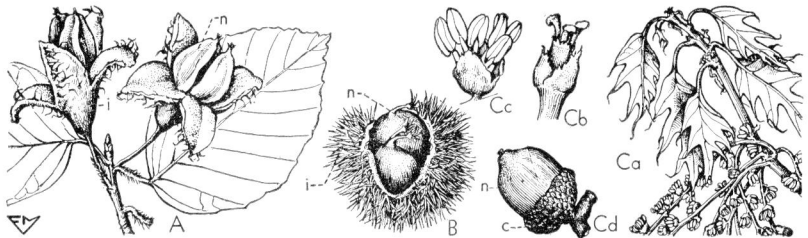

**Fig. 110.** FAGACEAE. A, *Fagus sylvatica*: fruiting branch, × ½. B, *Castanea mollissima*: fruit, × ½. C, *Quercus borealis*: Ca, twig with staminate catkins, × ½; Cb, pistillate flower, × 3; Cc, staminate flower, × 3; Cd, acorn, × ½. (c cup, i involucre, n nut.) (From L. H. Bailey, *Manual of cultivated plants*, The Macmillan Company, 1949. Copyright 1924 and 1949 by Liberty H. Bailey.)

Deciduous or evergreen monoecious trees and shrubs (dioecious in *Nothofagus*); leaves alternate, simple, cleft, lobed, or entire, deciduously stipulate; flowers functionally unisexual (rarely bisexual), apetalous, the staminate solitary (*Nothofagus*, *Fagus*), in pendulous heads, or in catkinlike racemes, the pistillate solitary or few in clusters; staminate flowers with a 4–6–7 imbricately lobed perianth of tepals, stamens 4–40, filaments filiform, distinct, the anthers basifixed, 2-celled, the cells confluent, dehiscing longitudinally, the flower solitary on a bract or in cymules, a rudimentary pistil usually present; pistillate flower solitary within an involucre of generally many adnate and imbricated bracteoles or in a 2–3-flowered cymule with or without an involucre, pistil 1, the ovary inferior and adnate to which is a 4–6-lobed perianth of tepals, 3–6-loculed with styles and carpels as many as the locules, the placentation axile, the ovules 2 in a locule but 1 aborting, both basal or nearly so; fruit a 1-seeded nut subtended or enveloped by a cupule or involucre which may or may not be muricate, bristly, or spiny; in some genera (as *Castanea*) 3 nuts are present in the involucre; seed with a large embryo and no endosperm.

A dominant family of 6 genera and about 600 species, mostly of temperate and subtropical regions of the northern hemisphere, but with *Nothofagus* (about 14 spp.) restricted to the antarctic regions of the southern hemisphere, and *Pasania* (about 100 spp.) pantropical. The family is represented in the United States by 1 Californian species of *Lithocarpus* (a genus of about 100 spp., mostly in southeastern Asia), about 60 of the more than 300 species of *Quercus*, 1 of the 10 species of *Fagus*, 2 of the 10 species of *Castanea*, and (on the Pacific coast) by 2 of the 30 species of *Castanopsis*.

Members of the family are best recognized by the inferior tricarpellate ovary, and by the characteristic fruit which is always subtended or enveloped by a cupule or involucre.

Most available evidence supports the view that the Fagaceae and Betulaceae are more closely related to one another than to other amentiferous families. Wettstein, Rendle, and Hutchinson retained them in the Fagales, whereas Bessey aligned them with the Betulaceae and Juglandaceae and treated the 3 to have been derived from sapindaceous ancestors. Hjelmquist (1948) segregated the families as 2 orders, Betulales and Fagales, and retained both within the Amentiferae—a taxon he accepted as an advanced phyletic unit. As a result of studies of floral morphology and anatomy, Berridge (1914) concluded the Fagaceae to have been derived from epigynous rosaceous ancestors. By a reinterpretation of her data, augmented by original anatomical findings, Tippo (1938) contended there was no serious barrier to interpreting them to have been derived from 3-carpellate stocks that were ancestral also to the present-day Hamamelidaceae. He placed both the Fagaceae and Betulaceae in the Fagales, with the Betulaceae as the more primitive.

Economically the family is of importance for the lumber produced by its members throughout its range. The species of *Quercus* are sources of many kinds of oak lumber, from *Fagus* is obtained beech lumber, and from *Castanea* chestnut timber. Tannic acid is extracted from the insect galls of oak species, and commercial cork is obtained from the bark of *Quercus suber*. Species of *Fagus, Castanea,* and *Quercus* are cultivated for their edible fruits. In this country there are cultivated for ornament more than 65 spp. of *Quercus,* 7 spp. of *Castanea,* 5 of *Castanopsis,* 5 of *Lithocarpus,* 8 of *Nothofagus,* and 2 of *Fagus*.

*LITERATURE:*

BERRIDGE, E. M. The structure of the flower of the Fagaceae, and its bearing on the affinities of the group. Ann. Bot. 28: 509–526, 1914.

CAMUS, A. Les chênes. Monographie du genre *Quercus*. 3 vols., 619 pl. Paris, 1934–1949.

DYAL, S. C. A key to the species of oaks of eastern North America based on foliage and twig characters. Rhodora, 38: 53–63, 1936.

FRIESNER, R. C. Chromosome numbers in ten species of *Quercus,* with some remarks on the contributions of cytology to taxonomy. Butler Univ. Bot. Stud. 1: 77–103, 1930.

HJELMQUIST, H. Studies on the floral morphology and phylogeny of the Amentiferae. Bot. Notiser, Suppl. 2: 1–171, 1948.

LANGDON, L. M. Ontogenetic and anatomical studies of the flower and fruit of the Fagaceae and Juglandaceae. Bot. Gaz. 101: 301–327, 1939.

MULLER, C. H. Oaks of Trans-Pecos, Texas. Amer. Midl. Nat. 24: 703–728, 1940.

———. The problem of genera and subgenera in the oaks. Chron. Bot. 7: 12–14, 1942.

SAX, H. J. Chromosome numbers in *Quercus.* Journ. Arnold Arb. 11: 220–223, 1930.

SCHWARTZ, O. Entwurf zu einem natürlichen System der Cupuliferen und der Gattung *Quercus* L. Notizbl. Bot. Gart. Berlin, 13: 1–22, 1936.

TILLSON, A. H. and MULLER, C. H. Anatomical and taxonomic approaches to subgeneric segregation in American *Quercus*. Amer. Journ. Bot. 29: 523–529, 1942.
TRELEASE, WM. The American Oaks. Mem. Nat. Acad. Sci. 20: 1–255, 1924.

## Order 13. URTICALES

An order treated by Engler and Diels to be composed of the families Ulmaceae, Rhoipteleaceae, Moraceae, and Urticaceae, and distinguished from other orders of presumed alliance by the usually 2-carpellate unilocular superior ovary with a single ovule and the androecium of typically few to several stamens. Studies by Withner (1941) indicate that the Rhoipteleaceae belong in the Juglandales, and those by Tippo (1940) make it clear that the Eucommiaceae belong here and not in the Rosales. Bessey placed these 3 families within the Malvales, and Hallier included them in his Terebinthales. Bechtel (1921) found the members of the order to possess a combination of primitive and specialized characters, and concluded that it probably was "not far removed from primitive entomophilous ancestors." This conclusion was amplified further by Bechtel's opinion that the taxon is a natural order made up of the 3 families as classified by Engler, but that "the natural position . . . is at the culmination of a distinct line of descent from a protoangiospermous plexus from which also the ranalian line descended." The order is now generally accepted as phyletically advanced and as the terminal taxon of an evolutionary line from which the Fagales also have been derived.

### ULMACEAE. ELM FAMILY

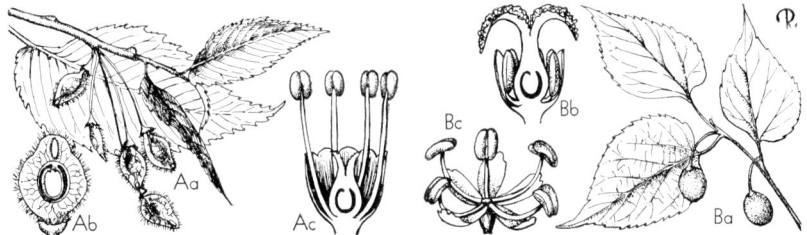

**Fig. 111.** ULMACEAE. A, *Ulmus americana*: Aa, fruiting branch, × ½; Ab, fruit, vertical section, × 1; Ac, flower, vertical section, × 4. B, *Celtis occidentalis*: Ba, fruiting branch, × ½; Bb, flower, vertical section, × 4; Bc, staminate flower, × 6. (From L. H. Bailey, *Manual of cultivated plants*, The Macmillan Company, 1949. Copyright 1924 and 1949 by Liberty H. Bailey.)

Monoecious (or bisexually flowered) trees or shrubs, sap watery; leaves simple, alternate, caducously stipulate, often obliquely based, petioled; flowers bisexual (*Ulmus*) or unisexual by abortion of opposing sex (most other genera), zygomorphic, solitary, cymose, or in axillary fasciculate aggregations, borne on twigs of previous season (Ulmoideae) or of current season (Celtidoideae); floral parts obscurely spirally arranged, perianth of 4–8 more or less connate sepals,[50] imbricate, campanulate, herbaceous; the stamens erect in bud, arising from the hypanthium, usually the same number and opposite the sepals or calyx lobes, the filaments distinct, the anthers 2-celled, dehiscing longitudinally; pistil 1, the ovary

---
[50] The perianth components were designated sepaloid leaves by Rendle, sepals by Hutchinson, and tepals by others. Bechtel (1921) demonstrated the perianth cup when present to be the "fused bases of floral envelopes and stamens." This cup (an hypanthium), characteristic of the Ulmoideae, is absent or scarcely apparent in the Celtidoideae. Bechtel's finding that the perianth represents a single and outer whorl of parts justifies its designation as a calyx of distinct or basally connate sepals.

superior, 2-carpelled, usually unilocular, the ovule 1, apical, anatropous, and pendulous from locule apex, the styles 2, linear, stigmatose along upper inner surface; fruit a broadly winged samara whose single seed contains a straight embryo with flat cotyledons (Ulmeae), or a drupe whose seed contains a curved embryo with folded or rolled cotyledons and no endosperm.

A family of about 15 genera and more than 150 spp. distributed throughout much of the northern hemisphere and more particularly in the tropics and subtropics. Species of 3 genera are indigenous to North America: *Ulmus* is represented by 5–6 of its 18–20 spp. (none west of the Rocky Mountains), *Celtis* by perhaps 5 of its 70-odd spp., and *Planera* by a single species in the southeast.

The family is composed of 2 well-marked subfamilies (Ulmoideae and Celtidoideae) distinguishable from the related families of Urticaceae and Moraceae by the presence of watery sap, the usually oblique leaves with deciduous stipules, the flowers generally bisexual, the stamens erect in bud, and the fruit a samara or drupe whose seed generally lacks endosperm.

The position of the family is one of primitiveness within the Urticales, and Bechtel (1921) considered the usually bisexually flowered *Ulmus* to be the primitive genus. Sax (1933) found chromosome numbers to indicate closer relationships with some of Betulaceae than with others of the Amentiferae but conceded that the great diversity in morphological and anatomical characters would preclude any close phyletic affinity between these taxa. The family represents a case of reduction from ancestral types that were much more complex, and the present anemophilous condition undoubtedly has been derived from an ancestral entomophilous condition.

The family is not of major economic importance. *Ulmus* (the elm) is a source of lumber used to a limited extent in furniture making, the mucilaginous bark of *U. fulva* (slippery elm) is utilized in the manufacture of medicinal troches. Seeds of some species of *Celtis* are edible, and the fragrant wood of *Planera Abelica* is known in the cabinet trade as false sandalwood. Several genera and species are grown domestically as ornamentals.

*LITERATURE:*

BECHTEL, A. R. The floral anatomy of the Urticales. Amer. Journ. Bot. 8: 386–410, 1921.
FERNALD, M. L. and SCHUBERT, B. Studies in the British herbaria. The type of *Celtis occidentalis*, L. Rhodora, 50: 155–162, 1948.
SAX, K. Chromosome numbers in *Ulmus* and related genera. Journ. Arnold Arb. 14: 82–84, 1933.

## MORACEAE.[51] MULBERRY FAMILY

Deciduous or evergreen, monoecious or dioecious trees or shrubs (herbaceous in *Dorstenia*), with milky latex (except in *Humulus* and *Cannabis*); leaves alternate (rarely opposite), simple, often with 3–5 basal palmate veins, entire, serrate or lobed, the stipules 2, small and lateral or each pair forming a cap over the bud and leaving a cylindrical scar; flowers unisexual, minute, regular, the inflorescence basically cymose but often much modified,[52] the perianth in 2 whorls generally

[51] The name *Moraceae* Lindl. (1846), antedated by *Artocarpaceae* Horaninov (1834), has been proposed (1945) for conservation.

[52] In *Morus* the inflorescence of each sex is condensed into a pendulous amentiferous structure, in *Maclura, Cudrania,* and *Broussonetia* the pistillate inflorescence is reduced to a globose head, in *Dorstenia* the pedicels and peduncles have become coalesced and dorsiventrally compressed into a laminate receptacle over whose ventral surface the minute sessile flowers are densely disposed, and in *Ficus* the receptacle has by involution developed into a hollow fleshy axis (termed a *syconium*) bearing the flowers over the inner surface.

## DIVISION IV. EMBRYOPHYTA SIPHONOGAMA 463

of 4 (2-6) free or more or less connate segments (sepals or tepals of authors); the staminate flower usually with stamens equal in number (generally 4 or reduced to 1-2 in *Ficus, Morus, Dorstenia,* and *Artocarpus*) and opposite the perianth segments, filaments distinct, the anthers 2-celled, versatile, dehiscing longitudinally; the pistillate flower with or without a minute generally 4-lobed biseriate perianth, the pistil 1, the ovary superior to inferior, basically 2-carpelled with 1 usually aborting, 1-loculed, the ovule 1 usually anatropous and mostly pendulous, the styles filiform and mostly 2 (only 1 in *Maclura*) with same number of stigmas; fruit basically a drupe and often aggregated (*Morus*) or connate together with perianths and axes (*Artocarpus*), or an achene within a fleshy receptacle (*Ficus*); seed usually with endosperm, the embryo mostly curved.

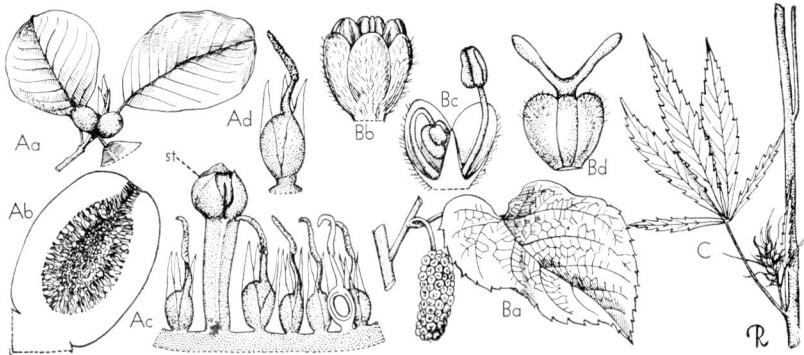

**Fig. 112.** MORACEAE. A, *Ficus altissima*: Aa, fruiting branch, × ⅒; Ab, flowering inflorescence, vertical section, × 1; Ac, section of "receptacle" bearing pistillate and staminate flowers, × 5; Ad, pistillate flower, × 6. B, *Morus alba*: Ba, fruit and leaf, × ½; Bb, staminate flower, × 6; Bc, same, vertical section, × 6; Bd, pistillate flower, × 12. C, *Cannabis sativa*: pistillate inflorescence and leaf, × ¼. (From L. H. Bailey, *Manual of cultivated plants,* The Macmillan Company, 1949. Copyright 1924 and 1949 by Liberty H. Bailey.)

A family of about 73 genera and over 1000 species, mostly of pantropical distribution. The 3 genera with species indigenous to the United States are: *Morus* (mulberry), *Maclura* (Osage orange), *Humulus* (hop). Other genera represented by naturalized or escaped species are *Broussonetia* (paper mulberry), and *Cannabis* (hemp). The family is divided into 4 subfamilies or tribes: Moroideae (stamens incurved in bud, leaves folded in bud), Artocarpoideae (stamens straight, leaves convolute in bud, ovule amphitropous and apical), Conocephaloideae (stamens straight, ovule orthotropous and basal or apical), and Cannaboideae (stamens short and straight, ovule apical and anatropous, herbs with watery sap). By some authors (Rendle, Hutchinson) the Cannaboideae is treated as a separate family (Cannabinaceae), a view supported also by recent morphological studies.

The family is characterized by the presence of milky latex and/or usually 2 stigmas and the usually pendulous single ovule. Removal of *Humulus* and *Cannabis* as representing a separate family strengthens the distinguishing value of the latex character.

Ample evidence has been presented (Bechtel, Tippo) that the Moraceae belong within the Urticales (Bentham and Hooker treated all genera of the order as composing a single family, the Urticaceae); that the family stands advanced over the Ulmaceae, as evidenced by the greater reduction in perianth parts and in number of stamens; and that the Urticaceae phyletically are higher than it. It was the opinion of Bechtel that within the family, *Maclura* is more advanced than *Morus*, and that *Humulus* and *Cannabis* form a distinct taxon within the family that is the more highly advanced.

Economically the family is important for the many edible fruits produced: figs (*Ficus*), mulberries (*Morus*), breadfruit and jack fruit (*Artocarpus*); the fibers of hemp (*Cannabis*) are used for cordage and the drug marijuana is obtained from the staminate flowers of the same plant; hops (*Humulus*) are grown for their fruit used in flavoring beer; genera cultivated domestically as ornamentals include the figs (*Ficus*, with 44 of its approximately 600 spp.), *Cecropia*, with 1 of its 50 spp., fustic (*Chlorophora*), *Cudrania*, with 2 of its 4 spp., the pickaback plant (*Dorstenia*, with 1 of its about 125 spp.).

*LITERATURE:*

BECHTEL, A. R. The floral anatomy of the Urticales. Amer. Journ. Bot. 8: 386–410, 1921.
CONDIT, I. J. The structure and development of flowers in *Ficus Carica*, L. Hilgardia, 6: 443–481, 1932.
——. The fig. 222 pp. Waltham, Mass., 1947.
NAKAI, T. *Morus alba* and its allies in the herbaria of Linnaeus, Thunberg, and others. Journ. Arnold Arb. 8: 234–238, 1927.
SATA, N. A monographic study of the genus *Ficus* from the point of view of economic botany. Contr. Inst. Hort. & Econ. Bot. Taihoku Univ. 32: Parts I, II, 1–405; Parts III, IV, 1–289, 1944.
STANDLEY, P. C. The Mexican and Central American species of *Ficus*. Contr. U. S. Nat. Herb. 20: 1–35, 1917.
TIPPO, O. Comparative anatomy of the Moraceae and their presumed allies. Bot. Gaz. 100: 1–99, 1938.

## URTICACEAE. NETTLE FAMILY

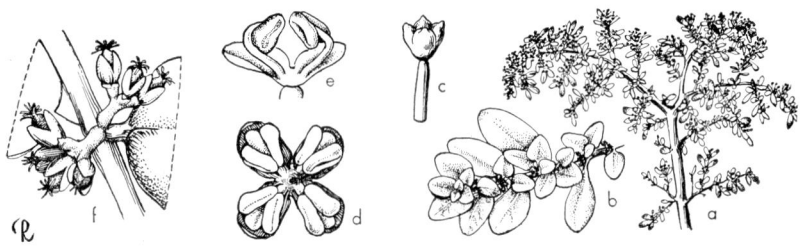

**Fig. 113.** URTICACEAE. *Pilea microphylla*: a, flowering branch, × ½; b, same, × 3; c, staminate flower-bud, × 5; d, staminate flower, face view, × 15; e, same, vertical section, × 15; f, pistillate flowers, × 5. (From L. H. Bailey, *Manual of cultivated plants*, The Macmillan Company, 1949. Copyright 1924 and 1949 by Liberty H. Bailey.)

Monoecious or dioecious fibrous herbs, or infrequently subshrubs or small trees, rarely climbing vines, herbage with stinging hairs in a few genera (*Urtica, Laportea, Urera*), cystoliths usually present in epidermal cells, the sap watery; leaves alternate or opposite, simple, stipulate (except in *Parietaria*); the inflorescence basically a bracteate cyme, sometimes modified into a head by loss of pedicels, or reduced to a single flower, the cymes usually borne on a simple axis arising in upper leaf axils with axis very short or elongate (erect or pendent and catkinlike); flowers unisexual (sometimes imperfectly bisexual in *Parietaria*), regular, minute, mostly green, the perianth present or absent, when present usually biseriate, of 4–5 distinct or connate parts undifferentiated into calyx and corolla; staminate flowers mostly with 4 (occasionally 3–5) stamens, opposite the perianth segments, the filaments distinct, bent inwards in bud and springing back elastically at anthesis, releasing pollen in a sudden burst, the anthers 2-celled dehiscing longitudinally, a rudimentary pistil often present; pistillate flowers with single pistil, the ovary superior or inferior (free or adnate to perianth), 1-loculed, 1-carpelled, the ovule solitary, seeming basal and falsely orthotropous by a shifting in position from the locule apex, the style 1, the stigma 1 and often a brushlike tuft, scalelike staminodes often present at base of pistil; fruit an achene or drupe, often enclosed by the persistent perianth; seed with straight embryo surrounded by oily endosperm.

A family of about 42 genera and nearly 600 species, mostly tropical and subtropical, with nearly 40 per cent of its components in the New World. The largest genus is *Pilea* with about 200 mostly South American species. Five genera are represented in our native flora: *Urtica*, by 3 of 35 spp.; *Boehmeria*, by 5 of 80 mostly tropical spp.; *Pilea* (*Adicea*), by 3 of over 200 tropical spp.; *Laportea* (*Urticastrum*), by 1 of 45 mostly tropical spp.; *Hesperocnide*, by 1 of 3 spp.; and *Parietaria*, by 2 of 7–8 spp. The family is widely distributed over the country but is less common in the southwest and is represented only by 3 genera and 8 spp. on the Pacific coast.

The Urticaceae are distinguished from other families of the order by the stinging hairs (when present), the unicarpellate ovary terminated by a single style, the solitary basal orthotropous ovule, and the cymose inflorescences on short axillary shoots.

Within the Urticales the family is concluded to be the most highly advanced, as evidenced by the complete loss of a second carpel, the shifting of the ovule from a terminal to a basal position (Bechtel, 1921), and the development of the herbaceous habit from the arborescent (the latter the more characteristic of the order).

Economically the family is of little importance. Ramie (*Boehmeria nivea*) is a commercial source of fiber used in cordage, species of *Pilea* and of *Pellionia* are grown under glass as novelties, and young herbage of many temperate species may serve as a source of edible greens.

*LITERATURE:*

BECHTEL, A. R. The floral anatomy of the Urticales. Amer. Journ. Bot. 8: 386–410, 1921.
HERMANN, F. J. The perennial species of *Urtica* in the United States east of the Rocky Mountains. Amer. Midl. Nat. 35: 773–778, 1946.
KILLIP, E. P. The Andean species of *Pilea*, Contr. U. S. Nat. Herb. 26: 475–530, 1939.
SELANDER, S. *Urtica gracilis* Ait. in Fennoscandia. Svensk. Bot. Tidskr. 41 (2): 264–282, 1947.
WEDDELL, H. A. Monographie de la famille des Urticées. Arch. Mus. Hist. Nat. (Paris) 9: 1–592, 1856–1857.
———. Considerations générales sur la famille des urticées suivie de la description des tribus et des genres. Ann. Sci. Nat., Ser. IV, 7: 307–396, 1857.
WINKLER, H. Urticaceae. Engl. Bot. Jahrb. 57: 501–608 (1920–1922).

## Order 14. PODOSTEMALES [53]

A monotypic order having the characters of the family Podostemaceae. Engler treated the taxon as the primitive suborder of the Rosales until the last revision of the system (1936), when it was reinterpreted to be of an even more primitive

---

[53] The name of the order was given as Podostemonales by Engler and Diels and most predecessors. Sprague has pointed out (Kew Bull. 1933, p. 46) that since the type genus is *Podostemum* the correct name of the family is Podostemaceae, a conserved name (vice Podostemonaceae). For this reason the name Podostemales is adopted.

alliance, and the terminal order of the line containing the Fagales and Urticales (from which it differed in part by the numerous stamens and absence of endosperm). This view has not been held by most botanists, who consider it a highly reduced aquatic and presumably with rosalian affinities.

## PODOSTEMACEAE.[53] RIVER WEED FAMILY

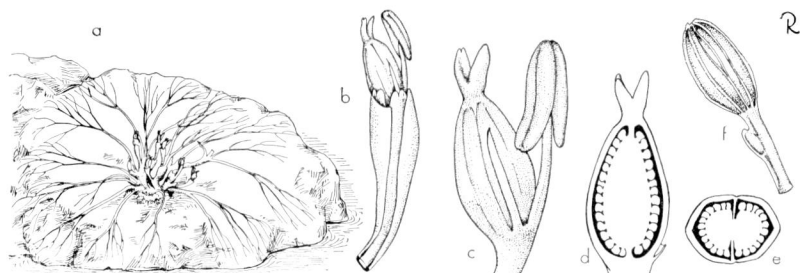

**Fig. 114.** PODOSTEMACEAE. *Oserya Coulteriana*: a, plant in flower on exposed rock, × ½; b, inflorescence, × 3½; c, flower, habit, × 8; d, pistil, vertical section, × 8; e, ovary, cross-section, × 8; f, fruit, × 6. (Drawn from material, courtesy H. E. Moore, Jr.)

Herbaceous mosslike perennial aquatics, usually attached to stones by adhesive-secreting polymorphic, often thalluslike, photosynthetic, creeping roots; growing mostly in running water and often under waterfalls; primary axis small, branching into a dorsiventral thalluslike structure with filiform to laminate segments, secondary leaf-bearing branches usually produced, milky latex present in many species; leaves alternate, usually simple, linear to lamellate, often basally sheathing; flowers usually produced only when plants are exposed by low water, bisexual, zygomorphic, very minute, fragrant, entomophilus, solitary and terminal on an elongated pedicel or cymose, perianth undifferentiated into calyx and corolla, sometimes of 2–3 distinct or basally connate segments (bracts?) or the parts wholly united and then spathe-like with an inner whorl of minute bractlets, the bud enclosed in a capsulelike structure (spathella); the stamens 1–4 or numerous, hypogynous, in 1–several whorls, the filaments basally connate, arising from ventral side of pistil, the anthers typically 4-celled at anthesis, introrse, 1 staminode on each side of the stamens; pistil 1, the ovary superior, sessile or briefly stalked, usually 2–3-loculed and -carpelled, the placentation axile (sometimes the ovary 1-loculed with free-central placentation), the ovules numerous and anatropous, integuments 2, the styles 2–3, filiform or very short, often strongly papillate, usually distinct, the stigmas as many as styles; fruit a septicidal capsule; seeds numerous, minute, no endosperm present.

A moderate-sized family of 43 genera and about 140 species, distributed pantropically and with a few extensions into north and south temperate regions of all continents. It is represented in this country by the indigenous *Podostemum ceratophyllum* that ranges from western Quebec southward to Georgia and southwestern Arkansas. Fassett (1939) treated *P. abrotanoides* as a form of *P. ceratophyllum* lacking any disjunctive distribution. The genus *Oserya* (4 spp.) is South American, except for *O. Coulteriana* of Mexico.

The members are identified readily by their occurrence in streams of running water where they are attached to rocks, by their mosslike or lichenlike appearance, and their minute flowers cymose or solitary on long pedicels from a usually spathe-like envelope.

The phylogenetic position of the family was considered by Warming (1890) and later by Engler to be close to the Saxifragaceae. More recently and on the basis of the embryology, Maheshwari (1945, pp. 31–32) has concluded that it is "almost certain that the Podostemaceae are much reduced apetalous derivatives of the Crassulaceae." Hutchinson (1926) placed the family, together with the Hydrostachyaceae, in the order Podostemonales with the notation (p. 13) "possibly very much reduced apetalous types of Saxifragales, with peculiar habit, but position altogether problematical."

The plants are of no economic importance. The morphology of their vegetative parts is highly variable and their life history is unusual and provides an interesting study (see Willis, 1902, and Arber, 1920).

*LITERATURE:*

ARBER, A. Water plants. 436 pp. Cambridge, England, 1920. [Cf. pp. 112–122, 327–332, for Podostemaceae.]
ENGLER, A. Podostemonaceae. In Engler and Prantl, Die natürlichen Pflanzenfamilien, ed. 2, Bd. 18a: 3–68, 1930.
FASSETT, N. C. *Podostemum* in North America. Rhodora, 41: 525–529, 1939.
HAMMOND, B. L. Development of *Podostemon ceratophyllum*. Bull. Torrey Bot. Club, 64: 17–36, 1937.
MAGNUS, W. Embryology of the Podostemonaceae. Flora, n.s. 5: 275–336, 1913.
MAHESHWARI, P. The place of angiosperm embryology in research and teaching. Journ. Indian Bot. Soc. 24: 25–41, 1945.
NASH, G. V. Podostemonaceae. North Amer. Flora, 22: 3–6, 1905.
WARMING, E. Familien Podostemonaceae. Kongel. Dansk. Videnskab. Selsk. Skrifter, Sjette Raekke. 2: 77–130, 1882.
———. Podostemonaceae. In Engler and Prantl, Die natürlichen Pflanzenfamilien, III (2a): 1–22, 1890.
WENT, F. A. F. C. Morphological and histological peculiarities of the Podostemonaceae. Proc. Int. Cong. Plt. Sci. Ithaca (1926), 1: 351–358, 1925 [1929].
WILLIS, J. C. On the dorsiventrality of the Podostemonaceae with reference to current views on evolution. Ann. Bot. 16: 593–594, 1902.

## Order 15. PROTEALES

A monotypic order with the characters of its 1 family Proteaceae. The cyclic flowers are mostly 2-merous, bisexual or unisexual, the stamens opposite the petaloid perianth segments and usually adnate to them by the filaments, and the seed lacks endosperm. Hallier accepted Engler's view of primitiveness for the order and considered it derived from his Proberberideae. Bessey expressed doubt as to its relationship, and placed it in the Sapindales as phyletically more advanced than amentiferous families also assigned to the Rosales. Rendle, following Engler, placed the order between the Urticales and Santalales, noting that it was difficult to associate it phyletically with other orders. Hutchinson (1948) considered it a terminal taxon derived from stocks ancestral to the Thymelaeaceae. These views are divergent and are evidence of need of much further study on the phylogeny of the order. For the present it can only be noted that the order is not basically primitive, but cannot yet be allied closely with any other existing order.

### PROTEACEAE. PROTEA FAMILY

Trees and shrubs, rarely herbaceous, occasionally dioecious; leaves alternate, rarely opposite or whorled (*Macadamia*), simple, entire or pinnately cut, estipulate; inflorescence a usually showy bracteate head, spike, or raceme; flowers bisexual (unisexual by abortion in *Leucadendron*), actinomorphic or zygomorphic. the perianth uniseriate, consisting of a petaloid hypogynous tetramerous and mostly gamosepalous calyx, valvate but variously split on opening of the bud; stamens 4, opposite the calyx lobes, the filaments usually adnate to calyx tube or lobes and not distinct, the anthers 2-celled, distinct, dehiscing longitudinally; the pistil 1, with or without basal scales or disc, usually on a gynophore, the ovary superior, 1-loculed, 1-carpelled, the ovules 1–many, placentation parietal or the solitary ovule pendulous, the style 1 and slender, the stigma 1, often bulbous; fruit a follicle, achene, samara, or drupe, in *Banksia* the follicles borne in a large woody conelike fruiting inflorescence; seeds without endosperm, sometimes winged.

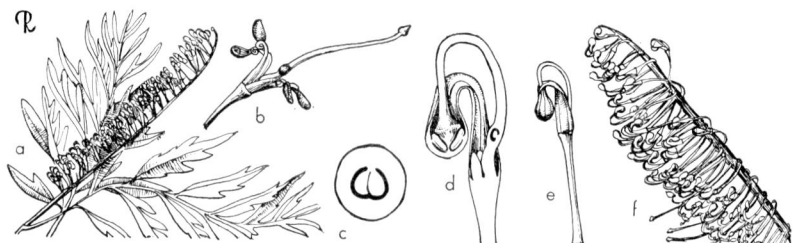

**Fig. 115.** PROTEACEAE. *Grevillea robusta*: a, inflorescence and leaf, × ¼; b, flower, × 1; c, ovary, cross-section, × 10; d, flower (before anthesis), vertical section, × 2; e, same, habit, × 1; f, inflorescence, × ½. (From L. H. Bailey, *Manual of cultivated plants,* The Macmillan Company, 1949. Copyright 1924 and 1949 by Liberty H. Bailey.)

A dominant family of mainly the drier regions of the southern hemisphere, composed of about 55 genera and 1200 species. Of this number about 15 genera and 475 species occur in South Africa (*fide* Phillips), and many genera and about 700 species are native to Australia. The family is not represented in the northern hemisphere (Abyssinia the most northern occurrence) except when cultivated or naturalized from cultivation.

The Proteaceae are distinguished by the usually 4 stamens opposite the same number of perianth segments and the filaments adnate to them, by the parietal placentation, and by the flowers often aggregated into heads and enveloped by large densely hairy or showy bracts.

The family is important domestically for its ornamentals, and about 20 genera and 100 species are offered in the American trade. They are cultivated primarily in California. The Queensland nut (*Macadamia*) is popular in Hawaii for its fruit, an edible delicacy, and the species is grown to a small extent in southern California.

### Order 16. SANTALALES

The Santalales are distinguished from allied taxa by the flowers usually unisexual (often bisexual in Olacaceae), the perianth parts smaller and opposite the adnate stamens, and the seed with endosperm.

Engler and Diels treated them as composed of 2 suborders and 7 families as follows (families preceded by an asterisk not treated in this text):

Santalineae  
    Olacaceae      *Grubbiaceae  
    *Opiliaceae      Santalaceae  
    *Octoknemataceae      *Myzodendraceae  
Loranthineae  
    Loranthaceae

Bessey placed the families of the order high up in his Celastrales and derived them from rosaceous ancestors. Hutchinson likewise treated them as derived from the Celastrales. Rendle considered them allied to Proteales, "differing mainly in

## DIVISION IV. EMBRYOPHYTA SIPHONOGAMA

the inferior ovary." Results of morphological studies by Smith (1942) were in accord with views of Engler and of Schellenberg that the Olacaceae are more primitive than the Loranthaceae and that both belong in the Santalales. Schellenberg (1942) concurred with most of Engler's internal classification of the order but rejected the view that the taxon could contain families of parasites and at the same time be a relatively primitive dicot. (For references, see under Santalaceae.)

## OLACACEAE. OLAX FAMILY

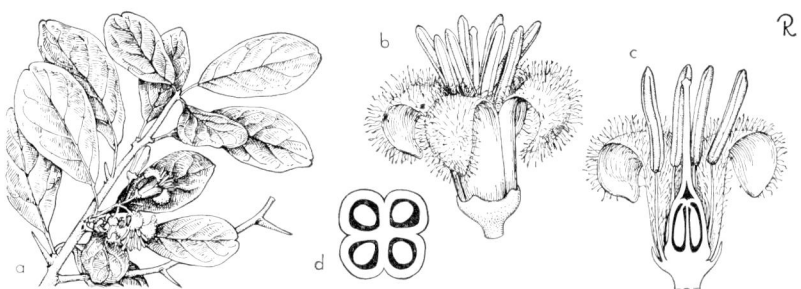

**Fig. 116.** OLACACEAE. *Ximenia americana*: a, flowering branch, × ½; b, flower, habit, × 2; c, same, vertical section, × 2; d, ovary cross-section, × 4. (b and c redrawn from Britton and Sargent respectively.)

Trees, shrubs, or vines; leaves usually alternate, simple and mostly entire, estipulate; inflorescence cymose or a thyrse; flowers usually bisexual (the plants sometimes polygamodioecious), the perianth biseriate, the sepals 4–6, imbricate or open in bud, the petals 4–6, distinct or variously connate, valvate; stamens 4–12, opposite the petals when of same number, distinct or rarely monadelphous, some occasionally antherless staminodes, the anthers 2-celled, dehiscing longitudinally or by porelike apical slits; disc present, often annular; pistil 1, the ovary superior (sometimes seemingly inferior by adnation to the surrounding disc), the carpels and locules 3–4 or sometimes the locule 1 by incomplete septation, the placentation typically axile (free-central from locule apex when unilocular), the ovules solitary in each locule, pendent, the style 1 with a 2–5-lobed stigma; fruit a berry or drupe, often surrounded by the persistent enlarged calyx; the seed with copious endosperm.

A primarily pantropical family of about 25 genera and 150 species. *Olax* (35 spp.) of the Old World tropics is one of the largest genera. Two genera are represented by species indigenous to peninsular Florida, *Schoepfia* (15 spp.) by *S. chrysophylloides*, and *Ximenia* (5 spp.) by *X. americana*.

The Olacaceae are distinguished from other families of the order by the stamens usually twice as many as the petals or more, the ovary superior, and the ovules freely pendulous from the placenta.

The family is of no known economic importance in the United States.

## SANTALACEAE. SANDALWOOD FAMILY

Trees, shrubs, or herbs, a few genera monoecious or dioecious, some parasitic on roots or on branches of trees; leaves simple, mostly opposite, sometimes alternate, entire, foliaceous or reduced to scales, estipulate; the inflorescence various, usually a raceme, spike, head, dichasium, or the flowers solitary in leaf axil; flowers minute, regular, bisexual or unisexual, the perianth uniseriate, sepaloid or petaloid in character and the segments valvate and distinct or basally connate, usually 4-5-lobed; stamens of same number and opposite to the perianth segments, attached at perianth base or on its cup, the filaments short, the anther 2-celled, basifixed, dehiscing vertically; pistil 1, the ovary mostly inferior and embedded in receptacular tissues or superior and borne on or surrounded by a nectar-secreting receptacular disc which may have lobes projecting between the stamens,[54] 1-loculed, the carpels 3-5 (rarely 2), the ovules 1-5 but usually 3 (only 1 maturing), borne pendulously on ovary floor, the integument one or none, the placentation basal, the style 1 and terminal, the stigma 1 and capitate or 3-5-lobed; fruit an achene or drupe; seed without testa, the embryo straight, endosperm fleshy.

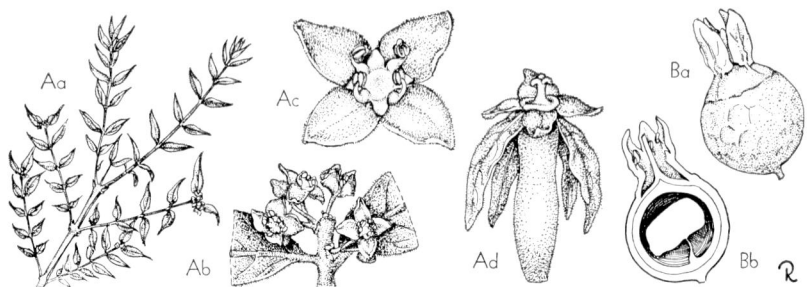

**Fig. 117.** SANTALACEAE. A, *Buckleya distichophylla*: Aa, flowering branch, × ⅙; Ab, staminate inflorescence, × 2; Ac, staminate flower, × 8; Ad, pistillate flower, × 3. B, *Commandra umbellata*: Ba, ovary vertical section, × 3; Bb, same, cross-section, × 3.

A family of 26 genera widely distributed throughout temperate and tropical regions. The number of species as recognized by various authors varies from 250 (Engler and Gilg) to 600 (Rendle). It is represented in the United States by 4 genera: *Comandra*, distributed over most of the country by 1 or more of the 4 indigenous species (a fifth species native to Europe); *Buckleya*, by 1 of its 5 species (others in eastern Asia), a dioecious shrub of North Carolina and Tennessee; *Nestronia*, a monotypic endemic shrub (polygamodioecious) of southeastern U. S.; *Pyrularia*, a dioecious shrub or tree with 1 of its 2 species (the other in the Himalayas) in mountainous woods from Pennsylvania to Georgia.

The members of the family are distinguished by the placental situation, the ovule integument reduced to 1 or absent, the 3-5-carpellate ovary, and the seed without seed coat. It differs from the allied *Loranthaceae* in the plants being non-parasitic or parasitic only on roots of a host, and in the ovary usually with 3 ovules.

The Santalaceae undoubtedly are most closely allied to the Loranthaceae (the families were united by Baillon), and because of the 4-5-merous flower and less complicated floral and fruit morphology is the less advanced. The Santalaceae were divided into 3 tribes by Engler and Diels on the basis of ovary position and the character of the disc.

Economically the family is of little domestic importance, and only *Buckleya* and *Pyrularia* are cultivated (infrequently) as novelties. In the tropics and subtropics,

---

[54] In some genera (as *Santalum*) an hypanthiumlike cup surrounds the pistil. It has been shown by Smith (1942) that, contrary to the usual situation, this is not an hypanthium of foliar parts but is receptacular in origin, with the perianth segments and stamens originating from or near its rim.

the aromatic and sweet-scented sandalwood (*Santalum album*) is prized for cabinet-making and for use in perfumery. The sweet flesh of fruit of some genera is edible.

*LITERATURE:*

RAO, L. N. Studies in the Santalaceae. Ann. Bot. n.s. 6: 151–175, 1942. [Morphological.]

SCHELLENBERG, G. Über Systembildung und über die Reihe der Santalales. Berlin Deutsches Bot. Ges. 50a: 136–145, 1932. Festschrift.

SMITH, F. H. Floral anatomy of the Santalaceae and some related forms. Oregon State Monogr. Stud. Bot. no. 5: 1–93, 1942.

SMITH, F. H. and SMITH, E. C. Anatomy of the inferior ovary of *Darbya*. Amer. Journ. Bot. 29: 464–471, 1942.

## LORANTHACEAE. MISTLETOE FAMILY

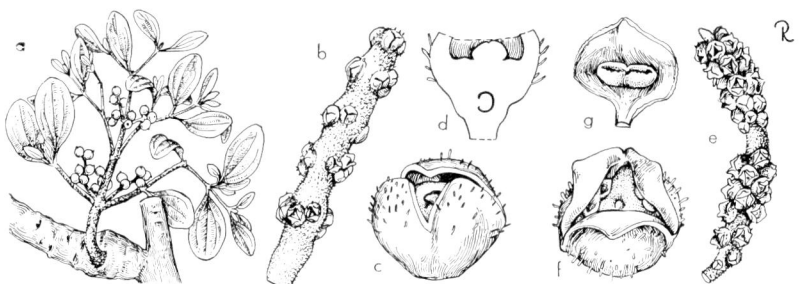

**Fig. 118.** LORANTHACEAE. *Phoradendron flavescens*: a, fruiting plant on host, × ⅙; b, pistillate flowers, × 3; c, pistillate flower, habit, × 15; d, same, vertical section (perianth partially excised), × 15; e, staminate flowers, × 2; f, staminate flower, × 15; g, sepal and stamen, × 15.

Herbs or shrubs parasitic on tree branches (rarely erect terrestrial trees, as *Nuytsia* of Australia), attached to the host by modified roots (haustoria); stems usually dichotomously branched, nodes generally swollen; leaves usually persistent and leathery, mostly opposite or whorled, rarely alternate, simple, entire, thick and green or reduced to connate scales, estipulate; flowers solitary or inflorescence a panicle, raceme, spike, bisexual or unisexual (when so, the plants dioecious), actinomorphic, green or (in many tropical elements) brightly colored, mostly small, the parts borne on a cup-shaped receptacle, the perianth biseriate, the 2 whorls similar and 2–3-merous, both green and sepallike or both large, brilliantly colored and petaloid (no apparent differentiation into calyx and corolla), the parts free or united (when united, the tube often split), the Loranthoideae with a weakly toothed or erose rim below the perianth (a calyculus) sometimes interpreted as calyx; stamens equal in number to perianth lobes or parts and borne on them or at their base, the anthers 2-celled (or 1-celled by confluence) or transversely multiloculate, dehiscing longitudinally or by transverse slits or by terminal pores (in staminate flowers, a rudimentary pistil may be present); pistil 1, the ovary inferior (seemingly embedded in receptacle), obscurely 1-loculed, probably 3–4-carpelled, the ovules not differentiated, the sporogenous cells arising from 3–4 points over a relatively large central placental area on the base of the ovary, the placentation basal, in some (as *Viscum*) the ovary cavity scarcely evident macroscopically, the style simple or absent, the stigma 1 often sessile; fruit a 2–3-seeded berry or a drupe, often viscid; seed without testa, the endosperm abundant, the embryos often 2–3 within each seed and when paired are usually united by cotyledons.

Primarily a tropical family, of about 30 genera and 1100 or more species, but one which extends into temperate zones of both hemispheres. *Loranthus*, a tropical and mostly African genus, the largest with about 500 species, and *Phoradendron*, an American genus, second largest with about 135 species. The Loranthaceae are represented in this country by the scaly-leaved *Arceuthobium* (a primarily American genus of about 12 spp. which is parasitic on conifers), with 7 spp. in the Pacific coast states, an eighth in the northeast, and the evergreen *Phoradendron* (parasitic on numerous genera) with 8 spp. in the west and a ninth in the east, extending as far north as New Jersey.

The family is distinguished by its usually aerial parasitic habit, the cup-shaped receptacle, the inferior ovary lacking a clearly defined locule, and the absence of distinct ovules.

The genera represent 2 rather natural groups usually designated subfamilies: Loranthoideae with about 10 genera, characterized by the presence of a calyculus below the more obvious perianth, and the Viscoideae whose genera lack a calyculus. The indigenous *Phoradendron* and *Arceuthobium* both belong to the Viscoideae.

Economically the family is domestically important only as the source for the mistletoe of yuletide popularity. The mistletoe of the American trade is *Phoradendron flavescens*, but that of Europe is *Viscum album*.

*LITERATURE:*

ALLARD, H. A. The eastern false mistletoe (*Phoradendron flavescens*); when does it flower? Castanea, 8: 72–78, 1943.

DANSER, B. H. A new system for the genera of Loranthaceae-Loranthoideae, with a nomenclator for the Old World species of this subfamily. Verh. Kon. Akad. Wetensch. Amsterdam, Afd. Naturk. Tweede Sect. 29: 1–128, 1933.

DOWDING, E. Floral morphology of *Arceuthobium americanum*. Bot. Gaz. 91: 42–54, 1931.

GILL, L. S. *Arceuthobium* in the United States. Trans. Conn. Acad. Arts and Sci. 32: 111–245, 1935.

KRAUSE, K. Loranthaceae. Engl. Bot. Jahrb. 57: 464–495, 1920–1922.

SCHAEPPI, H. and STEINDL F. Blütenmorphologische und embryologische Untersuchungen an Loranthoideen. Vierteljahr Naturf. Ges. Zurich, 87: 301–372, 1942.

TRELEASE, W. The genus *Phoradendron*. A monographic revision. 224 pp., 245 fig. Urbana, 1916.

## Order 17. ARISTOLOCHIALES

This order was treated by Engler as composed of the families Aristolochiaceae, Rafflesiaceae, and Hydnoraceae. The order is characterized by the perianth primarily uniseriate and petaloid, the ovary usually inferior, the placentation axile with 3–6 locules or parietal with as many carpels in a single locule, the stamens numerous and free from the perianth.

Bessey treated them as the most advanced taxon within his Myrtales, an order derived from the Rosales. Hallier accepted the order as here circumscribed (plus the Balanophoraceae) and derived it from his Lardizabaleae. Wettstein considered them as derived from myrtaceous and annonaceous ancestors, and included them within his order Polycarpicae, a taxon to which he assigned the Magnoliaceae as the primitive family. Hutchinson accepted them as a separate order (to which he added the Nepenthaceae) and considered it to represent "probably reduced Berberidales by way of Menispermaceae . . ." Undoubtedly the order is more advanced than the Engler system indicated and its apparent simplicity one of reduction. The views of Hutchinson and Wettstein (that it has descended from ranalian or magnoliaceous ancestral stocks) are the more widely accepted.

## ARISTOLOCHIACEAE.[55] BIRTHWORT FAMILY

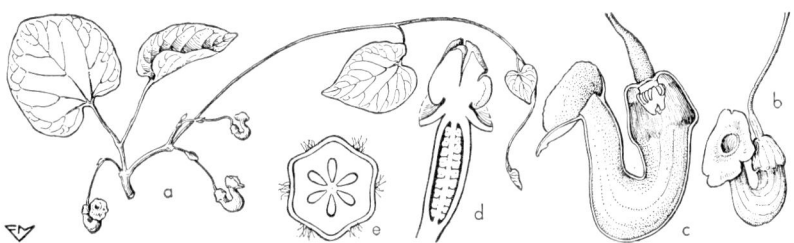

**Fig. 119.** ARISTOLOCHIACEAE. *Aristolochia durior*: a, flowering branch, × ⅙; b, flower, × ½; c, flower, perianth excised, × 1; d, same, in vertical section, perianth removed, × 2; e, ovary, cross-section, × 6. (From L. H. Bailey, *Manual of cultivated plants*, The Macmillan Company, 1949. Copyright 1924 and 1949 by Liberty H. Bailey.)

Low herbs, or more commonly woody usually scandent shrubs; leaves alternate, simple, entire, petioled, estipulate; flowers bisexual, actinomorphic (*Asarum*) or zygomorphic (*Aristolochia*), solitary, in axillary clusters, or racemose, the perianth usually a lurid petaloid gamosepalous calyx, variously 3-lobed or unilateral, often bizarrely colored and fetid, occasionally (in *Asarum*) an inner whorl of 3 minute teeth (vestigial corolla?) present; stamens 6–36, free or adnate to style and producing a column or gynostemium, the filaments short and thick when present, the anthers free or adnate to style, 2-celled, dehiscing longitudinally; pistil 1, the ovary inferior (rarely half inferior), of mostly 4–6 locules (sometimes incompletely so), the carpels 4–6, the placentation axile, each locule with several to many anatropous ovules, the style 1, short and stout, the stigmas as many as carpels; fruit a septicidal capsule, often dehiscing basally (parachutelike); seeds variable in form, the embryo minute, the endosperm copious.

A primarily tropical family with a few members throughout most temperate regions, composed of 6 genera and about 400 species. Two genera are represented by species indigenous to North America: *Asarum* (incl. *Hexastylis*) by perhaps 6 of its 15 species, with 1 in the Pacific coast states and the others in the east; *Aristolochia* (an essentially tropical genus of about 300 spp.) by about 7 spp. over much of the country but not common.

The family is distinguished by the usual adnation of stamens to style, the generally 6-loculed more or less inferior ovary, and the flowers lacking a corolla (or essentially so), but the calyx usually enlarged and petaloid and often of bizarre trumpetlike or bell-like shapes.

Hutchinson (1948), who placed the family as a terminal derivative from the Ranales via herbaceous members of the Berberidaceae, pointed out that relationship, not obvious from study of temperate Aristolochiaceae, becomes more evident from examination of such genera as *Saruma* with its conspicuous petals and separate pistils and its ranalian wood anatomy. It is presumed that the more or less woody climbing members of the family may have been derived from the herbaceous.

Economically the family is of slight importance for its several ornamental species of Dutchman's-pipe, pelican flower, or birthwort (*Aristolochia*), and wild ginger (*Asarum*).

*LITERATURE:*

PEATTIE, D. C. How is *Asarum* pollinated? Castanea, 5: 24–29, 1940.

---

[55] The name Asaraceae Link was accepted by J. K. Small for this family, but Aristolochiaceae (de Jussieu, 1830), is the conserved name for the taxon.

## RAFFLESIACEAE.[56] RAFFLESIA FAMILY

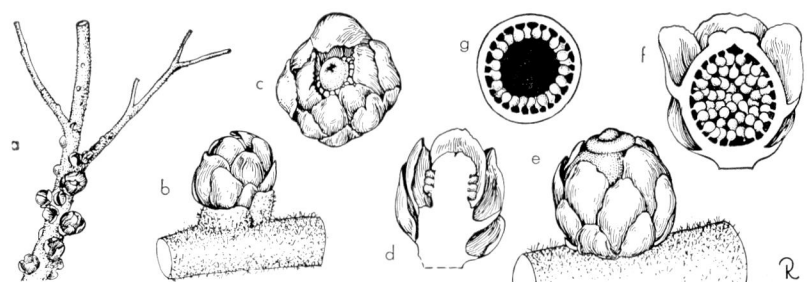

**Fig. 120.** RAFFLESIACEAE. *Pilostyles Thurberi*: a, branch of host with flowering staminate plants, × 1½; b, staminate plant on host, × 5; c, staminate flower, habit, × 8; d, same, vertical section, × 6; e, pistillate plant on host, × 1; f, pistillate flower, vertical section, × 1; g, ovary, cross-section, × 1. (e–g, after Torrey.)

Usually dioecious or monoecious fleshy herbs, parasitic on roots and branches of various hosts; the vegetative body thalloid or reduced to myceliumlike tissues that invade the host; leaves usually scalelike; flowers minute to large (*Rafflesia Arnoldii* of Malaya produces a flower to 20–36 in. across and weighing to 20 pounds, the largest flower known), solitary, unisexual, calyx of 4–10 distinct or basally connate segments, sometimes petaloid, corolla absent; staminate flowers with an indefinite number of stamens, the anthers sessile, usually 2-celled, dehiscing by slits or apical pores; pistillate flower with a single pistil, the ovary inferior to half inferior, unilocular, the carpels 4–6–8, the placentation parietal, the ovules numerous, the integument 1, the style 1 or none, the stigma discoid, capitate or of many lobes or surfaces; fruit a berry, the seed with endosperm.

A primarily subtropical family of 7 genera and 27 species, mostly in the Old World, but *Apodanthes* occurring in South America and *Pilostyles* in Mexico with 2 species extending northward from Mexico and Baja California into New Mexico and southern California. The plants are of no economic importance.

*LITERATURE:*

SOLMS-LAUBACH, H. VON. Rafflesiaceae. In Engler, Das Pflanzenreich, 5 (IV. 75): 1–19, 1901.

### Order 19. POLYGONALES [57]

An order containing the single family, Polygonaceae, characterized by the usually bisexual flowers possessing a superior unilocular uniovular 2–4-carpellate ovary and the fruit a nutlet or achene. Bessey included the family as an advanced taxon of his Caryophyllales, while Hallier, recognizing similar affinities, included both taxa in his Centrospermae. The position accorded the order parallels that of Engler. Hutchinson considered it and his order Chenopodiales as "reduced degraded types of Caryophyllales," with it and the Chenopodiales representing the termination of a line descended from the Ranales via caryophyllaceous ancestors. Undoubtedly the Polygonales are from caryophyllaceous stocks.

---

[56] The name Rafflesiaceae R. Br. (1845) has priority over Cytinaceae Hook. f. (1873).
[57] Order 18, the Balanophorales (containing the single family, Balanophoraceae), has no representatives known to occur in this country.

# POLYGONACEAE. BUCKWHEAT FAMILY

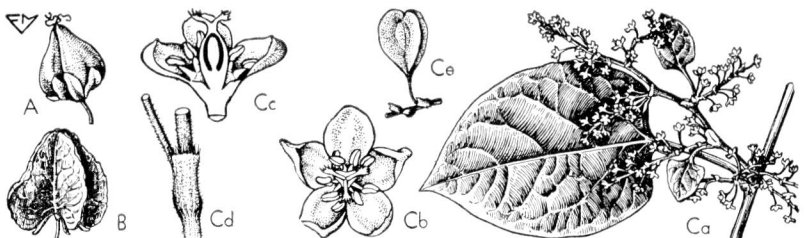

**Fig. 121.** POLYGONACEAE. A. *Fagopyrum sagittatum*: fruit, × 2. B, *Rumex Patientia*: fruit, × 2. C, *Polygonum cuspidatum*: Ca. portion of inflorescence, × ½; Cb, flower, face view, × 5; Cc, flower, vertical section, × 5; Cd, stem, node and ochrea, × ½; Ce, fruit, × 1. (From L. H. Bailey, *Manual of cultivated plants*. The Macmillan Company, 1949. Copyright 1924 and 1949 by Liberty H. Bailey.)

Herbs, shrubs, or rarely trees, sometimes twining; stems often with swollen nodes, occasionally geniculate; leaves alternate (rarely opposite), simple, usually (except in Eriogoneae) with a sheathing stipular growth (ochrea) at petiole base; flowers usually bisexual (when unisexual, the plants monoecious or dioecious), actinomorphic, basically in cymes or cymules that often are disposed in racemes, panicles spikes or heads, the perianth biseriate with usually 3, 4, 5, or 6 distinct undifferentiated tepals [58] (2 whorls of 3 each, the basic situation, the 5-tepal condition representing usually a fusing of 1 of the inner and 1 of the outer whorls), the inner sometimes enlarged or modified with hooks, spines, wings, or tubercles, the tepals often persistent, enlarged, and membranous in fruit; the stamens mostly 6-9 in basically 2 series (the 6 outer often introrse, the 3 inner extrorse), the filaments free or basally adnate, the anthers 2-celled, dehiscing longitudinally; pistil 1, subtended by an annular (often lobed) nectar-secreting glandular disc, the ovary superior, sessile, compressed or 3-angled, 1-loculed (sometimes falsely 3-loculed), the carpels 2, 3, or 4, ovule 1 and seemingly basal, the style 1, the stigmas 2-4; fruit a flat angled or winged achene; seed with usually a curved embryo and copious mealy endosperm.

A family of about 32 genera and 800 or more species, mostly of temperate distribution primarily in the northern hemisphere. *Polygonum* occurs on all continents, but most genera are of restricted distribution, and several are localized endemics. In this country the family is represented by 14 genera, 4 monotypic, and restricted to California and/or adjoining states (*Pterostegia, Gilmania, Nemacaulis, Hollisteria*), the others primarily of western American arid lands and include *Lastarriaea, Chorizanthe* (50 spp.—35 in southwest), *Oxytheca* (9 spp.), *Eriogonum* (perhaps 150 spp.), and *Emex*, an adventive monoecious Mediterranean plant. Genera of more general distribution in this country include *Polygonum* (200 spp., about 70 indigenous), *Rumex* (150-30), *Polygonella* (5-3), *Brunnichia* (1 spp. here, a second in west Africa), *Coccoloba* (125 of tropical American spp., 2 indigenous), and *Oxyria* (1 sp. here, the other Himalayan). Some American authors (notably Small, Rydberg and their devotees) have accepted the subdivisions of

---

[58] Wide divergence of opinion is represented in the literature as concerns the floral descriptions for the family. The evidence and conclusions of Laubengayer (1937), Lundblad (1922), and Geitler (1929) have been considered in the drafting of the description of this family. Literature dealing with the morphology and anatomy of members of the family was reviewed by Vautier (1949), whose own researches were essentially in accord with the findings by Laubengayer.

*Polygonum* and treated its species as representing several genera (including *Avicularia, Persicaria, Bistorta, Tovara,* and *Tiniaria*).[59]

The family is characterized by the presence of the nodal ochreas (or in their absence, by the involucrate heads), the unilocular ovary with its solitary basal ovule, and the 1-seeded fruit with a usually S-shaped embryo. The arctic and subarctic *Koenigia* (also Himalayan) is the only genus outside the Eriogoneae not possessing an ochrea.

In some genera the perianth parts clearly are biseriate while in others they seemingly are spirally arranged, but it was believed by Laubengayer (1937) that the "fundamental plan is trimerous and whorled," and he concluded that "the apparent spiral condition, which is found in the perianth of some forms, is fundamentally a whorled one as definitely shown by anatomical study. This condition is made possible by the fusing of an outer tepal with an inner one." He was of the opinion that the base of the ovary locule was not the base of the ovary, and that the lower third is filled in with ovarian tissue, with the result that the solitary seemingly basal ovule is in fact a terminal ovule persisting from unknown ancestral forms that had a multiovulate free-central type of placentation, and that the present apparent funiculus is actually "a free central placenta which is greatly reduced."[60] From these data Laubengayer concluded the family to be allied to and phyletically more advanced than the Caryophyllaceae. Hutchinson has also indicated a similar derivation, as did Bessey.

The Polygonaceae are not an economically important family. Domestically the buckwheat (*Fagopyrum*) and rhubarb (*Rheum*) are important for food. The remainder are mostly ornamental or noxious weeds. Notable among the ornamentals are the mountain-rose vine (*Antigonon*), the silver-lace vine (*Polygonum Aubertii*), sacaline, a coarse perennial (*Polygonum sachalinense*), and the sea grape (*Coccoloba Uvifera*).

*LITERATURE:*

BRENCKLE, J. F. Notes on *Polygonum*. Phytologia, 2: 402–406, 1948.
EMBERGER, L. La structure de la fleur des Polygonacées. C. R. Acad. Sc. 208: 313–317, 1939.
GEITLER, L. Zur Morphologie der Blüthen von *Polygonum*. Oesterr. Bot. Zeit. 78: 229–241, 1929.
GOODMAN, G. J. A revision of the North American species of *Chorizanthe*. Ann. Mo. Bot. Gard. 21: 1–102, 1934.
HEDBERG, O. Pollen morphology in the genus *Polygonum* L. s.lat. and its taxonomic significance. Svensk Bot. Tidsk. 40: 371–404, 1946.
JOSHI, A. C. The nature of the ovular stalk in Polygonaceae and some related families. Ann. Bot. n.s. 2: 957–959, 1933.
LAUBENGAYER, R. A. Studies in the anatomy and morphology of the polygonaceous flower. Amer. Journ. Bot. 24: 329–343, 1937.
LUNDBLAD, H. Über die baumechanischen Vorgänge bei dem Entstehen von Anomomeri bei homochlamydeischen Blüten. Lund. 1922.
MITRA, G. C. The origin, development and morphology of the ochrea in *Polygonum orientale* L. Journ. Indian Bot. Soc. 24: 191–200, 1945.

[59] Hedberg (1946) found the pollen morphology of these generic segregates to be sufficiently distinct to enable him to classify the species as belonging to one of the following 7 genera: *Koenigia, Persicaria, Polygonum* (s. str.), *Pleuropteropyrum, Bistorta, Tiniaria, Fagopyrum*. Keys based on pollen types and on gross morphological characters of the plant were provided.

[60] In an opposing opinion, based on studies of the presumed related multiovulate genera of Amaranthaceae, Joshi (1933) held that a true funiculus is present in the basal ovule of Polygonaceae and that "the unusual vascular supply of the funiculus is to be explained by the fact that the ovule in the Polygonaceae terminates the floral axis."

PERDRIGEAT, C. A. Anatomie comparée des Polygonacées et ses rapports avec la morphologie et la classification. Act. Soc. Linn. Bordeaux, 55: 1–91, 1900.
RECHINGER, K. H. JR. Vorarbeiten zu einer Monographie der Gattung *Rumex* I. Bot Centralbl. Beih. 49: 1–132, 1932.
———. The North American species of *Rumex*. Field Mus. Nat. Hist., Bot. Ser. 17: 1–151, 1937.
SMALL, J. K. A monograph of the North American species of *Polygonum*. Mem. Bot. Columbia College, 1: 1–183, 1895.
STANFORD, E. E. The amphibious group of *Polygonum*, subgenus *Persicaria* I. Rhodora 27: 109–112; II, *op. cit.* 125–130; III, *op. cit.* 146–152; IV, *op. cit.* 156–166, 1925.
STEWART, A. N. The Polygoneae of eastern Asia. Contr. Gray Herb. 88: 3–119, 1930.
STOKES, S. G. The genus *Eriogonum*, a preliminary study based on geographic distribution. San Francisco, Cal., 1936.
VAUTIER, S. La vascularisation florale chez les Polygonacées. Candollea, 12: 219–343, 1949.
WHEELER, L. C. *Polygonum Kelloggii* and its allies. Rhodora, 40: 309–317, 1938.
WODEHOUSE, R. P. Pollen grains in the identification and classification of plants. VI. Polygonaceae. Amer. Journ. Bot. 18: 749–764, 1931.

## Order 20. CENTROSPERMAE

An order containing the Chenopodiaceae, Amaranthaceae, Nyctaginaceae, Phytolaccaceae, *Gyrostemonaceae, *Achatocarpaceae, Aizoaceae, Portulacaceae, Basellaceae, and Caryophyllaceae. (Family names preceded by an asterisk [*] are not treated in this text.) In the Centrospermae, the perianth is typically biseriate, the embryo generally coiled or curved, and the superior ovary typically unilocular. Bessey designated it the Caryophyllales derived from ranalian ancestors; Hallier accepted Engler's name but greatly expanded its circumscription and derived it largely from his ranalian Podophylleae; Wettstein followed Engler but included also the Cactaceae, as allied to the Aizoaceae; and Hutchinson recognized these families as representing 2 orders, the primitive Caryophyllales (derived from ranalian stocks) and the Chenopodiales (much advanced over and derived from the Caryophyllales).

## CHENOPODIACEAE. GOOSEFOOT FAMILY

Predominantly halophytic annual or perennial herbs, shrubs, or rarely small trees (in *Haloxylon*), sometimes with fleshy nodal or jointed nearly leafless stems; leaves usually alternate, rarely opposite (in *Salicornia, Nitrophila*), simple, very fleshy and terete in some or reduced to scales in others, estipulate; flowers often bracteate, bisexual, or unisexual and the plants dioecious (as in *Grayia*) or monoecious (*Sarcobatus*), minute, greenish, mostly actinomorphic, usually in small dense dichasial or unilateral cymes. the perianth uniseriate, typically of 5 (varying from 2–5) connate sepals (sometimes absent in staminate flowers), usually persisting in fruit; petals absent; stamens as many as calyx lobes and opposite them, inserted on a disc on the calyx or hypogynous, the filaments usually distinct, the anthers 2-celled, incurved in bud, dehiscing longitudinally; pistil 1, the ovary usually superior or (in *Beta*) inferior, 1-loculed, 2–3-carpelled, the ovule solitary, erect or suspended from a basal funiculus, campylotropous, the styles 1–3, the stigmas mostly 2–3; fruit an indehiscent nutlet, several often aggregated together by connation of somewhat fleshy calyces; seed with a peripheral or coiled embryo surrounding the endosperm (the latter little or none in *Salsola, Sarcobatus*, and *Suaeda*).

A family of about 102 genera and 1400 species, of world-wide distribution but with centers in xerophytic and halophytic areas especially on the prairies and plains of North America, the pampas of South America, the shores of the Red, Caspian, and Mediterranean seas, the central Asiatic basin, the South African karroo, and the salt plains of Australia. The family is represented in the United States by about 22 genera. Several (as *Salsola, Salicornia, Suaeda*) are indicators of saline habitats and are cosmopolitan in distribution. About 18 genera are represented in the western states (endemic genera being *Nitrophila, Aphanisma, Mono-lepis,* and *Allenrolfea*). The largest genera include *Chenopodium* (represented by 30 of its about 100 species) and *Atriplex* (by about 6 of its 36 species).

The Chenopodiaceae are distinguished from related families by the fleshy habit (when present), the absence of scarious bracts, the 1-loculed, 1-ovuled, and 2-3-carpelled ovary, and the embryo surrounding the endosperm or spirally coiled.

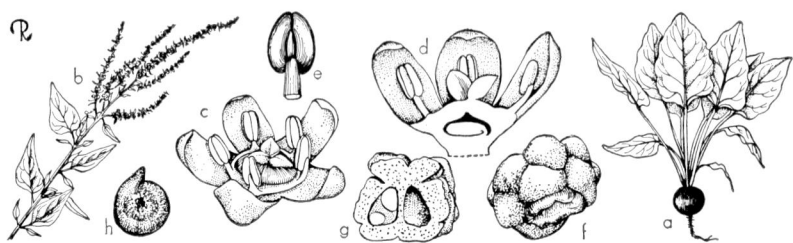

**Fig. 122.** CHENOPODIACEAE. *Beta vulgaris*: a, garden beet foliage and root, reduced; b, flowering stem, × 1/20; c, flower, × 10; d, same, vertical section, × 10; e, anther, dorsal side, × 20; f, fruit, × 3; g, same, vertical section, × 3; h, seed, × 3. (c-e after LeMaout and Decaisne.) (From L. H. Bailey, *Manual of cultivated plants,* The Macmillan Company, 1949. Copyright 1924 and 1949 by Liberty H. Bailey.)

The family was divided by Ulbrich (1934) into the Cyclobeae (embryo annular or conduplicate) and Spirolobeae (embryo coiled spirally), and thence into 8 subfamilies and 14 tribes. The American genus *Nitrophila,* and the genera of the Chenopodioideae and the Salicornioideae were included in the more primitive Cyclobeae, and those of the subfamilies Sarcobatoideae, Suaedoideae, and Salsoloideae were included in the Spirolobeae and in that phyletic sequence.

Economically the family is of minor importance, with the garden beet (*Beta vulgaris*) a commercial source of sugar (sucrose) and a root vegetable, followed in importance by the potherbs spinach (*Spinacia oleracea*) and Swiss chard (*Beta vulgaris* var. *Cicla*). Oil of wormwood, a vermifuge, is obtained from seeds of *Chenopodium anthelminticum*. A few members are grown as ornamentals, notably cypress spurge (*Kochia*).

*LITERATURE:*

AELLEN, P. Beiträge zur Systematik der *Chenopodium*-arten Amerikas. Fedde, Repert. Spec. Nov. Reg. Veg. 26: 31–64, 1929.
AELLEN, P. and JUST, T. Key and synopsis of the American species of the genus *Chenopodium.* Amer. Midl. Nat. 30: 47–76, 1943.
HALL, H. M. and CLEMENTS, F. E. The phylogenetic method in taxonomy. The genus *Atriplex.* Carnegie Inst. Wash. Pub. 326: 235–346, 1923.
LORZ, A. Cytological investigations on five chenopodiaceous genera with special emphasis on chromosome morphology and somatic doubling in *Spinacia.* Cytologia, 8: 241–276, 1937.
STANDLEY, P. C. Chenopodiaceae. North Amer. Flora, 21: 1–93, 1916.
ULBRICH, E. Chenopodiaceae. In Engler and Prantl, Die natürlichen Pflanzenfamilien, ed. 2, Bd. 16c: 377–584, 1934.

## AMARANTHACEAE.[61] AMARANTH FAMILY

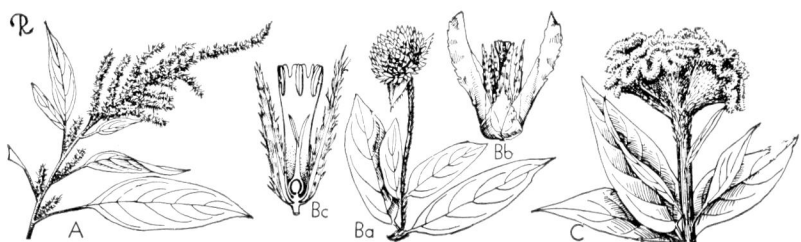

**Fig. 123.** AMARANTHACEAE. A, *Amaranthus caudatus*: flowering branch, reduced. B, *Gomphrena globosa*: Ba, flowering branch, × ½; Bb, flower with involucre, × 2; Bc, flower, vertical section, × 3. C, *Celosia argentea* var. *cristata*: flowering branch, × ⅙. (From L. H. Bailey, *Manual of cultivated plants*, The Macmillan Company, 1949. Copyright 1924 and 1949 by Liberty H. Bailey.)

Annual or perennial herbs (rarely shrubs), trees, or vines; leaves alternate or opposite, simple, generally entire, estipulate; flowers bisexual or less often unisexual (the plants then polygamodioecious or strictly dioecious), actinomorphic, each subtended typically by a membranous to scarious persistent bract and 2 similar bractlets (bracteoles of some authors), solitary racemose or spicate, perianth uniseriate, a calyx of 3–5 free or partially connate more or less dry and membranous sepals; stamens generally 5, opposite sepals or calyx lobes, the filaments usually connate for part or all their length into a membranous tube, lobes or petaloid enations (simple or fringed) may alternate with the anthers, each anther 4-celled at anthesis (Amaranthoideae) or 2-celled (Gomphrenoideae), dehiscing by longitudinal slits; pistil 1, the ovary superior, unilocular, 2–3-carpelled, the ovule solitary and basal or (in Celosieae) several on a seemingly single basal funiculus, campylotropous, the styles 1–3 with stigmas of same number and variable in form; fruit a circumscissile capsule (*Celosia*), or more frequently a utricle or nutlet, and rarely a drupe or berry; seed with embryo enveloping the abundant mealy endosperm.

A family of 64 genera and about 800 species, most abundant in tropical regions and especially tropical America and Africa. Nearly one-third of the genera are monotypic, with 5 indigenous in the New World. Twenty genera occur in the New World, of which 13 are in the United States, primarily in the south and southeast. Only 2 genera (*Amaranthus, Tidestromia*) occur in the Pacific coast states, and 6 genera (*Amaranthus, Acnida, Celosia, Iresine, Froelichia, Gomphrena*) are within the *Gray's manual* region of the northeast. The largest genera of the family include the American *Alternanthera* (170 spp.), the Australian *Ptilotus* (100 spp.), the primarily American *Gomphrena* (90 spp.), the pantropical *Celosia* (60 spp.), and nearly cosmopolitan *Amaranthus* (50 spp.).

The Amaranthaceae are distinguished by the presence of usually scarious bracts, a perianth, and by the usually connate filaments. The dense or congested inflorescence usually provides a superficial clue to identity.

The amaranths have been presumed to be among the more primitive members of the order, but recent studies of the bracts and bractlets provide evidence that the basic inflorescence is a dichasium of 3 flowers, of which 2 usually have been

---

[61] In much of the literature the name of the family appears as Amarantaceae (and that of the type genus as *Amarantus* vice *Amaranthus*); however, Sprague (Kew Bull. pp. 287–288, 1928) indicated the correct spelling of the type genus to be *Amaranthus*, and that of the family Amaranthaceae.

lost and only the bractlets remain. By this view each bracteated flower is interpreted to represent a relic of an ancestral dichasium. The views of Hutchinson and others that this family is not primitive but has evolved from caryophyllaceous ancestors seem to agree with existing data.

Economically the family is of little importance. It contributes few food products (herbage of several members is edible as greens) and only a few ornamentals, as Prince's feather (*Amaranthus*), globe amaranth (*Gomphrena*), and a red-foliaged form of *Iresine*. Several genera contain noxious weeds, notably *Amaranthus, Iresine,* and *Acnida.*

*LITERATURE:*

JOSHI, A. C. Contribution to the anatomy of the Chenopodiaceae and Amaranthaceae. Journ. Indian Bot. Soc. 10: 213–265, 1931.

SCHINZ, H. Amaranthaceae. Engler and Prantl, Die natürlichen Pflanzenfamilien, ed. 2, Bd. 16c: 7–85, 1934.

STANDLEY, P. C. The application of the generic name *Achyranthes.* Journ. Wash. Acad. Sci. 5: 72, 1915.

———. The North American tribes and genera of Amaranthaceae. Journ. Wash. Acad. Sci. 5: 391–396, 1915.

———. Amaranthaceae. North Amer. Flora, 21: 95–169, 1917.

STUCHLIK, J. Zur Synonymik der Gattung *Gomphrena.* Fedde, Repert. Spec. Nov. 11: 36–151, 1912–1913.

ULME, E. B. and BRAY, W. L.. A preliminary synopsis of the North American species of *Amaranthus.* Bot. Gaz. 19: 267–313, 1899.

## NYCTAGINACEAE.[62] FOUR-O'CLOCK FAMILY

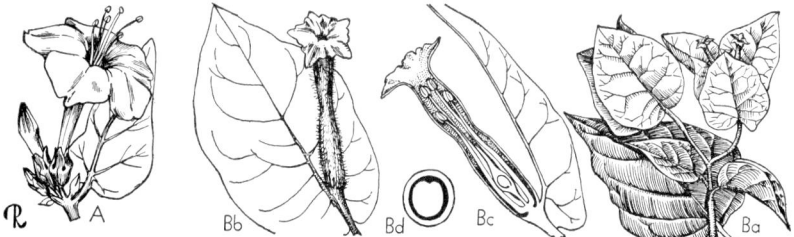

**Fig. 124.** NYCTAGINACEAE. A, *Mirabilis Jalapa*: flowering branch tip, × ½. B. *Bougainvillea spectabilis*: Ba, flowering branch, × ¼; Bb, flower and subtending bract, × 1; Bc, same, vertical section, × 1; Bd, ovary, cross-section, × 3. (From L. H. Bailey, *Manual of cultivated plants,* The Macmillan Company, 1949. Copyright 1924 and 1949 by Liberty H. Bailey.)

Herbs or (in the tropics) shrubs or trees; leaves usually opposite, simple, entire, estipulate; flowers bisexual or rarely unisexual (plants then monoecious or dioecious), actinomorphic, the inflorescence cymose, usually 2–5 foliaceous and often highly colored bracts subtending each flower (each bract fertile and flower-bearing in *Oxybaphus*), the perianth uniseriate, composed of a petaloid calyx of typically 5 connate sepals (seemingly a corolla), the calyx plicate or contorted in bud; corolla absent; stamens 1–30 (even in the same species), hypogynous, distinct or the filaments connate basally in a tube, unequal, the anthers 2-celled, dehiscing longitudinally; pistil 1, the ovary superior, unilocular, 1-carpelled, the ovule 1, basal, inverted (anatropous or campylotropous), the style 1 and slender, the

---

[62] The name Nyctaginaceae Lindl. (1836) is conserved over Allioniaceae Reichb. (1828).

stigma 1; fruit an achene, sometimes enveloped by the persistent calyx which may be variously modified and thereby facilitate dissemination; seed with straight or curved embryo and little or much endosperm.

A family of 28 genera and about 250 species, distributed mostly in tropics and subtropics of both hemispheres but only 2 genera indigenous to the Old World. It is restricted in this country mostly to the southern and Pacific regions (*Oxybaphus*, with 5 indigenous species, the only genus in the northeast), where it is represented in the Pacific states by 8 genera and 28 species (only *Hermidium* a monotypic endemic), and in the south by 7 of these and 2 other genera. The largest genera of the family are *Mirabilis* (60 spp.) and *Abronia* (33 spp.), 14 genera are monotypic.

Members of the family are distinguished by the often colored sepaloid bracts and the petaloid calyx, and by the unicarpellate ovary. However, there is no way by which it may be determined macroscopically that the petaloid calyx is not a corolla, and the distinction is of morphological rather than of taxonomic significance. For this reason, keys to families treating the Nyctaginaceae "bring out" the family with those families possessing gamopetalous corollas.

Heimerl (1934) recognized the family as composed of 5 tribes—the first 4 with ovaries glabrous and stamens more or less connate basally, the Mirabileae with a straight embryo and large cotyledons, the shrubby Pisoneae, and the herbaceous Boldoeae, and Colignoneae, all with the curved embryo and small linear cotyledons (leaves opposite in Colignoneae and alternate in Boldoeae), and the fifth, the Leucastereae with a hairy ovary and distinct filaments.

Economically the Nyctaginaceae are of little domestic importance except for such ornamentals as the garden annuals, four-o'clock (*Mirabilis*) and sand verbena (*Abronia*), and the woody subtropical vine *Bougainvillea*.

*LITERATURE:*

HEIMERL, A. Studien über einige Nyctaginaceae in l'Herb. Delessert. Ann. Conserv. Jard. Bot. Genève 5: 177–186, 1901.

———. Nyctaginaceae. Engler and Prantl, Die natürlichen Pflanzenfamilien, ed. 2, Bd. 16c: 86–134, 1934.

MAHESHWARI, J. Contributions to the morphology of *Boerhaavia diffusa*. Journ. Indian Bot. Soc. 9: 42–61, 1930.

STANDLEY, P. C. The Allioniaceae of the United States with notes on Mexican species. Contr. U. S. Nat. Herb. 12: 303–389, 1909.

———. Allioniaceae. North Amer. Flora, 21: 171–254, 1918.

## PHYTOLACCACEAE.[63] POKEWEED FAMILY

Herbs, shrubs, or trees (sometimes lianous); leaves alternate, simple, entire, stipules absent or minute; flowers bisexual or rarely unisexual (plants then monoecious), actinomorphic (zygomorphic in *Anisomeria*), small, cymose or racemose, the perianth mostly uniseriate, composed of a calyx of 4–5 usually more or less connate persistent sepals, the petals absent (staminodes petaloid in the tropical American *Stegnosperma*); stamens 3–many (varying within the same species), often borne on an hypogynous disc in 1 or more but commonly 2 whorls (the outer whorl petaloid in *Stegnosperma*) alternate or opposite the calyx lobes, the filaments distinct or basally connate, the anthers 2-celled, dehiscing longitudinally; gynoecium of 1–16 distinct or connate pistils, when 1 the pistil is usually 2–many-carpelled and -loculed (unilocular and uniovulate in *Rivina*), when pistils are 2–more each is 1-loculed and 1-carpelled, the ovary superior (inferior only in

[63] The name Phytolaccaceae Lindl. (1836) is conserved over Petiveriaceae Link (1829).

*Agdestis*), when 2–more carpelled the placentation axile, the locules uniovulate, when unilocular the ovules basal, campylotropous or amphitropous, the style none or short, the stigmas as many as carpels, usually linear to filiform (peltate in *Rivina*); fruit very variable depending on gynoecial situation, and may be a berry, drupe, schizocarp, utricle, or achene; seed often arillate, with abundant endosperm enveloped by the embryo.

A family of about 17 genera and 125 species, largely of the American tropics and subtropics. *Phytolacca* (ca. 35 spp.) is the largest genus. Three genera have species indigenous to this country: the pokeberry *Phytolacca* (*P. americana*) from Maine to Minnesota and southward, the pigeonberry (*Rivina humilis*) from Florida to Texas, and *Petiveria,* a monotypic genus whose northern limits are in southern Florida.

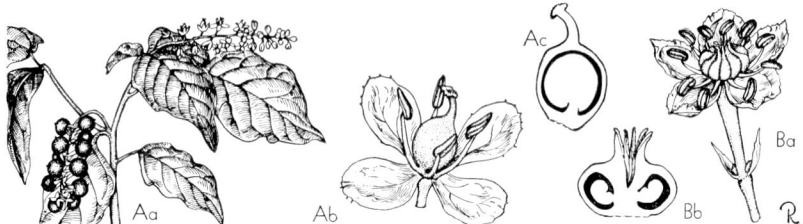

**Fig. 125.** PHYTOLACCACEAE. A, *Rivina humilis*: Aa, flowering and fruiting branch, × ½; Ab, flower, × 4; Ac, pistil, vertical section, × 8. B, *Phytolacca americana*: Ba, flower habit, × 5; Bb, pistil, vertical section, × 5. (Ba after Schnizlein, Bb after Baillon.)

Members of the family are distinguished from those of related families by the often multicarpellate pistil with a single ovule in each locule, the nonpetaloid minute perianth, and (in domestic indigens) by the berrylike fruit with colored juices or pulp.

The family is divided into 5 tribes: *Stegnosperma* and *Agdestis,* each in tribes by themselves, are characterized by the presence of petals (or petaloid staminodes) in the former and of an inferior ovary in the latter; the Rivineae (9 genera) are characterized by unicarpellate pistils; the unigeneric Barbeuieae have a bicarpellate ovary and the Euphytolacceae have a multicarpellate ovary.

Economically the family is of little significance. Two species of *Phytolacca* and *Rivina humilis* are grown for ornament, as also are species of *Agdestis, Ercilla,* and *Petiveria.* Young shoots of *Phytolacca americana* are a source of edible "greens" or potherbs.

*LITERATURE:*

HEIMERL, A. Phytolaccaceae. In Engler and Prantl, Die natürlichen Pflanzenfamilien, ed. 2, Bd. 16c: 135–164, 1934.
WALTER, H. Phytolaccaceae. In Engler, Das Pflanzenreich, 39 (IV. 83): 1–154, 1909.
WILSON, P. Petiveriaceae, North Amer. Flora, 21: 257–266, 1932.

## AIZOACEAE.[64] MESEMBRYANTHEMUM FAMILY

Annual or perennial herbs, or low shrubs; leaves alternate or more commonly opposite or in false whorls, simple (pinnate in *Acthephyllum*), fleshy or reduced to scales, stipulate or more often not so; flowers bisexual, actinomorphic, solitary,

---

[64] This family is sometimes designated by the names Ficoideae Juss. (1789), Mesembryaceae Lindl. (1836), or Tetragoniaceae Reichb. (1827), but Aizoaceae Braun (1864) has been proposed for conservation.

DIVISION IV. EMBRYOPHYTA SIPHONOGAMA    483

in axile dichasia or terminal in monochasia, the perianth uniseriate, composed of a calyx of 5-8 connate herbaceous green sepals, free or adnate to the ovary, the corolla none (the apparent petals are petaloid staminodes);[65] stamens basically 5 but usually many (by splitting), the outermost ones often sterile and petaloid, the filaments distinct or basally connate into bundles or a short monadelphous sheath, the anthers small, 2-celled, dehiscing longitudinally; pistil 1, the ovary superior or inferior, 1-5-loculed (occasionally to 20-loculed), the carpels basically 3-5, the placentation typically axile, but may be parietal or basal, the ovules mostly many (seldom 1), anatropous or campylotropous, the style 1 or absent, the stigmas 2-20 and usually radiating; fruit a loculicidal capsule (dehiscence septicidal or transverse in some), sometimes leathery and tardily dehiscent or berrylike; seed with a large embryo enveloping the mealy endosperm.

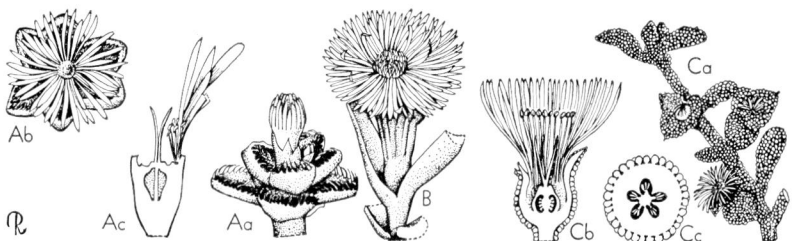

Fig. 126. AIZOACEAE. A, *Faucaria tigrina*: Aa, plant in flower, × ⅜; Ab, flower fully open, top view, × ⅜; Ac, flower, vertical section, × ¾. B, *Carpobrotus edulis*: flowering branch, × ¼. C, *Cryophytum crystallinum*: Ca, flowering branch, × ¼; Cb, flower, vertical section, × ½; Cc, ovary, cross-section, × 1. (All redrawn: A from Flowering Plants of South Africa; B from Curtis' Botanical Magazine; C from Reichenbach, Icones.) (From L. H. Bailey, *Manual of cultivated plants*, The Macmillan Company, 1949. Copyright 1924 and 1949 by Liberty H. Bailey.)

A large family, particularly of South Africa. The number of genera now recognized to be more than 100 and the species about 600.[66] The New World genera number 8-9, of which 7 occur in many parts of this country but primarily in the southwestern and Pacific areas. They are, together with numbers of total and indigenous or naturalized species: *Mollugo* (12-2, widely distributed), *Geocarpon* monotypic in Missouri, *Glinus* (10-2), *Cypselea* a monotypic West Indian genus naturalized from Florida to California, *Trianthema* (15-1), *Sesuvium* (5-3) from the coast of New York to Florida and California, *Tetragonia* (1, native of New Zealand and east Asia), and *Cryophytum* (45-3) along the coast of southern California and believed to have been "probably introduced . . . by natural means before the advent of white men" (usually treated in regional floras as *Mesembryanthemum*).

The members of the family are recognized by the combination of fleshy leaves,

---

[65] The petaloid staminodes are not grossly distinguishable from true petals and keys to the family or its subdivisions often treat the taxa as if petaloid.

[66] The South African taxa have been studied intensely during the last quarter century, and the Linnaean genus *Mesembryanthemum* is now divided among 100 separate genera or more; cf. Brown (1920), Bolus (1928-1938), Pax and Hoffmann (1934, pp. 198-221 for keys and synonymy), and Jacobsen (1938). Acceptance of this segregation was deferred by Pax and Hoffmann (1934) because of unavailability of adequate material. They treated the family as composed of 24 genera. Phillips (1926, pp. 232-249) accepted 52 genera, and many botanists now accept the more liberal views of Brown and of Nels.

5-8-lobed calyx, numerous stamens, often many petaloid showy staminodia, and the multicarpellate ovary with the ovules usually many on each placenta.

Members of the Aizoaceae are divided into subfamilies and tribes, on the basis of the position of the ovary, the type of placentation, and the character of the ovule. Buxbaum (1944) considered them (and the Cactaceae) to belong in the Centrospermae and to have been derived from stocks ancestral to the Phytolaccaceae, a view presented in substance much earlier by Hallier.

Members of the family are of domestic economic importance largely for their ornamentals. Species of about 80 genera are cultivated as novelties. One species, New Zealand spinach (*Tetragonia expansa*), is cultivated to a limited extent as a source of table greens.

*LITERATURE:*

BOLUS, H. M. L. Notes on *Mesembryanthemum* and some allied genera. Parts I, II, III Cape Town, 1928–1938.

BROWN, N. E. New and old species of *Mesembryanthemum*, with critical notes. Journ. Linn. Soc. Bot. (Lond.) 45: 53–140, 1920.

———. *Mesembryanthemum* and allied genera. Journ. Bot. 66: 75–80, 265–268, 322–329 1928; *op. cit.* 67: 17–20, 1929.

BUXBAUM, F. Untersuchungen zur Morphologie der Kakteenblüte. 1. Das Gynoecium Bot. Arch. (Leipzig), 45: 190–247, 1944.

DE VOS, M. P. Cytological studies in genera of *Mesembryanthemeae*. Beitr. Sukkulentenk. u. Pflege, 1943: 1–160, 71 figs., 1943.

JACOBSEN, H. Verzeichnis der Arten der Gattung *Mesembryanthemum* L. Fedde Repert. Spec. Nov. Beihefte 106: 1–198, 1938; Suppl. 1, II. Nachtr. 1–34, 1939.

PAX, F. and HOFFMANN, K. Aizoaceae. Engler and Prantl, Die natürlichen Pflanzenfamilien, 2, Bd. 16c: 179–233, 1934.

PHILLIPS, E. P. The genera of South African flowering plants. Cape Town, 1926.

WILSON, P. Tetragoniaceae. North Amer. Flora, 21: 267–277, 1932.

WULF, H. C. Cytological research of the Mesembryanthemeae. Cactus & Succ. Journ. Great Britain 10: 42–43, 1948.

## PORTULACACEAE.[67] PURSLANE FAMILY

Annual or perennial herbs or suffrutescent shrubs; leaves alternate or opposite, often rosulate, usually fleshy, simple, stipules scarious or setaceous (none only in *Claytonia*); flowers cymose, racemose, or solitary, bisexual, actinomorphic, often showy, perianth biseriate,[68] the calyx composed of 2 green herbaceous sepals (several in *Lewisia* and *Grahamia*), distinct or basally connate, often caducous, imbricated, the corolla of 4–6 petals (sometimes 2 in *Calyptridium* or 3 in *Montia*), distinct or connate basally; the stamens as many as and opposite the petals or 2–4 times as many (by splitting), rarely fewer (in Calyptridiinae), distinct, occasionally borne on corolla, the anthers 2-celled, introrse, dehiscing longitudinally; pistil 1, the ovary superior (half inferior in *Portulaca*), 1-loculed, the carpels 2–3, placentation basal, the ovules several to many or solitary, campylotropous, the styles

---

[67] The name Portulacaceae has been conserved over other names for this family.

[68] Morphologically, the perianth is interpreted to be uniseriate. By this view its calyx is considered an involucre, its sepals as bracts, and the apparent corolla as a perigone composed of tepals. Evidence for this view is that the stamens (when the same number) are opposite the tepals, and the floral anatomy is in harmony with it. This interpretation was accepted by Hoffmann and Pax, but most systematists (including Rendle, Fernald, Small, Abrams, Britton, and Bailey) perhaps for reasons of convenience have treated the perianth as biseriate and as composed of calyx and corolla. Taxonomically, it seems best to continue to refer to these parts as sepals and petals.

and stigmas 2–5; fruit a capsule, circumscissile (*Portulaca, Lewisia*) or loculicidal by 2–3 apical valves, rarely an indehiscent nut; seed with copious endosperm surrounded by embryo.

A family of 16 genera (Abrams accepted 20) and more than 500 species. The Pacific coast states constitute 1 center of distribution for the family, and southern South America is another. Nine genera occur indigenously in the western states. Of them (together with their number of total and indigenous species), *Talinum* (50–6), *Claytonia* (20–5), and *Portulaca* (125–1) occur in the south and southeast with the last 2 extending into the northeast. The genera with domestic representatives restricted to the west and southwest include: *Calandrinia* (150–5), *Calyptridium* (4–4), *Lewisia* (18–13), *Montia* (50–15), *Spraguea* (2–2), and *Talinopsis* (1–1). The African *Anacampseros* (60 spp.) is the only other large genus of the family.

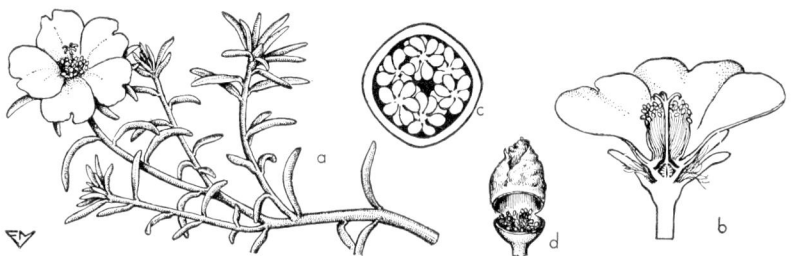

**Fig. 127.** PORTULACACEAE. *Portulaca grandiflora*: a, flowering branch, × ½; b, flower, vertical section, × 1; c, ovary, cross-section, × 4; d, capsule, × 1. (From L. H. Bailey, *Manual of cultivated plants,* The Macmillan Company, 1949. Copyright 1924 and 1949 by Liberty H. Bailey.)

Members of the Portulacaceae are distinguished generally by the 2–5-styled unilocular ovary containing usually several campylotropous ovules on a basal central placenta, by the usually 2 sepals, and the fleshy leaves.

The Portulacaceae are closely related to the Basellaceae, the genera *Portulacaria* of South Africa and *Philippiamra* of Chile standing intermediate between them. They are of close affinity to the Aizoceae, and authors are not agreed as to which family some of the genera belong. Within the Portulacaceae, those genera with few stamens are probably more primitive than those characterized by a presumed splitting of stamens into many.

Economically the family is domestically important for the ornamentals selected from it, notably rose moss (*Portulaca grandiflora*), and many species of *Talinum, Lewisia, Calandrinia*. Herbage of *Portulaca oleracea* is used to a limited extent as a potherb and salad green.

*LITERATURE:*

CRETE, P. Développement de l'embryon chez le *Calandrinia procumbens* Moris. Remarques embryographiques sur quelques Portluacacées Compt. Rend. Paris Acad. Sci. 227: 81–83, 1948.

PAX, F. and HOFFMANN, K. Portulacaceae. In Engler and Prantl, Die natürlichen Pflanzenfamilien, ed. 2, Bd. 16c: 234–262, 1934.

POELLNITZ, K. VON. Zur Kenntnis der Gattung *Portulaca*. Fedde, Repert, Spec. Nov. 50: 107–112, 1941.

RYDBERG, P. A. Portulacaceae. North Amer. Flora 21: 279–336, 1932. [The genera *Talinum, Talinaria, Talinopsis, Portulaca* by Percy Wilson.]

## BASELLACEAE. BASELLA FAMILY

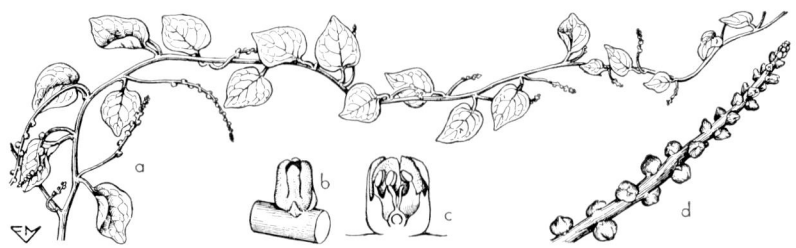

Fig. 128. BASELLACEAE. *Basella rubra*: a, flowering branch, × ⅛; b, flower, side view, × 2; c, flower, vertical section, × 3; d, inflorescence, × ½. (From L. H. Bailey, *Manual of cultivated plants,* The Macmillan Company, 1949. Copyright 1924 and 1949 by Liberty H. Bailey.)

Climbing perennial herbs, sometimes suffrutescent, herbage often fleshy; leaves alternate, simple, fleshy, petiolate, estipulate; flowers in spikes, racemes, or panicles, each flower subtended by an involucre of 2 bracts, bisexual, actinomorphic, the perianth uniseriate, represented by a calyx of 5 free or basally connate sepals, often colored, imbricate, persistent in fruit, the corolla none; stamens 5, opposite sepals and adnate to their bases, the filaments distinct, the anthers 2-celled, dehiscing longitudinally; pistil 1, the ovary superior, 1-loculed, 3-carpelled, the ovule solitary, campylotropous, the placentation basal, the style 1, the stigmas usually 3; fruit a drupe enveloped by the annular or spirally twisted embryo.

A family of 5 genera and about 22 species, mostly of tropical America and the West Indies. One species, *Basella alba,* is native to the Old World (Asia). *Boussingaultia* has 2 of its 12 species in southern and southeastern United States. *Anredera* has 1 species in southern Texas that extends south to northwestern South America.

The family is readily identifiable by its scandent succulent character, its bisexual apetalous flowers whose fleshy persistent calyx envelopes the drupe in fruit.

The Basellaceae are related most closely to the Portulacaceae, and probably are an advanced terminal offshoot of the latter. By Bentham and Hooker they were united with the Chenopodiaceae, but differ from that family by the biseriate perianth. Within the family the genera fall into 2 more or less natural groups, one with the embryo spirally twisted and stamens straight in the bud (as in *Basella*) and the other with the embryo annular and the stamens curved in bud (as in *Boussingaultia* and *Anredera*).

The family is of little domestic importance. The Madeira vine (*Boussingaultia*) is cultivated as an ornamental, while Malabar (*Basella*) produces edible shoots eaten as spinach. In tropical America the tuberous starchy root of *Ullucus tuberosus* is eaten as a potato substitute.

*LITERATURE:*

ULBRICH, E. Basellaceae. Engler and Prantl, Die natürlichen Pflanzenfamilien, ed. 2, Bd. 16c: 263–271, 1934.

## CARYOPHYLLACEAE.[69] PINK FAMILY

Annual or perennial herbs, infrequently suffrutescent shrubs; stems characteristically with swollen nodes; leaves opposite (rarely alternate), simple, mostly linear to lanceolate, often connected basally by a transverse line or by a shortly

---

[69] The older name Alsinaceae, Wahlenb. (1824) is displaced by the conserved name of Caryophyllaceae, Reichb. (1828).

connate-perfoliate base (*Dianthus*), decussate, the stipules scarious or more often absent; flowers bisexual or infrequently unisexual (the plants then dioecious), actinomorphic, mostly in determinate simple to complex dichasial inflorences or solitary and terminal, the perianth usually biseriate, the calyx of typically 5 basally connate sepals (distinct in Alsinoideae, and sepals or lobes sometimes 4), rarely subtended by bracts (*Dianthus*), the corolla with typically same petal number as calyx lobes, the petals distinct and free, often differentiated into claw and limb, sometimes minute or none; stamens in 1–2 whorls, usually of same number as petals or twice as many, filaments distinct or briefly connate, the anthers 2-celled, dehiscing longitudinally, petaloid staminodes sometimes present; pistil 1, the ovary superior, 1-loculed with free-central placentation or basally 3–5-loculed with axile placentation in lower third or less and free-central placentation above (in some Silenoideae), or ovules basal when solitary, the carpels 2–5, the ovules solitary to many, the styles and stigmas 2–5 (rarely 1), as many as the carpels; fruit a capsule dehiscing apically by valves, teeth, or circumscissilely, or a utricle or achene; seed with embryo usually curved around the firm or hard endosperm (embryo straight in *Dianthus*, endosperm soft in *Agrostemma*).

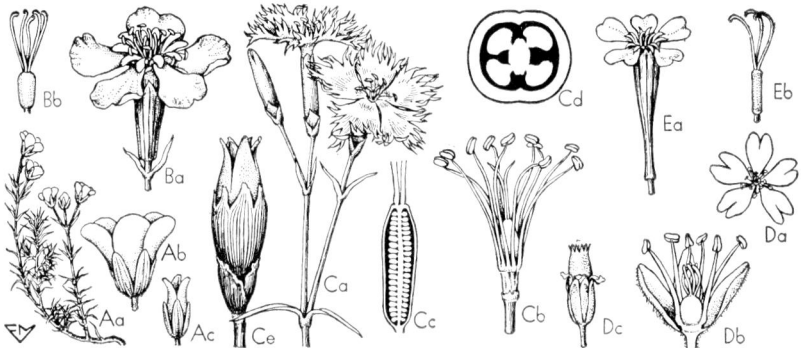

**Fig. 129.** CARYOPHYLLACEAE. A, *Arenaria laricifolia*: Aa, flowering branch, × ¼; Ab, flower, side view, × 1; Ac, capsule, × 1. B, *Lychnis Viscaria*: Ba, flower, × 1; Bb, pistil, × 1. C, *Dianthus plumarius*: Ca, flowering branch, × ½; Cb, flower, less perianth, × 1; Cc, ovary, vertical section, × 2; Cd, ovary, cross-section, × 5; Ce, capsule, × 1. D, *Cerastium tomentosum*: Da, flower, face view, × ½; Db, same, side view, less corolla, × 2; Dc, capsule, × 1. E, *Silene Armeria*: Ea, flower, × 1; Eb, pistil, × 1½. (From L. H. Bailey, *Manual of cultivated plants*, The Macmillan Company, 1949. Copyright 1924 and 1949 by Liberty H. Bailey.)

A family of about 80 genera and 2100 species (incl. Illecebraceae), primarily of the north temperate regions but with a few genera of the south temperate regions and higher altitudes of mountains in the tropics. The Mediterranean area is the principal center of distribution for the family. The largest genera include *Silene* (500 spp.), *Dianthus* (350 spp.), *Arenaria* [including *Minuartia, Moehringia, Merckia*] (160 spp.), and *Stellaria, Cerastium, Lychnis*, and *Gypsophila* (each of about 100 spp.). Many genera (as *Silene, Lychnis, Spergula, Spergularia* (*Tissa*), *Sagina, Arenaria, Paronychia, Scleranthus*) are widespread throughout the colder vegetated regions of the northern hemisphere, while others (as *Colobanthus, Lyallia*) occupy a similar position in the southern hemisphere. In the United States, about 13 genera are represented by indigenous species, and 9 others by species that are widely naturalized or adventive (often as noxious weeds). Among the indigens, *Arenaria, Sagina, Silene, Stellaria*, and *Cerastium* occur in general over much of the country, while *Polycarpon* occurs mostly in the warmer areas, with *Herniaria, Scopulophila, Achyronychia,* and *Loeflingia* primarily in 1 or more of the Pacific coast states, and *Stipulicida* monotypic in the southeast. The more important and widespread adventive genera include *Holosteum, Agrostemma, Dianthus, Spergula,* and *Saponaria.*

The Caryophyllaceae (with which the Illecebraceae are here included) are distinguished by the combination of opposite leaves associated with leaf bases connate and sheathing or connected by a line, the 2–5-carpelled unilocular ovary with freecentral placentation (at least apically so), and the 1-chambered capsule (rarely 3-chambered as in some *Lychnis,* spp.) dehiscing by valves or teeth.

Botanists have presented many views to account for the evolution of the Caryophyllaceae. With few exceptions these more or less fit 1 of 2 basic concepts: (1) that of Eichler (1875), who considered the family to have originated from the Phytolaccaceae by conversion of the outer stamen whorl to petals and of the outer carpel whorl to stamens, a view accepted in principle by Pax, Wettstein, and by Rendle; or (2) that of Wernham (1911), who considered the family to have evolved from ranalian ancestors and to have been the source of origin for the Primulaceae and such reduced families as the amaranths and chenopods, a view accepted in principle by Bessey, Hutchinson, and others. It has been suggested also (Dickson, 1936) that the family may have been derived from the Geraniaceae, but the findings of Thompson (1942) did not support this view. The Wernham theory, on the basis of current information, seems more nearly to represent the probable origin and relationships of the family.

Economically the family is important for the large number of ornamentals available from it. Notable among them are the florists' carnation (*Dianthus Caryophyllus*) and perhaps 75 other species of the genus, baby's-breath (*Gypsophila*), catchfly (*Silene*), maltese cross (*Lychnis*), sandwort (*Arenaria*), and mouse-ear chickweed (*Cerastium*).

*LITERATURE:*

DICKSON, J. Studies in floral anatomy. III. An interpretation of the gynoecium in the Primulaceae. Amer. Journ. Bot. 23: 385–393, 1936.
FERNALD, M. L. *Arenaria,* I-VI. Rhodora, 21: 1–22, 1919.
HITCHCOCK, C. L. and MAGUIRE, B. A revision of the North American species of *Silene.* Univ. Wash. Publ. Biol. 13: 1–73, 1947.
LUDERS, H. Systematische Untersuchungen über die Caryophyllaceae mit einfachem Diagram. Engler's Bot. Jahrb. 40 (91): 1–38, 1907.
MAGUIRE, B. Studies in the Caryophyllaceae. Bull. Torrey Bot. Club, 73: 326, 1946.
———. Studies in the Caryophyllaceae—1. A synopsis of the North American species of *Arenaria,* Sect. *Eremogone* Fenzl. Bull. Torry Bot. Club, 74: 38–56, 1947.
———. Studies in the Caryophyllaceae. II. *Arenaria Nuttallii* and *Arenaria filiorum,* section *Alsine.* Madroño, 8: 258–263, 1946.
MATTFELD, J. Biologische und morphologische Blütenformen bei den Caryophyllaceen. Notizbl. Bot. Gard. Berl. 14: 470–482, 1939.
PAX, F. Zur Phylogenie der Caryophyllaceae. Engl. Bot. Jahrb. 61: 223–241, 1927.
PAX, F. and HOFFMANN, K. Caryophyllaceae. In Engler and Prantl, Die natürlichen Pflanzenfamilien, ed. 2, Bd. 16c: 275–364, 1934.
ROBINSON, B. L. Caryophyllaceae. In Gray, Synop. Flora, 1: 208–255, 1897.
ROHWEDER, H. Beiträge zur Systematik und Phylogenie der Gattung *Dianthus.* Engler, Bot. Jahrb. 66: 249–368, 1934.
———. Ueber die Bedeutung der Karyologie der Caryophyllaceae. Lilloa, 9: 203–210, 1943.
ROSSBACH, R. P. *Spergularia* in North and South America. Rhodora, 42: 158–193, 203–213, 1940.
SIMMLER, G. Monographie der Gattung *Saponaria.* Denkschr. Akad. Wiss. Math. Naturw. (Wien), 85: 434–509, 1910.

THOMPSON, B. F. The floral morphology of the Caryophyllaceae. Amer. Journ. Bot. 29: 333–349, 1942.
WERNHAM, H. F. Floral evolution; with particular reference to the sympetalous dicotyledons. III. The Pentacyclidae. New Phytologist, 10: 145–159, 1911.
WILLIAMS, F. N. A revision of the genus *Silene*. Journ. Linn. Soc. Bot. 32: 1–196, 1896.
———. A revision of the genus *Arenaria*. Journ. Linn. Soc. Bot. 33: 326–437, 1898.

## Order 21. RANALES

The Ranales, as here circumscribed, are characterized by the floral parts typically spirally arranged (cyclic in advanced members), numerous, distinct, the perianth often not differentiated into calyx and corolla, and the gynoecium apocarpous and multicarpellate (i.e., composed of many unilocular unicarpellate pistils).

This order was considered by Engler and Diels to comprise 19 families distributed among 4 suborders as follows:

Nymphaeineae
  Nymphaeaceae
  Ceratophyllaceae
Trochodendrineae
  Trochodendraceae
  Cercidiphyllaceae
Ranunculineae
  Ranunculaceae
  Lardizabalaceae
  Berberidaceae
  Menispermaceae

Magnoliineae
  Magnoliaceae
  *Himantandraceae (So. Pacific)
  Calycanthaceae
  *Lactoridaceae (Juan Fernandez)
  Anonaceae
  *Eupomatiaceae (Austral., N. Guinea)
  Myristicaceae
  *Gomortegaceae (Chile)
  Monimiaceae
  Lauraceae
  Hernandiaceae

Families whose names are preceded above by an asterisk are not accounted for in the treatments that follow, but to the Trochodendrineae have been added the families Eupteleaceae and Tetracentraceae, and to the Magnoliineae have been added (or accepted) the families Winteraceae, Degeneriaceae, Illiciaceae, and Schisandraceae.

Most botanists now accept the view that the Magnoliineae and Trochodendrineae are better treated as 1 order, the Magnoliales, and the remaining 2 suborders treated as the order Ranales. Of the 2, the Magnoliales may be the more primitive. The Ranales (*sensu* Engler and Diels) are believed by many to be among the most primitive living angiosperms and to be the taxon (or the derivative of a taxon) from which most other living angiosperms (or their ancestors) may have evolved. Reasons supporting this view are given in the chapters on phylogeny and on systems of classification in Part I of this text.

## NYMPHAEACEAE. WATER-LILY FAMILY

Aquatic, annual (*Euryale*) or more commonly perennial herbs, the stem cauline (*Cabomba*) or generally rhizomatous, erect (*Victoria*) or creeping (*Nymphaea*); leaves alternate, simple mostly capillarily dissected and immersed (in *Cabomba*), usually floating or emersed (*Nelumbo*), peltate or falsely so, dimorphic in *Cabomba*, usually smooth (prickly beneath in *Victoria* and *Euryale*), mostly long petioled (sessile in *Cabomba*), milky latex often present; flowers solitary and long-peduncled, often showy, bisexual, actinomorphic, often fragrant, the perianth basically trimerous, biseriate (Cabomboideae) or spirally arranged; calyx with sepals 3 (Cabomboideae), 4–5 (Nymphaeoideae), or indefinite (Nelumboideae), distinct, usually green but often as large as petals (larger than petals and yellow in *Nuphar*); corolla polypetalous, the petals 3 (Cabomboideae) to indefinite, usually showy (small and scalelike in *Nuphar*), often with dorsal nectaries, the innermost petals being petaloid staminodia in Nymphaeoideae; stamens 3-6 (Cabomboideae) and cyclic, or more commonly very numerous and acyclic, introrse, the filament often extending as a sterile appendage beyond anther sacs, the anthers 2-celled, dehiscing longitudinally; gynoecium multipistillate (apocarpous) (Cabomboideae and Nelumboideae) or the pistil 1 and syncarpous (Nymphaeoideae), when apocarpus the unicarpellate pistils 3–many, borne on or enveloped by the torus (the ovary superior), unilocular with 1–several ovules on a parietal placenta, when syncarpous the ovary usually superior (inferior in *Victoria*, *Euryale*), with 5–35 locules and carpels, the ovules many per locule, anatropous, the placentation parietal or lamellate, the style 1 or absent, the stigmas 1 and discoid or 5–35 and often radiate; fruit a follicle (Cabomboideae), and aggregate of indehiscent nutlets (Nelumboideae), or a leathery tardily dehiscent berry; seeds with straight embryo, the endosperm starchy (absent in *Nelumbo*).

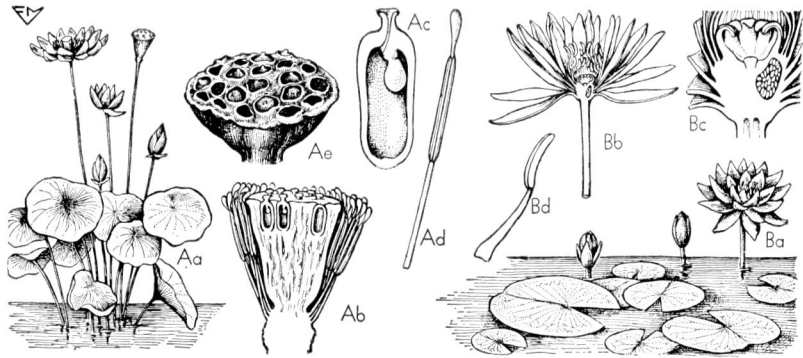

**Fig. 130.** NYMPHAEACEAE. A, *Nelumbo nucifera*: Aa, flowering plant, much reduced; Ab, flower, vertical section, less perianth, × ½; Ac, pistil, vertical section, × 2; Ad, stamen, × 1; Ae, pod, × ⅕. B, *Nymphaea odorata*: Ba, flowering plant, much reduced; Bb, flower, vertical section, × ¼; Bc, same, less perianth, × 1; stamen, × 1. (From L. H. Bailey, *Manual of cultivated plants,* The Macmillan Company, 1949. Copyright 1924 and 1949 by Liberty H. Bailey.)

A family of 8 genera and about 90 species, in fresh waters of most of the earth. Three genera are tropical: *Euryale* and *Barclaya* in Asia, and *Victoria* in South America. *Brasenia* is indigenous to all continents except Europe. *Nymphaea* (*Castalia*) and *Nuphar* occur throughout most of the northern hemisphere, while *Cabomba* is a New World genus. *Nelumbo* has one species indigenous to eastern North America and another in the region from India to Australia.

The family is distinguished by the aquatic habit, the usually long-petioled peltate or pseudopeltate leaves (latex-producing), and the long-peduncled flowers with ovules disposed on parietal or lamellate placentae.

Most American botanists accept the Nymphaeaceae to be composed of 3 well-defined subfamilies as noted above. Bessey (1915) treated each as a distinct family, placed the Cabombaceae and the Nelumbaceae in the Ranales, and removed the Nymphaeaceae (*sensu stricto*) to the Rhoeadales on the basis of the syncarpous pistil. Small (1933) also recognized them to comprise 3 families, but placed all in the Ranales. Hutchinson (1926) accepted the Cabombaceae, but retained the other 2 subfamilies within his concept of Nymphaeaceae. On the basis

## DIVISION IV. EMBRYOPHYTA SIPHONOGAMA

of the gynoecial conditions, it is clear that the genera concerned represent either 1 family (Nymphaeaceae) or 3 separate families (*sensu* Bessey).

Species of all genera except *Barclaya* are in the commercial trade as garden or conservatory ornamentals. Formerly the large rhizomes of *Nelumbo pentapetala* were important to the American Indian as a source of starchy food.

*LITERATURE:*

CONARD, H. S. The waterlilies. A monograph of the genus *Nymphaea*. Carnegie Inst. Wash. 292 pp., 1905.

———. Waterlilies; monocots or dicots? Amer. Bot. 42: 104–107, 1936.

FERNALD, M. L. The name of the American lotus. Rhodora, 36: 23–24, 1934.

GLEASON, H. A. The preservation of well-known binomials *Nelumbo lutea* vice *N. pentapetala*. Phytologia, 2: 201–212, 1947.

LYON, H. L. Observations on the embryogeny of *Nelumbo*. Minn. Bot. Studies, 2: 643–655, 1901.

MILLER, G. S. and STANDLEY, P. C. The North American species of *Nymphaea*. Contr. U. S. Nat. Herb. 16: 63–109, 1912.

SMALL, J. K. The water-lilies of the United States. Journ. N. Y. Bot. Gard. 32: 117–121, 1931.

TAYLOR, H. J. The history and distribution of yellow *Nelumbo*, water chinquapin or American lotus. Proc. Iowa Acad. Sci. 34: 119–124, 1927.

## CERATOPHYLLACEAE. HORNWORT FAMILY

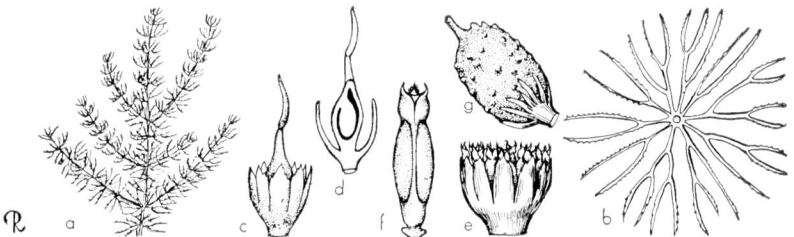

Fig. 131. CERATOPHYLLACEAE. *Ceratophyllum demersum*: a, vegetative branch, × ¼; b, whorl of leaves, × 1; c, pistillate flower, × 12; d, same, vertical section, × 12; e, staminate flower, × 12; f, stamen, × 25; g, fruit, × 15. (c–g after Eichler.) (From L. H. Bailey, *Manual of cultivated plants,* The Macmillan Company, 1949. Copyright 1924 and 1949 by Liberty H. Bailey.)

Rootless, submerged, perennial aquatic monoecious plants; leaves whorled, dichotomously divided, sessile, the filiform or linear segments with serrulate margins, estipulate; flowers unisexual, actinomorphic, minute, and inconspicuous, each solitary at a node; the staminate with a perianth of 10–15 sepals basally connate into a gamosepalous calyx, the stamens 10–20, spirally arranged on a flat receptacle, with very short filaments, the anthers erect, linear-oblong, 2-celled, dehiscing longitudinally, a thickened, often colored connective projecting beyond them; pistillate flowers with similar perianth, the gynoecium of a single pistil, the ovary superior, 1-loculed, 1-carpelled, the ovule 1, pendulous, anatropous, the placentation parietal, the style slender and acute, the stigma 1 and undifferentiated from style; fruit a nut, terminated by the persistent style; seed with large straight embryo and no endosperm.

The family, represented by a single genus (*Ceratophyllum*) and 3 species, is of cosmopolitan distribution and is represented in this country by *C. demersum*, common in shallow stagnant or slow-moving waters.

The family is recognized by the whorled dichotomously dissected serrulate leaves and by the unisexual flowers solitary in the axil of one leaf of a whorl. The characteristic of a single lateral branch produced at a node is also of value in identifying the family.

The Ceratophyllaceae undoubtedly are of close affinity with the Nymphaeaceae, as is evidenced by the whorled elements of the androecium, by the typical ranalian unicarpellate pistil, and by the pendulous parietal ovule.

Economically the family is of little importance, although the plant is of value in wild-life conservation practices as a protective cover for freshly laid spawn and young fish fry. The plants are sold for use in pools and aquaria.

*LITERATURE:*

JONES, E. N. The morphology and biology of *Ceratophyllum demersum.* Univ. Iowa Stud. Bot. 13: 11–46, 1931.

MUENSCHER, W. C. Fruits and seedlings of *Ceratophyllum.* Amer. Journ. Bot. 27: 231–233, 1940.

PEARL, R. Variation and differentiation in *Ceratophyllum.* Carnegie Inst. Wash. Publ 58, 1907.

## TROCHODENDRACEAE. TROCHODENDRON FAMILY

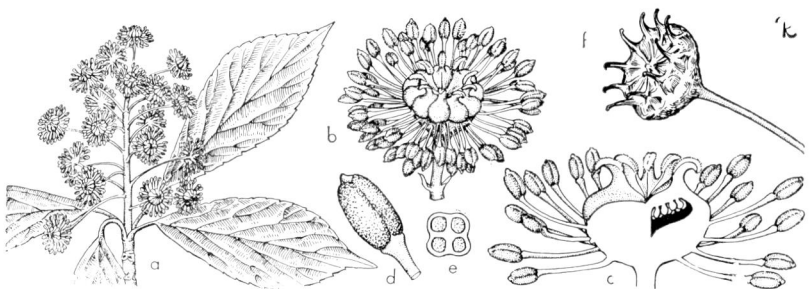

**Fig. 132.** TROCHODENDRACEAE. *Trochodendron aralioides*: a, flowering branch, × ⅜; b, flower, habit, × 1½; c, same, vertical section, × 2; d, anther, habit, × 6; e, same, cross-section, × 6; f, fruit, × 1½. (a–e redrawn from Curtis' Bot. Mag.)

Trees; leaves pseudoverticillate, simple, pinnate-veined, often subpersistent, petioled, estipulate; inflorescence a racemelike pleiochasium, terminal at inception, soon appearing axillary; flowers bisexual, actinomorphic, the pedicels usually bearing 2–5 minute bracteoles on or just below torus, the perianth none (bracteoles sometimes considered perianth remnants); stamens numerous, in 3 or 4 whorls on an expanded torus, the filaments long and slender, the anthers 2-celled (4-sporangiate), dehiscing longitudinally; gynoecium of (4–) 6–11 sessile pistils in a single whorl, these laterally concrescent, the ovary 1-loculed, with 2 rows of numerous ovules near the ventral suture, the style conduplicate and deeply canaliculate ventrally, apically stigmatic; fruit a follicetum composed of laterally coalescent follicles, dehiscing loculicidally on its upper surface; seeds dependent in 2 rows, mixed with sterile ovules, with basal and apical winglike projections, the endosperm waxy, the embryo minute, ellipsoid.

A unigeneric family (*Trochodendron*) of Japan and Formosa represented by the single species *T. aralioides* (interpreted by some authors to be of 2 or 3 species).

The family is distinguished from its relatives by the persistent pinnate-veined leaves, the bisexual bracteolate flowers (lacking perianth) with sessile concrescent pistils, and the fruit a group of follicles cohering to form a follicetum.

The classification of *Trochodendron* has been interpreted variously by many authors. Early botanists allied it with the Magnoliaceae, and in 1888 Prantl described the family Trochodendraceae, placing in it also *Euptelea* and *Cercidiphyllum*. Engler and Diels (1936) considered the family more primitive than the Cercidiphyllaceae, and to contain *Trochodendron* and *Euptelea*. These genera, together with *Eucommia* and *Tetracentron*, are now considered by Smith (1945) to compose 5 separate families. This opinion follows the important studies of van Tieghem (1900). Bessey (1915) accepted Trochodendraceae as a family of the Ranales, but failed to cite any authority for it or to indicate his concept of its circumscription. Hutchinson included *Euptelea* within the family and placed it in his Magnoliales, assigning the Eucommiaceae to the Hamamelidales. Wettstein (1935) recognized 4 of the genera as separate families, but retained *Tetracentron* in the Magnoliaceae. Details of the interrelationships of these families are provided by Smith (1945).[70]

*Trochodendron aralioides* is cultivated domestically to a limited extent as an ornamental in the southeastern part of the country, where it attains a height of 40 feet or more.

*LITERATURE:*

BAILEY, I. W. and NAST, C. G. Morphology and relationships of *Trochodendron* and *Tetracentron*. Journ. Arnold Arb. 26: 143–154, 267–276, 1945.
CROIZAT, L. *Trochodendron, Tetracentron*, and their meaning in phylogeny. Bull. Torrey Bot. Club, 74: 60–76, 1947.
SMITH, A. C. A taxonomic review of *Trochodendron* and *Tetracentron*. Journ. Arnold Arb. 26: 123–142, 1945.

## TETRACENTRACEAE. TETRACENTRON FAMILY

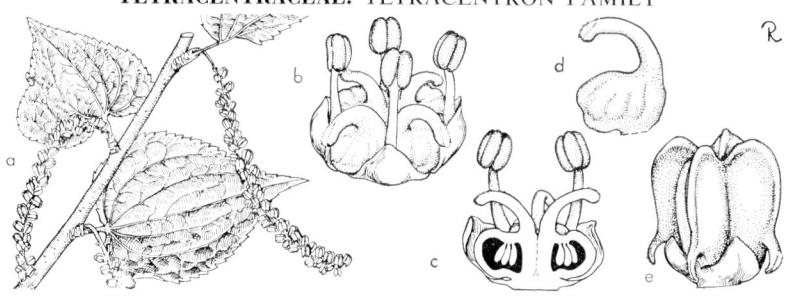

**Fig. 133.** TETRACENTRACEAE. *Tetracentron sinense*: a, flowering branch, × ¼; b, flower, habit, × 8; c, same, vertical section, × 8; d, single pistil, × 10; e, fruit, × 3.

A unigeneric family (*Tetracentron sinense* of south-central China and adjacent Burma being the only species) allied only to Trochodendraceae, differing in having the leaves solitary, palmate-nerved, and with stipular flanges; the inflorescence spikelike, the flowers numerous and arranged in clusters of 4, the flowers with 4 persistent imbricate sepals, the torus flattened and inconspicuous; the stamens 4 and opposite the sepals; the pistils 4 with a pronounced ventral overgrowth, the ovules 6 or fewer and borne on the placentae only at the middle of each locule; the folliceta with a strong ventral development; and the seeds with a more spongy and winglike outer integument.

[70] See also Croizat, L. in Bull. Torrey Bot. Club, 74: 60–76, 1947.

The single species, a hardy tree 15–90 feet high, is cultivated to a limited extent as an ornamental.

The genus, known to science for slightly more than half a century, has been treated by most authors (including Wettstein, 1935) as a member of the Magnoliaceae, but it has not received the consideration of most phylogenists (e. g., Hutchinson, Engler, and Diels). The family name, in a French form, dates from the morphological study of van Tieghem (1900), but it was apparently properly latinized as recently as 1945 (by Smith). Papers by Nast and Bailey and by Smith, cited under Trochodendraceae, also discuss Tetracentraceae; according to these authors the 2 families are closely related but have no immediate allies.

*LITERATURE:*

TIEGHEM, P. VAN. Sur les dicotylédones du groupe des Homoxylées. Journ. de Bot. 14: 259–297, 330–361, 1900.

## EUPTELEACEAE. EUPTELEA FAMILY

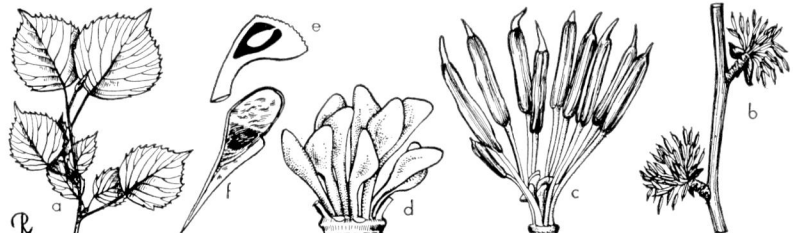

**Fig. 134.** EUPTELEACEAE. *Euptelea polyandra*: a, vegetative branch, × ⅙; b, flowering branch, × ½; c, flower, × 3; d, gynoecium, × 9; e, ovary, vertical section, × 10; f, fruit, × 2. (From L. H. Bailey, *Manual of cultivated plants,* The Macmillan Company, 1949. Copyright 1924 and 1949 by Liberty H. Bailey.)

Hardy deciduous trees or shrubs; twigs with imbricately scaled buds, the terminal bud aborted, replaced by the distal axillary bud; leaves alternate, simple, pinnate-veined, petioled, estipulate; inflorescence appearing before leaves, composed of 6–12 single flowers borne in axils of bracts around growing point and subsequently lateral, the flowers bisexual; perianth none; stamens numerous, in a single whorl on an expanded flattened receptacle, the filaments filiform or slightly flattened, subequal to anthers in length, the anthers 2-celled, dehiscing longitudinally, basifixed, becoming twisted; gynoecium of 6–18 stalked pistils, the latter free, borne in a single whorl just within the stamens, the ovary with a ventral or distal stigmatic margin covered with tangled sticky processes, unilocular, the ovules mostly 1–3, attached to ventral edge of the locule; fruit small stipitate samaras, clustered, these essentially circumalate with the ventral edge indented; seed with small embryo and copious oily granular endosperm.

A unigeneric family (*Euptelea*) of 2 species, one indigenous to Japan (*E. polyandra*) and a second to China and northeastern India. Elements of the latter have been treated to represent as many as 5 different species, but all were considered conspecific by Smith (1946).

The family is characterized by a combination of morphological and anatomical characters, including deciduous leaves, bisexual flowers with stipitate pistils, and by the fruits a cluster of samaras. It has been treated as a member of the Trochodendraceae, but Nast and Bailey (1946) and Smith (1946) considered it to be a very distinct and isolated member of the Ranales, possibly worthy of subordinal rank. The following is a partial quotation from Nast and Bailey (p. 191):

... the family Eupteleaceae exhibits evidences of general ranalian affinities, but does not appear to be closely related to any specific surviving family of the ranalian complex. It obviously cannot be placed in close proximity to those woody ranalian families ... which are characterized by having monocolpate and derived types of pollen [pollen tricolpate in *Eupteleaceae*] and no aromatic secretory cells. Although it appears to belong in the category of ranalian families having tricolpate and derived types of pollen and no aromatic secretory cells, it cannot be placed in close proximity to any of them ... the family Cercidiphyllaceae ... is not closely related to the Eupteleaceae. Nor does the latter family form a natural compact grouping with the Schisandraceae.

Both species of *Euptelea* are cultivated domestically to a limited extent as ornamentals.

*LITERATURE:*

NAST, C. G. and BAILEY, I. W. Morphology of *Euptelea* and comparison with *Trochodendron*. Journ. Arnold Arb. 27: 186–192, 1946.
SMITH, A. C. A taxonomic review of *Euptelea*. Journ. Arnold Arb. 27: 175–185, 1946.

## CERCIDIPHYLLACEAE. CERCIDIPHYLLUM FAMILY

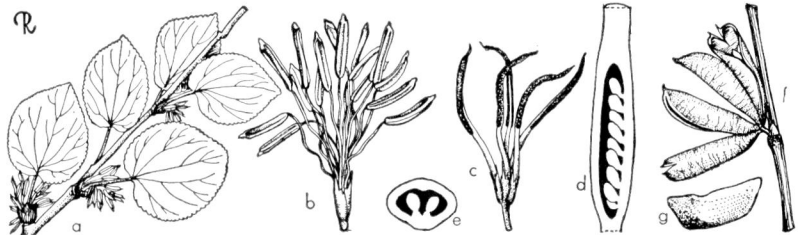

**Fig. 135.** CERCIDIPHYLLACEAE. *Cercidiphyllum japonicum*: a, branch with staminate flowers, × ½; b, staminate flower, × 2; c, pistillate flowers, × 1; d, ovary, vertical section, × 10; e, ovary, cross-section, × 10; f, follicles, × 1; g, seed, × 5. (From L. H. Bailey, *Manual of cultivated plants,* The Macmillan Company, 1949. Copyright 1924 and 1949 by Liberty H. Bailey.)

The family consists of 1 species; a dioecious deciduous tree; branches bearing short shoots (spurs); leaves dimorphic, usually opposite and pinnate-veined on the elongated terminal twigs, solitary (alternate) and palmate-veined on the short shoots, simple, stipules present and caducous; flowers unisexual, lacking any perianth,[71] appearing solitary but actually decussately disposed in highly reduced congested inflorescences, produced before or with the leaves, each flower subtended by a bract; staminate flowers sessile, stamens 8–13, filaments filiform, the anthers 2-celled, basifixed, dehiscing longitudinally; pistillate inflorescence (the "flower" of authors) pedunculate, the flowers 2–6, sessile, each subtended by a bract, the gynoecium a single pistil, the ovary superior, 1-loculed, 1-carpelled, the ovules many in 2 rows, attached to the ventral edge of the locule (e.g., placentation is parietal), the style linear, bearing 2 stigmatic ridges (representing a single stigma

---

[71] Hutchinson *et al.* have stated that a calyx is present. It has been held by many authors, and established by Swamy and Bailey, that the apparent calyx is composed of 2 pairs of decussate bracts. Likewise, the "flower" associated with *Cercidiphyllum* in the literature has been demonstrated (1949) to be an inflorescence of usually 4 or more apetalous bracteate flowers.

but an extension of the 2 sterile edges of the carpel); fruit a follicle, the seed winged, the endosperm copious and firm to firm-fleshy.

A unigeneric family of China and Japan represented by the single species (*Cercidiphyllum japonicum*) (the Chinese form probably is only varietally distinct). Fossil evidence indicates it to be the remaining member of a once widespread genus having many species.

The family is unlike any other extant angiosperm. It is distinguished by a combination of characters: dimorphic leaves of palmate or pinnate venation, short shoots, unisexual flowers dioeciously disposed, the pistillate inflorescence of 4–6 apetalous flowers each with a multiovulate pistil, and the staminate inflorescence of several bracteate apetalous staminate flowers.

The family long was included within the Magnoliaceae, but the recent data presented by Swamy and Bailey refuted this concept and reinterpreted the "flower" to be an inflorescence of bracteate flowers, a view supported by the abaxial position of the pistils whereby the ventral suture of each faces outwards, and this condition is not known to be present in multipistillate apocarpous gynoecia. Swamy and Bailey rejected the phyletic views of earlier workers and concluded that the family was not of close affinity with any other family and that it did not belong in any existing order of angiosperms, unless the Ranales be recognized as a useful repository for primitive relic dicotyledonous plants.

The tree is important in Asia for its lumber and as an ornamental. Domestically it is grown extensively as a hardy ornamental, and is prized for its ovoid symmetry in form and large size.

*LITERATURE:*

HARMS, H. Zur Kenntnis der Gattung *Cercidiphyllum*. Mittiel. Deutsch. Dendrol. Ges. 26: 71–87, 1918.

SWAMY, G. L. and BAILEY, I. W. The morphology and relationships of *Cercidiphyllum*. Journ. Arnold Arb. 30: 187–210, 1949.

## RANUNCULACEAE. BUTTERCUP FAMILY

Annual or perennial herbs, occasionally shrubs or vines (*Clematis* spp.) or very rarely trees (*Paeonia*); leaves mostly alternate (opposite in *Clematis* and *Ranunculus* spp.), usually compound (entire in *Caltha, Coptis, et al.*) and most generally palmate (pinnate in *Xanthorrhiza, Actaea*), mostly estipulate (rudimentary stipules in some *Thalictrum*); flowers typically bisexual (unisexual in some spp. of *Thalictrum*, the plants then monoecious or dioecious), actinomorphic or (in the Delphineae) zygomorphic, solitary, in determinate cymose inflorescences or racemose to paniculate perianth biseriate and differentiated into calyx and corolla or not so, segments distinct and free, variable in number, the corolla often modified and less petaloid than the often showy sepals, nectiferous glands usually present (covered or enveloped by a scale in *Ranunculus*); the stamens usually many, spirally arranged, hypogynous, distinct, the anthers 2-celled, basifixed, dehiscing longitudinally; the gynoecium of 3–many (very rarely only 1) distinct spirally disposed simple pistils (compound and multicarpellate in *Nigella*), the ovary superior, 1-loculed, 1-carpelled, the ovules 1–many, anatropous, the placentation parietal along ventral suture, the style and stigma 1 (5–12 in *Nigella*); fruit typically a follicle, sometimes an achene (*Ranunculus*), berry (*Actaea*), or rarely a capsule (*Nigella*); seed with minute embryo and a copious generally watery-fleshy endosperm.

A moderately large family, chiefly of the cooler temperate regions of the earth and especially of the northern hemisphere, composed of about 35 genera and perhaps 1500 species. About 20 genera and nearly 300 species are indigenous to

Fig. 136. RANUNCULACEAE. A, *Trollius europaeus*: Aa, flowering branch, × ½; Ab, basal leaf-blade, × ½; Ac, flower, vertical section, × ½; Ad, pistil, vertical section, × 3; Ae, petal, × 2. B, *Thalictrum rugosum*: Ba, portion of inflorescence, × ½; Bb, node and cauline leaf, × ½; Bc, staminate flower, × 2. C, *Delphinium elatum*: Ca, flower, × 1; Cb, flower, vertical section, × 1. D, *Aconitum Carmichaelii* var. *Wilsonii*: Da, flower, side view, × ½; Db, flower, vertical section, × ½. E, *Aquilegia flabellata*: flowering branch, × ½. F, *Nigella hispanica*: flower and leaf, × 1. G, *Helleborus niger*: Ga, flower, face view, × ½; Gb, flower, vertical section, × ½. (p petal, pi pistil, s sepal.) (From L. H. Bailey, *Manual of cultivated plants*, The Macmillan Company, 1949. Copyright 1924 and 1949 by Liberty H. Bailey.)

this country, with 4 additional genera represented by naturalized or adventive species. Notable among the genera represented in this country are the following (together with the number of total and indigenous species): *Ranunculus* [72] (250–96), *Aquilegia* (70–29), *Clematis* [72] (200±–25), *Thalictrum* (90–14), *Delphinium* (250–30), and *Anemone* [72] 100±–25). Genera with one or a few indigenous species include: *Myosurus* (6), *Cimicifuga* (5), *Isopyrum* (4), *Caltha* (3), *Actaea* (3), *Coptis* (3), *Hepatica* (2), and the following with one indigenous species each *Trautvetteria*, *Anemonella* (*Syndesmon*), *Trollius*, *Hydrastis*, *Paeonia*, *Xanthorhiza*. Most of the genera have representatives well distributed over the mesophytic regions of the country. A few are of more restricted distribution: *Hydrastis*, *Xanthorhiza*, and *Anemonella* to the eastern part of the country, *Paeonia* (primarily of Asia) and *Aquilegia* largely to the Rocky Mountain region, *Delphinium* mostly to the Pacific coast states, and *Clematis* to the more southerly parts of the country.

The Ranunculaceae are distinguished from related families by the usually herbaceous texture, the leaves often divided or compound, the flowers mostly bisexual with reduced or modified petals, the stamens mostly numerous and spirally arranged, the gynoecium of generally several to many pistils, and the ovary unicarpellate.

Contrary to the phyletic position assigned it by Engler, this family is now generally accepted to be among the most primitive of dicots—a conclusion published long ago by de Jussieu (1783), accepted by Bentham and Hooker, emphasized by Bessey (1915), by Hallier, and by Hutchinson (1948), and generally supported by current phylogenetic studies. Whether it is more primitive than the Winteraceae and Degeneriaceae [73] remains to be determined. For a detailed presentation of the components of this family and their relationships, see Rendle (1938, pp. 138–149).

The family is accepted generally as a natural taxon. However, as suggested by Worsdell (1908) and emphasized by Corner (1946), there is strong evidence to justify the segregation of *Paeonia* as a separate unigeneric family (Paeoniaceae) removing it from the Ranales to a position near the Dilleniaceae of the Parietales. Phylogenists have treated *Paeonia* as comprising 1 of the 3 tribes of the buttercup family (i.e., Helleboreae, Anemoneae, and Paeonieae). The Helleboreae are characterized by the multiovulate pistil, and the Anemoneae [74] by the uniovulate pistil.

Generic distinctions are admittedly weak among the Ranunculaceae, as indicated by the lack of agreement as to the generic limits of *Anemone*, *Pulsatilla*, *Hepatica*, *Clematis*, and *Ranunculus*. The wide variations of perianth conditions, while of significance in distinguishing genera, serve to emphasize the significance of the essential organs to any phyletic study of the family.

Economically the family is of domestic importance for the large number of ornamentals available from its components, represented by 27 genera and about 280 species. Among the more important genera are *Anemone*, *Delphinium*, *Aquilegia*, *Helleborus*, *Thalictrum*, *Paeonia*, *Ranunculus*, and *Trollius*. Roots of the golden seal (*Hydrastis*) continue to be in demand by the crude drug trade. From

---

[72] Small, following Rydberg, accepted the segregation of *Ranunculus* into 3 genera, *Batrachium*, *Halerpestes*, and *Ranunculus*. Rydberg recognized, in addition, the genera *Cyrtorhyncha*, *Beckwithia*, and *Coptidium*. By application of the same generic concepts, the genus *Clematis* was considered by one or both of these authors to be represented by the genera *Clematis*, *Viorna*, *Viticella*, and *Atragene*. Many authors accept the segregation of *Pulsatilla* from *Anemone* (pronounced àn-e-mo-ne, not aye-nèm-o-nee) and some treat *Hepatica* as not generically distinct from *Anemone*.

[73] Representatives of these 2 South Pacific families do not occur as indigens and perhaps only rarely as exotics in this country.

[74] Within this tribe belong the Clematideae treated as an independent tribe by Robinson and Fernald in *Gray's manual*, ed. 7 (1908).

monkshood (*Aconitum*) are obtained febrifuges important in internal medicine; the plant contains strong narcotics that are exceedingly poisonous when parts are eaten.

*LITERATURE:*

BAILEY, I. W. Origin of the angiosperms: need for a broadened outlook. Journ. Arnold Arb. 30: 64–70, 1949.

BENSON, L. Pacific States *Ranunculi*. Amer. Journ. Bot. 23: 26–33, 169–176, 1936.

———. The North American subdivisions of *Ranunculus*. Amer. Journ. Bot. 27: 799–807, 1940.

———. North American *Ranunculi*—I. Bull. Torrey Bot. Club, 68: 157–172; II, 477–490; III, 640–659, 1941; IV op. cit. 69: 298–316; V, 373–386, 1942.

———. The relationship of *Ranunculus* to the North American floras. Amer. Journ. Bot. 29: 491–500, 1942.

———. A treatise on the North American *Ranunculi*. Amer. Midl. Nat. 40: 1–261, 1948.

BOIVIN, B. American *Thalictra* and their old world allies. Rhodora, 46: 337–377; 391–445; 453–487, 1944.

CORNER, E. J. H. Centrifugal stamens. Journ. Arnold Arb. 27: 423–437, 1946.

DAVIS, K. C. Native and cultivated *Ranunculi* of North America and segregated genera. Minn. Bot. Studies, 2: 459–507, 1900.

DREW, W. B. North American representatives of *Ranunculus*, sect. *Batrachium*. Rhodora, 38: 1–47, 1936.

ERICKSON, R. O. Taxonomy of *Clematis* section *Viorna*. Ann. Mo. Bot. Gard. 30: 1–62, 1943.

EWAN, J. A synopsis of the North American species of *Delphinium*. Univ. Colo. Stud. D. 2: 55–244, 1945.

FERNALD, M. L. The North American species of *Anemone* Sect. *Anemonanthea*. Rhodora, 30: 180–188, 1928.

———. What is *Actaea alba?* Rhodora 42: 260–265, 1940.

———. Virginian botanizing under restrictions. Part 1. Field studies of 1942 and 1943. Rhodora, 45: 407–411, 1943.

FERNALD, M. L. and SCHUBERT, B. G. Studies of American types in British herbaria. III. A few of Philip Miller's species. Rhodora, 50: 186, 1948.

GLEASON, H. A. *Actaea alba* versus *Actaea pachypoda*. Rhodora, 46: 146–148, 1944.

GREGORY, W. C. Phylogenetic and cytological studies in the Ranunculaceae. Trans. Amer Phil. Soc. N.S. 31: 443–521, 1941.

HUTCHINSON, J. Contributions towards a phylogenetic classification of flowering plants. Kew Bull. 1923, p. 81.

HUTH, E. Revision der kleineren Ranunculaceen-Gattungen *Myosurus, Trautvetteria, Hamadryas, Glaucidium, Hydrastis, Eranthis, Coptis, Anemonopsis, Actaea, Cimicifuga,* und *Xanthorrhiza*. Engl. Bot. Jahrb. 16: 278–324, 1892.

LANGLET, O. Über Chromosomenverhältnisse und Systematik der Ranunculaceae. Svensk Bot. Tidskr. 26: 381–400, 1932.

MUNZ, P. A. *Aconita cultorum*. Gentes Herb. 6: 463–506, 1945.

———. *Aquilegia*, the cultivated and wild columbines. Gentes Herb. 7: 1–150, 1946.

PAYSON, E. B. The North American species of *Aquilegia*. Contr. U. S. Nat. Herb. 20: 133–157, 1918.

POLUNIN, N. Supplementary notes on arctic and boreal species in Benson's "North American Ranunculi." Bull. Torrey Bot. Club 71: 246–253, 1944.

SAUNDERS, A. P. and STEBBINS, G. L. Cytogenetic studies in *Paeonia*. I. The compatibility of the species and the appearance of the hybrids. Genetics, 23: 65–82, 1938.

STEBBINS, G. L. JR. Cytogenetic studies in *Paeonia*. II. The cytology of the diploid species and hybrids. Genetics, 23: 83–1'0, 1938.

———. The western American species of *Paeonia*. Madroño, 4: 252–260, 1938.

———. Notes on some systematic relationships in the genus *Paeonia*. Univ. California Publ. Bot. 19: 245–266, 1939.

STERN, F. C. A study of the genus *Paeonia*. 155 pp., 15 col. pl. London, 1946.

ULBRICH, E. Über die systematische Gliederung und geographische Verbreitung der Gattung *Anemone* L. Engl. Bot. Jahrb. 37: 172–334, 1905–1906.

WODEHOUSE, R. P. Pollen grains in the identification and classification of plants. VII. The Ranunculaceae. Bull. Torrey Bot. Club, 63: 495–514, 1936.
WORSDELL, W. C. The affinities of *Paeonia*. Journ. Bot. (Lond.), 46: 114–116, 1908.

## LARDIZABALACEAE. LARDIZABALA FAMILY

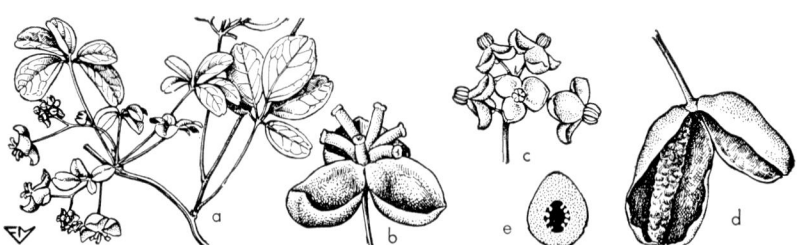

**Fig. 137.** LARDIZABALACEAE. *Akebia quinata*: a, flowering branch, × ¼; b, pistillate flower, × 1; c, staminate flower, × 1; d, fruit, × ¼; e, ovary, cross-section, × 6. (From L. H. Bailey, *Manual of cultivated plants*, The Macmillan Company, 1949. Copyright 1924 and 1949 by Liberty H. Bailey.)

Mostly monoecious or dioecious twining shrubs (only *Decaisnea* of the Orient an erect shrub); leaves alternate, palmately compound (pinnate only in *Decaisnea*), estipulate; flowers mostly unisexual (bisexual in *Decaisnea*), actinomorphic, racemose or solitary, trimerous, the perianth biseriate (corolla absent in some genera and/or simulated by small petaloid nectaries), the calyx of 3–6 distinct often petaloid sepals; corolla (when present) of 6 smaller distinct petals, the staminate flowers with 6 distinct or basally connate stamens, the anthers 2-celled, distinct, basifixed, extrorse, dehiscing longitudinally, nectaries usually present between filaments and perianth, the pistillate flowers with gynoecium of 3–15 free distinct and divergent pistils, the ovary superior, 1-loculed, 1-carpelled, the ovules many (infrequently solitary), the placentation parietal, the stigma 1, oblique, subsessile; fruit a berry, usually splitting longitudinally at maturity; seed with small straight embryo and a copious firm-fleshy endosperm.

A family of 7 genera and 20 species native in the Himalayas, China, Japan, and Chile (only *Boquila* and *Lardizabala* in Chile). Represented in the United States by cultigens as the ornamental hardy vine *Akebia* (2 spp.) and in warmer parts of the country, by the monotypic *Decaisnea,* by *Lardizabala* (1–2 spp.) and *Stauntonia* (2 spp.).[75] The fruits of all species are edible.

*LITERATURE:*

GAGNEPAIN, F. Révision des Lardizabalées asiatiques de l'herbier du Muséum. Bull. Mus. Hist. Nat. (Paris), 14: 64–70, 1908.

## BERBERIDACEAE. BARBERRY FAMILY

Perennial herbs or shrubs; rootstocks sometimes of creeping rhizomes or tubers; leaves alternate or basal, simple or pinnately compound, deciduous or less commonly persistent (*Mahonia, Nandina*), mostly estipulate and with petioles basally

---

[75] The rarely cultivated *Sargentodoxa cuneata*, placed in this family by many authors, was segregated by Hutchinson (1926, p. 100) into the unigeneric Sargentodoxaceae because "the family in many ways combines in its two sexes the characters of the Lardizabalaceae and Schizandraceae, with the male flowers similar to the former, and the gynaecium of the latter." This view was accepted by Stapf and by Rehder.

## DIVISION IV. EMBRYOPHYTA SIPHONOGAMA

dilated; flowers bisexual, actinomorphic, solitary, in axillary cymes, a false raceme, or thyrse, the perianth biseriate, the 4–6 petals and like number of sepals similar or clearly differentiated, segments distinct and free, hypogynous, often caducous, 1–2 whorls of often petaloid nectaries sometimes between corolla and androecium, the stamens 4–18, distinct, generally in 2 whorls, those of the outer whorl opposite the petals, the anthers 2-celled, basifixed, mostly dehiscing by flaplike valves recurving from theca base, infrequently by longitudinal splits (*Podophyllum*); pistil 1, the ovary superior, 1-loculed (2-loculed in some *Epimedium* spp.), 2–3-carpelled,[76] the ovules few and basal (rarely 1) or many on a single parietal placenta (seemingly basal in some species), ascending or erect, anatropous, the style short and thick or absent, stigma 1; fruit usually a berry, follicular (as in *Jeffersonia*), or the biovulate ovary pericarp rupturing and withering after fertilization, with the drupelike naked seeds maturing independently (in *Caulophyllum*); seed with a small embryo (long in *Berberis*) and copious fleshy endosperm (horny in *Caulophyllum*), sometimes arillate.

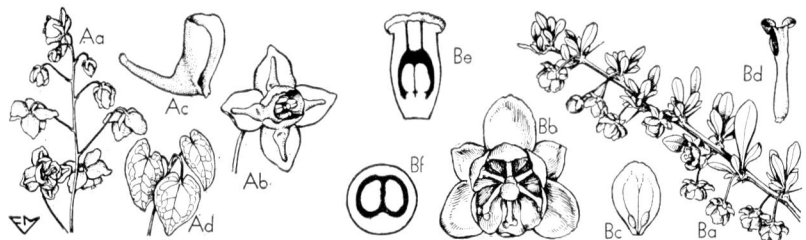

**Fig. 138.** BERBERIDACEAE. A, *Epimedium versicolor*: Aa, inflorescence branch, × ½; Ab, flower, × 1; Ac, petal, × 2; Ad, leaf, × ⅓. B, *Berberis Thunbergii*: Ba, flowering branch, × ½; Bb, flower, × 2; Bc, petal with 2 basal glands, × 2; Bd, stamen, × 4; Be, pistil, vertical section, × 5; Bf, ovary, cross-section, × 8. (From L. H. Bailey, *Manual of cultivated plants,* The Macmillan Company, 1949. Copyright 1924 and 1949 by Liberty H. Bailey.)

A family of 10–12 genera and about 200 species, in the north temperate regions (some spp. of *Berberis* extend into southern South America). Seven genera are represented in this country by about 17 indigenous species. Most of these occur in the Pacific coast states (*Mahonia* 10 spp.,[77] *Vancouveria* 3 spp., *Achlys* 1 sp.) while in the eastern part of the country occur *Jeffersonia* (1 sp., a second native in Manchuria), *Diphylleia* (1 sp., a second in Japan), *Podophyllum* (1 sp., 4 others in Asia), *Caulophyllum* (1 sp., a second in eastern Asia). *Berberis*, a large, primarily Asiatic genus, of about 175 species, is represented by 1 naturalized noxious species (*B. vulgaris*) and another (*B. canadensis*) native in eastern U.S.

The Berberidaceae are distinguished from related families by the stamens biseriate with the outer opposite the petals, the valvate anther dehiscence, and the

---

[76] The ovary of present-day members of the family is usually clearly unilocular and is seemingly 1-carpelled with parietal placentation. However, Chapman (1936) demonstrated that these members probably have evolved from proranalian ancestors that had flowers with 3 spirally arranged simple pistils that fused to form a 3-loculed ovary with axile placentation. It was hypothesized further that 2 carpels were suppressed and their placentae moved to one side of the ovary, and the locules lost by compression; this resulted in a unilocular condition, derived from a tricarpellate ovary.

[77] On the basis of evidence obtained from only 2 species of *Mahonia* and 42 species of *Berberis,* Dermen (1931) concluded that cytogenetic data do not justify the recognition of these 2 taxa as separate genera.

unipistillate gynoecium. In most genera (except *Podophyllum*) the stamen number (equal to number of petals) is distinctive.

There is general agreement that the genera of this family are not closely related to one another and that they are separable in 2 distinct subfamilies; one with *Berberis, Caulophyllum,* and *Leontice* having been derived from a 3-carpelled ancestor and the second containing the other genera by virtue of presumed derivation from a 2-carpelled ancestor. It was the opinion of Chapman (1936) that "it is doubtful whether any existing families may be related as the immediate predecessors of the Berberidaceae," and she pointed out that this family and the Ranunculaceae arose by parallel evolution from a proranalian complex. On the basis of carpellary anatomy, she expressed doubt that the Berberidaceae could have been the progenitors of the Papaveraceae or of the order Rhoeadales.

Economically the family is of domestic importance for members of ornamental value (about 100 species of 9 genera are offered in the trade). Fruits of *Podophyllum* are edible (preserves and beverages) but leaves and roots are poisonous. *Berberis* fruits likewise are edible, and plants of the naturalized *B. vulgaris* are the obligate host of the aecidial stage of wheat rust.

*LITERATURE:*

ABRAMS, L. R. The Mahonias of the Pacific states. Phytologia, 1: 89–94, 1934.

AHRENDT, L. W. A. A survey of the genus *Berberis* in Asia. Journ. Bot. 79: Suppl. 1–80, 1941.

CHAPMAN, M. Carpel anatomy of the Berberidaceae. Amer. Journ. Bot. 23: 340–348, 1936.

DERMEN, H. A study of chromosome number in two genera of Berberidaceae: *Mahonia* and *Berberis.* Journ. Arnold Arb. 12: 281–287, 1931.

HIMMELBAUER, W. Die Berberidaceen und ihre Stellung im System. Denk. Akad. Wiss. Math. Naturw. (Wien), 89: 733–796, 1914.

MAURITZON, J. Zur Embryologie der Berberidaceen. Acta Hort. Gotoburg, 11: 1–17, 1936.

SCHMIDT, E. Untersuchungen über Berberidaceen. Beih. Bot. Centralbl. 45: 329–396, 1928.

STEARN, W. T. *Epimedium* and *Vancouveria* (Berberidaceae), a monograph. Journ. Linn. Soc. (Bot.) Lond. 51: 409–535, 1937–1938.

## MENISPERMACEAE. MOONSEED FAMILY

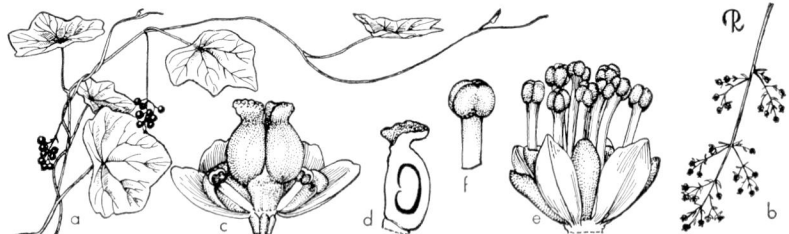

Fig. 139. MENISPERMACEAE. *Menispermum canadense*: a, fruiting branch, × ⅙; b, pistillate inflorescence, × ½; c, pistillate flower (with staminodes), × 6; d, pistil, vertical section, × 8; e, staminate flower, × 6; f, anther, × 12. (From L. H. Bailey, *Manual of cultivated plants,* The Macmillan Company, 1949. Copyright 1924 and 1949 by Liberty H. Bailey.)

Mostly twining woody dioecious vines, rarely erect shrubs or small trees; leaves alternate, persistent or deciduous, simple (rarely trifoliolate in few tropical spp.), mostly entire or occasionally palmately lobed, mostly palmately veined, petiolate, estipulate; flowers unisexual, minute, greenish, generally actinomorphic, the perianth mostly many-seriate, calyx and usually the corolla present, the sepals and petals 2–3-merous (mostly 6), distinct; the staminate flowers with usually 6 stamens (sometimes 3, or the number indefinite), opposite petals when of same number, distinct or variously connate (markedly monadelphous in some genera), the anthers 4-celled or falsely so, dehiscing longitudinally; pistillate flowers with or without staminodes, the gynoecium of usually 3–6 distinct sessile or stipitate pistils (sometimes solitary, rarely many), the ovary superior, 1-loculed, the ovules 2 aborting to 1, anatropous, the placentation parietal, the carpel 1, the style very short or none, the stigma terminal, capitate or discoid, entire or lobed; the fruit a drupe, or achene; seed with or without fleshy endosperm (present in *Menispermum, Cocculus, Calycocarpum*), usually curved.

A family of about 70 genera and 400 species, distributed largely throughout paleotropic regions. Species of a few genera extend into the eastern Mediterranean region and eastern Asia, but none is indigenous to Europe. Four species of 3 genera are native in the United States: *Menispermum canadense* as far north as Manitoba and Quebec, *Cocculus carolinus* (*Cebatha*) from Arizona to Florida and north to Virginia, *C. diversifolius* in southern Texas, and *Calycocarpum Lyonii* from eastern Texas to Florida north to Kansas and Illinois.

The family may be distinguished by the dioecious character (monoecious in *Albertisia*), the lianous habit (not so in some spp. of *Cocculus et al.*), the usually 3-merous floral situation, the double whorl of sepals, and the curved seed.

The Menispermaceae have been considered in the past as a part of the Berberidaceae, the Magnoliaceae, or the Annonaceae. DeCandolle recognized it as distinct, a view upheld by Eichler and successors. Most contemporary authors have treated it as of close affinity with the Berberidaceae and the Lardizabalaceae. It was divided by Diels into 8 tribes, and these and their genera are separated primarily on characters of the fruit and seed.

Economically the family is of little domestic importance. A few species of *Menispermum, Cocculus,* and *Cissampelos* are grown for ornament.

*LITERATURE:*

Diels, L. Menispermaceae. In Engler, Das Pflanzenreich, 46 (IV. 94): 1–345, 1910.

## MAGNOLIACEAE. MAGNOLIA FAMILY

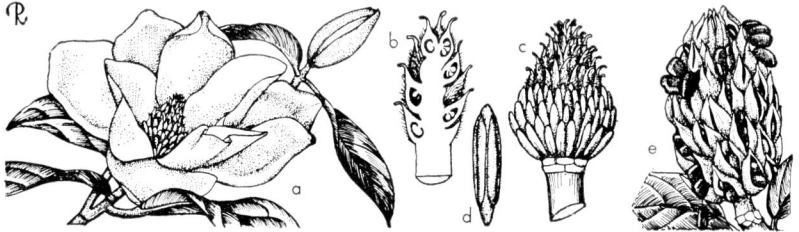

**Fig. 140.** Magnoliaceae. *Magnolia grandiflora*: a, flowering branch, × ¼; b, gynoecium, vertical section, × ½; c, flower, less perianth, × ½; d, stamen, × 2; e, fruit, × ¼. (Redrawn from Sargent.) (From L. H. Bailey, *Manual of cultivated plants,* The Macmillan Company, 1949. Copyright 1924 and 1949 by Liberty H. Bailey.)

Deciduous or evergreen trees or shrubs; leaves alternate, simple, mostly entire, petioled, pinnately veined, the stipules usually present and then enclosing the young bud, early deciduous and leaving a circular scar; flowers bisexual (except *Kmeria*) actinomorphic, often large and showy, usually solitary, terminal or axillary, the lower bud often subtended by a bracteate spathaceous sheath, the perianth of cyclic or spirally disposed distinct sepals and petals (not always differentiated), the sepals often 3, the petals 6 to many; stamens numerous, hypogynous, distinct, spirally disposed on the basal portion of the floral axis (the androphore), the anthers 2-celled, dehiscing longitudinally, introrse; the gynoecium sessile or borne on a gynophore, of many (rarely 2 or 3) pistils spirally arranged on a usually elongated axis, the ovary 1-loculed, 1-carpelled, placentation parietal, the ovules 1–several, style 1, stigma 1; fruit a follicle, samara, or berry; the seeds with minute embryo and a copious watery-fleshy endosperm, often suspended from the follicle by an elongated funiculus.

A family of 10 genera and over 100 species,[78] primarily of temperate regions of the northern hemisphere, with distributional centers in eastern Asia, Malaysia, and eastern North America south to the West Indies and Brazil. In this country it is represented by 8 species of *Magnolia* (incl. *Tulipastrum*) and the monotypic *Liriodendron*. *Magnolia* (about 77 spp.), mostly Asiatic, is the largest genus; *Michelia* (15 spp.) occurs in tropical Asia from Java to China; and *Talauma* (15 spp.), mostly of southeastern Asia, is represented by 4 species indigenous to tropical America from Mexico to Brazil.

The Magnoliaceae are trees or shrubs characterized by the presence of bisexual flowers (except for the monotypic *Kmeria*), an androecium of numerous spirally arranged stamens, and a gynoecium with usually numerous simple pistils spirally arranged on an elongated axis or torus.

The Magnoliaceae have often been considered to be the most primitive extant family of the order. Many botanists (including Hallier, Hutchinson, *et al.*) also believe the family to be one of the most primitive taxa of living angiosperms, and in defense of this belief Hallier (1905) compared the elongated floral axis bearing numerous spirally disposed pistils with the sporophyll-bearing axis of the Bennettitales. The cytological findings of Whitaker (1933) emphasize the validity of the segregation of the families Winteraceae, Trochodendraceae, Illiciaceae, Schisandraceae, and Eupteleaceae from the Magnoliaceae. Smith (1945) mentions that the Magnoliaceae are relatively highly specialized both vegetatively and florally, casting some doubt on the assumption of the primitive nature of the family, and implying that such groups as Winteraceae, etc., may be at least as primitive. Bailey, Nast, and Smith (1943) consider the Magnoliaceae to be closely allied only to 2 small Australasian-Pacific families, the Himantandraceae and Degeneriaceae.

Economically the family is important here for the 20 species of *Magnolia* cultivated as ornamentals and for the tulip-tree wood (*Liriodendron*) used in furniture and cabinet work. Three Asiatic species of *Michelia* and one of *Talauma* are grown for ornament in southern parts of the country. The wood of several species of *Magnolia* is used in cabinet work.

*LITERATURE:*

BAILEY, I. W., NAST, C. G. and SMITH, A. C. The family Himantandraceae. Journ. Arnold Arb. 24: 190–206, 1943.

DANDY, J. E. The Genera of Magnolieae. Kew Bull., pp. 257–264, 1927.

FRITEL, P. H. Remarques sur quelques espèces fossiles du genre *Magnolia*. Bull. Soc. Geol. France, 13: 277–292, 1913.

[78] Note that the above circumscription of the family excludes such taxa as the Schisandreae, Illicieae, and Tetracentreae, as outlined by Dalla Torre and Harms. The genera comprising these are now considered to represent 4 families: Schisandraceae, Illiciaceae, Winteraceae, and Tetracentraceae. This view, stated by Dandy (1927), Hutchinson, and others, reduces the number of genera and species credited to the Magnoliaceae from that given in most earlier works.

Good, R. D'O. The past and present distribution of the Magnolieae. Ann. Bot 39: 409 430, 1925.
Howard, R. A. The morphology and systematics of the West Indian Magnoliaceae. Bull Torr. Bot. Club, 75: 335–357, 1948.
McLaughlin, R. P. Systematic anatomy of the woods of the Magnoliales. Trop. Woods, 34: 3–38, 1933.
Maneval, W. E. The development of *Magnolia* and *Liriodendron,* including a discussion of the primitiveness of the Magnoliaceae. Bot. Gaz. 57: 1–31, 1914.
Millais, J. G. Magnolias. 251 pp. London, 1927.
Smith, A. C. Geographical distribution of the Winteraceae. Journ. Arnold Arb. 26: 48–59, 1945.
Whitaker, T. W. Chromosome number and relationship in the Magnoliales. Journ. Arnold Arb. 14: 376–385, 1933.

## ILLICIACEAE. ILLICIUM FAMILY

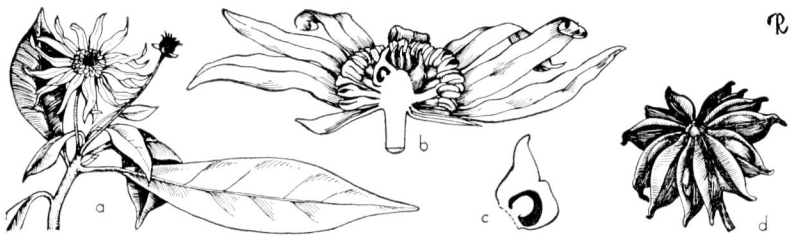

**Fig. 141.** Illiciaceae. *Illicium floridanum*: a, flowering branch, × ⅓; b, flower, vertical section, × 1; c, pistil, × 2; d, fruits, × 1. (From L. H. Bailey, *Manual of cultivated plants,* The Macmillan Company, 1949. Copyright 1924 and 1949 by Liberty H. Bailey.)

Shrubs or small trees; leaves persistent, alternate (often clustered at distal nodes), simple, entire, often aromatic, estipulate; flowers bisexual, actinomorphic, axillary near twig tips, solitary or in 2's or 3's, rarely borne on complex glomerules, perianth usually multiseriate and undifferentiated into calyx and corolla, the parts (tepals) 7–33, becoming centripetally larger or the innermost ones transitional toward stamens; stamens usually numerous (rarely as few as 4), 1–several seriate, the anthers 2-celled, basifixed, dehiscing longitudinally, the connective sometimes enlarged and extending beyond the thecae; gynoecium of 7–15 (rarely to 21) pistils, these whorled, distinct and free, the ovary superior, 1-loculed, 1-carpelled, the ovule borne ventrally (parietal) near base of locule, anatropous, the style tapering, conduplicate, its ventral surface distally stigmatic; fruit a folliceturm of 1-seeded ventrally dehiscent follicles; seed with minute embryo and copious watery-fleshy endosperm.

The family was first definitely established by Smith (1947), although previously *Illicium* had been regarded as distinct from the Magnoliaceae proper, in one degree or another, by numerous students. With the Schisandraceae, it forms a sharply delimited group in the order. It is distinguished readily from the Schisandraceae by its nonscandent habit, bisexual flowers, short (rather than greatly modified) receptacle, comparatively few pistils in a single whorl, vascularized conduplicate style, single ovule, and fruit a folliceturm composed of dehiscent distinct follicles.

Most taxonomists have treated this group as representing a tribe of the Magnoliaceae (Engler, Rendle, Bentham and Hooker), often linking the genus with *Drimys* and its relatives, while others (including Hutchinson, 1926) have placed it in the Winteraceae,[79] a disposition rejected by Bailey and Nast (1945) on the basis of differences in pollen morphology, stem anatomy, carpel morphology, and karyology. Smith (1947) observed that "*Illicium* . . . has no close allies other than *Schisandra* and *Kadsura*. The three genera will probably be treated by future phylogenists as composing a suborder of the Ranales . . ." Contrasting the Illiciaceae with the Schisandraceae, the same author concluded that ". . . the two groups . . . have become specialized in different ways and have each retained certain primitive characteristics. To say that one is more primitive than the other seems impossible."

Except for the fact that *Illicium verum* provides a volatile oil of commercial value, members of the family are of little domestic importance aside from their ornamental value in the warmer parts of the country, where 4 species are available in the trade.

*LITERATURE:*

BAILEY, I. W. and NAST, C. G. The comparative morphology of the Winteraceae. VII. Summary and conclusions. Journ. Arnold Arb. 26: 37–47, 1945.

SMITH, A. C. Taxonomic notes on the Old World species of Winteraceae. Journ. Arnold Arb. 24: 119–164, 1943.

———. The families Illiciaceae and Schisandraceae. Sargentia, 7: 1–224, 1947.

## SCHISANDRACEAE.[80] SCHISANDRA FAMILY

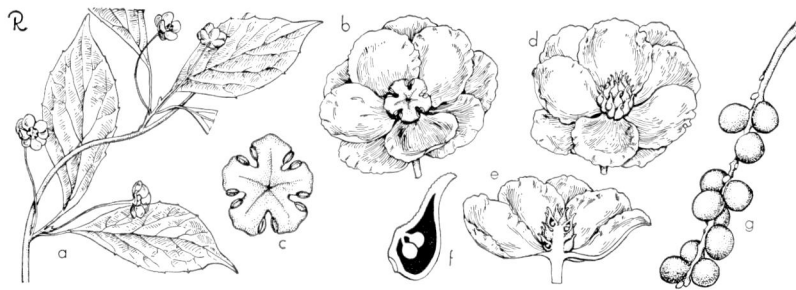

**Fig. 142.** SCHISANDRACEAE. *Schisandra coccinea*: a, flowering branch, × ½; b, staminate flower, × 2; c, androecium, top view, × 4; d, pistillate flower, habit, × 2; e, same, vertical section, × 2; f, pistil, vertical section, × 6; g, fruiting peduncle, × ½. (b–f, adapted from Gray.)

Clambering or twining woody, monoecious or dioecious vines; leaves alternate, simple, deciduous (or tardily so), estipulate; flowers unisexual, actinomorphic; the perianth of few to many essentially undifferentiated segments (tepals), in 2–many series, the receptacle often much modified; staminate flowers with an androecium

---

[79] The Winteraceae, a primitive ranalian family of predominantly southern hemisphere distribution, composed of 6 genera and about 90 species, is known in the United States only by 1 or 2 rarely cultivated species of *Drimys*. The family is of particular morphological, phyletic, and phytogeographical significance, and interested students should consult the findings and opinions of Smith (1943, 1947) and Bailey and Nast (1945).

[80] Rehder has pointed out (Journ. Arnold Arb. 25: 129–131, 1944) that the original spelling of the name on which this family name was based is *Schisandra* Michx. (1803) and that it was changed illegitimately by de Candolle in 1817 to *Schizandra*. Rehder demonstrated further that there was no orthographic basis for this change in spelling and that the original spelling, *Schisandra*, must be retained.

DIVISION IV. EMBRYOPHYTA SIPHONOGAMA    507

of 4–80 stamens (variable in composition among the species), the filaments at least basally connate into a modified column, the anthers 2-celled, basifixed, dehiscing longitudinally, connectives variously or not modified; gynoecium composed of 12–300 distinct and free acyclic pistils, the torus conical to cylindrical (*Schisandra*) or obovoid to ellipsoid (*Kadsura*), the ovary superior, 1-loculed, 1-carpelled, conduplicate, the ovules mostly 2-5, anatropous, ventrally attached and pendulous (parietal), the ovary not wholly closed along ventral margins, and the 2 parallel edges each bearing an erose or ciliate stigmatic crest, the distal end of which is usually projected into an unvascularized pseudostyle; fruit aggregate, baccate, composed of a highly modified torus and sessile drupelike indehiscent carpels; seed usually 1–5, with a small embryo and copius endosperm (probably water-fleshy).

Two genera (*Schisandra* and *Kadsura*) and 47 species compose the family. Only a single species of the family (*Schisandra glabra*) is American; all others are indigens of eastern and southeastern Asia and Malaysia. The American representative is a monoecious (or perhaps often dioecious) high-climbing vine, occurring with great rarity mostly on steep slopes of ravines in a few counties of South Carolina, Georgia, northwestern Florida, Alabama, Arkansas, Tennessee, and Louisiana.

The Schisandraceae are readily separated from the closely related Illiciaceae as noted under that family.

Economically the family is of slight domestic importance. One species of *Kadsura* and 3 of *Schisandra* are cultivated to a limited extent as ornamental vines.

*LITERATURE:*

SMITH, A. C. The families Illiciaceae and Schisandraceae. Sargentia, 7: 1–224, 1947.

## CALYCANTHACEAE. CALYCANTHUS FAMILY

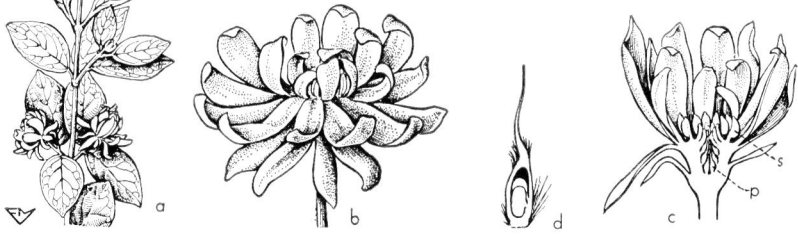

**Fig. 143.** CALYCANTHACEAE. *Calycanthus floridus*: a, flowering branch, × ¼; b, flower, × 1; c, flower, vertical section, × 1 (p pistil, s stamen); d, pistil, vertical section, × 5. (From L. H. Bailey, *Manual of cultivated plants*, The Macmillan Company, 1949. Copyright 1924 and 1949 by Liberty H. Bailey.)

Deciduous or evergreen shrubs; leaves opposite, simple, entire, short-petioled, estipulate; flowers bisexual, actinomorphic, perigynous, solitary, axillary, fragrant, the perianth spirally disposed and composed of numerous imbricated showy tepals, undifferentiated into sepals and petals, borne on the outer rim of a thickened cuplike receptacle; stamens 5–30, inserted on the receptacle rim, filaments short and distinct, the anthers laterally extrorse, 2-celled, the innermost ones sometimes sterile; gynoecium with about 20 distinct and free pistils, borne on the inside of the receptacle, the ovary superior, 1-loculed, 1-carpelled, the ovules 1-2, anatropous, ascending, the placentation parietal, the style linear to filiform, the stigma 1;

fruit an achene, 1-seeded (by abortion when ovary 2-ovuled), enclosed within the enlarged and fleshy receptacle; seed with large embryo and no endosperm.

A family of 2 genera and 6 species. *Calycanthus,* a deciduous genus characterized by stamens 10–30, has 1 species native to California and probably 3 in the southeastern United States. *Chimonanthus (Meratia),* characterized by stamens 5, the species evergreen or deciduous, has 2 species in China and Japan.

Members of the family are readily distinguished by the combination of opposite leaves, the spirally arranged perianth, and the gynoecium of numerous simple pistils situated within a cuplike receptacle.

The Calycanthaceae resemble the Magnoliaceae and the Annonaceae in the spiraled arrangement of perianth parts and in the usually numerous stamens and pistils. However, because neither the androecial nor gynoecial elements are clearly disposed spirally and because of the presence of the cuplike receptacle (not an hypanthium), the opposite leaves and the large embryo, some authors (including Pollard, 1908, and Hutchinson) have placed the family as closely allied to Rosaceae. Until further evidence is available, it is probable that the majority of botanists may retain it in the Magnoliales rather than in the Rosales.

The family is of no domestic economic importance except for the ornamental value of its members. Species of both genera are available in the trade and are prized for the aromatic fragrance of the flowers.

*LITERATURE:*

POLLARD, C. L. Calycanthaceae. North Amer. Flora, 22: 237–238, 1908.
SAX, K. Chromosome behavior in *Calycanthus.* Journ. Arnold Arb. 14: 279–282, 1933.

## ANNONACEAE.[81] CUSTARD-APPLE FAMILY

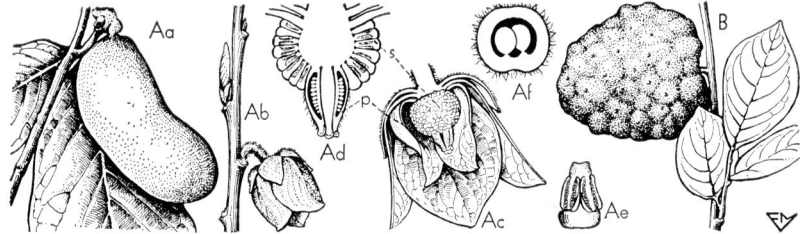

**Fig. 144.** ANNONACEAE. A, *Asimina triloba*: Aa, fruit, × ¼; Ab, flowering branch, × ½; Ac, flower, perianth excised, × 1; Ad, same, vertical section, less perianth, × 2; Ae, stamen, × 5; Af, ovary, cross-section, × 10. B, *Annona Cherimola*: fruit, × ⅙. (p pistil, s stamens.) (From L. H. Bailey, *Manual of cultivated plants,* The Macmillan Company, 1949. Copyright 1924 and 1949 by Liberty H. Bailey.)

Trees, shrubs, or vines with aromatic wood and foliage; leaves alternate, simple, entire, deciduous or persistent, estipulate; flowers bisexual (unisexual and plants dioecious in *Stelechocarpus* of Malaya), actinomorphic, hypogynous, the perianth usually triseriate, the outer whorl a calyx of 3 (rarely 2) basally connate or distinct valvate sepals, the 2 inner whorls a corolla of usually 6 distinct similar or dissimilar petals, imbricate or valvate in each whorl, the axis usually extending and enlarging beyond point of perianth attachment; stamens numerous, distinct, spirally arranged, the filament thickened and short, the anthers 4-celled at anthesis,

---

[81] The spelling Annonaceae has been conserved over that of Anonaceae on the grounds that the correct spelling of the type genus is *Annona* (not *Anona*). For discussion, cf. Kew Bull. p. 344, 1928.

extrorse, terminated or overtopped by an enlarged variously shaped connective; gynoecium of few to numerous distinct pistils (basally connate in *Monodora* of Africa), the ovary superior, 1-loculed and 1-carpelled (except in *Monodora*), the ovules 1–many, anatropous, typically parietal but sometimes seemingly basal, the style 1 and very short or absent, the stigma 1; fruit a berry or the maturing pistils becoming connate and adnate to the floral axis to form single fleshy aggregate fruit (*Annona*); seed large, the embryo small, the endosperm very large and ruminate.

A family of about 80 genera and 850 species, occurring paleotropically, but with relics of earlier distributions remaining in some temperate regions. In this country it is represented by 8 species of *Asimina* (*Pityothamnus*) the pawpaw, in the southeast and northward through the Mississippi valley to western New York and Nebraska, and by 1 species of *Annona* (*A. glabra*) in peninsula Florida. The larger genera of the family include: *Uvaria*, Old World tropics, 100 spp.; *Annona*, tropical America, 90 spp.; *Polyalthia*, paleotropics, mostly Asiatic, 70 spp.; *Xylopia*, pantropical, 60 spp.; *Guatteria*, Mexico to South America, 60 spp.; *Oxymitra*, Asiatic tropics, 50 spp.; *Artabotrys*, pantropics, 40 spp.

The Annonaceae are distinguished by the absence of stipules, the spirally arranged stamens and their distinctive usually enlarged connectives, multipistillate gynoecium, and the ruminate endosperm of the seeds.

There is general agreement that this family is a derivative from magnoliaceous stocks. Hutchinson treated it (together with the Eupomatiaceae) as comprising the Annonales, noting that the latter was related to but more advanced than the Magnoliales. Engler and Diels subdivided it into 5 tribes, with *Annona* in the Xylopieae and *Asimina* (treated as a congeneric with *Uvaria*) in the Uvarieae.

Economically the family is of considerable importance throughout the tropics of the world as a source of edible fruit, but few of its members are of importance domestically. The pawpaw (*Asimina*) is grown to a limited extent in temperate regions for its edible fruit and in warmer parts of the country are grown the soursop, cherimoya, sweetsop, custard apple, and ilama (all spp. of *Annona*). Other genera grown in warm parts of the country include *Cananga* and *Rollinia*.

*LITERATURE:*

ASANA, J. J. and ADATIA, R. D. Contributions to the embryology of the Anonaceae. Bombay U. J., Sect. B, Biol. Sci. n.s. 16(3): 7–21, 1947.
DIELS, L. Die Gliederung der Anonaceen und ihre Phylogenie. Sitzungsb. preuss. Akad. Wiss. 1932: 77–85.
FRIES, R. E. Revision der Arten einiger Anonaceen-Gattungen. I–V. Acta Hort. Bergiani, 10: 1–341, 1931; *op. cit.* 12: 1–220, 289–577, 1939.

## MYRISTICACEAE. NUTMEG FAMILY

Dioecious or monoecious evergreen trees or shrubs with usually aromatic wood and foliage; inner bark often exuding a reddish liquid; leaves alternate, simple, entire, coriaceous, often with pellucid dots, estipulate; flowers unisexual, actinomorphic, racemose, corymbose, fasciculate or capitate, inconspicuous; perianth limited to a gamosepalous saucer- to funnel-shaped and usually 3-lobed calyx; the staminate flowers with 2–30 stamens, the filaments connate in a solid column, the anthers 2-celled, extrorse, distinct or connate, dehiscing longitudinally; the pistillate flowers unipistillate, ovary superior, 1-loculed, 1 carpelled, the ovule solitary, parietal but sometimes seemingly basal, anatropous, the style 1 and short (or absent), the stigma 1; fruit a dehiscent drupe; seed partially or completely enveloped by an often brightly colored laciniate or subentire fleshy aril, the embryo small, the endosperm copious and ruminate.

The family was long considered to be unigeneric (*Myristica*) and to consist of about 100 species grouped in well-marked sections, until divided by Warburg (1897) into 15 genera that were interpreted to be represented by 260 species. The latter view is the more generally accepted. These genera, accepted also by Smith and Wodehouse (1937), are widely distributed; 5 occur in the New World tropics, 6 in Africa, and 4 in Asia. The northern limits for American representatives are in southern Mexico and the Lesser Antilles.

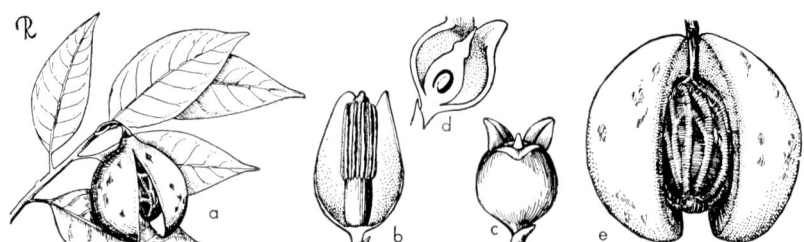

**Fig. 145.** MYRISTICACEAE. *Myristica fragrans*: a, fruiting branch, × ¼; b, staminate flower, perianth removed to show androecium, × 2; c, pistillate flower, × 1, d, same, vertical section, × 1; e, fruit showing aril (mace) about seed, × ½. (From L. H. Bailey, *Manual of cultivated plants,* The Macmillan Company, 1949. Copyright 1924 and 1949 by Liberty H. Bailey.)

The Myristicaceae are distinguished from related families by the unisexual flowers, the monadelphous stamens, and the arillate seed.

The characteristics of the family are distinctive and set it well apart from other families. Bessey and Hallier placed it in the Ranales (or Annonales); Wettstein concurred and indicated it to be closely related to the Annonaceae; Hutchinson included it in his Laurales (segregating the Annonaceae as the Annonales) with its nearest allies found among the Lauraceae. Smith and Wodehouse (1938) agreed with Hutchinson. Joshi (1946) on the basis of studies of pollen and embryology disagreed with Hutchinson's views, noting that his evidence "appears to favor Wettstein's view about the close relationship of the two families Myristicaceae and Anonaceae."

Economically the family is important as the source of nutmeg of commerce (*Myristica fragrans*), obtained from the seed, and for mace obtained from the dried aril. This species is grown to a limited extent in southern Florida. The seeds of some of the American species yield a wax of limited commercial use.

*LITERATURE:*

GARRATT, G. A. Systematic anatomy of the woods of the Myristicaceae. Trop. Woods, 35: 6–48, 1933.

JOSHI, A. C. A note on the development of pollen of *Myristica fragrans* Van Houtten and the affinities of the family Myristicaceae. Journ. Indian Soc. Bot. 25: 139–143, 1946.

SMITH, A. C. and WODEHOUSE, R. P. The American species of Myristicaceae. Brittonia, 2: 393–510, 1938.

WARBURG, O. Monographie der Myristicaceen. Nova Acta Acad. Leop.-Carol. 68: 1–680, 1897.

## MONIMIACEAE. MONIMIA FAMILY

Evergreen, sometimes dioecious or monoecious, trees or shrubs (rarely vines); leaves mostly opposite (sometimes whorled or alternate), simple, coriaceous, aromatic, estipulate; flowers bisexual in most genera or unisexual in others, actino-

## DIVISION IV. EMBRYOPHYTA SIPHONOGAMA

morphic, perigynous, solitary or cymose, variable in structure, the perianth inconspicuous (absent in a few genera), of 1 whorl and calyciform when parts few (4–8) or biseriate and inner segments petaloid when numerous, borne on rim of convex to deeply urceolate hypanthium (perianth tube or receptacle of some authors); stamens few to numerous in 1–2 series or scattered over inner hypanthium surface in staminate flowers, the filaments short, often flattened, frequently with lateral glandular appendages, the anthers 2-celled, basifixed, dehiscing longitudinally or transversely or by 2 flaplike upturning valves; gynoecium of many distinct pistils (rarely reduced to 1), the ovary superior, 1-loculed, 1-carpelled, the ovule 1, erect or pendulous, anatropous, the placentation parietal, the style 1, short or linear, the stigma 1 and terminal, staminodia sometimes present in pistillate flowers; fruit an achene or drupe enclosed by the hypanthium (the latter sometimes fleshy) or the pistils separate in fruit; seed with small embryo, the endosperm copious, oily, not ruminate.

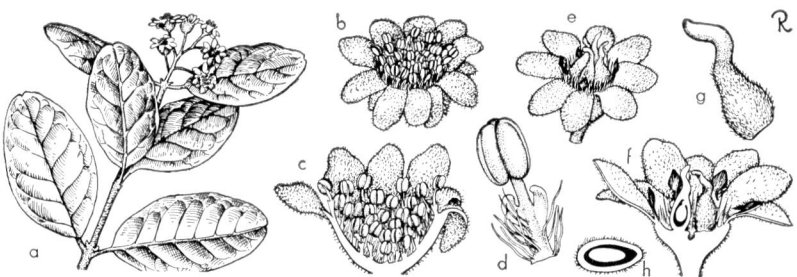

Fig. 146. MONIMIACEAE. *Boldea boldus*: a, flowering branch, × ½; b, staminate flower, habit, × 2; c, same, vertical section, × 3; d, stamen, × 10; e, pistillate flower, habit, × 2; f, same, vertical section, × 3; g, pistil, × 6; h, ovary, cross-section, × 10.

A family of about 32 genera and 350 species, chiefly of the southern hemisphere and these primarily in Australia, Polynesia, Madagascar, and Oceania. Range extensions project into South America, north to Mexico (*Mollinedia*), and into tropical Africa. The largest genus is *Siparuna*, of tropical America, with about 125 species. No species are indigenous to this country.

The Monimiaceae are considered closely related to the Calycanthaceae, as evidenced by the floral hypanthium and fruit, but are distinguished from them by the achene, the copious endosperm, the stamens often dehiscing transversely or valvately, and the plants sometimes dioecious or monoecious.

The family is of considerable phylogenetic interest, and interpretation of its position in the phyla has been varied, for it has both magnoliaceous and lauraceous characters which are countered in part by the very inconspicuous and generally unisexual flowers. Its present and previous distribution is more akin to that of the Lauraceae. Its approach to an apetalous condition and the presence of an hypanthium together with valvate stamens is suggestive of rosalian affinities. The morphological evidence seems stronger for alignment within the Laurales of Hutchinson than within the Magnoliales, a view upheld by the wood anatomy studies of Garratt (1934).

Economically the family is of little domestic importance. A few species each of 3 genera (*Hedycarya, Laurelia,* and *Boldea*) are cultivated primarily for ornament in southern Florida and California. The 2 latter produce edible fruits, and wood of many members of the family is of commercial value in regions of nativity, and Chilean boldo wood (*Boldea boldus*) sometimes reaches our markets as a rarity for cabinet work.

*LITERATURE:*

Garratt, G. A. Systematic anatomy of the woods of the Monimiaceae. Trop. Woods, 39: 18–44, 1934.
Pax, F. Monimiaceae. In Engler and Prantl, Die natürlichen Pflanzenfamilien, III(2): 94–105, 1897.
Perkins, J. R. and Gilg, E. Monimiaceae. In Engler, Das Pflanzenreich, 4(IV.101): 1–67, 1911.

## LAURACEAE. LAUREL FAMILY

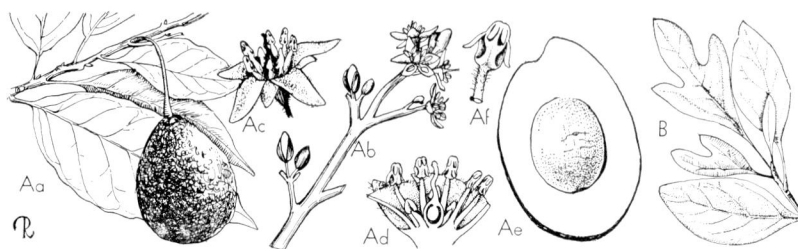

Fig. 147. Lauraceae. A, *Persea americana*: Aa, fruiting branch, × ⅙; Ab, twig with flowers, × ½; Ac, flower, × 1; Ad, same, vertical section, × 2; Ae, stamen (basal portion of filament removed), × 5; Af, fruit, vertical section, × ¼. B, *Sassafras albidum*: foliage, × ¼. (After L. H. Bailey, *Manual of Cultivated Plants,* The Macmillan Company, 1949.)

Mostly evergreen (deciduous in temperate regions), sometimes dioecious, trees or shrubs (*Cassytha,* a twining parasitic perennial herb with leaves reduced or absent), bark and foliage usually aromatic; leaves usually alternate, occasionally opposite or subopposite, simple, usually entire, mostly penninerved, usually punctate and coriaceous, estipulate; flowers in usually axillary, occasionally subterminal, panicles, spikes, racemes, or umbels; generally bisexual, sometimes unisexual, actinomorphic, mostly 3-merous, small, greenish yellowish or white, the perianth biseriate, of usually 6 basally connate usually undifferentiated sepallike segments, deciduous or persistent, the tube usually persisting as a cupule at base of fruit; the androecium typically of 4 whorls of 3 stamens each, adnate to perianth tube, the innermost usually reduced to staminodes (sometimes only a single whorl remains functional, and one or more whorls of staminodes may be absent), the filaments usually free, rarely those whorls united, the third whorl usually bearing a pair of usually sessile and distinct basal glandular protuberances (occasionally connate and disc-forming), the anthers basifixed, 2-celled or 4-celled at anthesis, those of the 2 outer whorls mostly introrse, the inner third whorl extrorse, dehiscing by flaplike valves opening upwards (when 4-celled, the valves superposed, except in *Nectandra* where the cells are arranged in an arc); pistil 1, the ovary usually superior, 1-loculed, the ovule solitary, anatropous, pendulous, the placentation parietal, the style 1, stigma 1, occasionally 2–3-lobed; fruit a drupe or berry usually surrounded at base by enlarged and often persistent perianth tube seated on an enlarged receptacle or pedicel; seed with large straight embryo, the endosperm absent.

A family of 45 genera and about 1100 species of forest trees and shrubs of mostly tropical southeastern Asia (Australia to Japan) and paleotropical America (species of 2 genera extend to Canada); a few members occur in Africa, and *Laurus nobilis* extends to the Mediterranean region of southern Europe. The family is represented here by indigenous species of 9 genera, all, with the exception of 1, in eastern parts of the country and mostly in the southeast: *Umbellularia,* a monotypic genus, occurs in California and Oregon; *Sassafras albidum* occurs east of the Mississippi from Florida to Ontario (a second species *S. tzumu,* indigenous to central China); *Lindera* (*Benzoin*), with about 60 species of temperate and subtropical Asia, is represented by 2 species, chiefly in eastern North America, (*L. Benzoin*) extending from Maine to Ontario south to Florida and Texas, and *L. melissaefolia* from North Carolina to Florida and west to Missouri; *Litsea,* the pond spice, by one of its nearly 200 mostly Asiatic species; *Persea* (60 spp.), chiefly an American genus, has 3 spp. native in the southeastern parts as far north as Virginia; *Octea* (250 spp.), a large pantropical genus, has 1 species native in peninsular Florida; *Nectandra* (about 100 spp.) and *Licaria* (about 50 spp.) both tropical American genera, each with a single species native in peninsular Florida; *Cassytha,* a genus of somewhat more than 25 spp., mostly of tropical Pacific range, has 1 pantropic species in Florida.

The Lauraceae are distinguished by the small, undifferentiated perianth, the trimerous stamens in several whorls, the valvate anther dehiscence, and the drupaceous fruit whose single seed lacks endosperm.

The family was divided by Pax (1889) into the subfamilies Perseoideae (anthers 4-celled dehiscing by 4 valves) and the Lauroideae (anthers usually 2-celled dehiscing by 2 valves). The genera *Cryptocarya, Lindera, Laurus,* and *Cassytha* belong to the presumably more advanced Lauroideae, with other genera in the Perseoideae. Hutchinson placed the family (together with the Monimiaceae, Hernandiaceae, and Myristicaceae *et al.*) in his Laurales, an order he considered as reduced from perhaps winteraceous ancestors of the Magnoliales.

Economically the family is important for the aromatic oils that are responsible for the fragrance of many of its members: the avocado (*Persea americana*), cinnamon and camphor (*Cinnamomum*), benzoin (*Lindera*), sassafras (*Sassafras*), and many fragrant woods used in cabinet work. Avocado growing is a major fruit industry in southern states. Species of about 7 genera of the family are cultivated domestically for ornament.

*LITERATURE:*

ALLEN, C. K. Studies in the Lauraceae, VI. Preliminary survey of the Mexican and Central American species. Journ. Arnold Arb. 26: 280–434, 1945.

KOSTERMANS, A. J. G. H. Revision of the Lauraceae, I-III. Meded. Bot. Mus. Utrecht, 25: 12–50, 1936; 42: 500–604, 1937; 43: 46–119, 1938.

PAX, F. Lauraceae. In Engler and Prantl, Die natürlichen Pflanzenfamilien, III(2): 106–126, 1891.

REHDER, A. The American and Asiatic species of *Sassafras.* Journ. Arnold Arb. 1: 242–245, 1920. [cf. also Fernald, in Rhodora, 38: 179, 1936; *op cit.* 47: 141, 1945.]

## HERNANDIACEAE. HERNANDIA FAMILY

Trees and shrubs (rarely lianous), monoecious in some genera; leaves alternate, palmately compound or simple, large, estipulate; flowers bisexual or unisexual, actinomorphic, in axillary corymbs or large thyrses (paniculate cymes), the perianth usually biseriate, the 4–8 segments of each series calyciform and sepallike; stamens 3–5, opposite outer sepals, the anthers 2-celled, dehiscing longitudinally or by 2 lateral valves, the glandlike staminodes often present in 1–2 whorls outside the stamens; pistil 1, the ovary inferior, 1-loculed, 1-carpelled, the ovule solitary, pendulous, the placentation probably parietal, the style 1, stigma 1; fruit basically an achene, incompletely enveloped by the expanded inflated receptacle, sometimes a 2–4-winged samara; seed with straight embryo and no endosperm.

A family of 4 genera and about 25 species of almost pantropical distribution. *Hernandia*, a monoecious genus of 14 species, is the largest. Other genera include *Gyrocarpus, Illigera,* and *Sparattanthelium*. None is native to this country.

The family lacks natural affinities and homogeneity, and of the few characters that bond all elements together or distinguish the family from others are the combination of inferior uniovulate ovary, the usually monoecious condition, the biseriate calyx, and the often winged samara. When present, the valvate anthers and glandlike staminodes are distinctive.

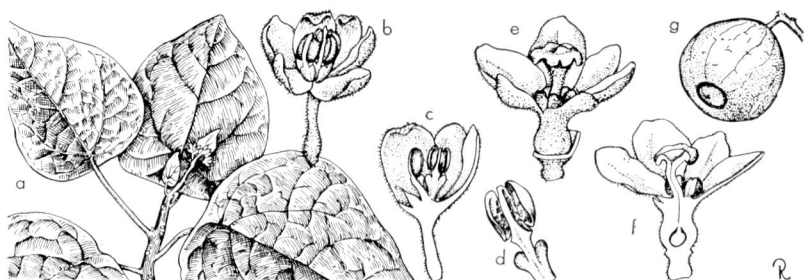

**Fig. 148.** HERNANDIACEAE. *Hernandia ovigera*: a, flowering branch, × ⅓; b, staminate flower, habit, × 2; c, same, vertical section, × 2; d, stamen, × 4; e, pistillate flower, habit, × 2; f, same, vertical section, × 2; g, fruit, × ⅜.

The Hernandiaceae have been accepted as a family by Wettstein, Engler and Diels, and by Hutchinson, each of whom placed it as near to but advanced over the Lauraceae. Bentham and Hooker and Hallier retained it within the Lauraceae. It is likely that its genera will be further segregated into additional families.

Species of *Hernandia* are cultivated for ornament and as specimen trees in southern Florida.

## Order 22. RHOEADALES

The Rhoeadales, as here circumscribed, are characterized by their usually herbaceous habit, the flowers bisexual and hypogynous with typically cyclic perianth parts and androecium, the gynoecium syncarpous with 2–many carpels and the placentation prevailingly parietal.

This order was considered by Engler and Diels to be composed of 7 families distributed among 5 suborders as follows (names of families not treated in this text are preceded by an asterisk):

    Rhoeadineae
      Papaveraceae
    Capparidineae
      Capparidaceae
      Cruciferae
      *Tovariaceae (trop. Amer.)

    Resedineae
      Resedaceae
    Moringineae
      Moringaceae
    Bretschneiderineae
      *Bretschneideraceae (southwest China)

The order has generally been accepted to include the families listed above, although not always of the same interrelationships (Hallier included the Moringaceae in the Leguminosae). Norris (1941) considered the Resedaceae and the Capparidaceae the most primitive families of the order, and noted that the Cruciferae, Fumariaceae, and Papaveraceae were "derived by subsequent parallel

# DIVISION IV. EMBRYOPHYTA SIPHONOGAMA

evolution from a common ancestral group somewhat resembling the existing Resedaceae and Capparidaceae." Similar views were held by Puri. The present placental situation is prevailingly parietal, but as Puri has shown for the Moringaceae, Cruciferae, and Capparidaceae this probably was derived from the axile type.

*LITERATURE:*

NORRIS, T. Torus anatomy and nectary characteristics as phylogenetic criteria in the Rhoeadales. Amer. Journ. Bot. 28: 101–113, 1941.
PURI, V. (for references, see under Capparidaceae, Cruciferae, and Moringaceae.)

## PAPAVERACEAE. POPPY FAMILY

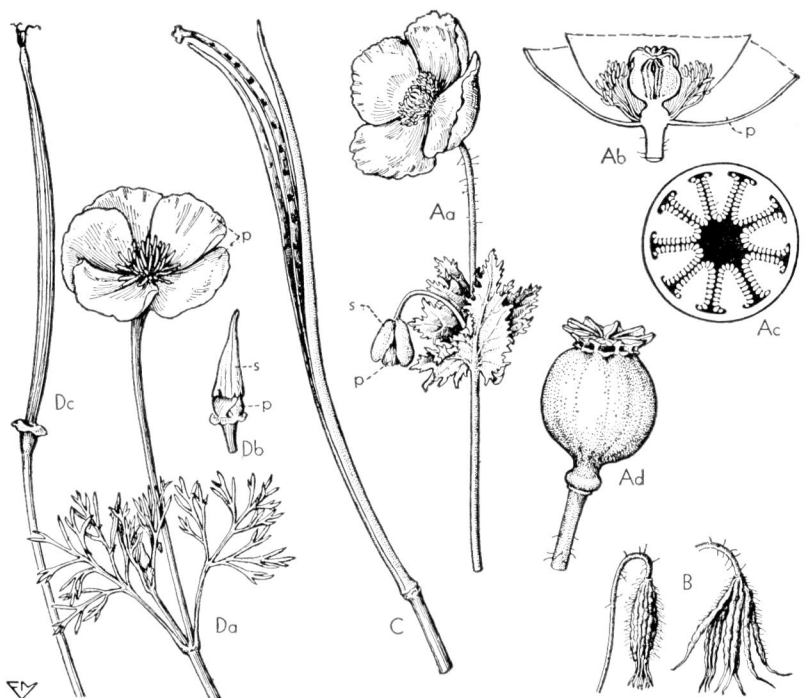

**Fig. 149.** PAPAVERACEAE. A, *Papaver somniferum*: Aa, flowering branch, × ¼; Ab, flower, vertical section, × ½; Ac, ovary, cross-section, × 2; Ad, capsule, × ½. B, *Platystemon californicus*: fruits, × 1. C, *Glaucium flavum*: capsule, × ½. D, *Eschscholzia californica*: Da, flowering branch, × ½; Db, bud, × ½; Dc, capsule, × ½. (s sepal, p petal.) (From L. H. Bailey, *Manual of cultivated plants*, The Macmillan Company, 1949. Copyright 1924 and 1949 by Liberty H. Bailey.)

Herbaceous annuals or perennials, rarely shrubs (*Dendromecon*) and very rarely trees (*Bocconia* a tree to 30 ft), sap usually milky or colored (watery in *Eschscholzia, Hunnemannia, et al.*); leaves alternate (uppermost ones whorled in Platystemoneae), entire to pinnately or palmately cleft, estipulate; flowers mostly solitary (paniculate in *Macleaya*), bisexual, actinomorphic, showy, the perianth biseriate or 3-seriate, the calyx of 2–3 distinct sepals (connate and calyptrate in

*Eschscholzia*), usually caducous, the corolla of 4–6 or 8–12 petals (absent in *Macleaya*), the petals in 1–2 whorls (rarely in 3), distinct, imbricate, often crumpled in bud; stamens numerous in several whorls (only 4 stamens in *Pteridophyllum, Hypecoum*), hypogynous, the filaments often alate and petaloid, the anthers 2-celled, dehiscing longitudinally; gynoecium of a single pistil (pistils several and coherent in *Platystemon*), the ovary superior, 1-loculed (rarely several-loculed by connation of intruding placentae or 2-loculed by a spurious septum), 2–many carpels, the ovules numerous on each placenta (solitary and basal in *Bocconia*), anatropous or campylotropous, the style usually 1 or obsolete (several in *Platystemon*), the stigmas as many as the carpels and alternate or opposite the placentae; fruit a capsule (follicular in *Platystemon*), dehiscing by pores or valves, rarely indehiscent; seed with minute embryo and copious oily or mealy endosperm.

A family of 28 genera and about 250 species, mostly of the subtropic and temperate regions of the northern hemisphere, with centers of distribution in western North America and eastern Asia (12 genera are strictly American, and 9 are Asiatic). The family is rare in the southern hemisphere. *Glaucium* (20 spp.) and *Papaver* (90 spp.) are primarily of the Mediterranean region, although the latter is represented by a few species in Asia and America. Other large genera include: *Meconopsis* (Asia, 45 spp.), *Hypecoum* (China, 15 spp.) and *Argemone* (Calif. to Mexico, 10 spp.).[82] The following genera, together with their number of species (*vide* Fedde), occur in the United States: [83] *Meconella* (including *Hesperomecon*) 13, *Romneya* 2, *Stylomecon* 1, *Arctomecon* 3, *Sanguinaria* 1, *Stylophorum* 1 (2 spp. in eastern Asia), *Chelidonium* 1, *Canbya* 2, and *Papaver* 1 (*P. californicum*).

Members of the Papaveraceae, as here delimited, are readily distinguished by the combination of actinomorphic bisexual flowers with caducous calyx, the usually crumpled corolla, the stamens generally numerous in several whorls, the unilocular compound ovary with parietal placentation, and the usually capsular fruit dehiscing by pores or flaplike valves. The presence of colored or milky sap is a reliable character when present.

There has been no serious challenge to the relative position of the family within the Rhoeadales. Undoubtedly *Platystemon* with its distinct pistils represents a primitive element, but the gynoecial situation, as shown by Arber (1933), is not fundamentally different from that encountered in other genera of the family. Basically, the ovary of the other genera is a compound structure derived from the marginal fusion of a single whorl of carpels, followed by the intrusion of the parietal placentae (each placenta composed of the margins of 2 adjacent carpels).[84] The division of the family into 3 subfamilies was accepted by Fedde (1936) on the basis of the following differences: Hypecoideae, flowers actinomorphic, stamens 4, carpels 2; Papaveroideae, flowers actinomorphic, stamens numerous, carpels 2–many; and Fumarioideae, flowers zygomorphic, stamens diadelphous in

---

[82] There is marked difference of opinion as to the number of species composing the Pacific coast genera; Fedde (1936) recognized the genera *Eschscholzia, Platystemon,* and *Dendromecon* to be composed of 120, 60, and 20 species respectively, whereas Abrams (1944) counted the species of those genera to be 12, 1, and 2 respectively. It appears that Fedde may have been influenced by the writings of E. L. Greene, who applied binomials freely to taxa not generally accepted as species, and that the concepts of Abrams may more closely approach the actual condition.

[83] Excluding genera already accounted for above.

[84] See papers by Dickson (1935) and Saunders (1937) for opposing views wherein it was held that the papaveraceous ovary consists of 2 whorls of carpels, that is that there are twice as many carpels as there are placentae, with half the number fertile and the other half sterile and comprising the median region of each of the fertile carpels. The views were rejected by Arber, Eames, Puri, and others as incompatible with the floral anatomy.

# DIVISION IV. EMBRYOPHYTA SIPHONOGAMA

3's, carpels 2. American authorities currently accept the segregation of the Fumarioideae as a distinct family (Fumariaceae), a view followed in this text.

The Papaveraceae are of economic importance for the opium of commerce, obtained from the sap of unripe capsules of *Papaver somniferum* (the alkaloid, morphine, is not present in the seeds sometimes ground and used as flour). Of secondary importance for their ornamental value are scores of species from about 20 genera, especially the Oriental poppy (*Papaver orientale*), Iceland poppy (*P. nudicaule*), bush poppy (*Dendromecon*), California poppy (*Eschscholzia*), celandine poppy (*Stylophorum*), Welsh and blue poppy (*Meconopsis*), prickly poppy (*Argemone*), tulip poppy (*Hunnemannia*), and plume poppy (*Macleaya*).

*LITERATURE:*

ARBER, A. Studies in flower structure IV. On the gynaeceum of *Papaver* and related genera. Ann. Bot. n.s. 2: 649–664, 1938.
DICKSON, JEAN. Studies in floral anatomy. II. The floral anatomy of *Glaucium flavum* with reference to other members of the Papaveraceae. Journ. Linn. Soc. Lond. (Bot.) 50: 175–224, 1935.
FEDDE, F. Papaveraceae-Hypecoideae et Papaveroideae. In Engler, Das Pflanzenreich, 40 (IV. 104): 1–430, 1909.
———. Papaveraceae. In Engler and Prantl, Die natürlichen Pflanzenfamilien, ed. 2, Bd. 17b: 1–145, 1936.
HUTCHINSON, J. *Bocconia* and *Macleaya*. Kew. Bull. 275–282, 1920.
———. Contributions towards a phylogenetic classification of flowering plants. Kew Bull. 161–168, 1925.
PRAIN, D. A revision of the genus *Chelidonium*. Bull. Herb. Boissier, 3: 570–587, 1895.
SAUNDERS, E. R. Floral Morphology. 2 vols. Cambridge, 1937, 1939.
SOUÉGES, R. Embryogénie des Papaveracées. Developpement de l'embryon chez le *Roemeria violacea* Medic. (*R. hybrida DC.*) Paris, Acad. des Sci. Compt. Rend. 226: 979–981, 1948.
TAYLOR, G. and COX, E. H. M. An account of the genus *Meconopsis*. 130 pp. London, 1934.

## FUMARIACEAE. FUMITORY FAMILY

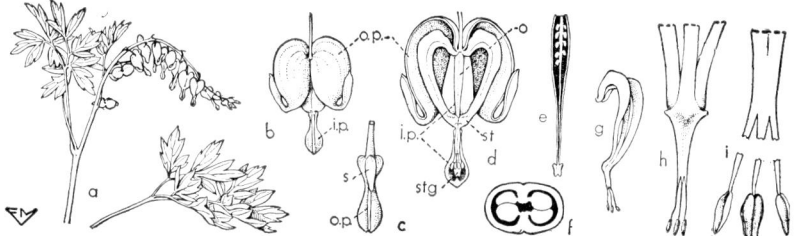

Fig. 150. FUMARIACEAE. *Dicentra spectabilis*: a, portion of plant in flower, × ⅛; b, flower, habit (less calyx), × ½ (i.p. inner petal, o ovary, o.p. outer petal, s sepal, st stamen, stg stigma); c, bud, showing calyx, × 1; d, flower, one outer petal removed, × ¾; e, pistil, vertical section, × 1; f, ovary, cross-section, × 8; g, three connate stamens, × ¾; h, same, distal end, × 2; i, same, showing nature of anthers, × 4. (After L. H. Bailey, *Manual of Cultivated Plants,* The Macmillan Company, 1949.)

Herbaceous plants with watery sap, sometimes lianous; leaves alternate, in basal rosettes or cauline, rarely subopposite, usually much-divided or dissected; flowers bisexual, transversely zygomorphic, usually racemose, the perianth triseriate, the calyx of 2 minute caducous sepals; corolla of 4 more or less coherent and sometimes basally connate petals in 2 whorls, one (as in *Corydalis*) or both of the 2 outer petals usually basally saccate or spurred, the inner ones narrower, crested and united over the anthers; stamens 6,[85] three on each side of the pistil, the filaments somewhat winged and coherent to connate for much of their length, 1 or 2 nectar glands usually present at base of androecium; pistil 1, the ovary superior, 1-loculed, 2-carpelled, the placentation parietal, the ovules 2–many, the style 1 and slender, stigma 1, sometimes 2-lobed, stigmatic surfaces 2, 4, or 8; fruit a transversely septate capsule dehiscing by valves, or an indehiscent 1-seeded nut (*Fumaria*); seeds with minute embryo and copious soft watery-fleshy endosperm, the cotyledons 1 (some spp. of *Corydalis*) or 2.

A family of 19 genera and about 425 species (*vide* Fedde) distributed mostly in the Old World and primarily in temperate Eurasia. Four genera (with a total of 7 spp.) occur in South Africa. Three genera have indigenous representatives in this country: *Adlumia* (monotypic) a biennial vine of the northeast with tripinnate leaves; *Dicentra* (*Bicuculla*), a genus of perhaps 300 species of north temperate regions and South Africa, by 4 spp. in the Pacific states and 4 additional species in the eastern states, and *Corydalis* widely distributed with 10 indigenous species. Aside from *Corydalis*, the only other large genera are *Fumaria* (50 spp.) and *Rupicapnos* (30 spp.), both primarily of the Mediterranean region. Only *Dactylicapnos* (8 spp.) is exclusively Asiatic.

The circumscription of this family has been variously interpreted. Most European botanists (including Engler and Diels) have treated it as a subfamily of the Papaveraceae. Hutchinson recognized it as a family but included with it the Hypecoideae, here retained in the Papaveraceae. Fedde, the most recent monographer of the group, placed the Fumariaceae within the Papaveraceae. For purposes of convenience, it seems desirable to treat it as a distinct family (excluding the Hypecoideae), distinguished by the zygomorphic corolla, distinctive androecium of 6 stamens, and the closed flowers with coherent to partially connate petals.

The family is of little economic importance except for the few ornamental species, bleeding heart (*Dicentra officinalis*) the most important.

*LITERATURE:*

FEDDE, F. Papaveraceae, Unterfamilie III. Fumarioideae. In Engler and Prantl, Die natürlichen Pflanzenfamilien. ed. 2, Bd. 17b: 121–145, 1936.

HUTCHINSON, J. The genera of Fumariaceae and their distribution. Kew Bull. 97–115, 1921.

NORRIS, T. Torus anatomy and nectary characteristics as phylogenetic criteria in the Rhoeadales. Amer. Journ. Bot. 28: 101–113, 1941.

OWNBEY, G. B. Monograph of the North American species of *Corydalis*. Ann. Mo. Bot. Gard. 34: 187–259, 1947.

## CAPPARIDACEAE. CAPER FAMILY

Herbs, shrubs, or trees, sometimes lianous, without latex; leaves alternate (rarely opposite), simple or palmately compound (frequently unifoliolate), the stipules minute, glandular to spinose, or absent; flowers bisexual or unisexual (the plants then monoecious, as in *Podandrogyne*), actinomorphic or more often zygomorphic,

---

[85] The various interpretations of the androecial situation in the Fumariaceae (*sensu* Hutchinson) have been reviewed by Norris (1941, pp. 108–109). In those genera possessing an androecium of 6 anthers (2 dithecal and 4 monothecal), the stamens are arranged in 2 groups of 3 each on opposite sides of the pistil, with 1 monothecal stamen on each side of a dithecal stamen. Norris believed that the stamens are basically in 2 whorls, the 4 monothecal in the inner and the 2 dithecal stamens in the outer whorl; he hypothesized that the present situation was derived from ancestral forms having 8 monothecal stamens, the 2 present outer dithecals representing the fusion of 4 monothecals. Likewise, the 4 dithecal stamens now present in genera such as *Hypecoum* are believed to represent fusion of all 8 ancestral monothecals into 4 pairs.

generally in racemes, bracteate but lacking bracteoles, the perianth usually biseriate (corolla wanting in a few genera), the sepals 4–8 and usually 4, distinct or basally connate or wholly connate (the calyx when calyptrate dehiscing transversely at anthesis and falling as a hood), the petals 4–many (sometimes absent), equal or the 2 posterior ones the larger, clawed or sessile, distinct (connate in *Emblingia*); stamens 4–many (when many, the androecium derived from tetrandrous condition by splitting of the 4 primordia and then many filaments often lacking anthers), the anthers 2–4-celled, dehiscing longitudinally, sometimes the androecium and gynoecium borne on an internodal development termed the androgynophore; pistil 1, usually borne on a short or long gynophore, the ovary superior, 1-loculed (sometimes falsely bilocular), the carpels typically 2 but sometimes 4 (the primitive number), the placentation typically parietal, the ovules few to many on each placenta, campylotropous, the style 1, short to filiform and elongate, the stigma usually 2-lobed or capitate; fruit a capsule dehiscing by valves or an often elongate or torulose berry, sometimes an indehiscent nut; seeds usually reniform, the embryo curved, the endosperm present and fleshy.

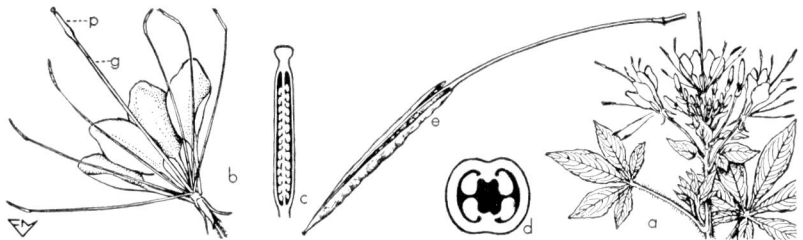

**Fig. 151.** CAPPARIDACEAE. *Cleome spinosa*: a, flowering branch × ⅙; b, flower, × ½ (g gynophore, p pistil); c, pistil, vertical section, × 2; d, ovary, cross-section, × 10; e, capsule, × ½. (From L. H. Bailey, *Manual of cultivated plants*, The Macmillan Company, 1949. Copyright 1924 and 1949 by Liberty H. Bailey.)

The Capparidaceae are composed of 46 genera and about 700 species, distributed paleotropically in both hemispheres. Three of the genera, including the 2 largest, are pantropical (*Capparis*, 350 spp., *Cleome*, 200 spp., and *Crataeva*, 20 spp.). Fifteen genera are restricted to Africa and account for one-third of the species, 3 genera are Australian, 15 are indigenous to the New World, and the others are Eurasian. The 7 genera with species indigenous to this country are: *Cleome, Cleomella, Polanisia* (united with *Cleome* by Pax and Hoffmann, 1936), *Isomeris Wislizenia, Oxystylis,* and *Cristatella*. All but *Cristatella* (which occurs from Texas northward to Nebraska) are widespread in distribution, but are more common in dry arid habitats.

Members of the family may be recognized by the presence of the gynophore (the androgynophore is distinctive when present), the usually zygomorphic flowers with 6 to many stamens, and the uniloculate ovary with parietal placentation. It is distinguished from the Cruciferae by the nontetradynamous stamens, the 1-loculed ovary, and the usually zygomorphic flower.

Phylogenists are agreed that the Capparidaceae and Cruciferae are 2 closely related families, and evidence seems to indicate that either each has been derived from a common ancestor, or that the Cruciferae may have arisen from a primitive member of the Capparidaceae. Both have an apparently 2-carpelled ovary and parietal placentation. Puri (1945) was of the opinion that this has been derived from a quadrilocular ovary that was 4-carpelled and had axile placentation.

Economically the family is important for capers used in seasoning foods (capers, the dried flower buds of *Capparis spinosa* of the Mediterranean). The spider flower (*Cleome spinosa*) is a popular garden annual. Other genera whose species are cultivated domestically for ornament are *Capparis, Gynandropsis,* and *Polanisia.*

LITERATURE:

ORR, M. Y. Observations on the structure of the seed in the Capparidaceae and Resedaceae. Notes Roy. Bot. Gard. Edinb. 12: 259–260, 1921.

PAX, F. and HOFFMANN, K. Capparidaceae. In Engler and Prantl, Die natürlichen Pflanzenfamilien. ed. 2, Bd. 17b: 146–223, 1936.

PAYSON, E. B. A synoptical revision of the genus *Cleomella.* Univ. Wyoming Publ. in Sci. 1: 29–46, 1922.

PURI, V. Studies in floral anatomy. III. On the origin and orientation of placental strands. Proc. Nat. Acad. Sci. India, 15(3): 74–91, 1945.

———. Studies in floral anatomy. VI. Vascular anatomy of the flower of *Crateva religiosa,* with special reference to the nature of the carpels in the Capparidaceae. Amer. Journ. Bot. 37: 363–370, 1950.

RAGHAVAN, T. S. Studies in the Capparidaceae, I. The life history of *Cleome Chelidonii,* L. f. Journ. Linn. Soc. Lond. (Bot.) 51: 43–72, 1937.

———. Studies in Capparidaceae, II. The floral anatomy and some structural features of the Capparidaceous flower. Journ. Linn. Soc. Lond. (Bot.) 52: 239–257, 1939.

STOUDT, H. N. The floral morphology of some of the Capparidaceae. Amer. Journ. Bot. 28: 664–675, 1941.

WOODSON, R. E. JR. *Gynandropsis, Cleome,* and *Podandrogyne.* Ann. Mo. Bot. Gard. 35: 139–146, 1948.

## CRUCIFERAE.[86] MUSTARD FAMILY

Annual, biennial, or perennial herbs, rarely subshrubs, sap watery, forked or stellate unicellular hairs commonly present; leaves alternate or rarely subopposite, simple, estipulate; flowers bisexual, actinomorphic (zygomorphic in *Iberis*), typically racemose, usually ebracteate, the perianth triseriate, calyx of 4 distinct sepals in 2 whorls, the corolla of 4 distinct usually clawed petals (absent in a few genera), diagonally disposed (cruciform), the androecium typically of 6 stamens in 2 whorls (stamens 16 in the Asiatic *Megacarpaea*), usually tetradynamous, the pair of the outer whorl in a transverse plane alternating with the 2 longer pairs of the inner whorl in the median plane, filaments of each inner pair sometimes connate, those of outer pair sometimes winged and/or toothed, 1, 2, or 4 basal glands often present, the anthers 2-celled (rarely 1-celled), dehiscing longitudinally; pistil 1, the ovary superior, usually 2-loculed by means of a false but complete septum, sessile (rarely stipitate), 2- (or 4-) carpelled,[87] the placentation parietal, the ovules usually few to many, borne on periphery of ovary wall in angle formed

---

[86] Some authors adopt for this family the alternative name Brassicaceae Lindl. (1836) or Raphanaceae Horaninov (1847).

[87] The carpellary situation has been interpreted variously. The views of Puri (1945) provide a partial interpretation of the problems raised by the complex floral morphology, and those by Eames and Wilson provide another interpretation. In essence, Puri has contended that the present-day 2-loculed situation with parietal placentation has resulted from a 4-loculed ancestral type with axile placentation. Furthermore he has given an explanation for the evolution of the false septum (replum) that separates the 2 locules. It should be noted that these views of 1945 represent a revised opinion by Puri and not that held by him in 1941. Eames and Wilson (1930) held that the modern crucifer ovary has been derived from 4 carpels, 2 fertile ones in the inner whorl and 2 sterile ones representing an outer whorl; and that the replum is the fused walls of the 2 fertile carpels with their ovules now in the locules of the sterile carpels. See Fig. 152 L and M.

by union of false septum and ovary wall, the style 1 or obsolete, the stigmas 2, their commissural face usually at right angles to the septum (except in *Moricanda, Matthiola*, spp. of *Thelypodium, Streptanthus,* and others); fruit usually valvately dehiscent and then a silique (elongated) or silicle (short), sometimes an indehiscent 1-few-seeded nut; seed with large embryo and little or no endosperm, the cotyledons accumbent (edges to the radicle), incumbent (dorsal side to radicle), or convolute (dorsal side folded over radicle).

A large family of 350 genera and about 2500 species [88] distributed primarily in the northern hemisphere and particularly in its cooler regions. Ten genera are cosmopolitan and nearly so, and include *Draba* 270 spp., *Cardamine* 130 spp., *Lepidium* 130 spp., *Sisymbrium* 80 spp., and *Thlaspi* 60 spp. The genera *Arabis* 100 spp., *Erysimum* 80 spp., and *Barbarea* 12 spp. are widely distributed in the northern hemisphere. The remaining genera are distributed numerically somewhat as follows: Asia 107, the Mediterranean region 48, United States and Canada 48, Central and South America 42, North Africa 33 (exclusive of genera of general Mediterranean distribution), South Africa 8, Europe (exclusive of the Mediterranean) 27, Australia and New Zealand 17, Mexico 8. Within the United States, 35 genera are restricted to the western states, 7 to Texas and adjoining states; only 1 (*Leavenworthia*) is indigenous only to the eastern part of the country.

The family is characterized (with exceptions as noted above) by the presence of 4 sepals, 4 petals diagonally disposed, the usually tetradynamous stamens, the single sessile and seemingly bilocular ovary with parietal placentation, the distinctive fruit (silique or silicle), and the cotyledons in 1 of 3 basic positions with reference to the radicle.

The phyletic position of the Cruciferae is accepted generally as belonging in this order. However, there is disagreement as to whether it has been derived from papaveraceous ancestors (as held by Bentham and Hooker and by Hutchinson, 1948) or from capparidaceous ancestors. On the basis of the androecial and gynoecial morphology and anatomy, the preponderance of evidence favors the capparidaceous alliance.[89]

Economically the family is of considerable importance for the food crops, weeds, and ornamentals that belong to it. Among the important food crops are cabbage, cauliflower, broccoli, rutabaga, kohlrabi, turnips, and Brussels sprouts (all from *Brassica*), radish (*Raphanus*), and watercress (*Nasturtium*). Weeds of economic significance include mustards (*Brassica, Barbarea*), shepherd's-purse (*Capsella*) and peppergrass (*Cardaria, Lepidium*). Two condiments, mustard (*Brassica*) and horse-radish (*Armoracia*) belong here. Among the ornamentals (about 50 genera are cultivated domestically) are stocks (*Matthiola*), rocket (*Hesperis*), candytuft (*Iberis*), wallflower (*Cheiranthus, Erysimum*), honesty (*Lunaria*), sweet alyssum (*Lobularia*), basket-of-gold (*Alyssum*), rock cress (*Arabis*).

*LITERATURE:*

ARBER, A. Studies in floral morphology. I. On some structural features of the cruciferous flower. New Phytologist, 30: 11–46, 1931. II. On some normal and abnormal crucifers, with a discussion on teratology and atavism; *op. cit.* 317–354, 1931.

BAILEY, L. H. The cultivated Brassicas. Gentes Herb. 1: 53–108, 1922; *op. cit.* 2: 211–267, 1930.

———. Certain noteworthy Brassicas. Gentes Herb. 4: 319–330, 1940.

[88] Data as to size, number, and distribution of genera taken from Schulz (1936).

[89] For the abundant literature on the subject of the morphology and phylogeny of the Cruciferae, the student is referred to the papers of Arber, Eames, Eggers, Hayek, Puri, and Saunders.

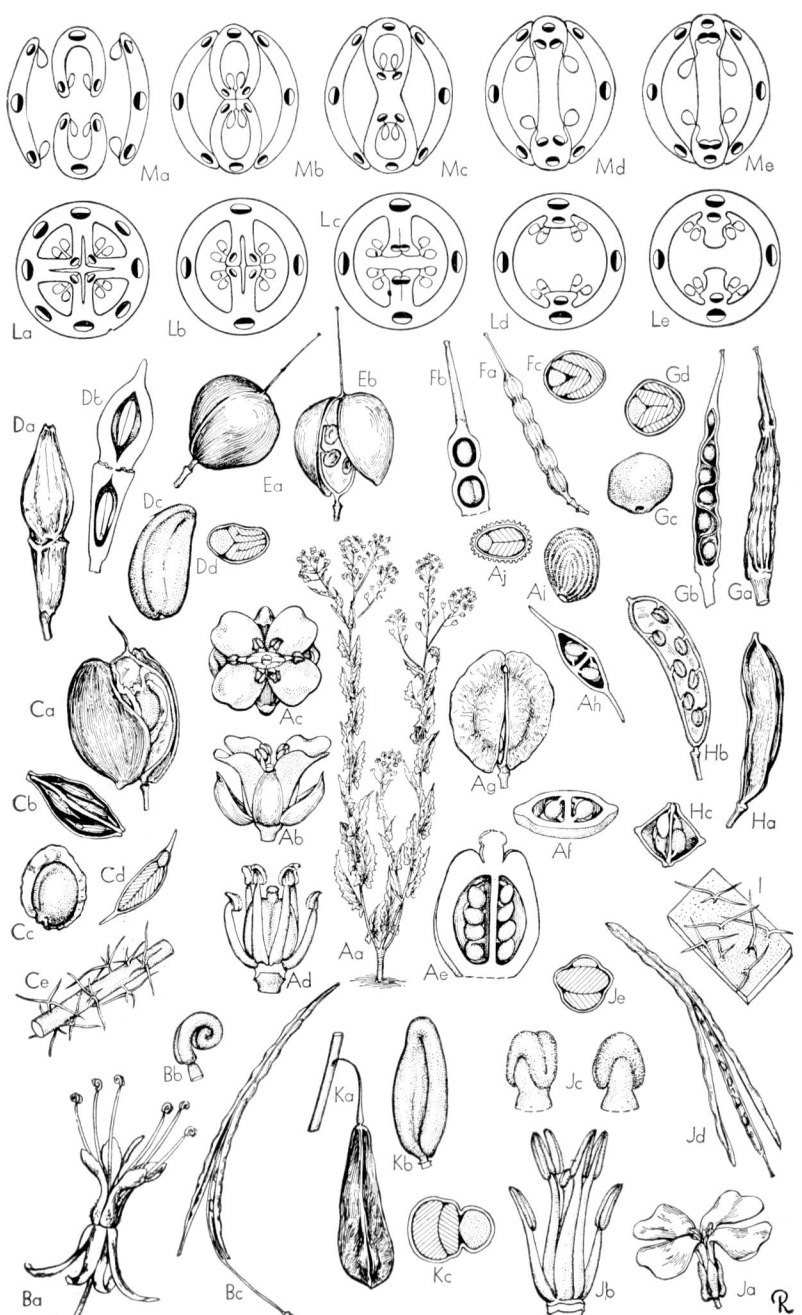

522

BALDWIN, J. T. JR. Chromosomes of Cruciferae II. Cytogeography of *Leavenworthia.* Bull. Torrey Bot. Club, 72: 367–378, 1945.
DETLING, LEROY E. The Pacific coast species of *Cardamine.* Amer. Journ. Bot. 24: 70–76, 1937.
———. A revision of the North American species of *Descourania.* Amer. Midl. Nat. 22: 481–520, 1939.
EAMES, A. J. and WILSON, C. L. Carpel morphology in the Cruciferae. Amer. Journ. Bot. 15: 251–270, 1928.
———. Crucifer carpels. Amer. Journ. Bot. 17: 638–656, 1930.
EGGERS, O. Über die morphologische Bedeutung des Leitbundelverlaufes in der Blüten der Rhoeadalen und über des Diagramm der Cruciferen und Capparidaceen. Archiv. Wiss. 24: 14–58, 1935.
FERNALD, M. L. *Draba* in temperate northeastern America. Rhodora, 36: 241–261, 285–305, 314–344, 353–371, 392–404, 1934. [Reprinted as Contrib. Gray Herb. no. 105.]
HAYEK, AUGUST VON. Entwurf eines Cruciferen-Systems auf phylogenetischer Grundlage. Bot. Centralbl., Beih. 27 (127): 127–355, 1911.
HITCHCOCK, C. L. The genus *Lepidium* in the United States. Madroño, 3: 265–320, 1936
———. A revision of the Drabas of western North America. Univ. Wash. Publ. Biology, 11: 1–132, 1941.
HOPKINS, M. *Arabis* in eastern and central North America. Rhodora, 39: 63–98, 106–148, 155–186, 1937. [Reprinted as Contrib. Gray Herb. no. 116.]
JANCHEN, E. Das System der Cruciferen. Oesterr. Bot. Zeits. 91: 1–28, 1942.
JARETZKY, R. Untersuchungen über Chromosomen und Phylogenie bei einigen Cruciferen. Jahrb. Wiss. Bot. 68(1): 1–45, 1928. [For same title, see also Jahrb. Wiss. Bot. 76: 485–527, 1932.]
MANTON, I. Introduction to the general cytology of the Cruciferae. Ann. Bot. 46: 509–556, 1932.

---

Fig. 152. CRUCIFERAE. A, *Thlaspi arvense*: Aa, habit of plant, × 1/6; Ab, flower, habit, × 4; Ac, flower, top view, × 4; Ad, flower, less perianth, × 5; Ae, pistil, vertical section transverse to replum, × 10; Af, ovary, cross-section, × 10; Ag, silicle, side view, × 1; Ah, silicle, cross-section, × 1½; Ai, seed, × 5; Aj, seed, cross-section (cotyledons accumbent). B, *Stanleya pinnata*: Ba, flower, habit, × 1; Bb, anther, × 4; Bc, silique, × 1. C, *Alyssum saxatile*: Ca, silicle (dehiscing), × 5; Cb, silicle, cross-section, × 5; Cc, seed, winged, × 5; Cd, seed, cross-section, × 8; Ce, stellate hairs, × 2. D, *Cakile maritima*: Da, fruit, habit, × 1½; Db, fruit, vertical section, × 1½; Dc, seed, × 3; Dd, seed, cross-section, × 3. E, *Lesquerella ovalifolia*: Ea, silicle, habit, × 2; Eb, silicle, dehiscing, × 2. F, *Raphanus raphanistrum*: Fa, silique, habit, × 1; Fb, silique, vertical section of distal end, × 2; Fc, seed, cross-section (cotyledons conduplicate), × 4. G, *Brassica arvensis*: Ga, silique, habit, × 1½; Gb, silique, vertical section, × 1½; Gc, seed, × 6; Gd, seed, cross-section, × 6. H, *Descourainia pinnata*: Ha, silique, habit, × 4; Hb, silique, vertical section, × 4; Hc, silique, cross-section, × 8. I, *Draba rupestris*: leaf section with mostly forked hairs, × 4. J, *Hesperis matronalis*: Ja, flower, habit, × 1; Jb, flower, less perianth, × 3; Jc, stigma, side and edgewise views, × 8; Jd, silicle, dehiscing, × ½; Je, seed, cross-section (cotyledons incumbent), × 8. K, *Isatis tinctoria*: Ka, fruit, habit, × 1½; Kb, seed, × 6; Kc, seed, cross-section (cotyledons incumbent), × 12. L, cross-sectional diagrams to illustrate the *Puri theory* (1945) for hypothetical derivation of crucifer type ovary: La, 4-loculed (and -carpelled) presumed ancestral type with axile placentation as in some Moringaceae, note inverted placental strands; Lb, reduction to bilocular and bicarpellate condition, with axile placentation, no ovule reduction; Lc, each axile placenta with components has split and fused to adjacent halves of intermediate (and "lost") carpels, note fusion of inverted placental strands; Ld, carpel margins shortened and placental strands moved peripherally, close to median marginal strand; Le, sterile tissue intruding from each peripheral and parietal placenta toward center producing, on fusion, a "false septum" or replum (as in Af or Cb). M, cross-sectional diagrams to illustrate the *Eames and Wilson theory* (1930) for the derivation of crucifer type ovary: Ma, two whorls of two open fertile carpels; Mb, the two inner carpels closed and connate, with the two outer carpels open and appressed against the inner; Mc, the two outer carpels sterile by loss of ovules, the inner two with smaller locules by contraction, the ventral bundles retreating from their marginal position; Md, the locules of inner carpels lost by compression, the ovules pushed outside into the locules of the sterile carpels, the inverted ventral bundles near the dorsal bundle; Me, same, the replum thinner and ventral bundles fused. (A-K from L. H. Bailey, *Manual of cultivated plants*, The Macmillan Company, 1949. Copyright 1924 and 1949 by Liberty H. Bailey.)

Payson, E. B. Monograph of the genus *Lesquerella*. Ann. Mo. Bot. Gard. 8: 103–236, 1921.
——. A monographic study of *Thelypodium* and its immediate allies. Ann. Mo. Bot. Gard. 9: 233–324, 1922.
——. Species of *Sisymbrium* native to North America north of Mexico. Wyo. Univ. Publ. Bot. 1: 1–27, 1922.
Puri, V. Studies in floral anatomy. I. Gynaecium constitution in the Cruciferae. Proc. Indian Acad. Sci. 14: 166–187, 1941.
——. Studies in floral anatomy. III. On the origin and orientation of placental strands. Proc. Nat. Acad. Sci., India 15: 74–91, 1945.
Rollins, R. C. The cruciferous genus *Stanleya*. Lloydia, 2: 109–127, 1939.
——. The cruciferous genus *Physaria*. Rhodora, 41: 392–415, 1939.
——. A monographic study of *Arabis* in western North America. Rhodora, 43: 289–325, 348–411, 425–481, 1941. [Reprinted as Contr. Gray Herb. no. 138.]
——. Some generic relatives of *Capsella*. Contrib. Dudley Herb. 3: 185–198, 1941.
——. Generic revisions in the Cruciferae: *Sibara*. Contr. Gray Herb. 165: 133–143, 1947.
Schulz, O. E. Monographie der Gattung *Cardamine*. Engler Bot. Jahrb. 32: 280–623, 1903.
——. *Draba*. In Engler, Das Pflanzenreich, Heft 89: 1–343, 1927.
——. Cruciferae. In Engler and Prantl, Die natürlichen Pflanzenfamilien, ed. 2, Bd. 17b: 227–658, 1936.
Smith, F. H. Some chromosome numbers in the Cruciferae. Amer. Journ. Bot. 25: 220–221, 1938.
Sun, Von Gee. The evaluation of taxonomic characters of the cultivated Brassicas with a key to species and varieties. I-II. Bull. Torrey Bot. Club, 73: 244–281, 370–377, 1946.
Zohary, M. Follicular dehiscence in Cruciferae. Lloydia, 11: 226–228, 1948 [1949].

## RESEDACEAE. MIGNONETTE FAMILY

**Fig. 153.** Resedaceae. *Reseda odorata*: a, flowering branch, × ⅛; b, inflorescence, × ½; c, flower, × 3; d, same, vertical section, × 7; e, ovary, cross-section, × 7; f, capsule, × 1. (From L. H. Bailey, *Manual of cultivated plants*, The Macmillan Company, 1949. Copyright 1924 and 1949 by Liberty H. Bailey.)

Annual or perennial herbs, suffrutescent or (in the north African *Ochradenus* and *Randonia*) shrubs, sap watery; leaves alternate, simple or pinnately divided, stipules glandlike and minute; flowers bisexual or infrequently unisexual (in some spp. of several genera, the plants then usually monoecious), zygomorphic, in racemes or spikes, perianth biseriate, the sepals (and petals) usually 4–8, valvate in bud, distinct, the petals sometimes lacking or only 2, the 2 posterior ones usually much the larger and often laciniately lobed; stamens 3–40, usually borne on a unilaterally expanded disc, the anthers 2-celled, dehiscing longitudinally, introrse; pistil 1, the ovary superior, 1-loculed, the carpels 2–6, distinct or marginally connate, usually open at the top and each with its own stigma, the ovules numerous, anatropous, the placentation parietal, the stigmas as many as carpels, styles none; fruit a capsule or berry; seeds reniform with curved embryo and the endosperm present and fleshy.

A family of 6 genera and about 70 species, of which about 60 species belong to the primarily Mediterranean *Reseda*. Several species of *Reseda* are adventive from southern Europe in waste places of much of this country. *Oligomeris*, with flowers having only 2 petals and no disc, is primarily African, but 1 species (*O. linifolia*) is widely distributed as an indigen of somewhat arid and saline habitats of the warmer parts of the northern hemisphere and in this country extending northward from Mexico into western Texas west to southern California. The remaining 4 genera occur in southern Europe, North Africa, and southwestern Asia.

Members of the family are usually distinguished by the combination of characters as represented by the glandular stipules, the zygomorphic flowers, the often lobed petals, and the open unilocular ovary each of whose terminal lobes (carpel tips) is terminated by a minute stigma, and by the capsule usually open apically

The Resedaceae are undoubtedly a phyletically advanced family within the order, as treated by Engler, Rendle, and Bessey. Hutchinson, following Bentham and Hooker, aligned them with the Violaceae. The presence of a third and innermost seed coat in species of both Resedaceae and Capparidaceae was concluded by Orr (1921) to provide added evidence of the close relationship between these 2 families. Arber (1942) concluded that despite the open character of the pistil apex, the presence of pollen-tube conveying canals from stigmas to placentas provided evidence that the gynoecium is angiospermous.

Economically the family is of little importance. Mignonette (*Reseda odorata*) long has been cultivated for the fragrance of its flowers, and dyer's weed (*R. luteola*) was once much grown as a source of a yellow dye used in the textile industry.

*LITERATURE:*

ARBER, A. Studies in flower structure VII. On the gynaecium of *Reseda*, with a consideration of paracarpy. Ann. Bot. n.s. 6: 43–48, 1942.
BOLLES, F. Resedaceae. In Engler and Prantl, Die natürlichen Pflanzenfamilien, ed. 2, Bd. 18a: 659–692, 1930.
EIGSTI, O. J. Cytological studies in the Resedaceae. Bot. Gaz. 98: 363–369, 1926.
HENNIG, L. Beiträge zur Kenntnis der Resedaceen-Blüte und Frucht. Planta, 9: 507–563, 1929.
ORR, M. Y. Observations on the structure of the seed in the Capparidaceae and Resedaceae. Notes Roy. Bot. Gard. Edinb. 12: 259–260, 1921.

## MORINGACEAE. MORINGA FAMILY

Deciduous trees; leaves alternate, 2–3 pinnately compound with opposite pinnae, the stipules and stipels reduced to basal glands or none; flowers bisexual, zygomorphic, in hairy axillary cymose panicles, the perianth biseriate, the sepals 5 and reflexed, borne on a very short hypanthium [90] that envelops a saucerlike disc,

[90] The flower is described throughout most of the literature as having a short calyx tube and 5 calyx lobes. It was shown by Puri (1942) that the "floral cup" is in part receptacular and in part appendicular: the basal and posterior portions are an invagination of the receptacle and the rim and anterior portion represent the fused bases of sepals, petals, and stamens. The "cup" is of the *Rosa* type, and for taxonomic purposes is better designated an hypanthium.

In his study of the carpellary situation, Puri showed the 2 rows of ovules on each placenta to represent those of the 2 ventral margins of the same carpel that had become folded inward, rather than 1 row from each of 2 adjoining carpel margins as is typical of normal parietal placentation. For this reason, and because of the "solidifying" of the carpels, he designated the placentation extracarpellary, and the cavity within the ovary an ovarian chamber, instead of a locule; distinctions of morphological rather than of taxonomic significance.

unequal and imbricated, the petals 5, distinct, borne on the hypanthium, the 2 posterior ones smaller and reflexed with laterals ascending and the anterior one larger; androecium of 5 functional declinate stamens of unequal length alternating with an outer whorl of 3–5 filiform or setiform staminodes, borne on disk rim, the filaments distinct, the anthers 1-celled, dehiscing longitudinally, forming a head through which the style protrudes at anthesis; pistil 1, the ovary superior, stipitate, usually curved and villous, 1-loculed, 3-carpelled, the ovules numerous, in 2 rows on each of 3 parietal placentae, anatropous, pendulous, the style 1 and slender, the stigma 1 and truncate; the fruit an elongated siliquelike triquetrous capsule, 3-valved, rostrate; seeds many, large, ovate, 3-winged or wingless, the embryo straight, the endosperm absent.

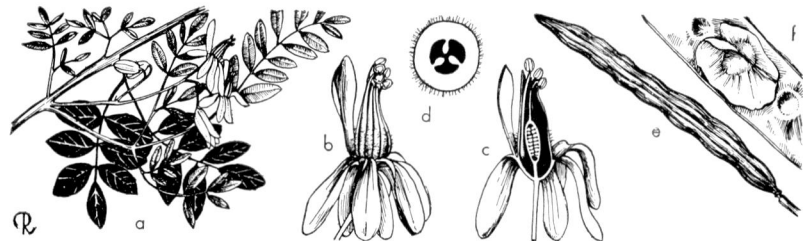

Fig. 154. MORINGACEAE. *Moringa oleifera*: a, flowering branch, × ½; b, flower, × 1; c, same, vertical section, × 1; d, ovary, cross-section, × 5; e, fruit, × ⅙; f, seed on capsule-valve, × ½. (From L. H. Bailey, *Manual of cultivated plants*, The Macmillan Company, 1949. Copyright 1924 and 1949 by Liberty H. Bailey.)

A unigeneric family (*Moringa*), considered by Philips (1926) to have 4 species and by Pax (1936) and Puri (1942) to have 10 species, native in the Old World tropics; adventive in the New World tropics.

Members of the family are distinguished readily by the alternate decompound leaves with opposite pinnae, the pentamerous zygomorphic flowers with 5 stamens and 3–5 alternating staminodia borne on a short hypanthium, and the long 3-valved fruit with often winged seeds.

Bessey placed the family in the Rhoeadales. Wettstein did likewise, treating it as doubtfully near the Resedaceae. Hutchinson included it with the Capparidaceae in his Capparidales, while Datta and Mitra (1947) considered it most closely related to the Violaceae (accepting Hutchinson's concept and disposition of the Violales). On the basis of new carpellary evidence, presumably not known to Datta and Mitra, Puri (1942) considered the family to be characterized in part by its complex carpellary situation, known elsewhere only in other families of the Rhoeadales (Cruciferae and Capparidaceae).

Economically the family is important domestically for the horse-radish tree (*Moringa oleifera*), cultivated for ornament and its edible fruits in southern Florida and southern California. The roots are a source of an edible condiment. Oil of ben, a nondrying oil, is obtained from the plant.

*LITERATURE:*

DATTA, R. M. and MITRA, J. N. The systematic position of the family Moringaceae based on a study of *Moringa pterygosperma* Gaertn. (*M. Oleifera* Lam.). Journ. Bombay Nat. Hist. Soc. 47: 355–357, 1947.

PAX, F. Moringaceae. Engler and Prantl, Die natürlichen Pflanzenfamilien, ed. 2, Bd. 17b: 693–698, 1936.

PHILIPS, E. P. The genera of South African flowering plants. Cape Town, 1926.

PURI, V. Studies in floral anatomy. II. Floral anatomy of the Moringaceae with special reference to gynaecium constitution. Proc. Nat. Inst. Sci. India, 3: 71–88, 1942.

## Order 23. SARRACENIALES

An order of insectivorous plants; herbs with usually alternate leaves, insectivorous by means of "pitchers" or sensitive usually glandular-viscid hairs, and actinomorphic flowers with uniseriate to biseriate perianths.

The 3 families (Sarraceniaceae, Nepenthaceae, and Droseraceae) are considered by most to compose a natural taxon (Hallier reduced the Nepenthaceae to tribal status within the Sarraceniaceae), although Hutchinson transferred the Nepenthaceae to the Aristolochiales.

### SARRACENIACEAE. PITCHER-PLANT FAMILY

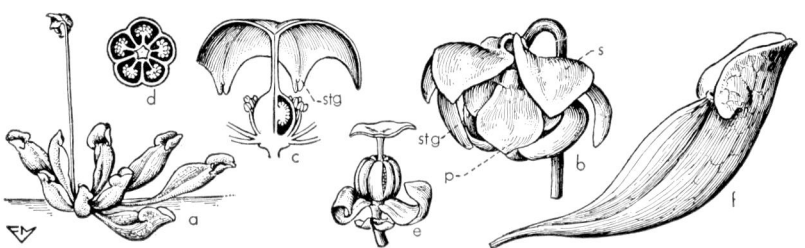

**Fig. 155.** SARRACENIACEAE. *Sarracenia purpurea*: a, flowering plant, × ¹/₁₂; b, flower, × ¹/₄; c, same, vertical section, less perianth, × ¹/₃; d, ovary, cross-section, × 1; e, capsule, × ¹/₄; f, leaf, × ¹/₄. (p petal, s sepal, stg stigma.) (From L. H. Bailey, *Manual of cultivated plants*, The Macmillan Company, 1949. Copyright 1924 and 1949 by Liberty H. Bailey.)

Herbaceous perennials; leaves rosulate, tubular, often alate, with usually a small terminal lamina, the tube often retrorsely hairy within; flowers bisexual, actinomorphic, solitary, and scapose or (in *Heliamphora*) racemose, the perianth biseriate, the sepals 4–5, distinct, imbricate, hypogynous, persistent, often colored and showy, the petals 5 or absent, distinct, hypogynous; stamens numerous but not clearly cyclic nor spiraled, distinct, hypogynous, the anthers 2-celled, dehiscing longitudinally; pistil 1, the ovary superior, 3–5-loculed and -carpelled and usually with as many lobes, the placentation axile, the ovules numerous on each placenta, anatropous, the style 1 and apically briefly lobed or the apex much expanded and peltate with as many lobes as carpels and each lobe tip stigmatic beneath; fruit a loculicidal capsule; seeds minute, with small embryo and soft-fleshy endosperm.

A family of 3 genera and 14 species of bog plants, of which 9 spp. belong to *Sarracenia* (eastern North America), 4 to *Heliamphora* (northern South America), and a single species of *Darlingtonia* (montane meadows of northern California and adjoining Oregon).

Members of the family are distinguished by the basal rosettes of tubular leaves, the persistent often colored sepals, the numerous stamens and the often umbrella-like style apex.

The Sarraceniaceae are of no significant economic importance. Several species of *Sarracenia* are listed in the trade. *Darlingtonia* is offered as a novelty, but is of difficult culture. Plants of the family are of special interest because of their insectivorous character. The pitchers secrete an enzyme that mixes with accumulated rain water and hastens the disintegration of insects enmeshed in the retrorse hairs.

*LITERATURE:*

EDWARDS, H. *Darlingtonia californica* Torrey. Proc. Calif. Acad. Sci. 6: 161–166, 1876.
HECHT, A. The somatic chromosomes of *Sarracenia*. Bull. Torrey Bot. Club, 76: 7–9, 1949.

UPHOF, J. C. T. Sarraceniaceae. In Engler and Prantl, Die natürlichen Pflanzenfamilien, ed. 2, Bd. 17b: 704–727, 1936.
WALCOTT, M. V. Illustrations of North American Pitcher-plants. 34pp., 15 col. pl. Smithsonian Inst., Washington, D. C., 1935.

## NEPENTHACEAE. NEPENTHES FAMILY

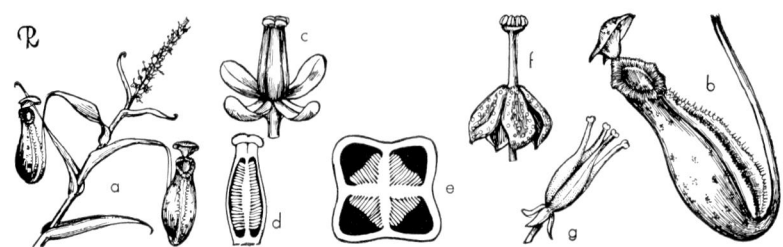

**Fig. 156.** NEPENTHACEAE. *Nepenthes*: a, flowering branch, × ⅛; b, leaf-tip pitcher, × ⅙; c, pistillate flower, × 3; d, pistil, vertical section, × 4; e, ovary, cross-section, × 10; f, staminate flower, × 2; g, capsule, × 3. (From L. H. Bailey, *Manual of cultivated plants*, The Macmillan Company, 1949. Copyright 1924 and 1949 by Liberty H. Bailey.)

Dioecious, herbaceous, suffrutescent, or shrubby plants, often climbing by their leaves; leaves alternate, composed of a petiole, a winged or expanded portion followed by a constricted often coiled zone (tendril), terminated by a pendant often highly colored urceolate to cylindrical pitcher (ascidium), that has a recurved fluted rim and a variously margined lid (operculum); flowers unisexual, actinomorphic, small and greenish, racemose to paniculate, the perianth of 2 dimerous whorls, the sepals 3–4, usually distinct, nectiferous; staminate flowers with 4–24 stamens, the filaments monadelphous in a column, the anthers distinct, 2-celled, dehiscing longitudinally; pistillate flowers with a single pistil composed of a superior 3–4-loculed and -carpelled ovary, the placentation axile, the ovules many, the style 1 and stout to obsolete, terminated by a discoid stigma; fruit an elongated leathery loculicidally dehiscent capsule; seeds numerous, filiform, ascending, imbricated, the embryo straight, surrounded by fleshy endosperm.

A unigeneric family of about 60 species, primarily of Borneo but extending through the Old World tropics to southern China and Australia. None is native in the New World.

The genus *Nepenthes* is easily distinguished by its dioecious character, the pendant showy pitchers, and the monadelphous staminate flowers.

Many species and hybrids are cultivated domestically under glass as novelties. The pitchers secrete enzymes with digestive properties that attack insects which fall into the usually water-containing vessels and drown.

*LITERATURE:*

HARMS, H. Nepenthaceae. In Engler and Prantl, Die natürlichen Pflanzenfamilien, ed. 2, Bd. 17b: 728–765, 1936.

## DROSERACEAE. SUNDEW FAMILY

Annual or perennial glandular herbs or rarely subshrubs, mostly bog plants (*Aldrovanda* a submerged aquatic); leaves alternate, usually in basal rosettes, often circinate in bud, both surfaces (except in *Dionaea*) generally covered with viscid stalked glands responsible for trapping small insects; flowers bisexual, actinomorphic, hypogynous, the inflorescence determinate, racemose to paniculate, the

# DIVISION IV. EMBRYOPHYTA SIPHONOGAMA

perianth biseriate, the sepals 4–5, mostly briefly and basally connate, persistent, imbricate, the petals 5, distinct, convolute; stamens 5–20 in 1 or more pentamerous whorls, distinct or rarely the filaments basally connate, the anthers 2-celled, the cells sometimes divergent and with broad connective, dehiscing longitudinally, extrorse, the pollen in tetrads; pistil 1, the ovary superior sometimes nearly half inferior, 1-loculed with 3–5 carpels and as many parietal placentae or (in *Dionaea* and *Drosophyllum*) the several ovules situated at the base of the unilocular ovary (in some spp. the ovary 3–5-loculed with axile placentation), the ovules few to many on each placenta, the styles 3–5, distinct, often forked or branched, the stigmas as many as stylar tips; fruit a loculicidal capsule; seeds numerous, the embryo straight, the endosperm crystalline-granular.

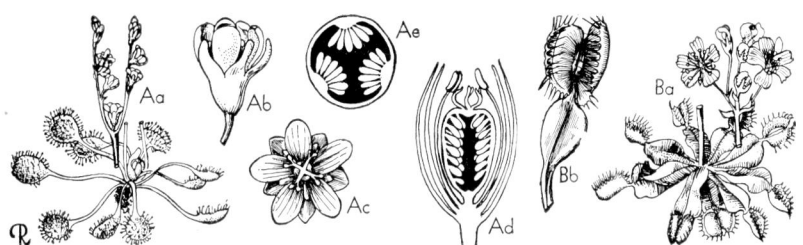

Fig. 157. DROSERACEAE. A, *Drosera rotundifolia*: Aa, plant in flower, × ½; Ab, flower, side view, × 2; Ac, same, face view, × 2; Ad, same, vertical section, × 12; Ae, ovary, cross-section, × 15. B, *Dionaea muscipula*: Ba, plant in flower, × ¼; Bb, leaf, × ½. (From L. H. Bailey, *Manual of cultivated plants*, The Macmillan Company, 1949. Copyright 1924 and 1949 by Liberty H. Bailey.)

A family of 4 genera and perhaps 90 species. Three of the genera are monotypic: *Drosophyllum* (of the western Mediterranean), *Dionaea* (of North and South Carolina),[91] and *Aldrovanda* (widespread in the Old World). *Drosera*, with perhaps 85–88 species and most abundant in Australia, is represented by a few species in most regions of all continents. In this country, the almost cosmopolitan *D. rotundifolia* occurs in all but the southwestern section of the country. The more boreal *D. intermedia* (*D. longifolia*) occurs across the northern part of the country but in the east extends south to Florida. Five other species occur in the eastern part of the country.

The family is distinguished by the insectivorous character of its rosulate leaves, the bisexual primarily 5-merous flowers, and by the determinate inflorescence (often mistaken to be a true raceme or panicle).

Many authors (incl. Wettstein) placed the Droseraceae in the Parietales near Violaceae and Ochnaceae on the basis of the parietal placentation. Others followed Lindley, who placed them in the Saxifragaceae because of the determinate inflorescence, the often half-inferior ovary, and similarities of androecium and perianth as well as of ovule morphology. In many respects, such as the insectivorous and aquatic habit of *Aldrovanda*, and the glandular hairs, the family is similar to the Lentibulariaceae, a similarity due to parallel development rather than phyletic relationship.

The family is of little or no economic importance. Venus'-flytrap (*Dionaea*) is sold in the trade as a novelty. The leaves of *Drosera* yield a violet dye, once but no longer of commercial importance.

---

[91] Accepted by J. K. Small as constituting a separate family, Dionaeaceae.

*LITERATURE:*

DIELS, L. Droseraceae. In Engler, Das Pflanzenreich, 26 (IV. 112): 1–136, 1906.

———. Droseraceae. In Engler and Prantl, Die natürlichen Pflanzenfamilien, ed. 2, Bd. 17b: 766–784, 1936.

MANSFELD, R. Zur Nomenclatur der Farn- und Blüten-Pflanzen Deutschlands. III. Fedde Repert. 46: 121, 1939. [For nomenclature of *D. longifolia* L.]

WYNNE, F. E. *Drosera* in eastern North America. Bull. Torrey Bot. Club, 71: 166–174, 1944.

## Order 24. ROSALES

The Rosales are characterized by the flowers generally cyclic, typically pentamerous, hypogynous to epigynous (perigyny common), the androecium commonly of many whorls, the gynoecium apocarpous to syncarpous but the styles generally distinct, and integuments typically 2.

The order was considered by Engler and Diels to be composed of 17 families distributed among 2 suborders as follows: [92]

Saxifragineae
  Crassulaceae
  *Cephalotaceae (Australia)
  Saxifragaceae
  Pittosporaceae
  *Byblidaceae (Australia)
  *Brunelliaceae (So. Amer.)
  Cunoniaceae
  *Myrothamnaceae (So. Afr.)
  *Bruniaceae (So. Afr.)
  Hamamelidaceae
  *Roridulaceae (So. Afr.)
  Eucommiaceae

Rosineae
  Platanaceae
  Crossosomataceae
  Rosaceae
  *Connaraceae (pantropic)
  Leguminosae

Students familiar with earlier editions of the Engler system will note the transfer by Diels of the Podostemaceae from the Rosales to an order by themselves immediately preceding the Proteales.

## CRASSULACEAE.[93] ORPINE FAMILY

Annual or perennial herbs, shrubs, or rarely scandent, succulents; leaves opposite, whorled, or alternate, mostly persistent, usually simple and entire, fleshy, estipulate; flowers bisexual (rarely unisexual, the plants then usually dioecious), actinomorphic the inflorescence cymose, of lateral cymes, or a monochasium, bracteate; perianth of calyx and corolla, the sepals 4–30 (3), generally distinct or nearly so (connate in some *Kalanchoë*), the petals of same number as sepals, distinct (connate in some spp. of *Kalanchoë* and *Cotyledon*); stamens typically in 2 whorls (1 whorl sometimes absent), usually as many or twice as many as petals, distinct (rarely basally connate), generally hypogynous unless petals basally connate and then usually borne on the corolla tube, the anthers 2-celled, dehiscing longitudinally, introrse; gynoecium of 3–more distinct or basally connate pistils (usually as

---

[92] Families whose names are preceded by an asterisk are not accounted for in the treatments that follow.

[93] The name *Crassulaceae* DC. (1805) has been conserved over *Sedaceae* Necker (1770) *nomen subnudum*. For discussion of the latter name, see Barnhart in Bull. Torrey Bot. Club, 22: 13, 1895.

# DIVISION IV. EMBRYOPHYTA SIPHONOGAMA

many as petals), each subtended by an hypogynous scalelike nectiferous gland (petaloid in *Monanthes,* absent in *Greenovia*), the ovary superior, 1-loculed, 1-carpelled, the ovules many or rarely few, the placentation parietal, the style 1, usually linear, the stigma 1; fruit a follicle; seeds with straight embryo and usually with the endosperm fleshy and scant, rarely none.

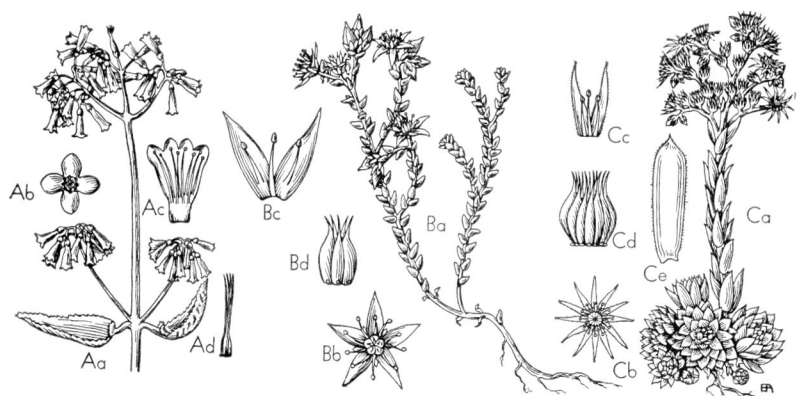

**Fig. 158.** CRASSULACEAE. A, *Kalanchoë Daigremontiana*: Aa, inflorescence, × ⅙; Ab, flower, face view, × ½; Ac, corolla expanded with stamens, × ½; Ad, pistils, × ½. B, *Sedum acre*: Ba, habit in flower, × ½; Bb, flower, face view, × 1; Bc, two petals and stamens, × 2; Bd, pistils, × 2. C, *Sempervivum tectorum*: Ca, habit in flower, × ¼; Cb, flower, face view, × ½; Cc, two petals with stamens, × 1; Cd, pistils, × 2; Ce, cauline bract, × 1. (From L. H. Bailey, *Manual of cultivated plants,* The Macmillan Company, 1949. Copyright 1924 and 1949 by Liberty H. Bailey.)

A family of wide geographical distribution although almost entirely absent from Australia and Oceania, and only a few representatives of 3–4 genera occur in South America. The criteria for the demarcation of genera are few and weak, resulting in the number of genera for the family being interpreted variously from a few (Kuntze) to 33 (Berger) and the species from 500 (Abrams) to 1300 (Berger). The plants for the most part inhabit the drier parts of the earth, but principally south-central Asia, the Mexican highlands, South Africa, and the Mediterannean region, with many endemics occurring on the Canary and a few on the Madeira Islands. In this country they are most abundant in arid region of the west and southwest. The largest genera include *Crassula* (about 250 spp., mostly in South Africa), *Sedum* (350 spp., mostly of temperate and northern boreal regions), *Kalanchoë* (125 spp., incl. *Bryophyllum*), *Echeveria* (80 spp.), *Aeonium* (38 spp.), and *Sempervivum* (about 30 spp. of Eurasia). Britton and Rose (1905) considered the North American species to represent 25 genera (16 of them relegated to synonymy by Berger, and 6 included under *Sedum* by Fröderström). Abrams (1944) recognized 9 genera in the Pacific states, but of these most authors combine *Tillaea* and *Tillaeastrum* under *Crassula;* and *Sedella* (properly *Parvisedum*) *Gormania,* and *Rhodiola* under *Sedum.* Most American authors reject Berger's inclusion of *Graptopetalum* under *Sedum,* and treat it as a valid genus. By these views, the family is represented in this country by the genera *Crassula, Diamorpha, Dudleya, Echeveria, Graptopetalum, Hasseanthus, Lenophyllum, Parvisedum* (*Sedella*), *Sedum,* and *Villadia.*

The family is characterized in general by the gynoecium composed of usually as many pistils as there are petals, by a like number of stamens in each whorl, by the presence of a scalelike gland at the base of each pistil, and by the usually fleshy or succulent character of stems and foliage.

Mauritzon (1933) on the basis of embryogeny, considered the Crassuloideae to stand apart from a taxon composed of the Cotyledonoideae, Sempervivoideae, Sedoideae, and Echeverioideae, and considered the Kalanchoideae to be a third taxon that was phyletically somewhat intermediate between the other 2. Baldwin (1938) gave cytological evidence in support of earlier views that *Bryophyllum* was not generically distinct from *Kalanchoë*. On the basis of morphological studies, Quimby concluded the family to be a natural grouping, provided that the genus *Penthorum* was excluded.[94]

Members of the family are currently of domestic importance only as ornamentals, and many are grown as novelties by fanciers of succulent plants. They represent about 18 genera and 400 species.

*LITERATURE:*

ABRAMS, L. Crassulaceae. In Illustated Flora of the Pacific states, 2: 330–346, 1944.
BALDWIN, J. T. JR. *Kalanchoë:* the genus and its chromosomes. Amer. Journ. Bot. 25: 572–579, 1938.
———. Certain cytophyletic relations of the Crassulaceae. Chron. Bot. 5: 415–417, 1939.
BERGER, A. Crassulaceae. Engler and Prantl, Die natürlichen. Pflanzenfamilien, ed. 2, Bd. 18a: 352–483, 1930.
BRITTON, N. L. and ROSE, J. N. Crassulaceae. North Amer. Flora, 22: 7–94, 1905
CLAUSEN, R. T. Studies in the Crassulaceae—III. *Sedum*, subgenus *Gormania*, section *Eugormania*. Bull. Torrey Bot. Club, 69: 27–40, 1942.
CLAUSEN, R. T. and UHL, C. H. The taxonomy and cytology of the subgenus *Gormania* of *Sedum*. Madroño, 7: 161–180, 1944.
CLAUSEN, R. T., MORAN, R., and UHL, C. H. The taxonomy and cytology of *Hasseanthus*. Desert Plt. Life, 17: 69–83, 1945.
ENGLER, A. Saxifragaceae. Die natürlichen Pflanzenfamilien, ed. 2, Bd. 18a: 74–226, 1930.
FRÖDERSTRÖM, H. The genus Sedum L., I. Acta Horti Gotoburg. 5: 1–75, 1930; II. *op. cit.* 6: 5–111, 1931.
HAMET, R. Monographie du genre *Kalanchoë*. Bull. Herb. Boissier, Ser. 2, 7: 869–900, 1907; *op. cit.* 8: 17–48, 1908.
MAURITZON, J. Studien über die Embryologie der Familien Crassulaceae und Saxifragaceae. Håkan Ohlssons, Lund, 1933.
MORAN, R. Delimitation of genera and subfamilies in the Crassulaceae. Desert Plt Life, 14: 125–128, 1942.
———. The status of *Dudleya* and *Stylophyllum*. *Op. cit.* 14: 149–157, 1942.
———. A revision of *Dudleya*, subgenus *Stylophyllum* I. *Op. cit.* 14: 190–192, 1942; II. 15: 9–14; III, 24–28; IV, 15: 40–44; V, 56–60, 1943.
POELLNITZ, K. VON. Zur Kenntnis der Gattung *Echeveria* DC. Fedde Repert. Spec. Nov. 39: 193–270, 1936.
PRAEGER, R. L. An account of the genus *Sedum* as found in cultivation. Journ. Roy. Hort. Soc. 46: 1–314, 1921.
———. An account of the *Sempervivum* group. 107 fig. London, 1932.
QUIMBY, M. W. The floral morphology of the Crassulaceae. Unpbl. Ph.D. thesis, Cornell Univ. 1939.
SCHÖNLAND, S. Crassulaceae. Engler and Prantl, Die natürlichen Pflanzenfamilien, Bd. 3 (2a): 23–38, 1890.
———. Materials for a critical revision of Crassulaceae. (The South African species of the genus *Crassula* L.) Trans. Roy. Soc. South Afr. 17: 151–293, 1929.
TILLSON, A. H. The floral anatomy of the Kalanchoideae. Amer. Journ. Bot. 27: 595–600, 1940.
UHL, C. H. Cytotaxonomic studies in the sub-families Crassuloideae, Kalanchoideae, and Cotyledonoideae of the Crassulaceae. Amer. Journ. Bot. 35: 695–706, 1948.

[94] *Penthorum* was considered the primitive genus of the Saxifragaceae by Engler (1930), by Berger, and by Wettstein. J. K. Small accepted it as a distinct family, Penthoraceae.

# SAXIFRAGACEAE. SAXIFRAGE FAMILY

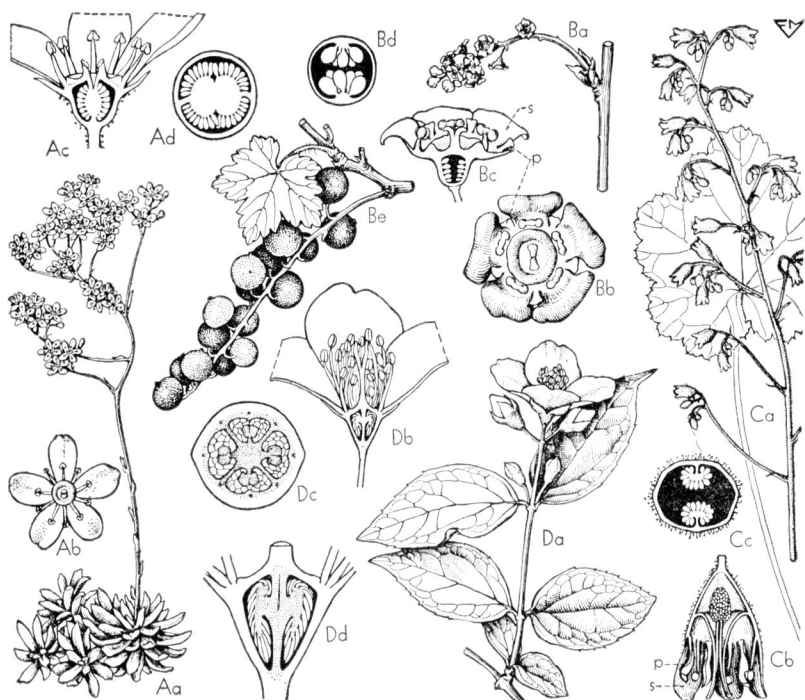

**Fig. 159.** SAXIFRAGACEAE. A, *Saxifraga Macnabiana*: Aa, plant in flower, × ¼; Ab, flower, face view, × 1; Ac, same, vertical section, perianth excised, × 2; Ad, ovary, cross-section, × 3. B, *Ribes sativum*: Ba, inflorescence, × ½; Bb, flower, face view, × 3; Bc, same, vertical section, × 3; Bd, ovary, cross-section, × 8; Be, fruit, × ½. C, *Heuchera sanguinea*: Ca, inflorescence and basal leaf, × ½; Cb, flower, vertical section, × 2; Cc, ovary, cross-section, × 4. D, *Philadelphus inodorus*: Da, flowering branch, × ½; Db, flower, vertical section, petals excised, × 1; Dc, ovary, cross-section, × 3; Dd, ovary, vertical section, × 3. (p petal, s sepal.) (From L. H. Bailey, *Manual of cultivated plants*, The Macmillan Company, 1949. Copyright 1924 and 1949 by Liberty H. Bailey.)

Herbs, shrubs, or small trees; leaves alternate or sometimes opposite (Hydrangeoideae, *Bauera*), simple or compound, usually deciduous, mostly estipulate; flowers usually bisexual and actinomorphic (unisexual in some *Ribes*, zygomorphic in some *Saxifraga*, neuter in some *Hydrangea*), basically in cymose inflorescences, but sometimes racemose to paniculate (cf. Rendle, pp. 320–321), the perianth biseriate, hypanthium sometimes present, the sepals mostly 4–5, often gamosepalous, sometimes highly colored and petaloid, the petals usually of same number, borne on the receptacle or on an hypanthium, sometimes smaller than calyx or absent; stamens usually of same number and alternate with the petals or twice as many and the outer ones opposite them (only 3 in *Tolmiea*), distinct, the anthers usually 2-celled (1-celled in *Leptarrhena*), dehiscing longitudinally, staminodes or nectar glands often present between the stamens or coalesced in a thalamus; gynoecium of a single pistil with axile (2–5-loculed) or parietal (1-loculed) placentation (2 basal ovules in *Eremosyne*), or composed of 2–5 separate pistils, often basally connate, the styles and stigmas as many as carpels, the ovary superior to inferior (often half inferior), sometimes situated within an hypanthium, the ovules anatropous and several to many on each placenta (the latter often intruding); fruit a capsule or berry; seed with a small embryo and abundant mostly soft-fleshy endosperm, rarely winged (*Sullivantia*).

The Saxifragaceae, as delimited above by Englerian concepts, are represented by about 80 genera and 1200 species. More genera (30) occur in this country than elsewhere, with 6 others restricted to South America and 4 extending from Mexico to Chile. Eight genera are in both eastern Asia and in North America, and other genera are in Australia and New Zealand (9), South Africa (6), Oceania (5), Australasia (3), Europe (1). Of the genera having species indigenous to this country, 16 are exclusively western, and only 2 are restricted to eastern areas. The larger genera of the family (with total number of species) include: *Saxifraga* 325, *Ribes* 140, *Hydrangea* 80, *Chrysosplenium* 85, *Philadelphus* 70, *Escallonia* 50, *Polyosma* 50, *Parnassia* 45, *Deutzia* 70, *Heuchera* 30, and *Astilbe* 25.

The family is difficult to separate from related families, especially from the Rosaceae. It differs from the latter in the more abundant endosperm, its estipulate or rarely stipulate leaves, and (among genera of both families having multipistillate gynoecia) by the constantly few pistils and few stamens. It is allied to the Crassulaceae, but the latter differ in the floral parts of a regular numerical plan and the pistils subtended by scalelike glands.

Engler (1930) treated the family as composed of 15 subfamilies. Both Hutchinson and Small treated several of these as families.[95] Rendle accepted the earlier Engler view that the Saxifragaceae consisted of 7 rather than 15 subfamilies. Hutchinson restricted the Saxifragaceae to the herbaceous members and transferred the primarily woody representatives to his Cunoniales. Bessey recognized as families the Saxifragaceae, Grossulariaceae, and Hydrangeaceae. He retained the 3 within his Rosales, but as more advanced than the roseaceous or leguminaceous taxa.

The family is economically important for the many ornamentals belonging to it. Notable among them are the saxifrage, mock orange (*Philadelphus*), coral-bells (*Heuchera*), hydrangea, escallonia, deutzia, astilbe, and bergenia. Currants and gooseberries (*Ribes* spp.) are bush fruits of minor importance.

*LITERATURE:*

BERGER, A. A taxonomic review of currants and gooseberries. Tech. Bull., N. Y. State Agr. Exp. Sta. 109: 1–118, 1924.
BRITTON, N. L. Iteaceae. North Amer. Flora, 22: 181, 1905.
COVILLE, F. V. and BRITTON, N. L. Grossulariaceae. North Amer. Flora, 22: 193–225, 1908.
DANDY, J. E. The genera of Saxifragaceae. Kew Bull. 100–118, 1927.
ENGLER, A. Saxifragaceae. In Engler and Prantl, Die natürlichen Pflanzenfamilien, ed. 2, Bd. 18a: 74–225, 1930.
ENGLER, A. and IRMSCHER, E. *Saxifraga*. In Engler, Das Pflanzenreich, 67(IV.117.I): 1–448, 1916; *op. cit.* 69 (IV.117.II): 449–709, 1919.
JANCZEWSKI, E. DE. Monographie des grosseilliers, *Ribes* L. Mém. Soc. Genève, 35: 199–516, 1907.
JOHNSON, A. M. Revision of the North American species of the Sect. *Boraphila*, genus *Saxifraga*. Univ. Minn. Stud. Biol. Sci. 4: 109, 1923.
———. Studies in *Saxifraga* I–II. Amer. Journ. Bot. 14: 323–326, 1923; *op. cit.* 18: 797–802, 1931.

[95] Hutchinson recognized, as segregates, the Escalloniaceae, Greyiaceae, Grossulariaceae, and Hydrangeaceae. Small treated the Saxifragaceae of the southeastern states as composed of the families Penthoraceae, Parnassiaceae, Saxifragaceae, Iteaceae, and Hydrangeaceae. Britton and Rydberg also accepted the Grossulariaceae as a distinct family.

LAKELA, O. A monograph of the genus *Tiarella* L. in North America. Amer. Journ. Bot. 24: 344–351, 1937.
ROSENDAHL, C. O. Die nordamerikanischen Saxifraginae und ihre Verwandtschafts-Verhältnisse in Beziehung zu ihre geographischen Verbreitung. In Engler, Bot. Jahrb. 37: Beibl. 83: 1–87, 1905.
———. Revision of the genus *Mitella*, with a discussion of geographical distribution and relationships. In Engler, Bot. Jahrb. 50: suppleb., 375–397, 1914.
ROSENDAHL, C. O., BUTTERS, F. K., and LAKELA, O. A monograph of the genus *Heuchera*. Minn. Stud. Pl. Sci. 2: 1–180, 1936.
RYDBERG, P. A. Penthoraceae. North Amer. Flora, 22: 75, 1905.
———. Parnassiaceae. North Amer. Flora, 22: 77–80, 1905.
SKOVSTED, A. Cytological studies in the tribe Saxifrageae. Dansk. Bot. Arkiv. 8: 1–52, 1934.
SMALL, J. K. and RYDBERG, P. A. Saxifragaceae. Hydrangeaceae. North Amer. Flora, 22: 81–178, 1905.

## PITTOSPORACEAE. PITTOSPORUM FAMILY

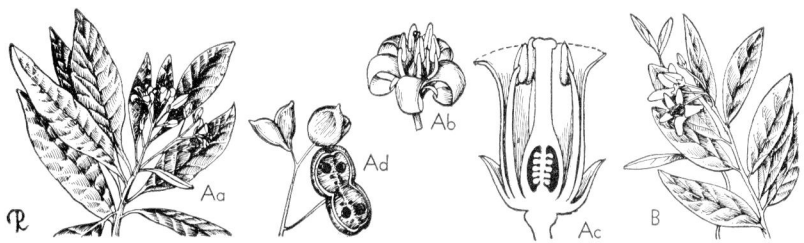

**Fig. 160.** PITTOSPORACEAE. A, *Pittosporum undulatum*: Aa, flowering plant, × ¼; Ab, flower, × 1; Ac, same, vertical section, × 2 (after Schnizlein); Ad, fruiting branch, × ½. B, *Sollya fusiformis*: flowering branch, × ½. (From L. H. Bailey, *Manual of cultivated plants*, The Macmillan Company, 1949. Copyright 1924 and 1949 by Liberty H. Bailey.)

Trees, shrubs, or woody climbers, sometimes spiny; leaves simple, alternate or whorled, often leathery, estipulate; flowers typically bisexual, actinomorphic (zygomorphic in *Cheiranthera*), solitary, cymose, or paniculate, the perianth usually campanulate or rotate, the sepals 5, distinct or basally connate, the petals 5, sometimes basally connate or coherent, imbricate; stamens 5, alternate with petals, distinct (monadelphous in *Marianthus*), hypogynous, the filaments shorter than anthers, the anthers lanceolate to linear, 2-celled, dehiscing (mostly introrsely) by pores or less frequently by longitudinal splits; pistil 1, the ovary superior, 2–5-loculed and carpelled with axile placentation or 1-loculed and 2-carpelled with parietal placentation, the ovules anatropous in 2 ranks, style 1 and short, the stigmas usually as many as carpels or coalesced and the number obscured; fruit a loculicidal capsule or berry; seed with abundant firm-fleshy endosperm and a minute linear embryo.

A family of 9 genera and 200 species, distributed throughout the warmer regions of the Old World and absent from the Americas. *Pittosporum*, the largest genus (160 spp.), occurs on all continents of that hemisphere but more particularly in Oceania and Australasia.

Members of the family are recognized by their leathery alternate or whorled leaves, and by the flower having the combination of short filaments and often poricidal anthers, multiovulate placentae, a 2–5-carpelled ovary, and a usually simple style.

The phyletic position of the family seems close to the Saxifragaceae and particularly to the genus *Escallonia* and its relatives.

Economically the family is known domestically for species of *Sollya* and *Pittosporum* which are grown as ornamentals in the warmer parts of the country.

LITERATURE:

PRITZEL, E. Pittosporaceae. In Engler and Prantl, Die natürlichen Pflanzenfamilien, ed. 2, Bd. 18a: 265–288, 1930.

## CUNONIACEAE. CUNONIA FAMILY

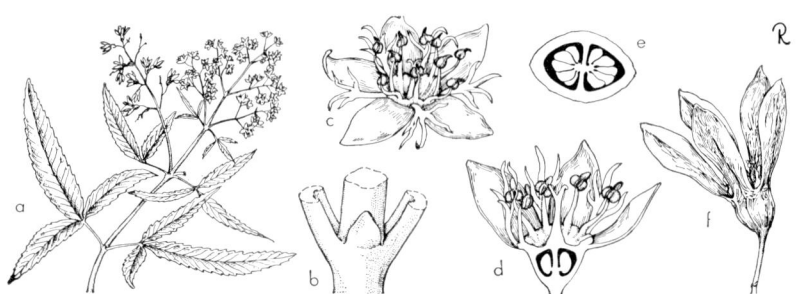

**Fig. 161.** CUNONIACEAE. *Ceratopetalum gummiferum*: a, flowering branch, × ⅓; b, node with stipule, × 5; c, flower, habit, × 3; d, same, vertical section, × 4; e, ovary, cross-section, × 10; f, fruit, × 1½.

Trees or shrubs; leaves pinnately or trifoliolately compound or rarely simple, opposite or whorled, margins often glandular serrate, stipules present and sometimes large and connate; flowers bisexual or sometimes unisexual (the plants then dioecious), actinomorphic, small, solitary capitate or paniculate, the perianth biseriate, the sepals 3–6 (mostly 4–5) sometimes basally connate, the petals 3–5, mostly smaller than the sepals and often lacking, sometimes basally connate; stamens mostly numerous, sometimes few and then opposite the sepals, usually on an annular nectiferous glandular disc, the filaments distinct, the anthers short and 2-celled, dehiscing longitudinally; gynoecium of 1–5 unicarpellate pistils or more commonly of a single 2–5-carpelled pistil, the ovary superior, the carpels and locules 2–5 and placentation axile (when ovary is compound), the ovules few to numerous in 2 rows on recurved intruding carpellary margins on each placenta, the styles and stigmas distinct, as many as carpels; fruit a capsule or nut; seed with small embryo and abundant endosperm.

A family of 26 genera and about 240 species (*Weinmannia*, distributed from Mexico and the West Indies to Chile and from the Philippines to New Zealand, has 126 spp.). The family is restricted almost exclusively to the southern hemisphere, and mostly to Oceania and Australasia. Three genera occur in South America and 2 in South Africa. None is known to be indigenous to this country.[96]

Members of the family are allied to the Saxifragaceae but differ in the usually treelike habit, the leaves always opposite or whorled, and the carpellary margins of each placental zone recurved or distinct.

The family is of little domestic importance. Species of *Ackama, Callicoma, Ceratopetalum,* and *Weinmannia* are grown for ornament in California. Lightwood (*Ceratopetalum apetalum*) is an important timber tree in New South Wales.

LITERATURE:

ENGLER, A. Cunoniaceae. In Engler and Prantl, Die natürlichen Pflanzenfamilien, ed. 2, Bd. 18a: 229–262, 1930.

[96] *Lyonothamnus* of California, and treated by Britton (North Amer. Flora, 22: 180, 1905) as belonging to this family is here included within the Rosaceae.

# HAMAMELIDACEAE.[97] WITCH-HAZEL FAMILY

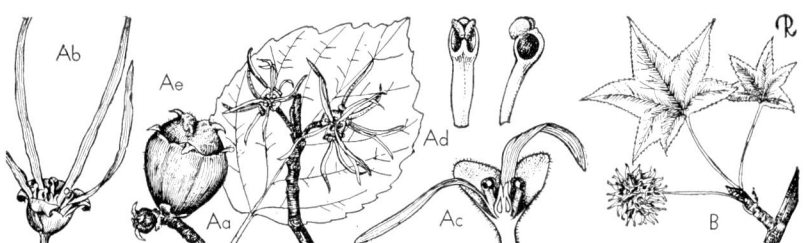

**Fig. 162.** HAMAMELIDACEAE. A, *Hamamelis virginiana*: Aa, twig in flower and leaf, × ½; Ab, flower, × 1; Ac, flower, vertical section, × 1; Ad, stamen, face and side view, × 6; Ae, fruit, × ½. B, *Liquidambar Styraciflua*: twig in fruit, × ¼. (From L. H. Bailey, *Manual of cultivated plants*, The Macmillan Company, 1949. Copyright 1924 and 1949 by Liberty H. Bailey.)

Deciduous or evergreen trees or shrubs, often with stellate hairs; leaves alternate, simple, glandular-toothed to palmately lobed, stipulate; flowers bisexual or unisexual (the plants then monoecious [*Liquidambar*] or dioecious), actinomorphic or less often zygomorphic, axillary, capitate, or spicate, sometimes subtended by highly colored bracts, the perianth present or absent, the calyx of mostly 4–5 basally connate sepals, the tube more or less adnate to the ovary, the corolla (sometimes absent) of 4–5 distinct petals seemingly borne on the calyx, imbricate or valvate; stamens 2–8 in a single whorl; perigynous, distinct, the anthers 2-celled, dehiscing longitudinally or by upturning valves, the connective often exserted; pistil 1, the ovary half inferior to inferior (rarely superior), 2-loculed, the 2 carpels diverging and often separating apically, the placentation axile, the ovules 1 or more in each locule, pendulous, anatropous, the styles and stigmas 2; fruit a loculicidal capsule (infrequently septicidal), often with a woody to leathery exocarp and bony endocarp; seed sometimes winged, the large and straight embryo enveloped by a thin fleshy endosperm.

A family of 23 genera and about 100 species, mostly of Asia (17 genera), with 2 genera of Africa and 1 of Australia. *Hamamelis* (6 spp.), *Liquidambar* (4 spp.), and *Fothergilla* (4 spp.) have species indigenous to eastern North America, with species of the first 2 occurring also in eastern Asia. *Corylopsis* (22 spp.) of eastern Asia is the largest genus and *Dicoryphe* (14 spp.) of Madagascar is next in size. Eleven genera are monotypic.

The Hamamelidaceae are distinguished by the bicarpellate, bilocular ovary, whose diverging apices are each terminated by a slender recurved-tipped style, and by the 1 to many pendulous ovules. The stellate indumentum is characteristic when present, as is the usually woody or bony capsule.

The families Hamamelidaceae, Platanaceae, and Myrothamnaceae were treated by Wettstein and have been generally accepted to compose the order Hamamelidales. Hutchinson (1948) considered this order to terminate a line of evolutionary ascent and to have been derived from stocks ancestral to the Rosales and derived from the Magnoliales. A similar origin was postulated by Tippo (1938), but he considered, to the contrary, that the Hamamelidales were not an evolutionary terminus, but that they (or their ancestral forms) gave rise to the Casuarinaceae, the Fagales, and the Urticales. Current data seem to support these views more

---

[97] The name *Hamamelidaceae* Lindl. (1846) has been conserved over *Altingiaceae* Hayne (1830) and *Parrotiaceae* Horaninov (1834). For discussion of *Altingiaceae*, see Barnhart in Bull. Torrey Bot. Club, 22: 14, 1895.

closely than they do those of Engler, Rendle, or Warming, who held that the Hamamelidaceae are allied more nearly to the Saxifragaceae and the Cunoniaceae of the Rosales.

Economically the family is important domestically for the witch-hazel extract prepared from *Hamamelis* bark and used as a liniment, and species of sweet gum or storax (*Liquidambar*), winter hazel (*Corylopsis*), *Fothergilla*, *Hamamelis*, and *Loropetalum* are grown for ornament.

*LITERATURE:*

BRITTON, N. L. Hamamelidaceae. North Amer. Flora, 22: 185–187, 1905.
HALLIER, H. Über den Umfang, die Gliederung und die Verwandtschaft der Familie der Hamamelidaceen. Beih. Bot. Centralbl. 14: 247–260, 1903.
HARMS, H. Hamamelidaceae. In Engler and Prantl, Die natürlichen Pflanzenfamilien, ed. 2, Bd. 18a: 303–345, 1930.
HORNE, A. S. A contribution to the study of the evolution of the flower with special reference to the Hamamelidaceae, Caprifoliaceae, and Cornaceae. Trans. Linn. Soc. Bot. (Lond.) 8: 239–304, 1914.
TIPPO, O. Comparative anatomy of the Moraceae and their presumed allies. Bot. Gaz. 100: 1–99, 1938.
TONG, KOE-YANG. Studien über die Familie der Hamamelidaceae mit besonderer Berücksichtigung der Systematik und Entwicklungsgeschichte von *Corylopsis*. 72 pp. 1931. [Republished in Bull. Dept. Biol. Sun Yatsen Univ. 2: 1–72, 1931. Cf. Lignan Sci. Journ. 12: 186–189, 1933, for detailed review.]
WILSON, P. Altingiaceae. North Amer. Flora, 22: 189, 1905.

## EUCOMMIACEAE. EUCOMMIA FAMILY

Fig. 163. EUCOMMIACEAE. *Eucommia ulmoides*: a, habit in fruit, × ½; b, staminate inflorescence, × ½; c, staminate flower, × 1; d, pistillate inflorescence, × ½; e, pistillate flower, × 1; f, same, vertical section, × 2. (From L. H. Bailey, *Manual of cultivated plants*, The Macmillan Company, 1949. Copyright 1924 and 1949 by Liberty H. Bailey.)

A unigeneric family of a single species, a deciduous dioecious elmlike tree, characterized in part by presence of the rubber latex and lamellate pith; leaves alternate, simple, petioled, serrate, estipulate; flowers unisexual, actinomorphic, solitary in bract axils at twig bases or adventitious, lacking perianth; staminate flowers pedicelled, the stamens 4–10, the filaments very short, distinct, the anthers mucronate, linear, 2-celled, dehiscing longitudinally; pistillate flowers short-pedicelled, bracteate, the pistil 1, the ovary unilocular, 2-carpelled (1 aborts), bifid at apex, the placentation basically axile, the ovule 1 in each locule, anatropous, pendulous, the styles 2, short and reflexed, inner surface of each style stigmatic; fruit 1-seeded, a samara; seed with large straight embryo, the endosperm abundant.

The phyletic position of the family was studied by Tippo (1940), who held that ". . . the Eucommiaceae do not belong in the Hamamelidales near the Hamamelidaceae, but that they are closer to the Ulmaceae, in the Urticales." This view was held also by Wettstein (1935). Conclusions identical to these were reached independently by Varossieau (1942) without benefit of knowledge of Tippo's earlier findings. Harms (1930) placed the family in the Rosales near the Hamamelidaceae but conceded its true relationship could be within the Urticales, an order now believed generally to be a highly advanced derivative of the Hamamelidaceae or of stocks ancestral to them.

Only the 1 species, *Eucommia ulmoides,* of China is known. It is cultivated in temperate regions as an ornamental.

*LITERATURE:*

HARMS, H. Eucommiaceae. In Engler and Prantl, Die natürlichen Pflanzenfamilien, ed. 2, Bd. 18a: 348–351, 1930.
PARKIN, J. *Eucommia ulmoides.* Kew Bull. 1921: 177–185, 1921.
TIPPO, O. The comparative anatomy of the secondary xylem and the phylogeny of the Eucommiaceae. Amer. Journ. Bot. 27: 832–838, 1940.
VAROSSIEAU, W. W. On the taxonomic position of *Eucommia ulmoides* Oliv. Blumea, 5: 81–92, 1942.

## PLATANACEAE. PLANE-TREE FAMILY

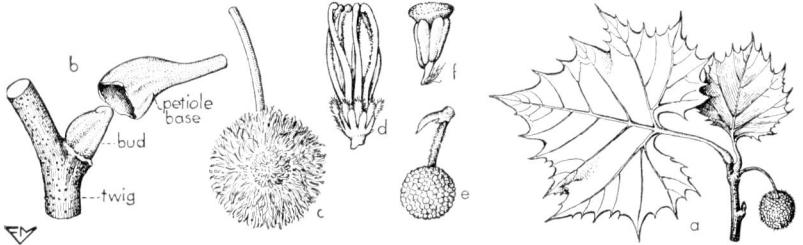

**Fig. 164.** PLATANACEAE. *Platanus occidentalis*: a, twig with fruit, × ⅙; b, subpetiolar bud; c, pistillate inflorescence, × 2; d, pistillate flower, × 4; e, fruit, × ¼; f, staminate flower, × 4. (From L. H. Bailey, *Manual of cultivated plants,* The Macmillan Company, 1949. Copyright 1924 and 1949 by Liberty H. Bailey.)

A unigeneric family of monoecious trees; bark of branches exfoliating, forming furrows on old trunks, stellate hairs common on vegetative parts; leaves alternate, simple, palmately 3–9-lobed, petiole dilated basally and enclosing the axillary bud, the stipules membranous, caducous, and encircling the twig; flowers unisexual, in 1–several dense globose heads that are distributed along a pendulous peduncle;[98] staminate flowers several to many in the male head, each composed of a calyx cup, 3–7 tridentate and minute petals alternating with the whorl of 3–7 stamens, occasional rudimentary pistils present in a few flowers of the head, the stamens distinct, the filaments very short or obsolete, the anthers long, 2-celled, dehiscing longitudinally, with large apically peltate connective, small staminodes often present; pistillate flowers often not markedly differentiated, hypogynous, the perianth present, the calyx cupular and 3–5 lobed or of 3–5 distinct sepals, the petals usually absent (present in *P. racemosa*), staminodes present, usually in a whorl of 3–4 around the outer pistils, the pistils mostly 5–9, the ovary superior, uniloculate, 1-carpelled, incompletely closed at apex, the placentation parietal, the ovules 1–2, pendulous, orthotropous, the style 1 and linear, usually recurved, the stigma extending most of the inner face of the style length; fruit a 1-seeded linear quadrangular achene or rarely follicular; seed with the embryo straight and the endosperm a thin fleshy layer.

[98] The findings and opinions of Boothroyd (1930) were drawn on freely in the preparation of the following description of floral parts and organization.

The single genus (*Platanus*) is composed of perhaps 8 species,[99] indigenous to many regions of the northern hemisphere, exclusive of Africa. Four species occur in North America: 2 in the southwest, a third in the east, and the fourth restricted to Mexico.

The family is distinguished by the subpetiolar axillary buds and exfoliating bark, and by the monoecious character associated with unisexual flowers aggregated in globose heads on pendulous peduncles.

Opinions differ as to the nature of the much-reduced inflorescence and flowers in members of the family. Boothroyd concluded the inflorescence of *P. racemosa* (a raceme of globose heads) to be primitive among present-day species, and postulated it to have been derived by reduction from ancestral stocks having open panicles. Brouwer (1924) concluded the monocephalic inflorescence (as in *P. occidentalis*) to be primitive, and the racemose type to have been derived from it by the subdivision of 1 head into several, a view contradicted by Boothroyd's anatomical evidence. Some authors (including Hutchinson) accept the view of Brouwer that the *Platanus* flower consists only of a pair of organs, in the one a pistil and a scale, or in the other a stamen and its scale. Macroscopic examination of the components of a head do not support this latter view.

The phyletic position of the Platanaceae, on the basis of apparent similarities of inflorescence, simplicity of pistil, and variability in number of floral parts, long was considered to be within the Urticales. It is the present consensus (Boothroyd, Schönland, Wettstein, Tippo, *et al.*) that the Platanaceae are of closer affinity to the Rosaceae and Hamamelidaceae than to other taxa.

Most species of the genus are cultivated domestically for ornament and are known as sycamore, plane tree, and button-ball tree.

*LITERATURE:*

BOOTHROYD, L. E. The morphology and anatomy of the inflorescence and flower of the Platanaceae. Amer. Journ. Bot. 17: 678–693, 1930.

BRETZLER, E. Studien über die Gattung *Platanus* L. Verh. Leop-Carol. Akad. Naturf. 77: 115–226, 1899.

BROUWER, J. Studies in Platanaceae. Rec. Trav. Bot. Néerlandais 21: 369–382, 1924.

GLEASON, H. A. Platanaceae. North Amer. Flora, 22: 227–229, 1908.

GRIGGS, R. F. Characters and relationships of Platanaceae. Bull. Torrey Bot. Club, 36: 369–395, 1909.

TIPPO, O. Comparative anatomy of the Moraceae and their presumed allies. Bot. Gaz. 100: 1–99, 1938.

## CROSSOSOMATACEAE. CROSSOSOMA FAMILY

Shrubs or small trees, bark rough; leaves alternate, often clustered on short shoots, simple, leathery, entire; flowers bisexual, actinomorphic, solitary, terminal on short shoots, showy, the perianth biseriate, the sepals 5, persistent, basally connate, the petals 5, white or purplish, deciduous, imbricate, orbicular to oblong; stamens usually 15–50 in 3–4 whorls, distinct, arising from a disc lining the hypanthium, the anthers oval to oblong, 2-celled, dehiscing longitudinally; gynoecium of 3–5 distinct stipitate pistils, each pistil with an elongate unilocular unicarpellate ovary, the ovules several to many, 2-ranked on a single parietal placenta, the styles and stigmas 2 or styles obsolete; fruit a follicle; seed globose to reniform, enclosed by a fimbriate aril, the embryo arcuate, the endosperm scant.

A single American genus (*Crossosoma*) of 4 species, 2 in California and 2 in Arizona.

[99] Bretzler (1924) recognized a single polymorphic species, a view now considered the extreme in conservatism.

The plants are distinguished by the bisexual flowers of 3–5 pistils and an androecium of many stamens, all borne on the rim or sides of a short hypanthium, and by the glossy seeds enveloped by a multifingered aril.

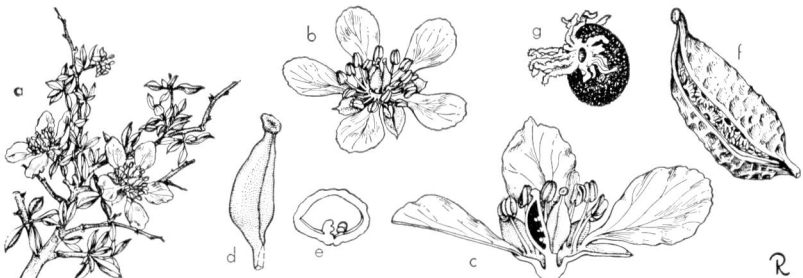

**Fig. 165.** CROSSOSOMATACEAE. *Crossosoma Bigelowii*: a, flowering branch, × ½; b, flower, × 1; c, same, vertical section, × 2; d, pistil, × 3; e, ovary, cross-section, × 6; f, follicle, × 2; g, seed with fimbriate aril, × 5.

Most botanists have accepted the view that the Crossosomataceae belong in the Rosales near the Rosaceae or Pittosporaceae. Hutchinson (1926) placed them in the Dilleniales (an order considered by him as intermediate between the Magnoliales and Rosales). Lemesle (1948) proposed their transfer to a position intermediate between the Paeoniaceae and the Ranunculaceae, a view given indirect support by Corner's opinion that the Paeoniaceae (on the basis of their centrifugal stamens) should be considered closely allied with the Dilleniaceae.

The family is of no known economic importance.

*LITERATURE:*

LEMESLE, R. Position phylogénétique de *l'Hydrastis canadensis* L. et du *Crossosoma californica* Nutt., d'après les particularités histologiques du xylème. Paris Acad. des Sci. compt. rend. 227: 221–223, 1948.

SMALL, J. K. Crossosomataceae. North Amer. Flora, 22: 231–232, 1908.

## ROSACEAE. ROSE FAMILY

Trees, shrubs, or herbs, often thorny, sometimes climbing; leaves alternate (rarely opposite), simple or pinnately compound, the stipules usually present, sometimes caducous or adnate to petiole; flowers usually bisexual or infrequently unisexual (the plants then dioecious, as in *Aruncus*), actinomorphic (zygomorphic in *Hirtella*) in various types of determinate or indeterminate inflorescences, the perianth biseriate or one or both series absent, generally perigynous, basal portions usually adnate into an hypanthium, the calyx typically of 5 basally connate sepals, the petals usually present and arising from hypanthium rim (absent in *Alchemilla, Sanguisorba, Acaena, et al.*), imbricate (convolute in *Gillenia*), the hypanthium often bearing a nectiferous glandular disc; stamens commonly numerous (sometimes definite and then usually 5 or 10) and in 1–several whorls of 5 stamens each, perigynous around the gynoecium, arising from the hypanthium, distinct (monadelphous in *Chrysobalanus*), the anthers small, 2-celled, dehiscing longitudinally (rarely transversely or by pores); gynoecium of usually 1 compound pistil or of many simple pistils cyclically or spirally disposed, situated usually within an hypanthium or the hypanthium adnate to a compound ovary, the ovary superior or inferior, when compound composed of 2–5 locules and carpels with placentation axile (ovules few to several in each locule) and terminated by as many styles (or

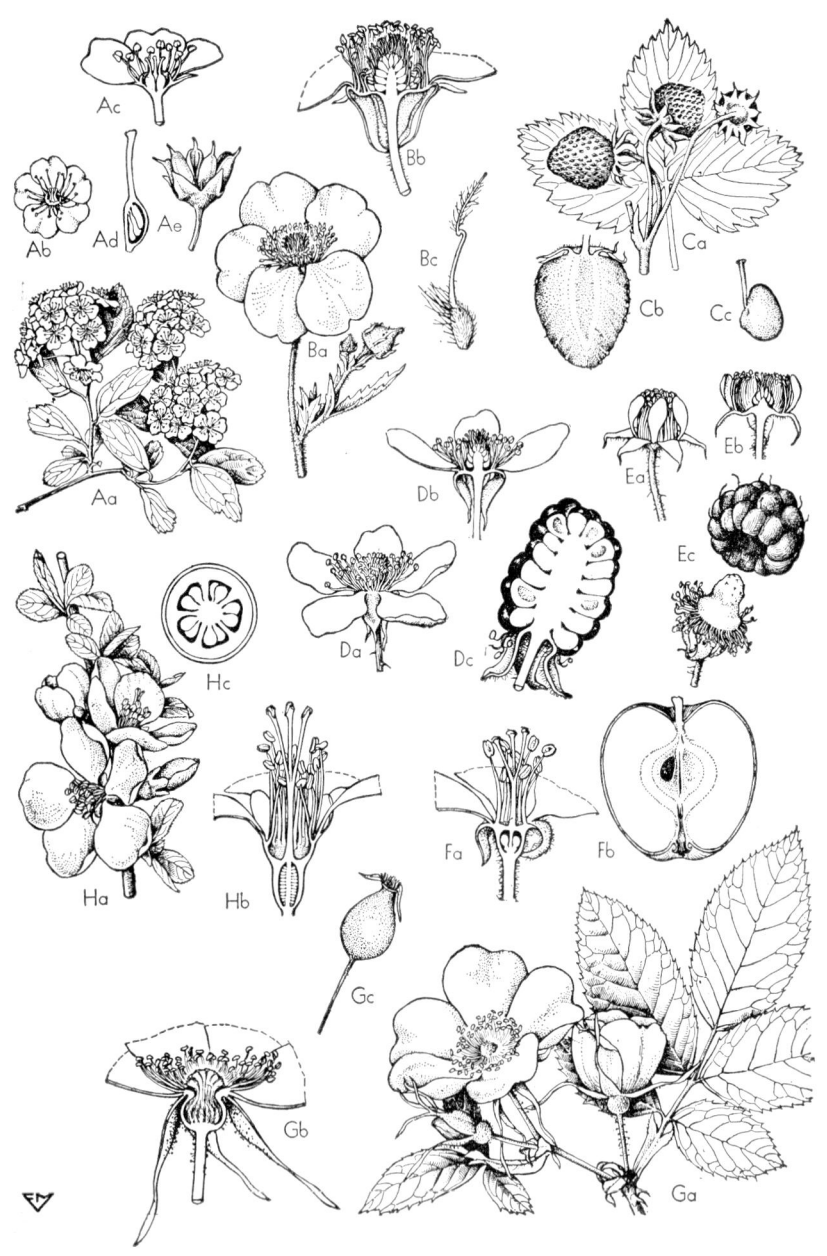

style branches) or stigmas as carpels; fruit an achene, follicle, pome, drupe, or aggregation of drupelets; seed with small embryo and usually without endosperm.

A large family of perhaps 115 genera and 3200 species, distributed over most of the earth and abundant in eastern Asia, North America, and Europe. About 50 genera have species indigenous to this country; 38 genera occur in the Pacific states, and one-half are restricted to that area or extend only southeasterly into Mexico, 3 genera are restricted to the southeastern states, and about 30 genera occur in the northeastern section of the country. The largest genera of the family (together with approximations of total and indigenous species) include: *Potentilla* (300–50), *Rubus* (600–1000, with perhaps 200–400 indigenous), *Rosa* (150–50), *Crataegus* (300–1000 with perhaps 75–200 indigenous), *Prunus* (150–175 with 25–30 indigenous), and *Spiraea* (75–12).

The family often is subdivided into several subfamilies or tribes. The evidence seems to favor their recognition as subfamilies. The 6 subfamilies are distinguished by characters used in the following key:

1. Fruit a follicle or capsule (an achene in *Holodiscus*) ................ *Spiraeoideae*
1. Fruit indehiscent, dry or fleshy.
   2. Ovary inferior, of 2–5 carpels .................................... *Pomoideae*
   2. Ovary superior.
      3. Gynoecium of a single pistil (rarely 2–5), the fruit a drupe or a berry.
         4. Style terminal or essentially so, ovule(s) pendulous, flowers actinomorphic ........................................................ *Prunoideae*
         4. Style basal, ovules erect, flowers zygomorphic or somewhat so .......................................................... *Chrysobalanoideae*
      3. Gynoecium of mostly 10 or more pistils (if few or one, the fruit dry and indehiscent).
         5. Pistils basally connate and basally adnate to hypanthium .... *Neuradoideae*
         5. Pistils distinct and free from hypanthium walls (usually unicarpellate in Cercocarpae) ....................................................... *Rosoideae*

Most of these subfamilies have been treated as independent families by various authors,[100] a view rejected by most phylogenists. The Rosaceae family as a unit is a more natural assemblage than are some other large families (e.g., Saxifragaceae, Leguminosae). Its members are characterized by the usual presence of stipules, the general pentamerous plan of the flower (exclusive of the gynoecium), the presence of an hypanthium in most genera (adnate to the ovary when the latter is inferior), and the near absence of endosperm.

The family is of considerable economic importance in temperate regions. This is especially true of its fruit-producing members from which are obtained the apple

[100] The Spiraeoideae as the Spiraeaceae, the Pomoideae as the Pomaceae or Malaceae, the Rosoideae as the Poteriaceae, the Prunoideae as the Amygdalaceae, and Chrysobalanoideae as the Chrysobalanaceae.

---

**Fig. 166.** ROSACEAE. A, *Spiraea Vanhouttei*: Aa, flowering branch, × ½; Ab, flower, face view, × 1; Ac, flower, vertical section, × 2; Ad, pistil, vertical section, × 6; Ae, follicles, × 2. B, *Geum bulgaricum*: Ba, flower, × ½; Bb, same, vertical section, × 1; Bc, achene, × 2. C, *Fragaria chiloensis*: Ca, fruit and basal leaf, × ¼; Cb, fruit, vertical section, × ½; Cc, nutlet with persistent style, × 5. D, *Rubus roribaccus*: Da, flower, × nearly 1; Db, same, vertical section, × 1; Dc, fruit, vertical section, × 1. E, *Rubus occidentalis*: Ea, flower, × 1; Eb, flower, vertical section, × 1; Ec, fruit separated from torus, × 1. F, *Malus sylvestris*: Fa, flower, vertical section, corolla excised, × 1; Fb, fruit, vertical section, × ⅓. G, *Rosa canina*: Ga, flowering branch, × ½; Gb, flower, vertical section, × 1; Gc, hip, × ½. H, *Chaenomeles lagenaria*: Ha, flowering branch, × ½; Hb, flower, vertical section, × 1; Hc, ovary, cross-section, × 4. (From L. H. Bailey, *Manual of cultivated plants*, The Macmillan Company, 1949. Copyright 1924 and 1949 by Liberty H. Bailey.)

(*Malus*), pear (*Pyrus*),[101] quince (*Cydonia*), cherry, plum, prune, peach, nectarine, apricot, almond, sloeberry (spp. of *Prunus*), loquat (*Eriobotrya*), blackberry, raspberry, loganberry (spp. of *Rubus*), stawberry (*Fragaria*), and medlar (*Mespilus*). Many ornamental trees and shrubs occur within the family, notably the spiraea (*Spiraea*), ninebark (*Physocarpus*), pearlbush (*Exochorda*), cotoneaster (*Cotoneaster*), firethorn (*Pyracantha*), hawthorn (*Crataegus*), photinea (*Photinia*), flowering quince (*Chaenomeles*), mountain ash (*Sorbus*), snow wreath (*Neviusia*), kerria (*Kerria*), jetbead (*Rhodotypos*), Japanese cherry (*Prunus*), rose (*Rosa*), and shrubby cinquefoils (*Potentilla*).

*LITERATURE:*

BAILEY, L. H. A monograph of the genus *Rubus* in North America. Gentes Herb. 5: 1–932, 1941–45.
———. Species *Batorum*—Addenum I. Studies in *Rubus*. Gentes Herb. 7: 193–349, 1947; Addenum II, *op. cit.* 481–526, 1949.
———. The *Pyrus-Malus* puzzle. Gentes Herb. 8: 40–43, 1949.
CAMP, W. H. The *Crataegus* problem. Castanea, 7: 51–55, 1942.
EINSET, JOHN. Chromosome studies in *Rubus*. Gentes Herb. 7: 181–192, 1947.
ERLANSON, E. Q. Cytological conditions and evidences for hybridity in North American, wild roses. Bot. Gaz. 87: 443–506, 1929.
———. American wild roses. Amer. Rose Annual, 83–90, 1932.
———. Experimental data for a revision of the North American wild roses. Bot. Gaz. 96: 197–259, 1934.
FERNALD, M. L. Minor transfers in *Pyrus*. Rhodora, 49: 229–233, 1947. [Summarizes reasons for uniting *Malus, Sorbus,* and *Aronia* with *Pyrus.*]
GUSTAFSSON, ÅKE. The genesis of the European blackberry flora. Lunds Univ. Arssk. N. F. Avd. II. 39 (6): 1–199, 1943.
HERRING, P. Classification of roses. Dansk. Bot. Ark. 4 (9): 1–24, 1925.
JONES, G. N. A synopsis of the North American species of *Sorbus*. Journ. Arnold Arb. 20:1–43, 1939.
———. American species of *Amelanchier*. Illinois Biol. Monogr. 20: 1–126, 1946.
MCVAUGH, R. The status of certain anomalous native crabapples in eastern United States. Bull. Torrey Bot. Club, 70: 418–429, 1943.
MASON, S. North American species of *Prunus*. Bull. U. S. Dept. Agr. 179: 1–75, 1915.
MEYER, K. Kulturgeschichte und systematische Beiträge zur Gattung *Prunus*. Fedde, Rep. Spec. Nov. Beih. 22: 1–64, 1923.
NIELSEN, E. L. A taxonomic study of the genus *Amelanchier* in Minnesota. Amer. Midl. Nat. 22: 160–206, 1939.
PALMER, E. J. Synopsis of North American *Crategi*. Journ. Arnold Arb. 6: 5–128, 1925.
———. *Crataegus* in the northeastern and central United States and adjacent Canada. Brittonia, 5: 471–490, 1946.
RICKETT, H. W. The inflorescence of *Crataegus*. Bull. Torrey Bot. Club, 70: 489–495, 1943.
RUBTSOV, G. A. Geographical distribution of the genus *Pyrus* and trends in its evolution. Amer. Nat. 78: 358–366, 1944.
RYDBERG, P. A. Rosaceae. North Amer. Flora, 22: 239–533, 1908.
———. Notes on Rosaceae: I, *Opulaster, Spiraea, Petrophytum, Luetkea, Aruncus*. Bull. Torrey Bot. Club, 35: 535–542, 1908; II, *Schizonotus, Chamaebatiaria, Sericotheca, Horkelia, Horkeliella, Ivesia, Comarella,* and *Stellariopsis*. *Op. cit.* 36: 397–407, 1909; III-V, *Potentilla. Op. cit.* 37: 375–386, 487–502, 1910; 38: 79–89, 1911; *Argentina, Comarum, Duchesnea, Fragaria, Sibbaldia, Sibbaldiopsis, Dasiphora, Drymocallis. Op. cit.* 38: 351–367, 1911; VII, *Alchemilla, Aphanes, Sanguisorba, Poteridium, Poterium, Acaena, Agrimonia, Adenostoma, Coleogyne. Op. cit.* 41: 319–332, 1914; VIII, *Dryas, Geum, Sieversia, Cowania, Fallugia, Kuntzia* vs. *Purshia, Chamaebatia,*

---

[101] Generic limits cannot be drawn sharply between many of the apparent genera of the family, and by some authors the genera *Malus, Pyrus, Sorbus,* and *Aronia* are considered to comprise a single genus, *Pyrus*. For opinions in support of one or the other of these views cf. references to papers by Bailey (1949) and Fernald (1947).

Cercocarpus. *Op. cit.* 41: 483–503, 1914; IX, *Rubacer, Rubus. Op. cit.* 42: 117–160, 1915; X, *Rubus* hybrids. *Op. cit.* 42: 463–479, 1915; XI, Roses of California and Nevada. *Op. cit.* 44: 65–84, 1917.

SAX, K. The origin of the Pomoideae. Proc. Amer. Soc. Hort. Sci. 30: 147–150, 1934.

WIEGAND, K. M. The genus *Amelanchier* in eastern North America. Rhodora, 14: 117–161, 1912.

WOLF, T. Monographie der Gattung *Potentilla*. Bibliot. Bot. 16 (71): 1–714, 1908.

## LEGUMINOSAE.[102] PEA FAMILY

Herbs, shrubs, or trees; leaves mostly alternate, compound (mostly pinnately so, but sometimes palmately compound) or simple by suppression of leaflets, stipulate or infrequently estipulate; flowers bisexual, commonly zygomorphic (actinomorphic in the Mimosoideae), perianth biseriate, the calyx gamosepalous and 5-lobed, the corolla typically of 5-petals (rarely absent or reduced to a single petal), distinct or the 2 anterior ones basally connate (for descriptive details, see under respective subfamilies), a hypanthium sometimes produced (as in *Arachis*); stamens mostly 10 (sometimes numerous in Mimosoideae), distinct, monadelphous, or diadelphous, the anthers 2-celled, dehiscing by longitudinal slits or infrequently by pores, sometimes with an apical deciduous gland; pistil 1, the ovary superior, 1-loculed, 1-carpelled, the placentation parietal along ventral suture, the ovules 2-many in 2 alternating rows on a single placenta, amphitropous, anatropous or infrequently campylotropous, pendulous or ascending, the style and stigma 1; fruit usually a legume or loment, sometimes follicular, or indehiscent or tardily so, the basal portion sometimes sterile and stalklike and termed the peg (*Arachis*); seed with a usually leathery testa, funiculus sometimes produced into a more or less fleshy aril or callosity, the endosperm none or very scant and hard and glassy.

The Leguminosae, as here defined, generally are considered to be one of the 3 largest families of angiosperms,[103] represented by about 550 genera and perhaps 13,000 species. The family is cosmopolitan in distribution, and is represented in this country by about 115 genera, of which nearly 35 are naturalized from Old World or tropical American sources. Approximately 44 genera are indigenous over a large part of the country; about 20 of these are restricted to the southeast, perhaps a dozen to the south-central region, 8–10 to the Pacific coast region, but none to the northeast.

The family as circumscribed above is divided into 3 subfamilies, distinguished as follows:

1. Flowers actinomorphic, calyx and corolla valvate in bud ............ *Mimosoideae*
1. Flowers zygomorphic, perianth segments predominately imbricate in bud.
   2. Corolla caesalpinaceous, aestivation imbricate-ascending, the posterior petal innermost, the petals typically 5 and distinct ................ *Caesalpinioideae*
   2. Corolla papilionaceous, aestivation imbricate-descending, the posterior petal outermost, the two anterior petals (forming the keel) often basally connate. ................................................................ *Lotoideae*

These 3 subfamilies are treated as distinct families by many botanists (the Mimosaceae, the Caesalpiniaceae, and the Papilionaceae respectively). Authors who accept them as separate families treat the trio as a single order. The relative

---

[102] The name Leguminosae has been conserved for the family (*sensu latiore*). According to the Rules (ed. 3) "those who regard the *Papilionaceae* as constituting an independent family may use that name, although it is not formed in the prescribed manner."

[103] The Compositae are the largest family, the Orchidaceae and Leguminosae compete for second place.

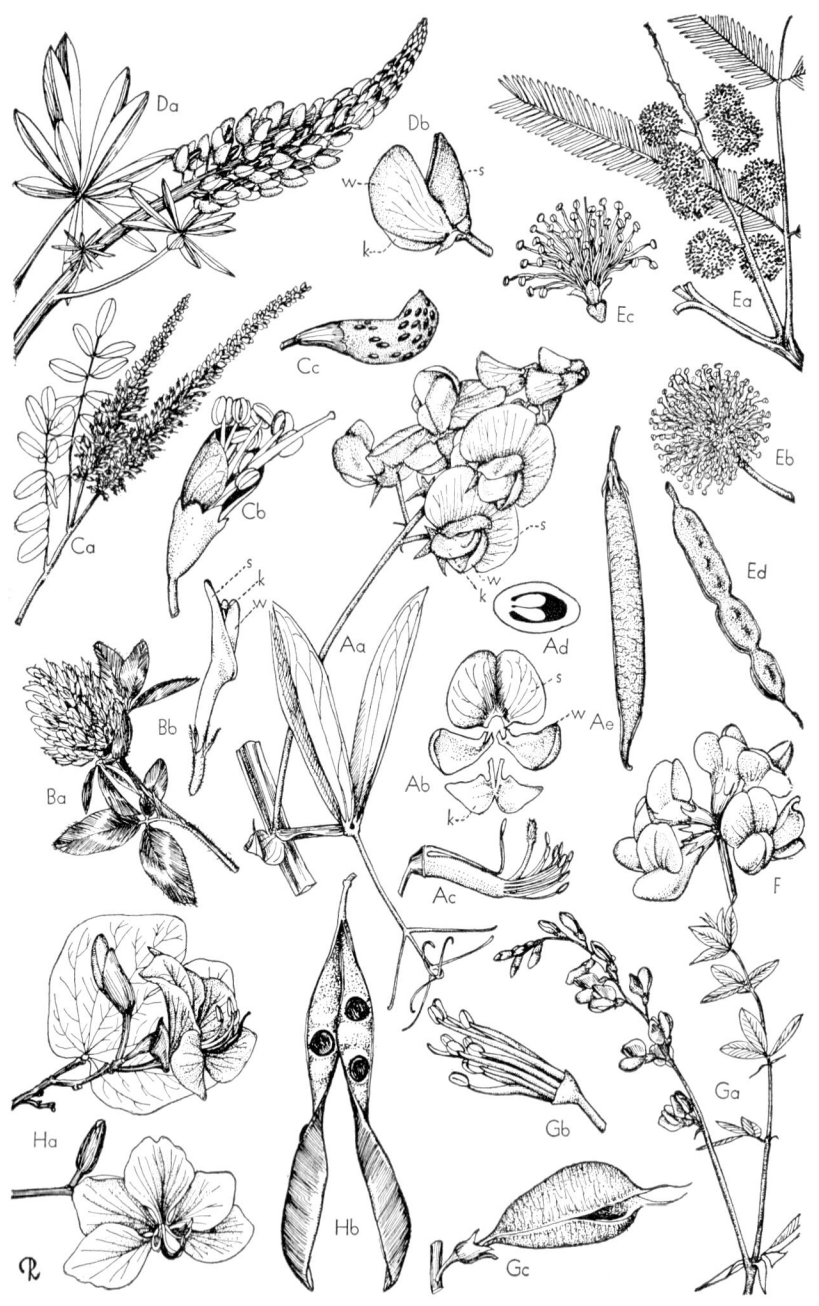

546

DIVISION IV. EMBRYOPHYTA SIPHONOGAMA 547

positions of the taxa remain essentially unchanged in either case [104] and while characters that distinguish one subfamily from the other appear to be more apparent than the characters that bond them together (those of the gynoecium), many botanists, perhaps with undue conservatism and adherence to tradition, have held the Leguminosae to comprise a single family rather than 3 or 4 families. Senn (1943) pointed out that comparative studies of wood anatomy produced "no sharp lines separating the three subfamilies" and that adequate cytological data were lacking for the support or denial of the elevation of each of these to the rank of family. Martin (1946) found seeds of Lotoideae to be distinct morphologically from those of other taxa of the family. The Mimosoideae are accepted to be the most primitive of the 3 taxa, and the Lotoideae the most advanced.

The Mimosoideae are composed of about 40 genera divided among 5 tribes and are almost exclusively tropical or subtropical in distribution. Indigenous genera (together with approximate number of total and indigenous species) include: *Calliandra* (150–5), *Acacia* (350–10), *Prosopis* (10–3), *Pithecellobium* (100–2), *Mimosa* (350–2), and *Schrankia* (10–5).

The Caesalpinioideae are represented by about 135 genera distributed among 9 tribes and, like the Mimosoideae, the majority of the genera are of paleotropical distribution. By conservative views, the indigenous species represent less than a dozen genera, among which are *Cassia* (450–30), *Cercis* (7–2), *Gleditsia* (6–2), *Gymnocladus* (1), *Cercidium* (10–2), *Hoffmanseggia* (20–4), and *Caesalpinia* (35–1).

The Lotoideae [105] with about 375 genera is the largest subfamily and to it belong the majority of legumes of temperate regions of both the northern and southern hemispheres. The subfamily is considered composed of 10 tribes. It is characterized by the usually gamosepalous calyx and the papilionaceous corolla. The latter is a corolla of 5 unequal petals, the posterior petal is outermost and is designated the standard (vexillum), the lateral pair of similar petals is distinct, each usually long-clawed, and together they comprise the wings (alae); the 2 innermost petals are closely appressed, usually coherent or connate along their adjoining margins, and generally envelop the stamens and pistil; they comprise the keel (carina) of the corolla. The stamens, enclosed within the keel, are usually 5 or 10, commonly either monadelphous or diadelphous, and are distinct in only a few genera. Among the more widespread indigenous genera are *Astragalus* (1500–250), *Lupinus* (225–120), *Trifolium* (275–65), *Crotalaria* (250–15), *Hosackia* (50–45),

---

[104] In addition to recognizing as distinct the 3 families mentioned above, some authors consider the unigeneric and wholly American tribe Kramerieae of the Caesalpinioideae to be a fourth taxon deserving family status (Krameriaceae). This view, accepted by the "Britton school," was rejected by Hallier and by Hutchinson, who excluded *Krameria* from the legumes and placed it as derived from or within the Polygalaceae.

[105] In much of the literature this taxon, when treated as a subfamily, has been designated the Papilionatae or the Papilionoideae. For explanation of the validity of the name Lotoideae, see Rehder (1945).

---

**Fig. 167.** LEGUMINOSAE. A, *Lathyrus latifolius*: Aa, flowering branch, × ½; Ab, perianth expanded, × ½; Ac, flower, less perianth, × 1; Ad, ovary, cross-section, × 8; Ae, fruit, × ½. B, *Trifolium pratense*: Ba, flowering branch, × ½; Bb, flower, × 3. C, *Amorpha fruticosa*: Ca, flowering branch, × ¼; Cb, flower, ×4; Cc, fruit with calyx, × 2. D, *Lupinus polyphyllus*: Da, flowering branch, × ¼; Db, flower, × 1. E, *Acacia decurrens*: Ea, flowering branch, × 1; Eb, inflorescence, × 4; Ec, flower, × 5; Ed, fruit, × ½. F, *Lotus corniculatus*: inflorescence, × 1. G, *Baptisia australis*: Ga, flowering branch, × ¼; Gb, flower, less perianth, × 1; Gc, fruit, × ½. H, *Bauhinia variegata*: Ha, flowering branches, × ¼; Hb, fruit (dehisced), × ¼. (k keel, s standard, w wing.) (From L. H. Bailey, *Manual of cultivated plants*, The Macmillan Company, 1949. Copyright 1924 and 1949 by Liberty H. Bailey.)

*Desmodium* (175–35), *Tephrosia* (160–35), *Dalea* (150–50), *Lespedeza* (60–25), and *Baptisia* (20).

Economically the Leguminosae are one of the most important families of flowering plants. They provide many articles of food, fodder, dyes, gums, resins, oils,[106] and in addition to this, members of over 140 genera are grown domestically for ornament. Outstanding among this vast assemblage of food products are the garden peas (*Pisum*), lentils (*Lens*), peanut (*Arachis*), yam bean (*Pachyrhizus*), beans (*Phaseolus*), cowpeas (*Vigna*), velvet beans (*Stizolobium*), and soybean (*Glycine*); fodder and forage plants include clover (*Trifolium*), alfalfa (*Medicago*), soybean (*Glycine*), lupine (*Lupinus*), vetch (*Vicia*), bird's-foot trefoil (*Lotus*), and sweet clover (*Melilotus*); outstanding ornamentals include wisteria (*Wisteria*), sweet pea, (*Lathyrus*), lupine (*Lupinus*), redbud (*Cercis*), orchid tree (*Bauhinia*), royal poinciana (*Delonix*), wattles (*Acacia*), broom (*Cytisus, Genista*), senna (*Cassia*), and albizia (*Albizia*).

LITERATURE:

BARNEBY, R. C. Pugillus Astragalorum. I, Leafl. W. Bot. 3: 97–114, 1944; II, Proc. Calif. Acad. Sci. ser. 4, 15: 147–170, 1944; III-VI, Leafl. W. Bot. 4: 49–63, 65–147, 228–238, 1944–1946; 5: 1–9, 1947. VII, Amer. Midl. Nat. 37: 421–516, 1947. VIII-IX, Leafl. W. Bot. 5: 25–35, 82–89, 1947–1948.

BRITTON, N. L. and ROSE, J. N. Mimosaceae. North Amer. Flora, 23: 1–194, 1928.

———. Krameriaceae. *Op. cit.* 23: 195–200, 1930.

———. Caesalpiniaceae. *Op. cit.* 23: 201–349, 1930. [For Fabaceae, see under Rydberg.]

BURKART, A. Materiales para una monographia del genéro *Prosopsis* (Leguminosae). Darwiniana, 4: 51–128, 1940.

———. Las Leguminosas argentinas silvestres y cultivadas. i-xix, 1–590. Buenos Aires, 1943.

CAPITAINE, L. Étude analytique et phytogeographique du groupe des Legumineuses. Bull. Geogr. Bot. 23 (A): 1–500, 1913. [Incl. keys to genera.]

DORMER, K. J. Vegetative morphology as a guide to the classification of the Papilionatae. New Phytologist, 45: 145–161, 1946.

EASTWOOD, A. A tentative key to the small-flowered lupines of the western United States. Leafl. West. Bot. 4: 217–223, 251–254, 1946.

FASSETT, N. C. The leguminous plants of Wisconsin. 157 pp. Madison, 1939.

FERNALD, M. L. The genus *Oxytropis*, in northeastern America. Rhodora, 30: 137–155, 1928.

FOX, WM. B. The Leguminosae of Iowa. Amer. Midl. Nat. 34: 207–230, 1945.

HOPKINS, M. *Cercis* in North America. Rhodora, 44: 93–211, 1942.

JONES, M. E. Revision of North American species of *Astragalus*. 330 pp., Salt Lake City, 1923.

KRUKOFF, B. A. The American species of *Erythrina*. Brittonia, 3: 205–337, 1939.

LARISSEY, M. M. Monograph of the genus *Baptisia*. Ann. Mo. Bot. Gard. 27: 119–244, 1940.

MACBRIDE, J. F. A revision of the genus *Astralagus*, subgenus *Homalobus*, in the Rocky Mts. Contr. Gray Herb. 65: 28–39, 1922.

MARTIN, A. C. The comparative internal morphology of seeds. Amer. Midl. Nat. 36: 513–660, 1946.

MOORE, J. A. The vascular anatomy of the flower in the papilionaceous Leguminosae, I-II. Amer. Journ. Bot. 23: 279–290, 349–355, 1936.

PALMER, E. J. Conspectus of the genus *Amorpha*. Journ. Arnold Arb. 12: 157–197, 1931.

PIERCE, W. P. Cytology of the genus *Lespedeza*. Amer. Journ. Bot. 26: 736–744, 1939.

PIPER, C. V. Studies in American Phaseolineae. Contrib. U. S. Nat. Herb. 22: 663–701, 1926.

REHDER, A. Notes on some cultivated trees and shrubs, II. Leguminosae subfam. Lotoideae. Journ. Arnold Arb. 26: 477, 1945.

---

[106] For detailed accountings of these products, see those by Bois, Hill, Holland, and Robbins as cited in Chap. XIV.

ROLLINS, R. C. Studies in the genus *Hedysarum* in North America. Rhodora, 42: 217–239, 1940.
RYDBERG, P. A. Fabaceae. North Amer. Flora, 24: 1–462, 1919–1929.
―――. Genera of North American Fabaceae. I-VII. Amer. Journ. Bot. 10: 485–498, 1923; *op. cit.* 15: 195–203, 425–432, 584–595, 1928; *op. cit.* 16: 197–206, 1929; *op. cit.* 17: 231–238, 1930.
SCHUBERT, B. G. *Desmodium*: Preliminary studies, I-II. Contr. Gray Herb. 129: 1–31, 1940; *op. cit.* 136: 78–115, 1941.
―――. The *Hedysarum* of Sesse and Mocino. Contr. Gray Herb. No. 161: 19–25, 1946.
SENN, H. A. Chromosome number relationships in the Leguminosae. Bibliographia Genetica, 12: 175–337, 1938.
―――. Cytological evidence on the status of the genus *Chamaecrista* Moench. Journ. Arnold Arb. 19: 153–157, 1938.
―――. North American species of *Crotalaria*. Rhodora, 41: 317–367, 1939.
―――. The relation of anatomy and cytology to the classification of the Leguminosae. Chron. Bot. 7: 306–308, 1943.
SOUÈGES, R. Embryogénie des Papilionacées. Paris Acad. Sci. Compt. Rend. 226: 609–611, 761–764, 2101–2103, 1948.
SMITH, C. P. Species lupinorum. Nos. 1–31, 1938–1948. [incomplete.]
WHEELER, L. C. *Astragalus* versus *Oxytropis*. Leafl. West. Bot. 2: 209–210, 1939.
WIGGINS, I. L. Taxonomic notes on the genus *Dalea* Juss. Contr. Dudley Herb. 3: 41–64, 1940.
WOOD, C. E. JR. The American barbistyled species of Tephrosia (Leguminosae). Rhodora, 51: 193–231, 233–302, 305–364, 369–384, 1949.

## Order 26. GERANIALES [107]

The Geraniales are characterized by the stamens typically twice as many as sepals in 2 whorls or the outer whorl missing, and the ovules pendulous with a ventral raphe and the micropyle pointing upwards or erect with a dorsal raphe and the micropyle pointing downwards. In addition, the ovary is syncarpous, the styles often persistent in fruit, and the seeds generally lack endosperm.

The order was treated by Engler and Diels to be composed of the following 6 suborders and 21 families:[108]

Geraniineae
  Oxalidaceae
  Geraniaceae
  Tropaeolaceae
  Linaceae
  Erythroxylaceae
  Zygophyllaceae
  *Cneoraceae
  Rutaceae
  Simaroubaceae
  Burseraceae
  Meliaceae
  *Akariaceae

Malpighiineae
  Malpighiaceae
  *Trigoniaceae
  *Vochysiaceae
Polygalineae
  Tremandraceae
  Polygalaceae
Dichapetalineae
  *Dichapetalaceae
Tricocceae
  Euphorbiaceae
  *Daphniphyllaceae
Callitrichineae
  Callitrichaceae

Bessey was the only other phylogenist to include all of the above families within the order. Wettstein, Rendle, and Hallier considered them to comprise 3 orders. Hutchinson distributed the families of the Geraniineae among 4 orders (Geraniales, Rutales, Meliales, and Malpighiales), and the other families among 5 other orders

---

[107] Order 25, the Pandales, is composed of a single west African species, the dioecious *Panda oleosa*.
[108] Families whose names are preceded by an asterisk are not accounted for in the treatment that follows.

as discussed below. The accumulation of evidence from all fields of botany indicates that the Geraniales (*sensu* Engler) are not a natural taxon and that they may yet be accepted as comprising 4 or 5 more or less disjunctive orders.

## OXALIDACEAE. OXALIS FAMILY

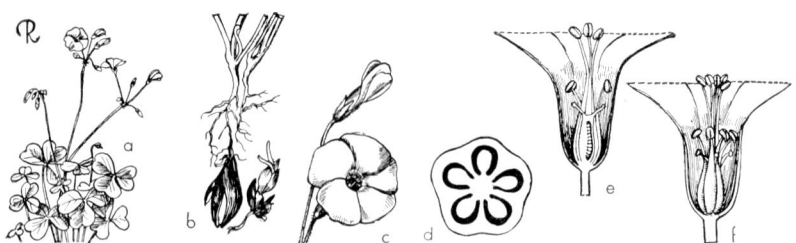

**Fig. 168.** OXALIDACEAE. *Oxalis Bowiei*: a, flowering plant, × ⅛; b, stems showing subterranean bulbs, × ¼; c, flower and bud, × ½; d, ovary, cross-section, × 10; e, flower, vertical section, × 2; f, flower, perianth removed, × 2. (From L. H. Bailey, *Manual of cultivated plants,* The Macmillan Company, 1949. Copyright 1924 and 1949 by Liberty H. Bailey.)

Herbs (sometimes suffrutescent) or shrubs, rarely arborescent (as in *Averrhoa*), often producing fleshy rhizomes or bulblike tubers; leaves alternate, pinnately or palmately compound or simple by suppression of leaflets, petioled, the leaflets folded back in bud and at night, estipulate; flowers bisexual (cleistogamous flowers sometimes produced), actinomorphic, solitary umbellate cymose or racemose, the perianth biseriate (apetalous when cleistogamous), the calyx of 5 lobes or segments, imbricate, the corolla of 5 petals, distinct or sometimes basally connate, contorted in bud; stamens 10 (5 sometimes reduced to staminodes), basally connate, hypogynous, in 2 series with those of outer series opposite the petals, the anthers 2-celled, introrse, dehiscing longitudinally; pistil 1, the ovary superior, 5-loculed, 5-carpelled, the placentation axile, the ovules 1 or more in each locule, anatropous, the styles 5 and distinct, persistent, the stigmas terminal on each style and usually capitate, sometimes shortly divided; fruit a loculicidal capsule, or rarely a berry; seed sometimes arillate (the aril separating from testa elastically and expelling seeds explosively from capsule), the embryo straight, enveloped by soft-fleshy endosperm.

A family of 7 genera and about 1000 species, mostly pantropical, with numbers decreasing as ranges extend north and south into temperate zones. *Oxalis* (about 850 species) is represented domestically by about 25 species,[109] and representatives of the genus occur over much of the country.

Members of the Oxalidaceae are distinguished from those of related families by the shortly monadelphous stamens, the 5 distinct styles, the single quinquilocular pistil, the usually palmately compound leaves, and the characteristic fruit dehiscence of those members having arillate seeds.

Most authors (including Hallier, Bessey, Wettstein, Hutchinson, and Rendle) agree that the family belongs in the Geraniales, and treat it as having affinities with the Geraniaceae (with which they were united by Bentham and Hooker).

Economically, members of the family are of little importance. Two species of the Asiatic genus *Averrhoa* are large trees cultivated for ornament in warm parts of

---

[109] By J. K. Small, the southeastern members of the genus were treated as representing 4 genera: *Oxalis, Bolboxalis, Ionoxalis,* and *Xanthoxalis.*

# DIVISION IV. EMBRYOPHYTA SIPHONOGAMA

the country and the gooseberrylike fruits of the cultivated Carambola (*A. Carambola*) are edible. Over 30 species of *Oxalis* are offered in the trade as ornamentals.

*LITERATURE:*

HANKS, L. T. and SMALL, J. K. Oxalidaceae. North Amer. Flora, 25: 25–58, 1907.
KUNTH, R. Oxalidaceae. In Engler, Das Pflanzenreich, 95: (IV. 130) 1–481, 1930.
——. Oxalidaceae. In Engler and Prantl, Die natürlichen Pflanzenfamilien, ed. 2, Bd. 19a: 11–42, 457, 1931.

## GERANIACEAE. GERANIUM FAMILY

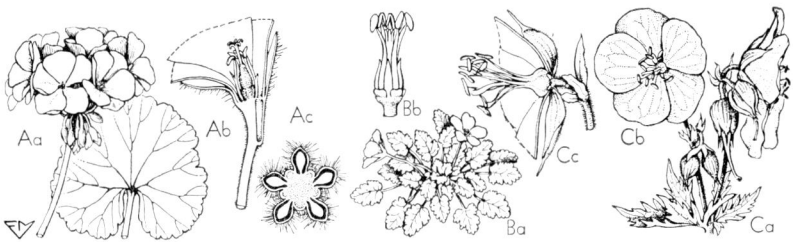

Fig. 169. GERANIACEAE. A, *Pelargonium zonale*: Aa, inflorescence and leaf-blade, × ¼; Ab, flower, perianth in vertical section, × 1; Ac, ovary, cross-section, × 5. B, *Erodium chamaedryoides*: Ba, plant in flower, × ½; Bb, flower, less perianth, × 3. C, *Geranium pratense*: Ca, inflorescence, × ½; Cb, flower, face view, × ½; Cc, flower, perianth in vertical section, × 1. (From L. H. Bailey, *Manual of cultivated plants*, The Macmillan Company, 1949. Copyright 1924 and 1949 by Liberty H. Bailey.)

Mostly herbaceous plants, sometimes suffrutescent or shrubby, the stems often fleshy; leaves alternate or opposite, compound or if simple then lobed or divided, rarely glabrous, the venation mostly palmate, stipules present; flowers bisexual, actinomorphic or zygomorphic (*Pelargonium* and sometimes *Erodium*), the inflorescence cymose or umbellate (rarely reduced to a 1-flowered peduncle), bracteate, determinate, the perianth biseriate, the calyx of 5 usually distinct and imbricated sepals (4 in *Vivania*, 8 in *Dirachma*), the corolla of 5 distinct petals (rarely 8, 4, 2, or none), nectiferous glands usually alternating with the petals; the stamens typically 5–15, in 1–3 whorls of 5 each with androecia of 1 or 2 whorls sometimes with stamens reduced to antherless or scalelike staminodes, sometimes basally connate (rarely connate in 5 bundles of 3 each as in *Monsonia*), the anthers 2-celled, dehiscing longitudinally; pistil 1 and 3–5-lobed, the ovary superior, the locules and carpels typically 3–5 (8 in *Dirachma*), the placentation axile, the ovules usually 1–2 in each locule (many in *Balbisia*), pendulous, anatropous, the styles 3–5, slender (beaklike), stigmas of same number and ligulate (rarely capitate); fruit capsular, dehiscing septicidally (loculicidal in *Vivania*) into as many 1–2- or many-seeded usually dehiscent "mericarps" as carpels, the styles usually adhering to the ovarian beak and the basal portion recurving elastically and sometimes spirally (*Erodium*); seed with plicate cotyledons incumbent, the embryo mostly curved, endosperm usually none (present in *Biebersteinia*).

A family of 11 genera and about 850 species widely distributed over temperate and subtropical regions of the northern and southern hemispheres. The genera *Geranium* (375–21) and *Erodium* (75–3) are represented domestically by widely distributed indigens. Other genera include *Pelargonium* (250 spp.), *Monsonia*, and *Sarcocaulon* of the Old World, and *Vivania* (28 spp.), and *Balbisia* (6 spp.) of South America.

The family is characterized by the typically 5-merous flowers that produce a beaked or lobed fruit distinctive by its usually elastic dehiscence, the separation of the "mericarps," and the usual absence of endosperm.

Economically the family is important primarily for the florist's geranium (*Pelargonium zonale*) and for other species and hybrids of the genus grown for aromatic foliage with its aromatic oils and for flowers. A few species of crane's-bill (*Geranium*) and stork's-bill (*Erodium*) are cultivated as garden ornamentals.

*LITERATURE:*

FERNALD, M. L. *Geranium carolinianum* and allies of northeastern North America. Rhodora, 37: 295–301, 1935.

JONES, G. N. and F. F. A revision of the perennial species of *Geranium* of the United States and Canada. Rhodora, 45: 5–26, 32–53, 1943.

KNUTH, R. Geraniaceae. In Engler, Das Pflanzenreich, 53 (IV. 129): 1–640, 1912.

———. Geraniaceae. In Engler and Prantl, Die natürlichen Pflanzenfamilien, ed. 2, Bd. 19a: 43–66, 1931.

MOORE, H. E. JR. A revision of the genus *Geranium* in Mexico and Central America. Contr. Gray Herb. 146: 1–108, 1943.

## TROPAEOLACEAE. NASTURTIUM FAMILY

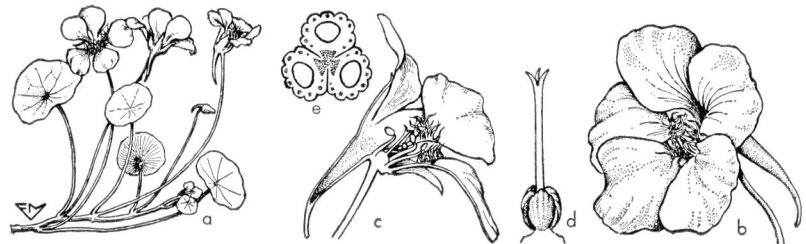

**Fig. 170.** TROPAEOLACEAE. *Tropaeolum majus*: a, flowering branch, × ⅙; b, flower, × ½; c, flower, vertical section, × ½; d, pistil, × 2; e, ovary, cross-section, × 5. (From L. H. Bailey, *Manual of cultivated plants,* The Macmillan Company, 1949. Copyright 1924 and 1949 by Liberty H. Bailey.)

Somewhat succulent herbs with watery acrid sap, mostly prostrate, twining, or clambering; leaves alternate, simple (infrequently pinnately compound), peltate, sometimes lobed or dissected, estipulate; flowers bisexual, zygomorphic and spurred, solitary and axillary (rarely in umbels), the perianth biseriate, the calyx bilabiate, of 5 distinct sepals, the dorsal one produced into a spur, the corolla of 5 distinct (usually clawed) petals, imbricated, the upper 2 petals differing in shape from the lower 3 (usually smaller and situated in the opening of the spur); stamens 8 in 2 whorls, distinct, somewhat perigynous, unequal, declinate, the anthers 2-celled, dehiscing longitudinally; pistil 1, the 3-lobed ovary superior, 3-loculed, 3-carpelled, the placentation axile, each locule with a single pendulous anatropous ovule, the style 1 and apical, the stigmas 3 and linear; fruit a 3-seeded schizocarp, each mericarp separating from the short axis and remaining indehiscent, usually rugose or furrowed; seed with straight embryo, usually lacking endosperm.

A unigeneric family (*Tropaeolum*) of about 50 species, distributed mostly in mountainous regions from Mexico south to central Chile and Argentina. None is indigenous to this country.

The family, now accepted by most authors, once was included within the Geraniaceae from which it differs by the distinct stamens and the fruit a schizocarp splitting into three 1-seeded mericarps. Unlike the Geraniaceae, no beak is produced on the top of the *Tropaeolum* ovary.

Eight species are cultivated domestically for ornament, notably the nasturtium (*T. majus*) and the canary-bird flower (*T. peregrinum*).

LITERATURE:

BUCHENAU, F. Tropaeolaceae. In Engler, Das Pflanzenreich, 10 (IV. 131): 1–36, 1902.
FERENHOLTZ, H. Tropaeolaceae. In Engler and Prantl, Die natürlichen Pflanzenfamilien, ed. 2, Bd. 19a: 67–82, 1931.
NASH, G. V. Tropaeolaceae. North Amer. Flora, 25: 89–91, 1910.

## LINACEAE. FLAX FAMILY

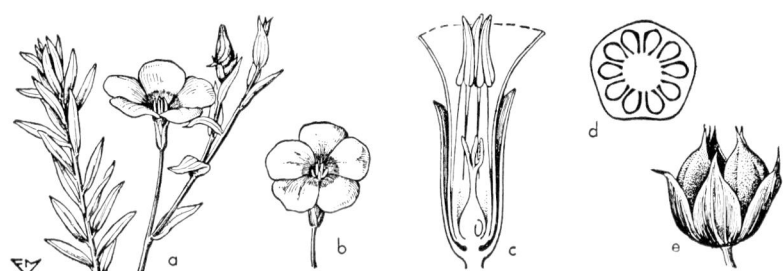

Fig. 171. LINACEAE. *Linum grandiflorum*: a, flowering and sterile branch, × ½; b, flower, × ½; c, flower, vertical section, petals excised, × 3; d, ovary, cross-section, × 8; e, capsule, × 2. (From L. H. Bailey, *Manual of cultivated plants*, The Macmillan Company, 1949. Copyright 1924 and 1949 by Liberty H. Bailey.)

Mostly herbs, sometimes shrubs; leaves alternate or opposite, rarely whorled, simple, entire, stipules present or absent; flowers bisexual, actinomorphic, the inflorescence a dichasial cyme or cincinnus (sometimes appearing racemose), the perianth biseriate, the calyx of 5 distinct or basally connate sepals (rarely 4-parted with 3-fid lobes), imbricate, the corolla usually of 5 petals (rarely 4), the petals contorted in bud, distinct, often clawed and when so the claw naked or crested, early deciduous; the stamens 5 and alternate with petals if 10 or more, the filaments basally connate to form a ring outside of which may be nectiferous glands, sometimes the 5 stamens alternated with 5 or 10 toothlike staminodes, the anthers 2-celled, introrse, dehiscing longitudinally; pistil 1, the ovary superior, the carpels and locules 5 (rarely 3–4) or falsely 10-loculed by the intrusion of carpel midribs, the placentation axile, the ovules typically 2 in each locule, pendulous, anatropous, the styles as many as ovary locules, distinct, filiform, each terminated by a capitate stigma; fruit a septicidal capsule or drupe (in some Hugonieae) surrounded by the persistent calyx; seed with a usually straight embryo, the endosperm copious (*Linum*), scant or absent.

A family of 9 genera and 200 species, of cosmopolitan distribution, and primarily of temperate regions of both northern and southern hemispheres. The Linaceae are represented in this country by about 35 indigenous species of *Linum* (including *Cathartolinum* and *Hesperolinon*), the largest genus, composed of about 140 species. Other genera include *Radiola* (1), *Reinwardtia* (2), *Hugonia* (20 spp.)

The family is distinguished by the generally 5-merous flowers, with the contorted corolla composed of distinct and usually clawed fugacious petals, and by the shortly connate filaments and septicidally dehiscent capsule.

Bessey and Rendle included it within the Geraniales, as did Wettstein. Hallier transferred it to the Guttales as a derivative from the Ochnaceae and considered it, or stocks directly ancestral to it, to be the progenitor of several evolutionary lines including those giving rise to the Sapotales, the Scrophulariaceae, and others of the Tubiflorae, the Passiflorales, and probably the Polygalales (see Fig. 14).

The family is important for the flax plant (*Linum usitatissimum*) that is widely cultivated as a fiber plant (from whose fibers linen cloth is made) and for the seeds (the source of linseed oil). About 25 species of *Linum* and one of *Reinwardtia* are grown domestically as ornamentals.

*LITERATURE:*

NESTLER, H. Beiträge zur systematichen Kenntnis der Gattung *Linum*. Beih. Bot. Centralbl. 50 (2): 497–551, 1933.

SMALL, J. K. Linaceae. North Amer. Flora, 25: 67–87, 1907.

WINKLER, H. Linaceae. In Engler and Prantl, Die natürlichen Pflanzenfamilien, ed. 2, Bd. 19a: 82–130, 1931.

## ERYTHROXYLACEAE. COCA FAMILY

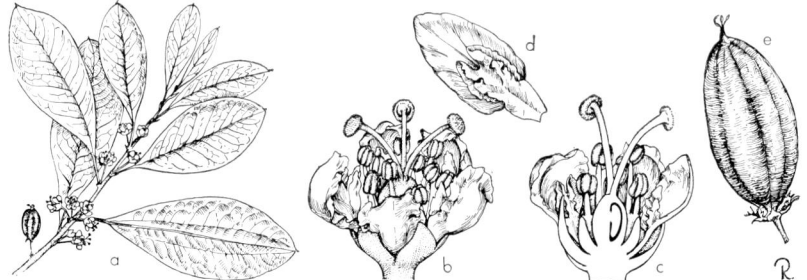

**Fig. 172.** ERYTHROXYLACEAE. *Erythroxylon Coca*: a, flowering branch, $\times$ ½; b, flower habit, $\times$ 5; c, same, vertical section, $\times$ 5; d, petal, ventral side, $\times$ 3; e, fruit, $\times$ 3.

Shrubs or small trees; leaves alternate (opposite in *Aneulopha*), simple, entire to crenate, stipulate (intrapetiolar); flowers bisexual, actinomorphic, inconspicuous, and solitary or fasciculate in leaf axils or the inflorescence a thyrse, the perianth biseriate, the calyx campanulate, sepals 5 (rarely 6), distinct and usually persistent, imbricate or valvate, the corolla somewhat rotate, the petals 5, distinct, convolute or imbricate, with bifid ligulate appendages or with callosities on inner face; stamens usually 10 in 2 whorls, more or less basally connate into a tube and externally glandular, the anthers 2-celled, dehiscing longitudinally; pistil 1, the ovary superior, 3-carpelled, 3-loculed but usually with only 1 locule developing in fruit, the placentation axile, the fertile locule with 1–2 functional ovules, the ovules pendulous, anatropous, the styles 3 and distinct or basally connate, each terminated by a clavate or capitate stigma or the stigmatic surface obliquely depressed; fruit basically a berry, sometimes, drupaceous; seed with a straight embryo, endosperm present and fleshy or rarely absent.

A family of 3 genera (*Erythroxylon* 200 spp., *Nectaropetalum* 4 spp., and *Aneulophus* 1 sp.), largely of the American tropics. None of them is indigenous to this country, and while the northern limits of distribution extend into Mexico and Cuba, the species reach their greatest development in South America. A few species are African.

The Erythroxylaceae are distinguished by the monadelphous stamens, the appendages on the petals, and the drupaceous fruit. They were included in the Linaceae by Bentham and Hooker and by Hallier, but subsequent workers (including Bessey) have considered them a separate family and usually as a part of the Geraniales (Hutchinson transferred them to his Malpighiales).

Economically the family is of greatest importance for the cocaine extracted from the leaves of coca [110] (*Erythroxylon Coca*). This shrub grows to a height of 12–15 feet and is also grown for ornament or as a novelty in warmer parts of the country.

*LITERATURE:*
BRITTON, N. L. Erythroxylaceae. North Amer. Flora, 25: 59–66, 1907.
SCHULZ, O. E. Erythroxylaceae. In Engler and Prantl, Die natürlichen Pflanzenfamilien, ed. 2, Bd. 19a: 130–143, 1931.
TIEGHEM, P. VAN. Structure et affinités des Erythroxylacées. Un nouvel exemple de cristarque. Bull. Mus. Hist. Nat. Paris, 9: 287–295, 1903.

## ZYGOPHYLLACEAE.[111] CALTROP FAMILY

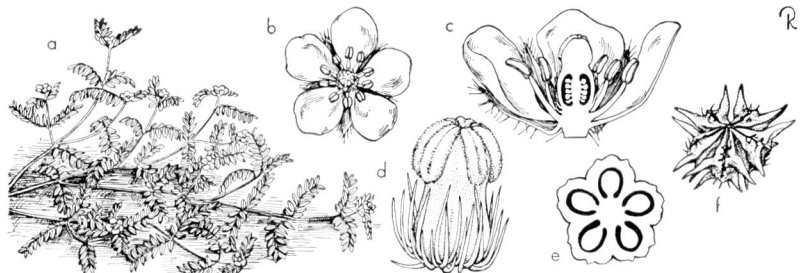

**Fig. 173.** ZYGOPHYLLACEAE. *Tribulus terrestris*: a, flowering stems, × ¼; b, flower, face view, × 2; c, same, vertical section, × 4; d, gynoecium, × 8; e, ovary, cross-section, × 10; f, fruit, × ½.

Mostly herbs and shrubs, rarely trees (*Guaiacum* spp.); branches often jointed at nodes; leaves opposite (alternate in few genera), often fleshy to coriaceous, mostly pinnately compound (occasionally simple or 2-foliolate), persistent stipules present and usually leathery, hairy, fleshy, or spinescent; flowers bisexual (unisexual and the plants dioecious in *Neoleuderitzia*), actinomorphic or rarely zygomorphic, solitary paired or in cymes, the perianth biseriate, the whorls imbricate or rarely valvate, the calyx persistent, sepals usually 5 (4), distinct or rarely basally connate, the corolla rarely absent, the petals as many as sepals, distinct, the disc usually present, convex or depressed, rarely annular; stamens hypogynous, in 1, 2, or 3 whorls of 5 stamens each, often unequal, outer ones usually opposite petals, the filaments distinct, naked or with basal scales, the anthers 2-celled, introrse, dehiscing longitudinally; pistil 1, the ovary superior, usually furrowed, angled, or winged, sessile or rarely on short gynophore, usually 4–5-loculed and -carpelled, placentation axile, the ovules usually 2–many (rarely 1) on each placenta, pendulous, the style 1, angular or furrowed, the stigma usually 1 and simple (rarely styles 4–5 and distinct or the stigma discoid); fruit a loculicidal or septicidal capsule, rarely a drupaceous berry; seed with straight or curved embryo and the endosperm copious and hard or none.

[110] Not to be confused with cocoa or cocas, a product of the fruit of *Theobroma Cacao* of the Sterculiaceae.
[111] The name Zygophyllaceae has been conserved over all earlier names applied to this family, including that of Nitrariaceae Lindley (1830).

A family of about 27 genera and nearly 200 species, primarily pantropical with ranges extending into temperate regions of northern and southern hemispheres. The members are especially abundant in the drier areas of the Mediterranean regions. The family is represented domestically by 1 indigenous species of *Guaiacum* in southern Florida, by 6 of *Kallstroemia* that occur across the southern parts of the country, and by 1 each of *Fagonia, Larrea, Porlieria,* and *Peganum* in the southwest. *Tribulus* is represented by the widely naturalized species of *T. terrestris.* The largest genera are *Zygophyllum* (100 spp.), mostly African and Australian, and *Fagonia* (40 spp.), predominantly of the Old World.

The Zygophyllaceae are characterized by the usually pinnate or 2-foliolate leaves and the paired persistent stipules, by the flowers with a disc and the distinct stamens bearing basal scales, and by the usually 4–5-loculed ovary terminated by a single style. Most phylogenists have retained them in the Geraniales, except Hutchinson, who included them in his Malpighiales.

Economically the family is important for lignum vitae (*Guaiacum officinale*), the hardest and most dense of woods. This species and others of the genus and of *Zygophyllum, Tribulus,* and *Larrea* are grown domestically for ornament, mostly in the warmer regions.

*LITERATURE:*

DESCOLE, H. R., O'DONNELL, C. A., and LOURTEIG, A. Revisión de las Zigofilaceas Argentinas. Lilloa, 5: 257–352, 1940.

ENGLER, A. Zygophyllaceae. In Engler and Prantl, Die natürlichen Pflanzenfamilien, ed. 2, Bd. 19a: 144–184, 1931.

VAIL, A. M. and RYDBERG, P. A. Zygophyllaceae. North Amer. Flora, 25: 103–116, 1910.

## RUTACEAE.[112] RUE FAMILY

Herbs, shrubs, and trees, glandular-punctate and often strong-smelling; leaves alternate or opposite, simple or palmately or pinnately compound (often heathlike in xerophytic members), sometimes reduced to spines, estipulate; flowers bisexual (sometimes unisexual and the plants then dioecious in some Zanthoxyleae), actinomorphic (zygomorphic in a few genera of Ruteae and Cuspariinae), in varying types of inflorescence, the perianth typically biseriate, usually imbricate, the sepals 3–5 and distinct or basally connate, the corolla usually present, hypogynous or perigynous, the petals 3–5, rarely none, usually distinct (gamopetalous in *Correa* and in the Cuspariinae), the disc present, situated between stamens and ovary, varying from an apparent receptacular swelling to a cuplike structure; the stamens 3–10, or more, basically in 2 whorls, those of the outer ones usually opposite the petals (the disc itself may represent vestigial remains of a third and innermost whorl), all stamens attached at base or rim of disc, some occasionally reduced to staminodes, distinct or rarely basally connate, rarely adnate to petals, usually straight and unequal, sometimes declinate, the anthers 2-celled, introrse, dehiscing longitudinally, sometimes with basal appendages, the connective often with glandular apex; gynoecium of usually a single pistil, but sometimes the carpels weakly connate or only basally or apically connate, the ovary superior, usually deeply lobed, typically 4–5-loculed and -carpelled with placentation axile,[113] the ovules 1–2 or several in each locule, collateral or superposed, the styles as many as carpels and distinct or connate and seemingly 1, each containing a continuous stylar canal, stigma 1; fruit a valvate capsule, or a leathery-rind berry (hesperidium), or

[112] The name Rutaceae has been conserved over other names for this family. Synonyms include Aurantiaceae, Pteleaceae, and Zanthoxylaceae.

[113] In *Feronia*, of the Aurantioideae, the ovary is unilocular with typically 5 parietal placentae.

separating into mericarps, or sometimes a winged berry or drupe or samara; seed with a large straight or bent embryo, the endosperm present and fleshy or absent.

A family of about 140 genera and 1300 species, widely distributed in the temperate and tropical regions of northern and southern hemispheres, but most numerous in South Africa and Australia. Genera with species indigenous to this country include *Ptelea* (10 spp.), *Zanthoxylum* (150–4), and *Amyris* (15–2). The first 2 genera extend as far north as Minnesota and Ontario, the indigenous species of the third are restricted to southern Florida. Genera having naturalized species include *Ruta, Glycosmis, Triphasia, Poncirus,* and *Citrus*.

**Fig. 174.** RUTACEAE. A, *Citrus Aurantium*: Aa, twig in flower, × ½; Ab, flower, vertical section, × 1. B, *Dictamnus albus*: Ba, flowering branch, × 1/10; Bb, flower, × ½; Bc, same, vertical section, × ½; Bd, ovary, vertical section, × 3; Be, same, cross-section, × 3. C, *Ptelea trifoliata*: fruiting branch, × ½. (From L. H. Bailey, *Manual of cultivated plants,* The Macmillan Company, 1949. Copyright 1924 and 1949 by Liberty H. Bailey.)

The Rutaceae are distinguished by the presence of translucent pellucid dots in the foliage, the lobed ovary elevated on a disc, and the outer stamens usually opposite the petals. The development of oil glands producing an aromatic oil is a characteristic feature of the family.

The family, together with Simaroubaceae and Burseraceae, was segregated by Hutchinson into his Rutales, a view held earlier by Rendle (and in principle by Wettstein (who called the order the Terebinthales), who included also the Meliaceae. Hallier attached special phyletic significance to the family, deriving it from stocks ancestral to the Berberidaceae and considering it to be the progenitor of his Terebinthales (an order in which he placed the amentiferous families as the advanced taxa). The narrower circumscriptions by Rendle, Wettstein, and Hutchinson are supported by morphological and anatomical features, particularly by the predominantly woody character, the usual presence of oil ducts or glands, and the disc formations around the base of the ovary. The disc in members of the Aurantioideae was concluded by Tillson and Bamford (1938) to be a vestigial third whorl of stamens.

The family contains many members of economic importance. Notable among them are the citrus fruits: orange, grapefruit, tangerine, lime, citron, and lemon (all from *Citrus*), and kumquat (*Fortunella*). Ornamentals include the cork tree (*Phellodendron*), hop tree (*Ptelea*) prickly ash (*Zanthoxylum*), dittany (*Dictamnus*), common rue (*Ruta*), trifoliolate orange (*Poncirus*), orange jessamine (*Murraya*), and Cape chestnut (*Calodendrum*).

## LITERATURE:

ENGLER, A. Rutaceae. In Engler and Prantl, Die natürlichen Pflanzenfamilien, ed. 2. Bd. 19a: 187–359, 458–459, 1931.

GREENE, E. L. The genus *Ptelea* in western and southwestern United States and Mexico. Contr. U. S. Nat. Herb. 10: 49–78, 1906.

MOORE, J. A. Floral anatomy and phylogeny in the Rutaceae. New Phytologist, 35: 318–322, 1936.

RECORD, S. J. and HESS, R. W. American woods of the family Rutaceae. Tropical Woods, 64: 1–28, 1940.

SWINGLE, W. T. A new taxonomic arrangement of the orange subfamily, Aurantioideae. Journ. Washington Acad. Sci. 28: 530–533, 1938.

———. The botany of *Citrus* and the wild relatives of the orange subfamily. In The Citrus Industry, 1: 129–474, 1943.

TANAKA, TYOZABURO. The taxonomy and nomenclature of Rutaceae-Aurantioideae. Blumea, 2: 101–110, 1936.

TILLSON, A. H. and BAMFORD, R. The floral anatomy of the Aurantioideae. Amer. Journ. Bot. 25: 780–793, 1938.

WILSON, P. Rutaceae. North Amer. Flora, 25: 173–224, 1911.

———. Notes on Rutaceae—V. Species characters in *Ptelea* and *Taravalia*. Bull. Torrey Bot. Club, 38: 295–297, 1911.

## SIMAROUBACEAE.[114] QUASSIA FAMILY

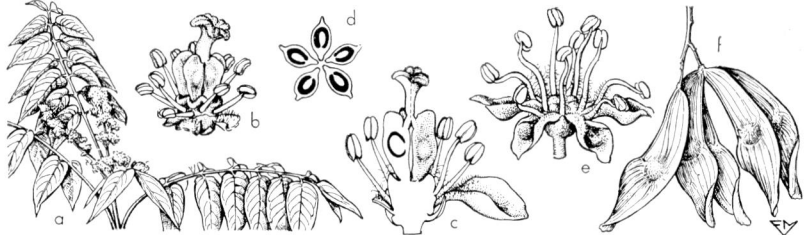

**Fig. 175.** SIMAROUBACEAE. *Ailanthus altissima*: a, flowering branch, much reduced; b, pistillate flower (perianth partially excised), × 4; c, same, vertical section, × 5; d, gynoecium, cross-section through the ovaries, × 6; e, staminate flower, habit, × 3; f, fruit, × ½. (From L. H. Bailey, *Manual of cultivated plants*. The Macmillan Company, 1949. Copyright, 1924 and 1949 by Liberty H. Bailey.)

Shrubs or trees; leaves alternate, pinnately compound (rarely simple, as in *Suriana*), lacking pellucid dots, stipules usually absent (plants essentially leafless in *Holacantha*); flowers unisexual by abortion (plants mostly dioecious) or less frequently bisexual, actinomorphic, in axillary racemes, compound panicles or cymose spikes, perianth biseriate, the calyx gamosepalous, lobes 3–8, imbricate or valvate, corolla polypetalous or sometimes the petals absent; the stamens as many as petals or twice as many (rarely numerous), often appendaged by scales at base, borne on or at base of disc, distinct, the anthers 2-celled, dehiscing longitudinally; gynoecium on a short broad gynophore (disc), comprised of 2–5 uniloculate simple pistils (rarely only 1) or the pistils basally connate (or connate by styles) into a lobed 2–8-loculed and -carpelled ovary with axile placentation, the ovules solitary or in pairs (rarely more) in each locule, the styles 2–8, distinct or connate, some-

---

[114] This family name is based on that of its type genus *Simarouba*, established and so spelled by Aublet (1775) to commemorate the barbaric Carib name for plants of the genus. The spelling was changed by de Candolle (1811) to *Simaruba*, but as pointed out by Sprague (Kew Bull. 1929, p. 243) the original spelling must be used. Simaroubaceae (not Simarubaceae) is the correct spelling of the family name.

times wanting; fruit a capsule, schizocarp, or samara, rarely a berry or drupe; seed with straight or curved embryo, the endosperm scant or none.

A family of mostly pantropical distribution, of about 32 genera and 200 species, and a few representatives extending into temperate regions. It is represented in the warmer areas of the southeastern part of this country by indigenous species of the genera *Castela* [*Castelaria*] (9-1), *Holocantha* (2-1), *Suriana* (1-1), *Simarouba* (6-1), *Picramnia* (40-1), *Alvaradoa* (5-1), and by the widely naturalized subtropical to temperate Asiatic genus, *Ailanthus* (15-1). Cronquist (1945) treated the New World indigens as members of 12 genera.

The family was excluded from the Geraniales by Wettstein, Rendle, Hutchinson, and Hallier, and segregated with others as a member of the Rutales (Terebinthales *fide* Hallier) for reasons given above under Rutaceae. It differs from the Rutaceae by the absence of pellucid dots in the foliage, the predominance of unisexual flowers, and (when present) by the scales on the filament bases.

Economically the family is important for various bitters prepared from the bitter principle present in the bark of most members, and domestically for a few ornamentals: Chinese tree of heaven (*Ailanthus altissima*) and 1 shrubby half-hardy species of *Picrasma*.

*LITERATURE:*

CRONQUIST, A. Studies in the Simaroubaceae, I. The genus *Castela*. Journ. Arnold Arb. 25: 122-128, 1944; II. The genus *Simarouba*. Bull. Torrey Bot. Club, 71: 226-234, 1944. III. The genus *Simaba*. Lloydia, 4: 81-92, 1944; IV. Resume of the American genera. Brittonia, 5: 128-147, 1944. [Cf. also, *op. cit.* 5: 469-470, 1945].

ENGLER, A. Simarubaceae. In Engler and Prantl, Die natürlichen Pflanzenfamilien, ed. 2, Bd. 19a: 359-405, 1931.

SMALL, J. K. Simaroubaceae. North Amer. Flora, 25: 227-239, 1911.

WEBBER, I. E. Systematic anatomy of the woods of the Simaroubaceae. Amer. Journ Bot. 23: 577-587, 1936.

## BURSERACEAE.[115] BURSERA FAMILY

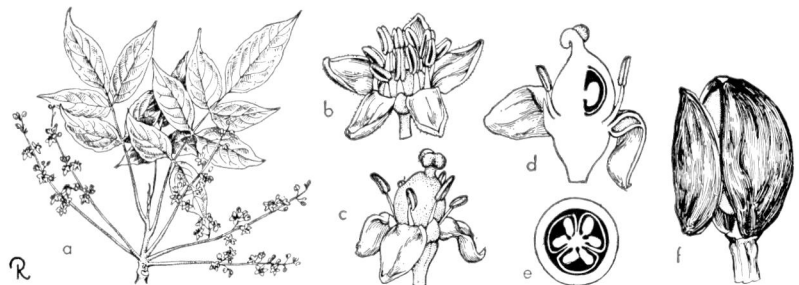

Fig. 176. BURSERACEAE. *Bursera Simaruba*: a, flowering branch, × ⅓; b, staminate flower, habit, × 4; c, pistillate flower, habit, × 4; d, pistillate flower, vertical section, × 5; e, ovary, cross-section, × 12; f, fruit, × 2.

Deciduous shrubs or large trees, aromatic oil- or resin-secreting; leaves alternate (rarely opposite), usually pinnately compound or decompound or infrequently reduced to a single leaflet, rachis often winged, estipulate; flowers bisexual or uni-

[115] Sometimes known as the torchwood family. The name Burseraceae has been conserved over other names for the family.

sexual (plants then polygamodioecious), actinomorphic, minute, usually solitary or in panicles, the perianth biseriate, calyx and corolla imbricate or valvate, the sepals 3–5, more or less basally connate, the petals 3–5 (rarely absent) and alternate with sepals, usually distinct; disc present, annular to cup-shaped, rarely absent, sometimes adnate to the calyx; stamens in 1–2 whorls, of same or typically double the number of petals, hypogynous, distinct, sometimes unequal, the outer ones opposite the petals, the anthers 2-celled, dehiscing longitudinally; pistil 1, the ovary superior, 2–5-loculed and -carpelled, the placentation axile, the ovules 2 (rarely 1) and collateral on each placenta, the style 1 or none, the stigma lobes usually as many as carpels; pistillate flowers (when present) often with staminodes; fruit a 1–5-seeded berry [116] with or without a valvate epicarp, or sometmes a tardily dehiscent capsule; seed with straight or rolled embryo, no endosperm present.

A family of about 20 genera and 500–600 species whose greatest development is in tropical America and northeastern Africa. It is represented in this country by 3 species of *Bursera* (*Elaphrium*): *B. Simaruba,* the gumbo limbo, that extends northward into southern Florida, and 2 species (*B. odorata* and *B. microphylla*) that extend from Mexico into Arizona and adjoining southwestern areas. Among the largest genera are *Bursera* (60 spp. in tropical America), *Commiphora* (90 spp. northern Africa), *Canarium* (90 spp. tropical Asia and Africa), and *Protium* (60 spp. in tropical America).

The family is distinguished from the Rutaceae and Simaroubaceae by the presence of resin ducts or chambers in the bark, by the distinct stamens and the short single style, and by the usual absence of punctate pellucid glands.

Members of the family are of little importance domestically. However, they are very rich in gums and resins of considerable value in world markets, notably frankincense and myrrh of biblical fame, from *Boswellia* and *Commiphora,* respectively. A resin, copal, used as a cement and varnish in Mexico, is obtained from native species of the elephant tree (*Bursera*). A few species of *Bursera* and *Garuga* are cultivated domestically in warm regions as ornamentals.

*LITERATURE:*

ENGLER, A. Burseraceae. In Engler and Prantl, Die natürlichen Pflanzenfamilien, ed. 2, Bd. 19a: 405–457, 1931.
LAM, H. J. Studies in phylogeny. Blumea, 3: 114–158, 1938.
ROSE, J. N. Burseraceae. North Amer. Flora, 25: 241–261, 1911.
WEBBER, I. E. Systematic anatomy of the woods of the Burseraceae. Lilloa, 6: 441–465, 1941.

## MELIACEAE.[117] MAHOGANY FAMILY

Shrubs or trees, the wood often scented; leaves usually alternate, pinnately compound, decompound, or rarely simple, lacking pellucid dots, estipulate; flowers bisexual (rarely unisexual, the plants then polygamodioecious), actinomorphic, often in cymose panicles, the calyx usually imbricate (rarely valvate) and small, the sepals 4–5, usually basally connate, corolla contorted or imbricate, the petals 4–5 (rarely 3–8), distinct or connate, or adnate to the staminal tube and then valvate; the stamens 8–10, rarely 5 or numerous, hypogynous, mostly monadelphous by the connate filaments (distinct in *Cedrela*), a disc usually present and between stamens and ovary, the anthers 2-celled, dehiscing longitudinally; pistil 1, the ovary superior, usu-

---

[116] Throughout the literature the baccate fruit of this family is designated a drupe or as drupaceous with 1–5 stones. There seems no morphological basis for treating it other than a berry that on occasion may be 1-seeded by abortion.

[117] The name Meliaceae has been conserved over other names for this family.

ally 2–5-loculed and -carpelled (rarely 1–many-loculed), the placentation axile, the ovules mostly 2 and paired (seldom more, but in *Swietenia* are 12, and in other genera sometimes 1) on each placenta, pendulous, anatropous, the style 1 or none, the stigma often capitate or discoid, lobed or not so; fruit a berry, capsule, or rarely a drupe; seeds often winged, endosperm fleshy or none.

Primarily a pantropical family of 50 genera and about 800 species, of which a single species (*Swietenia Mahoganii*) is indigenous northward into southern Florida. The Asiatic chinaberry tree (*Melia Azedarach*) is naturalized in southern parts of the country. The largest genera include *Cedrela* (100 spp., American tropics), *Trichilia* (200 spp., tropical America and Africa), and *Guarea* (100 spp. America and Africa).

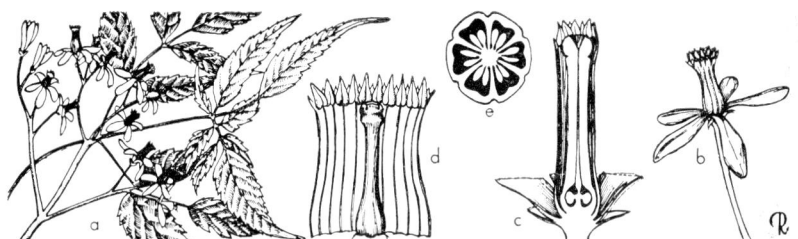

Fig. 177. MELIACEAE. *Melia Azedarach*: a, flowering branch, × ½; b, flower, × 1; c, same, vertical section, × 3; d, pistil and expanded androecium, × 2; e, ovary, cross-section, × 6. (c-e adapted from Schnizlein.) (From L. H. Bailey, *Manual of cultivated plants*, The Macmillan Company, 1949. Copyright 1924 and 1949 by Liberty H. Bailey.)

The family is distinguished from related taxa by the peculiar staminal tube, the discoid or capitate stigmas, and the usually winged seeds. It is separated from the Burseraceae by the lack of resin-producing ducts.

This taxon was included by Hallier in his Terebinthales and considered to have been derived from the Rutaceae. Bessey retained it in the Geraniales, while Wettstein allied it with the Burseraceae and Polygalaceae. Hutchinson segregated it as the only family of his Meliales, stating that it differed from his Rutales primarily in the "leaves usually not gland-dotted and the stamens connate into a tube."

The Meliaceae are of little domestic importance. The chinaberry tree (*Melia Azedarach*) is grown widely in dooryards in the south. Mahogany (*Swietenia*) is cultivated to a limited extent as an ornamental, 1 species of the shrubby *Turraea* is grown in California, and the West Indian cedar (*Cedrela odorata, C. Toona*) is cultivated for ornament in warmer parts of the country. Much of the mahogany lumber of commerce is obtained from *Swietenia*, but the African mahogany is from *Khaya senegalensis*, also of this family.

*LITERATURE:*

CANDOLLE, C. DE. Meliaceae. In de Candolle, Monogr. Phan. 1: 399–752, 1878.

HARMS, H. Meliaceae. In Engler and Prantl, Die natürlichen Pflanzenfamilien, ed. 2, Bd. 19b: 1–172, 1940.

JULIANO, J. B. Studies on the morphology of the Meliaceae. Philippine Journ. Agr. 23: 253–266, 1934.

KRIBS, D. A. Comparative anotomy of the woods of Meliaceae. Amer. Journ. Bot. 17: 724–738, 1930.

PANSHIN, A. J. Comparative anatomy of the woods of the Meliaceae, subfamily Swietenioideae. Amer. Journ. Bot. 20: 638–668, 1933.

WILSON, P. Meliaceae. North Amer. Flora, 25: 263–296, 1924.

## MALPIGHIACEAE.[118] MALPIGHIA FAMILY

Trees or shrubs, usually lianous, sometimes with stinging hairs and the unicellular hairs often variously branched or medianly attached; leaves usually opposite, sometimes alternate or ternate, simple, often with petiolar glands and jointed petioles, stipules generally present and variable; flowers mostly bisexual, showy, variously arranged, usually bracteate, cleistogamous flowers often present, mostly actinomorphic, or obliquely zygomorphic, the calyx imbricate (rarely valvate), the sepals 5, mostly distinct, some or all with large sessile or stalked glands, the corolla convolute, the petals 5, unequal, distinct, clawed, fringed, or toothed; stamens typically 10 in 2 whorls, some or half often reduced to staminodes, hypogynous, usually basally connate, the anthers very diverse, 2-celled, introrse, often with an enlarged connective, dehiscing longitudinally; pistil 1, the ovary superior, typically of 3 (rarely 2, 4, or 5) locules, carpels, and lobes, the placentation axile, the ovule solitary in each locule, pendulous, semianatropous, styles usually distinct and then as many as carpels, rarely connate, the stigmas entire or minutely lobed; fruit a samara, schizocarp, capsule, berry, or rarely a drupe; seed with a large embryo, endosperm none.

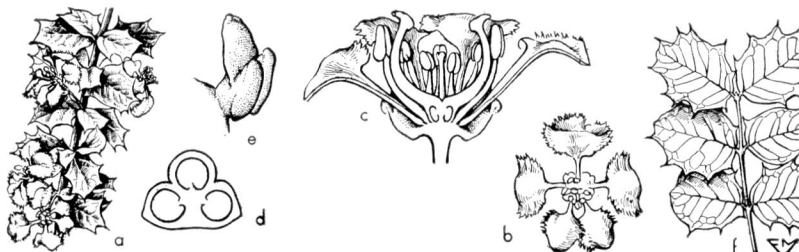

**Fig. 178.** MALPIGHIACEAE. *Malpighia coccigera*: a, flowering branch, × ½; b, flower, face view, × 1; c, flower, vertical section, × 3; d, ovary, cross-section, × 10; e, sepal with basal glands, × 4; f, twig with leaves, × ½. (From L. H. Bailey, *Manual of cultivated plants*, The Macmillan Company, 1949. Copyright 1924 and 1949 by Liberty H. Bailey.)

A tropical family of perhaps 60 genera and 850 species, primarily of the American tropics and subtropics. Seven species of 5 genera are indigenous in the warmer regions of this country; *Byrsonima* (100–1) in southern Florida, *Malpighia* (35–1) and *Thryallis* (8–1) in Texas, and *Aspicarpa* (13–3) and *Janusia* (12–1) from western Texas to southern Arizona. About 16 genera have species indigenous to Mexico, of which *Banisteria* (100 spp.) is one of the largest in the family.

Distinguishing characters of the family include the medifixed unicellular hairs (sometimes stinging), the often lianous habit, the glandular calyx and prominently clawed petals, the peculiar anthers, and the often winged or lobed fruits.

The family has been accepted by most botanists as allied to the Geraniales. Hutchinson placed it (together with the Erythroxylaceae) in his Malpighiales and derived the latter from tiliaceous stocks, while Hallier included it in the Polygalales.

Members of the family are of little domestic importance. The Barbadoes cherry (*Malpighia glabra*) and the related *M. coccigera* are cultivated as ornamental shrubs in warm parts of the country, as is also the yellow-flowered vine *Stigmaphyllon ciliatum*.

---

[118] The name Malpighiaceae has been conserved over other names for this family.

# DIVISION IV. EMBRYOPHYTA SIPHONOGAMA

*LITERATURE:*
NIEDENZU, F. Malpighiaceae. In Engler, Das Pflanzenreich, Hefte 91, 93, 94 (IV. 141): 1–870, 1928.
SMALL, J. K. Malpighiaceae. North Amer. Flora, 25: 117–171, 1910.

## TREMANDRACEAE.[119] TREMANDRA FAMILY

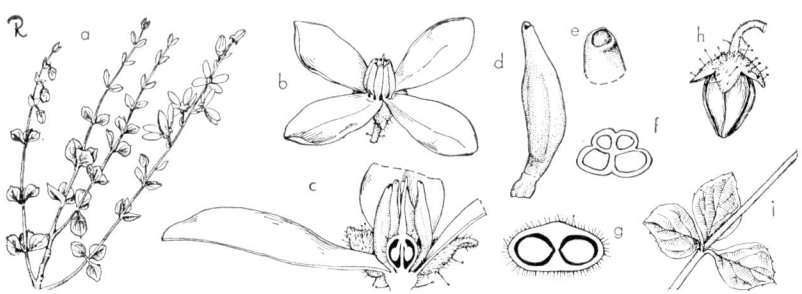

**Fig. 179.** TREMANDRACEAE. *Tetratheca ciliata*: a, flowering branch, × ½; b, flower, habit, × 1½; c, same, vertical section (perianth partially excised), × 3; d, stamen, × 7; e, anther tip with single pore, × 20; f, anther at anthesis, cross-section, × 10; g, ovary, cross-section, × 10; h, bud, showing calyx, × 2; i, stem with whorl of leaves, × ¾.

Suffrutescent herbs or small shrubs, often with glandular hairs, stems sometimes winged; leaves opposite, whorled, or alternate, simple, often heathlike, estipulate; flowers bisexual, actinomorphic, solitary in axils, the calyx valvate, the sepals 4–5 (rarely 3), distinct, the corolla induplicate-valvate, the petals of same number as sepals and alternate with them, distinct; a lobed glandular disc sometimes between stamens and corolla; stamens 8–10 (rarely 6), usually in 2 whorls, hypogynous, distinct, the anthers 2–4-celled, dehiscing by a transverse terminal valve or more or less prolonged into a beak with terminal pores; pistil 1, the ovary superior, 2-loculed and -carpelled, the placentation axile, the ovules 1–2 (rarely 3) in each locule, anatropous, pendulous, the style and stigma solitary, simple; fruit a compressed loculicidal or septicidal capsule; seeds often hairy, with an arillike appendage from the chalaza, embryo small and straight, endosperm copious.

A family of 3 genera and about 30 species, of southern and western Australia. One species of *Tetratheca,* a dwarf shrub, is cultivated in warmer parts of this country.

The Tremandraceae were retained by Bessey in the Geraniales. Wettstein placed them (together with the Polygalaceae and the Rutales *sensu* Hutchinson) in his Terebinthales, but neither Rendle nor Hallier accounted for them in their classifications. Hutchinson placed them in his Pittosporales. This diversity of opinion reflects a lack of information prerequisite to any conclusive views on the phyletic relationships of the taxon.

## POLYGALACEAE.[120] MILKWORT FAMILY

Herbs, shrubs, or small trees, sometimes climbing or twining (chlorophyll-less saprophytes in the Malayan *Epirrhizanthes*); leaves usually alternate, sometimes opposite or whorled, simple, sometimes scalelike, mostly estipulate or with small stipular glands; flowers bisexual, zygomorphic, each subtended by a bract and 2

---

[119] The name Tremandraceae has been conserved over other names for the family.
[120] The name Polygalaceae has been conserved over other names for this family.

bractlets, solitary spicate or racemose, the calyx zygomorphic, persistent, imbricate, the sepals typically 5 (4–7) with the 2 lower united or the sepals distinct with the 2 inner sepals largest and often winged or petaloid (similar to the wing petals of a papilionaceous flower), the corolla basically of 5 distinct hypogynous petals but usually only 3 present (the 2 upper and lower median), generally more or less basally adnate to the androecium, the lower median petal often concave with or without a fringed crest (keel); stamens basically 10 in two 5-merous whorls but usually only 8 (rarely 3–7), monadelphous in a split sheath to beyond the middle (rarely distinct), the anthers basifixed, usually confluently 1-celled, dehiscing by an apical or subterminal pore (rarely 2-celled and each splitting lengthwise), the pollen usually distinctive; an annular intrastaminal disc sometimes present on receptacle within the staminal whorl; pistil 1, the ovary superior, the carpels and locules usually 2 (sometimes 1, 3, or 5), the placentation axile, the ovule solitary on each placenta (rarely 2–6), pendulous, anatropous, the style 1, the stigmas (or stigma lobes) as many as carpels; fruit usually a loculicidal capsule, rarely a nut, samara, or drupe; seed often hairy, with a conspicuous micropylar aril or callosity, the embryo straight, axial, the endosperm soft fleshy (rarely absent).

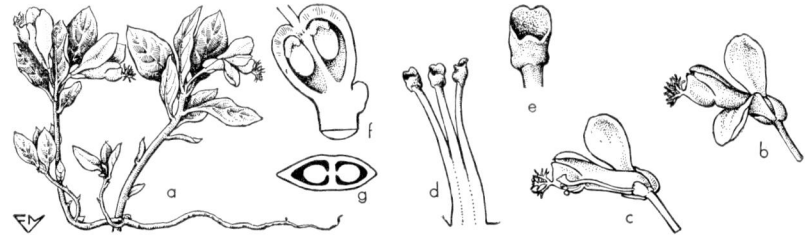

**Fig. 180.** POLYGALACEAE. *Polygala paucifolia*: a, flowering plant, × ½; b, flower, × 1; c, same, vertical section, × 1; d, stamens, × 5; e, anther, × 15; f, ovary, vertical section, × 5; g, same, cross-section, × 8. (From L. H. Bailey, *Manual of cultivated plants*, The Macmillan Company, 1949. Copyright 1924 and 1949 by Liberty H. Bailey.)

A family of about 10 genera and 700 species, widely distributed, except in New Zealand and arctic regions of Asia and North America. The genus *Polygala* (475 spp.) is represented in this country by 45 indigenous species. Other North American genera are *Monnina* (80 spp.) represented by a single species in New Mexico and Arizona, and *Securidaca* (*Elsota*) extending from Mexico southward.

The family is characterized by the peculiar androecial structure, the modified perianth, and the usually biloculate ovary. The flowers superficially resemble the papilionaceous flowers of Leguminosae, but the similar parts are not homologous.

The Polygalaceae were included in the Geraniales by Bessey, in the Polygales by Hallier, in the Terebinthales by Wettstein, as "of doubtful position" by Rendle, and in the Polygalales (*sensu strictu*, not of Hallier) by Hutchinson. Small [121] also accepted the order Polygalales and included in it (for North America) the families Vochyaceae, Polygalaceae, and Dichapetalaceae; all characterized generally by the 2 inner sepals petaloid and winglike, the zygomorphic corollas, and the tendency toward reduction and cohesion of stamens. The taxonomic validity of the order Polygalales (*sensu* Hutchinson) is more readily established than is its phyletic position.

The family is of slight domestic economic importance. About 20 species of *Polygala*, and 1 or a few of *Securidaca* and *Comesperma* are cultivated as ornamentals.

[121] Small. J. K. Polygalales. In North Amer. Flora. 25: 299. 1924.

*LITERATURE:*

BLAKE, S. F. A revision of the genus *Polygala* in Mexico, Central America, and the West Indies. Contr. Gray Herb. 47: 1–122, 1916.

———. Polygalaceae. North Amer. Flora, 25: 305–379, 1924.

CHODAT, R. Monographia polygalacearum. Mem. Soc. phys. hist. nat. Genève. vols. 30–31 et Suppl., 1891–93.

HOLM, T. Morphology of North American species of *Polygala*. Bot. Gaz. 88: 167-185, 1929. [Primarily vegetative morphology.]

PENNELL, F. W. On the typification of Linnaean species as illustrated by *Polygala verticillata*. Rhodora, 41: 378–384, 1939.

WHEELOCK, W. E. The genus *Polygala* in North America. Mem. Torrey Bot. Club, 2: 109–152, 1891.

## EUPHORBIACEAE.[122] SPURGE FAMILY

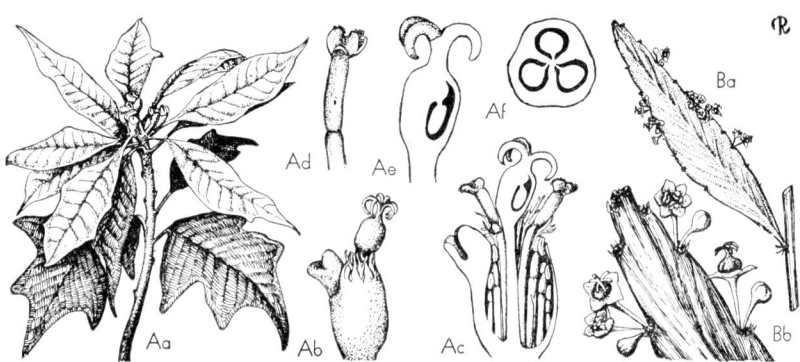

Fig. 181. EUPHORBIACEAE. A, *Euphorbia pulcherrima*: Aa, flowering branch, × ¼; Ab, cyathium, × 1; Ac, same, vertical section, × 2; Ad, staminate flower, × 6; Ae, pistillate flower, vertical section, × 3; Af, ovary, cros-section, × 5. B, *Xylophylla angustifolia*: Ba, stem bearing flowering phyllodium, × ½; Bb, phyllodium tip with inflorescences, × 2. (From L. H. Bailey, *Manual of cultivated plants*, The Macmillan Company, 1949. Copyright 1924 and 1949 by Liberty H. Bailey.)

Monoecious or occasionally dioecious herbs, shrubs, or trees, often with milky juice, sometimes fleshy and cactuslike; leaves mostly alternate, sometimes opposite or whorled, simple or variously compound, usually stipulate (sometimes reduced to hairs, glands, or spines); flowers unisexual, sometimes much reduced by suppression of parts, mostly actinomorphic, variously disposed but the inflorescence usually determinate, both calyx and corolla present or the latter or both absent, perianth valvate or imbricate, sepals and petals usually distinct, usually 5-merous; staminate flowers with stamens usually as many or twice as many as petals (when corolla is present) or reduced to 1 (as in *Euphorbia*),[123] distinct or monadelphous (filaments partially connate in *Ricinus*), anthers 2-celled (rarely 3-4-celled at anthesis), dehiscing longitudinally or transversely (as in *Euphorbia*) (rarely by apical pores), intrastaminal disc usually present in multistaminate flowers (sometimes differentiated into glands), pistillode frequently present; pistillate flowers with or without

---

[122] The name Euphorbiaceae has been conserved over other names for the family.

[123] For explanation of the floral situation in *Euphorbia* and other related genera, cf. Wheeler in Bull. Torrey Bot. Club, 63: 449, 1936, and Rhodora, 43: pl. 655, 1941; Rendle, ed. 2, 2: 256–265, 1938, Haber 1925, and in Willis, J. C. Dictionary of Flowering Plants and Ferns, ed. 6, pp. 256–259, 1931.

staminodes, often pedicillate (*Euphorbia*), the ovary superior, 3-loculed and -carpelled (rarely 2–4), placentation axile, the ovules 1 or less commonly 2 and collateral in each locule, pendulous, anatropous, the micropyle usually carunculate, the styles 3, distinct or basally connate, each often 2-lobed, stigmas 3 or 6 and linear or broadened, often papillate or dissected into filiform segments; fruit usually a schizocarp capsule splitting often elastically into three 1-seeded cocci that dehisce ventrally; seeds with straight or bent embryo and copious soft-fleshy to fleshy endosperm.

A large family of 283 genera and about 7300 species,[124] of almost cosmopolitan distribution, mainly of the tropics but extending also into the temperate regions of northern and southern hemispheres. Two major centers of distribution are tropical America and Africa. Fifteen genera have more than 100 species each, and the largest include *Euphorbia* (over 1600 spp.), *Croton* (700), *Phyllanthus* (480), *Acalypha* (430), *Glochidion* (280), *Macaranga* (240), *Manihot* (160), *Jatropha* (150), and *Tragia* (140). In this country, about 25 genera are represented by a total of about 225 indigenous species, with the greatest number of genera (16) and indigenous species (about 80) occurring in the southeastern region.

The family is distinguished by the milky sap (when present), the unisexual flowers, the ovary superior and usually trilocular, placentation axile, and the ovules collateral, pendulous with ventral raphe and a usually carunculate micropyle.

Hallier included the Euphorbiaceae in his Passionales. Wettstein, Rendle, and Hutchinson each isolated them in an order by themselves (termed the Tricoccae by the first 2, who included also the Callitrichaceae, and termed the Euphorbiales by the latter). The most constant character for the order (*sensu* Engler) is the superior, tricarpellate, trilocular ovary, and each locule with 1 or 2 pendulous ovules with usually ventral raphe. The Euphorbiaceae were considered by Pax and Hoffmann to be composed of 4 subfamilies: the Phyllanthoideae, with cotyledons broader than the radicle, and ovules 2 in each locule; the Crotonoideae (to which belong most North American genera) differ in ovule solitary in each locule; the Porantheroideae (Australian), with radicle as wide as cotyledons and biovulate locules, and the Australian Ricinocarpoideae with uniovulate locules.

The Euphorbiaceae are of considerable economic importance from the world viewpoint, since products of the family include rubber (*Hevea*), tung oil (*Aleurites Fordii*), castor oil (*Ricinus*), and cassava and tapioca (*Manihot*). Of these only tung oil is produced commercially in this country. Many members of the family are grown domestically as ornamentals, notably the poinsettia and crown of thorns (*Euphorbia* spp.) croton (*Codiaeum* spp.), Otaheite gooseberry (*Phyllanthus*), and the castor-bean plant (*Ricinus*).

*LITERATURE:*

BULLOCK, A. A. *Pedilanthus* vs. *Tithymalus*. Kew Bull. 1938: 468–470.

CROIZAT, L. De Euphorbio antiquorum atque officinarum, 1–127, 1934.

———. On the classification of *Euphorbia*. I. How important is the cyathium? Bull. Torrey Bot. Club, 63: 525–531, 1936.

———. On the phylogeny of the Euphorbiaceae and some of their presumed allies. Rev. Univ. 25: 205–220, 1940.

———. A study of *Manihot* in North America. Journ. Arnold Arb. 23: 216–225, 1942.

———. Peculiarities of the inflorescence in the Euphorbiaceae. Bot. Gaz. 103: 771–779, 1942.

———. New and critical Euphorbiaceae chiefly from the southeastern United States. Bull. Torrey Bot. Club, 69: 445–460, 1942.

———. The family Euphorbiaceae: when and by whom published. Amer. Midl. Nat. 30: 808–809, 1943.

[124] These figures, and those given as totals for particular genera of the family, are taken from Pax and Hoffmann (1931).

FERGUSON, A. M. Crotons of the United States. Annual Rept., Mo. Bot. Gard. 12: 33–73, 1901.
HABER, J. M. The anatomy and morphology of the flower of Euphorbia. Ann. Bot. 39: 657–707, 1925.
JABLONSZKY, E. Euphorbiaceae-Phyllanthoideae-Bridelieae. In Engler, Das Pflanzenreich, 65 (IV. 147. VIII): 1–98, 1915.
JANSSONIUS, H. H. A contribution to the natural classification of the Euphorbiaceae. Trop. Woods, 19: 8–10, 1929.
MOYER, L. S. Species relationships in *Euphorbia* as shown by the electrophoresis of latex. Amer. Journ. Bot. 21: 293–313, 1934.
NORTON, J. B. S. North American species of *Euphorbia* section *Tithymalus*. Annual Rept., Mo. Bot. Gard. 11: 85–144, 1900.
PAX, F. Euphorbiaceae-Jatropheae. In Engler, Das Pflanzenreich, 42 (IV. 147): 1–148, 1910; Euphorbiaceae-Adrineae, *op. cit.* 44 (IV. 147. II): 1–111, 1910; Euphorbiaceae-Cluytieae, *op. cit.* 47 (IV. 147. III): 1–124, 1911; Euphorbiaceae-Gelonieae, *op. cit.* 52 (IV. 147. IV): 1–41, 1912; Euphorbiaceae-Hippomaneae, *op. cit.* 52 (IV. 147. V): 1–319, 1912; Euphorbiaceae-Acalypheae-Chrozophorinae, *op. cit.* 57 (IV. 147. VI): 1–142, 1912; Euphorbiaceae-Acalypheae-Mercurialinae, *op. cit.* 63 (IV. 147. VII): 1–473, 1914.
———. Phylogenie der Euphorbiaceae. Engler, Bot. Jahrb. 59: 129–182, 1924.
PAX, F. and HOFFMANN, K. Euphorbiaceae-Acalypheae. In Engler, Das Pflanzenreich, 68 (IV. 147. IX, XI): 1–134, 1919; Euphorbiaceae-Dalechampieae, *op. cit.* 68 (IV. 147. XII): 1–59, 1919; Euphorbiaceae-Phyllanthoideae-Phyllantheae, *op. cit.* 85 (IV. 147. XVI–XVII): 1–231, 1924.
———. Euphorbiaceae. In Engler and Prantl, Die natürlichen Pflanzenfamilien, ed. 2, Bd. 19c: 1–251, 1931.
PERRY, B. A. Chromosome number and phylogenetic relationships in the Euphorbiaceae. Amer. Journ. Bot. 30: 527–543, 1943.
———. Chromosome number relationships in the genus *Euphorbia*. Chron. Bot. 7: 413, 414, 1943.
RECORD, S. J. The American woods of the family Euphorbiaceae. Trop. Woods, 54: 7–40, 1938.
SCHOUTE, J. C. On the aestivation in the cyathium of *Euphorbia fulgens*, with some remarks on the morphological interpretation of the cyathium in general. Rec. Trav. Bot. Néerl. 34: 168–181, 1937.
WHEELER, L. C. *Euphorbia* in the Pacific states. Bull. So. California Acad. Sci. 35: 127–147, 1936.
———. Revision of the *Euphorbia polycarpa* group of the southwestern U. S. and adjacent Mexico; a preliminary treatment. Bull. Torrey Bot. Club, 63: 397–416, 1936; 429–450, 1936.
———. Typification of the generic synonyms of *Pedilanthus*. Contrib. Gray Herb. 124: 43–74, 1939.
———. *Euphorbia* subgenus *Chamaesyce* in Canada and the United States exclusive of southern Florida. Rhodora, 43: 97–154, 168–205, 223–286, 1941.
———. The genera of living *Euphorbieae*. Amer. Midl. Nat. 30: 456–503, 1943.
WHITE, A., DYER, R. A. and SLOANE, B. L. The succulent Euphorbieae, 2 vols. Pasadena, 1941.

## CALLITRICHACEAE. WATER-STARWORT FAMILY

Monoecious annual herbs, aquatic or terrestrial, stems slender and delicate; leaves opposite (the upper ones often rosulate in aquatic species), entire, estipulate; flowers unisexual, actinomorphic, solitary in leaf axils (rarely flowers of both sexes in same axil), perianth absent, each flower subtended by 2 hornlike bracteoles; staminate flower comprised of a single stamen, the anther 2-celled dehiscing laterally and longitudinally; pistillate flower composed of a single pistil, the ovary superior, 2-lobed, -carpelled, and -loculed (or 4-loculed by false septa), the placentation axile, the ovules 1 in each locule, pendulous, anatropous, the styles 2, filiform.

papillose; fruit a schizocarp splitting at maturity into 2 or 4 mericarps, the lobes usually dorsally keeled or winged; seed with straight embryo and fleshy endosperm.

A unigeneric family (*Callitriche*) of almost cosmopolitan distribution, composed of perhaps 26 species. The number of species has been interpreted variously; some authors consider all taxa to be variants of 1 or 2 species (a view contradicted by cytological data), but Hegelmaier (1864) accepted 24 species and Jörgensen (1923, 1925) recognized 44 species. By liberal interpretation there are about 10 species indigenous to this country.

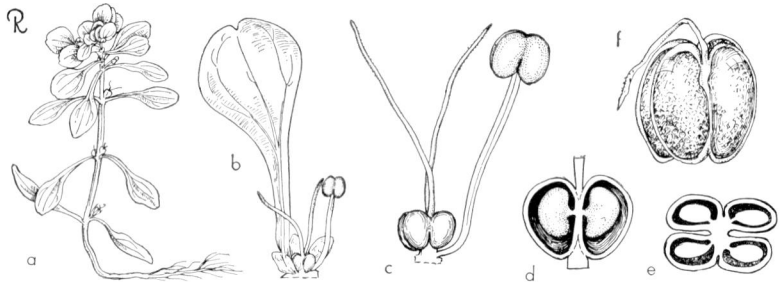

**Fig. 182.** CALLITRICHACEAE. *Callitriche heterophylla*: a, plant in flower, × 1; b, leaf, with flower and bracts, × 5; c, flower, bracts removed, × 12; d, ovary, vertical section, × 25; e, ovary, cross-section, × 25; f, fruit, × 15.

The phyletic position of the family is obscured by the extreme reduction of parts and of the vascular anatomy. Early botanists, including Hegelmaier and Bentham and Hooker, included it in the Haloragaceae. Eichler, Baillon, Rendle, Wettstein, and Pax considered it an advanced member of the Euphorbiales. Hallier included it as a primitive member of his Guttales, derived from Linaceae. Hutchinson included it in the Lythrales, together with Haloragaceae and Onagraceae.

Economically the family is of no importance.

*LITERATURE:*

BAILLON, H. Recherches sur l'organogénie du *Callitriche*. Bull. Soc. Bot. Fr. 5: 337–341. 1858.
HEGELMAIER, F. Monographie der Gattung *Callitriche*. Stuttgart, 1664.
———. Zur Systematik von *Callitriche*. Verh. Bot. Verein Brandenburg, 9: 1–40, 1867.
HYLANDER, N. Studien über nordische Gefasspflanzen. Upsala Univ. Årsskrift, 1945, [Cf. pp. 234–236 for taxonomic discussion of genus.]
JÖRGENSEN, C. A. Studies on Callitrichaceae. Bot. Tidsskr. 38 (2): 81–122, 1923.
———. Frage der systematischen Stellung der Callitrichaceae. Jahrb. wiss. Bot. 64: 440–442, 1925.
PAX, F. and HOFFMANN, K. Callitrichaceae. In Engler and Prantl, Die natürlichen Pflanzenfamilien, ed. 2, Bd. 19c: 236–240, 1931.

## Order 27. SAPINDALES [125]

Usually woody plants, differing from the Geraniales in the ovules pendulous with the dorsal raphe and micropyle upward or erect with ventral raphe and micropyle downward.

[125] This order derives its name from the type family Sapindaceae, and it from its type genus *Sapindus*. At one time the family Anacardiaceae was known by the older name Terebinthaceae and the order was designated as the Terebinthales.

# DIVISION IV. EMBRYOPHYTA SIPHONOGAMA

The order was treated by Engler and Diels as composed of 11 suborders and 23 families as follows (family names preceded by an asterisk not treated in this text):

| | | |
|---|---|---|
| Buxineae | Celastrineae | Icacinineae |
| Buxaceae | Cyrillaceae | Icacinaceae |
| Empetrineae | *Pentaphyllacaceae | *Aetoxicaceae |
| Empetraceae | *Corynocarpaceae | Aceraceae |
| Coriariineae | Aquifoliaceae | Hippocastanaceae |
| Coriariaceae | Celastraceae | Sapindaceae |
| Limnanthineae | Hippocrataceae | Sabiineae |
| Limnanthaceae | *Salvadoraceae | Sabiaceae |
| Anacardiineae | *Stackhousiaceae | Melianthineae |
| Anacardiaceae | Staphyleaceae | Melianthaceae |
| | | Didiereineae |
| | | *Didiereaceae |
| | | Balsaminineae |
| | | Balsaminaceae |

The families of the Sapindales (*sensu* Engler) have been interpreted by all other phylogenists to belong in from 3 to 6 orders. All have included the Limnanthaceae in the Geraniales (Gruinales), and all except Wettstein included the Balsaminaceae in the same order. There was general agreement that the Sapindales should be restricted to include the Sapindaceae, Hippocastanaceae, Sabiaceae, Melianthaceae, Aceraceae, and Anacardiaceae, with the Celastraceae, Aquifoliaceae, and Hippocrataceae composing the Celastrales (a view rejected by Hallier, who included them in his Guttales). The alliance of some families assigned by Engler to this order has been interpreted very diversely (especially the Buxaceae, Empetraceae, and Coriariaceae) as indicated in the discussions that follow here under each of these families.

## BUXACEAE. BOXWOOD FAMILY

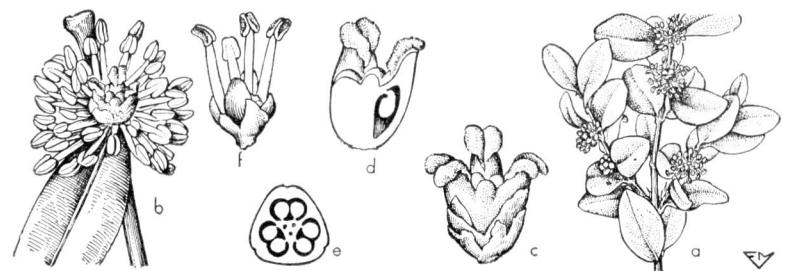

**Fig. 183.** BUXACEAE. *Buxus sempervirens*: a, flowering branch, × ½; b, inflorescence and flowers, × 2; c, pistillate flower, × 3; d, same, vertical section, × 3; e, ovary, cross-section, × 5; staminate flower, × 3. (From L. H. Bailey, *Manual of cultivated plants*, The Macmillan Company, 1949. Copyright 1924 and 1949 by Liberty H. Bailey.)

Monoecious or dioecious evergreen herbs, shrubs, or trees; leaves alternate or more often opposite, simple (not linear), leathery, estipulate; flowers unisexual, actinomorphic, bracteate, spicate or racemose, inconspicuous, perianth composed only of a calyx or sometimes wanting, sepals usually 4, basally connate (in the pistillate flower, the calyx sometimes 4–12-parted); staminate flowers typically with 4 stamens opposite calyx lobes or stamens numerous, anthers 2-celled, dehiscing by valves or longitudinal slits, pistillode sometimes present, pistillate flowers fewer than staminate and sometimes solitary, the pistil 1, the ovary superior, typically 3-loculed and -carpelled (rarely the carpels 2 or 4), sometimes deeply lobed at least apically, placentation axile, the ovules usually 2 in each locule, collateral, pendulous with a dorsal raphe, anatropous, the styles as many as carpels, simple, basally connate or distinct and divergent; fruit a loculicidal capsule dehiscing elastically, or indehiscent and berrylike; seed glossy black, usually with a caruncle, embryo straight, endosperm fleshy.

A family of 6 genera and 30–60 species, indigenous to the tropics and subtropics of especially the Old World. *Buxus* (25–40 spp.) is the largest genus. *Pachysandra* (4 spp.), primarily an Asiatic genus, has 1 species (*P. procumbens*) indigenous to mountainous areas of West Virginia south to western Florida and Louisiana. The monotypic *Simmondsia* (a dioecious shrub) extends north from Mexico into the southwestern parts of this country.

The Buxaceae are distinguished from related families by the absence of milky sap, the perianth represented only by a calyx, the 3-loculed ovary with typically 2 collateral and pendulous ovules with dorsal raphes in each locule, and the glossy black carunculate seeds.

The family was included in the Euphorbiaceae by Bentham and Hooker, was treated as a distinct family but in the same order (Tricoccae) by Wettstein and by Rendle, was placed in the Hamamelidales by Hutchinson and in the Celastrales by Bessey.

Economically, the family is of importance for its ornamentals: boxwood (*Buxus*), Japanese spurge (*Pachysandra*), and *Sarcococca*, and the hard fine-grained wood of boxwood is used in commerce.

*LITERATURE:*

Müller, J. Buxaceae. In de Candolle, Prodromus, 16 (1): 7–23, 1869.
Pax, F. Buxaceae. In Engler and Prantl, Die natürlichen Pflanzenfamilien, III (5): 130–136, 1890.

## EMPETRACEAE.[126] CROWBERRY FAMILY

**Fig. 184.** Empetraceae. *Empetrum nigrum*: a, fruiting branch, × ½; b, twig (of *Corema Conradii*), × 2; c, staminate flower, × 4; d, pistillate flower, × 8; e, same, vertical section, × 8. (From L. H. Bailey, *Manual of cultivated plants,* The Macmillan Company, 1949. Copyright 1924 and 1949 by Liberty H. Bailey.)

Small, sometimes dioecious, evergreen, pulvinate, heathlike shrubs; leaves alternate, deeply grooved beneath, linear, estipulate; flowers usually bisexual (occasionally unisexual), axillary or congested in terminal heads, actinomorphic, small, apetalous, sepals 2–6, sometimes petaloid, bracteate, or absent, imbricate, in 2 whorls (the inner whorl interpreted as petals by some authors), stamens 2–4, hypogynous, distinct, the anthers 2-celled, dehiscing longitudinally, no disc present, pistil 1 (rudimentary in staminate flowers), ovary superior, 2–9-loculed and -car-

[126] The name Empetraceae has been conserved over other names for this family.

pelled, placentation axile, the ovule solitary in each locule, anatropous or nearly campylotropous (amphitropous), the style 1, short, variously lobed, fringed, or divided, the stigmatic branches as many as carpels; fruit a fleshy or dry berry containing 2 or more 1-seeded pyrenes; seed with long straight embryo, endosperm present.

A family of 3 genera and 8 species of mountainous regions of North and South America, extending to the arctic and the antarctic, and in the circumpolar regions of the northern hemisphere. *Empetrum* and *Corema* are native in the Rocky Mountains, in associated mountain ranges of the west, and the Adirondacks and Alleghenies in the east, also in relic floras from New Jersey to Michigan. The monotypic *Ceratiola* (*C. ericoides*) occurs only in the Atlantic coastal plain from Florida to South Carolina and westward to Mississippi.

The ericaceous habit associated with the apetalous condition, the few stamens, the often flabellate style branches and fruit are characteristic of the family.

Hallier and Wettstein included the family within the Bicornes (primitive Gamopetalae), Hutchinson and Rendle each placed it in the Celastrales, and Bessey included it in the Sapindales. The alliance of the family with the sapindaceous or celastraceous taxa depends largely on the phyletic significance of the erect ovule and its ventral raphe. Wettstein rejected this character as of no major importance, and based his views of alliance (as an advanced and perhaps degenerate taxon) with the Ericaceae on embryological and endosperm affinities, and on the compatibilities indicated by serological studies.

The family is of little economic importance. Species of *Empetrum* and *Corema* are cultivated to a limited extent, and with difficulty, as ornamentals or novelties.

*LITERATURE:*

FERNALD, M. L. and WIEGAND, K. M. The genus *Empetrum* in North America. Rhodora, 15: 211–217, 1913.
GOOD, R. O'D. The genus *Empetrum*. Journ. Linn. Soc. Bot. 47: 489–523, 1927.
PAX, F. Empetraceae. In Engler and Prantl, Die natürlichen Pflanzenfamilien, III (5): 123–127, 1895.

## CORIARIACEAE.[127] CORIARIA FAMILY

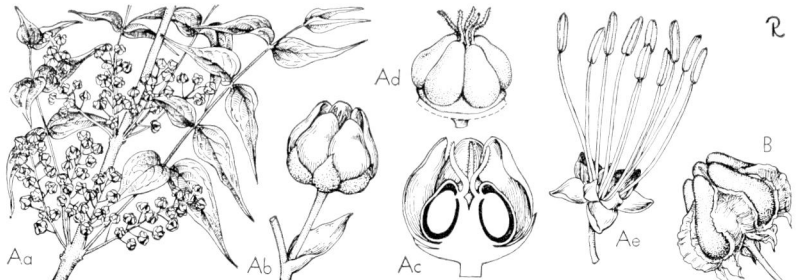

**Fig. 185.** CORIARIACEAE. A, *Coriaria japonica*: Aa, flowering branch, × ⅓; Ab, pistillate flower, habit, × 2; Ac, same, vertical section, × 3; Ad, gynoecium, habit, × 3; Ae, staminate flower, × 2. B, *Coriaria myrtifolia*: fruit, × 2. (Ab-Ae, redrawn from Curtis' Bot. Mag.: B, from Baillon.)

Suffrutescent perennial herbs or shrubs with angular twigs; leaves opposite or whorled, simple, entire, estipulate; flowers bisexual or sometimes unisexual, actinomorphic, axillary or racemose, perianth biseriate, sepals 5 and imbricate, petals 5,

[127] The name Coriariaceae has been conserved over other names for this family.

smaller than the sepals, keeled within, fleshy, persistent and enlarging in fruit and adhering to the fruiting carpels to produce a pseudodrupe, the stamens 10, distinct or those opposite the petals adnate to petal keel, hypogynous, anthers exserted, 2-celled, dehiscing longitudinally, gynoecium of 5–10 distinct pistils, the ovary superior, uniloculate, 1-carpelled, the ovule solitary, pendulous from a parietal placenta, anatropous, the style terminal, linear, stigmatic along ventral side, usually divergent, those of adjoining pistils distinct; fruit an achene seemingly drupaceous by accrescence of the persistent fleshy petals; seed with straight embryo and scant endosperm.

A unigeneric family (*Coriaria*) of about 10 widely distributed species. A few are indigenous to South America, and one extends north into Mexico. None is native in this country. Centers of distribution include eastern Asia, the Mediterranean region, New Zealand, and Chile.

The Coriariaceae are not closely related to any other family, the Empetraceae perhaps the closest allies, and represent a relic of ancient stocks. Wettstein considered the family to be of questionable phyletic relationship and placed it, together with the Cyrillaceae, between Hippocastanaceae and Balsaminaceae in the Terebinthales. Bessey included it in the same order (as Sapindales) and Hutchinson treated it as a distinct order allied to his Dilleniales and Pittosporales.

The plants are of no domestic importance, although 1–2 species are cultivated as ornamentals in warmer parts of the country. The foliage and fruits are reported to be very poisonous, and in Mexico the fruits are reportedly used as a dog poison. Dyes from the fruits have been used to prepare indelible inks.

*LITERATURE:*

ENGLER, A. Coriariaceae. In Engler and Prantl, Die natürlichen Pflanzenfamilien, III (5): 128–129, 1890.

GOOD, R. O'D. The geography of the genus *Coriaria*. New Phytologist, 29: 170–198, 1930.

## LIMNANTHACEAE. LIMNANTHUS FAMILY

**Fig. 186.** LIMNANTHACEAE. *Limnanthes Douglasii*: a, flowering branch, × ½; b, flower, face view, × 1; c, same, vertical section, × 2; d, pistils, × 5; e, gynoecium, cross-section, × 4. (c after Engler & Prantl, d-e from LeMaout & Decaisne.) (From L. H. Bailey, *Manual of cultivated plants,* The Macmillan Company, 1949. Copyright 1924 and 1949 by Liberty H. Bailey.)

Annual herbs; leaves alternate, pinnately dissected, estipulate; flowers bisexual, actinomorphic, solitary, and usually on long axillary peduncles, perianth biseriate, the sepals 3- or 5-merous, valvate, distinct or nearly so, persistent, the petals 3 or 5, contorted, withering-persistent, distinct, usually clawed, often alternating with as many nectiferous glands; the stamens twice as many as the petals, in 2 whorls, distinct, the outer ones alternating with petals and often with basal glands, the anthers 2-celled, dehiscing longitudinally; gynoecium a single pistil of 3–5 distinct

## DIVISION IV. EMBRYOPHYTA SIPHONOGAMA

ovaries and the styles gynobasically connate into a common structure with as many branches as ovaries (stigmas capitate in *Limnanthes*), the ovary superior, each one uniloculate, 1-carpelled, placentation basal-parietal, the ovule solitary, anatropous, ascending; fruit a 1-seeded achene (nutlet); seed with a straight embryo and no endosperm.

A family of 2 genera (*Limnanthes* and *Floerkea*) and 8 species, all indigenous to North America and predominantly of the Pacific regions, with 7 species occurring in California. Only the monotypic *Floerkea* occurs in the northeastern part of the country and then extends only as far south as Kentucky and Delaware. The plants generally are of aquatic habitats or their environs.

The family has been included within the Geraniaceae, but is now accepted to be distinct, on the basis of the ovule ascending and with a single integument, and the absence of endosperm in the seed. Superficially the 2 families are readily distinguished by difference in fruit type and dehiscence, and the contorted petals and gynobasic style of Limnanthaceae are distinctive.

The family is of no significant importance. One species of meadowfoam (*Limnanthes Douglasii*) is cultivated to a limited extent as an ornamental annual of gardens.

*LITERATURE:*

RUSSELL, A. M. A comparative study of *Floerkea proserpinacoides* and allies. Contrib. Bot. Lab. Univ. Penn. 4: 401–418, 1920.
RYDBERG, P. A. Limnanthaceae. North Amer. Flora, 25: 97–100, 1910.

## ANACARDIACEAE.[128] CASHEW FAMILY

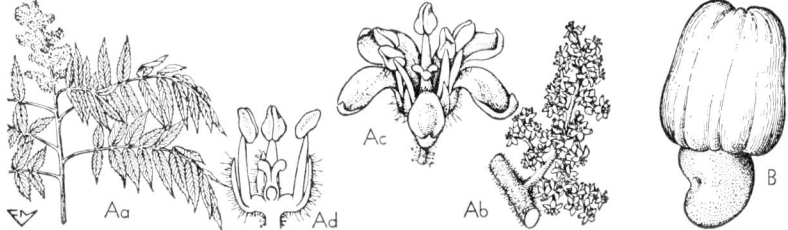

**Fig. 187.** ANACARDIACEAE. *Rhus typhina*: Aa, flowering branch, × 1/16; Ab, segment of inflorescence, × 1/2; Ac, perfect flower, × 3; Ad, same, less petals, vertical section, × 4. B, *Anacardium occidentale*: fruit, × 1/2. (From L. H. Bailey, *Manual of cultivated plants*. The Macmillan Company, 1949. Copyright 1924 and 1949 by Liberty H. Bailey.)

Trees or shrubs, usually with resinous bark; leaves alternate (opposite in *Dobinea*), simple trifoliolate or pinnate, estipulate or the stipules obscure; flowers typically bisexual but usually unisexual by reduction, usually actinomorphic, small, in terminal or axillary panicles, the perianth usually biseriate and mostly imbricate, the sepals 3–5, basally connate and sometimes (together with other floral parts) adnate to the gynoecium, the petals 3–5 or none, distinct or rarely basally connate, usually hypogynous, a short hypanthium sometimes present; stamens typically 10 in 2 whorls, rarely more and usually fewer (*Anacardium* has 1 stamen and 6–9 staminodes; *Rhus* has 5 stamens), distinct or infrequently basally connate, arising from or under the edge or rim of an annular intrastaminal disc (sometimes pro-

---

[128] The name Anacardiaceae Lindl. (1830) is conserved over Terebintaceae Juss. (1789), "Terebinthaceae" of authors, Spondiaceae Kunth (1924), *et al.*

duced into a gynophore), the anthers 2-celled, dehiscing longitudinally; pistil 1, the ovary superior, usually unilocular and 3-carpelled but functionally 1-carpelled, rarely 5-loculed and -carpelled (*Buchanania*) with a single functional ovule, the placentation basically axile, ovules solitary in each locule, when unilocular the ovule anatropous and parietal, or seemingly basal, the style usually 1 (occasionally 2–6 and widely divergent), the stigmas generally as many as carpels; fruit usually a drupe with a resinous mesocarp; seed with curved embryo, the endosperm scant or absent.

A family of 73 genera and about 600 species, of both hemispheres but extending into the north temperate areas of Eurasia and this continent. In the United States *Rhus, Toxicodendron,* and *Schmaltzia* (combined by some authors as a single genus, *Rhus*) are of widespread distribution from the Pacific to the Atlantic, and are represented by about 20 indigenous species; the monotypic *Malosma* occurs in California, indigenous species of *Cotinus* extends from Texas to Tennessee (along a former shore of the gulf), and 1 species of the West Indian *Metopium* reaches southern Florida. The largest genera of the family include *Rhus* (50 spp. mostly pantemperate), *Searsia* (50 spp. in Africa), *Schinus* (30 spp. in South America), *Schmaltzia* (40 spp. in North America), *Toxicodendron* (30 spp., north temperate zone), *Semecarpus* (40 spp. in Malaysia), and *Mangifera* (30 spp. in Indomalaysia).

The family is generally distinguished from related families by the combination of intrastaminal disc, presence of resin ducts, the usually unilocular ovary, and drupaceous fruit.

The Anacardiaceae have been included in the Sapindales by all phylogenists. Hallier considered the family (sub-Terebinthaceae) to have been ancestral to most of the amentiferous taxa (as Juglandaceae, Leitneriaceae, Fagaceae, Urticaceae, etc.) and also to the Aceraceae, and to have been derived from the Rutaceae. Hutchinson treated it as one of the more advanced members of the Sapindales.

The family is important for the edible nuts (seeds) of the cashew (*Anacardium occidentale*) and pistachio (*Pistacia vera*), for the edible aromatic flesh of the fruit of the mango (*Mangifera indica*), mombin (*Spondias* spp.) and Kaffir plum (*Harpephyllum caffrum*), for resins, oils, and lacquers obtained from the varnish tree (*Toxicodendron vernicifera*) and mastic tree (*Pistacia Lentiscus*), for a few ornamentals, notably the smoke tree (*Cotinus Coggygria*), and sumac (*Rhus* spp.), and for much of the commercial supply of tannic acid (*Schinopsis* spp., of South America).

*LITERATURE:*

BARKLEY, F. A. Studies in the Anacardiaceae. I-II, Ann. Mo. Bot. Gard. 28: 263–264, 499–500, 1937; V, Amer. Midl. Nat. 24: 680, 1940.

———. A monograph of the genus *Rhus* and its allies in North and Central America including the West Indies. Ann. Mo. Bot. Gard. 24: 265–498, 1937.

———. *Schmaltzia.* Amer. Mdl. Nat. 24: 647–665, 1940.

———. A key to the genera of the Anacardiaceae. Amer. Midl. Nat. 28: 465–474, 1942.

———. *Schinus.* Brittonia, 5: 160–198, 1944.

BARKLEY, F. A. and BARKLEY, E. D. A short history of *Rhus* to the time of Linnaeus. Amer. Midl. Nat. 19: 265–333, 1938.

BARKLEY, F. A. and REED, M. J. Studies in the Anacardiaceae IV. Amer. Midl. Nat. 22: 209–211, 1939.

———. *Pseudosmodingium* & *Mosquitoxylum.* Amer. Midl. Nat. 24: 666–679, 1940.

ENGLER, A. Anacardiaceae. In de Candolle, Monogr. Phaner. 4: 171–546, 1883.

———. Anacardiaceae. In Engler and Prantl, Die natürlichen Pflanzenfamilien, III (5): 138–178, 1895.

HEIMSCH, C. Wood anatomy and pollen morphology of *Rhus* and allied genera. Journ. Arnold Arb. 21: 279–291, 1940.

MCNAIR, J. B. The geographical distribution of poison sumac (*Rhus Vernix* L.) in North America. Amer. Journ. Bot. 12: 393–397, 1925.

MUKHERJI, S. A monograph of the genus *Mangifera* L. Lloydia, 12: 73–136, 1949.
RECORD, S. J. American woods of the family Anacardiaceae. Trop. Woods, 60: 11–45, 1939.
SWEET, H. R. and BARKLEY, F. A. A most useful plant family, the Anacardiaceae. Bull. Mo. Bot. Gard. 24: 216–229, 1936.

## CYRILLACEAE.[129] CYRILLA FAMILY

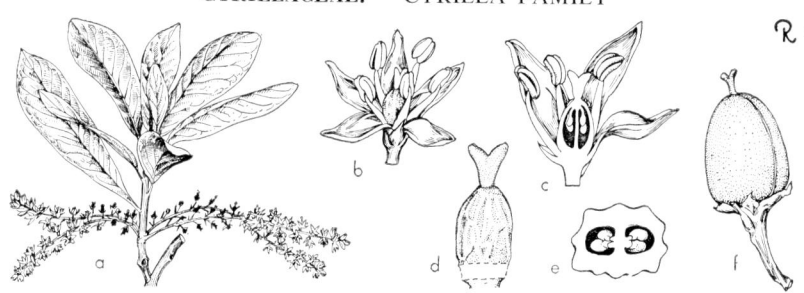

**Fig. 188.** CYRILLACEAE. *Cyrilla racemiflora*: a, flowering branch, × ½; b, flower, habit, × 5; c, flower, vertical section, × 6; d, pistil, habit, × 8; e, ovary, cross-section, × 15; f, fruit, × 5.

Deciduous or evergreen shrubs or small trees; leaves alternate, simple, mostly coriaceous, estipulate; flowers bisexual, actinomorphic, in racemes, the sepals 5, basally connate, imbricate, or rarely valvate, persistent and often enlarged in fruit, the petals 5, distinct or basally connate, imbricate or contorted; stamens 10 in 2 whorls of 5 each or the inner whorl lacking or reduced to staminodes, hypogynous, the filaments dilated and distinct, the anthers 2-celled, dehiscing longitudinally; pistil 1, the ovary superior, 2–4-loculed and -carpelled, the placentation axile, the ovules usually 1 or 2 (rarely 4) collaterally disposed on each placenta, pendulous, anatropous, the style 1 and short or nearly obsolete, the stigmas 2 and linear-ovate; fruit a dehiscent capsule or a leathery to fleshy drupaceous berry, often angled or winged; seed with small straight embryo, the endosperm fleshy.

An American family of 3 genera (*Cyrilla*, *Cliftonia*, and *Costaea*). The first 2 are monotypic indigens of southeastern United States, and *Costaea* has 3 species extending from Cuba to Brazil and Colombia.

The family is set apart from related families by the racemose inflorescence, the petals basally short-connate, the several-celled few-seeded ovary, and by the capsular or winged and drupaceous fruit.

It is of little economic importance. Leatherwood (*Cyrilla racemosa*) and the buckwheat tree (*Cliftonia monophylla*) are cultivated as ornamental shrubs prized for the fragrant white flowers and showy autumn coloration of the foliage.

*LITERATURE:*

UPHOF, TH. Cyrillaceae. In Engler and Prantl, Die natürlichen Pflanzenfamilien, ed. 2, Bd. 20b: 1–12, 1942. [Not seen.]

## AQUIFOLIACEAE. HOLLY FAMILY

Evergreen or deciduous trees or shrubs, leaves alternate, simple, coriaceous, stipules minute or none; flowers bisexual or unisexual (the plants then dioecious or polygamodioecious), actinomorphic, small and greenish, solitary to few in fascicles or axillary cymes, the sepals 3–6, more or less basally connate, imbricate, the petals

---

[129] The name Cyrillaceae has been conserved over other names for this family.

4–9, distinct or scarcely connate basally, hypogynous, imbricate; the stamens 4–9, distinct, alternating with and sometimes adhering basally to petals, the anthers 2-celled, dehiscing longitudinally, no disc present (sterile anther-bearing staminodes usually present in pistillate flowers, a rudimentary pistil present in most staminate flowers); pistil 1, the ovary superior, 3–many-loculed, the carpels as many as locules, the placentation axile, the ovules 1–2 on each placenta, pendulous, anatropous, the style 1 and terminal or absent, the stigma lobed or capitate; fruit a berry with usually 4 pyrenes; seed with minute straight embryo and a copious fleshy endosperm.

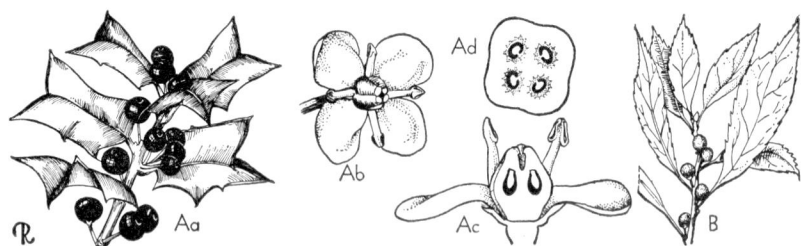

**Fig. 189.** AQUIFOLIACEAE. A, *Ilex cornuta*: Aa, twig with fruit, × ½; Ab, flower habit. × 2; Ac, flower, vertical section, × 3; Ad, ovary, cross-section, × 6. B, *Ilex verticillata*: fruiting branch, × ⅜. (From L. H. Bailey, *Manual of cultivated plants*, The Macmillan Company, 1949. Copyright 1924 and 1949 by Liberty H. Bailey.)

A family of 3 genera (*Ilex, Nemopanthus, Byronia*) and about 300 species. *Ilex* is a widely distributed genus of perhaps 295 species, represented domestically by about 15 indigenous species; *Nemopanthus* is a monotypic genus of northeastern North America, and *Byronia* a Polynesian and Australian genus of 3 species. *Ilex* is common to eastern United States and Asia, with its chief center of world distribution in Central and South America.

The family is distinguished by the absence of such characters as conspicuous stipules and interstaminal disc, and by the presence of determinate axillary inflorescences and the polygamodioecious character of the plants.

Most phylogenists have treated the Celastrineae of Engler as an order distinct from the Sapindales, and have included the Aquifoliaceae in it. Hallier included this and the next family in his Guttales, a view not shared by most contemporary botanists.

Economically the family is important for the white hard wood of *Ilex* spp. and for the ornamental value of a number of species of this genus of which about 30 spp. and innumerable hybrids and clones are cultivated in this country.

*LITERATURE:*

LOESENER, T. Monographia Aquifoliacearum. Nov. acta acad. Caes. Leop. Carol. 78: 1–598, 1901; *op. cit.* 89: 1–313, 1908.

———. Über die Aquifoliaceen, besonders über *Ilex*. Mitt. deutsch. dendr. Ges. 28: 1–66, 1919.

———. Aquifoliaceae. In Engler and Prantl, Die natürlichen, Pflanzenfamilien, ed. 2, Bd. 20b: 38–86, 1942. [Not seen.]

TRELEASE, W. Revision of the North American Ilicineae and Celastraceae. Trans. Acad. St. Louis, 5: 343, 357, 1892.

## CELASTRACEAE. STAFF-TREE FAMILY

Trees or shrubs, often climbing or twining; leaves alternate or opposite, simple, deciduous or persistent, stipules small and caducous or absent; flowers bisexual, or sometimes functionally unisexual (the plants then polygamodioecious), actinomorphic, small and greenish, cymose, the sepals 4–5, basally connate, mostly imbricate, the petals 4–5 (rarely absent), imbricate (rarely valvate); stamens 4–5 (rarely 10) and alternate with petals, arising from or below the rim of an annular disc, distinct, the anthers 2-celled, dehiscing longitudinally; disc present and often adnate to the ovary; pistil 1, the ovary superior (sometimes seemingly inferior by adnation of disc), 2–5-loculed and -carpelled, the placentation axile, the ovules usually 2 on each placenta, usually erect and anatropous, the style 1 and short, the stigma usually capitate or obscurely 2–5-lobed; fruit a loculicidal capsule, berry, samara, or drupe; seed usually covered by a bright-colored pulpy aril, the embryo generally enveloped by the endosperm.

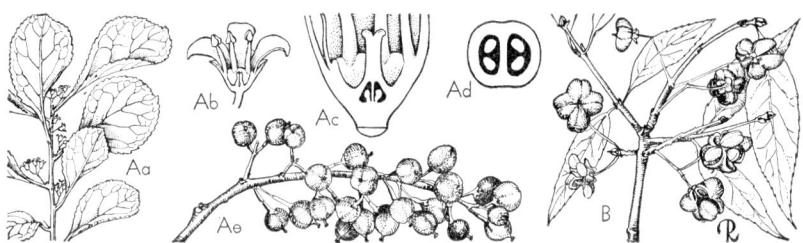

Fig. 190. CELASTRACEAE. A, *Celastrus orbiculatus*: Aa, flowering branch, × ¼; Ab, staminate flower, vertical section, × 2; Ac, pistil of pistillate flower, vertical section, × 6; Ad, ovary, cross-section, × 20; Ae, fruiting inflorescence, × ½. B, *Euonymus Maackii*: fruiting branch, × ½. (From L. H. Bailey, *Manual of cultivated plants*, The Macmillan Company, 1949. Copyright 1924 and 1949 by Liberty H. Bailey.)

A family of 45 genera and about 500 species, widely distributed except in the arctic, and represented in this country by 10 genera and about 20 species. In the eastern half of the country are indigenous species of *Euonymus, Pachystima* and *Celastrus*, with *Maytenus, Rhacoma, Gyminda,* and *Schaefferia* having indigenous species restricted to the southeast, while in the southwest and west are *Mortonia, Forsellesia, Canotia*,[130] *Euonymus,* and *Pachystima*.

The Celastraceae are distinguished by the cymose inflorescences of small greenish flowers, by the ovary usually surrounded by the disc, situated inside the point of stamen attachment and often adnate to the stamens, and by the seed usually enveloped by a bright-colored aril. Several closely related families differ in their having ovaries with uniovulate locules whereas in Celastraceae there are usually 2 ovules in each locule.

The family is of slight economic importance. Species of 8 genera are cultivated domestically for ornament. These genera include: *Elaeodendron, Catha, Gymnosporia,* and *Maytenus* grown in southern areas and *Celastrus, Euonymus, Tripterygium,* and *Pachystima* in cooler temperate areas.

*LITERATURE:*

CROIZAT, LEON. A study in the Celastraceae. Siphonodonoideae subf. nov. Lilloa, 13: 31–43, 1947.
ENSIGN, M. A revision of the celastraceous genus *Forsellesia (Glossopetalon).* Amer. Midl. Nat. 27: 501–511, 1942.

[130] *Canotia* was included in the Koeberliniaceae by Barnhart (No. Am. Fl. 25: 101–102, 1910), a view accepted by Jepson (1936), but Engler and Diels (1936) treated the type genus, *Koeberlinia*, as the type of the subfamily Koeberlinioideae of the Capparidaceae. Kearney and Peebles (1942) retained *Canotia* in the Celastraceae, observing that it was "an anomalous plant in this family."

HARRIS, J. A. Correlation in the inflorescence of *Celastrus scandens*. Ann. Rep. Mo. Bot. Gard., 20: 116–122, 1909.
LOESENER, T. Celastraceae. In Engler and Prantl, Die natürlichen Pflanzenfamilien, ed. 2, Bd. 20b: 87–197, 1942. [Not seen.]
LUNDELL, C. L. Revision of the American Celastraceae I. *Wimmeria, Microtropis,* and *Zinowiewia*. Contr. Univ. Michigan Herb. No. 3: 1–46, 1939.
MCNAIR, G. T. Comparative anatomy within the genus *Euonymus*. Univ. Kansas Sci. Bull. 19: 221–260, 1930.
SPRAGUE, T. A. The correct spelling of certain generic names: 6. *Euonymus* or *Evonymus*. Kew Bull. 1928: 294–296.

## HIPPOCRATEACEAE. HIPPOCRATEA FAMILY

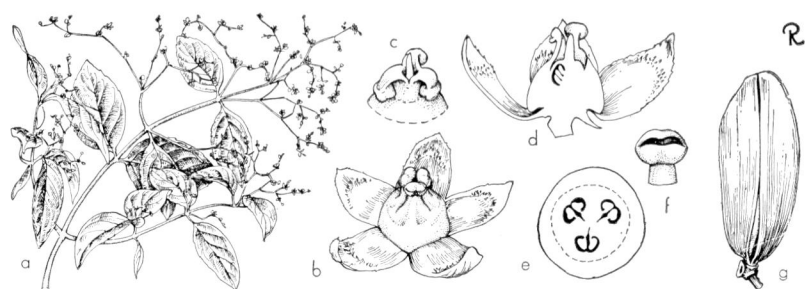

**Fig. 191.** HIPPOCRATEACEAE. *Hippocratea volubilis*: a, flowering branch, × ¼; b, flower, habit (stamens erect), × 5; c. section of flower (stamens recurved), × 5; d, flower, vertical section, × 6; e, ovary, cross-section, × 10; f, anther, × 12; g, fruit, × ½.

Trees, shrubs, or vines; leaves usually opposite, simple, minutely stipulate or estipulate; inflorescence thyrsoid, cymose racemose, paniculate or fasciculate; flowers bisexual, actinomorphic, usually small, bracteolate; flowers often with abundant mucilaginous fibers; perianth biseriate, the sepals usually 5, distinct or nearly so, imbricate, persistent, the petals usually 5, distinct, imbricate or valvate, suberect or rotate; stamens usually 3 (rarely 2, 4, or 5), the filaments often dilated and briefly connate basally, inserted within the disc, this usually annular and continuous but rarely discontinuous and forming staminiferous pockets, the anther cells 2 or confluent and seemingly 1, usually dehiscing transversely; pistil 1, the ovary superior and the disc sometimes adnate to its base and concealing it, 3-loculed and -carpelled (locules rarely 5), the placentation axile, the ovules 2–14 in each locule, usually collateral or 2-ranked, anatropous, the style usually short and trifid (rarely absent), the stigmas obscure or obvious, usually 3, entire or bifid; fruit a schizocarp, a 3-valved capsule, or a berry; seeds compressed or angular or winged in capsular fruits, no endosperm present.

A pantropical and subtropical family of about 18 genera and about 225 species, of which 12 genera and 115 species are American. The monotypic *Hippocratea* (*H. volubilis*) extends northward into southern Florida and is the only member of the family indigenous in the United States. The larger tropical American genera include: *Tontelea* (30 spp.), *Salacia* (29 spp.), and *Cheiloclinium* (20 spp.).

Distinguishing characters for the family are found in the combination of predominantly lianous habit, usually opposite leaves, the (usually) 3 stamens arising from the base of the tricarpellate ovary and inside the disc, and the often winged or angular seeds.

The Hippocrateaceae have been accepted as a distinct family of the Celastrales (or Celastrineae) by all except Hallier, who combined them with the Celastraceae. Smith (1940) acknowledged the affinity of the family with the Celastraceae, but pointed out numerous differences between the 2, especially in the relationship of androecium to the disc, the anther dehiscence, and in the fruit and seed.

The members of the family are of no consequential domestic economic importance.

*LITERATURE:*

LOESENER, T. Hippocrateaceae. In Engler and Prantl, Die natürlichen Pflanzenfamilien 3(5): 222–230, 1896.

———. Hippocrateaceae. In Engler and Prantl, Die natürlichen Pflanzenfamilien, ed. 2, Bd. 20b: 198–231, 1942. [Not seen.]

MIERS, J. On the Hippocrateaceae of South America. Trans. Linn. Soc. Bot. 28: 319–432, 1872. [Includes all in the New World.]

SMITH, A. C. The American species of Hippocrateaceae. Brittonia, 3: 341–555, 1940.

———. Notes on Old World Hippocrateaceae. Amer. Journ. Bot. 28: 438–443, 1941.

## STAPHYLEACEAE. BLADDERNUT FAMILY

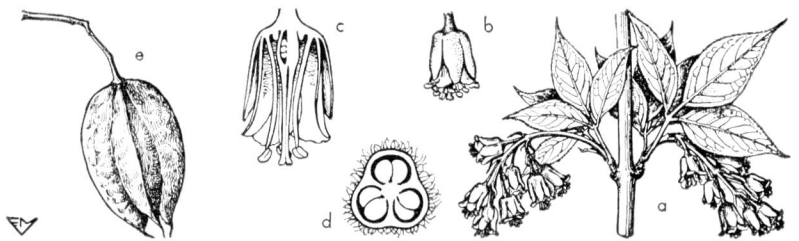

Fig. 192. STAPHYLEACEAE. *Staphylea trifolia*: a, flowering branch, × ½; b, flower, × 1; c, same, vertical section, × 2; d, ovary, cross-section, × 6; e, fruit, × ½. (From L. H. Bailey, *Manual of cultivated plants*, The Macmillan Company, 1949. Copyright 1924 and 1949 by Liberty H. Bailey.)

Shrubs and trees; leaves mostly opposite, pinnately compound (some are trifoliolate), stipulate; flowers usually bisexual, actinomorphic, in drooping racemes or panicles, 5-merous, the sepals 5, the petals 5, rising from or below the hypogynous disc; stamens 5, alternating with the petals, distinct, inserted near the outside of a large cup-shaped disc, the anthers 2-celled, dehiscing longitudinally; pistil 1, the ovary superior, 2–3-loculed and -carpelled, the placentation axile, the ovules numerous in usually 2 rows on each placenta, anatropous, usually ascending, the styles as many as carpels and usually distinct; fruit a membranous often lobed inflated capsule dehiscing apically, infrequently a berry; seeds few, sometimes arillate, with a straight embryo and abundant endosperm.

A family of 6 genera and about 24 species, of temperate regions in the northern hemisphere, a few extending into northern South America. Two of the 7 species of *Staphylea* are indigenous to the United States, *S. trifolia* from Quebec to Georgia and west to Minnesota and Oklahoma, and *S. Bolanderi* of California. This genus and *Turpinia* (10 spp.), of Mexico southward, is represented by species in both Asia and the Americas.

The Staphyleaceae were united by Bentham and Hooker with the Sapindaceae, but are distinguished from them by the cuplike intrastaminal disc, the more numerous ovules, the abundant endosperm, and straight embryo. They were included in the Celastrales by Wettstein, Rendle, and Bessey, but are placed in the Sapindales (*sensu strictu*) by Hutchinson. Hallier transferred them to his Rosales, and believed them to be allied to Cunoniaceae and Saxifragaceae. On the basis of cytological evidence Foster (1933) concluded them to have had a common origin with the Aceraceae.

The members of the family are of little economic importance. A few species of *Staphylea, Euscaphis,* and *Turpinia* are cultivated domestically for ornament.

*LITERATURE:*

FOSTER, R. C. Chromosome number in *Acer* and *Staphylea.* Journ. Arnold Arb. 14: 386–393, 1933.
KRAUSE, J. Staphyleaceae. In Engler and Prantl, Die natürlichen Pflanzenfamilien, ed. 2, Bd. 20b: 255–321, 1942. [Not seen.]

## ICACINACEAE. ICACINA FAMILY

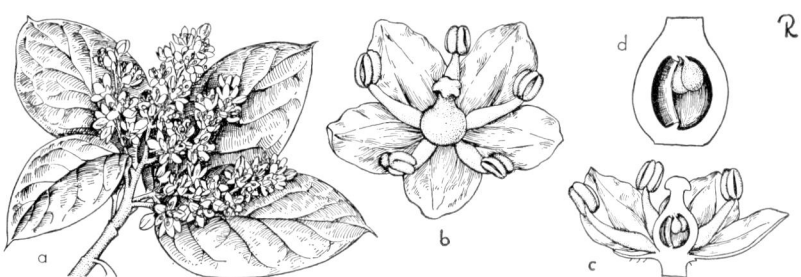

**Fig. 193.** ICACINACEAE. *Villaresia mucronata*: a, flowering branch, × ⅜; b, flower, habit, × 3; c, same, vertical section, × 3; d, ovary, vertical section, × 6. (Redrawn from Curtis' Bot. Mag.)

Trees or shrubs, often vines (rarely herbaceous); leaves usually alternate, simple, often coriaceous, estipulate; flowers bisexual (rarely unisexual by abortion), actinomorphic, usually in panicles, the calyx 4–5-lobed, small, the petals 4–5, usually distinct or sometimes connate, valvate; stamens as many as petals and alternate with them, distinct, the anthers 2-celled, usually introrse, sometimes 4-lobed, dehiscing longitudinally; pistil 1, the ovary superior, 1-loculed, mostly 3- or 5-carpelled (all except 1 carpel usually lost by suppression or compression), the placentation apical, the usually 2 ovules pendulous from locule apex, anatropous, the style 1, the stigmas usually 3 (rarely 2 or 5); fruit usually a drupe (1-seeded), rarely dry and winged (a samara); seed with straight or curved embryo, the endosperm usually present.

A family of pantropical distribution, composed of about 38 genera and 225 species, and probably more abundant in the southern than in the northern hemisphere. None of the genera is represented by species indigenous to the United States, although species of 3 genera extend north into Mexico.

The family is recognized by the woody, often lianous, character, the usually alternate estipulate leaves, and the 1-loculed 3–5-carpelled ovary with usually 2 pendulous ovules.

A few species of *Pennantia* and *Villaresia* are cultivated domestically as ornamentals.

*LITERATURE:*

ENGLER, A. Icacinaceae. In Engler and Prantl, Die natürlichen Pflanzenfamilien, III(5): 233–257, 1893.

## ACERACEAE. MAPLE FAMILY

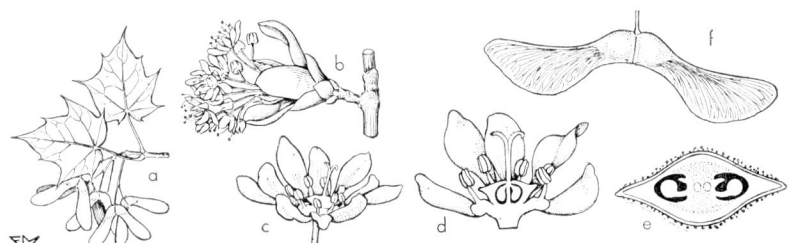

Fig. 194. ACERACEAE *Acer platanoides*: a, fruiting branch, × ⅙; b, inflorescence, × ½; c, flower, habit, × 1½; d, same, vertical section, × 2; e, ovary, cross-section, × 5; f, fruit, × ½. (From L. H. Bailey, *Manual of cultivated plants,* The Macmillan Company, 1949. Copyright 1924 and 1949 by Liberty H. Bailey.)

Trees or shrubs; leaves opposite, simple with usually palmate venation or pinnately compound, sap often milky, estipulate; flowers bisexual or more commonly unisexual (the plants monoecious, dioecious, or polygamodioecious), actinomorphic, in corymbs, racemes or panicles, the perianth usually biseriate, the sepals 4–5, distinct or basally connate, imbricate, the petals 4–5 or none, distinct, imbricate; disc usually present (extrastaminal or intrastaminal), usually flat but sometimes lobed or divided or reduced to teeth; stamens 4–10, mostly 8, distinct, usually arising from edge of disc (rudimentary ovary often present in staminate flowers), the anthers 2-celled; pistil 1, the ovary superior, 2-loculed and -carpelled, usually compressed at right angles to the septum, the placentation axile, the ovules 2 in each locule, orthotropous to anatropous, the styles 2, divergent and distinct or basally connate each with a terminal stigma; fruit a samaroid schizocarp splitting into 2 one-winged mericarps; seed lacking endosperm.

A family of 2 genera (*Acer, Dipteronia*) and about 150 species, with all but 2 species belonging in *Acer*. The maples are indigenous mostly to mountainous or upland regions of the northern hemisphere, with about 15 species indigenous to the United States. *Dipteronia,* native in central China, differs from *Acer* in the fruit winged around its periphery.

The Aceraceae are distinguished from related families, notably the Sapindaceae, with which they were combined by Bentham and Hooker, by the leaves opposite and usually palmate, the actinomorphic flowers, and the fruit a schizocarp or a samara.

Economically the family is important for maple lumber (especially from *A. saccharum*), for maple syrup and sugar manufactured from the sap, and for the 60-odd species cultivated domestically for ornament.

*LITERATURE:*

ANDERSON, E. and HUBRICHT, L. The American sugar maples I. Phylogenetic relationships, as deduced from a study of leaf variation. Bot. Gaz. 100: 312–323, 1938.
DANSEREAU, P. and LAFOND, A. Introgression des caractères de l'*Acer Saccharophorum* K. Koch et de l'*Acer nigrum* Michx. Contr. Inst. Bot. Univ. Montréal No. 37: 15–31, 1941.
GLEASON, H. A. The preservation of well-known binomials. Phytologia, 2: 201–212, 1947.
HALL, B. A. The floral anatomy of the Aceraceae. Ph.D. Thesis, Cornell, 1947.
KELLER, A. C. *Acer glabrum* and its varieties. Amer. Midl. Nat. 27: 491–500, 1942.
PAX, F. Monographie der Gattung *Acer*. Engler, Bot. Jahrb. 6: 287–374, 1885; *op. cit.* 7: 177–263, 1885–86; (Nachtrage) *op. cit.* 11: 72–83, 1889.
———. Aceraceae. In Engler, Das Pflanzenreich, 8 (IV. 163): 1–89, 1902.
ROUSSEAU, J. Histoire de la nomenclature de l'*Acer saccharophorum* K. Koch (*A. saccharum* Marshall) depuis 1753. Contr. Inst. Bot. Univ. Montréal, 1940.

## HIPPOCASTANACEAE.[131] HORSE-CHESTNUT FAMILY

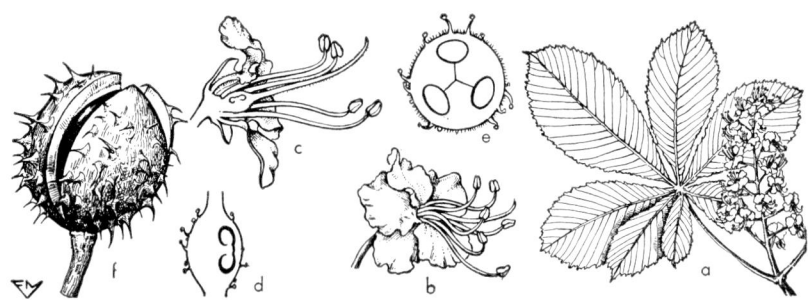

**Fig. 195.** HIPPOCASTANACEAE. *Aesculus Hippocastanum*: a, twig with leaf and inflorescence, × 1/10; b, flower, × 1; c, same, vertical section, × 1; d, ovary, vertical section, × 2; e, ovary, cross-section, × 4; f, fruit, × 1/2. (From L. H. Bailey, *Manual of cultivated plants*, The Macmillan Company, 1949. Copyright 1924 and 1949 by Liberty H. Bailey.)

Trees or shrubs; leaves opposite, palmately compound with 3–9 leaflets, estipulate; flowers bisexual (sometimes the upper ones unisexual and staminate), zygomorphic, in terminal thyrses, the sepals 4–5, basally connate (distinct in *Billia*), imbricate, the petals 4–5, distinct, unequal, clawed; extrastaminal disc present, often 1-sided; stamens 5–9, distinct, hypogynous, the anthers 2-celled, dehiscing longitudinally; pistil 1, the ovary superior, 3-loculed and -carpelled, the placentation axile, the ovules 2 in each locule, the style and stigma 1; fruit a usually 1-loculed 1-seeded leathery loculicidal capsule whose husk dehisces by 3 valves; seeds large and nonarillate, the embryo curved and endosperm absent.

A family of 2 genera (*Aesculus, Billia*) and perhaps 24 species, native mostly in North and South America. About 5 species of *Aesculus* are indigenous to the United States where the genus is of widespread distribution. The evergreen *Billia* (2 spp.) is indigenous from Colombia north into Mexico.

The Hippocastanaceae are included in the Sapindaceae by many authors (including Hallier and Hutchinson), from which they differ in the palmately compound leaves, the large flowered thyrse, and the leathery capsule with a very large usually solitary seed. Its distinction from the Sapindaceae has been strengthened by the removal of the monotypic Chinese *Bretschneidera* (included in Hippocastanaceae by Pax *et al.*) as a separate family, Breitschneideraceae, which is presumed to be of close affinity with the Moringaceae.

Economically the family is perhaps most important for the horse-chestnut tree (*Aesculus Hippocastanum*), a native of northern Greece, and cultivated throughout temperate parts of the world as an ornamental. Several of the American species of buckeye (*A. glabra, parviflora*, and *californica*) also are grown as ornamentals. Hybrids are produced readily within the genus.

*LITERATURE:*

BUSH, B. F. Notes on *Aesculus* species. Amer. Midl. Nat. 12: 19–26, 1930.
HOAR, C. S. Chromosome studies in *Aesculus*. Bot. Gaz. 84: 156–170, 1927.
PAX, F. Hippocastanaceae. In Engler and Prantl, Die natürlichen Pflanzenfamilien, III(5): 273–276, 1895.
WIGGINS, I. L. The lower California buckeye, *Aesculus Parryi* A. Gray. Amer. Journ. Bot. 19: 406–410, 1932.

[131] The name Hippocastanaceae (Torrey and Gray, 1838), antedated by Paviaceae Horaninov (1834) and Aesculaceae Lindl. (1836), has been proposed (1945) for conservation.

## SAPINDACEAE.[132] SOAPBERRY FAMILY

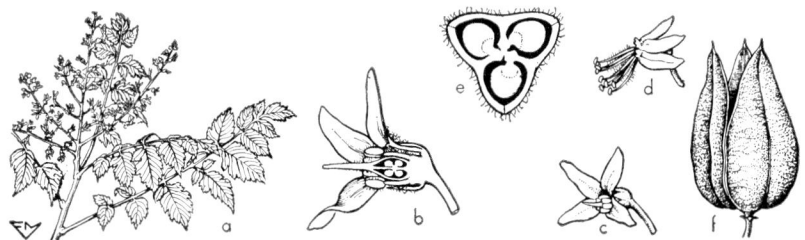

Fig. 196. SAPINDACEAE. *Koelreuteria paniculata*: a, flowering branch, × ¹⁄₁₀; b, perfect flower, vertical section, × 2; c, same, habit, × 1; d, staminate flower, × 1; e, ovary, cross-section, × 10; f, capsule, × ½. (From L. H. Bailey, *Manual of cultivated plants*, The Macmillan Company, 1949. Copyright 1924 and 1949 by Liberty H. Bailey.)

Trees or shrubs, or sometimes tendril-producing vines, rarely herbs; leaves usually alternate, simple or more commonly pinnately compound, stipulate only in climbing species; flowers bisexual or unisexual (polygamodioecism the usual condition, the apparently bisexual flowers generally unisexual in function), actinomorphic or commonly zygomorphic, minute, in racemose to paniculate unilateral cymes, the sepals usually 5, and distinct (in actinomorphic flowers seemingly 4 by connation and then associated with a loss of 1 petal in the corolla), mostly imbricate, the petals 5 or absent (except as noted above), distinct, equal or unequal, often with scaly or hair-tufted nectaries on lower inner side, imbricate; extrastaminal disc present and glandular; stamens typically 10 in 2 whorls of 5 each (often reduced to 8, 5, or 4) are rarely many, arising from inside the disc, distinct, often hairy, the anthers 2-celled, pistillode present in staminate flowers; pistil 1, the ovary superior, usually 3-loculed and -carpelled (rarely the locule number 1, 2, or 4), the placentation mostly axile (sometimes parietal), the ovules 1–2 on each placenta, ascending, the style mostly 1 (rarely 2–4) and 1–3 lobed; fruit very variable as to type; seed often arillate, the embryo mostly curved, endosperm wanting.

A family of 130 genera and about 1100 species, primarily pantropical in distribution and abundant in Asia and America. One tree (*Sapindus Saponaria*, the U. S. element sometimes distinguished as *S. Drummondii*) extends from the American tropics as far north as Kansas, and a shrub (*Dodonaea viscosa*) extends into Arizona. In the southern extremities of Florida there occurs a single species each of *Talisia, Exothea, Hypolate,* and *Cupania,* these representing outlying stations of West Indian representatives. The ubiquitous tropical American balloon vine (*Cardiospermum Halicacabum*) is naturalized as far north as Delaware and west to Kansas. Among the large genera of the family are the lianous *Serjania* and *Paullinia* with about 350 species each, and *Cupania* with about 35 species of mostly timber-producing trees.

The Sapindaceae are characterized by the usually pinnate leaves, the springlike circinately coiled tendrils of lianous genera, the small usually polygamodioecious flowers, the scale- or gland-appendaged petals, the unilateral extrastaminal disc, the typically tricarpellate ovary, and the usually arillate seed.

The members of the family are of minor domestic economic importance. Trees of the varnish tree (*Koelreuteria paniculata*) and the shrub *Xanthoceras sorbifolia* are cultivated as hardy ornamentals. In warmer regions the lychee (*Litchi chinensis*) is a tree much prized for the soft fleshy edible aril of its seed and for the seeds themselves (edible when roasted). Ornamentals in warm regions include the longan

---

[132] The name Sapindaceae has been conserved over other names for this family.

(*Euphoria*), Spanish lime or genip (*Melicocca*), Mexican buckeye (*Ungnadia*), and soapberry (*Sapindus*).

*LITERATURE:*

RADLKOFER, L. Sapindaceae. In Engler and Prantl, Die natürlichen Pflanzenfamilien, III(5): 277–366, 1895.

———. Sapindaceae. In Engler, Das Pflanzenreich, 98a-h (IV. 165. I-VIII): 1–1539, 1931–1934.

## SABIACEAE.[133] SABIA FAMILY

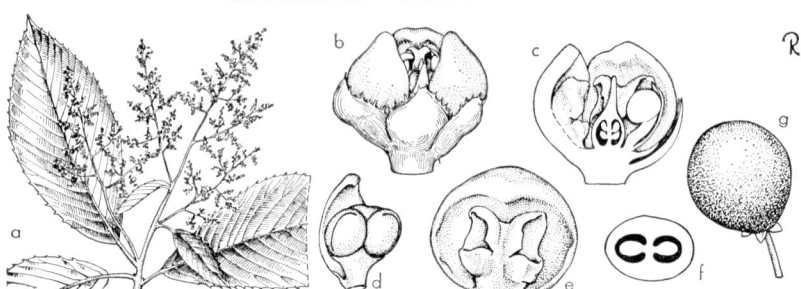

**Fig. 197.** SABIACEAE. *Meliosma myriantha*: a, flowering branch, × ¼; b, flower, habit, × 12; c, same, vertical section, × 12; d, stamen, undehisced, × 15, e, same, dehisced, × 15; f, ovary cross-section, × 25; g, fruit, × 5.

Trees, shrubs, or vines; leaves alternate, simple or pinnate, estipulate; flowers usually bisexual (sometimes the plants polygamodioecious), zygomorphic, often in panicles, the sepals 3–5, distinct or basally connate, imbricate, the petals 4–5, sometimes basally connate, the outer ones often broad and imbricate, with the inner 2 much reduced; annular disc small; stamens 3–5, distinct, opposite petals or opposite only the outer ones, free or adnate to petals, all or only 2 fertile (remainder reduced to staminodes), the anthers 2-celled, the connective usually thickened; pistil 1, the ovary superior, 2-loculed and -carpelled (derived from a 3-carpellate condition yet typical of some members), the placentation axile, the ovules usually 2 on each placenta, pendulous horizontal or ascending, semianatropous, the styles 2 but often connate, the stigmas or stigmatic tips 2; fruit a berry, sometimes dry and leathery; seed with large embryo, the endosperm scant or none.

A family of 4 genera (*Sabia, Meliosma, Ophiocaryon, Phoxanthus*) and about 90 species, mostly of tropical eastern Asia, but *Meliosma* (60 spp.) is occasional throughout the tropics, and in this hemisphere extends northward into Mexico (4 spp.).

The family is distinguished by the combination of 4–5-merous flowers whose stamens are opposite the petals and with often a thickened connective between or beneath the anther cells. Phylogenists are agreed that it is closely allied to the Sapindaceae.

Species of *Meliosma* are cultivated as ornamental trees in the warmer parts of this country.

*LITERATURE:*

WARBURG, O. Sabiaceae. In Engler and Prantl, Die natürlichen Pflanzenfamilien, III(5): 367–374, 1895.

[133] The name Sabiaceae has been conserved over other names (as Millingtoniaceae) for this family.

DIVISION IV. EMBRYOPHYTA SIPHONOGAMA 585

## MELIANTHACEAE. MELIANTHUS FAMILY

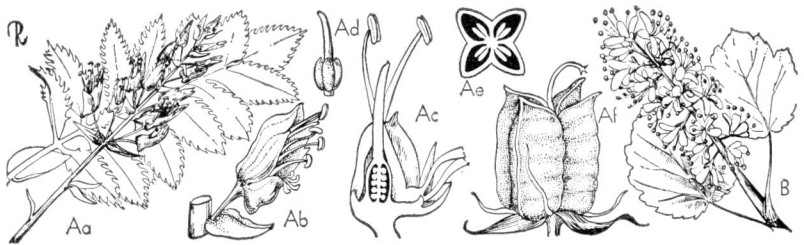

Fig. 198. MELIANTHACEAE. A, *Melianthus major*: Aa, inflorescence, × ¼ and leaf, × ⅛; Ab, flower, × 1; Ac, same, vertical section, × 2; Ad, pistil, × ½; Ae, ovary, cross-section, × 6; Af, fruit, × 1. B, *Greyia Sutherlandii*: flowering branch, × ¼. (From L. H. Bailey, *Manual of cultivated plants*, The Macmillan Company, 1949. Copyright 1924 and 1949 by Liberty H. Bailey.)

Trees, shrubs, or infrequently suffrutescent herbs; leaves alternate, pinnately compound or infrequently simple, stipules intrapetiolar; flowers bisexual or bisexual with also staminate and pistillate flowers dioeciously disposed, zygomorphic, in terminal or axillary racemes, the pedicels twisting 180° by time of anthesis, the sepals 5 or 4 by connation, distinct or basally connate, unequal, imbricate, the petals 4–5, distinct, clawed, unequal, imbricate; disc present, extrastaminal, crescent-shaped or annular with 10 projections; stamens 4–5, or 10, distinct or shortly basally connate, alternating with petals, often declinate, the anthers 2-celled, dehiscing longitudinally; pistil 1, the ovary superior, 4–5-loculed, and -carpelled (1-loculed with 5 parietal placentae in *Greyia*), usually deeply lobed, the placentation axile, the ovules 1–several on each placenta, erect or pendulous, anatropous, the style 1, the stigma 4–5-lobed (sometimes truncate, capitate, or toothed); fruit a loculicidal or apically dehiscing capsule; seeds sometimes arillate, the embryo straight, endosperm present.

A family of 2–3 genera (*Melianthus, Greyia*,[134] *Bersama*) and 38 species, all African.

The Melianthaceae were considered by Bentham and Hooker and by Hallier to be a part of the Sapindaceae, from which they differ in the flowers turning halfway around before anthesis, in the stamens often basally connate and typically 4, and in the seeds with copious endosperm.

Three species of the evergreen honey bush (*Melianthus*) and one of *Greyia* are cultivated domestically in warm regions as decorative shrubs and trees.

LITERATURE:

GÜRKE, M. Melianthaceae. In Engler and Prantl, Die natürlichen Pflanzenfamilien, III(5): 374–383, 1895.

## BALSAMINACEAE. BALSAM FAMILY

Herbaceous plants (sometimes aquatics) or suffrutescent, often somewhat succulent, rarely epiphytic; leaves alternate, opposite, or in whorls of 3's, simple, usually estipulate; flowers bisexual, zygomorphic, solitary or several together on axillary

---

[134] Hutchinson (1926) segregated *Greyia* (3 spp.) as a distinct and new family (Greyiaceae) on the basis of the leaves simple, stipules absent, and the ovary 1-loculed with 5 intrusive parietal placentae. He transferred it from the Sapindales to the Cunoniales (near Escalloniaceae). This change leaves the Melianthaceae a much more natural taxon.

peduncles, often resupinate, spurred, nodding, pentamerous, the calyx zygomorphic, the sepals 3–5, imbricate, often petaloid, the posterior very large and saclike and gradually prolonged backward into a tubular nectiferous spur, the petals 5, alternate with the sepals, distinct or connate and appearing as if 3, the lower ones larger than the upper ones; stamens 5, hypogynous, syngenesious or nearly so, the filaments flattened and closely covering the ovary (and often the style) like a hood or sheath, the anthers 2-celled, coherent to connate; pistil 1, the ovary superior, 5-loculed and -carpelled, the placentation axile, the ovules 3 to many on each placenta, anatropous, pendulous, the style 1, short or obsolete, the stigmas 1–5; fruit a fleshy 5-valved capsule, explosively dehiscing; the valves coil elastically, and on splitting the tension forcibly distributes the seeds; sometimes the fruit a berry; seeds with straight embryo and no endosperm.

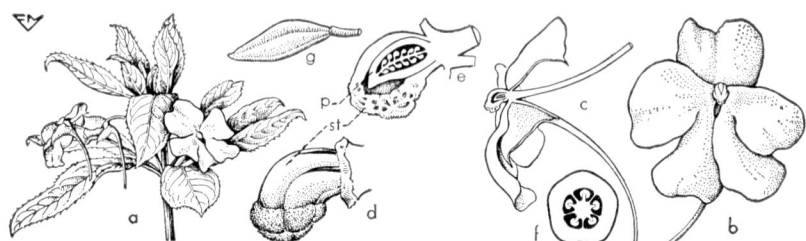

**Fig. 199.** BALSAMINACEAE. *Impatiens glandulifera*: a, flowering branch, × ⅙; b, flower, × ½; c, same, vertical section, × ½; d, flower less perianth, × 2; e, same, vertical section, × 2; f, ovary, cross-section, × 5; g, undehisced capsule, × ½. (st stamen, p pistil.) (From L. H. Bailey, *Manual of cultivated plants*, The Macmillan Company, 1949. Copyright 1924 and 1949 by Liberty H. Bailey.)

A family of 2 genera (*Impatiens, Hydrocera*) and about 450 species of which about 420 belong to *Impatiens*. The family is widely distributed but is most abundant in the tropics of Asia and Africa. In the New World it is absent from South America, and only 2 species of *Impatiens* are indigenous to the United States, both in the eastern half.

The family is best distinguished from allied taxa by the peculiar androecial situation, especially by the coherence or connation of the anthers about the ovary and stigma; the elastic dehiscence of the usually succulent capsule is also distinctive. The nectiferous spur of the flower has caused some botanists to ally the family with or close to the Geraniaceae or Tropaeolaceae, but in the Balsaminaceae it is strictly an outgrowth of the calyx, whereas in the former there is evidence that receptacular tissues are involved in the spur formation. Most phylogenists (except Wettstein) have in the past treated the family as a member of the Geraniales (Gruinales), but most current opinions (except Hutchinson's) do not concur.

Economically the members of the family are of little importance. Several species of *Impatiens* are cultivated domestically as ornamentals, notably the garden balsam (*I. Balsamina*), and the Himalayan balsam (*I. glandulifera*, of which the name *I. Roylei* is a synonym).

*LITERATURE:*

CARROLL, F. B. The development of the chasmogamous and the cleistogamous flowers of *Impatiens fulva*. Contrib. Bot. Lab. Univ. Pennsylvania, 4: 144–184, 1920.
RYDBERG, P. A. Balsaminaceae. North Amer. Flora, 25: 93–96, 1910.
WARBURG, O. and REICHE, K. Balsaminaceae. In Engler and Prantl, Die natürlichen Pflanzenfamilien, III(5): 383–392, 1895.

## Order 28. RHAMNALES

Usually woody plants, the flowers often unisexual and/or apetalous, differing from the Geraniales and the Sapindales in the stamens in a single whorl, as many as the sepals and alternating with them (opposite the petals); the ovary usually with 1-2 ascending ovules. The order is distinguished from the polypetalous orders that follow in the low stamen number and the presence of a disc surrounding or subtending the ovary.

Two families compose the order: Rhamnaceae and Vitaceae. These families were included within the Celastrales by Bessey (Celastrineae of Engler), but have been accepted as a separate order by Wettstein, Rendle, and Hallier, each of whom restricted it to these 2 taxa, and who indicated affinities with the Rosales. Hutchinson also accepted the order Rhamnales, but included in it these families plus the Elaeagnaceae and the Heteropyxidaceae, and treated the order as closely allied to the Celastrales.

### RHAMNACEAE.[135] BUCKTHORN FAMILY

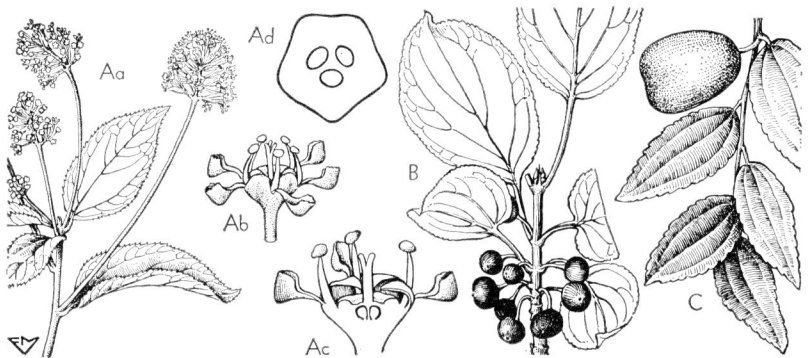

**Fig. 200.** RHAMNACEAE. A, *Ceanothus americanus*: Aa, flowering branch, × ½; Ab, flower, × 4; Ac, same, vertical section, × 5; Ad, ovary, cross-section, × 10. B, *Rhamnus cathartica*: fruiting branch, × ½. C, *Ziziphus Jujuba*: twig with fruit, × ½. (From L. H. Bailey, *Manual of cultivated plants,* The Macmillan Company, 1949. Copyright 1924 and 1949 by Liberty H. Bailey.)

Trees or shrubs, erect or climbing (by hooks, tendrils, or twining stems), rarely herbs, sometimes suffrutescent; leaves mostly alternate, simple, usually stipulate; flowers bisexual (rarely unisexual and the plants then mostly monoecious), actinomorphic, small and greenish, mostly in axillary corymbs or cymose inflorescences, perigynous, a hypanthium usually present, the sepals 5 (rarely 4), the lobes valvate, the petals of same number as sepal lobes (occasionally absent) and alternate with them, often concave, frequently clawed; stamens as many as petals and opposite them, enclosed by the petals on emergence, arising from outside an intrastaminal disc that lines or rims the hypanthium, the anthers 2-celled, dehiscing longitudinally; pistil 1, the ovary superior (or seemingly inferior by adnation to it of the intrastaminal disc), 2–4-loculed and -carpelled, the placentation basal, the ovules 1 (rarely 2) in each locule, anatropous, the style 1 or 2; fruit a berry (drupaceous), capsule, or rarely samaroid; seeds with embryo large and straight, the endosperm copious or scant.

---

[135] The name Rhamnaceae R. Br. (1827) is conserved over Frangulaceae DC. (1805).

A family of almost cosmopolitan distribution, composed of about 45 genera and 550 species. *Rhamnus*, the largest genus with perhaps 90 species, is represented by 12 species indigenous to the United States. Other genera having indigenous species include *Ceanothus* (83–69) best represented in the west, but with 4 species in the east, *Berchemia* (10–1), *Krugiodendron* (1–1), *Reynosia* (9–1), *Colubrina* (15–5), *Gouania* (40–1), and *Sageretia* (10–2) exclusively or predominately in the southwest, while in the southwest and west are also species of *Condalia* (12–5) and *Adolphia* (2–1).

The Rhamnaceae are distinguished by the simple unlobed leaves, the markedly perigynous flowers, the stamens opposite the usually strongly concave petals, and the basal ovules.

Species belonging to about 16 genera are cultivated in this country as ornamentals. Notable among them are buckthorns (*Rhamnus*), tea bush (*Ceanothus*), jujube (*Zizyphus*), supplejack (*Berchemia*), raisin tree (*Hovenia*), Jerusalem thorn (*Paliurus*), and species of *Pomaderris, Reynosia, Noltea,* and *Spyridium*. The fruits of the jujube are edible and tasty. The purgative known as *cascara sagrada* is obtained from the Californian *Rhamnus Purshiana*.

*LITERATURE:*

GORHAM, R. P. The known distribution of the buckthorns in the Maritime Provinces. Acadian Nat. 1: 118–124, 1944.

HOWELL, J. T. Studies in *Ceanothus*, I–V, Leafl. West. Bot. 2: 159–165, 202–208, 228–240, 259–262, 285–289, 1939–1940.

LAMMERTS, W. E. Origin and description of new *Ceanothus* hybrids. Journ. Calif Hort. Soc. 9: 121–125, 1948.

LAUTERBACH, C. Rhamnaceae. Engler, Bot. Jahrb. 57: 326–353, 1920–1922.

MCMINN, H. E. The importance of field hybrids in determining species in the genus *Ceanothus*. Proc. Cal. Acad. IV, 25: 323–356, 1944.

RECORD, S. J. American woods of the family Rhamnaceae. Trop. Woods, 58: 6–24, 1939.

VAN RENSSELAER, M. and MCMINN, H. E. Ceanothus. 308 pp. Santa Barbara, Cal., 1942.

WEBERBAUER, A. Rhamnaceae. In Engler and Prantl, Die natürlichen Pflanzenfamilien, III(5): 393–427, 1895.

WOLF, C. B. The North American species of *Rhamnus*. Monogr. Rancho Santa Ana Bot. Gard. Bot. Ser. 1: 1–136, 1938.

## VITACEAE.[136] GRAPE FAMILY

Mostly climbing shrubs with tendrils (stem growth sympodial; the tendril represents the main axis that has been subordinated by the more vigorous growth of the axillary branch in the opposing leaf axil); seldom erect shrubs or small trees, the nodes often swollen or jointed; leaves alternate (the lower ones sometimes opposite), simple or pinnately or palmately compound, pellucid punctate dots frequently present, the stipules petiolar or none; flowers minute, bisexual or unisexual (the plants then usually monoecious), actinomorphic, numerous in cymose inflorescences arising opposite a leaf, the sepals 4–5 (rarely 3–7), distinct or basally connate, the petals of same number as sepals, minute or obsolete, flat, distinct (connate in *Leea*) or (as in *Vitis*) apically connate, separating from each other at the base, and the corolla early deciduous as a cap; disc evident, annular or lobed; stamens as many as petals and opposite them, somewhat perigynous, arising from base of disc, the anthers distinct or connate, 2-celled, dehiscing longitudinally; pistil 1, the ovary

[136] The name Vitaceae Lindl. (1836) has been proposed for conservation (cf. Rehder, Journ. Arnold Arb. 26: 278, 1945) and is antedated by several little-known names. Some authors (incl. Hutchinson) have accepted the name Ampelidaceae Lowe (1868) for the family, a name based on Ampelideae Kunth (1821).

superior, 2- (rarely 3–6–) loculed and -carpelled, the placentation axile, the ovules 1–2 on each placenta, anatropous, the style 1 and short, the stigma discoid or capitate; fruit a berry; seeds with straight embryo and copious endosperm.

A family of 11 genera and about 600 species, widely distributed in the tropics and subtropics with ranges extending into the north and south temperate regions. *Cissus*, the largest genus, has about 300 species, almost wholly in the tropics and southern hemisphere. The family is represented in the United States by indigenous species of *Vitis* (50–30) with 2 species on the west coast and the others mainly in the south and east, *Cissus* (300–4) and *Ampelopsis* (15–2) mostly in the southeast, and *Parthenocissus* (10–2) of the eastern part of the country.

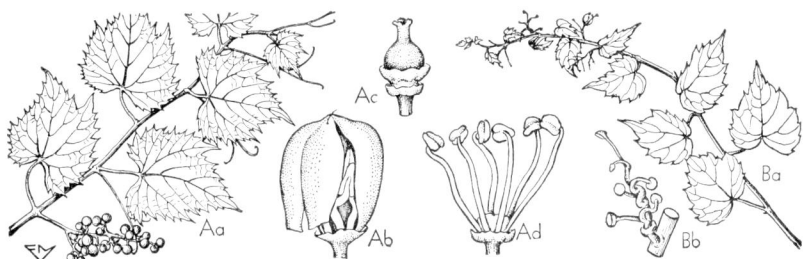

Fig. 201. VITACEAE. A, *Vitis riparia*: Aa, fruiting branch, × 1/6; Ab, bud opening, × 8; Ac, pistillate flower, × 5; Ad, staminate flower, × 5. B, *Parthenocissus tricuspidata*: Ba, sterile branch, × 1/4; Bb, stem with disc-tipped tendrils, × 1. (From L. H. Bailey, *Manual of cultivated plants*, The Macmillan Company, 1949. Copyright 1924 and 1949 by Liberty H. Bailey.)

The Vitaceae are characterized by the climbing habit, the terminal buds developing into apparently lateral tendrils, the inflorescences opposite a leaf at the node, the few stamens opposite the petals, the usually 2-loculed ovary with axile placentation, the capitate or discoid stigma, and the fruit a berry.

Economically the family is most important for the wine grape (*Vitis vinifera*) and for other species grown extensively for their edible fruit from which are obtained wines and raisins. The Boston ivy (*Parthenocissus tricuspidata*) and Virginia creeper (*P. quinquefolia*) are cultivated as ornamental vines prized for suitability for growing on stone or masonry walls. Several species of *Cissus* are cultivated as desirable house plants.

*LITERATURE:*

BAILEY, L. H. The species of grapes peculiar to North America. Gentes Herb. 3: 151–241, 1934.
BIOLETTI, F. T. Outline of ampelography for the vinifera grapes in California. Hilgardia, 11: 227–293, 1938.
DORSEY, M. J. Pollen development in *Vitis* with special reference to sterility. Minnesota Agr. Exp. Sta. Bull. 144: 1–60, 1914.
GILG, E. Vitaceae (Ampelidaceae). In Engler and Prantl, Die natürlichen Pflanzenfamilien, III(5): 427–456, 1895.
OBERLE, G. D. A genetic study of variations in floral morphology and function in cultivated forms of *Vitis*. N. Y. State Agr. Exp. Sta. Tech. Bull. 250: 3–63, 1938.
PLANCHON, J. E. Monographie des Ampélidées vraies. In de Candolle, Monogr. Phaner. 5: 305–654, 1887.
REGEL, E. Conspectus specierum generis vitis regiones americae borealis, chinae borealis et Japoniae habitantium. Act. Hort. Petrop. 2: 289–319, 1873.
SAX, K. Chromosome counts in *Vitis* and related genera. Proc. Soc. Hort. Sci. 1929: 32–33. 1930.

# SELECTED FAMILIES OF VASCULAR PLANTS

## Order 29. MALVALES [137]

Predominantly woody plants of generally pantropical and subtropical distribution, frequently stellate-pubescent, and tissues often mucilage-producing; characterized by the usually bisexual and actinomorphic cyclic flowers with mostly pentamerous perianths, the calyx valvate, and the stamens usually numerous (or in more than 1 whorl), the ovary multicarpellate, placentation usually axile and derived from a parietal condition (by deep intrusion of placentae) still encountered in some members (see Tiliaceae).[138]

The order was treated by Engler and Diels to be composed of the following 4 suborders and 8 families (family names preceded by an asterisk are not treated in this text):

| | | |
|---|---|---|
| Elaeocarpineae | Malvineae | Scytopetalineae |
| Elaeocarpaceae | Tiliaceae | *Scytopetalaceae |
| Chlaeneae | Malvaceae | |
| *Chlaenaceae | Bombacaceae | |
| | Sterculiaceae | |

Most phylogenists have followed Eichler's concept of the order as adopted by Engler. Hutchinson, however, restricted the Malvales to include only the Malvaceae, and placed the Tiliaceae, Bombacaceae, and Sterculiaceae in a new order, the Tiliales (the Elaeocarpaceae were included by him in the Tiliaceae). Of the 2, the Malvales were considered by him to be the more advanced and to represent "a fixed type of the Tiliales, and whence little or no further evolution has proceeded." Bessey expanded Eichler's concept of Malvales to include also the Balanopsidaceae, Ulmaceae, Moraceae, and Urticaceae.

## ELAEOCARPACEAE. ELAEOCARPUS FAMILY

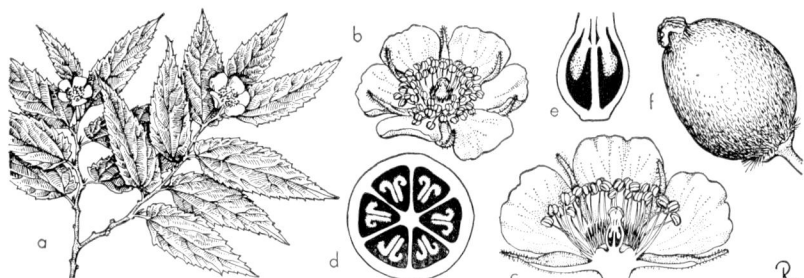

**Fig. 202.** ELAEOCARPACEAE. *Muntingia Calabura*: a, flowering branch, × ¼; b, flower, × 1; c, same, vertical section, × 1½; d, ovary, cross-section, × 3; e, same, vertical section, × 6; f, fruit, × 1.

Trees or shrubs; leaves simple, entire, alternate, or opposite, stipulate, slime cells lacking; flowers usually bisexual, actinomorphic, in racemes, panicles, or dichasia, hypogynous, lacking any involucre, perianth biseriate, the sepals 4–5, distinct or connate, valvate, the petals 4–5 or often none, distinct (rarely basally connate), often incised or hairy, usually valvate; stamens many, distinct, arising from a disc,

---

[137] This order was first named Columniferae by Eichler, a name adopted by Wettstein, and modified by Hallier to Columniferes.

[138] Phylogenetic considerations of these taxa were presented by Edlin (New Phytologist, 34: 1–20, 122–143, 1935). For critical notes, see also *op. cit.* 35: 93, 1936.

the anthers mostly 2-celled, dehiscing by 2 terminal pores; intrastaminal disc present and sometimes developed into an androphore; pistil 1, the ovary superior, 2-many-loculed (rarely 1-loculed), the placentation axile, the ovules 2-many in each locule, anatropous, pendulous, the style 1 and mostly simple or shortly lobed; fruit a capsule or drupaceous; seed with straight embryo and copious endosperm.

A family of 8 genera and 125 species, of pantropical distribution, often united with the Malvaceae from which it differs by the absence of slime cells beneath the bark, the frequent lack of corolla, the petals usually valvate when present, the stamens arising from the disc, and the often baccate fruit. No members of the family are indigenous in the United States (a few species of *Muntingia* and *Sloanea* extend northward into Mexico).

A few species of *Aristotelia, Crinodendron, Elaeocarpus,* and *Muntingia* are cultivated as ornamentals in the warmer parts of the southern states.

*LITERATURE:*

Mauritzon, J. Zur Embryologie der Elaeocarpaceae. Ark. Bot. 26A: 1–8, 1934.
Schumann, K. Elaeocarpaceae. In Engler and Prantl, Die natürlichen Pflanzenfamilien, III(6): 1–8, 1895.

## TILIACEAE.[139] LINDEN FAMILY

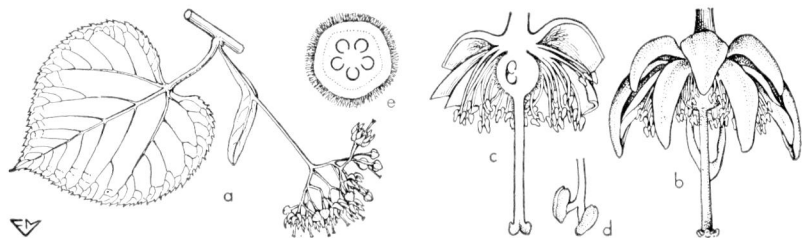

**Fig. 203.** Tiliaceae. *Tilia americana*: a, twig with leaf, bract and inflorescence, × ¼; b, flower, × 2; c, same, vertical section, × 2; d, anther, × 4; e, ovary, cross-section, × 4 (From L. H. Bailey, *Manual of cultivated plants,* The Macmillan Company, 1949. Copyright 1924 and 1949 by Liberty H. Bailey.)

Trees or shrubs, rarely herbaceous (as in *Corchorus*), pubescence mostly of branched hairs; leaves alternate (rarely opposite), simple, commonly deciduous, often oblique, stipulate; flowers bisexual (rarely unisexual, the plants then monoecious), actinomorphic, cymose, the sepals 5 or rarely 3–4, distinct or basally connate, usually valvate, the petals as many as sepals or lacking, sometimes sepaloid, distinct, stamens 10–numerous, hypogynous, distinct or briefly connate at base or in fascicles of 5–10 each, the anthers 2-celled, dehiscing by apical pores or by longitudinal slits; pistil 1, the ovary superior, 2–10-loculed and -carpelled, the placentation axile,[140] the ovules 1–several in each locule, ascending or pendulous, more or less anatropous, the style 1 and stigmas usually as many as locules; fruit fleshy or dry (dehiscent or indehiscent), of various types; seed with straight embryo, the endosperm present and varying from scant to copious.

[139] The name Tiliaceae has been conserved over other names for this family.
[140] Weibel (1945) has shown that the axile placentation in this family has been derived from parietal and is not a constant character. In *Sparmannia* the ovary is initially unilocular with 5–6 deeply intruded placentae that tardily become coherent or connate at the center, while in others (as *Belotia* spp.) the placentation may be basally axile and distally parietal, and in a few (*Goethalsia, Mollia*) the ovary remains unilocular with the placentation parietal.

A family of mostly woody plants, containing 41 genera and about 400 species, mostly of tropical distribution with a few extensions into temperate regions. *Tilia*, the dominant representative in this country, contains several polymorphic species resulting in varying opinions as to their number (18–65 for the genus) and perhaps only 6–8 indigenous to this country. Numerous hybrids are known. In warmer parts of the United States are indigenous species of *Triumfetta* (70–2) and *Corchorus* (40–3). The largest genus is the pantropical *Grewia* (160 spp.).

The Tiliaceae are distinguished by the nearly distinct stamens, the 2-celled anthers, and the typically cymose inflorescences. In some genera (*Tilia*) the inflorescence arises from a short shoot situated in the axil of a foliage leaf, with a conspicuous membranous bract adnate to the peduncle or primary axis of the inflorescence.

Economically the family is of domestic importance for *Tilia* spp., whose lumber is known as basswood or whitewood and whose trees known also as lindens are grown for ornament and shade. Species of *Grewia, Sparmannia, Corchorus, Corchoropsis,* and *Entelea* are grown for ornament in warm parts of the country.

LITERATURE:

BURRET, M. Beiträge zur Kenntnis der Tiliaceen. Notizb. Bot. Gart. Berlin, 9: 592–880, 1926.
BUSH, B. F. The glabrate species of *Tilia*. Bull. Torrey Bot. Club, 54: 231–248, 1927.
———. The Mexican species of *Tilia*. Amer. Midl. Nat. 11: 543–560, 1929.
CHATTAWAY, M. M. Anatomical evidence that *Grewia* and *Microcos* are distinct genera. Trop. Woods, 38: 9–11, 1934.
DERMEN, H. Chromosome numbers in the genus *Tilia*. Journ. Arnold Arb. 13: 49–51, 1932.
ENGLER, V. Monographie der Gattung *Tilia*. 159 pp. Berlin, 1909.
SARGENT, C. S. Notes on North American Trees III. *Tilia*. Bot. Gaz. 66: 421–438, 494–511, 1918.
SCHUMANN, V. Tiliaceae. In Engler and Prantl, Die natürlichen Pflanzenfamilien, III(6): 8–30, 1895.
WEIBEL, R. La placentation chez les Tiliacées. Candollea, 10: 155–177, 1946.

## MALVACEAE.[141] MALLOW FAMILY

Herbs, shrubs, or trees, the vesture often lepidote or stellate, sap often mucilaginous; leaves alternate, simple, entire or variously lobed, usually palmately veined, stipulate; flowers bisexual (rarely unisexual and the plants then generally dioecious as in *Napaea*), actinomorphic, basically in cymes of cincinni but often solitary in axils, the perianth typically biseriate but the calyx frequently subtended by an involucre (epicalyx) of distinct or connate bracts (sometimes interpreted as an involucel of bracteoles or as stipules), the sepals 5, distinct or basally connate, valvate, the petals 5, distinct, often basally adnate to the androecium, convolute or less commonly imbricate; stamens very numerous, in 2 whorls (the outer whorl usually absent), hypogynous, monadelphous, the filaments apically distinct, the anthers 1-celled, reniform, dehiscing longitudinally, the pollen grains distinctive, usually spiny and large; pistil typically 1, the ovary superior, 2–many-loculed and -carpelled with the locules usually in a ring or infrequently superposed (as in *Malope* or *Kitaibelia*), the placentation axile, the ovules 1–many in each locule, anatropous, ascending to pendulous (as in Sidineae), the style 1 and apically branched or as many as carpels, the stigmas as many or twice as many as carpels and often capitate to discoid or introrsely decurrent; fruit typically a loculicidal capsule or the mature carpels of the ovary separating from one another and from the axis, sometimes a berry (*Malvaviscus*) or a samara; seed often pubescent or

[141] The name Malvaceae has been conserved over other names for this family.

## DIVISION IV. EMBRYOPHYTA SIPHONOGAMA

comose (as in *Gossypium*), the embryo straight or curved, the endosperm mostly present and often oily.

The family is composed of about 82 genera and 1500 species, distributed over most of the earth and particularly abundant in the American tropics. The members indigenous to the United States are distributed among 27 genera and number nearly 200 species. Of them, *Hibiscus* (200–20), *Malvastrum* (75–8), *Sida* (180–16), and *Abutilon* (100–11), are rather widespread, while only *Napaea* (2–1) is restricted to the northeast, *Cienfuegosia* (20–1), *Malachra* (6–1), and *Urena* (3–1)

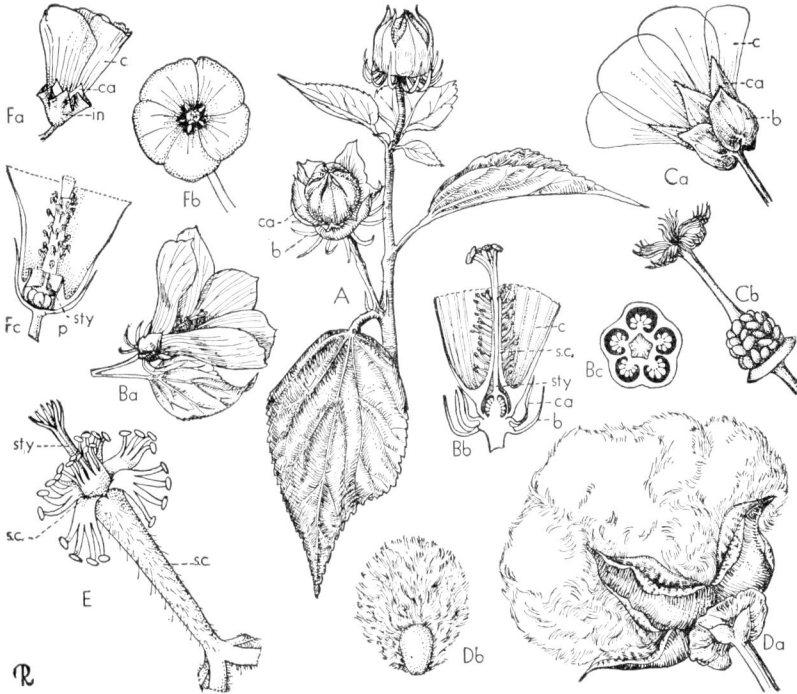

**Fig. 204.** MALVACEAE. A, *Hibiscus palustris*: fruiting branch, × ⅜. B, *H. Moscheutos*: Ba, flower, × ¼; Bb, same, vertical section, perianth partially excised, × ½; Bc, ovary, cross-section, × 2. C, *Malope trifida*: Ca, flower habit, × ½; Cb, same, gynoecium, × 4. D, *Gossypium hirsutum*: Da, boll, × 1; Db, seed, × ½. E, *Anoda cristata*: Fa, flower, side view, × ½; Fb, same, face view, × ½; Fc, partial vertical section, × 1. (b bract, c corolla, ca calyx, in involucre, p pistil, s.c. staminal column, sty style.) (From L. H. Bailey, *Manual of cultivated plants*, The Macmillan Company, 1949. Copyright 1924 and 1949 by Liberty H. Bailey.)

to the southeast, and *Anoda* (10–7), *Horsfordia* (4–2), *Sidalcea* (20–20), *Sphaeralcea* (250–30), and *Lavatera* (20–1) to the southwest and western parts of the country. *Gayoides* (1–1) occurs from the southeast west to Arizona, and *Iliamna* (4–4) from the southwest to Illinois and West Virginia. *Gossypium* (40–2) occurs from Florida to Arizona. A few other genera (as *Althaea, Thespesia,* and *Malva*) are represented by species adventive or naturalized from other countries.

The Malvaceae are related to the Sterculiaceae, Bombacaceae, and Tiliaceae, from which they differ in the 1-celled anthers (sometimes so in Bombacaceae) and the markedly monadelphous stamens. The pollen of most malvaceous plants is distinctive in that it is spiny and the grains are large. The presence of an involucre and the convolute corolla are diagnostic when present.

Economically the family is of greatest importance for the cotton of commerce, obtained from the woolly coma of the seeds and for the oil and pulp (meal) obtained from the seeds. Fruits of okra (*Hibiscus esculentus*) are cooked and eaten. Notable among the 30 genera whose species are grown domestically for ornament are the hollyhock (*Althaea*), poppy mallow (*Callirhoë*), rose of Sharon (*Hibiscus*), and *Malvaviscus, Malva, Kitaibelia,* and *Sidalcea.*

*LITERATURE:*

BAKER, E. G. Synopsis of genera and species of Malveae. Journ. Bot. (Lond.) 28: 15–18, 140–145, 207–213, 239–243, 339–343, 367–371, 1890; 29: 49–53, 164–172, 362–366, 1891; 30: 71–78, 136–142, 235–240, 290–296, 324–332, 1892; 31: 68–76, 212–217, 267–273, 334–338, 361–368, 1893. [Reprinted, 124 pp.; 1894.]

EASTWOOD, A. The shrubby Malvastrums of California with descriptions of new species and a key to the known species. Leafl. West. Bot. 1: 213–220, 1936.

FERNALD, M. L. *Hibiscus Moscheutos* and *H. palustris.* Rhodora, 44: 266–278, 1942.

FRIES, R. E. Zur Kenntnis der süd- und zentralamerikanischen Malvaceenflora. Svensk Vet. Akad. Handl. III. 24: 1–37, 1947.

GANDOGER, M. Le genre *Sida* (Malvacées). Bull. Soc. Bot. France, 71: 627–633, 1924.

HANSON, H. C. Distribution of the Malvaceae in southern and western Texas. Amer. Journ. Bot. 8: 192–206, 1921.

HOCHREUTINER, B. P. G. Revision du genre *Hibiscus.* Ann. Cons. Jard. Genève, 4: 23–191, 1900.

———. Monographia generis *Anodae.* Ann. Cons. Jard. Bot. Genève, 20: 29–68, 1916.

HOWARD, R. A. *Atkinsia* gen. nov., *Thespesia,* and related West Indian genera of the Malvaceae. Bull. Torrey Bot. Club, 76: 89–100, 1949.

HUTCHINSON, J. B. Notes on the classification and distribution of genera related to *Gossypium.* New Phytol. 46: 123–147, 1947.

HUTCHINSON, J. B., SILOW, R. A. and STEPHENS, S. G. The evolution of *Gossypium* and the differentiation of the cultivated cottons. London, 1947.

JANDA, C. Die extranuptialen Nektarien der Malvaceen. Österreich. Bot. Zeitschr. 86: 81–130, 1937.

KEARNEY, T. H. The North American species of *Sphaeralcea* subgenus *Eusphaeralcea.* Univ. California Publ. Bot. 19: 1–128, 1935.

———. Type of the genus *Malvastrum.* Leafl. West. Bot. 5: 23, 24, 1947.

KESSELER, E. VON. Observations on chromosome number in *Althaea rosea, Callirhoë involucrata,* and *Hibiscus coccineus.* Amer. Journ. Bot. 19: 128–130, 1932.

LANG, C. H. Investigations of the pollen of the Malvaceae, with special reference to the inclusions. Journ. Royal Microscop. Soc. III. 57: 75–102, 1937.

LEWTON, F. L. The value of certain anatomical characters in classifying the *Hibisceae.* Journ. Washington Acad. Sci. 15: 165–172, 1925.

MOLBY, E. E. The preliminary study of the epidermal appendages of the Mallow family. Trans. Illinois Acad. Sci. 23: 169–173, 1931.

MORTON, C. V. The correct names of the small-flowered mallows. Rhodora, 39: 98–99, 1937.

ROUSH, EVA M. F. A monograph of the genus Sidalcea. Ann. Mo. Bot. Gard. 18: 117–244, 1931.

SCHERY, R. W. Monograph of *Malvaviscus.* Ann. Mo. Bot. Gard. 29: 183–244, 1942.

SCHUMANN, K. Malvaceae. In Engler and Prantl, Die natürliche Pflanzenfamilien, III (6): 30–53, 1895.

SKOVSTED, A. Chromosome numbers in the Malvaceae. I. Journ. Genetics, 31: 263–296, 1935.

WEBBER, I. E. Systematic anatomy of the woods of the Malvaceae. Trop. Woods 38: 15–36, 1934.

WEBBER, J. M. Cytogenetic notes on *Sphaeralcea* and *Malvastrum*. Science II. 81: 639–640, 1935.

———. Chromosomes in *Sphaeralcea* and related genera. Cytologia, 7: 313–323, 1936.

WIGGINS, I. L. A resurrection and revision of the genus *Iliamna* Greene. Contr. Dudley Herb. 1: 213–229, 1936.

## BOMBACACEAE. BOMBAX FAMILY

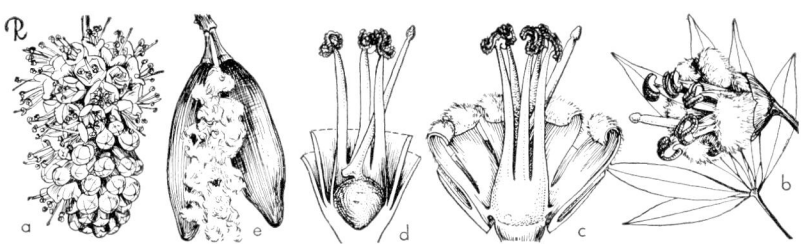

Fig. 205. BOMBACACEAE. *Ceiba pentandra*: a, inflorescence, × ¼; b, flower, × 1 and leaf, × ¼; c, flower, perianth partially excised, × 1; d, same, stamens removed in part, × 1; e, fruit, × ½ (From L. H. Bailey, *Manual of cultivated plants*, The Macmillan Company, 1949. Copyright 1924 and 1949 by Liberty H. Bailey.)

Trees, the trunks often tall and unusually thick; leaves alternate, simple or palmately compound, deciduous, often with slime cells and the vesture of stellate hairs or peltate scales (lepidote), stipules caducous; flowers bisexual, large and showy, commonly bracteate, often appearing before the leaves, actinomorphic (rarely slightly zygomorphic), solitary or fasciculate in leaf axils or situated opposite a leaf, the perianth sometimes subtended by an involucre, the sepals 5, distinct or basally connate, valvate, the petals 5 or occasionally absent, contorted in bud; stamens 5–many, distinct or monadelphous, the anthers often 1-celled, dehiscing longitudinally, the pollen grains smooth; staminodes often present; pistil 1, the ovary superior, 2–5-loculed and -carpelled, the placentation axile, the ovules 2 or more in each locule, erect, anatropous, the style 1, capitate or lobed, the stigmas 1–5; fruit a loculicidal capsule (sometimes an indehiscent pod) or berrylike; seed smooth, sometimes embedded in a pithlike tissue or in a woolly proliferation of the pericarp, occasionally arillate, the endosperm scant to absent.

A tropical, and primarily American, family of 22 genera and 140 species, none of which is indigenous to the United States. The larger genera include *Bombax* (60 spp.), *Ceiba* (20 spp.), *Durio* (15 spp.), and *Adansonia* (10 spp.).

The family is closely allied to the Malvaceae and the Sterculiaceae, from which it is distinguished by the smooth pollen grains, and the pericarp of the fruit (pithy to woolly). Staminodes are more common here than in either of the other 2 families.

Economically the family is important as the source of kapok (fruits of *Ceiba*), and balsa wood (*Ochroma*). Species of a few genera are cultivated for ornament in the warmer parts of this country, notably of the baobab tree (*Adansonia*), the cotton tree (*Bombax*), the guinea chestnut (*Pachira*), and the Brazilian floss-silk tree (*Chorisia*).

*LITERATURE:*

PITTIER, H. On *Gyranthera* and *Bombacopsis*, with a key to the American genera of Bombaceae. Journ. Washington Acad. Sci. 16: 207–214, 1926.

SCHUMANN, K. Bombacaceae. In Engler and Prantl, Die natürlichen Pflanzenfamilien. III (6): 53–68, 1895.

## STERCULIACEAE.[142] STERCULIA FAMILY

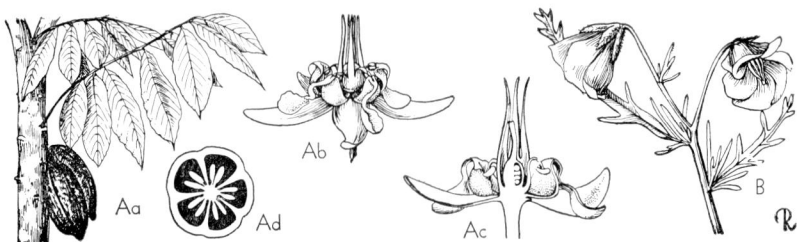

**Fig. 206.** STERCULIACEAE. A, *Theobroma Cacao*: Aa, trunk with fruits and leafy branches, much reduced; Ab, flower, × 2; Ac, same, vertical section, × 2; Ad, ovary, cross-section, × 8. B, *Mahernia verticillata*: flowering branch, × 1. (From L. H. Bailey, *Manual of cultivated plants*, The Macmillan Company, 1949. Copyright 1924 and 1949 by Liberty H. Bailey.)

Trees, shrubs (sometimes lianous), or herbs; leaves alternate, simple, entire, or infrequently palmately lobed or compound, vesture often stellate, stipules caducous; flowers generally bisexual (unisexual and the plants monoecious in *Sterculia* and *Cola*), actinomorphic or less often zygomorphic, pentamerous, usually axillary in various inflorescence types and often cauliflorous, the perianth often uniseriate, the sepals 3–5, basally connate, valvate, the petals small and reduced or lacking, hypogynous, sometimes adnate to base of androecium, contorted in bud; stamens in 2 whorls monadelphously connate into a single tube or all distinct, those of the outer whorl reduced to staminodes or wanting, the anthers 2-celled, dehiscing longitudinally, rarely cohering apically; pistil 1, the ovary superior, 4–5-loculed and -carpelled (rarely the locules 10–12), the placentation axile, the ovules 2–several in each locule, ascending or horizontal, anatropous, the styles as many as carpels and distinct or more commonly variously connate; fruit leathery or fleshy (rarely woody), dehiscent or indehiscent, sometimes the carpels splitting into cocci; seed with straight or curved embryo and abundant endosperm.

A pantropical and subtropical family of about 50 genera and 750 species, with 7 genera represented by species indigenous in the southern belt of the country from Florida to California. *Ayenia* (10–4) is domestically the most widespread, occurring in arid areas over much of the south. In the southeast also are *Melochia* (*Moluchia*, *Riedlea*) (50–3), and *Waltheria* (35–1), the latter extending westward into Arizona. *Hermannia* (150–1), primarily an African genus, extends northward from Mexico into Arizona. The monotypic *Fremontodendron* (*Fremontia*) occurs from central Arizona to California and southward.

The family is distinguished by the usually monadelphous stamens that differ from those of the Malvaceae in being 2-celled, and is separated from the allied Bombacaceae by the seeds with copious endosperm, the absence of a pithy or woolly pericarp, and the androecium of typically 5 fertile stamens alternated with as many staminodes.

Economically the family is important as the source of cacao or chocolate produced from the fermented seeds (beans) of the tropical American tree (*Theobroma Cacao*). Species of a few genera are cultivated domestically in warm regions as ornamentals, notably of the bottle tree (*Brachychiton*), the African cola (*Cola*), Dombeya (*Dombeya*), Chinese parasol tree (*Firmiana*), flannelbush (*Fremontodendron*), honey bell (*Mahernia*), Sterculia (*Sterculia*), and Thomasia (*Thomasia*).

[142] The name Sterculiaceae Lindl. (1830) is conserved over Buttneriaceae R. Br. (1814) and "Buettneriaceae" of authors.

*LITERATURE:*

Gazet du Chatelier, G. La structure florale des Sterculiacées. Comp. Rend. Acad. Sci. (Paris), 210: 57–59, 1940.
Harvey, M. A revision of the genus *Fremontia*. Madroño, 7: 100–110, 1943.
Mildbraed, J. Sterculiaceae. Engl. Bot. Jahrb. 62: 347–367, 1928–1929.

## Order 30. PARIETALES

A large order whose families are characterized in general by the flowers having a usually biseriate and pentamerous perianth with the calyx imbricated, the corolla usually present, the stamens as many as the petals or more, the ovary commonly of 3 carpels and unilocular with parietal placentation and numerous ovules, and endosperm usually present. Available evidence indicates that the order is not a phylogenetic taxon, and the realignment of the families into several orders is to be expected. In some instances this may result in transfer of families to existing orders and in others in the establishment of new orders. Indications of evidence in support of these realignments are to be found in the discussion given below and under the families concerned.

The Parietales were subdivided by Engler and Diels into 10 suborders, and treated as composed of 31 families, as follows (names of families not treated in this text are preceded by an asterisk):

| Theineae | Tamaricineae | Turneraceae |
|---|---|---|
| Dilleniaceae | Elatinaceae | *Malesherbiaceae |
| Actinidiaceae | Frankeniaceae | Passifloraceae |
| Eucryphiaceae | Tamaricaceae | *Achariaceae |
| *Medusagynaceae | Cistineae | Papayineae |
| Ochnaceae | Cistaceae | Caricaceae |
| *Strasburgeriaceae | Bixaceae | Loasineae |
| *Caryocaraceae | Cochlospermineae | Loasaceae |
| *Marcgraviaceae | Cochlospermaceae | Datiscineae |
| *Quiinaceae | Flacourtiineae | Datiscaceae |
| Theaceae | Canellaceae | Begoniineae |
| Guttiferae | Violaceae | Begoniaceae |
| (Hypericaceae) | Flacourtiaceae | Ancistrocladineae |
| *Dipterocarpaceae | Stachyuraceae | *Ancistrocladaceae |

Bessey placed the last 3 suborders in his Loasales, the Tamaricineae in his Caryophyllales, the Dilleniaceae (including Actinidiaceae) in the Ranales, and the remainder in his Guttiferales. Wettstein followed the Engler pattern in general, but segregated the Theineae as the Guttiferales and included the Canellaceae in his Polycarpicae (Magnoliales). Rendle followed most of Wettstein's treatment, but retained the Canellaceae within the Parietales and transferred the Datiscaceae and Begoniaceae into the Peponiferae (Cucurbitales) with the Cucurbitaceae. Hallier placed all the Theineae and Tamaricineae in his Guttales (together with many other families as shown in Fig. 14) except the Dilleniaceae (in his Ranales) and the Actinidiaceae (included in the Clethraceae); the Canellaceae he allied (as did Wettstein) with the magnoliaceous families, the Violaceae with the Polygales, the Datiscaceae and Begoniaceae with the Peponiferae, and the Flacourtiaceae, Turneraceae, Passifloraceae, and Caricaceae in his Passionales. Hutchinson distributed the families among 10 different orders (as indicated under each family) which were not the equivalent of any of Engler's suborders.

## DILLENIACEAE.[143] DILLENIA FAMILY

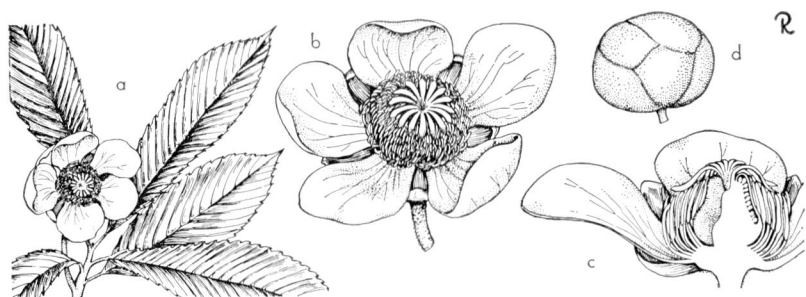

**Fig. 207.** DILLENIACEAE. *Dillenia indica*: a, flowering branch, × ⅙; b, flower, habit, × ⅜; c, same, vertical section (perianth partially excised), × ½; d, flower bud, × ⅓. (Redrawn from Baillon.)

Trees or shrubs, sometimes lianous, rarely herbaceous; leaves alternate, simple, venation seemingly parallel, the stipules caducous, alate and adnate to petiole or lacking; flowers bisexual or unisexual (the plants then dioecious or less commonly monoecious), rarely large and showy, actinomorphic, the sepals 5, imbricate, persistent, the petals 5 or fewer, imbricate, often crumpled in bud; stamens numerous, distinct or variously basally fasciculate, hypogynous, mostly persistent, the anthers 2-celled, dehiscing by apical pores or longitudinally; gynoecium usually of several distinct pistils (rarely 1), the ovary superior, 1-loculed and -carpelled, the placentation parietal but sometimes appearing basal, the ovules 1 or more, each with 2 integuments, erect, the styles as many as pistils, distinct; fruit a follicle or berrylike; seeds with copious fleshy endosperm, the embryo minute, the pericarp usually provided with a more or less lacerated aril.

As circumscribed above (the Actinidiaceae treated as a separate family), the Dilleniaceae are composed of 11 genera and 275 species, mostly of Australasia and tropical America. None is represented by species indigenous to the United States although species of *Tetracera* (35 spp.), *Davilla* (35 spp.), and *Curatella* (2 spp.) extend northward into Mexico. *Hibbertia* of Australia, the largest genus, has about 110 species. Other genera include *Doliocarpus* (20 spp.), *Dillenia* (20 spp.), and *Wormia* (35 spp.).

The Dilleniaceae (*sensu stricto*) are distinguished by the gynoecium composed of separate pistils, the numerous stamens in bundles, the parietal placentation, the seeds with endosperm, an aril and a minute straight embryo. It is separated from the Actinidiaceae by the character of the stamens (distinct or adnate to the petals in Actinidiaceae), the pistils distinct and not connate, the seeds with 2 integuments, and the embryo small and not large. Evidence is accumulating to establish the family as much more primitive than was allowed by Engler, and it is probable that its alliance is with or near the ranalian taxa (as done by Bessey, Hallier, and Hutchinson).

Economically the family is of little domestic importance. A few species of *Dillenia* and *Hibbertia* (incl. *Candollea*) are cultivated for ornament in warm parts of the country.

*LITERATURE:*

CROIZAT, L. Notes on the Dilleniaceae and their allies: *Austrobaileyeae* subfam. nov. Journ. Arnold Arb. 21: 397–404, 1940.

DIELS, L. Dilleniaceae. Engl. Bot. Jahrb. 57: 436–459 (1920–1922).

[143] The name Dilleniaceae has been conserved over other names for this family.

Gilg, E. Dilleniaceae. In Engler and Prantl, Die natürlichen Pflanzenfamilien, III (6): 100–128, 1895.

Gilg, E. and Werdermann, E. Dilleniaceae. In Engler and Prantl, Die natürlichen Pflanzenfamilien, ed. 2, Bd. 21: 7–36, 1925.

## ACTINIDIACEAE. ACTINIDIA FAMILY

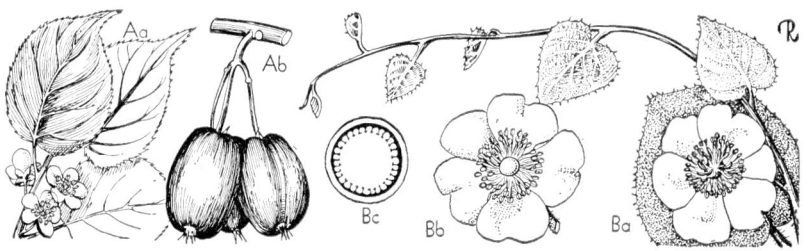

**Fig. 208.** Actinidiaceae. A, *Actinidia arguta*: Aa, flowering branch (staminate), × ½; Ab, fruit, × ½. B, *A. chinensis*: Ba, flowering branch, perfect flower, × ½; Bb, staminate flower, × ½; Bc, ovary, cross-section, × 3. (From L. H. Bailey, *Manual of cultivated plants*, The Macmillan Company, 1949. Copyright 1924 and 1949 by Liberty H. Bailey.)

Trees shrubs or woody vines; leaves alternate, simple, estipulate; flowers bisexual or unisexual (the plants then dioecious), usually fascicled, cymose, or paniculate, the perianth biseriate, pentamerous, imbricate; stamens 10–many, distinct or adnate to the base of the petals, hypogynous, the anthers 2-celled, versatile, dehiscing by apical pores or longitudinally; pistil 1, the ovary superior, locules and carpels 3–5 or more (sometimes the septa scarcely meeting at ovary centers), the placentation axile, the ovules with a single integument, numerous in each locule, anatropous, the styles usually as many as carpels, distinct or connate, generally persistent; fruit a berry (dehiscent at full maturity) or leathery capsule; seeds nonarillate, with usually large embryos and abundant endosperm.

A family of 4 primarily tropical genera (*Actinidia, Saurauia, Clematoclethra, Sladenia*) and about 285 species, primarily of Asiatic distributions. *Saurauia* (250 spp.) is composed of trees and shrubs, and has about one-third of its species in the American tropics, with its most northern occurrence here being in Mexico. *Actinidia*, a genus of about 22 species of woody vines, is restricted to the subtropics and temperate regions of eastern Asia.

The family, treated by most authors as a component of the Dilleniaceae, was recognized by Tieghem as distinct from the latter family as early as 1899, and he separated the 2 on the basis of the single multiloculate pistil, the axile placentation, the stamens distinct or in fascicles adnate to the base of each petal, the ovules with single integuments, and the nonarillate seed containing a usually large embryo.

Hutchinson recognized this taxon as distinct from the Dilleniaceae, but restricted the lianous and dioecious Actinidiaceae to the single genus *Actinidia* and recognized the Saurauiaceae as a separate family containing only the genus *Saurauia*. Hallier included both *Actinidia* and *Saurauia* within the Clethraceae of the Ericales. Hutchinson placed the Dilleniaceae in his Dilleniales (together with the Crossosomataceae) and the Actinidiaceae in his Theales.

Economically the family is of domestic importance for the tara vine (*Actinidia arguta*) cultivated (as are other species of the genus) as an ornamental and for its exotic edible fruit. *Clematoclethra*, a hardy vine, is cultivated to a limited extent.

*LITERATURE:*

Dunn, Stephen T. A revision of the genus *Actinidia*. Journ. Linn. Soc. Bot. (Lond.) 39 394–410, 1911.

GILG, E. and WERDERMANN, E. Actinidiaceae. In Engler and Prantl, Die natürlichen Pflanzenfamilien, ed. 2, Bd. 21: 36–47, 1925.

LECHNER, S. Anatomische Untersuchungen über die Gattungen *Actinidia, Saurauia, Clethra,* und *Clematoclethra* mit besonderer Berücksichtigung ihrer Stellung im System. Beih. Bot. Centralbl. 32: 431–467, 1915.

TIEGHEM, P. VAN. Sur les genres Actinidie et Sauravie, considérées comme types d'une famille nouvelle, les Actinidiacées. Journ. Bot. (Paris), 13: 170–174, 1899.

## EUCRYPHIACEAE. EUCRYPHIA FAMILY

**Fig. 209.** EUCRYPHIACEAE. *Eucryphia cordifolia*: a, flowering branch, × ⅜; b, flower, habit, × ¾; c, same, vertical section, × 1; d, ovary, cross-section, × 7; e, bud with caducous calyptrate calyx, × 1½; f, fruit, × 1½; g, seed, × 3.

Evergreen trees or shrubs, often resinous; leaves opposite, simple or pinnately compound, stipules minute and connate; flowers bisexual, actinomorphic or nearly so, showy white, solitary in axils, the sepals 4, imbricate, cohering at apex and separating at base at anthesis, and thus calyptrately deciduous, the petals 4, imbricate or convolute, distinct; stamens numerous, distinct, hypogynous, arising in many whorls from a thin disc that peripherally seemingly is produced into tubular excrescences (staminodes?), the anthers 2-celled, small, dehiscing longitudinally; pistil 1, the ovary superior, the locules and carpels 5–12 (rarely to 18), the placentation axile, the ovules several in each locule, pendulous, the styles as many as carpels, slender; fruit a woody or leathery capsule, the carpels of the ovary separating by dehiscence along ventral sutures into distinct follicular units (actually boat-shaped valves); seeds winged, compressed, the embryo embedded in the endosperm.

A unigeneric family (*Eucryphia,* with 4 spp.) indigenous to Chile and Australia. Three species (*E. glutinosa*) and 2 hybrids are cultivated domestically as half-hardy small trees.

The family is of interest because of its disjunctive distribution. Hallier included it within the Theaceae and considered it to have been derived from the Malvales. Wettstein accepted it as a distinct family, allied to both the Dilleniaceae and Ochnaceae. Hutchinson treated it as an intermediate member of his Guttiferales, an order he interpreted as an advanced hypogynous type of Theales.

*LITERATURE:*

FOCKE, W. O. Eucryphiaceae. In Engler and Prantl, Die natürlichen Pflanzenfamilien, III (6): 129–131, 1895.

GILG, E. Eucryphiaceae. In Engler and Prantl, Die natürlichen Pflanzenfamilien, ed. 2, Bd. 21: 47–50, 1925.

## OCHNACEAE. OCHNA FAMILY

Trees or shrubs, rarely herbaceous; leaves alternate, simple (rarely pinnately compound), leathery, stipulate; flowers bisexual, usually actinomorphic, in panicles racemes or cymes, the sepals 4–5 (rarely 10), distinct or basally connate, usually imbricate, the petals 4–5 (rarely 10), distinct, contorted or imbricate in bud; stamens 5, 10 or many, distinct, hypogynous, sometimes with 1–3 series of staminodes, occasionally borne on an elongated androphore, the anthers 2-celled, basifixed, each cell dehiscing by a terminal pore or longitudinally; pistil 1, the ovary superior, entire or deeply lobed, borne on an enlarged toral disc or gynophore, usually 2–5-loculed and -carpelled (rarely 10–15-loculed), the placentation axile (rarely parietal and the ovary then 1-loculed with intrusive placentas), the ovules 1–many in each locule, erect or rarely pendulous, the style typically 1, the stigmas 1–5; fruit baccate and borne usually on an enlarged torus, the carpels separating into distinct fleshy cocci, rarely a capsule; seed with a usually straight embryo, the endosperm present or absent.

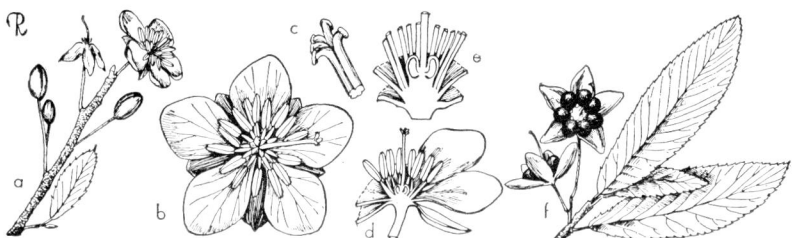

**Fig. 210.** OCHNACEAE. *Ochna multiflora*: a, flowering branch, × ½; b, flower, × 1; c, stigma, × 5; d, flower, vertical section, × 1; e, ovary, × 4; f, fruiting branch, × ½. (From L. H. Bailey, *Manual of cultivated plants,* The Macmillan Company, 1949. Copyright 1924 and 1949 by Liberty H. Bailey.)

A family of 21 genera and 375 species of pantropical distribution, but especially abundant in northeastern South America. The northern limits of the family in the New World are in Mexico, where 2 species of *Ouratea* (200 spp.) are indigenous. Other large genera include *Ochna* (90 spp., all Old World), *Sauvagesia* (18 spp. mostly Brazil) and *Luxemburgia* (12 spp. Brazil).

The family is distinguished by the stamens distinct and usually 1–3 times the number of petals, the single usually deeply lobed ovary, the gynophore, the generally poricidal anthers, and the carpels usually separating into baccate cocci in fruit.

Economically the family is of little domestic importance. One species (*Ochna multiflora*) is cultivated as an ornamental to an appreciable extent in warm parts of the country.

*LITERATURE:*

GILG, E. Ochnaceae. In Engler and Prantl, Die natürlichen Pflanzenfamilien, ed. 2, Bd. 21: 53–87, 1925.

## THEACEAE.[144] TEA FAMILY

Tree or shrubs; leaves alternate, simple, leathery or membranaceous, usually persistent, estipulate; flowers usually bisexual (unisexual and plants dioecious in *Eurya*), actinomorphic, generally solitary in axils or occasionally fasciculate, peri-

[144] This family is currently treated in the literature under the names Theaceae and Ternstroemiaceae. The latest authority to hold to the latter name was Rehder (Bibliography of Cultivated Trees and Shrubs, 1949). Correctness of the name Theaceae has been set forth by Sprague (Journ. Bot. Lond. 61: 17–19, 83–85, 1923) and challenged by Fawcett and Rendle (*op. cit.* 61: 52, 85–86, 1923). It is probable that the solution may lie in the future conservation of one name or the other.

anth parts often spirally arranged, frequently subtended by a pair of bracts, the sepals 5 (rarely more), distinct, usually persistent, imbricate, the petals 5 (rarely 4 or more), distinct or very shortly basally connate, contorted or imbricate; stamens numerous and rarely 15 or less, distinct, or basally monadelphous or fascicled in 5 bundles that are opposite and often basally adnate to the petals, the anthers 2-celled, dehiscing longitudinally (rarely poricidally); pistil 1, the ovary superior, sessile, mostly 3–5-loculed and -carpelled, the placentation axile, the ovules usually 2–more in each locule (rarely 1), anatropous, the styles as many as carpels and distinct to wholly connate; fruit usually a loculicidal capsule (the central column often persisting after dehiscence, subligneous, sometimes fleshy and indehiscent; seed nonarillate, with a large straight or curved embryo, the endosperm scant or absent.

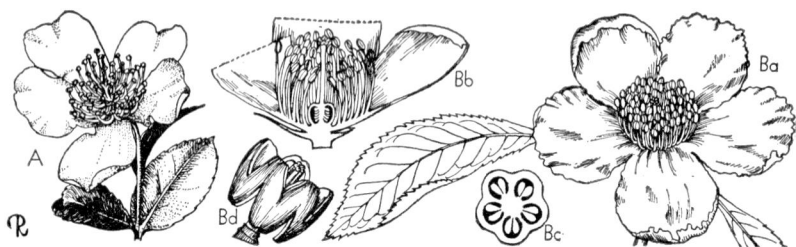

Fig. 211. THEACEAE. A, *Camellia Sasanqua*: flower, × ⅜. B, *Franklinia alatamaha*: Ba, flowering branch, × ½; Bb, flower, vertical section (petals excised), × ½; Bc, ovary, cross-section, × 1; Bd, capsule, × ½. (Ba-Bc redrawn from Sargent.) (From L. H. Bailey, *Manual of cultivated plants*, The Macmillan Company, 1949. Copyright 1924 and 1949 by Liberty H. Bailey.)

A tropical and subtropical family of about 30 genera and 500 species, of which 200 species from 10 genera occur in the New World, where 5 genera are endemic. A number of genera and species extend into temperate regions; in this country are 1 or more indigenous species of *Gordonia* (32 spp.) north to Virginia, the endemic and monotypic *Franklinia* of southeast Georgia now known only in cultivation, *Stewartia* (including *Malachodendron*) (6 spp.) of eastern Asia and eastern North America north to Virginia. The larger and mostly Asiatic genera include: *Eurya* (80 spp.), *Ternstroemia* (135 spp.), *Adinandra* (70 spp.), and *Camellia* (incl. *Thea*) (100 spp.).

The family is distinguished from related families by the several whorls of numerous stamens often fasciculate and the bundles adnate to the subtending petals, the spiral arrangement of the perianth parts and, in genera where present, the involucral bracts that intergrade into the sepals.

Economically the family is most important for the tea plant of commerce (*Camellia sinensis*). Domestically, many genera contribute significant ornamentals of warm climates, notably *Camellia*, loblolly bay (*Gordonia*), *Franklinia*, *Stewartia*, *Ternstroemia*, *Cleyera*, and *Eurya*.

*LITERATURE:*

AIRY-SHAW, H. K. Notes on the genus *Schima* and on the classification of the Theaceae-Camellioideae. Kew Bull. 1936: 496–500.

DIELS, L. Theaceae. Engl. Bot. Jahrb. 57: 431–435, 1920–1922.

FERNALD, M. L. Must all rare plants suffer the fate of *Franklinia?* Journ. Franklin Inst. 226: 383–397, 1938.

JENKINS, C. F. The historical background of Franklin's tree. Pennsylvania Mag. Hist. and Biogr. 58: 193–208, 1933.

HARPER, F. and LEEDS, A. N. A supplementary chapter on *Franklinia alatamaha.* Bartonia, 19: 1–13, 1938.
MELCHOIR, H. Theaceae. In Engler and Prantl, Die natürlichen Pflanzenfamilien, ed. 2, Bd. 21: 109–154, 1925.
RECORD, S. J. American woods of the family Theaceae. Trop. Woods, 70: 23–33, 1942.
SPRAGUE, T. A. The correct spelling of certain generic names. Kew Bull. 1928: 362–363. [*Stewartia* vs. *Stuartia.*]

For revisionary treatments of Asiatic and tropical American genera, see Kobuski, C. E. Studies in Theaceae, I–XIX. Journ. Arnold. Arb. 1935–50.

## GUTTIFERAE.[145] GARCINIA FAMILY

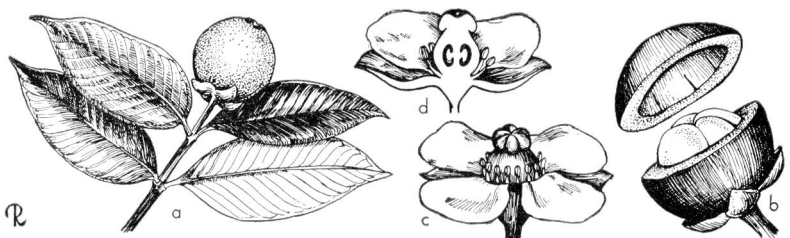

**Fig. 212.** GUTTIFERAE. *Garcinia Mangostana*: a, fruiting branch, × ¼; b, fruit with end of rind removed, × ⅓; c, flower, × ½; d, same, vertical section, × ½. (From L. H. Bailey, *Manual of cultivated plants,* The Macmillan Company, 1949. Copyright 1924 and 1949 by Liberty H. Bailey.)

Trees or shrubs, sap resinous, oil glands present; leaves opposite or whorled, rarely alternate, simple, estipulate; flowers usually unisexual, sometimes bisexual on the same plant and functionally polygamodioecious, rarely all bisexual, actinomorphic, the sepals 2–10 (or more), decussate or imbricated, the petals 4–12, susually imbricate, subvalvate, or contorted; stamens few or numerous, hypogynous, distinct or variously united, often collected into bundles or phalanges, the anthers 2-celled, dehiscing longitudinally; pistillode often present in staminate flowers; pistil 1, the ovary superior, 1-many-loculed, carpels 3 or 5 or as many as the locules, placentation axile basal, or infrequently parietal, the ovules 1–many on each placenta, anatropous, styles as many as locules or carpels and usually connate, the stigmas as many as carpels or locules, often peltate or radiating; fruit often capsular, sometimes a berry or drupaceous; seed with large embryo and no endosperm, often arillate.

As here interpreted (excluding the Hypericaceae) the family is almost exclusively tropical in distribution, and composed of about 35 genera and 400 species (over 200 spp. in *Clusia*). No species is indigenous to the United States, but species of *Clusia, Calophyllum, Mammea,* and *Rheedia* extend from South and Central America into Mexico and the West Indies.

The Guttiferae and Hypericaceae are closely related to one another (the former the more primitive) and many phylogenists (including Engler and Wettstein) have united them, for their allegedly characteristic anatomical differences lose force when comparison is made with the subfamilies Calophylloideae and Clusioideae of

---

[145] The name Guttiferae has been conserved for this family, with the exception that the alternative name Clusiaceae may be used by those preferring the ending -aceae. Engler and Diels (1936) treated this family as composed of 5 subfamilies, including the Hypericoideae, but an increasing number of botanists accept the view followed here that the Hypericaceae deserve recognition as a separate family.

the Guttiferae. The family seems to have been derived from the Theaceae or from stocks ancestral to them, a view accepted by Vestal, Hutchinson, and others.

Economically the family is of little domestic importance. Except for *Rheedia,* the genera mentioned above and also *Garcinia* are cultivated in the warmer parts of the country for ornament or for their edible fruit. The mangosteen (*Garcinia Mangostana*) is one of the most highly prized edible fruits of the tropics, but requires warmer temperatures for fruit production than are available in this country; a similar situation exists for the mammee apple (*Mammea americana*).

*LITERATURE:*

ENGLER, A. Guttiferae. In Engler and Prantl, Die natürlichen Pflanzenfamilien, ed. 2, Bd. 21: 154–237, 1925.
JULIANO, VESQUE. Guttiferae. In de Candolle, Monogr. Phaner. 8: 1–669, 1893.
LAUTERBACH, C. Guttiferae. Engl. Bot. Jahrb. 58: 1–49, 1922–23.
LOTT, H. J. Nomenclatural notes on *Hypericum*. Journ. Arnold Arb. 19: 149–152, 1938; 19: 279–290, 1938.
PLANCHON, J. F. and TRIANA, J. Mémoire sur la famille des Guttifères. Ann. sci. nat. ser. 4, 13: 306–376, 1860; *op. cit.* 14: 227–367, 1860.
VESTAL, P. A. The significance of comparative anatomy in establishing the relationship of the Hypericaceae to the Guttiferae and their allies. Philippine Journ. Sci. 64: 199–256, 1937.

## HYPERICACEAE. HYPERICUM FAMILY

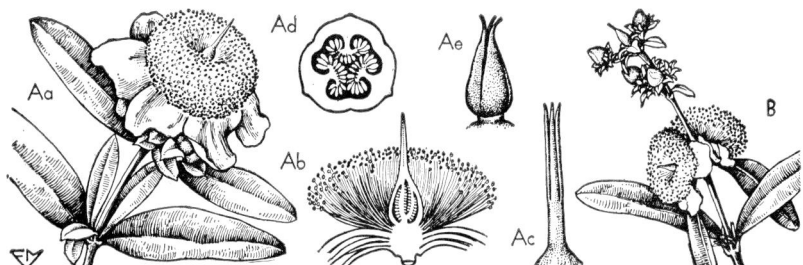

Fig. 213. HYPERICACEAE. A, *Hypericum frondosum*: Aa, flowering branch, × ½; Ab, flower, vertical section, × 1; Ac, style and stigma, × 2; Ad, ovary, cross-section, × 3; Ae, capsule, × 1. B, *H. prolificum*: flowering branch, × ½. (From L. H. Bailey, *Manual of cultivated plants*, The Macmillan Company, 1949. Copyright 1924 and 1949 by Liberty H. Bailey.)

Trees, shrubs, or herbs (rarely woody lianes), sap resinous; leaves opposite or whorled, simple, often pellucid- or black-punctate, estipulate; flowers bisexual, actinomorphic, cymose; perianth biseriate, the sepals 4–5, imbricate, often basally connate, outer ones usually smaller (rarely 4 with the 2 outer much larger), the petals as many as sepals, sessile or clawed (claw, when present, often with nectar pit or groove), distinct, imbricated or contorted; stamens numerous, hypogynous, usually fascicled in 3–5 (rarely 6–8) bundles the members of which are often more or less connate by the filaments, rarely the stamens monadelphous, the anthers 2-celled, dehiscing longitudinally; pistil 1, the ovary superior, 3–5-loculed and -carpelled (rarely unilocular with 3–5 carpels), the placentation usually axile (rarely parietal), the ovules numerous, anatropous, the styles 3–5, distinct or basally connate (sometimes coherent until after anthesis), linear; fruit a capsule or berry; seed without endosperm, the embryo straight or curved.

A family of about 8 genera and 350 species with *Hypericum* (300 spp.) the largest genus (generic segregates accepted by Small, Rydberg, *et al.* include

*Crookea, Sarothra, Sanidophyllum, Triadenum*). *Hypericum* is represented by about 40 species indigenous to the United States. The closely related *Ascyrum* (5–7 spp.) has about 4 species indigenous to the country. *Vismia* (32 spp.) of tropical America extends northward into Mexico.

The Hypericaceae are best distinguished from related families by the combination of pellucid or black dotted opposite leaves, the fascicled stamens, the 3–5-loculed ovary and essentially distinct styles. Studies by Vestal (1937) have shown that they are advanced over the Guttiferae, and that within the family the tropical woody genera are the more primitive. He found *Hypericum* to be, anatomically, a homogeneous taxon lacking bases for generic segregations.

Economically the plants are of little importance. Some of the shrubby and a few of herbaceous St.-John's-wort (*Hypericum* spp.) are cultivated for ornament and other species are noxious weeds or poisonous when eaten by livestock. St.-Andrew's-cross (*Ascyrum hypericoides*) is an evergreen shrub in warm regions and used there as a landscape subject.

*LITERATURE:*

FERNALD, M. L. The varieties of *Hypericum* sect. *Elodea* (*H. virginicum*). Rhodora, 38: 433–436, 1936.
KELLER, R. and ENGLER, A. Guttiferae subfam. Hypericoideae. In Engler and Prantl, Die natürlichen Pflanzenfamilien, ed. 2, 21: 174–188, 1925.
SAMPSON, A. W. and PARKER, K. W. St. Johnswort on range lands of California. California Agr. Exp. Sta. Bull. 503: 1–48, 1930.
SVENSON, H. K. Plants of the Southern United States. III. Woody species of *Hypericum*. Rhodora, 42: 8–19, 1940.
VESTAL, P. A. The significance of comparative anatomy in establishing the relationship of the Hypericaceae to the Guttiferae and their allies. Philippine Journ. Sci. 64: 199–256, 1937.

## ELATINACEAE.[146] WATERWORT FAMILY

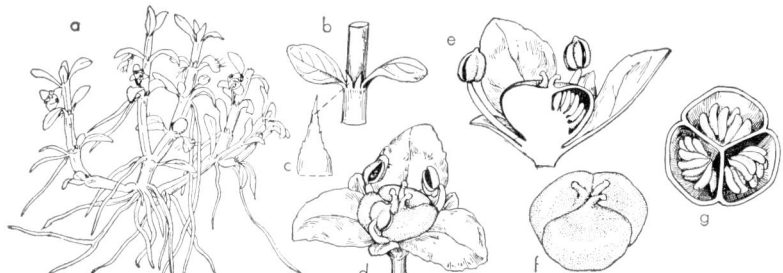

Fig. 214. ELATINACEAE. *Elatine americana:* a, flowering plant, $\times$ 1½; b, node with leaves, $\times$ 3; stipule, $\times$ 15; d, flower, habit, $\times$ 12; e, same, vertical section, $\times$ 12; f, pistil, $\times$ 15; g, ovary, transverse view, looking into basal half, $\times$ 15. (Drawn from material, courtesy S. J. Smith.)

Annual or perennial aquatic herbs or suffrutescent perennials of similar habitats; leaves opposite or whorled, simple, stipulate; flowers minute, bisexual, actinomorphic or zygomorphic, solitary or in small axillary dichasia, the perianth biseriate, persistent, the sepals 3–5, distinct or basally connate, the petals same number as sepals, distinct, hypogynous; stamens in 1–2 whorls of 3–5 stamens each, distinct, hypogynous, the anthers 2-celled, dehiscing longitudinally; pistil 1, the ovary superior, 3–5-loculed and -carpelled, the placentation axile, the ovules numerous, in 2

---

[146] The name Elatinaceae has been conserved for this family.

or more rows on each placenta, anatropous, the styles 3–5, distinct and short; fruit a septicidal capsule; seeds with straight or curved embryo and lacking endosperm.

A family of 2 genera (*Elatine* 10 spp., *Bergia* 20 spp.) of cosmopolitan distribution. *Elatine* is widespread in fresh-water habitats of temperate and warm regions with about 4 spp. indigenous to the United States, and *Bergia* is almost exclusively tropical and subtropical in distribution, with 1 species (*B. texana*) occurring from Texas to Missouri and west to California.

The Elatinaceae are readily distinguished among aquatics by their opposite or whorled leaves and paired stipules, by the axillary often cymose flowers that are 3–5-merous, and the numerous axillary ovules.

The family is of no known economic importance.

*LITERATURE:*

FASSETT, N. C. Notes from the herbarium of the University of Wisconsin — XVII. *Elatine* and other aquatics. Rhodora, 41: 367–377, 1939.

FERNALD, M. L. The genus *Elatine* in eastern North America. Rhodora, 19: 10–15, 1917.
———. *Elatine americana* and *E. triandra*. Rhodora, 43: 208–211, 1941.

GAUTHIER, R. and RAYMOND, M. Le genre *Elatine* dans le Québec. Contr. Inst. Bot. Univ. Montréal, 64: 29–35, 1949.

NIEDENZU, F. Elatinaceae. In Engler and Prantl, Die natürlichen Pflanzenfamilien, ed. 2, 21: 270–276, 1925.

## FRANKENIACEAE.[147] FRANKENIA FAMILY

**Fig. 215.** FRANKENIACEAE. *Frankenia grandifolia:* a, flowering branch, × ⅓; b, inflorescence, × 1; c, flower, habit, × 3; d, petal, ventral side, × 4; e, flower, less perianth, × 4; f, ovary cross-section, × 15; g, ovary, vertical section, × 10. (g, after Torrey.)

Perennial herbs or small shrubs; leaves opposite, decussate, simple, entire, the pairs joined at base by a ciliated line, estipulate; flowers bisexual, actinomorphic, bracteate, in terminal or axillary cymes or solitary, the perianth biseriate, the sepals 4–7, forming a tubular short-lobed calyx, the petals of same number, distinct, hypogynous, imbricated, long-clawed, each with a ligular scale at base of a spreading limb and decurrent down the claw; stamens usually 6 (4–7) in 2 whorls, hypogynous, distinct or shortly connate at base, often didynamous, the anthers 2-celled, extrorse, versatile, dehiscing longitudinally; pistil 1, the ovary superior, unilocular, the carpels 2–4, the placentae parietal, usually fertile basally, the ovules 2–3 to many, anatropous, ascending or the long funiculus recurved, the style 1 and slender, with 3 included branches or stigmas; fruit a loculicidal capsule enclosed within the persistent calyx; seeds with straight embryo and a mealy endosperm.

[147] The name Frankeniaceae has been conserved for this family.

A family of 4 genera and 34 species, occurring on all continents but not widely distributed except in the Mediterranean region, represented in the United States by 2 indigenous species of *Frankenia* in saline habitats of southern California and a third extending from Texas to Colorado.

The family is distinguished by its decussate heathlike foliage, the appendaged petals, and the long-funiculate ovules on parietal placentae. The affinity of the plants for saline or arid habitats also is characteristic.

The Frankeniaceae were placed in his Guttales by Hallier, who interpreted them to have been derived from the Linaceae via the Tamaricaceae and to be the most advanced family of the order; Rendle, following Wettstein, included them with the Parietales (an order differing from the Parietales of Engler by the removal of the Theineae as the Guttiferales); Bessey included all 3 families of the Tamaricineae in his Caryophyllales, and Hutchinson placed the Elatinaceae with the Caryophyllales but segregated the Tamaricaceae and Frankeniaceae as his Tamaricales.

The family is of little economic importance. Plants of the alkali heath (*Frankenia grandifolia*) and of a few Old World species of *Frankenia* are cultivated to a limited extent as novelties.

*LITERATURE:*

GUNDERSEN, A. The Frankeniaceae as a link in the classification of dicotyledons. Torreya, 27: 65–71, 1927.

NIEDENZU, F. Frankeniaceae. In Engler and Prantl, Die natürlichen Pflanzenfamilien, ed. 2. 21: 276–281, 1925.

## TAMARICACEAE.[148] TAMARISK FAMILY

**Fig. 216.** TAMARICACEAE. *Tamarix parviflora*: a, flowering branch, × ½; b, sterile twig with leaves, × 2; c, flower habit, × 4; d, flower, less perianth, × 10; e, ovary, vertical section, × 10. (From L. H. Bailey, *Manual of cultivated plants,* The Macmillan Company, 1949. Copyright 1924 and 1949 by Liberty H. Bailey.)

Small, heathlike trees or shrubs with slender flexuous branches, halophytic or xerophytic; leaves subulate to scalelike, alternate, appressed, estipulate; flowers minute, bisexual, actinomorphic, ebracteate, solitary (Reaumurieae) or in dense spikelike racemes (Tamariceae), hypogynous, 4–5-merous, the perianth biseriate, imbricated, the sepals and petals distinct, the latter persisting in fruit, the corolla and androecium arising from a fleshy nectiferous disc; stamens as many or twice as many as petals, distinct or basally connate, the anthers 2-celled, dehiscing longitudinally; pistil 1, the ovary superior; unilocular, usually 3–4-carpelled, the placentation parietal or (in *Tamarix*) reduced to a basal-parietal position, the ovules 2–many on each placenta, ascending, anatropous, on short funiculi, the styles as many as carpels, distinct or basally connate or sometimes absent and the 3–4 stigmas sessile; fruit a capsule, sometimes becoming falsely and incompletely

[148] The name Tamaricaceae has been conserved for this family

multiloculate; seeds densely bearded all over or at distal end, rarely winged, the embryo straight, the endosperm absent (Tamariceae) or present (Reaumurieae).

A family of 4 genera and about 100 species, mostly of the Mediterranean region and in central Asia. None is indigenous to the United States, but 1 species of Tamarix (*T. gallica*) is naturalized extensively in dry or saline habitats from South Carolina across the south to California. See under Frankeniaceae for notes on presumed phyletic relationships.

The plants of this family are readily distinguished by the sinuous branchlets, the appressed minute subulate leaves, and the hairy seeds.

Several species of tamarisk (*Tamarix*) are cultivated domestically as ornamental shrubs prized for their feathery verdure and pink bloom. One species of false tamarisk (*Myricaria germanica*) is grown to a limited extent as an ornamental.

*LITERATURE:*

NIEDENZU, F. Tamaricaceae. In Engler and Prantl, Die natürlichen Pflanzenfamilien, ed. 2, 21: 282–289, 1925.

## CISTACEAE.[149] ROCK-ROSE FAMILY

**Fig. 217.** CISTACEAE. *Helianthemum nummularium:* a, flowers, × ½; b, stem node, × 1; c, pistil, ovary in vertical section, × 4; d, ovary, cross-section; e, capsules, × 1; f, flower, vertical section, × 1. (From L. H. Bailey, *Manual of cultivated plants,* The Macmillan Company, 1949. Copyright 1924 and 1949 by Liberty H. Bailey.)

Herbs or shrubs, the hairs often stellate; leaves mostly opposite, simple, stipulate or not so; flowers bisexual, actinomorphic, solitary to cymose or in cymose racemes, the perianth biseriate (cleistogamous flowers occur in some genera), the sepals 3–5, convolute, distinct, sometimes unequal (the 2 outer generally smaller), the petals 5, rarely 3 or none, distinct, convolute (the convolutions of calyx and corolla in opposite directions) caducous or ephemeral; stamens numerous, hypogynous, borne on an elongated and disclike projection of the receptacle, centrifugal (basipetal), distinct, the anthers introrse, 2-celled, dehiscing longitudinally; pistil 1, the ovary superior, unilocular (sometimes falsely 5–10-loculed by intrusion of placentae), the carpels 3 or 5–10, the placentation parietal often on intruded placentae, the ovules 2–many on each placenta, orthotropous (rarely anatropous), funiculi well developed, the style 1 and stigmas 3–5 or connate and 1; fruit a leathery or woody capsule, loculicidally dehiscent; seeds small, angular, the embryo not straight, the endosperm present.

A family of 8 genera (some botanists accept only 3–4) and about 175 species, widely distributed in the warmer parts of the northern hemisphere but especially abundant in the Mediterranean region. In the United States it is represented by *Helianthemum* (2 of its 80 spp. in southern California), and in the eastern states

[149] The name Cistaceae has been conserved for this family.

by *Crocanthemum* [150] (by perhaps 9 of its 25 spp.), *Lechea* (by perhaps 12 of its 15 spp.), and *Hudsonia* by its 3 spp.

The Cistaceae are distinguished from related families (as the Violaceae, Bixaceae, and Hypericaceae) by the usually opposite leaves, the perianth series reversely convolute, the caducous or ephemeral petals, the numerous stamens, and the seed with copious endosperm.

The Cistineae (treated by Engler to contain the Cistaceae and Bixaceae) were included within the Guttiferales by Bessey, and retained in the Parietales by Wettstein and by Rendle. Hutchinson segregated these 2 families (plus the Cochlospermaceae, Flacourtiaceae, and Canellaceae) into his order Bixales.

Economically the family is of slight domestic importance. A few species and hybrids of *Cistus* are cultivated for ornament in the warmer parts of the country, and a few of *Helianthemum* throughout the country. *Hudsonia* is infrequently introduced into the rockery.

*LITERATURE:*

DANSEREAU, P. M. Monographie du genre *Cistus*. Boissiera, Fasc. 4, 1939.
FERNALD, M. L. *Crocanthemum*: has it really stable generic characters? Rhodora, 609–616, 1941.
GROSSER, W. Cistaceae. In Engler, Das Pflanzenreich, 14 (IV. 193): 1–161, 1903.
HODGDON, A. R. A taxonomic study of *Lechea*. Rhodora, 40: 29–69, 87–131, 1938.
JANCHEN, E. Bemerkungen zu der Cistaceen-Gattung *Crocanthemum*. Österr. bot. Zeitschr. 71: 266–270, 1922.
———. Cistaceae. In Engler and Prantl, Die natürlichen Pflanzenfamilien, ed. 2, 21: 289–313, 1925.
SCHREIBER, B. O. The genus *Helianthemum* in California. Madroño, 5: 81–85, 1939.

## BIXACEAE.[151] BIXA FAMILY

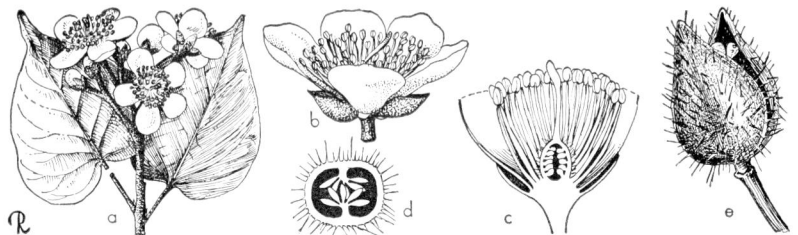

**Fig. 218.** BIXACEAE. *Bixa Orellana:* a, flowering branch, × ⅓; b, flower habit, × ½; c, same, vertical section, × ½; d, ovary, cross-section, × 2; e, fruit, × ½. (From L. H. Bailey, *Manual of cultivated plants,* The Macmillan Company, 1949. Copyright 1924 and 1949 by Liberty H. Bailey.)

Shrubs or small trees with reddish sap; leaves alternate, simple, palmately veined, stipulate; flowers bisexual, actinomorphic, paniculate, the sepals 5 (rarely 4), imbricated, distinct, deciduous, the petals 5, large and colored, imbricated and twisted in bud, distinct; stamens numerous, hypogynous, distinct, the anthers 2-celled, dehiscing longitudinally or rarely (as in *Bixa*) by 2 apical pores; pistil 1, the ovary superior, unilocular or falsely and incompletely bilocular by placental intrusion, the carpels 2, the placentation parietal, the ovules numerous, anatropous, funiculi well developed, the style slender with 2 stigmas; fruit a loculicidal capsule, often echinate without, 2-valved; seeds with brilliant red fleshy testa, the endosperm present.

[150] For views on the presence and absence of taxonomic validity of these generic segregates from *Helianthemum,* cf. Janchen (1925) and Fernald (1941).
[151] The name Bixaceae has been conserved over other names for this family.

A unigeneric family containing the monotypic *Bixa*, of mostly tropical America, widely naturalized pantropically but not indigenous. For many years the family was interpreted to contain 4 genera (see Warburg, 1895), but largely on the evidence submitted by van Tieghem all but *Bixa* were segregated to comprise the Cochlospermaceae. This restricted concept of Bixaceae has been accepted by almost all phylogenists (Hallier merged Bixaceae *sensu latiore* with the Tiliaceae), and the Cochlospermaceae are accepted as taxonomically valid.

The Bixaceae are readily distinguished by the colored sap and by the red fleshy testa of the seeds.

Economically the family is important for its 1 species, Arnotto (*Bixa Orellana*), grown in warm regions for its showy fruit and for dyes made from the fleshy testa enveloping the seeds.

*LITERATURE:*

PILGER, R. Bixaceae. In Engler and Prantl, Die natürlichen Pflanzenfamilien, ed. 2, 21: 313–315, 1925.
TIEGHEM, P. VAN. Sur les Bixacées, les Cochlospermacées et les Sphaerosepalacées. Journ. Bot. (Paris), 14: 33–42, 1900.
WARBURG, O. Bixaceae. In Engler and Prantl, Die natürlichen Pflanzenfamilien, III (6) 307–314, 1895.

## COCHLOSPERMACEAE. COCHLOSPERMUM FAMILY

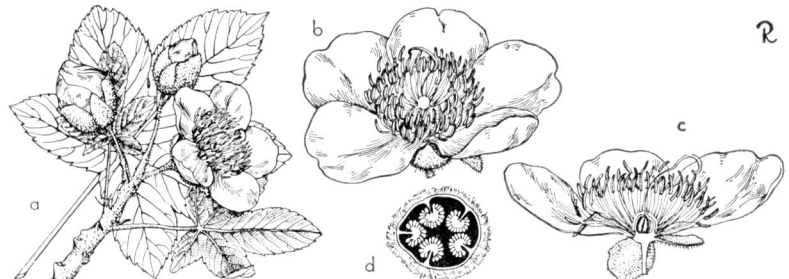

Fig. 219. COCHLOSPERMACEAE. *Cochlospermum vitifolium*: a, flowering branch superposed on leaf, both, × ¼; b, flower habit, × ½; c, same, vertical section, × ½; d, ovary, cross-section, × 3.

Trees, shrubs, or rhizomatous herbs, sap orange or reddish; leaves alternate, simple with palmate lobes and veins, stipulate; flowers showy, usually appearing before the leaves, bisexual, actinomorphic or slightly zygomorphic, the sepals and petals 4–5, deciduous, and imbricated, both distinct; stamens numerous, distinct, equal or unequal (those of one side with longer declinate filaments), the anthers 2-celled, linear, dehiscing by terminal porelike slits; pistil 1, the ovary superior, unilocular, the carpels 3–5, the placentae parietal and often intruding, and the ovary then falsely or basally 3–5-loculed, the ovules numerous, the style 1, the stigmas minute and dentate, as many as carpels; fruit a large capsule with thick outer and thin inner wall, valves 3–5; seed with curved embryo and an oily endosperm.

A family of 3 genera and 25 species, of tropical distribution. *Amoreuxia* (4 spp.) is represented by 2 species indigenous to southern Arizona, and one species of *Cochlospermum* (*Maximilianea*) [152] (15 spp.) extends northward into Mexico.

[152] The name *Cochlospermum* Kunth (1822) is conserved over *Maximilianea* Mart. (1819) not *Maximiliana* Mart. 1824? a genus of Palmae.

# DIVISION IV. EMBRYOPHYTA SIPHONOGAMA

The third genus, *Sphaerosepalum* of Madagascar, is sometimes treated as a unigeneric family.

The family differs from the Bixaceae primarily in the palmately lobed or divided leaves and the endosperm oily instead of granular or bony. The large yellow flowers are showy, and *Cochlospermum vitifolium* is cultivated as an ornamental tree in southern California.

*LITERATURE:*

BLAKE, S. F. The American species of *Maximilianea* (*Cochlospermum*). Journ. Wash. Acad. Sci. 11: 125–132, 1921.

PILGER, R. Cochlospermaceae. In Engler and Prantl, Die natürlichen Pflanzenfamilien, ed. 2, 21: 316–320, 1925.

## CANELLACEAE.[153] WILD CINNAMON FAMILY

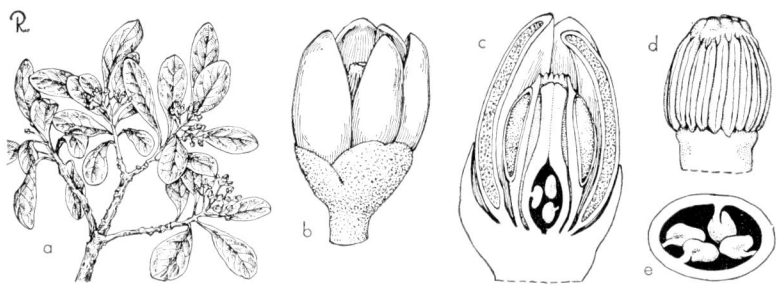

**Fig. 220.** CANELLACEAE. *Canella Winterana*: a, flowering branch, × ¼; b, flower, side view, × 4; c, same, vertical section, × 7; d, androecium, habit, × 7; e, ovary, cross-section, × 10.

Trees, possessing aromatic oils; leaves alternate, simple, entire, coriaceous, glandular-punctate or -dotted, estipulate; flowers bisexual, actinomorphic, solitary, cymose, or racemose, the perianth biseriate, often subtended by 3 persistent imbricated bracts, the sepals 4–5, distinct, imbricated, thick, the petals 4–5 or none, distinct or basally connate, imbricated; stamens 20 or less, monadelphous, the anthers extrorse, 2-celled, dehiscent by longitudinal valves; pistil 1, the ovary superior, unilocular, carpels 2–5, the placentae parietal and not intruding, the ovules 2–many on each placenta, semianatropous, the style solitary and thick, stigmas or their lobes as many as carpels; fruit a berry; seeds with oily endosperm and a straight or slightly curved embryo.

A tropical family of 5 genera and 11 species of disjunctive distribution; *Canella* (2 spp.) occurring from Venezuela to southern Florida (*C. Winterana*), *Cinnamodendron* (3 spp.) in tropical America, *Plenodendron*, a monotypic genus of Puerto Rico, *Cinnaosma* (2 spp.) in Madagascar, and *Warburgia* (3 spp.) in tropical Africa.

The family is distinguished by the stamens connate by their filaments into a tube that almost or completely envelops the pistil, by the parietal placentation, and the fruit a berry.

Economically the family is of little importance. The leaves and bark of the wild cinnamon (*Canella Winterana*) are used in medicine and as a condiment.

---

[153] This family, known in much of the literature as the Winteranaceae, had its name taken from the type genus *Winterana* L. (1759). However, the name *Canella* P. Br. ex Swartz (1791) has been conserved, and *Winterana* is a *nomen rejiciendum*, and the valid name for the family is Canellaceae.

The tree is cultivated to a limited extent in southern Florida as an ornamental prized for its purple flowers and black berries.

LITERATURE:

GILG, E. Canellaceae. In Engler and Prantl, Die natürlichen Pflanzenfamilien, ed. 2, 21: 323–329, 1925.

## VIOLACEAE.[154] VIOLET FAMILY

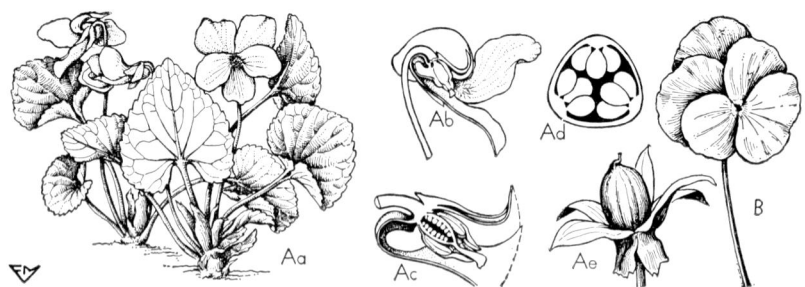

**Fig. 221.** VIOLACEAE. A, *Viola papilionacea*: Aa, flowering plant; Ab, flower, perianth partially removed, × 1; Ac, flower, vertical section, × 2; Ad, ovary, cross-section, × 4; Ae, capsule, × 1. (From L. H. Bailey, *Manual of cultivated plants*, The Macmillan Company, 1949. Copyright 1924 and 1949 by Liberty H. Bailey.)

Shrubs or herbs (usually perennial), rarely climbing (as in *Anchieta*); leaves alternate (opposite in *Hybanthus*), simple (sometimes lobed or divided), the stipules minute or leafy; flowers bisexual, zygomorphic or actinomorphic (sometimes cleistogamous), solitary or the inflorescence various, the sepals 5, distinct or nearly so, usually persistent, imbricated, the petals 5, the lowermost often spurred and larger than others, imbricated or contorted; stamens 5, hypogynous or slightly perigynous, closely connivent around the pistil, the anthers 2-celled, introrse, one of them often spurred, dehiscence by longitudinal slits; pistil 1, the ovary superior, unilocular, carpels 3–5, the placentae parietal, the ovules 1–2 or numerous on each placenta, anatropous, the style 1 with stigma of varying shapes; fruit a loculicidal capsule (sometimes dehiscing explosively) or a berry; seeds winged in some woody lianes, the embryo straight and a copious fleshy endosperm present.

A family of about 16 genera and 850 species, of wide distribution and occurring on all continents. The Violaceae are represented in the United States by indigenous species of 2 genera, by perhaps 60 of the 400 species of *Viola*, these occurring in all but the most arid regions of the country, and by 2 of the 80 species of *Hybanthus* (including *Cubelium*) that are indigenous from Arizona north to Kansas and Colorado, and a third species (*H. concolor*) eastward to Georgia and north to Ontario and New York. The tropical woody *Rinorea* (260 spp.) (not indigenous to North America) has actinomorphic flowers and all stamens alike and ecalcarate.

The Violaceae are distinguished by the 5-merous flowers (but the gynoecium often tricarpellate), the stamens basally coherent and introrse, and (in the Violeae) by the zygomorphic corolla. In *Viola* and others, the stamens possess a fingerlike curved nectar-secreting horn that projects backward from the connective of each of the two lower (abaxial) anthers into the spur of the lower petal. Also in *Viola*, and in *Hybanthus*, apetalous cleistogamous flowers are produced in midsummer on short pedicels near the ground, after the normal flowering season is past.

Hallier included the family as a primitive member of his Polygalines, and Bessey

[154] The name Violaceae has been conserved for this family.

retained it in the Guttales. Rendle, following Wettstein, included it in the Parietales. Hutchinson included it in his Violales (together with Resedaceae), interpreting this order (and his Polygalales) to be advanced and terminal taxa derived from ranalian stocks via the Rhoeadales.

Economically the family is important domestically for the English violet (*V. odorata*) of the florist's trade. About 120 species of *Viola* are offered in the domestic trade for cultivation as ornamentals.

*LITERATURE:*

ARNAL, C. Recherches morphologiques et physiologiques sur la fleur des Violacées. 262 pp. Dijon, 1945.

BAIRD, V. B. Wild violets of North America. Berkeley, Cal., 1942.

BAKER, M. S. Studies in western violets. I–III, Madroño, 3: 51–56, 232–238, 1935–36; *op. cit.* 218–231, 1940; IV–V, Leafl. West Bot. 5: 141–147, 173–177, 1949; VI, Madroño, 10: 110–128, 1949.

BAMFORD, R. and GERSHOY, A. Studies in North American violets. II. The cytology of some sterile $F_1$ violet hybrids. Vermont Agr. Exp. Sta. Bull. 325: 1–53, 1930. [1931.]

BECKER, W. Violae Mexicanae et Centrali-Americanae. I. Fedde, Repert. Spec. Nov. 19: 392–400, 1924. II, *op. cit.* 20: 1–12, 1924.

BOLD, H. C. and GERSHOY, A. Studies in North American violets. IV. Chromosome relations and fertility in diploid and tetraploid species hybrids. Vermont Agr. Exp. Sta. Bull. 378: 1–35, 1934.

BRAINERD, E. *Viola palmata* and its allies. Bull. Torrey Bot. Club, 37: 581–590, 1911.

———. The caulescent violets of the southeastern United States. Bull. Torrey Bot. Club, 38: 191–198, 1911.

———. Violets of North America. Bull. Vermont Agr. Exp. Sta. 224: 1–172, 1921.

———. Some natural violet hybrids of North America. Bull. Vermont Agr. Exp. Sta. 239: 1–205, 1924.

CLAUSEN, J. Studies on the collective species *Viola tricolor* L. I–II. Bot. Tidsskrift, 37: 205–221, 1921; *op. cit.* 37: 363–416, 1922.

———. Chromosome number and relationship of some North American species of *Viola*. Ann. Bot. 43: 741–764, 1929.

DAVIS, H. A. and DAVIS, T. The violets of West Virginia. Castanea, 14: 53–86, 1949.

EXELL, A. W. The phylogeny of Violaceae. Journ. Bot. (Lond.) 63: 330–333, 1925.

GERSHOY, A. Studies in North American Violets—I. General considerations. Bull. Vermont Agr. Exp. Sta. 279: 1–18, 1928.

———. Studies in North American violets. III. Chromosome numbers and species characters. Bull. Vermont Agr. Exp. Sta. 367: 1–91, 1934.

HOLM, T. Comparative studies of North American violets. Beih. Bot. Centralb. 50: 135–182, 1932.

MCCULLOUGH, H. M. Studies in soil relations of species of violets. Amer. Journ. Bot. 28: 934–941, 1941.

MELCHOIR, H. and BECKER, W. Violaceae. In Engler and Prantl, Die natürlichen Pflanzenfamilien, ed. 2, 21: 329–377, 1925.

MILDBRAED, J. Violaceae. Engl. bot. Jahrb. 62: 368–375, 1928–1929.

SPOTTS, A. M. The violets of Colorado. Madroño, 5: 16–27, 1939.

TODD, E. E. A short survey of the genus *Viola*. Journ. Roy. Hort. Soc. 55: 223–243 (1930); 57: 212–229 (1932).

## FLACOURTIACEAE. FLACOURTIA FAMILY

Trees or shrubs, rarely climbing; leaves alternate (rarely opposite), 2-ranked, simple, coriaceous, persistent, stipules caducous; flowers generally bisexual (sometimes unisexual, the plants monoecious or dioecious, as in *Pangium*), actinomorphic mostly in lateral or terminal cymose inflorescences, the sepals 2–15, equal, usually distinct, imbricated, sometimes the calyx undifferentiated from the corolla, the petals when present usually equal in number to sepals, sometimes more numerous, with or without an opposite basal scale, imbricated; stamens generally numerous, hypogynous, distinct or sometimes in bundles alternating with sepals (opposite petals), the anthers 2-celled, dehiscing longitudinally, often attenuate or appendaged; a disc often present between stamens and pistil; pistil 1, the ovary superior (rarely half inferior to inferior), unilocular, the carpels 2–10, the placentae parietal and sometimes much intruded, the ovules numerous on each placenta, of various types, the style 1 or as many as carpels and distinct; fruit a loculicidal (rarely indehiscent) capsule or berry; seeds often conspicuously arillate, with a straight embryo and the endosperm mostly abundant.

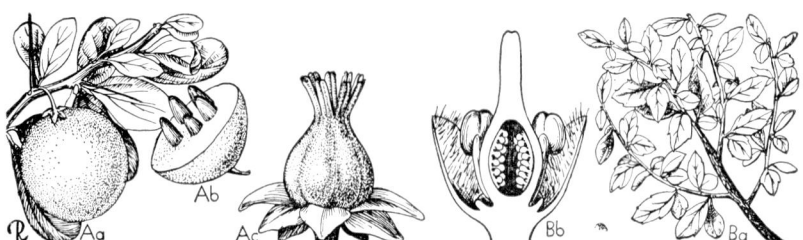

Fig. 222. FLACOURTIACEAE. A, *Dovyalis caffra*: Aa, fruiting branch, × ¼; Ab, fruit, terminal half removed exposing seeds, × ¼; Ac, pistillate flower, × 3. B, *Azara microphylla*: Ba, vegetative branch, × ⅓; Bb, flower, vertical section, × 6. (From L. H. Bailey, *Manual of cultivated plants*, The Macmillan Company, 1949. Copyright 1924 and 1949 by Liberty H. Bailey.)

A family of 84 genera and about 850 species, of pantropical and subtropical distribution. No species is indigenous to this country, but 10 genera and about 25 species extend northward into Mexico.

The family is distinguished from related taxa by the numerous stamens, the variously modified disc and often enlarged receptacle, and the often undifferentiated perianth. Hutchinson included it in his Bixales, Hallier placed it in his Passionales, and others retained it in either the Guttiferales or Parietales.

Economically the family is of little domestic importance. Chaulmugra oil (from *Gymnocardia odorata*) of medicinal value is imported from India. Domestically, species of 7 genera are cultivated for ornament (including *Azara, Berberidopsis, Carrierea, Idesia,* and *Xylosma*), and species cultivated for their edible fruits include the Kei apple and Ceylon gooseberry (*Dovyalis* spp.) and the ramontchi (*Flacourtia indica*). *Taraktogenos* is grown here in warm parts for the production of a medicinal oil.

*LITERATURE:*

GILG, E. Flacourtiaceae. In Engler and Prantl, Die natürlichen Pflanzenfamilien, ed. 2, 21: 377–456, 1925.

RECORD, S. J. American woods of the family Flacourtiaceae. Trop. Woods, 68: 40–57, 1941.

## STACHYURACEAE. STACHYURUS FAMILY

Shrubs or small trees; leaves alternate, simple, deciduous, minutely stipulate; flowers bisexual or sometimes unisexual, actinomorphic, subtended by a pair of basally connate bracteoles, in axillary racemes appearing before the leaves, the sepals and petals 4, distinct, imbricated; stamens 8, distinct, the anthers 2-celled, dehiscing longitudinally; pistil 1, the ovary 4-locular,[155] the placentation appearing

---

[155] Morphologically the ovary in this family is interpreted to be unilocular with intruding parietal placentae that meet in the center. For purposes of identification it is treated as 4-locular.

axile, the ovules numerous, the style 1 with a usually 4-lobed capitate-peltate stigma; fruit a berry, often leathery; seeds with straight embryo and an abundant endosperm.

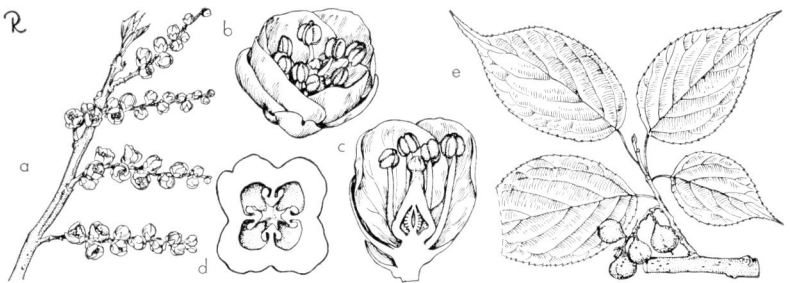

**Fig. 223.** STACHYURACEAE. *Stachyurus praecox*: a, flowering branch, × ½; b, flower, × 3; c, same less perianth, × 6; d, ovary, crosss-ection, × 8; e, fruit, × 1.

A unigeneric family (*Stachyurus*) of 5-6 species indigenous to eastern and central Asia. *S. praecox*, of Japan, occasionally cultivated in this country as an ornamental shrub.

The family was once united with the Theaceae, from which it differs in the fewer stamens, the unbranched style, the morphologically parietal placentation.

*LITERATURE:*
GILG, E. Stachyuraceae. In Engler and Prantl, Die natürlichen Pflanzenfamilien, ed. 2, 21: 457–459, 1935.

## TURNERACEAE.[156] TURNERA FAMILY

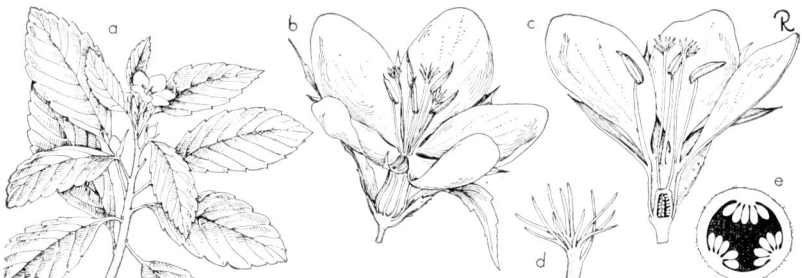

**Fig. 224.** TURNERACEAE. *Turnera ulmifolia*: a, flowering branch, × ¼; b, flower, habit, × 1; c, same, vertical section, × 1; d, stigma, × 5; e, ovary, cross-section, × 10.

Herbs, shrubs, or trees; leaves alternate, simple, entire or lobed, often basally biglandular, stipules, small or none; flowers bisexual, actinomorphic, often bibracteolate, axillary, solitary or fasciculate, hypanthoid, the sepals 5, borne on an hypanthium, imbricated, deciduous, the petals 5, on hypanthium, distinct, clawed, contorted; the stamens 5, opposite the sepals, distinct, borne on the hypanthium, the anthers 2-celled, dehiscing longitudinally; pistil 1, the ovary superior, unilocular, carpels 3, the placentae parietal, somewhat intruded, each with numerous ovules, the styles 3, linear or flattened, apically fringed; fruit a loculicidal capsule; seeds arillate, the embryo straight and endosperm fleshy or bony.

[156] The name Turneraceae has been conserved for this family.

A tropical American family of 6 genera and about 110 species. *Turnera*, the largest genus with 60 species, has 1 species indigenous in Texas. *T. ulmifolia* of the West Indies and southward is naturalized in Florida. Three genera and 4 species are indigenous to Mexico.

The family is distinguished by the presence of the hypanthium, the tricarpellate ovary, the fimbriate-tipped stigmas, and arillate seeds.

The species are of no known domestic economic importance.

*LITERATURE:*

GILG, E. Turneraceae. In Engler and Prantl, Die natürlichen Pflanzenfamilien, ed. 2, 21: 459–466, 1925.

## PASSIFLORACEAE.[157] PASSION-FLOWER FAMILY

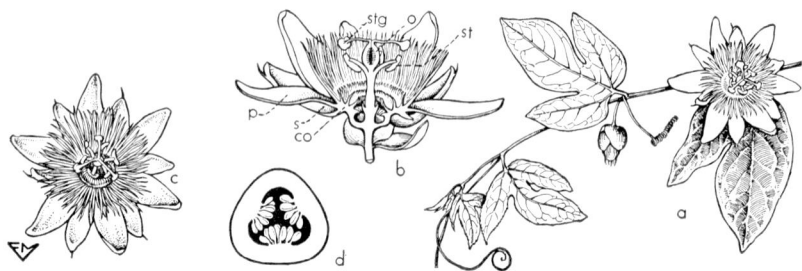

**Fig. 225.** PASSIFLORACEAE. *Passiflora caerulea*: a, flowering branch, × ⅙; b, flower, vertical section, × ⅓ (co corona, o ovary, p petal, s sepal, st stamen, stg stigma); c flower, face view, × ¼; d, ovary, cross-section, × 2. (From L. H. Bailey, *Manual of cultivated plants,* The Macmillan Company, 1949. Copyright 1924 and 1949 by Liberty H. Bailey.)

Shrubs, or herbs, often lianous with axillary tendrils; leaves alternate, simple or compound, stipulate; tendrils opposite the leaves, often corresponding to the terminal flower of a dichasium or first flower of a monochasium; flowers bisexual (rarely unisexual and the plants then monoecious or, in *Adenia,* dioecious), actinomorphic, axillary and usually in pairs, bracteate, the sepals 5 (4), distinct or basally connate, often petaloid or fleshy, imbricated, persistent, the petals 5 (4) or wanting, distinct or briefly basally connate, often smaller than sepals, imbricated; the corona fleshy, usually concave to cup-shaped, situated between perianth and androecium;[158] the stamens 5 or more, usually opposite the petals, arising from

[157] The name Passifloraceae has been conserved for this family.

[158] The gross morphology of the corona of the Passifloraceae, explained below, is based largely on the recent studies by Puri (1948). The corona is interpreted broadly to comprise (1) the hypanthium and (2) all accessory structures situated between the perianth parts and the stamens. The *hypanthium* (floral cup or receptacle) is of appendicular origin and does not represent an expansion and concavity of the receptacle or torus. The 1 or 2 whorls of usually filamentous sterile structures just within the petals are the *radii,* the usually much shorter whorl(s) within the radii are the *pali,* the next inside paii is a usually delicate membrane, the *operculum,* which generally closes the nectary chamber from the outside; these 3 parts of the corona (the radii, pali, and operculum) are believed to represent enations from the perianth parts and are not staminodes. Within the operculum there is a more or less annulate nectiferous region of which a conspicuous rim or ring is termed the *annulus.* At the base of the androgynophore (or sometimes "moved down" onto the hypanthium) is a minute and often lobed rim or cup termed the *limen* (probably staminodal in nature). In some species there is a distinct nodal swelling above the limen, the *trochlea.*

DIVISION IV. EMBRYOPHYTA SIPHONOGAMA       617

corona base or from receptacle rim or hypogynous from a gynophore apex (thereby forming an androgynophore), distinct or basally connate, anthers 2-celled, dehiscing longitudinally, staminodes present in some genera; pistil 1, often raised on a gynophore or more commonly an androgynophore, ovary superior, unilocular, carpels 3–5, placentae parietal and often broadly intruding (derived from ancestral axile placentae), the ovules numerous on each placenta, anatropous, styles as many as carpels, distinct or all connate, stigmas 3–5, often capitate or discoid; fruit a berry or a loculicidal capsule (rarely indehiscent); seed with straight embryo and fleshy endosperm.

A predominantly tropical American family of 11 genera and about 600 species. *Passiflora* (including *Tacsonia*) has about 400 species, of which about 7 are indigenous in the southeastern part of this country and one (*P. lutea*) extends north into Missouri and Pennsylvania. A few members of the family occur in New Zealand, Africa, Madagascar, and Asia.

The family is distinguished by the combination of the usually climbing habit, the 1-flowered peduncles often in pairs, the variously modified corona, the gynandrophore, and the often mucilaginously pulpy aril of the seeds.

Economically members of the family are of domestic importance as ornamentals and for the edible fruit of the several species of Granadilla (*Passiflora edulis, P. laurifolia, P. quadrangularis, P. ligularis,* and others). More than 20 species of *Passiflora* are cultivated here for the ornamental value of the vines and the showy unusual flowers, *P. caerulea* the most common.

*LITERATURE:*

HARMS, H. Passifloraceae. In Engler and Prantl, Die natürlichen Pflanzenfamilien, ed. 2, 21: 470–507, 1925.

KILLIP, E. P. The American species of Passifloraceae. Field Mus. Nat. Hist. Bot. 19: 1–613, 1938.

PURI, V. Studies in floral anatomy. IV. Vascular anatomy of the flower of certain species of Passifloraceae. Amer. Journ. Bot. 34: 562–573, 1947.

———. Studies in floral anatomy. V. On the structure and nature of the corona in certain species of the Passifloraceae. Journ. Indian Bot. Soc., pp. 130–149, 1948.

## CARICACEAE. CARICA FAMILY [159]

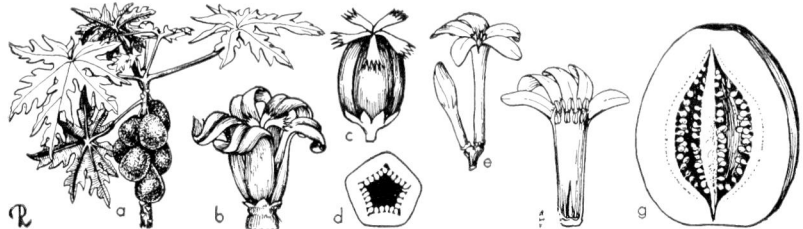

Fig. 226. CARICACEAE. *Carica Papaya*: a, crown of plant in fruit, much reduced; b, pistillate flower, × ⅓; c, pistil, × 1; d, ovary, cross-section, × 1; e, staminate flower, × ½; f, same, expanded, × ½; g, fruit, vertical section, × 1/20. (From L. H. Bailey, *Manual of cultivated plants,* The Macmillan Company, 1949. Copyright 1924 and 1949 by Liberty H. Bailey.)

Small soft-wooded often dioecious or monoecious trees, sap milky, trunks rarely branched, foliage in a terminal crown; leaves alternate; palmately lobed (rarely entire), large, mostly long-petioled, estipulate; flowers unisexual or bisexual, usually

[159] Sometimes known as the pawpaw family but that name may better be reserved for *Asimina* of the Anonaceae. It is known also in the tropics as the papaya family.

of four types: [160] (1) *staminate,* sessile, on staminate plants in clusters on pendant racemes 3–10 dm or more long, stamens 10 in 2 series, sessile, the pistil rudimentary rarely functional; (2) *pistillate,* subsessile, on pistillate plants, solitary or in few-flowered corymbs, in leaf axils, corolla gamopetalous, about 3 cm long, ovary large, globose, the 5 stigmas sessile and fimbriate, fruits globose to pyriform or ovoid; (3) *long fruited* type, similar to type 2 but the inflorescence multiflowered, about 6–10 cm long, composed of 5–6 flowered corymbs, corolla gamopetalous, goblet-shaped, stamens 10, sessile at base of petals, ovary usually functional and when so the fruit cucumber-shaped, mostly 3–5 dm long; (4) *polygamous,* flowers of 2 sorts, one with the 10 sessile stamens all at throat of corolla, petals connate into an elongated tube, the other with only 5 stamens, long-filamented, attached near base of ovary, corolla tube very short or scarcely apparent; calyx very small in all types, gamosepalous and 5-lobed; petals 5, connate or distinct as noted above; stamens usually 10 in 2 whorls, the inner one often lacking, the anthers 2-celled, dehiscing longitudinally; pistil 1, the ovary superior, unilocular, carpels usually 5, the placentae parietal and often intruding and producing a falsely 5-loculed chamber, the ovules numerous, the styles 5, broadly cuneate to fan-shaped, simple or fimbriate; fruit a large berry; seeds with straight embryo and a fleshy endosperm.

A predominantly tropical American family of 3 genera and 27 species. *Carica,* with about 24 species, is American, with most species in the Andes (one, *C. Papaya,* now adventive pantropically) and the other 2 genera indigenous to tropical Africa.

Members of the family are readily distinguished by the stiff unbranched thin-barked trunks crowned by the large leaves (giving a palmlike appearance), by the milky sap, the monoecious or dioecious character, the gamopetalous corollas bearing 5 or 10 stamens, and the unusual floral polymorphism.

Economically the family is important for the highly prized edible fruit, Papaya, available in an increasing number of commercial variants.

*LITERATURE:*

HARMS, H. Caricaceae. In Engler and Prantl, Die natürlichen Pflanzenfamilien, ed. 2, 21: 510–522, 1925.

HEILBORN, O. Taxonomical studies on *Carica.* Svensk. Bot. Tidsk. 30 H (3): 217–224, 1936.

HOFMEYR, J. D. J. The genetics of *Carica papaya.* Chron. Bot. 6: 245–247, 1941.

POPE, W. T. Papaya culture in Hawaii. Hawaii Agr. Exp. Sta. Bull. 61: 1–40, 1930.

## LOASACEAE.[161] LOASA FAMILY

Herbs or shrubs, sometimes lianous; hairs often rough or scabrous, sometimes stinging and of peculiar multicellular structures; leaves opposite or alternate, simple or pinnately divided or fid, estipulate; flowers bisexual, actinomorphic, mostly perigynous (i.e., the receptacle usually extending above top of inferior ovary), the perianth parts in 4's or 5's, the sepals imbricated, persistent, the petals flat or boat-shaped, distinct (*Mentzelia*) or connate (*Eucnide, Sympetaleia*), induplicate-valvate, often alternated with an inner series of petaloid staminodes or nectar scales; stamens numerous, distinct or more commonly connate in bundles opposite the petals, or all filaments basally connate in a low ring or short cylinder, the anthers usually 2-celled (1-celled in *Sympetaleia*), dehiscing longitudinally, the outer whorl sometimes sterile and petaloid; pistil 1, the ovary inferior or almost completely so, the locules 1–3, the carpels 3–7, the placentation usually parietal or sometimes axile, the ovules solitary to numerous on each placenta, anatropous, integument 1, the style and stigma 1; fruit a loculicidal capsule, the valves usually spirally twisted and convex; seed with straight embryo, the endosperm present.

A predominantly American family of 15 genera and about 250 species, most abundant in western South America, with the Old World representative a single monotypic genus (*Kissenia*) occurring from Arabia to southwest Africa. It is represented in the United States by about 30 indigenous species of 5 genera, all in the western provinces extending east to Illinois and Florida. *Mentzelia* (including *Bartonia*) by 44 of 61 species, *Eucnide* by 1 of 8, *Sympetaleia* by 1 of 2, *Petalonyx* by 4 of 5, and *Cevallia* by a single species.

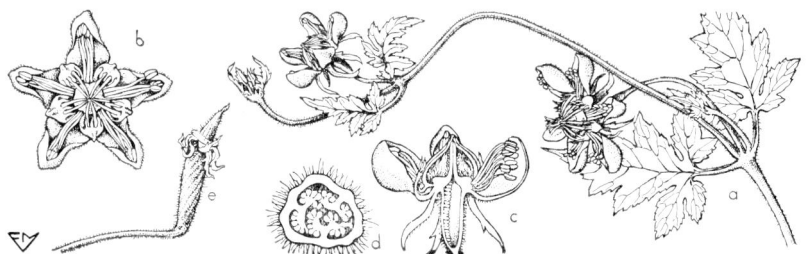

**Fig. 227.** LOASACEAE. *Caiophora lateritia*: a, flowering branch, × ½; b, flower, face view, × 1; c, flower, vertical section, × 1; d, ovary, cross-section, × 4; e, capsule, × ½. (From L. H. Bailey, *Manual of cultivated plants*. The Macmillan Company, 1949. Copyright 1924 and 1949 by Liberty H. Bailey.)

The family is characterized by its peculiarly shaped multicellular and usually stinging hairs (pagodalike, bulbously based with retrorse apex, or simple harpoonlike), its variety of staminode types, the often boat-shaped or concave petals, and the often spirally arranged capsule valves.

Economically the members of the family are of little importance. A few species each of *Blumenbachia*, *Caiophora*, *Eucnide*, *Loasa*, and *Mentzelia* are cultivated for ornament or as novelties.

*LITERATURE:*

DARLINGTON, J. A monograph of the genus *Mentzelia*. Ann. Mo. Bot. Gard. 21: 103-226, 1934.
GILG, E. Loasaceae. In Engler and Prantl, Die natürlichen Pflanzenfamilien, ed. 2, 21: 522-543, 1925.

## DATISCACEAE.[162] DATISCA FAMILY

Perennial herbs or small trees, mostly dioecious; hairs often lepidote-scaly; leaves alternate, simple or pinnately compound or divided, estipulate; flowers unisexual or rarely bisexual, actinomorphic, in axillary fascicles, or spicate to racemose; staminate flowers with calyx 3-9-lobed, the petals 8 and distinct or absent, the stamens variable in number (4-25?), often opposite the calyx lobes, distinct, the filaments often short, the anthers 2-celled, dehiscing longitudinally, a pistillode sometimes present; pistillate flowers (often some with a few stamens) with calyx tube adnate to ovary; staminodes often present; pistil 1, the ovary inferior, unilocular (sometimes open at apex), the carpels 3, the placentae parietal, the ovules numerous on each placenta, anatropous, the styles 3 and bifid; fruit a capsule, dehiscing apically and opening at the top between the persistent styles; seeds with a straight oily embryo and a scant endosperm.

[162] The name Datiscaceae has been conserved for this family.

A family of 3 genera (*Datisca, Tetrameles,* and *Octomeles*) and 4 species. *Datisca* has 1 species (*D. glomerata*) extending from Mexico into the Coast Ranges of California, and a second species in Asia. The other 2 genera are monotypic, the first a tree of India to Java, and the second occurring in the Philippines and East Indies.

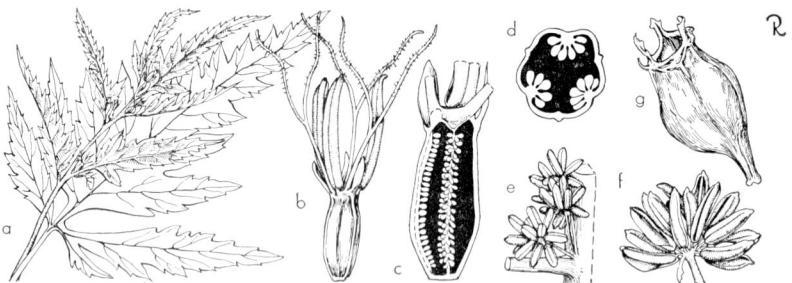

**Fig. 228.** DATISCACEAE. *Datisca glomerata*: a, pistillate branch, × ½; b, pistillate flower, × 3; c, ovary, vertical section, × 6; d, same, cross-section, × 9; e, staminate inflorescence, × 2; f, staminate flower, × 2; g, fruit, × 3.

The family is distinguished by the typically unisexual flowers and dioecious habit, an inferior ovary with parietal placentation, the nonvalvate capsule with an apical opening, and the oily embryo.

The family is of little domestic importance. Durango root (*Datisca glomerata*) is occasionally grown as a novelty, and *D. canabina* is planted for ornament.

*LITERATURE:*

GILG, E. Datiscaceae. In Engler and Prantl, Die natürlichen Pflanzenfamilien, ed. 2, 21: 543–547, 1925.

## BEGONIACEAE.[163] BEGONIA FAMILY

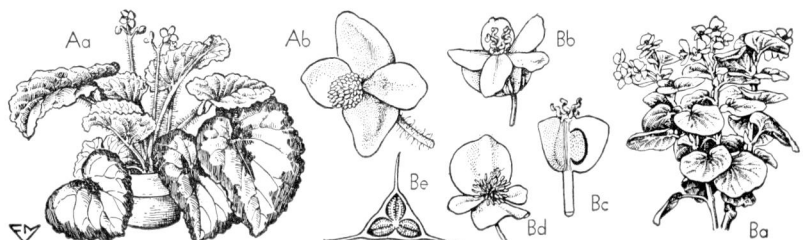

**Fig. 229.** BEGONIACEAE. A, *Begonia Rex-cultorum*: Aa, flowering plant, × 1/10; Ab, staminate flower, × ½. B, *B. semperflorens*: Ba, flowering branch, × ⅛; Bb, pistillate flower, × ½; Bc, same, vertical section, × ½; Bd, staminate flower, × ½; Be, ovary, cross-section, × 1. (From L. H. Bailey, *Manual of cultivated plants,* The Macmillan Company, 1949. Copyright 1924 and 1949 by Liberty H. Bailey.)

Mostly erect, creeping or acaulescent succulent monoecious herbs or low shrubs, stems somewhat jointed; leaves alternate, mostly 2-ranked, simple, mostly palmately nerved and often so lobed, petioled, the base oblique, the stipules caducous or early deciduous; flowers unisexual, zygomorphic or actinomorphic, the perianth of staminate flowers of 2 valvate petaloid sepals and 2 usually smaller valvate petals

[163] The name Begoniaceae has been conserved for this family.

DIVISION IV. EMBRYOPHYTA SIPHONOGAMA                                        621

and that of pistillate flowers of 2–many similar imbricated petaloid tepals (undifferentiated into calyx and corolla); stamens numerous, in many whorls, distinct or briefly connate basally, the anthers 2-celled, basifixed, a connective often exserted, dehiscing longitudinally (rarely by pores); pistil 1, the ovary inferior (half inferior in *Hillebrandia*), often 1–3 (6) winged, the locules 3 (unilocular with 5 parietal placentae in *Hillebrandia*), the carpels typically 3 (2 or 5), the placentation usually axile, the ovules numerous on simple or lobed placentae, anatropous, the styles 2–5, distinct or basally connate, the stigmas strongly papillose on all sides and often twisted; fruit a loculicidal, bony capsule or rarely a berry; seeds with straight oily embryo and no endosperm.

A family widely distributed in the tropics, with greatest development in northern South America. The 5 genera are *Begonia* (about 800 spp.), *Hillebrandia* (monotypic) Hawaii, *Begoniella* (3 spp.), Colombia, *Semibegoniella* (3 spp., Ecuador), *Symbegonia* (monotypic, New Guinea). No indigenous species occur in the United States, but a number of species are indigenous to Mexico.

The family is readily distinguished by the unisexuality of the flowers, the zygomorphic staminate flowers, the numerous stamens in many whorls, the typically inferior angled or winged ovary with forked or lobed placentae, the usually twisted stigmas, and the seeds with oily embryo and the lack of endosperm.

Economically the family is important for the many species of *Begonia* cultivated domestically as ornamentals, over 130 being currently offered in the American trade.

*LITERATURE:*

BUGNON, P. Valeur morphologique de l'ovaire infere chez les *Begonia*. Bull. Soc. Linn. Normandie, Ser. 7, 9: 7–25, 1926.
IRMISCHER, E. Begoniaceae. In Engler and Prantl, Die natürlichen Pflanzenfamilien, ed. 2, 21: 548–588, 1925.
SMITH, L. B. and SCHUBERT, B. G. Studies in the Begoniaceae—II Mexico, Central America. Contr. Gray Herb. 161: 26–29, 1946.
———. Some Mexican Begonias. Contr. Gray Herb. 165: 90–94, 1947.

## ORDER 31. OPUNTIALES

The Opuntiales of Engler are the cacti, and contain the single family Cactaceae. The order was accepted also by Rendle and (as the Cactales) by Wettstein but, as noted in the following account of the Cactaceae, the family has been included in other orders by most phylogenists and by several it has been treated as allied more closely to the Centrospermae than to the Parietales.

## CACTACEAE.[164] CACTUS FAMILY

Fleshy, herbaceous, or woody plants, stems simple or cespitose, many forms branched and treelike, often greatly enlarged and cylindrical, flattened or fluted, frequently constricted and jointed, the sap watery, or milky (in species of Coryphanthae; leaves alternate, simple, flat and leaflike (but fleshy) in *Pereskia* and *Pereskiopsis* but cylindric, scalelike, or absent in other genera, usually with clusters of spines and sometimes bristles in the axils (areoles), the spines of an areole often radiating with a central one porrect (glochids present in the tribe Opuntieae); flowers usually solitary, sometimes clustered, bisexual (unisexual in the series Stenopetalae of the Opuntieae), actinomorphic (sometimes zygomorphic by the curvature of perianth tube), the hypanthium adnate to the ovary or free from it, perianth weakly differentiated into sepals and petals, the latter very numerous (rarely 8–10), the stamens numerous, arising spirally or in groups from the inner

[164] The name *Cactaceae* Lindl. (1836) is conserved over *Opuntiaceae* HBK (1823).

face of the hypanthium, anthers 2-celled, dehiscing longitudinally; pistil 1, the ovary typically inferior (sometimes sunken or embedded in the stem, but stalked in *Pereskia*), unilocular, the carpels 3-many, the placentae parietal, the ovules numerous, anatropous, style 1 (as many as carpels in *Pereskia*), the stigmas as many as carpels and radiating; fruit a berry, often glochidiate, spiny, or bristly; seeds with straight or curved embryo, the endosperm usually lacking, sometimes viscid (*Rhipsalis*).

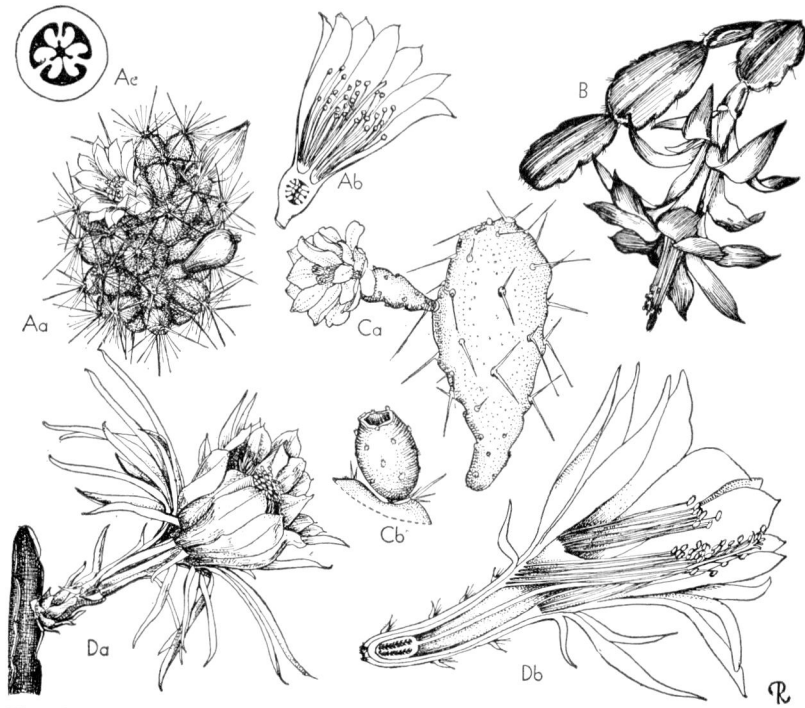

**Fig. 230.** CACTACEAE. A, *Mammillaria*: Aa, plant in flower, × 1; Ab, flower, vertical section, × 2; Ac, ovary, cross-section, × 5. B, *Schlumbergera Russellianus*: flowering branch, × 1. C, *Opuntia compressa*: Ca, flowering branch, × ¼; Cb, fruit, × ¼. D, *Nopalxochia Ackermannii*: Da, flower, × ¼; Db, same, vertical section, × ⅖. (From L. H. Bailey, *Manual of cultivated plants*, The Macmillan Company, 1949. Copyright 1924 and 1949 by Liberty H. Bailey.)

Cacti are native in the U.S. except Maine, New Hampshire and Vermont. The only genus alleged to have species in the Old World (West Africa and Ceylon) is *Rhipsalis,* and most botanists now reject the opinion that these are indigenous there. Indigenous species extend from Patagonia to British Columbia in the west and to Cape Cod in the east (including the West Indies), with the center of distribution in the dry areas of Mexico and large numbers of species extending southward into Chile and Argentina and northward through Arizona and into adjoining states. Some cacti are found in almost every state in the United States.

There is currently a lack of agreement on the generic limits within the family, and opinions of cactus specialists differ as to the number of genera. This is particularly true as regards the South American cacti. A current figure places the total number of genera at about 120 with 37 in the United States. Schumann (1899) accepted 21 genera for the family, Britton and Rose (1923) recognized about 100, and Parish (1936) perhaps ultraconservatively reduced these to 26. If the genera established for the South American cacti are accepted as taxonomically valid the total number for the family is in excess of 150. The fleshy character of cacti makes it difficult to prepare adequate and abundant herbarium specimens. This, plus the paucity of morphological differentiation in floral and reproductive characters, and an absence of basic cytological differences, has resulted in the establishment of many genera and species almost exclusively on vegetative features. Conservative cactologists, placing more reliance on differences of reproductive rather than on vegetative structures, have rejected the high number of genera resulting from the "splitting" alleged to have been practiced by Britton and Rose and their successors, and consider the genera in this country to be 5–8. Regardless of the generic concept accepted, there is reasonable agreement that the number of species for the family is between 1200 and 1800. Reference should be made to the works by Britton and Rose, Small, and Kearney and Peebles for data on the distribution of members of the family in the United States.

The Cactaceae are characterized, in general, by the fleshy habit, the spines or glochids arranged in areoles, the flowers solitary and with an undifferentiated perianth of very numerous segments basally fused to form an hypanthium, the numerous stamens arranged spirally or in clusters and the glochidiate spiny or bristly berry.

Engler placed the Cactaceae in the Opuntiales, an order he considered to have been derived from the Parietales. Mez, on the basis of the floral plan and serological data, considered them as allied to the Loasaceae and within the Parietales. Bessey and Hutchinson placed them near the Cucurbitaceae. However, as pointed out by Wettstein, by Hallier, by Buxbaum (1948), by Maheshwari (1945), and by Martin (1946), evidence is available from the embryology, floral morphology, and anatomy to suggest their transfer to a position within or near the Centrospermae. Buxbaum (1944) further derived them from the Phytolaccaceae and considered them to represent an example of parallel evolution with the Aizoaceae.

The family is composed of 3 subfamilies: (1) Pereskioideae, with flat somewhat fleshy leaves, spines, and flowers in panicles (*Pereskia* and *Maihuenia*); (2) Opuntioideae, succulents with usually flattened jointed stems (pads), small cylindrical to subulate caducous leaves, glochidiate areoles, and rotate flowers (*Opuntia, Pereskiopsis,* and *Nopalea,* the best-known genera); and (3) Cereoideae, succulents, leaves reduced to minute scales, areoles without glochidia, flowers (except in *Rhipsalis*), funnelform or salverform (includes most other genera).

Economically the Cactaceae are of domestic importance as ornamentals and are cultivated extensively in the open or under glass in all parts of the country. About 130 genera and over 1200 binomials are listed currently in the trade literature as cultivated in this country. The fruit of the Indian fig or prickly pear (*Opuntia* spp.) is edible and a product common in Mexican markets.

*LITERATURE:*

BACKEBERG, C. Blätter für Kakteenforschung, 1934–1938.
———. Zur Geschichte der Kakteen. Cactaceae, Zweiter Teil, 1942.
———. Eine neue Sippe: Lobiviae. Kakteenkunde, 11–19, 1943.
BACKEBERG, C. and KNUTH, F. M. Kaktus-ABC. Copenhagen, 1935.
BAXTER, E. M. California Cactus. Los Angeles, Cal., 1935.
BEARD, E. C. Some chromosome complements in the Cactaceae and a study of meiosis in *Echinocereus papillosus.* Bot. Gaz. 99: 1–21, 1937.
BENSON, L. The cacti of Arizona. Bull. Univ. Ariz. Biol. Sci. 4: 1–134, 1940.
———. A revision of some Arizona Cactaceae. Proc. Cal. Acad. IV, 25: 245–268, **1944**.
BERGER, A. Kakteen, 1–346, 1929.

BOISSEVAIN, C. H. and DAVIDSON, C. Colorado cacti. 72 pp. Pasadena, Cal., 1941.
BORG, J. Cacti. 419 pp. New York, 1937.
BRAVO, H. Las cactaceas de Mexico. Mexico, D. F., 1937.
BRITTON, N. L. and ROSE, J. N. The Cactaceae. 4 vols. Washington, D. C. 1919–1923.
BUXBAUM, F. Untersuchungen zur Morphologie der Kakteenbüte. Bot. Arch. (Leipzig), 45: 190–247, 1944.
CASTELLANOS, A. and LELONG, H. V. Los géneros de las Cactaceae Argentinas. An. Mus. Argentino Cien. Nat. Publ. 86: 383–419, 1938.
CLOVER, E. U. Vegetational Survey of the Lower Rio Grande Valley, Texas. Madroño, 4: 41–66, 77–100, 1937.
———. The Cactaceae of Southern Utah. Bull. Torrey Bot. Club, 65: 397–412, 1938.
———. Cacti of the Canyon of the Colorado River and Tributaries. Bull. Torrey Bot. Club, 68: 409–419, 1941.
CRAIG, R. T. The *Mammillaria* handbook, with descriptions, illustrations, and key to the species of the genus *Mammillaria* of the Cactaceae. Pasadena, Cal., 1945.
CROIZAT, L. A check list of Colombian and presumed Colombian Cactaceae. Caldasia, 9: 337–355, 1944.
ENGELMANN, G. Synopsis of the Cactaceae of the Territory of the United States and adjacent regions. (Proc. Amer. Acad. iii), Cambridge, 1856.
ENGLER, A. Historische Entwicklung der Ansichten über die systematische Stellung der Reihe Opuntiales. In Engler and Prantl, Die natürlichen Pflanzenfamilien, ed. 2, 21: 592–594, 1925.
HASELTON, S. E. *Epiphyllum* handbook. Pasadena, Cal., 1946.
JUST, T. The use of embryological formulas in plant taxonomy. Bull. Torrey Bot. Club, 73: 351–355, 1946.
MAHESHWARI, P. The place of angiosperm embryology in research and teaching. Journ. Indian Bot. Soc. 24: 25–41, 1945.
MARSHALL, W. T. and BOCK, T. M. Cactaceae, with illustrated keys of all tribes, sub-tribes and genera. 227 pp. Pasadena, Cal., 1941.
MARTIN, A. C. The comparative internal morphology of seeds. Amer. Midl. Nat. 36: 513–660, 1946.
SMALL, J. K. Chronicle of the cacti of eastern North America. Journ. N. Y. Bot. Gard. 36: 1–11, 25–36, 1935.
VAUPEL, E. Cactaceae. In Engler and Prantl, Die natürlichen Pflanzenfamilien, ed. 2, 21: 594–651, 1925.

## ORDER 32. **MYRTIFLORAE** [165]

The Myrtiflorae were characterized by Engler as a taxon whose components showed a transition from perigyny (in the more primitive) to epigyny; the stem tissues in general contain an internal phloem (intraxylary phloem), the leaves are more often opposite than alternate, the flowers are cyclic, and the development of an hypanthium (sometimes wholly adnate to the ovary) is distinctive.

The order was subdivided by Engler and Diels into 4 suborders, and treated as composed of 23 families, as follows (names of families not treated in this text are preceded by an asterisk):

| | | |
|---|---|---|
| Thymelaeineae | *Sonneratiaceae | Melastomaceae |
| *Geissolomataceae | *Crypteroniaceae | Hydrocaryaceae |
| *Penaeaceae | Punicaceae | Onagraceae |
| *Oliniaceae | *Lecythidaceae | Haloragaceae |
| Thymelaeaceae | Rhizophoraceae | Hippuridineae |
| Elaeagnaceae | Nyssaceae | Hippuridaceae |
| Myrtineae | Alangiaceae | *Thelygonaceae |
| Lythraceae | Combretaceae | Cynomoriineae |
| *Heteropyxidaceae | Myrtaceae | *Cynomoriaceae |

[165] The name Myrtiflorae was used by Eichler and adopted by Engler and associates, and by Rendle. Wettstein adopted the conventional ending -ales and designated the order the Myrtales, as also did Bessey and (with much narrower circumscription) Hutchinson.

# DIVISION IV. EMBRYOPHYTA SIPHONOGAMA

Wettstein accepted the order to be of the same circumscription as did Eichler and Engler, but included the Hydrocaryaceae within the Onagraceae and separated the Gunneraceae as a family distinct from the Haloragaceae. Bessey also accepted the Eichler-Engler views for the most part, but retained the Nyssaceae within the Cornaceae (of the Umbelliflorae), and transferred the Thymelaeineae to his Celastrales. Rendle followed the Engler treatment, but accepted the Nyssaceae and Alangiaceae as separate families of the Umbelliflorae and noted (p. 423) that they were "regarded as more nearly allied to the Myrtiflorae." Hallier included the Thymelaeineae in his Daphnales, restricted the Myrtiflorae (as the Myrtines) to include, among the families here treated, the Lythraceae, Lecythidaceae, Rhizophoraceae, Combretaceae, Myrtaceae, and Melastomaceae. He placed the Alangiaceae in his Santalales, the Nyssaceae in the Cornaceae, the Onagraceae (including the Hydrocaryaceae) in his Polygalines, and the Haloragaceae (including the Hippuridaceae) in his Ranales. Hutchinson distributed these families among 5 orders: the Thymelaeaceae in the Thymelaeales, Elaeagnaceae in the Rhamnales, the Lythraceae, Punicaceae, Onagraceae (including Hydrocaryaceae), and Haloragaceae (including Hippuridaceae) in his Lythrales, the Nyssaceae and Alangiaceae in the Umbelliflorae, and the balance of those treated here in his Myrtales.

## THYMELAEACEAE.[166] MEZEREUM FAMILY

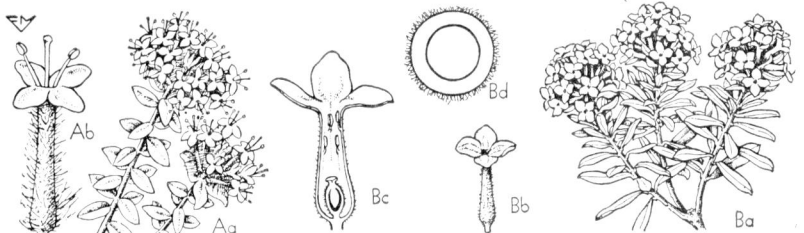

**Fig. 231.** THYMELAEACEAE. A, *Pimelea ferruginea*: Aa, flowering branch, × ½; Ab, flower, × 2. B, *Daphne Cneorum*: Ba, flowering branch, × ½; Bb, flower, × 1; Bc, same, vertical section, × 2; Bd, ovary, cross-section, × 10. (From L. H. Bailey, *Manual of cultivated plants*, The Macmillan Company, 1949. Copyright 1924 and 1949 by Liberty H. Bailey.)

Trees or shrubs, rarely herbs; leaves alternate or opposite, simple, persistent or deciduous, entire, estipulate; flowers bisexual or unisexual (when unisexual the plants mostly dioecious), actinomorphic, terminal or axillary, sometimes solitary, mostly racemose to umbellate, the sepals 4–5, imbricated, petaloid, usually connate and developed into a tube [167] with spreading lobes, the petals scalelike, 4–12 (or absent), arising usually from near the mouth of the tube; stamens as many as the sepals and alternating with them, or twice as many, or reduced to 2, perigynous, the anthers 2-celled, introrse, dehiscing longitudinally; an hypogynous nectiferous disc (annular, cupular, or of scales) often present; pistil 1, the ovary superior, the locules 1 or less commonly 2, the carpels typically 1, the ovule anatropous, solitary and pendulous (when biloculate, the ovary has 1 ovule in each locule), the style 1 or none and stigma typically discoid; fruit a drupe or nut, rarely a capsule or berry; seeds with straight embryo, the endosperm copious or absent.

[166] The name Thymelaeaceae Meiss. (1857) is conserved over Daphnaceae St. Hil (1805), Thymelaceae Lindl. (1836) *et al.*

[167] It is probable that this so-called calyx tube is a true hypanthium of appendicular origin as is evidenced by the reduced petals and the stamens borne on it.

A family of 40 genera and about 500 species, of nearly cosmopolitan distribution (absent in coldest regions), with centers of concentration in South Africa, Australia, the Mediterranean region, and the steppes of central and western Asia. It is represented in the United States by the 2 indigenous species of *Dirca*, *D. occidentalis* of California and *D. palustris* of eastern North America. The European *Daphne Mezereum* has become naturalized in parts of the northeastern area. The larger genera of the family include *Gnidia* (90, Africa), *Pimelea* (80, New Zealand, Australia to New Guinea), *Daphne* (50, Eurasia), and *Thymelea* (20, Mediterranean region).

The Thymelaeaceae are distinguished by the seemingly uniseriate gamosepalous perianth, the perigynous stamens of definite number, the petals reduced to hypanthial appendages, and the superior typically unilocular and uniovulate ovary.

Hutchinson (1948) reversed his earlier views, that the affinities of the family were with the Caryophyllales via the Geraniales, and stated, "I now consider the Thymelaeaceae to be apetalous relatives of the Bixaceae, especially of the exotic family Flacourtiaceae." Bessey included the family in his Celastrales. Hallier treated it as the primitive taxon of his Daphnales, an order of 4 families that he considered to have been derived from the Linaceae.

In this country, a dozen or more species of shrubs representing the genera *Dais*, *Daphne*, *Dirca* (leatherwood), *Edgeworthia* (paper bush), and *Pimelea* (rice flower) are cultivated for ornament, especially in the warmer parts.

*LITERATURE:*

DOMKE, W. Untersuchungen über die systematische und geographische Gliederung der Thymelaeaceen nebst einer Neubeschreibung ihrer Gattungen. Bibl. Bot. 27 (111): 1–151, 1934.

GILG, E. Thymelaeaceae. In Engler and Prantl, Die natürlichen Pflanzenfamilien, III (6b): 216–245, 1894.

## ELAEAGNACEAE.[168] OLEASTER FAMILY

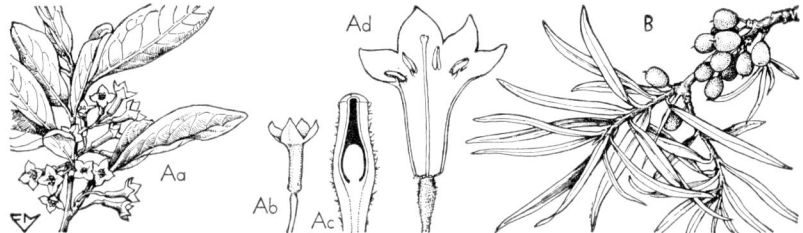

Fig. 232. ELAEAGNACEAE. A, *Elaeagnus umbellata*: Aa, flowering branch, × ½; Ab, flower, × 1; Ac, ovary, vertical section, × 5; Ad, flower, calyx expanded, × 2. B, *Hippophaë rhamnoides*: fruiting branch, × ½. (From L. H. Bailey, *Manual of cultivated plants*, The Macmillan Company, 1949. Copyright 1924 and 1949 by Liberty H. Bailey.)

Mostly erect shrubs, seldom trees, vegetative parts densely covered in part with silvery, brownish, or golden colored lepidote or stellate hairs, branches sometimes spiny; leaves alternate, opposite, or whorled, simple, entire, estipulate; flowers bisexual or unisexual (the plants then dioecious as in *Shepherdia* or *Hippophaë*, or polygamodioecious as in *Elaeagnus*), actinomorphic, axillary, solitary or in clusters or racemose, the perianth seemingly uniseriate and developed into an hypanthium (the tubular receptacle of authors), saucer-shaped to tubular, the lobes 4 (rarely

[168] The name Elaeagnaceae has been conserved for this family.

2 or 6), valvate; stamens as many or twice as many as lobes, perigynous, arising from the hypanthium, a lobed perigynous disc usually present; pistil 1, the ovary superior, unilocular, 1-carpelled, the ovule solitary, basal, anatropous, the integuments 2, the style and stigma 1; fruit a dry and indehiscent achene enveloped by the fleshy persistent perianth and hence drupaceous; seed with straight embryo and scant or no endosperm.

A family of 3 genera and about 45 species (of which 40 belong to *Elaeagnus*); mostly steppe and rock plants, chiefly of southern Asia, Europe, and North America. The family is represented in the United States by 2 indigenous species of *Shepherdia* occurring over much of the boreal part of the continent, and 1 of *Elaeagnus* of similar distribution. Two Japanese species of *Elaeagnus* are naturalized in the southeast. Nelson (1935) rejected the validity of generic distinctions and treated all species as belonging in *Elaeagnus*.

The Elaeagnaceae are readily distinguished by the silvery to golden-brown lepidote or stellate indumentum that densely covers the twigs, lower side of leaves, and hypanthia, and by the achenes enveloped by the fleshy persistent perianth.

Most phylogenists have considered the Elaeagnaceae to be closely allied to the Thymelaeaceae and have treated the 2 families similarly as noted under the Myrtiflorae (p. 625). However, Hutchinson (1948) included the Elaeagnaceae in his Rhamnales, believing them to be "a more advanced group than and related to the Rhamnaceae."

Economically the family is important domestically for the several shrubs cultivated for ornament, notably the buffalo berry (*Shepherdia argentea*), the oleaster (*Elaeagnus angustifolia*) and related species, and the sea buckthorn (*Hippophaë rhamnoides*).

*LITERATURE:*

GILG, E. Elaeagnaceae. In Engler and Prantl, Die natürlichen Pflanzenfamilien, III (6b): 246–251, 1894.

NELSON, A. The Elaeagnaceae — a mono-generic family. Amer. Journ. Bot. 22: 681–683, 1935.

SERVETTAZ, C. Monographie des Eléagnacées. Beih. Bot. Centralb. 25 (2): 1–420, 1909

## LYTHRACEAE.[169] LOOSESTRIFE FAMILY

Herbs, shrubs, or trees (rarely spinescent, as in *Lawsonia* spp.); leaves usually opposite or whorled, simple, mostly entire, stipules minute or commonly absent; flowers bisexual, generally actinomorphic (occasionally zygomorphic, as in *Cuphea*), cleistogamous flowers present in some genera, perigynous, an hypanthium present and sometimes subtended by an epicalyx of connate pairs of bracts (as in *Lythrum*), the sepals commonly 4, 6, or 8, valvate, seemingly marginal on the hypanthium (receptacle or calyx tube of most authors), the petals of same number as sepals or absent (as in *Peplis* or *Rotala*), seemingly distinct and arising from rim or upper inner surface of hypanthium, alternate with sepals, crumpled in bud, imbricated, often fugacious; stamens usually twice as many as petals, in 2 whorls, those of outer whorl alternate with petals and emerging from hypanthium some distance below them, occasionally 1 whorl missing or stamens reduced to 1 (rarely the stamens very numerous), the filaments usually unequal in length, the anthers 2-celled, introrse, dorsifixed, dehiscing longitudinally; pistil 1, the ovary superior, sessile or short-stipitate, the locules and carpels usually 2–6 (rarely the locule 1), the placentation typically axile (septa sometimes disappearing in upper part of ovary), the ovules several to numerous on each placenta, anatropous, ascending.

[169] The name Lythraceae Lindl. (1836) is conserved over Ammanniaceae Horaninov (1834) and Salicariaceae Desvaux (1827).

the style 1 (heterostyly occasional), the stigma usually discoid or capitate; fruit a variously dehiscing capsule; seeds with a straight embryo, the endosperm lacking.

There are about 23 genera and 475 species known for the family; generally distributed, but abundant in the American tropics. Most of the tropical elements are woody shrubs (trees in *Lagerstroemia* and *Lafoensia*) while those of temperate regions are suffrutescent or herbaceous. The Lythraceae are represented in the United States by 1 or more indigenous species of *Lythrum* (25-7, widespread), *Ammannia* (20-4, widespread but more common to the south), *Rotala* (35-2, widespread), *Didiplis* (monotypic, Minnesota to Texas and Florida), *Decodon* (monotypic, eastern and central), *Cuphea* (*Parsonsia*) (200-6, Kansas to Georgia and southward, 1 species to New England), and *Heimia* (2-1, western Texas).

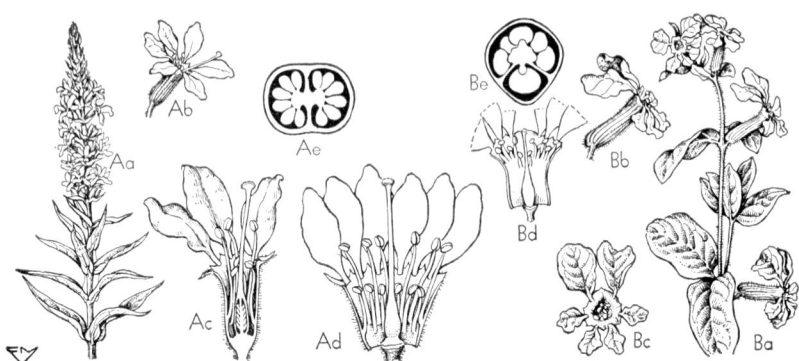

**Fig. 233.** LYTHRACEAE. A, *Lythrum Salicaria*: Aa, flowering branch, × ¼; Ab, flower, × 1; Ac, same, vertical section, × 2; Ad, flower, perianth expanded, × 2; Ae, ovary, cross-section, × 12. B, X *Cuphea purpurea*: Ba, flowering branch, × ⅓; Bb, flower, side view, × ½; Bc, flower, face view, × ½; Bd, flower (less petals), hypanthium expanded, × ½; Be, ovary, cross-section, × 4. (From L. H. Bailey, *Manual of cultivated plants*, The Macmillan Company, 1949. Copyright 1924 and 1949 by Liberty H. Bailey.)

The family is distinguished by the presence of the hypanthium, the superior ovary, the crumpled corolla, the often unequal stamens of typically twice the number of sepals, and the seed without endosperm.

Economically the family is of importance for members of ornamental value, notably the Crepe myrtle (*Lagerstroemia indica*) widely cultivated in warmer parts of the country, henna or mignonette tree (*Lawsonia inermis*) grown in California and much of the south, a dozen species of *Cuphea* as garden annuals, and the spiked or purple loosestrife (*Lythrum Salicaria*).

*LITERATURE:*

KOEHNE, E. Lythraceae. In Engler and Prantl, Die natürlichen Pflanzenfamilien, III (7): 1-16, 1898.

———. Lythraceae. In Engler, Das Pflanzenreich, 17 (IV. 216): 1-326, 1903. [See also, Nachträge, in Engler Bot. Jahrb. 41: 74-110, 1907.]

## PUNICACEAE. POMEGRANATE FAMILY

Shrubs or small trees, sometimes spiny, estipulate; twigs caducously 4-winged; leaves mostly opposite or fasciculate, simple, eglandular; flowers bisexual, actinomorphic, terminal and solitary or cymose, usually perigynous, a tubular to urceolate hypanthium present (calyx tube or receptacle of authors) and adnate to ovary, the

## DIVISION IV. EMBRYOPHYTA SIPHONOGAMA 629

calyx lobes (sepals) 5-8, valvate, fleshy, the corolla lobes (petals) 5-7, imbricated, emerging from edge of hypanthium, crumpled in bud; stamens very numerous, emerging in many whorls from within upper half or more of hypanthium, the filaments distinct and nearly equal, the anthers 2-celled, dorsifixed, dehiscing longitudinally; pistil 1, the ovary inferior, the locules and carpels generally 8-12 (3 in some spp.), in early ontogeny of ovary the locules in 2 concentric whorls and the placentation axile, but as development progresses the outer series of mostly 5-9 locules is carried upward and ultimately superposed over (or alternated with) the originally "inner" series (of usually 3 locules) and the placentation of the upper series transposed to a parietal position, the ovules numerous, anatropous, the style and stigma 1; fruit a berry with persistent calyx lobes; seeds with a fleshy testa that forms the pulp within the fruit, the embryo straight, the endosperm absent.

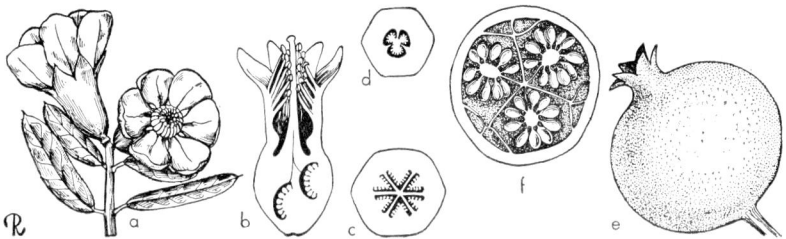

Fig. 234. PUNICACEAE. *Punica Granatum*: a, flowering branch, × ½; b, flower, vertical section (less calyx), × 1; c, ovary, cross-section, upper part, × 1; d, same, lower part, × 1; e, fruit, × ½; f, fruit, cross-section, lower half, × ½. (From L. H. Bailey, *Manual of cultivated plants*, The Macmillan Company, 1949. Copyright 1924 and 1949 by Liberty H. Bailey.)

A unigeneric family of 2 species (*Punica Granatum* the pomegranate, and *P. Protopunica*) a native of semitropical Asia but widely naturalized pantropically. Early authors (including Bentham and Hooker and Hallier) included the genus in the Lythraceae. The pomegranate is cultivated in warm parts of this country for its edible fruits and as a decorative ornamental.

*LITERATURE:*
NIEDENZU, F. Punicaceae In Engler and Prantl, Die natürlichen Pflanzenfamilien, III (7): 22-25, 1898.

## RHIZOPHORACEAE.[170] MANGROVE FAMILY

Shrubs or trees, mostly of tropical shores, nodes swollen; leaves usually opposite, coriaceous and persistent, stipules caducous or none; flowers bisexual (rarely unisexual by abortion, the plants then monoecious), actinomorphic, axillary, solitary or cymose, perigynous to epigynous, the sepals 3-14, more or less basally connate, valvate, persistent, the petals of same number as sepals and often smaller, usually fleshy or coriaceous, distinct and often clawed, frequently lacerate or emarginate, convolute or inflexed in bud; stamens 2-4 times as many as sepals, usually in 1 series, often in pairs opposite the petals, situated on the outer edge of a lobed perigynous or epigynous disc, the filaments mostly very short, the anthers introrse, typically 4-celled at anthesis (pollen sacs numerous in *Rhizophora*), dehiscence generally by longitudinal ruptures; pistil 1, the ovary variable in position depending on degree of adnation of perianth (i.e., superior, half inferior, or completely inferior), the locules and carpels 2-4 (rarely unilocular by suppression of septa o-

---
[170] The name Rhizophoraceae has been conserved over other names for this family.

carpels and locules sometimes 5), the placentation axile, the ovules usually 2 from each placenta, pendulous, anatropous with micropyle upward and outward, the style usually 1, the stigma with mostly as many lobes as carpels; fruit usually a berry, somewhat juicy, terminated by the persistent calyx, rarely dehiscent (rarely a dry dehiscent capsule or a drupe); seeds with straight and often green embryo, endosperm usually present, the seed (in some genera) germinating within the fruit while latter remains attached to twig.

A family of about 17 genera and 70 species. Several of the genera (*Rhizophora* 8 spp., *Ceriops* 2 spp., *Bruguiera* 6 spp., *et al.*) comprise the characteristic mangrove vegetation of muddy tidal flats and shore lines throughout the paleotropic regions of the world. In this country *Rhizophora Mangle* is the only indigenous species, and is restricted to muddy shores and to the everglades of peninsular Florida.

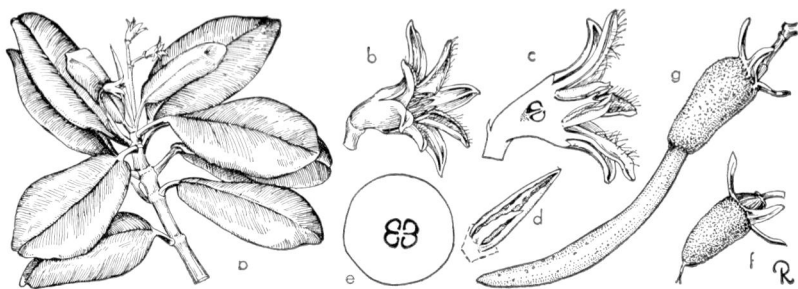

**Fig. 235.** RHIZOPHORACEAE. *Rhizophora Mangle*: a, flowering branch, × ½; b, flower, habit, × 1½; c, flower, vertical section, × 2; d, stamen, after anthesis, × 3; e, ovary, cross-section, × 4; f, fruit, × ½; g, germinating fruit, × ½.

The family is distinguished by the coriaceous leaves, the stamens numerous and with anthers 4-chambered or the anther sacs numerous, the ovules typically 2 on each placenta, the chlorophyllaceous cotyledons of the embryo and the seed often germinating within the fruit.

It is of no domestic importance, but elsewhere tannin is made from bark and foliage, and the wood is used for piling and underwater construction.

*LITERATURE:*

EGLER, F. E. The dispersal and establishment of red mangrove, *Rhizophora*, in Florida. Carib. Forester. 9: 299–310, 1948.

MARCO, H. F. Systematic anatomy of the woods of the Rhizophoraceae. Trop. Woods 44: 1–20, 1935.

SALVOZA, F. M. *Rhizophora*. Nat. and Appl. Sci. Bull. Univ. Philippines, 5: 179–237, 1936. [A monographic treatment.]

SCHIMPER, A. F. W. Rhizophoraceae. In Engler and Prantl, Die natürlichen Pflanzenfamilien, III (7): 42–56, 1898.

## NYSSACEAE. NYSSA FAMILY

Trees or shrubs, mostly dioecious; leaves alternate, simple, deciduous, estipulate; flowers minute, bisexual, or more commonly unisexual (when so the plants often polygamodioecious), actinomorphic, axillary, in umbellate, capitate, spicate, or racemose inflorescence (inflorescence subtended by 2 showy bracts in *Davidia*), the perianth adnate to ovary, the calyx lobes 5 and toothlike or obsolete, the petals 5 or more or absent, imbricate; stamens 5–10 (12), typically biseriate, exserted, the anthers 2-celled, introrse, dehiscing longitudinally, dorsifixed or basifixed; central fleshy disc usually present; pistil 1, the ovary inferior, unilocular and uniovulate (*Nyssa*) or 6–10-loculed with each placenta axile and uniovulate (*Davidia*), the ovule pendulous, anatropous, the style 1, often reflexed or coiled, sometimes cleft, the stigma typically 1; fruit a 1-seeded drupe or the stone 5–6-seeded; seed with straight embryo and scant endosperm.

A family of 3 genera and 8 species of eastern North America and Asia. *Nyssa* (6 spp.) is indigenous to the United States and to Asia, extending from the length of the Atlantic coast west to Illinois and Texas. The other genera are monotypic: *Davidia* in China, and *Camptotheca* in China west into Tibet.

Fig. 236. NYSSACEAE. *Nyssa sylvatica*: a, fruiting branch, × ½; b, staminate flower, × 6; c, same, vertical section, × 6; d, bisexual flower, × 4; e, same, vertical section, × 4. (From L. H. Bailey, *Manual of cultivated plants*, The Macmillan Company, 1949. Copyright 1924 and 1949 by Liberty H. Bailey.)

The Nyssaceae are distinguished by the estipulate alternate leaves, the flowers solitary when bisexual, and when unisexual the staminate usually racemose or capitate, by the stamens usually twice as many as the petals, and the fruit drupaceous. The uniovulate ovary is distinctive when present.

The family was retained in the Myrtiflorae alliance by Wettstein and Bessey, as well as by Rendle, who noted that there was evidence that it was more closely allied to the Cornaceae with which it was united by Hallier. Hutchinson included it in the Umbelliflores, and Rickett (1945) treated it as one of 2 families comprising the Cornales, an order close to and more advanced than the Umbellales. The view that the family has evolved from the Cornaceae, or stocks ancestral to it, was supported also by the cytological findings of Dermen (1932).

The family is domestically important for the pepperidge or sour gum tree (*Nyssa sylvatica*) cultivated to a limited extent as an ornamental prized for the usually brilliant coloration of the autumn foliage, and for the dove tree (*Davidia involucrata*) of northern China.

*LITERATURE:*

DERMEN, H. Cytological studies of *Cornus*. Journ. Arnold Arb. 13: 410–415, 1932.
MIRANDA, F. El género *Nyssa* en México. Anales Inst. Biol. México 15: 369–374, 1945.
RICKETT, H. W. Nyssaceae. North Amer. Flora, 28B: 313–316, 1945.
UPHOF, J. C. T. Die amerikanischen *Nyssa*-arten. Mitteil. deutsch. dendr. Gesell. 43: 2–16, 1931.
WANGERIN, W. Nyssaceae. In Engler, Das Pflanzenreich, 41 (IV. 220a): 1–20, 1910.

## ALANGIACEAE. ALANGIUM FAMILY

Trees or shrubs, sometimes spiny; leaves alternate, simple, entire or lobed, estipulate; flowers bisexual, actinomorphic, bracteate, with jointed pedicels in axillary cymes, the perianth biseriate and adnate to ovary, the calyx lobes 4–10, toothlike or obsolete, the petals as many as calyx lobes, valvate, linear to lorate, sometimes basally coherent, the corolla tubular and later spreading, more or less villous within; stamens as many as petals or 2–4 times as many, distinct, arising from an enlarged disc, the anthers 2-celled, dehiscing longitudinally; pistil 1, the ovary inferior, commonly 1- (rarely 2-) loculed, the carpels 2–3, the ovule solitary, pendulous, anatropous, the style and stigma 1, or of 2–3 lobes; fruit a drupe, usually crowned by the calyx lobes and disc; seed with straight embryo and copious endosperm.

A unigeneric family (*Alangium*) of about 22 species in the tropics and subtropics of the Old World. One or 2 species of *Alangium* (*Marlea*) are cultivated domestically as half-hardy shrubs.

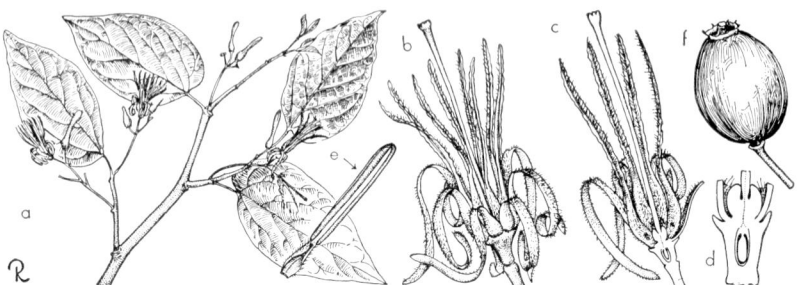

**Fig. 237.** ALANGIACEAE. *Alangium chinense*: a, flowering branch, × ½; b, flower, habit, × 1½; c, same, vertical section, × 1½; d, ovary, vertical section, × 4; e, undehisced stamen, ×2; f, fruit, × 2.

The family is closely related to the Nyssaceae-Cornaceae alliance, and is distinguished from those taxa by the articulated pedicels, the valvate petals, the corolla somewhat villous within, and the ovary mostly with a single uniovulate locule.

The Alangiaceae were not treated by Bessey, and by Hallier they were included in the Santalales (perpetuating an older view held by Reichenbach and by Spach). Hutchinson accepted the family and included it in his Umbelliflorae. Its retention by Engler and Diels, with the Nyssaceae, here in the Myrtiflorae probably reflects the influence of the earlier views held by Wangerin (1910).

*LITERATURE:*

WANGERIN, W. Alangiaceae. In Engler, Das Pflanzenreich, 41 (IV. 220b): 1–24, 1910.

## COMBRETACEAE.[171] COMBRETUM FAMILY

Trees or shrubs, often lianous, spinescent in some genera; leaves alternate or less commonly opposite, simple, estipulate; flowers bisexual (rarely unisexual), actinomorphic or occasionally zygomorphic, bracteate, spicate, racemose or paniculate, the perianth typically biseriate, the parts fusing to form an hypanthium (receptacle or calyx tube) that is adnate to ovary, the calyx lobes (sepals) 4–5 (8), persistent, valvate, the corolla lobes typically of same number or absent, imbricated or valvate, small; stamens usually 2–5 or twice as many as calyx lobes and biseriate, the anthers versatile, 2-celled, dehiscing longitudinally; pistil 1, the ovary inferior, unilocular, mostly with as many ribs or angles as calyx lobes and alternate with them, the ovules 2–6, anatropous, pendulous on long funiculi from locule apex, the

[171] The name Combretaceae has been conserved over other names for this family.

micropyle turned upwards and out, the style slender and solitary with a capitate or no obvious stigma; fruit leathery and drupaceous, 1-seeded (by ovule abortion), often winged; seed without endosperm.

A pantropical family of 18 genera and 500 species. The largest genera are *Combretum* (370 spp.) and *Terminalia* (200 spp.). It is represented in the United States (in peninsular Florida) by *Conocarpus* (monotypic), *Bucida* (3 spp., West Indies), and *Laguncularia* (monotypic). *Terminalia Catappa* of Oceania is naturalized in southern Florida.

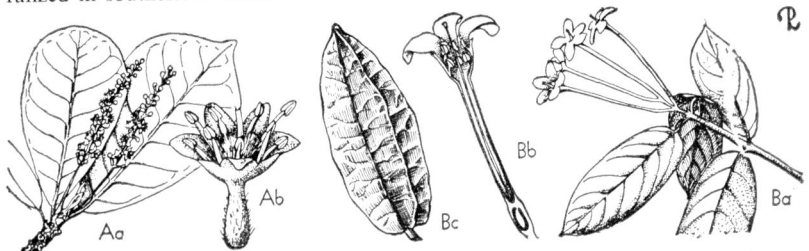

**Fig. 238.** COMBRETACEAE. A, *Terminalia Catappa*: Aa, flowering branch, × ⅙; Ab, flower, × 5. *Quisqualis indica*: Ba, flowering branch, × ⅙; Bb, flower, vertical section, × ½; Bc, fruit, × ¾. (From L. H. Bailey, *Manual of cultivated plants*, The Macmillan Company, 1949. Copyright 1924 and 1949 by Liberty H. Bailey.)

The Combretaceae are distinguished by the ovules 4–6 in the single locule, and all suspended from the locule apex by slender funiculi. When present, the wings on the indehiscent drupaceous fruit also are distinctive.

The family is of little domestic economic importance. The tropical or Indian almond (*Terminalia Catappa*) is cultivated in warmer parts of the country for ornament and for its edible nuts; about 4 species of vines from *Combretum* and *Quisqualis* are cultivated for ornament, and trees of the black olive (*Bucida Buceras*) are cultivated as a novelty in Florida.

*LITERATURE:*

BRANDIS, D. Combretaceae. In Engler and Prantl, Die natürlichen Pflanzenfamilien, III (7): 106–130, 1898.

EXELL, A. W. The Genera of Combretaceae. Journ. Bot. (Lond.) 69: 113–128, 1931.

## MYRTACEAE. MYRTLE FAMILY

Shrubs or trees; leaves usually opposite, simple, mostly entire, coriaceous, pellucid-dotted, estipulate; flowers bisexual, actinomorphic, commonly cymose, racemose, or paniculate, solitary in a few genera, mostly epigynous, the sepals usually 4–5 (mostly inconspicuous or absent in *Eucalyptus*), distinct or basally connate, the petals usually 4–5, generally distinct (connate, forming an operculum in *Eucalyptus*), imbricated; stamens numerous (rarely few), sometimes in fascicles opposite the petals, generally distinct, the anthers usually versatile (basifixed in *Calothamnus*), 2-celled, introrse, dehiscing longitudinally or sometimes apically, the connective often conspicuous and gland-tipped; pistil 1, the ovary usually inferior, sometimes nearly half inferior, the locule 1 with 3–several strongly intruded parietal placentae or more commonly the latter centrally connate resulting in the locules as many as carpels and the placentation axile, the ovules 2–many on each placenta, obliquely pendulous, anatropous or campylotropous, the style and stigma 1; fruit a berry or loculicidal capsule, rarely drupaceous or nutlike; seeds usually few, the embryo variously shaped, the endosperm scant or none.

A family of about 80 genera and 3000 species, almost entirely tropical distribution with 2 principal centers of concentration, tropical America for the berry-fruited Myrtoideae, and Australia for the capsular Leptospermoideae. *Eugenia*, once considered to have upwards of 800 species, is interpreted by some to be largely American with most of its Asiatic members assigned to *Syzygium* and a few smaller genera. The family is represented in this country by about a dozen indigenous species (mostly in Florida) that belong to *Eugenia* (including *Anamomis, Mosiera*) (600–9), *Psidium* (110–1), and *Calyptranthes* (80–2). One species of the Australian *Melaleuca* (100 spp.) is naturalized in peninsular Florida. *Psidium* (the guava) of the American tropics is naturalized pantropically, and in many places the species have become noxious weeds. *Eucalyptus* of Australia (600 spp.) contains some of the world's tallest trees.

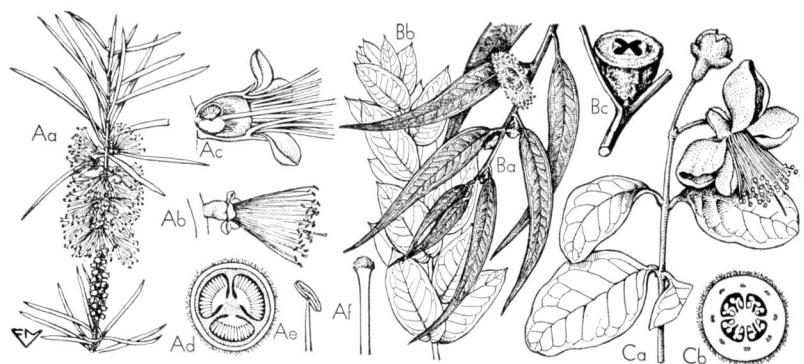

Fig. 239. MYRTACEAE. A, *Callistemon speciosus*: Aa, flowering branch, × ⅙; Ab, flower, × ½; Ac, flower, vertical section, stamens excised, × 1½; Ad, ovary, cross-section, × 4; Ae, anther, × 4; Af, end of style with stigma, × 4. B, *Eucalyptus Globulus*: Ba, flowering branch, × ⅙; Bb, branch with juvenile foliage, × ⅙; Bc, fruit, × ½. C, *Feijoa Sellowiana*: Ca, flowering branch, × ½; Cb, ovary, cross-section, × 3. (From L. H. Bailey, *Manual of cultivated plants*, The Macmillan Company, 1949. Copyright 1924 and 1949 by Liberty H. Bailey.)

The Myrtaceae are distinguished by the glandular-punctate leaves, the great number of stamens (due perhaps to splitting) whose anther connectives are often gland-tipped, the inferior ovary with axile or deeply intruding parietal placentae, and the seed testa modified into wings, or of varying textures as horny, leathery, or membranous.

Economically the family is of considerable importance throughout the world, but of limited importance in the United States. From the world viewpoint, it is important for the edible fruit of the guava (*Psidium*), the rose apple (*Syzygium Jambos*), the jaboticaba (*Myrciaria cauliflora*), the Surinam cherry (*Eugenia uniflora*), for spices such as the clove (dried flower buds of *Syzygium aromaticum*), the allspice (unripe berry of *Pimenta dioica*), oil of bay rum (*Pimenta racemosa*). Species of a number of genera are cultivated domestically as ornamentals in warmer areas, notably *Eucalyptus* (about 90 spp. cultivated in California), the bottle brushes (*Callistemon* and *Melaleuca*), the Brisbane box (*Tristania*), *Feijoa Sellowiana*, the Australian tea tree (*Leptospermum*), the lilly-pilly (*Acmena Smithii*), and the myrtle of classical literature (*Myrtus communis*).

LITERATURE:

ATCHISON, E. Chromosome numbers in the Myrtaceae. Amer. Journ. Bot. 34: 159–164, 1947.

# DIVISION IV. EMBRYOPHYTA SIPHONOGAMA

BERRY, E. W. The origin and distribution of the family Myrtaceae. Bot. Gaz. 59: 484-490, 1915.
BLAKELY, W. F. A key to the eucalypts, with descriptions of 500 species and 138 varieties. 339 pp. Sydney, 1934.
BURRET, M. Myrtaceenstudien. I, Notizbl. Bot. Gart. Berlin, 15: 479–550, 1941; II, Fedde Repert. Spec. Nov. 50: 50–60, 1941. [Pertinent to American genera.]
DADSWELL, H. E. and INGLE, H. D. The wood anatomy of the Myrtaceae I. A note on the genera *Eugenia, Syzygium, Acmena* and *Cleistocalyx*. Trop. Woods, 90: 1–7, 1947.
DIELS, L. Myrtaceae. Engl. Bot. Jahrb. 57: 356–426 (1920–22).
GAGNEPAIN, F. Classification des *Eugenia*. Bull. Soc. Bot. France, 64: 94–103, 1917.
HENDERSON, M. R. The genus *Eugenia* (Myrtaceae) in Malaya. The Garden's Bull., Singapore 12 (1): 1–293, 1949. [Rejects transfer of species from *Eugenia* to *Syzygium*.]
INGHAM, N. D. Eucalyptus in California. Calif. Agric. Exp. Sta. Bull. 196: 29–112, 1908.
MAIDEN, J. H. A critical revision of the genus *Eucalyptus*. 8 vols. Sidney, 1903–1929.
MERRILL, E. and PERRY, L. The Myrtaceae of China. Journ. Arnold Arb. 19: 191–247 (1938). [Includes notes on tropical American members.]
MUELLER, F. VON. Eucalyptographia, a descriptive Atlas of the Eucalypts of Australia and the adjoining islands. 10 vols. Melbourne, 1879–84.
NIEDENZU, F. Myrtaceae. In Engler and Prantl, Die natürlichen Pflanzenfamilien, III (7): 57–105, 1898.
SCHNEIDER, C. L. Natural establishment of *Eucalyptus* in California. Madroño, 10: 31, 32, 1948.
WALTHER, E. A key to the species of *Eucalyptus* grown in California. Proc. Cal. Acad. Sci. IV. 17: 67–87, 1928.

## MELASTOMACEAE.[172] MELASTOMA FAMILY

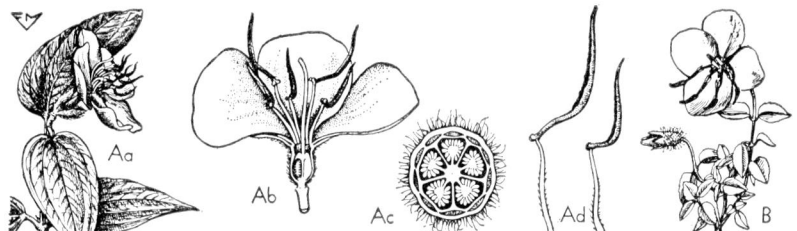

**Fig. 240.** MELASTOMACEAE. A, *Tibouchina semidecandra*: Aa, flowering branch, × ¼; Ab, flower, vertical section, × 1; Ac, ovary, cross-section, × 2; Ad, dimorphic stamens, × 1. B, *Schizocentron elegans*: flowering branch, × ½. (From L. H. Bailey, *Manual of cultivated plants,* The Macmillan Company, 1949. Copyright 1924 and 1949 by Liberty H. Bailey.)

Herbs, shrubs, or trees; erect, climbing, or epiphytic; leaves commonly opposite and decussate (one of a pair often smaller than the other), rarely alternate by abortion of one of a pair, simple, usually with 3–9 palmate more or less parallel veins closely and transversely anastomosing, oil glands lacking, estipulate; flowers bisexual, actinomorphic (slightly zygomorphic as to androecium), mostly 4–5-merous, cymose, the perianth biseriate, perigynous or epigynous, lacking any corona, the sepals 5 (3–6), valvate or connate into a calyptralike hood, the petals mostly 5, distinct, imbricate or convolute; the stamens mostly twice as many as

---

[172] The spelling Melastomataceae was used in the list of *nomina familiarum conservanda* of ed. 3 of the Rules, but (as noted therein by Weatherby) if Haloragaceae may be derived from *Haloragis,* the name Melastomaceae derived from *Melastoma* also seems acceptable.

petals in 2 whorls (rarely as many as petals by loss of 1 whorl and rarely more than twice as many), the filaments distinct, often geniculate, inflexed in bud, the anthers typically 2-celled (sometimes seemingly unithecate or 4-celled as in *Conostogia* and some spp. of *Miconia*), introrse, basifixed, each anther dehiscing usually by a single pore (rarely longitudinally or by a pore for each theca), connectives often variously appendaged; pistil 1, the ovary superior or more commonly inferior by adnation of hypanthium, locules and carpels usually 4–14 (rarely unilocular or carpels 2, 6 or more), the placentation axile (rarely seemingly basal), the ovules usually very numerous on each placenta, anatropous, the style and stigma 1 and simple; fruit a loculicidal capsule or berry; seeds with minute embryo and lacking endosperm.

A large pantropical family of about 150 genera and 4000 species of which about 3000 are American. In parts of Brazil, where the species are abundant, the family forms a characteristic component of the vegetation. It is represented in this country only in the east and southeast, where occur about 10 indigenous species of *Rhexia* along the Gulf and northward in decreasing numbers along the Atlantic coast to Maine, a second and West Indian genus *Tetrazygia* (16 spp.) is represented by a single species in southern Florida. The largest genera of the family include *Miconia* (900 or more), *Tibouchina* (200+), *Leandra* (200±), *Clidemia* (160±), *Medinilla* (160), *Memecylon* (130), and *Sonerila* (70).

The family is distinctive and readily identified by the leaf venation and the stamen morphology.

Economically the family is of little domestic importance. A few species of *Rhexia, Tibouchina, Heterocentron,* and *Medinella* are cultivated for ornament.

*LITERATURE:*

COGNIAUX, A. Mélastomacées. In de Candolle, Monographiae Phanerogamarum, 7: 1–1256, 1891.

GLEASON, H. A. The relationships of certain myrmecophilous Melastomes. Bull. Torrey Bot. Club, 58: 73–85, 1931.

KRASSER, F. Melastomataceae. In Engler and Prantl, Die natürlichen Pflanzenfamilien, III (7): 130–199, 1898.

[For recent treatments with keys, see those by Gleason in Flora of Surinam, Flora of Panama, etc.]

## TRAPACEAE.[173] WATER-CHESTNUT FAMILY

A unigeneric family (*Trapa*) of 3 Old World aquatic species (Flerov, 1925, recognized 11 species), of which 1 (*T. natans*) is naturalized to an increasing extent in this country. Many botanists have retained the genus in the Onagraceae, but *Trapa* differs from members of that family in that the plants are floating annuals whose rosulate leaves have inflated petioles, and the flowers are characterized by the absence of an inner whorl of stamens, the ovary is half inferior, always bilocular with a single pendulous ovule in each locule (1 of the 2 ovules often aborts), and the fruit is a 1-seeded top-shaped drupe whose fleshy pericarp is early deciduous and covers a large 2–4-horned stony endocarp (pyrene), the seed has 2 cotyledons markedly unequal in size (one very large, the other minute and scale-like). The horns of the characteristic pyrene (fruit of most authors) represent the persistent sepals and may be apically barbed.

Engler and Diels placed this family ahead of the Onagraceae in their schema, but evidence obtained from the morphology of the flower and fruit strongly indi-

---

[173] The oldest name for this family is Trapaceae, first validly published by Dumortier in 1829. [Analyse des famillies, pp. 36, 144.] The later name, Hydrocaryaceae (1893), has been used, and in addition to having been superfluous when published, there is no genus *Hydrocarya*.

cates it to be a taxon derived by reduction from the Onagraceae. In addition to the differences of gross morphology that separate the Hydrocaryaceae from the Onagraceae, there are ample embryological (see Just, 1946) and cytological data to support the segregation.

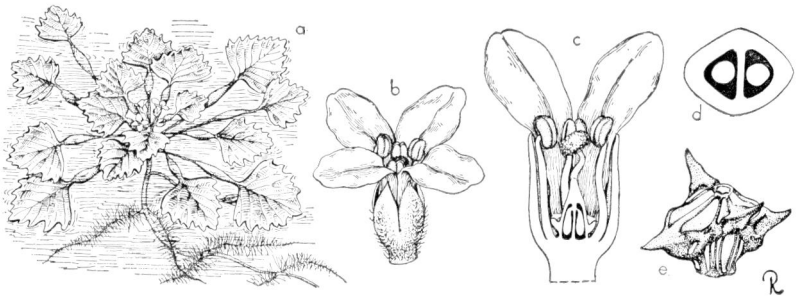

**Fig. 241.** HYDROCARYACEAE. *Trapa natans*: a, plant in flower, × ¼; b, flower, habit, × 2; c, same, vertical section, × 3; d, ovary, cross-section, × 6; e, fruit, less fleshy exocarp, × ½.

*LITERATURE:*

FLEROX, A. F. De genere *Trapa* L. Bull. Jard. Bot. Rép. Russe, 24: 13–45, 1925. [Text in Russian with German summary.]
GAMS, H. Die Gattung *Trapa* L. Pflanzenareale, 1: 39–41, 1927.
JUST, T. The use of embryological formulas in plant taxonomy. Bull. Torrey Bot. Club 73: 351–355, 1946.
RAIMANN, R. Hydrocaryaceae. In Engler and Prantl, Die natürlichen Pflanzenfamilien, III (7): 223–226, 1898.
VASIL'EV, V. N. Systematics and biology of the genus *Trapa* L. Sovet. Bot. 15: 343–345, 1947. [In Russian.]

## ONAGRACEAE.[174] EVENING-PRIMROSE FAMILY

Plants mostly herbs (occasionally aquatics), rarely shrubs (*Fuchsia*) or trees (*Hauya*); leaves alternate or opposite, simple, the stipules none or present and caducous (as in *Fuchsia, Circaea*); flowers solitary in axils or spicate to racemose (paniculate in *Fuchsia* spp.), bisexual, actinomorphic or infrequently zygomorphic (as in *Lopezia*), typically tetramerous, the perianth biseriate (corolla lacking in *Ludwigia* spp.) forming an hypanthium adnate to the ovary, the sepals (or calyx lobes) 4 or rarely 2–3–5 (6 in *Jussiaea* spp.), distinct or the hypanthium extending much above the ovary, valvate, persistent (as in *Jussiaea* and *Ludwigia*) or more commonly deciduous, the petals (or corolla lobes) mostly 4, sometimes 2 or more or small or none, mostly clawed, convolute or imbricate; stamens usually the same number as corolla lobes or in 2 whorls and twice as many (1 fertile stamen and 1 staminode in *Lopezia*), when biseriate those of outer whorl alternate with corolla lobes, distinct, arising from or near hypanthium rim, the anthers typically 2-celled, dehiscing longitudinally, sometimes each cell transversely divided into 2 (*Circaea*)

---

[174] The Rules of Nomenclature provide that a family name is taken from that of its type genus. The type genus of this family is *Oenothera* L. but the name Oenotheraceae Warming (1879) is antedated by Onagraceae Dumortier (1829). Engler and others adopted the name Oenotheraceae for the family, and a few have used Epilobiaceae Horaninov (1834). No name has been conserved to date for the family. For discussion of this nomenclatural situation, see Sprague in Journ. Bot. (Lond.) 60: 72, 1922.

or several (*Clarkia, et al.*) false cells by transverse plates, versatile or innate, the pollen grains of different genera of distinctive morphology; pistil 1, the ovary inferior, the locules and carpels typically 4 (rarely 2 or 5), the placentation axile (septa sometimes incomplete), the ovules 1–many on each placenta, mostly horizontal or ascending, anatropous, the style 1 and slender, the stigma usually capitate, sometimes notched (*Circaea*) or with radiate branches (*Oenothera* and *Epilobium* spp.); fruit usually a loculicidal capsule, a berry (*Fuchsia*), a bristly 1–2-seeded nutlet (*Circaea*) or a 1-seeded smooth nut (*Gaura*); seeds comose (*Epilobium, Zauschneria*) or glabrous, the endosperm lacking, the embryo straight or nearly so

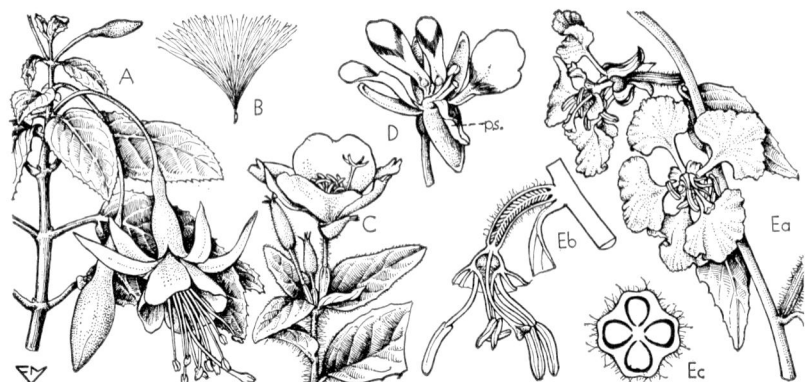

Fig. 242. ONAGRACEAE. A, *Fuchsia hybrida*: flowering branch, × ½. B, *Epilobium angustifolium*: comose seed, × ¼. C, *Oenothera pilosella*: flowering branch, × 1. D, *Lopezia coronata*: flower, × 1 (p.s. petaloid staminodium). E, *Clarkia elegans*: Ea, flowering branch, × ½ (note doubling of corolla); Eb, flower, vertical section, less perianth, × 1; Ec, ovary, cross-section, × 4. (From L. H. Bailey, *Manual of cultivated plants*, The Macmillan Company, 1949. Copyright 1924 and 1949 by Liberty H. Bailey.)

A family of about 20 genera and 650 species, of world-wide distribution, especially in temperate regions of the New World but also abundant in South America. Genera having species indigenous to this country (together with approximate number of total species) include *Oenothera* (including *Raimannia, Kneiffia, Hartmannia, Pachylophus, Sphaerostigma, Anogra, Lavauxia, Galpinsia*, and *Chylismia*) 200, *Epilobium* 200, *Jussiaea* 40, *Ludwigia* 30, *Gaura* 18, *Godetia* 15, *Lopezia* 14, *Boisduvalia* 10, *Gayophytum* 9, *Clarkia* 7, *Zauschneria* 5, *Hauya* about 8.

The Onagraceae are distinguished by the flowers having typically the numerical plan of 2 or 4, and the inferior multiovulate ovary terminated by an hypanthium from whose rim emerge sepals, petals, and stamens.

The family was included within the Myrtiflorae (or Myrtales) by Bessey, Wettstein, and Rendle. Hallier included it as one of the primitive genera in his Polygalines (an order he treated as derived from the Linaceae), and Hutchinson included it within his Lythrales (an order considered to have been derived from the Caryophyllales). It was subdivided by Raimann into 8 tribes, based primarily on fruit characters, of which the following are represented by genera native in this country: Jussieae, Epilobieae, Hauyeae, Onagreae (with 6 subtribes), Gaureae, Fuchsieae, and Circeae (members of the Lopezieae do not extend northward from Mexico into this country). Most American botanists have retained the genus *Trapa* in this family, but recent studies have shown it to be distinct (see Maheshwari, 1945) and it is segregated here as the Hydrocaryaceae.

DIVISION IV. EMBRYOPHYTA SIPHONOGAMA 639

Economically the family is not of major importance. Species of most genera are cultivated domestically as ornamentals. The predominantly tropical American genus *Fuchsia* (100 spp.) provides a wide assortment of ornamental shrubs grown in the open in warm regions.

*LITERATURE:*

BAEHNI, C. and BONNER, C. E. B. La vascularisation des fleurs chez les *Lopezieae* (Onagracées). Candollea, 11: 305–322, 1948.

———. La vascularisation du tube floral chez les Onagracées. Candollea, 12: 345–359, 1949.

BHADURI, P. N. Cytological studies in the genus *Gaura*. Ann. Bot. 5 (17): 1–14, 1941. [See also, *op. cit.* 6: 230–244, 1942.]

BONNER, C. E. B. The floral vascular supply in *Epilobium* and related genera. Candollea, 11: 277–303, 1948.

CLELAND, R. E. Cyto-taxonomic studies on certain Oenotheras from California. Proc. Amer. Philos. Soc. 75: 339–429, 1935.

———. Analysis of wild American races of *Oenothera* (*Onagra*). Genetics, 25: 636–644, 1940.

———. The problem of species in *Oenothera*. Amer. Nat. 78: 5–28, 1944.

DAVIS, B. M. The history of *Oenothera biennis*. Linnaeus, *Oenothera grandiflora* Solander, and *Oenothera Lamarckiana* of DeVries in England. Proc. Amer. Philos. Soc. 65: 349–378, 1926.

———. The segregation of sulfur and dwarf from crosses involving *Oenothera franciscana* and certain hybrid derivatives. Proc. Amer. Philos. Soc. 77: 99–160, 1937.

FERNALD, M. L. The identity of *Circaea latifolia* and the Asiatic *C. quadrisulcata*. Rhodora, 17: 222–224, 1915.

———. The identity of *Circaea canadensis* and *C. intermedia*. Rhodora, 19: 85–88, 1917.

———. The identities of *Epilobium lineare, E. densum* and *E. ciliatum*. Rhodora, 46: 377–386, 1944.

FOSTER, R. C. The rediscovery of *Riesenbachia* Presl. Contr. Gray Herb. 155: 60–62, 1945. [Shows flowers of the genus to have a corolla, contrary to earlier misconceptions.]

GATES, R. R. Early historico-botanical records of the Oenotheras. Proc. Iowa Acad. Sci. 17: 85–124, 1910.

———. Some phylogenetic considerations on the genus *Oenothera*, with descriptions of two new species. Journ. Linn. Soc. Lond. (Bot.) 49: 173–197, 1933.

HAUSSKNECHT, C. Monographie der Gattung *Epilobium*. 318 pp. Jena, 1884.

HILEND, M. A revision of the genus *Zauschneria*. Amer. Journ. Bot. 16: 58–68, 1929.

HITCHCOCK, C. L. Revision of the North American species of *Godetia*. Bot. Gaz. 89: 321–361, 1930.

JOHANSEN, D. A. A proposed phylogeny of the Onagraceae based primarily on number of chromosomes. Proc. Nat. Acad. Sci. 15: 882–885, 1929.

———. Studies on the morphology of the Onagraceae. VIII. *Circaea pacifica*. Amer. Journ. Bot. 21: 508–510, 1934.

LÉVEILLÉ, H. Monographie du genre *Oenothera*. Bull. Acad. Internat. Geogr. Bot. 17: 257–332, 1908.

———. Iconographie du genre *Epilobium*. 328 pp. Le Mans, 1910–11.

MAHESHWARI, P. The place of angiosperm embryology in research and teaching. Journ. Indian Bot. Soc. 24: 25–41, 1945.

MUNZ, P. A. Studies in Onagraceae. I, A revision of the subgenus *Chylismia* of the genus *Oenothera*. Amer. Journ. Bot. 15: 223–240, 1928; II, Revision of North American species of the subgenus *Sphaerostigma*, genus *Oenothera*, Bot. Gaz. 85: 233–270, 1928; III. Revision of the North American species of the subgenus *Taraxia* and *Eulobus* of the genus *Oenothera*. Amer. Journ. Bot. 16: 246–257, 1929; IV, Revision of the subgenera *Salpingia* and *Calyophis* of the genus *Oenothera*. *Op. cit.* 16: 702–715, 1929; V, The North American species of the subgenera *Lavauxia and Megapterium* of the genus *Oenothera*. *Op. cit.* 17: 358–370, 1930; VI, The subgenus *Anogra* of the genus *Oenothera*. *Op. cit.* 18: 309–327, 1931; VII, The subgenus *Pachylophus* of the genus *Oenothera*. *Op. cit.* 18: 728–738, 1931; VIII, The subgenera *Hartmannia* and *Gauropsis* of the genera *Oenothera* and *Gayophytum*. Op. cit. 19: 755–778, 1932; IX, The subgenus *Raimannia* of the genus *Oenothera*. Op. cit. 22: 645–663, 1933; X, The subgenus *Kneiffia* and miscellaneous other species of *Oenothera*. Bull. Torrey Bot. Club, 64: 287–306, 1937; XI, A revision of the genus *Gaura*. Op. cit. 65: 105–122, 211–228, 1939; XII, A revision of the New World species of *Jussiaea*. Darwiniana, 4: 179–284, 1942; XIII, The American species of *Ludwigia*. Bull. Torrey Bot. Club, 71: 152–165, 1944.

———. A revision of the genus *Boisduvalia* (Onagraceae). Darwiniana, 5: 124–152, 1941.
———. A revision of the genus *Fuchsia* (Onagraceae). Proc. Cal. Acad. 25: 1–138, 1943.
———. The *Oenothera Hookeri* group. El Aliso, 2: 1–47, 1949.
MUNZ, P. A. and HITCHCOCK, C. L. A study of the genus *Clarkia*, with special reference to its relationship to *Godetia*. Bull. Torrey Bot. Club, 56: 181–197, 1929.
RAIMANN, R. Onagraceae. In Engler and Prantl, Die natürlichen Pflanzenfamilien, III (7): 199–223, 1893.
SMITH, J. D., and ROSE, J. N. A monograph of the Hauyeae and Gongylocarpeae, tribes of the Onagraceae. Contr. U. S. Nat. Herb. 16: 287–298, 1913.
SPRAGUE, T. A. and RILEY, L. A. M. A recension of *Lopezia*. Journ. Bot. 62: 7–16, 1924.
———. Notes on *Raimannia* and allied genera. Kew Bull. 1921: 198–201.

## HALORAGACEAE.[175] WATER MILFOIL FAMILY

**Fig. 243.** HALORAGACEAE. *Myriophyllum exalbescens*: a, flowering branch, × 1; b, staminate flower, × 5; c, pistillate flower, × 15; d, same, vertical section, × 10. (From L. H. Bailey, *Manual of cultivated plants*, The Macmillan Company, 1949. Copyright 1924 and 1949 by Liberty H. Bailey.

Diverse appearing aquatic or terrestrial herbs, rarely suffrutescent; leaves opposite or alternate (often within the same genus), sometimes whorled, very variable in size, pectinate when submersed, estipulate or (in *Gunnera* spp.) the stipules reduced to scales or an ochreate sheath; flowers usually unisexual (the plants monoecious or sometimes polygamomonoecious), actinomorphic, often minute, mostly subtended by a pair of bracteoles, solitary or in axillary clusters or corymbose to paniculate, the perianth biseriate, uniseriate or none, adnate to ovary, when present, the calyx limb 2–4-lobed or none, the petals (when present) 2–4 and deciduous, distinct, usually larger than calyx lobes, imbricated or valvate; stamens typically 4 or 8, when in 2 whorls those of outer whorl opposite the petals, distinct, filaments short, the anthers basifixed, 2-celled, dehiscing longitudinally; pistil 1, the ovary inferior, the locules 1–4, the carpels 4, the placentation axile, each locule with a single pendulous anatropous ovule, or when unilocular the ovule solitary, parietal, and pendulous, the styles as many as locules, the stigmas often plumose; fruit a small nut or drupaceous, sometimes winged; seed with a straight cylindrical or obcordate embryo and with endosperm.

[175] This is the spelling adopted in Appendix II of ed. 3 of the Rules, as opposed to the older spellings of Halorrhagaceae or Haloragidaceae. For discussion see Sprague in Kew Bull. 1928, p. 354. The circumscription of the family in this text excludes *Hippuris*, here segregated as the Hippuridaceae.

A family of 8 genera and about 100 species, widely distributed on all continents. The largest genera include *Haloragis* with 60 spp. and *Gunnera* with 30 spp., both mostly of the southern hemisphere, and *Myriophyllum* 40 spp., cosmopolitan. It is represented in the United States by indigenous species of *Proserpinaca* (2 in the east), and *Myriophyllum* (1 species widespread *M. brasiliense*, 3 in the east, and 3 in the west).

The Haloragaceae are distinguished by the usually monoecious plants, the predominantly unisexual flowers, the stamens 4 or 8, and the inferior ovary with a single ovule in each locule. The aquatic members have pectinately pinnate submerged leaves and the terrestrial have usually very large leaves (2–7 ft across) arising from a short stout stem.

The family has been considered to be much reduced from (i.e., advanced over) the Onagraceae. Hallier rejected this view and included it in his Ranales as a derivative of the Nymphaeaceae. Hutchinson placed it as advanced over the Onagraceae in his Lythrales (including in it also *Hippuris*). Wettstein segregated *Gunnera* as a separate and more advanced family, the Gunneraceae, a view supported by the floral morphology, embryology, and gross morphology.

Economically the plants are of little importance. Water milfoil (*Myriophyllum*) is of importance in limnological conservation practices and as an aquarium subject. Gunneras are gigantic herbs cultivated in warm areas to a limited extent for foliage effects.

*LITERATURE:*

BALDWIN, J. T. JR. Chromosomes of *Proserpinaca*. Rhodora, 41: 584, 1939.
FERNALD, M. L. and GRISCOM, L. *Proserpinaca palustris* and its varieties. Rhodora, 37: 177–178, 1935.
KNUPP, N. D. The flowers of *Myriophyllum spicatum* L. Proc. Iowa Acad. Sci. 18: 61–73, 1911.
PETERSEN, O. G. Halorrhagidaceae. In Engler and Prantl, Die natürlichen Pflanzenfamilien, III (7): 226–237, 1889. [Incl. Hippuris.]
SCHINDLER, A. K. Halorrhagaceae. In Engler, Das Pflanzenreich, 23 (IV. 225): 1–133, 1905.
SEIDELIN, A. Hippuridaceae, Halorrhagidaceae and Callitrichaceae. Meddelser øm Grönland, 36: 297–332, 1910.

## HIPPURIDACEAE. HIPPURIS FAMILY

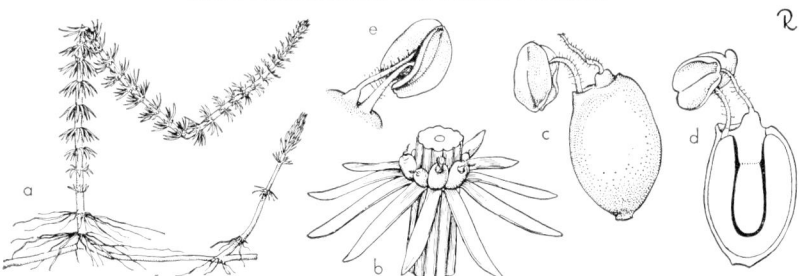

**Fig. 244.** HIPPURIDACEAE. *Hippuris vulgaris*: a, plant (stem normally erect), × ¼; b, node with flowers in leaf axils, × 2; c, flower, habit, × 12; d, same, vertical section, × 12; e, top of flower showing style between anther cells, × 15.

A family recognized by Engler and Diels (following Schindler, 1904) as composed of the single submerged aquatic species, *Hippuris vulgaris*. Plants of the family occur in the cool temperate regions of both the northern and southern hemispheres of the New World (naturalized in parts of the Old World). Most

American botanists treat the genus as composed of 3 species, and include it within the Haloragaceae. The description of the latter family as given above is circumscribed *sensu* Engler and Diels and does not admit *Hippuris*.

The family is similar to the Haloragaceae only in occupying an aquatic habitat. It differs in the leaves linear, entire, mostly in whorls of 6 or more, the submerged ones longer and more flaccid than the aerial, the stems sympodial (not monopodial), the flowers mostly bisexual, the perianth none, the androecium reduced to a single stamen, the ovary unilocular and the single pendulous ovule, the style 1 and filiform, stigmatic down one side, situated in a groove formed by the lobes of the single anther.

Hallier followed Bentham and Hooker and included both *Hippuris* and *Callitriche* (of Callitrichaceae) in the Haloragaceae, but placed the family in his Ranales and considered it to have been derived from the Nymphaeaceae; Bessey accepted the Hippuridaceae as distinct from Haloragaceae, as also did Wettstein and Rendle. Hutchinson included *Hippuris* within the Haloragaceae but retained *Callitriche* in a family by itself. The evidence seems to justify recognition of the Hippuridaceae as a family distinct from the Haloragaceae.

*LITERATURE:*

JUEL, H. O. Studien über die Entwicklungsgeschichte von Hippuris vulgaris. Nova Acta R. Soc. Upsaliensis, Ser. 4, 2(11), 1911.

SCHINDLER, A. K. Die Abtrennung der Hippuridaceen von den Halorrhagaceen. Engler's Bot. Jahrb. 34. Beibl. 77, 1904.

## Order 33. UMBELLIFLORAE [176]

The Umbelliflorae are characterized by a tendency of the flowers to be in determinate umbels (simple or compound), by the simplification (by reduction) of floral parts, by epigyny, and by the carpels reduced to 2, and each uniovulate.

The order was treated by Engler and associates to contain 3 families: Araliaceae; Umbelliferae; Cornaceae.

This circumscription of the order was adopted by Rendle. It was accepted by Wettstein, who amplified it to include the Garryaceae, and whose views were followed by Bessey except that the latter included both *Garrya* and *Nyssa* within his concept of the Cornaceae. Hallier recognized only 2 families, the Cornaceae and the Umbelliferae (the Araliaceae included in the Umbelliferae). Hutchinson (1948) revised his earlier views (1926), and restricted his Umbellales to include only the Umbelliferae, with the Cornaceae included in his Cunoniales (the latter a derivative of his Rosales) and presumed to be descended from the Philadelphaceae or from stocks ancestral to them. He treated the Araliaceae as a separate order, the Araliales, derived from the Cornaceae. Rickett (1945) recognized 2 orders: the Umbellales in which he placed the Araliaceae and Umbelliferae, and the Cornales which contained the Cornaceae and the Nyssaceae. If the latter view is accepted, the Old World Alangiaceae would also be placed in the Cornales.

## ARALIACEAE.[177] GINSENG FAMILY

Herbs, shrubs, or trees, sometimes lianous, occasionally dioecious; stems solid, pithy, often prickly; leaves alternate (rarely opposite), simple or more commonly palmately, ternately, or pinnately compound or decompound, the hairs often

[176] The name Umbelliflorae was used by Eichler and adopted by Engler and associates. Bessey, in an effort to terminate all orders uniformly with the ending -ales adopted Umbellales and was followed in this by Hallier, Hutchinson, and Rickett.

[177] The name Araliaceae Vent. (1799) has been conserved over Hederaceae Gisecke (1792) and other names for this family.

stellate, the stipules usually present and modified as a membranous border of petiole base or liguliform; flowers bisexual, or more commonly unisexual (the plants then dioecious or polygamodioecious), actinomorphic, minutely bracteate, small and often greenish, in heads or umbels (the latter sometimes racemose, corymbose, umbellate, or paniculate); perianth biseriate and mostly pentamerous, the calyx cupuliform or inconspicuous, adnate to ovary and represented usually by 5 minute teeth or a seamlike rim, the petals 5–10 (rarely 3), broad at base, arising from disc, caducous (falling separately or as a calyptrate cap), usually valvate; stamens usually as many as petals and alternate with them (rarely numerous), distinct, arising from disc, the anthers 2-celled, didymous, dorsifixed, dehiscing longitudinally; nectiferous epigynous disc present, covering ovary top and usually confluent with style bases; pistil 1, the ovary inferior, usually with 2–15 locules and carpels (sometimes only 1), the placentation axile, the ovule solitary in each locule, pendulous, anatropous, the raphe ventral, the styles as many as carpels and distinct and recurved or connate into a single cone or column, sometimes absent and the stigmas then sessile; fruit a berry (rarely a drupe), the ovary locules sometimes separating in fruit as mericarps or pyrenes; seeds with a small embryo in one end, the endosperm copious, sometimes ruminate.

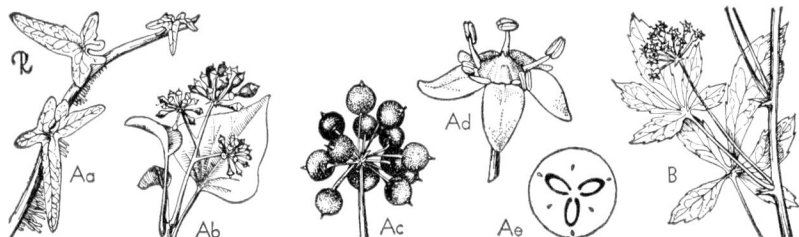

**Fig. 245.** ARALIACEAE. A, *Hedera Helix* var. *pedata*: Aa, sterile stem with leaves and aërial roots, × ½; Ab, adult foliage with young fruiting inflorescence, × ¼; Ac, fruit, × ½; Ad, flower habit, × 2; Ae, ovary, cross-section, × 4. B, *Acanthopanax Sieboldianus*: flowering branch, × ½. (From L. H. Bailey, *Manual of cultivated plants*, The Macmillan Company, 1949. Copyright 1924 and 1949 by Liberty H. Bailey.)

A family of about 65 genera and more than 800 species, primarily tropical. The 2 great centers of distribution are the Indo-Malayan region and tropical America, each mostly with genera peculiar to themselves. The family is represented in the United States by the following 3 genera (together with number of total and indigenous species); *Aralia* (over 30–6), *Panax* (6–2), *Oplopanax* (*Echinopanax*) (1). *Hedera Helix* (indigenous to Europe) is naturalized in several areas from Maryland southward.

The Araliaceae are distinguished by the usually umbellate inflorescence, the typically 5-merous flower, the inferior ovary with each locule uniovulate, and the fruit a berry.

Economically the family is of little domestic importance. Many variants of English ivy (*Hedera Helix*) are cultivated as evergreen vines climbing by aerial roots; 6 other species also are cultivated similarly. Species of a number of genera are grown as ornamental shrubs or small trees, notably of *Acanthopanax*, *Aralia*, *Polyscias*, and *Kalopanax*. The pith of the rice-paper plant (*Tetrapanax papyriferus*) is the source of Chinese rice paper, and the plant is grown here as a novelty. Ginseng roots, used in medicinal preparations, are obtained from *Panax quinquefolius*.

*LITERATURE:*

HARMS, H. Araliaceae. In Engler and Prantl, Die natürlichen Pflanzenfamilien, III (8): 1–62, 1894.
LI, HUI-LIN. The Araliaceae of China. Sargentia, 2: 1–134, 1942.
SMITH, A. C. Araliaceae. North Amer. Flora, 28B: 3–41, 1944.

## UMBELLIFERAE.[178] CARROT FAMILY

Plants mostly biennial or perennial herbs, occasionally suffrutescent, rarely shrubs; stems usually with a large pith that shrinks or dries at maturity with the internode becoming hollow; leaves alternate (rarely opposite, as in *Apiastrum* spp.), often basal, sometimes heteromorphic, usually pinnately or palmately compound (or decompound), infrequently simple, the petioles usually sheathing; inflorescence an interdeterminate umbel (the pedicels lost by reduction and the inflorescence capitate in a few genera), sometimes simple (the pedicels then termed primary rays) or more commonly compound (each primary ray, or peduncle, terminated by an umblet whose pedicels are termed secondary rays), the umbel often subtended by an involucre of distinct simple or compound bracts (the umblets often by an involucel of bractlets), sometimes caducous, persistent in fruit or deciduous; flowers bisexual (sometimes unisexual: the latest umblets staminate in many genera, polygamomonoecism exists when a few longer-pedicelled staminate flowers occur with a majority of bisexual ones as in *Astrantia,* all unisexual and the plant monoecious as in *Echinophora* of Eurasia or the plants dioecious as in *Arctopus* of Africa); the flowers actinomorphic, the perianth biseriate, the calyx adnate to the ovary and only the 5 lobes distinct and often reduced to teeth, or obsolete enations, the corolla of 5 distinct petals, usually apically inflexed; stamens 5, inflexed in bud but ultimately spreading, alternate with petals, arising from an epigynous disc, the anthers 2-celled, basi- or dorsifixed, dehiscing longitudinally; pistil 1, the ovary inferior, bilocular, 2-carpelled, the placentation axile, the ovule solitary in each locule, anatropous, pendulous, the raphe ventral and integument one, the styles 2, often swollen and spreading at base (the *stylopodium*), the stigmatic tip of each style scarcely differentiated; fruit a schizocarp composed of 2 mericarps coherent and dehiscing by their faces (commissure), flattened dorsally (parallel to the commissural face) or laterally (at right angles to the commissural face) or the schizocarp terete, each mericarp with typically 5 primary ribs (rarely intercalated by 4 secondary ribs) distinguished as lateral (1 down or adjoining each margin) dorsal (the median rib) and intermediate (1 between each lateral and the dorsal), the ribs thin or corky and filiform to winged (sometimes spinescent), oil tubes (*vittae*) present or obsolete in the intercostal spaces (the intervals between the ribs), or under the ribs, and on the commissural surface, each mericarp 1-seeded and usually suspended after dehiscence by a slender wiry stalk or carpophore;[179] seed with minute embryo, the endosperm firm but watery-fleshy and abundant.

A family of about 200 genera and 2900 species, mostly of the northern hemisphere, and represented in this country by 75 genera and 380 species of which 51 genera have a total of 318 indigenous species (24 genera are represented domes-

---

[178] Some authors, as allowed by Art. 23 of the Rules (ed. 3), reject the name Umbelliferae for this family and substitute for it an alternative name, Apiaceae Lindl. (1836) or Ammiaceae Presl *ex* Britton and Brown (1913).

[179] In most works it has been implied that the carpophore was the product of the splitting of a prolongation of the receptacle or floral axis. However, it was shown (Jackson, 1933) that only a small basal portion is axial and the remainder is appendicular and formed from the "ventral portion of the two carpels." The carpophore is not, for the most part, a separate structure morphologically distinct from the carpels but is an integral part which, in ripe fruit, often becomes free from the rest of each mericarp and appears as a special organ.

**Fig. 246.** UMBELLIFERAE. A, *Apium graveolens*: Aa, inflorescence and leaf-blade, × ½; Ab, fruit, side and face view, × 10. B, *Eryngium planum*: Ba, flowering branch, × ½; Bb, flower, × 3. C, *Trachymene caerulea*: Ca, flowering branch, × ½; Cb, flower, × 3; Cc, same, vertical section, × 3; Cd, ovary, cross-section, × 10. D, *Foeniculum vulgare*: Da, fruit, × 3; Db, same, cross-section, × 6. E, *Levisticum officinale*: Ea, inflorescence and leaf-blade, × ½; Eb, fruit, face view, × 3; Ec, same, side view, × 3. F, *Daucus Carota*: Fa, inflorescence and leaf-blade, × ½; Fb, umblet, × 4; Fc, fruit, side and face view, × 4. (From L. H. Bailey, *Manual of cultivated plants,* The Macmillan Company, 1949. Copyright 1924 and 1949 by Liberty H. Bailey.)

tically by about 62 adventive or naturalized species). Species of the family are distributed over much of the country although very few are aquatics. Nearly half of the native genera occur only in the western part of the country. The largest genera (together with number of total and indigenous species) include *Pimpinella* (200), *Eryngium* (200-28), *Azorella* (100), *Hydrocotyle* (80-5), *Bupleurum* (75-1). Native genera widespread in this country include *Sanicula* (17 spp.), *Cicuta* (7 spp.), *Thaspium* (3 spp.), and *Eryngium* (28 spp.).

The Umbelliferae are distinguished by the aromatic herbage, the usually sheathing petioles, the typically umbellate inflorescence, the pentamerous perianth and androecium, a 2-carpellate bilocular inferior ovary, and the fruit a schizocarp.

The family is accepted generally as a natural taxon readily separated from related families by the character of the fruit. It has been considered to be an advanced family in the order and, because of its fruit and prevalently herbaceous habit, to be more highly developed than the Araliaceae.

Economically the family is important for products used as food, condiments, and ornamentals. Some members possess resins or alkaloids in lethally poisonous quantities when the parts (especially roots and fruits) are eaten. Others are cultivated domestically for ornament. Among these economically important plants are: for food, carrot (*Daucus*), parsnip (*Pastinaca*), celery (*Apium*), and parsley (*Petroselinum*); for flavoring, anise (*Pimpinella*), caraway (*Carum*), dill (*Anethum*), chervil (*Anthriscus*), fennel (*Foeniculum*), lovage (*Levisticum*); as poisonous plants, water hemlock (*Cicuta*), poison hemlock (*Conium*), fool's parsley (*Aethusa*); for ornament, blue laceflower or didiscus (*Trachymene*), goutweed (*Aegopodium*), angelica (*Angelica*), sea holly (*Eryngium*), and cow parsnip (*Heracleum*).

*LITERATURE:*

BAUMANN, M. G. *Myodocarpus* and die Phylogenie der Umbelliferen-Frucht. Umbelli-floren-Studien I. Schweiz. Bot. Gesell. Ber. 56: 13–112, 1946.

COULTER, J. M. and ROSE, J. N. Monograph of the North American Umbelliferae. Contr. U. S. Nat. Herb. 7: 9–256, 1900. [For Supplement, see *op. cit.* 12: 441–451, 1909.]

DRUDE, O. Umbelliferae. In Engler and Prantl, Die natürlichen Pflanzenfamilien, III (8): 63–250, 1897–98.

HARDIN, E. The flowering and fruiting habits of *Lomatium*. State Coll. Washington Res. Stud. 1: 15–27, 1929.

HÅKANSSON, A. Studien über die Entwicklungsgeschichte der Umbelliferen. Lunds Univ. Årsskr. II(2): 18 (7): 1–120, 1923.

JACKSON, G. A study of the carpophore of the Umbelliferae. Amer. Journ. Bot. 20: 121–144, 1933.

MATHIAS, M. E. Monograph of *Cymopterus*, including a critical study of related genera. Ann. Mo. Bot. Gard. 17: 213–474, 1928.

———. The genus *Hydrocotyle* in northern South America. Brittonia, 2: 201–237, 1936.

———. A revision of the genus *Lomatium*. Ann. Mo. Bot. Gard. 25: 225–297, 1938.

MATHIAS, M. E. and CONSTANCE, L. A synopsis of the North American species of *Eryngium*. Amer. Midl. Nat. 25: 361–387, 1941.

———. New combinations and new names in Umbelliferae. I–II. Bull. Torrey Bot. Club, 68: 121–124, 1941; *op. cit.* 69: 244–248, 1942.

———. New North American Umbelliferae. I–II. Bull. Torrey Bot. Club, 69: 151–155, 1942; *op. cit.* 70: 58–60, 1943.

———. A synopsis of the American species of *Cicuta*. Madroño, 6: 145–151, 1942.

———. Umbelliferae. North Amer. Flora, 28B (1): 43–160, 1944; 161–295, 1945.

MURLEY, M. R. Fruit key to the Umbelliferae in Iowa, with plant distribution records. Iowa State Coll. Journ. Sci. 20: 349–364, 1946.

WOLFF, H. Umbelliferae-Apioideae-Bupleurum, Trinia et reliquae Ammineae heteroclitae. In Engler, Das Pflanzenreich, 43 (IV. 228): 1–214, 1910; Umbelliferae-Saniculoideae. *Op. cit.* 61 (IV. 228): 1–305, 1913; Umbelliferae-Apioideae-Ammineae-Carinae. Ammineae novemjugatae et genuinae. *Op. cit.* 90 (IV. 228): 1–398, 1927.

## CORNACEAE. DOGWOOD FAMILY

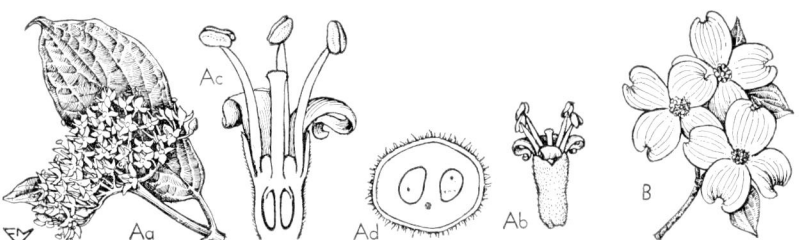

Fig. 247. CORNACEAE. *Cornus stolonifera*: Aa, inflorescence, × ½; Ab, flower, × 2; Ac, flower, vertical section, × 4; Ad, ovary, cross-section, × 8. B, *C. florida*: inflorescences with bracts, × ¼. (From L. H. Bailey, *Manual of cultivated plants*, The Macmillan Company, 1949. Copyright 1924 and 1949 by Liberty H. Bailey.)

Trees, shrubs, or suffrutescent subshrubs (woody lianes in *Griselinia* spp.); leaves opposite or less commonly alternate, simple, sometimes persistent (*Aucuba*, et al.), usually petiolate, estipulate (branched-ciliate stipules in *Helwingia*); flowers small, bisexual or unisexual (the plant then usually monoecious or polygamodioecious, dioecious in *Aucuba*), actinomorphic, in cymes or panicles (rarely racemes), sometimes with large showy foliaceous bracts (as in *Cornus florida*, *C. canadensis*, et al.), the perianth biseriate, adnate to ovary, the sepals 4–5, the petals 4–5 (sometimes absent), distinct, mostly valvate; stamens of same number as petals and alternate with them, distinct, the filaments short, the anthers 2-celled, dehiscing laterally (rarely introrsely), basifixed or dorsifixed; pistil 1, the ovary inferior, the locules and carpels 2–4 (rarely the locule 1), the placentation typically axile (parietal in *Aucuba*), the ovule 1 in each locule, anatropous, pendulous, with 1 integument; the style 1 or several, arising from an epigynous glandular ring or disc, the stigmas as many as styles and subcapitate; fruit typically a drupe, sometimes a berry (*Aucuba*, *Griselinia*) or rarely a syncarp; seed with small embryo in copious endosperm.

A family of wide distribution, of about 10 genera and 90 species of tropical and temperate plants. Most species occur in the temperate regions of North America and Asia, but others occur in tropics and subtropics of South America and New Zealand (*Griselinia* 6 spp.), in New Zealand (*Corokia* 3 spp.), Indo-Malaysia (*Mastixia* 25 spp.), Madagascar (*Kaliphora* with 1 and *Melanophylla* with 3 spp.), continental South Africa (*Curtisia* 1 spp.), and temperate Asia (*Aucuba* 3, *Helwingia* 3, and *Torricellia* 3 spp.). *Cornus* (about 45 spp.) is chiefly an Asiatic and American genus. The family is represented in the United States by 15 indigenous species of *Cornus*.

The Cornaceae are distinguished as woody or subligneous plants, with the perianth and androecium 4–5-merous, the ovary inferior, and the fruit a fleshy indehiscent drupe or berry.

The family has been included in the Umbelliflorae by most botanists except Hutchinson (1948) who transferred them, on the basis of their woody habit and stem anatomy, to his Cunoniales as derivatives of his Philadelphaceae and "particularly [from] the genus *Broussiasia* from the Pacific Islands." He pointed out also that the Cornaceae and Araliaceae differed from the Umbelliferae in the petals commonly valvate rather than imbricate in bud. Several authorities have interpreted the American species of *Cornus* as belonging to 4 separate genera: the large bracted arborescent species (sect. Cynoxylon), represented by *C. florida*, as *Benthamidia*; the large bracted suffrutescent or seemingly herbaceous species (sect. Arctocrania), represented by *C. canadensis*, as *Chamaepericlymenum*; the nonbracteate species with flowers in capitate heads that appear before the leaves (sect.

Tanycrania), represented by *Cornus mas* (an exotic species), and *C. sessilis,* as *Macrocarpium*; and the remainder retained in *Cornus.* This view has been rejected by Rickett (1942, 1945). Dermen (1932) concluded from cytological evidence that *Cornus mas* was the primitive type of the genus, and that the *C. florida* and *C. alternifolia* lines were evolved from it as parallel derivatives. He also concluded that the Nyssaceae were derivatives from the Cornaceae, a view now accepted by most botanists.

Members of the family are of domestic importance chiefly as ornamentals and for which use species of *Aucuba, Cornus, Corokia, Griselinia,* and *Helwingia* are cultivated. Fruits of the cornelian cherry (*Cornus mas*) are edible.

*LITERATURE:*

DERMEN, H. Cytological studies of *Cornus.* Journ. Arnold Arb. 13: 410–415, 1932.
HARA, HIROSHI. The nomenclature of the flowering dogwood and its allies. Journ. Arnold Arb. 29: 111–115, 1948.
HORNE, A. S. The polyphyletic origin of the Cornaceae. Proc. Brit. Assoc. Adv. Sci. Portsmouth, p. 585, 1911.
HUTCHINSON, J. Neglected generic characters in the family Cornaceae. Annals of Bot. 6: 83–93, 1942.
RICKETT, H. W. *Cornus amomum* and *Cornus candidissima.* Rhodora, 36: 269–274, 1934.
———. The names of *Cornus.* Torreya, 42: 11–14, 1942.
———. *Cornus asperifolia* and its relatives. Amer. Midl. Nat. 27: 259–261, 1942.
———. *Cornus stolonifera* and *Cornus occidentalis.* Brittonia, 5: 149–159, 1944.
WANGERIN, W. Cornaceae. In Engler, Das Pflanzenreich, 41 (IV. 229): 1–110, 1910.
WILKINSON, A. M. Floral anatomy of some species of *Cornus.* Bull. Torrey Bot. Club, 71: 276–301, 1944.

## Order 34. DIAPENSIALES

An order composed of the single family Diapensiaceae, closely allied to the Ericales and distinguished from them by the pollen never in tetrads, the gynoecium tricarpellate (true also of Clethraceae, of the Ericales), the stamens basically in 2 whorls with 1 whorl reduced to staminodes, the integuments 2 (basally connate), and in differences in embryology.

The family was included within the Ericales until 1924, when it was separated as a distinct order by Engler and Gilg. Prior to that time it was treated by Engler and associates as advanced over the Ericaceae. Hallier included it in his Bicornes as the most advanced taxon of that order, a view also accepted in principle by Wettstein. Diels considered the family to have been derived from saxifragaceous ancestors (or their relatives) in the Rosales.

## DIAPENSIACEAE.[180] DIAPENSIA FAMILY

Low-growing evergreen shrubs, sometimes suffrutescent herbs (*Galax*); leaves alternate, somewhat imbricated or reniform and petiolate, simple, seemingly estipulate; flowers bisexual, actinomorphic, hypogynous, the perianth biseriate, 5-merous, the calyx 5-lobed or the sepals distinct, imbricated, persistent; corolla gamopetalous, the lobes 5, imbricated; no disc present; stamens 5, epipetalous and alternate with the lobes or hypogynous, distinct or basally connate in a ring, often alternating with staminodes, the anthers usually 2-celled, dehiscing longitudinally or (in *Pyxidanthera*) transversely; pistil 1, the ovary superior, the locules and carpels 3, the placentation axile, the ovules few to numerous, anatropous or amphitropous, the style 1, the stigma 3-lobed; fruit a loculicidal 3-valved capsule; seeds with straight or slightly curved embryo and an abundant endosperm.

A family of 6 genera and 10 species of the cooler and arctic regions of the

[180] The name Diapensiaceae has been conserved over other names for this family.

northern hemisphere. The family occurs in the New World mostly in the eastern mountainous regions of this country where it is represented by indigenous species of 4 genera (all but the first are monotypic endemics): *Diapensia lapponica* of circumpolar distribution extends southward into high elevations of the White Mountains of New England and the Adirondacks of New York; *Shortia galacifolia,* rare, in the mountains of the southeast; *Pyxidanthera barbulata,* in sandy pine barrens of New Jersey to North Carolina; and *Galax aphylla* in open woods from Virginia to Georgia.

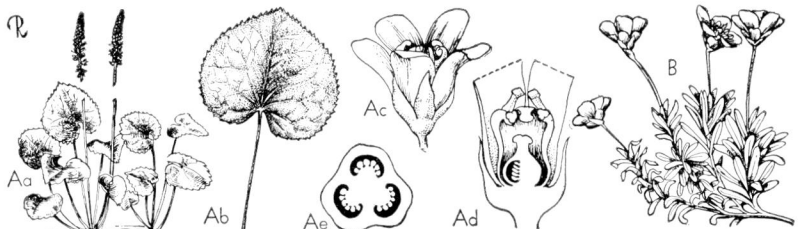

Fig. 248. DIAPENSIACEAE. A, *Galax aphylla*: Aa, habit in flower, × ⅛; Ab, basal leaf, × ¼; Ac, flower, × 2; Ad, same, vertical section, × 3; Ae, ovary, cross-section, × 8. B, *Diapensia lapponica*: flowering branch, × ½. (From L. H. Bailey, *Manual of cultivated plants,* The Macmillan Company, 1949. Copyright 1924 and 1949 by Liberty H. Bailey.)

The Diapensiaceae are distinguished by the combination of a 5-merous perianth and androecium and trilocular ovary, the adnation of the stamens on the corolla, the pollen not in tetrads, and by the absence of a disc.

Economically the members of the family are of little importance. Leaves of *Galax aphylla* are collected in the wild and preserved for decorative uses in the florist industry. Plants of *Diapensia, Pyxidanthera,* and *Shortia* are infrequently cultivated for ornamental purposes.

*LITERATURE:*

BALDWIN, J. T. JR. *Galax*: The genus and its chromosomes. Journ. Hered. 32: 249–254, 1941.
DIELS, L. Diapensiaceen-Studien. Bot. Jahrb. 50 (Suppl.): 304–330, 1914.
DRUDE, O. Diapensiaceae. In Engler and Prantl, Die natürlichen Pflanzenfamilien, IV (1): 80–84, 1889.
EVANS, W. E. A revision of the genus *Diapensia,* with special reference to the Sino-himalayan species. Notes, R. B. G., Edinb., 15: 209–236, 1927.
JENKINS, C. F. Asa Gray and his quest for *Shortia galacifolia.* Arnoldia, 2: 13–28, 1942.
PRINCE, A. E. *Shortia galacifolia* in its type locality. Rhodora, 49: 159–161, 1947.

## Order 35. ERICALES [181]

The Ericales are characterized by the flowers generally pentamerous, the petals distinct and free (in the primitive taxa) or more commonly basally connate, the stamens obdiplostemonous and inserted at the edge of a usually hypogynous nectiferous disc, pollen often in tetrads, placentation axile, the ovules numerous and each with a single integument.

The order was divided by Engler and Diels into 2 suborders, and treated as composed of 4 families, as follows:

        Ericineae                     Epacridineae
            Clethraceae                Epacridaceae
            Pyrolaceae
            Ericaceae

[181] Sometimes known also as the Bicornes.

Various authors (including Rydberg in North American Flora) have included the Lennoaceae in this order but, for reasons given in the discussion accompanying it, that family has been transferred to the Tubiflorae, immediately preceding the Boraginaceae.

## CLETHRACEAE. PEPPERBUSH FAMILY

**Fig. 249.** CLETHRACEAE. *Clethra alnifolia*: a, flowering branch, × ½; b, flower, × 2; c, petal with stamens, × 2; d, anthers, × 10; e, flower, vertical section, × 4; f, ovary, cross-section, × 8; g, capsule, × 2. (From L. H. Bailey, *Manual of cultivated plants*, The Macmillan Company, 1949. Copyright 1924 and 1949 by Liberty H. Bailey.)

Tall shrubs or low trees (*Schizocardia* a large tree); leaves alternate, simple, estipulate; flowers bisexual, actinomorphic, hypogynous, very fragrant or spicily aromatic, in terminal racemes or panicles (axillary in *Schizocardia*), the calyx 5-lobed, imbricate, persistent around the fruit, the corolla polypetalous, saucer-shaped, the petals 5 and distinct; no disc present; stamens 10 (12), in 2 whorls, distinct, hypogynous, the filaments pubescent or glabrous, the anthers 2-celled, extrorse, sagittate, inverted or inflexed in bud, dehiscing by apical pores, the pollen in single grains; pistil 1, the ovary superior, the locules and carpels 3 (5 in *Schizocardia*), mostly 3-lobed, the placentation axile, the ovules numerous, anatropous, borne on placental intrusions, the style 1 and stigma 3-lobed; fruit a 3-valved loculicidal capsule whose septa separate from the central column; seeds trigonous or flattened, sometimes winged, the embryo short, cyclindrical, the endosperm fleshy.

Two genera are in the family: *Clethra* with about 30 species, mostly American or paleotropic (a few are Asiatic and some occur in the Madeira Islands), extending from Maine to Brazil, and the arborescent monotypic *Schizocardia*, native in British Honduras. *Clethra* is represented in this country by 3 indigenous species: *C. alnifolia* along the coastal plain, *C. acuminata* in the mountains of Virginia to Georgia, and *C. tomentosa* on the coastal plain from North Carolina to Alabama.

The Clethraceae are distinguished from other families of the order by the polypetalous corolla, the temporarily inverted anthers (in bud), the pollen grains single and not in tetrads, and the 3-locular ovary (5-loculed in *Schizocardia*).

The opinion of Asa Gray, that the family represented only a tribe of the Ericaceae, was followed by Bentham and Hooker and many American botanists. Klotzsch's conclusion that the taxon deserved recognition as a family was accepted by Drude and subsequently by most phylogenists. Hallier treated it as the primitive taxon of his Bicornes and derived it from the Ochnaceae, as did also Wettstein. Bessey considered it the primitive family of his Ericales.

A few species of sweet pepperbush or white alder (*Clethra*) are cultivated as ornamentals in many parts of the country and are prized for their fragrant summer bloom.

DIVISION IV. EMBRYOPHYTA SIPHONOGAMA 651

LITERATURE:
BRITTON, N. L. Clethra. North Amer. Flora, 29: 3–9, 1914.
DRUDE, O. Clethraceae. In Engler and Prantl, Die natürlichen Pflanzenfamilien, IV (1): 1–2, 1889.
SMITH, A. C. and STANDLEY, P. C. Schizocardia, a new genus of trees of the family Clethraceae. Trop. Woods, 32: 8–11, 1932.

## PYROLACEAE.[182] PYROLA FAMILY

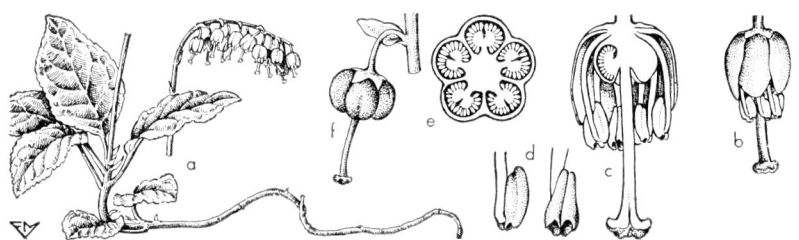

Fig. 250. PYROLACEAE. Ramischia secunda: a, flowering plant, × ½; b, flower, × 2; c, same, vertical section, × 3; d, anthers, side and ventral view, × 5; e, ovary, cross-section, × 5; f, capsule, × 2. (From L. H. Bailey, Manual of cultivated plants, The Macmillan Company, 1949. Copyright 1924 and 1949 by Liberty H. Bailey.)

Perennial herbs with creeping scaly rootstocks or chlorophyll-less fleshy saprophytes, sometimes briefly suffrutescent; leaves alternate, nearly opposite or in false whorls, simple, foliaceous or scalelike, persistent or deciduous, mostly coriaceous and toothed, estipulate; flowers bisexual, actinomorphic or nearly so, bracteate, the calyx of 5 distinct or briefly connate persistent sepals, the corolla waxy, the petals 4–5, distinct, hypogynous; disc present or absent; stamens 8–10, distinct, hypogynous, the filaments often dilated basally, the anthers dorsifixed, 2-celled, the thecae (cells) slightly separated by a connective, and each produced into a tubelike apex with a terminal pore and the pollen in tetrads (opening by slits and the pollen grains single in Monotropoideae, often reflexed after anthesis so that the pore becomes distal; pistil 1, the ovary superior, 5-lobed, the locules and carpels 5 (sometimes the locules with incomplete septation at upper levels), the placentation typically axile with the placenta intruded into each locule and bifurcate, the ovules numerous in each locule, anatropous, the style and stigma 1; fruit a loculicidal capsule; seed with a loose testa (termed an aril by some authors), minute embryo and abundant endosperm.

A family of 10 genera and about 32 species indigenous to the boreal and temperate regions of the northern hemisphere. It is represented by 2 subfamilies, the Pyroloideae with green foliaceous leaves and the Monotropoideae, lacking chlorophyll and with scalelike leaves. In this country and Canada the Pyroloideae are represented by the monotypic Ramischia and Moneses, and Chimaphila (4 spp., 1 in Japan and 3 in America). The Monotropoideae are represented by indigenous species of Allotropa (monotypic in the northwest), Monotropa (including Hypopitys) (4–3), Pterospora (monotypic, North America), Sarcodes (monotypic, Oregon to California and Nevada), Monotropsis (3, in the southeast), Pleuricospora (2, Pacific coast), and Newberrya (3 to 5 spp. from Washington to California).

The family is distinguished from the Ericaceae, to which it is closely related, by

[182] The name of this family was spelled "Pirolaceae" by Engler and associates. For the correctness of the spelling Pyrolaceae, see Sprague, T. A., Kew Bull. 1923: 359–360.

the herbaceous habit, the corolla of distinct petals, and the uniformly loculicidal capsule. In this regard it should be noted that the genera of Ericaceae sect. Ledeae also have polypetalous corollas and that in the Ericoideae of the latter family the capsule may be loculicidal.

Many botanists (including Hutchinson) have held the view that the Pyrolaceae are not sufficiently distinct from the Ericaceae to be treated as a separate family. Others (including Small, Rydberg) have not only recognized the Pyrolaceae as a taxon distinct from Ericaceae, but have accepted the segregation of the saprophytic chlorophyll-less Monotropoideae as a separate family, the Monotropaceae. Copeland (1947) rejected the view that the Pyrolaceae and Monotropaceae represented separate families distinct from Ericaceae, a view held also by Henderson (1920), and treated the first as a tribe of Arbutoideae and the second as a subfamily presumed derived from the Andromedeae. Wodehouse (cf. Smith and Standley under Clethraceae) pointed out that the single pollen grains of Monotropoideae were unlike any other ericaceous pollen and held that the subfamily, unlike the Pyroloideae, was distinct from the Ericaceae.

The family is of no economic importance except for the few species cultivated for ornament: pipsissewa or prince's pine (*Chimaphila*), shinleaf (*Pyrola*), and one-flowered pyrola (*Moneses*).

*LITERATURE:*

ANDRES, H. Piroleen Studien. Beiträge zur Kenntnis der Morphologie, Phytogeographie und allgemeinen Systematik der Pirolaceae. Verh. Bot. Ver. Brand. 56: 1–76, 1914. [Cf. also, Fedde Repert. Spec. Nov. 19: 209–224, 1923.]

CAMP, W. H. Aphyllous forms in *Pyrola*. Bull. Torrey Bot. Club, 67: 453–465, 1940.

COPELAND, H. F. The structure of *Monotropsis* and the classification of the Monotropoideae. Madroño, 5: 105–119, 1939.

———. Further studies on Monotropoideae. Madroño, 6: 97–119, 1941.

———. Observations on the structure and classification of the Pyroleae. Madroño, 9: 65–102, 1947.

DRUDE, O. Pirolaceae. In Engler and Prantl, Die natürlichen Pflanzenfamilien, IV(1): 3–11, 1889.

FERNALD, M. L. Transfers in *Pyrola*. Rhodora, 43: 167, 1941.

HENDERSON, M. W. A comparative study of the structure and saprophytism of the Pyrolaceae and Monotropaceae with reference to their derivation from the Ericaceae. Contr. Bot. Lab. Univ. Penn. 5: 42–109, 1920.

RYDBERG, P. A. Pyrolaceae. North Amer. Flora, 29: 21–32, 1914.

SMALL, J. K. Monotropaceae. North Amer. Flora, 29: 11–18, 1914.

## ERICACEAE.[183] HEATH FAMILY

Mainly shrubs, occasionally suffrutescent perennial herbs, trees, or rarely trailing or scrambling vines to 20 meters long; leaves alternate, sometimes opposite or whorled, simple, often coriaceous and persistent, estipulate; flowers bisexual, actinomorphic or slightly zygomorphic (as in *Rhododendron* spp.), solitary in axils or in axillary or terminal clusters, racemes, or panicles; perianth biseriate, parts of each series usually more or less connate, calyx typically 4–7 lobes (sepals sometimes distinct), usually persistent, the corolla of 4–7 sometimes distinct but usually connate petals, often funnelform campanulate or urceolate in form, convolute or imbricated; the stamens as many or more commonly twice as many as petals or corolla lobes, arising from the base of a disc (hypogynous in Rhododendroideae except for transition to epigynous within *Gaultheria* (*Chiogenes*), epigynous in Vaccinioideae), distinct, the filaments sometimes flattened or dilated and basally coherent to connate especially in the tropical Vaccinioideae and sometimes forming a

[183] The name Ericaceae DC (1805) is conserved over Rhodoraceae Vent. (1799).

## DIVISION IV. EMBRYOPHYTA SIPHONOGAMA

tube, straight or S-curved, the anthers 2-celled, the thecae often saccate and basally bulbous, frequently appendaged, each theca dehiscing introrsely by a terminal pore or chink or longitudinally, pollen grains in tetrads; pistil 1, the ovary superior (in Rhododendroideae excepting certain spp. of *Gaultheria*) or inferior (in Vaccinioideae), locules and carpels 4–10, typically 5, the placentation axile, the ovules usually numerous in each locule, anatropous, the style usually 1, conical to filiform (rarely with as many branches as carpels), the stigma simple; fruit a capsule or a berry (when capsular, the fruit may be baccate by its enclosure within the fleshy persistent and sometimes adnate calyx, as in *Gaultheria*); seed usually small, with straight embryo and fleshy endosperm.

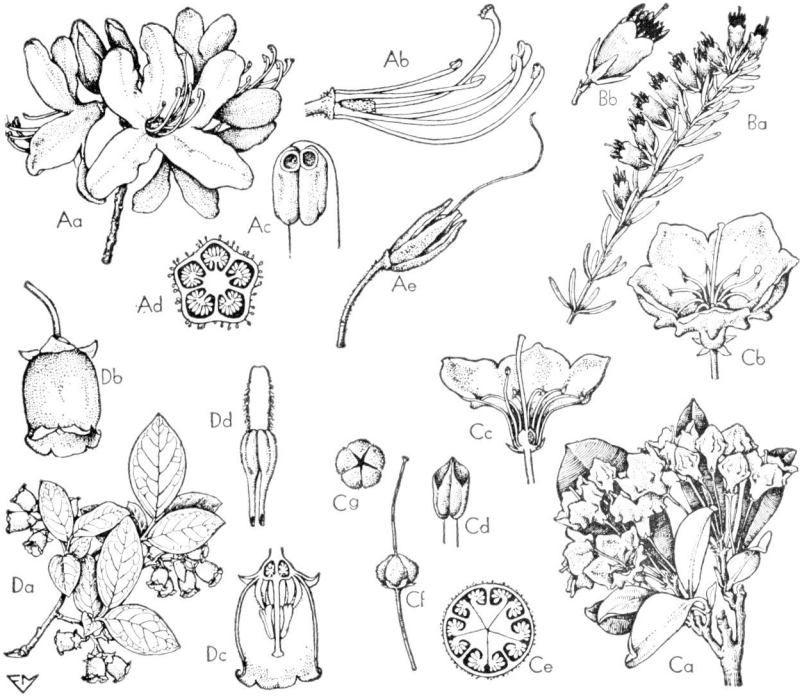

**Fig. 251.** ERICACEAE. A, *Rhododendron Vaseyi*: Aa, inflorescence, × ½; Ab, flower, less perianth, × 1; Ac, anther, × 5; Ad, ovary, cross-section, × 5; Ae, capsule, × 1. B, *Erica mediterranea*: Ba, flowering branch, × 1; Bb, flower, × 2. C, *Kalmia latifolia*: Ca, flowering branch, × ½; Cb, flower, × 1; Cc, same, vertical section, × 1; Cd, anther, × 5; Ce, ovary, cross-section, × 6; Cf, capsule, side view, × 1; Cg, same, face view, × 1. D, *Vaccinium vacillans*: Da, flowering branch, × ½; Db, flower, × 2; Dc, same, vertical section, × 2; Dd, anther, × 5. (From L. H. Bailey, *Manual of cultivated plants*, The Macmillan Company, 1949. Copyright 1924 and 1949 by Liberty H. Bailey.)

The Ericaceae, as circumscribed above, are a family of about 70 genera and about 1900 species, very widely distributed on acid soils throughout temperate regions of northern and southern hemispheres and to a lesser extent in the subarctic, and from sea level to high elevations in the tropics. They are represented in the United States by about 25 genera. The larger genera of the family (together with approximate number of total and indigenous species) include *Erica* (650 spp., mostly South African), *Rhododendron* [184] (700–25 concentrated in the mountains of east and southeast Asia), *Gaultheria* (100–5, incl. *Chiogenes,* chiefly tropical American), *Vaccinium* (130–45, abundant in tropical and subtropical mountains), *Arctostaphylos* (55–30, mainly warm semideserts of North America), *Gaylussacia* (40–9, chiefly Brazilian), and *Leucothoë* (35–9).

The family is distinguished from allied groups by a combination of characters such as predominantly shrubs, the stamens distinct (except in many tropical Vaccinioideae), mostly twice as many as corolla lobes, and arising from a nectiferous disc (rarely adnate to corolla), and the ovary typically 4-more loculed.

Opinion is divided on the internal classification of the family. Botanists following tradition accept it as a single family as circumscribed above; others interpret it as 2 families, the Ericaceae (with ovary superior and fruit typically a capsule), and the Vacciniaceae (with ovary wholly or partially inferior and fruit typically a berry). The evidence in support of each view [185] is mainly subjective; until more objective evidence is forthcoming, the 2 subfamilies might well be treated as one.

According to Engler and Diels, following Drude, the Ericaceae are comprised of 4 subfamilies: (1) Rhododendroideae, with septicidal capsule and seed often winged or ribbed, the calyx deciduous, the stamens with erect or long adnate unappendaged anthers, the ovary superior; (2) Arbutoideae, with fruit a berry or loculicidal capsule, seed not winged, the calyx deciduous, the anthers appendaged and much folded, the ovary superior; (3) Vaccinioideae, with ovary inferior (otherwise the flowers superficially resembling the Arbutoideae); (4) Ericoideae, with ovary superior, fruit typically a loculicidal capsule or nut, the seeds round and not winged, calyx persistent, the anthers with short connective and the 2 cells apically spreading, often appendaged.

Economically the family is of domestic importance primarily for its ornamentals; notable exceptions are the cultivated blueberries and cranberries. Few are trees, notably the sorrel tree (*Oxydendrum*); the majority are evergreen or less commonly deciduous shrubs. Among them are rhododendrons and azaleas (*Rhododendron*), mountain laurel (*Kalmia latifolia*), leatherleaf (*Leucothoë*), andromeda (*Pieris*), heather (*Calluna*), and heath (*Erica*); a few are suffrutescent perennials or subshrubs, as the Mayflower or trailing arbutus (*Epigaea*), wintergreen (*Gaultheria procumbens*), and bearberry (*Arctostaphylos Uva-ursi*). The blueberry of commerce is the fruit of species of *Vaccinium*; the commercial cranberry is the fruit of *Vaccinium macrocarpon* (*Oxycoccus macrocarpus*). The name huckleberry, often used for low-growing species of *Vaccinium,* should be applied only to species of *Gaylussacia.* Most of the so-called briar pipes are made from burls of *Erica arborea* (or related species) of the Mediterranean region. Foliage of *Gaultheria Shallon* of the Pacific states is sold as "lemon leaf" in eastern (U.S.) markets.

*LITERATURE:*

ABRAMS, L. R. Notes on some type specimens of *Arctostaphylos.* Leafl. West. Bot. 1: 84–87, 1934.

———. The dwarf Gaultherias in California. Madroño, 2: 121–122, 1934.

ADAMS, J. E. A systematic study of the genus *Arctostaphylos* Adans. Journ. Elisha Mitchell Soc. 56: 1–62, 1940.

AIRY-SHAW, H. K. Studies in the Ericales: IV. Classification of the Asiatic species of *Gaultheria* Kew Bull. 1940: 306–330. [Includes American spp. and merges *Chiogenes* with *Gaultheria.*]

BALDWIN, J. T. JR. Chromosomes of *Kalmiopsis.* Rhodora, 40: 278–279, 1938.

[184] Copeland (1943) and Cox (1948), on the basis of anatomical and morphological evidence, recognize as taxonomically valid such genera as *Azalea, Azaleastrum, Tsusiophyllum,* etc.

[185] For discussions and evidence see papers by Camp, Sleumer, Cox, and Copeland.

———. Cytogeography of *Oxydendrum arboreum*. Bull. Torrey Bot. Club, 69: 134–136, 1942.
BOWERS, C. G. Rhododendrons and azaleas. Their origin, cultivation and development 549 pp., New York, 1936.
CAMP, W. H. Studies in the Ericales. I. The genus *Gaylussacia* in North America north of Mexico. Bull. Torrey Bot. Club, 62: 129–132, 1934.
———. Studies in the Ericales. IV. Notes on *Chimaphila, Gaultheria,* and *Pernetya* in Mexico and adjacent regions. Bull. Torrey Bot. Club, 66: 7–28, 1939.
———. *Phyllodoce* hybrids. New Flora and Silva, 47: 207–211, 1940.
———. Studies in the Ericales: A review of the North American Gaylussacieae; with remarks on the origin and migration of the group. Bull. Torrey Bot. Club, 68: 531–551, 1941.
———. On the structure of populations in the genus *Vaccinium*. Brittonia, 4: 189–204, 1942.
———. Survey of American species of *Vaccinium,* subgenus *Euvaccinium*. Brittonia, 4: 205–247, 1942.
———. A preliminary consideration of the biosystematy of *Oxycoccus*. Bull. Torrey Bot Club, 71: 426–437, 1944.
———. The North American blueberries with notes on other groups of Vacciniaceae. Brittonia, 5: 203–275, 1945.
COPELAND, H. F. A study, anatomical and taxonomic, of the genera of Rhododendroideae. Amer. Midl. Nat. 30: 533–625, 1944.
COX, H. T. Studies in the comparative anatomy of the Ericales. I. Ericaceae — subfamily Rhododendroideae. Amer. Midl. Nat. 39: 220–245, 1948.
DARROW, G. M., *et al.* Chromosome numbers in *Vaccinium* and related groups. Bull. Torrey Bot. Club, 71: 498–506, 1944.
DARROW, G. M. and CAMP, W. H. *Vaccinium* hybrids and the development of new horticultural material. Bull. Torrey Bot. Club, 72: 1–21, 1945.
DRUDE, O. Ericaceae. In Engler and Prantl, Die natürlichen Pflanzenfamilien, IV (1): 15–65, 1889.
EASTWOOD, A. A revision of the genera formerly included in *Arctostaphylos*. Leafl. West. Bot. 1: 97–100, 1934.
———. A revision of *Arctostaphylos* with key and descriptions. Leafl. West. Bot. 1: 105–127, 1934.
GOOD, R. D'O. The genera *Phyllodoce* and *Cassiope*. Journ. Bot. (Lond.), 64: 1–10, 1926.
HUME, H. H. Azaleas, kinds and culture. New York, 1948.
MATTHEWS, J. R. and KNOX, E. M. The comparative morphology of the stamen in the Ericaceae. Trans. and Proc. Roy. Soc. Edinb. 29: 248–281, 1926.
PECK, M. E. Native rhododendrons of the Pacific northwest. Rhododendron Yearb. 1947: 87–89.
REHDER, A. *Azalea* or *Loiseleuria*. Journ. Arnold Arb. 2: 156–159, 1921.
———. *Kalmiopsis,* a new genus of the Ericaceae from northwest America. Journ. Arnold Arb. 13: 30–34, 1932.
ROZANOVA, M. A. [A survey of the literature on the genera *Vaccinium* L. and *Oxycoccus* (Tourn.) Hill.] Bull. Appl. Bot. and Pl. Breed. VIII. 2: 121–186, 1934. [Title and text in Russian, cf. pp. 173–186, for English summary and the bibliography.]
SCHREIBER, B. O. The *Arctostaphylos canescens* complex. Amer. Midl. Nat. 23: 617–632, 1940.
SINCLAIR, J. The *Rhododendron* bud and its relation to the taxonomy of the genus. Notes Roy. Bot. Gard. Edinb. 19: 267–271, 1937.
SLEUMER, H. Ericaceae americanae novae vel minus cognitae III. Notizbl. Bot. Gart. Berlin, 13: 206–214, 1936. [Incl. key to *Leucothoë* spp.]
———. Vaccinioideen-Studien. Engl. Bot. Jahrb. 71: 375–510, 1941.
SMITH, A. C. The American species of Thibaudieae. Contr. U. S. Natl. Herb. 28: 311–547, 1932.
WHERRY, E. T. The American azaleas and their variations. Nat. Hort. Mag. 22: 158–166, 1943.
WILSON, E. H. and REHDER, A. A monograph of Azaleas. Publ. Arnold Arb. No. 9, 1–219, 1921.

## EPACRIDACEAE. EPACRIS FAMILY

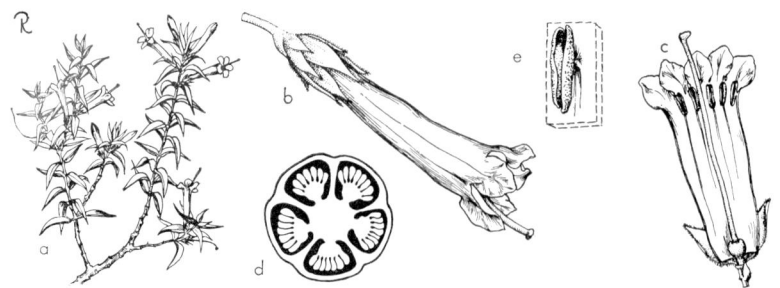

**Fig. 252.** EPACRIDACEAE. *Epacris impressa*: a, flowering branch, × ½; b, flower, × 2; c. same, perianth expanded, × 1½; d, ovary, cross-section, × 12; e, stamen, × 5.

Shrubs or small trees; leaves alternate, often crowded, simple, small and heath-like, usually stiff, estipulate; flowers bisexual (rarely unisexual, the plants then mostly monoecious), actinomorphic, bracteate, the calyx 4–5-lobed, persistent, the corolla gamopetalous, 4–5-lobed (rarely the lobes coherent and then the corolla opening transversely near the base of the tube), valvate or imbricate; stamens usually 5 (4), epipetalous or hypogynous, alternating with the corolla lobes, sometimes with alternating staminodes represented by clusters of hairs or glands, the anthers 1-celled at anthesis, dehiscing longitudinally; pistil 1, the ovary superior, often surrounded basally by an hypogynous glandular disc, the locules 1–10, the carpels 4–5, the placentation typically axile, the ovules solitary to many in each locule, the style 1, stigma capitate; fruit a 5-valved capsule or a drupe with a 1–5-seeded stone; seeds with straight embryo and fleshy endosperm.

A family of 23 genera and about 350 species, mainly Australasian, and none indigenous to this country. The largest genera are *Styphelia* (175 spp.) and *Epacris* (34 spp.). Species of *Cyathodes, Epacris* and *Leucopogon* are cultivated domestically to a small extent in the warmer parts.

The family is undoubtedly an advanced taxon of close alliance with the Ericaceae, and may be thought of as the Australasian counterpart of the family. It differs mostly in the single whorl of epipetalous stamens and the 1-celled anthers, but distinctions between the 2 families are weak.

### Order 36. PRIMULALES

The Primulales are characterized by the gamopetalous and generally pentamerous flowers, the unilocular superior ovary with free-central placentation, and the bi-integumented ovules. It is a small order containing only 3 families: Theophrastaceae, Myrsinaceae, and Primulaceae.

Curiously enough, Hallier retained them near his Bicornes and, as was true for the latter, derived them from the Ochnaceae (of his Guttales). Wettstein circumscribed them similarly and considered them allied to the Bicornes (Ericales). Bessey included the Plumbaginaceae and the Plantaginaceae in his Primulales, and derived the order from his Caryophyllales. A similar view, as concerns phylogeny, was held by Hutchinson. Recent morphological and anatomical studies strengthen the conclusion that the Primulales are advanced types derived from caryophyllaceous ancestors.

## THEOPHRASTACEAE. JOEWOOD FAMILY

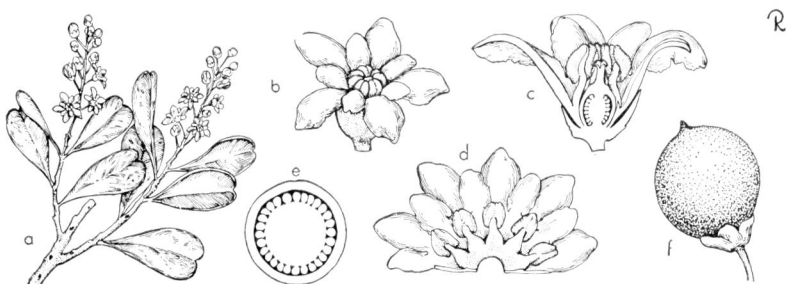

**Fig. 253.** THEOPHRASTACEAE. *Jacquinia keyensis*: a, flowering branch, × ⅓; b, flower, habit, × 2; c, same, vertical section, × 3; d, perianth, expanded, × 2; e, ovary, cross-section, × 8; f, fruit, × 1½.

Trees or shrubs; leaves usually alternate (opposite in *Jacquinia*), simple, persistent, sometimes spine-tipped (in *Jacquinia*), no resin ducts present, generally crowded, estipulate; flowers bisexual or unisexual (the plants then dioecious), actinomorphic, in racemes, corymbs, or panicles, the sepals 5, imbricated, distinct or nearly so, the corolla 5-lobed, gamopetalous, rotate, urceolate, or funnel-form; the androecium biseriate, the outer whorl reduced to 5 antesepalous petaloid staminodes, the inner of 5 functional epipetalous stamens with filaments distinct (connate in *Clavija*), the anthers 2-celled, dehiscing longitudinally (extrorse in *Jacquinia*); pistil 1, the ovary superior, unilocular, 5-carpelled, the placentation free-central, the ovules numerous, anatropous, the integuments 2 and immersed in a mucilaginous matrix, the style 1 and stigma entire to irregularly lobed; fruit a berry (rarely a 1-seeded drupe; seed with well developed embryo, endosperm copious.

A family of 4 genera and about 60 species of the American tropics and Hawaiian Islands. *Jacquinia* (25 spp.) is the only genus indigenous to the United States, and is represented in southern Florida by 1 species, *J. keyensis*.

The family has been combined with the Myrsinaceae, but it differs in anatomical characteristics, notably by the absence of resin ducts and the present of long strands of sclerenchymatous tissue beneath the leaf epidermis, and by the presence of staminodes and the large usually yellow or orange seeds (small and dark brown to black in Myrsinaceae).

The family is of no significant domestic importance, although *Clavija longifolia* is sometimes cultivated as an ornamental in the south.

*LITERATURE:*

MEZ, C. Theophrastaceae. In Engler, Das Pflanzenreich, 9 (IV. 236): 1–437, 1902.

## MYRSINACEAE.[186] MYRSINE FAMILY

Trees or shrubs; leaves mostly alternate, simple, usually persistent, coriaceous, glandular-punctate or with linear resin ducts, estipulate; flowers bisexual or unisexual (the plants then dioecious or polygamodioecious), actinomorphic, often glandular, bracteate (or bracteoles lost by reduction), small, fasciculate on scaly short shoots or spurs in leaf axils or paniculate, or corymbose or cymose, the sepals 4–6, distinct (in *Embelia*, *Heberdenia*) or more commonly basally connate,

[186] The name *Myrsinaceae* Lindl. (1836) has been conserved over *Ardisiaceae* Juss. (1810).

persistent, the petals usually connate, valvate or convolute, the corolla usually 4-6-lobed, rotate to salverform; stamens as many as corolla lobes and opposite them, usually epipetalous, distinct (monadelphous in some, syngenesious in *Ambylanthus* of India), the anthers 2-celled, dehiscing longitudinally or by apical slits or by pores, rarely transversely septate, usually longer than the filaments, staminodes absent; pistil 1, the ovary usually superior (inferior to half inferior in *Maesa*), the carpels and locules 4-6, the placentation axile or free-central (sometimes basal), the placentae much proliferated to form a cap over and around the ovules, the ovules solitary to several and uniseriate to multiseriate on each placenta, semianatropous to semicampylotropous, the style simple and short, the stigma simple or lobed; fruit a drupe with a fleshy exocarp and a stony endocarp containing 1-few seeds; the seeds with a cylindrical embryo and copious endosperm.

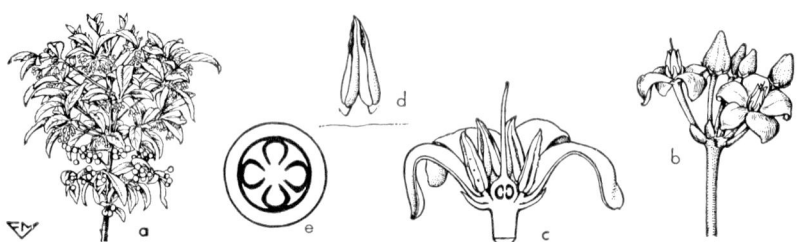

**Fig. 754.** MYRSINACEAE. *Ardisia crenata*: a, plant in flower and fruit, × ⅛; b, inflorescence, × 1; c, flower, vertical section, × 3; d, stamen, × 3; e, ovary, cross-section, × 8. (From L. H. Bailey, *Manual of cultivated plants*, The Macmillan Company, 1949. Copyright 1924 and 1949 by Liberty H. Bailey.)

A family of 32 genera and about 1000 species, with southern limits in New Zealand and South Africa, and northern limits in Japan, Mexico, and Florida. It is represented in the United States by 2 species indigenous to Florida, *Rapanea guaianensis* and *Icacorea paniculata*. Among the larger genera of the family are *Ardisia* (250 spp.), *Rapanea* (140 spp.), *Maesa* (100 spp.), and *Embelia* (60 spp.).

The Myrsinaceae are closely allied to the Primulaceae and differ from them in their woody habit, in the ovules generally buried in the proliferating placentae, and in the 1-few-seeded drupaceous fruit. See Theophrastaceae for distinctions separating them from that family.

Hallier treated them as distinct from but allied to the Theophrastaceae and to have been less highly advanced than the Primulaceae (the latter family connected to the Myrsinaceae by the Cyclamineae). Wettstein considered them to be the most advanced of the Primulales (because of frequency of dioecism, the transversely septate anthers in several genera, the half-inferior ovary of the Maesoideae, and polyembryony in *Ardisia*). Hutchinson placed the Myrsinaceae in an order by themselves (Myrsinales) and rejected, on the basis of woody vs. herbaceous texture, the majority opinion that the family was closely allied to the Primulaceae, and placed the two at widely divergent points in his classification.

The family is of little economic importance. A few species of 4 genera (*Ardisia, Maesa, Myrsine,* and *Suttonia*) are cultivated as ornamentals in warmer parts and under glass.

## PRIMULACEAE. PRIMULA FAMILY

Annual or perennial herbs, rarely suffrutescent; leaves mostly opposite or whorled, sometimes all basal, generally simple (pinnately dissected in the aquatic, *Hottonia*), often glandular-dotted or farinose; flowers bisexual, actinomorphic

# DIVISION IV. EMBRYOPHYTA SIPHONOGAMA

(zygomorphic in *Coris*), bracteate, typically 5-merous, the calyx 4–9-lobed, foliaceous, generally persistent, the corolla (absent in *Glaux*) gamopetalous with 4–9 imbricated lobes (polypetalous in *Pelletiera*), usually rotate to salverform (lobes reflexed in *Dodecatheon* and *Cyclamen*); the stamens as many as corolla lobes and opposite them, in a single whorl (the missing outer whorl sometimes represented by scalelike staminodes, as in *Soldanella, Samolus*), the anthers 2-celled, introrse, dehiscing longitudinally; pistil 1, the ovary superior (half inferior in *Samolus*), unilocular, the carpels typically 5, the placentation free-central, the ovules few to numerous, mostly semianatropous, the integuments 2, the style 1 (heterostyly common) and its stigma usually capitate; fruit usually a 5-toothed or -valved capsule (sometimes with 10 teeth), or a pyxis (as in *Anagallis* and *Centunculus*); seed with small straight embryo, endosperm abundant and hard to firm and semitransparent.

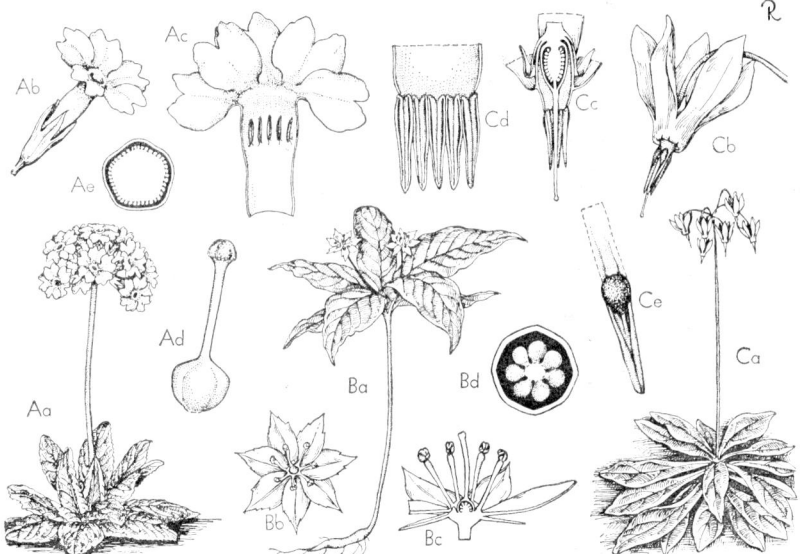

Fig. 255. PRIMULACEAE. A, *Primula denticulata*: Aa, plant in flower, × ¼; Ab, flower habit, × 1; Ac, perianth, expanded, × 1½; Ad, pistil, × 3; Ae, ovary, cross-section, × 5. B, *Trientalis borealis*: Ba, plant in flower, × ⅓; Bb, flower, face view, × 1; Bc, same, vertical section (perianth partially excised), × 2; Bd, ovary, cross-section, × 12. C, *Dodecatheon Meadia*: Ca, plant in flower, × ¾; Cb, flower, habit, × ¾; Cc, same, vertical section (perianth partially excised) × 1½.

A family of about 28 genera and nearly 800 species, widely distributed, occurring on all continents but most abundant in north temperate regions. Eleven genera are represented by indigenous species, and all except *Dodecatheon* and *Androsace* are primarily of eastern distribution. These genera are (together with number of total and indigenous species): *Primula* (400-6), *Lysimachia* (100-7), *Steironema* (5-5), *Androsace* (85-15), *Dodecatheon* (30-25), *Douglasia* (6-4), *Hottonia* (2-1), *Samolus* (10-5), *Trientalis* (3-1), *Glaux* (1-1), and *Centunculus* (1-1).

The Primulaceae are distinguished readily by the herbaceous habit, the usually gamopetalous corolla, the single whorl of stamens opposite the corolla lobes, the free-central placentation, the dehiscent capsule, and the usually numerous seeds

The evidence is reasonably conclusive that the Primulaceae and the Caryophyllaceae have fundamentally the same type of gynoecia and, as concluded by Douglas (1936) (and in essence by Dickson, 1936), ". . . the vascular pattern and the presence of locules at the base of the ovary point to the fact that the present much reduced flower of the Primulaceae has descended from an ancestor which was characterized by a plurilocular ovary and 'axial' placentation. This primitive flower might well be found in centrospermal stock, as Wernham, Bessey, and Hutchinson have suggested." This conclusion points to the artificiality of the taxa represented by the so-called Polypetalae and Gamopetalae, and gives impetus to the view that connation of perianth parts arose among the dicots many times by parallel or convergent evolution.

Economically the family is of domestic importance only for the ornamentals contributed by it. Species of most of the genera cited above are cultivated as also are species of *Cyclamen, Anagallis, Soldanella, Cortusa,* and *Omphalogramma.*

LITERATURE:

BRUUN, H. C. Cytological studies in *Primula* with special reference to the relation between the karyology and taxonomy of the genus. Symbol. Bot. Upsaliensis, 1: 1–239, 1932.

CONSTANCE, L. A revision of the genus *Douglasia.* Amer. Midl. Nat. 19: 249–259, 1938.

DICKSON, J. Studies in floral anatomy. III. An interpretation of the gynaeceum in the Primulaceae. Amer. Journ. Bot. 23: 385–393, 1936.

DOUGLAS, G. E. Studies in the vascular anatomy of the Primulaceae. Amer. Journ. Bot. 23: 199–212, 1936.

FASSETT, N. C. *Dodecatheon* in eastern North America. Amer. Midl. Nat. 31: 455–486, 1944.

FERNALD, M. L. The genus *Primula,* Sect. *Farinosae* in America. Rhodora, 30: 59–77, 1928.

———. *Steironema lanceolatum* and its varieties. Rhodora, 39: 438–442, 1937.

GOODMAN, G. J. and LEYENDECKER, P. J. The genus *Lysimachia* in Iowa. Proc. Iowa Acad. 49: 211, 212, 1942.

HANDEL-MAZZETTI, H. Die Subgenera, Sektionen und Subsektionen der Gattung *Lysimachia* L. Pflanzenareale, 2: 39–41, 1929.

PAX, F. Primulaceae. In Engler and Prantl, Die natürlichen Pflanzenfamilien, IV (1): 98–116, 1889.

PAX, F. and KNUTH, R. Primulaceae. In Engler, Das Pflanzenreich, 22 (IV. 237): 1–386, 1905.

ST. JOHN, H. Revision of certain North American species of *Androsace.* Mem. Victoria Museum, 126: 45–55, 1922.

SMITH, W. W. and FORREST, G. The sections of the genus *Primula.* Notes Edinb. Bot. Gard. 16: 1–47, 1928.

SMITH, W. W. and FLETCHER, H. R. The genus *Primula.* Trans. Bot. Soc. Edinb. 33: 122–181, 1941; *op. cit.* 209–294, 1942; Journ. Linn. Soc. Lond. 52: 321–335, 1942; Trans. Roy. Soc. Edinb. 50: 563–627, 1942; *op. cit.* 51: 1–69, 1943; Trans. Bot. Soc. Edinb. 33: 431–487, 1943; *op. cit.* 34: 55–158, 1944; Trans. Roy. Soc. Edinb. 61: 271–314, 1944; *op. cit.* 415–478, 1946; *op. cit.* 631–680, 1948; Trans. Bot. Soc. Edinb. 34: 402–468, 1948.

WILLIAMS, L. O. Revision of the Western Primulas. Amer. Midl. Nat. 17: 741–748, 1936.

## Order 37. PLUMBAGINALES

An order containing the single family Plumbaginaceae, and distinguished from presumably allied taxa by the pentamerous flowers (polypetalous or gamopetalous), the styles and/or stigmas 5, and the ovary unilocular with a single bi-integumented ovule (pendulous from a basal funiculus).

There is evidence that the Plumbaginales may have evolved from stocks ancestral to the Primulales, and most phylogenists (including Engler, 1924) have been of the opinion that they have close affinities with the caryophyllaceous taxa.

DIVISION IV. EMBRYOPHYTA SIPHONOGAMA 661

## PLUMBAGINACEAE.[187] LEADWORT FAMILY

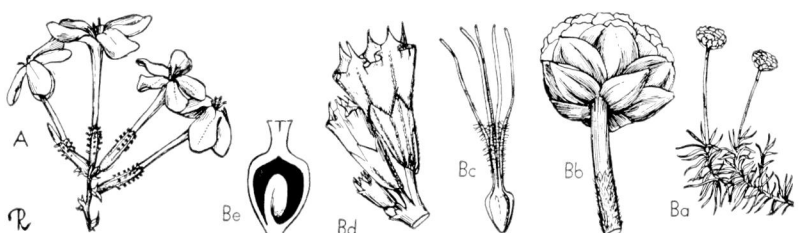

**Fig. 256.** PLUMBAGINACEAE. A, *Plumbago capensis*: inflorescence, × ½. B, *Armeria maritima*: Ba, flowering branch, × ¼; Bb, head showing involucre and sheath, × ½; Bc, pistil, × 4; Bd, cincinnus, × 3; Be, ovary, vertical section, × 8. (From L. H. Bailey, *Manual of cultivated plants,* The Macmillan Company, 1949. Copyright 1924 and 1949 by Liberty H. Bailey.)

Perennial herbs or shrubs, sometimes lianous; leaves alternate, often rosulate, estipulate; flowers bisexual, actinomorphic, bracteate, in cymules or cincinni (each reduced to a single flower in *Plumbago* that are capitulose (*Armeria*), spicate (*Acantholimon*), or usually racemose to paniculate (*Limonium*), the bracts sometimes forming an involucre; calyx 5-lobed (sometimes with smaller secondary lobes), gamosepalous, plicate, often 5–10-ribbed, -angled, or winged, the limb sometimes membranous or scarious and showy, often persistent, the corolla gamopetalous with the 5 lobes or segments extending almost to base (sometimes seemingly polypetalous), contorted and imbricated; the stamens 5, hypogynous (Plumbagineae) or perigynous (Staticeae), opposite the corolla lobes, introrse, the anthers 2-celled, dehiscing longitudinally, pollen grains dimorphic in some species of some genera; pistil 1, the ovary superior, unilocular, 5-carpelled, usually 5-lobed or -ribbed, the ovule solitary and pendulous from a basal funicle, anatropous, the integuments 2, the styles 5, opposite the sepals, distinct or basally connate, often hairy or glandular, the stigmas filiform, heterostyly sometimes present, papillations often dimorphic; fruit a utricle or tardily circumscissally dehiscent, often enclosed within the calyx; seed with straight embryo and firm crystalline-granular endosperm.

A family of 10 genera and about 300 species, mostly of semiarid regions of the Old World, especially the Mediterranean and central Asiatic regions. It is represented in the United States by a single polymorphic species of *Armeria* (*A. maritima*) of boreal and Pacific coast distribution and perhaps 3–5 polymorphic species of *Limonium* along the coasts. *Plumbago scandens* is native in Florida and from Texas to Arizona. The largest genera include *Limonium* (*Statice*) (150 spp.), *Acantholimon* (90 spp.), *Armeria* (40 spp.), and *Plumbago* (10 spp.).

The members of the family are distinguished by the 5-styled pistil and the unilocular and uniovulate ovary.

The family was placed by Hallier in the Centrospermae (from which it differs primarily in the anatropous ovule and straight embryo), by Wettstein and by Rendle in an order by itself (the Plumbaginales) close to the Primulales, by Hutchinson as one of 2 families in his Primulales. The evidence favors its separation in a distinct order, advanced over the Primulales, with both orders derived from the Centrospermae or their ancestors.

Economically the family is of little importance. A number of species of thrift (*Armeria*), and statice (*Limonium*), and one or more species of leadwort (*Cera-*

---

[187] The name Plumbaginaceae Lindl. (1836) is conserved over Armeriaceae Horaninov (1834).

*tostigma*), plumbago (*Plumbago*), and prickly thrift (*Acantholimon*) are cultivated for ornament.

*LITERATURE:*

BAKER, H. G. Relationships in the Plumbaginaceae. Nature, 161: 400, 1948.
——. Significance of pollen dimorphism in late-glacial *Armeria*. Nature (London), 161: 770–771, 1948.
——. Dimorphism and monomorphism in the Plumbaginaceae. I. A survey of the family. Ann. Bot. 12: 207–219, 1948.
BLAKE, S. F. *Limonium* in North America and Mexico. Rhodora, 18: 53–66, 1916.
——. *Statice* in North America. Rhodora, 19: 1–9, 1917.
BOISSIER, E. Plumbaginaceae. In de Candolle, Prodromus, 12: 617–696, 1848.
D'AMATO, F. Contributo all'embriologia delle Plumbaginaceae. Nuovo Giorn. Bot. Ital. N. S. 47: 349–382, 1940.
IVERSON, J. Blütenbiologische Studien, I. Dimorphie und Monomorphie bei *Armeria*. K. Danske Videnskae. Selskab. Biol. Meddelel. 15 (8): 1–39, 1940.
LAWRENCE, G. H. M. Armerias, native and cultivated. Gentes. Herb. 4: 391–418, 1940.
——. The genus *Armeria* in North America. Amer. Midl. Nat. 37: 757–779, 1947.
PAX, F. Plumbaginaceae. In Engler and Prantl, Die natürlichen Pflanzenfamilien IV (1): 116–125, 1889.
SPRAGUE, T. A. *Statice* and *Limonium*. Journ. Bot. 62: 267–268, 1924.
SUGIURA, T. Chromosome numbers in Plumbaginaceae. Cytologia, 10: 73–76, 1939.

## Order 38. EBENALES

A distinctive order of gamopetalously flowered plants characterized by the epipetalous stamens in usually 2–3 whorls, and the ovary basically septate with 1 to a few ovules on each of the axile placentae.

Engler and Diels treated the order as composed of 2 suborders and 7 families as follows (names of families not treated in this text are preceded by an asterisk):

Sapotineae
    Sapotaceae
    *Hoplestigmataceae

Diospyrineae
    Ebenaceae
    *Diclidantheraceae
    Symplocaceae
    Styracaceae
    *Lissocarpaceae

There is evidence that the order is not a natural taxon. Hallier accepted each of the suborders as orders, the first as the Sapotales and the second as the Santales (to which he added also the Loranthaceae, Olacaceae, and Alangiaceae). Rendle indicated that the suborders of Ebenales (*sensu* Engler) represented 2 divergent lines of development. Hutchinson likewise accepted 2 orders (Ebenales, Styracales) but on wholly different morphological grounds. It is to be expected that some of these families may be redistributed in future phylogenetic classifications. Copeland concurred with earlier authors that "the Theaceae are the living plants which most nearly represent the ancestry of the Ebenales." (Cf. under Styracaceae.)

## SAPOTACEAE.[188] SAPOTE FAMILY

Trees or shrubs, sap milky; leaves alternate (rarely opposite), simple, usually entire, coriaceous, sometimes stipulate; flowers bisexual, actinomorphic, bracteolate, solitary or more commonly cymose in leaf axils or on old stems, perianth triseriate, the sepals 4–12, imbricate, biseriate or spirally arranged, basally connate, the corolla gamopetalous, imbricated, the lobes usually as many as sepals, sometimes with lateral or dorsal appendages; the stamens epipetalous, distinct, typically

[188] The name Sapotaceae has been conserved over others for this family.

in 2 or 3 whorls of 4–5 each but usually only the inner whorl fertile (others reduced to staminodes or sometimes obsolete), the anthers 2-celled, dehiscing longitudinally, pistil 1, the ovary superior, the locules and carpels typically 4 or 5 (1–14), the placentation axile, each locule uniovulate, the ovule anatropous, integument 1, micropyle downwards, the style 1, often apically lobed; fruit a berry, often with a thin leathery to bony outer layer; seed with endosperm usually fleshy or none.

About 40 genera and 600 species of primarily tropical trees, common in the Old World and American tropics. Genera extending northward and into the United States (together with number of total and indigenous species) include *Bumelia* (26–6, to Illinois and Virginia), *Chrysophyllum* (50–2), *Dipholis* (14–1), *Pouteria* (*Lucuma*) (300+–1), *Mastichodendron* (8–1). Indigenous species of these genera, except the first, occur only in peninsular Florida, where the Sapodilla (*Achras Zapota*) also is a naturalized species.

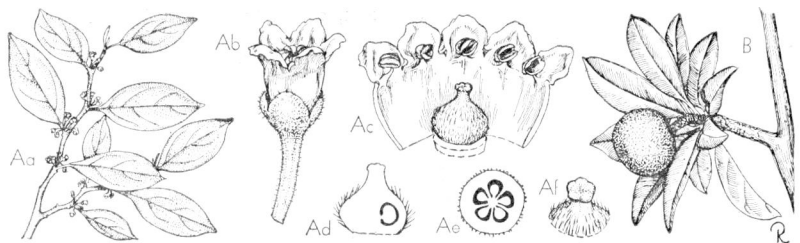

Fig. 257. SAPOTACEAE. A, *Chrysophyllum olivaeforme*: Aa, flowering branch, × ⅙; Ab, flower, habit, × 4; Ac, flower, perianth expanded, × 5; Ad, pistil, vertical section, × 6; Ae, ovary, cross-section, × 12; Af, ovary top with style and stigma, × 12. B, *Calocarpum Sapota*, fruiting branch, × ⅙.

The Sapotaceae (together with the Hoplestigmataceae of tropical Africa) differ from other families of the order by the completely septate superior ovary in which each locule contains a single ascending ovule and the ovule possesses but a single integument. In addition the family is distinguished by the presence of lactiferous ducts or sacs in the vegetative parts. Its phyletic position is not certain. The American genera are sharply defined, while the family as a whole is loosely knit and ill defined. Major taxonomic problems are encountered in the delimitation of its genera and species, and the nomenclature of these taxa has been badly confused.

The family is of considerable economic importance in world markets (chicle for chewing gum, is from *Achras Zapota*), but of little domestic importance. Guttapercha of commerce is obtained from the milky latex of several Old World genera (especially *Mimusops, Palaquium,* and *Payena*). Several members are cultivated pantropically for the edible fruits, notably for the sapodilla (*Achras Zapota, Manilkara zapotilla* a synonym), the sapote or marmalade plum (*Calocarpum Sapota, Pouteria mammosa* a synonym), the canistel or eggfruit (*Pouteria campechiana, Lucuma nervosa* a synonym), and the star apple (*Chrysophyllum Cainito*). (Note: the nomenclature of these species is unsettled, and authorities on the family are not in accord as to the identity or names to be used. There is no intention here to indicate one name as valid and its alleged synonym as invalid.)

*LITERATURE:*

BAEHNI, C. Mémoires sur les Sapotacées. I. Système de classification. Candollea, 7: 394–508. 1938.

CLARK, R. B. A revision of the genus *Bumelia* in the United States. Ann. Mo. Bot. Gard 29: 155–182, 1942.

CRONQUIST, A. Studies in the Sapotaceae — II. Survey of the North American genera. Lloydia, 9: 241–292, 1946.

———. Studies in the Sapotaceae. III. *Dipholis* and *Bumelia*. Journ. Arnold Arb. 26: 435–471, 1945.

———. Studies in the Sapotaceae — IV. The North American species of *Manilkara* Bull. Torrey Bot. Club, 72: 550–562, 1945.

———. Studies in Sapotaceae — VI. Miscellaneous notes. Bull. Torrey. Bot. Club, 73: 465–471, 1946.

ENGLER, A. Beiträge zur Kenntnis der Sapotaceae. Engl. Bot. Jahrb. 12: 496–525, 1890.

———. Sapotaceae. In Engler and Prantl, Die natürlichen Pflanzenfamilien, IV (1): 126–153, 1890.

GILLY, C. L. Studies in the Sapotaceae, II. The sapodillanispero complex. Trop. Woods, 73: 1–22, 1943.

LAM, H. J. On the system of the Sapotaceae with some remarks on taxonomical methods. Med. Bot. Mus. Herb. Utrecht, 65: 1939 et Rec. Trav. Bot. Neerl. 36: 509–525, 1939.

———. A tentative list of wild Pacific Sapotaceae, except those from New Caledonia. Blumea, 5: 1–46, 1942.

RECORD, S. J. American woods of the family Sapotaceae. Trop. Woods, 59: 21–51, 1939.

## EBENACEAE.[189] EBONY FAMILY

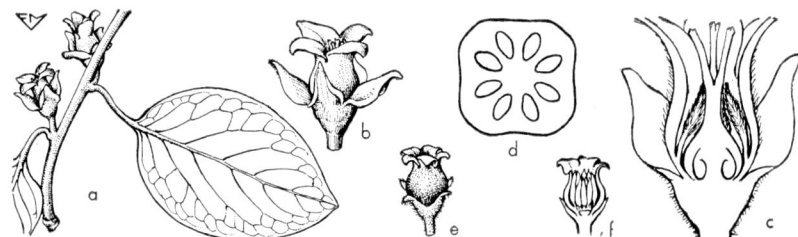

Fig. 258. EBENACEAE. *Diospyros virginiana*: a, flowering branch of pistillate plant, × ½; b, pistillate flower, × 1; c, same, vertical section, × 2; d, ovary, cross-section, × 4; e, staminate flower, × 1; f, same, vertical section, × 1. (From L. H. Bailey, *Manual of cultivated plants*, The Macmillan Company, 1949. Copyright 1924 and 1949 by Liberty H. Bailey.)

Generally dioecious trees or shrubs (heartwood often black, red, or green), lacking milky sap; leaves alternate, simple, entire, coriaceous, estipulate; flowers usually unisexual (plants infrequently monoecious), actinomorphic, 3–7-merous, axillary, solitary, or in small cymose inflorescences (staminate flowers mostly more abundant than pistillate), the calyx gamosepalous, persistent, the corolla gamopetalous, 3–7-lobed, hypogynous, urceolate, coriaceous, usually contorted and imbricated; stamens epipetalous or hypogynous, of same number as corolla lobes or 2–3 times as many (staminodes usually present in pistillate flower), distinct or united in pairs, the anthers 2-celled, introrse, dehiscing longitudinally; pistil 1, the ovary superior, the locules and carpels 2–16, the placentation axile, the ovules typically 2 (1) in each locule, pendulous and anatropous, the integuments 2, the styles and stigmas 2–8, styles distinct or basally connate; fruit a berry; seed with straight embryo and very hard, copious endosperm.

A widely distributed family of 5 genera and about 325 species. *Diospyros* (240 spp.) provides the only species indigenous to the United States (*D. virginiana*), native from Texas to Florida and north to Connecticut.

The Ebenaceae are distinguished from the Sapotaceae by the absence of milky

[189] The name Ebenaceae has been conserved over others for this family.

## DIVISION IV. EMBRYOPHYTA SIPHONOGAMA

sap, the unisexual flowers, the multilocular superior ovary, and by the ovules typically in pairs and always with 2 integuments. They differ from the Styracaceae, with which they are most closely allied, by the flowers usually unisexual and the ovary completely septate.

The family is of economic importance as a source of timber valued for cabinetmaking purposes, notably the black Macassar ebony (*Diospyros Ebenum*) [190] of India and the East Indies. Domestically it is of importance for the Japanese persimmon (*Diospyros Kaki*) and for the American persimmon (*D. virginiana*), the former a commercial crop produced in California and other warm areas.

*LITERATURE:*

BALDWIN, J. T. JR. and CULP, R. Polyploidy in *Diospyros virginiana* L. Amer. Journ. Bot. 28: 942–944, 1941.

GÜRKE, M. Ebenaceae. In Engler and Prantl, Die natürlichen Pflanzenfamilien, IV (1): 153–165, 1890.

HIERN, A. A monograph of the Ebenaceae. Trans. Cambridge Phil. Soc. 12: 27–300, 1873.

## SYMPLOCACEAE. SYMPLOCOS FAMILY

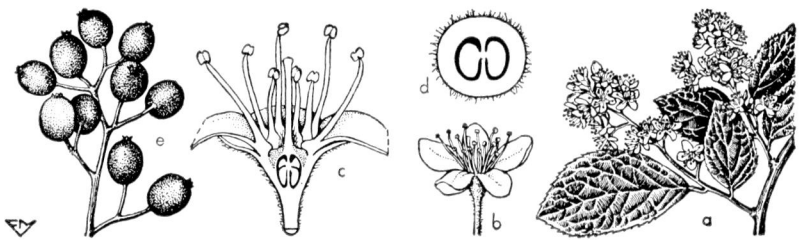

Fig. 259. SYMPLOCACEAE. *Symplocos paniculata*: a, flowering branch, × ½; b, flower, × 2; c, same, vertical section (corolla partially excised), × 4; d, ovary, cross-section. × 10; e, fruit, × 1. (From L. H. Bailey, *Manual of cultivated plants,* The Macmillan Company, 1949. Copyright 1924 and 1949 by Liberty H. Bailey.)

Trees or shrubs; leaves alternate, simple, coriaceous, estipulate; flowers bisexual (rarely some unisexual and the plant polygamodioecious), actinomorphic, in racemose or paniculate inflorescences, the calyx 5-lobed, imbricated, persistent, the corolla gamopetalous but divided nearly to base, the lobes 5–10, in 1 or 2 series, imbricated; stamens 4 to many (usually 12 or more), epipetalous, distinct or the filaments variously connate, the anthers globose, 2-celled, dehiscing longitudinally; pistil 1, the ovary inferior or half inferior, the locules and carpels 2–5, septation complete, the placentation axile, the ovules typically 2 in each locule, pendulous, anatropous, the integuments probably 2, the style 1, the stigma often capitate and 2–5-lobed; fruit a berry or drupaceous; seed with straight or curved embryo, the endosperm copious.

A unigeneric family of about 300 species of paleotropic Asia and America. About 8 species of *Symplocos* extend northward into Mexico; *S. tinctoria* is indigenous to the United States.

The Symplocaceae are distinguished from the closely allied Styracaceae by the inferior or half-inferior and completely septate ovary, and by the berrylike fruit with its persistent calyx.

A few Asiatic species of sweetleaf (*Symplocos*) are cultivated as hardy ornamental shrubs, most commonly *S. paniculata*.

---

[190] American ebony is the wood of *Byra Ebenus* (Leguminosae), perhaps better known as cocus wood or granadillo, and is streaked dark red in color.

## STYRACACEAE. STORAX FAMILY

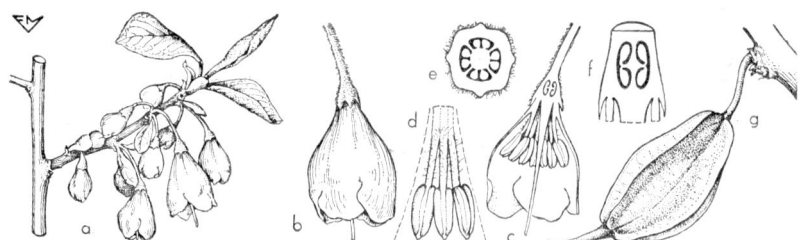

Fig. 260. STYRACACEAE. *Halesia carolina*: a, flowering branch, × ½; b, flower, × 1; c, same, vertical section, × 1; d, stamens, × 2; e, ovary, cross-section, × 4; f, ovary, vertical section, × 3; g, fruit, × 1. (From L. H. Bailey, *Manual of cultivated plants*, The Macmillan Company, 1949. Copyright 1924 and 1949 by Liberty H. Bailey.)

Small trees or shrubs, pubescence usually of stellate hairs or lepidote; leaves alternate, simple, soft-herbaceous to coriaceous, estipulate; flowers bisexual, actinomorphic, in racemose to paniculate cymose inflorescences, the calyx 4-5-cleft or -toothed, tubular, the corolla gamopetalous but lobed nearly to the base, the lobes usually 4-6, imbricate or valvate; stamens 8-12, in 1 whorl, epipetalous (rarely hypogynous), the filaments basally connate, the anthers 2-celled, oblong to linear, dehiscing longitudinally; pistil 1, the ovary superior (half inferior to inferior in *Halesia, Pterostyrax*), carpels mostly 3-5, usually unilocular above and 3-5-locular in lower half, the placentation axile below and parietal above, the ovules 2-8 on each placenta (all but 1-2 abort), anatropous, the integuments 1 or 2, the style 1, stigmas 1-5; fruit a drupe (the seeds within a stony endocarp) with a fleshy to dry and papery pericarp (when the latter, sometimes dehiscent); seeds with straight or slightly curved embryo and copious cellular endosperm.

A family of 8 genera and about 120 species, distributed in the warmer regions of South and Central America, southeastern United States, eastern Asia, and the Mediterannean region. It is represented in the southeastern part of the United States by 3 indigenous species each of *Styrax* and *Halesia*. *Styrax* (about 110 spp.) is the largest genus.

The Styracaceae are distinguished readily by the stellate pubescence, the stamens in a single whorl, the imperfectly septate ovary, the single style, and the fruit.

The family is important as the principal source of benzoin, a resin used in medicinal preparations (obtained from *Styrax Benzoin*). It is of domestic importance for a number of ornamental shrubs and small trees, notably silver bells (*Halesia*) and the half-hardy storax *Styrax*.

*LITERATURE:*

COPELAND, H. F. The *Styrax* of northern California and the relationships of the Styracaceae. Amer. Journ. Bot. 25: 771-780, 1938.
CORY, V. L. The genus *Styrax* in central and western Texas. Madroño, 7: 110-115, 1943.
PERKINS, J. Styracaceae. In Engler, Das Pflanzenreich, 30 (IV. 241): 1-111, 1907.

## Order 39. CONTORTAE [191]

Plants with commonly opposite simple or pinnately compound estipulate leaves, corolla gamopetalous and usually actinomorphic, the corolla lobes usually convolute, the stamens adnate to or near corolla base, the carpels 2 in number in each flower.

[191] This order was designated the Gentianales by Bentham and Hooker and others.

The order was subdivided by Engler and Diels into 2 suborders containing 6 families as follows (the Desfontaineaceae not treated in this text):

| Oleineae | Loganiaceae |
| Oleaceae | Gentianaceae |
| Gentianineae | Apocynaceae |
| Desfontaineaceae | Asclepiadaceae |

Wettstein, followed by Rendle, restricted the Contortae to contain only the suborder Gentianineae and segregated the Oleineae as the Oleales, considering the latter to have been derived from stocks allied to the Staphyleaceae, and the former to be advanced over the Tubiflorae. Hutchinson treated these families as composing 3 orders: Loganiales (Loganiaceae and Oleaceae), Apocynales (Apocynaceae and Asclepiadaceae), and Gentianales (Gentianaceae). Hallier included all within his Tubiflores, and derived the Oleaceae from the Scrophulariaceae and the others from the Linaceae of his Guttales. The Salvadoraceae, originally included in this order by Engler, were transferred by Engler and Gilg (1924) to the Sapindales.

## OLEACEAE.[192] OLIVE FAMILY

Trees or shrubs, sometimes lianous; leaves opposite (alternate in some *Jasminum* spp.), simple or pinnately compound, stipulate; flowers bisexual (unisexual in some *Fraxinus* spp., in *Phillyrea, et al.* and the plants then dioecious or polygamodioecious), actinomorphic, in axillary or terminal racemose, paniculate or thyrsiform inflorescences, the perianth biseriate (apetalous in some *Fraxinus* spp.), the calyx typically 4-lobed (rarely 4-15), valvate, rarely absent, the corolla gamopetalous (sometimes very deeply lobed or divided and seemingly polypetalous, or distinct as in *Fraxinus* spp.), typically 4-lobed (occasionally the lobes 6-12), mostly imbricate; the stamens typically 2 (4 in *Hesperelaea* and *Tessarandra*), distinct, the anthers 2-celled, the cells usually back to back, often apiculate by extension of the connective, dehiscing longitudinally; pistil 1, the ovary superior, the locules and carpels 2, the placentation axile, the ovules usually 2 in each locule (4-10 in *Forsythia, et al.*), anatropous, the style 1 or none, stigmas 1-2; fruit a berry (*Ligustrum*), drupe (*Olea, Jasminum*), loculicidal capsule (*Syringa, Forsythia*), a circumscissile capsule (*Menodora*), or a samara (*Fraxinus*); seed with straight embryo, the endosperm firm-fleshy.

A family of 22 genera and about 500 species, of temperate and paleotropical regions, notably Asia and the East Indies. It is represented domestically by several genera (together with number of total and indigenous species): *Fraxinus* (60-16), *Forestiera* (20-2), *Menodora* (18-3), *Osmanthus* (15-2), and *Chionanthus* (2-1, the other sp. in China). Only *Fraxinus* is indigenous to the extreme northeast, *Chionanthus* extends north to New Jersey, and the remainder are southern or western (*Forestiera* extends north into Illinois). Species of *Ligustrum* and *Jasminum* are naturalized in the southeast.

The Oleaceae are distinguished by the typically 2-merous flowers, the 2 anthers with cells back to back, and the 2-loculed superior ovary with generally 2 ovules each.

There are 2 subfamilies in the Oleaceae: Oleoideae and Jasminoideae, with all but 3 genera (*Jasminum, Menodora,* and *Nyctanthes*) in the Oleoideae. Hutchinson (1948) was of the opinion that the family is an unnatural assemblage, and that the dimerous androecium may have misled authorities into assuming a common origin for the taxa. He considered *Fraxinus* to be more closely allied to the Sapindaceae and *Ligustrum* to the Loganiaceae. On the basis of cytological studies, Taylor

---

[192] The name Oleaceae has been conserved over other names for this family.

(1945) proposed a revised classification of genera within the family, but made little attempt to relate it to other families.

The family is of considerable economic importance. The olive (*Olea*) is a source of food, and oil expressed from the fruit is of high value. Ash lumber (*Fraxinus*) is of value in cabinet work. Most of the other genera contribute important ornamentals, notably the lilac (*Syringa*), privet (*Ligustrum*), jasmine (*Jasminum*), golden bells (*Forsythia*), fringe tree (*Chionanthus*), fragrant olive (*Osmanthus*) and *Phillyrea*.

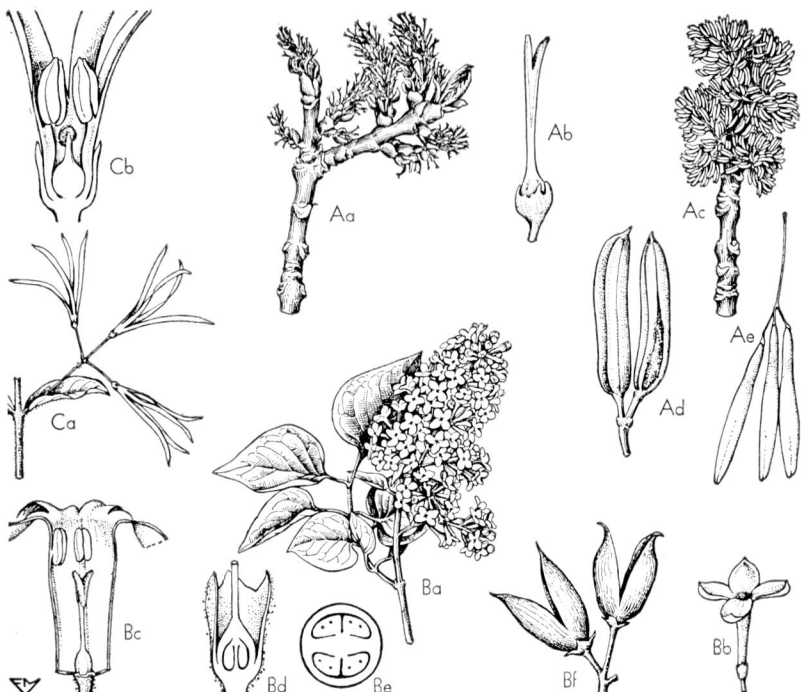

**Fig. 261.** OLEACEAE. A, *Fraxinus americana*: Aa, pistillate inflorescence, × ½; Ab, pistillate flower, × 4; Ac, staminate inflorescence, × ½; Ad, staminate flower, × 4; Ae, samaras, × ½. B, *Syringa vulgaris*: Ba, flowering branch, × ¼; Bb, flower, × 1; Bc, same, perianth expanded, × 2; Bd, ovary, vertical section, × 5; Be, same, cross-section, × 10; Bf, capsules, × 1. C, *Chionanthus virginica*: Ca, flowers, × ½; Cb, flower, perianth partially excised, × 5. (From L. H. Bailey, *Manual of cultivated plants,* The Macmillan Company, 1949. Copyright 1924 and 1949 by Liberty H. Bailey.)

*LITERATURE:*

ANDERSON, E. and WHELDEN, C. M. Studies in the genus *Fraxinus* II. Journ. Heredity, 27: 473–474, 1936.

KNOBLAUCH, E. Oleaceae. In Engler and Prantl, Die natürlichen Pflanzenfamilien, IV (2): 1–16, 1895.

LINGELSHEIM, A. Oleaceae-Oleoideae. In Engler, Das Pflanzenreich, 72 (IV. 243): 1–125, 1920.

MUNZ, P. A. and LAUDERMILK, J. D. A neglected character in western ashes (*Fraxinus*). El Aliso, 2: 49–62, 1949.

O'MARA, J. Chromosome Number in the Genus *Forsythia*. Journ. Arnold Arb. 11: 14–15, 1930.
REHDER, A. The genus *Fraxinus* in New Mexico and Arizona. Proc. Amer. Acad. Arts and Sci. 53: 199–212, 1917.
STEYERMARK, J. A. A revision of the genus *Menodora*. Ann. Mo. Bot. Gard. 19: 87–160, 1932.
TAYLOR, H. Cyto-taxonomy and phylogeny of the Oleaceae. Brittonia, 5: 337–367, 1945.
WESMAEL, A. Monographie des èspeces du genre *Fraxinus*. Bull. Soc. Roy. Bot. Belg. 31: 69–117, 1892.
WHELDEN, C. M. Studies in the genus *Fraxinus* I. A preliminary key to winter twigs for sections Melioides and Bumelioides. Journ. Arnold Arb. 15: 118–126, 1934.

## LOGANIACEAE.[193] LOGANIA FAMILY

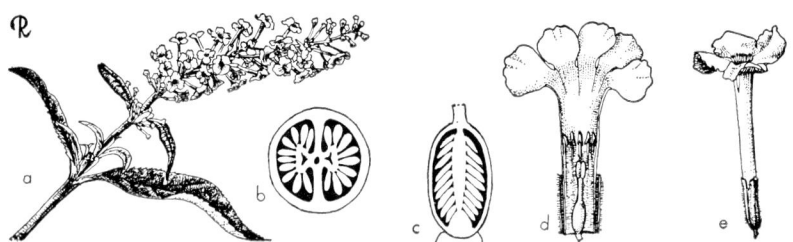

**Fig. 262.** LOGANIACEAE. *Buddleja Davidii*: a, inflorescence, × ½; b, ovary, cross-section, × 15; c, ovary, vertical section, × 8; d, flower, perianth expanded, × 2; e, flower, × 2. (From L. H. Bailey, *Manual of cultivated plants*, The Macmillan Company, 1949. Copyright 1924 and 1949 by Liberty H. Bailey.)

Herbs, shrubs, or trees (often lianous); leaves opposite (alternate in a few *Buddleja* spp.), simple, stipulate; flowers bisexual, actinomorphic, cymose to thyrsiform, usually bracteate and bracteolate, mostly 4–5-merous, the calyx 4–5-lobed or parted, the corolla gamopetalous, 4–5 or 10-lobed, aestivation various, the corolla-tube mouth often with crown of hairs; stamens epipetalous, as many as corolla lobes and alternate with them (rarely twice as many, or reduced to 1 as in *Usteria*), the anther cells 2 and side by side, dehiscing longitudinally; pistil 1, the ovary superior (half inferior in *Mitreola*), the locules and carpels typically 2 (unilocular in some *Strychnos* spp. and incompletely bilocular in *Fagraea*), the placentation axile (parietal in unilocular ovary), the ovules usually many, amphitropous or anatropous, the style 1 (2 with a single common stigma in *Cynoctonum*), stigmas 1–2 (rarely 4-cleft); fruit a septicidal capsule, the valves falling from the usually persistent axis, rarely a berry or drupe; seed sometimes winged, the embryo small and straight, the endosperm fleshy to bony.

A predominantly pantropical and warm temperate family of 32 genera and nearly 800 species, with half of the genera in the Old World, but well represented also in tropical America. Several genera are represented by species indigenous to this country. They are (together with number of total and indigenous species): *Spigelia* (35–3), herbs extending north to Kentucky and Missouri; *Gelsemium* (2–1), a vine occurring along the coast from Texas to Virginia; *Cynoctonum* (5–3), herbs of the southeastern coast north to Virginia; *Polypremum* (1–1), an annual herb from South America north to Maryland and Missouri; *Coelostylis* (2–1), an annual herb of Florida; and *Buddleja* (150–2), shrubs, Arizona to California. Other large genera in the family include *Strychnos* (200 spp.), *Nuxia* and *Fagraea* (30 spp. each), and *Logania* (21 spp.).

[193] The name Loganiaceae Horaninov (1834) is conserved over Spigeliaceae.

The Loganiaceae are distinguished from other families of the order by the opposite stipulate leaves, and by the bilocular superior ovary with typically axile placentation. An added and anatomical distinction is the presence of internal phloem in the majority of genera.

Most authors included the family in the Contortae, except Hallier, who placed it in his Tubiflorae, and Hutchinson, who segregated it (together with the Oleaceae) as the Loganiales. Engler was of the opinion that the family may be the most primitive of the order, with other families of the taxon having been derived from stocks ancestral to it. The circumscription of the family has been diversely interpreted (cf. Moore, 1947, for review) with one of the latest phylogenetic treatments (Klett, 1924) being in accord with that of Solereder (1895). Moore's phylogenetic views were that "Cytological data agree with taxonomic evidence that the Loganiaceae [are] an artificial group. Nevertheless, it is concluded that *Gelsemium*, *Polypremum*, and *Buddleja* are to be retained in the family."

The family is of economic importance as a source of the drug and poison strychnine (from seeds of *Strychnos Nux-vomica*), and curare poison is obtained from the bark of *Strychnos toxifera*. Many representatives of the family are cultivated domestically as ornamentals, notably species of the butterfly bush (*Buddleja*), Carolina or yellow jessamine (*Gelsemium*), pinkroot or Indian pink (*Spigelia*), Natal orange (*Strychnos spinosa*), and of *Logania* and *Geniostoma*.

*LITERATURE:*

KLETT, W. Umfang und Inhalt der Familie der Loganiaceen. Bot. Arch. 5: 312–338, 1924.
KRUKOFF, B. A. and MONACHINO, J. D. The American species of *Strychnos*. Brittonia, 4: 248–322, 1942.
MOORE, R. J. Cytotaxonomic studies in the Loganiaceae. I–III. Journ. Bot. 34: 527–538, 1947; *op. cit.* 35: 404–410, 1948; *op. cit.* 36: 511–516, 1949.
RECORD, S. J. American woods of the family Loganiaceae. Trop. Woods, 56: 9–13, 1938.
SOLEREDER, H. Loganiaceae. In Engler and Prantl, Die natürlichen Pflanzenfamilien, IV (2): 19–30, 1895.

## GENTIANACEAE.[194] GENTIAN FAMILY

Herbs, rarely subshrubs or shrubs, the plants sometimes saprophytic, branching often dichotomous; leaves opposite (alternate in *Menyanthes, et al.*), decussate, often basally connate, simple, estipulate; flowers bisexual (rarely polygamous), actinomorphic, usually showy, cymose, bracteate, and bracteolate or not so, the inflorescence dichasial or monochasial, usually 4–5-merous, the calyx gamosepalous, usually tubular, generally imbricate, the corolla gamopetalous, contorted (induplicate and valvate in Menyanthoideae) tubular rotate salverform or campanulate, scales or nectar pits often in corolla tube; stamens of same number as corolla lobes and alternate with them, epipetalous, distinct (syngenesious in *Voyria* and *Leiphaimos* spp.) usually versatile, the anthers 2-celled, introrse, dehiscing longitudinally (poricidal in *Exacum*), pollen variable and distinctive; annular disc present or absent; pistil 1, the ovary superior, usually unilocular with 2 usually intruded parietal placentae (bilocular with axile placentation in Exacineae), the carpels 2, the ovules usually numerous, anatropous, the style simple, stigma simple or 2-lobed; fruit a septicidal capsule; seed with a small embryo and a fleshy endosperm.

A family of 70 genera and about 800 species, of world-wide distribution and most abundant in temperate regions. Genera with species indigenous to the United States (together with approximate number of total and indigenous species) are: *Centaurium* (30–5), *Sabatia* (15–15), *Gentiana* (425–40), *Pleurogyne* (1–1), *Fra-*

[194] The name Gentianaceae has been conserved over other names for this family.

sera (8–8), *Swertia* (90–10), *Halenia* (25–2), *Bartonia* (5–5), *Eustoma* (5–3), *Lapithea* (2–2), *Microcala* (2–1), *Obolaria* (1–1), *Leiphaimos* (20–1), *Menyanthes* (1–1), and *Nymphoides* (20–2).

The Gentianaceae are usually divided into 2 subfamilies, the Gentianoideae (leaves opposite, corolla involute or imbricated, terrestrial plants), and Menyanthoideae (leaves alternate, corolla induplicate-valvate, marsh or aquatic plants). It is the opinion of some (including Wettstein) that these 2 taxa should be treated as separate families, a view supported by the morphological and anatomical findings of Lindsey (1938). The Gentianoideae are distinguished readily from related families by the opposite estipulate leaves, the typically superior unilocular bicarpellate multiovulate ovary, and parietal placentation developing to bilocular ovaries with axile placentation in advanced members.

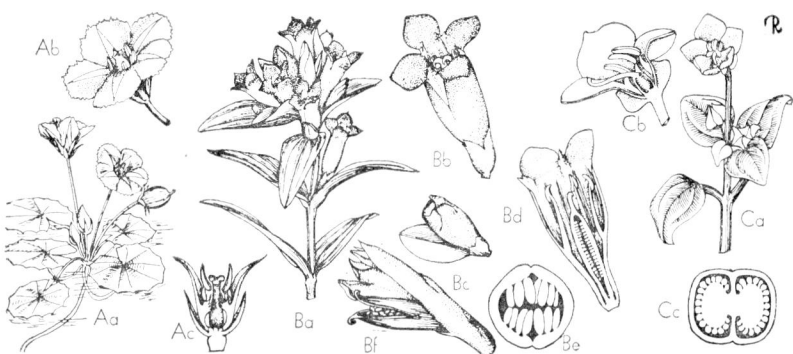

**Fig. 263.** GENTIANACEAE. A, *Nymphoides peltatum*: Aa, plant in flower, × ¼; Ab, flower, × ½; Ac, same, vertical section, × 1. B, *Gentiana cruciata*: Ba, flowering branch, × ½; Bb, flower, × 1; Bc, bud, × 1; Bd, flower, vertical section, × 3; Be, ovary, cross-section, × 5; Bf, capsule within calyx, × 1. C, *Exacum affine*: Ca, flowering branch, × ½; Cb, flower, vertical section, × 1; Cc, ovary, cross-section, × 5. (From L. H. Bailey, *Manual of cultivated plants*, The Macmillan Company, 1949. Copyright 1924 and 1949 by Liberty H. Bailey.)

The family is of little domestic economic importance except for the appreciable number of genera and species cultivated as ornamentals, of which the more important are gentians (*Gentiana*), prairie gentian (*Eustoma*), centaury (*Centaurium*), water snowflake (*Nymphoides*), buckbean or bogbean (*Menyanthes*), and *Exacum*.

*LITERATURE:*

ALLEN, C. K. A monograph of the American species of the genus *Halenia*. Ann. Mo. Bot. Gard. 20: 119–222, 1933.

CLAUSEN, R. T. Studies in the Gentianaceae: *Gentiana*, section *Pneumonanthe*, subsection *Angustifoliae*. Bull. Torrey Bot. Club. 68: 660–663, 1941.

GILG, C. Beiträge zur Kenntnis der Gentianaceen-Gattung *Curtia* Cham. et Schlecht. Notizbl. Bot. Gart. Berlin, 14: 66–93, 1938–39.

———. Beiträge zur Morphologie und Systematik der Gentianoideae-Gentianeae-Erythraeinae. Notizbl. Bot. Gart. Berlin, 14: 417–430, 1939.

GILG, E. Gentianaceae. In Engler and Prantl, Die natürlichen Pflanzenfamilien, IV (2): 50–108, 1895.

KUSNEZOW, N. I. Subgenus *Eugentiana* Kusnez. generis *Gentiana* Tournef. Act. Hort. Petrop. 15: 1–507, 1896–1904.

LINDSEY, A. A. Anatomical evidence for the Menyanthaceae. Amer. Journ. Bot. 25: 480–485, 1938.

———. Floral anatomy in the Gentianaceae. Amer. Journ. Bot. 27: 640–651, 1940.

McCoy, R. W. On the embryology of *Swertia carolinensis*. Bull. Torrey Bot. Club, 76: 430–439, 1949.
Maguire, B. Great Basin plants, VI. — Notes on *Gentiana*. Madroño, 6: 151–153, 1942.
Rork, C. L. A cytotaxonomic investigation of the genus *Gentiana* and related genera. Cornell Univ. Abs. Theses 1945: 180–183, 1946.
Wilkie, D. Gentians. 187 pp. London, 1936.

## APOCYNACEAE.[195] DOGBANE FAMILY

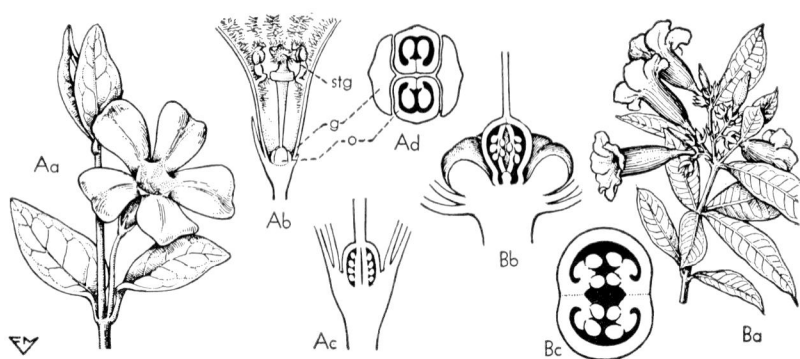

**Fig. 264.** Apocynaceae. A, *Vinca minor*: Aa, flowering branch, × 1; Ab, flower, perianth in vertical section, × 2; Ac, ovaries, vertical section, × 4; Ad, same, cross-section, × 10. B, *Allamanda neriifolia*: Ba, flowering branch, × ¼; Bb, ovary, vertical section, × 3; Bc, same, cross-section, × 10. (g gland, o ovary, stg stigma.) (From L. H. Bailey, *Manual of cultivated plants,* The Macmillan Company, 1949. Copyright 1924 and 1949 by Liberty H. Bailey.)

Trees, shrubs, or herbs, sap usually milky; leaves opposite and decussate, sometimes alternate or whorled, simple, entire, mostly estipulate; flowers bisexual, actinomorphic (sometimes weakly zygomorphic), solitary or racemose to cymose, bracteate and bracteolate, typically pentamerous, the calyx 5-lobed (rarely 4-lobed), imbricated, often glandular (squamellate) within, the corolla gamopetalous, contorted in bud, usually salverform or funnelform, the tube often appendaged within; the stamens as many as corolla lobes and alternate with them, epipetalous, distinct, the anthers 2-celled at anthesis, introrse, often sagittate, free or adherent by viscid exudates to the stigma or "clavuncle," dehiscing longitudinally, an apical connective often produced, the pollen granular or rarely in tetrads; pistil 1 and composed basically of 2 distinct unicarpellate ovaries terminated by a single thickened distally stigmatic clavuncle, each ovary superior, to subinferior, usually unilocular and unicarpellate with the placentation parietal in each ovary,[196] the ovules few to many, anatropous or campylotropous, style usually 1, the stigma or "clavuncle" massive and variable in form; fruit a follicle, capsule, berry or drupaceous; seed naked, comose or with a paper wing, occasionally arillate, endosperm straight, fleshy to firm-fleshy.

A family of about 300 genera and 1300 species, of almost cosmopolitan distribution, but most abundant pantropically. Of these, 46 genera and 203 species are indigenous or naturalized in North America, with only 9 genera and 33 species

---
[195] The name Apocynaceae Lindl. (1836) is conserved over Plumariaceae Horaninov (1834).

[196] In some genera (incl. *Ambelania, Couma, Lacmellea*) the ovary is syncarpous by connation of the 2 ovaries, resulting in a bilocular bicarpellate ovary with axile placentation (unilocular and parietal in *Allamanda*). In the primitive Pleiocarpeae, the gynoecium is composed of 5–8 carpels.

indigenous to this country (18 genera and 52 species are Mexican). Only *Amsonia, Apocynum,*[197] and *Trachelospermum* with 17, 7, and 1 spp. respectively, are widespread in this country. *Angadenia, Urechites, Echites,* and *Vallesia* are represented by 1–2 species each in southern Florida, while the monotypic *Cycladenia* occurs in northern California and 2 species of *Macrosiphonia* are native in the southwest.

The Apocynaceae are closely related to the Asclepiadaceae, and differ in the presence of a single style, the absence of a corona, the pollen grains distinct or in tetrads (not in pollinia), the stamens free from the stigma, and in the absence of translators connecting the contents of adjoining anther sacs.

The family was subdivided by Asa Gray into 2 tribes, raised by Schumann to the subfamilies Plumeroideae and Echitoideae. Hallier included the Asclepiadaceae with the Apocynaceae (as did Demeter, 1922), and derived them from the Linaceae. Bessey, and all other contemporary botanists, accepted the 2 families as distinct and included both in his Gentianales, an order he derived from geraniaceous stocks. Hutchinson (1948) included the Apocynaceae as the only family of his Apocynales, and considered them as derived from stocks ancestral to the Loganiaceae.

Economically the family is of importance domestically for its ornamentals, notably amsonia (*Amsonia*), oleander (*Nerium*), periwinkle (*Vinca*), Natal plum (*Carissa*), allamanda (*Allamanda*), frangipani (*Plumeria*), yellow oleander (*Thevetia*), crepe-jasmine (*Ervatamia*), and Chilean jasmine (*Mandevilla*). Of these only *Amsonia* and *Vinca* are hardy in the north. Many members of the family are poisonous when vegetative parts and fruits are eaten.

*LITERATURE:*

BOKE, N. H. Development of the perianth in *Vinca rosea* L. Amer. Journ. Bot. 35: 413–423, 1948.

DEMETER, K. Vergleichende Asclepideen-Studien. Flora, 115: 130–176, 1922.

MONACHINO, J. A resume of the American Carisseae (Apocynaceae). Lloydia, 9: 293–309, 1946.

SCHUMANN, K. Apocynaceae. In Engler and Prantl, Die natürlichen Pflanzenfamilien, IV (2): 109–189, 1895.

WOODSON, R. E. JR. Studies in the Apocynaceae. I. A critical study of the Apocynoideae (with special reference to the genus *Apocynum*). Ann. Mo. Bot. Gard. 17: 1–212, 1930; II, A revision of the genus *Stemmodenia*. op. cit. 15: 341–378, 1928; III, A monograph of the genus *Amsonia*. op. cit. 15: 379–434, 1928; IV, The American genera of Echitoideae. op. cit. 20: 605–790, 1933; 21: 613–623; 22: 153–306; 23: 169–438, 1936; V, A revision of the Asiatic species of *Trachelospermum* Lem. Sun-yatsenia, 3: 67–105, 1936; VI, *Kibatalia* and its immediate generic affinities. Philippine Journ. Sci. 60: 205–229, 1936; VII, An evaluation of the genera *Plumeria* L. and *Himatanthus* Willd. Ann. Mo. Bot. Gard. 25: 189–224, 1938.

———. Apocynaceae. North Amer. Flora, 29: 103–192, 1938.

———. Observations on the inflorescence of Apocynaceae (with special reference to the American genera of Echitoideae). Ann. Mo. Bot. Gard. 22: 1–48, 1935.

WOODSON, R. E. JR. and MOORE, J. A. The vascular anatomy and comparative morphology of apocynaceous flowers. Bull. Torrey Bot. Club, 65: 135–166, 1938.

## ASCLEPIADACEAE.[198] MILKWEED FAMILY

Perennial herbs, shrubs, or rarely small trees, sometimes fleshy or cactuslike, generally with milky sap, usually lianous; leaves caducous or vestigial in some succulent representatives, opposite or whorled, rarely alternate, simple, generally

---

[197] Woodson (1930) restricted *Apocynum* to North America, segregating the Eurasian species among the *Trachomitum* (2 spp.) and *Poacynum* (3 spp.).

[198] The name Asclepiadaceae Lindley (1847) has been conserved over other names, including Stapeliaceae Horaninov (1834) and Vincaceae Horaninov (1834).

entire, stipules present and minute; inflorescence one of 2 types, determinate and a dichasial or monochasial cyme, or racemose or umbelliform; flowers bisexual, actinomorphic, typically pentamerous (except the gynoecium), the perianth biseriate; calyx of distinct or basally connate sepals, imbricate or open; corolla 5-lobed, the lobes contorted or less commonly valvate, the corolla tube often terminated by a morphologically highly variable faucal annulus;[199] stamens 5, rarely distinct and usually adnate or adherent to the gynoecium to produce a gynostegium (comparable

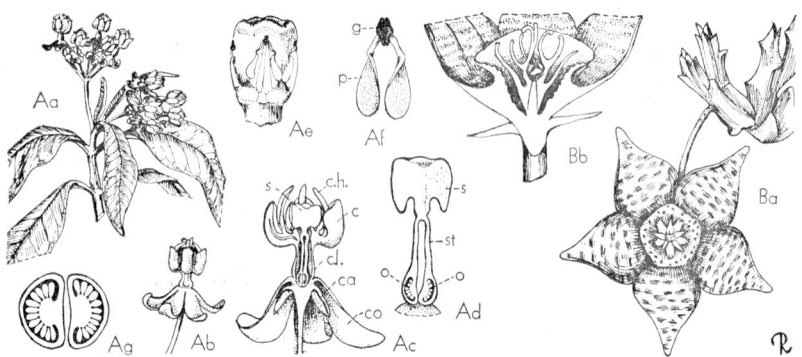

**Fig. 265.** ASCLEPIADACEAE. A, *Asclepias curassavica*: Aa, habit in flower, × ½; Ab, flower, × 1; Ac, same, vertical section, × 2; Ad, gynoecium, vertical section, × 4; Ae, stigma and anthers, × 4; Af, anther, × 10; Ag, ovaries, cross-section, × 10. B, *Stapelia variegata*: Ba, branch in flower, × ¼; Bb, flower, vertical section, perianth excised, × ½. (c, corona hood, cl column, c.h. corona horn, ca calyx, co corolla, g "gland," o ovary, p pollinium, s stigma, st style.) (From L. H. Bailey, *Manual of cultivated plants*, The Macmillan Company, 1949. Copyright 1924 and 1949 by Liberty H. Bailey.)

to the column in Orchidaceae), the anthers 2-celled and biloculate (4-celled in *Secamone*), in the Cynanchoideae (to which belong all New World species) the pollen grains at maturity agglutinated within each anther sac into a sac-shaped pollinium, the pollinia united in pairs, each pollinium bearing a translator arm and the translator arms of each pair "themselves joined by a roughly sagittate body called the gland";[200] in the less specialized Old World Periplocoideae, the pollen

---

[199] Terminology used for the so-called coronal parts of the asclepiadaceous flowers follows that adopted by Woodson (1941, pp. 198-200), who recognized the "corona" of most authors to be composed of 1 or more of the 3 following elements: (1) a *faucal annulus* arising from the corolla tube, (2) a fleshy true *corona* consisting of "various elaborations or enations of the staminal filaments only," or (3) an apical *sterile appendage* arising from each of the anthers. In much of the literature, the presence of any 2 of these components is referred to as a double corona and of all 3 as a triple corona.

[200] The translator arm is sometimes termed the *retinaculum* or *connective*, and the gland the *corpusculum*. As pointed out by Woodson (1941, p. 194), "A pair of pollinia, therefore, consists of the contents of adjacent anther cavities of contiguous anthers. The translators, with their glands, are formed between the neighboring anthers." It is believed that the translators are formed from the "solidified secretion of special glandular cells located upon the stigma head" and "are molded when still in the liquid state by the available spaces between the young anthers," and the uniting "gland" (not a true gland in the morphological sense) is always 2-parted.

is granular but the grains united in tetrads with each of the 5 translator arms concavely spoon- or cornucopia-shaped and ending below in an adhesive disc, the pollen tetrads at maturity falling into the spoon or horn. The gynoecium consists of a single bicarpellate pistil composed typically of 2 distinct nearly superior ovaries (slightly inferior, becoming so during development), 2 styles, and a common single 5-lobed often much-enlarged stigma; each ovary unilocular, 1-carpellate, the ovules anatropous and pendulous, imbricated in several series on a ventral and parietal placenta; the enlarged stigma (discoid, conical, or beaked) is nonreceptive except for the 5 longitudinal strips of glandular stigmatic surface on the thickened edge or the lower side exposed between the contiguous anthers; fruit a follicetum of 2 follicles (commonly 1 aborts); seeds (with few exceptions) with a tufted micropylar coma of long silky hairs, the embryo large, the endosperm thin and small.

A primarily pantropical family (especially of South America) with only a few genera occurring in temperate regions of the northern and southern hemispheres. The taxonomic status of the family is in need of thorough revision, and opinions differ as to the number of genera involved. Rendle (1925) placed it at 280, Willis (1931) at 320, but if Woodson's more critical and considered view (1941) that the North American species represent only 9 genera is proved the most correct, the number for the family may not exceed 75–100 at most. The number of species for the family is perhaps 1800. Woodson (1941) re-evaluated the genera of this country (reducing the number by about one-third) to be as follows: *Asclepias*, the largest North American genus (including *Acerates, Asclepiodora, Podostigma, Solanoa, Gomphocarpus* of American auth.), *Cynanchum* (including *Mellichampia, Metastelma, Basistelma,* and *Astephanus* of American auth.), *Matelea* (including *Gonolobus* and *Vincetoxicum* of most American auth.), *Gonolobus* (including *Lachnostoma* of North American auth.), and *Sarcostemma* (including *Philibertia* and *Funastrum*).

The Asclepiadaceae are most closely allied to the Apocynaceae, from which they differ in the modification of the stamens associated with the presence of translators, and the presence of the gynostegium, an alliance accepted by phylogenists (cf. discussion under Apocynaceae).

Economically the family is important domestically for the "down" of low quality obtained from the seed, and for a few genera grown as ornamentals, notably the butterfly weed (*Asclepias tuberosa*), bloodflower (*A. curassavica*), blue milkweed (*Oxypetalum caeruleum*), wax plant (*Hoya carnosa*), carrion flower (*Stapelia* spp.), and species of *Huernia, Periploca, Araujia,* and *Ceropegia*. The rubber vine (*Cryptostegia grandiflora*) has been grown as a commercial source of natural rubber and as an ornamental. Some species of *Asclepias* are important as livestock poisons; the sap of *Matelea* has been used as an arrow poison, while that of the Ceylon milk plant (*Gymnema lactiferum*) is used as a food for man.

LITERATURE:

BROWN, R. On the "Asclepiadeae" a natural order of plants separated from the Apocineae of Jussieu. Mem. Wern. Soc. Edinb. 1: 12–78, 1809 [1808–10].
PERRY, L. M. *Gonolobus* within the Gray's Manual range. Rhodora, 40: 281–287, 1938.
ROTHE, W. Ueber die Gattung *Marsdenia* R. Br. und die Stammpflanze der Condurangorinde. Engl. Bot. Jahrb. 52: 354–434, 1915.
SCHUMANN, K. Asclepiadaceae, in Engler and Prantl, Die natürlichen Pflanzenfamilien, IV (2): 189–306, 1895.
WHITE, A. and SLOANE, B. L. The Stapelieae. 3 vols. Pasadena, 1937.
WOODSON, R. E. JR. The North American Asclepiadaceae. I. Perspective of the genera. Ann. Mo. Bot. Gard. 28: 193–244, 1941.
——— Some dynamics of leaf variation in *Asclepias tuberosa*. Ann. Mo. Bot. Gard. 34: 253–432, 1947.

## Order 40. TUBIFLORAE

A large order of primarily herbaceous plants with gamopetalous corollas, characterized by the floral parts usually in 4 isomerous whorls, or with an oligomerous gynoecium and (when zygomorphic) androecium, the stamens epipetalous, hypogynous, and the ovules each with a single integument.

Engler and Diels recognized the order to be composed of the following 8 suborders and 23 families (only the Columelliaceae are not treated in this text):

Convolvulineae
  Convolvulaceae
  Polemoniaceae
  Fouquieriaceae
Lennoineae
  Lennoaceae
Boragineae
  Hydrophyllaceae
  Boraginaceae
Verbenineae
  Verbenaceae
  Labiatae

Solanineae
  Nolanaceae
  Solanaceae
  Scrophulariaceae
  Bignoniaceae
  Pedaliaceae
  Martyniaceae
  Orobanchaceae
  Gesneriaceae
  Columelliaceae
  Lentibulariaceae
  Globulariaceae

Acanthineae
  Acanthaceae
Myoporineae
  Myoporaceae
Phrymineae
  Phrymaceae

Wettstein accepted the Tubiflorae with much the circumscription as given them by Engler (except that the Fouquieraceae were placed in the Parietales, and the Plantaginaceae were included here), and Rendle did likewise (except that the Convolvulaceae were segregated as the Convolvulales, and that neither the Fouquieraceae nor Lennoaceae were treated). Other phylogenists distributed the families among 4 or more orders as indicated in the following treatments. Rendle gave an explanation of the phylogenetic derivation of members of this order (pp. 487–490), but the taxon may ultimately be shown not to be homogeneous; the families concerned appear to represent several more or less unrelated orders.

## CONVOLVULACEAE.[201] MORNING-GLORY FAMILY

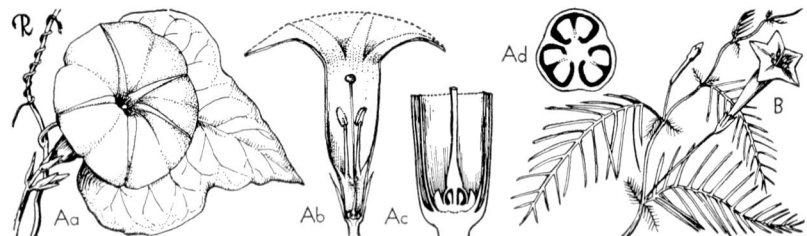

**Fig. 266.** CONVOLVULACEAE. A, *Ipomoea Leari*: Aa, flowering branch, × ¼; Ab, flower, vertical section, × ½; Ac, ovary, vertical section, × 4; Ad, ovary, cross-section, × 20. B, *Quamoclit pennata*: flowering branch, × ½. (From L. H. Bailey, *Manual of cultivated plants*, The Macmillan Company, 1949. Copyright 1924 and 1949 by Liberty H. Bailey.)

Erect or twining herbs or shrubs or small trees, sap usually milky, often turf-forming or thorny when shrubby, some genera are yellow almost leafless twining parasites (as *Cuscuta*); leaves alternate, simple (entire, lobed, or pinnately divided to pectinate), estipulate; inflorescence an axillary dichasium (racemose or paniculate in *Porana*) or the flowers solitary and axillary, the peduncle jointed; flowers

[201] The name Convolvulaceae has been conserved over other names for this family.

bisexual (rarely unisexual by abortion and the plants then dioecious), actinomorphic, often showy, generally pentamerous (except the gynoecium), bracteate, with bracts (usually in pairs) sometimes forming an involucre, sometimes cleistogamous (*Dichondra*); the calyx of 5 usually distinct sepals, imbricate, persistent, the corolla 5-lobed or entire, often funnelform or salverform, rarely with appendages within (*Cuscuta*), plaited or twisted in bud, hypogynous; stamens 5, distinct, epipetalous at base of corolla tube, alternate with lobes, the anthers 2-celled, dorsifixed, dehiscing longitudinally, usually introrse (pollen characters useful in separating some genera); intrastaminal disc usually present, ring- or cup-shaped, usually lobed; pistil 1, the ovary superior, entire or deeply 2-lobed (in *Dichondra*), typically bilocular and 2-carpelled with axile placentation (sometimes 1-4 locules, the latter by false septa, and rarely 3-5 carpelled), the ovules 1-2 in each locule, erect, anatropous, sessile; style usually filiform and simple (styles sometimes 2, as in *Cressa, Cuscuta*, or absent in *Erycibe*), the stigma terminal and capitate or 2 distinct stigmas varying in type; fruit usually a loculicidal capsule (indehiscent or sometimes fleshy, as in *Argyreia*, some *Cuscuta* spp.); seed smooth or hairy, the embryo large, with folded emarginate or bilobed cotyledons (embryo filiform and lacking apparent cotyledons in *Cuscuta*), surrounded by a hard cartilaginous endosperm.

A family of about 50 genera and 1200 or more species, primarily of tropics and subtropics with ranges extending into north and south temperate regions, and particularly abundant in tropical America and tropical Asia. *Ipomoea*, the largest genus, has about 400 mostly lianous species, *Convolvulus* of more temperate distribution has about 200 species, and *Cuscuta* the third largest with about 120 species. The family is represented in the United States by indigenous species of the following genera: *Breweria, Calonyction, Convolvulus, Cressa, Cuscuta, Dichondra, Evolvulus, Ipomoea, Jacquemontia, Pharbitis*.

The family is distinguished by milky sap or latex usually present, by the presence of bicollateral vascular strands, the plaited corolla, the erect sessile ovules with axile placentation, and the folded cotyledons.

Some authors place *Dichondra* and *Cuscuta* in separate families (Dichondraceae, Cuscutaceae), but by the more conservative view they are retained in Convolvulaceae, with the family divided into the Convolvuloideae and Cuscutoideae (the former subdivided into 5 tribes). Both Wettstein and Rendle segregated the family as a separate order (Convolvulales), and Wettstein considered them to be allied to the Malvales or Geraniales, while Rendle followed Engler's views on their affinity. Bessey included them in his Polemoniales, and Hutchinson treated them as the more advanced of 2 families comprising his Solanales.

Economically the family is important in the United States for the sweet potato (*Ipomoea Batatas*), for numerous noxious weeds, and for several ornamentals, notably the common morning glory (*Ipomoea purpurea*), the wood rose (*I. tuberosa*), moonflower (*Calonyction aculeatum*), cypress vine (*Quamoclit pennata*), and Christmas vine (*Porana paniculata*).

*LITERATURE:*

ABRAMS, L. R. Notes on the types of some Californian species of *Convolvulus*. Contr. Dudley Herb. 3: 351-373, 1946.
ALLARD, H. A. The direction of twist of the corolla in the bud, and twining of the stems on Convolulaceae and Dioscoreaceae. Castanea, 12: 88-94, 1947.
DEAN, H. L. An addition to bibliographies of the genus *Cuscuta*. Univ. Iowa Stud. Nat. Hist. 17: 191-197, 1937.
FOGELBERG, S. O. The cytology of *Cuscuta*. Bull. Torrey Bot. Club, 65: 631-645, 1938.
HOUSE, H. D. Synopsis of the Californian species of *Convolvulus*. Muhlenbergia, 4: 49-56, 1908.
———. The North American species of the genus *Ipomoea*. Ann. New York Acad. Sci. 18: 181-263, 1908.

KING, J. R. and BAMFORD, R. The chromosome number in *Ipomoea* and related genera. Journ. Heredity, 28: 279–282, 1937.
PETER, A. Convolvulaceae. In Engler and Prantl, Die natürlichen Pflanzenfamilien, IV (3a): 1–40, 375–377, 1897.
SMITH, B. E. A taxonomic and morphological study of the genus *Cuscuta* in North Carolina. Journ. Elisha Mitchell Sci. Soc. 50: 283–302, 1934.
VAN OOSTROOM, S. J. A monograph of the genus *Evolvulus*. Meded. Bot. Mus. Herb. Rijksuniv. 14: 1–267, 1934.
WOLCOTT, G. B. Chromosome numbers in the Convolvulaceae. Amer. Nat. 71: 190–192, 1937.
YUNCKER, T. G. Revision of the North American and West Indian species of *Cuscuta*. Illinois Biol. Monogr. 6: n. 2, 3: 1–141, 1921.
———. The genus *Cuscuta*. Mem. Torrey Bot. Club, 18: 113–331, 1932.

## POLEMONIACEAE. PHLOX FAMILY

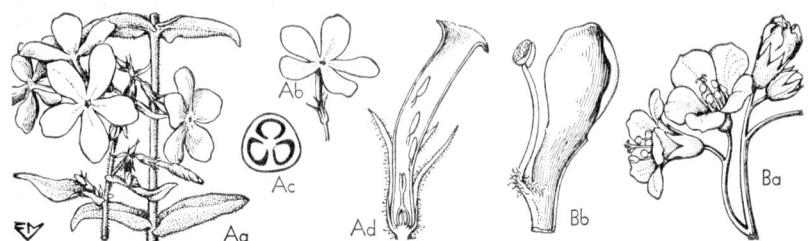

Fig. 267. POLEMONIACEAE. A, *Phlox divaricata*: Aa, inflorescence, × ½; Ab, flower, × ½; Ac, ovary, cross-section, × 10; Ad, flower, vertical section, corolla partially excised, × 2. B, *Polemonium reptans*: Ba, flowers, × 1; Bb, segment of corolla with stamen, × 3. (From L. H. Bailey, *Manual of cultivated plants*, The Macmillan Company, 1949. Copyright 1924 and 1949 by Liberty H. Bailey.)

Annual to perennial herbs, rarely shrubs, small trees, or twining vines, sap not colored; leaves usually alternate or the lower or all opposite, entire or palmately or pinnately divided (pinnately compound in *Cobaea*), estipulate; inflorescence usually cymose, corymbose to capitate, the flowers rarely solitary and axillary; flowers bisexual, actinomorphic or weakly zygomorphic (subbilabiate) in *Loeselia* and *Bonplandia*, usually showy, the calyx 5-lobed, imbricate or valvate, the corolla contorted in bud, salverform to rotate (campanulate in *Cobaea*) hypogynous, 5-lobed, the tube usually well developed; stamens 5, epipetalous and arising from corolla tube at various and unequal heights, the anthers 2-celled, dehiscing longitudinally; intrastaminal disc usually present; pistil 1, the ovary superior, 3-locular (carpels rarely 2 or 4), the placentation axile, the ovules 1–many on each placenta, sessile, anatropous, the style filiform, terminal, terminated by 3 branches or stigmas (rarely 2); fruit a loculicidal capsule (septicidal in *Cobaea*), rarely indehiscent; seeds sometimes with mucilaginous coat (*Collomia, Gilia* spp.), the endosperm usually copious and fleshy to firm-fleshy.

An American family of about 13 genera and 265 species, with the majority in western United States (a few spp. of *Phlox* and *Polemonium* occur in eastern Asia). Generic lines have been confused and subject to various interpretations (see Mason, 1945). The following genera are native to this country: *Collomia, Gilia, Eriastrum* (*Hugelia*), *Gymnosteris, Langloisia, Leptodactylon, Linanthus* (including *Linanthastrum*) *Loeselia, Navarretia, Phlox, Polemonium* (including *Polemoniella*), and *Siphonella*; all are represented in the western part of the country, and the genera *Gilia, Phlox,* and *Polemonium* occur in the eastern part of the country. The genus *Cobaea* extends from Mexico to Chile and *Cantua* from Peru to Chile, with *Gilia* extending down the Andes and to Argentina.

Members of the family are characterized by the absence of milky latex, the gamosepalous calyx, the typically tricarpellate ovary, and the usually numerous ovules and seeds. The closest affinity of the family seems to be with the Convolvulaceae.

The taxonomy of the family and its members, especially the phylogenetic aspects, has been particularly perplexing to botanists. Hallier treated it as derived probably from the Linaceae, Bessey from the boraginaceous stocks, while Wettstein followed the Englerian view, and Rendle (following a suggestion of Wettstein) considered it to have been derived from sympetalous relatives of the Rosales. Hutchinson placed it in his Polemoniales and derived them from the Geraniales, with the Ranales ancestral to both. On the basis of the floral morphology, Dawson (1936) concluded that the Polemoniaceae are closely related to both the Caryophyllaceae and the Geraniaceae and that the tricarpellate condition may have had its origin with stocks ancestral to modern caryophyllaceous plants at a time prior to the reduction of a trilocular ovary to one with free-central placentation. She interpreted the polemoniaceous genera *Cantua* and *Cobaea* to be primitive in the family, and concluded the basal disc around the ovary to be the remains of a lost whorl of stamens.

Economically the Polemoniaceae are important for a few ornamentals grown as annuals, biennials, perennials, or vines. Notable among them are: *Phlox, Polemonium, Gilia, Cobaea,* and *Linanthus,* all grown as garden flowers.

*LITERATURE:*

BRAND, A. Polemoniaceae. In Engler, Das Pflanzenreich, 27 (IV. 250): 1–203, 1907.
CONSTANCE, L. and ROLLINS, R. C. A revision of *Gilia congesta* and its allies. Amer. Journ. Bot. 23: 433–440, 1936.
CRAIG, T. A revision of the subgenus *Hugelia* of the genus *Gilia* (Polemoniaceae). Bull. Torrey Bot. Club, 61: 385–396, 1934.
DAVIDSON, J. F. The present status of the genus *Polemoniella* Heller. Madroño, 9: 58–60, 1947.
———. The genus *Polemonium* [Tournefort] L. Univ. Calif. Publ. Bot. 23: 209–282, 1950.
DAWSON, M. L. The floral morphology of the Polemoniaceae. Amer. Journ. Bot. 23: 501–511, 1936.
FLORY, W. S. A cytological study on the genus *Phlox*. Cytologia, 6: 1–18, 1934.
———. Chromosome numbers in the Polemoniaceae. Cytologia Fujii, 1: 171–180, 1937.
GRAY, A. Revision of North American Polemoniaceae. Proc. Amer. Acad. 8: 247–282, 1870.
KEARNEY, T. H. and PEEBLES, R. H. *Gilia multiflora* Nutt. and its nearest relatives. Madroño, 7: 59–63, 1943.
MASON, H. L. Notes on Polemoniaceae. Madroño, 6: 200–205, 1942.
———. The genus *Eriastrum* and the influence of Bentham and Gray upon the problem of generic confusion in Polemoniaceae. Madroño, 8: 65–91, 1945.
PETER, A. Polemoniaceae. In Engler and Prantl, Die natürlichen Pflanzenfamilien, IV (3a): 40–54, 1897.
WHERRY, E. T. Phlox. Bartonia. 11: 5–35, 1929; 12: 24–53, 1931; 13: 18–37, 1931; 15: 14–26, 1933; 16: 38–45, 1935.
———. The eastern long-styled Phloxes, I. Bartonia, 13: 18–37, 1932; II, *op. cit.* 14: 14–26, 1932.
———. *Polemonium* and *Polemoniella* in the eastern States. Bartonia, 17: 5–12, 1936.
———. A provisional key to the Polemoniaceae. Bartonia, 20: 14–17, 1940.
———. The genus *Polemonium* in America. Amer. Midl. Nat. 27: 741–760, 1942.
———. The minor genus *Polemoniella*. Amer. Midl. Nat. 31: 211–215, 1944.
———. Two linanthoid genera. Amer. Midl. Nat. 34: 381–387, 1945.
———. Supplementary notes on the genus *Polemonium*. Amer. Midl. Nat. 34: 376–380, 1945.
———. The *Phlox carolina* complex. Bartonia, 23: 1–9, 1945.
WHITEHOUSE, E. Annual *Phlox* species. Amer. Midl. Nat. 34: 388–401, 1945.

## FOUQUIERIACEAE. CANDLEWOOD FAMILY

Shrubs or small trees, armed with stout petiolar spines; leaves alternate, simple, small, somewhat fleshy, obovate in high phyllotaxy, or on short shoots in spine axils, the primary leaves soon deciduous with petioles persisting and becoming spinescent, estipulate; flowers showy, bisexual, actinomorphic, in terminal panicles, the sepals 5, unequal, strongly imbricated, the petals 5, connate into a tube with short spreading lobes, imbricated in bud; stamens 10–17, in 1–2 whorls, hypogynous, with a stiffish, often reddish, basal portion puberulent on the outside and developed into interstaminal tooth on the rim or with an interstaminal ligule on the inside, the anthers 2-celled, versatile, dehiscing longitudinally; pistil 1, the ovary superior, unilocular, 3-carpelled, the placentae parietal and intruded, the ovules 4–6 on each placenta, the styles 3, connate to about the middle; fruit a 3-valved capsule; seeds winged or marginally hairy with a straight embryo and scant but firm-fleshy endosperm.

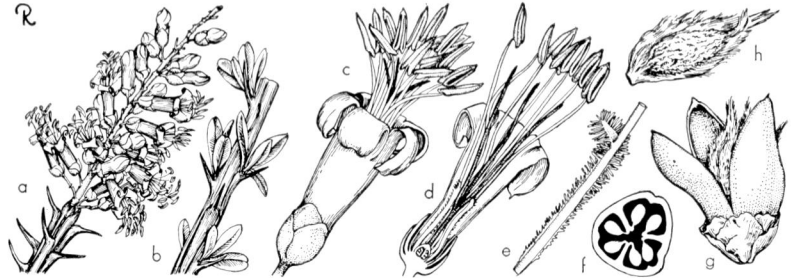

**Fig. 268.** FOUQUIERIACEAE. *Fouquieria splendens*: a, flowering branch, × ⅜; b, sterile branch section, × ⅜; c, flower, × 1½; d, flower, vertical section, × 1½; e, filament base, × 3; f, ovary, cross-section, × 8; g, fruit, × 1; h, seed, × 1½.

A unigeneric family (*Fouquieria*) of 3 xerophytic species in Mexico, one of which (*F. splendens*) extends northward to Texas, Arizona, and southern California. Nash (1903) treated it as composed of 6 species, with a seventh sometimes segregated as the monotypic *Idria*.

The phyletic position of the Fouquieriaceae is unsettled. In earlier Englerian works, and by Wettstein, it was included in the Parietales, and was advanced to its present position by Engler and Gilg (1924). Bessey included it in his Ebenales and Hutchinson in his Tamaricales.

The Fouquieriaceae are characterized by the thorny stout stems, the early deciduous leaves, the gamopetalous flowers with stamens basally hairy, and the partially connate styles.

The family is of little economic importance. The coachwhip cactus or ocotillo (*F. splendens*) is grown in the southwest as a hedge subject, and in California to a limited extent in landscape work in arid areas.

*LITERATURE:*

KHAN, R. The ovule and embryo-sac of *Fouquieria*. Proc. Nat. Inst. Sci. India, 9: 253–256, 1943.

NASH, G. V. Revision of the family Fouquieriaceae. Bull. Torrey Bot. Club, 31: 449–459, 1903.

SCOTT, F. M. Some features of the anatomy of *Fouquieria splendens*. Amer. Journ. Bot. 19: 673–678, 1932.

## LENNOACEAE.[202] LENNOA FAMILY

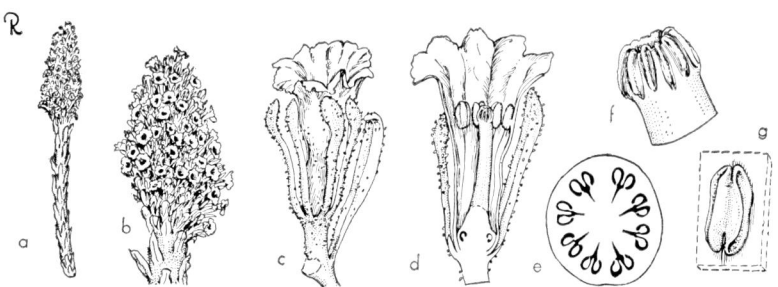

**Fig. 269.** LENNOACEAE. *Pholisma arenarium*: a, plant in flower, × ⅙; b, inflorescence, × ⅓; c, flower and bract, × 3; d, flower, vertical section, × 4; e, ovary, cross-section, × 12; f, stigma, × 12; g, stamen, × 12.

Parasitic, herbaceous, fleshy plants lacking chlorophyll; root parasites, stems simple or branched; leaves reduced to short scales; flowers bisexual, actinomorphic or sometimes slightly zygomorphic, in a compact thyrse or a peltately expanded head (in *Ammobroma*), the pedicels sometimes seemingly embedded in the fleshy stem, the sepals 5–10, distinct or nearly so, linear to subulate, puberulent or plumose, the corolla gamopetalous, 5–6-lobed (8 lobes in *Lennoa*), imbricated, tubular to funnelform (salverform in *Lennoa*); stamens 5–10 in 1 whorl (2 whorls in *Lennoa*), epipetalous, the filaments very short and adnate to corolla tube for most of their length, and emerging from it below the mouth, the anther cells 2 and parallel (basally divergent in *Lennoa*), introrse, the pollen grains single; disc absent; pistil 1, the ovary superior, the carpels 6–14, the number of apparent locules twice as many as carpels by false septation of the loculus, placentation axile, the ovule 1 in each apparent locule, nearly horizontal, anatropous, the style 1, solid and simple, the stigma subcapitate or peltate, crenate; fruit a fleshy capsule, tardily and irregularly circumscissile, the seeds seemingly enclosed by a bony endocarp formed from the locule wall, and pyrenelike, the embryo globose with undifferentiated cotyledons and radicle, endosperm present.

A family of 3 genera and 4 species, restricted to southwestern United States and Mexico. It is represented in this country by *Pholisma* and *Ammobroma* (both monotypic), with the former (*P. arenarium*) in southern and Baja California and the latter (*A. Sonorae*) in southwestern Arizona and southeastern California extending southward into Mexico. Both plants are small herbs rising for a few inches above the usually sandy soil, the stems orange or brown in color with minute violet to purple flowers. *Lennoa* (with 2 spp.) is indigenous primarily to Mexico where it is parasitic, mostly on composites, and Blake (1926) reported it from Colombia.

The position of the family has been interpreted variously. Bessey, Engler, and Hutchinson included it with the Ericales, and allied it with the Monotropoideae, placing it (as did Rydberg) as allied to the saprophytic Monotropoideae of the Pyrolaceae. Engler and Gilg and Wettstein, following Hallier (1912, 1923), advanced it to a position near the Hydrophyllaceae and the Boraginaceae, a view supported by the morphological studies of Sussenguth (1927) and Copeland (1935), and by the embryological views of Maheshwari (1945). This transfer to the Tubiflorae is based on the pollen being in single grains, the carpels each with

[202] The name Lennoaceae has been conserved over other names for this family.

2 horizontal ovules, the presence of pyrenes, and the lack of differentiation in the embryo.

The plants are of no economic importance.

*LITERATURE:*

COPELAND, H. F. The structure of the flower of *Pholisma arenarium*. Amer. Journ. Bot. 22: 366–383, 1935.
DRUDE, O. Lennoaceae. In Engler and Prantl, Die natürlichen Pflanzenfamilien, IV (1): 12–15, 1891.
HALLIER, H. Über die Lennoeen. Beih. Bot. Centralbl. 40 (2): 1–19, 1923.
MAHESHWARI, P. The place of angiosperm embryology in research and teaching. Journ Indian Bot. Soc. 24: 25–41, 1945.
RYDBERG, P. A. Lennoaceae. North Amer. Flora, 29: 19–20, 1914.
SUSSENGUTH, K. Über die Gattung *Lennoa*. Flora, 122: 264–305, 1927.

## HYDROPHYLLACEAE.[203] WATERLEAF FAMILY

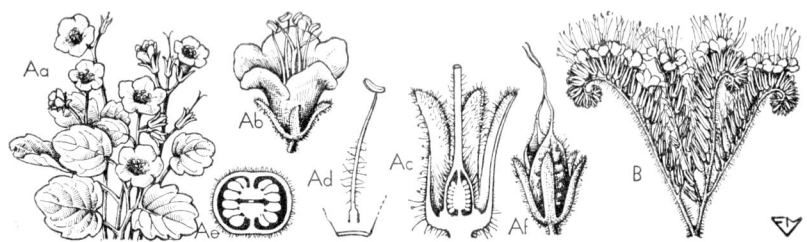

**Fig. 270.** HYDROPHYLLACEAE. *Phacelia campanularia*: Aa, flowering branches, × ½; Ab, flower, × 1; Ac, same, vertical section, × 3; Ad, stamen, × 1½; Ae, ovary, cross-section, × 6; Af, capsule, × 1. B, *Phacelia tanacetifolia*: inflorescence, × ½. (From L. H. Bailey, *Manual of cultivated plants,* The Macmillan Company, 1949. Copyright 1924 and 1949 by Liberty H. Bailey.)

Mostly annual or perennial herbs (shrubby in *Eriodictyon*), often scabrid hairy, glandular, or bristly; leaves alternate or opposite, often in basal rosettes, entire or pinnately divided (rarely palmately divided); inflorescence cymose, often helicoid or circinate, sometimes 2 or more ebracteate cymes in dichasia or umbellate inflorescences, or the flowers solitary and axillary; flowers bisexual, actinomorphic, usually pentamerous (the African *Codon* has pleiomerous flowers); calyx 5-lobed (sepals almost distinct in some genera), imbricate, the sinuses often with appendages; corolla mostly 5-lobed, imbricate or infrequently contorted, rotate, campanulate, or funnelform; stamens mostly 5, epipetalous, arising from base of corolla, alternate with petals, the filaments equal or not so, the anthers 2-celled, dehiscing longitudinally, versatile, introrse, often alternating with or subtended by scalelike or hairy appendages; hypogynous staminodes present or absent; pistil 1, the ovary superior (half inferior in *Nama* spp.), typically unilocular with 2 parietal more or less intruding placentae (sometimes meeting in center, or the ovary appearing as if bilocular), the carpels 2, the ovules often numerous (4 in *Hydrophyllum*), pendulous, anatropous, or amphitropous, the micropyle pointing upward and outward, the styles 1 and somewhat divided or 2, stigma capitate; fruit mostly a loculicidal capsule dehiscing by 2 or rarely 4 valves, sometimes indehiscent; seeds sometimes carunculate, variously pitted, reticulate, sculptured, or muricate, the

[203] The name Hydrophyllaceae has been conserved over Hydroleaceae and other older names.

endosperm copious or thin, cartilaginous or soft- to firm-fleshy, the embryo small and straight.

A widely distributed family occurring on all continents except Australia. It is particularly abundant in western North America and extends south to the Strait of Magellan. It is represented by about 20 genera and 265 species, of which 15 genera occur in this country, with all except *Hydrophyllum, Phacelia, Nama, Nemophila, Ellisia,* and *Hydrolea* restricted to the west and southwest. *Phacelia,* the largest genus, has about 130 species.

The Hydrophyllaceae are distinguished from the related families of Polemoniaceae and Boraginaceae by a combination of characters, notably the imbricate aestivation of the perianth, the usually numerous ovules borne on the 2 parietal placentae, and the tendency toward helicoid cymes.

Phylogenists are agreed that the family is closely allied to the Convolvulaceae, the Polemoniaceae, and Boraginaceae. Bessey included them in his Polemoniales, Wettstein and Rendle placed them in the suborder Boragineae of the Tubiflorae, and Hutchinson restricted the Polemoniales to be composed of only the Polemoniaceae and the Hydrophyllaceae. Hallier included it within the Boraginaceae and removed the latter from the Tubiflorae, treating it as the primitive taxon of his Campanulales.

Economically the family is of little importance aside from the few genera (as *Nemophila, Wigandia* and *Phacelia*) grown to a limited extent as garden ornamentals.

*LITERATURE:*

ABRAMS, L. and SMILEY, F. J. Taxonomy and distribution of *Eriodictyon*. Bot. Gaz. 60: 115–133, 1915.

BRAND, A. Hydrophyllaceae. In Engler, Das Pflanzenreich, 59 (IV. 251): 1–210, 1913.

CAVE, M. S. and CONSTANCE. L. Chromosome numbers in the Hydrophyllaceae. Univ. Cal. Publ. Bot. 18: 205–216, 1942; 293–298, 1944; 449–465, 1947; 23: 363–382, 1950.

CHANDLER, H. P. A revision of the genus *Nemophila*. Bot. Gaz. 34: 194–215, 1902.

CHITTENDEN, R. J. and TURRILL. W. B. Taxonomic and genetical notes of some species of *Nemophila*. Kew Bull. 1926: 1–12, 1926.

CONSTANCE, L. The genus *Eucrypta* Nutt. Lloydia, 1: 143–152, 1938.

———. The genera of the tribe Hydrophylleae of the Hydrophyllaceae. Madroño, 5: 28–33, 1939.

———. The genus *Pholistoma* Lilja. Bull Torrey Bot. Club, 66: 341–352, 1939.

———. The genus *Ellisia*. Rhodora, 42: 33–39, 1940.

———. The genus *Nemophila* Nutt. Univ. Cal. Publ. Bot. 19: 341–398, 1941.

———. The genus *Hydrophyllum* L. Amer. Midl. Nat. 27: 710–731, 1942.

———. A revision of *Phacelia* subgenus *Cosmanthus* (Hydrophyllaceae). Contr. Gray Herb. 168: 1–48, 1949.

DUNDAS, F. W. A revision of the *Phacelia californica* group (Hydrophyllaceae) from North America. Bull Soc. Cal. Acad. Sci. 33: 152–168, 1934.

EASTWOOD, A. Small-flowered species of the genus *Nemophila* from the Pacific Coast Bull. Torrey Bot. Club, 28: 137–160, 1901.

GRAY, A. Conspectus of the North American Hydrophyllaceae. Proc. Amer. Acad. 10: 312–332, 1875.

HOWELL, J. T. Studies in *Phacelia*—a revision of species related to *P. pulchella* and *P. rotundifolia*. Amer. Midl. Nat. 29: 1–26, 1943.

———. A systematic study of *Phacelia humilis* and its relatives. Amer. Midl. Nat. 30: 6–29, 1943.

———. A revision of *Phacelia* section *Militzia*. Proc. Cal. Acad. IV, 25: 357–376, 1944

———. Studies in *Phacelia*. I–IV. Leafl. Western Bot. 3: 95–96, 117–120, 190–191, 1941–42; *op. cit.* 4: 150–152, 1945.

———. Studies in *Phacelia*—Revision of species related to *P. Douglasii, P. linearis* and *P. Pringlei*. Amer. Midl. Nat. 33: 460–494, 1945.

———. A revision of *Phacelia* sect. *Euglypta*. Amer. Midl. Nat. 36: 381–411, 1947.

Peter, A. Hydrophyllaceae. In Engler and Prantl, Die natürlichen Pflanzenfamilien, 4 (3b): 54–71, 1895.

Voss, J. W. A revision of the *Phacelia crenulata* group for North America. Bull. Torrey Bot. Club, 64: 81–96, 1937.

## BORAGINACEAE.[204] BORAGE FAMILY

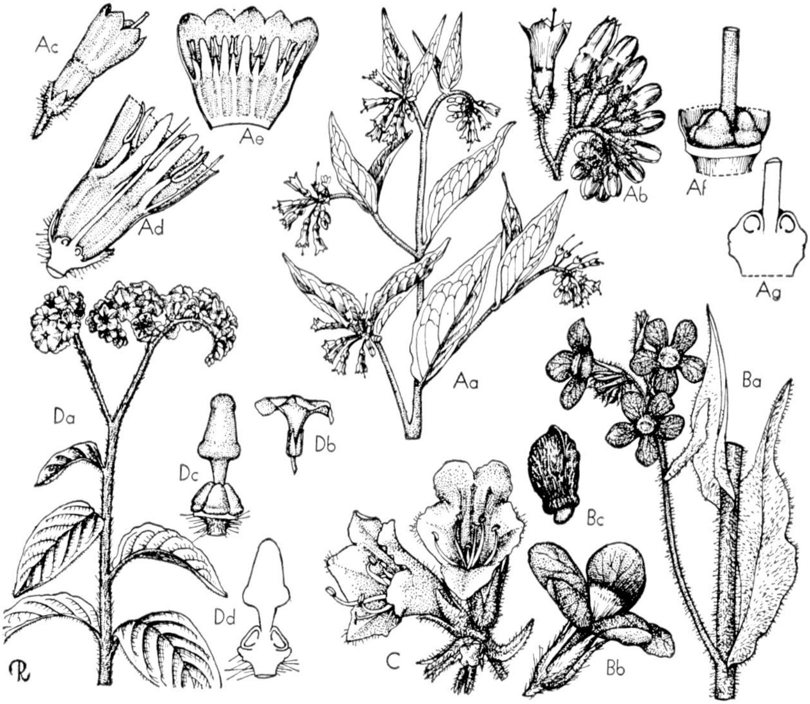

**Fig. 271.** Boraginaceae. A, *Symphytum asperum*: Aa, flowering branch, × ¼; Ab, cymule, × 1; Ac, flower, × 1; Ad, same, vertical section, × 2; Ae, corolla expanded, × 1; Af, ovary, habit, × 4; Ag, same, vertical section, × 4. B, *Anchusa azurea*: Ba, portion of flowering branch, × ½; Bb, flower, × 1; Bc, nutlet, × 5. C, *Echium plantagineum*: flowers, × ½. D, *Heliotropium arborescens*: Da, flowering branch, × ½; Db, flower, × 1; Dc, pistil, × 3; Dd, same, vertical section, × 3. (From L. H. Bailey, *Manual of cultivated plants,* The Macmillan Company, 1949. Copyright 1924 and 1949 by Liberty H. Bailey.)

Herbs, shrubs, or trees, or rarely lianous (*Cordia, Tournefortia* spp.), usually scabrous or hispid hairy and sometimes glabrous; leaves with cystoliths, generally alternate, the lowermost sometimes opposite, simple, usually entire, estipulate; inflorescence determinate, usually composed of 1 or more scorpioid or helicoid cymes that uncoil as the flowers open, and may be glomerate-racemose or spicate, at times loosely cymose or with solitary flowers bracteolate or not so; flowers mostly bisexual, actinomorphic or rarely zygomorphic (*Lycopsis, Echium*), hypogynous; sepals 5, distinct or basally connate, imbricate or rarely valvate, calyx

[204] The name Boraginaceae Lindley (1836) is conserved over older names, and is based on *Borago,* formerly spelled *Borrago* by many authors. For discussion of correctness of the spelling *Borago,* see Sprague (1928).

sometimes irregular; corolla 5-lobed, imbricate or contorted in bud, rotate, salverform, funnelform, or campanulate, the corolla tube with folds or sometimes terminated or partially closed by faucal appendages (scales); stamens 5, epipetalous, equal or less commonly unequal, dorsally appendaged (as in *Borago*) or not so, alternate with corolla lobes, the anthers 2-celled, dehiscing longitudinally, basifixed or basally dorsifixed, introrse; annular nectiferous disc present or absent; pistil 1, the ovary superior, bilocular but becoming falsely 4-locular at maturity, the carpels 2, placentation at times seemingly basal but actually axile, the ovules usually 4 (2 in each carpel), or fewer by abortion, anatropous, erect, inverted, or nearly horizontal, micropyle facing upwards (the ovules curved downwards or descending and micropyle facing downwards in the Heliotropioideae, *Harpagonella, Barthriosperma,* and others); the style 1, gynobasic or (in the Heliotropioideae and Ehretioideae) terminal, generally simple, the stigma mostly simple, capitate (bilobed in some genera, as *Anchusa* or *Pulmonaria*, or 4 in *Cordia* or 2 in *Echium*); fruit of 4 nutlets (2 in *Cerinthe*) or a 1–4-seeded nut or a drupe as in Cordioideae, Heliotropioideae, and Ehretioideae, variously sculptured and vestured or glabrous and glossy; seeds usually without endosperm, or the latter fleshy and usually scant when present (most abundant in Heliotropioideae).

A family of wide distribution, composed of about 100 genera and 2000 species. Of this number about 19 are represented in the United States by indigenous species, and 9 others by naturalized or adventive species. The larger North American genera, primarily western in distribution, include *Cryptantha, Plagiobothrys, Lithospermum,* and *Mertensia*. The family long has been divided into the subfamilies Cordioideae, Ehretioideae, Heliotropioideae, and Boraginoideae, with the latter subdivided into 5 tribes. Most taxonomic treatments are based on fruit or nutlet characters, and fruiting material is almost essential to critical identification.

The Boraginaceae are distinguished from Hydrophyllaceae, Labiatae, Verbenaceae, and other families by the leaves predominantly alternate, terete stems, the usually circinate cymose inflorescence, the usually actinomorphic corollas, frequently with faucal appendages, and the characteristic fruit usually with an erect embryo.

While the majority of botanists accept the family in a broad or more or less conservative sense, others have accepted the elevation of each of the subfamilies to family status (as Cordiaceae, Ehretiaceae, and Heliotropiaceae). Recent phylogenists (Bessey, Hallier, Wettstein, Hutchinson) have interpreted the family in the broad sense, and it is usually the authors of local floras who lack knowledge of the family from the world viewpoint who have accepted the segregate taxa. Johnston, in his monographic studies, and Brand, have treated it as composed of several subfamilies rather than of small microfamilies. In her studies of the floral anatomy in this family, Lawrence (1937) concluded that the subfamilies were best treated as components of the one family; and that "without doubt the Boraginoideae with their deeply cut nutlets . . . are the most highly evolved." Hallier rejected the view that the family belonged in the Tubiflorae, circumscribed it to include the Lennoaceae and Hydrophyllaceae, and treated it as the primitive taxon of his Campanulales, considering it to have been derived directly from the Annonaceae or from stocks ancestral to them. Hutchinson indicated (1948, p. 129) that the woody subfamilies should be kept distinct, as Ehretiaceae.

The Boraginaceae are of slight economic importance. A number of species of about 30 genera are cultivated to a limited extent for ornament, notably heliotrope (*Heliotropium*), Virginia bluebells (*Mertensia*), forget-me-nots (*Myosotis*), lungwort (*Pulmonaria*), borage (*Borago*), alkanet (*Anchusa*), honeywort (*Cerinthe*), hounds' tongue (*Cynoglossum*), comfrey (*Symphytum*), viper's bugloss (*Echium*), and *Cordia*.

*LITERATURE:*

BRAND, A. Borraginaceae—Borraginoideae—Cynoglosseae. In Engler, Das Pflanzenreich, 78 (IV. 252): 1–183, 1921.

———. Borraginaceae—Borraginoideae—Cryptantheae. In Engler, Das Pflanzenreich, 97 (IV. 252): 1–236, 1931.

GEITLER, L. Vergleichend-zytologische Untersuchungen an *Myosotis*. Jahrb. Bot. 83: 707–724, 1936.

GÜRKE, M. Borraginaceae. In Engler and Prantl, Die natürlichen Pflanzenfamilien, IV (3a): 71–131, 1897.

JOHNSTON, I. M. Studies in the Boraginaceae. I, Contr. Gray Herb, n.s. 68: 43–80, 1923; II, *op. cit.* 70: 1–61, 1924; III, *op. cit.* 73: 42–78, 1924; IV, *op. cit.* 74: 3–114, 1925; V, *op. cit.* 75: 40–49, 1925; VI, *op. cit.* 78: 1–118, 1927; VII, *op. cit.* 81: 1–83, 1928; VIII, *op. cit.* 92: 1–95, 1930; IX, Contr. Arnold Arb. 3: 1–102, 1932; X, Journ. Arnold Arb. 16: 1–67, 1935; XI, *op. cit.* 145–205, 1935; XII, *op. cit.* 18: 1–25, 1937; XIII, *op. cit.* 20: 375–402, 1939; XIV, *op. cit.* 21: 48–66, 1940; XV, *op. cit.* 336–355, 1940; XVI, *op. cit.* 29: 227–241, 1948; XVII, *op. cit.* 30: 85–110, 1949; XVIII, *op. cit.* 30: 111–138, 1949.

LAWRENCE, J. R. A correlation of the taxonomy and the floral anatomy of certain of the Boraginaceae. Amer. Journ. Bot. 24: 433–444, 1937.

MACBRIDE, J. F. The true Mertensias of western North America. Contr. Gray Herb. 48: 1–20, 1916.

———. A revision of the North American species of *Amsinckia*. Contr. Gray Herb, n.s. 49: 1–16, 1917.

MOORE, J. A. Morphology of the gynobase in *Mertensia*. Amer. Midl. Nat. 17: 749–752, 1936.

RECORD, S. J. and HESS, R. W. American woods of the family Boraginaceae. Trop. Woods, 67: 19–33, 1941.

SMITH, S. G. Cytology of *Anchusa* and its relation to the taxonomy of the genus. Bot. Gaz. 94: 394–403, 1932.

SPRAGUE, T. A. The correct spelling of the generic name: *Borago* vs. *Borrago*. Kew Bull. pp. 288–292, 1928.

WADE, A. E. Notes on the genus *Myosotis*. Journ. Bot. (Lond.) 80: 127–129, 1944.

WILLIAMS, L. O. A monograph of the genus *Mertensia* in North America. Ann. Mo. Bot. Gard. 24: 17–159, 1937.

## VERBENACEAE.[205] VERBENA FAMILY

Herbs, shrubs, or trees, the stems or twigs often quadrangular; leaves usually opposite or whorled, mostly simple, sometimes palmately or pinnately compound (as in *Vitex*) estipulate; inflorescence variable, usually determinate and of one or another modification of a dichasial cyme, mostly bracteolate; flowers typically bisexual or polygamous by abortion, zygomorphic or rarely actinomorphic; perianth generally pentamerous (tetramerous in *Physopsis*), calyx typically 5-lobed or toothed (rarely with 6–8 teeth or lobes), persistent; corolla usually with as many lobes as calyx, the lobes mostly unequal, usually salverform or rarely campanulate, sometimes 2-lipped, imbricate in bud; stamens primitively 5 (as in *Tectona* and *Geunsia*), commonly 4 and didynamous (the fifth represented by a staminode or not so) or infrequently diandrous (as in *Oxera*) with 3 staminodes, anthers 2-celled, dehiscing longitudinally, introrse; pistil 1, ovary superior, usually with as many lobes as locules, carpels mostly 2 (4 in *Duranta*, 5 in *Geunsia*), locules as many as carpels or twice as many (by false septation), placentation axile (sometimes parietal in the undeveloped ovary), ovules usually solitary in each apparent locule, mostly erect, anatropous, the micropyle facing downward, style 1 and terminal (rarely more or less depressed into the ovary apex), stigma lobes usually as many as carpels; fruit generally a drupe with as many pyrenes as ovules in the

---

[205] The name Verbenaceae Persoon (1806) has been conserved over older names, as Pyrenaceae Vent (1799).

ovary, or of nutlets (*Verbena*), or a 2–4-valved capsule (*Avicennia*); seed with a straight embryo and the endosperm usually absent (present and fleshy in *Avicennia, Stilbe,* and *Chloanthes*).

The Verbenaceae are predominantly a tropical or subtropical family, although *Verbena* (ca. 230 spp.) extends into temperate regions of the New World, and a few species in cooler parts of the Old World. The family is composed of about 98 genera and 2614 or more species, the larger genera including *Vitex* (269 spp.), *Clerodendrum* [206] (383 spp.) *Verbena* (231 spp.) and *Lantana* (155 spp.). Genera with species indigenous to this country include: *Verbena, Phyla, Stylodon,* and *Callicarpa* extending into the cooler regions, and *Avicennia, Citharexylum, Duranta, Lantana, Priva,* and *Stachytarpheta* in the southern and southeastern extremities. *Verbena, Lippia, Bouchea, Aloysia, Lantana,* and *Tetraclea* represent the family in the western part of the country. *Clerodendrum, Aegiphila, Vitex,* and *Premna* have escaped from cultivation in the southern parts.

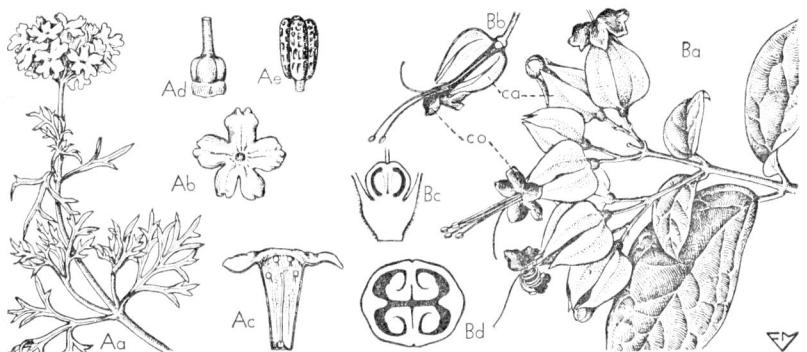

**Fig. 272.** VERBENACEAE. A, *Verbena bipinnatifida*: Aa, flowering branch, × ½; Ab, flower, face view, × ¾; Ac, flower, corolla expanded, × 1; Ad, ovary, × 8; Ae, nutlets, × 3. B, *Clerodendrum Thomsoniae*: Ba, flowering branch, × ½; Bb, flower, vertical section, × ½; Bc, ovary, vertical section, × 3; Bd, same, cross-section, × 6. (ca calyx, co corolla.) (From L. H. Bailey, *Manual of cultivated plants*, The Macmillan Company, 1949. Copyright 1924 and 1949 by Liberty H. Bailey.)

The verbena family is generally accepted as belonging within the Tubiflorae and of close affinity to the Labiatae, although Bessey separated the Labiatae and Verbenaceae (on the basis of corolla zygomorphy and gynoecial characters) as a distinct order, the Lamiales. Hallier retained it within the Tubiflorae and derived it from the Scrophulariaceae. Hutchinson broadened its circumscription to include the Phrymaceae, and (as of 1926) included it within his Lamiales (an order much expanded over that accepted by Bessey), but later (1948) segregated them as the Verbenales. The latter were considered by him to be wholly unrelated to the Labiatae and to have been derived from rubiaceous stocks. Moldenke, following Endlicher, Eichler, and others, accepted Avicenniaceae as a family distinct from Verbenaceae on the basis of its wood anatomy, articulate branches, imbricate scalelike prophylls, the free-central placentation, and the pendant orthotropous ovules. He

---

[206] Authorities differ in usage as to the correct spelling of this generic name. Rehder (1949) *et al.* have used Clerodendron, but Linnaeus, the author of the name, used *Clerodendrum* in *Species plantarum* (p. 637, 1753) and in *Genera plantarum*, ed. 5 (no 707, 1754). The spelling was changed by Adanson (1763) and so adopted by Bentham and Hooker (1876). However, the Rules (ed. 3) provide no authority for changing the original spelling employed by Linnaeus.

also placed *Phryma* in the Phrymaceae, *Campylostachys, Eurylobium, Euthystachys,* and *Stilbe* in the Stilbaceae, and *Congea, Sphenodesme,* and *Symphorema* in the Symphoremaceae.

The family is closely allied to the Labiatae and is distinguished from them by the undivided ovary, the terminal style, and the usually nonverticillate inflorescence. It differs from the terminal-styled members of the Boraginaceae by the micropyle facing the opposite direction from that assumed by the ovule, and in the leaves usually opposite rather than alternate.

Economically the family is perhaps most important for teak lumber (*Tectona grandis*) of East India. A number of genera contain important ornamentals; notable among them are: *Callicarpa, Caryopteris, Clerodendrum, Duranta, Holmskoldia, Lantana, Petrea, Verbena,* and *Vitex.*

LITERATURE:

BAKHUIZEN VAN DEN BRINK, R. C. Revisio generis Avicenniae. Bull. Jard. Bot. Buitenzorg, III, 3: 199–223, 1921.
BEALE, G. H. The genetics of *Verbena* I. Journ. Genetics, 40: 337–358, 1940.
BRIQUET, J. Verbenaceae. In Engler and Prantl, Die natürlichen Pflanzenfamilien, IV (3b): 132–182, 1897.
DERMEN, H. Cytological study and hybridization in two sections of *Verbena*. Cytologia, 7: 160–175, 1936.
JUNELL, S. Zur Gynäceummorphologie und Systematik der Verbenaceen und Labiaten nebst Bemerkungen über ihre Samenentwicklung. Symbolae Bot. Upsal. 1(4): 1–219, 1934.
KOBUSKI, C. E. A revision of the genus *Priva*. Ann. Mo. Bot. Gard. 13: 1–34, 1926.
MOLDENKE, H. N. A monograph of the genus *Timotocia*. Fedde Repert. Spec. Nov. 39: 129–153, 1936.
———. A monograph of the genus *Callicarpa* as it occurs in America and in cultivation. Fedde Repert. Spec. Nov. 40: 38–131, 1936.
———. A monograph of the genus *Cornutia*. Fedde Repert. Spec. Nov. 40: 153–205, 1936.
———. A monograph of the genus *Priva*. Fedde Repert. Spec. Nov. 41: 1–76, 1936.
———. An alphabetical list of invalid and incorrect scientific names proposed in the Verbenaceae and Avicenniaceae. 59 pp. New York, 1942. Privately published.
———. The recorded common and vernacular names of Verbenaceae and Avicenniaceae arranged according to genera and species. Phytologia, 2: 89–123, 1944.
MOLDENKE, H. N. and A. L. A brief historical survey of the Verbenaceae and related families. Plant Life, 2: 13–98, 1946.
PERRY, L. M. A revision of the North American species of *Verbena*. Ann. Mo. Bot. Gard. 20: 239–362, 1933.
PITTIER, H. The middle American and Mexican species of *Vitex*. Contr. U. S. Natl. Herb. 20: 483–487, 1922.
RECORD, S. J. and HESS, R. W. American woods of the family Verbenaceae. Trop. Woods, 65: 4–21, 1941.

## LABIATAE.[207] MINT FAMILY

Predominantly annual or perennial herbs, sometimes shrubs, and rarely trees (*Hyptis* spp.), or lianous (*Scutellaria* spp.), herbage usually with aromatic oils, stems and twigs usually quadrangular; leaves opposite or whorled, simple to pinnately dissected and compound, estipulate; inflorescence usually composed of axillary pairs of dichasial or circinate cymes that form an apparent whorl (verticil), or the flowers solitary in each axil (*Scutellaria*), or sometimes the primary axis with shortened internodes and the verticils congested into a contiguous series

---

[207] The name Labiatae is one of 8 family names not ending in -aceae, and this name has been conserved subject to the provision that the alternative name Lamiaceae may be used.

**Fig. 273.** LABIATAE. A, *Salvia splendens*: Aa, inflorescence, × ½; Ab, flower, vertical section, × 1; Ac, ovary, × 3; Ad, same, vertical section, × 3; Ae, nutlets, × 2. B, *Stachys grandiflora*: Ba, inflorescence, × ¼; Bb, flower, × ½; Bc, same, corolla expanded, × 1. C, *Molucella laevis*: whorl of flowers, × ½. D, *Teucrium lucidum*: Da, inflorescence, × 1; Db, flower, × 2. E, *Monarda didyma*: Ea, inflorescence, × ½; Eb, flower, × 1. F, *Mentha spicata*: Fa, inflorescence, × ½; Fb, flower, × 3. G, *Nepeta Faassenii*: Ga, inflorescence, × ½; Gb, flower, × 2. (From L. H. Bailey, *Manual of cultivated plants*, The Macmillan Company, 1949. Copyright 1924 and 1949 by Liberty H. Bailey.)

or head (as in *Lamium, Prunella, Hyptis,* or *Monarda*), the flowers bracteolate or not, and each cyme usually subtended by a foliaceous bract that may exceed the cyme; flowers bisexual, zygomorphic (actinomorphic in *Mentha* and *Elsholtzia* or nearly so); calyx persistent, imbricate, typically 5-lobed, sometimes bilabiate (the lobes occasionally obsolete and seemingly 2, usually with 5, 10, 13, or 15 conspicuous ribs; corolla typically 5-lobed, bilabiate (upper lip seemingly obsolete in *Teucrium*), the lower lip typically 3-lobed and often concave; stamens 2 or 4, epipetalous, usually didynamous with anterior pair usually the longer, a staminode rarely present, distinct (monadelphous in *Coleus*), the anthers 2-celled, dehiscing longitudinally, the connective often much developed (as in *Salvia*) and the anterior cell reduced or absent; hypogynous nectiferous disc often present between stamens and ovary; pistil 1, the ovary superior, 4-lobed, carpels 2, locules 2 or seemingly 4 by intrusion of ovary wall, the placentation basal and derived from axile type, the ovules 4, anatropous, erect, the micropyle facing downward, the style 1 and gynobasic, arising from the central depression of the lobes, shortly 2-branched, the stigmas minute at branch tips; fruit composed of typically 4 nutlets (rarely the pericarp fleshy), distinct or cohering in pairs, enclosed within the persistent calyx; seed with scant fleshy endosperm that is often absorbed by the developing embryo.

The Labiatae are a large family of about 200 genera and 3200 species, of cosmopolitan distribution, but whose center is chiefly in the Mediterranean region, where they form a dominant part of vegetation. Some of the subfamilies are localized in distribution, as the Prostantheroideae in Australia and Tasmania, the Prasioideae in Malaya, India, and China, and the Catopherioideae in Central America. Most of the North American genera belong to the more cosmopolitan subfamilies Stachydoideae and Ajugoideae. The family is widespread throughout this country, where it is represented by about 48 genera, of which the larger include *Salvia, Pycnostachys, Scutellaria, Stachys, Monarda, Monardella.*

The mint family is readily distinguished from all other families except the Verbenaceae by the distinctive gynoecium. The Labiatae subfamilies Ajugoideae and Prostantheroideae resemble the Verbenaceae in that the style is terminal and not gynobasic as in other mints, likewise several verbenaceous genera have a nearly gynobasic style. The presence of these and other intergrading characters makes it difficult if not impossible to separate all members of one family from the other by any single character or combination of characters, and in most reliable keys artificial characters are used. The Boraginaceae differ from the Labiatae consistently by the technical character of the ovule (micropyle pointing upward and the raphe directed outward in Boraginaceae, and the micropyle inferior and raphe inward in the mints).

Hallier, Wettstein, and Rendle followed the Englerian view that the family belonged in the Tubiflorae. Bessey separated it and the Verbenaceae into his Lamiales, largely on the basis of corolla zygomorphy and the gynoecial situation. Hutchinson (1926) did likewise at first but later (1948) restricted the Lamiales to the Labiatae, segregated the Verbenaceae as the Verbenales, and considered the 2 as unrelated (the former in his Herbaceae and the latter in his Lignosae).

Economically the family is of importance as a source of volatile aromatic essential oils and garden ornamentals. Among the more important essential oils are sage (*Salvia*), lavender (*Lavandula*), rosemary (*Rosmarinus*), mint (*Mentha* spp.), patchouly (*Pogostemon*). In addition to many of the above, others serve as important culinary herbs prized for the flavor or aroma imparted to foods, especially pot marjoram (*Origanum*), hyssop (*Hyssopus*), pennyroyal (*Hedeoma pulegioides*), basil (*Ocimum*), thyme (*Thymus*), and savory (*Satureja*). Hoarhound (*Marrubium*) is used in medicinal preparations and confections. The principal ornamentals include: salvia (*Salvia*), bugloss (*Ajuga*), lion's-ear (*Leonotis*).

## DIVISION IV. EMBRYOPHYTA SIPHONOGAMA

dragonhead (*Dracocephalum*), false dragonhead (*Physostegia*), Oswego tea (*Monarda*), skullcap (*Scutellaria*), and species of *Nepeta, Stachys, Teucrium, Thymus, Coleus, Lavandula, Pycnanthemum*.

*LITERATURE:*

BENTHAM, G. Labiatarum genera et species etc. London, 1832–1836.
BRIQUET, J. Labiatae. In Engler and Prantl, Die natürlichen Pflanzenfamilien, IV (3a): 183–375, 1897.
BUSHNELL, E. P. Cytology of certain Labiatae. Bot. Gaz. 98: 356–362, 1936.
CHAYTON, D. A. A taxonomic study of the genus *Lavandula*. Journ. Linn. Soc. Lond. (Bot.) 51: 153–204, 1937.
DUNN, S. T. A key to the Labiatae of China. Notes Roy. Bot. Gard. Edinb. 6: 127–208, 1915.
EPLING, C. Monograph of the genus *Monardella*. Ann. Mo. Bot. Gard. 12: 1–106, 1925.
———. Notes on Linnaean types of American Labiatae. Journ. Bot. 67: 1–12, 1929.
———. Preliminary revision of American *Stachys*. Fedde Repert. Spec. Nov. Beih. 80: 1–75, 1934.
———. Notes on *Monarda*: the subgenus *Cheilyctis*. Madroño, 3: 20–31, 1935.
———. The California salvias. A review of *Salvia*, section *Audibertia*. Ann. Mo. Bot. Gard. 25: 95–188, 1938.
———. *Scylla, Charybdis* and Darwin. Amer. Nat. 72: 547–561, 1938.
———. A revision of *Salvia*: subgenus *Calosphace* I. Fedde Repert. Spec. Nov. Beih. 110(1): 1–160, 1938; II, *op. cit.* 112(2): 161–380, 1939; and in Publ. Univ. Cal. L. A. in Biol. Sci. 2: 1–383, 1940.
———. A note on the occurrence of *Salvia* in the new world. Madroño, 5: 34–37, 1939.
———. Supplementary notes on American Labiatae I, Bull. Torrey Bot. Club, 37: 509–534, 1940; II, *op. cit.* 68: 552–568, 1941; III, *op. cit.* 71: 484–497, 1944; IV, *op. cit.* 74: 512–518, 1947.
———. The American species of *Scutellaria*. Cal. Pub. Bot. 20: 1–146, 1942.
———. A synopsis of the tribe Lepechinieae (Labiatae). Brittonia, 6: 352–364, 1948.
FERNALD, M. L. A synopsis of the Mexican and Central American species of *Salvia*. Contr. Gray Herb. 19: 489–556, 1900.
———. The geographic segregation of *Monarda fistulosa* and its var. *mollis*. Rhodora, 46: 494–496, 1944.
GRANT, E. and EPLING, C. A study of *Pycnanthemum* (Labiatae). Cal. Pub. Bot. 20: 195–240, 1943.
HERMANN, F. J. Diagnostic characteristics in *Lycopus*. Rhodora, 38: 373–375, 1936.
HOWELL, J. T. The genus *Pogogyne*. Proc. Cal. Acad. Soc. 20: 105–128, 1931.
JORGENSEN, C. A. Cytological and experimental studies in the genus *Lamium*. Hereditas, 9: 126–136, 1927.
JUNELL, S. Zur Gynäceummorphologie und Systematik der Verbenaceen und Labiaten nebst Bemerkungen über ihre Samenentwicklung. Symbolae Bot. Upsala, (1)4: 1–210, 1934.
———. Die Samenentwicklung bei einigen Labiaten. Svensk Bot. Tidsk. 31: 67–110, 1937.
LEONARD, E. C. The North American species of *Scutellaria*. Contr. U. S. Natl. Herb. 22: 703–748, 1927.
LEWIS, H. A revision of the genus *Trichostema*. Brittonia, 5: 276–303, 1945.
MCCLINTOCK, E. A proposed retypification of *Dracocephalum* L. Leafl. West. Bot. 5: 171, 172, 1949.
———. A review of the genus *Monarda* (Labiatae). Univ. Pub. Bot. 20: 147–194, 1942.
MCCLINTOCK, E. and EPLING, C. A revision of *Teucrium* in the New World, with observations on its variation, geographical distribution and history. Brittonia, 5: 491–510, 1946.
RECHINGER, K. H. fil. Monographische Studie über *Teucrium* Sect. *Chamaedrys*. Bot. Arch. (Leipzig), 42: 335–420, 1941.
RONNIGER, K. Beiträge zur Kenntnis der Gattung *Thymus* I. Fedde Repert. Spec. Nov. 20: 331–332, 334–336, 1924.

RUTTLE, M. L. Cytological and embryological studies on the genus *Mentha*. Gartenbauwissenschaft, 4: 428–468, 1931.
STEWART, S. R. *Mentha arvensis* and some of its North American variations. Rhodora, 46: 331–335, 1944.
STEWART, W. S. Chromosome numbers of California Salvias. Amer. Journ. Bot. 26: 730–732, 1939.

## NOLANACEAE. NOLANA FAMILY

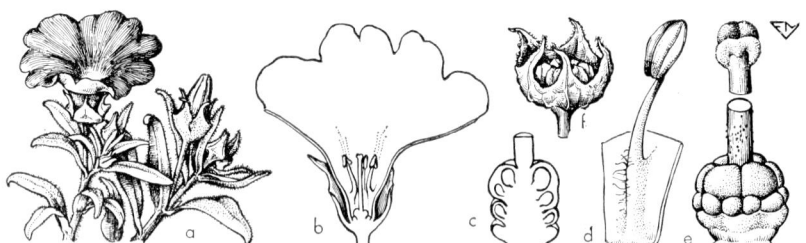

Fig. 274. NOLANACEAE. *Nolana paradoxa*: a, flowering branch, × ½; b, flower, vertical section, × 1; c, ovary, vertical section, × 5; d, stamen, on section of corolla-tube, × 3; e, pistil, × 5; f, fruit, × 1. (From L. H. Bailey, *Manual of cultivated plants*, The Macmillan Company, 1949. Copyright 1924 and 1949 by Liberty H. Bailey.)

Herbs or small shrubs (mostly strand plants); leaves alternate or becoming paired toward stem apex, simple, mostly fleshy or semisucculent, estipulate; flowers solitary in leaf axils, bisexual, actinomorphic; the calyx 5-cleft; corolla gamopetalous, hypogynous, 5-lobed, plicate in bud, campanulate to funnelform; stamens 5, usually equal or sometimes unequal, alternating with corolla lobes and arising from near the base, the anthers 2-celled, dehiscing longitudinally; disc present, often lobed; pistil 1, the ovary superior, often radially or transversely lobed, the carpels typically 5, the locules as many as carpels or (by transverse lobing) 2–3 times as many, the placentation axile, the ovules anatropous, the style 1 and terminal, stigma of 2 or more lobes; fruit a 1–7-seeded mericarp, sometimes incorrectly termed a nutlet, the endocarp usually stony; seed with a curved embryo and copious endosperm.

A family of 2 genera and over 60 species native in Chile and Peru, mostly along the shore line. *Nolana*, the largest genus, has about 57 species, and *Alona* has 6 species.

The family is anomalous, and numerous morphological characters have led some botanists to align it with one or another family. Bentham and Hooker and Hutchinson included it with the Convolvulaceae on the basis of corolla form, the twisted aestivation, and pentamerous ovary. The mericarpous fruit is suggestive of affinities to Boraginaceae, but the strongest affinity appears to be with the Solanaceae, with which it was united by Duval and others, but from which it differs by its peculiar nutletlike fruit.

Wettstein and Rendle accepted the family and included it in the Tubiflorae. Bessey placed it, together with the Solanaceae, in his Polemoniales, and both Hallier and Hutchinson treated the genus as belonging in the Solanaceae.

One or two species of *Nolana* are cultivated domestically as garden annuals (especially *N. paradoxa*).

*LITERATURE:*

JOHNSTON, I. M. A study of the Nolanaceae. Contr. Gray Herb. 112: 1–83, 1936.
WETTSTEIN, R. Nolanaceae. In Engler and Prantl, Die natürlichen Pflanzenfamilien, IV (3b): 1–4, 1895.

## SOLANACEAE.[208] NIGHTSHADE FAMILY

Fig. 275. SOLANACEAE. A, *Nierembergia hippomanica* var. *violacea*: Aa, flowering branch, × ½; Ab, flower, × ½; Ac, same, perianth partly excised, × 2; Ad, flower, face view, × ½. B, *Lycopersicon esculentum*: Ba, flowering branch, × ½; Bb, flower, × 1; Bc, flower, vertical section, corolla partly excised, × 2; Bd, ovary, cross-section, × 5. C, *Schizanthus pinnatus*: Ca, flower, × 1; Cb, same, perianth expanded, × 1. D, *Browallia viscosa*: Da, flower, face view, × ½; Db, flower, perianth excised in vertical section, × 1. E, *Nicotiana alata* var. *grandiflora*: Ea, flower, vertical section, × ½; Eb, bud, partially expanded, × ½; Ec, capsule, × ½. (From L. H. Bailey, *Manual of cultivated plants*, The Macmillan Company, 1949. Copyright 1924 and 1949 by Liberty H. Bailey.)

Herbs, shrubs, trees, often lianous or creeping, the stems with bicollateral vascular strands; leaves alternate or becoming opposite at or near the inflorescences, simple or rarely pinnatisect, estipulate; inflorescence typically an axillary cyme or combination of cymes, sometimes helicoid [209]; flowers bisexual, actinomorphic or slightly zygomorphic, hypogynous; calyx 5-lobed or -parted (sometimes 4-6-lobed), persistent, often enlarging in fruit; corolla gamopetalous, rotate to tubular, typically 5-lobed, variously shaped but rarely bilabiate (as in *Schizanthus*), usually plicate or convolute (rarely valvate); stamens epipetalous, alternating with corolla lobes, usually unequal, primitively 5, sometimes 4 and didynamous or not so, or only 2 (*Schizanthus*), when less than 5 the "lost" stamen often represented by a staminode, the anthers 2-celled (1 cell sometimes undeveloped, as in *Browallia*),

[208] The name Solanaceae has been conserved over other names for the family.
[209] See Rendle, 2: 516–517, for details.

dehiscing longitudinally or sometimes conically connivent and dehiscing poricidally (as *Solanum*), the connective sometimes enlarged (*Cyphomandra*); hypogynous disc usually present and apparent; pistil 1, ovary superior, carpels 2, typically 2-loculed (or 3–5 by false septation) rarely unilocular (as in *Henoonia,* which also has a single ovule, or unilocular apically as in *Capsicum,* the placentation axile, the ovules numerous (few in *Cestrum*) on intruding placentae, anatropous or somewhat amphitropous; style 1, stigma 2-lobed; fruit a berry (sometimes enclosed within an inflated persistent calyx as in *Physalis*), or septicidal capsule; seeds smooth or pitted, embryo embedded in the fleshy and semitransparent endosperm.

A large family of about 85 genera and in excess of 2200 species, distributed primarily in tropical America and South America (where there are 38 endemic genera). Some of the larger genera include *Solanum* (about 1500 spp.), *Cestrum* (250 spp.), *Lycium* (100 spp.), *Physalis* (100 spp.), *Nicotiana* (100 spp.), and *Cyphomandra* (30 spp.). Genera represented in this country by indigenous species include *Physalis, Solanum,* and *Chamaesaracha* in the northeastern and middle states, *Capsicum* and *Petunia* in peninsular Florida, and *Lycium, Margaranthus, Saracha, Nicotiana,* and *Oryctes* in the southwestern and western states. Species of *Nicandra, Hyoscyamus,* and *Datura* are naturalized or adventive over much of the United States and probably indigenous in the southwest.

The Solanaceae were interpreted by Wettstein as probably of a polyphyletic origin, as evidenced by alliance with several families. It is most closely related to the Scrophulariaceae, and is best distinguished from them by the actinomorphic corolla, the typically 4 or 5 stamens, the usually plicate corolla, and the invariable anatomical character of bicollateral vascular strands. The family was divided by Wettstein into the tribes: Nicandreae, Solaneae, Datureae, Cestreae, and Salpiglossideae.

The family was included by Bentham and Hooker and by Bessey within the Polemoniales (separated by Bessey from his Scrophulariales by the actinomorphic corolla), whereas Hallier considered it to be the primitive member of the Tubiflorae and (together with the Scrophulariaceae) to have been derived probably from the Linaceae. Hutchinson included it as the primitive taxon of his Solanales, together with the Convolvulaceae, an order ancestral to his Personales.

The Solanaceae are a family of considerable economic importance, and are the source of food plants as the potato and eggplant (*Solanum* spp.), tomato (*Lycopersicon*), strawberry tomato (*Physalis*), and red pepper (*Capsicum*); the fumitory, tobacco (*Nicotiana* spp.); such drug plants as henbane (*Hyoscyamus*), belladonna and atropine (*Atropa*), and stramonium (*Datura*) and ornamentals from many genera including *Petunia, Salpiglossis, Schizanthus, Lycium, Solanum, Streptosolen, Cestrum, Datura, Solandra, Browallia, Nierembergia,* and *Brunfelsia.*

LITERATURE:

BAEHNI, C. L'ouverture du bouton chez les fleurs des Solanées. Candollea, 10: 399–494, 1946.
BARTLETT, H. H. The purple-flowered Androcerae (*Solanum*) of Mexico and southern U. S. Contr. Gray Herb. 36: 627–629, 1909.
BITTER, G. Solana Nova vel Minus Cognita. I–XIV. Fedde Repert. Sp. Nov. vols. 10–13, 1912–1914.
———. Die Gattung *Lycianthes*. Abhandl Nat. Ver. Bremen, 24: 292–520, 1919–1920.
COMES, O. Monographie du genre *Nicotiana.* Paris, 1899.
DUNAL, M. F. Solanaceae. In de Candolle, Prodomus, 13(1): 1–690, 1852.
FERNALD, M. L. A revision of the Mexican and Central American Solanums of the subsection Torvaria. Contr. Gray. Herb. 19: 557–562, 1900.
FRANCEY, P. Monographie du genre *Cestrum* L. Candollea, 6: 46–398, 1935; *op. cit.* 7: 1–132, 1936.

FRIES, R. E. Die Arten der Gattung *Petunia*. K. Svensk Vetensk. Handl. vol. 46, 1911
GOODSPEED, T. H. [See Chapter VIII, p. 189, for references on *Nicotiana*.]
HAWKES, J. G. Potato Collecting Expeditions in Mexico in South America. II. Systematic Classification of the Collections. Imperial Bureau of Plant Breeding and Genetics. [pp. 1–142]. Cambridge, England, 1944.
HITCHCOCK, C. L. A monographic study of the genus *Lycium* of the Western Hemisphere. Ann. Mo. Bot. Gard. 19: 179–364, 1932.
LAMM, R. Cytogenetic studies in *Solanum*, Sect. *Tuberarium*. Dissert. 1–128, Lund, 1944.
MURRAY, M. A. Carpellary and placental structure with Solanaceae. Ph.D. Thesis. Univ. Chicago, 1946. [Not seen.]
RYDBERG, P. A. The North American species of *Physalis* and related genera. Mem. Torrey Bot. Club, 4: 297–372, 1896.
SAFFORD, W. E. Synopsis of the genus *Datura*. Journ. Wash. Acad. Sci. 11: 173–189, 1921.
SCHULZ, O. E. Solanacearum genera nonnulla, in Urban. Symb. Antill. 6: 140–279, 1909.
WETTSTEIN, R. Solanaceae. In Engler and Prantl, Die natürlichen Pflanzenfamilien, IV (3b): 4–39, 1895.

## SCROPHULARIACEAE.[210] FIGWORT FAMILY

Mostly herbs or small shrubs, sometimes lianous (*Maurandia, Rhodochiton*), a few are chlorophyll-less parasites (*Hyobanche*) and others are chlorophyll-containing parasites or saprophytes (*Melampyrum, Pedicularis, Castilleja*); leaves usually alternate or opposite, rarely whorled (*Veronicastrum*), deciduous or infrequently persistent (*Hebe*), simple, entire, or pinnately lobed or incised, estipulate; inflorescence variable, determinate or indeterminate, rarely umbellate, bracts and bracteoles usually present (brightly colored in *Castilleja*); flowers bisexual, typically zygomorphic, or nearly actinomorphic; calyx usually deeply 4–5 lobed or divided, imbricate or valvate; corolla gamopetalous, the tube sometimes very short (as *Veronica*) or conspicuous (*Penstemon* or *Digitalis* spp.), often bilabiate, sometimes personate (as *Linaria*), the lobes usually 4–5 (rarely 6–8) and imbricate, one or more anterior petals sometimes produced into a spur (*Linaria, Diascia*) or basally gibbous to saccate (*Antirrhinum*) or the limb developing into 2 unequal inflated lips (*Calceolaria*); stamens epipetalous but arising from the extreme base of corolla tube, distinct (sometimes coherent in pairs) sometimes 5 (*Verbascum* and *Capraria*), commonly 4 and didynamous (fifth stamen sometimes represented by a filamentous staminode, as in *Penstemon* or reduced to a scale in some *Scrophularia* spp.), sometimes only 2 (*Veronica, Calceolaria*), the anthers 2-celled (one cell sometimes larger than the other, or in *Buchnera* one cell obsolete), dehiscing longitudinally or rarely poricidally (in *Seymeria*), introrse; nectiferous annular or unilateral disc usually present and often lobed; pistil 1, the ovary superior, the locules and carpels 2 (rarely unilocular), the placentation axile, the ovules numerous, anatropous, on usually enlarged placentae, the style 1 and terminal, the stigma bilobed or 2 (as in *Gratiola*); fruit typically a capsule, septicidal, occasionally loculicidal (spp. of Buchnereae and Euphrasieae) or sometimes poricidal (*Antirrhinum*), rarely a berry (*Leucocarpus, Halleria*) or dry indehiscent capsule (*Hebenstretia*); seeds smooth or variously surfaced, angled or winged, the endosperm soft-fleshy to fleshy.

A large family of about 210 genera and nearly 3000 species, of cosmopolitan distribution and represented on all continents. The largest genera are *Pedicularis* (probably 600 spp.), *Calceolaria* (500), *Penstemon* (250), *Verbascum* (250), *Linaria* (150), *Mimulus* (150), *Veronica* (excl. *Hebe*, 150), *Hebe* (140), and *Castilleja* (100). Of these all are of the northern hemisphere except *Hebe* (Austral-

---

[210] The name Scrophulariaceae has been conserved over other earlier names for the family, and the name Rhinanthaceae as used by J. K. Small and others is invalidated.

asia) and *Calceolaria* (western South America). The larger genera in this country include *Penstemon, Mimulus, Veronica, Gerardia, Chytra, Castilleja,* and *Pedicularis.*

The Scrophulariaceae are separated from the Solanaceae by the nonplicate usually zygomorphic corolla, the collateral (not bicollateral) vascular strands, and the reduction (usually) of the posterior stamen; from the Gesneriaceae and Orobanchaceae by the usually bilocular ovary; and from the Pedaliaceae and Bignoniaceae by the presence of endosperm.

**Fig. 276.** SCROPHULARIACEAE. A, *Antirrhinum majus*: Aa, flowers, × ½; Ab, flower, vertical section, perianth partially excised, × 1. B, *Calceolaria crenatiflora*: Ba, portion of inflorescence, × ½; Bb, flower, vertical section, × 1. C, *Torenia Fournieri*: Ca, flowering branch, × ½; Cb, flower, side view, × ½; Cc, flower, perianth expanded and partially excised, × ½. D, *Veronica longifolia*: Da, inflorescence, × ½; Db, flower, × 2; Dc, same, vertical section, × 3; Dd, capsule, × 2; De, ovary, cross-section, × 10. E, *Penstemon laevigatus*: Ea, inflorescence, × ½; Eb, flower, vertical section, × 1. (From L. H. Bailey, *Manual of cultivated plants*, The Macmillan Company, 1949. Copyright 1924 and 1949 by Liberty H. Bailey.)

The family was considered by Wettstein, following Bentham (1846), to be composed of 3 subfamilies (Pseudosolaneae, Antirrhinoideae, Rhinanthoideae) and 10 tribes. Rendle and Hutchinson accepted elevation of the African tribe Selagineae (including *Hebenstretia* and *Selago*) to family status, the Selaginaceae. Hallier considered the Scrophulariaceae to be one of the primitive families of the Tubiflorae, including in it such related taxa as the Globulariaceae, Lentibulariaceae,

Plantaginaceae, and Selaginaceae. Bessey treated it as an advanced component of his Scrophulariales, derived perhaps from the Bignoniaceae. Pennell (1935) rejected the earlier view that the nearly actinomorphic and 5-stamened flower of *Verbascum* was primitive in the family (indicating presumed alliance to the Solanaceae), considered the primitive corolla (of Scrophulariaceae) to have been zygomorphic, and transferred the Verbasceae to "a more advanced position, influenced by its stigma, seeds, elaborate anthers, etc." In doing so he noted that here was an example of ". . . the recovery of a structure which had become previously obsolete; that the suppressed posterior stamen could appear again is a remarkable bit of evolutionary information."

The Scrophulariaceae are not a particularly important family, and except for the drug plant *Digitalis,* the plants are valued primarily as garden ornamentals. Notable among the latter are the snapdragons (*Antirrhinum*), speedwells, (*Veronica*), slipperflower (*Calceolaria*), beardtongues (*Penstemon*), monkey flower (*Mimulus*), Kenilworth ivy (*Cymbalaria*), coral plant (*Russelia*), wishbone flower (*Torenia*), and foxgloves (*Digitalis*).

*LITERATURE:*

BENTHAM, G. Scrophulariaceae. In de Candolle, Prodromus, 10: 186–586, 1846.
CLAUSEN, J., KECK, D. D. and HIESEY, W. M. *Penstemon*: a study in cytotaxonomy and transplanting. Carnegie Inst. Wash. Publ. no. 520: 260–295, 1940.
CLOKEY, I. W. and KECK, D. Reconsideration of certain members of *Pentstemon* subsection *Spectabiles.* Bull. So. Cal. Acad. Sci. 38: 8–13, 1939.
DILL, F. E. Morphology of *Veronicastrum virginicum.* Trans. Kansas Acad. 44: 158–163, 1941.
EASTWOOD, A. Studies in *Castilleja.* I. *Castilleja* in the Marble Mountains, Siskiyou County, California. Leafl. West. Botany, 2: 241–245, 1940.
———. Some apline Castillejas from the High Sierra of California. Amer. Midl. Nat. 30: 40–46, 1943.
———. Synopsis of the Mexican and Central American species of *Castilleja.* Contr. Gray Herb. 36: 564–591, 1909.
FISCHER, J. Zur Entwicklungsgeschichte und Morphologie der Veronicabluete. Zeitschr. Bot. 12: 113–161, 1920.
GLEASON, H. A. Specific names in *Gratiola.* Phytologia, 2: 503, 504, 1948.
GRANT, A. G. A monograph of the genus *Mimulus.* Ann. Mo. Bot. Gard. 11: 99–388, 1924.
HARLE, A. Die Arten und Formen der *Veronica,* Sektion *Pseudolysimachia* Koch auf Grund systematischer und experimenteller Untersuchungen. Biblioth. Bot. 104: 1–35, 1932.
KECK, D. D. Studies in *Penstemon.* I. Univ. Calif. Publ. Bot. 16: 367–428, 1932; II, Madroño, 3: 200–219, 1936; III, *op. cit.* 248–250, 1936; IV, Bull. Torrey Bot. Club, 64: 357–381, 1937; V. Amer. Midl. Nat. 18: 790–829, 1937; VI, Bull. Torrey Bot. Club, 65: 233–255, 1938; VII, Amer. Midl. Nat. 23: 594–616, 1940; VIII, *op. cit.* 33: 128–206, 1945.
LIMPRICHT, W. Studien über die Gattung *Pedicularis.* Fedde Repert. Spec. Nov. 20: 161–265, 1924.
MILLSAPS, V. The structure and development of the seed of *Paulownia tomentosa* Steud. Journ. Elisha Mitchell Sci. Soc. 52: 56–75, 1936.
MINOD, M. Contribution à l'étude du genre *Stemodia.* Bull. Soc. Bot. Genève, II. 10: 155–252, 1918.
MUNZ, P. A. The Antirrhinoideae-Antirrhineae of the New World. Proc. Cal. Acad. Sci. IV, 15: 323–397, 1926.
MUNZ, P. A. and JOHNSTON, I. M. The Penstemons of Southern California. Bull. So. Calif. Acad. Sci. 23: 21–40, 1924.
MURBECK, S. Monographie der Gattung *Celsia.* Acta Univ. Lund. n.s. 22 (1): 1–239, 1925.
———. Monographie der Gattung *Verbascum.* Acta Univ. Lund. n.s. 29 (2): 1–630, 1933; Nachträge [Suppl.] in *op. cit.* 32: 1–45, 1936.

PENNELL, F. W. Studies in the Agalinanae, a subtribe of the Rhinanthaceae. Bull. Torrey Bot. Club, 40: 119–130, 401–439, 1913.
———. Scrophulariaceae of the southeastern United States. Proc. Acad. Nat. Sci. Phila. 71: 224–291, 1920.
———. Scrophulariaceae of the central Rocky Mountain states. Contr. U. S. Nat. Herb. 20: 313–381, 1920.
———. "Veronica" in North and South America. Rhodora, 23: 1–22, 29–41, 1921. [Cf. also *op. cit.* 34: 149–151, 1932.]
———. Scrophulariaceae of the West Gulf States. Proc. Acad. Nat. Sci. Phila. 73: 450–535, 1921.
———. The genus *Afzelia*; a taxonomic study in evolution. Proc. Acad. Nat. Sci. Phila. 77: 335–373, 1926.
———. *Agalinis* and allies in North America. Proc. Acad. Nat. Sci. Phila. 80: 339–449, 1928; 81: 111–249, 1929.
———. Genotypes of the Scrophulariaceae in the first edition of Linné's "Species Plantarum". Proc. Acad. Nat. Sci. Phila. 86: 9–26, 1930.
———. Revision of *Synthyris* and *Besseya*. Proc. Acad. Nat. Sci. Phila. 85: 77–106, 1934.
———. *Pedicularis* of the group *Bracteosae*. Bull. Torrey Bot. Club, 61: 441–448, 1934.
———. The Scrophulariaceae of eastern temperate North America. Monog. Acad. Nat. Sci. Phila. 1: 1–650, 1935.
———. *Castilleja* in Alaska and northwestern Canada. Proc. Acad. Nat. Sci. Phila. 86: 517–540, 1935.
———. Scrophulariaceae of Trans-Pecos Texas. Proc. Acad. Nat. Sci. Phila. 92: 289–308, 1941.
PENNELL, F. W. and WHERRY, E. T. The genus *Chelone* of eastern North America. Bartonia, 10: 12–23, 1929.
SCHLECHTER, R. Scrophulariaceae. Engl. Bot. Jahrb. 59: 99–117, 1924–1925.
SOO, R. VON. Systematische Monographie der Gattung *Melampyrum*. I. Fedde Repert. Spec. Nov. 23: 159–176, 1926; II, *op. cit.* 385–397, 1927; III, *op. cit.* 24: 124–193, 1927.
SRINIVASAN, V. K. Morphological and cytological studies in the Scrophulariaceae. II. Floral morphology and embryology of *Angelonia grandiflora* C. Morr. and related genera. Journ. Indian Bot. Soc. 19: 197–222, 1940.
STIEFELHAGEN, H. Systematische und pflanzengeographische Studien zur Kenntnis der Gattung *Scrophularia*. Engler, Bot. Jahrb. 44: 406–496, 1910.
WETTSTEIN, R. Scrophulariaceae. Engler and Prantl, Die natürlichen Pflanzenfamilien, IV (3b): 39–107, 1895.

## BIGNONIACEAE.[211] BIGNONIA FAMILY

Trees or shrubs, often climbing or twining vines, rarely herbs (*Incarvillea*) or suffrutescent (*Eccremocarpus* spp.); leaves opposite, decussate, rarely alternate, simple or pinnately compound (terminal leaflet then often reduced to a tendril), often glandular (between petiole bases, in leaf-vein axils) [212], estipulate; inflorescence usually of dichasial cymes compounded into a monochasium, bracts and bracteoles produced; flowers bisexual, zygomorphic, showy, the calyx with 5 teeth or lobes, sometimes bilabiate (spathelike in *Spathodea*) or truncate, the corolla campanulate or funnelform, usually imbricate, lobes or teeth 5 or more, sometimes bilabiate; stamens epipetalous, typically 4 and didynamous with a posterior staminode, sometimes 2 (*Catalpa*) and with or without 3 staminodes, the anthers 2-celled, the cells usually widely divergent and then seemingly one above the other, coherent or distinct, dehiscing longitudinally; hypogynous disc present; pistil 1, the ovary superior, 2-carpeled, typically bilocular with axile placentation or some-

[211] The name Bignoniaceae has been conserved over other earlier names for the family.
[212] See Seibert (1948) for details on presence and taxonomic utility of glands on vegetative and reproductive parts of especially lianous members of the family.

times unilocular with 2 bifid parietal placentae (as *Eccremocarpus, Kigelia*), the ovules numerous, anatropous, usually erect with micropyle pointing downward, the style terminal and simple, the stigma 2-lobed; fruit usually a 2-valved septicidal or loculicidal capsule, fleshy and indehiscent in a few genera (as *Parmentiera, Kigelia, Crescentia*); seed in capsular fruits usually abundant, winged, and compressed (not winged in fleshy fruits), sometimes comose (as *Chilopsis*), the endosperm lacking.

A primarily tropical family of many genera (about 110) but relatively few species (about 750), particularly abundant in northern South America. A few genera occur in tropical Africa and Madagascar, a few in Asia, and only 2 have species in both the Old and New World (*Catalpa, Campsis*). *Catalpa* extends northward into Missouri and Indiana; *Bignonia* and *Campsis* (*Tecoma*) are each represented in the southern and eastern states by a single indigenous species. *Chilopsis linearis* is an indigen of the southwest and the only member of the family native to the west coast states (California). *Stenolobium stans* (*Tecoma*) occurs along the coastal plain from Florida to Texas and *Crescentia* occurs in southern Florida.

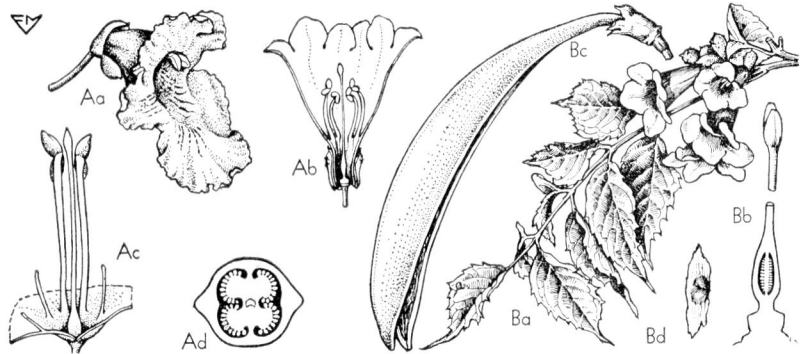

**Fig. 277.** BIGNONIACEAE. A, *Catalpa bignonioides*: Aa, flower, × ½; Ab, flower, perianth expanded, ¼; Ac, pistil with two stamens, × 1; Ad, ovary, cross-section, × 4. B, *Campsis radicans*: Ba, flowering branch, × ⅛; Bb, pistil with ovary in vertical section, × 1; Bc, capsule, × ⅓; Bd, seed, ½. (From L. H. Bailey, *Manual of cultivated plants*, The Macmillan Company, 1949. Copyright 1924 and 1949 by Liberty H. Bailey.)

The Bignoniaceae are distinguished from other related families by the absence of endosperm in the seeds, the structure of the fruit, the often conspicuously winged seeds, and in being predominantly tropical vines. The identification of genera and species of the families is difficult and both flowers and fruit are almost essential to critical determinations. Schumann subdivided the family into 4 tribes (Bignonieae, Tecomeae, Eccremocarpeae, and Crescentieae) on the basis of locular condition of the ovary, the type of fruit and the nature of its septum, and the seeds.

Wettstein considered the family as advanced over but allied to the Scrophulariaceae and Gesneriaceae, Bessey followed Engler's classification as to the relative position of this family, and Hallier included them in his Tubiflorae as derived from the Scrophulariaceae. Hutchinson placed them in his Personales intermediate between the Gesneriaceae and Pedaliaceae.

Members of the family are of economic importance both for lumber and as cultivated ornamentals. West Indian boxwood (*Tabebuia*) is prized for its lumber, and the catalpa (mostly *Catalpa speciosa*) is prized for fence-post material. Among the more notable ornamentals are several trees, mostly tropical: South African tulip tree (*Spathodea*), the sausage tree (*Kigelia*), roble blanco (*Tabebuia*), calabash tree (*Crescentia*), and *Jacaranda* and the hardy Paulownia of China (*Paulownia*) long but apparently erroneously placed in the Scrophulariaceae (cf. Campbell, 1930); the more common vines include the hardy or half-hardy trumpet vine (*Campsis*), trumpet flower (*Bignonia capreolata*), and *Eccremocarpus scaber*; the tender vines such as cats-claw (*Doxantha Unguis-cati*), Cape honeysuckle (*Tecomaria*), the Australian bower plant (*Pandorea*), and the ubiquitous *Pyrostegia ignea*; a few semihardy to tropical shrubs include the flowering willow (*Chilopsis*), and yellow bells (*Stenolobium*). Several species of *Incarvillea* are hardy perennials.

*LITERATURE:*

BLAKE, S. F. On the type species of *Bignonia*. Journ. Bot. (Lond.), 61: 191–193, 1923. [With reply by T. A. Sprague.]

BUREAU, L. E. Monographie des Bignoniacées. Paris, 1864.

CAMPBELL, D. H. The relationships of *Paulownia*. Bull. Torrey Bot. Club, 57: 47–50, 1930.

MELCHIOR, H. Beitrag zur Systematik und Phylogenie der Gattung *Tecoma*. Ber. deutsch. bot. Ges. 59: 18–31, 1941.

PICHON, M. Sur le centre de dispersion des Bignoniacées. Bull. Soc. Bot. France, 93: 121–123, 1946.

RECORD, S. J. and HESS, R. W. American timbers of the family Bignoniaceae. Tropical Woods, 63: 9–38, 1940.

SCHUMANN, K. Bignoniaceae. In Engler and Prantl, Die natürlichen Pflanzenfamilien, IV (3b): 189–252, 1895.

SEIBERT, R. J. The use of glands in a taxonomic consideration of the family Bignoniaceae. Ann. Mo. Bot. Gard. 35: 123–136, 1948.

SMITH, E. C. Chromosome behavior in *Catalpa hybrida* Spaeth. Journ. Arnold Arb. 22: 219–221, 1941.

SPRAGUE, T. A. On the type-species of *Bignonia*. Journ. Bot. (Lond.) 61: 192–193, 1923.

## PEDALIACEAE.[213] PEDALIUM FAMILY

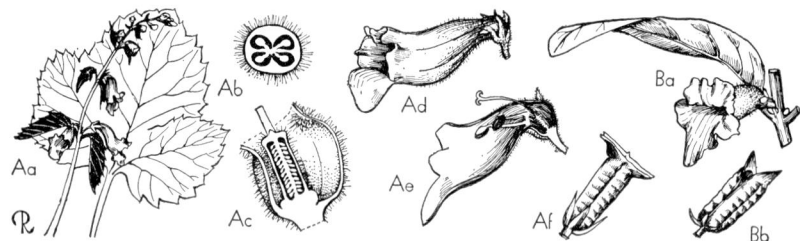

Fig. 278. PEDALIACEAE. A, *Ceratotheca triloba*: Aa, inflorescence, × ⅛, and leaf, × ¼; Ab, ovary, cross-section, × 5; Ac, ovary and calyx, vertical section, × 2; Ad, flower, × ½; Ae, flower, vertical section, × ½; Af, capsule, × ½. B, *Sesamum indicum*: Ba, flower, × ½; Bb, capsule, × ½. (From L. H. Bailey, *Manual of cultivated plants*, The Macmillan Company, 1949. Copyright 1924 and 1949 by Liberty H. Bailey.)

Annual or perennial herbs or rarely shrubs; leaves opposite or the uppermost sometimes alternate, simple, entire or lobed, estipulate; flowers usually solitary in axils or in simple axillary dichasia, bisexual, zygomorphic; calyx of usually 5 basally connate sepals (sometimes 4); corolla usually broadly tubular, 5-lobed

[213] The name Pedaliaceae has been conserved over other names for the family.

## DIVISION IV. EMBRYOPHYTA SIPHONOGAMA

and somewhat bilabiate; stamens epipetalous, distinct, 4 and didynamous (only 2 in *Trapella*), the fifth (posterior) represented by a small staminode, the anthers 2-celled, dehiscing longitudinally, introrse; pistil 1, the ovary mostly superior (inferior in *Trapella,* an Asiatic aquatic), 2-carpeled, the locules 2 or (by false septation) 4, the placentation axile, the ovules 1 to many on each placenta, anatropous, the style 1 and slender, the stigmas 2; fruit a loculicidal capsule or nut, often spiny or with wings, hooks, or thorns; seeds smooth, with a thin fleshy endosperm and a small straight embryo.

A family of 16 genera and about 50 species, mostly maritime or desert plants of Old World tropics and subtropics, with some naturalized in New World tropics. The largest genus is *Sesamum* (16 spp.) of Africa and India, and *Sesamum indicum* is naturalized from Florida to Texas and *Ceratotheca triloba* is naturalized in Florida.

The family is distinguished from related families by the usually 4-loculed or incompletely 4-loculed bicarpellate ovary, the axile placentation, and the usually beaked or barbed fruits.

Economically, *Sesamum indicum*, known as benne, is important for the sesame oil expressed from the seeds, and used as an acceptable substitute for olive oil, especially in European countries. Over 3,000,000 acres were devoted to the culture of the plant in India within the last decade. The plant is grown to a limited extent as a crop for home consumption in the southern states and has become naturalized in many areas. *Ceratotheca triloba* is cultivated domestically as an ornamental to a limited extent.

## MARTYNIACEAE. MARTYNIA FAMILY

**Fig. 279.** MARTYNIACEAE. *Proboscidea Jussieui:* a, flowering branch, × ½; b, flower, side view, × ½; c, same, vertical section, × ½; d, anthers, face view, × 1½; e, ovary, cross-section × 6; f, fruit, × ¼. (From L. H. Bailey, *Manual of cultivated plants,* The Macmillan Company, 1949. Copyright 1924 and 1949 by Liberty H. Bailey.)

Stout annual or perennial herbs, viscid-pubescent; leaves opposite, becoming alternate toward branch ends, undulate or lobed, estipulate; inflorescence terminal, racemose; flowers bisexual, zygomorphic; calyx spathaceous or of 5 distinct sepals, sometimes subtended by 1 or 2 bracts often becoming thick and fleshy at maturity; corolla gamopetalous, the tube basally cylindrical and campanulate or infundibular above, often ventricose and oblique, somewhat bilabiate, always 5-lobed; stamens epipetalous, distinct, usually 4 and didynamous with a posterior staminode, or 2 with the second pair forming staminodes (in *Martynia*), the anthers 2-celled, divergent, dehiscing longitudinally, those of each stamen pair coherent before anthesis; annular disc present; pistil 1, the ovary superior, bicarpellate, unilocular, the placentation parietal, with the two placentae winged and often cohering to form false septa, the ovules few to many, anatropous, the style 1, slender, the stigma with 2 flat sensitive lobes; fruit a horned capsule with fleshy viscid-hairy deciduous exocarp and a woody endocarp crested on a median line above and sometimes below, the single proboscislike persistent style splitting at fruit maturity into 2 hornlike processes; seeds sculptured, somewhat compressed, the embryo straight, endosperm none.

A family of 5 genera and 16 species, native to the New World tropics. *Proboscidea* has 4 species indigenous from Louisiana westward to California.

The Martyniaceae were included in the Pedaliaceae by Bentham and Hooker, Hallier, and Hutchinson, and by some earlier authors were included in the Bignoniaceae. They were treated as a family by Bessey, Rendle, and Wettstein. Members of the family are distinguished by their characteristic fruits and are separated from the Bignoniaceae and Pedaliaceae by the parietal placentae, and from the Gesneriaceae by the viscid pubescence, the superior ovary, the large calyx, and the inflorescence.

Members of the family are known generally as unicorn plants, without regard for genus or species distinction. Much of the material of this family is misidentified, and determinations should be checked against Van Eseltine's revision (1929). Material of *Ibicella lutea* and *Proboscidea Jussieui* (*Martynia louisiana*) are grown for ornament or for their novel fruit. In addition to the ornamental value of the plants, the young fruits are pickled and eaten in some areas.

*LITERATURE:*

MAYBERRY, M. W. *Martynia louisiana* Mill., an anatomical study. Trans. Kan. Acad. 50: 164–171, 1947.

SHEAD, A. C. The viscid substance covering the leaves and stems of *Martynia*. Proc. Oklahoma Acad. Sci. 4: 18, 1925.

STAPF, O. Martyniaceae. In Engler and Prantl, Die natürlichen Pflanzenfamilien, IV (3b): 265–269, 1895.

VAN ESELTINE, G. P. A preliminary study of the Unicorn Plants. N. Y. State Agr. Exp. Sta. Tech. Bull. 149: 1–41, 1929.

## OROBANCHACEAE.[214] BROOMRAPE FAMILY

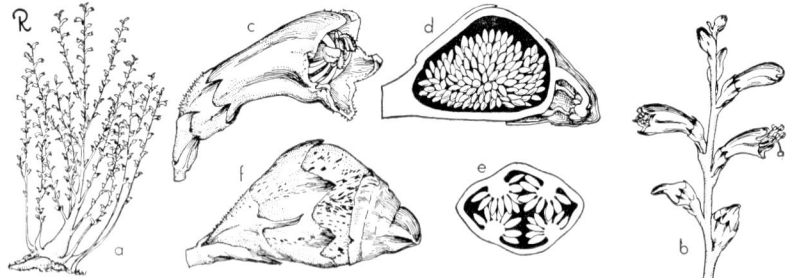

**Fig. 280.** OROBANCHACEAE. *Epifagus virginiana*: a, plant in flower, × 1/10; b, flowering branch tip, × 1; c, flower, habit, × 3; d, flower, vertical section, × 4; e, ovary, cross-section, × 5; f, fruit, × 4.

Annual or perennial somewhat fleshy herbs, root parasites, and commonly lacking chlorophyll or seemingly so; leaves alternate, scalelike; flowers solitary in a leaf or bract axil, usually subtended by a pair of bracteoles arising from the pedicel (the axis much condensed and spicate in *Boschniakia*), bisexual, zygo-

[214] The name Orobanchaceae has been conserved over earlier names for this family.

morphic (in *Epifagus* the upper ones usually sterile and the lower ones perfect and cleistogamous); calyx 2–5-lobed or divided or spathaceous, the segments open or valvate; corolla straight or arcuate, usually bilabiate or obliquely so, 5-lobed, imbricate, the 2 adaxial lobes innermost; stamens epipetalous, distinct, alternate with the lobes, 4, didynamous, the posterior stamen lacking or reduced to a small staminode, the anthers 2-celled, dehiscing longitudinally, often coherent in pairs, occasionally one half anther undeveloped and sterile; pistil 1, the ovary superior, unilocular (basally bilocular in *Christisonia* of tropical Asia), the carpels 2 and median (rarely 3), the placentation parietal with the placentae as many as carpels by connation of ventral margins or seemingly twice as many by the branches of each intruded placenta having reflexed and folded back against the ovary wall (sometimes the branches meeting in center of ovary), the ovules numerous, anatropous, the style 1 and slender, the stigma terminal, usually 2–4-lobed; fruit a loculicidal capsule, often enveloped by the persistent calyx, usually 2-valved and leathery; seeds minute, usually pitted or roughened, the embryo undifferentiated, the endosperm soft-fleshy and oily.

A family of about 13 genera and 140 species, mostly of north temperate regions but primarily of the warm temperate parts of the Old World. *Orobanche*, with about 90 species, is the largest genus. The family is represented in the United States by the monotypic *Epifagus* (*Leptamnium*) of eastern United States, *Conopholis* occurring from Maine to Michigan and south to Florida, and *Boschniakia* of northwestern America and northeastern Asia. *Orobanche* (including *Aphyllon* and *Thalesia*) occurs over much of the country and is represented by about 8 indigenous species and several that are adventive from Europe.

The members of the family are readily recognized by the lack of green coloration (stems usually in shades of red-brown–tan–purple, sometimes whitish), and the character of root parasitism. In addition, the flowers are distinctive by the parietal placentation, the usually intruded and spreading or T-shaped placentae, and the consistently 4 didynamous stamens.

Some phylogenists have derived the family from the Gesneriaceae (on the basis of the unilocular ovary with parietal placentation), and others have derived it from the Scrophulariaceae. The evidence, as reviewed by Boeshore (1920), strongly favors the latter view.

The members of the family are of no significant economic importance. In previous times some of them have been used medicinally for their alleged therapeutic properties.

*LITERATURE:*

ACHEY, D. AM. A revision of the sect. *Gymnocaulis* of the genus *Orobanche*. Bull. Torrey Bot. Club, 60: 441–450, 1933.
BECK-MANNAGETTA, G. Orobanchaceae. In Engler and Prantl, Die natürlichen Pflanzenfamilien, IV (3b): 123–132, 1895.
―――. Orobanchaceae. Engler. Das Pflanzenreich, 96 (IV. 261): 1–348, 1930.
MUNZ, P. A. The North American species of *Orobanche*, sect. *Myzorrhiza*. Bull. Torrey Bot. Club, 57: 611–623, 1930.
PUGSLEY, H. W. Notes on *Orobanche*. Journ. Bot. (Lond.) 78: 105–116, 1940.

# GESNERIACEAE.[215] GESNERIA FAMILY

Herbs, shrubs, or rarely trees; a few are lianous, others are epiphytic; leaves opposite or in basal rosettes, those of a pair equal in size or one much reduced and stipulelike, usually decussate, occasionally whorled, rarely alternate (due to complete reduction of the opposing leaf), usually herbaceous and hairy (coriaceous in

[215] The name Gesneriaceae has been conserved over earlier names for this family.

epiphytes); flowers showy, solitary in leaf axils or in cymose inflorescences, bisexual, zygomorphic (actinomorphic in *Ramonda* [216]); calyx 5-lobed, the lobes sometimes basally connate, valvate or rarely imbricate; corolla typically 5-lobed, rotate to campanulate or infundibularform-salverform, mostly bilabiate, imbricate; stamens 4 and didynamous or 2 (5 in *Ramonda, Sinningia* spp.), a staminode often present, epipetalous, the anthers coherent in pairs or in some genera connate (stamens then syngenesious), rarely distinct, 2-celled, dehiscing longitudinally; annular nectiferous disc conspicuous, sometimes lobed, or represented by 5 distinct glands; pistil 1, the ovary superior or half inferior or wholly inferior, unilocular, the carpels 2, the placentation parietal, with placentae often intruded and bifid, sometimes meeting in center or nearly so and the ovary appearing bi- or quadri-loculate, the ovules numerous, anatropous, the style 1 and slender, the stigma often bilobed; fruit usually a loculicidal capsule (septicidal in *Ramonda*), rarely fleshy and berrylike (as in *Cyrtandra*); seeds numerous, minute, usually with abundant endosperm (little or none in the Cyrtandroideae).

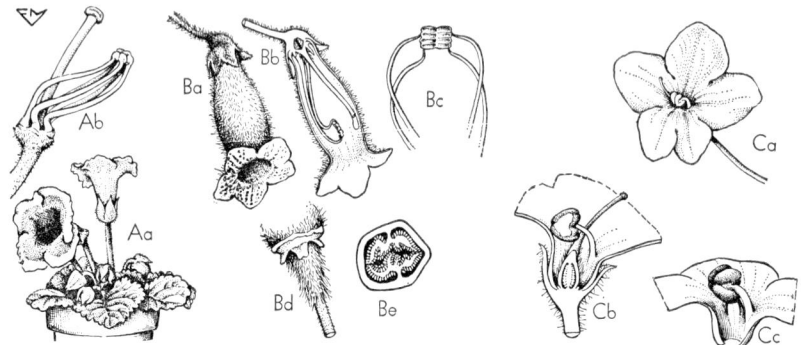

**Fig. 281.** GESNERIACEAE. A, *Sinningia speciosa*: Aa, plant in flower, × ⅙; Ab, flower, less perianth, × ½. B, *Kohleria hirsuta*: Ba, flower, × ½; Bb, same, vertical section, × ½; Bc, stamens, × 1; Bd, glandular disc at top of ovary, × 1½; Be, ovary, cross-section, × 3. C, *Saintpaulia ionantha*: Ca, flower, × ½; Cb, same, vertical section, corolla partially excised, × 1½; Cc, stamens, × 1½. (From L. H. Bailey, *Manual of cultivated plants*, The Macmillan Company, 1949. Copyright 1924 and 1949 by Liberty H. Bailey.)

The Gesneriaceae are a large family of 85 genera and about 1200 species, of primarily tropical and subtropical distribution of both hemispheres, with 2 genera (*Haberlea* and *Ramonda*) native to temperate parts of Europe. None of the family is native to the United States although a dozen or more genera extend northward into Mexico.

The family is closely allied to the Scrophulariaceae, Bignoniaceae, and probably to the Orobanchaceae, and as a taxon, is separated from them with difficulty. In general, it may be distinguished from the Scrophulariaceae by the unilocular ovary and parietal placentation, from the Bignoniaceae by the same character or from the unilocular bignoniads by the differences in the fruit and seed (siliquelike and winged respectively in the Bignoniaceae), and from the Orobanchaceae by the non-parasitic habit.

Economically, members of the family are important primarily as ornamentals grown in the open in warm climates or under glass in the cooler regions (except

---

[216] Throughout much of the literature the spelling of this name appears as *Ramondia*. However, as pointed out by Lawrence (Gentes Herb. 8: 68, 1949) the original spelling was *Ramonda* and should be retained.

species of *Ramonda* and *Haberlea* which are prized as rock garden subjects in temperate regions. The most important ornamentals include: gloxinia (*Sinningia*), African violet (*Saintpaulia*), Cape primrose (*Streptocarpus*), and species of *Achimenes, Smithiantha, Kohleria* (*Isoloma*), *Aeschynanthus* (*Trichosporum*), and *Episcia*.

*LITERATURE:*

BEDDOME, R. H. Gesneriaceae; with annotated list of the genera and species which have been introduced into cultivation. Journ. Roy. Hort. Soc. (Lond.), 33: 74–100, 1908.

CLARK, C. B. Cyrtandreae (*Gesneracearum tribus*). In de Candolle, Monograph. Phan. 5: 1–303, 1883.

FRITSCH, K. Gesneriaceae. In Engler and Prantl, Die natürlichen Pflanzenfamilien, IV (3b): 133–185, 1895.

———. Die Keimpflanzen der Gesneriaceen, pp. 188. Jena, 1904.

———. Beiträge zur Kenntnis der Gesnerioideae. Engler Bot. Jahrb. 50: 392–439, 1913.

HANSTEIN, J. Die Gesneraceen des kön. Herbariums und der Gärten zu Berlin. Linnaea, 26: 145–216, 1853; *op. cit.* 27: 693–785, 1854; *op. cit.* 29: 497–592, 1857–1858; *op. cit.* 34: 225, 462, 1865–1866.

MORTON, C. V. A Revision of *Besleria*. Contr. U. S. Nat. Herb. 26: 395–474, 1939.

———. The West Indian species of *Columnea*. The West Indian species of *Alloplectus*, and a revision of *Cremosperma*. Contr. U. S. Nat. Herb. 29: 1–35, 1944.

———. Las especies sudamericanas del género *Monopyle*. Revista Unversitaria (Univ. Cuzco, Peru), 87: 98–116, 1945.

OERSTED, A. S. Centralamericas Gesneraceen. 78 pp. Copenhagen, 1858.

RIDLEY, H. N. The Gesneraceae of the Malay Peninsula. Journ. Straits Branch Roy. Asiatic Soc. no. 43: 1–92, 1905.

STANDLEY, P. C. and MORTON, C. V. Gesneriaceae (of Costa Rica). Field Mus. Publ. Bot. 18: 1137–1187, 1938.

URBAN, I. Enumeratio Gesneriacearum (of West Indies) in Urban, Symb. Antill. 2: 344–388, 1901.

## LENTIBULARIACEAE.[217] BLADDERWORT FAMILY

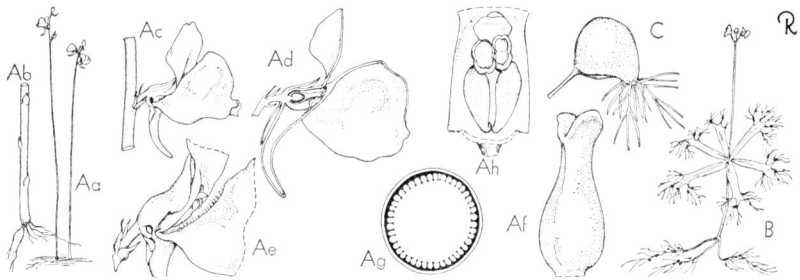

**Fig. 282.** LENTIBULARIACEAE. A, *Utricularia cornuta*: Aa, flowering plant (in water), × ⅛; Ab, stem, basal portion, × 1; Ac, flower, side view, × 1; Ad, flower, vertical section, × 1½; Ae, flower, basal portion, showing palate, × 2; Af, pistil, habit, × 6; Ag, ovary, cross-section, × 10; Ah, flower, perianth excised, showing stamens and pistil, × 1. B, *U. inflata*: habit in flower, showing rosette of floating leaves with bladders, × ⅛. C, *U. neglecta*: habit of bladder, × 2. (C, after Engler.)

Annual and perennial herbs, aquatics (and then often rootless), or plants usually of wet places, predominantly insectivorous; leaves alternate or in basal rosettes, often dimorphic in aquatic representatives with submerged leaves usually finely

---

[217] The name Lentibulariaceae has been conserved, and the name Pinguiculariaceae as adopted by J. K. Small is a *nomen rejiciendum* unless *Pinguicula* is segregated as a separate family.

divided, and bearing insectivorous bladders of complex structure,[218] and aerial leaves composing a floating rosette or reduced to scalelike enations or absent; in some terrestrial genera (as the tropical *Genlisea*) dimorphism is represented by rosettes of foliage leaves and tubular or pitcherlike insectivorous leaves appressed into the ground; flowers bracteate, pedicels of some each with a pair of bracteoles, in scapose racemes or solitary on a scape (*Pinguicula*), bisexual, zygomorphic, the calyx 2-5-lobed or divided, the segments open or imbricate, the corolla gamopetalous, 5-lobed, imbricate, bilabiate, the lower lip saccate or spurred, often personate, with the palate very variable in form; stamens 2, with 2 staminodes sometimes produced, arising from extreme base of corolla tube, the anthers 1-celled (the theca sometimes partially medianly constricted), dehiscing longitudinally; no disc produced; pistil 1, the ovary superior, unilocular, the carpels 2, the placentation free-central, the ovules numerous (reduced to 2 in *Biovularia*) on or often sunken into a globose placental mass, anatropous, the style 1 or commonly obsolete with the 2-lobed stigma sessile; fruit a capsule, dehiscing by 2-4 valves, circumscissilely, or by an irregular splitting; seeds minute with a poorly differentiated embryo and no endosperm.

A widely spread family of about 5 genera [219] and 260 species, occurring on all continents and its members comprising an important element of aquatic and marsh vegetation. The largest genera (together with number of species) are *Utricularia* ca. 200, *Pinguicula* 32, and *Genlisea* 12. By conservative generic concepts the family is represented in the United States by indigenous species of only the first 2 of the above-mentioned genera, with most of the species occurring in appropriate habitats over much of the region east of the Rockies (3 species of *Utricularia* are native to California but are infrequent or rare).

The Lentibulariaceae superficially resemble some of the Scrophulariaceae, but are distinguished from most of the latter by the stamens 2, and from all by the free-central placentation. The presence of the tiny bladders on aquatic representatives, or other forms of insect traps, is also distinctive.

The members of the family are of little economic importance. A few species of butterwort (*Pinguicula*) are offered in the domestic trade as ornamentals, and species of bladderwort (*Utricularia*) are grown in aquaria.

*LITERATURE:*

BARNHART, J. H. Segregation of genera in Lentibulariaceae. Mem. N. Y. Bot. Gard. 6: 39–64, 1916.
HEIDE, F. Lentibulariaceae (*Pinguicula*). Meddelelser øm Grönland, 36: 441–481, 1912.
KAMIENSKI, F. Lentibulariaceae. In Engler and Prantl, Die natürlichen Pflanzenfamilien, IV (3b): 108–123, 1895.
LLOYD, F. E. The range of structural and functional variation of the "traps" of *Utricularia*. Proc. Fifth Internat. Bot. Congr. Cambridge, 1930: 450–451, 1931.
———. The range of structural and functional variety in the traps of *Utricularia* and *Polypompholyx*. Flora, 126: 303–328, 1932.
———. *Utricularia*. Biol. Rev. 10: 72–110, 1935.
———. The carnivorous plants. 352 pp. Waltham, Mass., 1942.
LLOYD, F. E. and TAYLOR, G. Some new species of *Utricularia*. Contr. Gray Herb. 165: 82–90, 1947.

[218] See Lloyd (1942, pp. 213–267) for details of structure function and taxonomic value. He has given also a review of the pertinent literature and an extensive bibliography.

[219] Barnhart (1916) gave a very excellent summary of generic segregation in the family, recognizing it to be composed of 17 genera, but pointed out that it was not as important that the taxonomic validity of his genera be accepted as it was that the taxa represented by them be recognized as natural assemblages and be given subgeneric or sectional rank if rejected as genera. The majority of American taxonomists have not accepted his segregates of *Utricularia* as representing distinct genera.

LUETZELBURG, P. VON, Beiträge zur Kenntnis der Utricularien. Flora, 100: 145–212, 1910.
MERL, E. M. Beiträge zur Kenntnis der Utricularien und Genliseen. Flora, 108: 127–200, 1915.
PRINGSHEIM, N. Zur Morphologie der Utricularien. Berlin, 1869.
ROSSBACH, G. B. Aquatic Utricularias. Rhodora, 41: 113–128, 1939.

## GLOBULARIACEAE. GLOBULARIA FAMILY

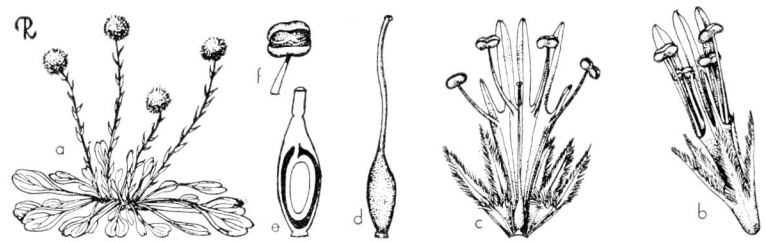

**Fig. 283.** GLOBULARIACEAE. *Globularia Aphyllanthes*: a, flowering plant, × 10; b, flower, × 4; c, flower, perianth expanded, × 4; d, pistil, × 6; e, ovary, vertical section, × 10; f, anther, × 10. (From L. H. Bailey, *Manual of cultivated plants,* The Macmillan Company, 1949. Copyright 1924 and 1949 by Liberty H. Bailey.)

Perennial herbs or shrubs; leaves alternate, simple, estipulate; inflorescence a head subtended by a multibracteate involucre; flowers bisexual, zygomorphic, usually minute, on a scaly receptacle, the calyx tubular, 5-lobed, actinomorphic or weakly bilabiate, the corolla bilabiate (lower lip often much reduced and 2-lobed), 4–5-lobed; stamens epipetalous, distinct, 4 and didynamous or sometimes 2, arising from near top of corolla tube and alternate with lobes, the anthers 2-celled but becoming confluent at anthesis and dehiscing by a single longitudinal slit, versatile; annular or glandular disc often present; pistil 1, the ovary superior, unilocular, the carpels 2, the ovule solitary, pendulous from locule apex, anatropous, the style 1 and filiform, the stigma capitate or briefly 2-lobed; fruit a 1-seeded nutlet enclosed within the persistent calyx; seed with straight embryo surrounded by a fleshy endosperm.

A small family, restricted to the Old World and primarily to the Mediterranean region, composed of 3 genera and about 23 species. *Globularia* (19 spp.) is the largest genus, and several species are cultivated in this country as hardy perennials.

The family was merged with the Selaginaceae by Bentham and Hooker, with the Scrophulariaceae by Hallier, but retained as distinct by Engler, Bessey, Rendle, Wettstein, and Hutchinson. The latter interpreted it to be the primitive family of his Lamiales, while Wettstein derived it from the Scrophulariaceae.

## ACANTHACEAE.[220] ACANTHUS FAMILY

Perennial armed or unarmed herbs or shrubs, rarely trees; some are lianes, xerophytes, aquatics, or mesophytes; leaves opposite and decussate, simple, estipulate; cystoliths (appearing as protuberances or streaks) are common on vegetative parts (none in *Aphelandra*); inflorescence usually a dichasial cyme (sometimes axillary and congested as the verticils of Labiatae) or developing into a monochasium, or the flowers racemose or solitary; flowers bisexual, zygomorphic, bracteate with bracteoles common and often conspicuous or even involucrate, the calyx deeply 4–5-lobed or sometimes much reduced (as in *Thunbergia*), contorted or

[220] The name Acanthaceae has been conserved over earlier names for this family.

imbricate, the corolla typically 5-lobed, usually bilabiate, upper lip usually erect and bifid (absent in *Acanthus* and others); stamens usually 4 and didynamous or 2 (rarely 5, as in *Pentstemonacanthus*), epipetalous, distinct (sometimes connate in pairs), staminodes sometimes present, the anthers exceedingly variable in position and form, 2-celled and sometimes the cells separated by the connective or 1-celled, sometimes spurred, the pollen diversely surfaced, dehiscence longitudinal; annular or glandular nectiferous disc present; pistil 1, the ovary superior, the carpels and locules 2, the placentation axile, the ovules 2 or more in each locule, anatropous, in 2 rows, the style 1 and slender, the stigmas 2 and variously shaped, the posterior often the smaller; fruit usually a loculicidal capsule (a drupe in a few genera), often elastically dehiscent with valves recurving from persistent central column (character of capsule and dehiscence varies with different genera); seeds with testa of various types (as mucilaginous, scaly, hairy, or with an indurated funicle), embryo large, the endosperm usually absent.

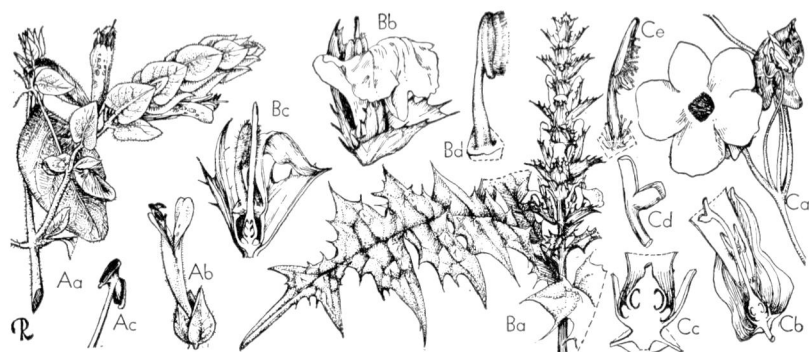

**Fig. 284.** ACANTHACEAE. A, *Beloperone guttata*: Aa, flowering branch, × ½; Ab, flower, × ½; Ac, anther, × 2. B, *Acanthus montanus*: Ba, inflorescence with leaf, × ⅛; Bb, flower with bract, × 1; Bc, same, vertical section, × 1; Bd, stamen, × 2. C, *Thunbergia alata*: Ca, flower, × ½; Cb, same, vertical section (corolla-limb excised), × 1; Cc, ovary, vertical section, × 3; Cd, stigma, × 4; Ce, stamen, × 3. (From L. H. Bailey, *Manual of cultivated plants*, The Macmillan Company, 1949. Copyright 1924 and 1949 by Liberty H. Bailey.)

The Acanthaceae are a large pantropical family of about 240 genera and over 2200 species, and have 4 centers of distribution: Indo-Malaya, Africa, Brazil, and Central America northward into Mexico. Relatively few genera and species extend into the United States; 3 genera occur in the northeast (*Justicia* [*Dianthera*], *Ruellia*, *Dyschoriste*). These and 6 others occur in the southeastern quarter (*Thunbergia*, *Hygrophila*, *Elytraria* [*Tubiflora*], *Stenandrium*, *Dicliptera*, [*Diapedium*] and *Yeatsia*). In addition, *Carlowrightia*, *Anisacanthus*, *Tetramerium*, *Dicliptera*, *Siphonoglossa*, *Jacobinia*, and *Beloperone* have species indigenous to the southwestern states (only *Beloperone californica* occurs in California).

From related families the Acanthaceae are distinguished by a number of characters, notably the usual presence of cystoliths in vegetative parts, the presence and development of floral bracts and bracteoles, the usually bilabiate corollas associated with the bilocular ovary, the generally bivalvate elastically dehiscing capsules, and usually by the curved retinacula supporting the seeds. The anthers and stamens provide many characters diagnostic of groups of genera.

## DIVISION IV. EMBRYOPHYTA SIPHONOGAMA

Botanists have consistently retained the Acanthaceae as a distinct taxon, and the majority are agreed that it is derived from the Scrophulariaceae or stocks ancestral to them. Hutchinson considered it the most advanced taxon of his Personales, and Bessey previously had adopted a similar view by placing it at the top of his Scrophulariales.

The family is of little domestic importance. Species of a few genera, mostly tropical, are cultivated as ornamentals. Of these the following may be the more common: bear's-breech (*Acanthus*), *Ruellia*, *Pseuderanthemum*, *Aphelandra*, clock vine (*Thunbergia*), *Jacobinia*, *Fittonia*, shrimp plant (*Beloperone*), *Justicia*, and cardinal's guard (*Pachystachys*).

*LITERATURE:*

FERNALD, M. L. *Ruellia* in the eastern United States. Rhodora, 47: 1–38, 47–63, 69–90, 1945.
HARTMANN, A. Zur Entwicklungsgeschichte und Biologie der Acanthaceen. Flora, 116: 216–258, 1923.
KOBUSKI, C. E. A monograph of the American species of the genus *Dyschoriste*. Ann. Mo. Bot. Gard. 15: 9–90, 1928.
PENFOUND, W. T. The biology of *Dianthera americana* L. Amer. Midl. Nat. 24: 242–247, 1940.
THARP, B. C. and BARKLEY, F. A. The genus *Ruellia* in Texas. Amer. Midl. Nat. 42: 1–86, 1949.

## MYOPORACEAE.[221] MYOPORUM FAMILY

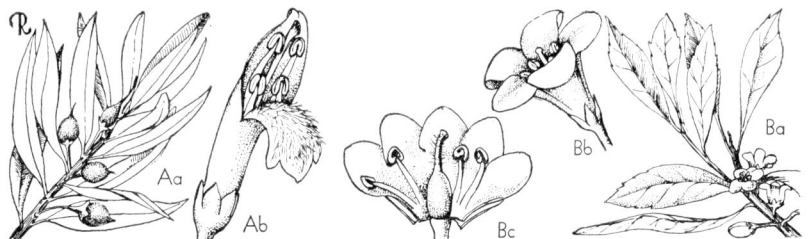

**Fig. 285.** MYOPORACEAE. A, *Bonita daphnoides*: Aa, fruiting branch, × ¼; Ab, flower, × 1½. B, *Myoporum laetum*: Ba, flowering branch, × ¼; Bb, flower, × 3; Bc, same, perianth expanded, × 3. (From L. H. Bailey, *Manual of cultivated plants*, The Macmillan Company, 1949. Copyright 1924 and 1949 by Liberty H. Bailey.)

Shrubs or infrequently trees, herbage often with stellate, glandular, or plumose hairs; leaves alternate or rarely opposite, simple, entire, estipulate; inflorescence a fasciculate cyme or the flowers solitary in leaf axils; flowers bisexual, zygomorphic or rarely actinomorphic, the calyx 5-lobed or -fid, persistent, the corolla usually 5-lobed, imbricate, sometimes bilabiate; stamens epipetalous, alternate with corolla lobes, 4 and didynamous (the fifth sometimes represented by a staminode) or rarely same number as corolla lobes, the anthers 2-celled, the cells often divergent and apically confluent, dehiscing longitudinally; pistil 1, the ovary superior, bilocular or the locules 3–10 by false septation, the carpels 2, the ovules superposed in pairs and 2–8 in each locule, pendulous, anatropous, the style 1 and simple, stigma 1; fruit a berry or drupe; seed with scant or no endosperm.

A family, mostly of the Old World (except for a few in Pacific Islands), of 5 genera and about 110 species, chiefly Australian and neighboring islands. The

[221] The name Myoporaceae has been conserved over earlier names for this family.

larger genera are *Myoporum* (about 30 spp.) and *Pholidia* (60 spp.) None is American.

Several species of *Myoporum* and *Bontia* are cultivated as ornamental shrubs in the warmer parts of the country.

## PHRYMACEAE. LOPSEED FAMILY

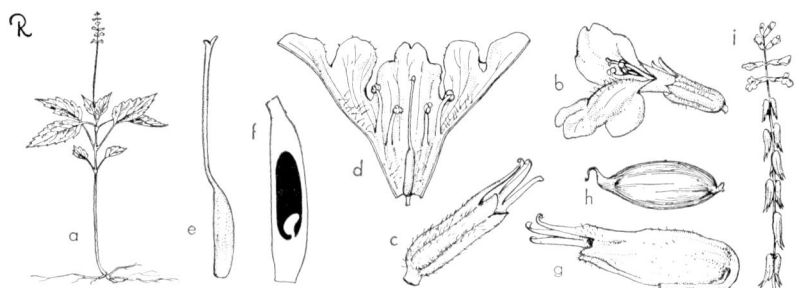

**Fig. 286.** PHRYMACEAE. *Phryma leptostachya*: a, plant in flower, × 1/12; b, flower, side view, × 3 c, calyx, × 5; d, flower, corolla expanded, × 3; e, pistil, × 6; f, ovary, vertical section, × 10; g, fruiting calyx, × 4; h, fruit, × 4; i, fruiting inflorescence, × 1/2.

A family of somewhat doubtful taxonomic status, composed of a single monotypic genus (*Phryma Leptostachya*), and allied to the Verbenaceae from which it differs chiefly in the ovary possessing a solitary erect orthotropous ovule. In addition the plant, a perennial herb, is characterized by the flowers solitary in opposing axils disposed in elongated slender terminal spikelike racemes and becoming strongly reflexed in fruit; the calyx bilabiate; the corolla is bilabiate with lower lip 3-lobed and much larger than the upper; the stamens 4 and didynamous; ovary bicarpellate, unilocular and uniovulate; the fruit a 1-seeded nutlet enclosed by the persistent calyx; the seeds without endosperm and the cotyledons convolute.

*Phryma* is indigenous to eastern North America and northeastern Asia. Some authors have recognized the Asiatic populations to comprise 3 species while others treat them as conspecific with the American plant.

Bentham and Hooker, Hallier, and Hutchison treated the genus as belonging in the Verbenaceae. Engler, Bessey, and Wettstein and most American botanists have accepted Phrymaceae as a distinct family, and Rendle did not account for them.

*Phryma* is of no known economic importance.

*LITERATURE:*

HOLM, T. *Phryma leptostachya* L., a morphological study. Bot. Gaz. 56: 306–318, 1913.

## Order 41. PLANTAGINALES

This order is composed of only the single family, Plantaginaceae, and the phyletic position assigned it by various authors is given in the discussion under the family.

## PLANTAGINACEAE.[222] PLANTAGO FAMILY

Herbs or rarely branched subshrubs; leaves all basal or nearly so, commonly alternate or rarely opposite, sometimes much reduced, venation often seemingly parallel, the bases often sheathing, estipulate; inflorescence bracteate but without

[222] The name Plantaginaceae has been conserved over other names for this family.

bracteoles, scapose, capitate or spicate; flowers usually bisexual, actinomorphic, the calyx tubular and 4-toothed (sometimes deeply divided), membranous, the corolla scarious, typically 4-lobed or -toothed (sometimes 3-lobed), imbricate; stamens epipetalous and alternate with corolla lobes, 4 and equal or rarely 1–2, exserted, the anthers 2-celled, large, versatile, dehiscing longitudinally; pistil 1, the ovary superior, the locules 1–4 (usually 2), the carpels 2, the ovules 1 or more in each locule, semianatropous, the placentation usually axile (free-central or basal in unilocular ovary); the style 1, filiform, bifid, bearing stigmatic hairs; fruit a circumscissile capsule or a bony nut; seeds with small straight embryo enveloped by a fleshy endosperm.

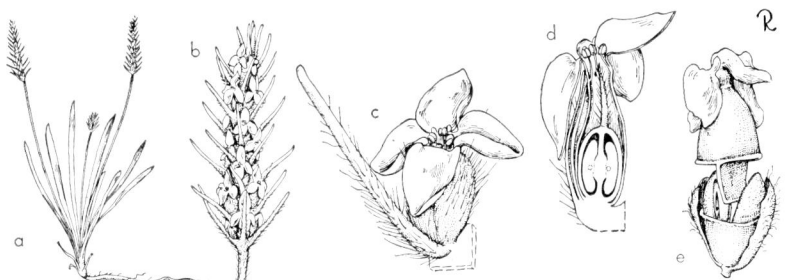

**Fig. 287.** PLANTAGINACEAE. *Plantago aristata*: a, plant in flower, × ⅙; b, inflorescence, × 1; c, flower, habit, with bract, × 6; d, same, vertical section, × 6; e, capsule, dehiscing, × 5.

A family of 3 genera: *Plantago* a cosmopolitan genus of about 200 spp., *Litorella* of Europe and the antarctic with 2 species, and the monotypic Andean *Bougueria*. *Plantago* is widespread through this country, where it is represented by about 30 species, of which 4 are naturalized from Europe.

The Plantaginaceae are readily distinguished by their usually rosulate foliage and apparent parallel venation, the spicate or capitate inflorescences on stout or wiry scapes, the 4-merous flowers (gynoecium bicarpellate) with membranous corollas, the stamens much exserted on wiry filaments, and producing an abundance of dry powdery pollen.

Wettstein and Hallier included the family in the Tubiflorae, as allied to the Scrophulariaceae. Bessey placed it in his Primulales as derived from or related to the Plumbaginaceae. Hutchinson (1948) retained it in the Plantaginales and derived the order from primulaceous ancestors.

The Plantaginaceae are of importance primarily as a source of Psyllium seeds (*Plantago Psyllium*) whose mucilaginous seed coats have made them of reputed laxative value. A number of species are noxious lawn weeds.

*LITERATURE:*

PILGER, R. Plantaginaceae. In Engler, Das Pflanzenreich, 102 (IV. 269): 1–466, 1937.

### Order 42. RUBIALES

An order of largely paleotropical distribution, whose plants are characterized by usually opposite leaves and cymose inflorescences; the gamopetalous flowers vary from actinomorphic to zygomorphic, the ovary is inferior, and the anthers are always distinct.

Engler and Diels included the following 5 families in the order, of which all except Dipsacaceae have species indigenous to this country: Rubiaceae; Adoxaceae: Dipsacaceae; Caprifoliaceae; Valerianaceae.

The order has been accepted as a natural taxon by all recent taxonomists, except Hutchinson, who restricted the Rubiales to contain the Rubiaceae and Caprifoliaceae and transferred the remaining families to his Asterales, a view he modified later (1948) by placing the 3 latter families in a new order, the Valerianales, and treating the Rubiales and Valerianales as very unrelated taxa. He derived the former from loganiaceous ancestors, and the latter from the Saxifragaceae, indicating (p. 204) that they "have had a very different phylogenetic history from the Compositae, and that the supposed affinity is due to convergent evolution."

## RUBIACEAE.[223] MADDER FAMILY

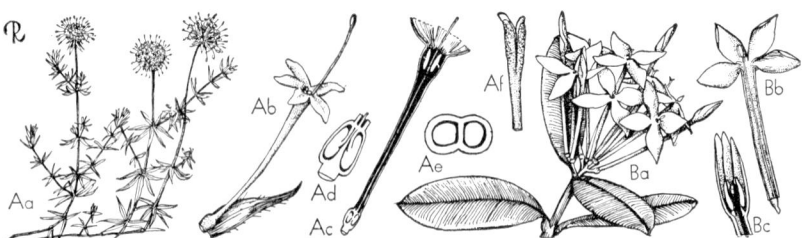

**Fig. 288.** RUBIACEAE. A, *Crucianella stylosa*: Aa, flowering stems, × ⅛; Ab, flower with subtending bract, × 1½; Ac, same, vertical section( corolla-lobes excised), × 2; Ad, ovary, vertical section, × 5; Ae, ovary, cross-section, × 8; Af, stigma, × 4. B, *Ixora coccinea*: Ba, flowering twig, × ½; Bb, flower, perianth expanded, × ½; Bc, opening bud tip, vertical section, showing stamens, × 1. (From L. H. Bailey, *Manual of cultivated plants,* The Macmillan Company, 1949. Copyright 1924 and 1949 by Liberty H. Bailey.)

Tress or shrubs, sometimes lianous, infrequently herbs; leaves opposite or whorled, simple, entire or rarely toothed, the stipules present and interpetiolar or intrapetiolar, sometimes foliaceous and not distinguishable from the leaves (as in the Galieae) or reduced to glandular setae (as in *Pentas*), distinct or connate; inflorescence basically a dichasial cyme (only the central flower present in some, as *Gardenia*), the dichasia sometimes aggregated into globose heads (the flowers becoming basally adnate, as in *Morinda* or *Sarcocephalus*); flowers bisexual, usually actinomorphic or rarely zygomorphic and somewhat bilabiate (as in *Henriquezia*), calyx 4–5-lobed, the lobes or segments open in aestivation, sometimes becoming enlarged in fruit (as in *Nematostylis*), the corolla gamopetalous, usually salverform, rotate or funnelform, 4–5-lobed (rarely 8–10), aestivation various and providing sectional characters; stamens as many as corolla lobes and alternate with 1em, epipetalous on the corolla tube, the anthers 2-celled, dehiscing longitudinally, introrse, usually distinct; pistil 1, the ovary inferior (rarely superior, as in *agamea,* or half inferior, as in *Synaptanthera*), the carpels 2 or more, the locules sually 2 with axile or seemingly basal placentation, sometimes several (or 1 with arietal placentation in *Gardenia*), the ovules usually numerous in each locule (uniovulate in *Pavetta* with ovule sunken in the fleshy funiculus), the style 1 and slender, often 2-branched, the stigmas usually linear, 1 on each style branch, or solitary and 2-lobed; fruit a loculicidal or septicidal capsule or indehiscent and separating into 1-seeded segments (*Galium*), a fleshy berry in some genera (*Coffea, Mitchella*); seeds sometimes winged, the endosperm usually copious and fleshy or rarely cartilaginous (none present in the Guettardeae).

[223] The name Rubiaceae has been conserved over other names for this family.

The Rubiaceae are a large pantropical and subtropical family of nearly 400 genera (of which nearly half are monotypic) and 4800 to 5000 species. Species and genera of the tribes Galieae, Anthospermeae, and Oldenlandieae are predominantly herbaceous, and extend into temperate zones (*Nertera* extends from the equator to Cape Horn, and *Galium* from the equator to the arctic). The family is especially abundant in northern South America. About 50 genera occur indigenously in Mexico, and 14 in the United States. Notable among the latter are *Houstonia, Galium,* and *Cephalanthus* in the cooler parts, and *Hedyotis* (*Oldenlandia*), *Diodia, Pentodon, Pinckneya,* and *Bouvardia* primarily in the warmer areas. Other genera in the southwest include *Kelloggia* and *Crusea*.

The Madder family is closely allied to the Caprifoliaceae, and there is no single character distinguishing them. In general, the presence of stipules in the Rubiaceae and the usual lack of them in Caprifoliaceae is a good field character.

The Rubiaceae are of economic importance primarily for several tropical crops, notably coffee (*Coffea*), quinine (*Cinchona*), and ipecac (*Cephaelis*). In addition to these a number of ornamentals grown in this country include gardenia (*Gardenia*), madder (*Rubia*), bead plant (*Nertera*), partridgeberry (*Mitchella*), and species of *Galium, Asperula, Ixora, Bouvardia, Manettia, Coprosma,* and *Serissa*.

*LITERATURE:*

FAGERLIND, F. Embryologische, zytologische und bestäubungs-experimentelle Studien in der Familie Rubiaceae nebst Bemerkungen über einige Polyploiditätsprobleme. Acta. Horti Bergiani, 11: 195–470, 1937.
HILEND, M. and HOWELL, J. T. The genus *Galium* in southern California. Leafl. West. Bot. 1: 145–168, 1935.
HOMEYER, H. Beiträge zur Kenntnis der Zytologie und Systematik der Rubiaceen. Bot. Jahrb. von Engl. 67: 237–263, 1935–1936.
JOSHI, A. C. A note on the morphology of the ovule of Rubiaceae with special reference to cinchona and coffee. Curr. Sci. 7: 236–237, 1938.
JOVET, P. Aux confins des Rubiacées et des Longaniacées. Not. Syst. [Paris] 10: 39–53, 1941.
STANDLEY, P. C. Rubiaceae. North Amer. Flora, 32: 1–300, 1918–1934 [incomplete].

## CAPRIFOLIACEAE.[224] HONEYSUCKLE FAMILY

Shrubs, sometimes lianous (*Lonicera* spp.), rarely herbaceous (*Triosteum* and *Sambucus Ebulus*) or suffrutescent (*Linnaea*); leaves opposite (connate-perfoliate in *Lonicera* spp.), usually simple, rarely pinnately compound (*Sambucus*), usually estipulate (stipulate in *Sambucus*) or the stipules reduced to nectiferous glands (as in *Viburnum, Leycesteria*); inflorescence cymose or modifications of it (racemose monochasium in *Diervilla*), usually bracteolate in some 2-flowered cymes, the adjoining ovaries basally or wholly adnate (*Kolkwitzia*); flowers usually bisexual (sterile or neutral "flowers" produced in some *Viburnum* spp.), actinomorphic or zygomorphic, the calyx 5-lobed or -toothed, usually small, the corolla gamopetalous, typically 5-lobed, variable in form, often bilabiate, rotate, or salverform, lobes imbricate; stamens epipetalous and alternate with corolla lobes, distinct, typically 5 or 4 (by suppression of posterior stamen, as in *Linnaea* or *Dipelta*), the anthers 2 celled, dehiscing longitudinally, generally introrse (extrorse in *Sambucus*); pistil 1, the ovary inferior, the carpels [225] usually 3–5, the locules 1–5, the placentation typically axile with an approach toward parietal, the ovules usually 1 in each locule (numerous in *Leycesteria*), pendulous, the style 1, slender

[224] The name Caprifoliaceae has been conserved over other names for this family.
[225] For explanation and interpretation of carpellary situation in several genera of this family, see especially the papers by Wilkinson (1948–1949).

or obsolete, the stigmas as many as carpels and distinct or united; fruit a berry or drupe; seeds with usually a small straight embryo, the endosperm copious and soft to watery-fleshy (firm-fleshy in *Viburnum*).

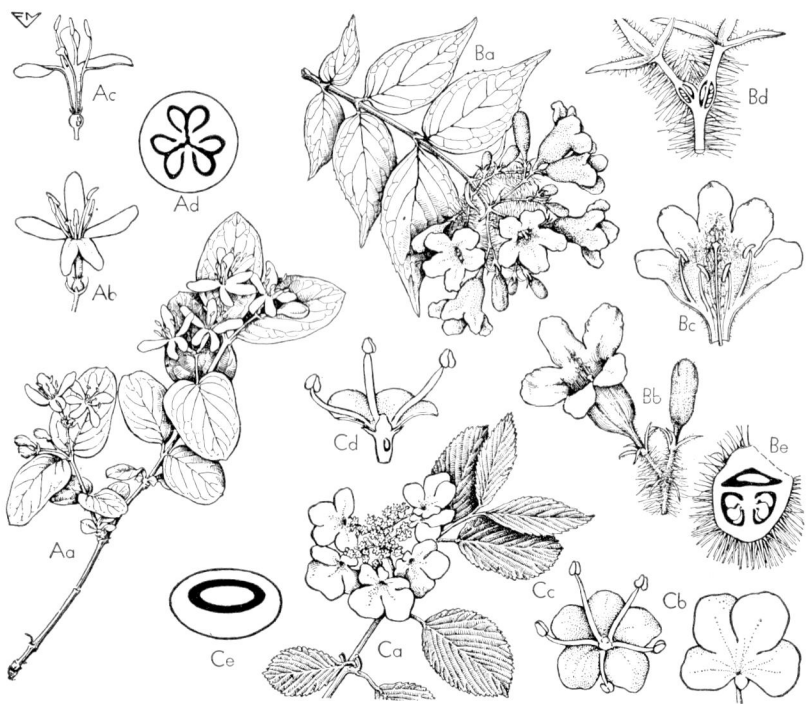

**Fig. 289.** CAPRIFOLIACEAE. A, *Lonicera tatarica*: Aa, flowering branch, × ½; Ab, flower, × 1; Ac, same, vertical section, × 1; Ad, ovary, cross-section, × 10. B, *Kolkwitzia amabilis*: Ba, flowering branch, × ½; Bb, flower and bud, × 1; Bc, corolla expanded, × 1; Bd, ovaries, vertical section, × 2; Be, same, cross-section, × 10. C, *Viburnum tomentosum*: Ca, flowering branch, × ¼; Cb, sterile flower, × ½; Cc, perfect flower, × 3; Cd, same, vertical section, × 3; Ce, ovary, cross-section. (From L. H. Bailey, *Manual of cultivated plants,* The Macmillan Company, 1949. Copyright 1924 and 1949 by Liberty H. Bailey.)

A family of about 18 genera [226] and 275 species, primarily of the northern hemisphere (spp. of *Sambucus* and *Viburnum* occur in South America, and *Alseuosmia* is endemic in New Zealand) and particularly in eastern Asia and eastern North America. The largest genera include *Lonicera* (180 spp.), *Viburnum* (120 spp.), *Abelia* (30 spp.), and *Sambucus* (20 spp.). All except *Abelia* have species

[226] The monotypic genus *Adoxa* was retained in the Caprifoliaceae by Bentham and Hooker and in the 7th edition of Gray's Manual. Most phylogenists have treated it as a separate family, the Adoxaceae. Its phyletic position is yet in considerable doubt, but the family differs from the Caprifoliaceae in the absence of any apparent calyx, the splitting of each stamen almost to the filament base (anthers 1-celled), the ovary half inferior, the style 3–5-parted, and the fruit a dry berry containing 3–5 cartilaginous nutlets.

indigenous to the United States, as do also the genera *Symphoricarpos, Diervilla, Linnaea, Triosteum.*

The family is readily distinguished from most other families (except Rubiaceae) by the inferior ovary, opposite leaves, and multicarpellate ovary. The difficulty of separating it from Rubiaceae has been indicated under the latter family.

Economically the family is important primarily for a number of hardy ornamental shrubs, although wine made from the ripened fruit of *Sambucus* is popular in many localities. Most of the genera have species of ornamental value, and over 100 species, representing 13 genera, are currently offered by the American trade. Notable among these genera are the honeysuckles (*Lonicera*), coralberry or snowberry (*Symphoricarpos*), elderberry (*Sambucus*), beauty bush (*Kolkwitzia*), bush honeysuckle (*Diervilla*), twinflower (*Linnaea*), and *Viburnum, Abelia, Leycesteria,* and *Weigela.* The species *Lonicera japonica* has become naturalized in much of eastern United States and is a noxious weed.

*LITERATURE:*

AIRY-SHAW, H. K. A revision of the genus *Leycesteria*. Kew Bull. 1932: 241–245.
ARNAL, C. Étude sur les pistilis à ovaire infère. I. *Sambucus nigra.* Bull. Soc. Bot. de France, 95: 60–66, 1948.
FERNALD, M. L. Century of additions to flora of Virginia. Rhodora, 43: 647–652, 1941.
FRITSCH, K. Caprifoliaceae. In Engler and Prantl, Die natürlichen Pflanzenfamilien, IV (4): 156–169, 1897.
GRAEBNER, P. Die Gattung *Linnaea.* Engler, Bot. Jahrb. 29: 120–145, 1900.
HÖCH, F. Zur systematischen Stellung von *Sambucus.* Bot. Centralbl. 51: 233–234, 1892.
HORNE, A. S. A contribution to the study of the evolution of the flower, with special reference to the Hamamelidaceae, Caprifoliaceae, and Cornaceae. Trans. Linn. Soc. Lond. 8: 239–309, 1914.
HULL, E. D. Notes on the coral berry *(Symphoricarpos orbiculatus).* Rhodora, 49: 117–119, 1947.
KECK, D. D. *Lonicera* and *Symphoricarpos* in southern California. Bull. So. Calif. Acad. Sci. 25: 66–73, 1926.
JONES, G. N. A monograph of the genus *Symphoricarpos.* Journ. Arnold Arb. 21: 201–252, 1940.
MORTON, C. V. Mexican and Central American species of *Viburnum.* Contr. U. S. Natl. Herb. 26: 339–366, 1933.
REHDER, A. Synopsis of the genus *Lonicera.* Rept. Mo. Bot. Gard. 14: 27–232, 1903.
RYDBERG, P. A. The North American twinflowers. Torreya, 1: 52–54, 1901.
SAX, K. and KRIBS, D. A. Chromosomes and phylogeny in Caprifoliaceae. Journ. Arnold Arb. 11: 141–153, 1930.
SCHWERIN, F. VON. Revisio generis *Sambucus.* Mitt. deutsch. dend. Ges. 1920: 194–231.
SVENSON, H. K. Plants of the southern United States, I. *Viburnum dentatum.* Rhodora, 42: 1–6, 1940.
WILKINSON, A. M. The floral anatomy and morphology of some species of *Cornus* and of the Caprifoliaceae. Cornell U. Abs. Theses, 1945: 184–187, 1946.
———. Floral anatomy and morphology of some species of the tribe Lonicereae of the Caprifoliaceae. Amer. Journ. Bot. 35: 261–271, 1948.
———. Floral anatomy and morphology of some species of the tribes Linnaeaceae and Sambuceae of the Caprifoliaceae. Amer. Journ. Bot. 35: 365–371, 1948.
———. Floral anatomy and morphology of some species of the genus *Viburnum* of the Caprifoliaceae. Amer. Journ. Bot. 35: 455–464, 1948.
———. Floral anatomy and morphology of *Triosteum* and of the Caprifoliaceae in general. Amer. Journ. Bot. 36: 481–489, 1949.

## VALERIANACEAE. VALERIAN FAMILY

Annual or perennial herbs, rarely subshrubs; leaves in basal rosettes or opposite, often pinnately much divided (at least the cauline ones), estipulate, the bases often sheathing; inflorescence a many-flowered compound dichasial cyme or monochasium, sometimes condensed and capitate, bracteate and usually bracteolate; flowers bisexual or unisexual (the plants then usually dioecious, as *Valeriana* spp.), irregular [227] (almost regular in *Patrinia*), the calyx usually tardily developing, represented by an epigynous ring or rarely 2–4-toothed (as in *Fedia*) which develops in fruit (the subtending bracteoles sometimes mistaken for the calyx), the corolla usually tubular and 5-lobed, imbricate, often basally spurred or saccate, sometimes bilabiate (*Centranthus*); stamens epipetalous and alternating with corolla lobes, varying in number (4 in *Patrinia*, 3 in *Valeriana* and *Valerianella*, 2 in *Fedia*, 1 in *Centranthus*), the anther cells 2-celled, dehiscing longitudinally; pistil 1, the ovary inferior, basically 3-loculed but 2 locules usually suppressed (presence evident in *Valeriana* and *Fedia*) and sterile, the carpels 3, the ovule solitary, pendulous, anatropous, the placentation basically axile but seemingly parietal, the style 1 and slender, the stigma simple or of 2–3 branches or lobes; fruit an achene, the calyx often developing into a winged, awned or plumose pappus; seed with straight embryo, the endosperm absent.

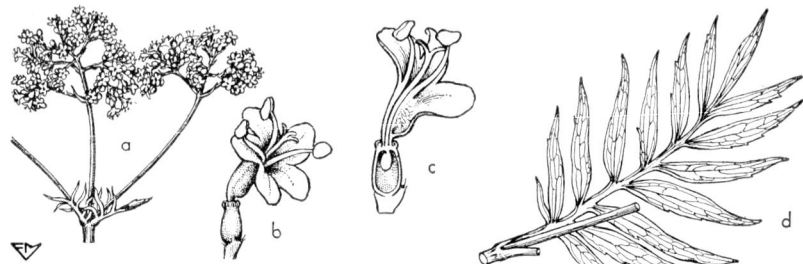

**Fig. 290.** VALERIANACEAE. *Valeriana officinalis*: a, portion of inflorescence, × ½; b, flower, × 3; c, same, vertical section, × 4; d, leaf, × ¼. (From L. H. Bailey, *Manual of cultivated plants*, The Macmillan Company, 1949. Copyright 1924 and 1949 by Liberty H. Bailey.)

A family of 10 genera and about 370 species, mostly of the north temperate region except for occurrences in the Andes of South America. *Valeriana* (about 210 spp.) is the largest genus and is represented in the United States by about 6 species. *Valerianella* has 19 (of 64) species indigenous to this country. *Plectritis* has about 4 species in this country and 4 in Chile.

The Valerianaceae are readily identified by the irregular often spurred or saccate flowers, the reduction in stamen number, the tricarpellate ovary and its 2 abortive ovules and locules, the inferior ovary terminated by an annulate calyx, and by the development of the calyx on the 1-seeded nonendospermous fruit.

Economically the family is of domestic importance only for a few ornamentals, notably red valerian (*Centranthus ruber*), common valerian or garden heliotrope (*Valeriana officinalis*), corn salad (*Valerianella*), and African valerian (*Fedia eriocarpa*). *Valeriana officinalis* is the source of a drug used to some extent in the treatment of some cardiac ailments.

*LITERATURE:*

DYAL, S. C. *Valerianella* in North America. Rhodora, 40: 185–212, 1938. Cf. also pp. 465–467.

[227] The term irregular is used here in contrast with zygomorphic, for in the Valerianaceae the flower usually has no single plane of symmetry.

# DIVISION IV. EMBRYOPHYTA SIPHONOGAMA

## DIPSACACEAE. TEASEL FAMILY

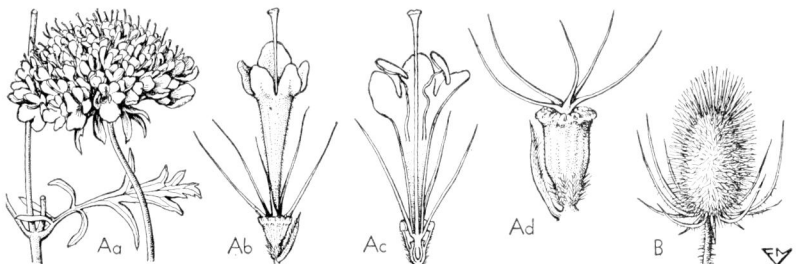

Fig. 291. DIPSACACEAE. A, *Scabiosa atropurpurea*: Aa, flowering branch, × ½, and leaf, × ¼; Ab, flower, × 2; Ac, same, vertical section, × 2; Ad, fruit, × 2. B, *Dipsacus sylvestris*: fruiting inflorescence, × ¼. (From L. H. Bailey, *Manual of cultivated plants*, The Macmillan Company, 1949. Copyright 1924 and 1949 by Liberty H. Bailey.)

Annual, biennial, or perennial herbs, rarely shrubs (in *Scabiosa* spp.); leaves opposite or rarely whorled, estipulate; inflorescence a dense involucrate head or an interrupted spike; flowers bisexual, zygomorphic (rarely irregular), each flower enveloped by an epicalyx formed by connation of two subtending bracteoles, the calyx small and variable, cuplike or divided into 5–10 pappuslike segments, the corolla gamopetalous, 4–5-lobed, imbricate; stamens epipetalous at the base of corolla tube, alternate with the lobes, usually 4 (sometimes 2–3), distinct, the anthers 2–celled, dehiscing longitudinally, introrse; pistil 1, the ovary inferior, unilocular, the carpels 2 with one suppressed, the ovule solitary, pendulous from locule apex, anatropous, the style 1 and filiform, the stigma simple (lateral in *Dipsacus*) or 2-lobed; fruit an achene, enclosed within the epicalyx, often crowned by the persistent calyx; seed with straight embryo, the endosperm fleshy.

An Old World family of 9 genera and about 160 species. None is indigenous to the United States, but some species of *Dipsacus* and *Scabiosa* are naturalized. The family is primarily of eastern Mediterranean and Balkan areas, extending eastward across the steppes of Russia into India.

It is of importance for the teasel (a noxious biennial weed) and for a number of ornamentals, notably species of *Scabiosa, Cephalaria, Morina,* and *Pterocephalus*. The ripened fruiting inflorescences of *Dipsacus fullonum* are used in fulling cloth (i.e., raising the nap), and the plants are grown to a limited extent for this purpose.

*LITERATURE:*

JURICA, H. S. Development of head and flower of *Dipsacus sylvestris*. Bot. Gaz. 71: 138–145, 1921.

PHILLIPSON, W. R. Studies in the development of the inflorescence II. The capitula of *Succisa pratensis* Moench and *Dipsacus fullonum* L. Ann. Bot. (Lond.) n. s. 11: 285–297, 1947.

## Order 43. CUCURBITALES [228]

This order was circumscribed by Engler and Diels to contain only the Cucurbitaceae. The order was considered by them as allied to the Campanulales because of the coherent to connate androecial parts, and to differ by the presence of unisexual flowers, the typically tricarpellate ovary, and the seed lacking any endosperm. For phylogenetic considerations, see under Cucurbitaceae family below.

[228] Sometimes designated as the Peponiferes (Hallier) or Peponiferae (Rendle).

## CUCURBITACEAE. GOURD FAMILY

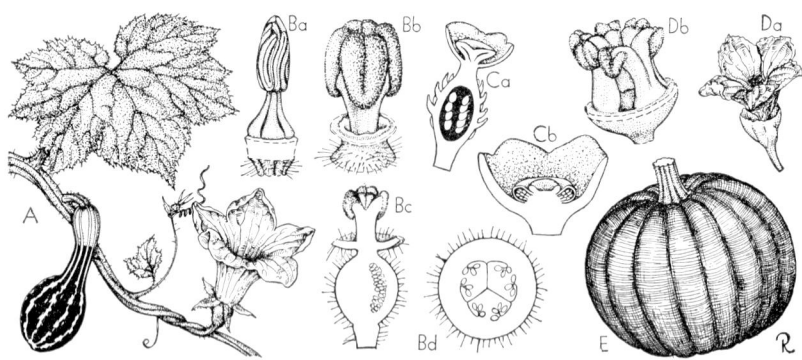

**Fig. 292.** CUCURBITACEAE. A, *Cucurbita Pepo* var. *ovifera*: branch with fruit and pistillate flower, × ⅛. B, *Cucurbita maxima*: Ba, staminate flower, less perianth, × ½; Bb, pistillate flower less perianth and ovary, × 1; Bc, pistillate flower, vertical section, less perianth, × ½; Bd, ovary, cross-section, × 1. C, *Cyclanthera explodens*: Ca, pistillate flower, vertical section, × 3; Cb, staminate flower, vertical section, × 6. D, *Mormordica Balsamina*: Da, staminate flower, habit, × ½; Db, same, less perianth, × 2; E, *Cucurbita Pepo*: fruit, much reduced. (Adapted in part from L. H. Bailey, *Manual of cultivated plants*, The Macmillan Company, 1949. Copyright 1924 and 1949 by Liberty H. Bailey.)

Climbing or prostrate annual or infrequently perennial mostly monoecious or dioecious herbs, rarely suffrutescent (*Dendrosicyos,* a small tree on Socotra), stems often 5-angled; leaves alternate, usually palmately or pinnately 5-lobed or divided, spirally coiled tendrils usually present and situated at the upper side of the petiole base, where they may be simple or variously forked or branched (no tendrils in *Ecballium*),[229] no stipules produced; the inflorescence axillary, determinate, and usually cymose or modifications of cymose, sometimes reduced to a solitary flower; flowers unisexual (very rarely bisexual, as in *Schizopepon* of Japan), actinomorphic, very variable in structure, the calyx lobes 5, the corolla gamopetalous (rarely polypetalous, as in *Fevillea*), the petals or lobes 5, campanulate to rotate or salverform; staminate flower with primitively 5 biloculate distinct stamens alternating with the petals (as in *Fevillea*), but the androecium usually highly modified in 1 of 4 types, in all of which the anthers dehisce by a single longitudinal split, as (1) four of the stamens coherent by basal portions of the filaments into 2 pairs, with the fifth stamen standing apart (as in *Thladiantha*); (2) a similar situation, but the coherent filaments having become connate and the androecium seemingly of two 2-celled anthers and one 1-celled anther (as in *Bryonia, Momordica, Citrullus*); (3) a situation more complex than the second is that represented by the spiral twisting of the anthers associated with their cohesion into a central column and the filaments all connate except sometimes at the extreme base (as in *Cucurbita*), and (4) the stamens strictly monadelphous

---

[229] The morphology of the tendril has been interpreted variously: Braun (1876) believed it a highly modified bracteole that had dropped in position from the peduncle or pedicel to the side of the leaf; Engler (1904) concluded it to be a modified stipule, pointing out that in some genera the tendril was reduced to a spine; Müller (1887) and Hagerup (1930) independently presented evidence in support of the view that the tendril was a combination of a branch (the stiff basal portion) and a highly modified foliage leaf (the upper twining portion). Of these interpretations, the last is currently the most generally accepted.

with the thecae of the anthers in 2 horizontal linklike rings around the edge of the peltate mass of connective and filament tissue, the thecal "links" dehiscing seemingly transverely by a single suture (*Cyclanthera*); rudimentary ovary often present in staminate flowers; pistillate flower with a single pistil, the ovary inferior, the carpels primitively 5, usually 3 or sometimes 4, the locule usually 1 with parietal placentation, or sometimes 3 with axile placentation,[230] the ovules usually numerous (solitary and apically parietal in *Sechium*), and anatropous with 2 integuments, the style 1 (rarely 3), with as many branches or stigma lobes as carpels (sometimes each lobe bifid or notched); fruit basically a berry with soft or hardened pericarp (as in many gourds), often termed a pepo, dehiscent (often explosively so as in *Ecballium* or *Cyclanthera*) or more commonly indehiscent; seeds many (solitary and germinating within the fruit in *Sechium*), with straight embryo having large cotyledons and no endosperm, the seedcoat comprised of many layers, with the outermost sometimes extraovulate and derived from carpellary tissues.

The Cucurbitaceae are represented by about 100 genera and 850 species, primarily of pantropical and subtropical distributions, and are about equally divided between the Old and New Worlds, with range extensions into the temperate northern and southern hemispheres. None of the genera is particularly large, the larger including *Cayaponia* (70 spp.), *Melothria* (60 spp.), *Gurania* (49 spp.), *Cyclanthera* (40 spp.), *Trichosanthes* (25 spp.), and *Sicyos* (35 spp.). The family is represented in the United States by 1 species each of *Echinocystis*, *Melothria*, and *Sicyos* in much of the eastern half, by 1 or more species of *Cayaponia*, *Cucurbita*, and *Momordica* in the south and southeast, and of *Apodanthera*, *Brandegea*, *Cucurbita*, *Cyclanthera*, *Echinopepon*, *Iberillea*, *Marah*, *Maximowiczia*, *Sicyos*, *Sicyosperma*, and *Tumamoca* in the southwest and western areas.

The gourd family is distinguished by the prostrate or scandent habit, the tendril-bearing herbaceous stems, the unisexual flowers with the inferior ovary, with generally 3 parietal placentae which almost completely fill the locule, and the unusual modifications of the androecium.

The Cucurbitaceae were subdivided by Pax (1889) into 5 subfamilies (Fevilleae, Melothrieae, Cucurbiteae, Sicyoideae, and Cyclanthereae), with the first subdivided into 4 tribes, the second into 5 tribes, and the third into 4 tribes.

The determination of the phyletic position of this family requires much more study of the floral anatomy, embryology, cytology, and comparative morphology. Pax and Engler and Wettstein followed the view of Eichler that, on the basis of corolla, androecium, and ovary characters, the family was allied to the Campanulaceae. The view of Robert Brown, adopted by Bentham and Hooker, was that the inferior ovary with its parietal placentation and ovule characters indicated the nearest relative to be the Passifloraceae. Rendle, and later Hutchinson, adopted in principle the view of Hallier that the Cucurbitaceae, Begoniaceae, and Datiscaceae comprised a separate order (Cucurbitales) derived from the Passifloraceae. Bessey included the cucurbits (together with the begonias) in his Loasales, a taxon derived from the more advanced Rosales. Vuillemin (1923) grouped the Begoniaceae and Cucurbitaceae together, and then attempted to demonstrate their alliance with the apetalous orders (by interpreting the cucurbit corolla to be an inner petaloid upfolding of the calyx). From all these views, and the evidence on which they are based, it seems probable that the Cucurbitaceae will be removed

---

[230] The placental situation is often complicated by the intrusion and reflection of the carpel margins, together with an increase of fleshy placental region to the extent that the intruding portions may seemingly meet (in parietal situations) or actually become connate (in axile situations). The situation is interpreted to be parietal unless connation has taken place.

from the Campanulales and be treated as a terminal and advanced taxon of a polypetalous order.

Economically the family is domestically important as source of food and ornament. Prominent among the former are pumpkin and squash (marrow) (*Cucurbita*), cucumber, gherkin, and muskmelon (*Cucumis*), and watermelon and citron (*Citrullus*). Species of about 16 other genera are grown as ornamentals (mostly as gourds), including Chinese watermelon (*Benincasa*), ivy gourd (*Coccinea*), squirting cucumber (*Ecballium*), calabash gourd (*Lagenaria*), dishcloth gourd (*Luffa*), balsam apple (*Momordica*), chayote (*Sechium*), cassabanana (*Sicana*), bur cucumber (*Sicyos*), and the snake gourd (*Trichosanthes*).

*LITERATURE:*

BAILEY, L. H. The domesticated Cucurbitas. Gentes Herb. 2: 63–115, 1929.
———. Three discussions in Cucurbitaceae. Gentes Herb. 2: 175–180, 1930.
———. Species of *Cucurbita*. Gentes Herb. 6: 267–322, 1943.
———. Jottings in the Cucurbitae. Gentes Herb. 7: 449–477, 1948.
BRAUN, A. Morphologie der Cucurbitaceen-ranke. Sitzb. ver. deutsch. Naturf. und Arz. Hamburg. 1876: 101–110.
COGNIAUX, A. Cucurbitaceae. In de Candolle, Monogr. Phan. 3: 325–1008, 1881.
———. Cucurbitaceae. Fevilleae et Melothrieae. In Engler, Das Pflanzenreich, 66 (IV. 275, I): 1–277, 1916.
COGNIAUX, A. and HARMS, H. Cucurbitaceae. Cucurbiteae-Cucumerimae. In Engler, Das Pflanzenreich. 88 (IV. 275. II): 1–246, 1924.
HAGERUP, O. Vergleichende morphologische und systematische Studien über die Ranken und andere vegetative Organe der Cucurbitaceen und Passifloraceen. Dansk. Bot. Arkiv. 6(8): 1–104, 1930.
HEIMLICK, L. F. The development and anatomy of the staminate flower on the cucumber. Amer. Journ. Bot. 14: 227–237, 1927.
JUDSON, J. E. The morphology and vascular anatomy of the pistillate flower of the cucumber. Amer. Journ. Bot. 16: 69–86, 1929.
———. The floral development of the staminate flower of the cucumber. Papers Michigan Acad. Sci. 9: 163–168, 1929.
MÜLLER, E. G. O. and PAX, F. Cucurbitaceae. In Engler and Prantl, Die natürlichen Pflanzenfamilien, IV (5): 1–39, 1889.
NAUDIN, C. Observations relatives à la nature des vrilles et à la structure de la fleur chez les Cucurbitacées. Ann. Sci. Nat. Bot. Sér. IV, 4: 5–19, 1855.
RUTTLE, M. L. Chromosome number in the genus *Cucurbita*. Tech. Bull., N. Y. Agr. Exp. Sta. 186: 1–12, 1931.
WHITAKER, T. W. Cytological and phylogenetic studies in the Cucurbitaceae. Bot. Gaz. 94: 780–790, 1933.
———. American origin of the cultivated cucurbits. Ann. Mo. Bot. Gard. 34: 101–111, 1947.
WHITAKER, T. W. and BOHN, G. W. The taxonomy, genetics, production and uses of the cultivated species of *Cucurbita*. Econ. Bot. 4: 52–81, 1950.
WHITAKER, T. W. and CARTER, G. F. Critical notes on the origin and domestication of the cultivated species of *Cucurbita*. Amer. Journ. Bot. 23: 10–15, 1946.

## Order 44. CAMPANULATAE

The order Campanulatae is characterized by the perianth and androecium typically 5–merous but the number of carpels generally fewer (except in some Campanulaceae), the stamens in a single whorl and the bithecal anthers often coherent to connate, and the ovary often unilocular with a single ovule (except Campanulaceae).

Engler and Diels placed the following 6 families in the order: Campanulaceae, Brunoniaceae, Calyceraceae, Goodeniaceae, Stylidaceae, Compositae.

## CAMPANULACEAE.[231] BELLFLOWER FAMILY

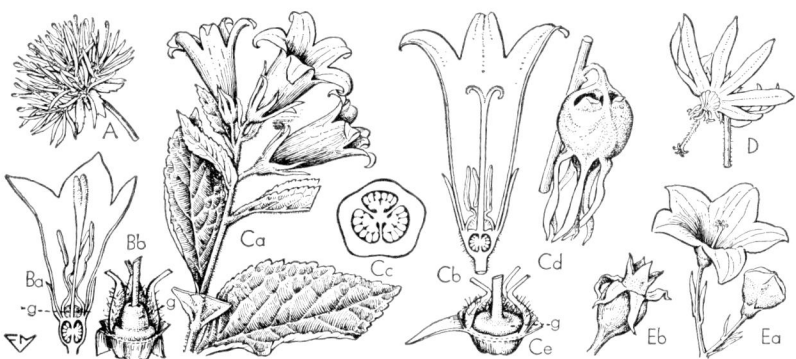

Fig. 293. CAMPANULACEAE subfamily *Campanuloideae*. A, *Jasione perennis*: inflorescence, × 1. B, *Adenophora polymorpha*: Ba, flower, vertical section, × ½; Bb, detail of gland above ovary, × 2. C, *Campanula latifolia*: Ca, flowering branch, × ½; Cb, flower, vertical section, × 1; Cc, ovary, cross-section, × 3; Cd, capsule, × about 1; Ce, detail of gland above ovary, × 2. D, *Michauxia campanuloides*: flower, × ½. E, *Platycodon grandiflorum*: Ea, flower and bud, × ½; Eb, capsule, × ½. (g gland.) (From L. H. Bailey, *Manual of cultivated plants*, The Macmillan Company, 1949. Copyright 1924 and 1949 by Liberty H. Bailey.)

Annual or perennial herbs or subshrubs, rarely arborescent (trees to 30 feet high in *Clermontia* spp.), sap watery or milky; leaves alternate or rarely opposite (whorled in *Ostrowskia* and in *Siphocampylus* spp.), simple, estipulate; inflorescence basically a determinate dichasial or monochasial cyme, and seemingly racemose or thyrsiform (superficially paniculate), sometimes the flowers in involucrate heads or solitary in leaf axils, usually bracteate and often bracteolate; flowers bisexual, actinomorphic to zygomorphic, the calyx lobes 3–10 and usually 5, imbricate or valvate, the corolla actinomorphic and campanulate or tubular to strongly bilabiate, often split down 1 side when zygomorphic, valvate, lobes or segments variable in number but usually 5 (usually 6 in *Canarina* and *Michauxia*, or 3–4 in *Wahlenbergia* and some *Edrianthus*), clearly gamopetalous in most genera but sometimes seemingly polypetalous and the petals distinct (as in *Phyteuma*, *Michauxia*), sometimes absent (apetalous cleistogamous flowers present in *Legenere*, *Heterocodon*, and *Specularia* spp.); stamens as many as corolla lobes or petals and alternate with them, distinct or variously coherent or connate, the filament bases often expanded, forming a dome-shaped chamber over the nectiferous epigynous disc (the latter enlarged and glandlike in *Adenophora*), epipetalous at the extreme base of corolla or more commonly seemingly free from the corolla, the anthers 2-celled, dehiscing longitudinally, introrse, distinct, coherent or connate (connation complete or connate anthers split wholly or partially down 1 side); pistil 1, the ovary inferior (half inferior in *Wahlenbergia, Diastatea, Lobelia* spp. and *Edrianthus* or superior in *Cyananthus* of Asia), 3–10-lobed, the carpels typically 5 or 2 (in Lobelioideae), the locules 2, 3, 5, or rarely 10 (by false septation), the placentation axile (ovary unilocular with 2 parietal placentae in *Downingia, Legenere*, and *Howellia* of western North America, in *Apetahia* of Tahiti, and sometimes so, with deeply intruded placentae, in spp. of tropical American *Sipho-*

---

[231] The name Campanulaceae has been conserved over other names for the family. The Campanulaceae are treated here to include also (as a subfamily) the taxon separated by some authors as the Lobeliaceae.

*campylus*), the ovules numerous and anatropous (rarely few and seemingly basal, as in *Merciera* and *Siphocodon* of Africa), the style 1 and slender, sometimes with 2–5 branches, the stigmas (or stigma lobes) usually 2–5; fruit a capsule dehiscing apically by slits (Wahlenberginae, and *Lobelia* and relatives), circumscissilely (as in *Lysipomia*), or more commonly by apical or basal pores, sometimes a berry (as in *Canarina* and *Centropogon*); seeds with a small, usually straight embryo and a copious fleshy endosperm.

The Campanulaceae (including the Lobelioideae) are a family of about 60 genera and 1500 species, widely distributed over the earth, and mostly in temperate and subtropical regions, with most of the tropical representatives occurring at the higher elevations. The larger genera include (together with an approximate number of species) *Campanula* 230, *Lobelia* 225, *Siphocampylus* 200, *Centropogon* 200, *Wahlenbergia* 70, *Phyteuma* 40, *Cyanea* 50, and *Lightfootia* 40. Genera with species indigenous to this country belong to the Campanuloideae and Lobelioideae. Those of the Campanuloideae include *Campanula, Triodanis, Heterocodon,* and *Githopsis,* and those of the Lobelioideae include *Nemacladus, Parishella, Downingia, Howellia, Porterella, Legenere, Lobelia,* and *Hippobroma.* Only *Campanula, Triodanis,* and *Lobelia* are widespread throughout the country, *Hippobroma* occurs only in southern Florida as a northern extension from the West Indies, and the remaining genera occur primarily in the southwestern and Pacific coast states.

Some authors, notably the American, have treated the Campanulaceae as a family distinct from the Lobeliaceae. However, most monographers and students of these taxa (including McVaugh), and most phylogenists (except Hutchinson) have treated them as comprising 2 of the 3 subfamilies of the Campanulaceae (Campanuloideae, Cyphioideae, and Lobelioideae). The dominant genus in the first and third of these subfamilies (*Campanula* and *Lobelia,* respectively) contrast strongly with one another, but when all genera are taken into consideration, the differences and distinctions become less significant. For a review of the situation see McVaugh (1945, pp. 13–17).

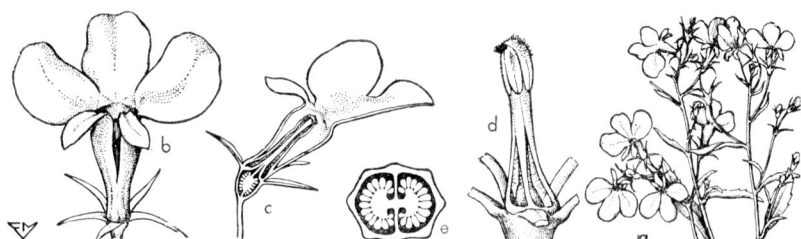

**Fig. 294.** Campanulaceae subfamily *Lobelioideae. Lobelia Erinus*: a, flowering branch, × ½; b, flower, × 2; c, same, vertical section, × 2; d, flower, less perianth, × 4; e, ovary, cross-section, × 8. (From L. H. Bailey, *Manual of cultivated plants,* The Macmillan Company, 1949. Copyright 1924 and 1949 by Liberty H. Bailey.)

The Campanulaceae are distinguished from other allied families by the usually inferior ovary with typically axile placentation and many ovules on 2–5 placentae, and by the stamens frequently united by coherence or connation of filaments or anthers.

The family is economically of domestic importance only for the relatively large number of ornamentals. About 120 species of bellflowers (*Campanula*) are cultivated, as also are about 20 species of *Lobelia,* 10 of *Wahlenbergia,* of *Edrianthus,* and of *Codonopsis,* and one or a few species of *Platycodon, Ostrowskia, Michauxia, Adenophora, Specularia, Sympyandra, Jasione, Pratia,* and *Phyteuma.*

## LITERATURE:

BEDDOME, R. H. An annotated list of the species of *Campanula*. Journ. Roy. Hort. Soc. 32: 196–221, 1907.
DE CANDOLLE, A. Campanulaceae. In de Candolle, Prodromus, 7 (2): 414–496, 1839.
GUINOCHET, M. Recherches de taxonomie expérimentale sur la flore des Alpes et de la région mediterranéene occidentale. II, Sur quelques formes du *Campanula rotundifolia* L. *sens. lat.* Bull. Soc. Bot. Fr. 89: 70–75, 153–156, 1942.
MCVAUGH, R. Studies in the taxonomy and distribution of the eastern North American species of *Lobelia*. Rhodora, 38: 241–263, 276–298, 305–329, 346–362, 1936.
———. A key to the North American species of *Lobelia* (Sect. *Hemipogon*). Amer. Midl. Nat. 24: 681–702, 1940.
———. A revision of "Laurentia" and allied genera in North America. Bull. Torrey Bot. Club, 67: 787–798, 1940.
———. A monograph on the genus *Downingia*. Mem. Torrey Bot. Club, 19, no. 4, pp. 1–57, 1941.
———. Campanulales: Campanulaceae, Lobelioideae. North American Flora, 32A: 1–134, 1943.
———. The genus *Triodanis* Rafinesque, and its relationships to *Specularia* and *Campanula*. Wrightia, 1: 13–52, 1945.
———. Generic status of *Triodanis* and *Specularia*. Rhodora, 50: 38–49, 1948.
ROCK, J. F. A monographic study of the Hawaiian species of the tribe Lobelioideae. Bernice P. Bishop Mus. 1919.
SCHÖNLAND, S. Campanulaceae. In Engler and Prantl, Die natürlichen Pflanzenfamilien, IV (5): 40–70, 1889.
WIMMER, F. E. Campanulaceae-Lobelioideae, Teil I. In Engler, Das Pflanzenreich, 106 (IV. 276b): 1–260, 1943.

## GOODENIACEAE.[232] GOODENIA FAMILY

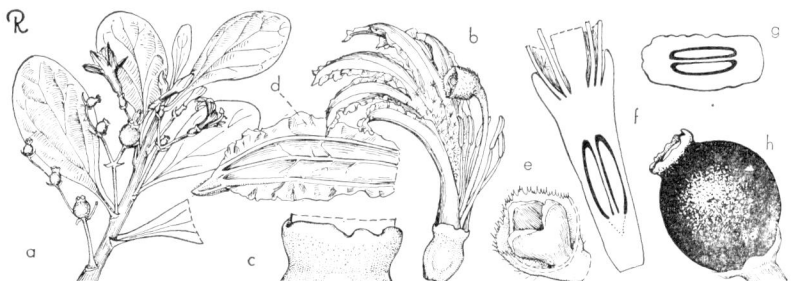

**Fig. 295.** GOODENIACEAE. *Scaevola Plumeri*: a, flowering branch, × ⅜; b, flower, habit, side view, × 1½; c, top of ovary and calyx, × 5; d, corolla-lobe, × 3; e, stigma with indusiumlike cup, × 5; f. flower (basal portion), vertical section, × 5; g, ovary, cross-section, × 5; h, fruit, × 4.

Perennial herbs or small shrubs, lacking milky sap; leaves alternate or rarely opposite or basal, simple, estipulate; inflorescence determinate, cymose or racemose or sometimes capitate, paniculate or solitary; flowers bisexual, usually zygomorphic, the calyx 5-lobed, the corolla usually 5-lobed, bilabiate or rarely 1-lipped, valvate or induplicate; stamens 5, alternate with corolla lobes, free or rarely shortly adnate to corolla base, distinct or the anthers coherent and forming a cylinder

---

[232] The name Goodeniaceae has been conserved over other names for the family. It is to be noted that J. K. Small (Man. S. E. Flora, p. 1925, 1933) placed the genus *Scaevola* (included here in the Goodeniaceae) in the Brunoniaceae, but following Engler and Diels the latter family is held to contain only the monotypic genus *Brunonia* of Australia (flowers actinomorphic in heads), and is treated separately in this text.

around the style, sometimes syngenesious, the anthers 2-celled, dehiscing longitudinally, introrse, pistil 1, the ovary inferior (occasionally half inferior or wholly superior, as in *Velleia*), the carpels 2, the locule 1 or 2, the ovules 1, 2, or many, anatropous, ascending, the placentation basal or axile, the style 1 and filiform, the stigma simple or 2–3-branched and surrounded by a subtending indusiumlike cup, fruit usually a capsule dehiscing by valves, or a berry drupe or nut; seed with a straight embryo and a fleshy endosperm.

A primarily Australasian family of 10–12 genera and about 300 species, with species of *Scaevola* pantropically distributed along the coasts of both hemispheres. *Scaevola Plumieri*, a suffrutescent subshrub, is indigenous on sand dunes of peninsular Florida. The Goodeniaceae are accepted generally to be closely allied to the Lobelioideae of the Campanulaceae, and differ from them primarily in the complete absence of a milky latex, in the more complex anatomy of cambial tissue, and by the development of the indusiate pollen-collecting cup that subtends the stigmas.

The family is of no known economic importance.

## BRUNONIACEAE. BRUNONIA FAMILY

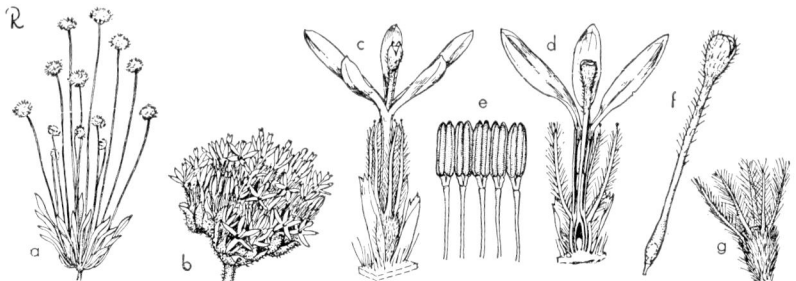

**Fig. 296.** BRUNONIACEAE. *Brunonia australis*: a, habit, × ⅛; b, inflorescence, × 1½; c, flower, with subtending bracts, × 4; d, same, vertical section, × 4; e, upper portion of androecium showing syngenesious anthers, × 5; f, pistil, × 6; g, fruit, × 3.

A family of Australia, represented by the monotypic genus *Brunonia*, a perennial herb about a foot high with spatulate leaves in basal rosettes, closely allied to the Goodeniaceae, from which it differs in the flowers actinomorphic or very nearly so, arranged in a capitate head, the perianth and androecium 4 or 5-merous, stamens 5 and syngenesious, the ovary superior, unilocular, and apparently unicarpellate, with a single basal anatropous ovule, the style simple, the fruit a small nut and the seed lacking endosperm.

Earlier botanists accepted it as a subfamily of the Goodeniaceae, but most recent workers elevated it to the rank of family.

The single species *Brunonia australis,* known as blue pincushion, is cultivated in this country as a hardy perennial.

## STYLIDIACEAE.[233] STYLIDIUM FAMILY

Perennial herbs or small shrubs, lacking milky sap; leaves usually in basal rosettes and grasslike, estipulate; inflorescence mostly scapose, racemose or corymbose; flowers bisexual or unisexual (the plants then usually monoecious), usually zygo-

---

[233] The name Stylidiaceae has been conserved over other names for the family. The generic name *Stylidium* Swartz (1805) was conserved over *Candollea* Labill (1805), and since this is the type genus of the family, the name of the latter was changed from Candolleaceae to Stylidiaceae.

morphic, the calyx usually 5- (sometimes 4-9-) lobed or the lobes connate into a somewhat bilabiate cup, the odd lobe posterior, the corolla usually 5-lobed, the lobes usually unequal with the anterior lobe larger or smaller than others (sometimes termed the labellum); stamens 2 (3 in some spp.), adnate to the style and usually forming a column or gynostemium (homologous to that of the Orchidaceae) or distinct and free (as in *Donatia* of Chile), free from corolla, the anther cells 2, often divergent, extrorse, dehiscing longitudinally; pistil 1, the ovary inferior, primitively bilocular with axile placentation, the carpels 2 but sometimes the posterior carpel abortive and the ovary unilocular with a single parietal placenta, or the septum is lost and free-central placentation exists, the ovules numerous, anatropous, the stigmas 2, terminal and above the adnate anthers; fruit usually a capsule (rarely indehiscent); seed with minute embryo and a fleshy endosperm.

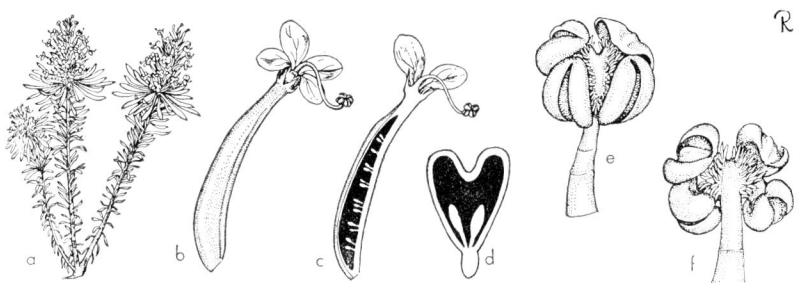

**Fig. 297.** STYLIDIACEAE. *Stylidium adnatum*: a, flowering branch, × ¼; b, flower, habit, × 3; c, same, vertical section, × 3; d, ovary cross-section, × 12; e, tip of gynandrium with two stamens and bilobed stigma, × 18; f, same, as seen from below, × 18.

A family of 3 or 6 genera and about 125 species of Australia and New Zealand. A few species of *Forstera* (*Phyllachne*) and *Stylidium* (*Candollea*) are cultivated in the United States, especially in California.

The family is of special phyletic interest because of the peculiar gynostemium produced in the Stylidioideae, and its members are distinguished from other related families by the reduction in stamen number, the usual adnation of stamens to the style, and the extrorse anthers. The abortion of one of the two carpels is distinctive when present, as also is the loss of the dividing septum and resultant free-central placentation. The stem anatomy of caulescent species differs from that of most dicots in that secondary growth results from the formation of new bundles outside the primary cylinder, a condition somewhat similar to that present in some woody monocots (as *Dracaena* or *Yucca*).

## CALYCERACEAE.[234] CALYCERA FAMILY

Annual or perennial herbs, sometimes suffrutescent; leaves alternate and cauline or in basal rosettes, entire or pinnately lobed, estipulate; inflorescence an involucrate head; flowers usually bisexual, actinomorphic or zygomorphic, the calyx 4-6-angled and toothed, or the lobes sometimes foliaceous, the corolla 4-6-lobed, valvate; stamens 4-6 and alternate with corolla lobes, arising from near mouth of corolla tube, the filaments often basally or wholly connate, the anthers 2-celled, dehiscing longitudinally, introrse, distinct or basally coherent to connate around the style; pistil 1, the ovary inferior, the carpel seemingly 1, the ovary unilocular,

[234] The name Calyceraceae has been conserved over other names for the family.

with a solitary ovule anatropous and pendulous, the style 1 and slender, the stigma capitate; fruit an achene, crowned by the persistent calyx lobes, adjoining achenes sometimes connate; seed with straight embryo, the endosperm scant to copious.

A tropical American family of 3 genera and about 60 species, of which *Acicarpha tribuloides* is reported to be naturalized in "fields and roadsides of northern Florida."

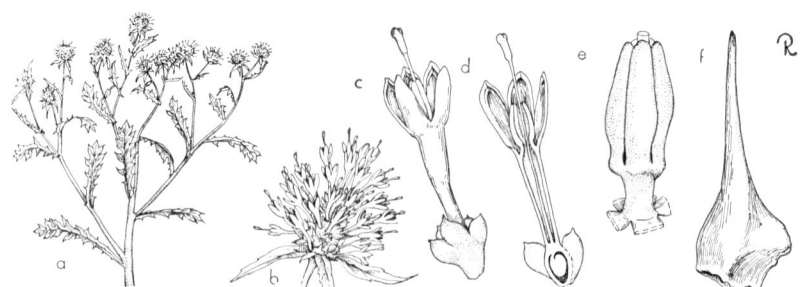

**Fig. 298.** CALYCERACEAE. *Acicarpha tribuloides*: a, flowering branch, × ½; b, inflorescence, × 2; c, flower, habit, × 8; d, same, vertical section, × 8; e, androecium, × 25; f, fruit, × 5.

The family is closely allied to the Compositae, from which it differs in the variable number of perianth lobes, the anthers more or less distinct, the ovule pendulous, and the stigma undivided.

It is of no known economic importance.

## COMPOSITAE.[235] COMPOSITE FAMILY

Herbs, shrubs, or less commonly trees or climbers, the sap sometimes milky; leaves usually alternate, opposite, or rarely whorled (as in *Eupatorium* spp.), simple or pinnately or palmately lobed, divided or compound, frequently in basal rosettes, sometimes heathlike, needlelike, or reduced to scales in some xerophytes, frequently decurrent or sometimes auriculate, estipulate; the primary inflorescence indeterminate, a capitulum of usually few to many flowers (reduced to a single flower in *Echinops* and a few others), ebracteate or bracteate, the bracts often chaffy and scarious, deciduous or persistent, the receptacle variously shaped, usually flattened concavely or convexly, sometimes conical to columnar, ordinarily subtended by an involucre of 1–more series of distinct or variously connate bracts (phyllaries); flowers bisexual or unisexual (the plants then usually monoecious, or sometimes dioecious), the calyx generally considered to be represented by a pappus[236] or seemingly absent, the pappus very diverse in form and type; corolla

---

[235] The name Compositae is one of 8 family names not terminated by the conventional ending -aceae. It is a conserved name, with the provision that an alternative name ending in -aceae may be used in its place, as Asteraceae. For details of variations of gross morphology, see Rendle, 2: 586–611, 1925.

[236] In most of the literature the pappus is accepted to be a reduced and highly modified calyx, but Koch (1930) has suggested that in some genera at least (as in the Heliantheae) the calyx and corolla have fused into a single petaloid structure, and that the pappus represents enations (often trichomelike) from the ovary. Small (1917–1918) held this to be the general situation in the majority of composites.

**Fig. 299.** COMPOSITAE. A, *Aster tanacetifolius*: Aa, flowering branch, × ½; Ab, disc-flower, × 2. B, *Chrysanthemum rubellum*: Ba, flowering branch, × ½; Bb, ray-flower, × 1; Bc, same, corolla partially excised, × 2; Bd, disc-flower, × 3; Be, same, corolla expanded to show syngenesious stamens, × 3. C, *Cichorium Intybus*: Ca, flowering branch, × ½; Cb, ray-flower, corolla partially excised, × 2. D. *Gerbera Jamesonii*: Da, head, × ½; Db, ray-flower, corolla partially excised, × 2; Dc, disc-flower, × 2. E, *Helianthus annuus*: Ea, flowering branch, × ¼; Eb, segment of receptacle bearing disc-flowers, ray-flowers and interfloral bracts, × 2; Ec, disc-flower, × 2. (From L. H. Bailey, *Manual of cultivated plants,* The Macmillan Company, 1949. Copyright 1924 and 1949 by Liberty H. Bailey.)

gamopetalous and derived by connation of 5 petals, usually represented by one of three types, (1) tubular or discoid corolla, 5-lobed with a conspicuous tube and usually a short limb (the latter split and somewhat spreading in some *Centaurea* spp.), (2) ligulate or ray corolla, usually lorate with 3-5 apical teeth (teeth sometimes absent) and a very short tube, (3) bilabiate corolla, modified from a tubular corolla and having a 3-lobed (or toothed) upper lip and 2 slender usually recurved lower lips; ligulate or ray corollas (the so-called strap-shaped "petals") may be restricted to the periphery of the radiate head or may cover the entire receptacle (the head then *ligulate*). When peripheral they usually are neutral or pistillate, when comprising a ligulate head they are usually bisexual (sometimes unisexual and the plants dioecious); tubular or discoid corollas may occupy the entire receptacle (the head then discoid) or all except the periphery, and are usually bisexual (sometimes unisexual and the plants then monoecious or dioecious); stamens 5, epipetalous, syngenesious (anthers sometimes only coherent), the anthers introrse, 2-celled, dehiscing longitudinally, forming a cylinder around the style, often variously appendaged at apex or base of anther sac, the filaments distinct and coiled before anthesis (expanding and straightening as the anthers are forced upward by the elongating stigmas and style); pistil 1, the ovary inferior, unilocular, with a single basal anatropous ovule, the carpels 2, the style 1 and slender, usually 2-branched or the branches bifid, the stigmas 2, surfaces often much restricted, variously shaped; fruit an achene (drupaceous in *Chrysanthemoides* of South Africa), often compressed or obcompressed, crowned or appendaged with any one of a wide assortment of persistent or deciduous pappus types or the pappus lacking, sometimes enveloped by a persistent bract; seed with a large straight embryo and no endosperm.

The Compositae are the largest family of vascular plants, and the genera are estimated to number about 950 and the species probably 20,000. They are distributed over most of the earth and in almost all habitats. The greater proportion are herbaceous, although about 2 per cent are trees or shrubs (mostly tropical representatives). There is no modern treatment of the family as it occurs in this country, and an estimate places the genera present to number about 230, of which about 20 are represented by species widely naturalized from Europe or tropical America (as *Inula, Galinsoga, Anthemis, Chrysanthemum, Cichorium, Matricaria, Cotula, Lapsana, Tussilago, Arctium, Carduus, Silybum, Tragopogon, Taraxacum, Sonchus*, and others). Some genera are widespread over much of the country (as *Solidago, Bidens, Helianthus, Erigeron, Aster, Xanthium, Coreopsis, Senecio,* and *Hieracium*), while the indigenous species of others are restricted to particular regions; as, for example, *Chrysogonum* in Virginia and Pennsylvania southward to Florida; *Balduina, Brintonia, Chrysoma, Marshallia, Polypteris* and *Stokesia* in the south and southeast; *Baileya, Cosmos, Engelmannia, Hemizonia, Laphamia, Palafoxia, Sanvitalia, Tithonia,* and *Zinnia* in the southwest (some representing northern range extensions from Mexico); and *Apargidium, Calycadenia, Centromadia, Holocarpha, Lasthenia, Madia, Rigiopappus, Haplopappus* (*Stenotus*), and *Whitneya* in 1 or more of the Pacific coast states or mostly so. A few genera are represented mostly in the subarctic or colder temperate areas, as *Arnica, Antennaria,* and *Petasites*.

The family is generally readily recognized by the inflorescence an involucrate head, the 5-lobed gamopetalous corollas, the usual presence of a pappus, the inferior bicarpellate uniloculate ovary with a single basal ovule, and the 5 syngenesious stamens. The fruit an achene with a seed lacking endosperm is also distinctive.

Cassini's classification of the family, and adopted by Bentham and Hooker, is generally followed (with various modifications), and according to him it is composed of 2 primary subdivisions: (1) the Tubiflorae with 12 tribes all characterized

**Fig. 300.** COMPOSITAE. A, *Tagetes patula*: flowering branch, × ½; B, *Calendula officinalis*: Ba, flowering branch, × ½; Bb, anthers, × 8; Bc, ray-flower, × 1; Bd, disc-flower, corolla expanded, × 2; Be, seeds, × 2. C, *Ageratum Houstonianum*: Ca, flowering branch, × ½; Cb, anthers, × 8; Cc, disc-flower, × 3; Cd, same, corolla expanded, filaments excised, × 3. D, *Centaurea montana*: Da, flowering branch, × ½; Db, ray-flower, × 1; Dc, disc-flower, × 1; Dd, style and syngenesious stamens of disc-flower, × 2. E, *Echinops exaltatus*: Ea, flowering branch, × ½; Eb, 1-flowered inflorescence, × 1; Ec, flower, × 1. (From L. H. Bailey, *Manual of cultivated plants*, The Macmillan Company, 1949. Copyright 1924 and 1949 by Liberty H. Bailey.)

by the corolla of the disc flowers tubular or bilabiate (not ligulate), and by the absence of lactiferous vessels; and (2) the Liguliflorae with a single tribe (the Cichorieae) characterized by the flowers all ligulate and anastomosing lactiferous vessels present. The tribes of the Tubiflorae (together with a few representative North American genera) are as follows:

1. *Vernonieae:* flowers all alike and tubular, never yellow, anthers sagittate or tailed. *Vernonia, Elephantopus, Stokesia.*
2. *Eupatorieae:* differs from above in anthers blunt at base. *Eupatorium, Mikania, Ageratum, Liatris, Piqueria.*
3. *Astereae:* all or only the central flowers tubular, the disc flowers commonly yellow, anthers blunt at base. *Aster, Solidago, Erigeron, Baccharis, Bellis, Chrysopsis.*
4. *Inuleae:* as in Astereae but anthers tailed. *Anaphalis, Antennaria, Gnaphalium, Inula, Pluchea.*
5. *Heliantheae:* as in Astereae but the style branches each with crown of hairs below the stigma, receptacle with chaffy bracts. *Ambrosia, Bidens, Cosmos, Coreopsis, Helianthus, Layia, Heliopsis, Iva, Madia, Parthenium, Ratibida, Rudbeckia, Silphium, Tithonia, Xanthium.*
6. *Helenieae:* as in Heliantheae but the receptacle bractless. *Baileya, Eriophyllum, Gaillardia, Helenium, Tagetes.*
7. *Anthemideae:* as in Helenieae but the involucral bracts with membraneous tips and edges, pappus none or abortive. *Anthemis, Achillea, Chrysanthemum, Artemisia, Tanacetum, Matricaria.*
8. *Senecioneae:* as in 6 or 7, but the pappus hairy. *Arnica, Petasites, Cacalia, Erechtites, Senecio, Tussilago.*
9. *Calenduleae:* ray flowers usually pistillate and disc flowers usually staminate, pappus none, receptacle bractless. All Old World genera, as *Calendula, Dimorphotheca, Osteospermum.*
10. *Arctotideae:* style thickened or hairy below bifurcation. Mostly South African, as *Arctotis, Gazania, Venidium.*
11. *Cynareae:* as in 10, receptacle mostly bristly, the thistles. *Arctium, Carduus, Centaurea, Cirsium.*
12. *Mutisieae:* ray, and usually disc, flowers bilabiate. *Perezia, Trixis,* and Old World or tropical American genera as *Gerbera* and *Mutisia,* respectively.

Some American authors (notably Britton and students) have treated the family as a separate order (Carduales), and have subdivided it into 3 families: Ambrosiaceae, the nonsyngenesious genera of Heliantheae in 2 tribes, Carduaceae (established to contain tribes 1–12 above except for the nonsyngenesious genera of Heliantheae) in 19 tribes, and the Cichoriaceae, the equivalent of the Liguliflorae above. A number of authors have used the name Asteraceae in lieu of Compositae for the family in the broad sense.

Much has been written on the presumed phylogeny and biology of the family, and the student is referred to the treatments in Small, Bentham, Lavialle, Koch, and Willis as cited in the bibliography below. There are some botanists who hold that the family may be of a polyphyletic origin, but all are agreed that it is the most advanced taxon of dicots.

Economically the family is of considerable importance, but when viewed in relation to its size the importance is considerably less than that of such families as the grasses, legumes, or rosaceous plants. Members important as sources of food to man include lettuce (*Lactuca*), globe artichoke (*Cynara*), endive (*Cichorium*), salsify (*Tragopogon*), and chicory (*Cichorium*). The contact insecticide pyrethrum is obtained from *Chrysanthemum coccineum.* The red dye safflower is from *Carthamus tinctorius.* The pollen of ragweed (*Ambrosia artemisiifolia*) is a primary cause of hayfever. Many members are noxious weeds, and others are used to a limited extent in medicinal or patented preparations. Economically a large number of genera (in excess of 215) are offered in the American trade as ornamentals. Among the more important of these are perennial asters (*Aster*),

DIVISION IV. EMBRYOPHYTA SIPHONOGAMA 731

garden or China aster (*Callistephus*), chrysanthemum (*Chrysanthemum*), coreopsis, cosmos, dahlia, gerbera, strawflowers (*Helichrysum, Helipterum, Xeranthemum*), sunflower (*Helianthus*), *Doronicum*, cineraria (*Senecio*), edelweiss (*Leontopodium*), stevia of florists (*Piqueria*), marigolds (*Tagetes*), zinnia, globe thistle (*Echinops*), and wormwoods (*Artemisia*).

*LITERATURE:*

BABCOCK, E. B. The genus *Crepis*. I, The taxonomy, phylogeny, distribution, and evolution of *Crepis*, Univ. Cal. Pub. Bot. 21: 1–198, 1947. II, Systematic treatment, *op. cit.* 22: 199–1030, 1947.

BABCOCK, E. B., STEBBINS, G. L. JR. and JENKINS, J. A. Chromosomes and phylogeny in some genera of the Crepidinae. Cytologia Fujii, 1: 188–210, 1937.

BENTHAM, G. Notes on the classification, history, and geographical distribution of Compositae. Journ. Linn. Soc. (Bot.) Lond. 13: 335–577, 1873.

BLAKE, S. F. A revision of the genus *Viguiera*. Contr. Gray Herb. 54: 1–205, 1918.

———. Asteraceae described from Mexico and the southwestern United States by M. E. Jones, 1908–1935. Contr. U. S. Nat. Herb. 29: 117–137, 1945.

CHUTE, H. M. The morphology and anatomy of the achene. Amer. Journ. Bot. 17: 703–723, 1930.

CLAUSEN, J., KECK, D. D. and HIESEY, W. M. Experimental studies on the nature of species. II. Plant evolution through amphiploidy and autoploidy, with examples from the Madiinae. Publ. 564. Carnegie Inst. of Washington, 1945. III. Environmental responses of climatic races of *Achillea*. *Op. cit.* Publ. 581, 1948.

CONSTANCE, L. A systematic study of the genus *Eriophyllum*. Univ. Cal. Bot. Pub. 18: 69–136, 1937.

CRONQUIST, A. Revision of the western North American species of *Aster* centering about *Aster foliaceus* Lindl. Amer. Midl. Nat. 29: 429–468, 1943.

———. Notes on Compositae of the northeastern United States. I, Inuleae. Rhodora, 47: 182–184, 1945; II, Heliantheae and Helenieae. *Op. cit.* 396–403, 1945; III, Inuleae and Senecioneae. *Op. cit.* 48: 116–125, 1946; IV, *Solidago*. *Op. cit.* 49: 69–79, 1947.

———. Revision of the North American species of *Erigeron* north of Mexico. Brittonia, 6: 121–302, 1947.

FERNALD, M. L. Dwarf Antennaria of northeastern America. Rhodora, 26: 95–102, 1924.

———. VI, Studies in *Solidago*. VII, Memoranda on *Antennaria*. VIII, Varieties of *Gnaphalium obtusifolium*. Rhodora, 37: 201–239, 1936.

———. Notes on *Hieracium*. Rhodora, 45: 317–325, 1943.

———. Key to *Antennaria* of the "Manual Range." Rhodora, 47: 221–235, 239–247, 1945.

———. Transfers and animadversions on *Artemisia*. Rhodora, 47: 247–256, 1945.

———. Studies on North American plants. [*Helianthus*]. Rhodora, 48: 74–80, 1946.

FRIESNER, R. C. The genus *Solidago* in northeastern North America. Butler Univ. Bot. Studies, 3: 1–64, 1933.

GAISER, L. O. The genus *Liatris*. Rhodora, 48: 165–183, 216–263, 273–326, 331–382, 393–412, 1946.

———. Chromosome studies in *Liatris*. 1. Spicatae and Pycnostachyae. Amer. Journ. Bot. 36: 122–135, 1949.

GLEASON, H. A. Carduaceae: Vernonieae. North Amer. Flora, 33: 47–101, 1922.

GOOD, R. D'O. Some Evolutionary Problems presented by Certain Members of the Compositae. Journ. Bot. 69: 299–305, 1931.

HALL, H. M. Compositae of Southern California. Univ. Cal. Pub. Bot. 3: 1–302, 1907.

———. *Baeria, Lasthenia, Monolopia*. North Amer. Flora, 34: 76–82, 1914–1915.

———. The genus *Haplopappus*. A phylogenetic study in the Compositae. Publ. 389. Carnegie Inst. Washington, 1928.

HALL, H. M. and CLEMENTS, F. E. The experimental method in taxonomy. The North American species of *Artemesia, Crysothamnus,* and *Atriplex*. Publ. 326. Carnegie Inst. Washington, 1923.

HEISER, C. B. JR. Hybridization between the sunflower species *Helianthus annuus* and *H petiolaris*. Evolution, 1: 249–262, 1947.

HOFFMANN, O. Compositae. In Engler and Prantl, Die natürlichen Pflanzenfamilien, IV (5): 87–387, 1894.
JOHANSEN, D. A. Cytology of the tribe Madinae, family Compositae. Bot. Gaz. 95: 177-208, 1933.
KECK, D. A. Studies upon the taxonomy of the Madinae. Madroño, 3: 4–18, 1935.
KOCH, M. F. Studies in the anatomy and morphology of the Composite flower. I–II. Amer. Journ. Bot. 17: 938–952, 995–1010, 1930.
LARSEN, E. L. *Astranthium* and related genera. Ann. Mo. Bot. Gard. 20: 23–44, 1933.
LAVIALLE, P. Recherches sur le développment de l'ovaire en fruit chez les Composées. Ann. Sci. Nat. (Bot.) ser. 9, 15: 39–152, 1912.
MAGUIRE, B. A monograph of the genus *Arnica*. Brittonia, 4: 386–527, 1943.
MALTE, M. O. *Antennaria* of arctic America. Rhodora, 36: 101–117, 1934.
MATTFELD, J. Compositae. Engl. Bot. Jahrb. 62: 386–451, 1928–1929.
POPHAM, R. A. A key to the genera of the Compositales of northeastern North America. Ohio Biol. Survey Bull. 7: 103–129, 1941.
ROBINSON, B. L. A generic key to the Compositae-Eupatorieae. Contr. Gray Herb. 42: 430–437, 1913.
———. A monograph of the genus *Brickellia*. Mem. Gray Herb. 1: 1–151, 1917.
RYDBERG, P. A. Carduaceae. North Amer. Flora, 33: 1–45, 1922 [Incl. family description and key to 19 tribes]; Carduaceae: Heleneae, Tageteae, Anthemideae, Liabeae, Neurolaeneae, Senecioneae (pars.). *Op. cit.* 34: 1–75, 83–359, 1914–1927.
SHARP, W. M. A critical study of certain epappose genera of the Heliantheae-Verbesininae of the natural family Compositae. Ann. Mo. Bot. Gard. 22: 51–152, 1935.
SHERFF, E. E. Revision of the genus *Coreopsis*. Field Mus. Nat. Hist. Bot. Ser. 11: 279–475, 1936.
———. The genus *Bidens*. Field Mus. Nat. Hist. Bot. Ser. 16: 1–709, 1937.
SHINNERS, L. H. The genus *Aster* in Wisconsin. Amer. Midl. Nat. 26: 398–420, 1941; The genus *Aster* in Nova Scotia. Rhodora, 45: 344–351, 1943; The genus *Aster* in West Virginia. Castanea, 10: 61–74, 1945.
———. Revision of the genus *Krigia* Schreber. Wrightia, 1: 187–206, 1947.
SMALL, J. The origin and development of the Compositae. New Phytol. 16: 159–174, 1917; *op. cit.* 17: 13–37, 69–92, 114–142, 1918. [Includes an extensive bibliography.]
STEBBINS, G. L. JR. Cytology of *Antennaria*. I. Normal species. Bot. Gaz. 94: 134–151, 1932.
WATSON, E. E. Contributions to a monograph of the genus *Helianthus*. Mich. Acad. Sci. 9: 305–475, 1928.
WIEGAND, K. M. *Eupatorium purpureum* and its allies. Rhodora, 22: 57–70, 1920.
———. *Aster lateriflorus* and some of its relatives. Rhodora, 30: 161–179, 1928.
———. *Aster paniculatus* and some of its relatives. Rhodora, 35: 16–38, 1933.
WIEGAND, K. M. and WEATHERBY, C. A. The nomenclature of the verticillate *Eupatoria*. Rhodora, 39: 297–306, 1937.
WILLIS, J. C. Compositae, in A dictionary of the flowering plants and ferns. ed. 6, pp. 164–168, 1931.
WODEHOUSE, R. P. Pollen grain morphology in the classification of the Anthemideae. Bull. Torrey Bot. Club, 53: 479–485, 1926.
ZAHN, K. H. Compositae—*Hieracium*. In Engler, Das Pflanzenreich, 76, 77, 79, 82 (IV 280): 1–1705, 1921–1923.

# APPENDIX I

# SUGGESTED SYLLABUS FOR ELEMENTARY COURSE IN TAXONOMY

Many schools offer a single course in plant taxonomy as an elective to students majoring in general biology or arts and sciences, and who previously may have completed only an introductory course in biology or botany. Few of these students may become professional botanists, and a taxonomy course to meet their needs must be on a relatively elementary level. In this connection it is recognized that this textbook includes material beyond the interest and grasp of that student. On the other hand, effort has been made to introduce each phase of the subject at an elementary level and to give it depth by including also supporting evidence, supplementary material, and references to detailed studies that are primarily of interest to the advanced student.

The syllabus that follows is provided as a possible aid to the instructor in organizing an introductory 1-term course in taxonomy. It is only a suggestive outline. It represents a composite opinion obtained from several teachers of elementary taxonomy courses in the United States, and it is based on the presumption that the course will consist of two weekly lectures for a 15-week term, together with integrated laboratory periods. In the preparation of this syllabus, cognizance has been taken of the monotony that may result from an extended series of lectures on selected families, and it is believed that this may be alleviated in part by the interspersion of lectures on other taxonomic subjects. No laboratory syllabus is included, since its subject matter will depend on the season of the year, availability of material, and accessibility of areas for field study. However, experience has shown that some of the lecture topics are best covered when coordination exists between the lecture and laboratory. In these instances, suggested laboratory topics are indicated. Parenthetical numbers in leads of the syllabus refer to pertinent text pages.

Lecture I
  A. Definitions
    1. Taxonomy (3); 2. Identification (3–4); 3. Nomenclature (4); 4. Classification (4–5).
  B. Objectives of the course
    1. Significance of taxonomy (6–7); 2. Relationships to other botanical disciplines (7–8); 3. Acquisition of a familiarity with (a) terminology, (b) literature, (c) use of keys, (d) botanical nomenclature, (e) bases of plant relationship, (f) selected families of vascular plants.
  C. Plant classification
    1. Necessity for (10–12).
    2. Historical review of
      (a) Pre-Linnaean systems (14–18), (b) Linnaeus' systems and other contribution*

(18–26), (c) natural systems (26–33), (d) presumed phylogenetic systems (33–38), (e) present trends (38–39).
Suggested laboratory exercises
   Study of vegetative structures and their terminology as used in identification (57).
   Keys (use and construction of) for vegetative material (228).
   Identification of woody material by aid of keys (such as Muenscher's Keys to Woody Plants).
Lecture II. Evolution and units of classification
  A. Evolution and its significance to taxonomy (92–94, 172)
  B. Units of classification, the components of an evolutionary schema (217)
  C. Species
     1. Concepts of (50); 2. Species name a binomial (194); 3. Subdivisions of: (a) subspecies (55), (b) varietas (55), (c) forma (56).
  D. Genus
     1. Concepts of (48–49); 2. Names are nouns (206); 3. Subdivisions of (50).
  E. Family (46–47, 206)
  F. Order (46, 205)
  G. Class (46)
  H. Division (44)
Suggested laboratory exercises
   Demonstration in laboratory or in the field of material to illustrate several species (with a few variants) of (a) 1 genus, (b) of 2 genera of same family, (c) and of species representative of 1 or 2 other families (to illustrate family characteristics).
   Emphasis to be placed on (a) similarities of relationships; (b) the species as a biological unit comprised of 2 or more intraspecific units, of which 1 is the nomenclatural type of the species; and (c) names (and their endings) applied to the units.
Lecture III. Pteridophyta—the ferns
  A. Characteristics (334)
  B. Morphological structures and their significance.
     1. Spore; 2. Sporangium; 3. Sorus; 4. Indusium; 5. Fronds (sterile and fertile, similar and dimorphic).
  C. Classification (335, 101)
  D. Selected families, emphasizing artificiality of
     1. Polypodiaceae (349); 2. Hydropterideae (352); 3. "Fern allies" (335–341).
     Note: for each family summarize distinguishing characteristics, relationships, world-wide distribution, local representatives, economic importance.
Lectures IV and V. Gymnospermae
  A. Characteristics of the subdivision (355, 102)
  B. Orders, their characteristics, and important families
     1. Cycadales (356); 2. Ginkgoales (357); 3. Taxaceae (360); 4. Coniferae (364); 5. Gnetales or Ephedrales (368)
  C. Selected families (see note under Pteridophyta)
Suggested laboratory exercise to follow Lecture V (in anticipation of Lecture VI):
   (a) Gross morphology of reproductive structures and essential terminology (note starred [*] terms in Glossary), (b) structure of typical angiosperm flower( components of perianth, androecium, and gynoecium) (65–72), (c) types of ovaries and their origins (73), (d) evolution of inferior ovary (79–81), (e) variations of androecial situations (70–71), (f) basic fruit types (follicle, achene, capsule, berry, drupe, nut) (85–87).
Lecture VI.—Angiospermae
  A. The angiosperms
     1. Characteristics and size (370); 2. Origins and possible affinities (103); 3. Phylogeny of (104); 4. Classification guiding principles of: (a) Engler (34), (b) Hutchinson (36, 133), (c) Bessey (36, 125).
  B. Dicotyledoneae and Monocotyledoneae
     1. Characteristics of each (371, 438); 2. Relative primitiveness of the two (107).
  C. Ranales
     1. Characteristics (489); 2. Reasons for presumed primitiveness (127, 135); 3. Composition of (families) and why.

APPENDIX I 735

Lectures VII through IX
  Explanation of selected families, as indicated under III–D above with coordinated laboratory exercises.
Lecture X. Nomenclature
  A. Rules of nomenclature
    1. What are they? (200); 2. To whom do they apply? 3. The need for Latin names (192).
  B. Systems of nomenclature (194–195)
    1. Polynomial; 2. Trinomial; 3. Binomial.
  C. Codes of nomenclature
    1. Paris code (196); 2. Rochester (American code) (197); 3. Vienna code (198); 4. Cambridge code (199).
  D. Significant features of current rules
    1. A plant may have only one valid name; 2. Explain *nomina generica conservanda* (205); 3. Requirements of legitimate names (207–210); 4. Citation of authors' names (209); 5. Retention of epithets when taxa are remodeled, divided, or combined (210–212); 6. Conditions for rejection of a name (213); 7. Orthography of names (216).
  Suggested laboratory exercises
    (a) Study of sheets of problems comprising actual examples, to illustrate the application of important Articles of the Rules.
Lectures XI through XII—Selected families
  Each to be taken up individually as noted under III-D.
  Students should be encouraged to make keys that will enable them to separate families one from another.
Lecture XIV. Biosystematics
  Note: some instructors may prefer to substitute here a lecture on plant geography and ecology (cf. pp. 141–165).
  A. Scope and objectives of the subject (169–172)
  B. Evolution
    1. Lamarckism, Darwinism, Mutation theory, significance of chromosomes and genes (172); 2. Review essentials of meiosis to establish significance of chromosome numbers to biosystematics; 3. Polyploidy (174), what it is and how achieved; 4. Biosystematic categories and their significance:
      (a) ecotype (176), (b) ecospecies (177), (c) cenospecies (178), (d) comparium (178).
    Importance to taxonomy (181–183).
    Limitations (186).
Lectures XV through XVII—Selected families
Lecture XVIII—Field and herbarium techniques
  A. Nature and significance
  B. Collecting procedures
    1. Use of vasculum (234); 2. Plant presses and drying techniques (235); 3. Field notes (269).
  C. Mounting and preserving dried plants (243–257).
  D. Organizing material in the herbarium (257–260).
Lectures XIX through XXII—Selected families
Lecture XXIII—Literature of taxonomic botany
  A. Types of literature. Explain what is meant by:
    1. A serial or journal (308); 2. A review serial (307); 3. A monograph and a revision (304); 4. A manual and a flora (275, 288).
  B. Important indices and their significance
    1. Index Kewensis (286); 2. Index Filicum (287); 3. Gray Herbarium Card Index (287); 4. Index Londinensis (288).
  C. Important floras and manuals of the United States and areas covered by each, to be selected from list (291).
Lectures XXIV through XXVII—Selected Monocot Families
Lecture XXVIII—Phylogeny
  Note: Some instructors prefer to substitute here, or as the last lecture, one on either (1) famous plant explorers and their contributions, or (2) taxonomists important in

the development of the science in this country. In the first group there might be included Marco Polo, Thunberg, von Humboldt, Bonpland, Spruce, Douglas, Forrest, David, E. H. Wilson, and Fairchild. In the second group, there should be included Walter, Muhlenberg, Bartram, Michaux, Sullivant, Torrey, Gray, Eaton, Rafinesque, Pursh, Nuttall, Engelmann, Howell, Britton, Rydberg, Small, Jepson.
A. Significance to taxonomy.
B. Evidence used in phylogenic classifications (97–101)
   1. Paleobotanical; 2. Anatomical; 3. Morphological; 4. Phytogeographical; 5. Physiological; 6. Cytogenetical.
C. Phylogenetic trends in the pteridophytes (101)
D. Phylogenetic trends in the gymnosperms (102)
E. Phylogenetic trends in the angiosperms (103, 108)

Lectures XXIX and XXX. Selected Monocot Families.

## APPENDIX II

# ILLUSTRATED GLOSSARY
# OF TAXONOMIC TERMS

This glossary attempts to include the taxonomic terms used in the text and to include with each a definition according to the best current authority or information. Many of the terms are illustrated. Twenty-two figures, each of related terms, are a part of the glossary. References to figures illustrating terms are indicated parenthetically. As a guide to the beginning student, certain terms believed essential to a minimum working vocabulary are preceded by an asterisk (\*). References to glossaries and botanical dictionaries are provided on pp. 317–319.

*Abaxial.* The side of an organ away from the axis or center of the axis; dorsal.
*Abortive.* Defective; barren; imperfectly developed.
*Abrupt.* Changing suddenly rather than gradually, as a leaf that is narrowed quickly to a point, not tapering; also pinnate leaf that has no terminal leaflet.
\**Acaulescent.* Stemless, or apparently stemless; sometimes the stem is subterranean or protrudes only slightly; a descriptive rather than a morphological term.
*Accessory Fruit.* A fruit, or assemblage of fruits, conspicuous by fleshy parts not part of the pistil, as the strawberry, whose fleshy receptacle is soft and edible and whose fruits (achenes) are embedded in its surface (Fig. 321, k).
*Accumbent.* Lying against and face to face, as cotyledons (Fig. 317, Db).
*Acephalous.* Headless.
\**Achene* (akene). A small dry indehiscent one-seeded fruit with tight thin pericarp (Figs. 321 l, 320 b, c).
\**Acicular.* Needle-shaped (Fig. 305 b).
*Acorn.* The fruit of the oak (*Quercus*) and composed of a nut and its cup or cupule.
*Acropetal.* Arising or developing in a longitudinal plane from a lower toward a more apical position, the opposite of basipetal. *See* Centripetal.
*Actinomorphic.* Regular, symmetrical (Fig. 9 e).
\**Acuminate.* Said of an acute apex whose sides are somewhat concave and taper to a protracted point. (Fig. 307 d).
\**Acute.* Sharp, ending in a point, the sides of the tapered apex essentially straight or slightly convex (Fig. 307 e).
*Acyclic.* Arranged in spirals, not in whorls (Fig. 4 Aa).
*Adaxial.* The side toward the axis; ventral.
*Adherent.* A condition existing when two dissimilar organs or parts touch each other connivently but are not grown or fused together. *See* Coherent and Adnate.
\**Adnate.* Grown to, organically united with another part, as stamens with the corolla; the fusion of unlike parts. *See* Connate.

*Adventitious Buds.* Buds appearing on occasion, rather than resident in regular places and order, as those arising about wounds.

*Aestivation.* The arrangement of the perianth or its parts in the bud. Vernation is leaf arrangement in the bud.

*Aggregate Fruit.* One formed by the coherence of pistils that were distinct in the flower, as blackberry (*Rubus*) (Fig. 321 g).

*Alate.* Winged (Fig. 322 c).

*Albumen.* Starchy or other nutritive material accompanying the embryo; commonly used in the sense of endosperm, for the material surrounding the embryo.

*\*Alternate.* Any arrangement of leaves or other parts not opposite or whorled; placed singly at different heights on the axis or stem (Fig. 303 a).

*Ament.* Catkin (Fig. 109).

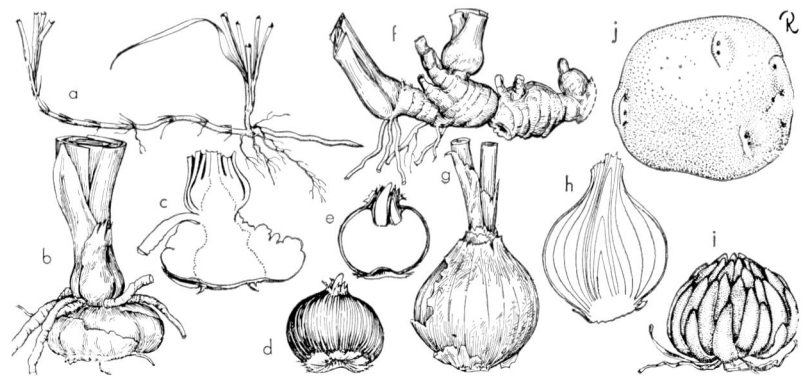

**Fig. 301.** Rootstocks: a, stoloniferous rhizome; b, corm, with membranous tunic; c, same, vertical section; d, corm with fibrous tunic; e, same, vertical section; f, rhizome; g, bulb, tunicated; h, same, vertical section; i, bulb, scaly; j, tuber.

*Amphiploid (Amphidiploid).* A type of polyploid characterized by the addition of both sets of chromosomes from each of 2 species.

*Amphitropous.* Said of an ovule whose stalk (funiculus) is curved about it so that the ovule tip and stalk base are near each other (Fig. 319 Ab).

*Ampulla.* A bladder, as in *Utricularia*.

*Anastomosing.* Netted; interveined; said of leaves marked by cross veins forming a network; sometimes the vein branches meeting only at the margin.

*Anatropous.* Said of an ovule that is reversed, one whose opening (micropyle) is close to the point of funiculus attachment (Fig. 319 Ac).

*\*Androecium.* The male element or household; the stamens as a unit of the flower (Fig. 317).

*Androgynophore.* An axis or stalk bearing both stamens and pistil above the point of perianth attachment (see in Fig. 225 b).

*Androphore.* A stalk bearing the androecium.

*Anemophilous.* Wind-pollinated; the opposite of entomophilous.

*Annual.* Of one season's duration from seed to maturity and death.

*Annular.* In a ring or arranged in a circle.

*Anterior.* Front; on the front side; away from the axis; toward the subtending bract.

*\*Anther.* The pollen-bearing part of the stamen (Figs. 4, 317), borne at the top of the filament or sometimes sessile. See Pollen Sac.

*Antheridium.* In Cryptogams, the organ corresponding to an anther or male organs in flowering plants.

*Antheriferous.* Anther-bearing.

*Anthesis.* Flowering; strictly, the time of expansion of a flower when pollination takes place, but often used to designate the flowering period; the act of flowering.

\**Apetalous.* Without petals; petals missing.

*Apex* (pl. *Apices*). The tip or distal end.

*Aphyllous.* Leafless.

*Apiculate.* Terminated by an *apicula*, a short, sharp, flexible point (Fig. 307 i).

*Apocarpous.* With carpels separate, not united; frequently applied to a gynoecium of separate pistils. See Syncarpous (Figs. 5 a, 134 d).

*Apomict.* In general, a plant produced without fertilization. See p. 183.

*Aposepalous.* Having the sepals distinct from one another, the calyx being composed of separate elements; polysepalous.

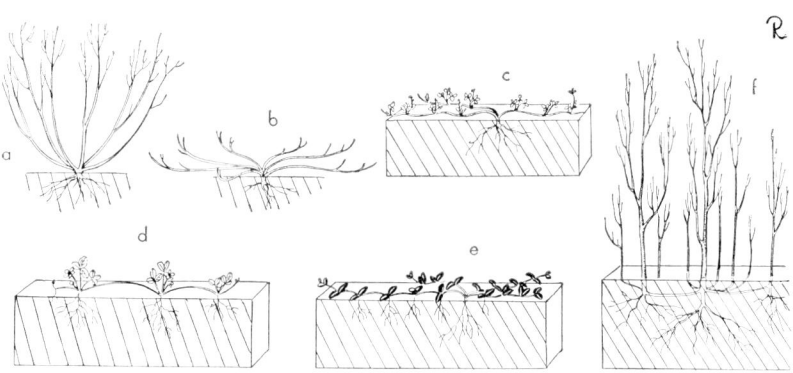

**Fig. 302.** Stem habit types: a, ascending; b, branches decumbent; c, procumbent; d, stoloniferous; e, repent; f, soboliferous.

*Appendage.* An attached subsidiary or secondary part, as a projecting part or a hanging part or supplement (Fig. 317 h, i).

*Appressed.* Closely and flatly pressed against; adpressed.

*Arachnoid.* Cobwebby by soft and slender entangled hairs; also spiderlike.

*Arborescent.* Of treelike habit.

*Archegonium.* The female or egg-containing sex organ of higher cryptogams.

*Arcuate.* Curved or bowed.

*Areole.* The open space formed by anastomosing veins; a small pit or raised spot, often bearing a tuft of hairs, glochids, or spines (Fig. 310 h).

*Aril* (*Arillus*). An appendage or an outer covering of a seed, growing out from the hilum or funiculus; sometimes it appears as a pulpy covering.

*Arillate.* Provided with an aril.

\**Aristate.* Bearing a stiff bristlelike awn or seta; tapered to a very narrow, much-elongated apex (Fig. 307 b).

*Armature.* Any covering or occurrence of spines, barbs, hooks, or prickles on any part of the plant.

*Armed.* Provided with any kind of strong or sharp defense, as of thorns, spines, prickles, barbs.

*Articulate.* Jointed; provided with nodes or joints, or places where separation may naturally take place.

*Ascending.* Rising up; produced somewhat obliquely or indirectly upward.
*Ascidium.* A cup- or pitcher-shaped organ (as in *Nepenthes* leaves).
*Asexual.* Sexless; without sex.
*Assurgent.* Ascending, rising.
*Attenuate.* Showing a long gradual taper, applied to bases or apices of parts (Fig. 307 n).
*Auricle.* An ear-shaped part or appendage, as the projections at the base of some leaves and petals (Fig. 307 t).
*Autoploid.* A polyploid in which each of the 3 or more chromosome sets has been derived from the same species.
*Awl-Shaped.* Narrow and sharp-pointed; gradually tapering from base to a slender or stiff point.
*Awn.* A bristlelike part or appendage (Fig. 322 c).

**Fig. 303.** Leaf arrangements: a, alternate; b, alternate, distichous; c, opposite; d, opposite, decussate; e, whorled; f, fascicled; g, imbricated; h, cauline; i, rosulate (basal); j, equitant; k, cross-section through equitant arrangement.

*Axil.* Upper angle that a petiole or peduncle makes with the stem that bears it.
*Axile.* Belonging to the axis. *See also* Placentation.
*Axillary.* In an axil.
*Axis.* The main or central line of development of any plant or organ; the main stem.
*Baccate.* Berrylike; pulpy or fleshy.
*Banner.* The uppermost petal of a papilionaceous corolla; the standard or vexillum.
*Barbed.* Said of bristles or awns provided with terminal or lateral spinelike hooks that are bent backward sharply (Fig. 310 l).
*Basifixed.* Attached or fixed by the base, as an ovule or anther that is affixed to its support by its bottom rather than by its side (Fig. 4 Da).
*Basipetal.* Developing in a longitudinal plane from an apical or distal point toward the base, the opposite of acropetal. *See* Centrifugal.
*Beak.* A long, prominent, and substantial point; applied particularly to prolongations of fruits and pistils (Fig. 322).
*Beard.* A long awn or bristlelike hair, as in some grasses; a tuft, line, or zone of pubescence, as in some corollas.
\*Berry. Pulpy, indehiscent, few- or many-seeded fruit; technically, the pulpy fruit resulting from a single pistil, containing 1 or more seeds but no true stone, as the tomato or grape (Fig. 32 d).

APPENDIX II                                                                 741

*Biciliate.* Provided with 2 cilia or elaters, as in motile spores.
*Biennial.* Of 2 seasons' duration from seed to maturity and death.
*Bifid.* Two-cleft, as apices of some petals or leaves.
*Bifurcate.* Forked, as some Y-shaped hairs, stigmas, or styles (Fig. 318 f).
*Bilabiate.* Two-lipped, often applied to a corolla or calyx; each lip may or may not be lobed or toothed (Fig. 316 c, d).
*Biotype.* A genotypic race.
*Biovulate.* Containing 2 ovules.
*Biseriate.* In 2 whorls or cycles, as a perianth comprised of a calyx and a corolla.
\*Bisexual.* Having both sexes present and functional in the one flower.
*Bladdery.* Inflated; the walls thin like the bladder of an animal.
\*Blade.* The expanded part of a leaf or petal.
*Bole.* A strong unbranched caudex; the trunk.

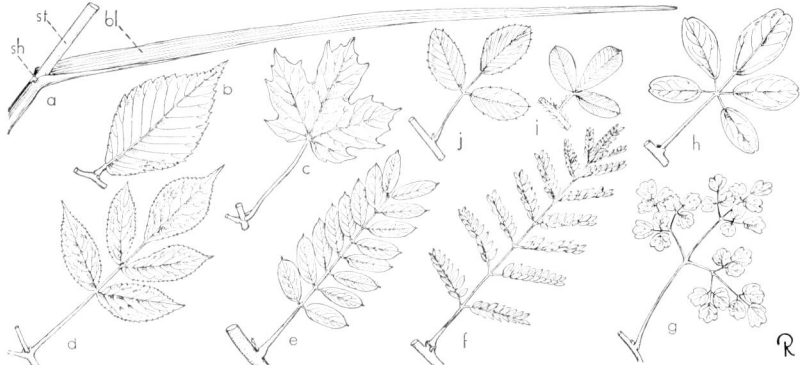

**Fig. 304.** Leaf parts and types: a, leaf sessile, parallel venation (bl blade, sh sheath, st stem); b, leaf alternate and petioled, pinnate venation; c, leaves opposite and petioled, palmate venation; d, leaf compound, odd-pinnate; e, leaf compound, even-pinnate; f, leaf decompound (2-pinnate); g, leaf decompound, ternate; h, leaf compound, palmate; i, leaf trifoliolate and palmate; j, leaf trifoliolate and pinnate.

*Bostryx.* A helicoid cyme (Fig. 313 c).
*Bract.* A much-reduced leaf, particularly the small or scalelike leaves in a flower cluster or associated with the flowers; morphologically a foliar organ.
*Bracteate.* Bearing bracts.
*Bracteole.* A secondary bract; a bractlet.
*Bractlet.* Bract borne on a secondary axis, as on the pedicel or even on the petiole; bracteole.
*Bristly.* Bearing stiff strong hairs or bristles.
\*Bulb.* A thickened part in a resting state and made up of scales or plates on a much-shortened axis.
*Bulbel.* Bulb arising from the mother bulb.
*Bulbil.* A small bulb, usually in leaf axils or sinuses.
*Bulblet.* Little bulb produced in a leaf axil, inflorescence, or other unusual area.
*Bullate.* Blistered or puckered, as the leaf of a Savoy cabbage (Fig. 310 d).
*Bush.* A low and thick shrub, without distinct trunk.
\*Caducous.* Falling off early, or prematurely, as the sepals in some plants.
*Calcarate.* Spurred (Fig. 316 d).
*Caliciform.* Calyxlike.

*Callosity.* A thickened raised area, often paler in color; the presence of a callus.
*Callus.* A hard prominence or protuberance; in a cutting or on a severed or injured part the roll of new covering tissue.
*Calyculus.* A simulated calyx composed of bracts or bractlets.
*Calyculate.* Calyxlike; bearing a part resembling a calyx; particularly, furnished with bracts against or underneath the calyx resembling a supplementary or outer calyx.
*Calyptra.* A hood or lid; particularly the hood or cap of the capsule of a moss, the lid in fruit of eucalyptus, or the form of the calyx as in some papaveraceous flowers.
*\*Calyx.* The outer whorl of floral envelopes, composed of the sepals; the latter may be distinct, or connate in a single structure, sometimes petaloid as in some ranunculaceous flowers.

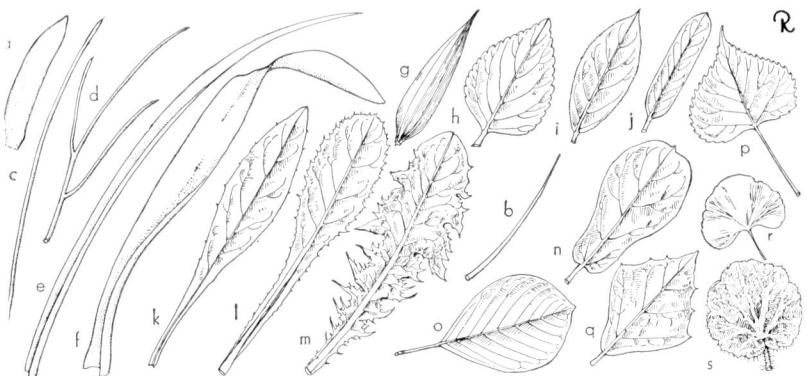

**Fig. 305.** Leaf form or outline: a, subulate; b, acicular; c, filiform; d, filiform segments; e, linear; f, lorate; g, lanceolate; h, ovate; i, elliptic; j, oblong; k, oblanceolate; l, spatulate; m, runcinate; n, pandurate; o, obovate; p, deltoid; q, rhombate (rhomboid); r, reniform; s, orbicular.

*Calyx Tube.* The tube of a gamosepalous calyx; sometimes employed for hypanthium, which see.
*\*Campanulate.* Bell-shaped (Fig. 315 b).
*Campylotropous.* An ovule curved by uneven growth so that its axis is approximately at right angles to its funiculus (stalk) (Fig. 319 Ad).
*Canaliculate.* With a channel or groove.
*Canescent.* Gray-pubescent and hoary, or becoming so.
*Cap.* A convex removable covering of a part, as of a capsule; in the grape, the cohering petals falling off as a cap.
*Capillary.* Hairlike; very slender.
*Capitate.* Headed; in heads; formed like a head; aggregated into a very dense or compact cluster (Fig. 318 b).
*Capitulum.* A dense inflorescence comprised of an aggregation of usually sessile flowers (Fig. 313, e, f).
*Capsular.* Pertaining to a capsule; formed like a capsule.
*\*Capsule.* A dry fruit resulting from the maturing of a compound ovary (of more than one carpel), usually opening at maturity by one or more lines of dehiscence (Fig. [loculicidal] 320 f; [poricidal] 320 g; [septicidal] 320 e).
*Carinal.* Keeled, especially when the keel contains other parts of the flower.

# APPENDIX II

*Carinate.* Keeled; provided with a projecting central longitudinal line or ridge on the dorsal or under surface.

*Carpel.* One of the foliar units of a compound pistil or ovary; a simple pistil has one carpel. See Pistil (Fig. 6 a, m). A foliar, usually ovule-bearing unit of a simple ovary, 2 or more combined by connation in the origin or development of a compound ovary; a female- or mega-sporophyll of an angiosperm flower. For discussion, see pp. 72–75.

*Carpellate.* Possessing or comprised of carpels.

*Carpophore.* The stalk of a sporocarp; a prolongation of the receptacle (axis, stele) that projects above the point of perianth attachment and supports the gynoecium; or, in the Umbelliferae, a wiry stalk (primarily of carpellary origin) that supports each half (carpel) of the dehiscing fruit. See Gynophore, (Fig. 320 k).

**Fig. 306.** Leaf or petal margins: a, entire; b, undulate; c, crenate; d, serrate, e, serrulate; f, double serrate; g, dentate; h, denticulate; i, incised; j, lacerate; k, laciniate; l, lobed; m, cleft; n, parted; o, pinnatifid; p, pectinate; q, palmate (palmatifid); r, pedate; s, revolute (cross-section in diagram); t, crispate; u, ciliate.

*Cartilaginous.* Tough and hard but not bony; gristly.

*Caryopsis.* An achene; sometimes restricted to the fruit of grasses and distinguished from the achene of Compositae by its origin from a superior ovary.

*\*Catkin.* A scaly-bracted, usually flexuous spike or spikelike inflorescence of cymules; ament; prominent in willows, birches, oaks (Fig. 109 Aa).

*Caudate.* Bearing a taillike appendage, as spadices of some aroids or some leaf apices (Fig. 307 c).

*Caudex.* Stem.

*\*Caulescent.* More or less stemmed or stem-bearing; having an evident stem above ground.

*\*Cauline.* Pertaining or belonging to an obvious stem or axis, as opposed to basal or rosulate (Fig. 303 h).

*Centralium.* A central lengthwise cavity as found in seeds of some palms.

*Centrifugal.* Developing or progressing more or less in a transverse plane from the center toward the periphery (outside), the opposite of centripetal. Actually, in flowers, it may be synonymous with basipetal.

*Centripetal.* Developing from the outside toward the center or axis, as stamens of *Ranunculus*; sometimes synonymous with acropetal; the opposite of centrifugal

*Centrum.* The center of a solid.

*Cernuous.* Drooping.

*\*Cespitose, Caespitose.* Matted; growing in tufts; in little dense clumps; said of low plants that make tufts or turf of their basal growth.

*\*Chaff.* A small, thin, dry and membranous scale or bract; in particular, the bracts in the flowerheads of composites.

*Chalaza.* The basal part of an ovule where it is attached to the funiculus.

*Chartaceous.* Of papery or tissuelike texture and not usually green in color.

*Choripetalous.* Having separate and distinct petals; polypetalous.

*\*Ciliate.* Fringed with hairs; bearing hairs on the margin (Fig. 306 u).

*Cincinnus.* A helicoid cyme; sometimes restricted to those characterized by short internodes, as in *Limonium* or *Armeria*.

*Cinereous.* Ash-colored; light gray.

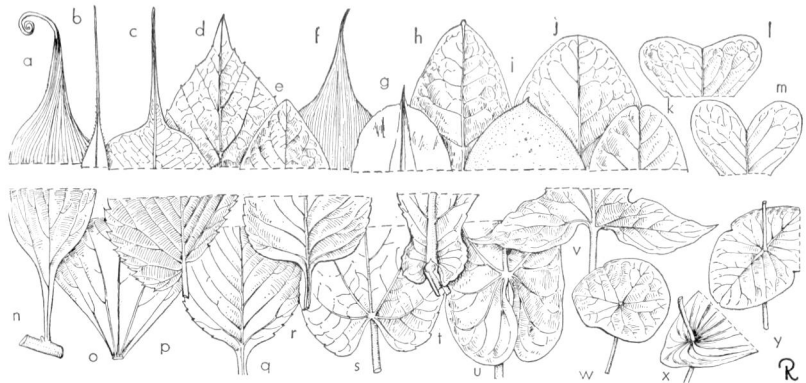

**Fig. 307.** Leaf or petal bases and apices: a, cirrhous; b, aristate; c, caudate; d, acuminate; e, acute; f, cuspidate; g, mucronate; h, mucronulate; i, apiculate; j, obtuse; k, retuse; l, emarginate; m, obcordate; n, attenuate; o, cuneate; p, oblique; q, obtuse; r, truncate; s, cordate; t, auriculate; u, sagittate; v, hastate; w, peltate; x, perfoliate; y, connate-perfoliate.

*Circinate.* Rolled coilwise from top downward, as in unopened fern fronds; with the apex nearest the center of the coil.

*\*Circumscissile.* Opening or dehiscing by a line around the fruit or anther, the valve usually coming off as a lid (Fig. 320 d).

*Cirrhous.* Tendrillike, as a leaf with a slender coiled apex (Fig. 307 a).

*Cladophyll.* A flattened foliaceous stem having the form and function of a leaf, but arising in the axil of a minute, bractlike, often caducous, true leaf. Example, *Ruscus* and *Asparagus*.

*Clambering.* Vinelike, forming a mat or canopy over the undergrowth, often without the aid of tendrils or twining stems.

*Clasping.* Partly or wholly surrounding stem.

*Clavate.* Club-shaped; said of a long body thickened toward the top, like a baseball bat (Fig. 310 g).

*\*Claw.* The long narrow petiolelike base of the petals or sepals in some flowers.

*\*Cleft.* Divided to or about the middle into divisions, as a palmately or pinnately cleft leaf (Fig. 306 m).

*Cleistogamous Flowers.* Small, closed, self-fertilized flowers, as in some violets and in many other plants; they are mostly on or under the ground.

*Clone. A group of individuals resulting from vegetative multiplication; any plant propagated vegetatively and therefore presumably a duplicate of its parent. Originally spelled clon, but changed to clone by its coiner (H. J. Webber) for reasons of philology and euphony; proposed in application to horticultural varieties.

Coccus (plural Cocci). A berry; in particular, one of the parts of a lobed, sometimes leathery or dry fruit with 1-seeded cells.

Coherent. Descriptive of two or more similar parts or organs of the same series touching one another more or less adhesively but not fused. See Connate.

Columella. The carpophore of the mericarpous fruits of Umbelliferae, sometimes restricted to the basal unbifurcated portion.

Column. Body formed of union of stamens, style, and stigmas in orchids, or of stamens, as in mallows (Fig. 98 Ac, Bc).

Coma. The leafy crown or head, as of many palm trees; a tuft of hairs, as in Asclepias.

Commissure. The place of joining or meeting; as the face by which one mericarp joins another, or the faces of appressed stigmas or style branches.

*Comose. Bearing a tuft or tufts of hair (Fig. 322 q).

Composite. Compound; said of an apparently simple or homogeneous organ or structure made up of several really distinct parts.

*Compound. Of 2 or more similar parts in one organ.

*Compound Leaf. A leaf of 2 or more leaflets; in some cases (Citrus) the lateral leaflets may have been lost and only the terminal leaflet remains. *Ternately compound* when the leaflets are in 3's; *palmately compound* when 3 or more leaflets arise from a common point to be palmate (if only 3 are present they may be sessile); *pinnately compound* when arranged along a common rachis (or if only 3 are present at least the terminal leaflet is stalked); *odd-pinnate* if a terminal leaflet is present and the total number of leaflets for the leaf is an odd number; *even-pinnate* if no terminal leaflet is present and the total is an even number.

*Compound Pistil (or ovary). A pistil produced by the connation of 2 or more carpels. The number of cells or locules within the ovary may or may not indicate the number of carpels. An ovary having more than 1 complete cell or locule is almost always compound, but many 1-celled ovaries are also compound. A pistil having a 1-celled ovary, but more than 1 placenta or more than 1 style or more than 1 stigma, or any combination of these duplications, may be presumed to be compound so far as taxonomic considerations are concerned.

Compressed. Flattened, especially flattened laterally. See Obcompressed.

Conduplicate. Folded together lengthwise (Fig. 319 Ca).

Cone. A dense and usually elongated collection of flowers or fruits comprising usually sporophylls and bracts on a central axis, the whole forming a detachable homogeneous fruitlike body; some cones are of short duration, as the staminate cones of pines, and others become dry and woody persistent parts.

Confluent. Merging or blending together.

Congeneric. Belonging to the same genus.

*Connate. United or joined; in particular, said of like or similar structures joined as 1 body or organ. See Adnate.

Connate-Perfoliate. Said of opposite, sessile leaves, connate by their bases, the axis seemingly passing through them (Fig. 307 y).

*Connective. The filament or tissue connecting the 2 cells of an anther, particularly when the cells are separated.

Connivent. Coming together or converging, but not fused; coherent; the parts often arching (treated by some authors as synonymous with adherent).

*Contiguous.* Touching without fusion; used irrespective of whether the parts are like or unlike.

*Contorted.* Twisted; convolute (in aestivation).

*Convolute. Said of floral envelopes in the bud when one edge overlaps the next part (petal or sepal or lobe) while the other edge or margin is overlapped by a preceding part; rolled, the margins overlapping.

*Cordate. Heart-shaped; with a sinus and rounded lobes at the base, and ovate in general outline; often restricted to the basal portion rather than to the outline of the entire organ (Fig. 307 s).

*Coriaceous. Of leathery texture, as a leaf of *Buxus*.

*Corm. A solid bulblike part of the stem, usually subterranean, as the "bulb" of *Crocus* and *Gladiolus*.

*Cormel.* A corm arising vegetatively from a mother corm.

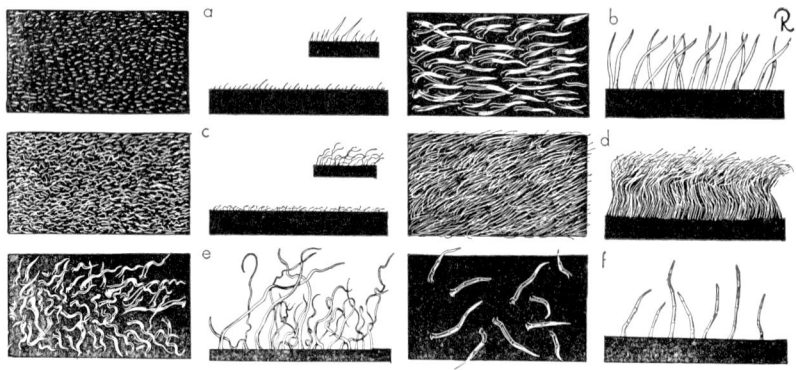

**Fig. 308.** Vesture types (surface view at left, sectional view at right): a, puberulous; b, velutinous; c, tomentose; d, woolly; e, villous; f, pilose (a and c with enlarged inserts of sectional view, all others drawn at same scale).

*Corniculate.* Bearing or terminating in a small hornlike protuberance or process.

*Corolla. Inner circle or second whorl of floral envelopes; if the parts are separate they are petals and the corolla is said to be polypetalous; if not separate, they are teeth, lobes, divisions, or are undifferentiated, and the corolla is said to be gamopetalous or sympetalous (Figs. 315, 316).

*Corolliform.* Appearing as if a corolla, as the calyx, in Nyctaginaceae.

*Corona. Crown, coronet; any appendage or extrusion that stands between the corolla and stamens, or on the corolla, as in some amaryllids, or that is the outgrowth of the staminal part or circle, as in the milkweeds.

*Corymb. Short and broad, more or less flat-topped indeterminate inflorescence, the outer flowers opening first (Fig. 313 a).

*Costa.* A rib, the midvein of a simple leaf; or less commonly the rachis of a pinnately compound leaf.

*Costapalmate.* Said of a palmate palm leaf whose petiole continues through the blade as a distinct midrib, as in the palmetto.

*Cotyledon.* Seed leaf; the primary leaf or leaves in the embryo; in some plants the cotyledon always remains in the seed coats and in others (as bean) it emerges on germination.

*Cotyloid.* Concave or cup-shaped, as the receptacle of *Calycanthus*.

*Creeper.* A trailing shoot that takes root mostly throughout its length; sometimes applied to a tight-clinging vine.

APPENDIX II 747

*Cremocarp.* A dry, dehiscent, 2-seeded fruit of the umbellifer family, each half a mericarp borne on a hairlike carpophore; a schizocarp.
*\*Crenate.* Shallowly round-toothed or obtusely toothed, scalloped (Fig. 306 c).
*Crested.* With elevated and irregular or toothed ridge.
*Crispate.* Curled, an extreme form of undulate (Fig. 306 t).
*Crown.* Corona; also that part of the stem at the surface of the ground; also a part of a rhizome with a large bud, suitable for use in propagation.
*Crownshaft.* A trunklike extension of the bole formed by the long, broad, overlapping petiole bases of some palms.
*Cryptogam.* A plant reproducing by spores instead of by seeds, as ferns, mosses, algae.
*\*Culm.* The stem of grasses and bamboos, usually hollow except at the swollen nodes.

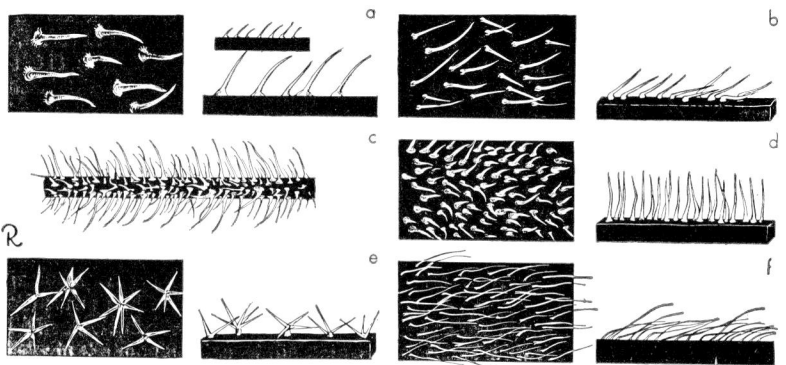

**Fig. 309.** Vesture types (surface view at left, sectional view at right): a, scabrous; b, strigose; c, hispid; d, hirsute; e, stellate; f, sericeous.

*Cultigen.* Plant or group known only in cultivation; presumably originating under domestication; contrast with indigen.
*Cultivar.* A variety or race that has originated and persisted under cultivation, not necessarily referable to a botanical species.
*\*Cuneate.* Wedged-shaped; triangular, with the narrow end at point of attachment, as the bases of leaves or petals (Fig. 307 o).
*Cupule.* Cuplike structure at base of some fruits (as in some palms) formed by the dry and enlarging floral envelopes.
*\*Cuspidate.* With an apex somewhat abruptly and sharply concavely constricted into an elongated, sharp-pointed tip (Fig. 307 f).
*Cyathium.* A type of inflorescence characteristic of *Euphorbia*; the unisexual flowers condensed and congested within a bracteate envelope from which they emerge at anthesis (Fig. 181, Ab, Ac).
*Cyclic.* Whorled, the opposite of spiralled (Figs. 4a, 8b).
*Cymba.* A woody, durable, boatlike spathe or spathe valve that encloses the inflorescence, opens and persists, as in many palms. *See* **Manubrium**.
*Cymbiform.* Boat-shaped.
*\*Cyme.* A broad, more or less flat-topped, determinate flower cluster, with central flowers opening first (Fig. 313 b).
*Cymule.* Diminutive of cyme, usually few-flowered.
*Cystoliths.* Intercellular concretions, usually of calcium carbonate.

*Deciduous.* Falling at the end of one season of growth or life, as the leaves of nonevergreen trees.

*Declinate.* Bent downward or forward, the tips often recurved.

*Decompound.* More than once compound (Fig. 304 f).

*Decumbent.* Reclining or lying on the ground, but with the end ascending.

*Decurrent.* Extending down and adnate to the stem, as the leaf base of *Verbascum*, many composites.

*Decussate.* Opposite leaves in 4 rows up and down the stem alternating in pairs at right angles (Fig. 303 d).

*Deflexed.* Reflexed.

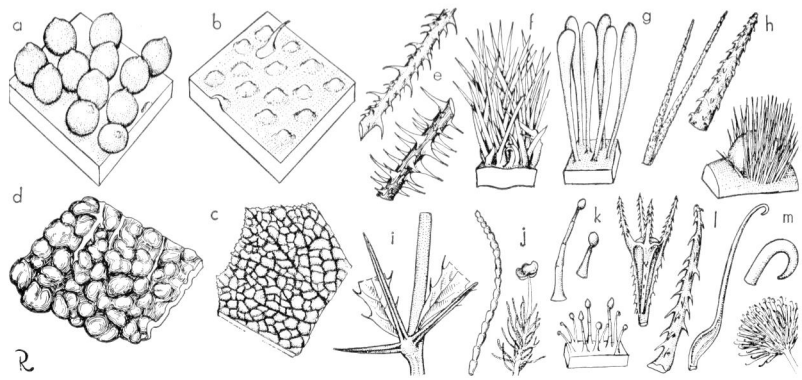

**Fig. 310.** Surface and trichome types: a, papillate (*Cryophytum* leaf); b, muricate (*Delosperma* leaf); c, rugose (*Begonia*); d, bullate (Savoy Cabbage); e, prickles (*Rosa* and *Rubus*); f, echinate (*Castanea* involucre); g, clavellate (*Iris* beard); h, glochidiate (*Opuntia*), glochids and an areole; i, spines (*Berberis* node), not a true trichome; j, moniliform (*Tradescantia* filament); k, glandular (*Solenomelos*); l, retrorsely barbed (*Bidens* fruit); m, uncinate (*Arctium* involucre).

*Dehiscence.* The method or process of opening of a seed pod or anther: *loculicidally* dehiscent when the split opens into a cavity or locule (Fig. 320 f), *septicidally* when opening at point of union of septum or partition to the side wall (Fig. 320 e), *circumscissilely* when the top valve comes off as a lid (Fig. 320 d), *poricidally* when opening by means of pores whose valves are often flaplike (Fig. 320 g).

*Deliquescent.* Having the primary axis or stem much-branched above, as in an elm or banyan tree; the opposite of excurrent; said of perianth parts that quickly become semiliquid, as *Eichornia, Tradescantia.*

*Deltoid.* Triangular; deltalike (Fig. 305 p).

*Dentate.* With sharp, spreading, rather coarse indentations or teeth that are perpendicular to the margin (Fig. 306 g).

*Denticulate.* Minutely or finely dentate (Fig. 306 h).

*Depressed.* More or less flattened endwise or from above; pressed down.

*Determinate.* Said of an inflorescence when the terminal (or central) flower opens first and axis prolongation is thereby arrested. Example, a cyme.

*Diadelphous.* In 2 sets as applied to stamens when the androecium is comprised of two bundles or clusters. In many legumes represented by 9 stamens in one bundle and a solitary stamen in the second (Fig. 317 g).

*Dialypetalous.* Polypetalous, the corolla being composed of separate and distinct petals.

*Diandrous.* Having two stamens, as in *Veronica*.

\*Dichasium. A determinate inflorescence represented by a false dichotomy with the first flower to open situated between 2 lateral flowers (Figs. 312 e, 313 b).

\*Dichotomous. Forked, in 1 or more pairs.

*Diclesium.* An achene enclosed within a free but persistent perianth envelope, as in *Mirabilis*.

*Diclinous.* Unisexual; requiring 2 flowers to represent both sexes.

\*Didynamous. With 4 stamens, in 2 pairs of 2 different lengths (Fig. 317 a).

*Diffuse.* Loosely branching or spreading; of open growth.

*Digitate.* Handlike; compound with the members arising from one point, as the leaflets of horse chestnut. See Palmate.

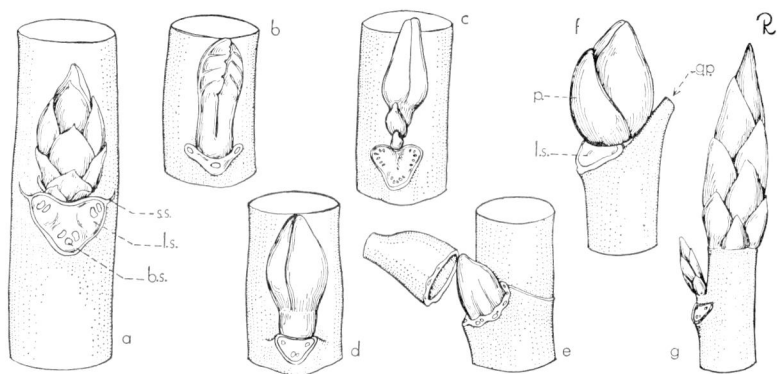

**Fig. 311.** Bud types and bud-scale arrangements: a, axillary bud with imbricate scales; b, naked bud; c, buds (three) superposed; d, stalked bud with valvate scales; e, sub-petiolar bud; f, pseudo-terminal bud; g, terminal bud. (b.s. bundle scar, g.p. growing point abscission scar, l.s. leaf scar, p. pseudo-terminal bud, s.s. stipule scar.)

*Dimorphic.* Occurring in 2 forms, as in ferns with sterile foliaceous fronds and fertile fronds (or segments) that are not leaflike; or having juvenile and adult foliage types, as in *Eucalyptus* or *Hedera*.

\*Dioecious. Having staminate and pistillate flowers on different plants; a term properly applied to a taxonomic unit, not to flowers.

*Diplostemonous.* Having the stamens in 2 whorls, those of the outer whorl alternate with the petals, and those of the inner whorl opposite the petals.

\*Disc. A more or less fleshy or elevated development of the receptacle or of coalesced nectaries or staminodes about the pistil; receptacle in the head of Compositae; a flattened extremity, as on tendrils of Virginia creeper. The term is an Anglicization of the Latin *discus*, and is sometimes spelled disk, and usually so when with reference to the Compositae.

*Discoid.* Having only disc flowers, as in some Compositae; or, in reference to stigma, disc-shaped (Fig. 318 c).

*Disk Flowers.* The tubular flowers in the center of heads of most Compositae, as distinguished from the ray flowers.

*Dissected.* Divided into many slender segments.

*Distichous.* Two-ranked, with leaves, leaflets or flowers on opposite sides of a stem and in the same plane (Fig. 303 b).

*Distinct. Separate; not connate nor coherent with parts in the same series; the petals of a polypetalous flower are distinct. *Compare* Free.

*Diurnal.* Opening only during hours of daylight.

*Divaricate.* Spreading very far apart; extremely divergent.

*Divergent.* Spreading broadly, but less so than when divaricate.

*Divided. Separated to very near the base.

*Dorsal. Back; relating to the back or outer surface of a part or organ, as the lower side of a leaf; the opposite of ventral.

*Dorsifixed. Attached by the back; often, but not necessarily, versatile, as anthers in *Lilium* (Fig. 4 Da).

*Dorsiventral.* Flattened and provided with a definite dorsal and ventral surface; laminate.

*Double.* Said of flowers that have more than the usual or normal number of floral envelopes, particularly of petals; full.

**Fig. 312.** Inflorescence types: a, scapose (flower solitary); b, spike; c, raceme; d, panicle; e, dichasium; f, thyrse.

*Double-serrate.* With coarse serrations bearing minute teeth on their margins (Fig. 306 f).

*Downy.* Covered with very short and weak soft hairs.

*Drupe. A fleshy 1-seeded indehiscent fruit with seed enclosed in a stony endocarp (a pyrene); stone fruit (Fig. 321 a).

*Drupelet.* One drupe of a fruit composed of aggregate drupes, as in the raspberry, the so-called seed being a pyrene.

*E-* or *Ex-*. In Latin-formed words usually denoting, as a prefix, that parts are missing, as: estipulate, without stipules; estriate, without stripes.

*Ebracteate.* Without bracts.

*Echinate.* With stout, bluntish prickles (Figs. 310 f, 322 o).

*Eciliate.* Without cilia.

*Elater.* A ribbonlike band which, in *Equisetum,* is hygroscopic, somewhat clavate, and an aid in dispersal of the spores to which 4 are attached.

*Elliptic. Oval in outline, being narrowed to rounded ends and widest at or about the middle (Fig. 305 i).

*Elongate.* Lengthened; stretched out.

*Emarginate. With a shallow notch at the apex (Fig. 307 l).

*Embryo.* The rudimentary plant in the seed, usually developing from a zygote.

*Embryotega.* A disclike callosity on the seed coat in species of Commelinaceae, Flagellariaceae, and Mayacaceae.
*Enation.* An epidermal outgrowth.
*Endarch.* Said of a type of xylem whose development is toward the center or the axis, that is, centripetal; the opposite of exarch.
*Endemic.* Native or confined naturally to a particular and usually restricted area or region; biologically a relic of once wide distribution.
*Endocarp.* The inner layer of the pericarp or fruit wall.
*Endosperm.* The starch- and oil-containing tissue of many seeds; often referred to as the albumen.
*\*Ensiform.* Sword-shaped.
*\*Entire.* With a continuous margin; not in any way indented; whole (may or may not be hairy or ciliate).

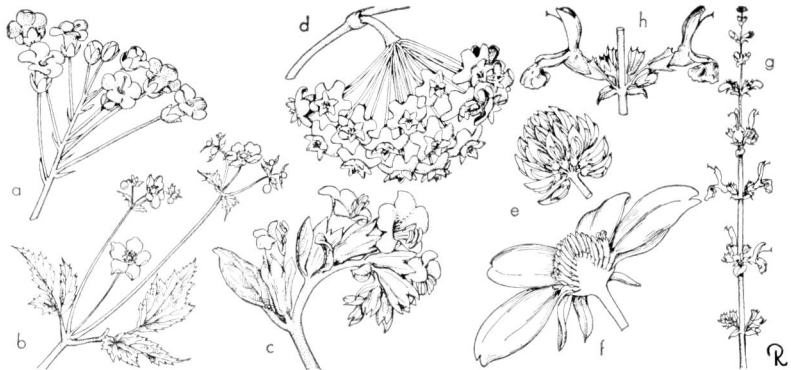

**Fig. 313.** Inflorescence types (somewhat schematic). a, corymb; b, cymes (dichasia) (*Anemone*); c, helicoid cyme (cincinnus); d, umbel (*Hoya*); e, capitulum (*Trifolium*); f, capitulum, vertical section to show conical torus (*Helianthus*); g, inflorescence of verticillate cymes (*Salvia*); h, "verticel" (*Salvia*).

*Entomophilous.* Insect-pollinated.
*\*Ephemeral.* Persisting for 1 day only, as flowers of spiderwort.
*Epi-.* A Greek prefix signifying on or upon.
*Epibiotic.* A species which is a survival of a lost flora; a near-endemic.
*\*Epigynous.* Borne on or arising from the ovary; used of floral parts when ovary is inferior and flower not perigynous. The term is not applicable to the ovary itself (Fig. 8 F, G).
*\*Epipetalous.* Borne on or arising from the petals or corolla.
*Epiphyte.* A plant growing on another or on some other elevated support.
*\*Equitant.* Overlapping in 2 ranks, as the leaves of *Iris* (Fig. 303 j, k).
*\*Erose.* Said of a margin when appearing eroded or gnawed or of a jaggedness too small to be fringed or too irregular to be toothed.
*\*Estipulate.* Without stipules.
*Eusporangiate.* Said of fern sporangia that have developed from a group of cells that first divided periclinally to produce inner (sporogenous-forming) and outer (sterile, sporangium-forming) layers of cells. *See* Leptosporangiate.
*Even-pinnate.* *See* Compound Leaf.
*Evergreen.* Remaining green in its dormant season; sometimes applied to plants that are green throughout the year; properly applied to plants and not to leaves, but due to the persistence of leaves.

*Excurrent.* Extending beyond the margin or tip, as a midrib developing into a mucro or awn; or, descriptive of the habit of a plant with a continuous unbranched axis, as *Picea* or *Abies.* The opposite of deliquescent.

*Exfoliate.* To peel off in shreds, thin layers or plates, as bark from a tree trunk.

*Exine.* The outer coat of a pollen grain (extine).

*Exocarp.* The outer layer of the pericarp or fruit wall.

*\*Exserted.* Sticking out; projecting beyond, as stamens from a perianth; not included.

*Exstipulate.* Without stipules.

*\*Extrorse.* Looking or facing outward; said of anther dehiscence and best determined by a cross section of an undehisced anther.

*Eye.* The marked center of a flower; a bud on a tuber, as on a potato; a single-bud cutting.

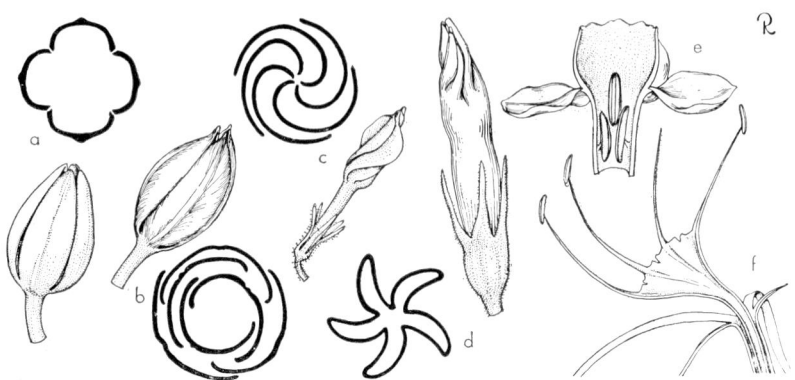

**Fig. 314.** Aestivation and corona types: a, valvate aestivation (*Clematis paniculata*), habit of bud and diagram of perianth arrangements; b, imbricate aestivation (*Tulipa*), habit of flower and diagram; c, convolute aestivation (*Phlox*), habit of opening bud and diagram; d, plicate aestivation (*Nicotiana*), habit of opening bud and diagram; e, corona from perianth, vertical section (*Narcissus*); f, corona from stamen filaments, vertical section (*Hymenocallis*).

*\*Falcate.* Sickle-shaped.

*Falls.* Outer whorl or series of perianth parts of an iridaceous flower, often broader than those of inner series and, in some *Iris,* drooping or flexuous.

*Farinaceous.* Containing starch, or starchlike materials; sometimes applied to a surface covered with a mealy coating, as leaves of some *Primula* spp.; farinose.

*Fasciated.* Much flattened; an abnormal or sometimes teratological widening and flattening of the stem.

*Fascicle.* A condensed or close cluster, as of flowers, or of most pine leaves (Fig. 303 f).

*\*Fasciculate.* Congested in close clusters or in bundles, with or without subtending bracts, as leaves of most *Pinus.*

*Fastigiate.* With branches erect and more or less appressed, as in Lombardy poplar.

*Feminine.* Pistillate (in higher plants).

*Fenestrate.* Perforated with openings or with translucent areas.

*Fertile.* Said of pollen-bearing stamens and seed-bearing fruits.

*Fertilization.* The union of 2 gametes resulting in a zygote.

*Fetid.* Having a disagreeable odor.

*-fid.* Combining form denoting cleft; cut halfway to the middle.

*\*Filament.* Thread; particularly the stalk of the stamen, terminated by the anther.

*Filiform.* Threadlike; long and very slender (Figs. 305 c, 318 d).

*\*Fimbriate.* Fringed.

*\*Flabellate.* Fanlike. Example, leaf of *Phoenix, Ginkgo.*

*Flaccid.* Limp, floppy.

*Flexuous.* Having a more or less zigzag or wavy form; said of stems of various kinds; withy.

*\*Floccose.* Covered with tufts of soft woolly hairs that usually rub off readily.

*Florets.* Individual flowers, especially of composites and grasses; also other very small flowers that make up a very dense form of inflorescence.

*Floricane.* The flowering and fruiting stem, especially of a bramble (*Rubus*).

*Floriferous.* Flower-bearing.

*\*Flower.* An axis bearing 1 or more pistils or 1 or more stamens or both: when only the former, it is a *pistillate flower,* when only the latter a *staminate flower,* when both are present it is a *perfect flower* (i.e., bisexual or hermaphroditic). When this perfect flower is surrounded by a perianth representing 2 floral envelopes (the inner envelope the corolla, the outer the calyx), it is a *complete flower.*

*Foliaceous.* Leaflike; said particularly of sepals and calyx lobes and of bracts that in texture, size, or color look like small or large leaves.

*Follicetum.* An aggregate of follicles (usually distinct) representing the product of an apocarpous multipistillate gynoecium.

*\*Follicle.* Dry dehiscent fruit opening only on the dorsal (front) suture and the product of a simple pistil (Fig. 320 a).

*Foveolate.* Pitted, the pit (*foveola*) solitary or not so.

*\*Free.* Not adnate or adherent to other organs; as petals free from the stamens or calyx; or, the veinlets (as in ferns) not united. Sometimes, however, the word is used in the sense of *distinct*.

*Free-central. See* Placentation.

*Frond.* Leaf of fern; sometimes used in the sense of foliage; especially of palms or other compound leaves. Used by Linnaeus for the leaves of palms.

*\*Fruit.* The ripened ovary (pistil) with the adnate parts; the seed-bearing organ.

*Fruticose.* Shrubby or shrublike in the sense of being woody.

*\*Fugacious.* Falling or withering away very early.

*Funiculus.* The stalk by which an ovule is attached to the ovary wall or placenta.

*\*Funnelform.* With tube gradually widening upward and passing insensibly into the limb, as in many flowers of *Convolvulus*; infundibuliform (Fig. 315 c).

*Furcate.* Forked.

*Furrowed.* With longitudinal channels or grooves.

*Fusiform.* Spindle-shaped; narrowed both ways from a swollen middle.

*Galea.* A helmet, as 1 sepal in an aconite flower.

*Gametophyte.* The generation that bears the sex organs; in ferns, a thalluslike minute or small body bearing archegonia and antheridia; in the angiosperms reduced to the 3-nucleate pollen tube and the 8-nucleate (or -celled) embryo sac and contents.

*\*Gamopetalous.* With a corolla of 1 piece, the petals united, at least at the base, the corolla removable as a single structure; sympetalous (Figs. 93 h, 315).

*Gamophyllous.* With leaves, or foliar units, connate by their edges.

*Gamosepalous.* With a calyx whose sepals are marginally connate, in whole or in part.

*Gemma.* An asexual propagule sometimes appearing as, but not homologous with, a vegetative bud.

*Geniculate.* Bent, like a knee.
*Genotype.* A type (plant or population) determined by genetical characters.
*Gibbous.* Swollen on 1 side, usually basally, as in a snapdragon corolla.
*Glabrate.* Nearly glabrous, or becoming glabrous with maturity or age.
*\*Glabrous.* Not hairy; often incorrectly used in the sense of *smooth* (which see).
*Gland.* Properly, a secreting part or prominence or appendage, but often used in the sense of a glandlike body.
*Glandular.* Having or bearing secreting organs, or glands (Fig. 310 k).
*Glandular-Pubescent.* With glands and hairs intermixed, or hairs terminated by pinheadlike glands.
*Glandular-Punctate.* See Punctate.
*Glaucescent.* Slightly glaucous.
*\*Glaucous.* Covered with a bloom, or whitish substance that rubs off.

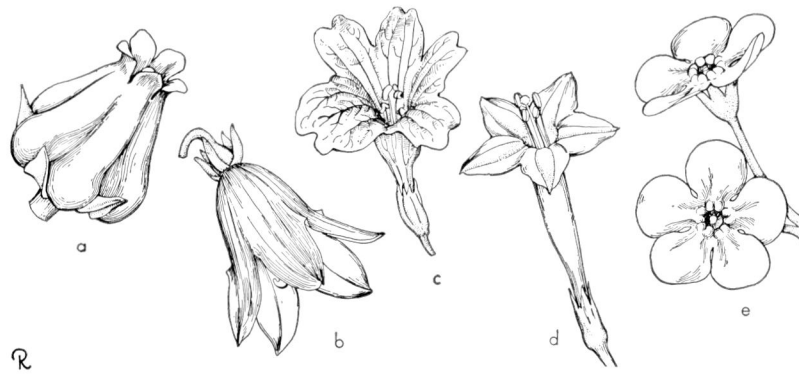

**Fig. 315.** Corolla types. a, urceolate (*Pieris*); b, campanulate (*Campanula*); c, funnelform (*Salpiglossis*); d, salverform (*Quamoclit*); e, rotate (*Brunnera*).

*Glochid.* A minute barbed spine or bristle, often in tufts, as in many cacti (Fig. 310 h).
*Glomerate.* In dense or compact cluster or clusters.
*Glume.* A small chafflike bract; in particular, one of the 2 sterile bracts at the base of most grass spikelets.
*Glutinous.* Sticky.
*Granular, Granulose.* Covered with very small grains; minutely or finely mealy.
*\*Gynoecium.* The female element of a flower; a collective term employed for the several pistils of a single flower when referred to as a unit; when only 1 pistil is present, pistil and gynoecium are synonymous.
*Gynophore.* Stipe of an ovary prolonged within the perianth.
*Gynostemium.* The column in an orchid flower, formed by adnation of stamens and the style and stigma.
*Haft.* The narrow constricted portion of an organ or part.
*Halophyte.* A plant tolerant of various mineral salts in the soil solution, usually of sodium chloride.
*\*Hastate.* Having the shape of an arrowhead, but with the basal lobes pointed or narrow and standing nearly or quite at right angles; halberd-shaped (Fig. 307 v).
*Hastula.* Terminal part of petiole on upper surface of leaf blade of a palmate-leaved palm, sometimes called a ligule.

APPENDIX II 755

*Haustoria.* The absorbing organs (often rootlike) of parasitic plants.
*\*Head.* A short dense spike; capitulum (Fig. 313 e).
*Heart-Shaped.* Cordate; ovate in general outline but with 2 rounded basal lobes; has reference particularly to the shape of the base of a leaf or other expanded part.
*\*Helicoid Cyme.* A sympodial determinate inflorescence whose lateral branches develop from the same side; a bostryx; often erroneously designated as a scorpioid cyme (Fig. 313 c).
*Hemi-.* In Greek compounds, signifying half.
*Herb.* Plant naturally dying to the ground; without persistent stem above ground; lacking definite woody firm structure.
*\*Herbaceous.* Not woody; dying down each year; said also of soft branches before they become woody.

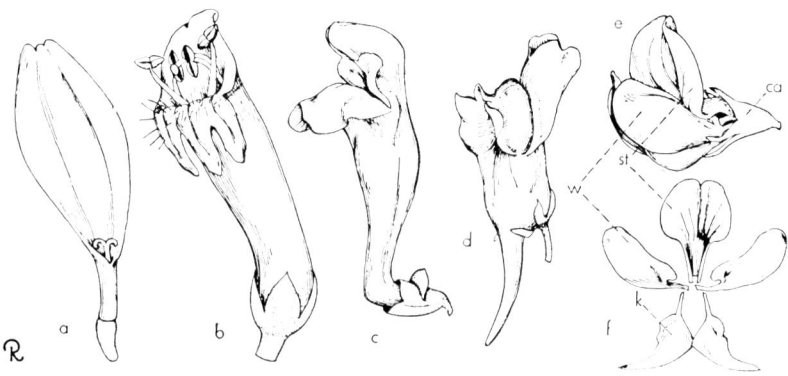

**Fig. 316.** Corolla types. a, ligulate or ray type (*Helianthus*); b, tubular or disc-type (also bilabiate) (*Penstemon*); c, bilabiate (base geniculate) (*Scutellaria*); d, bilabiate and personate (base spurred or calcarate) (*Linaria*); e, papilionaceous, habit, side view; f, papilionaceous, petals spread (*Lotus*). (k, keel petals; w, wing petals; st, standard or *vexillum*.)

*Herbage.* Vegetative parts of plant.
*Hermaphroditic.* Bisexual.
*Hetero-.* In Greek composition, signifying various, or of more than 1 kind or form; as heterophyllous, with more than 1 kind or form of leaf.
*Heterogamous.* With 2 or more kinds or forms of flowers.
*Heterogeneous.* Lacking in uniformity in kind of part or organ.
*Heterosporous.* Producing 2 kinds of spores, representative of 2 sexes, as in *Isoetes, Salvinia*, and other ferns.
*Hilum.* In the seed, the scar or mark indicating the point of attachment.
*Hippocrepiform.* Horseshoe-shaped.
*\*Hirsute.* With rather rough or coarse hairs.
*Hirtellous.* Softly or minutely hirsute or hairy.
*\*Hispid.* Provided with stiff or bristly hairs (Fig. 309 c).
*Hispidulous.* Somewhat or minutely hispid.
*Hoary.* Covered with a close white or whitish pubescence.
*Homo-.* In Greek compounds, signifying alike or very similar.
*Homochlamydeous.* With a perianth of tepals, undifferentiated into calyx and corolla.
*Homogeneous.* Uniform as to kind; the opposite of heterogeneous.

*Homosporous.* Producing spores of one kind only, as in Polypodiaceae, Osmundaceae.

*Horsetail. Equisetum,* or any member of the sphenopsid group.

*Husk.* An outer covering of some fruits (as *Physalis* or *Juglans*), usually derived from the perianth or involucre.

*\*Hyaline.* Translucent when viewed in transmitted light, or transparent.

*Hybrid.* A plant resulting from a cross between parents that are genetically unlike; more commonly, in descriptive taxonomy, the offspring of 2 different species or their infraspecific units.

*Hygroscopic.* Capable of expanding or contracting on presence or absence of water or water vapor.

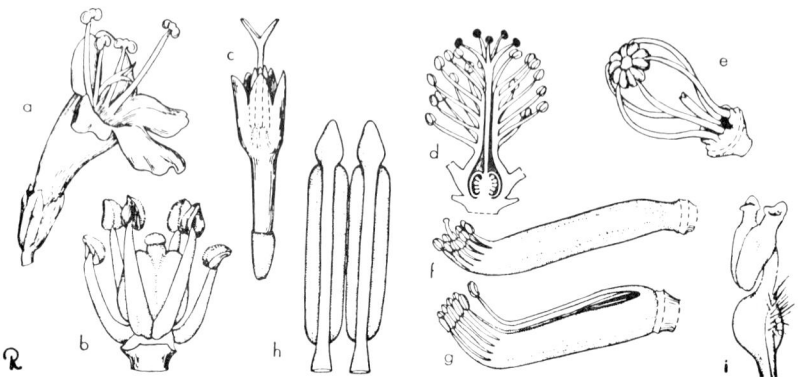

**Fig. 317.** Androecial types. a, stamens didynamous (*Origanum*); b, stamens tetradynamous (*Thlaspi*); c, stamens syngenesious (*Aster*); d, stamens monadelphous (*Hibiscus*); e, stamens syngenesious (*Sinningia*); f, stamens monadelphous (*Desmodium*); g, stamens diadelphous (*Lathyrus*); h, stamens with distal appendage (*Inula*); i, stamen with basal appendage (*Chimaphila*).

*\*Hypanthium.* The cuplike receptacle derived usually from the fusion of floral envelopes and androecium, and on which are seemingly borne calyx, corolla, and stamens (as in fuchsia or plum); once generally accepted to have been formed solely by the enlargement or depression of the torus; literally "beneath the flower"; the fruitlike body (as the rose hip) formed by enlargement of the cuplike structure and bearing the achenes (the true fruits) on its upper and inner surface; sometimes erroneously termed the calyx tube (which see). (Fig. 8 B, G).

*Hypocotyl.* The axis of an embryo below the cotyledons which on seed germination develops into the radicle.

*Hypocrateriform.* See Salverform.

*\*Hypogynous.* Borne on the torus, or under the ovary; said of the stamens or petals; not applicable to the ovary (Fig. 8 A).

*\*Imbricated.* Overlapping, as shingles on a roof (Fig. 303 g).

*Imparipinnate.* Unequally pinnate; odd-pinnate; with a single terminal leaflet.

*Incised.* Cut; slashed irregularly, more or less deeply and sharply; an intermediate condition between toothed and lobed (Fig. 306 i).

*Included.* Not protruded, as stamens not projecting from the corolla; not exserted.

*Incumbent.* Said of cotyledons that, within the seed, lie face to face with the back of one against the hypocotyl; anthers are incumbent when turned inward (Fig. 319 Ba).

*Indehiscent.* Not regularly opening, as a seed pod or anther.
*Indeterminate.* Said of those kinds of inflorescence whose terminal flowers open last, hence the growth or elongation of the main axis is not arrested by the opening of the first flowers.
*Indigen.* Indigenous inhabitant; a native.
*Indumentum.* A rather heavy hairy or pubescent covering.
*Induplicate.* Rolled or folded inwards.
*Indurated.* Hardened, usually ontogenetically.
*Indusium.* The epithelial excrescence that, when present, covers or contains the sporangia of a fern when the latter are in sori.
*Inferior.* Beneath, lower, below; as an inferior ovary, one that seemingly is below the calyx leaves (Fig. 8 F, G).

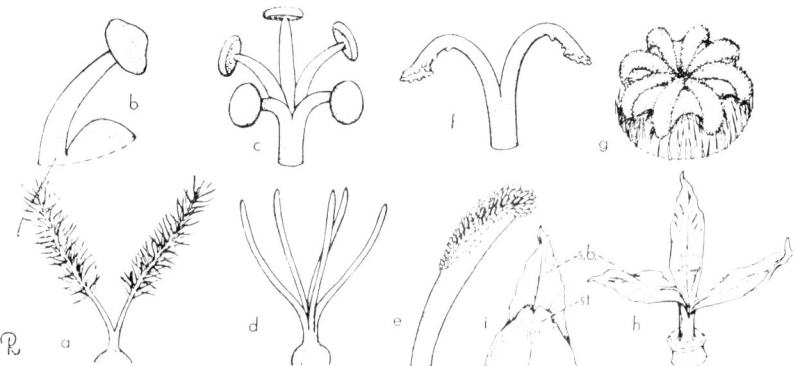

**Fig. 318.** Style and stigma types. a, stigmas plumose (*Gramineae*); b, stigma capitate (*Alchemilla*); c, stigmas discoid (*Hibiscus*); d, style branches filiform (*Armeria*, hairs omitted); e, stigma linear (*Kitaibelia*, a single branch); f, style bifurcate (*Arnica*); g, stigma branches radiate (*Papaver*); h, style branches petaloid (*Iris*); i, stigma transverse (*Iris*). (s.b., style branch; st. stigma.)

*Inflated.* Blown up; bladdery.
*Inflorescence.* Mode of flower bearing; technically less correct but much more common in the sense of a flower cluster (Figs. 312, 313).
*Infra-.* In combinations, signifying below.
*Infrafoliar.* Below the leaves.
*Infraspecific.* Referring to any unit of classification below the species level.
*Infundibular.* Funnel-shaped (Fig. 315 c).
*Insectivorous.* Insect-catching, as *Dionaea* or *Sarracenia*.
*Inserted.* Attached; as a stamen growing on the corolla.
*Integument.* The covering of an organ; the outer envelope of an ovule.
*Inter-.* In composition, signifying between, particularly between closely related parts or organs.
*Interfoliar.* Among the leaves.
*Internode.* The part of an axis between 2 nodes.
*Interrupted.* Not continuous; in particular, referring to the interposition of small leaflets or segments between others.
*Intrafoliar.* Within the leaves, as inflorescences in some palms.
*Introrse.* Turned or faced inward or toward the axis, as an anther whose line of dehiscence faces toward the center of flower.
*Inverted.* Turned over; end for end; top side down.

*Involucel.* A secondary involucre; small involucre about the parts of a cluster.
*\*Involucre.* One or more whorls of small leaves or bracts (phyllaries) standing close underneath a flower or flower cluster (Figs. 110, 299).
*\*Involute.* Rolled inward or toward the upper side; said of a flat body, as a leaf. *See* Revolute.
*\*Irregular flower.* A flower having some parts different from other parts in same series, and incapable of being divided into 2 equal halves (Fig. 9 y). *See* Zygomorphic.
*Jointed.* With nodes, or points of real or apparent articulation.
*Jugum.* A pair, as of leaflets.
*Karyotype.* The chromosomal complex characteristic of a group of allied plants, associated with both morphology and number of chromosomes.
*Keel.* The two front united petals of a papilionaceous flower.

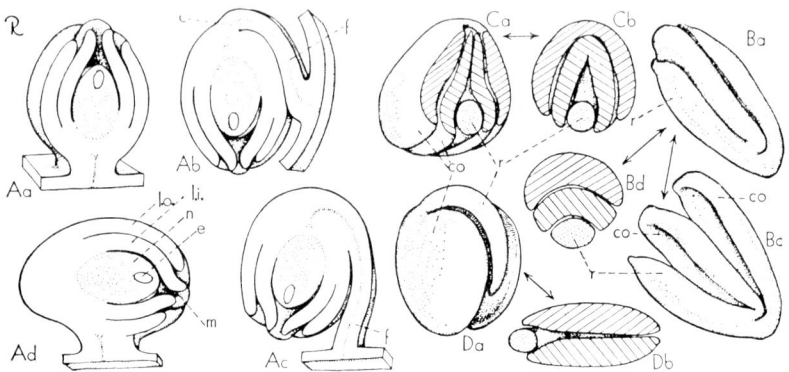

**Fig. 319.** Ovule and cotyledon types: Aa, ovule orthotropous; Ab, ovule amphitropous; Ac, ovule anatropous; Ad, ovule campylotropous. B, Cotyledons incumbent: Ba, embryo with cotyledons folded against radicle; Bb, same, separated; Bc, same, cross-section. C, Cotyledons conduplicate: Ca, cross-section near distal end of embryo; Cb, same, median. D, Cotyledons accumbent: Da, embryo, habit; Db, same, cross-section (c chalaza, co cotyledon, e egg, f funiculus, Ii inner integument, Io outer integument, m micropyle, n nucellus, r radicle.)

*Keeled.* Ridged like the bottom of a boat.
*\*Labellum.* Lip; a perianth part, particularly the lip of an orchid flower (Fig. 98 Ad, Ba).
*Labiate.* Lipped; or, a member of the Labiatae.
*Lacerate.* Torn; irregularly cleft or cut (Fig. 306 j).
*Laciniate.* Slashed into narrow pointed lobes (Fig. 306 k).
*Lactiferous.* Producing or bearing latex (milky sap).
*Lacuna.* A cavity, hole, or gap.
*Lageniform.* Gourd-shaped.
*Lamellate.* Provided with many finlike blades or cross partitions. *See* Placentation.
*\*Lamina.* A blade or expanded portion.
*\*Lanate.* Woolly, with long, intertwined, curly hairs.
*\*Lanceolate.* Lance-shaped; much longer than broad; widening above the base and tapering to the apex (Fig. 305 g).
*Lanuginose.* Woolly or cottony; downy, the hairs somewhat shorter than in lanate.
*Lanulose.* Very short-woolly.
*Lateral.* On or at the side.

*Latex.* Milky sap.
*Lax.* Loose, the opposite of congested.
*Leaf Stalk.* The stalk of a leaf; petiole.
\**Leaflet.* One part of a compound leaf; secondary leaf.
*Legume.* Simple fruit dehiscing on both sutures, and the product of a simple unicarpellate ovary (Fig. 320 j).
*Lemma.* In grasses, the flowering glume, the lower of the 2 bracts immediately enclosing the flower (Fig. 75 lb).
*Lenticular.* Lens-shaped.
*Lepidote.* Surfaced with small scurfy scales.
\**Lianous.* Vinelike.
\**Ligneous.* Woody.
\**Ligulate.* Strap-shaped, as a leaf, petal, or corolla (Fig. 316 a).

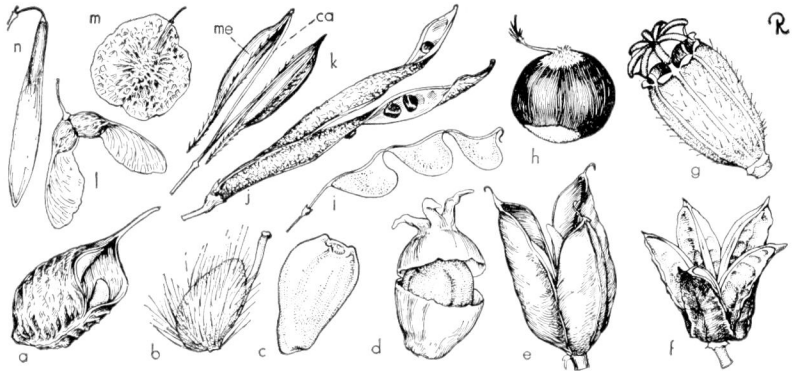

**Fig. 320.** Fruit types (dry): a, follicle (*Helleborus*); b, achene (*Potentilla*); c, achene (Compositae); d, pyxis (circumscissile); e, capsule (septicidal); f, capsule (loculicidal); g, capsule (poricidal); h, nut; i, loment; j, legume; k, schizocarp; l, samara (*Acer*); m, samara (*Ptelea*); n, samara (*Fraxinus*). (ca carpophore, me mericarp.)

*Ligule.* A strap-shaped organ or body; particularly, a strap-shaped corolla, as in the ray flowers of composites; also a projection from the top of the sheath in grasses and similar plants (Fig. 75 c).
\**Limb.* The expanded flat part of an organ; in particular, the expanding part of a gamopetalous corolla.
\**Linear.* Long and narrow, the sides parallel or nearly so, as blades of most grasses (Figs. 305 e, 318 e).
*Lineate.* Lined; bearing thin parallel lines.
*Lingulate.* Tongue-shaped.
\**Lip.* One of the parts in an unequally divided corolla or calyx; these parts are usually 2, the upper lip and the lower lip, although 1 lip is sometimes wanting; the upper lip of orchids is by a twist of the ovary made to appear as the lower; a labium.
\**Lobe.* Any part or segment of an organ; specifically, a part of petal or calyx or leaf that represents a division to about the middle (Fig. 306 l).
*Lobule.* A small lobe.
\**Locule* (*Loculus*). Compartment or cell of an ovary, anther, or fruit, a descriptive term lacking morphological meaning.

*Loculicidal.* Dehiscence on the back, more or less midway between the partitions, into the cavity (Fig. 320 f).

*Lodicule.* One of 2 or 3 scales appressed to the base of the ovary in most grasses, believed to be rudiments of ancestral perianth parts (Fig. 75 A)

*Loment.* A leguminous fruit, contracted between the seeds, the 1-seeded segments separating at fruit maturity (Fig. 320 i).

*\*Lorate.* Strap-shaped; often also flexuous or limp or with apex not acuminate or acute (Fig. 305 f).

*Lycopsid.* Any one of the Lycopsida, to which *Lycopodium* belongs.

*Lyrate.* Pinnatifid, but with an enlarged terminal lobe and smaller lower lobes.

*Macrospore.* The larger of the 2 kinds of spores, as in *Selaginella* and related plants; a megaspore.

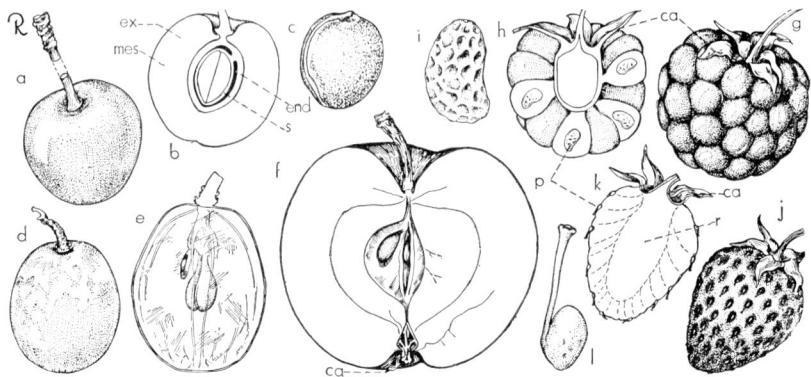

**Fig. 321.** Fruit types (fleshy): a, drupe (*Prunus*); b, same, vertical section; c, pyrene or "pit" of drupe; d, berry (*Vitis*); e, same, vertical section; f, pome (*Malus*), vertical section; g, aggregate fruit of drupelets (*Rubus*); h, same, vertical section; i, pyrene from *Rubus* drupelet; j, accessory fruit (*Fragaria*); k, same, vertical section; l, achene from *Fragaria* fruit. (ca calyx, end endocarp [bony wall of pyrene], ex exocarp, mes mesocrap, p pyrene [the so-called "pit" or "seed"], r receptacle, s seed.)

*Manubrium.* The long, thin, more or less cylindrical base of certain cymbas or palm spathes.

*Marcescent.* Withering, but the remains persisting.

*Marginal Placentation.* See Placentation.

*Masculine.* Staminate (in higher plants).

*Megasporangium.* A sporangium containing only megaspores (Fig. 41 Ad).

*Megaspore.* The larger of 2 spore sizes produced by heterosporous ferns; the spore that on germination gives rise to the female gametophyte; 1 of 4 (megaspores) produced within the ovule by 2 divisions of the megasporocyte.

*Megasporophyll.* A sporophyll that bears megaspores, often produced in the axil of a bract; a carpel.

*\*Membranous, Membranaceous.* Of parchmentlike texture.

*Mericarp.* See Schizocarp (Fig. 320 k).

*Meristem.* Undifferentiated tissue whose cells are capable of developing into various organs or tissues; that of a growing point is *promeristem*.

*-merous.* In composition, referring to the numbers of parts; as flowers 5-merous, in which the parts of each kind or series are 5 or in 5's.

*Micropyle.* The opening between the integuments into an ovule (foramen).

*Microsporangium.* The microspore-containing case; an anther sac.

*Microspore.* The smaller of the 2 kinds of spores in such pteridophytes as *Selaginella*; in angiosperms, each microsporocyte divides twice to produce 4 microspores (pollen grains).

*Microsporophyll.* A sporophyll bearing microsporangia; in angiosperms, the anther.

\**Midrib.* The main rib of a leaf or leaflike part, a continuation of the petiole.

\**Monadelphous.* Stamens united in one group by connation of their filaments, as in some Leguminosae and all mallows (Figs. 317 d, f, 204 Bb, C).

*Moniliform.* Constricted laterally and appearing beadlike (Fig. 310 j).

*Monocephalic.* One-headed or scapose, as in dandelion.

*Monochasium.* A cyme reduced to single flowers on each axis (the laterals of the dichasium having been lost by reduction).

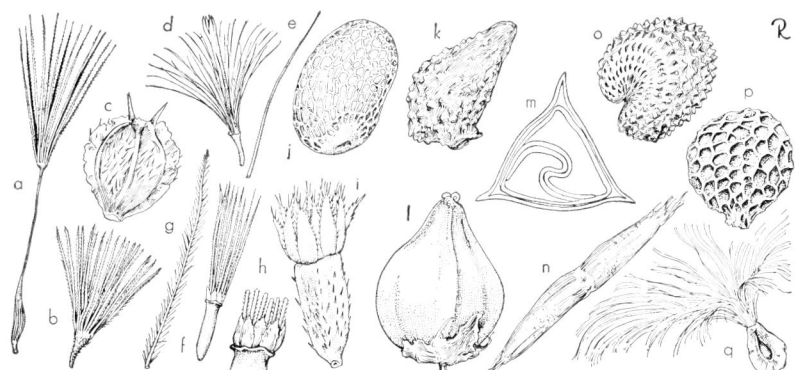

**Fig. 322.** Achene and seed types: a, achene, beaked (*Tragopogon*); b, achene, plumose pappus; c, achene, winged and awned; d, achene, capillary pappus; e, same, detail of capillary trichome; f, achene with double pappus, of scales and plumose trichomes; g, same, detail of plumose trichome; h, same, detail of outer pappus scales (trichomes excised); i, achene with pappus of scales; j, seed with reticulate coat (testa); k, seed with muricate surface; l, triquetrous smooth seed (*Fagopyrum*), m, same, cross-section, embryo curved; n, seed, winged; o, seed with rows of tubercles; p, seed with alveolate or honeycombed surface, pitted; q, winged seed with coma.

*Monochlamydous.* Having a perianth of a single series, as in Phytolaccaceae.

*Monocolpate.* Said of pollen grains with a single groove.

\**Monoecious.* With staminate and pistillate flowers on the same plant, as in corn. See Dioecious (Fig. 9 d).

*Monogynous.* With one pistil.

*Monopetalous.* Gamopetalous.

*Monophyletic.* Derived from a single ancestral line, as opposed to polyphyletic.

*Monotypic.* In reference to a genus, composed of a single species.

*Motile.* Self-propelling, as spores or sperms, by means of cilia or elaters.

*Mucro.* A short and sharp abrupt spur or spiny tip.

\**Mucronate.* Terminated abruptly by a distinct and obvious mucro (Fig. 307 g).

*Mucronulate.* Diminutive of mucronate (Fig. 307 h).

*Multicarpellate.* Referring to a compound ovary, formed by the union of 2 or more carpels.

*Multiciliate.* With many cilia.

\**Multiple Fruit.* One formed from several flowers into a single structure having a common axis, as in mulberry.

*Muricate.* Rough, due to presence of many minute spiculate excrescences on the epidermis (Figs. 310 b, 322 k).

*Muriform.* With bricklike markings, pits, or reticulations, as on some seed coats and achenes.

*Naked Flower.* One with no floral envelopes (perianth).

*Navicular.* Boat-shaped, as in glumes of most grasses.

*Nectary.* A nectar-secreting gland, often appearing as a protuberance, scale, or pit.

*Neutral Flower.* A sterile flower composed of a perianth without any essential organs; formerly sometimes applied to staminate flowers.

*Nocturnal.* Said of flowers that open at night and close during the day.

*\*Node.* A joint where a leaf is borne or may be borne; also, incorrectly, the space between 2 joints, which is properly an internode.

*Nodose.* Knobby, knotty.

*Novirame.* A flowering or fruiting shoot arising from a primocane, sometimes encountered in blackberries.

*\*Nut.* An indehiscent 1-celled and 1-seeded hard and bony fruit, even if resulting from a compound ovary (Fig. 320 h).

*Nutlet.* A small or diminutive nut; nucule.

*Ob-.* A Latin prefix, usually signifying inversion, as *obconical,* inversely conical, cone attached at the small point; *oblanceolate,* inversely lanceolate, with the broadest part of a lanceolate body above the middle; *obovate,* inverted ovate; *obovoid,* ovoid, but attached at the smaller end.

*Obcompressed.* Flattened at right angles to the primary plane or axis, as achenes of some composites that are flattened at right angles to the radius of the receptacle.

*\*Obcordate.* Deeply lobed at the apex; the opposite of cordate (Fig. 307 m).

*Obdiplostemonous.* With the stamens in 2 alternating whorls, those of the outer whorl opposite the petals.

*Oblanceolate.* The reverse of lanceolate, as a leaf broader at the distal third than at the middle and tapering toward the base (Fig. 305 k).

*\*Oblique.* Slanting; unequal-sided (Fig. 307 p).

*\*Oblong.* Longer than broad, and with the sides nearly or quite parallel most of their length (Fig. 305 j).

*\*Obovate.* The reverse of ovate, the terminal half broader than the basal (Fig 305 o).

*\*Obovoid.* Said of a terete solid that is obovate in outline.

*Obsolescent.* Nearly obsolete; applied to a nonfunctional part or organ scarcely small enough to be vestigial.

*Obsolete.* Not evident or apparent; rudimentary, vestigial.

*\*Obtuse.* Blunt, rounded (Fig. 307 j).

*Ocrea.* A nodal sheath formed by fusion of 2 stipules, as in many Polygonaceae.

*Odd-pinnate.* See Compound Leaf (Fig. 304 d).

*Oligo-.* In Greek compounds, signifying few, as: *oligandrous,* with few stamens; *oligospermous,* with few seeds.

*Ontogeny.* The developmental cycle of an individual; an ontogenetic situation being one that developed during the current generation; the opposite of phylogeny.

*Operculate.* With a cap or lid.

*Operculum.* A lid or cover produced by circumscissile dehiscence.

*\*Opposite.* Two at a node, on opposing sides of an axis (Fig. 303 c, d).

*\*Orbiculate.* Circular or disc-shaped, as leaf of *Nelumbo* (Fig. 305 s).

*Ortho-.* In Greek compounds, signifying straight, as an *orthotropous* ovule or seed,

an erect straight seed with the micropyle at the apex and hilum at the base (Fig. 319 Aa).
*Ovary. Ovule-bearing part of a pistil. When borne above the point of attachment of perianth and stamens, or surrounded by an hypanthium that is not adnate to it, it is a *superior ovary*; when below attachment of these floral envelopes and adnate to them, it is an *inferior ovary*; when intermediate, it is a *half-inferior or subinferior ovary*.
*Ovate. With an outline like that of hen's egg, the broader end below the middle (Fig. 305 h).
Ovoid. A solid that is oval (less correctly ovate) in flat outline.
Ovulate. Bearing ovules; said of gymnospermous megasporophylls, where the ovules are naked and not enclosed in a pistil.
*Ovule. The body which, after fertilization, becomes the seed; the egg-containing unit of the ovary.
Palate. In personate corollas, a rounded projection or prominence of the lower lip, closing the throat or very nearly so.
Palea, Palet. In the grass flower, the upper of the 2 enclosing bracts, the lower one being the lemma (Fig. 75 Ib).
Paleobotany. The study of plants of previous geological periods, especially as evidenced by fossils, casts, impressions, or other preserved parts.
Palman. Undivided part of a palmate leaf between the petiole and segments of the blade, of particular pertinence to leaves of fan palms.
*Palmate. Lobed or divided or ribbed in a palmlike or handlike fashion; digitate, although this word is usually restricted to leaves compound rather than to merely ribbed or lobed (Fig. 306 q).
*Palmatifid. Cut about halfway down in a palmate form.
Pandurate. Fiddle-shaped; obovate with a marked concavity along each basal side (Fig. 305 n).
*Panicle. An indeterminate branching raceme; an inflorescence in which the branches of the primary axis are racemose and the flowers pedicellate (Fig. 312 d).
Paniculate. Resembling a panicle.
Papilionaceous Corolla. Butterflylike, pealike flower, with a standard, wings, and keel. (Fig. 316 e, f).
Papillate (Papillose). Bearing minute pimplelike protuberances (papillae) (Fig. 310 a).
Pappus. Peculiar modified outer perianth series of composites, borne on the ovary (persisting in fruit), being plumose, bristlelike, scales, or otherwise; once commonly accepted as a modified calyx, but now believed in some genera of Compositae to represent a modification of the corolla.
*Parietal. Borne on the walls within a simple or compound ovary, or on intrusions of the wall that form incomplete partitions or false septa within the ovary; a descriptive term applied to the location of ovules or a placental zone, and lacking morphological significance. In a simple ovary the placenta is usually solitary and is parietally situated; by some authors this form of parietal placentation is designated *marginal* or *ventral placentation* because each placenta usually represents the zone of fusion of 2 ventral margins of the same carpel. As for compound ovaries with parietal placentation, some authors treat those types derived from axile placentation as *falsely parietal* and as phylogenetically more advanced than parietal or axile placentation.
*Parted. Cut or cleft not quite to the base (Fig. 306 n).
Patent. Spreading.
Pectinate. Comblike or pinnatifid with very close narrow divisions or parts; also

used to describe spine conditions in cacti when small lateral spines radiate like comb teeth from areole (Fig. 306 p).

*Pedate.* Said of palmately lobed or divided leaf of which the 2 side lobes are again divided or cleft (Fig. 306 r).

*\*Pedicel.* Stalk of 1 flower in a cluster.

*\*Peduncle.* Stalk of a flower cluster, or of a solitary flower when that flower is the remaining member of an inflorescence.

*Peg.* Used to designate the stalk on which are seemingly borne the developing ovaries and fruits in certain genera such as *Eucommia, Arachis* and others, but which is formed from the sterile basal portion of the ovary and hence is not a gynophore.

*\*Pellucid.* Clear, almost transparent in transmitted light.

*\*Peltate.* Attached to its stalk inside the margin; peltate leaves are usually shield-shaped (Fig. 307 w).

*Pendulous.* Drooping, hanging downward.

*Penninerved.* With nerves arising along the length of a central midrib; pinnately nerved.

*Pentamerous.* Five-merous, the parts in 5's or multiples of 5.

*Perennate.* Lasting the whole year through; self-renewing by lateral shoots from the base.

*Perennial.* Of 3 or more season's duration.

*\*Perfoliate.* Descriptive of a sessile leaf or bract whose base completely surrounds the stem, the latter seemingly passing through the leaf (Fig. 307 x).

*Pergamentaceous.* Parchmentlike.

*\*Perianth.* The two floral envelopes considered together; a collective term for the corolla and calyx; perigone.

*Pericarp.* The wall of a ripened ovary, that is, the wall of a fruit; sometimes used loosely to designate a fruit.

*Perigone.* The perianth, more commonly used when the parts are not as scarcely differentiated into calyx and corolla and are then termed tepals.

*Perigynium.* The so-called flask or papery sheath that envelops the achene in *Carex.*

*\*Perigynous.* Borne or arising from around the ovary and not beneath it, as when calyx, corolla and stamens arise from the edge of a cup-shaped hypanthium; such cases are said to exhibit perigyny (Fig. 8 B).

*Periphery.* The outer wall or margin; the inner side or face of the ovary wall as opposed to the faces of its septa.

*Persistent.* Remaining attached; not falling off.

*Personate.* Said of a 2-lipped corolla the throat of which is closed by a palate, as in toadflax (Fig. 316 d).

*Perulate.* Scale-bearing, as most buds.

*\*Petal.* One unit of the inner floral envelope or corolla of a polypetalous flower, usually colored and more or less showy.

*Petaloid.* Petallike; in color and shape resembling a petal (Fig. 318 h).

*\*Petiole.* Leaf stalk.

*Petiolule.* Stalk of a leaflet.

*Phanerogam.* A seed plant or spermatophyte, as opposed to a cryptogam.

*Phenotype.* A type (plant or population) determined by its appearance, as opposed to a genotype.

*Phloem.* A complex tissue of the vascular anatomy containing sieve tubes, phloem parenchyma, and other elements such as fibers, stone cells, companion cells, etc.; the inner bark of a woody plant; the outer of the 2 tissues of a vascular strand (the xylem the inner one).

*Phyllary.* A bract of the involucre, as in the Compositae.

*Phylloclade.* A branch, more or less flattened, functioning as a leaf, as in Christmas cactus.

*Phyllodium.* Leaflike petiole with no blade, as in some acacias and other plants.

*Phyllotaxy.* The arrangement of leaves or floral parts on their axis; generally expressed numerically by a fraction, the numerator representing the number of revolutions of a spiral made in passing from one leaf past each successive leaf to reach the leaf directly above the initial leaf, and the denominator representing the number of leaves passed in the spiral thus made.

*Phylogeny.* The evolutionary development of a population or of parts or organs of members of a given population.

*Phylum.* One of the primary divisions of the plant or animal kingdom, whose members are assumed to have a common ancestry.

\**Pilose.* Shaggy with soft hairs (Fig. 308 f).

\**Pinna.* A primary division or leaflet of a pinnate leaf.

\**Pinnate.* Feather-formed; with the leaflets of a compound leaf placed on either side of the rachis (Fig. 304).

\**Pinnatifid.* Cleft or parted in a pinnate (rather than palmate) way (Fig. 306 o).

*Pinnatisect.* Cut down to the midrib in a pinnate way.

\**Pinnule.* A secondary pinna or leaflet in a pinnately decompound leaf.

\**Pistil.* A unit of the gynoecium, comprised of ovary, style (when present), and stigma. It may consist of 1 or more carpels; when of 1 carpel, it is a *simple pistil*; when of 2 or more carpels, it is a *compound pistil*. See Carpel and Ovary.

\**Pistillate.* Having pistils and no functional stamens; female (Fig. 9 b).

*Pistillode.* A rudimentary or vestigial pistil present in some staminate flowers.

*Pith.* The soft spongy central cylinder of most angiosperm stems, composed mostly of parenchyma tissue.

*Pitted.* Having little depressions or cavities.

\**Placenta.* A place or part in the ovary where ovules are attached; a location or zone; in a simple ovary the placenta is usually a single zone formed by the connation of 2 ventral margins of the 1 carpel (rarely these margins intrude and recurve, the placentae then 2 in number); in a compound ovary with parietal placentation each placenta is usually formed by connation of adjoining ventral margins of carpels and the number of placentae equals the carpel number, or these margins may intrude and recurve with the number of placentae then twice that of the carpels; when the placentation is free-central the placentae lose their individuality and cannot be counted. Unlike the situation in mammals, where the placenta is an organ, the placenta in angiosperm ovaries is not a tissue histologically distinct from adjoining tissues; in some instances it may be borne on an intrusion or proliferation of carpellary tissue. The term, morphologically and taxonomically, is descriptive and does not connote phylogenetic significance.

\**Placentation.* The arrangement of ovules within the ovary. Several types are recognized: *parietal placentation,* which see; *axile placentation,* wherein the ovules are borne at or near the center of a compound ovary on the axis formed by the union and fusion of the septa (partitions) and usually in vertical rows (in 2-celled ovaries they are borne at or near the center and on the cross partition or on a proliferation of it, often filling the locules, as in most Solanaceae, Scrophulariaceae); *free-central placentation,* wherein the ovules are borne on a central column with no septa present, as in Primulaceae; *basal placentation,* wherein the ovules are few or reduced to 1 and borne at the base of the ovary, the ovule when solitary often filling the cavity, as in Labiatae, Plumbaginaceae; *lamellate placentation,* wherein the ovules are borne on platelike lamellae within the ovary, as in *Nuphar,* a modification of parietal placentation.

*Plexus.* A network, often of anastomosing veins.

*Plicate.* Folded, as in a fan, or approaching this condition.

\**Plumose.* Plumy; featherlike; with fine hairs, as the pappus of some composites (Figs. 322 b, e, 318 a).

*Pod.* A dehiscent dry fruit; a rather general uncritical term, sometimes used when no other more specific term is applicable, as for the fruit of *Nelumbo.*

\**Pollen.* Spores or grains borne by the anther, containing the male element (gametophyte).

*Pollen Sac.* The microsporangium, containing the pollen; in most angiosperms each anther is comprised of 4 pollen sacs, 2 in each lobe or half of the anther, the tissues separating them disintegrating prior to anthesis and the resulting anther seemingly 2-celled or biloculate.

*Pollination.* The transfer of pollen from the dehiscing anther to the receptive stigma; the act of pollinating (germination of the pollen grain is a stage intermediate between pollination and fertilization).

*Pollinium.* A coherent mass of pollen, as in orchids and milkweeds (Figs. 98 Dd, 265 Af).

*Polyandry.* The production of an indefinite number of stamens in an androecium.

*Polygamodioecious.* Said of a species that is functionally dioecious but has a few flowers of the opposite sex or a few bisexual flowers on all plants at flowering time.

*Polygamous.* Bearing unisexual and bisexual flowers on the same plant.

*Polygonal.* Many-angled.

\**Polypetalous.* With a corolla of separate petals, as opposed to gamopetalous (Figs. 5 c, 9 c).

*Polyphylesis.* A situation representing a polyphyletic origin; that is, the result of evolution along two or more lines, having different origins.

*Polyploid.* A plant with a chromosome complement of more than 2 sets of the monoploid number.

*Polymorphic.* Represented by 2 or more forms, as a species of many closely related infraspecific taxa; very variable as to habit or some morphological feature.

*Polysepalous.* Having a calyx of separate sepals, as opposed to synsepalous or gamosepalous.

*Pome.* Fruit of apple, pear, quince, and related rosaceous genera (Fig. 321 f).

*Pore.* A small more or less round aperture.

*Poricidal. See* Dehiscence (Fig. 4 Cd).

*Porrect.* Said of cactus spines when the central spines of an areole stand perpendicular to the surface; arrect.

*Posterior.* At or toward the back; opposite the front; toward the axis; away from the subtending bract.

*Praemorse.* Appearing as if the end had been chewed or bitten off, very coarsely erose.

\**Prickle.* A small and weak spinelike body borne irregularly on the bark or epidermis (Fig. 310 e).

*Primocane.* The first season's shoot or cane of a biennial woody stem, as in many brambles.

*Primordium.* A group of undifferentiated meristematic cells, usually of a growing point, capable of differentiating into various kinds of organs or tissues.

*Procumbent.* Trailing or lying flat, but not rooting (Fig. 302 c).

*Proliferous.* Bearing offshoots or redundant parts; bearing other similar structures on itself.
*Prophyll.* A bracteole.
*Prostrate.* A general term for lying flat on the ground.
*Protandrous.* Said of a flower whose anthers mature and release their pollen before the stigma of the same flower is receptive.
*Protogynous.* A flower whose stigma is receptive to pollen before pollen is shed from anthers of the same flower.
*Prothallus.* The gametophyte stage or generation of pteridophytes, a cellular and usually flattened thalluslike structure on the ground, bearing the sexual organs, as the antheridia and archegonia.
*Pruinose.* Having a bloom on the surface.
*Pseud-, Pseudo-.* In combination means false, not genuine, not the true or the typical.
*Pseudobulb.* The thickened or bulbiform stems of certain orchids, being solid and borne above ground.
*Pseudoterminal Bud.* Seemingly the terminal bud of a twig, but actually the uppermost lateral bud with its subtending leaf scar on one side and the scar of the terminal bud often visible on opposite side. Example, *Castanea*.
\**Puberulent.* Minutely pubescent; the hairs soft, straight, erect, scarcely visible to the unaided eye (Fig. 308 a).
\**Pubescent.* Covered with short soft hairs; downy.
*Pulvinate.* Cushion-shaped.
*Pulviniform.* Having the shape of a pulvinus; pad- or cushion-shaped.
*Pulvinus.* A minute gland or swollen petiole (petiolule) base responsive to vibrations and heat, as in leaves of sensitive plant (*Mimosa*)
\**Punctate.* With translucent or colored dots or depressions or pits.
*Pungent.* Ending in a stiff sharp point or tip; also, acrid (to the taste).
*Pustular.* Blistery, usually minutely so.
*Pyrene, Pyrena.* The nutlet in a drupe; a seed and the bony endocarp, as a cherry or peach pit or the seed of a raspberry drupelet (Fig. 321 i).
*Pyriform.* Pear-shaped.
*Pyxis.* A capsule dehiscing circumscissilely, the top coming off as a lid (Fig. 320 d).
\**Raceme.* A simple, elongated, indeterminate inflorescence with pedicelled or stalked flowers (Fig. 321 c).
*Racemose.* Having flowers in racemelike inflorescences that may or may not be true racemes.
*Rachilla, Rhachilla.* A diminutive or secondary axis, or rachis; in particular, in the grasses and sedges the axis that bears the florets.
\**Rachis.* Plural, rachides or rachises. Axis bearing flowers or leaflets; petiole of a fern frond.
*Radiate.* Standing on and spreading from a common center; also, having ray flowers, as in the Compositae (Fig. 318 g).
*Radical.* Arising from the root or its crown; said of leaves that are basal or rosulate.
*Radicle.* The embryonic root of a germinating seed.
*Ramiform.* Branching.
*Rank.* A vertical row; leaves that are 2-ranked are in 2 vertical rows, and may be alternate or opposite.
*Raphe.* That portion of the funiculus of an ovule that is adnate to the integument, usually represented by a ridge, present in most anatropous ovules, diagnostic in *Sarracenia*.

*Raphides.* Needlelike crystals of calcium oxalate, as in the tuber of *Arisaema*, and in vegetative parts of many plants.

*Ray.* Outer modified floret of some composites, with an extended or straplike part to the corolla; also a branch of an umbel or umbellike inflorescence (Fig. 316 a).

*\*Receptacle.* Torus; the more or less enlarged or elongated end of the stem or flower axis on which some or all of the flower parts are borne; sometimes the receptacle is greatly expanded, as in the Compositae.

*Reclinate, Reclining.* Bent down or falling back from the perpendicular.

*Recurved.* Bent or curved downward or backward.

*Reflexed.* Abruptly recurved or bent downward or backward.

*\*Regular Flower.* A flower with the parts in each series or set so arranged as to be vertically divisible into equal halves by 2 or more planes.

*\*Reniform.* Kidney-shaped (Fig. 305 r).

*Repand.* Weakly sinuate.

*Replum.* Partition between the two loculi of cruciferous fruits.

*Resinous.* Containing or producing resin, said of bud scales when coated with a sticky exudate of resin (as in *Aesculus* spp.).

*Resupinate.* Twisted 180°, as the ovary (and flower) of most orchids; upside down.

*\*Reticulate.* Netted (Fig. 322 i).

*Retrorse.* Bent or turned over backward or downward (Fig. 310 l).

*\*Retuse.* Notched slightly at a usually obtuse apex (Fig. 307 k).

*\*Revolute.* Rolled backward; with margin rolled toward lower side. *See* Involute (Fig. 306 s).

*Rhizoid.* A rootlike structure in function and general appearance but not so in anatomy (Fig. 80 a).

*Rhizomatous.* Producing or possessing rhizomes.

*\*Rhizome.* Underground stem; rootstock; distinguished from a root by presence of nodes, buds, or scalelike leaves.

*Rhizophore.* A leafless stem that produces roots, as in *Selaginella*.

*Rhombate, Rhomboidal.* Shaped like a rhomboid (Fig. 305 q).

*\*Rib.* In a leaf or similar organ, the primary vein; also, any prominent vein or nerve.

*Rootstock.* Subterranean stem; rhizome.

*\*Rosette.* An arrangement of leaves radiating from a crown or center and usually at or close to the earth, as in *Taraxacum* (dandelion).

*Rosulate.* In rosettes (Fig. 303 i).

*Rostellum.* A small beak.

*Rostrate.* Having a beak or beaklike projection.

*\*Rotate.* Wheel-shaped; said of a gamopetalous corolla with a flat and circular limb at right angles to the short or obsolete tube (Fig. 315 e).

*Rotund.* Nearly circular; orbicular inclining to be oblong.

*Rudimentary.* Imperfectly developed and nonfunctional. *See* Vestigial.

*\*Rugose.* Wrinkled, usually covered with wrinkles, the venation seeming impressed into the surface (Fig. 310 c).

*Ruminate.* Mottled in appearance; applied to a surface or tissue showing dark and light zones of irregular outline.

*Runcinate.* Coarsely serrate to sharply incised with the teeth pointing toward the base, as in *Taraxacum* (Fig. 305 m).

*Runner.* A slender trailing shoot taking root at the nodes

*Rushlike.* Resembling rushes (*Juncus, Typha, Scirpus,* and others).

*Saccate.* Bag-shaped, pouchy.

*Sagittate.* Like an arrowhead in form; triangular, with the basal lobes pointing downward or concavely toward the stalk (Fig. 307 u).
*Salverform.* Said of a gamopetalous corolla with the slender tube and an abruptly expanded flat limb, as that of the phlox; hypocrateriform.
*Samara.* Indehiscent winged fruit, as of the maple (*Acer*) and ash (*Fraxinus*) (Fig. 320 m, n).
Saprophyte. A plant (usually lacking chlorophyll) living on dead organic matter, as *Monotropa*.
Sarmentose. Producing long flexuous runners or stolons.
*Scabrous.* Rough; feeling roughish or gritty to the touch. (Fig. 309 a).
Scale. A name given to many kinds of small, mostly dry, and appressed leaves or bracts, often only vestigial.
Scandent. Climbing without aid of tendrils.
*Scape.* Leafless peduncle arising from the ground; it may bear scales or bracts but no foliage leaves and may be one- or many-flowered.
Scapose. Bearing flowers or inflorescence on a scape (Fig. 312 a).
*Scarious.* Applied to leaflike parts or bracts that are not green, but thin, dry, and membranaceous, often more or less translucent.
*Schizocarp.* A dry dehiscent fruit that splits into 2 halves, each half a mericarp, as in most umbellifers, or in *Acer* (Figs. 320 k, 246 Da).
Scorpioid. Said of a circinnately coiled determinate inflorescence in which the flowers are 2-ranked and borne alternately at the right and the left.
Scorpioid Cyme. A determinate inflorescence with the seemingly lateral flowers borne alternately on opposite sides of a pseudoaxis and sometimes appearing racemose. *See,* Helicoid Cyme.
Scutate. Like a small shield.
Secund. One-sided; said of inflorescences when the flowers appear as if borne from only 1 side (Fig. 98 Ca).
*Seed.* The ripened ovule; the essential part is the embryo, and this is contained within integuments.
*Segment.* One of the parts of a leaf, petal, calyx, or perianth that is divided but not truly compound.
*Sepal.* One of the separate parts of a calyx, usually green and foliaceous.
Sepaloid. Said of an involucre or corolla that simulates a calyx in appearance.
Septate. Partitioned; divided by partitions.
*Septicidal.* Dehiscence along or into the partitions (septa), not opening directly into the locule (Fig. 320 e).
*Septum (pl. septa).* A partition or cross wall.
Seriate. In series, usually in whorls or apparent whorls.
*Sericeous.* Silky (Fig. 309 f).
*Serrate.* Said of a margin when saw-toothed with the teeth pointing forward (Fig. 306 d).
*Serrulate.* Minutely serrate (Fig. 306 e).
*Sessile.* Not stalked; sitting (Fig. 304 a).
Seta. A bristle.
Setaceous. Bearing bristles.
Setiform. Bristle-shaped.
*Setose.* Covered with bristles.
Sheath. Any long or more or less tubular structure surrounding an organ or part.
Shrub. A woody plant that remains low and produces shoots or trunks from the base, not treelike nor with a single bole; a descriptive term not subject to strict circumscription.

*Sigmoid.* Said of a leaflet or segment that is curved sidewise in opposing directions; S-shaped.
*Siliceous.* Containing minute particles of silica.
*Silicle.* The short fruit of certain Cruciferae, usually not more than twice as long as wide.
*Silique.* The long fruit of certain Cruciferae.
*Silky.* Having a covering of soft appressed fine hairs; sericeous.
*Silvery.* With a whitish, metallic, more or less shining luster.
*Simple.* Said of a leaf when not compounded into leaflets, of an inflorescence when not branched.
*Sinus.* The space or recess between 2 lobes or divisions of a leaf or other expanded organ.
*Smooth.* Said of surfaces that have no hairiness, roughness, or pubescence, particularly of those not rough or scabrous. See Glabrous.
*Soboliferous.* Bearing or producing shoots from the ground, clump-forming; usually applied to shrubs or small trees, as spp. of *Syringa, Rhus,* and some palms (Fig. 302 f).
*Solitary.* Borne singly or alone.
*Sorus.* (*pl. sori*) A heap or cluster. The fruit dots or clusters of fruiting bodies of ferns usually located on the dorsal side of the frond.
*\*Spadix.* A thick, or fleshy spike of certain plants, as in members of the Araceae, surrounded or subtended by a spathe (Fig. 79 Ab, Bb).
*Spathaceous.* Spathelike.
*\*Spathe.* The bract or leaf surrounding or subtending a flower cluster or a spadix; it is sometimes colored and flowerlike, as in the calla. See Cymba (Fig. 79 Ba, Ea).
*Spathe Valves.* One or more herbaceous or scarious bracts that subtend an inflorescence or flower and generally envelop the subtended unit when in bud.
*Spatulate.* Spoon-shaped (Fig. 305 1).
*Sperm.* A male gamete or reproductive cell.
*Sphenopsid.* Any of a group of *Equisetum-* or horsetaillike plants; usually used for prehistoric ancestors or relatives of modern *Equisetum* spp.
*Spicate.* Spikelike.
*\*Spike.* A usually unbranched, elongated, simple, indeterminate inflorescence whose flowers are sessile, the flowers either congested or remote; a seemingly simple inflorescence whose flowers may actually be composite heads (*Liatris*), or other inflorescence types (*Phleum*) (Fig. 312 b).
*Spikelet.* A secondary spike; 1 part of a compound inflorescence which of itself is spicate; the floral unit, or ultimate cluster, of a grass inflorescence composed of flowers and their subtending bracts.
*\*Spine.* A strong and sharp-pointed woody body mostly arising from the wood of the stem (Fig. 310 i).
*Spinescent.* Terminated by a sharp spine or tip.
*Spinulose.* With small spines over the surface.
*Sporangiophore.* The stalk of a sporangium; a sporangium-bearing organ (peltate in *Equisetum*) (Fig. 39 Ad).
*\*Sporangium.* A spore case; a sac or body bearing spores; in most eusporangiate ferns it is composed of a stalk (sporangiophore), an annulus, and a capsule.
*\*Spore.* A simple reproductive body, usually composed of a single detached cell, and containing a nucleated mass of protoplasm (but no embryo) and capable of developing into a new individual; used particularly in reference to the pteridophytes and lower plants.

*Sporocarp.* A receptacle containing sporangia (as in *Salvinia, Marsilea*) (Fig. 53 c).
*Sporogenous.* Tissue capable of producing reproductive parts or organs.
*Sporophyll.* A spore-bearing leaf; a leaflike or foliaceous organ bearing reproductive parts or organs.
*Sporophyte.* In ferns and seed plants, the foliaceous vegetative plant, as opposed to the gametophyte.
*Spreading.* Standing outward or horizontally.
*Spur.* A tubular or saclike projection from a blossom, as of a petal or sepal; it usually contains a nectar-secreting gland.
*Squamate.* With small scalelike leaves or bracts; scaly.
*Squamose.* Covered with small scales, more coarsely so than when lepidote.
*Stalk.* The "stem" of any organ, as the petiole, peduncle, pedicel, filament, stipe.
*Stalked Bud.* One whose outer scales are attached above the base of the bud axis.
\**Stamen.* The unit of the androecium and typically composed of anther and filament, sometimes reduced to only an anther; the pollen-bearing organ of a seed plant (Fig. 317).
\**Staminate.* Having stamens and no pistils; male (Fig. 9 a).
\**Staminode (Staminodium).* A sterile stamen, or a structure resembling such and borne in the staminal part of the flower; in some flowers (as in *Canna* or Aizoaceae) the staminodes are petallike and showy.
*Standard.* The upper and broad, more or less erect petal of a papilionaceous flower; the narrow, usually erect or ascending unit of the inner series of perianth of an *Iris* flower as opposed to the broader, often drooping falls (Fig. 316 f).
\**Stellate.* Starlike; stellate hairs have radiating branches, or, when falsely stellate, are separate hairs aggregated into starlike clusters; hairs once or twice forked are often treated as stellate.
*Stem.* The main axis of a plant, leaf-bearing and flower-bearing as distinguished from the root-bearing axis.
*Sterile.* Lacking functional sex organs.
\**Stigma.* The part of the pistil that receives the pollen (Fig. 318).
*Stigmatic.* Pertaining to the stigma.
*Stipe.* The stalk of a pistil or other small organ when axile in origin; also, the petiole of a fern leaf.
*Stipel.* Stipule of a leaflet.
*Stipitate.* Borne on a stipe or short stalk.
\**Stipule.* A basal appendage of a petiole; the 3 parts of a complete leaf are blade, petiole, stipules (usually 2) (Fig. 102 b).
*Stolon.* A shoot that bends to the ground and takes root; more commonly, a horizontal stem at or below surface of the ground that gives rise to a new plant at its tip.
*Stomate.* The pore in a leaf (frequently the lower side) formed by the concavity of 2 sausagelike guard cells.
*Stone.* A large pyrene, which see (Fig. 321 c); *stone fruit,* a drupe or drupelet, especially as in Rosaceae.
*Striate.* With fine longitudinal lines, channels or ridges.
*Strict.* Straight and upright, little if at all branched, often rigid.
*Strigose.* With sharp, appressed straight hairs, stiff and often basally swollen (Fig. 309 b).
*Strobile.* Cone.
\**Strobilus.* A conelike structure containing the reproductive organs of one or both sexes, as in *Equisetum, Lycopodium, Cycas,* and the conifers.
*Stylar.* Within the style or stigma; pertaining to the style.

*Style.* More or less elongated part of the pistil between the ovary and the stigma (Fig. 318).

*Stylopodium.* A dislike enlargement at the base of the style, as in some umbellifers.

*Sub-.* As a prefix, usually signifying somewhat, slightly, or rather.

*Subherbaceous.* Herbaceous, but becoming woody as the season progresses.

*Subinferior.* Sometimes applied to a half-inferior ovary; somewhat inferior.

*Subpetiolar.* Under the petiole and usually enveloped by it (Fig. 164 b).

*Subshrub.* A suffrutescent perennial (the stems basally woody), or a very low shrub often loosely treated as a perennial.

*Subtend.* To stand below and close to, as a bract underneath a flower, particularly when the bract is prominent or persistent. The flower is in the axil of the bract.

*Subvalvate.* Incompletely or somewhat valvate.

*Subulate.* Awl-shaped, tapering from base to apex (Fig. 305 a).

*Succulent.* Juicy; fleshy; soft and thickened in texture.

*Suffrutescent (Suffruticose).* Pertaining to a low and somewhat woody plant; diminutively shrubby or fruticose; woody at base with herbaceous shoots produced perennially.

*Sulcate.* Grooved or furrowed lengthwise.

*Superior.* Said of an ovary that is free from the calyx or perianth. See Ovary.

*Supine.* Lying flat and with face upward.

*Suppressed.* Vestigial to the degree of not being evident superficially or macroscopically, but whose presence in ancestral forms may be indicated by other features, as anatomy.

*Suprafoliar.* Above the leaves.

*Suture.* A line or mark of splitting open; a groove marking a natural division or union; the lengthwise groove of a plum or similar fruit.

*Syconium.* The "fruit" of a fig (*Ficus*).

*Symmetrical.* Said of an actinomorphic flower that has the same number of parts in each series or circle, as 5 stamens, 5 petals.

*Sympetalous.* The petals united, at least at the base; see Corolla.

*Symphysis.* The connation of like parts from time of meristematic development, as petals in a gamopetalous corolla (not applied to ontogenetic fusion).

*Sympodial inflorescence.* A determinate inflorescence that simulates an indeterminate inflorescence, as a scorpioid cyme (compare latter with a helicoid cyme).

*Synandrium.* An androecium coherent by the anthers as in some aroids; when anthers are connate they are termed syngenesious.

*Syncarpous.* Having carpels united; applied to an ovary of 2 or more carpels; sometimes used when separate pistils within one flower are partially united. See Apocarpous (Fig. 5 c).

*Syngenesious.* Said of stamens connate by their anthers to form a cylinder about the style, as in the Compositae (Fig. 317 c, e).

*Synsepalous.* Gamosepalous; the sepals marginally connate, at least basally.

*Tailed.* Said of anthers having caudal appendages.

*Tapering.* Gradually becoming smaller or diminishing in diameter or width toward one end; not abrupt.

*Taxon.* (pl. *taxa*). A general term applied to any taxonomic element, population, or group irrespective of its classification level. See p. 53.

*Tendril.* A rotating or twisting threadlike process or extension by which a plant grasps an object and clings to it for support; morphologically it may be stem or leaf.

*Tepal.* A segment or unit of those perianths not clearly differentiated into typical corolla and calyx, as in tulip, onion, or pokeweed.

*Terete.* Circular in transverse section; imperfectly cylindrical because the object may taper one or both ways.

*Terminal.* At the tip, apical, or distal end.

*Ternate.* In 3's (Fig. 304 g).

*Terrestrial.* Of the ground; a land plant, as opposed to aquatics, epiphytes, or saprophytes.

*Testa.* Outer coat of a seed (developed from the integument).

*Tetrad.* A group of 4; a tetrahedron; in angiosperms, a four-celled pollen-mother cell.

*Tetradynamous.* An androecium of 6 stamens, 4 longer than the outer 2, as in most Cruciferae (Fig. 317 b).

*Tetrahedal.* Four-sided, as a 3-sided pyramid and its base.

*Tetramerous.* Four-merous.

*Tetrandrous.* With 4 stamens.

*Thalamus.* The receptacle of a flower; *Thalamiflorae,* a taxon whose floral parts are hypogynous, distinct, and separate from one another on the receptacle.

*Thallus.* A flat leaflike organ; in some cryptogams, the entire cellular plant body without differentiation as to stem and foliage.

*Theca.* The pollen sac of an anther.

*Throat.* The opening or orifice into a gamopetalous corolla, or perianth; the place where the limb joins the tube.

*Thyrse, thyrsus.* Compact and more or less compound panicle; more correctly, a paniclelike cluster with main axis indeterminate and the lateral axes determinate, as in most lilacs (Fig. 312 f).

*Tomentose.* With tomentum; densely woolly or pubescent; with matted soft woollike hairiness (Fig. 308 c).

*Tomentulose.* Somewhat or delicately tomentose.

*Torulose.* Twisted or knobby; irregularly swollen at close intervals.

*Torus.* Receptacle (Fig. 313 f).

*Trabecula.* A transverse partition, complete or incomplete, as in a sporangium of *Isoëtes.*

*Tree.* A woody plant that produces one main trunk or bole and a more or less distinct and elevated head.

*Tri-.* Three or 3 times.

*Triad.* In 3's.

*Trichome.* A hair or bristle.

*Tricolpate.* Three-grooved.

*Trifoliate.* Three-leaved, as in *Trillium.*

*Trifoliolate.* Having a leaf or leaves of 3 leaflets, as most clovers (Fig. 304 i, j).

*Trigonal.* Three-angled.

*Triquetrous.* Three-angled in cross section (Fig. 322 1).

*Triternate.* Three times 3; the leaflets or segments of a twice ternate leaf again in 3 parts.

*Truncate.* Appearing as if cut off at the end; the base or apex nearly or quite straight across (Fig. 307 r).

*Tuber.* A short congested part; usually defined as subterranean (as of a rootstock) although this is not essential.

*Tubercle.* A small tuber, or rounded protruding body.

*Tuberous.* Bearing or producing tubers.

*Tumid.* Swollen, inflated

*Tunic.* A loose membranous outer skin not the epidermis; the loose membrane about a corm or bulb.
*Tunicated.* Provided with a tunic as defined above; having concentric or enwrapping coats or layers, as bulb of onion.
*Turbinate.* Inversely conical; top-shaped.
*Turgid.* Swollen from fullness.
*Turion.* A young shoot or sucker, as an emerging stem of asparagus.
*Twig.* A young woody stem; more precisely the shoot of a woody plant representing the growth of the current season and terminated basally by circumferential terminal-bud scar.
\**Umbel.* An indeterminate, often flat-topped inflorescence whose pedicels and peduncles (rays) arise from a common point, resembling the stays of an umbrella; umbels are characteristic of the Umbelliferae and are there usually compound, each primary ray terminated by a secondary umbel (*umbellet*) (Fig. 313 d).
*Umbellate.* Umbelled; with umbels; pertaining to umbels.
*Umbelliform.* Umbellike; resembling an umbel.
*Umbo.* A conical projection arising from the surface.
*Uncinate.* Hooked obtusely at the tip (Fig. 310 m).
\**Undulate.* Wavy (up and down, not in and out), as some leaf or petal margins (Fig. 306 b).
*Unguiculate.* Narrowed into a petiolelike base; clawed.
*Unifoliolate.* Said of a compound leaf reduced to a single, and usually the terminal, leaflet.
*Unigeneric.* Said of a family composed of a single genus; monogeneric.
*Unijugate.* Applied to a compound leaf composed of 1 pair of leaflets.
*Unilateral.* One-sided.
*Unilocular.* Containing a single chamber or cell.
*Unisexual.* Of 1 sex; staminate only or pistillate only.
\**Urceolate.* Urn-shaped (Fig. 315 a).
*Utricle.* A small bladder; more commonly, a bladdery, 1-seeded, usually indehiscent fruit, as in some amaranths.
*Vaginate.* Sheathed.
*Vallecular.* Pertaining to the grooves between the ridges, as in fruits of Umbelliferae.
\**Valvate.* Opening by valves or pertaining to valves; meeting by the edges without overlapping, as leaves or petals in the bud.
\**Valve.* A separable part of a pod; the units or pieces into which a capsule splits or divides in dehiscing.
*Vascular.* Pertaining to the presence of vessels in the conducting tissues of the stele.
*Vasiform.* Of elongated funnel shape.
*Velamen.* A thin sheath or covering, usually parchmentlike.
*Velum.* A veil; a membranous indusium, as in *Isoëtes* (Fig. 43 e).
*Velutinous.* Clothed with a velvety indumentum composed of erect, straight, moderately firm hairs.
\**Venation.* Veining; arrangement or disposition of veins (Fig. 304).
\**Ventral.* Front; relating to the inner face or part of an organ; opposite the back or dorsal part.
*Ventricose.* With a 1-sided swelling or inflation, more pronounced than when gibbous.
\**Vernation.* The disposition or arrangement of leaves in the bud. *See* Aestivation.
*Verrucose.* Having a wartlike or nodular surface.
\**Versatile.* Hung or attached near the middle and usually moving freely, as an

anther attached crosswise on apex of filament and capable of turning (Fig. 4 Bb, Db).

*Verticil.* A whorl.

*Verticillate.* Arranged in whorls, or seemingly so.

**Verticillate Inflorescence.* One with the flowers in whorls about the axis, the whorls remote from one another (as in many salvias) or congested into head-like structures (catnip). Such whorls are false whorls, since they are actually sessile cymes arranged opposite one another in the axils of opposite bracts or leaves (Figs. 313 g, h, 3 p).

*Vesicle.* A small bladdery sac or cavity filled with air or fluid.

*Vessels.* Water-conducting cells of the xylem, usually standing end to end.

*Vestigial.* Imperfectly developed; said of a part or organ that was fully developed and functional in ancestral forms, or in an earlier or ancient generation, but is now a degenerate relic, usually smaller and less complex than its prototype.

*Vesture.* Anything on or arising from a surface causing it to be other than glabrous. Also vestiture.

*Vexillum.* The broad upper petal of a papilionaceous corolla; the standard or banner (Fig. 316 f).

**Villous.* Provided with long and soft, not matted, hairs; shaggy.

*Virgate.* Wandlike; long, straight, and slender.

**Viscid.* Sticky, or with appreciable viscosity.

*Voluble.* Twining.

**Whorl.* Three or more leaves or flowers at one node, in a circle (Fig. 303 e).

*Wing.* A thin, dry or membranaceous expansion or flat extension or appendage of an organ; also the lateral petals of a papilionaceous flower.

*Woolly.* Provided with long, soft, and more or less matted hairs; like wool; lanate.

*Xerophyte.* A plant of a dry arid habitat, such as the desert.

*Xylem.* The wood elements of a vascular cylinder; a water-conducting complex tissue of tracheids and parenchyma; the inner of 2 elements of a vascular strand. *See* Phloem.

**Zygomorphic.* Said of corollas when divisible into equal halves in one plane only, usually along an anterior-posterior line (Fig. 9 f). *See* Actinomorphic, Irregular.

# INDEX

NOTE: Generic names in parentheses are of taxa not accepted in this work as being taxonomically valid. Names of authors cited in the text are included in the index unless they appeared only in a bibliographic reference; an exception to this is that page references are not given to the repeated occurrence in Part Two of the names of Bentham and Hooker, Bessey, Engler, Hallier, Hutchinson, Rendle, and Wettstein.

Abaca cloth, 427
Abbe, E. C., 458
Abbe, E. C. and Earle, T. T., 451, 452
Abbreviations, periodical titles, 308–317; world herbaria, 273
*Abelia*, 714
*Abies*, 364
Abietineae, 364
*Abolboda*, 404
Abrams, L., 293, 377, 380, 386, 450, 484, 531
*Abronia*, 481
*Abutilon*, 593
*Acacia*, 546, 547, 548
*Acaena*, 541, 544
*Acalypha*, 566
Acanthaceae, described, 707–709
Acanthineae, 676
*Acantholimon*, 661, 662
*Acanthopanax*, 643
*Acanthus*, 708, 709
*Acer*, 581
Aceraceae, described, 581
(*Acerates*), 675
Achariaceae, 597
Achatocarpaceae, 477
*Achillea*, 730
*Achimenes*, 705
*Achlys*, 501
*Achras*, 663
(*Achrosanthes*), 436
*Achyronychia*, 488
*Acicarpha*, 726
*Ackama*, 536
*Acmena*, 634
*Acnida*, 479, 480
*Aconitum*, 497, 499
*Acorus*, 398
Acrotonae, taxonomy of, 436
*Actaea*, 496, 498
*Acthephyllum*, 482
*Actinidia*, 599
Actinidiaceae, described, 599–600

Actinomorphy, characteristics of, 84
(*Actinostachys*), 346
*Actinostrobus*, 366
Adamson, R. S. and Slater, T. M., 301
Adanson, M., 27, 42
*Adansonia*, 595
*Adenia*, 616
*Adenophora*, 721, 722
*Adenostoma*, 544
*Adiantum*, 349, 350, 351
*Adicea*, 465
*Adinandra*, 602
*Adlumia*, 518
*Adolphia*, 588
*Adoxa*, 714
Adoxaceae, 711; discussed, 714
Adventive plant, defined, 279
*Aechmea*, 406, 407
*Aegiphila*, 687
*Aegopodium*, 646
*Aeonium*, 531
*Aeschynanthus*, 705
Aesculaceae, 582
*Aesculus*, 582
*Aethusa*, 646
Africa, floras of, 301, 302
African cola, 596
African mahogany, 561
African valerian, 716
African violet, 705
*Afzelia*, 698
Agapantheae, 413, 415
*Agathis*, 362
Agavaceae, of Hutchinson, 413, **415**
Agavales, of Hutchinson, 415
*Agave*, 419, 420
Agavoideae, 419
*Agdestis*, 482
Age and area theory, 160–161
*Ageratum*, 727, 730
*Aglaonema*, 400
*Agrimonia*, 544
*Agrostemma*, 487

777

Agrostideae, 390
*Agrostis*, 390
*Ailanthus*, 559
Airplane photographs, sources of, 325; uses of, 268
Aizoaceae, described, 482–484
*Ajuga*, 690
Ajugoideae, 690
Akariaceae, 549
*Akebia*, 500
Alangiaceae, described, 631–632
*Alangium*, 632
Alaska, floras of, 292, 293
*Albertisia*, 503
Albertus Magnus, 15
*Albidella*, 383
*Albizia*, 548
*Alchemilla*, 541
Alder, white, 650
*Aldrovanda*, 528, 529
Alefeld, A., 320
*Aletris*, 419
*Aleurites*, 566
*Alfaroa*, 453
*Alisma*, 382, 383
Alismaceae, described, 382–383
Alismataceae, 382
Alismatales, 376
Alismateae, 383
Alkali heath, 607
Alkanet, 685
Allamanda, 672, 673
Allan, H. H., 174
Allen, C. K., 513
*Allenrolfea*, 478
Allieae, 413, 415
Allioideae, 415
Allioniaceae, 480
*Allium*, 413, 416
*Alloplectus*, 705
Allspice, 634
Almond, 544; Indian, 633; tropical, 633
*Alnus*, 457, 458, 459
*Alocasia*, 399, 400
*Aloe*, 413, 416
Aloin, 416
*Alona*, 692
*Alophia*, 424
*Alotropa*, 651
*Aloysia*, 687
Alpha taxonomy, 170, 171
*Alpinia*, 428
Alpiniaceae, 427
Alsinaceae, 486
*Alsophila*, 348, 349
Alston, A. H. G., 340
*Alstroemeria*, 419
Alstroemeriaceae, 419
*Alternanthera*, 479
Alternifoliae, families of, 127, 128

*Althaea*, 593, 594
*Althenia*, 377
Altingiaceae, 537
*Alvaradoa*, 559
*Alyssum*, 521, 522
Amarantaceae, 479
Amaranthaceae, described, 479–480
Amaranthoideae, 479
*Amaranthus*, 479, 480
(*Amarantus*), 479
Amarylleae, 418
Amaryllidaceae, described, 417–420; of Hutchinson, 413, 415, 419; of Pax, 419
Amaryllidoideae, 419
*Amaryllis*, 419, 420
*Ambelania*, 672
*Ambrosia*, 730
Ambrosiaceae, 730
*Ambylanthus*, 658
*Amelanchier*, 545
Amentaceae, 446
Amentiferae, 446
*Amentotaxus*, 362
American code of nomenclature, 199
American ebony, 665
American periodicals, list of, 309–313
American persimmon, 665
Ames, O., 436
(*Amesia*), 436
*Ammannia*, 628
Ammanniaceae, 627
Ammiaceae, 644
*Ammobroma*, 681
*Amomum*, 428
*Amoreuxia*, 610
*Amorpha*, 546, 548
*Amorphophallus*, 400
Ampelidaceae, 588
*Ampelopsis*, 589
Amphiapomict, 185
(*Amphiglottis*), 436
Amphiploidy, 175, 176
*Amsinckia*, 686
*Amsonia*, 673
Amsterdam Congress, 200
Amygdalaceae, 543
*Amyris*, 557
*Anacampseros*, 485
Anacardiaceae, 568; described, 573–575
Anacardiineae, 569
*Anacardium*, 573, 574
*Anacharis*, 385, 386
(*Anachelium*), 436
*Anagallis*, 659, 660
(*Anamomis*), 634
*Ananas*, 407
*Anaphalis*, 730
Anatomy, importance to phylogeny, 98, 107
*Anchieta*, 612
*Anchusa*, 684, 685

Ancistrocladaceae, 597
Ancistrocladineae, 597
Anderson, C. E., 415
Anderson, E., 99, 106, 165, 174, 269–270, 272
Anderson, E. and Abbe, E. C., 99
Andrews, H. N., Jr., 101, 104
Androecium, morphology of, 67–72
*Andromeda*, 654
Andromedeae, 652
*Andropogon*, 389
Andropogoneae, 387, 390
*Androsace*, 659
*Aneilema*, 408, 409
*Anemia*, 346
*Anemone*, 498
Anemoneae, 498
*Anemonella*, 498
Anemophilous flowers, 72
*Anemopsis*, 444
*Anethum*, 646
*Aneulopha*, 554
*Angadenia*, 673
*Angelica*, 646
*Angelonia*, 698
Angiopteridaceae, 344
Angiosperm anatomy, importance of, 97, 108
Angiospermae, a polyphyletic taxon, 106, 107; described, 370–371
Angiosperms, paleobotany of, 103–108; phylogeny of, 103–108; size of, 370, 371
*Anisacanthus*, 708
Anise, 646
*Anisomeria*, 481
*Annona*, 508, 509
Annonaceae, desribed, 508–509
Annotation of specimens, 272
Annotation slips, use of, 272
*Anoda*, 593
(*Anogra*), 638
Anonaceae, 508
*Anredera*, 486
Anthemideae, 730
*Anthemis*, 728, 730
*Antennaria*, 728, 730
Anther, dehiscence types, 71; morphology of, 71; position of, 71
*Antholyza*, 424
Anthophyta, 370
Anthospermeae, 713
*Anthriscus*, 646
*Anthurium*, 398, 400
*Antigonum*, 476
Antirrhinoideae, 696
*Antirrhinum*, 695, 696, 697
*Apargidium*, 728
*Apetahia*, 721
*Aphanes*, 544
*Aphanisma*, 478

*Aphelandra*, 707, 709
(*Aphyllon*), 703
Apiaceae, 644
*Apiastrum*, 644
*Apium*, 645, 646
*Aplectrum*, 436
Apocynaceae, described, 672–673
*Apocynum*, 673
*Apodanthera*, 719
*Apodanthes*, 474
Apomicts, 183; nomenclature of, 185, 201
Apomixis, 183–185
*Aponogeton*, 379
Aponogetonaceae, described, 379
Aponogetonales, 376
Apostasiaceae, 435
Apostasineae, 435
Appalachian highlands, 156
Appendicular theory of ovary position, 81
Apple, 543; Balsam, 720
Apricot, 544
*Apteria*, 432
Aquifoliaceae, described, 575–576
*Aquilegia*, 497, 498
*Arabis*, 521, 522
Araceae, described, 398–400
*Arachis*, 545, 548
*Aralia*, 643
Araliaceae, described, 642–643
*Araucaria*, 362; ovules enclosed, 370
Araucariaceae, described, 361–362
*Araujia*, 675
Arber, A., 65, 73, 107, 371, 387, 390, 467, 516, 521, 525
Arber, E. A. N. and Parkin, J., 98, 104, 134
Arbutoideae, 652, 654
*Arceuthobium*, 472
*Arceuthos*, 366
Archer, W. A., 246, 256
*Arctium*, 728, 730
*Arctomecon*, 516
*Arctopus*, 644
*Arctostaphylos*, 654, 655
Arctotideae, 730
*Arctotis*, 730
*Ardisia*, 658
Ardisiaceae, 657
Arecaceae, 394
Arecales, 394
*Arenaria*, 487, 488
*Arethusa*, 436
*Argemone*, 516, 517
(*Argentina*), 544
Argentina, flora of, 297, 298
*Argyreia*, 677
Arillatae, 425
*Arisaema*, 398, 399, 400
*Aristolochia*, 473
Aristolochiaceae, described, 473

Aristolochiales, described, 472
*Aristotelia*, 591
*Armeria*, 661
Armeriaceae, 661
*Armoracia*, 521
*Arnica*, 728, 730
Arnold Arboretum, 26
Arnold, C. A., 101, 104, 105, 106
Arnotto, 610
Aroideae, 398
*Aronia*, 544
Arrangement of material in the herbarium, 258–260
Arrow grass family, described, 379
Arrow poison, 675
Arrowroot, 430
Arrowroot family, described, 429
*Artabotrys*, 509
*Artemisia*, 730
Artichoke, globe, 730
Articulatae, described, 335
Artificial heat, to dry specimens, 242–243
Artificial systems of classification, 14–26
Artocarpaceae, 462
Artocarpoideae, 463
*Artocarpus*, 463, 464
Arum family, described, 398
*Aruncus*, 541
Asaraceae, 473
*Asarum*, 473
*Ascarina*, 446
Asclepiadaceae, described, 673–676
*Asclepias*, 674, 675
(*Asclepiodora*), 675
*Ascyrum*, 605
Ash, 668; mountain, 544; prickly, 557
Asia, floras of, 300, 301
*Asimina*, 509
*Asparagus*, 413, 416
*Asperula*, 713
*Aspicarpa*, 562
*Asplenium*, 349
(*Astephanus*), 675
*Aster*, 728, 729, 730; China, 731
Asteraceae, 726, 730
Astereae, 730
*Astilbe*, 534
*Astragalus*, 547
*Astranthium*, 732
*Astrantia*, 644
(*Atamasco*), 419
*Athrotaxis*, 365
*Athyrium*, 349
*Atkinsia*, 594
Atlantic coastal plain, 154
Atoxiaceae, 569
(*Atragene*), 498
*Atriplex*, 478, 731
*Atropa*, 694
Atropine, 694

Atwood, W. W., physiographic provinces, 154–157
*Aucuba*, 647
(*Auliza*), 436
Aurantiaceae, 556
Australia, floras of, 302
Australian bower plant, 700
Australian tea tree, 634
*Austrotaxus*, 360
Author citation, in nomenclature, 209, 210
Autoploidy, 175
Autumn crocus, 416
*Avena*, 391
*Averrhoa*, 550
*Avicennia*, 687
(*Aviculare*), 476
Avocado, 513
Axile placentation, defined, 75
*Ayenia*, 596
(*Azalea*), 654
(*Azaleastrum*), 654
*Azara*, 614
*Azolla*, 354
*Azorella*, 646

Babcock, E. B., 52, 159, 179, 181, 231
Babcock, E. B. and Stebbins, G. L., Jr., 185
Baby's-breath, 488
*Baccharis*, 730
Backer, C. A., 319
*Baeria*, 732
Bahama Islands, flora of, 297
Bailey, F. M., 302
Bailey, I. W., 98, 105, 504
Bailey, I. W. and Nast, C. G., 506
Bailey, I. W. and Sinnott, E. W., 48
Bailey, L. H., 216, 253, 256, 320, 326, 396, 450, 484, 544, 618
Bailey, L. H. and E. Z., 320
*Baileya*, 728, 730
Baillon, H., 289, 319
Baker, H. B., 143
Baker, H. G., 174
Balanophoraceae, 474
Balanophorales, 474
Balanopsidales, 451
*Balbisia*, 551
*Balduina*, 728
Baldwin, J. T., Jr., 532
Ballard, F., 249
Balloon vine, 583
Balsam, Canada, 364; of gardens, 586
Balsam apple, 720
Balsaminaceae, described, 585–586
Balsamineae, 569
Bambuseae, 387, 389, 390
Bancroft, H., 65
*Banisteria*, 562
*Banksia*, 468
*Baptisia*, 546, 548

INDEX  781

Barbadoes cherry, 562
*Barbarea*, 521
*Barberetta*, 417
Barbeuieae, 482
*Barclaya*, 490, 491
Barley, 391
Barnhart, J. H., 326, 530, 537, 706; biographical file, 326
Barrier factors in phytogeography, 149–150
*Barthriosperma*, 685
*Bartonia*, 671
(*Bartonia*), 619
Basal placentation, 78
*Basella*, 486
Basellaceae, described, 486
Basil, 690
*Basiphyllaea*, 436
(*Basistelma*), 675
Basitonae, taxonomy of, 436
Basket-of-gold, 521
Basswood, 592
Batidaceae, described, 455–456
Batidales, described, 455
(*Batrachium*), 498
Battandier, J. A. and Trabut, L., 301
Battiscombe, E., 301
*Bauera*, 533
Bauhin, J., 16, 17, 23
*Bauhinia*, 546, 548
Bay, J. C., 305
Bay rum, oil of, 634
Bayberry, 451
Bead plant, 713
(*Beadlea*), 436
Bean, W. J., 320
Bean, common, 548; velvet, 548; yam, 548
Bearberry, 654
Beardtongue, 697
Bear's-breech, 709
Beauty bush, 715
Bechtel, H. R., 461, 462, 463, 465
(*Beckwithia*), 498
Bedevian, A. K., 319
Beech, 460
Beech family, described, 459
Beer, birch, 459
Beet, 478
Beetle, A. A., 216
Beetles, control of in herbarium, 248–253
*Begonia*, 620, 621
Begoniaceae, described, 620–621
*Begoniella*, 621
Begoniineae, 597
*Belamcanda*, 424
Belk, E., 386, 390
Belladonna, 694
Bellair, G. and St.-Léger, L., 320
Bellflower, 722
Bellflower family, described, 721
*Bellis*, 730

*Beloperone*, 708, 709
*Belotia*, 591
Ben, oil of, 526
Benedict, R. C., 352
*Benincasa*, 720
Benne, 701
Bennettitales, alliance with Magnoliaceae, 504; phyletic importance of, 105
Benson, M. and Wellsford, E. J., 454
Bentham, G., 42, 116, 117, 118, 298, 302
Bentham, G. and Hooker, J. D., system of classification, 31, 32, 33, 36, 46, 114, 115–118, 289; see also note p. 777
(*Benthamidia*), 647
Benzoin, 513, 666
(*Benzoin*), 513
Berberidaceae, described, 500–502
*Berberidopsis*, 614
*Berberis*, 501, 502
*Berchemia*, 588
*Bergenia*, 534
Berger, A., 531
*Bergia*, 606
Bergmans, J. B., 321
Beridge, E. M., 460
*Bersama*, 585
*Besleria*, 705
Bessey, C. E., 36, 38, 39, 94, 95, 104, 106, 107, 114, 115, 131, 134, 136, 333; arrangement of families, 127–130; diagram of system, 126; system of classification, 36, 125–130; see also note, p. 776
*Besseya*, 698
*Beta*, 477, 478
*Betula*, 457, 458, 459
Betulaceae, described, 457–459
Betuleae, 457
Bews, J. W., 301, 390
Bibliographies of taxonomic literature, 305, 306
Bichloride of mercury, as an insecticide, 251, 252
Bicornes, 649, 656
(*Bicuculla*), 518
*Bidens*, 728, 730
*Biebersteinia*, 551
*Bignonia*, 699, 700
Bignoniaceae, described, 698–700
Bignonieae, 699
*Billbergia*, 406, 407
*Billia*, 582
Biochemistry, importance to phylogeny, 160
Biographical references, 326, 327
Biosystematic categories, 176–179
Biosystematics, aid in delimiting taxa, 181–182; apomixis, 183–185; crossing programs, 180–181; cytogenetic analysis, 180; defined, 169; endosperm as sterility barrier, 180; importance of environments,

179–180; infrageneric units, 181; limitations of, 186–188; methods, 179–181; relation to taxonomic interpretation, 181–183; role of environment, 179, 180; sterility barriers, 180, 182
Biosystematics and cytogenetics, 169–188
Biosystematics and modern taxonomy, 170–172
Biotype, defined, 178–179
*Biovularia*, 706
Birch, 459
Birch beer, 459
Birch family, described, 457
Bird-of-paradise flower, 427
Bird's-foot trefoil, 548
Birthwort, 473
Birthwort family, described, 473
(*Bistorta*), 476
*Bixa*, 609, 610
Bixaceae, described, 609–610
Black, J. M., 302
Black olive, 633
Black walnut, 455
Blackberry, 544
Bladdernut family, described, 579
Bladderwort, 706
Bladderwort family, described, 705
Blake, S. F., 204, 216, 681
Blake, S. F. and Atwood, A. C., 223, 273, 278, 288, 291
Blaser, H. W., 386, 393
*Blechnum*, 350
Bleeding heart, 518
(*Blephariglottis*), 436
*Bletia*, 436
*Bletilla*, 437
Bloodflower, 675
Bloodwort family, described, 416
Blue laceflower, 646
Blue milkweed, 675
Blue pincushion flower, 724
Blue poppy, 517
Blueberry, 654
*Blumenbachia*, 619
*Bocconia*, 515, 516
Bock, J., 15, 16
Bodmer, F., 192
*Boehmeria*, 465
Bogbean, 671
Bog-moss family, described, 402
Bois, D., 321
*Boisduvalia*, 638
Boissier, E., 42, 300
(*Bolboxalis*), 550
*Boldea*, 511, 512
Boldo wood, 512
Boldoeae, 481
Bolus, H. M. L., 483
*Bomarea*, 415, 419
Bombacaceae, described, 595

*Bombacopsis*, 595
*Bombax*, 595
*Bonplandia*, 678
Bonstedt, C., 321
*Bontia*, 710
Boothroyd, L. E., 539, 540
*Boquila*, 500
Borage, 685
Boraginaceae, described, 684–686
Boragineae, 676
*Borago*, 684, 685
Boreal floras, literature of, 266, 267
Borraginaceae, 684
Borrowing herbarium specimens, 270, 271
*Boschniakia*, 702, 703
*Bosenbergia*, 428
Boston ivy, 589
*Boswellia*, 560
Botanical glossaries, list of, 317–318
Botanical illustrations, 272, 273
Botanical libraries, catalogues of, 306, 307; list of, 330
*Botrychium*, 343
Bottle brush, 634
Bottle tree, 596
*Bouchea*, 687
*Bougainvillea*, 480, 481
*Bougueria*, 711
*Boussingaultia*, 486
*Bouvardia*, 713
*Bowenia*, 356, 357
Bowenioideae, 357
Bower, F. O., 342, 352
Bower plant, Australian, 700
Boxwood, West Indian, 699
Boxwood family, described, 569
*Brachychiton*, 596
Branch systems, types of, 65
*Brandegea*, 719
Brasil, floras of, 298
*Brassia*, 436
*Brassica*, 521, 522
Brassicaceae, 520
Braun, A., 34, 718
Breadfruit, 464
Breitschneideraceae, 514, 582
Breitschneiderineae, 514
Bretschneider, E., 326
*Bretschneidera*, 582
Bretzler, E., 540
*Breweria*, 677
*Brickelia*, 732
*Brintonia*, 728
Briquet, J., 220, 326
Brisbane box, 634
British Honduras, flora of, 297
Britten, J. and Boulanger, G. S., 326
Britton, N. L., 170, 197, 291, 295, 377, 386, 484; contributions to nomenclature, 197, 198; influence on nomenclature, 218, 219

Britton, N. L. and Millspaugh, C. F., 297
Britton, N. L. and Rose, J. N., 623
Broccoli, 521
Bromeliaceae, described, 405–407
Brongniart, A., 34; contributions of, 31
Brooks, J. S., 401
Broom, 548
Broomrape family, described, 702
*Broussiasia*, 647
*Broussonetia*, 462
Broun, M., 335
Brouwer, J., 540
*Browallia*, 693, 694
Brown, N. E., 483
Brown, R., 30, 31, 42, 435, 719
Bruckner, G., 409
*Bruguiera*, 630
Brunelliaceae, 530
(*Bruneria*), 401
Brunfels, O., 15, 48
*Brunfelsia*, 694
Bruniaceae, 530
*Brunnichia*, 475
*Brunonia*, 723, 724
Brunoniaceae, described, 724
*Brunsvigia*, 420
Brussels Congress, 199
Brussels sprouts, 521
*Bryonia*, 718
(*Bryophyllum*), 531
Bryophyta, 334
*Buchanania*, 574
Buchenau, F., 412
*Buchnera*, 695
Bucholz, J. T., 366
*Bucida*, 633
Buckbean, 671
*Buckleya*, 470
Buckthorn, sea, 627
Buckthorn family, described, 587
Buckwheat, 476
Buckwheat family, described, 475
Buckwheat tree, 575
*Buddleja*, 669, 670
Buettneriaceae, 596
Buffalo berry, 627
Bugloss, 690; viper's, 685
*Bumelia*, 663
*Bupleurum*, 646
Bur cucumber, 720
Burmann, J., 19
*Burmannia*, 432
Burmanniaceae, described, 431–432
Burmanniales, 431
Burnat, E., 327
*Burnatia*, 382
Bur-reed family, described, 374
Burret, M., 396
*Bursera*, 560

Burseraceae, described, 559–560
Burtt, Davy, J., 301
Bush Honeysuckle, 715
Bush poppy, 517
Butcher, R. W. and Strudwick, F. E., 299
Butomaceae, described, 383–384
Butomales, sensu Hutchinson, 385, 386
*Butomus*, 384
Buttercup family, described, 496
Butterfly bush, 670
Butterfly weed, 675
Butterwort, 706
Buttneriaceae, 596
Button-ball tree, 540
Buxaceae, described, 569–570
Buxbaum, F., 484, 623
Buxineae, 569
*Buxus*, 569, 570
Byblidaceae, 530
*Byra*, 665
*Byrsonina*, 562
*Byronia*, 576

Cabbage, 521
*Cabomba*, 489, 490
Cabombaceae, 490
Cabomboideae, 489, 490
*Cacalia*, 730
Cacao, 596
Cactaceae, described, 621–624
Cactales, 621
Cactus, coachwhip, 680
*Caesalpinia*, 547
Caesalpiniaceae, 545
Caesalpinioideae, 545, 547
Cain, S. A., 142, 158–160, 161, 162; principles of phytogeography, 159, 160
*Caiophora*, 619
*Cakile*, 522
Calabash gourd, 720
Calabash tree, 700
*Caladium*, 400
Calamitaceae, 335
Calamitales, 337
Calamites, era of, 145–146
*Calandrinia*, 485
*Calathea*, 430
*Calceolaria*, 695, 696, 697
(*Calceolus*), 436
*Caldesia*, 382
Calectasieae, 413, 415
*Calendula*, 727, 730
Calenduleae, 730
California, floras of, 293
California poppy, 517
*Calla*, 398, 400
Calla-lily of florists, 400
*Calliandra*, 547
*Callicarpa*, 687, 688
*Callicoma*, 536

*Callirhoë*, 594
*Callisia*, 408, 409
*Callistemon*, 634
*Callistephus*, 731
*Callitriche*, 568, 642
Callitrichaceae, described, 567–568
Callitrichineae, 549
*Callitris*, 366, 367
*Callitropsis*, 366, 367
Calloideae, 398
*Calluna*, 654
*Calocarpum*, 5P3. 773
*Calocasia*, 400
Calocasioideae, 398
*Calochortus*, 413
*Calodendrum*, 557
*Calonyction*, 677
Calophylloideae, 603
*Calophylum*, 603
*Calopogōn*, 436
*Calostemma*, 418
*Calothamnus*, 633
*Caltha*, 496, 498
Caltrop family, described, 555
*Calycadenia*, 728
Calycanthaceae, described, 507–508
*Calycanthus*, 508
Calyceraceae, described, 725–726
*Calydorea*, 424
*Calypso*, 436
*Calyptranthes*, 634
*Calyptridium*, 484, 485
Calyx, morphology of, 65–67
*Camassia*, 414
Cambridge Congress, 199, 200
*Camellia*, 602
Camp, W. H., 49, 243, 299, 327
Camp, W. H. and Gilly, C. L., 50, 51, 53, 176
*Campanula*, 721, 722
Campanulaceae, described, 721–723
Campanulales, 717
Campanulatae, described, 720
Campanuloideae, 722
Campbell, D. H., 143, 153, 158, 378
Camphor, 513
*Campylocentrum*, 436
*Campynema*, 419
*Campynemanthe*, 419
Campynematoideae, 419
*Campsis*, 699, 700
*Camptosorus*, 350
*Camptotheca*, 631
Canada balsam, 364
Canada, floras of, 293
*Cananga*, 509
*Canarina*, 721, 722
*Canarium*, 560
Canary-bird flower, 553
*Canbya*, 516

Candlewood family, described, 680
Candolle, see De Candolle
*Candollea*, 598
(*Candollea*), 725
Candolleaceae, 724
Candytuft, 521
*Canella*, 611
Canellaceae, described, 611–612
Canistel, 663
*Canna*, 429
Cannabinaceae, 463
*Cannabis*, 462
Cannaboideae, 463
Cannaceae, described, 428–429
*Canotia*, 577
*Cantua*, 678, 679
Cape chestnut, 557
Cape ponderweed, 379
Cape primrose, 705
Caper family, described, 518
Capers, 520
Capitalization of specific and trivial epithets, 202, 216, 219
Capparidaceae, alliance with Resedaceae, 525; described, 518–520
Capparidineae, 514
*Capparis*, 519, 520
*Capraria*, 695
Caprifoliaceae, described, 713–715
*Capsella*, 521
*Capsicum*, 694
Carambola, 551
Caraway, 646
Carbon disulfide, as an insecticide, 250–251
*Cardamine*, 521
*Cardamon*, 428
*Cardaria*, 521
Cardinal's guard, 709
*Cardiospermum*, 583
Carduaceae, 730
Carduales, 730
*Carduus*, 728, 730
*Carex*, 392, 393
*Carica*, 618
Caricaceae, 393; described, 617–618
Cariceae, 393
Caricoideae, 393
*Carissa*, 673
*Carlowrightia*, 708
*Carludovica*, 397, 398
Carnation, 488
Caroa, 407
Carolina Jessamine, 670
Carpel, defined, 72; morphology of, 72–75
Carpel polymorphism, theory of, 65
Carpels, determination of number in ovary, 79; open in angiosperms, 370
*Carpinus*, 457, 458
*Carrierea*, 614

INDEX 785

Carrion flower, 675
Carrot, 646
Carrot family, described, 644
(*Carteria*), 436
*Carthamus*, 730
Cartography, literature of, 323
*Cartonema*, 409
Cartonemaceae, 409
*Carum*, 646
*Carya*, 453, 454, 455
Caryocaraceae, 597
Caryophyllaceae, described, 486–489
*Caryoptis*, 688
Cascara sagrada, 588
Cashew, 574
Cashew family, described, 573
Cassabanana, 720
Cassava, 566
*Cassia*, 547, 548
Cassini, H., 729
*Cassiope*, 655
*Cassytha*, 512, 513
(*Castalia*), 490
*Castanea*, 459, 460
*Castanopsis*, 460
*Castela*, 559
(*Castelaria*), 559
*Castilleja*, 695, 696
Castor oil, 566
*Casuarina*, 443
Casuarinaceae, described, 442, 443
Casuarinales, 442
Catalogue of Linnaean herbarium, 328, 329
Catalogues of botanical libraries, 306, 307
*Catalpa*, 698, 699
Catchfly, 488
Categories of classification, 44–56
*Catha*, 577
(*Cathartolinum*), 553
*Catopsis*, 406
Cattail family, described, 373
*Cattleya*, 434, 437
Cauliflower, 521
*Caulophyllum*, 370, 501, 502
*Cayaponia*, 719
Caytoniales, phyletic importance of, 105
*Ceanothus*, 588
*Cecropia*, 464
Cedar, oil of, 367; West Indian, 561
*Cedrela*, 560, 561
*Cedrus*, 364
*Ceiba*, 595
Čelakovský, L. F., 60
Celandine poppy, 517
Celastraceae, described, 576–577
Celastrales, 578
Celastrineae, 569, 576
*Celastrus*, 577
Celery, 646

Cellophane, for mounting specimens, 245
Celosia, 478, 479
Celtidoideae, 461
*Celtis*, 461, 462
Cenospecies, concept of, 178
*Centaurea*, 727, 728, 730
*Centaurium*, 670
Centaury, 671
Centers of area, 164–165
Central America, floras of, 296, 297
Central lowlands of North America, 156
*Centranthus*, 716
Centrifugal androecia, 69–70
Centrifugal inflorescences, 61
Centripetal androecia, 69–70
Centripetal inflorescences, 61
*Centrogenium*, 436
Centrolepidaceae, 386, 402, 403
*Centromadia*, 728
*Centropogon*, 722
Centrospermae, described, 477
*Centunculus*, 659
*Cephaelis*, 713
*Cephalanthera*, 435
*Cephalanthus*, 713
*Cephalaria*, 717
Cephalotaceae, 530
Cephalotaxaceae, described, 362
*Cephalotaxus*, 362
*Cerastium*, 487
*Ceratiola*, 571
*Ceratopetalum*, 536
Ceratophyllaceae, described, 491–492
*Ceratophyllum*, 491, 492
Ceratopteridaceae, 351
*Ceratopteris*, 351, 352
*Ceratostigma*, 661–662
*Ceratotheca*, 701
*Ceratozamia*, 357
Cercidiphyllaceae, described, 495–496
*Cercidiphyllum*, 493, 495, 496
*Cercidium*, 547
*Cercis*, 547, 548
Cercocarpae, 543
*Cercocarpus*, 545
Cereoideae, 623
*Cerinthe*, 685
*Ceriops*, 630
*Ceropegia*, 675
Cesalpino, A., 16, 17
Cestreae, 694
*Cestrum*, 694
Ceylon gooseberry, 614
Ceylon milk plant, 675
*Cevallia*, 619
*Chaenomeles*, 542, 544
Chalk, L., 98
*Chamaebatia*, 544
*Chamaebatiaria*, 544
*Chamaecrista*, 549

*Chamaecyparis*, 366, 367
(*Chamaepericlymenum*), 647
*Chamaesaracha*, 694
Chamberlain, C. J., 368
Chamberlain, W., 323
Chapman, M., 501, 502
Chaudefaud, M., 104, 340
Chaulmugra oil, 614
Chayote, 720
Cheadle, V. I., 98, 107, 376, 383, 412
Cheeseman, T. F., 303
Cheeseman, T. F. and Hemsley, W. B., 303
*Cheilanthes*, 349
*Cheiloclinium*, 578
*Cheiranthera*, 535
*Cheiranthus*, 521
*Chelidonium*, 516
Chelone, 698
Chenopodiaceae, described, 477–478
Chenopodioideae, 478
*Chenopodium*, 478
Cherimoya, 509
Cherler, J. A., 16
Cherry, 544; Barbadoes, 562; Cornelian, 648; Surinam, 634
Chervil, 646
Chester, K. S., 100
Chestnut, 460; cape, 557
Chevalier, A., 451
Chickweed, 488
Chicle, 663
Chicory, 730
Chile, floras of, 298
Chilean jasmine, 673
*Chilopsis*, 699, 700
*Chimaphila*, 651
*Chimonanthus*, 508
China aster, 731
China fir, 366
Chinaberry tree, 561
Chinese parasol tree, 596
Chinese rice-paper, 643
Chinese watermelon, 720
Ching, R. C., 351, 352
(*Chiogenes*), 654, 655
*Chionanthus*, 667, 668
Chlaenaceae, 590
Chlaeneae, 590
*Chloanthes*, 687
Chloranthaceae, described, 445, 446
*Chloranthus*, 446
*Chlorideae*, 390
*Chlorophora*, 464
*Choananthus*, 418
Chocolate, 596
*Chorisia*, 595
*Chorizanthe*, 475
*Chosenia*, 447
Chouard, P. and Laumonnier, E., 321

Christensen, C., 287, 327, 342, 344, 347, 348, 349, 350, 351, 352, 354
*Christisonia*, 703
Christmas vine, 677
Christopherson, E., 303
Chromosome counts, limitations of, 187, 188
*Chrysalidocarpus*, 395
*Chrysanthemoideas*, 728
*Chrysanthemum*, 728, 729
Chrysobalanaceae, 543
Chrysobalanoideae, 543
*Chrysobalanus*, 541
*Chrysogonum*, 728
*Chrysoma*, 728
*Chrysophyllum*, 663
*Chrysopsis*, 730
*Chrysosplenium*, 534
(*Chylismia*), 638
*Chytra*, 696
Chufa, 394
*Cibotium*, 349
Cichoriaceae, 730
*Cichorium*, 728, 729, 730
*Cicuta*, 646
*Cienfuegosia*, 593
*Cimicifuga*, 498
*Cinchona*, 713
Cineraria, 731
*Cinna*, 399
*Cinnamodendron*, 611
*Cinnamomum*, 513
Cinnamon, 513; wild, 611
Cinnamon fern, 345
Cinnamon vine, 422
*Cinnaosma*, 611
Cinquefoil, 544
*Circaea*, 637, 638, 639
Circassian walnut, 455
*Cirsium*, 730
*Cissampelos*, 503
*Cissus*, 589
Cistaceae, described, 608–609
Cistineae, 597, 609
*Cistus*, 609
*Citharexylum*, 687
*Citron*, 557, 720
*Citrullus*, 718, 720
*Citrus*, 557
*Clarkia*, 638, 640
Class, a unit of classification, 45
Classification, artificial, 18; biosystematic, 169–170; categories of, 43, 44; defined, 4; history of, 13–41; natural, 93; natural vs. artificial, 13–14; principles of, 4
Classification system of, Bentham and Hooker, 31; Bessey, 35, diagram of, 126; Bock, 15; Eichler, 33; Engler, 34; de Candolle, 28; Hallier, 36, diagram of, 132; Hutchinson, 36, diagram of, 137;

INDEX 787

Jussieu, 28; Mez, 38; Rendle, 37; Theophrastus, 14; Tippo, 38, diagram of, 138; Wettstein, 35
Classification systems based on, form relationships, 26; habit, 14; phylogenetic concepts, 33
Classification systems, current, 114–139
Clausen, J., 165
Clausen, J., Keck, D. D. and Hiesey, W., 52, 159, 175, 176, 178, 179
Clausen, R. T., 55, 343
*Clavija*, 657
*Claytonia*, 484, 485
*Cleistes*, 436
Cleland, R. E., 181
*Clematis*, 496, 498
*Clematoclethra*, 599, 600
*Cleome*, 519, 520
*Cleomella*, 519
*Clermontia*, 721
*Clerodendrum*, 687, 688
*Clethra*, 600, 650
Clethraceae, described, 650–651
*Cleyera*, 602
*Clidemia*, 636
Clifford, G., 19, 20
*Cliftonia*, 575
Climbing fern family, described, 345
*Clinogyne*, 430
*Clintonia*, 414
*Clivia*, 418
Clock vine, 709
Clone, a taxonomic unit, 56, 217
Clove, 634
Clover, sweet, 548
Club moss family, described, 337
*Clusia*, 603
Clusiaceae, 603
Clusioideae, 603
Clusius, see L'Ecluse, C.
Cneoraceae, 549
Coachwhip cactus, 680
*Cobaea*, 678, 679
Coca, 555
Coca family, described, 554
Cocaine, 555
*Coccinea*, 720
*Coccoloba*, 475, 476
*Cocculus*, 503
Cochlospermaceae, described, 610–611
Cochlospermineae, 597
*Cochlospermum*, 610, 611
*Cocos*, 395, 396
*Cocothrinax*, 396
Cocus wood, 665
*Codiaeum*, 566
*Codon*, 682
*Codonopsis*, 722
(*Coeloglossum*), 436
*Coelogyne*, 437

*Coffea*, 712, 713
Coffee, 713
*Coix*, 389
*Cola*, 596
*Colchicum*, 413
*Coleogyne*, 544
*Coleus*, 690, 691
Colignoneae, 481
Collecting equipment, 234, 235
Collecting procedures, 234–235
Collins, J. F., 236
*Collomia*, 678
Colmeiro, M., 327
*Colobanthus*, 487
Color charts, sources of, 329–330
Color retention of plant materials, 255–256
*Colubrina*, 588
Columelliaceae, 676
Columniferae, 590
*Comarella*, 544
*Comarum*, 544
Combretaceae, described, 632–633
*Combretum*, 632
*Comesperma*, 564
Comfrey, 685
*Commandra*, 470
*Commelina*, 408, 409
Commelinaceae, described, 408–410
Commelinales, 409
Commelineae, 409
*Commiphora*, 560
Comparium, concept of, 178
Compilatory lists, preparation of, 276–277
Compositae, described, 726, 732
*Comptonia*, 450, 451
Conanthereae, 419
*Concocarpus*, 633
*Condalia*, 588
Coniferae, described, 359–360
Conifers, phylogeny of, 102–103
*Conium*, 646
Connaraceae, 530
Conocephaloideae, 463
*Conopholis*, 703
*Conostegia*, 636
Conostylideae, 417, 419
Conservatism of characters, defined, 99
Continental bridges, theory of, 144
Continental drift, theory of, 143, 144
Continental islands, 149
Continental land masses, 147
Continental platforms, islands of, 149; map of, 148
Continental shelf, 144
Contortae, described, 666–667
Convallariaceae, 416
Convolvulaceae, described, 676–678
Convolvulales, 677
Convolvulineae, 676
*Convolvulus*, 677

Conzatti, C., 296
Cooper, D. C. and Brink, R. A., 180
*Cooperia*, 418, 420
Copal, 560
Copeland, E. B., 344, 347, 348, 349, 351, 353, 354
Copeland, H. F., 24, 101, 652, 654, 681
*Coprosma*, 713
(*Coptidium*), 498
*Coptis*, 496, 498
Coral plant, 697
Coral-bells, 534
Coralberry, 715
*Corallorrhiza*, 436
*Corchoropsis*, 592
*Corchorus*, 591
Cordaites, importance of, 102
*Cordia*, 684, 685
Cordiaceae, 685
Cordilleran plateau, a physiographic province, 157
Cordilleran ranges, a physiographic province, 157
Cordioideae, 685
Core, E. L., 219
*Corema*, 570, 571
*Coreopsis*, 728, 730
*Coriaria*, 572
Coriariaceae, described, 571–572
Coriariineae, 569
*Coris*, 659
Cork bark, 460
Cork tree, 557
Cork-wood family, described, 451, 452
Corn, 391
Corn salad, 716
Cornaceae, described, 647–648
Cornelian cherry, 648
Corner, E. J. H., 70, 219, 498. 541
*Cornus*, 647, 648
*Cornutia*, 688
*Corokia*, 647
Corolla, morphology of, 66, 67
Corolliferae, 411
*Correa*, 556
Corrosive sublimate, as an insecticide, 251, 252
Corsiaceae, 432
*Cortusa*, 660
*Corydalis*, 517, 518
Corylaceae, 457
Coryleae, 457
*Corylopsis*, 537, 538
*Corylus*, 457, 458, 459
Corynocarpaceae, 569
Coryphananae, 621
*Cosmos*, 728, 730
Cosson, E., 327
Costa Rica, flora of, 297
*Costaea*, 575

Coste, H., 299
Costoideae, 428
*Costus*, 428
*Cotinus*, 574
*Cotoneaster*, 544
Cotton, 594
*Cotula*, 728
*Cotyledon*, 530
Cotyledonoideae, 532
Cotyloideae, families of, 129, 130
Cotype, defined, 205
Coulter, J. M., 293
*Couma*, 672
Coutinho, A. X. P., 299
Cow parsnip, 646
*Cowania*, 544
Cowpeas, 548
Cox, H. T., 654
Cranberry, 654
Crane's-bill, 552
*Cranichis*, 436
*Crassula*, 531
Crassulaceae, described, 530, 532; allied to Podostemaceae, 467
Crassuloideae, 532
*Crataegus*, 543, 544
*Crateva*, 519
Creeping Jennie, 338
*Cremosperma*, 705
Crepe-jasmine, 673
Crepe myrtle, 628
*Crescentia*, 699, 700
Crescentieae, 699
Cress, rock, 521; water, 521
*Cressa*, 677
*Crinum*, 419, 420
*Crinodendron*, 591
(*Criosanthes*), 436
*Cristatella*, 519
*Crocanthemum*, 609
Crocinae, 424
Crocoideae, 424
*Crocus*, 423, 424; autumn, 416; spring, 424
Croizat, L., 23, 43, 59
*Crookia*, 605
*Crossosoma*, 540, 541
Crossomataceae, described, 540–541
*Crotalaria*, 547
*Croton*, 566
Croton, 566
Crotonoideae, 566
Crowberry family, described, 570
Crowfoot, 338
Crown of thorns, 566
Cruciferae, described, 520–524
*Crusea*, 713
*Cryophytum*, 483
Cryptangieae, 393
*Cryptantha*, 685
Crypteroniaceae, 624

*Cryptocarya*, 513
*Cryptogamma*, 352
Cryptogams (vascular), phylogeny of, 101–102
*Cryptomeria*, 365, 366
*Cryptostegia*, 675
*Cryptostephanus*, 418
*Crysothamnus*, 732
Cuban hemp, 420
*Cubelium*, 612
Cucumber, 720; bur, 720; squirting, 720
*Cucumis*, 720
*Cucurbita*, 718
Cucurbitaceae, described, 718–720
Cucurbitales, 597, 717
Cucurbiteae, 719
*Cudrania*, 462, 464
Cultivated plants, literature of, 316, 317, 319, 321, 322
*Cunninghamia*, 365, 366
Cunoniaceae, described, 536
*Cupania*, 583
*Cuphea*, 627, 628
Cupressaceae, described, 366–368
Cupressineae, 364
Cupressoideae, 367
*Cupressus*, 366, 367
Curare, 670
*Curatella*, 598
*Curcuma*, 428
Curly grass family, described, 345
Currant, 534
*Cuscuta*, 676, 677, 678
Cuscutaceae, 677
Custard-apple, 509
Custard-apple family, described, 508
(*Cuthbertia*), 408
Cutler, H. C., 270
*Cyanaea*, 722
*Cyananthus*, 721
Cyanastraceae, 402
Cyanide gas as an insecticide, 249
*Cyanotis*, 408, 409
*Cyathea*, 348
Cyatheaceae, described, 348
*Cyathodes*, 656
Cycadaceae, described, 356–357
Cycadales, 355, 356
Cycadioideae, 357
*Cycas*, 356, 357
*Cycladenia*, 673
*Cyclamen*, 659, 660
Cyclanthaceae, described, 397–398
Cyclanthales, 397
*Cyclanthera*, 718, 719
Cyclanthereae, 719
Cyclobeae, 478
(*Cyclopogon*), 436
*Cydonia*, 544
*Cymbalaria*, 697

*Cymbidium*, 437
*Cymodocea*, 376, 377
Cymodoceae, 377
Cymodoceaceae, 377
*Cynanchum*, 675
*Cynara*, 730
Cynareae, 730
*Cynoctonum*, 669
*Cynodon*, 391
*Cynoglossum*, 685
Cynomoriaceae, 624
Cynomoriineae, 624
Cyperaceae, described, 392–394
Cyperales, 386
*Cyperus*, 392, 393
Cyphioideae, 722
*Cyphomandra*, 694
Cypress family, described, 366
Cypress spurge, 478
Cypress vine, 677
Cypripedilineae, 435
Cypripediloideae, 435
*Cypripedium*, 435, 436, 437
*Cypselea*, 483
*Cyrilla*, 575
Cyrillaceae, described, 575
*Cyrtandra*, 704
*Cyrtanthus*, 420
*Cyrtomium*, 351
*Cyrtopodium*, 436
(*Cyrtorhyncha*), 498
(*Cythera*), 436
Cytinaceae, 474
*Cytisus*, 548
Cytogenetics, analysis of data, 180; limitations of, 186–188
Cytogenetics and biosystematics, 169–188
Cytology, role in phylogeny, 99
Cytotaxonomy, 169

*Dacrydium*, 361
*Dactylicapnos*, 518
*Dactylis*, 390
Dade, H. A., 329
*Dais*, 626
*Dalea*, 548, 549
Dalla Torre and Harms, *Genera siphonogamarum*, described, 287
*Damasonium*, 382
*Danaea*, 344
Dandy, J. E., 386, 504
Daphnaceae, 625
*Daphne*, 626
Daphniphyllaceae, 549
*Darlingtonia*, 527
Darrah, W. C., 104, 106
Darwin, C., 8, 27, 33, 172
Darwin, E., 26
Darwinism, 172, 173; defined, 176
*Dasiphora*, 544

Dasypogoneae, 413, 415
Data for herbarium specimen labels, 246, 247, 248
Dates for publication, sources of, 327–328
*Datisca*, 620
Datiscaceae, described, 619–620
Datiscineae, 597
Datta, R. M. and Mitra, J. N., 526
*Datura*, 694
Datureae, 694
*Daucus*, 645, 646
*Davallia*, 350, 351
*Davidia*, 630, 631
*Davilla*, 598
Dawson, M. L., 679
Day lily, 416
Dayton, W. A., 366
DDT, as an herbarium insecticide, 252
Deam, C. C., 277, 294
DeBeer, C. R., 98
*Decaisnea*, 500
DeCandolle, A., 196, 289, 328; contribution to nomenclature, 195, 196
DeCandolle, A. P., 39, 42, 73, 116, 195, 289; *Prodromus*, 29, 30, 289; system of classification, 29
DeCandolle, A. and C., *Monographiae phanerogamarum*, described, 289
Decapitalization of specific epithets, 202, 216
*Decodon*, 628
Deep-freezing of flowers, 256
Deetz, C. H. and Adams, O. S., 323
Degener, O., 303
Degeneriaceae, 439, 489, 498, 504
De Jussieu, *see* Jussieu
*Delonix*, 548
Delphineae, 496
*Delphinium*, 497, 498
Demeter, K., 673
*Dendrobium*, 437
*Dendromecon*, 517
(*Dendropogon*), 407
*Dendrosicyos*, 718
*Dennstaedtia*, 350, 351
Dermen, H., 501, 631, 648
Descole, H., 297
*Descourania*, 522
Descriptions of taxa, preparation of, 272
Descriptive taxonomy, defined, 170
Deserts, as barrier factors, 150
Desfontainaceae, 667
Desfontaines, R., 29
*Desmodium*, 548
*Deutzia*, 534
DeVries, H., 8, 173
*Diamorpha*, 531
Diandrae, taxonomy of, 435, 436
(*Dianthera*), 708
*Dianthus*, 487, **488**

(*Diapedium*), 708
*Diapensia*, 649
Diapensiaceae, described, 648–649
Diapensiales, 648
(*Diaphoranthema*), 407
*Diascia*, 695
*Diastatea*, 721
*Dicentra*, 518
Dichapetalaceae, 549
Dichapetalineae, 549
Dichasium, in inflorescence phylogeny, 60, 61
*Dichondra*, 677
Dichondraceae, 677
*Dichorisandra*, 408
Dickason, F. G., 351
Dickson, J., 86, 488, 516, 660
Dickson, W. A., 303
Dicksoniaceae, described, 348–349
Diclidantheraceae, 662
*Dicliptera*, 708
*Dicoryphe*, 537
Dicotyledoneae, described, 438
*Dictamnus*, 557
Dictionaries, list of, 318, 319
Didiereaceae, 569
Didiereineae, 569
*Didiscus*, 646
*Dieffenbachia*, 400
Diels, L., 424, 530
*Diervilla*, 713, 715
Differentiation, Guppy's theory of, 161
*Digitalis*, 695, 697
*Dilatris*, 417
Dill, 646
*Dillenia*, 598
Dilleniaceae, 498; described, 598–599
Dillenius, J. J., 25
*Dimorphotheca*, 730
*Diodia*, 713
Dioecism, characteristics of, 83
*Dion*, 357
*Dionaea*, 528, 529
Dionaeaceae, 529
Dionioideae, 357
*Dioscorea*, 421
Dioscoreaceae, described, 421–422
Dioscoreae, 421
Diospyrineae, 662
*Diospyros*, 664, 665
*Dipelta*, 713
*Dipholis*, 663
Dipsacaceae, described, 717
*Dipsacus*, 717
Dipterocarpaceae, 597
*Dipteronia*, 581
*Dirachma*, 551
*Dirca*, 626
Discontinuous distributions, 163, 164
*Diselma*, 366, 367

Dishcloth gourd, 720
Dissections, permanent preparations of, 257
*Distichlis*, 389
Dittany, 557
Division, unit of classification, 44, 45
*Dobinea*, 573
Dobzhansky, T., 180
*Dodecatheon*, 659
*Dodonaea*, 583
Dogbane family, described, 672
Dogwood family, described, 647
*Doliocarpus*, 598
*Dombeya*, 596
*Donatia*, 725
*Doronicum*, 731
*Dorstenia*, 462, 463, 464
Douglas, G. E., 81, 660
*Dovyalis*, 614
*Downingia*, 721, 722
*Draba*, 521, 522
Dracaeneae, 419
Dracenioideae, 413, 415
*Dracocephalum*, 691
Dragonhead, 691; false, 691
*Drimys*, 505
*Drosera*, 529
Droseraceae, described, 528, 530
*Drosophyllum*, 529
Drude, O., 654
*Dryas*, 544
Drying plants, 235–237, 241–243
*Drymocallis*, 544
*Dryopteris*, 349, 351
*Duchesnia*, 544
Duckweed family, described, 400
*Dudleya*, 531
Durango root, 620
*Duranta*, 686, 687, 688
DuRietz, G. E., 51, 55, 159
*Durio*, 595
Dutchman's-pipe, 473
DuToit, A. L., 142, 143
*Dychoriste*, 708

Eames, A. J., 38, 65, 73, 80, 98, 101, 106, 107, 139, 334, 337, 340, 341, 344, 346, 347, 348, 349, 350, 354, 355, 516; classification of, 38
Eames, A. J. and Wilson, C. L., 520
Earth, amount below sea-level, 144; continental islands, 147–148; continents of, 147–148; evolution of, 142–144; island land masses of, 147–148; major mountain ranges, 148; vegetation areas of, 150, 153
Eastwood, A., 318
Eaton, A., 26
Ebenaceae, described, 664–665
Ebenales, described, 662
Ebony, American, 665; Macassar, 665

Ebony family, described, 664
*Ecballium*, 718, 719, 720
Eccremocarpeae, 699
*Eccremocarpus*, 698, 700
*Echeveria*, 531
Echeverioideae, 532
*Echinochloa*, 387
*Echinocystis*, 719
*Echinodorus*, 382
(*Echinopanax*), 643
*Echinopepon*, 719
*Echinophora*, 644
*Echinops*, 726, 727
*Echites*, 673
*Echium*, 684, 685
Economic botany, literature of, 319–322
Economic plants, literature of, 316, 317, 319–322
Ecospecies, 177, 178
Ecospecies and species, 182
Ecotype, 176, 177
Edelwiess, 731
*Edgeworthia*, 626
*Edrianthus*, 721, 722
Eel grass, 377
Effective publication, conditions of, 208
Eggers, O., 521
Eggfruit, 663
Eggplant, 694
Egypt, floras of, 302
Ehretiaceae, 685
Ehretioideae, 685
*Eichhornia*, 411
Eichhornieae, 411
Eichler, A. W., 33, 34, 95, 118, 427, 446, 487, 590; system of classification, 33, 34
Elaeagnaceae described, 626–627
*Elaeagnus*, 626, 627
Elaeocarpaceae, described, 590–591
Elaeocarpineae, 590
*Elaeocarpus*, 591
*Elaeodendron*, 577
(*Elaphrium*), 560
Elatinaceae, described, 605–606
*Elatine*, 606
*Elattosis*, 384
Elderberry, 715
*Eleocharis*, 392, 393
*Elephantopus*, 730
Elephant's Ear, 400
*Elettaria*, 428
*Eleusine*, 389
*Elisma*, 382
*Ellisia*, 683
Elm family, described, 461
(*Elodea*), 385
Elodeaceae, 386
*Elsholtzia*, 690
(*Elsota*), 564
*Elytraria*, 708

*Embelia*, 657
Embryological formulae, use of, 88
Embryology, importance to taxonomy, 87
Embryophyta Asiphonogama, 334
Embryophyta Siphonogama, 334; described, 354, 355
*Emex*, 475
Empetraceae, described, 570–571
Empetrineae, 569
*Empetrum*, 570, 571
Enantioblastae, 403
*Encephalartos*, 357
(*Encyclia*), 436
Encyclopedias, of cultivated plants, 319, 320, 321, 322
Endemism, 164
Endlicher, S., 31
Endosperm, as sterility barrier, 180
*Engelhardtia*, 453, 454, 455
*Engelmannia*, 728
Engler, A., 34–35, 39, 44, 94, 114, 534; see also note, p. 777
Engler, A. and Diels, L., 35, 290, 333; *Syllabus*, described, 290
Engler, A., and Prantl, K., 36, 290
English ivy, 643
Ensatae, 424
*Entelia*, 592
Entomophilous flowers, 72
Environment, importance in biosystematics, 179, 180
Epacridaceae, described, 656
Epacridineae, 649
*Epacris*, 656
*Ephedra*, 369
Ephedraceae, described, 368, 369
Ephedrine, 369
Epibiotics, 164
(*Epicladium*), 436
*Epidendrum*, 435, 436, 437
*Epifagus*, 702, 703
*Epigaea*, 654
Epigyny, significance of, 81
Epilobiaceae, 637
*Epilobium*, 368, 639
*Epimedium*, 501
*Epipactis*, 436, 437
(*Epipactis*), 436
*Epirrhizanthes*, 563
*Episcia*, 705
Epling, C., 99, 169, 272
Equisetaceae, described, 336–337
Equisetales, described, 336
Equisetinae, 335
*Equisetum*, 335, 336, 337
*Ercilla*, 482
Erdtman, G., 159
*Erechtites*, 730
*Eremosyne*, 533
*Eriastrum*, 678

*Erica*, 653, 654
Ericaceae, described, 652–655
Ericales, described, 649–650
Ericineae, 649
Erickson, R. O., 270
Ericoideae, 654
*Erigeron*, 728, 730
*Eriobotrya*, 544
Eriocaulaceae, described, 404, 405
Eriocaulales, 405
*Eriocaulon*, 405
*Eriodictyon*, 682
*Eriogonum*, 475
*Eriophorum*, 392
*Eriophyllum*, 730
*Erodium*, 551, 552
*Ervatamia*, 673
*Erycibe*, 677
*Eryngium*, 645, 646
*Erysimum*, 521
*Erythrina*, 548
*Erythrodes*, 436
*Erythronium*, 414
Erythroxylaceae, described, 554–555
*Erythroxylum*, 554
*Escallonia*, 534
Escalloniaceae, 534
*Eschscholzia*, 515, 516, 517
Ethylene dichloride, as an insecticide, 251
*Eucalyptus*, 633, 634
*Euchnide*, 618, 619
*Eucommia*, 493, 538, 539
Eucommiaceae, 461; described, 538–539
*Eucryphia*, 600
Eucryphiaceae, described, 600
*Eucrypta*, 683
Eufilicales, described, 344; see also Filicales
*Eugenia*, 634
*Eulophia*, 436
*Euonymus*, 577
Eupatorieae, 730
*Eupatorium*, 726, 730
*Euphorbia*, 565, 566
Euphorbiaceae, described, 565, 567
*Euphoria*, 584
Eupomatiaceae, 489
*Euptelea*, 493, 494, 495
Eupteleaceae, 564; described, 494–495
Europe, floras of, 298, 299, 300
*Eurya*, 601, 602
*Euryale*, 489, 490
*Eurycles*, 418
*Euscaphis*, 580
Eusporangiatae, described, 343
Eusporangiate fern, defined, 342
*Eustoma*, 671
*Eustylis*, 424
Evening-primrose family, described, 637
Evolution, definition and mechanics of, 172; of earth and continents, 142–144

# INDEX

Evolvulus, 677, 678
Exacineae, 670
Exacum, 670
Exochorda, 544
Exothea, 583
Experimental taxonomy, 169–188; limitations of, 186–188; methods in, 179–181; objectives of, 171–177; sterility barriers, 180

Fagaceae, described, 459–460
Fagales, described, 456–457
Fagerlind F., 105
Fagonia, 556
Fagopyrum, 475, 476
Fagraea, 669
Fagus, 459, 460
Fallugia, 544
False dragonhead, 691
False sandalwood, 462
False tamarisk, 608
Families, nomenclature of, 206
Families of vascular plants, 333–775
Family, a unit of classification, 46, 47
Farinosae, described, 402
Fassett, N. C., 99, 270, 466
Fawcett, W. and Rendle, A. B., 297
Fedde, F., 516, 518
Fedia, 716
Feijoa, 634
Fennel, 646
Fern family, described, 349
Ferns (true), described, 342
Fernald, M. L., 55, 164, 216, 242, 243, 267, 296, 377, 450, 457, 484, 544, 609
Fern-allies, 335
Feronia, 556
Festuca, 389
Festuceae, 390
Fevillea, 718
Fevilleae, 719
Ficoideae, 482
Ficus, 462, 463, 464
Field studies, 268–270; for floristic investigations, 277; mass collections, 269–270; notes to be taken, 268, 269, 278, 279
Field techniques, collecting equipment and procedures, 234–235; drying specimens, 241–243; Hodges' preservation method, 241; preparation of specimens, 235–237; Schultes' preservation method, 240; special methods for particular taxa, 256, 257; use of hydroxyquinoline sulfate, 241, 255; use of special aids, 239
Fig, 464; Indian, 623
Figwort family, described, 695
Filbert, 459
Filicales, in the Mesozoic, 145–147; described, 344 (as Eufilicales); phylogeny of, 101, 102

Filicinae, described, 342
Filmy fern family, described, 347, 348
Fimbristylis, 392
Firethorn, 544
Firmiana, 596
Fisher, M. J., 447
(Fissipes), 436
Fittonia, 709
Fitzroya, 366
Flacourtia, 614
Flacourtiaceae, described, 613–614
Flacourtiineae, 597
Flagellariaceae, 402
Flannelbush, 596
Flax family, described, 553
Floerkea, 573
Floras, evolution of, 163, features to be included in, 288; types of, 275
Floras of, Africa, 301, 302; Asia, 300, 301; Australia, 302; Europe, 298, 299, 300; Mexico and Central America, 296, 297; North America, 291–296; northeastern U. S., 295; Pacific islands, 303; South America, 297, 298; West Indies, 297; western U. S., 293–294; world, 288–290
Florin, R., 60, 102, 358, 360
Floristic areas of the earth, 147, 150–153
Floristic studies, 275–283; field work for, 277, 278; herbarium labels for, 248; physiographic areas, 297; procedures, 278; square degree as basis for area, 277; use of literature, 281, 282
Floristic zones, 150–153
Florist's geranium, 552
Flower, defined, 65, 370; evolutionary theories of, 66; morphology of, 64–88
Flower types, 81–85
Flowering plants, families of, 370–775
Flowering quillwort family, described, 380
Flowering quince, 544
Flowering rush family, described, 383
Flowering willow, 700
Fluviales, 375
Foeniculum, 645, **646**
Fogg, J. M., 256
Fokenia, 366, 367
Fool's parsley, 646
Forestiera, 667
Forget-me-not, 685
Forma, a unit of classification, 56
Formaldehyde, preserving freshly pressed material, 240
Formulas, of preserving solutions, 254, 255
Forsellesia, 577
Forskål, P., 25
Forsythia, 667, 668, **669**
Fortunella, 557

793

Fosberg, F. R., 55, 240, 241, 256
Foster, R. C., 580
*Fothergilla*, 537, 538
*Fouquieria*, 680
Fouquieriaceae, described, 680
Four-o'clock, 481
Four-o'clock family, described, 480
Foxglove, 697
*Fragaria*, 542, 544
Fragrant olive, 668
Frangipani, 673
Frangulaceae, 587
*Frankenia*, 607
Frankeniaceae, described, 606 607
Frankincense, 560
*Franklinia*, 602
*Frasera*, 670 671
*Fraxinus*, 667, 668, 669
Free-central placentation, 78
*Freesia*, 423, 424
Freezing flowers, 256
(*Fremontia*), 596
*Fremontodendron*, 596
*Freycinetia*, 374
Fringe tree, 668
*Fritillaria*, 413
Frizzell, D. L., 204
*Froelichia*, 479
Frog's-bit family, described, 385
Fruits, classification of, 85, 86, 87
Fuchs, L., 15
*Fuchsia*, 637, 638, 639, 640
*Fumaria*, 518
Fumariaceae, 517; described, 517, 518
Fumarioideae, 516
Fumigating herbaria, 250, 251
Fumitory family, described, 517
(*Funastrum*), 675
*Furcraea*, 420
Furtado, F. X., 396
Fusion and modification of perianth parts, 84 85

Gager, C. S., 231
Gagnepain, F., 368
*Gaillardia*, 730
Galantheae, 418
*Galanthus*, 420
*Galax*, 648, 649
*Gale*, 450, 451
(*Galeorchis*), 436
Galieae, 713
*Galinsoga*, 728
*Galium*, 712, 713
(*Galpinsia*), 638
Gamble, J. S. and Fisher, C. E. C., 301
*Garcinia*, 604
Garcinia family, described, 603
Garden heliotrope, 716
*Gardenia*, 712, 713

Garratt, G. A., 511
*Garrya*, 449
Garryaceae, alliance with Umbelliflorae, 642; described, 448–449
Garryales, described, 448
*Garuga*, 560
Gates, B., 243
*Gaultheria*, 652, 563, 564, 566
*Gaura*, 638
Gay, C., 298
*Gaylussacia*, 654, 655
*Gayoides*, 593
*Gayophytum*, 638
*Gazania*, 730
Gazetteers, list of, 325
Geissolomataceae, 624
Geitler, L., 475
*Gelsemium*, 669, 670
Genecology, 169
Genera, nomenclature of, 206
*Genera siphonogamarum*, of Dalla Torre and Harms, described, 287
Generic names, conservation of, 205
Genes, importance of, 173
Genesis and evolution, 142
*Geniostoma*, 670
Genip, 584
*Genista*, 548
*Genlisea*, 706
Genonomy, 169
Genotype, 178–179
Gentian, 671
Gentianaceae, described, 670–672
Gentianales, 673, 666, 667
Gentianineae, 667
Genus, a unit of classification, 48 50
*Geocarpon*, 483
Geography of vascular plants, 141–165
Geography, relationship to plant distributions, 147
Geologic eras, 145 147
Geologic maps, use and sources of, 268, 324, 325
Geologic periods, chart of, 146; discussion of, 145, 147
Geology and plant distributions, 142
Geomorphology, importance to taxonomy, 142 144
Geraniaceae, described, 551–552
Geraniales, described, 549
Geraniineae, 549
*Geranium*, 551, 552
Geranium, florist's, 552
Gerard, J., 15
*Gerardia*, 696
*Gerbera*, 729, 730
Gerth van Wijk, H. L., 319
Gesneriaceae, described, 703–705
*Geum*, 542, 544
*Geunsia*, 686

INDEX

Gherkin, 720
Gilbert-Carter, H., 318
*Gilia*, 678, 679
*Gillenia*, 541
Gillesieae, 413, 415
*Gilmania*, 475
Gilmour, J. S. L., 219
Ginger, wild, 473
Ginger family, described, 427
Ginger lily, 428
Ginger root, 428
*Ginkgo*, 357, 358
Ginkgoaceae, described, 357–359
Ginkgoales, 357
(*Ginkyo*), 358
Ginseng, 643
Ginseng family, described, 642
*Githopsis*, 722
*Gladiolus*, 423, 424
*Glaucium*, 515, 516
*Glaux*, 659
Gleason, H. A., 219, 398
*Gleditsia*, 547
*Gleichenia*, 347
Gleicheniaceae, described, 346–347
*Glinus*, 483
*Globba*, 428
Globe amaranth, 480
Globe artichoke, 730
Globe thistle, 731
*Globularia*, 707
Globulariaceae, described, 707
*Glochidion*, 566
Glossaries, list of, 317, 318
*Gloxinia*, 705
Glue, for mounting specimens, 245, 246
Glumiflorae, described, 386–387
*Glycine*, 548
*Glycosmis*, 557
*Glyptostrobus*, 365
*Gnaphalium*, 730
Gnetaceae, 368
Gnetales, described, 368; phyletic importance of, 104–105
*Gnetum*, 370
*Gnidia*, 626
*Godetia*, 638, 640
Godfery, M. J., 437
Goebel, K. I. E., 60
*Goethalsia*, 591
Goethe, J. W., 68, 73
Golden bells, 668
Golden seal, 498
Goldschmidt, R., 174
Gomortegaceae, 489
*Gomphocarpus*, 675
*Gomphrena*, 478, 479, 480
Gomphrenoideae, 479
Gondwana land, 143
*Gonolobus*, 675

Good, R. A., 142, 143, 153, 159, 162; theory of tolerance, 161, 162; vegetation areas of earth, 151–153
Goodeniaceae, described, 723–724
Goodspeed, T. H., 181
Goodspeed, T. H. and Bradley, M. V., 176, 178
*Goodyera*, 436
Gooseberry, 534; Ceylon, 614; Otaheite, 566
Goosefoot family, described, 477
*Gordonia*, 602
*Gormania*, 531
*Gossypium*, 593
*Gouania*, 588
Gourd family, described, 718
Goutweed, 646
*Grahamia*, 484
Graminales, 386
Gramineae, primitive characters of, 390
Granadilla, 617
Granadillo wood, 665
Grape family, described, 588
Grape, sea, 476
Grapefern family, described, 343
Grapefruit, 557
*Graptopetalum*, 531
Grass family, described, 398
Grasses, primitive characters of, 390
*Gratiola*, 695
Gray, A., 26, 42, 49, 54, 86, 267, 292, 318
Gray Herbarium card index, described, 287
Great Plains, a physiographic province, 156–157
Green coloration, preservation in liquid, 255, 256
Green, M. L., 200
Greene, E. L., 24, 170, 516
*Greenovia*, 531
Gregor, J. W., 176, 177, 178
Gregory, J. W., 144
Grew, N., 17
*Grewia*, 592
Grey, E., 321
*Greyia*, 477, 585
Greyiaceae, 534, 585
Griffen, F. J., Sherborn, C. D., and Marshall, H. S., 328
*Griselinia*, 647
Gronovius, J. F., 19
Grossulariaceae, 534
Ground pine, 338
Groundnut, 394; *see also* peanut
Grubbiaceae, 468
*Guaiacum*, 555, 556
*Guarea*, 561
Guatemala, flora of, 297
*Guattleria*, 509
Guava, 634
Guettardeae, 712

Gum, sour, 631
*Gunnera*, 640, 641
Gunneraceae, 641
Guppy, H. B., theory of differentiation, 161
*Gurania*, 719
Gustafsson, A., 52, 183, 184, 185
Guttales, 597
Guttapercha, 663
Guttiferae described, 603–604
Guttiferales, 597
*Guzmania*, 406, 407
*Gyminda*, 577
(*Gymnadenia*), 436
(*Gymnadeniopsis*), 436
*Gymnema*, 675
*Gymnocardia*, 614
*Gymnocladus*, 547
Gymnogrammeoideae, 352
Gymnospermae, described, 355–356; synopsis to orders of, 356
Gymnosperms, 370; ancestors of the angiosperms, 104; paleobotany of, 102–103; phylogeny of, 102, 103
*Gymnosporia*, 577
*Gymnosteris*, 678
Gynandreae, of Wettstein, 431
*Gynandropsis*, 520
Gynoecium, morphology of, 72–75
Gynostegium, 435
*Gypsophila*, 487, 488
*Gyranthera*, 595
*Gyrocarpus*, 514
(*Gyrostachys*), 436
Gyrostemonaceae, 477

*Habenaria*, 435, 436, 437
Haber, J. M., 565
*Haberlea*, 704, 705
*Habranthus*, 420
Haeckel, E., 389, 390
*Haemanthus*, 418, 419
Haemodoraceae, 419; described, 416–417
*Haemodorum*, 417
*Hagenbachia*, 417
Hagerup, O., 104, 718
*Halenia*, 671
(*Halerpestes*), 498
*Halesia*, 666
Hall, H. M., 159
*Halleria*, 695
Hallier, H., 36, 106, 107, 115, 136; diagram of system, 132; system of classification, 36, 130–133; *see also* note, p. 776
Hallock, F. A., 449
*Halophila*, 385, 386
Haloragaceae, 642; described, 640–641
*Haloragis*, 641
Halorhagidaceae, 640
*Haloxylon*, 477
Hamamelidaceae, described, 537–538

*Hamamelis*, 537, 538
Hansen's phytogeographic zones, 153
*Haplopappus*, 728, 732
Harms, H., 407, 539
*Harpagonella*, 685
*Harpephyllum*, 574
Harrington, H. D., 255, 256
*Harrisella*, 436
(*Hartmannia*), 638
Harvey, W. H. and Sonder, W., 301
*Hasseanthus*, 531
Hasselquist, F., 24, 25
*Hauya*, 637, 638
Hawaii, floras of, 303
Hawthorn, 544
Hayek, A., 299
Hazelnut, 459
Heat, as an insecticide, 252–253
Heath, 654
Heath family, described, 652
Heather, 654
*Hebe*, 695
*Hebenstretia*, 695
*Heberdenia*, 657
*Hechtia*, 406
*Heckeria*, 445
Hedberg, O., 476
*Hedeoma*, 690
*Hedera*, 643
Hederaceae, 642
*Hedrianthus:* see *Edrianthus*
Hedrick, U. P., 321
*Hedycarya*, 511
*Hedychium*, 428
*Hedyosmum*, 446
*Hedyotis*, 713
*Hedysarum*, 549
Hegelmaier, F., 568
Hegi, G., 299
Heilborn, O., 174
Heim, F., 452
Heimerl, A., 481
*Heimia*, 628
Heimsch, C., 98, 454, 455
Heimsch, C. and Wetmore, R. H., 454
Helenieae, 730
*Helenium*, 730
*Heleocharis:* see *Eleocharis*
*Heliamphora*, 527
Heliantheae, 730
*Helianthemum*, 608, 609
(*Helianthium*), 382
*Helianthus*, 728, 729, 730, 731
*Helichrysum*, 731
*Helicodiceros*, 400
Helicoid cyme, origin of, 61
*Heliconia*, 426
*Heliopsis*, 730
Heliotrope, 685; garden, 716
Heliotropiaceae, 685

Heliotropioideae, 685
*Heliotropium*, 684, 685
*Helipterum*, 731
Helleboreae, 498
*Helleborus*, 497, 498
*Helminthostachys*, 343
Helobiae, described, 375
*Helwingia*, 647
*Hemerocallis*, 415, 416
Hemiangiospermae, position of, 104
*Hemicarpha*, 392
*Hemitelia*, 348
*Hemizonia*, 728
Hemp, Cuban, 420; Manila, 427; Mauritian, 420
Hemsley, W. B., 296
Henbane, 694
Henderson, M. W., 652
Henequin, 420
Henna tree, 628
*Henoonia*, 694
*Henriquezia*, 712
*Hepatica*, 498
*Heracleum*, 646
Herbaceae, of Hutchinson, 59
Herbaceous versus woody habit, primitiveness of, 106–107
Herbalists, 15
Herbaria, 228–231; abbreviations of, 273; important in world, 231, 232; lost by fire, 260; significance of, 230–231
Herbarium, arrangement of material in, 259, 260
Herbarium beetles, control of, 248–253
Herbarium cases, organization of specimens in, 258, 260; types of, 257–258
Herbarium collections, essential field notes to be made, 269
Herbarium labels, maps printed on, 247, 273
Herbarium specimens, annotation of, 272; borrowing, 270, 271; labels for, 246, 247, 273
Herbarium studies, procedures, 270–273
Herbarium techniques and methods, 243–260
Herbarium techniques, adhesives for mounting specimens, 246; color retention of materials in fluid, 255; DDT as an insecticide, 252; deep-freezing of materials, 256; dissected preparations, 254–255, 257; fumigation, 250, 251; heat as an insecticide, 252, 253; heat in drying specimens, 242–243; housing bulky materials, 253–254; insect repellents, 253; insecticides, 249–253; labels for sheets, 246–248; liquid plastics, 246; liquid preservation of materials, 253–256; mounting specimens, 243–246; oxyquinoline sulfate as a preservative, 255; quick-freezing of material,
256; poisoning specimens, 251, 252; specimen preservation, 248–257
(*Herbertia*), 424
*Hermannia*, 596
*Hermidium*, 481
*Hernandia*, 514
Hernandiaceae, described, 513–514
*Herniaria*, 488
*Hesperelaea*, 667
*Hesperis*, 521, 522
*Hesperocnide*, 465
(*Hesperolinum*), 553
(*Hesperomecon*), 516
*Heteranthera*, 411
Heterantherae, 411
*Heterocentron*, 636
*Heterocodon*, 721, 722
Heteropyxidaceae, 624
Heterostylaceae, 380
(*Heterostylus*), 381
*Heuchera*, 533, 534
*Hevea*, 566
*Hewardia*, 424
*Hexalectris*, 436
*Hexastylis*, 473
*Hibbertia*, 598
*Hibiscus*, 593, 594
Hickory, 455
(*Hicoria*), 454
*Hieracium*, 728, 732
*Hierochloë*, 389
Hill, A. F., 321
*Hillebrandia*, 621
Himantandraceae, 489, 504
*Himatanthus*, 673
*Hippeastrum*, 419
*Hippobroma*, 722
Hippocastanaceae, described, 582
Hippocrataceae, described, 578–579
*Hippocratea*, 578
*Hippophaë*, 626, 627
Hippuridaceae, described, 641–642
Hippuridineae, 624
*Hippuris*, 641, 642
*Hirtella*, 541
History of classification, 13–41
Hitchcock, A. S., 200, 219, 328
Hjelmquist, H., 443, 446, 447, 448, 449, 451, 452, 453, 455, 456, 457, 458, 460
Hoarhound, 690
Hodge, W. H., preservation technique, 241
*Hoffmanseggia*, 547
Hofmeister, W., 30
Holland, J. H., 321
Holland, T. H., 143
*Hollisteria*, 475
Holly family, described, 575
Hollyhock, 594
*Holmskoldia*, 688
*Holocantha*, 558, 559

*Holocarpha*, 728
*Holodiscus*, 543
*Holosteum*, 488
Holotype, defined, 204
*Homalomena*, 398
Homonym Rule, 213
Honesty plant, 521
Honey bell, 596
Honeysuckle, 715
Honeysuckle family, described, 713
Honeywort, 685
Hooker, J. D., 31, 32, 33, 41, 49, 301; contributions to *Genera Plantarum*, 116, 117, 118
Hooker, W. J., 31, 42
Hop tree, 557
Hoplestigmataceae, 662
Hops, 464
Hordeae, 390
*Hordeum*, 391
*Horkelia*, 544
*Horkeliella*, 544
(*Hormidium*), 436
Hornwort family, described, 491
Horse-chestnut family, described, 582
Horse-radish, 521
Horse-radish tree, 526
Horsetail family, described, 336, 337
*Horsfordia*, 593
Horticultural plants, nomenclature of, 206
*Hosackia*, 547
*Hottonia*, 658
Hounds' tongue, 685
*Houstonia*, 713
*Houttuynia*, 444
*Hovenia*, 588
Howard, R. A., 252
*Howellia*, 721, 722
*Hoya*, 675
Hubbard, C. E., 390
Huckleberry, 654
*Hudsonia*, 609
*Huernia*, 675
(*Hugelia*), 678
*Hugonia*, 553
Hultén, E., 267, 292, 293
*Humulus*, 462, 464
*Hunnemannia*, 515, 517
Hutchinson, J., 34, 36–37, 38, 39, 58–59, 67, 94, 95, 96, 104, 106, 107, 114, 115, 290, 299, 333; diagram of system, 139; phyletic views, 59; system of classification, 36, 37, 133–139; *see also* note, p. 777
Hutchinson, J. and Dalziel, J. M., 302
Huxley, J., 173, 174
Hyacinth, 416
*Hybanthus*, 612
Hybrid, definition of, 174

Hybrids, fertility of, 174–175; types of, 174–176
Hydnoraceae, 472
*Hydrangea*, 533, 534
Hydrangeaceae, 534
Hydrangeoideae, 533
*Hydrastis*, 498
*Hydrastylus*, 424
Hydrocaryaceae, described, 636–637
*Hydrocera*, 586
Hydrocharitaceae, described, 385–386
*Hydrocleis*, 384
*Hydrocotyle*, 646
*Hydrolea*, 683
*Hydromystia*, 386
Hydrophyllaceae, described, 682–684
*Hydrophyllum*, 682, 683
Hydropteridales, 344; described, 352
Hydropteridineae, 352
Hydrostachyaceae, 446
Hydrostachyales, 446
*Hydrothrix*, 410
Hydroxyquinoline sulfate, as a preservative, 241, 255
Hyenales, 337
*Hygrophila*, 708
Hylander, N., 321, 327
*Hymenocallis*, 418, 419, 420
Hymenophyllaceae, described, 347, 348
*Hymenophyllum*, 348
*Hyobanche*, 695
*Hyoscyamus*, 694
Hypanthium, morphology of, 80, 81
Hypecoideae, 516, 518
*Hypecoum*, 516, 518
Hypericaceae, described, 604–605
*Hypericum*, 604, 605
*Hypolate*, 583
Hypoxidaceae, 419
Hypoxideae, 419
Hypoxidoideae, 419
*Hypoxis*, 419, 420
*Hyptis*, 688
Hyssop, 690
*Hyssopus*, 690

*Iberillea*, 719
*Iberis*, 520, 521
*Ibicella*, 702
(*Ibidium*), 436
Icacinaceae, described, 580
Icacinineae, 569
*Icacorea*, 658
Iceland poppy, 517
Identification of plants, 223–232
*Idesia*, 614
*Idria*, 680
Ilama, 509
*Ilex*, 576
*Iliamna*, 593

Illecebraceae (included with Caryophyllaceae), 488
Illegitimate names, rules pertaining to, 213
Illiciaceae, 504; described, 505-506
*Illicium*, 505
*Illigera*, 514
Illustrations, photographs, 272-273; types of, 272
*Impatiens*, 586
*Incarvillea*, 698, 700
*Index filicum*, description of, 287
*Index Kewensis*, notes concerning, 286, 287
*Index Londinensis*, description of, 288
Indexes of, maps, 325; plant names, 285-288
Indian almond, 633
Indian fig, 623
Indian pink, 670
Inferior ovary, defined, 79; derivation of, 79-81
Inflorescence, centrifugal, 61; centripetal, 61; defined, 59; diagrams of types, 63; evolution of, 59; morphology of types, 59-64; role of solitary flower, 60
Infrageneric categories, measured by biosystematics, 181, 182
Infraspecific categories, 53, 54
Insect control in herbaria, 248-253
Insecticides, use in herbarium, 249, 253
Interior highlands of North America, 156
International Rules of botanical nomenclature, beginnings of, 198-200; provisions of, 201, 202; *see also* Nomenclature
Interrupted fern, 345
Introduced plant, defined, 279
*Inula*, 728, 730
Inuleae, 730
*Ionopsis*, 436
(*Ionoxalis*), 550
Ipecac, 713
*Ipomoea*, 676, 677, 678
*Iresine*, 479, 480
Iridaceae, described, 422-425
Iridales, of Hutchinson, 424
Iridineae, 411
Iridoideae, 424
*Iris*, 422, 423
Iris family, described, 422
Irregular flower, characteristics of, 84
*Isatis*, 522
*Ischnosiphon*, 430
Isoetaceae, described, 341-342
Isoetales, 341
*Isoetes*, 341, 342
Isoetinae, described, 340
(*Isoloma*), 705
*Isomeris*, 519
Isophysideae, 424
*Isophysis*, 424
*isopyrum*, 498

*Isotria*, 436
Isotype, defined, 204
Iteaceae, 534
*Iva*, 730
*Ivesia*, 544
Ivy, Boston, 589; Engish, 643; Kenilworth, 697
Ivy gourd, 720
*Ixia*, 424
Ixiaceae, 422
Ixieae, 424
Ixioideae, 424
*Ixora*, 713

Jaboticaba, 634
*Jacaranda*, 700
Jackfruit, 464
Jack-in-the-pulpit, 400
Jackson, B. D., 20, 22, 224, 286, 306, 318
Jackson, G., 80-81, 644
*Jacobinia*, 708, 709
Jacobsen, H., 483
*Jacquemontia*, 677
*Jacquinia*, 657
Jamaica, flora of, 297
Janchen, E., 609
*Janusia*, 562
Japanese persimmon, 665
*Jasione*, 721, 722
Jasmine, 668; Chilean, 673
Jasminoideae, 667
*Jasminum*, 667, 668
*Jatropha*, 566
*Jeffersonia*, 501
Jenkin, T. J., 181, 182
Jerusalem thorn, 588
Jessamine, Carolina, 670; orange, 557
Jetbead, 544
Joewood family, described, 657
Johansen, D. A., 88
Johnston, I. M., 256, 685
Jones, G. N. and Meadows, E., 231, 260
Jones, M., 170
Jonker, F. P., 431, 432
Jordon, A., 179
Jorgensen, C. A., 568
Joshi, A. C., 476, 510
Juglandaceae, described, 453-455
Juglandales, described, 452-453
*Juglans*, 453, 454, 455
Jujube, 588
Julianales, 455
Juncaceae, described, 412-413; distinct from Glumiflorae, 386
Juncaginaceae, 379
Juncales, 386
Juncineae, 411
*Juncus*, 412
Juniperoideae, 367
*Juniperus*, 366, 367

*Jussiaea*, 637
Jussieu, Adrian de, 29
Jussieu, Antoine de, 28–29, 32, 42, 116, 498
Jussieu, B. de, 18, 20
Just, T., 43, 87, 103, 188, 355, 637
*Justicia*, 708, 709

*Kadsura*, 506, 507
Kaffir plum, 574
*Kalanchoë*, 530, 531
Kalanchoideae, 532
*Kaliphora*, 647
*Kallstroemia*, 556
Kalm, P., 24
*Kalmia*, 653, 654
*Kalmiopsis*, 655
*Kalopanax*, 643
Kanjilal, U. N., 301
Kearney, T. H. and Peebles, R. H., 293, 577, 623
Kei apple, 614
*Kelloggia*, 713
Kelsey, H. F., 193
Kenilworth ivy, 697
Kerosphaereae, taxonomy of, 436
*Kerria*, 544
*Keteeleria*, 364
Kew Rule, explained, 197
Key to, classes of Pteridophyta, 335; orders of Gymnospermae, 356; orders of dicots, 439–442; orders of monocots, 372; subfamilies of Rosaceae, 543, of Leguminosae, 545
Keys, construction of, 227, 228; testing of, 271; to woody plants, 57; types of, 225, 226; use in plant identification, 225
*Khaya*, 561
*Kibatalia*, 673
*Kigelia*, 699, 700
Kirk, J. W. C., 321
Kirk, T., 303
Kirouc, C., *see* Marie-Victorin, Fr.
*Kissenia*, 619
*Kitaibelia*, 592, 594
Klett, W., 670
*Kmeria*, 503, 504
(*Kneiffia*), 638
*Kniphofia*, 413
Kunth, C., 376, 421
Koch, M., 67, 726, 730
*Kochia*, 478
*Koeberlinia*, 577
Koeberliniaceae, 577
Koeberlinioideae, 577
*Koelreuteria*, 454, 583
*Koenigia*, 476
*Kohleria*, 704, 705
Kohlrabi, 521
*Kolkwitzia*, 713, 715

Komarov, V. L., 299
*Krameria*, 547
Krameriaceae, 547
Kramerieae, 547
Kranzlin, K., 429
Krause, K., 413, 415, 419
Kreyer, G. K., 182
*Krigia*, 732
Krock, T. O. B. N., 327
*Krugiodendron*, 588
Kumquat, 557
Kuntze, O., 531
*Kuntzia*, 544

Labels for herbarium specimens, 246–248
Labiatae, described, 688–692
*Lachnanthes*, 417
*Lachnocaulon*, 405
(*Lachnostoma*), 675
*Lacmellea*, 672
Lactoridaceae, 489
*Lactuca*, 730
*Laelia*, 437
*Lafoensia*, 628
*Lagenaria*, 720
Lagerberg, T. and Holmboe, J., 300
*Lagerstroemia*, 628
Laibach, F., 180
Lam, H. J., 53, 98, 105; Stachyosporous theory, 105, 106
Lamarck, J. B. A. P. M., 27, 28
Lamarckism, defined, 172
Lambrecht, H., 51
Lamellate placentation, 78
Lamiaceae, 688
Lamiales, 680, 687
*Lamium*, 689
*Langloisia*, 678
(*Languas*), 427
Lanjouw, J., 201
*Lantana*, 687, 688
*Laphamia*, 728
*Lapithea*, 671
*Laportea*, 464, 465
*Lapsana*, 728
Lardizabalaceae, described, 500
*Larix*, 364, 365
*Larrea*, 556
Lasègue, A., 328
Lasioideae, 398
*Lastarriaca*, 475
*Lasthenia*, 728, 732
*Lathyrus*, 546, 548
Lattice-leaf plant, 379
Laubengayer, R. A., 475, 476
Lauraceae, described, 512–513
Laurales, of Hutchinson, 513
Laurasia, 143
Laurel, mountain, 654
*Laurelia*, 511

# INDEX

Laurentian uplands, 156
Lauroideae, 573
*Laurus*, 512, 513
*Lavandula*, 690, 691
*Lavatera*, 593
(*Lavauxia*), 638
Lavender, 690
Lavialle, P., 730
Lawalree, A., 401
Lawrence, J. R., 685
*Lawsonia*, 627, 628
*Layia*, 730
Leadwort family, described, 661
*Leandra*, 636
Leatherleaf, 654
Leatherwood, 575, 626
*Leavenworthia*, 521
Leaves, origins of, 65
*Lechea*, 609
L'Ecluse, C., 15
Lecythidaceae, 624
*Leea*, 588
*Legenere*, 721, 722
Leguminosae, described, 545–549; synopsis to subfamilies of, 545
*Leiphaimos*, 670, 671
*Leitneria*, 452
Leitneriaceae, described, 451–452
Leitneriales, described, 451
Lemesle, R., 105, 541
*Lemna*, 400, 401
Lemnaceae, described, 400–402
Lemon, 557
Lemon leaf, 654
*Lennoa*, 681, 682
Lennoaceae, 650; described, 681–682
Lennoineae, 676
*Lenophyllum*, 531
*Lens*, 548
Lentibulariaceae, described, 705–706
Lentil, 548
*Leonotis*, 690
*Leontice*, 502
*Leontopodium*, 731
*Lepidium*, 521
Lepidodendrids, era of, 145–146
(*Leptamnium*), 703
*Leptarrhena*, 533
*Leptodactylon*, 678
*Leptopteris*, 345
*Leptospermum*, 634
Leptosporangiatae, described, 344
Leptosporangiate fern, defined, 342
*Lespedeza*, 548
*Lesquerella*, 522
Lettuce, 730
*Leucadendron*, 467
Leucastereae, 481
*Leucocarpus*, 695
Leucojaceae, 417

*Leucojum*, 420
*Leucopogon*, 656
*Leucothoë*, 654
*Levisticum*, 645, 646
*Lewisia*, 484, 485
*Leycesteria*, 713, 715
L'Heritier de Brutelle, C. L., 29
*Liatris*, 730
*Libertia*, 422
*Libocedrus*, 366, 367
*Licaria*, 513
Lid, J., 300
*Lightfootia*, 722
Lightwood, 536
Ligneous versus herbaceous characters, 58, 59
Lignosae, of Hutchinson, 59
Lignum vitae, 556
*Ligustrum*, 667, 668
Lilac, 668
*Lilaea*, 380, 381
Lilaeaceae, described, 380–381
Liliales, 411
Liliaceae, described, 413–416; of Hutchinson, 415
Liliiflorae, described, 411–412
Liliineae, 411
*Lilium*, 413, 414
Lilly-pilly, 634
Lily family, described, 413
Lime, 557; Spanish, 584
Limnanthaceae, described, 572–573
*Limnanthes*, 573
Limnanthineae, 569
*Limnobium*, 385, 386
*Limnocharis*, 384
Limnochariteae, 383
*Limnophyton*, 382
(*Limodorum*), 436
*Limonium*, 661
Linaceae, described, 553–554
(*Linanthastrum*), 678
*Linanthus*, 678, 679
*Linaria*, 695
Linden family, described, 591
*Lindera*, 513
Lindley, J., 42, 116, 529; system of classification, 31
Lindman, C. A. M., 300
Lindsey, A. A., 671
Link, H., 54, 59
*Linnaea*, 713, 715
Linnaeus, C., 13, 14, 39, 42, 48, 49, 54, 59, 178, 194, 195; binomial nomenclature, 23; biographical account, 18–21; catalogue of herbarium, 328, 329; herbarium, 25, 26; natural system of classification, 24; principles of nomenclature, 195; *Species plantarum*, 23; system of classification, 22, 23, 24

Linnaeus, C., *fil.*, 25
Linseed oil, 554
*Linum*, 553, 554
Lion's-ear, 690
*Liparis*, 436
*Lippia*, 687
*Liquidambar*, 537, 538
Liquids, as specimen preservatives, 253–256
*Liriodendron*, 504
Lissocarpaceae, 662
*Listera*, 436, 437
*Litchi*, 583
Literature, 284–332; abbreviations of periodical titles, 308–317; bibliographies of, 305–306; biographical references, 326–327; catalogues of botanical libraries, 306, 307; color charts, 329–330; cultivated plants, 316–317, 319–322; dictionaries, 318–319, economic plants, 316–317, 319–322; floras of North America, 291; gazetteers, 325; glossaries, 317–318; importance in floristic studies, 281, 282; maps, 323–326; Old World periodicals, 313–316; periodicals, 308–317; publication dates, 327–328; review serials, 307; type-specimen sources, 328, 329; *Union list of serials*, 309; use of in identification, 223, 224
*Lithocarpus*, 460
*Lithospermum*, 685
*Litorella*, 711
*Litsea*, 513
Little, E. L., 219
Lizard's-tail family, described, 444
Lloyd, F. E., 706
Loasaceae, described, 618–619
Loasales, 597
Loasineae, 597
L'Obel, M., 15
*Lobelia*, 721, 722
Lobeliaceae, 721, 722
Lobelioideae, 722
Loblolly bay, 602
*Lobularia*, 521
*Loeflingia*, 488
*Loeselia*, 678
Loesener, T., 428
Loganberry, 544
*Logania*, 669
Loganiaceae, described, 669–670
Lomandreae, 413, 415
Longan, 583
*Lonicera*, 713, 714, 715
Loosestrife, 628
Loosestrife family, described, 627
*Lopezia*, 637, 638
*Lophiola*, 417
*Lophotocarpus*, 382
Lopseed family, described, 710

Loquat, 544
Loranthaceae, described, 471–472
Loranthineae, 468
Loranthoideae, 472
*Loranthus*, 472
*Loropetalum*, 538
Lotoideae, 545, 547
Lotsy, P., 107, 390
Lotsy, P. and Godjin, W. A., 174
*Lotus*, 546, 548
Lovage, 646
Lowiaceae, 426
Lowioideae, 426
(*Lucuma*), 663
*Ludwigia*, 637
*Luetkea*, 544
*Luffa*, 720
*Lunaria*, 521
Lundell, C. E., 243
Lungwort, 685
Lupine, 548
*Lupinus*, 546, 547, 548
*Luxembergia*, 601
*Luzula*, 412
Luzuriagoideae, 415
*Lyallia*, 487
Lychee, 583
*Lychnis*, 487, 488
*Lycianthes*, 694
*Lycium*, 694
*Lycopersicon*, 693, 694
Lycopodiaceae, described, 337
Lycopodiales, 337
Lycopodiinae, described, 337
*Lycopodium*, 337, 338
Lycopsida, 337, 355; era of, 145–146
*Lycopsis*, 684
*Lycoris*, 419, 420
Lycosphens, 335
*Lygodium*, 346
*Lyonothamnus*, 536
Lyons, A. B., 319
(*Lysias*), 436
*Lysichiton*, 398
*Lysimachia*, 659
*Lysipomia*, 722
Lythraceae, described, 627–628
*Lythrum*, 627

*Macadamia*, 467
*Macaranga*, 566
Macassar ebony, 665
Macbride, J. F., 298
MacDaniels, L. H., 80, 243
Macdougal, T. A., 256
Mace, 510
*Macleaya*, 515, 516, 517
Macloskie, G., 298
*Maclura*, 462
*Macradenia*, 436

INDEX

(*Macrocarpium*), 648
*Macrosiphonia*, 673
*Macrozamia*, 357
McVaugh, R., 722
Madder, 713
Madder family, described, 712
Madeira vine, 486
*Madia*, 728, 730
*Maesa*, 658
Magnol, P., 28
*Magnolia*, 503, 504
Magnolia family, described, 503–505
Magnoliaceae, described, 503, 504; distinguished from Cercidiphylaceae, 496
Magnoliineae, 489
*Mahernia*, 596
Maheshwari, P., 87, 467, 623, 638, 681
Mahogany, African, 561
Mahogany family, described, 560
*Mahonia*, 500, 501
Maia, L. D'O., 415
*Maianthemum*, 414
Maillefer, F., 182, 243
Maiwald, V., 327
Malabar, 486
Malaceae, 543
*Malachodendron*, 602
*Malachra*, 593
*Malaxis*, 436
Malaysia, flora of, 303
Malesherbiaceae, 597
Mallow family, described, 592
*Malope*, 592, 593
*Malosma*, 574
Malpighi, N., 17
*Malpighia*, 562
Malpighiaceae, described, 562–563
Malpigiineae, 549
Maltese-cross, 488
*Malus*, 542, 544
*Malva*, 593, 594
Malvaceae, described, 592–595
Malvales, 590
*Malvastrum*, 593
*Malvaviscus*, 592, 594
Malvineae, 590
*Mammea*, 603
Mammee apple, 604
*Mandevilla*, 673
*Manettia*, 713
*Manfreda*, 419, 420
*Mangifera*, 574
Mango, 574
Mangosteen, 604
Mangrove family, described, 629, 630
*Manihot*, 566
Manila hemp, 427
(*Manilkara*), 663
Manning, W. E., 453, 454, 455
Manton, I., 181

Manuals, of cultivated plants, 319–322; *see also* Floras
Maple family, described, 581
Maps, air photographs, 325; geological, 324, 325; indexes to, 325; nautical charts, 324; on herbarium labels, 247, 273; soil, 325; sources of outline maps, 323; topographic, 323, 324; types of 323–326
*Marah*, 719
*Maranta*, 430
Marantaceae, described, 429–430
*Marattia*, 344
Marattiaceae, described, 344
Marattiales, described, 344
Marcgraviaceae, 597
*Margaranthus*, 694
Marginal placentation, 75
*Marianthus*, 535
Mariceae, 424
Marie-Victorin, Fr., 293
Marijuana, 464
Marjoram, pot, 690
Markgraf, F., 105, 368
(*Marlea*), 632
Marmalade plum, 663
Marrow, 720
*Marrubium*, 690
*Marsdenia*, 675
Marshall, H. S., 328
*Marshallia*, 728
*Marsilea*, 353
Marsileaceae, described, 352–353
Martin, A. C., 87, 98, 221, 623
Martius, K. F. P. von, 298
*Martynia*, 701
Martyniaceae, described, 701–702
Mason, H. L., 159, 164, 401
Mass collections, 269–270
Masters, M. T., 86
Mastic tree, 574
*Mastichodendron*, 663
*Mastixia*, 647
*Matelea*, 675
*Matricaria*, 728, 730
*Matthiola*, 521
*Maurandia*, 695
Maurilaun, K. von, 179
Mauritian hemp, 420
Mauritzon, J., 532
(*Maximilianea*), 610
*Maximowiczia*, 719
*Mayaca*, 402
Mayacaceae, described, 402–403
Maydeae, 387, 390
Mayflower, 654
*Maytenus*, 577
Meadowfoam, 573
*Meconella*, 516
*Meconopsis*, 516, 517
*Medinella*, 636

Medlar, 544
Medusagynaceae, 597
*Megacarpaea*, 520
Megasporophyll, 74
*Melaleuca*, 634
*Melampyrum*, 695, 698
*Melanophylla*, 647
Melastomaceae, described, 635-636
Melastomataceae, 635
*Melia*, 561
Meliaceae, described, 560-561
Meliales, 561
Melianthaceae, described, 585
Melianthineae, 569
*Melianthus*, 585
*Melicocca*, 584
*Melilotus*, 548
*Meliosma*, 584
(*Mellichampia*), 675
*Melochia*, 596
*Melothria*, 719
Melothrieae, 719
*Memecylon*, 636
Mendelism, 173
Mendle, G., 8, 173
Menispermaceae, described, 502-503
*Menispermum*, 503
*Menodora*, 667, 669
*Mentha*, 689, 690
*Mentzelia*, 618, 619
Menyanthaceae, 671
*Menyanthes*, 670, 671
Menyanthoideae, 670
(*Meratia*), 508
*Merciera*, 722
(*Merckia*), 487
Merrill, E. D., 245, 250, 253, 257, 303
Merrill, E. D. and Walker, E. H., 224, 306
*Mertensia*, 685
(*Mesadenus*), 436
Mesembryaceae, 482
*Mesembryanthemum*, 483
Mesembryanthemum family, described, 482
*Mespilus*, 544
*Metasequoia*, 365, 366
(*Metastelma*), 675
Metcalfe, C. R. and Chalk, L., 98
*Metopium*, 574
Mexican buckeye, 584
Mexico, floras of, 296
Meyer, A., 53
Mez, K. C., 37, 100, 623
Mezcal, 420
Mezereum family, described, 625
*Michauxia*, 721, 722
*Michelia*, 504
*Miconia*, 636
*Microcala*, 671
*Microcycas*, 357
(*Micropiper*), 445

Microspecies, 185
Microspermae, 411; described, 431
Mignonette tree, 628
Migrations, floral, 163
*Mikania*, 730
Miki, S., 376, 377, 385
Milk plant, Ceylon, 675
Milkweed family, described, 673
Milkwort family, described, 563
*Mimosa*, 547
Mimosaceae, 545
Mimosoideae, 545
*Mimulus*, 695, 696, 697
*Mimusops*, 663
Mint, 690
Mint family described, 688
(*Minuartia*), 487
Mirabileae, 481
*Mirabilis*, 480, 481
Missa, H., 25
Mistletoe, 472
Mistletoe family, described, 471
*Mitchella*, 712, 713
*Mitreola*, 669
Mock orange, 534
Modern taxonomy, objectives of, 171-172
(*Moehringia*), 487
Molasses, 391
Moldenke, H. N., 687
Molisch, serological contributions, 100
*Mollia*, 591
*Mollinedria*, 511
*Mollugo*, 483
*Molucella*, 689
(*Moluchia*), 596
Mombin, 574
*Momordica*, 718, 719, 720
Monandrae, taxonomy of, 435, 436
*Monanthes*, 531
*Monarda*, 689, 690, 691
*Monardella*, 690
*Moneses*, 651
Monimiaceae, described, 510-512
Monkey flower, 697
Monkshood, 499
*Monnina*, 564
Monochasium, in inflorescence phylogeny, 62
*Monochoria*, 411
Monocots versus dicots, primitiveness of, 107
Monocotyledoneae, described, 371-372; synopsis to orders of, 372
*Monodora*, 509
Monoecism, characteristics of, 83
Monograph, defined, 304
Monographs and revisions, bibliographies of, 305, 306; preparation of, 263-274
*Monolepis*, 478
*Monolopia*, 732

INDEX

*Monopyle*, 705
*Monotropa*, 651; preservation of in liquid, 255, 256
Monotropaceae, 652
Monotropoideae, 651
*Monotropsis*, 651
*Monsonia*, 551
*Monstera*, 400
Monsteroideae, 398
*Montia*, 484, 485
Moonflower, 677
Moonseed family, described, 502
Moore, C. and Betche, E., 303
Moore, H. E., Jr., 255
Moore, R. J., 670
Moraceae, described, 462–464
*Moraea*, 424
Moran, R., 49
*Moricanda*, 521
*Morina*, 717
*Morinda*, 712
*Moringa*, 526
Moringaceae, described, 525–527
Morning-glory family, described, 676
Moroideae, 463
Morphine, source of, 517
Morphological criteria in taxonomy, 51–88
Morphology, importance to phylogeny, 98–99
*Mortonia*, 577
*Morus*, 462, 463, 464
Moseley, M. F. Jr., 443
(*Mosiera*), 634
Moule, A. C., 358
Mountain ash, 544
Mountain laurel, 654
Mountain-rose vine, 476
Mounting specimens, 243–246
Mouse-ear chickweed, 488
Muenscher, W. C., 321
Muhlenberg, H. L., 26
Mulberry, 464
Mulberry family, described, 462–464
Müller, E. G. O., 718
*Muntingia*, 591
Munz, P. A., 293, 326
*Murdannia*, 408, 409
*Murraya*, 557
*Musa*, 426
Musaceae, described, 425–427
Muschler, R. 302
Muskmelon, 720
Musoideae, 426
Mustard, 521
Mustard family, described, 520
Mutations, 174
Mutiseae, 730
*Mutisia*, 730
*Mydocarpus*, 646

805

Myoporaceae, described, 709–710
Myoporineae, 676
*Myoporum*, 709, 710
*Myosorus*, 498
*Myosotis*, 685
*Myrica*, 450, 451
Myricaceae, described, 450–451
Myricales, described, 450
*Myricaria*, 608, 634
*Myriophyllum*, 641
*Myristica*, 510
Myristicaceae, described, 509–510
Myrothamnaceae, 530
Myrrh, 560
*Myrsina*, 658
Myrsinaceae, described, 657–658
Myrsinales, 658
Myrtaceae, described, 633–635
Myrtales, 624, 625
Myrtiflorae, 624–625
Myrtineae, 624
Myrtle, 634
Myrtle family, described, 633
*Myrtus*, 634
Myzodendraceae, 468

Nageli, C. W., 60
Najadaceae, 377; described, 378
Najadales, 375, 376
*Najas*, 378
Nakai, T., 447
*Nama*, 682, 683
*Nandina*, 500
*Napeae*, 592, 593
Napthalene flakes, as insect repellents, 253
Narcisseae, 419
*Narcissus*, 418, 419, 420
Narthecieae, 415
Nast, C. G., 454, 504
Nast, C. G. and Bailey, I. W., 494, 495
Nasturtium, 553
*Nasturtium*, 521; see also *Tropaeolum*
Nasturtium family, described, 552
Natal orange, 670
Natal plum, 673
Natural classification, defined, 93
Natural selection, role in evolution, 172–173
Natural systems of classification, 26–33
Naturalized plant, defined, 279
Nautical charts, sources of, 324
Naval stores, 364
*Navarretia*, 678
*Navia*, 407
Navioideae, 407
Neal, M. C., 322
*Nectandra*, 512, 513
Nectarine, 544
*Nectaropetalum*, 554
Nelumbaceae, 490

(*Nelumbium*), 491
*Nelumbo*, 489, 490, 491
Nelumboideae, 489, 490
*Nemacaulis*, 475
*Nemacladus*, 722
*Nemastylis*, 424
*Nematanthera*, 445
*Nematostylis*, 712
*Nemopanthus*, 576
*Nemostylis*, 424
*Neoglaziovia*, 407
*Neoleuderitzia*, 555
*Neomarica*, 424
(*Neottia*), 436
Neotype, defined, 204
Nepenthaceae, described, 528
*Nepeta*, 689, 691
*Nephrolepis*, 351
*Nerine*, 420
*Nerium*, 673
*Nertera*, 713
*Nestronia*, 470
Nettle family, described, 464
Neuradoideae, 543
Neutral flowers, defined, 83–84
*Neviusia*, 544
New World periodicals, list of, 310–313
New Zealand, floras of, 302, 303
New Zealand spinach, 484
*Newberrya*, 651
Nicandreae, 694
Nicholson, G., 322
*Nicotiana*, 693, 694
*Nidularium*, 407
*Nierembergia*, 693, 694
Nieuwland, J. A. and Slavin, A. D., 256
*Nigella*, 496, 497
Nightshade family, described, 693, 694
Ninebark, 544
Nitrariaceae, 555
*Nitrophila*, 477, 478
*Nivenia*, 422
*Nolana*, 692
Nolanaceae, described, 692
Nolineae, 419
*Noltea*, 588
Nomenclature, 192–222; American code, 199; Amsterdam Congress, 200; apomicts, 201; author citation, 209, 210; Britton's influence, 218, 219; Brussels Congress, 199; Cambridge Congress, 199, 200; capitalization of specific epithets, 202, 216; choice of names when combining two taxa, 212; choice of names when rank is changed, 212; codes of, 196–201; common names inadequate, 193, 194; conditions of valid publication, 208; conservation of species names, 200, 202, 206; conservation of names of taxa in higher categories, 205; cultivated plants, 206; dates for starting points, 205; dates of publication, 207; DeCandolle's contributions, 195–197; decapitalization of specific epithets, 202, 216, 219; designation of typical element, 202; distinguished from identification, 4; effective publication, 206–207; English names inadequate, 193; epithet defined. 208; family names, conservation of, 205; generic names conserved, 205; historical account of, 194–196; holotype defined, 204; homonym rule, 213; horticultural plants, 206; illegitimate names, 213; independent from zoological nomenclature, 202; infraspecific units, 211–212; isotype defined, 204; Kew Rule, 196, 196; kinds of types, 204, 205; Latin description required, 207; Latin language the basis, 192, 193; Linnaeus' Principles, 195; neotype defined, 204; *nomen rejiciendum*, 205; *nomina ambigua*, 214; *nomina confusa*, 215; *nomina dubia* rejected, 202; *nomina generica conservanda*, history of, 198–199; *nomina specifica conservanda* rejected, 200, 202; nothomorphs, 201; of cultivated plants, 206; of remodelled or divided taxa, 210; of taxa of changed rank, 212, 213; of united taxa, 212; original spelling retained, 216; orthographic errors, 216; paratype defined, 204; Paris Code, 196; Paris Convention, 198; principle of priority, 203; publication requirements, 208; purpose of, 202, 203; rejection of names, 213; retention of names on remodeling, 210, 211; Rochester Code, 197; sequence of classification units, 217; species names, 206. 211; Steudel's Nomenclator, 195, 196; Stockholm Congress, 201; superfluous names illegitimate, 213; syntype defined, 204; tautonomy examples, 215; tautonyms illegitimate, 215–216; taxon, adoption of the term, 201; the type method, 203–205; topotype defined, 205; type defined, 203; Type-basis Code, 199; units of classification, 217; valid publication, 208; Vienna Code, 198, 199
*Nomina ambigua*, 214
*Nomina confusa*, 215
*Nomina generica conservanda*, source of lists of, 205
*Nomina specifica conservanda*, rejected, 200, 202
*Nopalea*, 623
Norris, T., 514, 518
North American flora, described, 291, 292
Northeastern United States, floras of, 295
Northwestern United States, floras of, 293, 294

INDEX    807

*Nothofagus*, 459, 460
*Notholaena*, 350
Nothomorphs, nomenclature of, 201
*Nuphar*, 490
Nutmeg, 510
Nutmeg family, described, 509
Nuttall, T., 26
*Nuxia*, 669
*Nuytsia*, 471
Nyctaginaceae, described, 480-481
*Nyctanthes*, 667
*Nymphaea*, 489, 490
Nymphaeaceae, described, 489-491
Nymphaeineae, 489
Nymphaeoideae, 489, 490
*Nymphoides*, 671
*Nyssa*, 631
Nyssaceae, derivatives of Cornaceae, 648; described, 630-631

Oak, 460
Oats, 391
*Obolaria*, 671
Ocean basins, importance of, 147-149
*Ochna*, 601
Ochnaceae, described, 600-601
*Ochradenus*, 524
*Ocimum*, 690
Ocotillo, 860
*Octea*, 513
Octoknemataceae, 468
*Octomeles*, 620
*Odontoglossum*, 437
*Odontospermum*, 730
Odontostominae, 415
*Odontostomum*, 415
*Oenothera*, 638, 639, 640
Oenotheraceae, 637
Oil of, Bay Rum, 634; Ben, 526; betula, 459; cedar, 367; wormwood, 478
Okra, 594
Olacaceae, described, 469
Olax family, described, 469
Old World periodicals, list of, 313-316
(*Oldenlandia*), 713
Oldenlandieae, 713
*Olea*, 667, 668
Oleaceae, described, 667-669
Oleander, 673; yellow, 673
Oleaster, 627
Oleaster family, described, 626
Oleineae, 667
Oleoideae, 667
*Oligomeris*, 525
Olinaceae, 624
Olive, 668; black, 633; fragrant, 668
Olive family, described, 667
Oliver, D., 302
*Olsynium*, 424
Olyreae, 389

*Omphalogramma*, 660
Onagraceae, described, 637-640
*Oncidium*, 436
One-flowered pyrola, 652
O'Neill, H., 252
Onion, 416
*Onoclea*, 350
Ontogeny, contrasted with phylogeny, 92
*Onychium*, 352
*Ophiocaryon*, 584
Ophioglossaceae, described, 343
Ophioglossales, described, 343
*Ophioglossum*, 343
*Ophiopogon*, 415
Opiliaceae, 468
Ophrydoideae, 436
*Ophrys*, 437
*Oplopanax*, 643
Oppositifoliae, families of, 128, 129
*Opulaster*, 544
*Opuntia*, 623
Opuntiaceae, 621
Opuntiaeae, 621
Opuntiales, 621
Opuntioideae, 623
Orange, 557; Natal, 670; trifoliolate, 557
Orange jessamine, 557
hid tree, 548
Orchidaceae, described, 433-438; origin of, 431
Orchidales, 431
*Orchidanthera*, 426
*Orchis*, 436, 437
Order, a unit of classification, 46
Orders, nomenclature of, 205; of dicots, synopsis to, 439, 440, 441, 442
Oriental poppy, 517
*Origanum*, 690
Orobanchaceae, described, 702-703
*Orobanche*, 703
*Orontium*, 398, 400
Orpine family, described, 530
Orr, M. Y., 525
Orris root, 424
*Orthrosanus*, 422
*Oryctes*, 694
*Oryza*, 391
Oryzeae, 389
*Oserya*, 466
*Osmanthus*, 667, 668
*Osmunda*, 345
Osmundaceae, described, 345
*Ostenia*, 384
*Ostrowskia*, 721, 722
*Ostrya*, 457, 459
*Ostryopsis*, 457
Oswego tea, 691
Otaheite gooseberry, 566
*Ouratea*, 601
Outline maps, sources of, 323

Ovary, origin of, 72–75; simple versus compound, 74–75
Ovary positions, derivations of, 79, 80
Oxalidaceae, described, 550–551
*Oxalis*, 550
*Oxera*, 686
*Oxybaphus*, 480, 481
(*Oxycoccus*), 654, 655
*Oxydendrum*, 654
*Oxygyne*, 432
*Oxymitra*, 509
*Oxypetalum*, 675
Oxyquinoline sulfate, formula for use of, 255
*Oxyria*, 475
*Oxystylis*, 519
*Oxytheca*, 475
*Oxytropis*, 548
Ozenda, P., 65

*Pachira*, 595
(*Pachylophus*), 638
*Pachyrrhizus*, 548
*Pachysandra*, 570
*Pachystachys*, 709
*Pachystima*, 577
Pacific borderlands, a physio-graphic province, 157
Pacific coast states, floras of, 293
Pacific islands, floras of, 303
*Paeonia*, 496, 498
Paeoniaceae, 498; alliance with Crossosomataceae, 541
Paeonieae, 498
*Pagamea*, 712
*Palafoxia*, 728
*Palaquium*, 663
Paleobotany, of angiosperms, 103–108; of gymnosperms, 102–103; importance to phylogeny, 97, 98; the basis of phylogeny, 97–98
*Paliurus*, 588
Palm family, described, 394–395
Palmae, described, 394, 395
Palmales, 394
Panama, floras of, 297
Panama-hat-palm family, described, 397
*Panax*, 643
*Panda*, 549
Pandales, 440, 549
Pandanaceae, described, 373–374
Pandanales, described, 372
*Pandanus*, 374
*Pandorea*, 700
Pangaea, 143
*Pangium*, 613
Paniceae, 390
Panicle, in inflorescence phylogeny, 60, 61
*Panicum*, 389

Papaver, 515, 516, 517
Papaveraceae, alliance with Fumariaceae, 518; described, 515–517
Papaveroideae, 516
Papaya, 618
Papaya family, 617
Papayineae, 597
Paper bush, 626
*Paphiopedalum*, 434, 437
Papilionaceae, 545, 547
Papilionatae, 547
Pappus, morphological nature of, 67
Paradichlorbenzene, as an insecticide, 250, 253
Paratype, defined, 204
*Pariana*, 389
Parideae, 413, 415
Parietal placentation, 76, 77, 78
Parietales, 597
*Parietaria*, 464, 465
Paris code of nomenclature, 196
Paris Convention, 198
*Parishella*, 722
Parkeriaceae, described, 351–352
Parkin, J., 60
*Parmentiera*, 699
*Parnassia*, 534
Parnassiaceae, 534
*Paronychia*, 437
Parrotiaceae, 537
Parsley, 646
Parsnip, 646
(*Parsonsia*), 628
*Parthenium*, 730
*Parthenocissus*, 589
Partridgeberry, 713
*Parvisedum*, 531
*Pasania*, 460
*Paspalum*, 389
*Passiflora*, 617
Passifloraceae, described, 616–617
Passion-flower family, described, 616
Paste, for mounting specimens, 245, 246
*Pastinaca*, 646
Passionales, 566
Patchouly, 690
*Patrinia*, 711
*Paullinia*, 583
*Paulownia*, 700
*Pauridia*, 417
*Paurotis*, 396
Paviaceae, 582
Pawpaw, 509
Pawpaw family, 617
Pax, F., 417, 419, 488, 513, 526
*Payena*, 663
Pea, cow, 548; garden, 548; sweet, 548
Pea family, described, 545
Peach, 544
Peanut, 548

# INDEX

Pear, 544
Pearlbush, 544
Pecan, 455
Peck, M. E., 293
Pedaliaceae, described, 700–701
*Pedicularis*, 695, 696, 698
*Peganum*, 556
*Pelargonium*, 551, 552
Pelican flower, 473
*Pellaea*, 350
*Pelleteria*, 659
*Pellionia*, 465
*Peltandra*, 398
Pen and ink drawings, value of, 272, 273
Penaeaceae, 624
*Pennantia*, 580
Pennell, F. W., 55, 697
Pennyroyal, 690
*Penstemon*, 695, 696, 697
Pentaphyllaceae, 569
*Pentas*, 712
Penthoraceae, 532, 534
*Penthorum*, 532
*Pentodon*, 713
*Pentstemonacanthus*, 708
*Peperomia*, 445
*Peplis*, 627
Peponiferae, 717
Pepper, 445; red, 694
Pepper family, described, 444, 445
Pepperbush family, described, 650
Peppergrass, 521
Pepperidge tree, 631
(*Peramium*), 436
*Pereskia*, 621, 622, 623
Pereskioideae, 623
*Pereskiopsis*, 621, 623
*Perezia*, 730
Perianth, disposition of parts, 84; fusion and modification of, 84, 85; morphology of, 65–67
Perigyny, significance of, 81
Periodicals, abbreviations of titles, 308–317; American, 309; cultivated plants, 316, 317; in *Union list of serials*, 309; lists of, 308–317; Old World, 313–316; on economic plants, 316–317
*Periploca*, 675
Periplocoideae, 674
Periwinkle, 673
*Persea*, 512, 513
Perseoideae, 513
(*Persicaria*), 476
Peru, flora of, 298
(*Perularia*), 436
*Petalonyx*, 619
Petals, morphology of, 65–67
*Petasites*, 728, 730
*Petermannia*, 421
*Petiveria*, 482

Petiveriaceae, 481
*Petrea*, 688
*Petrophytum*, 544
Petrosaviaceae, 380, 413
Petrosavieae, 413
*Petroselinum*, 646
*Petunia*, 694
Pfeiffer, N. E., 342
Pfitzer, E., 435
*Phacelia*, 682, 683
*Phalaenopsis*, 434, 437
*Pharbitis*, 677
Phareae, 389
Pharmacognosy, importance to phylogeny, 100
*Pharus*, 387
*Phaseolus*, 548
*Phellodendron*, 557
*Phenakospermum*, 426
Phenotype, 179
Pherosphaeroideae, 361
*Philadelphus*, 533, 534
Philesiaceae, 413
(*Philibertia*), 675
*Philippiamra*, 485
Philippine islands, flora of, 303
Phillips, E. P., 483
*Phillyrea*, 667, 668
Philodendroideae, 398
*Philodendron*, 398, 400
(*Philotria*), 385
Philydraceae, 402
*Phleum*, 390
*Phlox*, 678, 679
Phlox family, described, 678
*Pholidia*, 710
*Pholisma*, 681, 682
*Pholistoma*, 683
*Phoradendron*, 471, 472
*Photinea*, 544
Photographs, limitations of, 272–273
*Phoxanthus*, 584
*Phryma*, 710
Phrymaceae, 687, 668; described, 710
Phrymineae, 676
*Phrynium*, 430
*Phyla*, 687
*Phyllachne*, 725
Phyllanthoideae, 566
*Phyllanthus*, 566
*Phyllitis*, 350
Phyllocladoideae, 361
*Phyllocladus*, 361
*Phyllodoce*, 655
*Phylloglossum*, 337
*Phyllospadix*, 377
Phylogenetic classification, defined, 93; types, 93
Phylogenetic considerations, in taxonomy, 92–108

Phylogenetic systems of classification, 33–39
Phylogeny, anatomy an integral part of, 98; cytological contributions to, 99; defined, 92; evidence from seed characters, 87; interrelationships with phytogeography, 141; paleobotanical contributions to, 97–98; pharmocological contributions to, 100, phytogeographic contributions to, 99; Sporne's Advancement Index, 95
Phylogeny of, angiosperms, 103–108; carpels and gynoecia, 72–75; floral structures, 64–88; Gnetales, 104–105; gymnosperms, 102–103; inflorescences, 59–64; ovary positions, 81; perianth parts, 66–67, 84–85; pteridophytes, 101–102; stamens, 67–72
Phylum, 44
*Phymatodes*, 350
*Physalis*, 694
Physiographic areas, significance of, 154, 157–158
Physiographic provinces of North America, 153–157
Physiographic units, as basis of floras, 277, 278
Physiology, significance to phylogeny of major units, 100–101
*Physocarpus*, 544
*Physopsis*, 686
*Physostegia*, 691
*Physurus*, 436
*Phytelephas*, 396
*Phyteuma*, 721, 722
Phytochemical studies and phylogeny, 100–101
Phytogeography, age and area, 160–161; barrier factors, 149–150; Cain's principles of, 158–160; centers of area, 164–165; cytogenetic criteria, 165; discontinuous distributions, 163, 164; distribution factors, 149–150; dynamic, 158–165; endemism, 164; epibiotics, 164; evolution of floras, 163; Good's floristic zones, 152–153; Good's theory of tolerance, 161–162; Guppy's theory of differentiation, 161; Hansen's floristic zones, 153; importance to taxonomy, 162, 163; migration and evolution, 163; polytopism, 164; role in phylogeny, 99–100; senescence, 165; significance of barriers, 149–150
*Phytolacca*, 482
Phytolaccaceae, described, 481–482
(*Piaropsis*), 411
*Picea*, 363, 364
Pichon, M., 383, 384, 409
Pickaback plant, 464
Pickerel-weed family, described, 410
*Picramnia*, 559

*Picrasma*, 559
*Pieris*, 654
*Pilea*, 465
Pilger, R., 60, 357, 359, 360, 364, 366, 367
*Pilostyles*, 474
*Pilularia*, 353
*Pimelea*, 626
*Pimenta*, 634
Pinaceae, described, 363, 364
*Pinckneya*, 713
Pine, 364; ground, 338; Princess, 338; umbrella, 366
Pine family, described, 363, 364
Pineapple family, described, 405
*Pinguicula*, 705, 706
Pinguiculariaceae, 705
Pink family, described, 486–487
Pinkroot, 670
*Pinus*, 363, 364
*Piper*, 445
Piper, C. V. 293
Piperaceae, described, 444, 445
Piperales, described, 443
Pipewort family, described, 404
Pipsissewa, 652
*Piqueria*, 730, 731
Pirolaceae, 651
Pistachio, 574
*Pistacia*, 574
*Pistia*, 399, 400
Pistioideae, 398
Pistil, defined, 72; morphology of, 72–75
*Pisum*, 548
Pita floja, 407
*Pitcairnia*, 406, 407
Pitcairnioideae, 405, 407
Pitcher-plant family, described, 527
*Pithecolobium*, 547
Pittendrigh, C. S., 407
Pittosporaceae, described, 535–536
*Pittosporum*, 535, 536
(*Pityothamnus*), 509
*Pityrogramma*, 351
*Placea*, 418
Placenta, defined, 74, 75
Placentation types, derivation of, 75–79
*Plagiobothrys*, 685
*Planera*, 462
Plane-tree family, described, 539
Plant distribution: *see* phytogeography
Plant geography: *see* phytogeography
Plant identification, 223–232
Plant kingdom, classification of, 324
Plantaginaceae, described, 710, 711
Plantaginales, described, 710
*Plantago*, 711
Plastics, use in mounting specimens, 246
Platanaceae, described, 539–540
(*Platanthera*), 436

INDEX

*Platanus*, 370, 540
*Platycarya*, 453, 455
*Platycerium*, 350, 351
*Platycodon*, 721, 722
(*Platypus*), 436
*Platystemon*, 515, 516
Platystemoneae, 515
*Plectritis*, 716
*Plenodendron*, 611
Pleuranthe, taxonomy of, 436
*Pleuricospora*, 651
*Pleurogyne*, 670
Pleuromeiales, 342
(*Pleuropteropyrum*), 476
*Pleurothallis*, 436
(*Plexia*), 436
*Pluchea*, 730
Plum, 544; Kaffir, 574; marmalade, 663; Natal, 673
Plumariaceae, 675
Plumbaginaceae, described, 661–662
Plumbaginales, described, 660
*Plumbago*, 661
Plume poppy, 517
*Plumeria*, 673
Plum-yew family, described, 362
*Poa*, 389
Poaceae, 387
*Poacynum*, 673
Poales, 386
*Podandrogyne*, 518, 520
Podocarpaceae, described, 361
Podocarpoideae, 361
*Podocarpus*, 361
*Podophyllum*, 501, 502
Podostemaceae, allied to Rosales, 530; described, 466
Podostemales, described, 465, 466
Podostemonaceae, 465
Podostemonales, 465
*Podostemum*, 465, 466
(*Podostigma*), 675
*Pogogyne*, 691
*Pogonia*, 436
*Pogostemon*, 690
Poinsettia, 566
Poison hemlock, 646
Poisoning herbarium specimens, 251–252
Pokeweed family, described, 481
*Polansia*, 519, 520
Polemoniaceae, described, 678–679
(*Polemoniella*), 678
*Polemonium*, 678, 679
Pollard, C. L., 508
Pollen grains, morphology of, 71; smear preparations, 71
Pollen sterility, role in phylogeny, 106
Pollinia, 71
Polunin, N., 216
*Polyalthia*, 509

*Polycarpon*, 488
Polychondreae, taxonomy of, 436
*Polygala*, 546
Polygalaceae, described, 563–565
Polygalales, 564
Polygalineae, 549
Polygonaceae, described, 475–477
Polygonales, described, 474
*Polygonatum*, 414, 416
*Polygonella*, 475
*Polygonum*, 475–476
Polynesia, floras of, 303
*Polyosma*, 534
Polyphylesis, 164
Polyploidy, 174, 175
Polypodiaceae, described, 349–351
*Polypodium*, 350
*Polypompholix*, 706
*Polypremum*, 669, 670
*Polypteris*, 728
*Polyrrhiza*, 436
*Polyscias*, 643
*Polystachya*, 436
*Polystichum*, 349, 350, 351
Polytopism, 164
Pomaceae, 543
*Pomaderris*, 588
Pomegranate, 629
Pomegranate family, described, 628
Pomoideae, 543
*Poncirus*, 557
Pondweed, Cape, 379
Pondweed family, described, 376
*Pontederia*, 411
Pontederiaceae, described, 410–411
Pontederieae, 411
*Ponthieva*, 436
Pool, R. J., 235, 386
Popenoe, W., 322
Poppy, blue, 517; bush, 517; california, 517; celandine, 517; Iceland, 517; oriental, 517; prickly, 517; Welsh, 517
Poppy family, described, 515
Poppy mallow, 594
*Populus*, 477, 448
*Porana*, 676, 677
Poranthoroideae, 566
*Porlieria*, 556
*Porterella*, 722
*Portulaca*, 484, 485
Portulacaceae, described, 484–485
*Portulacaria*, 485
*Posidonia*, 377
Posidonieae, 377
Post, G. E., 300
Pot Marjoram, 690
*Potamogeton*, 376, 377
Potamogetonaceae, described, 376–378
Potamogetonales, 376
Potamogetoneae, 377

Potato, 694
*Potentilla*, 543, 544
Poteriaceae, 543
*Poteridium*, 544
*Poterium*, 544
Pothoideae, 398
*Pothos*, 400
*Pouteria*, 663
Prairie gentian, 671
*Pratia*, 722
*Premna*, 687
Preparation of specimens, 235, 236, 237
*Prescottia*, 436
Preserving solutions, formulas for, 254, 256
Pressing plant materials, 238–240
Prickly ash, 557
Prickly pear, 623
Prickly poppy, 517
Prickly thrift, 662
Primitive versus advanced characters, 94, 95
Primitiveness of, monocots versus dicots, 107; woody versus herbaceous character, 106–107
*Primula*, 659
Primulaceae, described, 658–660
Primulales, described, 656
Primrose, Cape, 705
Prince's feather, 480
Prince's pine, 652
Princess pine, 338
Principes, described, 394
Pritzel, G. A., 306, 327
*Priva*, 687, 688
Privet, 668
*Probiscidea*, 701, 702
*Proserpinaca*, 641
*Prosopsis*, 547
Prostantheroideae, 690
Proteaceae, described, 467–468
Proteales, described, 467
*Protium*, 560
Prune, 544
*Prunella*, 690
Prunoideae, 543
*Prunus*, 543, 544
*Pseuderanthemum*, 709
*Pseudolarix*, 364
*Pseudophoenix*, 396
Pseudosolaneae, 696
*Pseudotsuga*, 363, 364
*Psidium*, 634
Psilophytales, era of, 145–146
Psilopsida, 355
Psilotaceae, described, 339–340
Psilotales, 339–340
Psilotineae, described, 339
*Psilotum*, 340
*Ptelea*, 557

Pteleaceae, 556
Pteridaceae, 349
*Pteridium*, 350
*Pteridophyllum*, 516
Pteridophyta, described, 334, 335; paleobotany of, 101–102; phylogeny of, 101, 102
Pteridospermae, 355; inflorescences of, 61; ovule arrangement, 61; phyletic importance of, 103–105
*Pteris*, 349
*Pterocarya*, 453, 455
*Pterocephalus*, 717
Pteropsida, 355
*Pterspora*, 651
*Pterostegia*, 475
*Pterostyrax*, 666
*Ptilotus*, 479
Publication dates, sources of, 327–328
Publication of botanical names, 206, 207
Pulle, A. A., 94, 114, 124, 298; system of classification, 122, 123
*Pulmonaria*, 685
Pulque, 420
(*Pulsatilla*), 498
Pumpkin, 720
*Punica*, 629
Punicaceae, described, 628–629
Puri, V., 514, 516, 519, 520, 521, 525, 526, 616
Pursh, F., 26
*Purshia*, 544
Purslane family, described, 484
*Puya*, 407
*Pycnanthemum*, 691
*Pycnostachys*, 690
*Pyracantha*, 544
Pyrenaceae, 686
Pyrethrum, 730
Pyrola family, described, 651
Pyrolaceae, described, 651–652
*Pyrolirion*, 420
Pyroloideae, 651
*Pyrostegia*, 700
*Pyrularia*, 470
*Pyrus*, 544
*Pyxidanthera*, 648, 649

*Quamoclit*, 676, 677
Quassia family, described, 558
Queensland nut, 468
*Quercus*, 459, 460
Quick-freezing of flowers, 256
Quiinaceae, 597
Quillwort family, described, 341–342
Quimby, M. W., 532
Quince, 544; flowering, 544
Quinine, 713
*Quisqualis*, 633
Quisumbing, E., 257

# INDEX 813

Race, a taxonomic unit, 56; edaphic and biotic, 177
Raceme, in inflorescence phylogeny, 61
*Radiola*, 553
Radish, 521
*Raffia*, 396
*Rafflesia*, 474
Rafflesiaceae, described, 474
Rafinesque, C. S., 49
Ragweed, 730
(*Raimannia*), 638
Raisin tree, 588
Raisz, E., 323
Ramie, 465
*Ramischia*, 651
*Ramonda*, 704, 705
Ramontchi, 614
Ranales, described, 489
*Ranalisma*, 382, 383
Randolph, L. F., 390
*Randonia*, 524
Ranunculaceae, described, 496–500
Ranunculineae, 489
*Ranunculus*, 496, 497
*Rapanea*, 658
Rapateaceae, 402; allied to Bromeliaceae, 407
Raphanaceae, 520
*Raphanus*, 521, 522
Raspberry, 544
Rastall, R. H., 143
*Ratibida*, 730
Rattenbury, J. A., 188
Raunkier, C., 158
*Rautanenia*, 382
*Ravenala*, 426
Ray, J., 17, 18, 27, 28
Reaumurieae, 607, 608
Receptacular theory, of ovary position, 79–80
Record, S. J., 98
(*Rectanthera*), 409
Red pepper, 694
Red squill, 416
Red valerian, 716
Redbud, 548
Redouté, P. J., 29
Redwood, 366
Reeves, R. G. and Bain, D. C., 294
Regel, E., 42
*Regnelledium*, 353
Rehder, A., 224, 306, 322, 326, 447, 450, 457, 500, 506, 547, 588, 601, 687
Reiche, K. F., 298
Reichenbach, H. G. L., 632
Reichert, E. T., 100
*Reinwardtia*, 553
Rendle, A. B., 37, 86, 95, 121, 326; system of classification, 37, 120, 121; *see also* note, p. 777

*Renealmia*, 428
Repellents for herbarium beetles, 253
Reproductive elements in taxonomy, 64–88
Reproductive organs, taxonomic importance of, 64
*Reseda*, 370, 525
Resedaceae, described, 524–525
Resedineae, 514
Restionaceae, 402, 403; in Glumiflorae, 386
Resurrection plant, 339
*Reussia*, 411
Review serials, list of, 307
Revision, defined, 304, 305
Revisions, preparation of, 264–274
*Reynosia*, 588
*Rhacoma*, 577
Rhamnaceae, described, 587–588
Rhamnales, 587
*Rhamnus*, 588
*Rhapidophyllum*, 396
*Rheedia*, 603, 604
*Rheum*, 476
*Rhexia*, 636
Rhinanthaceae, 695
Rhinanthoideae, 696
*Rhipsalis*, 622, 623
*Rhizophora*, 629, 630
Rhizophoraceae, described, 629, 630
*Rhodiola*, 531
*Rhodochiton*, 695
Rhododendroideae, 654
*Rhododendron*, 652, 653, 654
Rhodoraceae, 652
*Rhodotypos*, 544
Rhoeadales, described, 514
Rhoeadineae, 514
*Rhoeo*, 408, 409
Rhoipteleaceae, 453, 461
Rhubarb, 476
*Rhus*, 573, 574
*Rhynchospora*, 392
Rhynchosporaceae, 393
Rhynchosporoideae, 393
(*Rhynchophorum*), 445
*Ribes*, 533, 534
Rice, 391
Rice flower, 626
Rice-paper plant, 643
Ricinocarpoideae, 566
*Ricinus*, 565
Ricker, P. L., 236, 256
Rickett, H. W., 59–63, 634, 642, 648; on inflorescences, 59
Ridgway, R., 329
*Riedlea*, 596
*Riesenbachia*, 639
*Rigiopappus*, 728
*Rinorea*, 612
River weed family, described, 466

*Rivina*, 481
Rivinus, A. Q., 23
Robins, W. W., 322
Robinson, B. L., 170
Roble blanco, 700
Robyns, W., 302
Rochester code of nomenclature, 197
Rock cress, 521
Rock, J. F. C., 303
Rocket, 521
Rock-rose family, described, 608
Rocky Mountain province, 157
Rodway, L., 303
Roeper, J., 59
Rolfe, R. C., 435
*Rollinia*, 509
Rollins, R. C., 256
*Romneya*, 516
*Romulea*, 424
Roridulaceae, 530
Rosaceae, synopsis to subfamilies of, 543
Rosales, described, 530
*Rosmarinus*, 690
*Rosa*, 542, 543, 544
Rosaceae, described, 541, 543
*Roscoea*, 428
Rose, wood, 677
Rose apple, 634
Rose family, described, 541, 543
Rose moss, 485
Rose of sharon, 594
Rosemary, 690
Rosineae, 530
Rosoideae, 543
*Rotala*, 627
Royal poinciana, 548
*Roystonea*, 396
*Rubacer*, 545
Rubber, 566
Rubber vine, 675
*Rubia*, 713
Rubiaceae, described, 712–713
Rubiales, described, 711
*Rubus*, 542, 543, 544
Rudbeck, O., 18
*Rudbeckia*, 730
Rue, common, 557
Rue family, described, 556
*Ruellia*, 708, 709
Rules of nomenclature, *see* Nomenclature
Rum, 391
*Rumex*, 475
*Rupicapnos*, 518
*Ruppia*, 377
Ruscaceae, 413
Rush family, described, 412
Rusinae, 415
*Russelia*, 697
*Ruta*, 557,
Rutabaga, 521

Rutaceae, described, 556–558
Rutgers, F. L., 88
Rydberg, P. A., 170, 293, 294, 377, 386, 681
Rye, 391

*Sabal*, 396
*Sabatia*, 670
*Sabia*, 584
Sabiaceae, described, 584
Sabiineae, 569
Sacaline, 476
Saccado, P. A., 327
*Saccharum*, 391
Safflower dye, 730
Saffron, 424
Sage, 690
*Sageretia*, 588
*Sagina*, 487
*Sagittaria*, 382, 383
Sahni, B., 98, 105
*Saintpaulia*, 704, 705
Sake, 391
*Salacia*, 578
Salicaceae, described, 447–448
Salicales, described, 446
Salicariaceae, 627
*Salicornia*, 477, 478
Salicornioideae, 478
*Salix*, 447, 448
Salpiglossideae, 694
*Salpiglossis*, 694
Salsify, 730
*Salsola*, 477, 478
Salsoloideae, 478
Salvadoraceae, 569, 667
*Salvia*, 689, 690
*Salvinia*, 354
Salviniaceae, described, 353–354
*Sambucus*, 713, 714, 715
*Samolus*, 659
Sampaio, G., 300
Sand verbena, 481
Sandalwood, 471; false, 462
Sandalwood family, described, 469
Sandwort, 488
*Sanguinaria*, 516
*Sanguisorba*, 541, 544
*Sanicula*, 646
*Sanidophyllum*, 605
*Sanseviera*, 413
Santalaceae, described, 469–471
Santalales, described, 468
Santalineae, 468
*Santalum*, 471
*Sanvitallia*, 728
Sapindaceae, described, 583–584
Sapindales, 568, 569
*Sapindus*, 583, 584
Sapodilla, 663

# INDEX 815

Sapotaceae, described, 662–664
Sapote, 663
Sapotineae, 662
*Saracha*, 694
*Sararanya*, 374
Sarcobatoideae, 478
*Sarcobatus*, 477
*Sarcocaulon*, 551
*Sarcocephalus*, 712
*Sarcococca*, 570
*Sarcodes*, 651
Sarcopodiaceae, 368
*Sarcopus*, 368
*Sarcostemma*, 675
*Sargentodoxa*, 500
Sargentodoxaceae, 500
*Sarothra*, 605
*Sarracenia*, 527
Sarraceniaceae, described, 527–528
Sarraceniales, described, 527
*Saruma*, 473
*Sassafras*, 512, 513
*Satureja*, 690
Saunders, E. R., theory of carpel polymorphism, 65, 516, 521
*Saurauia*, 599, 600
Saurauiaceae, 599
Saururaceae, described, 444
*Saururus*, 444
Sausage tree, 700
*Sauvagesia*, 601
Savage, S., catalogue of Linnaean herbarium, 328, 329
Savory, 690
Sax, K., 462
*Saxegotheca*, 361
*Saxifraga*, 533, 534
Saxifragaceae, described, 533–535
Saxifragineae, 530
*Scabiosa*, 717
*Scaevola*, 723, 724
*Schaefferia*, 577
Schaffner, J. H., 294
Schellenberg, G., 469
Schery, R. W., 270
*Scheuchzeria*, 380
Scheuchzeriaceae, described, 379
*Schiekia*, 417
*Schinopsis*, 574
*Schinus*, 574
*Schisandra*, 506
Schisandraceae, 504; described, 506–507
*Schizaea*, 346
Schizaeaceae, described, 345–346
(*Schizandra*), 507
Schizandraceae, 506
*Schizanthus*, 693, 694
*Schizocardia*, 650
*Schizonotus*, 544
*Schizopepon*, 718

Schlechter, R., 435, 436, 437
*Schmaltzia*, 574
Schneider, C., 322
*Schoepfia*, 469
Schouw, J. F., 153
*Schrankia*, 547
Schultes, R. E., 240
Schwartz, O., 411
Schweinitz, L. D., 26
Sciadopityoideae, 366
*Sciadopitys*, 365, 366
*Scilla*, 413, 416
*Scindapsus*, 400
Scirpoideae, 393
*Scirpus*, 392, 393
Scitaminales, 425
Scitamineae, 411; described, 425
*Scleranthus*, 487
*Scleria*, 392
Sclerieae, 393
*Scoloyopus*, 413
*Scopulophila*, 488
Scorpioid cyme, origin of, 61
Scouring rush family, described, 336, 337
Screw-pine family, described, 373, 374
*Scrophularia*, 695
Scrophulariaceae, described, 695–697
*Scutellaria*, 688, 690, 691
Scytopetalaceae, 590
Scytopetalineae, 590
Sea buckthorn, 627
Sea grape, 476
Sea holly, 646
*Secale*, 391
*Secamone*, 674
*Sechium*, 719, 720
Section, a unit of classification, 50
*Securidaca*, 564
Sedaceae, 530
(*Sedella*), 531
Sedge family, described, 392
Sedoideae, 532
*Sedum*, 531
Seeds, taxonomic value of, 87
Seibert, R. J., 698
Selaginaceae, 696
Selagineae, 696
*Selaginella*, 339
Selaginellaceae, described, 338–339
Selaginellales, 338, 339
*Selago*, 696
*Semecarpus*, 574
*Semibegoniella*, 621
Sempervivoideae, 532
*Sempervivum*, 531
*Senecio*, 728, 730
Senecioneae, 730
Senescence, 165
Senn, H. A., 162, 165, 181
Senna, 548

Sepals, morphology of, 65–67
*Sequoia*, 365, 366
*Sequoiadendron*, 365, 366
(*Serapias*), 436
*Serenoa*, 396
Serials, lists of, 308–317
*Sericotheca*, 544
Series, a unit of classification, 50
*Serissa*, 713
*Serjania*, 583
Sesame oil, 701
*Sesamum*, 701
*Sesuvium*, 483
*Setaria*, 390
Sexual arrangements, in flowers, 82–84
*Seymeria*, 695
Sharp, A. J., 257
Shell ginger, 428
She-oak, 443
*Shepherdia*, 626, 627
Sherborn, C. D., 328
Shinleaf, 652
*Shortia*, 649
Shrimp plant, 709
Sial, 142
*Sibbaldia*, 544
*Sibbaldiopsis*, 544
*Sicana*, 720
Sicyoideae, 719
*Sicyos*, 719, 720
*Sicyosperma*, 719
*Sida*, 593
*Sidalcea*, 593, 594
Siebert, A. and Voss, A., 322
*Sieversia*, 544
*Silene*, 477, 488
*Silphium*, 730
Silver bells, 666
Silver-lace vine, 476
*Silybum*, 728
Sima, 142
*Simarouba*, 558, 559
Simaroubaceae, described, 558–559
(*Simaruba*), 558
Simarubaceae, 558
*Simmondsia*, 570
Simpson, G. G., 172
*Sinningia*, 704, 705
Sinnott, E. W., 106, 164
Sinnott, E. W. and Bailey, I. W., 106
*Siparuna*, 511
*Siphocampylus*, 721, 722
*Siphocodon*, 721, 722
*Siphonella*, 678
*Siphonoglossa*, 708
Sisal, 420
*Sisymbrium*, 521
Sisyrincheae, 424
*Sisyrinchium*, 422, 424
Skottsberg, C., 94, 114, 136; system of

classification, 124
Skullcap, 691
*Sladenia*, 599
Slipperflower, 697
Slippery elm, 462
Sloan, H., 25
*Sloanea*, 591
Sloeberry, 544
Small club moss family, described, 338–339
Small J., 726, 730
Small, J. K., 170, 296, 371, 375, 381, 386, 417, 427, 431, 484, 498, 529, 550, 564, 623, 652, 723
Smilacaceae, 413
*Smilax*, 413
Smith, A. C., 219, 230, 231, 493, 494, 504, 505, 506, 579
Smith, A. C. and Standley P. C., 652
Smith, A. C., and Wodehouse, R. P. 510
Smith, F. H., 469, 570
Smith, G. G., 243
Smith, G. H., 38, 139
Smith, J. E., 26
Smith, L. B., 407
Smith, W. W., 174
*Smithiantha*, 705
Smoke tree, 574
Snake gourd, 720
Snapdragon, 697
Snell, R. S., 393
Snow wreath, 544
Snowberry, 715
Soapberry, 584
Soapberry family, described, 583
Sodium silicate, for preserving dissections 257
Soil maps, 325; use of, 268
Solanaceae, described, 693–695
*Solandra*, 694
Solaneae, 694
Solanineae, 676
(*Solanoa*), 675
*Solanum*, 694
*Soldanella*, 659, 660
Solereder, H., 670
*Solidago*, 728, 730
*Sollya*, 536
*Sonchus*, 728
*Sonerila*, 636
Sonneratiaceae, 624
*Sorbus*, 544
*Sorghum*, 390
Sorrel tree, 654
Sour-gum tree, 631
Soursop, 509
South Africa, floras of, 301–302
South America, floras of, 297, 298
Southwestern U. S., floras of, 296
Soybean, 548

Spadiciflorae, of Wettstein, 431
Spanish lime, 584
Spanish moss, 407
*Sparattanthelium*, 514
Sparganiaceae, described, 374–375
*Sparganium*, 375
*Sparmannia*, 591
Spathiflorae, described, 398
(*Spathiger*), 436
*Spathodea*, 698, 700
Species, a unit of classification, 50, 51, 52, 53; delimited by biosystematic criteria, 181, 182
Species lectotypicae, 204
Species names, conservation of, 200, 202, 206; decapitalization of, 202, 216, 219
Specimens, herbarium: see under Herbarium Techniques
*Specularia*, 721, 722
Speedwell, 697
*Spergula*, 487
*Spergularia*, 487
Spermatophyta, 354–355
*Sphaeralcea*, 593
*Sphaerocephalum*, 611
(*Sphaerostigma*), 638
Sphenophyllales, 336
Sphenopsida, 337, 355
Spider flower, 520
Spiderwort family, described, 408
*Spigelia*, 669, 670
Spigeliaceae, 669
Spikelet, described, 387–389
Spinach, 478; New Zealand, 484
*Spinacia*, 478
*Spiraea*, 542, 543, 544
Spiraeaceae, 543
Spiraeoideae, 543
*Spiranthes*, 435, 436, 437
Spirolobeae, 478
*Spirodela*, 401
(*Spironema*), 409
Spondiaceae, 573
*Spondias*, 574
Sporne, K. R., 95
*Sporobolus*, 389
Sprague, T. A., 93, 200, 201, 204, 206, 479, 558, 601, 637, 640, 651
*Spraguea*, 485
Sprengel, C., 42, 45
Spurge family, described, 565
*Spyridium*, 588
Square degree, a basis of area for floras, 277
Squash, 720
Squirting cucumber, 720
Stachyosporous theory, summary of, 105–106
*Stachys*, 689, 690, 691
*Stachytarpheta*, 687

Stachyuraceae, described, 614–615
*Stachyurus*, 615
Stackhousiaceae, 569
Staff-tree family, described, 576
Stamen, defined, 67
Stamens, centrifugal, 69–70; morphology of, 67–72; types of, 71
Standley, P. C., 296, 297
*Stangeria*, 356, 357
Stangerioideae, 357
*Stanleya*, 522
*Stapelia*, 674, 675
Stapf, O., 500
*Staphylea*, 579, 580
Staphyleaceae, described, 579–580
Star apple, 663
Statice, 661
(*Statice*), 661
*Stauntonia*, 500
Stearn, W. T., 55, 326, 328, 329
Stebbins, G. L. Jr., 52, 164, 176, 178, 179, 184, 366
Steenis, C. G. G. J. van, 303
Steere, W. C., 216
*Stegnosperma*, 481
*Steironema*, 659
*Stelechocarpus*, 508
*Stellaria*, 487, 488
*Stellariopsis*, 544
Stelophyta, 355
Stemen, T. R. and Myers, W. S., 294
*Stemmodenia*, 673
*Stemodia*, 697
Stemonaceae, 411
*Stenandrium*, 708
*Stenolobium*, 699, 700
Stenomerideae, 421
*Stenophyllus*, 392
*Stenorrhynchus*, 436
*Stenotaphrum*, 391
(*Stenotus*), 728
*Sterculia*, 596
Sterculiaceae, described, 596–597
Sterility barriers, determination of, 180; significance of, 182, 183
Steudel's *Nomenclator botanicus*, 195, 196
Stevia, florist's, 731
Steyermark, J., 243
*Stewartia*, 602
*Stigmaphyllon*, 562
Stilbaceae, 688
*Stilbe*, 687
*Stizolobium*, 548
Stockholm Congress, 201
Stocks, 521
*Stokesia*, 728, 730
Storax, 538
Storax family, described, 666
Stork's-bill, 552
Stramonium, 694

Strasburgeriaceae, 597
Strawberry, 544
Strawberry tomato, 694
Strawflowers, 731
*Strelitzia*, 426
Strelitziaceae, 426
Strelitzioideae, 426
*Streptanthus*, 521
*Streptocarpus*, 705
*Streptosolen*, 694
Strobiloideae, families of, 128, 129
Strychnine, 670
*Strychnos*, 669, 670
(*Stuartia*), 603
Stylidiaceae, described, 724–725
*Stylidium*, 724, 725
*Stylodon*, 687
*Stylomecon*, 516
*Stylophorum*, 516, 517
*Styphelia*, 651
Styracaceae, described, 666
Styracales, 662
*Styrax*, 666
*Suaeda*, 477, 478
Suaedoideae, 478
Subdivision, a unit of classification, 45
Suborder, a unit of classification, 46
Subspecies, concepts of, 54, 55
Subtribe, a unit of classification, 47
Subtribes, nomenclature of, 206
*Succisa*, 717
Succulent material, housing in herbaria, 254, 255
Sugar, 391; from beets, 478
*Sullivantia*, 534
Sumac, 574
Sundew family, described, 528
Sunflower, 731
Superior ovary, defined, 79
Supplejack, 588
*Suriana*, 558, 559
Surinam cherry, 634
Sussenguth, K., 681
*Suttonia*, 658
Svenson, H. K., 23, 24
Swamp cypress: see *Taxodium*
Swamy, B. G. L., 435
Swamy, B. G. L. and Bailey, I. W., 495
Sweet alyssum, 521
Sweet clover, 548
Sweet gale family, described, 450
Sweet pea, 548
Sweet pepperbush, 650
Sweet potato, 677
Sweetleaf, 665
Sweetsop, 509
*Swertia*, 671
*Swietenia*, 561
Swiss chard, 478
Sycamore, 540

Syllabus, 733–736
*Symbegonia*, 621
*Symbryon*, 445
*Sympetaleia*, 618, 619
Symphoremaceae, 688
*Symphoricarpos*, 715
*Symphyandra*, 722
*Symphytum*, 684, 685
Symplocaceae, described, 665
*Symplocarpus*, 398
*Symplocos*, 665
Sympodial inflorescences, 162
Synanthae, described, 397
*Synaptanthera*, 712
(*Syndesmon*), 498
*Syngonanthus*, 405
Synopsis, defined, 225; to classes of Pteridophyta, 335; to orders of Dicotyledoneae, 439–442, Gymnospermae, 356, Monocotyledoneae, 372; to subfamilies of Leguminosae, 545, Rosaceae, 543
*Synthyris*, 698
Syntype, defined, 204
*Syringa*, 667, 668
Systematic botany, 2
Systems of classification, 114–139
Systems of classification, based on form relationship, 26–33; habit, 14–18; numbers or sexual arrangements, 18–26; phylogeny, 33–39
*Syzygium*, 634

*Tabebuia*, 699, 700
Taccaceae, 411
Tackholm, V. and Drar, M., 302
(*Tacsonia*), 617
*Tagetes*, 727, 730
*Taiwania*, 365
*Talauma*, 504
*Talinopsis*, 485
*Talinum*, 485
*Talisia*, 583
Tamacaceae, 421
Tamaricaceae, described, 607–608
Tamariceae, 607
Tamaricineae, 597
*Tamarix*, 607, 608
*Tamus*, 422
*Tanacetum*, 730
Tangerine, 557
Tannic acid, 574; source of, 460
Tapioca, 566
Tara vine, 599
*Taraktogenos*, 614
*Taraxacum*, 728
Taro, 400
Tasmania, flora of, 303
Tate, R., 303
Tautonomy, examples of, 215; illegitimate, 215–216

Taxaceae, described, 360
Taxales, 360
Taxodiaceae, described, 365–366
Taxodineae, 364
Taxodioideae, 366
*Taxodium*, 365, 366
Taxon, adoption of in Rules of nomenclature, 201; origin and definition of term, 53
Taxonomy, alpha, 171; descriptive, defined, 170; experimental, 171–172; interrelationships with allied sciences, 5; modern, 171–172; objectives of, 6–8; opportunities in, 10–12; principles of, 42–91; problems of, 8–10; scope of, 3–5; significance of, 7
*Taxus*, 360
Taylor, H., 667
Taylor, N., 377
Tea bush, 588
Tea family, described, 601
Teak, 688
Teasel family, described, 717
Techniques: *see* Herbarium Techniques
(*Tecoma*), 699
Tecomeae, 699
Tecophilaeaceae, 415, 419
*Tectona*, 686, 688
*Tenagocharis*, 384
Tepals, morphology of, 66
*Tephrosia*, 548
Tequila, 420
Terebintaceae, 573
Terebinthaceae, 568, 573
Terebinthales, 557, 568
*Terminalia*, 633
*Ternstroemia*, 602
Ternstroemiaceae, 601
*Tessarandra*, 667
Tetracentraceae, described, 493, 494
*Tetracentron*, 493
*Tetracera*, 598
*Tetraclea*, 687
*Tetraclinis*, 366
*Tetragonia*, 483, 484
Tetragoniaceae, 482
*Tetrameles*, 620
*Tetramerium*, 708
*Tetrapanax*, 643
*Tetratheca*, 563
*Tetrazygia*, 636
*Tetroncium*, 380
*Teucrium*, 689, 690, 691
*Thalassia*, 385, 386
(*Thalesia*), 703
*Thalia*, 430
*Thalictrum*, 496, 497, 498
Thallophyta, 334
*Thaspium*, 646
*Thea*, 602

Theaceae, described, 601–603
Theineae, 597
Thelygonaceae, 624
*Thelypodium*, 521
*Theobroma*, 596
Theophrastaceae, described, 657
Theophrastus, 14–15
*Thespesia*, 593
*Thevetia*, 673
*Thismia*, 432
Thismiaceae, 432
Thismieae, 432
Thistle, globe, 731
*Thladiantha*, 718
*Thlaspi*, 521, 522
Thomas, H. H., 98
*Thomasia*, 596
Thommen, E., 358
Thrift, 661
*Thrinax*, 395, 396
*Thryalis*, 562
Thujoideae, 367
*Thujopsis*, 366
Thunberg, C. P., 25
*Thunbergia*, 707, 708, 709
Thurniaceae, 402; in Glumiflorae, 386
Thyme, 690
Thymelaeaceae, described, 625–626
Thymelaeineae, 624
*Thymelea*, 626
*Thymus*, 690, 691
*Tibouchina*, 636
Tidestrom, I., 295
*Tidestromia*, 479
Tigrideae, 424
*Tigridia*, 424
*Tilia*, 591, 592
Tiliaceae, described, 591–592
Tiliales, 590
*Tillaea*, 531
*Tillaeastrum*, 531
*Tillandsia*, 406, 407
Tillandsioideae, 407
Tillson, A. H. and Bamford, R., 557
*Timotocia*, 688
*Tinantia*, 409
(*Tiniaria*), 476
Tippo, O., 98, 107, 114, 115, 138–139, 443, 454, 458, 460, 461, 463, 537, 538, 540; diagram of system, 138; system of classification, 38, 39
*Tipularia*, 436
(*Tissa*), 487
*Tithonia*, 728, 730
(*Tithymalus*), 566
*Tmesipteris*, 340
Tobacco, 694
*Todea*, 345
Tofieldieae, 415
*Tolmiea*, 533

Tomato, 694
Tontelea, 578
Topographic maps, kinds and sources of, 323, 324
Topotype, defined, 205; importance of, 268
Torch flower, 428
Torchwood family, 559
Torenia, 696, 697
Torrey, J., 26, 49
Torreya, 360
Torricellia, 647
Tournefort, J. P., 17, 18, 27, 42, 48
Tournefortia, 684
(Tovara), 476
Toxicodendron, 574
Trachelospermum, 673
Tracheophyta, 355
Trachomitum, 673
Trachymene, 645, 646
(Tradescantella), 408, 409
Tradescantia, 408, 409
Tradescantieae, 409
Tragia, 566
Tragopogon, 728, 730
Tragus, H., 15
Trailing arbutus, 654
Trapa, 636
Trapaceae, described, 636–637
Trapella, 701
Trautvetteria, 498
Traveler's-palm, 427
Traveler's-tree, 427
Trefoil, bird's-foot, 548
Tremandraceae, described, 563
Triadenum, 605
Trianthema, 483
Tribe, a unit of classification, 47; nomenclature of, 206
Tribulus, 556
Trichilia, 561
Trichomanes, 347, 348
Trichosanthes, 719, 720
(Trichosporum), 705
Trichostema, 691
Tricocceae, 549
Trientalis, 659
Trifoliolate orange, 557
Trifolium, 547
Triglochin, 379, 380, 381
Trigoniaceae, 549
Trillium, 413
Triodanis, 722
(Triorchis), 436
Triosteum, 713, 715
Triphasia, 557
Triphora, 436
Tripogandra, 408, 409
Tripterygium, 577
Tristania, 634

Triticum, 391
Triumfetta, 592
Trixis, 730
Trochodendraceae, 504; described, 492
Trochodendrineae, 489
Trochodendron, 492, 493
Trollius, 497, 498
Tropaeolaceae, described, 552–553
Tropaeolum, 552
Tropical almond, 633
Tropidia, 436
Trumpet flower, 700
Trumpet, vine, 700
Tsuga, 364
(Tsusiophyllum), 654
(Tubiflora), 708
Tubiflorae, 676
Tukey, H. B., 180
Tulip, 416
Tulip poppy, 517
Tulip tree, 700
Tulipa, 413
(Tulipastrum), 504
Tulipeae, 415
Tumamoca, 719
(Tumion), 360
Tung oil, 566
Turnera, 616
Turneraceae, described, 615–616
Turnip, 521
Turpentine, 364
Turpinia, 579, 580
Turraea, 561
Turreson, G., 164, 168, 169, 176, 185; biosystematic units of, 177
Turrill, W. B., 93, 98, 100, 106, 119, 170, 179, 182
Tussilago, 728, 730
Twinflower, 715
Type localities, importance of, 268
Type (nomenclatural), defined, 203–204
Type species, designation of, 202, 203–204
Type specimens, care of, 260; locations of, 328, 329
Type-basis code of nomenclature, 199
Type-method, in nomenclature, 203–205
Types, kinds of, 204–205; *see also under* Biosystematics, 203–205
Typha, 373
Typhaceae, described, 373
Typical element, nomenclature of, 202

Uhl, N. W., 376, 377, 380, 381, 385
Ulbrich, E., 478
Ullucus, 486
Ulmaceae, described, 461–462
Ulmoideae, 461
Ulmus, 461, 462
Umbel, in inflorescence phylogeny, 61, 62
Umbellales, 642

Umbelliferae, described, 644–646
Umbelliflorae, 642; alliance of Garryales, 448
Umbellularia, 513
Umbrella pine, 366
Umbrella plant, 393
*Ungnadia*, 584
Unicorn plant, 702
*Uniola*, 389
*Union list of serials*, described, 309
Unisexual flowers, morphology and disposition of, 83–84
United States, floras of, 293, 294, 295, 296
Units of classification, biosystematic, 176–179; explanation of, 44–56; nomenclature of, 217; nomenclatural considerations, 217; taxonomic considerations, 44–56
Uphof, T. J. C., 420
Urban, I., 327
*Urceolinia*, 418
*Urea*, 464
*Urechites*, 673
*Urena*, 593
*Urginea*, 416
*Urtica*, 464, 465
Urticaceae, described, 464, 465
Urticales, described, 461
*Urticastrum*, 465
*Usteria*, 669
*Utricularia*, 705, 706
*Uvaria*, 509

Vacciniaceae, 654
Vaccinioideae, 654
*Vaccinium*, 653, 654
*Vagaria*, 418
*Vanilla*, 436
Varossieau, W. W., 539
Vautier, S., 475
Vegetable hair, 407
*Valeriana*, 716
Valerianaceae, described, 715–716
Valerianales, 712
*Valerianella*, 716
Valid publication, conditions of, 208
*Vallesia*, 673
*Vallisneria*, 385, 386
Vallisneriaceae, 386
Vallisnerioideae, 386
*Vancouveria*, 501
Van Eseltine, G. P., 702
Van Tieghem, P., 493, 494, 610
Variety, a unit of classification, 55, 56
Varnish tree, 574, 583
Vascular cryptogams, phylogeny of, 101–102
Vegetation areas of earth, 150–153
Vegetative structures in classification, 57–64

*Velleia*, 724
Velloziaceae, 411, 424
Velvet beans, 548
Venetian turpentine, 364
*Venidium*, 730
Ventral placentation, 75
Venus flytrap, 529
*Veratrum*, 413
*Verbascum*, 695, 697
*Verbena*, 687, 688
Verbena, sand, 481
Verbenaceae, described, 686–688
Verbenales, 690
Verbenineae, 676
Verdoorn, F., 329, 340
Verdoorn, I. C., 257
*Verhuellia*, 445
Vernacular names, inclusion of in floras, 279
*Vernonia*, 730
Vernonieae, 730
*Veronica*, 695, 696, 697
*Veronicastrum*, 695
Verticillatae, described, 442
Vessel, defined, 370
Vessels, angiosperms lacking, 370
Vestal, P. A., 605
Vetch, 548
*Viburnum*, 713, 714, 715
*Vicia*, 548
*Victoria*, 489, 490
Vienna Code of nomenclature, 198, 199
Vierhapper, F., 412
*Vigna*, 548
*Villadia*, 531
*Villarsia*, 580
*Vinca*, 672, 673
Vincaceae, 673
*Vincetoxicum*, 675
*Viola*, 612, 613
Violaceae, alliance with Resedaceae, 525; described, 612–613
(*Viorna*), 498
Vipers' bugloss, 685
Virginia bluebells, 685
Virginia creeper, 589
Viscoideae, 472
*Viscum*, 471, 472
*Vismia*, 605
Visual aids, in taxonomic papers, 272
Vitaceae, described, 585–589
*Vitex*, 686, 687, 688
(*Viticella*), 498
*Vitis*, 588
*Vittaria*, 350
*Vivania*, 551
Vochysiaceae, 549
von Engeln, O. D., 142, 144
*Voyria*, 670
*Vriesia*, 407

*Wachendorffia*, 417
*Wahlenbergia*, 721, 722
Wallace, A. R., 8, 27, 33
Wallflower, 521
Walnut, 455
Walnut family, described, 453
Walter, T., 26
*Waltheria*, 596
Walton, J., 340
Wangerin, W., 632
Wappata, 383
Warburg, O., 510, 610
*Warburgia*, 611
Warming, E., 466
*Washingtonia*, 395, 396
Water clover family, described, 253, 353
Water fern family, described, 351
Water glass, for preserving dissections, 257
Water hemlock. 646
Water hyacinth, 411
Water milfoil family, described, 640
Water snowflake, 671
Water-chestnut family, described, 636
Watercress, 521
Waterleaf family, described, 682
Water-lily family, described, 489
Water-plantain family, described, 382
Watermelon, 720; Chinese, 720
Water-starwort family, described, 567
Waterwort family, described, 605
Wattles, 548
Wax plant, 675
Weatherby, C. A., 55, 351, 382, 635
Weatherwax, P., 390
Wegener, A., theory of continental drift, 142, 143
Wehrhahn, H. R., 322
Weibel, R., 591
*Weigela*, 715
*Weinmannia*, 536
Welsh Poppy, 517
Welwitschiaceae, 368
Wernham, H. F., 488, 660
West Indian boxwood, 699
West Indian cedar, 561
West Indies, floras of, 297
Wettstein, R. von, 94, 95, 101, 114; system of classification, 35–36, 121, 122
Wheat, 391
Wheeler, L. C., 565
Wherry, E. T., 335
Whiskey, 391
Whitaker, T. W., 504
White alder, 650
Whitewood, 592
*Whitneya*, 728
Widder, F., 358
*Widdringtonia*, 366

*Wiesneria*, 382
*Wigandia*, 683
Wild cinnamon, 611
Wild cinnamon family, described, 611
Wild ginger, 473
Wilkinson, A. M., 713
Willdenow, C. L., 26, 42
Willis, J. C., 318, 435, 467, 565, 675, 730; theory of age and area, 160–171
Willkomm, H. and Lange, J., 300
Willow family, described, 447
Wilson, C. L., 65, 67–71
Wilson, C. L. and Just, T., 81
Wilson, L. R., Downs, R. B., and Tauber, M. F., 330
Winge, Ø., 178
Winkler, H., 86, 426, 429
Winter hazel, 538
Winteraceae, 498, 504; described, 506
*Winterana*, 611
Winteranaceae, 611
Wintergreen, 654; commercial source of, 459
Wishbone flower, 697
*Wislizenia*, 519
*Wisteria*, 548
Witch-hazel family, described, 537
Withner, L. L., 461
*Witsenia*, 422
Wittstein, G., 319
Wodehouse, R. P., 652
Wood rose, 677
Woodson, R. E., Jr., 297, 409, 674, 675
Woody versus herbaceous characters, 58–59
Woody versus herbaceous habit, primitiveness of, 106–107
Wooton, E. O. and Standley, P. C., 294
*Wolffia*, 401
*Wolffiella*, 401
World floras, 288–290
World herbaria, abbreviations of, 273
*Wormia*, 598
Wormwood, 731; oil of, 478
Worsdell, W. C., 498
Wulff, E. V., 159–160, 162, 267; on phytogeography, 159

*Xanthium*, 728, 730
*Xanthoceras*, 583
*Xanthorrhiza*, 496, 498
Xanthorrhoraceae, 415
(*Xanthoxalis*), 550
*Xeranthemum*, 731
*Ximenia*, 469
*Xylopia*, 509
*Xylosma*, 614
Xyridaceae, described, 403–404
Xyridales, 404
*Xyris*, 403, 404

INDEX 823

Yam bean, 548
Yam family, described, 421
*Yeatsia*, 708
Yellow bells, 700
Yellow Jessamine, 670
Yellow oleander, 673
Yellow-eyed grass family, described, 403
Yew family, described, 360
Youngken, H. W., 451
Yucceae, 419

*Zamia*, 356, 357
Zamioideae, 357
*Zannichellia*, 377
Zanichelliaceae, 377
Zanichellieae, 377
*Zantedeschia*, 399, 400
Zanthoxylaceae, 556
*Zanthoxylum*, 557
*Zauschneria*, 638, 639
*Zea*, 389, 390, 391
*Zebrina*, 408, 409
Zephyrantheae, 418
*Zephyranthus*, 420
*Zeugites*, 387
*Zingiber*, 428
Zingiberaceae, described, 427–428
Zingiberales, 425
Zingiberoideae, 428
*Zinnia*, 728
*Zizyphus*, 588
*Zoisia*, 391
*Zostera*, 376, 377
Zosteraceae, 377
*Zosterella*, 411
Zygomorphy, characteristics of, 84
Zygophyllaceae, described, 555–556
*Zygophyllum*, 556